UNITED STATES DEPARTMENT OF THE INTERIOR, Douglas McKay, *Secretary*

FISH AND WILDLIFE SERVICE, John L. Farley, *Director*

FISHES OF THE GULF OF MAINE

BY HENRY B. BIGELOW AND WILLIAM C. SCHROEDER

First Revision

FISHERY BULLETIN 74

FISHERY BULLETIN OF THE FISH AND WILDLIFE SERVICE

VOLUME 53

[Contribution No. 592, Woods Hole Oceanographic Institution]

Reprint of First Revision, Published in 1953 by the U.S. Department of the Interior

Fishes of the Gulf of Maine
Fishery Bulletin 74, Fish and Wildlife Service

ISBN-10: 1-930665-60-1
ISBN-13: 978-1-930665-60-6

Library of Congress Control Number: 2002107351

THE BLACKBURN PRESS
P. O. Box 287
Caldwell, New Jersey 07006 U.S.A.
973-228-7077
www.BlackburnPress.com

CONTENTS

FISHES OF THE GULF OF MAINE

By HENRY B. BIGELOW and WILLIAM C. SCHROEDER

Museum of Comparative Zoology, Harvard University, and Woods Hole Oceanographic Institution

During the summer of 1912 the Bureau of Fisheries, with the cooperation of the Museum of Comparative Zoology of Harvard University, commenced an oceanographic and biological survey of the Gulf of Maine, with special reference to its fishes, to its floating plants and animals (plankton), to the physical and chemical state of its waters, and to the circulation of the latter. Cruises were made on the Fisheries schooner *Grampus* during the summers and autumns of 1912, 1913, 1914, 1915 and 1916, and during the winters and springs of 1913 and 1915. The work was interrupted by the war, but was resumed with a cruise of the Fisheries steamer *Albatross* in the late winter and spring of 1920, and was continued by the Fisheries steamer *Halcyon* during the winter and spring of 1920–21, and the summers of 1921 and 1922.

The first part of the general report, dealing with the fishes, was published in 1925, as Bulletin 40 (Pt. 1) of the United States Bureau of Fisheries;[1] subsequent parts describing the plankton of the offshore waters of the Gulf and the physical characteristics of its waters were published in 1926–27, as Part 2.

The preparation of the section on the fishes was assigned originally to W. W. Welsh, who had gathered a large body of original observations on the growth, reproduction, diet, and other phases of the lives of many of the more important species. The report was far advanced when it was interrupted by his untimely death, and H. B. Bigelow undertook to carry it to publication along the lines originally laid down. The new edition, entailing a general revision and the addition of much new material, has been prepared jointly by H. B. Bigelow and by W. C. Schroeder.

[1] The Bureau of Fisheries was transferred on July 1, 1939, from the Department of Commerce to the Department of the Interior, and on July 30, 1940, it was consolidated with the Bureau of Biological Survey to form the Fish and Wildlife Service.

210941—53——2

AREA COVERED

The term "Gulf of Maine" covers the oceanic bight from Nantucket Shoals and Cape Cod on the west, to Cape Sable on the east. Thus it includes the shore lines of northern Massachusetts, New Hampshire, Maine, and parts of New Brunswick and of Nova Scotia. The eastern and western boundaries adopted in this paper are 65° and 70° West longitude, respectively. Southern strays, or northern, which have no real status in the Gulf of Maine except by accident, are mentioned only briefly, or are relegated to footnotes. The Gulf of Maine has a natural seaward rim formed by Nantucket Shoals, by Georges Bank, and by Browns Bank. We have chosen the 150-fathom contour as the arbitrary offshore boundary, because this will include all of the species that are likely to be caught by commercial fishermen but will exclude almost the entire category of the so-called "deep-sea" fishes, which are numerous in the basin of the open Atlantic but are not constituents of the fauna of the Gulf of Maine, properly speaking.

The general oceanography of this area has been the subject of another report, but it may not be amiss to point out that the temperature of the Gulf and its fauna are boreal, and that its southern and western boundaries are the northern limit to common occurrence of many southern species of fishes and of invertebrates.

SCOPE OF THE WORK

Our aim has been a handbook for the easy identification of the fishes that occur in the Gulf of Maine, with summaries of what is known of the distribution, relative abundance, and more significant facts in the life history of each. The descriptions are as little technical as is compatible with scientific accuracy, and are limited chiefly to such of the external features of each kind of fish as may serve for identification in the field.

References to more detailed descriptions and synonymies are given to Bigelow and Schroeder (Fishes of the Western North Atlantic, Parts 1 and 2, 1948, 1953) for the cartilaginous fishes; also to Garman's beautiful plates for such of these as he pictured in his classic monograph, published in 1913, in vol. 36, of the Memoirs of the Museum of Comparative Zoology. References for the various species of bony fishes are to Jordan and Evermann's Fishes of North and Middle America (Bulletin 47, U. S. National Museum, 1896–1900, Parts 1–4), which still remains the only comprehensive work on the bony fishes of North America. Many of the illustrations have been borrowed from earlier publications, but some of them are original.

Keys are provided for all species as a further aid to identification.

In most cases the sizes of larval fish and eggs are given in millimeters (1 inch equals 25.4 mm.); the sizes of the larger fishes are in inches and feet; weights are in pounds.

The scientific nomenclature of the cyclostomes, of the elasmobranchs, and of the chimaeroids, follows Bigelow and Schroeder (Fishes of the Western North Atlantic, No. 1, Parts 1 and 2, 1948; 1953) that of the bony fishes follows Jordan, Evermann, and Clark's Check List of the Fishes and Fishlike Vertebrates of North and Middle America (Report, U. S. Commissioner of Fisheries for 1928 (1930), Part 2), unless otherwise noted. The families of bony fishes are arranged for the most part in the sequence employed by Jordan, Evermann, and Clark, except that the several families of luminescent fishes are grouped together, in the hope of making it easier for the nontechnical observer to identify such of them as may come to hand.

SOURCES OF INFORMATION

The literature dealing with the fishes of the Gulf of Maine begins with the earliest descriptions of New England. Captain John Smith, for instance, commented on the abundance of sturgeon, cod, hake, haddock, cole (the American pollock), cusks, sharks, mackerel, herring, cunners, eels, salmon, and striped bass, in his Generall Historie of Virginia, New England and the Summer Isles, published in 1616, while Wood in his New England's Prospect, 1634, gave much interesting information, some of which we quote hereafter.

The sea fishes of northern New England and of the Maritime Provinces had begun to attract scientific attention by the early part of the nineteenth century, and many local faunal lists have been published since then. The following are the most important of these, in chronological arrangement:

1850. Report on the sea and river fisheries of New Brunswick, within the Gulf of St. Lawrence and Bay of Chaleur, M. H. Perley, 137 pp., 1850. Fredericton, New Brunswick.

1853–1867. A history of the fishes of Massachusetts, David Humphreys Storer. Memoirs, American Academy of Arts and Sciences, New Series, vol. 5, pp. 49–92, 122–168, and 257–296; vol. 6, pp. 309–372; vol. 8, pp. 389–439; vol. 9, pp. 217–256, 39 pls. (Also in book form with supplement, 1867), Cambridge and Boston.

1879. A list of the fishes of Essex County, including those of Massachusetts Bay, George Brown Goode, and Tarleton H. Bean. Bulletin, Essex Institute, vol. 11, No. 1, pp. 1–38. Salem.

1884. Natural history of useful aquatic animals, George Brown Goode and associates, Section I, The Fisheries and Fishery Industries of the United States. Published jointly by the U. S. Fish Commission and the U. S. Bureau of the Census, 895 pp. Washington.

1908. Fauna of New England. 8. List of the Pisces, William C. Kendall. Occasional Papers, Boston Society of Natural History, vol. 7, No. 8, April 1908, pp. 1–52. Boston.

1914. An annotated catalogue of the fishes of Maine, William C. Kendall. Proceedings, Portland Society of Natural History, vol. 3, 1914, Part 1, pp. 1–198. Portland.

1922. The fishes of the Bay of Fundy, A. G. Huntsman. Contributions to Canadian Biology (1921), 1922, No. 3, pp. 1–24 (51–72). Ottawa.

These lists contain all the early published locality records of the rarer species, either first hand, or by reference to original sources, while the last two, with a paper by Gill,[2] and the first edition of the present book give complete bibliographies for the Canadian coasts of the Gulf and for the coasts of Maine and of Massachusetts. A similar list of the captures of deep water fishes along the outer part of the continental shelf is to be found in Goode and Bean's "Oceanic Ichthyology."[3]

The most pertinent extralimital lists are Smith's[4] and Sumner, Osburn and Cole's [5] lists of Woods

[2] Rept. U. S. Comm. Fish., (1904) 1905, pp. 163–188.
[3] Smithsonian Contribution to Knowledge, vol. 30, 1896.
[4] Bull. U. S. Fish Comm., Vol. 17, 1898, pp. 85–111.
[5] Bull. U. S. Bur. Fish., vol. 31, Pt. 2, 1913, pp. 549–794.

Hole fishes; Halket's [6] Checklist of the fishes of Canada and of Newfoundland, and Vladykov and McKenzie's The Marine Fishes of Nova Scotia.[7]

The literature dealing with the habits of the fishes of the Gulf of Maine is very extensive, for most of the important commercial species, and many of the others also, are common to both sides of the North Atlantic. Among general European manuals, Day's Fishes of Great Britain and Ireland,[8] Smitt's "Scandinavian Fishes," [9] and Ehrenbaum's summary of the many scattered accounts of the eggs and larvae of northern fishes [10] have been especially helpful.

A large amount of information as to local distribution and abundance of various fishes has been gleaned from unpublished material in the files of the U. S. Fish and Wildlife Service, as well as from the fishery statistics published by the Fisheries Branch, U. S. Fish and Wildlife Service (formerly the U. S. Bureau of Fisheries), by the Dominion of Canada, and by the Commonwealth of Massachusetts. The superintendents of the Woods Hole, Gloucester, and Boothbay hatcheries have supplied much valuable information, as have other members of the U. S. Fish and Wildlife Service. Among these, Leslie Scattergood has given many interesting pieces of information for Maine waters, while Howard Schuck has contributed authenticity to the account of the haddock. Dr. A. G. Huntsman has contributed his unpublished notes on the fishes of the Bay of Fundy and Gulf of St. Lawrence. Dr. A. H. Leim, Mr. R. A. McKenzie, and Dr. Vadim D. Vladykov have supplied us with pertinent information on certain species from the Nova Scotian-St. Lawrence River regions. The late Prof. J. P. McMurrich permitted the use of his unpublished plankton records, and a number of Newfoundland records were furnished by Drs. George W. Jeffers and E. Templeman.

The late W. F. Clapp has contributed many interesting notes gleaned during his experience as a fisherman before entering the scientific field. Harry Piers of the Provincial Museum of Halifax, has supplied interesting information on the occurrence of the blue shark. John Worthington has furnished us with pound-records for the Truro-Provincetown region covering a recent span of about fifteen years and has given us specimens of three species heretofore unreported in the Gulf of Maine. Benjamin H. Morrow has supplied interesting data from the vicinity of Sandwich, Mass. We have received much information about the striped bass in Nova Scotia from Major Howard Scott, through the kind offices of Henry Lyman. And we owe it to consultation with Dr. Å. Vedel Tåning of the Marine Biological Laboratory, Charlottenlund, Denmark, and the specimens contributed by Dr. C. E. Lucas of the Scottish Fisheries Laboratory, Aberdeen, that we have dared to reach a conclusion as to the relationship between the rosefish of our gulf and of north European waters. Francis Sargent, also of the Division of Marine Fisheries of Massachusetts, and Henry Lyman, editor of the Salt Water Sportsman, have been unfailing in their response to our many inquiries. Myvanwy Dick of the Harvard Museum of Comparative Zoology has been of assistance in the handling of certain of our study material and in the preparation of a number of illustrations. The illustrations of the hagfish and lamprey and most of those of the sharks, skates, rays, and chimaera are reprinted here through the courtesy of the Sears Foundation for Marine Research, publisher of the Fishes of the Western Atlantic, Memoir 1, Parts 1 and 2, in which the illustrations originally appeared. Claude Ronne of the Woods Hole Oceanographic Institution prepared many photographs from both original and published drawings, which were used to illustrate this book.

We owe a debt of gratitude, also, to the late Dr. Samuel Garman, who was ever ready with assistance until the time of his death, and to W. C. Adams, former director of the division of fisheries and game of the State of Massachusetts. We wish to express our hearty thanks to the many commercial fishermen and to the many salt water anglers of our acquaintance who have met our inquiries in the most cordial way and who have supplied us with a vast amount of first-hand information on the habits, distribution, and abundance of the commercial and game fishes, which could be had from no other source. The

[6] Checklist of the Fishes of the Dominion of Canada and Newfoundland 1913, 138 pp.

[7] Proc. Nova Scotia Inst. of Science, vol. 19, Pt. 1, 1935, pp. 17-113.

[8] The fishes of Great Britain and Ireland, by F. Day, Text vol. 1, CXII+336 pp., vol. 2, 388 pp., and atlas, 179 plates, 1880-1884. London and Edinburgh.

[9] A history of Scandinavian fishes. Second edition, vol. 1, 1892; vol. 2, 1895; 1,240 pp., 53 pls. Stockholm.

[10] Eier und Larven von Fischen. Nordisches Plankton, vol. I, 413 pp., 148 figs.; appeared in two parts as Lief. 4, 1905, and Lief. 10, 1919.

preparation of this book would have been out of the question without their help.

Finally, we have ourselves gathered a large body of data as to distribution, habits, spawning seasons, and like matters, through many years, at many localities, both inshore and on the offshore banks.

USE OF THE KEYS

The various fins and other structures mentioned in the keys are named in the accompanying outlines of a haddock and of a typical shark (fig. 1). A simple way to explain the use of the keys is to use the haddock as an example, running it down with the illustration at hand for reference.

Turning to Key A (p. 5), we find that our fish fits the second alternative under section 1, since it has bony jaws and pectoral fins, and is not shaped like an eel. This refers us to section 3.

There being only one gill opening on each side, we go from section 3 to section 5. As our fish does not have a tubular snout section 5 refers us to section 6, and this in turn to section 7, since neither the upper jaw nor the lower is greatly prolonged. Since the body is not square-cut close behind the dorsal and anal fins, but has a definite tail part, we proceed from section 7 to section 8, and from section 8 to section 11, for our fish has no sucking plate or disc, either on top of the head, or on the chest. Section 11 refers us in turn to section 12 because the tail fin is nearly symmetrical in outline. The anal fin being clearly and definitely separated from the caudal fin, we go from section 12 to section 13; and from section 13 to section 14, for our fish does not have any evident light-producing ("luminescent") spots either on its sides or on its head. Our fish does not have a fleshy fin or flap either in front of the ordinary dorsal fins or behind them, but all of its dorsal fins are supported by rays that are visible if held against the light. Consequently, we proceed from section 14 to section 18, and this refers us to section 22, there being no flaps or tags of skin on the sides of the head.[11] Our fish obviously does not lie flat on one side, i. e., it is not one of the flat fishes, which brings us to section 23, and

[11] There is a barbel on its chin, but this is very different in appearance from the skin flaps around the jaws that are characteristic of the few species that fall under the first alternative of section 18.

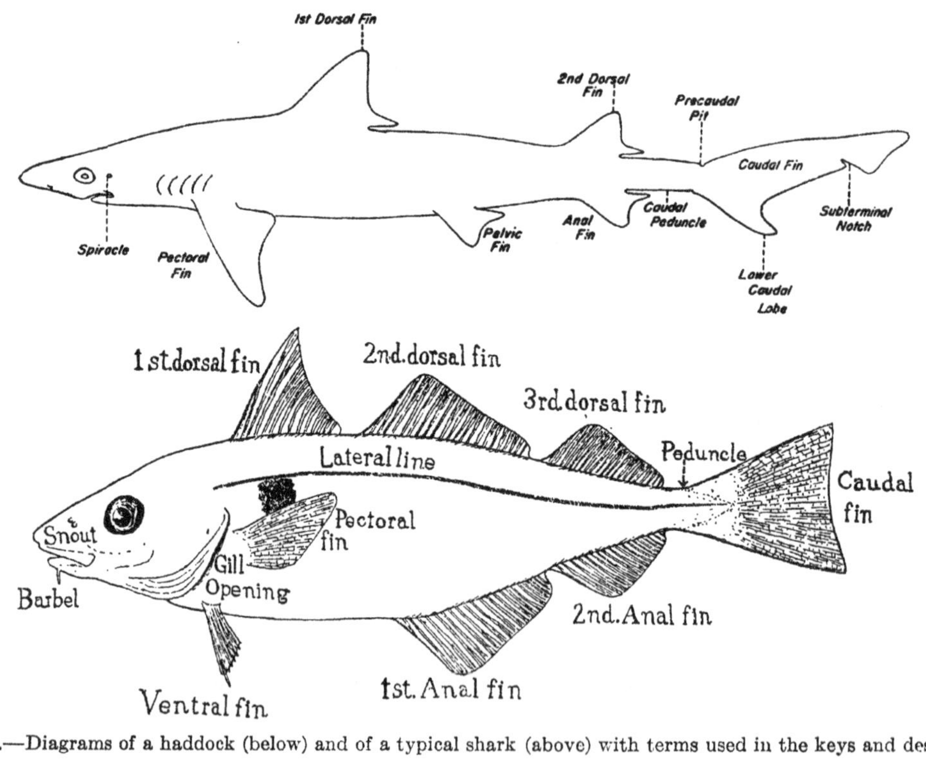

FIGURE 1.—Diagrams of a haddock (below) and of a typical shark (above) with terms used in the keys and descriptions.

this in turn carries us to Key E (p. 7) because it has three separate, well developed dorsal fins.

Since there are 3 dorsal fins and 2 anal fins, section 1 of Key E sends us to the key to the cod and silver hake families (p. 173). Turning to the first section of the latter we find that our fish fits the first alternative (3 dorsal fins and 2 anals), which refers it to section 2. And here the black lateral line and the dark blotch on each shoulder name it a haddock.

Any other Gulf of Maine species is to be named in the same way, starting with Key A, section 1, and following through the appropriate alternatives as they refer it from section to section.

KEY TO GULF OF MAINE FISHES

Key A

1. Mouth soft, with no firm jaws; no pectoral fins; form eel-like ---- 2
   Mouth has firm jaws; pectoral fins are present even if the form is eel-like ---- 3
2. Two separate fins on the back; no barbels on the snout ---- Lamprey, p. 12
   Only one fin on the back; with barbels on the snout ---- Hag, p. 10
3. Five gill openings on each side ---- 4
   Only one gill opening on each side ---- 5
4. General form cylindrical in all Gulf of Maine species; the forward edges of the pectoral fins are not attached to the sides of the head forward, past the gill openings; the gill openings are not confined to the lower surface; the upper margin of each orbit is free from the eyeball, as a free eyelid ---- Sharks, key, p. 16
   General form very flat, disclike; the forward edges of the pectoral fins are attached to the sides of the head forward past all of the gill openings; the gill openings are confined to the lower surface; the upper margin of each orbit is not free from the eyeball (no free eyelid) ---- Skates and Rays, key, p. 57
5. The bones of the head are fused in a tubular snout, with the mouth at its lip ---- refer to Key B, p. 6
   No tubular snout ---- 6
6. One or both jaws are prolonged as a bony sword or bill ---- refer to Key C, p. 6
   Neither jaw is greatly prolonged ---- 7
7. Body abruptly square-cut, close behind the very high dorsal and anal fins ---- refer to Sunfishes, key, p. 529
   Body with distinct tail part ---- 8
8. There is a sucking plate or disc, either on the top of the head or on the chest ---- 9
   There is no sucking disc or plate ---- 11
9. The sucking plate is on the top of the head ---- refer to Remora family, key, p. 485
   The sucking disc is on the chest ---- 10
10. General form is like a tadpole; the anal fin originates about as far back as the tips of the pectorals. refer to Sea snail family, key, p. 464
    General form is not like a tadpole, but is high arched, with longitudinal ridges; the anal fin originates far behind the tips of the pectorals ---- refer to Lumpfish family, key, p. 459
11. Tail like a shark, i. e., with the upper lobe much longer than the lower ---- Sturgeons, key, p. 81
    Tail with the upper and lower lobes of equal lengths, or nearly so ---- 12
12. No clear separation between the anal and the caudal fins, which together form one continuous fin (the anal portion may be either long or short) ---- refer to Key D, p. 6
    Anal and caudal fins are separated by a deep notch, or by a space ---- 13
13. Sides of body and head, or both, with luminescent spots or patches, easily seen if not damaged. refer to Luminescent fishes, key, p. 141
    No luminescent organs ---- 14
14. There is a fleshy ("adipose") fin, with neither rays nor spines, either in front of the rayed dorsal fin, or behind it ---- 15
    There is no fleshy ("adipose") fin, but both the dorsals (if there are two) are supported by rays or by spines that can be felt, if not seen ---- 18
15. The adipose fin is on the nape of the neck, in front of the dorsal fin ---- Tilefish, p. 426
    The adipose fin is behind the dorsal fin ---- 16
16. The dorsal fin extends nearly the entire length of the body ---- Lancetfish, p. 161
    The dorsal fin is short, standing about midway of the body ---- 17
17. Tail deeply forked ---- refer to Smelts and Argentine Key, p. 133
    Tail nearly square or only slightly forked ---- refer to Salmon key, p. 120
18. The head is fringed with fleshy tags or flaps ---- 19
    The head is not fringed with fleshy tags or flaps ---- 22
19. The pectorals are armlike ---- 20
    The pectorals are not armlike ---- 21

Key A—Continued

20. Body very broad and flat; mouth enormous — Goosefish, p. 532
    Body deep and flattened sidewise; mouth small — Sargassum fish, p. 541
21. The first (spiny) dorsal fin is longer than the second (soft-rayed dorsal); neither dorsal fin is fleshy — Sea raven, p. 454
    The first (spiny) dorsal fin is much shorter than the second (soft-rayed dorsal); both of the dorsals are thick and fleshy — Toadfish, p. 518
22. Fishes which lie flat on the one side, with both of their eyes on the other side; the upper side is dark, the lower side normally is pale — refer to Flatfish tribe key, p. 248
    Fishes which do not lie flat on one side — 23
23. Two or more separate and well-developed dorsal fins, each with continuous membrane — refer to Key E, p. 7
    Only one well-developed dorsal fin with continuous membrane (this, however, may be preceded by isolated spines or rays) — 24
24. Top of snout with several barbels or beards — Rockling (cod family in part), p. 234
    No barbels or beards on the top of the snout — 25
25. Jaws with very large canine tusks — refer to Wolffishes key, p. 503
    No large canine tusks in either jaw — 26
26. Dorsal fin soft-rayed throughout its length, except that it may be preceded by a few separate spines. — refer to Key F, p. 8
    At least the forward one-third of the dorsal fin is with stiff sharp rays or spines — refer to Key G, p. 9

Key B

Fishes with tubular snouts (from No. 5, p. 5).

1. Head is horselike; rear portion of trunk is slender, prehensile; no caudal fin — Sea horse, p. 315
   Head is not horselike; rear part of trunk is not prehensile; there is a caudal fin — 2
2. Body and head (measured from tip of snout) are only about 4 times as long as deep; the dorsal fin has a long, strong, saw-edged spine — Snipefish, p. 301
   Body and head (measured from tip of snout) are at least 25 times as long as deep; the dorsal fin does not have a large spine — 3
3. The snout is not longer than the dorsal fin; the anal fin is very small; no ventral fins; the caudal fin is rounded — Pipefishes, key, p. 312
   The snout is more than 6 times as long as the dorsal fin; the anal fin is about as large as the dorsal; ventral fins are present though small; the caudal fin is forked — Trumpetfish, p. 316

Key C

Fishes with bills or swords (from No. 6, p. 5)

1. Both of the jaws are elongated — 4
   Only one of the jaws is elongated — 2
2. Upper jaw elongated, as a sword — 3
   Lower jaw elongated — Halfbeak, p. 169
3. The sword is flattened dorso-ventrally, and is sharp-edged; the first dorsal fin is shorter than the sword forward of the eyes; no ventral fins — Swordfish, p. 351
   The sword is round-edged; the first dorsal fin is nearly twice as long as the sword — refer to Spearfishes or Marlins and Sailfish, key,[12] p. 358
4. The caudal fin is well developed — 5
   No caudal fin; the tip of the tail is whip-like — Snipe eel, p. 159
5. There are several finlets behind the dorsal and anal fins — Needlefish, p. 170
   No finlets behind the dorsal and anal fins — refer to Billfishes or Silver gars, key, p. 167

Key D

Bony fishes with snouts of ordinary form; with only one gill opening on each side, and with the anal fin continuous with the caudal fin around the tip of the tail (from No. 12, p. 5).

1. Only one dorsal fin — 2
   Two separate dorsal fins, the first much bigger than the second, but shorter — 7
2. Body band-shaped, the tail tapering to a whip-like tip — Cutlassfish, p. 350
   Body thick, eel-like; the vertical fins continuous around the tip of the tail in a broad band — 3
3. The dorsal fin is spiny from end to end — 4
   The dorsal fin is soft-rayed, at least for almost all its length — 5

[12] The sailfish would also come under this heading should one ever be taken in the Gulf of Maine. The distinctions between it and the spearfishes are given on page 358.

Key D—Continued

4. Mouth large and strongly oblique; there are no ventral fins ---------- Wrymouth, p. 500
Mouth small and horizontal; with small ventral fins ---------- Rock eel, p. 492
5. There are no ventral fins ---------- refer to Eel family key, p. 150
With small but distinct ventral fins, situated forward of the pectorals ---------- 6
6. The ventrals are situated behind the gill openings and are of ordinary form ---- refer to Eelpout family, key, p. 509
The ventrals are situated on the chin, well in front of the gill openings and are reduced to forked, barbel-like structures ---------- Cusk eel, p. 517
7. The ventral fins are situated below the points of origin of the pectorals; the skin is conspicuously scaly. refer to Grenadier family, key, p. 243
The ventral fins are situated far back, behind the tips of the pectorals; the skin is soft, without scales. Chimaera, p. 79

Key E

Bony fishes of ordinary form, with 2 or 3 well-developed dorsal fins and with the anal fin and the rearmost dorsal separated from the caudal fin. (from No. 23, p. 6).

1. Three dorsal fins and 2 anal fins ---------- refer to Cod family, key (in part), p. 173
Only 2 dorsal fins and 1 anal fin ---------- 2
2. With one or more small finlets between the second dorsal and anal fins and the caudal fin ---------- 3
No finlets between the second dorsal and anal fins and the caudal fin ---------- 4
3. With more than 3 dorsal finlets and 3 anal finlets ---------- refer to Mackerel family, key, p. 317
With only 2 dorsal finlets and 2 anal finlets ---------- Escolar, p. 349
With only 1 dorsal finlet and 1 anal finlet ---------- Mackerel scad (Pompano family, in part), p. 374
4. Head very broad; top and sides of head bony, with sharp spines or horns ---------- 5
Head not noticeably broad; sides of head have no spines or horns ---------- 7
5. First (lower) few rays of the pectoral fins are not separate from the remainder of the fin; the mouth is very large refer to Sculpin family, key, p. 440
First (lower) few rays of the pectoral fins are separate from the remainder of the fin; the mouth is not very large ---------- 6
6. Each of the first (lower) 2 or 3 rays of the pectoral fins have the form of a separate feeler; outline of tip of snout is concave; the first few spines of the first dorsal fin are not separate from the remainder of the fin. refer to Sea robin and Armored sea robin, key, p. 467
First (lower) few rays of the pectorals do not have the form of feelers, but are connected, one with the next, by membrane, as a separate fin; outline of tip of snout convex; the first few spines of the first dorsal are separate. Flying gurnard, p. 472
7. First spine of first dorsal fin is very much stouter than the other spines, and can be locked erect by the second spine; no ventral fins; skin of the sides is very hard ---------- Triggerfish, p. 520
First dorsal spine is not much stouter than the others and cannot be locked erect by the second spine; ventral fins are well-developed; skin of the sides is soft ---------- 8
8. The space between the two dorsal fins is nearly as long as the first dorsal fin, or longer; the ventral fins are situated behind the middle of the pectorals ---------- 9
There is little or no free space between the two dorsal fins; the ventrals are in front of the middle of the pectorals --- 11
9. Jaws long; teeth large and strong; anal with one spine ---------- Barracuda, p. 306
Jaws short; teeth weak ---------- 10
10. Anal fin is about as long as head (snout to gill openings) and has one weak spine - refer to Silverside family key, p. 302
Anal fin is only about half as long as head and has three stiff spines (only two spines in very young specimens). Mullet, p. 305
11. Caudal peduncle is extremely slender; the caudal fin is deeply forked ---------- Pompano family (in part) key, p. 371
Caudal peduncle is at least moderately deep and thick; the caudal fin is only moderately forked, at most ---------- 12
12. First dorsal fin is much lower than second dorsal ---------- 13
First dorsal fin is as high as the second dorsal fin, or higher ---------- 14
13. Anal fin is nearly as long as second dorsal fin ---------- Bluefish, p. 383
Anal fin is only about one half as long as second dorsal fin ---------- Rudderfish (Pompano family in part), p. 373
14. Body very thin through, flat sided, nearly two-thirds as deep as it is long to base of caudal fin; the back and also the ventral edge of the body are armed with bony plates; there is a finlet of three short spines in front of the anal fin ---------- John Dory, p. 297
Body stout, not more than one-third as deep as it is long; the sides are rounded; the back and lower surface are not armed with bony plates; there is no finlet in front of the anal fin ---------- 15

Key E—Continued

15. First dorsal fin soft-rayed; second dorsal fin at least 5 times as long as first dorsal. refer to Cod family key, in part, p. 173
First dorsal fin noticeably spiny; second dorsal fin as long as first dorsal 16
16. Second dorsal fin not much longer than the anal fin refer to Sea bass family key, in part, p. 389
Second dorsal fin is about twice as long as the anal fin refer to Weakfish family key, p. 417

Key F

Bony fishes with snouts of ordinary form; symmetrical tails; caudal fin distinct from the anal fin; neither with barbels on the top of the snout nor with canine tusks; and with only one well-developed dorsal fin; the latter is soft-rayed except that it may be preceded by a few short spines or by a series of hair-like rays without connecting membrane and that there may be an isolated spine on the top of the head (from No. 26, p. 6).

1. The rear parts of the dorsal fin and of the anal fin are broken up into series of almost separate finlets (fig. 191) 2
The rear parts of the dorsal and anal fins are not broken up into series of finlets 3
2. The forward parts of the dorsal and anal fins are very high and scythe-shaped; the pectorals are very long, reaching back considerably beyond the high part of the dorsal fin; there are no spines in front of the anal fin Sea bream, p. 361
The dorsal and anal fins are not very high and slope gradually rearward; the pectorals are small, their tips falling far short of the level of the front of the dorsal fin; the anal fin is preceded by two short stout spines. Leather jacket, p. 380
3. The mouth gapes back far beyond the eye refer to Anchovies key, p. 118
The mouth does not gape back much beyond the rear edges of the eyes, if that far 4
4. The whole of the anal fin is behind the rear end of the dorsal fin refer to Herring Tribe key, p. 85
Part or all of the anal fin is further forward than the rear end of the dorsal fin 5
5. There is a spine or a bristle-like rod on the top of the head over the eyes 6
There is no spine or bristle-like rod on the head over the eyes, but there may be a few short spines close in front of the dorsal fin 7
6. The spine on the top of the head is thick and very stiff and has no fleshy tab at its tip; mouth small; body stiff; fin rays slender, not fleshy refer to Filefish family, key, p. 521
The spine on the head is slender and flexible and has a fleshy tab or "bait" at its tip; body soft; mouth very large; fin rays thick and fleshy Deep-sea angler, p. 543
7. Form eel-like; snout sharp pointed Launce, p. 488
Form not eel-like; snout blunt 8
8. Dorsal fin originates on the head, about over the eyes Dolphin, p. 360
Dorsal fin originates far behind the eyes 9
9. Each ventral fin is represented by a single large stout spine refer to Stickleback key, p. 307
The ventral fins are of ordinary rayed type, or are lacking 10
10. The upper anterior profile of the head is conspicuously concave 11
The upper anterior profile of the head is more or less convex 12
11. The forward parts of the dorsal fin and of the anal fin are much higher than the rear parts, the first few rays of each being very much longer than the rays farther back Lookdown, p. 379
The dorsal and anal fins are only a little higher in front than rearward, the first few rays not being much longer than the rays farther to the rear Moonfish, p. 378
12. The forward rays of the dorsal and anal fins are very long and thread-like Thread-fin, p. 381
The forward rays of the dorsal and anal fins are not very long and thread-like 13
13. The entire body is armored with several rows of overlapping plates Alligator fish, p. 457
The body is not armored with overlapping plates 14
14. The skin is rough or prickly Refer to Puffers and Porcupine fishes, key, p. 526
The skin is smooth, though scaly 15
15. The front part of the dorsal fin is much higher than the rear part 16
The front part of the dorsal fin is not much higher than the rear part 17
16. The ventral fins are large and conspicuous Opah, p. 247
There are no ventral fins Refer to Butterfish and Harvest Fish, key, p. 363
17. The tail fin is conspicuously rounded 18
The tail fin is more or less deeply forked 19
18. The dorsal fin runs the whole length of the back from close behind the head to the caudal fin which it joins; there is a barbel on the chin Cusk, p. 238
The dorsal fin occupies only about one-third of the length of the back or less, and stands far to the rear; there is a considerable space between it and the caudal fin; there is no barbel on the chin Refer to Mummichog key, p. 162

Key F—Continued

19. The caudal peduncle is slender and has a conspicuous longitudinal keel on either side; the pectoral fins do not reach back as far as the point of origin of the dorsal fin ---------- Pilotfish, p. 372
   The caudal peduncle is deep and has no longitudinal keel; the pectoral fins reach back farther than the point of origin of the dorsal fin ---------- 20
20. There are 6–8 short detached spines, each with a small triangular fin membrane, on the back in front of the dorsal fin ---------- Barrelfish, p. 369
   There are no detached spines on the back in front of the dorsal fin ---------- 21
21. The ventral fins stand far behind the bases of the pectoral fins; the point of origin of the dorsal fin is little if any in advance of the anal fin; the pectoral fins (Gulf of Maine species) are very long, reaching back nearly to the base of the tail fin ---------- Flying fish, p. 172
   The ventral fins stand about under the base of the pectoral fins; the point of origin of the dorsal fin is far in advance of the anal fin; the pectoral fins are small, falling far short of the anal fin ---------- Black ruff, p. 370

Key G

Fishes as in Key F, except that at least the forward one-third of the single dorsal fin is spiny. There is no adipose fin behind the rayed dorsal nor fleshy flap in front of it (from No. 26, p. 6).

1. The body (tip of snout to base of caudal fin) is at least as deep as it is long ---------- Boarfish, p. 438
   The body is considerably longer than it is deep ---------- 2
2. The rear part of the dorsal fin is soft-rayed ---------- 3
   The whole length of the dorsal fin is spiny ---------- 8
3. Sides of head bony, with knobs or spines ---------- 4
   No knobs or spines on the sides of the head ---------- 5
4. Sides of head armed with conical spines; the spiny portion of the dorsal fin is at least as long as the soft part; the body is flattened sidewise ---------- Refer to Rosefish family, key, p. 430
   Sides of head with low rounded knobs only; the spiny portion of the dorsal fin is considerably shorter than the soft part; body tadpole-shaped ---------- Arctic sculpin (Sculpin family in part), p. 453
5. The ventral fins are much larger than the pectorals; the eyes are very large ---------- Short big-eye, p. 410
   The ventral fins are not larger than the pectorals; the eyes are not very large ---------- 6
6. The pectorals are sharply pointed; the body is much flattened sidewise ---------- Refer to Porgy family, key, p. 411
   The pectorals are rounded; the body is not much flattened sidewise ---------- 7
7. The rear (soft) part of the dorsal fin is nearly as long as the front (spiny) part; the anal fin is much higher than long ---------- Seabass (Seabass family in part), p. 407
   The rear (soft) part of the dorsal fin is less than half as long as the spiny (front) part; the anal fin is longer than high ---------- Refer to Cunner family, key, p. 473
8. The mouth is strongly oblique; there are no ventral fins ---------- Wrymouth, p. 500
   The mouth is not strongly oblique; ventral fins are present (very small in one species) ---------- Refer to Blenny fishes, key, p. 491

THE CYCLOSTOMES. CLASS AGNATHA

The lampreys are the most primitive of the true vertebrates, their skeletons being cartilaginous without any true bone, and their skulls hardly differentiated from the vertebral column which forms a simple notochordal sheath. They have no true jaws, no ribs, no shoulder or pelvic girdles, and no paired fins. They are eel-like in appearance, but are easily distinguishable from the true eels and, indeed, from most of the true fishes, by their peculiar jawless sucking mouth situated at the tip of the snout, and, further, from all Gulf of Maine eels by lacking pectoral fins.

THE HAGFISHES AND LAMPREYS. FAMILIES MYXINIDAE AND PETROMYZONIDAE

These two groups are easily distinguished, one from the other, by the fact that the hags have several barbels on the chin, that their mouths are not disc- or funnel-like, that they have only one continuous fin fold on the back and around the tail, and that their eyes are not visible externally, whereas the lampreys have no barbels, their mouths are disc- or funnel-like, their eyes are well developed after the larval stage is past, and they have one or more dorsal fins separate from the caudal fin.

Hagfish *Myxine glutinosa* Linnaeus 1758

Bigelow and Schroeder, 1948, p. 34.

Description.—The hag, like the lamprey, lacks paired fins and fin rays. Its skeleton is wholly cartilaginous, without bones, its mouth is jawless; and its skin is scaleless. It is easily recognized by its eel-like form; by its single finfold (a fold of skin, not a true fin) running right around the tail and forward on the lower surface of the body with no division into dorsal, caudal, and anal fins; by the single gill pore on each side, just forward of the origin of the ventral finfold; by its lipless mouth, star-shaped in outline when closed; by the single nasal aperture at the tip of the snout; by its peculiar barbels or "tentacles," two flanking the mouth on either side and four surrounding the nostril; and by the evertible tongue studded with rows of horny rasplike "teeth." We might also mention the series of mucous sacs on either side of the abdomen, and point out that the dorsal finfold originates about two-thirds of the distance back from snout toward tip of tail, and the ventral fin fold one-third the way back, with the vent piercing it.

Color.—Hags vary in color, perhaps to correspond with the color of the bottom, being grayish brown or reddish gray above, variously suffused, mottled, or piebald with darker or paler gray, with brown, or with bluish; they are whitish or pale gray below.

Size.—Gulf of Maine hags grow commonly to a length of about 1½ to 2 feet, with a maximum of 31 inches recorded off the coast of Maine.

Habits.—The hag is found chiefly if not exclusively where the bottom is soft mud, where (to judge from its actions during the brief time it survives in aquaria) it spends its time lying embedded in the clay or mud with the tip of the snout projecting. And it is at home only in comparatively low temperatures, cooler probably, than 50°, which confines it in summer to depths of 15 to 20 fathoms or more in the Gulf of Maine. It is not a true parasite, as has sometimes been suggested, their being no reason to believe it ever attacks living, uninjured fish, but is a scavenger.

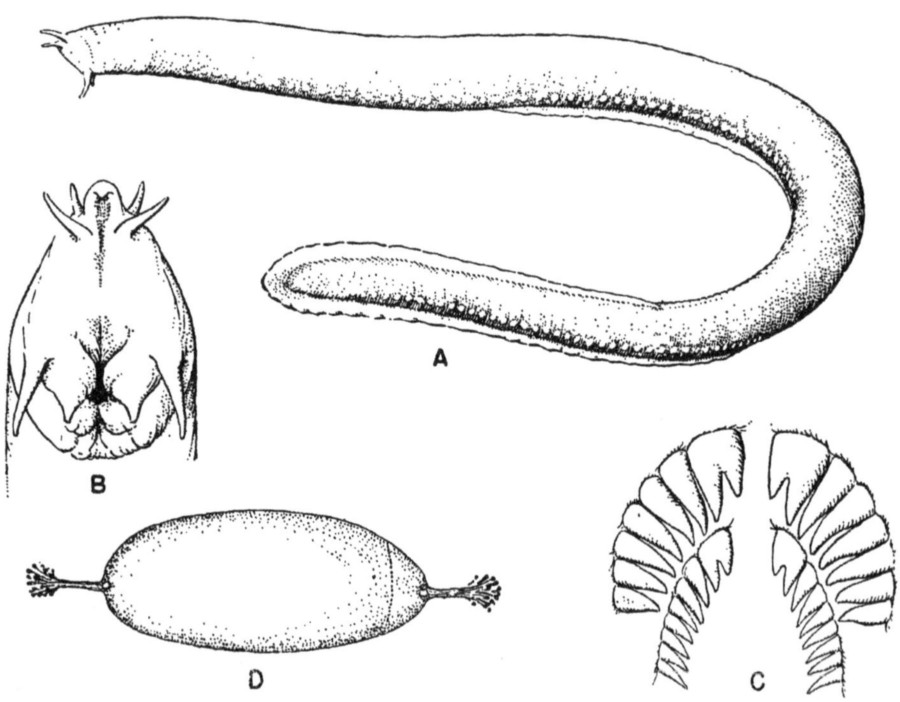

FIGURE 2.—Hagfish (*Myxine glutinosa*). A, adult, Gulf of Maine, from Bigelow and Schroeder, drawing by E. N. Fischer. B, lower view of head of same; C, tongue-teeth of same as seen from above, about 3 times natural size; D, egg, after Dean, about 2 times natural size.

Being blind, it doubtless finds its food by its greatly specialized olfactory apparatus. It feeds chiefly on fish, dead or disabled, though no doubt any other carrion would serve it equally well. And it is known to prey on marine annelid worms also, at least in Norwegian waters. It is best known for its troublesome habit of boring into the body cavities of hooked or gilled fishes, eating out the intestines first and then the meat, and leaving nothing but a bag of skin and bones, inside of which the hag itself is often hauled aboard, or clinging to the sides of a fish it has just attacked. In fact, it is only in this way, or entangled on lines, that hags ordinarily are taken or seen.

Being worthless itself, the hag is an unmitigated nuisance, and a particularly loathsome one owing to its habit of pouring out slime from its mucous sacs in quantity out of all proportion to its small size. One hag, it is said, can easily fill a 2-gallon bucket, nor do we think this any exaggeration.

In American waters the commercial fishes most often damaged by it are haddock and the hakes (*Urophycis*), these being the species most often fished for with long lines or with gill nets over the type of bottom the hag frequents. But it sometimes damages cod also, and European authors describe it as attacking ling (*Molva*) and other members of the cod tribe, herring, mackerel, sturgeon, and even mackerel sharks under similar circumstances.

Breeding habits.—The hag was formerly believed to be a functional hermaphrodite, with its single sex organ first developing sperm in the posterior portion, eggs later in the anterior portion. However, recent detailed studies of the sex organ appear to show that such is not the case, but that either the male portion of the common sex organ matures in a given individual with the female portion remaining rudimentary, or vice versa.[13]

It has long been known that the eggs are few in number (only 19 to 30 having been counted in any one female) and large (up to 25 mm. in length), and the horny shell has a cluster of anchor-tipped filaments at each end that make the eggs easy of identification. Until 1900 none had been found that certainly had been laid naturally. In that year, however, hag eggs were reported from the western part of Georges Bank and from the south coast of Newfoundland by Dean (1900);[14] from the neighborhood of the Faroe Islands by Jensen;[15] from Norway by Hjort;[16] off Morocco by Koefoed.[17] And they have been reported subsequently from the Bay of Fundy by Huntsman, from Frenchman Bay on the coast of Maine by Conel.[18] The eggs are deposited on bottom, where they stick firmly to fixed objects of one sort or another by their terminal filaments and by threads of slime.

The hag spawns throughout its range; also it spawns throughout the year, for females nearing ripeness and others nearly spent have been recorded for winter and spring, as well as summer and autumn, in one part of its range or another. The few eggs so far reported have been from depths of 50 to 150 fathoms, most of them trawled on mud, clay, or sand bottom.

We need only add that, to judge from their behavior in aquaria, the females cease to feed at the approach of sexual maturity, as many other fishes do. Newly hatched hags have never been seen, but inasmuch as the smallest yet described (about 2½ inches long), probably not long out of the egg, already resembled the adult in external appearance there is no reason to suppose that the hag passes through a larval stage greatly different from the adult.

General range.—Arctic seas, and both coasts of the north Atlantic; Murman Coast and northern Norway south regularly to the Irish Sea, and to Morocco as a stray in the East; northern part of Davis Strait, south to the latitude of Cape Fear, N. C., in the west. It is represented in the corresponding temperature-belt of the Southern Hemisphere by a form (or forms) resembling it so closely that it is doubtful whether any sharp line can be drawn between them.

Occurrence in the Gulf of Maine.—Apart from one record for the northern part of Davis Strait, the most northerly reports of the hag off the American coast are from southern Newfoundland and from the Grand Banks.[19] But it is generally distributed along outer Nova Scotia at appropriate depths. And it is only too common in the Gulf

[13] See Bigelow and Schroeder, Fishes Western North Atlantic, Pt. 1, ch. 2, 1948, pp. 35–36, for references.

[14] Mem. N. Y. Acad. Sci., vol. 2, Pt. 2, Art. 2, 1900.

[15] Viden. Meddel. Dansk naturhist. Forening, 1900, p. 1.

[16] Rept. Norwegian Fishery and Mar. Invest., vol. 1, 1900, No. 1, ch. 4, p. 75.

[17] Rept. Michael Sars North Atlantic Exped., Zool., vol. 4, No. 1, 1927, p. 18

[18] Science, N. Ser., vol. 75, 1932, pp. 19–20.

[19] It has not been reported for certain from West Greenland (so far as we can learn), from the outer coast of Labrador, or within the Gulf of St. Lawrence though it is to be expected in the deeper parts of the latter.

of Maine; perhaps it is not absent there from any considerable area of smooth bottom. Thus, it is abundant off the north end of Grand Manan; is reported from Passaamquoddy Bay and from various localities near Eastport; is to be found off-shore on muddy bottom all along the Maine coast; and is caught at times in considerable numbers on the Boon Island-Isles of Shoals fishing grounds and about Jeffreys Ledge, where we found it plentiful enough in the spring of 1913 to have gutted 3 to 5 percent of all the haddock in the gill nets. Fishermen report it as equally numerous in the deeper parts of Massachusetts Bay. On the offshore banks the hag is well known, and it has been trawled at various localities along the outer edge of the Continental Shelf off New England at depths of from 100 to 200 fathoms, and deeper. We ourselves took 11 large ones in one set of a Monaco deep-sea trap in 260 fathoms off Nantucket on July 9, 1908, and it has been taken in from 300 to 500 fathoms off Marthas Vineyard; as deep as 524 fathoms on the southeast slope of Georges Bank.

Sea lamprey *Petromyzon marinus* Linnaeus 1758

LAMPREY; SPOTTED LAMPREY; LAMPER; EEL-SUCKER

Bigelow and Schroeder, 1948, p. 46.

Description.—Lampreys are eel-like in appearance, but have a soft, cartilaginous skeleton. They lack paired fins but have well developed dorsal and ventral finfolds. In the adult the jaws are so rudimentary that apparently they are wanting; the mouth is a longitudinal slit when closed, but forms an elliptical disk at the tip of the snout when open, and is armed with many horny, hooked teeth arranged in numerous (11 to 12) rows, the innermost the largest. There are two dorsal finfolds, and there are seven open gill slits on each side, whereas the hag has only one gill pore on each side, and only one fin. The sea lamprey (the only member of its group known from our salt waters) can hardly be mistaken for any other fish, its eel-like appearance coupled with two dorsal fins and the jawless mouth placing it at a glance.

Color.—Small specimens (whether on their way downstream or in salt water) are white below and uniformly colored above, usually described as blackish blue, or as lead colored, and more or less silvery. But large specimens usually are olive brown above, or of varying shades of yellow-brown, green, red, or blue, mottled with a darker shade of the same color, or sometimes nearly black if the dark patches are confluent. The lower surface is whitish, gray, or of a pale shade of the same hue as the ground color of the back. During breeding season, the landlocked form takes on more brilliant hues, with the ground tint turning bright yellow.

Size.—The length at the time of transformation from the larval stage is about 4 to 8 inches (100–200 mm.). Sexually mature individuals, taken

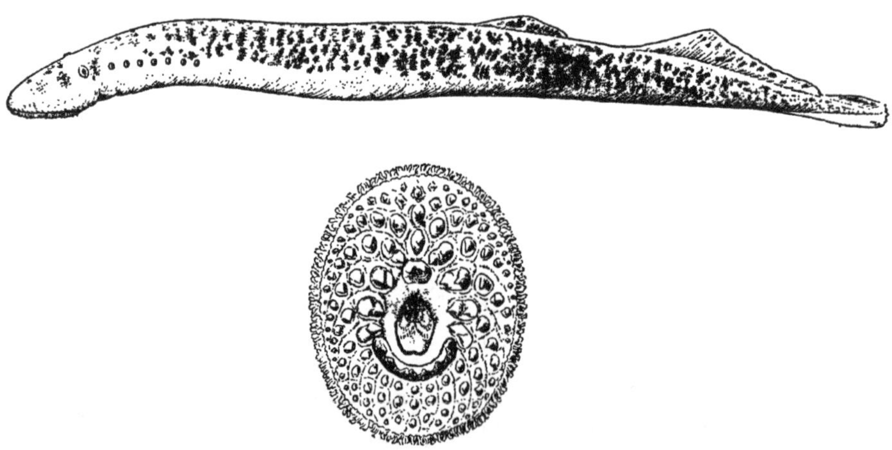

FIGURE 3.—Sea lamprey (*Petromyzon marinus*), about 18 inches long, Merrimac River; and open mouth disc of another Merrimac River specimen to show the arrangement of the horny teeth, about 0.9 times natural size. From Bigelow and Schroeder. Drawings by E. N. Fischer.

in American rivers, average 2 to 2½ feet long, up to a maximum of about 3 feet. One of 33 inches weighed 2¼ pounds.

Habits.—It has been known from early times that the sea lamprey breeds in fresh water. However, it does not enter all the streams within its range indiscriminately. As an illustration, we may cite outer Nova Scotia and the Bay of Fundy, where lampreys run in the St. Marys, Sackville, Annapolis, Shubenacadie, Petit Codiac, and St. Johns Rivers, but not in the Moser or Apple Rivers, although these last also are "salmon" rivers. Their requirements are a gravelly bottom in rapid water for their spawning beds, with muddy or sandy bottom in quiet water nearby, for the larvae.

In many small streams, and in larger ones if these are blocked by dams or high falls, they may spawn only a short distance upstream; even within the influence of the tide, although invariably in fresh water. But they are able to ascend falls, if these are not too steep and high, by clinging to the rocks by their oral discs and resting. And they may run upstream for very long distances in large rivers, as they did formerly in the Merrimac and probably still do in the St. Johns River. They are still to be found 200 miles or more from the sea in the upper tributaries of the Delaware and Susquehanna systems.

Since the breeding activities of the sea lamprey take place in fresh water, a brief account will suffice here. As the two sexes ripen, the males develop a strong ridge along the back, the females a crestlike fin between the anus and the caudal fin. Spawning, commencing when the temperature of the water is about 50° F. (10° C.) is completed by the time it has warmed to about 68°–70° (20°–21° C.), and a sea lamprey has been found to contain 236,000 ova. Working in pairs, sometimes with a second female assisting, they make depressions 2 to 3 feet in diameter and about 6 inches deep in the stream bed in stretches where the bottom is stony or pebbly, dragging the stones downstream in a pile with their suckerlike mouths. And they are able to move stones as large as one's fist. It is in these depressions that the eggs are deposited, not among the piles of discarded stones that have often been described as "nests." It seems that they all die after spawning; not only have they often been found dead, but their intestines atrophy, they are attacked by fungus, and they become so debilitated that recovery seems out of the question.

The larvae are different in appearance from the adults: blind, toothless, with mouths and fins of different shape. They continue in this state for a period estimated as 3 to 4 years, during most of which time they live in burrows in the mud or sand, or hide under stones. They are abundant in the mud of flats near the mouths of small tributary streams of river systems such as the Delaware and Susquehanna, where lampreys still breed in large numbers, and they subsist on minute organisms. At the end of this larval period, when they have grown to a length of 4 to 6 inches, they undergo transformation to the adult form and structure, an event occupying about two months, August to September or October. They run down to the sea in November or December, to live and grow there for one or two years, so that large ones, not yet mature, are to be found in salt water all the year round.

Little is known of the habits of the lampreys while they live in the sea further than that their mode of life centers around a fiercely predaceous nature. Judging from their land-locked relatives and from the occasions on which they have been found fastened to sea fish, they must be extremely destructive to the latter, which they attack by "sucking on" with their wonderfully effective mouths. The lamprey usually fastens to the side of its victim, where it rasps away until it tears through the skin or scales and is able to suck the blood. Its prey sucked dry, it abandons it for another. Probably lampreys are parasites and bloodsuckers pure and simple, for we cannot learn that anything but blood has been found in their stomachs, except fish eggs, of which lampreys are occasionally full.[20]

In salt water they have been found preying on mackerel, the various anadromous herrings, cod, haddock, American pollock (*Pollachius*), salmon, basking sharks, swordfish, hake (*Urophycis*), sturgeons and eels. Sometimes as many as three or four are fast at one time to a single shad, and they are said to be exceedingly aggressive in their attacks on other fishes. Occasionally they are found fast to driftwood, even to boats. When not clinging to anything they are strong, vigorous swimmers, progressing by an undulating motion.

[20] Goode, Fish. Ind. U. S., Sect. I, 1884, p. 677.

General range.—Atlantic coasts of Europe and of North America; from the west coast of Greenland to Florida in the western side of the Atlantic; from northern Norway to the Mediterranean in the eastern;[21] running up fresh rivers to breed, and landlocked in certain American lakes.

Occurrence in the Gulf of Maine.—No doubt the sea lamprey occurs along the whole coast line of the Gulf of Maine, for it is recorded in or at the mouths of numerous rivers and streams in Nova Scotia, New Brunswick, Maine, and Massachusetts; specifically in the St John, Annapolis, Petit Codiac, and Shubenacadie Rivers and from the St. Andrews region in salt water in the Bay of Fundy; from Eastport, Bucksport, Casco Bay, and the Presumpscott, Kennebec, and Penobscot Rivers in Maine; from the Merrimac River system; from the Exeter and Lamprey Rivers, tributaries of Great Bay, New Hampshire; and from the Parker River in northern Massachusetts.

Since lampreys never take the hook or are captured in nets except on rare occasions they are seldom seen in salt water; only when running up our rivers are they familiar objects. But they have been taken as far offshore as the seaward slopes of Banquereau, Sable Island, and LaHave Banks off Nova Scotia; on Browns Bank; in the deep gully between the latter and Georges Bank, and over the continental slope off Nantucket and off Marthas Vineyard.

Lampreys have long been known to run up New England rivers a little earlier in the spring than shad, perhaps beginning to work upstream as early as the beginning of April or even the end of March. In the rivers tributary to the Gulf of Maine the runs are at their peak during May and early June, with few, if any, entering later than that. The larvae have been reported by Doctor Huntsman as plentiful in the Shubenacadie (emptying into the Bay of Fundy) and no doubt they are to be found in the Merrimac system, in the Exeter River, and in other Gulf of Maine streams.

Abundance.—The construction of impassible dams has sadly reduced the numbers of lampreys in the larger rivers of New England. In the Merrimac, for example, once a famous lamprey river,[22] so few now succeed in surmounting the succession of dams that a recent survey yielded no evidence of any now having access to the upper reaches. Some lampreys, however, are said to breed in the river below the Lowell dam;[23] we have seen what resembled their "nests" in the Squannacook, a branch of the Nashua tributary to the Middle Merrimac, and they still continue numerous in some Gulf of Maine streams where they can reach suitable spawning grounds without too great difficulty. We may quote catches of up to 119 recently in the Shubenacadie, where larvae also have recently been reported in abundance,[24] and of more than 100 each on several occasions in the Exeter River,[25] where they are familiar spectacles, as they gather at the falls at Exeter, N. H. But we ought perhaps to caution the reader that while lampreys, like other anadromous fishes, may seem plentiful when condensed between the narrow bounds of a river's banks, their numbers as a whole do not rival those of the more abundant of the salt-water fishes.

Importance.—Lampreys were esteemed a great delicacy in Europe during the middle ages (historians tell us Henry I of England died of a surfeit of them) and considerable numbers were captured of old in the rivers of New England for human food, particularly in the Connecticut and Merrimac Rivers. But the lamprey fishery has been scarcely more than a memory for 40 years past except locally and in a small way for home consumption, or to supply the needs of biological laboratories. In the salt water of the Gulf of Maine the lamprey has never been of any commercial importance; the average fisherman might not see one in a lifetime, nor is there any sale for the few that are picked up by chance. But larvae are taken in considerable numbers for bait in the Susquehanna River, and perhaps elsewhere along the middle Atlantic coast.

---

[21] Also reported from "West Africa" by Günther, Cat. Fishes British Museum, vol. 8, 1870, p. 502.

[22] For an account of the lamprey fishery in New England during the first half of the 19th century, see Goode, Fish. and Fishery Ind. U. S., Sect. 1, 1884, p. 680.

[23] Bailey, Biol. Survey Merrimack Watershed, New Hampshire Fish and Game Dept., 1938, p. 155.

[24] Information gathered for us by Dr. A. G. Huntsman.

[25] Collected for the Biological Laboratory, Harvard University.

CARTILAGINOUS FISHES. CLASS CHONDRICHTHYES

The Shark and Skate Tribes, and the Chimaeroids

These are fishlike vertebrates with well-developed fins and teeth, and with 2 pairs of fins, one of them supported by the pectoral girdle, the other by the pelvic girdle. Their most distinctive character, as contrasted with the bony fishes (p. 80) is that their entire skeleton, including the skull, is cartilaginous, without any true bone, though it is partly calcified, especially in the vertebrae; the skull is far simpler than it is among the bony fishes; the gill filaments are attached throughout their lengths to the partitions between the gill openings instead of being free; and the rear portion of the digestive tract is modified into the so-called "spiral valve" by the development of a special fold from its lining layer, which only a few bony fishes have.

Fertilization is internal in all of them, and is effected by a pair of rodlike copulatory organs, each of which is developed from the inner edge of one of the two pelvic fins, and is supported by one or more cartilages.

The sharks and rays are usually looked upon as more primitive than the bony fishes.

SHARKS, TORPEDOES, SKATES, AND RAYS. SUBCLASS ELASMOBRANCHII

The most obvious external character by which all the sharks, skates, and rays are distinguishable from all of the bony fishes is that they have five or more gill openings on either side of the head, instead of only one. They recall the lampreys in this respect, but it is a commonplace that their jaws and teeth are extremely well-developed. Their skins are tough, and are studded in most of them with denticles (placoid scales), which are not homologous with the scales of bony fishes, for both dermis and epidermis take part in their formation, instead of the dermis alone. The teeth of the sharks and rays represent placoid scales that are modified and are embedded in the gums alone, not in the jaws. The fins are supported at their bases by segmented cartilaginous rods, supplemented in all of the sharks, and in some of the rays by numerous slender horny fibers further out, instead of by rays or spines of the sorts that are seen in the bony fishes. All of their fins are covered with the same leathery skin that clothes the body. Among sharks the tail is uneven ("heterocercal"), with the vertebral column extending out into its upper lobe, but it is whip-like in most of the skates and rays, with no definite caudal fin. The torpedo is an exception to this rule.

The modern representatives of the subclass may be grouped in two orders, the one (Selachii) to include all living sharks, the other (Batoidei) to include the sawfishes, the skates and the rays. They are separated one from the other by the following external differences, and there are skeletal differences between them as well:[26]

1. The gill openings are at least partly on the sides; the edges of the pectoral fins are not attached to the sides of the head in front of the gill openings; the upper edges of the orbits are free from the eyeballs, so that they form free eyelids................................Sharks, (p. 15).
The gill openings are entirely on the lower surface; the edges of the pectoral fins are attached to the side of the head in front of the gill openings; the upper edges of the orbits are attached to the eyeballs so that they do not form free eyelids........Sawfishes, skates and rays, (p. 57).

[26] For further discussion, see Bigelow and Schroeder, Fishes Western North Atlantic, Pt. 1, ch. 3, 1948, p. 64.

Sharks. Order Selachii

Sharks always are objects of interest, not only to fishermen and mariners but to seaside visitors generally, because of their evil appearance, their ferocity, the large size to which some of them grow, the destruction they wreak on fishermen's nets and lines as well as on the smaller fishes on which they prey, and because of the bad reputation certain kinds have earned as maneaters.

The Gulf of Maine is not particularly rich in sharks (very poor indeed compared with our southern coasts), for while the number of species actually recorded there is considerable (indeed any high-seas shark might stray thither) the little spiny dogfish alone is numerous in the sense in which this term is applied to the various commercial fishes. And only two of the larger species, the mackerel shark (*Lamna nasus*), and the blue shark (*Prionace glauca*), occur with us in numbers sufficient for one to be fairly sure of seeing them during a summer's boating off the coast north of Cape Cod.

With the larger sharks generally so scarce (the mackerel shark is harmless to anything larger than the fishes on which it feeds, and the blue shark is also harmless, although better armed), the danger of attacks on bathers is negligible in our Gulf. Indeed, not a single well-authenticated instance of the sort is on record [27] for the past 80 years for the coast north of Cape Cod, though the beaches are crowded every summer with vacationists. But as long as the white shark or man-eater (*Carcharodon carcharias*) does stray occasionally into the Gulf (p. 26), it is always remotely possible that we may be horrified some summer by the news of tragedies such as occurred on the New Jersey coast in July 1916, when several persons were killed or injured, presumably by a small shark of this species that was captured nearby a few days later,[28] and near Mattapoisett, on Buzzards Bay, Mass., on July 25, 1936, when a swimmer was fatally injured by a shark, species not determined.[29]

[27] In 1830 (an event often quoted) one Joseph Blaney, fishing from a small boat in Massachusetts Bay off Swampscott, Mass., was attacked by some fish that was seen to overset and sink his boat and, presumably, devoured him, for neighboring fishermen who hastened to his rescue found no trace of him. Whether his attacker was a large shark or a killer whale is an open question.

[28] Murphy and Nichols (Brooklyn Mus. Quart., vol. 3, 1916, No. 4, pp. 145–160) give a detailed account of this occurrence.

[29] See Gudger (Amer. Midland Natural., vol. 44, 1950, p. 714) for clinical details of this case.

All Gulf of Maine sharks give birth to young that are not only practically adult in structure but of relatively large size at birth, and there is a placental connection between mother and embryo in some, but not in others. Still other sharks lay eggs; this is true of the chain dogfish (*Scyliorhinus retifer*, p. 34), which is common out on the continental shelf from the offing of Cape Cod, southward, and of its immediate relatives; also of the heterodontids or Port Jackson sharks which are not represented in the Atlantic.

There is so little market for sharks in Gulf of Maine ports (attempts to introduce the dogfish as a food fish having failed so far) that the amounts landed in Maine and Massachusetts were only about 240,000 pounds in 1947, and about 309,500 pounds in 1949; they interest fishermen chiefly as nuisances because of the damage they do to nets and other gear, except that mackerel sharks are marketable.

It is possible to identify all the sharks so far known from the Gulf (and this includes all that are likely to occur there except strays) by the sizes and relative locations of the fins, and by such tooth characters as may be seen at a glance at the open mouth or easily felt with the finger (after the shark is dead!).

We have attempted in the following descriptions of the several species to include only such features as will tell what shark is at hand; for more minute particulars we refer the reader to our account of the sharks of the western North Atlantic (p. 2).

Key to Gulf of Maine Sharks

1. There is an anal fin ---------- 2
   There is no anal fin ---------- 16
2. Head greatly expanded sidewise, at level of eyes, in hammer- or shovel-form ---------- 3
   Head of ordinary shape, with rounded or pointed snout ---------- 4
3. Outline of front of head only slightly concave opposite nostrils if at all so; grooves (if any) from nostrils shorter than horizontal diameter of eyes; free tip of second dorsal fin is not longer than forward margin of the fin; rear margin of anal fin is only weakly concave; teeth near outer corners of mouth are rounded, without sharp cusps. Shovel head, p. 44
   Outline of front of head is deeply indented opposite each nostril; grooves from nostrils are more than twice as long as horizontal diameter of eye; free tip of second dorsal fin is considerably longer than front margin of the fin; rear margin of anal fin deeply concave; teeth near corners of mouth are like those near center of mouth, with sharp cusps ---------- Hammerhead, p. 45
4. Caudal peduncle (root of tail) is not widely expanded sidewise as a lateral keel on either side; upper lobe of caudal fin is much longer than lower lobe ---------- 8
   Caudal peduncle is widely expanded sidewise as a lateral keel on either side; lower lobe of caudal fin is nearly as long as upper lobe, suggesting the caudal fin of a mackerel or swordfish ---------- 5

5. Gill openings very large, the first pair nearly meeting below the throat; teeth tiny, many hundred in number; gill arches with numerous horny gill rakers directed inward-rearward ---------- Basking shark, p. 28
Gill openings, confined to sides of head; teeth large, few in number; gill arches do not have horny gill rakers ------ 6
6. Upper teeth broadly triangular, with serrate edges; anal fin is entirely behind second dorsal fin ---------- White shark, maneater, p. 25
Upper teeth with smooth-edged cusp, with or without a denticle on either side, at the base; anal fin is not entirely behind second dorsal fin ---------- 7
7. First two teeth from center in each jaw are similar to the succeeding teeth; origin of first dorsal fin is over or in front of inner corner of pectoral fin when latter is laid back; forward part of caudal fin has a small secondary lateral keel on each side, below the primary keel formed by the lateral expansion of the caudal peduncle. Mackerel shark, p. 20
First two teeth from center in each jaw are noticeably more slender and more flexuous than the succeeding teeth; origin of first dorsal fin is behind inner corner of pectoral fin when latter is laid back; forward part of caudal fin does not have a secondary longitudinal keel ---------- Sharp-nosed mackerel shark, mako, p. 23
8. Upper lobe of caudal fin is nearly or quite as long as head and body combined ---------- Thresher, p. 32
Upper lobe of caudal is less than one-half as long as head body combined ---------- 9
9. Second dorsal fin is nearly as high vertically as first dorsal fin ---------- 10
Second dorsal fin is less than one-half as high vertically as first dorsal fin ---------- 12
10. First dorsal fin is wholly or mostly forward of the origin of the pelvic fins ---------- 11
First dorsal fin is wholly posterior to bases of pelvic fins ---------- Chain dogfish, p. 34
11. Teeth high, narrow, sharp pointed, not in mosaic arrangement; snout conical; fifth gill openings well in front of pectoral fins ---------- Sand shark, p. 18
Teeth small, low, rounded, in mosaic arrangement; snout flat, broadly rounded in front; fifth gill openings are behind origins of pectoral fins ---------- Smooth dogfish, p. 34
12. Origin of first dorsal fin far behind inner corner of pectoral fin; upper surface brilliant blue in life. Blue shark, p. 38
Origin of first dorsal fin is over or anterior to inner corners of pectorals; ground color of upper surface is gray, brownish or dusky in life, not bright blue ---------- 13
13. Length of snout in front of mouth is not more than one-half as great as breadth of mouth; upper jaw has a furrow on either side extending from outer corner forward past level of eye; caudal peduncle with a low longitudinal keel on either side; upper and lower teeth are of shapes shown in figure 11; their margins coarsely serrate. Tiger shark, p. 37
Length of snout in front of mouth is more than two-thirds as great as breadth of mouth; furrows on upper jaw, if any, do not extend forward-inward as far as level of eye; caudal peduncle without longitudinal ridges; teeth are not of shape shown in figure 11, their margins either only very finely serrate or smooth ---------- 14
14. Outer corners of mouth have a short "labial furrow" extending inward-forward along each jaw; teeth are alike in the two jaws, directed sharply outward, margins of upper teeth smooth, as well as those of lower teeth. Sharp-nosed shark, p. 40
Outer corners of mouth have no labial furrow on lower jaw and upper labial furrow is so short as to be hardly noticeable; teeth directed only moderately outward, their margins only finely serrate; lowers noticeably more slender than uppers ---------- 15
15. Origin of first dorsal fin is about over inner corner of pectoral when latter is laid back; vertical height of first dorsal fin is less than distance from eye to first gill opening ---------- Dusky shark, p. 41
Origin of first dorsal is about over axil (armpit) of pectoral, its vertical height (after birth) is at least as great as distance from eye to third gill opening ---------- Brown shark, p. 43
16. Trunk much flattened dorso-ventrally; eyes on top of head; front margins of pectorals overlap the gill openings. Angel shark, note, p. 18
Trunk subcylindrical; eyes on side of head; front margins of pectorals do not overlap the gill openings ---------- 17
17. Each dorsal fin is preceded by a stout and conspicuous spine ---------- 18
Dorsal fin-spines either lacking, or are so nearly concealed in the skin that their presence can be detected by touch only ---------- 20
18. Upper teeth with 5 erect cusps; lower teeth with only one cusp, the successive cusps directed outward, forming a nearly continuous horizontal cutting edge all along the jaw ---------- *Etmopterus princeps*, p. 47
Upper and lower teeth are alike in shape ---------- 19
19. Upper teeth quadrangular as well as lower teeth, with one cusp directed outward, forming a nearly continuous horizontal cutting edge along each jaw ---------- Spiny dogfish, p. 47
Upper and also lower teeth each have 3 to 5 erect, triangular cusps ---------- Black dogfish, p. 51
20. First dorsal fin well in advance of pelvic fins; upper teeth noticeably different in shape from lower teeth ---------- 21
First dorsal fin stands over posterior part of bases of pelvic fins; upper are teeth similar to lower teeth in shape. Bramble shark, p. 56

21. Lower teeth erect, triangular, their edges serrate ---------- *Dalatias licha*, p. 55
Lower teeth quadrate, the cusp directed outward, forming a nearly continuous horizontal cutting edge; their outer margins deeply notched, the edges smooth ---------- 22
22. Dermal denticles rounded, overlapping, scale-like, entirely concealing the skin (fig. 20); each dorsal fin is preceded by a short spine, embedded nearly to its tip in the skin, but recognizable by touch ----- Portuguese shark, p. 52
Dermal denticles conical, only moderately close set, the skin visible between them; dorsal fins not preceded by spines ---------- Greenland shark, p. 53

NOTE.—Not yet known from the Gulf of Maine though reported from Marthas Vineyard.

THE SAND SHARKS. FAMILY CARCHARIIDAE

Outstanding characteristics of the sand sharks are that they have an anal fin; the two dorsal fins are without spines and are nearly equal in size; the rear end of the base of the first dorsal is over or in front of the origin of the pelvic fins; the anal fin is about as large as the dorsals; the upper lobe of the caudal fin is much longer than the lower, but occupies not more than one-third of the total length of the fish; there are no lateral keels on the caudal peduncle; the fifth gill openings are farther forward than the origins of the pectoral fins; and the teeth are slender and sharp-pointed.

Sand shark *Carcharias taurus* Rafinesque 1810

DOGFISH SHARK; GROUND SHARK

Bigelow and Schroeder, 1948, p. 100.
Garman, 1913, pl. 6, figs. 1–3.

Description.—The large size of the second dorsal fin, and of the anal as well (which is about equal to the first dorsal instead of much smaller) is of itself enough to distinguish this species from all other Gulf of Maine sharks. The fact that the first dorsal fin is located but little in front of the pelvics, and that the trunk seems crowded with fins of equal size, is a useful field mark. We may also point out that the pectoral fins are not much larger than the other fins—triangular rather than sickle-shaped; that the upper lobe of the tail is nearly one-third as long as head and body together and notched near its tip, with the lower lobe about one-third as long as the upper lobe; and that the head is flat above, the snout short, conical with rather sharp tip. The teeth also (alike in the two jaws) are diagnostic, being long, narrow, sharp-pointed, and smooth-edged, with one (rarely two) small spurs ("denticles") on either side near the base.

Size.—Most of the sand sharks that are caught in the northern part of their American range, from Delaware Bay to Cape Cod, are immature, of perhaps 4 to 6 feet. But adults up to 8 or 9 feet long are reported there from time to time, especially from the vicinity of Nantucket, where a commercial shark fishery yielded many of them in

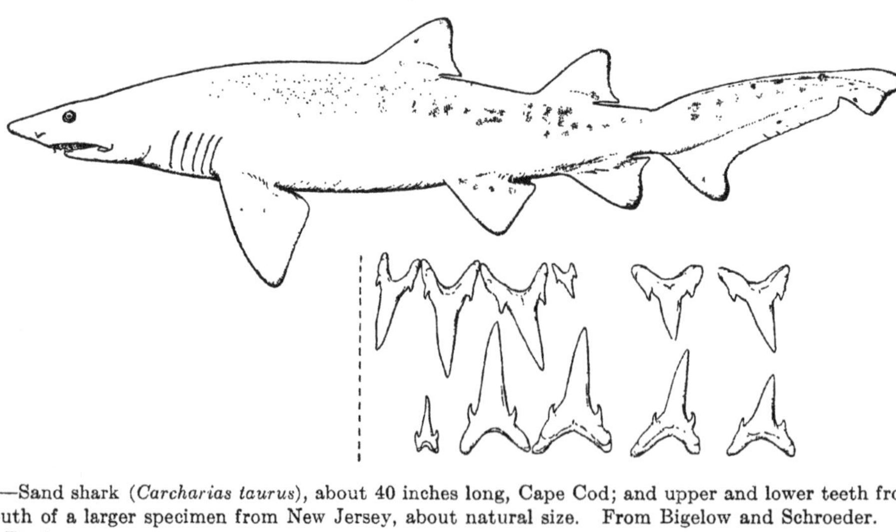

FIGURE 4.—Sand shark (*Carcharias taurus*), about 40 inches long, Cape Cod; and upper and lower teeth from front part of mouth of a larger specimen from New Jersey, about natural size. From Bigelow and Schroeder. Drawings by E. N. Fischer.

the early 1920's. And large ones, alone, have been reported from North Carolina, southward. The greatest recorded length is 10 feet 5 inches, from southwestern Florida. And the sand shark does not mature sexually until perhaps 7 feet long, or more. A weight of 250 pounds is recorded for one 8 feet 10 inches long, showing how much lighter a fish this is, length for length, than various other sharks.

Color.—Light gray-brown above, darkest along back, snout, and upper sides of pectorals, paling on the sides to grayish white on lower surface; sides of trunk rearward from pectorals variously marked with roundish to oval spots, of which there may be upwards of 100, varying in color from yellowish brown to ocher yellow. The rear margins of the fins are edged with black on some specimens, but not on others.

Habits and food.—Despite its trim appearance and voracious appetite, this is a comparatively sluggish shark, living mostly on bottom or close to it; more active and taking a bait more freely at night than by day. During its summer visits to the New England coast it holds so close to the coast that it has never been reported from Georges Bank, or from the outer part of the Continental Shelf. Most of those caught are from depths not greater than 1 to 5 fathoms, occasionally perhaps as deep as 10 fathoms, and many come right in to tide line along the beaches. They may sometimes be seen moving slowly to and fro at the surface, over bars, with dorsal and caudal fins showing above the water; and they sometimes enter the mouths of rivers. They capture great numbers of small fish, which are their chief diet, particularly menhaden, cunners, mackerel, skates, silver hake, flounders, alewives, butterfish, and south of Cape Cod, scup, weakfish, and bonito. Sand sharks have been seen surrounding and harrying schools of bluefish; they have even been known to attack nets full of bluefish, which gives a measure of their voracity. They also eat lobsters, crabs, and squid

Breeding.—The eggs of the sand shark are hatched within the parent and are retained there until the resultant young are ready for independent existence, but there is no placental connection between mother and developing embryo. It has recently been discovered that while a ripe female contains a large number of eggs, only two embryos develop as a rule, one in each oviduct; they are nourished (at least largely) by swallowing the unfertilized eggs [30] with which the stomach of the embyro becomes greatly distended. Females with large embryos have so far been reported only from Florida and from Louisiana, whereas others taken near Woods Hole have contained eggs only, making it likely that the small specimens that are so common along southern New England have come from a more southerly birthplace.

General range.—Coastal waters on both sides of the Atlantic; Maine to Florida and Brazil in the west; Mediterranean, tropical West Africa, Canaries, and Cape Verdes in the east; also South Africa; represented in Argentine waters and in the Indo-Pacific by close relatives.

Occurrence in the Gulf of Maine.—The sand shark is by far the most common of its tribe, next to the smooth and spiny dogfishes, along southern New England and at the westerly entrance to the Gulf of Maine. It is plentiful at Woods Hole from June to November, to be found anywhere in that region in shoal waters, even coming up to the wharves. At Nantucket, too, it is so abundant that shark fishing, with the sand shark as the chief objective, is a popular sport. The facts that a catch of about 1,900 sharks by three boats on Horseshoe Shoal, in Nantucket Sound, June to September 1918, was mostly of this species, as was another catch of 350 sharks, taken near Nantucket in the early 1920's, illustrate their numbers there. Scattered sand sharks are also caught along the outer beaches of Cape Cod by surf anglers (published records are for Monomoy, Chatham, and Provincetown) and there are enough of them along this stretch of beach in some summers (1951 was a case in point) for them to be a nuisance to anglers casting for striped bass in the surf at night.

In August 1947 we saw a large one at the surface pursuing a striped bass, that was being hauled aboard a fishing boat on a hand line, in the eastern side of Cape Cod Bay, where fishermen tell us that this is not an unusual happening. But this appears to be the northern boundary to their occurrence in any numbers, or with regularity. True, they are recorded at Cohasset, on the southern shore of Massachusetts Bay, where we caught one about 4 feet long, years ago in Boston Bay, and at Lynn, Mass. But so rarely does it stray north of Cape Ann that it has been reported only

[30] For an account of the embryos, see Springer, Copeia, 1948, No. 3, pp. 153-156.

twice from Casco Bay, and once from St. Andrews, New Brunswick, near the mouth of the Bay of Fundy, its most northerly known outpost, where one was taken in a weir in 1913.

In New England waters the sand shark occurs only as a summer visitor. The winter home of those that summer along the northeastern United States is not known, nor has any increase been noted in Florida waters (where they are taken at all times of year) coincident with their winter disappearance from the northern part of their range. Like various bony fishes they may move offshore, and perhaps southward, to escape winter chilling.

Importance.—There were commercial fisheries for the sand shark around Nantucket during the first quarter of the present century, but these were short lived, reputedly because of exhaustion of the local stock. And the sand shark is of no commercial importance on the New England coast at present. Westward from Cape Cod it is of some interest to anglers, who catch considerable numbers, both as objects of special pursuit, for it takes almost any natural bait readily, or incidentally while surf casting for better fish. But it is not plentiful enough in the Gulf of Maine to be worth fishing for.

There is no record of attacks by sand sharks on human beings in North American waters, though bathers often come close to them. Our own experience bears this out; in fact, it is looked upon as a harmless nuisance on the New England coast wherever it is plentiful enough to be familiar. But its relative (or relatives) of East Indian waters have a more sinister reputation.

MACKEREL SHARKS. FAMILY ISURIDAE

Sharks of this family are easily recognizable by the very firm half-moon-shaped (technically lunate) caudal fin, with lower lobe but little shorter than the upper, in combination with large awl-like or blade-shaped teeth, and with gill openings larger than any other Gulf of Maine shark except the basking shark. Their tail fins, in fact, recall the tails of such bony fishes as the mackerel tribe or the swordfish, in outline, likewise in firm texture, hence their common name. The basking shark also has a caudal fin and peduncle of this same sort, but its teeth are minute and very numerous, and its gill openings are so long that those of the two sides nearly meet on the lower surface of the throat.

Other diagnostic features are that they have an anal fin; that their caudal peduncle is expanded as a prominent longitudinal keel on either side; that their dorsal fins are not preceded by spines; and that the inner margins of their gill arches do not have horny gill rakers.

Mackerel shark *Lamna nasus* (Bonnaterre) 1788

PORBEAGLE; BLUE DOG (IN GULF OF MAINE)

Bigelow and Schroeder, 1948, p. 112.
Garman, 1911, pl. 6, figs. 4–6 (as *Isurus punctatus*).

This is a stout, heavy-shouldered shark, tapering in front to a pointed conical snout and behind to a very slim tail root. Its dorsal and pectoral fins are large; the former, originating a little rearward of the armpits of the pectorals, is triangular and about as high as it is long; the pectoral fins are only half as broad as long. The second dorsal and anal fins are very small indeed, and the pelvics but little larger. The second dorsal fin stands over the anal. There is a conspicuous transverse furrow or pit on the upper surface of the root of the tail, also one on the lower surface close in front of the origin of the caudal fin. The lower lobe of the caudal fin is two-thirds to three-fourths as long as the upper lobe, and there is a small secondary keel on the base of the caudal fin on either side, below and behind the rear end of the primary keel formed by the sidewise expansion of the caudal peduncle.

The teeth of the porbeagle are alike in the two jaws, slender, pointed, smooth-edged, and with a sharp denticle near the base on each side (young fish may not have these) which the mako lacks (p. 23).

The only Gulf of Maine sharks with which the porbeagle might be confused are the maneater (p. 25), or the mako (p. 23). And it is easily told from the former by its slender, smooth-edged teeth, as well as by the position of its second dorsal fin directly over the anal; from the mako by the shape of its teeth (*cf.* fig. 5 with fig. 6), each usually with a small basal denticle on either side, which the mako lacks; also by its stouter body and by the presence of the secondary longitudinal keel on the anterior part of its caudal fin.

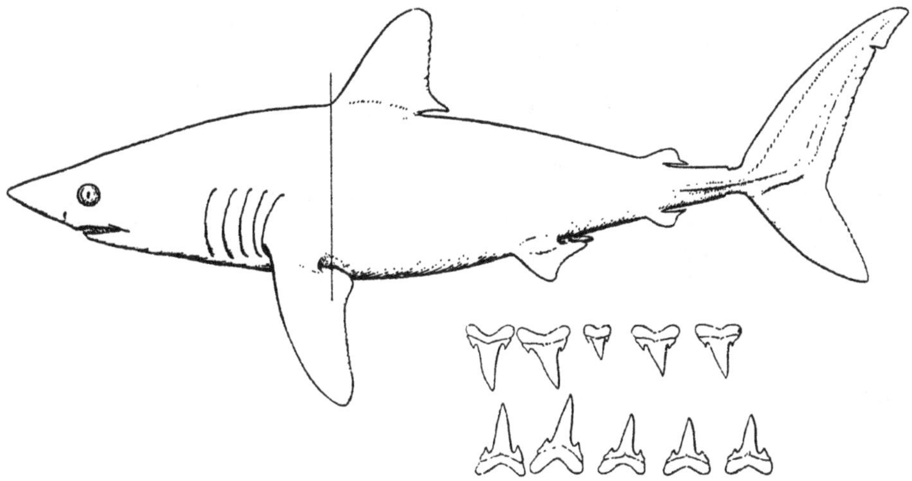

FIGURE 5.—Mackerel shark (*Lamna nasus*), about 37 inches long, Nahant, Massachusetts. Upper and lower first to fifth teeth from center of jaw of a larger specimen from Platts Bank, about 0.7 times natural size. From Bigelow and Schroeder. Drawings by E. N. Fischer.

Color.—Dark bluish gray to bluish black above, including the upper surfaces of the pectorals, changing abruptly, low down on the sides, to white below; lower surfaces of pectorals dusky to black on the outer one-half to one-third, more or less mottled white and dark toward their bases, and with the anterior and posterior edges narrowly rimmed with black; the anal is white or slightly dusky.

Size.—The common run of mackerel sharks in the Gulf of Maine are from 4 to 6 feet long, with few heavier than 200 pounds; thus 18 recently landed at Portland and Eastport, Maine,[31] averaged 4 feet 5 inches, the largest being about 8 feet long, the smallest 3 feet 7 inches.

Specimens longer than 7 to 8 feet are not common; only two longer than 8 feet have been recorded previously from the Gulf of Maine, one of which was 10 feet,[32] the largest recorded from either side of the North Atlantic. This shark has been said to reach a length of 12 feet. But the sizes of sharks often are overstated, unless actually measured, point to point, not around the curve of the body. Information as to the relationship between length and weight is restricted to a report of 305 pounds at 8 feet 3 inches, and of about 400 pounds at about 9 feet. One 3 feet long that we measured weighed 20 pounds.

[31] Scattergood, Copeia, 1949, No. 1, pp. 71-72.

[32] Hubbs, Copeia, No. 173, 1973, p. 101

Habits.—The whole mackerel-shark tribe lead a pelagic life, wandering about over the ocean in pursuit of the fishes on which they prey, and often uniting in small companies, though they can hardly be called gregarious. Like swordfish they spend much time at the surface on calm days, when their triangular back fins, followed by the tip of the caudal fin (the bluntness of the former and the wavy track of the latter identify the shark as such) may often be seen cutting through the water. We have sailed close to sharks probably of this species again and again, only to see them sound, just out of harpoon range, plainly visible at first but soon fading from sight as they swim downward.

The porbeagle has often been described as active and strong swimming. But it puts up only a very feeble resistance when hooked. We have never seen or heard of one jumping, as the mako often does (p. 24), nor is there any difficulty in landing one of 4 to 5 feet on an ordinary cod line. It is, in fact, as proverbial among fishermen for its sluggishness when hooked, as is the mako for its activity. While often seen "finning," many are caught close to the bottom, in depths down to 80 fathoms in the gill net fishery for ground fish that is carried on from Portland, Maine; some also on bottom on cod lines; how much deeper they may descend is not known.

Food.—In the Gulf of Maine the porbeagle feeds chiefly on mackerel and on the herring tribe; on butterfish; on ground fish, as cod, hake, cusk, rosefish, flounders, or other kinds available; and on squid. It has also the annoying custom of foraging on the cod and other fish that have been hooked on long lines and biting off the snoods. It is also known to prey on the spiny dogfish in the eastern Atlantic; probably in the Gulf of Maine also. But we find no record of its eating crustaceans of any kind.

Breeding.—The mackerel shark tribe are ovoviviparous; that is, the eggs are hatched within the maternal oviducts, but there is no placental connection between mother and young. The embryos, like those of the sand shark (p. 19), are nourished chiefly by swallowing the unfertilized eggs that lie nearby in the "uterus," and their stomachs become enormously swollen by the masses of yolk that are eaten in this way. Another interesting feature of the porbeagle embryo is that the upper lobe of its caudal fin is much longer at first than the lower lobe, the latter increasing in relative length with growth. The embryos also are very large at birth; young of 18, 19, and 24 inches have, for example, been found in a five-foot mother. Corresponding to their large size, gravid females contain only one to four young (0–2 in each oviduct).

General range.—Continental waters in both sides of the North Atlantic; southern Scandinavia, Orkneys and North Sea southward to the Mediterranean and northwest Africa in the east; northern coast of Newfoundland,[33] Newfoundland Banks and Gulf of St. Lawrence to New Jersey and perhaps to South Carolina in the west; represented in the northwest Pacific and in Australian-New Zealand waters by forms that are closely allied to it, but not identical.

Occurrence in the Gulf of Maine.—It has been known from the days of the earliest settlement that stout-shouldered, surface-swimming sharks of moderate size, with "mackerel" tails and slender, smooth-edged teeth are tolerably common in the Gulf of Maine; they are universally referred to by the fishing population as "mackerel sharks." During the first half of the last century only one such shark species was recognized in our waters. And while more recent researches have proved that two actually occur within the limits of the Gulf (this and the next described) the present species is the more northerly of the pair, and much the more frequently taken in the Gulf. Hence it is probable that most of the mackerel sharks that fishermen often see swimming lazily on the surface, and often catch, off the shores of northern New England, belong here.

Seemingly, the chief centers of population for the porbeagle in the western Atlantic are along outer Nova Scotia, and in the western side of the Gulf of Maine. Thus, while there are but two published records for it from the Newfoundland Banks, and one (besides verbal reports) in the Gulf of St. Lawrence, fishermen report it as the commonest large shark along the Atlantic coast of Nova Scotia. Apparently it tends to shun the cold waters of the Bay of Fundy, for it is recorded only twice from Passamaquoddy Bay, one in August 1900, the other on October 3, 1935.[34] But it is so plentiful farther west in the Gulf that incidental catches are on record of 19 that were taken in one night by six men on hand lines, and of about 150 taken by one crew during three weeks' cod fishing near Monhegan Island, Maine. We have ourselves hooked or sighted about one per three or four days' fishing, on the cod grounds in general in the western side of the Gulf, the majority near Platts Bank off Cape Elizabeth, but some also on Nantucket Shoals.[35] Certainly it is the most often seen of the larger sharks around the Isles of Shoals and near Cape Ann, and it has been characterized repeatedly as "common" in Massachusetts Bay.[36]

To the westward the porbeagle is described as not uncommon near Woods Hole (we have not seen it there). We saw a small one about 3 feet long taken in an otter trawl at 60 fathoms, off Marthas Vineyard, on February 20, 1950, by the *Eugene H*; and it has been reported on several occasions from Rhode Island waters. But it appears only as a stray off New York and to the southward.

Thus, the latitudinal range within which it occurs regularly off the American coast covers only something like 5°. And its on- and offshore range is correspondingly so narrow that no report

[33] One reported at Raleigh, on the Newfoundland side of the Strait of Belle Isle, July 1929, by Dr. W. G. Jeffers.

[34] Reported by McGonigle and Smith, Proc. Nova Scotia Inst. Sci., vol. 19, 1936, p. 160.

[35] Cod tagging cruises of the U. S. Bureau of Fisheries.

[36] Actually no sharks other than the spiny dogfish (p. 47) are "common" in the Gulf of Maine, in the sense that this term is applied to such fish as herring, cod, mackerel, and other species, but only as relative to other sharks of corresponding sizes.

of it has come to hand from Georges or Browns Banks, only one from the Nova Scotia slope off Sable Island, and two from the Grand Banks, as just noted. On the other hand, few come in-shore close enough to be picked up in pound nets or weirs.

All published records of mackerel sharks from the Gulf, and all that we have seen there, have been in the warm half of the year, and something like 70 percent of the landings of porbeagles on the coast of Maine are for August to November. But its presence in the Gulf in winter is proved by our receipt of a photograph of a porbeagle embryo, taken from a female caught in January, off Portland, Maine, in 1927. And it is also caught in winter as well as in summer in north European waters. Apparently it simply descends into deeper water during the winter to escape low surface temperatures, feeding little, else more of them would have been caught in the Gulf during the winter fishery with long lines for hake (*Urophycis*).

In the Gulf of Maine, females containing embryos have been taken in August (near Monhegan Island, Maine); in October (off Barnstable, Mass.); in November (off Portland, Maine); and in January (off Portland, Maine). But the fact that the largest embryos have been found in European seas in summer suggests that most of the young are not born until then.

Importance.—The liver oil of the porbeagle, mixed with other fish oils, was in demand for use in tanning leather during the first quarter of the 19th century. And it is interesting to read that as much as 11 gallons of oil has been obtained from the liver of a single shark 9 feet long.

This demand had almost entirely died before 1850 and has never revived. But a new demand has developed of late years for porbeagle meat, which resembles swordfish in taste as well as in appearance, resulting in landings for this purpose of about 46,000 pounds in 1944 on the coast of Maine, and of 71,600 pounds in 1945. Assuming an average weight of, say, 50 pounds, this corresponds to a commercial catch of about 900 to 1,400 sharks. There is no special fishery for porbeagles at present in the Gulf of Maine, or for any other sharks for that matter. About four-fifths of those brought in are taken in gill nets set on bottom for ground fish, and most of the sharks caught in this way are landed in Portland, Maine. The remainder are taken by seines, traps, weirs, hook and line or harpoons. And most of the porbeagles taken in these ways are discarded at sea.[37] The porbeagle is not "game" enough to be of any interest to sport-anglers.

Sharp-nosed mackerel shark *Isurus oxyrinchus* Rafinesque 1810

Atlantic mako

Bigelow and Schroeder, 1948, p. 124.

Description.—This shark resembles the common mackerel shark so closely that we need merely point out the points of difference. Most obvious of these is that while the first dorsal originates about above the armpits of the pectorals in the common mackerel shark, it stands over or behind the inner corner of the pectoral in the mako, and that the second dorsal originates a short distance in front of the anal. The teeth, too, differ rather noticeably in appearance, for while of the same awl-like type, those of the mako lack the lateral spurs or denticles that are characteristic of all but the smallest porbeagles, and those in the front part of the mouth are conspicuously flexuous in form. The mako, too, is more slender bodied; its snout is more narrowly conical; its upper and lower caudal lobes are more nearly equal in length; and the forward part of its caudal fin lacks the secondary lateral keels that are to be seen on the caudal fin of the porbeagle (*cf.* fig. 6 with fig. 5).

Color.—Deep blue-gray above when fresh-caught, appearing cobalt or ultramarine in the water, with gradual transition along the sides to snow-white below; but turning dark slate gray above soon after death (especially if preserved), and to bluish white or pale dirty gray below and on the lower surfaces of the pectorals.

Size.—The maximum length reported for a specimen of the Atlantic mako that was actually measured is about 12 feet,[38] though it has been said to grow to 13 feet. The largest western Atlantic specimen of which we find definite record, taken off St. Petersburg, Fla., was 10 feet 6 inches long, and one nearly as large (10 ft. 2 in.) was caught off New York Harbor many years ago. But the common run caught off the middle Atlantic United

[37] See Scattergood, Copeia, 1949, p. 70, for further details as to landings in Maine and methods of capture.

[38] 3.7 meters as calculated from the size of its jaws.

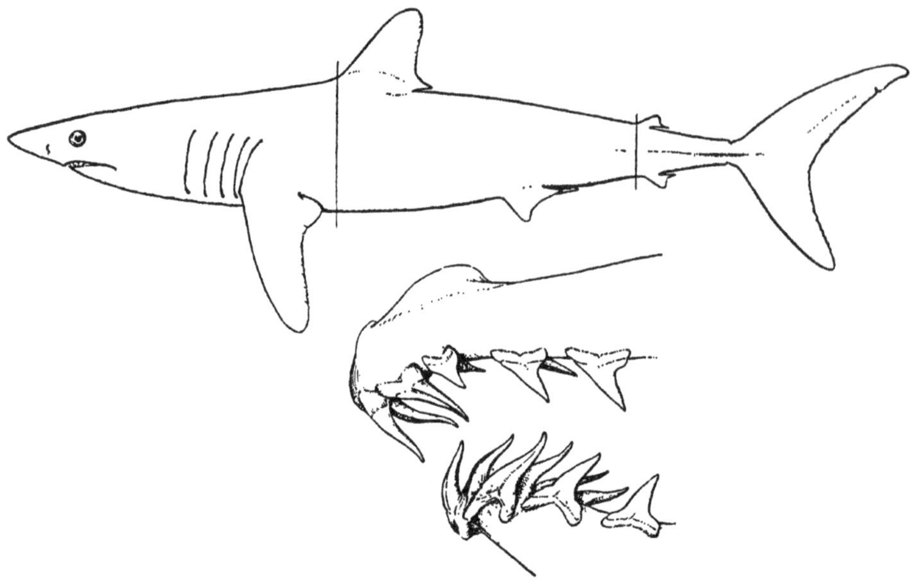

FIGURE 6.—Sharp-nosed mackerel shark, or Mako (*Isurus oxyrinchus*), about 64½ inches long, Maryland. Below, teeth in front of mouth of a large specimen, Cape Cod. From Bigelow and Schroeder. Drawings by E. N. Fischer.

States are perhaps 5 to 8 feet long. Males of about 6 feet are sexually mature (as indicated by the claspers). Recorded weights at different lengths are about 135 pounds at 6 feet, 230 pounds at 7 feet 8 inches; and about 300 pounds at 8 feet. The heaviest Atlantic mako caught on rod and reel of which we have found record was one of 786 pounds taken off Bimini, Bahamas, by Ernest Hemingway in 1936; the largest Pacific mako one of 798 pounds, taken by E. White-Wickham off New Zealand.[39]

Habits.—This is one of the most active and swift swimming of the sharks. In seas where it is more common than it is in our Gulf, it is often seen swimming at the surface, and it is famous for its habit of leaping clear of the water, not only when hooked, but under natural conditions. Seemingly it preys chiefly on schools of smaller fishes of the mackerel and herring tribes. But it also attacks larger fishes. A 730-pound mako, for example, that was harpooned near Bimini in the Bahamas, contained a 120-pound swordfish (*Xiphias gladius*) almost entire, while one weighing about 800 pounds, harpooned off Montauk, Long Island, was seen attacking a swordfish, and was found when landed to contain a large amount of its flesh.[40]

Young embryos of the mako, like those of the porbeagle (p. 22), have greatly dilated stomachs, being nourished on the unfertilized eggs that lie near them in the oviducts, and they are very large at birth, relative to the size of the mother.

General range.—This is an oceanic shark, of the tropical and warm-temperate belts of the Atlantic north and south, including the Mediterranean in the east and the Caribbean and Gulf of Mexico in the west. It is represented in the corresponding thermal belts of the Pacific and Indian Oceans by a close ally, the Pacific mako *Isurus glaucus*.

Occurrence in the Gulf of Maine.—The center of abundance for the mako lies in warmer seas to the southward of our Gulf. Considerable numbers journey northward, however, in summer along the continental shelf, as far as to the offing of southern New England, and a few are caught off Woods Hole. One of the earliest accounts of it in American waters was based partly on one from Cape Cod. During the past few summers we have heard repeatedly of makos seen jumping, or occasionally hooked near the northern end of

[39] A South African shark of 2,176 pounds, landed on rod and reel, and reported as a mako, is proved by the photograph of its teeth (London Illus. News, July 14, 1928, p. 53) to have been a maneater (*Carcharodon*).

[40] See Farrington (Field and Stream, vol. 47, Feb. 1943) for these instances of the mako attacking swordfish, and for other interesting notes on this shark.

Cape Cod, and in the summer of 1941 one about six feet long was landed on rod and reel in the southern side of Massachusetts Bay near Plymouth.[41] Thus stray individuals may be expected to visit the southern part of the Gulf in most summers, though we have never met it there ourselves. It has even been reported as far north as Seguin Island, Maine, but without convincing evidence that the shark in question was not a porbeagle.[42]

Importance.—The chief importance of the Atlantic mako, as of its Indo-Pacific relative, is as a game fish, because of its fast runs when hooked and of its habit of leaping. But it is not plentiful enough anywhere in the Gulf of Maine to be worth fishing for there especially.

Maneater *Carcharodon carcharias* (Linnaeus) 1758

WHITE SHARK

Bigelow and Schroeder, 1948, p. 134.
Garman, 1913, Pl. 5, figs. 5–9.

Description.—The maneater is of the general "mackerel shark" appearance, with firm lunate tail, the upper lobe only a little longer than the lower; and with triangular first dorsal of moderate size originating over the armpits of the pectorals, which are sickle shaped, and roughly twice as long as they are broad. The second dorsal and anal fins are very small, the former a little in advance of the latter; and the root of the tail bears a single well-marked keel on either side. The snout is conical, moderately pointed.

[41] Information from Dr. W. J. Mixter.

[42] Various early reports of it in the northern part of the Gulf seem to have referred, actually, to the porbeagle.

Unfortunately, there is no obvious field mark to distinguish a small maneater from a large porbeagle or from a large mako when seen swimming at any distance. Once captured, however, no confusion could arise, for instead of the slim catlike teeth of the porbeagle and of the mako, we find the maneater one of the best armed of all sharks; its teeth large and triangular, and similar in shape in the two jaws, except broadest in the upper, with nearly straight cutting edges and strongly serrated margins. As a precaution, any large active shark, upwards of 10 or 12 feet long, with the tail not long, out of ordinary proportions, should be looked upon with suspicion, for it might prove to be a maneater. If it were sluggish, resting with the dorsal fin high out of water, it would be no doubt a harmless basking shark (p. 28).

Color.—Maneaters up to 12 to 15 feet long are slaty brown or leaden gray above, sometimes almost black, shading more or less abruptly on the sides to dirty white below. There is a black spot in the armpit of each pectoral fin, and the lower surfaces of the pectorals are black toward their tips, usually with some black spots adjacent. The pelvics are white below, but olive along their anterior edges. Larger specimens (we have seen none) have been described as dun colored above or very pale leaden, and they may lack the black spot at the armpit of the pectoral fin.[43]

[43] Information from Stewart Springer, from large Florida specimens.

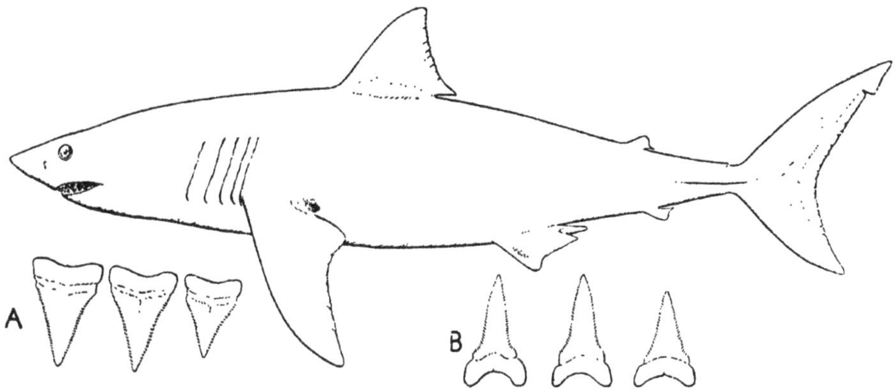

FIGURE 7.—Maneater (*Carcharodon carcharias*), Massachusetts, about 7 feet long. A, first three upper and B, first three lower teeth, from center of jaw, from a specimen about 8½ feet long, Woods Hole, about 0.6 times natural size. From Bigelow and Schroeder. Drawings by E. N. Fischer.

Size.—This is one of the largest of sharks. A gulf of Maine specimen about 3 feet long is the smallest, apart from embryos, that has been seen; one of about 5 feet the next smallest. So far as known it does not mature sexually until it has grown to a length of 12 to 14 feet. Among larger ones, from one place or another, the exact measurements for which have been reported, four have been between 14 and 16 feet long, two between 16 and 18 feet, and three between 19 and 21 feet. The largest on record was 36½ feet long;[44] the next largest about 30 feet, but perhaps not measured exactly.

Maneaters of a given length may vary widely in weight, because of variations in their condition. Thus one specimen 8 feet 2 inches long weighed only 342 pounds, but another of 8 feet 3 inches, weighed 600 pounds. Five, weighing between 910 and 1,000 pounds ranged from 9 feet 8 inches in length to 12 feet 6 inches. Three, of 13 to 13½ feet, weighed 1,291 to 1,344 pounds, but another, from South Africa of 13 feet 3 inches scaled 2,176 pounds, doubtless a very fat fish. A 15-foot 2-inch specimen weighed 1,720 pounds; and one of 21 feet, the largest that has been weighed so far, 7,100 pounds, its liver 1,005 pounds.[45]

Habits.—So few maneaters are seen that little is known of their way of life, apart from their voracity. Most of the records of them have been of specimens taken at or near the surface, and such specimens as visit our Gulf sometimes come very close inshore. Thus two specimens were seined close in, off Swampscott, at the northern entrance to Boston Harbor in 1939; one was harpooned in 1937 about 2 miles off Nantasket Beach, one of the most popular bathing resorts near Boston; another was harpooned about one-half mile off Cohasset, Mass., where the water is not over 20 feet deep; one in 10 feet of water in Provincetown Harbor, many years ago. Some have even been taken in fish traps close to the beach on Cape Cod and near Woods Hole; and in 1916 one was taken in the shallow water of Sandy Hook Bay, N. Y. On the other hand, the largest one that has been weighed yet was caught on a set line off the north coast of Cuba, at a depth of about 700 fathoms.

Nothing is known of its breeding habits, beyond the bare facts that it is ovoviviparous like others of the mackerel shark tribe.

The maneater is one of the most voracious of all the fish tribe, feeding indifferently on large prey and on small. Other sharks, 4 to 7 feet long and practically intact, have been found repeatedly in maneaters' stomachs; and a young sea lion of 100 pounds in one on the coast of California, while seals, sturgeons, and tuna have been found in maneaters no longer than 8 to 9 feet. In southern seas they are described as feeding regularly on sea turtles. But they also devour smaller fishes of whatever kinds are available, including small sharks and chimaeroids, also squids. When they come in on the fishing banks, they are known to take fish that they find hooked on long lines as porbeagles do (p. 22). Thus the mouth of one of 9 feet 8 inches, taken near Cohasset, Mass., and examined by us, carried several hooks with the snoods still attached, while its stomach contained a spiny dogfish (*Squalus acanthias*) that evidently had been torn off a hook. And a large Florida maneater, caught on a set line, contained 2 brown sharks (*Carcharhinus milberti*), 6 to 7 feet long, that had evidently been torn from hooks on the same set line on which the maneater was hooked. The maneater, like the Tiger shark, is not above feeding on slaughterhouse waste or other garbage.

General range.—This is an oceanic shark, widespread in the tropical and warm temperate belts of all oceans, including the Mediterranean. In the western side of the Atlantic it has been recorded as far north as St. Pierre Bank south of Newfoundland, and as far south as Brazil.[46]

Occurrence in the Gulf of Maine.—The maneater is usually looked on as a warm water shark, doubtless correctly so. None the less, it has been reliably reported from the southwestern part of the Gulf of Maine more often than it has from any other coastal sector of comparable length on the Atlantic coast of North America. At least 10, for example, were actually captured or were harpooned and lost in Massachusetts Bay alone during the period 1935 to 1948. We ourselves examined three of these, one that was netted at Swampscott; a female of 9 feet 8 inches weighing 980 pounds that was harpooned within half a mile of the land off Cohassett, in August 1940; one of about 3 feet, that was harpooned in July 1948

[44] This Australian specimen, the jaws of which are in the British Museum, is the basis for repeated statements that the maneater grows to 40 feet.

[45] For further details, see Bigelow and Schroeder, Fishes Western North Atlantic. Pt. 1, 1948, pp. 137–138.

[46] For details and references, see Bigelow and Schroeder, Fishes Western North Atlantic. Pt. 1, 1948, pp. 140–141.

near Boston Lightship, this last being the smallest that is on record to date (p. 26), and one about 14 feet long, weighing 1,050 pounds dressed, which sold for 10 cents a pound, was taken in a trap at North Truro on November 9, 1952.

Carrying the record back to earlier years, a 15-foot shark, taken at Monomoy Point at the elbow of Cape Cod in the autumn of 1928, appears to have been a maneater, and one of about 16 feet, taken in a trap at East Brewster, October 16, 1923, and identified by Dr. Samuel Garman, certainly was, while one of 7 feet 2 inches, taken in Massachusetts Bay, about 1910, was the basis of Garman's (1913, pl. 5, fig. 5) beautiful illustration. Earlier still, a 13-footer, taken at Provincetown, Cape Cod, in June 1848, was described by Storer as a new species, *atwoodi*, while two small ones were mentioned by him as taken by Massachusetts fishermen between 1820 and 1850. And Capt. Atwood reported seeing four, caught in mackerel nets at Provincetown many years ago.[47]

Proceeding northward, we find scattered records from the vicinity of Portland, Maine, most recently, a 13-footer caught in a gill net off Casco Bay in November 1931; one from Eastport, Maine, many years ago; a very large one (estimated as about 26 feet long) taken in a wier at Campobello Island, November 23, 1932 [48] it was suggested locally that it may have been the same specimen that had attacked a fishing boat off Digby, Nova Scotia, the preceding July (p. 27); one from Deer Island, New Brunswick, taken in a herring weir, August 24, 1949;[49] and one from Digby, on the Nova Scotian shore of the Bay of Fundy, July 2, 1932. And there are several reliable records for St. Margaret Bay on the outer coast of Nova Scotia, perhaps also for Halifax.

The most northerly positive record for it on the Atlantic coast of North America is for St. Pierre Bank, south of Newfoundland, where one attacked a fisherman in a dory many years ago, leaving in the sides of the boat pieces of its teeth, from which Dr. Garman was able to identify it.[50]

Westward and southward from the elbow of Cape Cod, we find nine or ten definite records for Nantucket and for the vicinity of Woods Hole (never more than two in any one year), with one of five feet (second smallest on record) netted at Sakonnet, Rhode Island, May 30, 1939. Maneaters are also reported occasionally near New York, notably one of about seven feet, taken in Sandy Hook Bay, July 1916, to which we recur below (p. 27).

Relation to man.—So few man-eaters visit our Gulf that they would deserve only the briefest mention were this not the only shark that is ever likely to attack human beings there. Strong and active, equipped as it is with a most terribly effective set of cutting teeth, it has borne an unsavory reputation as a man-eater from the earliest times, and it is probable that the 7-foot specimen listed earlier from South Amboy, Sandy Hook Bay, was the cause of the shark fatalities along the New Jersey beach in July 1916 (p. 16). A fatal attack on a swimmer at Mattapoisett, on Buzzards Bay, on July 25, 1936, may also have been by a man-eater, though in this case the shark was driven away without being identified.

This is also perhaps the only shark against which unprovoked attacks on small boats are proved by identification of their teeth, embedded in the wood. One such instance, from the Newfoundland Banks, was reported by Putnam [51] many years ago (p. 27). A recent local case is of a very large one that attacked a fishing boat in the Bay of Fundy off Digby Gut, Nova Scotia, July 2, 1932 and left in her keel or lower planking several of its teeth, by which it was identified.[52] Storer [53] wrote of a case where one (apparently the 13-foot specimen that he had described earlier as *atwoodi*) turned furiously on a boat, but was lanced to death and brought into Provincetown. And a 15-foot shark, probably this species to judge from the illustration of it that was published,[54] that was killed off Monomoy Point by two fishermen in November 1928, overturned their dory before it was subdued. And one of about 15 feet (similarly identified by teeth left in the planking) attacked a boat, from which it had been harpooned, in St. Margaret's Bay, Nova Scotia, on June 27, 1920.[55] Hence, so long as maneaters wander within

[47] Putnam, Bull. Essex Inst., vol. 6, 1874, p. 72.

[48] Piers, Proc. Nova Scotian Inst. Sci., vol. 18, 1934, p. 198.

[49] A female 12 feet, 8 inches long, weighing 1,299 pounds, reported by Scattergood, Trefethen, and Coffin, Copeia, 1951, p. 298.

[50] Putnam, Bull. Essex Inst., Salem, vol. 6, 1874, p. 72.

[51] Proc. Essex Inst. Salem, vol. 6, 1874, p. 72; teeth identified by Dr. S. Garman.

[52] Reported by Piers, Proc. Nova Scotia Inst. Sci., vol. 18, 1934, p. 198.

[53] Fishes of Mass., 1867, p. 248.

[54] Reported in Witman and Lee Co.'s Market Letter for Nov. 8, 1928; called to our attention by Dr. Lewis Radcliffe of the U. S. Bureau of Fisheries.

[55] For details of this occurrence, see Piers, Proc. Nova Scotia Inst. Sci., vol. 18, 1934, pp. 196–198.

our limits more often than had been realized previously, the possibility is always open of attacks on bathers along the Massachusetts shores of the Gulf.

Despite its ferocity, muscular strength and size, the man-eater does not put up so spectacular a resistance when hooked as does a mako, neither running so fast nor having the habit of jumping. Neither does it put up as strong a fight, pound for pound, as a tuna ordinarily does, or any of the swordfish tribe. Thus a 1,329-pound maneater was landed on rod and reel by an Australian angler in 53 minutes. One of 2,176 pounds, caught from the shore in South Africa, is the largest fish ever landed on rod and reel that has come to our notice.[56]

[56] London Illus. News, July 14, 1928, p. 53; photograph recorded as a mako but shown by its teeth to have been a maneater.

BASKING SHARKS. FAMILY CETORHINIDAE

Basking shark *Cetorhinus maximus* (Gunnerus) 1765

BONE SHARK

Bigelow and Schroeder, 1948, p. 147.

The basking shark resembles the mackerel sharks in the lunate shape of its caudal fin, with lower lobe nearly as long as upper; also in the presence of a noticeable lunate furrow above and one below on the root of the tail, and in the wide lateral expansion of the latter, forming a pronounced "fore and aft" keel on either side; also in the facts that the second dorsal fin and the anal fin are much smaller than the first dorsal, that its fifth gill opening is situated in front of the origin of the pectoral fin; in the position of the mouth on the under side of the head; and in the wide separation of the nostrils from the mouth. But the teeth of the basking shark are minute and very numerous (large and few in number in the mackerel sharks); its gill openings are so large that they extend right around the neck, with those of the first pair almost meeting below on the throat; and the inner margin of each gill arch bears a great number of horny, bristle-like rakers, directed inward-forward, that correspond to the rakers of various bony fishes in their position and in their function (see p. 30). It was the fancied resemblance of these rakers to the whalebone of the whalebone whales that suggested the vernacular name "bone shark" to the whalemen of olden times.

Corresponding to its feeding habits, the mouth of the basking shark is very large and widely distensible at the corners. The snout is short, conical, with rounded tip on large specimens. But it is much longer, relatively, on small ones,

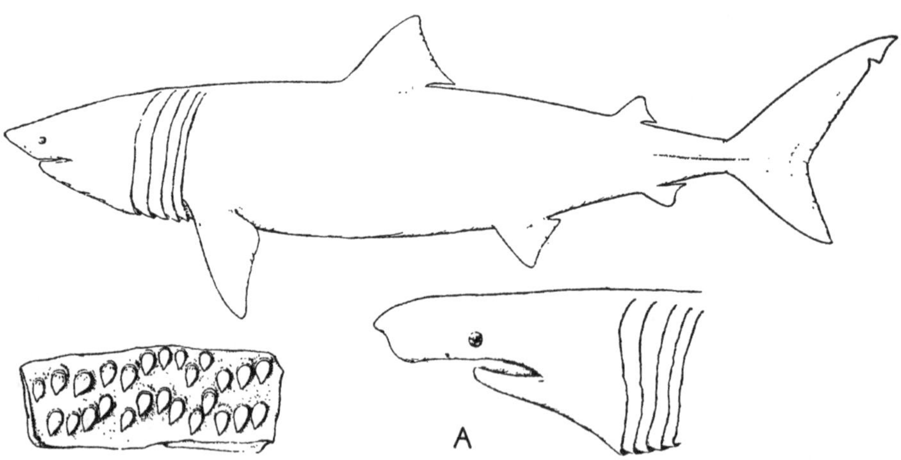

FIGURE 8.—Basking shark (*Cetorhinus maximus*), 26½-foot female, Marthas Vineyard. A, side view of head of 12-foot Long Island specimen; B, a group of the teeth of same, about 1.2 times natural size. From Bigelow and Schroeder. Drawings by E. N. Fischer.

projecting far beyond the mouth, obliquely truncate in front, terminating above in a sharp point, and with the head strongly compressed sideways abreast of the front of the mouth. This results in so bizarre an appearance that the young basking shark was thought at first to represent a separate species. A gradual transition takes place from the juvenile shape of head to the adult shape when a length of 12 to 16 feet has been reached. We need only note further that the triangular first dorsal fin stands midway between pectorals and pelvics; though not so high in proportion as that of the mackerel-shark tribe, it rises high in the air when a large basking shark lies awash on the surface, as is their habit, a convenient field mark (p. 29).

Color.—Upper surface grayish brown, slaty gray, or even almost black. The lower surface has been described repeatedly as white. But the Menemsha specimen described by Allen [57] was of a somewhat lighter shade below than above, without white markings, as was a Massachusetts Bay specimen recently examined by us; while one 14 feet long captured at West Hampton, L. I., on June 29, 1915 [58] had the belly as dark as the back, with a white patch underneath the snout in front of the mouth.

Size.—The basking shark rivals, though it does not equal, the whale shark of tropical seas in size. Reports that an occasional basking shark may reach a length of 50 feet probably are not an exaggeration, for the catch on the coast of Norway, for the period 1884 to 1905, included one of about 45 feet and three of about 40 feet, with the six next longest ranging between 36 feet and 30 feet 3 inches. The three longest for which we find definite measurements for the western Atlantic were of 32 feet 2 inches, 32 feet, and 30 feet 3 inches. But others up to 35 feet long have been credibly reported as killed near Eastport, Maine, many years ago; and one captured at Musquash Harbor, New Brunswick, near the mouth of the Bay of Fundy in 1851 was said to have been about 40 feet long. It is probable that they are at least 5 to 6 feet long when born, the three smallest so far reported having been between 5 feet 5 inches and about 8 feet 6 inches long. Matthews [59] concluded from studies of basking sharks taken near the Isle of Skye that fish up to 10 feet are in their first year, those of 15 feet in their second year. Males mature sexually at about 18 to 20 feet as indicated by the lengths of their claspers, females at about 20 to 23 feet; i. e., when 3 years old or perhaps 4, according to Mathews' estimate.

We find no exact weights for large basking sharks from the Atlantic. But 6,580 pounds for one of 28 feet, and 8,600 pounds for another of 30 feet, from Monterey, Calif., is doubtless a fair indication of what a fairly large one may be expected to weigh. Estimated weights for smaller ones, from the Pacific, are about 6,600 pounds at about 23 feet, 1,000 to 1,800 pounds at 13 to 15 feet, and 800 pounds at 8 feet 4 inches.[60] A young one, 12 feet long, killed off Digby, Nova Scotia, August 16, 1939, weighed 359 pounds, after it had bled, [61] and one almost 20 feet long, taken off Portland, Maine, in 1936, weighed 550 pounds, dressed.

Habits.—This is a sluggish, inoffensive fish, helpless of attack so far as its minute teeth are concerned. It spends much time sunning itself at the surface of the water, often lying with its back awash and dorsal fin high out of water, or on its side, or even on its back sunning its belly; sometimes it loafs along with the snout out of water, the mouth open, gathering its provender of plankton. They pay so little attention to boats that it is easy to approach one of them within harpoon range, and excellent motion pictures have been taken of them in Irish waters.[62] But they have also been seen jumping, perhaps to shake off parasites. Those seen in the Gulf of Maine are usually traveling singly. But they are known to congregate sometimes in loose schools which may include as many as 60 to 100 in the peak years of abundance for them in regions where they are more numerous than in the Gulf of Maine.[63] It is chiefly during the warm half of the year that basking sharks are encountered off the northeastern United States and in the northern part of their range in the opposite side of the Atlantic. It is likely that those that summer in the inshore parts of the Gulf simply withdraw in the fall, to pass the

[57] Bull. Boston Soc. Nat. Hist., No. 24, March 1921, p. 5.

[58] Described by Hussakof, Copeia, No. 21, 1915, pp. 25–27.

[59] Philos. Trans. Roy. Soc. London, Ser. B., vol. 234, 1950, pp. 247–316.

[60] For further details as to sizes of basking sharks, see Bigelow and Schroeder, Fishes, Western North Atlantic, Pt. I, 1948, pp. 151–152.

[61] Referred to by McKenzie, Proc. Nova Scotia Hist. Sci., vol. 20, 1940, p. 42.

[62] Shown in the film "Men of Arran."

[63] See Bigelow and Schroeder, Fishes Western North Atlantic, Pt. 1, 1948, pp. 153, 154, for details as to their centers of population and secular fluctuations in abundance in north European waters.

winter in deeper water where the temperature does not fall so low.

Next to its vast bulk and its curiously sluggish habit, the most interesting peculiarity of the basking shark is its diet, for it subsists wholly on tiny pelagic animals, which it sifts out of the water by means of its greatly developed gill rakers, exactly as plankton-feeders among fishes such as menhaden do, and whalebone whales with their baleen sieves. In several cases their stomachs have been found packed with minute Crustacea; this was true of the only western Atlantic specimen of which the stomach contents have been examined. And while digestion is so rapid that the food swallowed is soon reduced to a soupy mass, this usually is reddish, suggesting a crustacean origin.

All that is known of the breeding of the basking shark is that the structure of the internal sex organs of the female accords with the nourishment of the embryo within the maternal oviduct, that the ovary of a female, with empty oviduct contained something like 6 million immature ova instead of the few that are usual in sharks that bear "living" young, and that an embryo about a foot long was said, long ago, to have been taken from its mother.[64]

Basking sharks reported as "sea serpents" or as other "monsters".—The remains of basking sharks have been reported as "sea serpents" on several occasions; nor is this astonishing. "As the carcass of the shark rots on the shore, or is buffeted against the rocks, the whole of the gristly skeleton of the jaws and gill arches . . . as well as the pectoral and pelvic fins, is soon washed away,"[65] leaving only the cranium and the long backbone, with larger or smaller amounts of muscle, so frayed out as to suggest a hairy or bristly mane. As a recent instance from the Gulf of Maine, we may cite the newspaper and radio publicity, that was given, as a supposed sea serpent, to a basking shark skeleton, about 25 feet long, that beached near Provincetown on the outer shore of Cape Cod, in January 1937, that we examined.[66]

A more spectacular instance of the fanciful interpretation that is likely to be placed on any large stranded carcass that has decayed partially, was the famous "Animal of Stronsa," that came ashore on the island of that name in the Orkneys, in September 1808. It was pictured by an eyewitness as having three pairs of limbs, but the published illustration of its cranium, vertebrae, and pelvic skeleton[67] show that it was only the remains of some very large shark, probably a basking shark. It has also been suggested repeatedly that some of the stories of sea monsters of one sort or another may have been based on the dorsal and caudal fins of two or more basking sharks, swimming one behind another as they often do (we dare not touch further on the controversial subject of the "sea serpent").

General range.—This enormous fish, formerly thought to be an Arctic species, straying southward, is now known to be an inhabitant of the temperate-boreal zone of the North Atlantic.[68] It is represented in the corresponding thermal belts of the South Atlantic and of the North and South Pacific by a similar great shark (or sharks), whose exact relationship to the basking shark of the North Atlantic is still an open question.

The northern boundary of the normal range of the basking shark of the North Atlantic appears to follow the line of transition from waters of predominately Atlantic influence to those of Arctic origin. This, roughly, runs from the outer coast of Nova Scotia (1 record), and from southern Newfoundland (4 positive records) to western and southern Iceland, to the Orkney and Faroe Islands, and skirts the Norwegian coast to the North Cape, while basking sharks stray now and then to the Murman coast. To the southward, in the North Atlantic, they range as far as the Mediterranean and Morocco in the east, to North Carolina in the west.

Occurrence in the Gulf of Maine.—Before the coming of the white man this great shark seems to have been a regular inhabitant of the southern part of the Gulf of Maine. And tradition has it that large numbers were taken in Massachusetts waters, especially off the tip of Cape Cod, during the first half of the eighteenth century, for their liver oil which was then in demand for illuminating purposes. However, the local stock seems soon to have gone the same way as the local stock of the North Atlantic right whale; that is, into the try pot. And basking sharks seem never to have

[64] See Matthews, Philos. Trans. Roy. Soc. London, Ser. B, No. 612, vol. 234, 1950, pp. 347—366 for detailed account.

[65] Norman and Fraser, Giant Fishes, Whales and Dolphins, 1937, p. 21.

[66] For account and photograph, see Schroeder, New England Naturalist, No. 2, 1939, p. 1.

[67] Barclay, Mem. Wernerian Soc., Edinburgh, vol. 1, 1811, p. 418.

[68] It has long been realized that old tales of a tremendous whale-eating shark, on which Fabricius based his statement that the basking shark occurs in Greenland waters, were fiction.

visited the northeastern part of the Gulf in any numbers, there being only a few records for the vicinity of Eastport, Maine, and three from within the Bay of Fundy. At the present time the Gulf appears to harbor a sparse and fluctuating population, occasional members of which are encountered from time to time, here or there, but whether as immigrants into the Gulf from the open ocean is not known.

The list of specimens, the capture or stranding of which in the Gulf has come to our attention for the period 1908–1951 is as follows:

1908. One, 18 feet long, near Provincetown, taken in a fish trap; measured by J. Henry Blake.
1909. One, about 22 feet, in Provincetown Harbor.
1913. One, about 29 feet, Provincetown.
1925. One, about 29 feet, near Monhegan Island, Maine.
1931. Female, 12½ feet long, York Harbor, Maine.
1934. One, 29 feet, Whale Cove, Grand Manan Island, and one, 28 feet, Back Bay, Bay of Fundy.[69]
1936. Two off Portland, Maine; the first about 20 feet long, weighing 550 pounds dressed, about May 1; the second, much larger (reported as of about 40 ft.), August 2.
1939. Skeleton of one of about 25 feet, examined by us, found on the beach near Provincetown in January. One of about 25 feet, Yarmouth, Nova Scotia. One of 12 feet, Bay of Fundy off Digby Gut.[70]
1947. Female, about 13 feet long, examined by us, harpooned by W. T. Reid 3rd, near Boston Lightship, August 5th.
1949. A small one (size not recorded), near Rockport, Mass., September; identified from a good photograph by Miss D. E. Snyder of the Peabody Museum, Salem.
1951. One, 12 feet, near Bar Harbor, Maine, harpooned July 28.[71]

Occasional basking sharks also visit the shores of the southern coast of Massachusetts, westward from Cape Cod; one, for example, 12 to 14 feet long was taken at Menemsha on Marthas Vineyard, August 16, 1916; another of 20 feet 6 inches at that same locality on June 24, 1920;[72] one 20 feet 2 inches long was stranded in Hadleys Harbor, Naushon Island, July 1937; and one of 8 feet (among the smallest on record) was taken in a fish trap near Woods Hole on June 15, 1948.

Probably the basking shark is no more plentiful near shore in our Gulf in most years than the paucity of the recent records suggest, for popular interest in sharks is now so keen, as represented by newspaper publicity given to any unusual capture, that any well-grown one is apt to be seen in these frequented and hard-fished waters. We do not find evidence of any considerable incursion by them into coastal waters farther west since 1878, when 20, at least, were found dead in the fish traps near Woods Hole during the summer. And the only report that might be based on the basking shark on the offshore fishing banks that we have received from fishermen has been of a number of unusually large sharks of some sort, seen by Capt. Henry Klimm on the southeast part of Georges Bank during late June and early July 1947.

Importance.—The day of any regular fishery for the basking shark is long since past in New England waters, probably never to return. And no use is made there, nowadays, of the occasional specimens that are captured. But it may be of interest to point out that it was always hunted of old by the sperm whalers from New Bedford, for its liver oil was considered nearly or as good as sperm oil for illuminating purposes. Basking sharks are still the object of intermittent small vessel fisheries off the coast of Iceland, around the Orkneys, off western Ireland, and off southern Norway; also off Ecuador and Peru in the Pacific. And increasing numbers have been landed during the past few years in northern California, where they are considerably more plentiful than they are in the Gulf of Maine,[73] for fish meal and for the liver oil. The yield of oil per fish varies from about 80 gallons to about 200, occasionally to 400 gallons, with as much as 600 gallons reported. The liver of a 30-foot fish weighing 6,580 pounds, taken off Monterey, Calif., had a liver weighing 1,800 pounds, 60 percent of which was oil.[74] But, sad to say, it is very low in vitamin A.

The fishery, wherever carried on, is by harpoon. And basking sharks are so sluggish and so unsuspicious of a boat, large or small, that it usually is a simple matter to harpoon one that is seen at

[69] McKenzie, Proc. Nova Scotia Inst. Sci., vol. 20, 1939, p. 14.
[70] McKenzie, Proc. Nova Scotia Inst. Sci., vol. 20, 1939, p. 14.
[71] Personal communication from J. W. Burger.
[72] This specimen, mounted, in the New England Museum of Science and described by Allen (Bull., Boston Soc. Nat. Hist., No. 24, March 1921, pp. 3–10), served as chief basis for the illustration given here of the adult basking shark.

[73] According to MacGinitie (Science, N. Ser., vol. 73, 1931, p. 496), 21 basking sharks were landed in Monterey, Calif., between November 22, 1930 and February, 1931.
[74] MacGinitie, Science, N. Ser., vol. 73, May 1931, p. 496.

the surface. Once struck, however, a large one is likely to put up an astonishingly active and enduring resistance. We read, for example, of one of 35 to 38 feet harpooned by Capt. N. E. Atwood off Provincetown, Mass., about 1863, that towed the fishing smack all night, and broke loose finally.[75]

[75] Goode, Fish. Ind. U. S., 1884, Sect. 1, p. 669.

THRESHER SHARKS. FAMILY ALOPIIDAE

The threshers (several species are known) are peculiar among sharks for their enormously long tail fin. Their closest affinities in other respects are with the mackerel sharks.

Thresher *Alopias vulpinus* (Bonnaterre) 1758

THRASER; SWIVELTAIL; FOX SHARK

Bigelow and Schroeder, 1948, p. 167.
Garman, 1913, pl. 7, figs. 1–3.

Description.—The thresher is as easily distinguished from all other Gulf of Maine sharks by its long tail as the hammerhead is by its head, the upper caudal lobe being a little longer than the head and body of the fish together, curved much like the blade of an ordinary scythe, and notched near the tip, whereas the lower lobe measured along the front margin is hardly longer than the pelvic fins. We need merely point out in addition that the first dorsal fin (of moderate size and about as high as it is long) stands about midway between pectoral and pelvic fins; that the second dorsal fin and the anal are very small; that the pectoral fin is long and sickle shaped; and that the thresher is a stout-bodied shark with short snout and blunt, rounded nose. Its teeth are small, subtriangular with a single sharp cusp and are smooth edged. Those near the center of mouth are nearly symmetrical, but the successive teeth are increasingly oblique outward, with their outer margins increasingly concave.

Color.—Dark brown, blue-slate, slate gray, blue gray, leaden or even nearly black above, often with metallic luster, grading on the sides to white below, except that the snout and the lower surface of the pectorals are usually about as dark below as above, and that the sides near the pectorals may be more or less mottled with gray, the belly also. The iris is black or green.

Size.—Threshers vary considerably in size at birth, for while free living specimens have been reported as small as 46 inches, with many of 48 to 60 inches (some with umbilical scars still showing), one unborn embryo was 61 inches long. The state of development of the claspers of males, with the lengths (14 ft. 6 in. and about 15½ ft.) of females that have been found with embryos, makes it unlikely that they mature sexually until they are at least 14 feet long (tail included). Lengths up

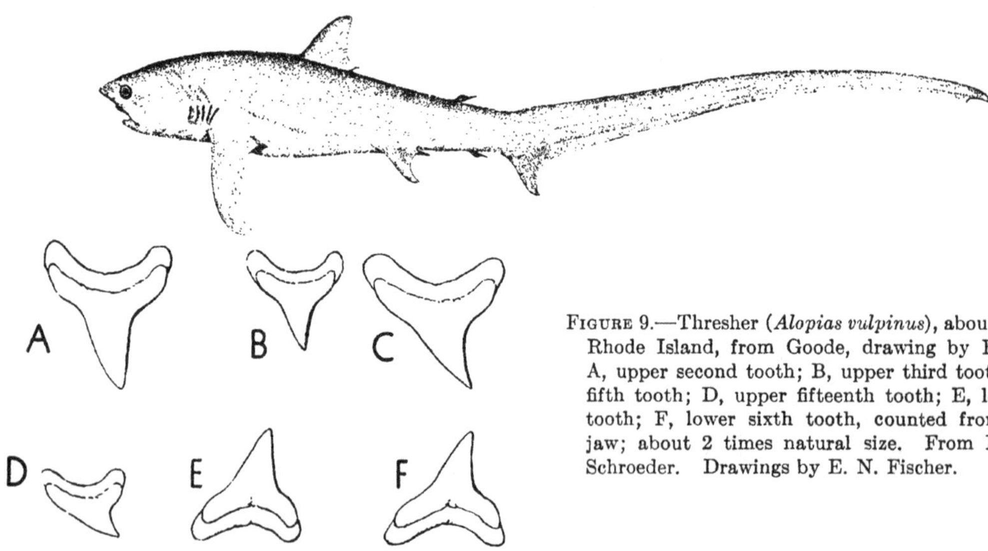

FIGURE 9.—Thresher (*Alopias vulpinus*), about 5 feet long, Rhode Island, from Goode, drawing by H. L. Todd. A, upper second tooth; B, upper third tooth; C, upper fifth tooth; D, upper fifteenth tooth; E, lower second tooth; F, lower sixth tooth, counted from center of jaw; about 2 times natural size. From Bigelow and Schroeder. Drawings by E. N. Fischer.

to 16 feet are usual;[76] the maximum length (tail included) is about 20 feet. Threshers are so largely tail that they are much lighter than many other sharks, length for length. The few actually weighed have ranged from about 300 to 320 pounds at about 10 feet, and 375 to 400 pounds at about 13 feet, to about 500 pounds at about 14½ feet. Perhaps 1,000 pounds is about the maximum to be expected for a very large one.

Habits.—The reports of threshers are mostly based on ones seen at the surface or caught either in nets set shoal, or in traps set close inshore. But a thresher has been hooked as deep as 35 fathoms in British waters.[77]

The thresher feeds chiefly if not exclusively on small schooling fishes; in American waters mostly on mackerel, menhaden, herring, and bluefish (*Pomatomus*); also on bonito and on squid. A pair of threshers often work in concert "herding" a school of fish, and it is to frighten its prey together that the enormously long, flail-like tail is employed. Allen [78] gives an interesting eyewitness account of a thresher pursuing and striking a single small fish with its tail.

The tale that the thresher leagues with the swordfish to attack whales is time honored, but has long since been relegated to the category of myth. And so weak toothed is this shark that the second part of the story (it makes a meal of its huge victim) is close to an impossibility. The thresher, we may add, does not harm human beings.

In American waters it is probable that threshers are born throughout its range, very small free living specimens having been caught off New England on the one hand, and off Florida on the other. The embryos do not develop a placental attachment with the mother, and either 2 or 4 have been reported in gravid females.

General range.—This is an oceanic shark of temperate and subtropical seas. In the Atlantic it is known from southern Ireland and the North Sea to Madeira and the Mediterranean in the east, and also from the Cape of Good Hope; from Nova Scotia and the Gulf of St. Lawrence to Cuba and the northern part of the Gulf of Mexico in the west, and again from southern Brazil and northern Argentina. Seemingly it does not occur in the equatorial belt of the Atlantic. But it does in the Pacific, where it is known from Oregon to Panama and Chile. Threshers of this same type are also found in the central and western Pacific and in the Indian Ocean. Whether the thresher of the eastern side of the Pacific is identical with that of the Indian Ocean remains to be determined.

Occurrence in the Gulf of Maine.—The thresher has often been seen off the southern coast of New England and in some numbers. Three about 16 feet long have been taken near Woods Hole, for example, in one trap in a single morning, and it has been classed as the commonest of the large sharks off Block Island. Scattered specimens also visit the Gulf of Maine in some years, though perhaps none in others. Thus two have been reported in print from Nantucket; we saw several large ones in Pollock Rip, off the southern angle of Cape Cod on August 4, 1913; it has been reported repeatedly on the coast of Massachusetts, as at Barnstable on Cape Cod Bay, where one about 10 feet long was taken in a trap on October 21, 1949, and from various localities in Massachusetts Bay (e. g. Boston Harbor and Nahant).

Records for it along the coast of Maine include the vicinity of Monhegan Island, east of Matinicus Island, the offing of Penobscot Bay where one weighing about 500 pounds (estimated) was caught in 1911 and another seen in 1911, in the vicinity of Eastport. It has also been taken in the cold waters of Passamaquoddy Bay; one for instance in a weir at Deer Island, August 28, 1936;[79] also in the Basin of Minas on the Nova Scotian shore of the Bay of Fundy. Occasionally a thresher is netted or seen off the outer coast of Nova Scotia. The most northerly record for it from our side of the Atlantic is for the Bay of Chaleur in the southern side of the Gulf of St. Lawrence. It is to be expected in Gulf of Maine waters only during the warm half of the year, perhaps May to October (April to late autumn for Woods Hole); in the cold season it altogether deserts our northern coasts for warmer seas.

Importance.—The thresher is not common enough in the Gulf of Maine to be of any importance to fishermen one way or another, or to play

[76] Several of that size have been taken in the traps at Woods Hole.

[77] There is another group of species of the genus, with very large eyes, that live at greater depths; for discussion of these, see Bigelow and Schroeder (Fish. Western North Atlantic, Pt. 1, 1948, pp. 162, 163).

[78] Science, N. Ser., vol. 58, 1923, pp. 31–32.

[79] Reported by McKenzie, Proc. Nova Scotia Inst. Sci., vol. 20, 1939, p. 14.

a practical role of any moment among the smaller fish. Further south, however, and wherever it is numerous in the Atlantic, it makes itself a pest, tangling and tearing mackerel nets as well as destroying and chasing away the more valuable fishes on which it feeds.

CAT SHARKS. FAMILY SCYLIORHINIDAE

Distinctive features of these little sharks are that they have five pairs of gill openings and an anal fin; that at least one-half of the base of the first dorsal fin is rearward of the point of origin of the pelvic fins; that the front margin of the nostrils does not bear a fleshy barbel; and that they lay eggs with horny shells and tendrils at the corners. Many species are known. The familiar spotted dogfishes of European seas (two species) fall in this group. And one species calls for mention here.

Chain dogfish *Scyliorhinus retifer* (Garman) 1881

Description.—The chain-like pattern of narrow black stripes with which the reddish-brown back and sides of this little shark are marked are so distinctive that there is no likelihood of confusing it with any other shark. We need only add that its first dorsal fin stands wholly behind the rear ends of the bases of its pelvic fins; that its second dorsal fin is about one-half as large in area as its first dorsal fin; that its tail fin is square-tipped and occupies only about one-fifth of the length of the fish; and that its teeth are similar in the two jaws, narrow-triangular with a small secondary cusp on either side.

Size.—The largest specimen measured so far was 17 inches long.

General range and occurrence in the Gulf of Maine.—The range of the chain dogfish is confined to the 40–125 fathom zone between the offings of Cape Lookout, North Carolina, and of Nantucket. It seems to be the most plentiful off Virginia, in the general offing of Chesapeake Bay, where considerable numbers are taken during the winter trawl fishing. They are caught now and then as far as the offing of Marthas Vineyard, and *Cap'n Bill II* trawled one, in July 1952, south of Nantucket Lightship, Lat. 40°02′ N; Long. 69°37′ W, at 75–90 fathoms which brings it within the arbitrary boundary of the Gulf of Maine.

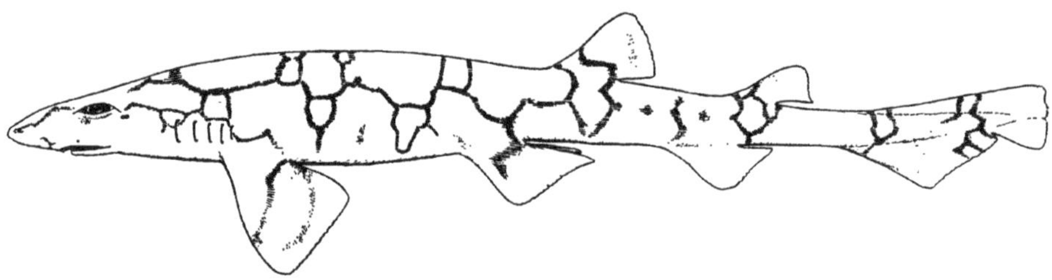

FIGURE 9A.—Chain dogfish (*Scyliorhinus retifer*), male, about 17 inches long, New Jersey. After Bigelow and Schroeder.

SMOOTH DOGFISHES. FAMILY TRIAKIDAE

These are rather small sharks, with two dorsal fins without spines, the second dorsal (in Atlantic species) nearly as large as the first, and they have an anal fin. The tail fin is very strongly asymmetrical, its lower anterior corner forming a low but rather definite lobe in some, but not in others. The teeth are small, with several rows in function imultaneously, flat, and pavement-like in some, but with three or four definite cusps in others. The eye has no nictitating ("winking") membrane, but only a longitudinal fold along the lower eyelid. They resemble the requiem sharks (Family Carcharhinidae, p. 36), except for the teeth, and for the lack of a nictitating membrane. Only one species is known from the Gulf of Maine, or is ever likely to be found there.

Smooth dogfish *Mustelus canis* (Mitchill) 1815

SMOOTH DOG; SMOOTH HOUND; GRAYFISH

Bigelow and Schroeder, 1948, p. 244.
Garman, 1913, pl. 4, figs. 6–9, as *Galeorhinus laevis*.

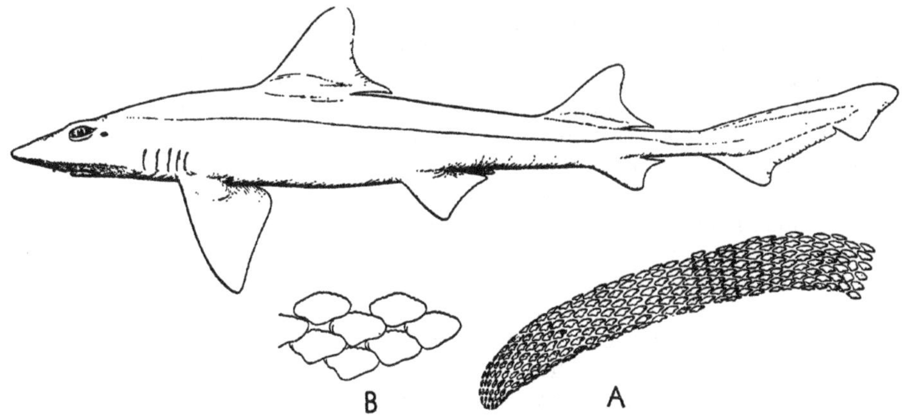

FIGURE 10.—Smooth dogfish (*Mustelus canis*), male, about 31 inches long, Woods Hole. A, tooth band of right-hand side of upper jaw, about 1.8 times natural size; B, teeth of another specimen, about 6 times natural size. From Bigelow and Schroeder. Drawings by E. N. Fischer.

Description.—The smooth dog is easily identified among Gulf of Maine sharks by having two large spineless dorsal fins, the second only a little smaller than the first, combined with low, flat, pavement-like teeth. So different, indeed, are its teeth from the awl-like or blade-like teeth of all our other sharks that a glance at the mouth is enough to separate this species from the young of any larger Gulf of Maine shark. In form this little shark is slender, flattened below, with tapering but blunt snout. Its first dorsal fin originates nearly over the hind angle of the pectorals. The second dorsal fin is about twice as large as the anal, over which its stands. The tail is of typical "shark" shape, i. e. with upper lobe much longer than lower. The hind margin of the upper lobe of the caudal is deeply notched near the tip; the lower caudal lobe is very small.

Color.—Upper surface grayish olive, slaty gray or brown, lower surface yellowish or grayish white. Newborn specimens have the upper part of the first dorsal fin edged with dusky gray; the apex of the second dorsal sooty edged or tipped, but with the rear edge white; the tail fin with a sooty blotch above near the tip, but white edged below. But these markings have mostly faded out by the time the little "dog" has grown to a length of two feet or so. Smooth dogs have a greater ability than most sharks to change shade to suit their surroundings, paling to a translucent pearly tint above white sand, but darkening on dark bottom.[80]

Size.—Smooth dogs range from about 11½ inches to about 14½ inches long when born. They mature sexually at about 3 feet, most of the mature females with young are between about 3 feet 3 inches and 4 feet 4 inches long; and a few grow to a length of about 5 feet.

Habits.—The smooth dog is most familiar as a shore fish and a bottom swimmer, commonly entering shoal harbors and bays, and even coming into fresh water. But fishermen also report them as far offshore as the "tile fish" grounds off southern New England and down to a depth of 80 to 90 fathoms. They reach the northern part of their range only as warm-season visitors; at Woods Hole they arrive sometime in May, to withdraw in late October or in November.

Food of the smooth dogfish consists chiefly of the larger Crustacea, and it is perhaps the most relentless enemy of the lobster, which had been eaten by no less than 16 percent of the fish examined by Field. Large crabs are likewise an important article in its diet, as are the smaller fishes. It has been estimated that 10,000 smooth dogfish, in Buzzards Bay, might devour more than 60,000 lobsters yearly, and perhaps one-fifth

[80] Experiments have shown that it requires only 1 to 2 hours for one to darken, but as much as 2 days to pale to the extreme; see Parker (Biol. Bull., vol. 66, 1934, p. 31).

million crabs, besides a great number of small fish (menhaden and tautog are the species most often found in smooth dogfish stomachs). And these figures are based on a sufficient number of observations of the stomach contents to serve as a general indication of the destructiveness of the smooth dogfish. They also feed on squid, especially in spring, and while they do not regularly take hard-shelled mollusks, razor clams have been found in the stomachs of several at Woods Hole. When kept in captivity they are constantly on the move, searching the bottom for food, which they find chiefly by the sense of smell though their sight is also keen.[81] Any crab that may be offered is soon found, seized, shaken to and fro, and eaten. And with packs of these sea hounds hunting over every square foot of our southern bays and sounds it is a wonder any of the larger crustaceans escape where dogfish are abundant. Field[82] also made the interesting observation that the smooth dogfish never molested healthy and active menhaden but soon devoured any sick or injured fish that might be in the same tank with them.

As this is not a characteristic Gulf of Maine fish, we need merely note that it is one of the sharks that develop a placental connection between the embryos and the mother. In other words, it is truly viviparous. The period of gestation appears to be about 10 months; off southern New England the young are born between early May and mid July. The number in a litter usually is between 10 and 20, but as few as 4 have been reported. A description of the unborn young is given by Fowler.[83]

General range.—Coastal waters of the western Atlantic, from Uruguay and southern Brazil, regularly to Cape Cod, and to Passamaquoddy Bay as a stray; also Bermuda.[84]

Occurrence in the Gulf of Maine.—The smooth dog is the second most numerous shark along the southern coast of New England, though falling far short of the spiny dogfish (p. 50). At Woods Hole, for example, pound-net catches varied during the summer of 1903 from 1 to 41, averaging about 7, and catches up to 100 have been reported from the vicinity at one time. Similarly, catches of 5 or 6 on a hand line are common in a few hours' fishing, with as many as 10 to 20 reported. But the elbow of Cape Cod and the region of Nantucket Shoals mark so definite a boundary to their dispersal eastward that while they have been reported from Provincetown, from various localities within Massachusetts Bay, and even from as far north as St. Andrews in the Bay of Fundy, where one was caught in July 1913, neither of us had ever seen one north of Cape Cod until September 21, 1951, when an angler (Ellery Sidney) showed us a female about 3 feet long that he had caught at Cohasset, while casting with an eel skin, for striped bass. So far as known its occasional incursions into the Gulf are sporadic, at least they have not been correlated with unusually warm summers or with the presence of other southern fishes. Neither has it been reported by fishermen from Georges or Browns Banks, nor was it detected there by the representatives of the Bureau of Fisheries during the trawling investigations of the years 1912 and 1913 (p. 60), or subsequently.

REQUIEM SHARKS. FAMILY CARCHARHINIDAE

This family, which includes a large number of species in tropical and temperate seas, is characterized by a head of normal shape, eye with a nictitating (winking) membrane, tail with the upper lobe considerably larger than the lower but not very long, 2 spineless dorsal fins, the first usually much larger than the second in most of the Atlantic species,[85] an anal fin, a caudal peduncle lacking lateral keels, and sharp, bladelike teeth with a single cusp. All bear "living" young;

[81] The senses of this shark have been studied by Parker (Bull., U. S. Bur. of Fish., vol. 29, 1911, pp. 43-57), and by Sheldon (Jour. Compar. Neurol. and Psychol., vol. 19, 1909, No. 3, p. 273).

[82] Rept. U. S. Comm. Fish., (1906), 1907, Spec. Pap. 6, pp. 14-16.

[83] Occas. Pap. Mus. Zool., Univ. Mich., No. 56, 1918, p. 15.

[84] Present indications are that several more or less isolated populations of this shark exist, with their areas of regular occurrence separated by wide gaps, where there is little or no intermingling. One of the best known is along the Atlantic coast, Cape Cod to North Carolina; another centers in the Gulf of Mexico-Caribbean region; a third is along southern Brazil and Uruguay. For further details, see Bigelow and Schroeder, Fishes Western North Atlantic, Part 1, 1948, pp. 250-251.

[85] The lemon shark (*Negaprion brevirostris*) of warmer waters, which has been known to stray to New Jersey, is an exception in this respect; its second dorsal is nearly as large as its first dorsal.

some have a placental connection between mother and embryo, but others do not.

Tiger shark *Galeocerdo cuvier* (LeSueur) 1822

LEOPARD SHARK

Bigelow and Schroeder, 1948, p. 266.

Description.—The tiger shark is characterized among the Atlantic members of its family by the forward position of its first dorsal fin (origin about over the arm pit of the pectorals), combined with a caudal peduncle with a low longitudinal ridge of skin on either side, besides a well-marked semilunar pit below as well as above; a very small second dorsal fin; a furrow, about as long as the snout along either side of the upper jaw; a very slender-tipped caudal fin with moderately large and pointed lower lobe; and large teeth alike in the two jaws, of very characteristic shape, with convex inner margins, deeply and conspicuously notched outer margins and strongly serrate edges (fig. 11).

Young tiger sharks are rather slender, but they become very heavy forward, with growth, though they continue tapering toward the tail. The first dorsal fin is high, triangular, and nearly as large as the pectorals, while the second dorsal is hardly one-third to one-fourth as high as the first and stands over the anal, which is of about equal size. The lower tail lobe is almost half as long as the upper, the rear margin of which is notched near the tip. The large size of the head, with very short, obtusely rounded front outline, and broad mouth occupying nearly four-fifths of the width of the head, with long grooves along the upper jaw, combined with the unique shape of its teeth, make the "tiger" easy to recognize among Gulf of Maine sharks.

Color.—Gray, or grayish brown, darkest on the upper surface. Young "tigers" up to 5 or 6 feet long, are more or less conspicuously spotted or barred with darker brown on the back and along the upper parts of the sides. But these markings fade with advancing age until large specimens are plain colored, or nearly so.

Size.—Tiger sharks are small at birth, corresponding to the large numbers in a litter, free living specimens having been reported only 18 to 19 inches long. By the time they mature they are among the larger sharks; but their size has often been overestimated. The majority of tigers caught in centers of abundance are less than 12 to 13 feet long, and the largest measured lately in the western Atlantic was one of about 18 feet, from Cuba. Repeated statements that the tiger grows to a maximum length of 30 feet have no reliable foundation, so far as we can discover.

A 4-foot specimen from Woods Hole weighed 25¾ pounds when taken from the water. Larger tigers vary widely in weight at given lengths depending on how fat they are and on the stage of development of the young in gravid females. Specimens from various localities have weighed 37 pounds at 5½ feet; 168 pounds at 6 feet; 366 to 718 pounds at 10 to 11 feet; 450 to 825 pounds at 11 to 12 feet; 630 to 1,324 pounds at 12 to 13

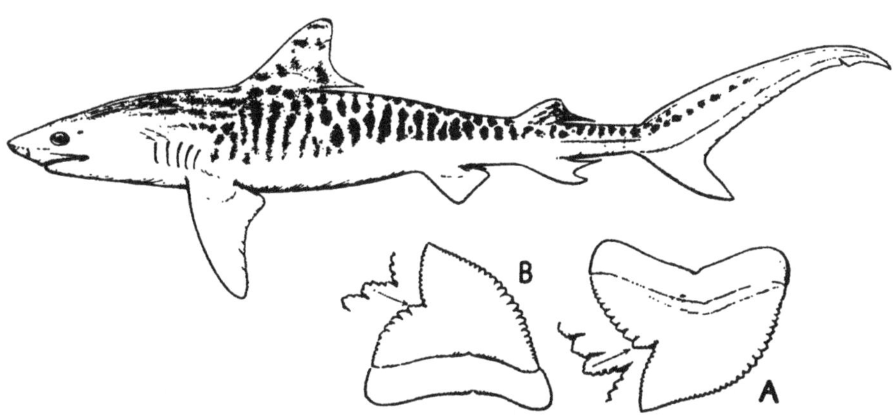

FIGURE 11.—Tiger shark (*Galeocerdo cuvier*), young male, about 49 inches long, Rhode Island. A, upper tooth, and B, lower tooth of larger specimen, enlarged. From Bigelow and Schroeder. Drawings by E. N. Fischer.

feet; and 1,028 to 1,395 pounds at 13 to 14 feet.[86]

Habits.—This voracious shark, with wide jaws and powerful teeth, preys upon the large sea turtles, other sharks, fish, and occasionally on invertebrates such as horseshoe crabs, crabs, conchs, whelks. It is proverbial for its habit of feeding on slaughter-house wastes or any other carrion. Remnants of squeteague, mackerel, hake, scup, menhaden, goosefish, and dogfish all have been found in stomachs of tiger sharks taken at Woods Hole.[87] There is no placental connection between mother and young, and the broods are very large, as many as 82 having been counted in a large female; but other litters as small as 10 to 14. In the West Indies it is much dreaded, whether or not with good cause.

General range.—Cosmopolitan in the warmer waters of all oceans; straying northward as far as Cape Cod on the American coast of the Atlantic.

Occurrence in the Gulf of Maine.—A few tiger sharks are taken in fish traps in the Woods Hole region every year, seldom before August or later than October although one was caught there July 20, 1951.[88] These specimens usually have been about 5 feet long, at most about 8 feet, and very rarely does a full-grown tiger shark stray so far from its tropical home. The tiger has not yet been recorded (on reliable evidence) from within the limits of the Gulf of Maine. It is included here because of the likelihood that a stray specimen may occasionally round the elbow of Cape Cod, or be encountered on the offshore Banks.[89]

[86] For further details and references, see Bigelow and Schroeder, Fishes Western North Atlantic, Pt. 1, 1948, p. 269.

[87] Bell and Nichols (Copeia, No. 92, March 1921, pp. 17–20) list the stomach contents of a number of tiger sharks caught off Morehead City, N. C.

[88] This shark was 8 feet, 3 inches long, taken in a pound net off Quisset Harbor, Buzzards Bay.

Blue shark *Prionace glauca* (Linnaeus) 1758

BLUE DOG

Bigelow and Schroeder, 1948, p. 282.
Garman, 1913, pl. 3, figs. 1–3 (as *Galeus glaucus*).

Description.—The blue shark is slender-bodied, thickest about its mid-length, and tapers toward head and tail (a shape usually named "fusiform"). Its snout is long with rounded tip. Its first dorsal fin is of moderate size, standing far back with the mid point of its base about midway between the inner corners of the pectorals (when these are laid back) and the points of origin of the pelvic fins. The second dorsal fin is less than one-half as high as the first, and is about equal in size to the anal over which it stands. The pectorals are narrow and very long, their tips reaching back nearly as far as the rear corner of the first dorsal. The lower lobe of the caudal fin (measured along its anterior edge) is about one-half as long as the upper lobe; the latter is conspicuously notched near the tip, and both of the lobes of the caudal fin are slender tipped.

The teeth are large, sharp-pointed, with serrate edges, and distinctive in shape. The uppers are so closely spaced that the bases of adjacent teeth

[89] The statement in the first edition that a tiger shark was once taken at Provincetown was an error. The original description of the specimen in question (Atwood, Proc. Boston Soc. Nat. Hist., vol. 10, 1865, p. 81) suggests that it was a mako (*Isurus oxyrinchus*).

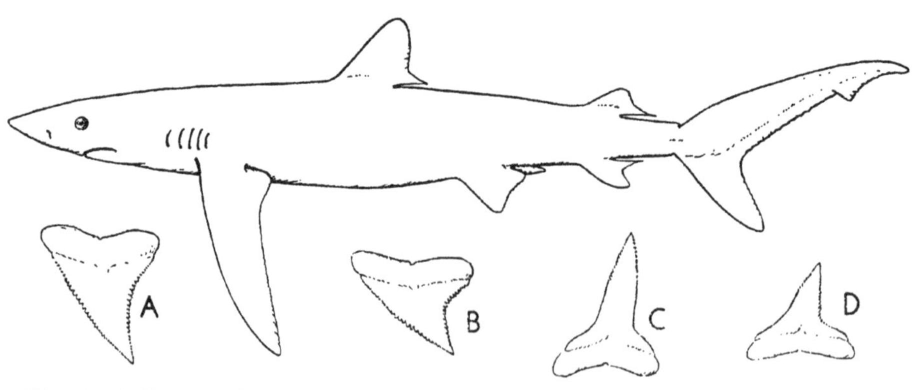

FIGURE 12.—Blue shark (*Prionace glauca*), male, about 7 feet 2 inches long, off Marthas Vineyard. A, third left-hand upper tooth, counted from mid-point of jaw; B, ninth left-hand upper tooth; C, third left-hand lower tooth; and D, eighth left-hand lower tooth; about 1.6 times natural size. From Bigelow and Schroeder. Drawings by E. N. Fischer.

overlap. The median upper tooth is nearly symmetrical, but those along the sides of the mouth have strongly convex outer margins, and deeply concave inner margins, while their points curve sharply outward toward the respective corner of the mouth. The lower teeth are narrower, more nearly symmetrical, and nearly erect.

Color.—Living specimens are dark indigo blue along the back, shading to a clear bright blue [90] along the sides; but this beautiful hue changes to a slaty or sooty gray soon after death. The lower surface is snow-white, but with the tips of the pectorals dusky and the anal fin partly sooty.

Size.—The usual length at birth seems to be between 1½ and 2 feet.[91] Blue sharks do not mature until they have grown to be 7 or 8 feet long, to judge from the sizes of the females that have been found with young; the longest we have handled was almost exactly 11 feet long. The fact that the greatest measured length so far reliably reported was only 12 feet 7 inches (3.83 meters) suggests that repeated characterizations of the blue shark as commonly growing to 15 feet are an exaggeration. If any grow to 20 feet, as is rumored, they must be giants of their kind.

Remarks.—The very long slender pectorals of the blue shark, combined with its long narrow snout, the position of its first dorsal fin far back, and its brilliant blue color, give it an aspect very different from that of the tiger shark (p. 37), of the sharp-nosed shark (p. 40), the dusky or brown sharks (pp. 41–43), or that of any other carcharhinid shark that might perhaps straggle to the Gulf of Maine.

Habits.—The blue shark is "encountered indifferently far out at sea and in continental waters, its wanderings no doubt directed chiefly by the search for food, though it may drift with ocean currents. It is frequently seen at the surface, swimming lazily with first dorsal fin and tip of caudal out of water, or basking in the sun. There is no reason to suppose that it ever descends to any great depth." [92] They sometimes follow sailing ships for days on end, to pick up scraps, and their habit of gathering when a sperm whale was killed, to feed on the carcass, was proverbial during the days of the sperm whale fishery.[93] But their normal diet is smaller fishes, of whatever kinds may be available. In northern waters herring, mackerel, spiny dogfish, and various others have been found in their stomachs. And we have several times seen a blue shark pick up a tagged cod, haddock or American pollock that we had put back in the water, on Georges Bank.

The blue shark is viviparous, that is to say, the embryo has a well developed placenta attached to the mother. As many as 28 to 54 young have been reported in a litter in the Mediterranean.

General range.—Cosmopolitan on the high seas in the warmer parts of all the oceans, including the Mediterranean; ranging northward to outer Nova Scotia and as a stray to the Banks of Newfoundland in the western side of the Atlantic; to England and Scotland in the east, with stray specimens reaching the Orkneys and southern Norway. This, we think, is by far the most numerous of the large, oceanic sharks; it is the one with which the sperm whalers were the most familiar; the one around which many of the superstitions about sharks have developed; and the one with which we have had to do most often.

Occurrence in the Gulf of Maine and along Nova Scotia.—Only one blue shark had been reported definitely from the Gulf of Maine in scientific literature, up to the time the first edition of this book was printed, though it was known to be rather common along outer Nova Scotia. But we have learned since then that it is a regular summer visitor to the southern and western parts of the Gulf, appearing occasionally in July, more often in August and September. In 1928, for example, we caught one on Stellwagen Bank on August 26, saw one over the northern end of Jeffreys Ledge on September 2, and caught four on Platts Bank on September 3, with others in sight from the vessel at nearly all times throughout the day. And many more have been seen or caught subsequently, on Platts Bank, in Massachusetts and Cape Cod Bays, where 18 were reported to us during the summer of 1935,[94] on Georges Bank where blue sharks, swimming at the surface, are a familiar sight in summer; and on Browns Bank. Two have also been re-

[90] "Sailor blue," as shown in Ridgeway's Color Standards and Color Nomenclature, 1912, p. 21.

[91] Embryos have been reported as long as about 17½ inches, and free-living specimens as small as 20–21 inches.

[92] Bigelow and Schroeder, Fishes Western North Atlantic, Pt. 1, 1948, p. 286.

[93] Nichols and Murphy (Brooklyn Mus. Sci. Bull., vol. 3, No. 1, 1916, p. 9) have given a graphic account of the blue shark as it was met with by whalers on the high seas.

[94] By J. R. Lowes, an experienced shark fisherman.

ported to us recently from the coast of Maine, a few miles east of Casco Bay.[95]

We have never heard of a blue shark in the northeastern corner of the Gulf, in the Bay of Fundy, nor along western Nova Scotia, whence they may be barred by colder surface waters. But fishermen are familiar with them off the outer coast of Nova Scotia, both offshore, and also near the coast at the times when the warm surface water presses inshore.

Blues were reported near Halifax, for instance, from time to time between August 15 and October 10, 1920, some coming close in to the entrance to the Harbor. And two specimens have been reported at Canso,[96] but whether the "blue dogs" described by local fishermen as common on the neighboring banks actually are this shark, or perhaps the porbeagle, seems doubtful. It has also been recorded from the southwest part of the Grand Bank of Newfoundland.[97]

Following westward from Cape Cod, we find many records of blues from the traps near Woods Hole, and they are often seen (or harpooned) on the continental shelf in the offing. Twenty-eight were counted 4 to 10 miles off Block Island for example, during one hour, and something like 150 to 200 during the day (13 of them were harpooned) on August 22, 1943.

Most of the blues that are seen or taken off our northern coast are medium sized or larger, though very small ones are taken from time to time.[98]

[95] By the late Walter H. Rich, who was long associated with the U. S. Bureau of Fisheries.

[96] Cornish, Contr. Canadian Biol. (1902–1905) 1907, p. 81.

[97] Rept. Newfoundland Fish. Res. Lab., 1935, p. 79.

[98] Robert Goffin reports one only 20 inches long, from Menemsha Bight, near Woods Hole, August 31, 1925; we have seen one of 21 inches, taken a few miles off Block Island, August 22, 1943; and F. D. Firth reports one 34⅛ inches long taken 65 miles southeast of Highland Light, Cape Cod, on October 23, 1930.

And for some obscure reason all but two of the adults seen in our Gulf, for which we have the pertinent information, have been males.

Commercial importance.—This shark is of no commercial value. A few are caught by anglers, mostly on natural bait, and a Blue will sometimes take an artificial lure; we hooked one off Boone Island, Maine, on a feather jig, tipped with pork rind. We have never had blues put up much resistance on a heavy hand line until hauled in to the side of the vessel, when they thrash about violently, but it is said that a large one will make long and powerful runs, if hooked on rod and reel.

The blue shark has always been looked on with contempt by the sperm whalers, who were more familiar with it than anyone else. We find no well-authenticated case of one attacking a swimmer. sailors' yarns to the contrary notwithstanding.

Sharp nosed shark *Scoliodon terrae-novae* (Richardson) 1836

Bigelow and Schroeder, 1948, p. 295.
Garman, 1913, pl. 2, figs. 1–4.

Description.—This little shark is separable from any other carcharhinid that has yet been reported from the Gulf of Maine or that is likely to be, by its upper and lower teeth which are perfectly smooth along the edges from tip to base, combined with a so-called "labial furrow" of considerable length running forward along each side of each jaw from the corner of the mouth toward the nostril. This last character, while not conspicuous, is a precise one.

The trunk is slender, highest about at the first dorsal fin, tapering both fore and aft. The snout varies rather widely in length and in bluntness at the tip. The point of origin of the first dorsal fin

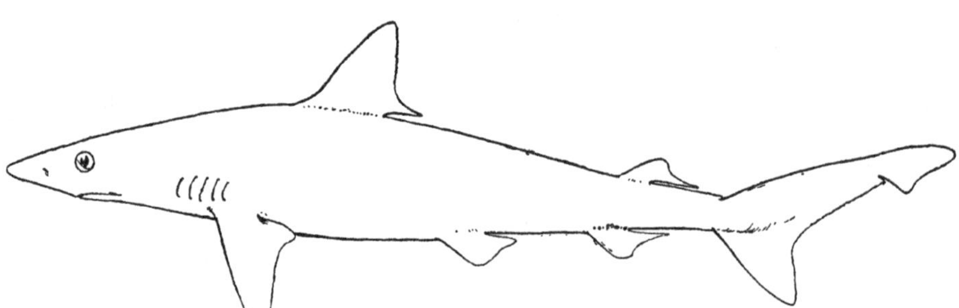

FIGURE 13.—Sharp-nosed shark (*Scoliodon terrae-novae*), female, about 31 inches long, from the Bahamas. From Bigelow and Schroeder. Drawing by E. N. Fischer.

is about over the inner corners of the pectorals when the latter are laid back; its height is about one-half as great as the distance from the tip of the snout to the level of the origin of the pectorals. The second dorsal is only about one-quarter as high as the first; its point of origin is about over the mid-point of the base of the anal fin; the anal is a little larger than the second dorsal. The tail fin occupies about one-quarter of the total length of the shark; its lower lobe (measured along the anterior edge) is a little less than one-half as long as the upper lobe, the rear edge of which is deeply notched near the tip. The pectoral fins are smaller relatively than in any other local species of this family, their length, armpit to tip, being only a little greater than the height of the first dorsal fin. The teeth are alike in shape in the two jaws, sharp-pointed and smooth edged; those in the center of the mouth are symmetrical and erect, but those along the sides have weakly concave inner margins, but deeply notched outer margins, and are increasingly oblique toward the corners of the mouth.

Color.—Brown to olive gray above, with the dorsal and caudal fins more or less dark edged; white below and along the rear margins of the pectorals.

Size.—Mature specimens are commonly between 26 and 30 inches long; a few grow to 36 inches.

General range.—Both sides of the tropical–subtropical Atlantic; Morocco to Cameroon and the Cape Verde Islands in the east; Uruguay to North Carolina in the west; occasional to Woods Hole, and as a stray to the Bay of Fundy.

Occurrence in the Gulf of Maine.—Our only reason for including this warm-water shark is that one was taken at Grand Manan Island,[99] at the mouth of the Bay of Fundy, in 1857.[1-2]

Early reports of it from Newfoundland were based on a misconception.

SHARKS OF THE GENUS *Carcharhinus*

The members of the genus *Carcharhinus* are set apart from other Atlantic members of the family Carcharhinidae by the following combination of characters: The mid-point of base of the first dorsal fin is at least as near to the level of the axils of the pectorals as to the level of the origin of the pelvics (separating them from the blue shark, p. 38); no labial furrows on lower jaw, and furrow on upper jaw reduced to a very short slit at the extreme corner of the mouth, directed outward (separating them from the tiger shark, p. 37, and from the sharp-nosed shark, p. 40); second dorsal fin much smaller than first dorsal (separating them from the lemon shark, p. 35, footnote 85); edges of upper teeth more or less finely serrate but without larger denticles near the base, and edges of lower teeth perfectly smooth, without lateral denticles (separating them from the tiger shark, p. 37, from the sharp-nosed shark, p. 40), and from *Paragaleus pectoralis*, a tropical shark that has been taken off southern New England.[3]

This is a warm-water group, fifteen species of which are known to inhabit the western side of the Atlantic, most of them resembling one another closely in general aspect. Only one of these (the dusky shark, described on p. 41) has yet been reported reliably from within the confines of the Gulf of Maine, while only one other (the brown shark, p. 43) is likely to be found there. If a stray *Carcharhinus* from offshore that does not agree with the following descriptions of one or other of these should be taken on Georges Bank, or on Nantucket Shoals east of the longitude of Cape Cod, we hope that its captor can identify it by means of the keys and descriptions of the genus that we have given in Part 1 of the Fishes of the Western North Atlantic.

Dusky shark *Carcharhinus obscurus* (LeSueur) 1818.

Bigelow and Schroeder, 1948, p. 382.

Description.—The combination of characters that place the dusky shark among the western Atlantic members of its genus are: Trunk about one-fifth as high at first dorsal fin as it is long to origin of the caudal fin, tapering both forward and rearward; snout broadly rounded in front, its length in front of the nostrils less than the distance between the nostrils; the front edge of the nostril is not expanded as a definite lobe; the midline of the back between the two dorsal fins has a low but definite ridge, a character which is very precise, though seemingly minor; the first dorsal fin is considerably smaller than in the brown shark

[99] This specimen, collected by A. E. Verrill, is in the Museum of Comparative Zoology.

[1-2] See Jordan and Evermann, Bull. 47, U. S. Nat. Mus., Pt. 1, 1896, p. 43, footnote.

[3] For description, see Bigelow and Schroeder, Fishes of the Western North Atlantic, Pt. 1, 1948, p. 276.

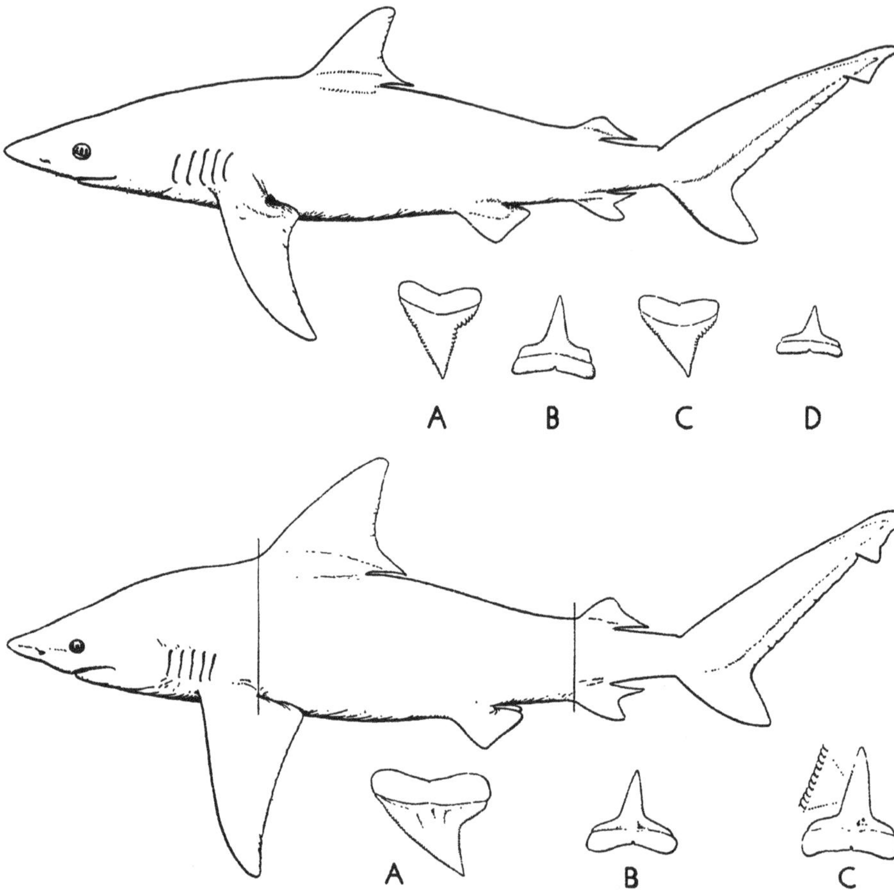

FIGURE 14.—Above: Dusky shark (*Carcharhinus obscurus*), female about 39 inches long, Woods Hole. A, third upper tooth; B, fourth lower tooth; C, ninth upper tooth; D, tenth lower tooth; about 2.4 times natural size. Below: Brown shark (*Carcharhinus milberti*), female, about 4 feet 10 inches long, from Woods Hole. A, ninth upper tooth; B, eighth lower tooth; C, third lower tooth; about 1.4 times natural size. From Bigelow and Schroeder. Drawings by E. N. Fischer.

(p. 43), with more deeply concave rear margin, its point of origin about over the inner corner of the pectoral (over the armpit of the pectoral in the brown shark); its apex is narrowly rounded. The free rear corner of the second dorsal fin is less than twice as long as the vertical height of the fin. The anal fin is a little longer, along the base, than the second dorsal and stands about under the latter. The caudal fin occupies between one-quarter and one-third of the total length of the shark, the lower caudal lobe (measured along its anterior edge) is about two-fifths as long as the upper lobe; and the upper lobe is noticeably slender toward its tip. The pectorals are about as long (from origin to tip) as the distance from the tip of the snout to the level of the first pair of gill openings, usually narrower, relatively, than in the brown shark, and sometimes more definitely sickle-shaped.

The upper teeth are broadly triangular; nearly erect toward the center of the mouth but weakly oblique toward its corners; their inner margins are nearly straight, the outer margins increasingly concave outward along the jaw. The lower teeth are erect, symmetrical, with narrow cusp on a broadly expanded base. Both the upper teeth and the lower are serrate along the edges, the lower the more finely so.

Color.—All the fresh caught specimens we have seen have been bluish or leaden gray on the back and upper part of the sides, including the pectorals, but this shark has also been described as pale gray above or even dirty white, perhaps over a white sand bottom. The trunk is white below, the pectorals grayish, darkening to sooty at their tips; the pelvics and anal fins grayish white.

Size.—The usual length at birth is a little more than three feet.[4] Adult dusky sharks so far measured have ranged from 10 feet 4 inches to 11 feet 8 inches in length, and they are said to grow to 14 feet, though perhaps not on very convincing evidence.

General range.—Western Atlantic, north to southern New England and to Georges Bank, south to southern Brazil, at least by name. A shark very closely allied to *obscurus* has been reported under that name in the eastern Atlantic, from Spain to Table Bay, South Africa, including Madeira, the Canaries, the Cape Verdes, Ascension Island, and St. Helena. But we have yet to learn its precise relationship to the *obscurus* of the western Atlantic.

Occurrence in the Gulf of Maine.—The dusky shark has been taken repeatedly off the coasts of New Jersey and of Long Island, N. Y.; also at Woods Hole, where we have handled 12 specimens during the past few summers, 6 of them in August 1944. But it so seldom strays to cooler waters farther east that only one shark has been recorded from Nantucket, and one from Georges Bank, that probably were of this species and not some other carcharhinid.[5] Thus it has no real place in the fauna of the Gulf.[6]

Brown shark *Carcharhinus milberti* (Müller and Henle) 1841

Sand bar shark

Bigelow and Schroeder, 1948, p. 368.
Garman, 1913, pl. 3, figs. 4–6 (as *Carcharinus platyodon*).

Description.—The brown shark differs from the dusky (only member of its genus that seems actually to have been taken within the Gulf) in the more forward position and larger size of its first dorsal fin, in its broader pectorals, and in its stouter trunk, heaviest forward (compare specimens in figure 14). Also, the anterior edge of its nostril is expanded as a low but definite triangular lobe, which is not the case in the dusky shark. Other characters (in combination) that mark it off from other members of this genus that might stray to the Gulf are: Mid-line of the back with a low ridge between the two dorsal fins; snout forward of a line connecting the front margins of the nostrils, considerably shorter than the distance between the nostrils; point of origin of second dorsal fin about over origin of anal fin, its free rear corner only a little longer than the height of the fin; apex of first dorsal fin angular; length of pectorals along anterior margin about as great as distance from tip of snout to level of second pair of gill openings; distance from rear tips of pelvic fins to origin of anal fin as long as base of anal fin, or longer, fifth gill openings longer than horizontal diameter of eye.

The teeth resemble closely those of the dusky shark (see figure 14).

Color.—Upper surface slate gray to brown; lower surface a paler tint of the same hue, or white; fins without any conspicuous black markings. When alive some of the dermal denticles are bright blue, at least on some specimens.

Size.—Sexual maturity is reached at a length of about 6 feet; maximum length about eight feet.[7]

General range.—Southern Brazil, Louisiana, both coasts of Florida, and northward along the Atlantic coast of the United States to southern New England; also the tropical-subtropical belt of the eastern Atlantic, and the Mediterranean, or represented there by an extremely close relative.[8]

Occurrence in the Gulf of Maine.—Next to the sand shark, this is the most numerous of the larger sharks along the coasts of New Jersey and of New York. Some visit the vicinity of Woods Hole, though so few that the number taken there in most summers probably is not greater than six or seven. It has not been reported as yet from

[4] Embryos have been reported up to 38 in. long (965 mm.), and a free living specimen of only 39 in. (993 mm.); see Bigelow and Schroeder, Fishes Western North Atlantic, Pt. 1, 1948, p. 387.

[5] Probably this species and not the brown shark because 11–12 feet long.

[6] In the first edition of this book, the dusky shark was said to have been taken at three localities within the Gulf. But one of these records, at least, was almost certainly based on a blue shark, and the others probably were (Bigelow and Schroeder, Fishes Western North Atlantic, Pt. 1, 1948, pp. 292, 368).

[7] Seven feet 10 inches is the greatest measured length that we have found recorded, with convincing evidence that the specimen actually was one of this species.

[8] If the eastern Atlantic–Mediterranean form is actually identical with the American, as seems to be the case, the specific name *milberti* of Müller and Henle, 1841, must be replaced by *plumbeus* proposed by Nardo in 1827 for the brown shark of the Adriatic.

within the limits of our Gulf, but is included here on the chance that a stray specimen may be taken, either on the outer coast of Cape Cod, on Nantucket Shoals, or on Georges Bank.

THE HAMMER-HEADED SHARKS. FAMILY SPHYRNIDAE

The peculiar hammer-shaped head, with eyes far apart, sufficiently characterizes the Gulf of Maine sharks of this family, which resembles the requiem sharks (p. 36) otherwise. Five species are known in the western Atlantic, all of them tropical-subtropical in nature. Two of these have been reported from our Gulf, but only as strays.

Shovelhead *Sphyrna tiburo* (Linnaeus) 1758

BONNET HEAD SHARK

Bigelow and Schroeder, 1948, p. 420.

Garman, 1913, pl. 1, figs. 4–6 (as *Cestracion tiburo*).

Description.—The peculiar shovel-shaped head of this shark is enough to distinguish it readily from any other shark known from the Gulf of Maine, except for the hammerhead, from which it is readily distinguished by the fact that its head is considerably narrower, is more rounded in front, and is not deeply indented opposite each nostril; that the posterior margin of its anal fin is only weakly concave, and that the outermost four or five of its lower teeth next each outer corner of its mouth are low and rounded, not blade-like. The eyes of the shovel-head shark, like those of the hammerhead, stand at either edge of the expanded head; the first dorsal fin originates a little behind the "armpit" of the pectoral, is somewhat higher than the pectorals are long, and is higher than long; the very small second dorsal fin originates a little behind the origin of the anal fin; the upper lobe of the tail is notably long (about one-third as long as the body of the fish) and deeply notched near the tip, the lower lobe is about one-third as long as the upper lobe. The anal fin is larger than the second dorsal fin, its posterior margin is only slightly concave; the pectorals are broadly triangular, their anterior margins about as long as the distance from the level of their own points of origin to the front of the mouth.

Color.—Gray or grayish brown above, and a paler shade of the same below; some are marked with a few small dark, roundish spots along the sides.

Size.—This shark is much smaller than the hammerhead, rarely exceeding 5 feet in length; it is said to reach 6 feet.

General range.—Tropical-warm temperate Atlantic; from southern Brazil to North Carolina, in the west, and as a stray to southern New England and Massachusetts Bay; tropical West Africa in the east; also from southern California

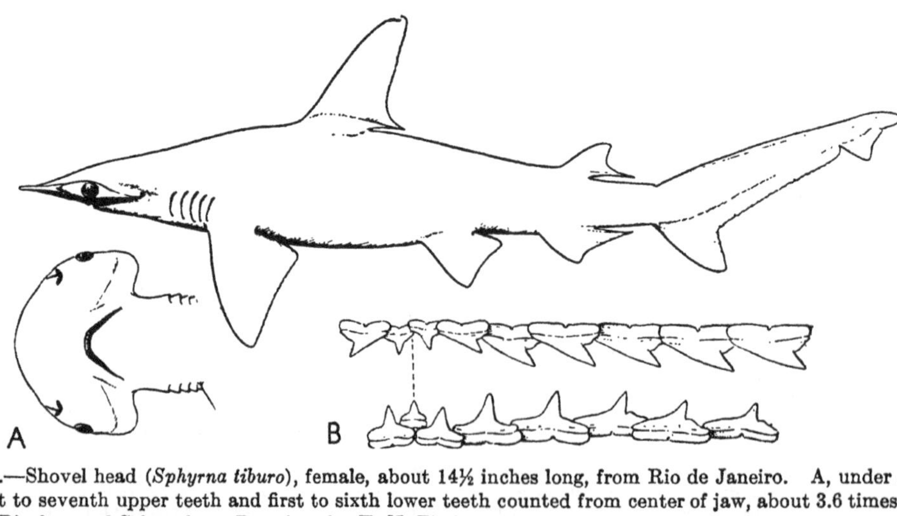

FIGURE 15.—Shovel head (*Sphyrna tiburo*), female, about 14½ inches long, from Rio de Janeiro. A, under side of head; B, first to seventh upper teeth and first to sixth lower teeth counted from center of jaw, about 3.6 times natural size. From Bigelow and Schroeder. Drawings by E. N. Fischer.

to Ecuador on the Pacific Coast of America, or represented there by a very close relative.[9]

Occurrence in the Gulf of Maine.—Our only reason for including the shovel-head here is that a stray specimen has been reported from Massachusetts Bay.[10] It has also been taken once at Newport, R. I., and a commercial shark fishery that was carried on in Nantucket Sound in the summer of 1918 is said to have yielded six of them.[11]

Common hammerhead *Sphyrna zygaena* (Linnaeus) 1758

Bigelow and Schroeder, 1948, p. 436.

[9] On this point, see Bigelow and Schroeder, Fishes of the Western North Atlantic, Pt. 1, 1948, p. 425, footnote 20. A shark has also been reported as *tiburo* from China and from the Philippines, but without convincing evidence as to its identity.

[10] By Garman, Mem. Mus. Comp. Zool., vol. 36, 1913, p. 161. Apparently the specimen is no longer in existence.

[11] Personal communication by R. H. Bodman, who operated this fishery.

Description.—The very differently shaped head of the hammerhead, the shape of its anal fin with much more deeply concave posterior margin, and the fact that the outermost four or five of its lower teeth on each side are blade-like, like those nearer the center of its mouth, are ready field marks to separate the hammerhead from the shovelhead (cf. fig. 16 with fig. 15). The anal fin, too, is only about as large as the second dorsal in the hammerhead (considerably larger than the second dorsal in the shovelhead). Otherwise the positions and shapes of the fins and the size and shape of the tail are much alike in the two species.

Color.—Leaden or brownish gray above, shading along the sides to pure or grayish white below; the tips and edges of the dorsal and caudal fins are more or less dusky; and the tips of the pectorals are black on some specimens.

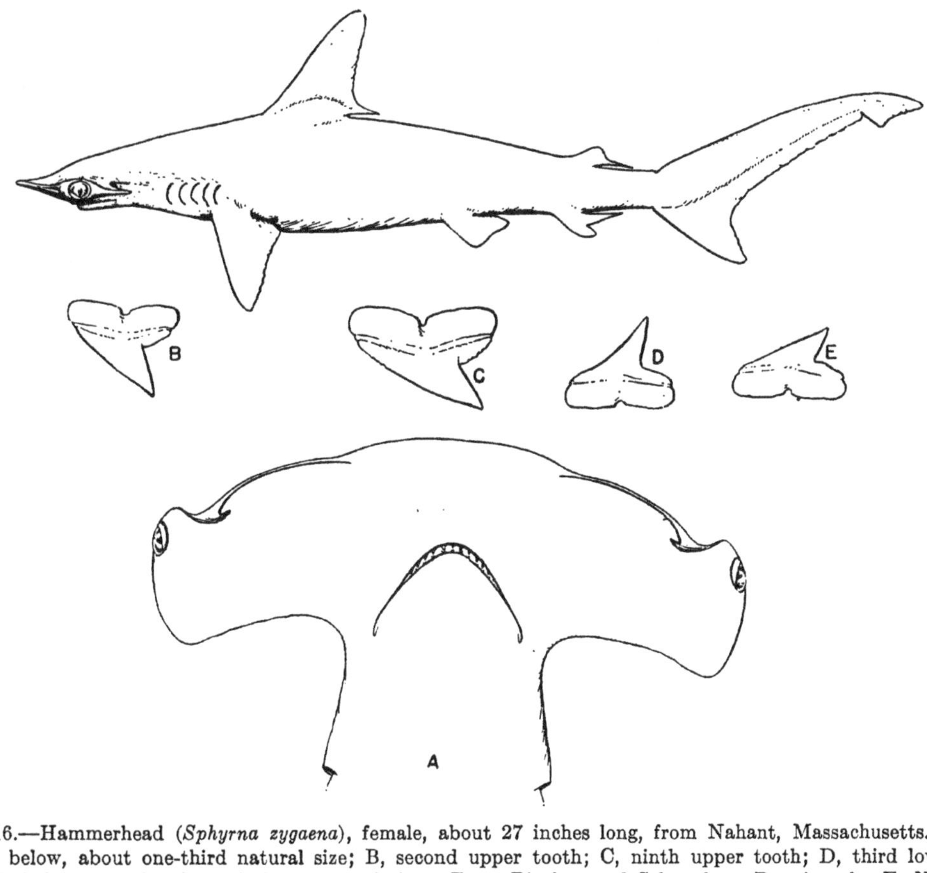

FIGURE 16.—Hammerhead (*Sphyrna zygaena*), female, about 27 inches long, from Nahant, Massachusetts. A, head from below, about one-third natural size; B, second upper tooth; C, ninth upper tooth; D, third lower tooth; E, ninth lower tooth; about 4 times natural size. From Bigelow and Schroeder. Drawings by E. N. Fischer.

Size.—It appears that hammerheads are commonly about 19 to 20 inches long when they are born; seemingly, they mature sexually at about 7 to 8 feet; they are often taken 9 to 11 feet long, and occasionally as long as 12 to 13 feet.[12] Most of those that visit southern New England are less than 6 to 7 feet long, some very small indeed.[13] In 1805, however, one of 11 feet was netted at Riverhead, L. I. And the fact that it contained parts of a man in its stomach has been chiefly responsible for the bad reputation of this species of hammerhead.

Two other large sharks closely related to the common hammerhead, the tropical hammerhead (*Sphyrna lewini* Griffith, 1834)[14] and the great hammerhead (*Sphyrna mokarran* Rüppell, 1835) [15] occur along the South Atlantic coast of the United States. The first of these, in particular, might stray as far as Cape Cod, as many tropical fishes do, for it has been recorded from the offing of Cape May, New Jersey. They resemble the common hammerhead closely in general appearance, but both of them may be distinguished from the latter by the fact that the front outline of their head is scalloped in the midline, not evenly rounded there as it is in the common hammerhead. For further accounts of them, see Bigelow and Schroeder.[16]

Habits.—Since hammerheads are an accidental visitor to the Gulf, we need only remark that they are pelagic in habit, often swimming with dorsal and caudal fins out of water, and are to be met with indifferently out at sea or near land. They feed chiefly on fish, including smaller sharks (including their own kind), and sting rays, the tail spines of which are sometimes found imbedded in their jaws. Like tiger sharks, they make themselves a pest in warmer latitudes where fisheries for sharks are carried on, by devouring those that they find entangled in the nets. As many as 30 to 37 embryos have been found in a gravid female, and the embryos do not develop any placental connection with the mother, so far as is known.

General range.—Widespread in the tropical to warm temperate belts of the Atlantic, of the Pacific, and probably of the Indian Ocean as well; north commonly to southern New England, straying to Massachusetts Bay and as far as Halifax, Nova Scotia.[17]

Occurrence in the Gulf of Maine.—Hammerheads (often in small schools) wander northward every summer, along the Atlantic seaboard; they are often to be seen basking at the surface (some harpooned) a few miles out, off Marthas Vineyard and Nantucket; and one is occasionally taken in one or another of the fish traps near Woods Hole. But the longitude of Cape Cod so sharply bounds their yearly dispersal that the only records from the Gulf of Maine, or from Nova Scotia waters, are of stray specimens from Chatham and Provincetown on the outer shores of the Cape; of one about 27 inches long from Nahant, in the inner part of Massachusetts Bay;[18] of two small ones recently from Casco Bay;[19] of one taken many years ago, off Brier I., on the Nova Scotian side of the Bay of Fundy;[20] of a 12-footer harpooned between Georges and Browns Banks in August 1928 by the sword fishing schooner *Doris M. Hawes;* of a small one caught in Halifax Harbor, Nova Scotia, in September 1932;[21] and of another about 21 inches long taken in a trap off Sambro Head, near Halifax, August 25, 1938.[22]

[12] The larger hammerheads that are sometimes reported probably are not this species, but the great hammerhead (*Sphyrna mokarran*, p. 46, note 16).

[13] Dozens of little ones, of about 2½ feet, have been seined on the outer shore of Long Island, N. Y., in August.

[14] The account of this species, in Bigelow and Schroeder, (Fishes of the Western North Atlantic, Pt. 1, 1948, p. 415) was as *diplana* Springer, 1941. But Fraser-Brunner (Rec. Austral. Mus., vol. 22, No. 3, 1950, pp. 213–214), has shown that it cannot be separated from the Indo-Pacific *S. lewini* of Griffith, 1834, a much older name.

[15] Tortonese has recently pointed out (Ann. Mag. Nat. Hist. Ser. 12, vol. 3, No. 36, 1950, p. 214) that the name *tudes* Valenciennes 1822 that has been applied commonly to the great hammerhead of the Atlantic actually belongs to a different species; consequently that the correct name of the great hammerhead is *mokarran* Rüppell, 1835, it being identical with that Indo-Pacific species.

[16] Fishes Western North Atlantic, Pt. 1, 1948, pp. 415, 428.

[17] For further details of distribution, see Bigelow and Schroeder, Fishes of the Western North Atlantic, Pt. 1, 1948, p. 442.

[18] This specimen, obtained many years ago by Louis Agassiz, is in the Museum of Comparative Zoology.

[19] Seen in the fish market at Portland, Maine, by the late Walter H. Rich.

[20] McKenzie, Proc. Nova Scotia Inst. Sci., vol. 20, 1939, p. 13.

[21] Vladykov, Proc. Nova Scotia Inst. Sci., vol. 19, Pt. 1, 1935. p. 8.

[22] McKenzie, Prov. Nova Scotia Inst. Sci., vol. 20, 1939, p. 13.

THE SPINY DOGFISHES. FAMILY SQUALIDAE

This group is characterized by the lack of an anal fin, combined with the presence of two dorsal fins, each of which is preceded by a fixed spine which is long and conspicuous in some, but so short in others that its presence can be detected only by touch. The teeth are alike in the two jaws in some, unlike in others.

Spiny dogfish *Squalus acanthias* Linnaeus 1758

DOGFISH; PIKED DOGFISH; GRAYFISH

Bigelow and Schroeder, 1948, p. 455.
Garman, 1913, pl. 14, figs. 1–4.

Description.—Any little gray or brownish shark, with a large sharp spine lying along the front margin of each dorsal fin, caught within the Gulf, or on the shoaler parts of the offshore fishing banks, is practically sure to be this "dog," of which there are thousands in the Gulf to every one shark of any other kind. One of its relatives, the black dogfish (p. 51), is a regular inhabitant of the deeper slopes of the offshore Banks that front the Gulf, where we also trawled more than 50 specimens of another relative *Etmopterus princeps* Collett 1904 during the summer of 1952. But there is no danger of confusing the common spiny-dog with either of these, for they are velvety black in color, the rear margins of their tail fins are indented near the tip, which is not the case in the spiny-dog, and each of their teeth, at least in the upper jaw (lower jaw as well in the black dogfish) has 3 to 5 sharp points, but only one point in the spiny dog.

This is a slender little shark, with flattened head and snout tapering to a blunt tip. Its first dorsal fin stands between pectorals and pelvics; its second dorsal fin is about two-thirds as large as the first; its pectorals form nearly an equilateral triangle; and its pelvics are well forward of its second dorsal fin. The dorsal fin spines lie close along the front margins of the two dorsals, the first not more than one-half as long, and the second nearly as long as the front margin of their respective fin, and they are very sharp. The spiny-dog has no anal fin, a lack separating it from all smooth-finned sharks known from the Gulf of Maine, except for the Greenland shark (p. 53), *Dalatias* (p. 55), and the bramble shark (p. 56). There is a low fold of skin on either side of the root of the tail back of the second dorsal fin, so small, however, that there is no danger of confusing it with the caudal keels of the mackerel-shark tribe. The teeth are small, their sharp points bent toward the outer corners of the mouth so that they form a nearly continuous cutting edge along each jaw.

Color.—The upper surface is slate colored usually, sometimes tinged brown, with a row of small

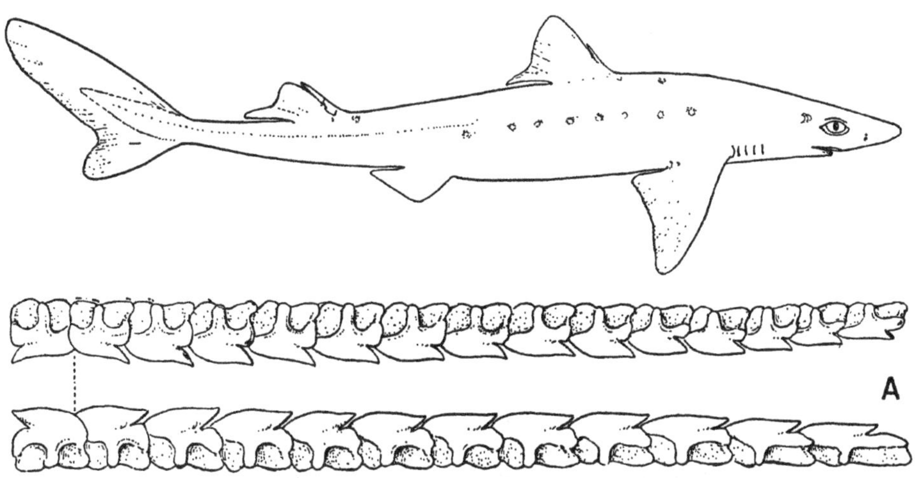

FIGURE 17.—Spiny dogfish (*Squalus acanthias*), female, 27 inches long; after Garman. A, upper and lower teeth, midpoint of mouth marked by the dotted line, about 3 times natural size. From Bigelow and Schroeder. Drawing by E. N. Fischer.

white spots on each side from the pectoral fin to abreast of the anal fin, and with a few other white spots in front of the first dorsal and behind it, also in front of the second dorsal fin. These spots are most conspicuous on small fish up to 12 or 14 inches long and they fade with growth until they disappear altogether in some specimens. The margins of the first and second dorsals, and of the caudal are more or less dusky at birth, but soon fade. The lower surface ranges from pale gray to pure white.

Size.—The majority are between 8⅝ and 13 inches long when born. Most of the adult males are from about 2 feet to a little less than 3 feet long; adult females are from a little less than 2½ feet to almost 3½ feet; maximum length about four feet. Mature females average 7 to 10 pounds, a few reach 15 pounds if very fat, and 20 pounds has been reported.

Habits.—Much has been written of the habits of the spiny dogfish, but nothing to recommend it from the standpoint either of the fishermen or of its fellow creatures in the sea. It is one of the more gregarious of our fishes, swimming in schools or packs. Swedish fishermen assert that young dogs school separately from their parents, and it is certain that fish of a size continue to associate together as they grow, the result being that any given school runs very even, consisting as a rule either of the very large mature females, or of medium-sized fish (either mature males or immature females), or of small immature fish of both sexes in about equal numbers.

Apart from their general seasonal migratory movements, dogfish are governed by the movements of the fishes on which they prey. And recent marking experiments have shown that some of them cover long distances in their wanderings, for two tagged near St. Johns, Newfoundland, in mid-July 1942 were recaught off Cape Ann,[23] one on November 23, 1943, the other on December 4 of that year,[24] while others from the same tagging experiment were caught within the Gulf of St. Lawrence.[25] Fortunately they seldom stay long in one place, but there is seldom, if ever, a time during the summer when they are not common on some part of the Gulf of Maine coast. So erratic are their appearances and disappearances that where one has good fishing today he may catch only dogfish tomorrow and nothing at all the day after, the better fish having fled these sea wolves and the latter departing in pursuit.

The dogfish use their back spines for defense, curling around in a bow and striking, which makes them hard to handle on the hook. It is probable, too, that the spines are slightly poisonous, general report to this effect being corroborated by the fact that the concave surfaces are lined with a glandular tissue resembling the poison glands of the venomous "weever" (*Trachinus draco*)[26] of Europe.

Voracious almost beyond belief, the dogfish entirely deserves its bad reputation. Not only does it harry and drive off mackerel, herring, and even fish as large as cod and haddock, but it destroys vast numbers of them. Again and again fishermen have described packs of dogs dashing among schools of mackerel, and even attacking them within the seines, biting through the net, and releasing such of the catch as escapes them. At one time or another they prey on practically all species of Gulf of Maine fish smaller than themselves, and squid are also a regular article of diet whenever they are found. Dogfish are also known to take worms, shrimps, and crabs. And when they first arrive at Woods Hole in May they are often found full of Ctenophores, being one of the few fish that eat these watery organisms. Often, too, they bite groundfish from the hooks of long lines, or take the baits and make it futile to fish with hook and line where they abound.

Fishermen are familiar with the fact that the female spiny dog bears "living" young (this has been known since the days of Aristotle). The eggs are large, well stored with yolk, and during early stages those in each oviduct (so-called "uterus") are contained in a horny capsule that breaks down later, leaving the embryos free in the "uterus," to which they have no placental attachment. The number in a litter is commonly 4 to 6; sometimes as many as 8 to 11, or as few as 2.

According to recent studies, the females carry their young for 18 to 22 months. Accordingly, the adult females caught in our Gulf contain either very early embryos, averaging only about three-fourths of an inch in length by September, or

[23] About 14 miles offshore.

[24] On Middle Ground about 25 miles off Cape Ann.

[25] Templeman, Fish. Res. Bull., Newfoundland Dept. Nat. Res., No. 15, 1944, pp. 67–69.

[26] Evans (Philos. Trans. Royal Soc., London, Ser. B, vol. 212, 1923, pp. 8, 27) describes the spines and gives clinical records of the effects of wounds inflicted by them.

much larger ones, 7 to 11 inches long by that month; i. e., nearly ready for birth. Similarly, we have taken females with embryos 9 to 10¼ inches long in November, on the Cholera Bank near New York Harbor. And it now seems established that most of the young are born on the offshore wintering grounds.[27] But dogfish so small as evidently to have been newborn are occasionally taken along southern New England and in the Gulf in early summer; also on Nantucket Shoals where the *Albatross II* trawled some of 10½ to 13 inches in August, showing that the season of production extends through the spring, or even into the summer as in 1905 when females taken off Gloucester in July gave birth to young on capture.[28]

General range.—Both sides of the North Atlantic, chiefly in the temperate and subarctic belt; also both sides of the northern Pacific;[29] and represented in the corresponding thermal belt of the southern hemisphere by a relative (or relatives) so close that it is doubtful whether they differ in any recognizable way from the spiny-dog of the north.

Occurrence in the Gulf of Maine. The spiny dogfish ("dogfish" or "dog" in common parlance) makes up for the comparative rarity of other sharks in the Gulf of Maine by its obnoxious abundance. To mention all the localities from which it has been reported there would be simply to list every seaside village and fishing ground from Cape Cod to Cape Sable. It is as familiar, too, on the offshore banks as it is along the coast; also along outer Nova Scotia, in the Gulf of St. Lawrence, on the Grand Banks, and along the east coast of Newfoundland to southeastern Labrador. There is no record of it from the North American coast north of Hamilton Inlet, but stray specimens have been taken along the southwest coast of Greenland.[30] To the southward, fishermen are familiar with it in season as far as Cape Lookout, N. C., and a few stray even to southern Florida and to Cuba.[31]

Dogfish are seasonal visitors on the coast, striking in about as early along New Jersey (March), and even on Georges Bank (March–April), as along North Carolina. In the inner parts of the Gulf of Maine the date of the first heavy run of dogfish varies widely from year to year and from place to place. We have not heard of them there before May. But the period of freedom may close as early as the last half of the month, in some years.

In 1903, for example, they had appeared as far north as Penobscot Bay by the middle of May. And while it is not until June that they usually arrive in numbers in the Massachusetts Bay region, it is sometimes impossible to set gill or drift nets anywhere between Cape Cod and Cape Elizabeth after the first days of that month, so numerous are they. In 1913 the first heavy run of dogfish struck Ipswich Bay on June 14, and they appeared there at about the same date in 1905, but there is much local variation in this respect. In 1903, for example, they did not appear until early July at Provincetown, though swarming a month earlier in Massachusetts Bay, in Ipswich Bay, and off Penobscot Bay. But in 1920 they appeared at Provincetown by May 25 to 26 when one set of mackerel traps caught 23 barrels of them, and another 21 barrels. They usually strike in all along the northern Maine and west Nova Scotia coasts by the end of June; but few are seen until late in July in Passamoquoddy Bay. They have been recorded as early as July 1 near Raleigh, on the Newfoundland side of the Strait of Belle Isle, but they are not caught in any numbers in the inner parts of the Gulf of St. Lawrence until well into July, and they have not been reported from southeastern Labrador until early in September.[32]

In the southern part of its range, from North Carolina to New York, the spiny dogfish is a spring and autumn transient only. West of Cape Cod (at Woods Hole, that is, and along Long Island)

[27] Females that we saw trawled off Block Island in 60–65 fathoms in late January 1950, gave birth to young on the deck of the vessel.

[28] McIntire, Rept. Comm. Fish. Game Massachusetts, (1905) 1906, p. 108.

[29] We have found no consistent differences between North Atlantic and North Pacific specimens. For further discussion of this point, and further details as to the occurrence of the spiny-dog in the two sides of the North Atlantic, see Bigelow and Schroeder (Fishes of the Western North Atlantic, Pt. 1, 1948, pp. 453, 463).

[30] Jensen (Selachians of Greenland, Mindeskr. Japetus Steenstrup, Pt. 2, No. 30, 1914, p. 7) lists several definite records of this species at Sukkertoppen and near Holsteinborg, West Greenland.

[31] Repeated reports of it as plentiful along eastern Florida seem to have referred to some other shark; the basis for similar reports from Cuba and Trinidad doubtless was the Cuban dogfish, *Squalus cubensis* Rivero.

[32] See Templeman (Res. Bull. 15, Newfoundland Dept. Nat. Res., 1941, pp. 56, 64) for dates of arrival around the coast of Newfoundland in different years.

they are transients mostly, passing north in spring and south in autumn, though some summer there; even considerable numbers in some years.[33] And it seems that most of them withdraw from Massachusetts Bay also during the warmest period, for few are taken there between June and September. But they continue present all summer along outer Cape Cod, and here and there throughout the northern and eastern parts of the Gulf, in varying abundance.

Most of the dogfish take their autumnal departure from the inner parts of the Gulf during October, few being caught on the coast north of Massachusetts Bay after November 1. But they sometimes stay later, as in 1903 (a big dogfish year), and again in 1942, when they were abundant along the outer shore of Cape Cod as late as the first week of November. Ordinarily none are caught within the Gulf of Maine north of Georges Bank in winter, but this has its exceptions. In 1913, for example, a few were caught 20 miles off Cape Ann on November 19 to 24, many near Boon Island from December 5 to 13, and on Jeffreys Ledge on December 11 and 12.

In 1882, schools were reported off Portsmouth, N. H., even as late as February, an exceptional event.

Dogfish appear earlier in spring and linger later into the winter on Georges Bank (fig. 18) than in the inner parts of the Gulf. It is safe to say that there are few there in March, the earliest definite record (obtained during the investigations of 1913, only year of record, being of 25 fish caught on the "winter cod ground" east of the shoals (long. about 67°, lat. about 41°40′) between the 20th and the 22nd, and of 46 from the same general region from the 27th to the 30th, while some are trawled there all summer. In 1913, a few were taken in November and in December; a few also on the southern part of the Bank (lat. about 41°, long. about 67°30′) on January 20 to 22 in 1914.

Apparently dogfish reach Browns Bank later than they do Georges, for none was taken there on April 14 in 1913, though they are only too plentiful there in summer. It is also likely that they depart earlier, although a few lingered as late as December 3 to 12 on Western Bank off Halifax in that year.

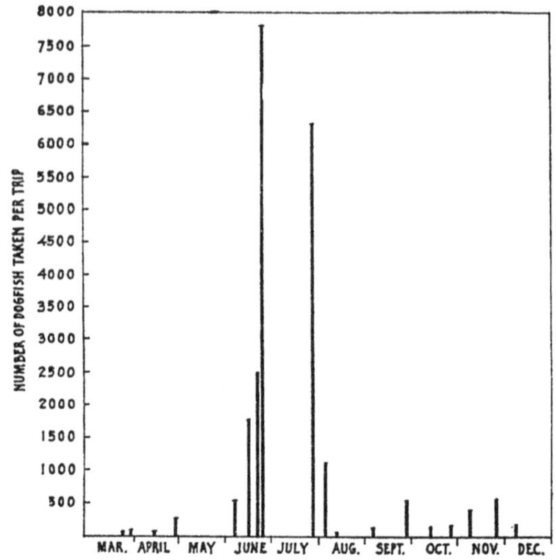

FIGURE 18.—Numbers of spiny dogfish caught on certain otter trawling trips to Georges Bank, during the different months of 1913.

It now seems certain that the spiny dogfish winter chiefly in deeper water offshore, for considerable numbers have been trawled at that season on the outer part of the continental shelf off Block Island, in 50 to 65 fathoms, where we saw several hundred (200 in one haul) trawled during the last week of January 1950; off New York in November and January;[34] also in February off the Middle Atlantic coast in 16 to 70 fathoms, south as far as the offing of Cape Hatteras. On the other hand, the fact that numbers of them have been found washed on shore in January on the southwest coast of Newfoundland suggests that some of those that summer in that general region may survive the winter in the deep trough of the Gulf of St. Lawrence. They are usually so thin when they reappear on the coast in spring as to suggest that they feed but little during the winter.

This is the only Gulf of Maine shark that even remotely rivals the important food fishes in numbers. Unfortunately, the statistics of the commercial landings for American waters do not afford any information in this regard. But spiny dogs must be plentiful indeed in our waters when they can sometimes be caught as fast as they can

[33] For details, see Bigelow and Schroeder, Fishes of the Western North Atlantic, Pt. 1, 1948, p. 464.

[34] Mr. Thomas Quast informs us that many were taken from the schooner *Victor*, long-lining for tile fish, on the outer edge of the continental shelf, off New York, during the second week of January 1928.

be hauled in; when a long line, with 1,500 hooks, has been known to bring in a dogfish on nearly every hook; and when an average trawl catch of 6,000 to 8,000 per trip was made on Georges Bank in 1913 during their season of abundance. At the time of the 1904 to 1905 peak it was estimated from recorded catches that at least 27,000,000 were being taken yearly off the coast of Massachusetts.[35]

More precise information from waters farther north is that 10,391,000 pounds, or 2 to 3 million individual dogfish, were caught in 1938, in Placentia Bay, Newfoundland, with no apparent effect on their numbers.[36] In short, they may be as plentiful in our Gulf as they are on the Cornish coast, where the record catch of 20,000 in a single haul was made many years ago.

Spiny dogfish appear to have been more numerous in the Massachusetts Bay region during the last quarter of the past century and during the early nineteen hundreds than they had been previously. At Woods Hole, on the contrary, they are said to have been much more plentiful before 1887 than they have been at any time since. To a certain extent, of course, reports of fluctuations in abundance from year to year must be discounted as reflecting the movements of the great schools that may visit one part of the coast one summer and another part the next, not a general alteration of the stock. But the many fishermen who reported to the Massachusetts Commissioners in 1905 were unanimously of the opinion that dogfish had multiplied steadily for 20 to 30 years past, and reports from British coasts were to the same effect. Perhaps the years 1904–1905 marked the apex of this wave of multiplication; at any rate dogfish were reported as distinctly less troublesome to the mackerel netters in 1913 than they had been previously. And little complaint has been made of them in late years.

But it is not safe to conclude from this that the stock is at a low ebb at present, for it was the hand- and long-line fishermen that suffered most from them; and it is only as they increase the amounts of trash fish dumped overboard that the dogfish bother the otter-trawlers.

Importance.—During the years when the ground fishery was chiefly by hook and line, fishing often was actually prevented by dogfish in Massachusetts and Ipswich Bays, unless cockles (*Polynices*) were used for bait, for dogfish do not take these. The general replacement of hook and line fishing by the otter trawl has put an end to widespread complaints on this score. But when schools of dogfish get into a net or seine, they so snarl the twine that disentanglement and repair may be the work of days. And it has been estimated that they may do some $400,000 worth of damage annually to fishing gear, and to fish caught by such gear, off the coast of Massachusetts alone, during their peaks of abundance there.

With the dogfish so plentiful and destructive, it is no wonder that serious efforts have been made to make them a source of revenue instead of a dead loss. And the dog is a far better food fish when fresh than is generally appreciated, as is evident by the large amounts landed in the fishing ports of northwestern Europe. But it has never been in any demand for the table, on our coasts, though it would offer a large supply of cheap food were a satisfactory method found for canning it. During their more recent periods of plenty various efforts have been made to utilize them on a large scale for fertilizer, and for liver oil (it compares favorably with cod for vitamin A, though it is much poorer in vitamin D), on the Atlantic coasts of the United States and Canada; however such developments have been short-lived. And dogfish have not been of sufficient value up to the present to compensate for a hundredth part of the damage they do.[37]

Black dogfish *Centroscyllium fabricii* (Reinhardt) 1825

Bigelow and Schroeder, 1948, p. 482.
Garman, 1913, pl. 10, figs. 5–8.

Description.—The notched margin of the upper tail lobe distinguishes this shark at a glance from the spiny dogfish, with which it agrees in having a long pointed spine at the front edge of each dorsal fin. It differs further from the common dogfish in that its dorsal spines are deeply grooved along each side, whereas in the "dog" they are rounded; in the location of the pelvic fins, the rear axils of

[35] Report, Comm. Fish and Game, Mass., (1906), 1907, p. 20.

[36] Templeman, Newfoundland Fish. Res. Bull., 15, 1944, p. 72.

[37] For further discussion of the damage done by dogfish and of their commercial possibilities, see Ann. Rept., Comm. Fish. Game Mass. (1905), 1906, pp. 97–169; Rept. U. S. Comm. Fish. (1902) 1904, pp. 228–229; Field, Doc. 622, Rept. U. S. Comm. Fish. (1906) 1907, pp. 21–23; Field, Bull. U. S. Bur. Fish., vol. 28, 1910, pp. 243–257; Mayor, Contr. Canad. Biol. (1918–1920) 1921, pp. 125–135; and Templeman, Newfoundland Fish Res. Bull. 15, 1944

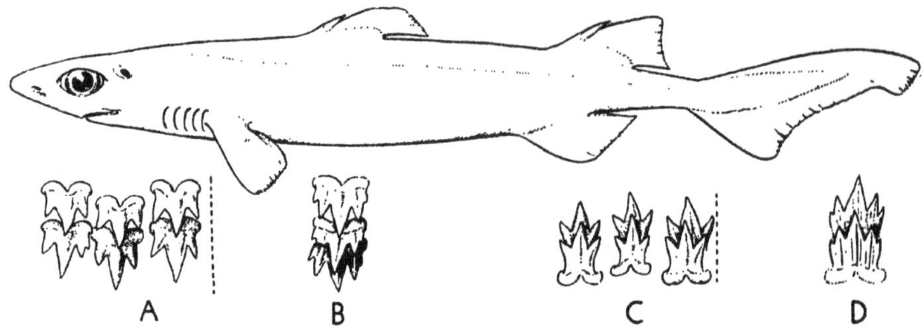

FIGURE 19.—Black dogfish (*Centroscyllium fabricii*), female, about 25 inches long, from the southeast slope of Georges Bank. A, first three upper teeth counted from center of jaw; B, twentieth upper tooth; C, first three lower teeth; D, lower sixteenth tooth; about 5 times natural size. From Bigelow and Schroeder. Drawings by E. N. Fischer.

which stand almost directly under the front origin of the second dorsal fin instead of some distance in front of the latter; in its small pectorals of rounded outline; in the shapes of its teeth, each of which has 3 or 5 sharp points; in its broad rounded snout; and in its very dark color. Like the spiny dogfish, it lacks an anal fin.

Size.—Adult specimens range from 2 to 3½ feet in length, that is, about the same size as the spiny dogfish.

Color.—Uniform dark brown to black, below as well as above.

Habits.—In West Greenland waters cephalopods, pelagic crustaceans, and medusae have been found in their stomachs, and females have been taken with embryos in February. Perhaps they are luminescent, for their skins bear minute deeply pigmented dots, suggesting the light organs of the brilliantly luminescent shark *Isistius brasiliensis.*

General range.—Northern North Atlantic; Faroe Bank, Faroe-Shetland Channel and Iceland in the east; West Greenland; Davis Strait; and outer slopes of the fishing banks in the west, southward to Georges Bank; chiefly deeper than 150 fathoms.

Occurrence in the Gulf of Maine. In the years when a long line fishery for halibut was carried on regularly, black dogfish were often caught along the slopes of the offshore Banks, from Grand to Browns and to the eastern part of Georges, if sets were made down to 200 fathoms or deeper. And while they dropped out of sight with the general abandonment of that fishery, no doubt they are as plentiful now as formerly, for we trawled about 100 of them, 6 to 24½ inches long, off southwestern Nova Scotia, at 290 to 580 fathoms, on the *Caryn* of the Woods Hole Oceanographic Institution, in June 1949. How far they range to the west and south, at the appropriate depths, is not known.[38]

Portuguese shark *Centroscymnus coelolepis* Bocage and Brito Capello, 1864

Bigelow and Schroeder, 1948, p. 494.

Garman, 1913, pl. 14, figs. 5–8.

Description.—This shark can be identified easily by the fact that while its general appearance (especially the absence of anal fin, the situation of its pelvics far back under the second dorsal, and its rather stout form and blunt snout) might lead a hasty observer to think he had caught a small Greenland shark; more careful examination, by touch if not by eye, would reveal a short spine close in front of each dorsal fin. The first dorsal fin is smaller than in any of our sharks except in the "Greenland," (p. 53), and in *Dalatias licha* (p. 55), the second dorsal is a little larger than the first, and the pelvics are larger than either of the dorsals. The tail is noticeably short and broad and the rear edge of its upper lobe is notched. The teeth are different in the two jaws; narrow, pointed, and of the seizing type in the upper; broader, oblong, with a notch on the outer side near the tip, and forming a continuous cutting edge in the lower. The dermal denticles are flat, scale-like, closely overlapping, and clothe the entire trunk.

Color.—Dark chocolate brown, belly as well as back and fins.

[38] Its range has been said to extend to New York, but without supporting evidence; and report of a young one from the Gulf of Mexico (Goode and Bean, Smithsonian Contrib. Knowledge, vol. 30, 1895, p. 11), probably was based on some other shark.

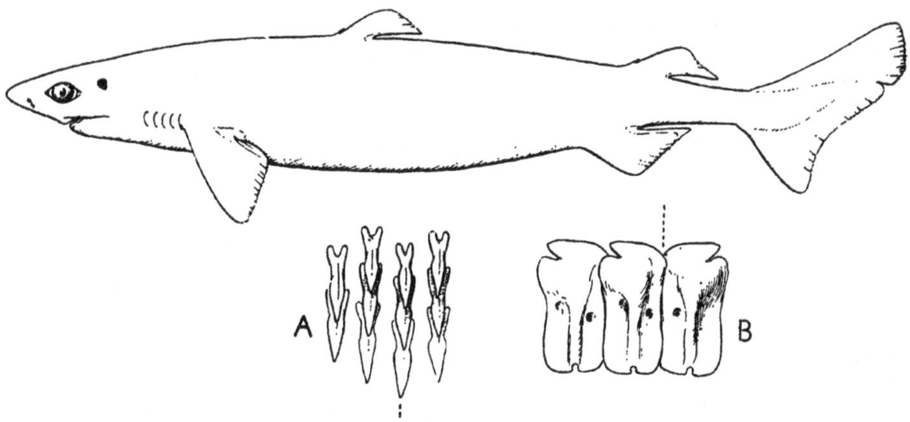

FIGURE 20.—Portuguese shark (*Centroscymnus coelolepis*), female about 42½ inches long, off Banquereau Bank. A, upper teeth, and B, lower teeth from center of mouth, about 3.4 times natural size. From Bigelow and Schroeder. Drawings by E. N. Fischer.

Size.—Adults measure from 3 to 3½ feet long, as they are caught. Garman records one 44 inches long taken off the coast of New England. About 9 inches is the smallest recorded.[39]

Habits.—Little is known of its habits beyond the fact that it is a deep-water species, and that it was caught regularly by Portuguese fishermen with hand lines, a fishery that Wright [40] described as follows:

> Some 600 fathoms of rope were let out, the first 30 or 40 fathoms of which had fastened to it at intervals of a fathom a series of small ropes, on each of which was a large hook baited with a codling. This fishing tackle remained below for about two hours, when they commenced to haul it in. When it arrived at the last few fathoms, they pulled in, one after another, five or six specimens from 3 to 4 feet long. The species was the *Centroscymnus coelolepis* Bocage and Capello. These sharks, as they were hauled into the boat, fell down into it like so many dead pigs.

Thirteen to 16 young have been found in females caught off Portugal.

General range.—This deep-water shark, originally discovered off Portugal, has since been taken at various other eastern Atlantic localities.[41] Definite records of it for the western Atlantic are from the slopes of the Nova Scotian Banks and of Georges, at depths of 180 to 250 fathoms, perhaps 15 to 20 specimens in all. But Goode and Bean's [42] old characterization of them as abundant on the Banks at 200 fathoms and deeper presents its local status more correctly, for fishermen long lining for halibut often caught one or two a trip in the deeper gullies between the offshore Banks.

THE GURRY SHARKS. FAMILY DALATIIDAE

The gurry sharks, like the spiny dogfishes, lack anal fins, but they have no spines in their dorsal fins. The teeth in the upper jaw are noticeably unlike those in the lower.

Greenland shark *Somniosus microcephalus* (Bloch and Schneider) 1801

SLEEPER SHARK; GURRY SHARK; GROUND SHARK

Bigelow and Schroeder, 1948, p. 516.
Garman, 1913, pl. 15, figs. 4–6.

Description.—The Greenland shark is notable for its small dorsal fins, without spines, the second dorsal being of about the same size as the first, and for small pectorals hardly larger than the pelvics, coupled with the absence of an anal fin and with a tail of more fish-like form than that of most other sharks except for the mackerel-shark tribe. Bearing these points in mind, particularly the absence of an anal fin and of dorsal spines, it cannot be confused with any shark common in our Gulf. And while it resembles the rare Portuguese shark in the sizes and relative situa-

[39] A male 228 mm. long, examined by us, in the U. S. National Museum from the continental edge south of Nantucket.

[40] Ann. Mag. Nat. Hist., Ser. 4, vol. 2, 1868, p. 426.

[41] Iceland; Faroe Bank; Madeira; Azores; Morocco; Cape Verde I.: For key to other species of the genus, see Bigelow and Schroeder, Fishes Western North Atlantic, P. 1, 1948, p. 494.

[42] Smithsonian Contrib. Knowledge, vol. 30, 1896, p. 14.

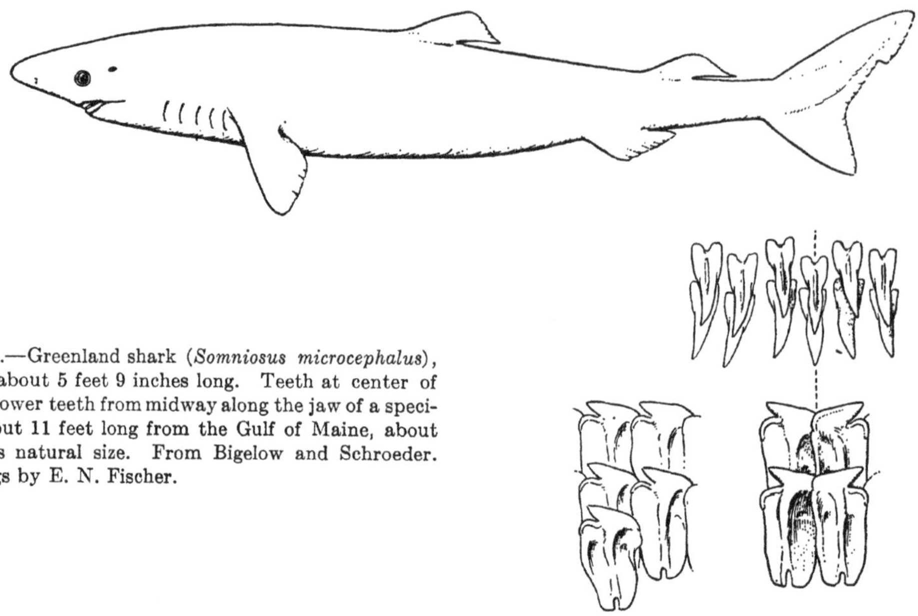

FIGURE 21.—Greenland shark (*Somniosus microcephalus*), female, about 5 feet 9 inches long. Teeth at center of mouth; lower teeth from midway along the jaw of a specimen about 11 feet long from the Gulf of Maine, about 1.8 times natural size. From Bigelow and Schroeder. Drawings by E. N. Fischer.

tions of its dorsal and anal fins, in its general form, and in its teeth, it is easily separable from the "Portuguese," both by lacking any trace of spines in its dorsal fins, by its thorn-like and loosely spaced dermal denticles, and by its more lunate tail. It also grows much larger than the Portuguese shark. We need only note, further, that while its upper teeth are narrow and awl-like, its lowers are broad, squarish, forming a nearly continuous cutting edge, with the single cusp directed sharply outward; that its gill openings are short and located low down on the sides of the neck; that its eyes are very small; and that it is stout shouldered, with blunt rounded snout, as Scoresby pictured it more than a century ago.[43]

Color.—Blackish, coffee brown, or ashy-, purplish-, or slate gray, below as well as above; changing to bluish gray if the epidermis is rubbed off, as is apt to happen when one is caught; the back and sides are marked with many indistinct dark crossbars on some specimens.

Size.—This is one of the larger sharks. It is said to grow to a length of 24 feet, but 21 feet is the largest of which we find definite record,[44] and 16- to 18-footers are unusual. One of 16½ feet was reported from the Grand Banks in 1934; one of 16 feet off Portland, Maine, in 1846; one of about 15 feet off Cape Ann in 1849; and another of about that same size was caught on a long line north of Cape Ann in February 1931. Perhaps 8 to 14 feet is a fair average for adults, that is not often exceeded among the hundreds caught annually off West Greenland and around Iceland. The 21-foot British specimen mentioned above was said to weigh about 2,250 pounds; two Gulf of Maine specimens, each about 11 feet long, weighed about 600 and 650 pounds, respectively.

Habits.—Off Greenland, and along the Labrador coast, the Greenland sharks tend to approach the surface in winter, often coming right up to the ice. But most of them withdraw in summer to 100 fathoms or deeper. And the few that visit our Gulf appear to hold rather closely to the bottoms of the deeper troughs, though a stray may come so close to the shore now and then, and into water so shoal as to blunder into a fish weir; one such event is on record for Passamaquoddy Bay.

This is one of the most sluggish of sharks, offering no resistance whatever when hooked, and it is entirely inoffensive to man.[45] But it is ex-

[43] Arctic Regions, 1820, vol. 2, pl. 15, figs. 3 and 4.

[44] Jenkins, Fishes British Isles, 1925, p. 325.

[45] Tales to the effect that it attacks Greenlanders in their kyaks are apparently mythical, and Doctor Porsild, Director of the biological station at Disko, said that the Eskimos do not fear it as they do the killer whale; nor is there any authentic instance on record of a shark attacking a human being near Iceland.

tremely rapacious. It devours any carrion eagerly, such as whale meat, blubber from whaling operations, or the carcasses of young seals that are left on the ice off the Newfoundland-Labrador coasts. And its habit of gathering when there has been a big killing of narwhals in Greenland waters is proverbial. Apart from carrion (which cannot be available except on rare occasions), its diet includes a wide variety of fishes, large and small. Seals are a favorite food, and in view of its sluggishness, it is somewhat astonishing that it should be able to capture prey as active as seals, halibut, and salmon. The specimen from Cape Cod Bay, mentioned above, contained half a dozen flounders and a large piece (with hide and hair) that had been bitten out of the side of a seal. It is also known to eat crabs, large snails, even medusae. Objects as large as an entire reindeer (without horns), a whole seal, a 3-foot cod, and a 39-inch salmon, found in Greenland shark stomachs, give some measure of their appetite. In line with this, they will bite on any fish or meat bait, the more putrid and ill smelling the better.

Large numbers of soft eggs, without horny capsules, ranging in size up to that of a goose egg, have been found repeatedly in female Greenland sharks, but never any embryos, suggesting that this may be an egg-laying species.[46]

General range.—Northern Atlantic, from Polar latitudes south to the North Sea and accidentally to the mouth of the Seine and perhaps to Portugal in the east; south regularly to Newfoundland and the northern part of the Gulf of St. Lawrence in the west, and less commonly to the Gulf of Maine. It is represented in the Mediterranean region, in the North Pacific, and in the sub-Antarctic by forms that appear to be distinct, though closely allied to it.[47]

Occurrence in the Gulf of Maine.—Although there is no reason to suppose that the Greenland shark ever appears in our Gulf save as a straggler from the north, its presence there has been signalized on a number of occasions. Two specimens, for example, were taken in the neighborhood of St. Andrews in 1915 (one caught in a weir and the other on a long line). It has been reported off Eastport; off Cape Elizabeth whence 6 were landed at Portland between 1925 and 1948;[48] on Jeffreys Ledge, where one of about 15 feet was caught on a long line, on February 16, 1931;[49] near Cape Ann; off Marblehead and Nahant; in Massachusetts Bay; off Barnstable in Cape Cod Bay; at Provincetown; and in Cape Cod Bay off the entrance to the Cape Cod Canal, where one between 10 and 11 feet long was taken by a trawler in April 1924, landed in Boston and identified by us.

Recorded captures in the Gulf include small specimens as well as large, and have been for all four seasons of the year, suggesting that when a Greenland shark does stray southward to the Gulf, it may survive there for years. The local records are distributed so widely as to show that an odd specimen is to be expected anywhere in the deeper parts of the Gulf. And rumor has it that they were more numerous in our waters in early colonial times when Atlantic right whales were still being killed in numbers off the Massachusetts coast.[50]

Commercial importance.—This shark is not plentiful enough in our Gulf to be even of potential value. But it has long supported a fishery off northern Norway, around Iceland, and in West Greenland waters, chiefly for its liver oil.[51] In Greenland the flesh is dried also for dog food, and to a small extent in Iceland for human consumption. But it produces an intoxicant poisoning if eaten fresh, though it is wholesome if dried.[52]

Dalatias licha (Bonnaterre) 1788

Bigelow and Schroeder, 1948, p. 502.

Description.—This shark resembles the Portuguese shark in the relative sizes and positions of its fins; also in its scales. But its dorsal fins do not have any trace of spines, while the serrate margins of its lower teeth, in combination with their triangular shape, mark it off from any other shark without an anal fin that is known yet from the North Atlantic. Its trunk is rather slender, its snout short and bluntly rounded, and the lower-anterior corner of its tail fin is not expanded as a definite lobe. Its upper teeth are slender, awl-

[46] The Mediterranean *Somniosus rostratus*, on the contrary, bears living young.

[47] For recent discussion of the species of *Somniosus*, see Bigelow and Schroeder, Fishes Western North Atlantic, Pt. 1, 1948, p. 515.

[48] Reported to us by the late W. W. Rich.

[49] This one was landed in Boston, where we saw it.

[50] When they gather to feed on whale, narwhal, and seal carcasses in their northern home, they may linger for a long time in the vicinity.

[51] The annual catch off West Greenland was around 32,000 during the first decade of the present century.

[52] For accounts, see Jensen, 1914 (Selachians of Greenland, Mindesk. Jap. Steenstrup, vol. 2, No. 30, 1914, p. 12); also Clark (Science, N. Ser., vol. 41, 1915, p. 795).

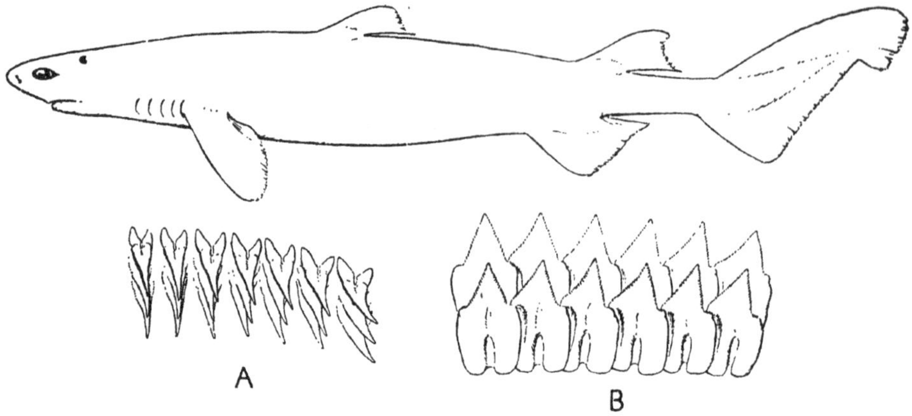

FIGURE 22.—Dalatias (*Dalatias licha*), female, 58 inches long, from Georges Bank. A, upper teeth and B, lower teeth from central part of mouth, about 1.5 times natural size. From Bigelow and Schroeder. Drawings by E. N. Fischer.

shaped, curving somewhat outward toward the corners of its mouth; but the lowers are erect, broadly triangular, with serrate edges.

Color.—Dark chocolate, cinnamon, or violet brown below as well as above; the upper surface sometimes with poorly defined blackish spots; the dorsal and pectoral fins with pale or whitish edges, the tail tipped with black.

Size.—Most of those caught are between 40 and 60 inches long; 72 inches is the longest recorded so far. The Gulf of Maine specimen illustrated in figure 22 was about 5 feet long and weighed 23½ pounds, gutted.

General range.—Eastern Atlantic, from tropical West Africa to the Irish Atlantic slope; recorded once from the American coast.

Occurrence in the Gulf of Maine.—Our only reason for mentioning this shark is that a female, about 5 feet long, was taken on the northern edge of Georges Bank on August 19, 1937 (fig. 22).[53]

THE BRAMBLE SHARKS. FAMILY ECHINORHINIDAE

The only living representative of this family (it is represented among the tertiary sharks) resembles the Greenland shark and its allies in lacking both anal fin and dorsal spines, but its teeth are alike in the two jaws.

Bramble shark *Echinorhinus brucus* (Bonnaterre) 1788

SPINY SHARK

Bigelow and Schroeder, 1948, p. 527.

Description.—The location of the first dorsal fin above the pelvics instead of about midway between the latter and the pectorals, and the very different shape of its tail fin (*cf.* fig. 23 with fig. 21), are the most conspicuous field marks separating this shark from the Greenland shark. *Brucus* also differs from the latter in that the teeth are alike in the two jaws, instead of unlike, and that the skin of its back and sides is sparsely strewn with large scales with either one or two sharp points.

[53] Recorded by Nichols and Firth, Proc. Biol. Biol. Soc. Wash., vol. 52, 1939, p. 85.

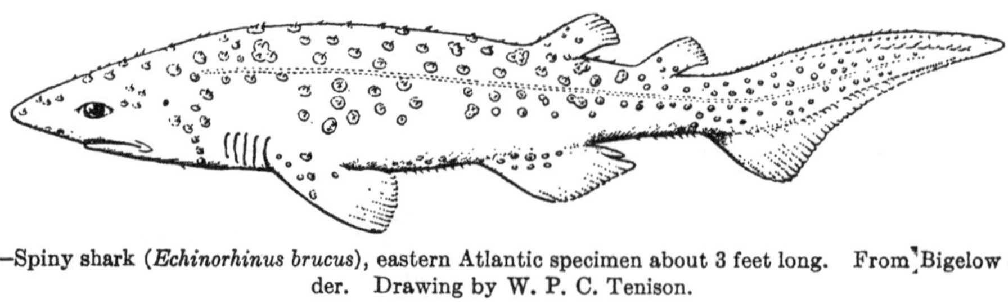

FIGURE 23.—Spiny shark (*Echinorhinus brucus*), eastern Atlantic specimen about 3 feet long. From Bigelow and Schroeder. Drawing by W. P. C. Tenison.

Color.—Described as dark gray, olive or brown above, with metallic reflections, and with or without darker blotches; as paler brown or gray to white below. The scales have been described as luminescent,[54] but there are no special luminous organs.

Size.—The largest of which we have found a record (a specimen from British waters) was 9 feet long. One 8 feet 4 inches long weighed about 300 pounds.

General range.—Eastern Atlantic (including the Mediterranean) from tropical West Africa to Ireland and the North Sea, and accidental in the western Atlantic; represented in South Africa; off California; in the Hawaiian, Japanese, and Australo-New Zealand regions, and in Arabian waters by forms that probably cannot be distinguished from *brucus* of the Atlantic.

Occurrence in the Gulf of Maine.—A single specimen of this little known shark came ashore at Provincetown in December 1878. This and one taken near Buenos Aires more recently [55] are the only records of it from the western Atlantic.

Torpedoes, Skates, and Rays. Order Batoidei

This tribe falls into four groups, so far as the Gulf of Maine fauna is concerned: first, the torpedoes (family Torpedinidae), with large caudal fin, interesting because provided with electric organs capable of giving a strong shock; second, the skates (family Rajidae), with very thin bodies, comparatively short tails without tail spines, and only a trace of caudal fin; third, the sting rays (families Dasyatidae and Rhinopteridae), with long whiplike tails armed with a stiff saw-edged spine (or spines); and fourth, the devil rays (Mobulidae) with two ear-like fins extending forward from the front of the head. Most of our common species belong to the second group.

Among torpedoes, skates, and rays, fertilization is internal as it is among sharks, and the modification of the posterior edges of the pelvic fins into rodlike semitubular claspers (the copulatory organs) distinguishes males and females at a glance. Some bear "living" young, ready for independent existence; others lay eggs.

[54] Cornish, Zoologist, Ser. 2, vol. 10, 1875, p. 4501.

[55] Berg, Com. Ictiol. Comm. Mus. Nac. Buenos Aires, vol. 1, No. 1, 1898, p. 10.

KEY TO GULF OF MAINE SKATES AND RAYS

1. The front of the head bears a pair of separate, ear-like fins, extending forward ---------- Devil ray, p. 77
   The front of head does not bear a pair of separate ear-like fins extending forward ---------- 2
2. There is a large triangular caudal fin, as well as two well developed dorsal fins on the tail ---------- Torpedo, p. 58
   There is no distinct caudal fin; the dorsal fins, if any, are very small ---------- 3
3. No long dorsal spine on tail ---------- Common skates 4
   There is a long saw-edged dorsal spine (or spines) on the tail ---------- 11
4. The upper surface of the disc is marked with conspicuous black rosettes ---------- Leopard skate, p. 66
   The markings on the upper surface of the disc are not in the form of black rosettes ---------- 5
5. There are no conspicuous thorns along the mid-dorsal zone of disc between the spiracles and the base of tail; the lower surface of disc is marked with black dots or dashes, marking the openings of the mucous pores.
   Barndoor skate, medium sized and large specimens, p. 61
   There are one or more rows of conspicuous thorns along the mid-dorsal zone of disc rearward from the spiracles; the lower surface of disc is not marked with black dots or dashes ---------- 6
6. There are no large thorns on the rear ¼–⅓ of tail ---------- Smooth-tailed or Prickly skate, p. 70
   There are one or more rows of large thorns along the rear part of tail as well as farther forward along it ---------- 7
7. There are no large thorns on upper side of disc between the spiracles and the level of axils of pectoral fins.
   Barndoor skate, very small specimens, p. 61
   The upper side of disc, rearward from spiracles, has more or fewer large thorns ---------- 8
8. The thorns of the midrow on the tail are much larger and more conspicuous than any other thorns on the tail, and not more than 9 or 10 in number ---------- Thorny skate, p. 72
   No one row of thorns along the tail is much larger or more conspicuous than the other thorns on the tail; there are at least 15 thorns in each of the rows along tail ---------- 9

9. There is only one row of large thorns along the midzone of the disc from the nape to the level of the axils of the pectoral fins; the first and second dorsal fins are separated by a definite space or at least by 1 or 2 thorns; the forward angle of the disc is less than 110°; the upper surface of the disc is marked with short dark bars as well as with roundish spots ---------- Brier skate, p. 65

   There are at least three rows of thorns along the midzone of the disc from the nape to the level of the axils of the pectoral fins; the first and second dorsal fins are not separated by a definite interspace or by a thorn or thorns; the forward angle of the disc is more than 125°; the upper surface is not marked with dark bars though it is variously spotted ---------- 10

10. Upper teeth in at least 72 series, most often 90–100; does not mature sexually until at least 26 inches long. Big skate, p. 63

    Upper teeth in not more than 66 series and usually less then 54; matures when only 18–20 inches long. Little skate, p. 67

11. There is a small dorsal fin on the upper side of the tail, in front of the spine (or spines); the crown of the head is high-domed, with the eyes and spiracles on the sides; there are only 7–9 series of teeth in the form of large flat grinding plates ---------- Cow nosed ray, p. 76

    There is no dorsal fin on the tail; the crown is low, flat, and with the eyes and spiracles on the upper surface; the teeth are in many series, in mosaic arrangement ---------- Sting ray, p. 74

THE TORPEDOES OR ELECTRIC RAYS. FAMILY TORPEDINIDAE

The trunk of the electric rays has the form of a flattened, roundish or oval disc, fleshier toward the margins than it is in other Gulf of Maine skates or rays, and the body is softer. The tail, too, is broader and shorter; there are one or two relatively larger dorsal fins on the tail, and the latter ends in a well-developed caudal fin also.

The most interesting feature of the electric rays is that they have two large electric organs, each of which occupies one side of the front part of the disc. In the only Gulf of Maine species the two organs together make up about one-sixth of the total weight of the fish.

Torpedo *Torpedo nobiliana* Bonaparte 1835

ELECTRIC RAY; NUMBFISH, CRAMPFISH

Bigelow and Schroeder, 1953, p. 96.
Garman, 1913, pl. 25, fig. 2, as *Narcacion nobilianus*.

Description.—No one would be apt to mistake a

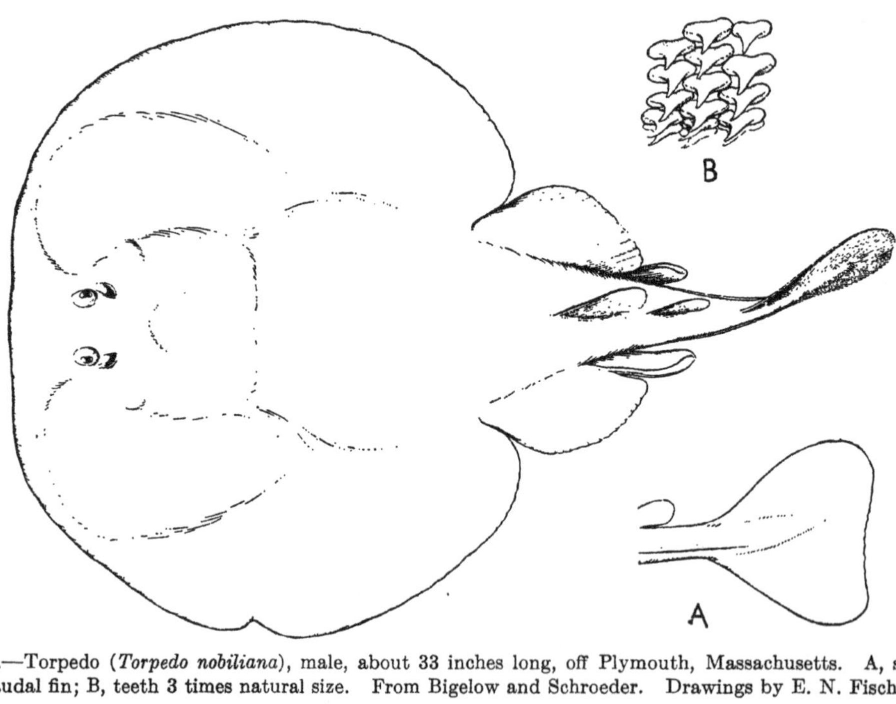

FIGURE 24.—Torpedo (*Torpedo nobiliana*), male, about 33 inches long, off Plymouth, Massachusetts. A, side view of caudal fin; B, teeth 3 times natural size. From Bigelow and Schroeder. Drawings by E. N. Fischer.

torpedo for any other Gulf of Maine skate or ray, the rounded outline of the disk and the large caudal fin identifies it at a glance. Furthermore, its skin is soft and naked, without the spines or thorns so characteristic of all our common skates. The disk is roughly subcircular, truncate in front, and somewhat broader than long. The eyes are very small and set far forward. The two dorsal fins, of which the first is the larger, stand on the forward end of the tail, the first, indeed, partly above the bases of the pelvic fins, and they are separated by an interspace nearly as long as the second dorsal fin. The tail fin is of ordinary fish form, triangular and nearly as long as it is deep. The tail is shorter than in the skates for it occupies only about two-fifths the total length of the fish, measured from the cloaca. The teeth are small, with sharp curved points, and are in about 60 series, with up to 7 rows exposed and functioning at one time.

Color.—Dark chocolate to purplish brown above, some with a few obscure darker spots; lower surface white except that the edges of disk, fins, and tail are of the same dark tint as the upper side.

Size.—Adult torpedoes are usually 2 to 5 feet long or a little longer, and heavy for their size. Specimens taken at Woods Hole average about 30 pounds, while most of those taken anywhere on our Atlantic coast weigh less than 75 pounds. But we have seen one only about 4 feet long from Chesapeake Bay that weighed about 100 pounds; one of 144 pounds was brought from Nantucket to the U. S. Fisheries Station at Woods Hole many years ago; and the heaviest taken near Provincetown were estimated long ago by a fisherman of keen observation as 170 to 200 pounds.

Habits.—The most interesting thing about the torpedo is its ability to give electric shocks of considerable strength to anyone touching it. The statement, even, has long been current that the shock from a large one in rested condition may be strong enough to throw a full grown man to the ground. And the story is told of a dog which was in the habit of wading on a Cape Cod beach in shoal water to catch flounders, but was so shocked by a torpedo that it ran away howling and could never be persuaded to go fishing again. In fact, this anecdote antedates the scientific naming of the New England torpedo. But shocks of a strength even approaching what is suggested by such reports are to be expected only from torpedos of the largest size in rested condition. The voltage recorded recently was 170 to 220 for one that had been kept in a live well. And the most we have felt ourselves from medium-sized torpedos lying on the dock at Woods Hole has been a slight benumbing sensation.

The torpedo, like others of its tribe, is a bottom fish. It is a fish eater. The stomach of one taken at Woods Hole contained a summer flounder (*Paralichthys dentatus*) about 14½ inches long. A 2-pound eel, a 1-pound flounder, plaice (*Pleuronectes platessa*), red mullet (*Mullus surmuletus*), a salmon weighing 4 or 5 pounds, and the remains of spotted dogfish (genus *Scyliorhinus*) have been found in the stomachs of British specimens. The wide distensibility of its jaws allows it to swallow fishes much larger than might be considered possible from the breadth of the mouth when closed. And it is generally believed that it stuns its prey by its electric shocks. Otherwise it is difficult to conceive how so sluggish a fish could capture such active prey.

It bears "living" young, but there is no placental connection between embryo and mother. And it seems that the young are born offshore, for the smallest torpedo yet recorded from American inshore waters (from New Jersey) was about 2 feet (610 mm.) long. And we doubt if it succeeds in producing young in the colder waters of our Gulf.

General Range.—Both sides of the North Atlantic [56] from southern Nova Scotia (La Have Bank), Bay of Fundy, and Georges Bank to North Carolina in the west; [57] and from northern Scotland to the Mediterranean, Azores, Madeira, and tropical West Africa in the east.

Occurrence in the Gulf of Maine.—The torpedo is more common south and west from Cape Cod than to the northward and eastward. But it strays past the elbow of the Cape often enough for it to be classed as a regular member of the Gulf of Maine fish fauna. The most northeasterly records for it are of one presumably of this species taken in St. Margarets Bay, Nova Scotia, some 30 years ago; one caught on a long line set for cod

[56] Comparison of American specimens with one from the North Sea revealed no differences.

[57] This torpedo is also reported from the Florida Keys and from Cuba, but on doubtful evidence.

on La Have Bank in 1890,[58] and from Eastport, Maine, at the mouth of the Bay of Fundy. It has also been taken at Williamsport, Maine; off Seguin Island where one was examined in 1880; at the mouth of Casco Bay; at Wood Island near Cape Elizabeth (1, in a trap, in 1894); near Cape Ann; off Plymouth in the southern side of Massachusetts Bay; near Provincetown; and on the outer coast of Cape Cod, so it would be no surprise to find it anywhere along the shores of the Gulf. It has been caught occasionally on Georges Bank;[59] there are records of long standing of torpedos off Nantucket and Marthas Vineyard, and they are caught yearly in Vineyard Sound and in Buzzards Bay.

Most of the reports of torpedoes within the Gulf have been based on single specimens. But it has been known for a long time that torpedoes are caught in much larger numbers in some years than in others. Thus they are said to have been unusually common near Provincetown in 1819 and for the next 4 or 5 years, when 60 to 80 were taken there yearly. Again in 1845 about a dozen came ashore or were caught otherwise near Provincetown. Any fluctuation, however, that may have taken place from year to year thereafter seems to have attracted no attention until the summer of 1896, when Dr. W. C. Kendall, of the U. S. Fish Commission collected several along the coast of Maine. The Massachusetts Bay specimen mentioned above, taken off Plymouth and now in the Harvard Museum of Comparative Zoology, is the only torpedo from the inner part of the Gulf of which we have heard since that time. But it is as likely to be found in the Gulf now as it ever was.

Importance.—The torpedo is of no commercial value nowadays, but its liver oil was considered equal to the best sperm for illuminating purposes before the use of kerosene oil was general. There is an old tale that its oil was a good cure for cramps if rubbed on externally, for stomach trouble if taken internally. And when one is landed on the dock at Woods Hole it is an object of interest to the workers at the Biological Laboratory because of its electric discharges.

SKATES. FAMILY RAJIDAE

Skates, with their disc-like outlines, thin as a shingle toward their outer edges, and with their rather long tails, are familiar objects along our shores. The outer edges of their pelvic fins are concave (convex in the sting rays), they have two very small dorsal fins on the rear part of the tail, but no distinct tail fin, and they lack the large tail spine that is so characteristic of the sting rays. The Gulf of Maine supports four species in abundance, while two others have been recorded on rare occasions.

The common skates look so much alike that fishermen seldom distinguish between them. For this reason we know very little about the individual differences in habits among the several species. All live chiefly on the bottom or close to it, spending much of the time partially buried in the mud or sand. They move through the water by undulations of the flexible pectoral fins, steering themselves with the tail. All are decidedly omnivorous, feeding largely on the larger Crustacea, such as shrimps, crabs, lobsters; on mollusks, worms, and to a greater or less extent on fish.

All the true skates lay large eggs with blackish or sea-green leathery shells, roughly oblong in outline, with a hollow tendril at each corner by which they become fastened to seaweeds or other objects. The empty eggshells, "mermaids purses," are familiar objects on our beaches among the flotsam along high water mark. While still in the egg the embryo skate develops temporary external gill filaments from the walls of the gill clefts, but these disappear completely before it hatches. Probably all our local skates spawn over a considerable part of the year, with incubation periods of several months up to a year or more.[60]

To give some idea of their abundance on the offshore banks we may note that the average number of skates (all species together) taken on Georges Bank, per trip of 4 to 7 days, on 25 trips by several trawlers, January to December 1913, was about 800, the largest catch 4,520, the

[58] Reported by G. F. O. Hansen, then second mate and later master of the U. S. Fish Commission schooner *Grampus*, who doubtless was acquainted with the torpedo at Woods Hole.

[59] The most recent record is of one 58 inches long, trawled on the southwest part in December 1930.

[60] Under aquarium conditions the incubation period for the little skate (*R. erinacea*) was 5 to 6 months (p. 69); and it ranged from 4½ to about 14¾ months for 6 common European skates; see Clarke, Jour. Marine Biol. Assoc. United Kingdom, vol. 12, No. 4, 1927, p. 587.

poorest 82. Again, on a trip to the northeastern part of the bank, September 1929, on the otter trawler *Kingfisher*, 37 hauls yielded from 0 to 105 skates per haul (total 495) and 42 trawl hauls by the *Eugene H*, fishing from Nantucket Lightship to the south-central part of Georges Bank in late June 1951 caught an average of 146 skates per haul (total, 6,130 skates), which works out at about 9 to 10 skates per acre.[61] Probably they are equally abundant on Browns Bank; certainly they are familiar enough there, but statistics are not available of the actual numbers caught. Skates are also plentiful inshore as appears from catches of about 1 skate to 33 fishes of all kinds on long lines, at various localities in the Gulf of Maine.[62]

In the Gulf of Maine, skates are only a nuisance for they bite the hook readily and often are caught in great numbers in otter trawls, most of them to be thrown back into the sea, the market demand for them being so small that the total landings reported for New England (Massachusetts and Maine) in 1947 was only 28,200 pounds; and 59,100 pounds for 1948. But some are now being landed in Maine for fish meal.[63] They are much more highly valued in northwestern Europe for food with landings for the years just preceding World War II, running around 90 to 100 million pounds.

Barn-door skate *Raja laevis* Mitchill 1817

Bigelow and Schroeder, 1953, p. 217.
Garman, 1913, pl. 22, fig. 2, as *R. stabuliforis*.

Description.—The barn-door skate is easily identified by its large size, its very pointed snout, and its smooth skin. The thorns along the mid-line of its back are comparatively small and run only from the hinder part of the disc back along the tail; the tail also has one or two rows of large, sharp spines (smaller on males than on females) along each side, besides the median row. There are small thorns on the snout also, sometimes below as well as above, and along the front edges of the pectoral fins. The male has a patch of erectile hooks on the outer part of each pectoral covering an area measuring 5 by 1¼ inches on one side, and 4½ by 1⅛ inches on the other in a specimen 52 inches long; otherwise the pectorals are smooth for the most part. The front angle of the disc is sharper than in our other skates, being more acute than a right angle, but the tip of the snout is blunt. The outer corners of the pectorals are angular and the disc as a whole is diamond or lozenge-shaped. The two dorsal fins are separated by a short interspace, with one or more spines, and the tip of the tail extends farther beyond the second dorsal fin than it does in most skates. The teeth of the female are flat and pavement-like, but those of adult males are provided with sharp slender cusps. Thirty to forty series of teeth have been counted in the upper jaw, 28 to 38 series in the lower jaw.

Color.—The barn-door skate like so many sea fish, varies in color. The upper surface is brown (as a rule usually of a distinctly reddish hue), variously marked with small scattered darker spots or blotches of varying size, and often with pale marblings or waterings; usually there is a large oval spot on the base of each pectoral fin, in line with the outer angle. The lower surface is not as uniformly pale as it is in most skates, its gray or white ground being shaded with darker toward the snout, and speckled on one-third grown specimens and larger, with black or dusky dots or short streaks that mark the mucous pores, a conspicuous feature.

Size.—The barn-door skate is our largest, growing to a length of 5 feet; it has been said to reach 6 feet though there is no definite record of one that large. One of 58 inches was 42 inches wide with a tail 27 inches long, and a female of 50 inches, taken by us, was 33¼ inches wide, with a 22-inch tail. Barn-door skates weigh about 4 to 6 pounds when 28 to 30 inches long, about 10 to 11 pounds at 36 inches, and about 19 to 21 pounds at 45 to 46 inches. Very small specimens are seldom taken.

Habits.—Barn-door skates are bottom fish. They prefer smooth to rocky ground, and we have caught them on very soft mud bottoms as well as on sand and gravel. The fact that the lower surface is more or less pigmented instead of white suggests that it hugs the bottom less closely than other skates, and it is a strong, active swimmer, as anyone will agree, who has landed a large one on a hand-line. Its usual depth range is from close to the tide line, down to about 100 fathoms. It is perhaps more plentiful at 25

[61] Three mile hauls with the trawl sweeping a strip about 35 feet wide.
[62] Examples are: 15 miles off Monhegan I., Maine, June 24–25, 1913, total fish caught, 5,463; skates 170. Twenty miles east of Cape Cod, Nov. 11, 1913; total fish caught 6,532, skates 202. Jeffreys Ledge, Dec. 11–12, 1913; total fish caught 3,996, skates 62.
[63] Scattergood, Copeia, 1950, p. 169.

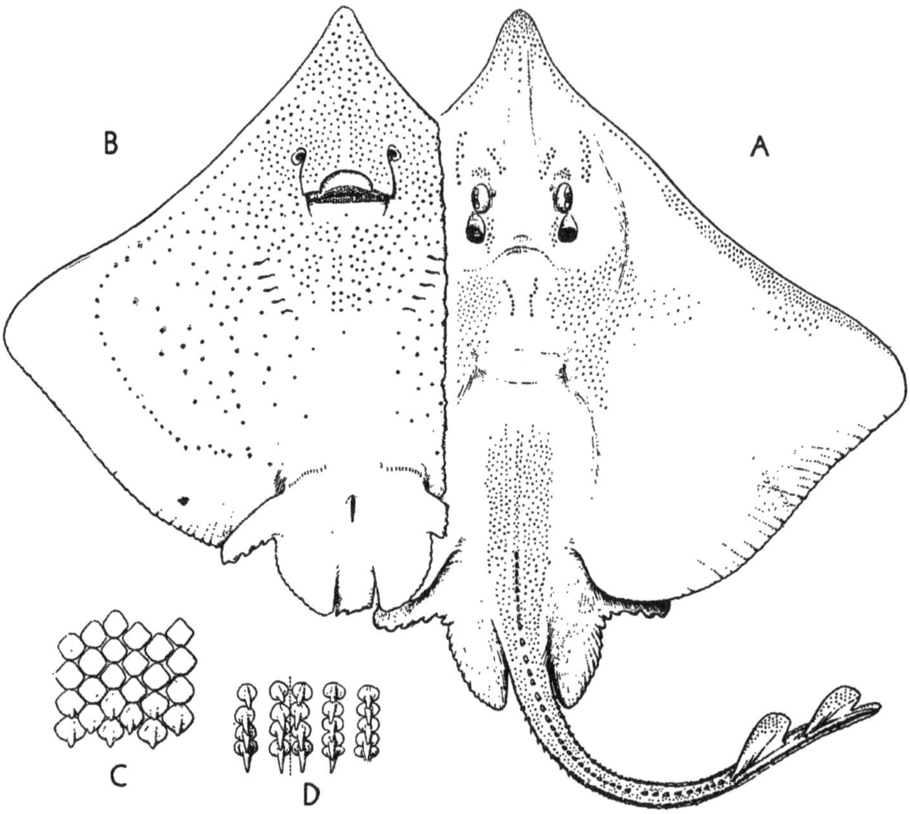

FIGURE 25.—Barn-door skate (*Raja laevis*). A, dorsal view of female, about 47 inches long, Massachusetts; B, ventral view of one of about 26½ inches to show the black markings; C, upper teeth from center of jaw of female 50 inches long; and D, upper teeth from center of jaw of male 52 inches long. B, C, and D from Nantucket Shoals. From Bigelow and Schroeder. Drawings by E. N. Fischer.

to 35 fathoms on Georges Bank and on Nantucket Shoals than deeper, judging from average catches of 32 per haul at 26 to 35 fathoms, 13 per haul at 36 to 49 fathoms, and 6 per haul at 50 to 75 fathoms in 42 trawl hauls by the *Eugene H*, late June 1951, fishing from Nantucket Lightship to the south-central part of Georges Bank. But the *Atlantis* found it widespread (though not numerous), as deep as 100 fathoms both in the open trough of our gulf and in the bowl west of Jeffreys Ledge during experimental trawling, in August 1936; and it has been reported as deep as 235 fathoms off Nantucket.

The temperature range of the barn-door skate is wider than that of the little skate (p. 67). They are found in the southern side of the Gulf of St. Lawrence in the icy-cold-bottom water on the banks, also, at lesser depths that warm in summer to 60° F. (16° C.) or more. In the Gulf of Maine, at one locality or depth or another, they are exposed to temperatures ranging from perhaps as low as 32° to as high as 64 to 68° and the upper limit must be considerably higher in the southern part of their range.

Garman has pointed out that the spines on the snout of this skate are usually worn smooth, as though used to dig in the mud or sand (very likely it thus obtains the bivalves that form part of its diet). It also feeds on worms, various crustaceans, particularly on large rock crabs and lobsters, shrimps, squid, and on fish. Probably it is more destructive to the latter than are any other of our skates thanks to its large size. Woods Hole records list spiny dogfish, alewives, herring, menhaden, butterfish, launce, cunners, tautog, sculpins, silver hake, hake, and flatfish among its foods. No doubt cod, haddock, and other fish, suffer to some extent from this skate on the offshore fishing grounds, for its European relative is a well-known enemy of the cod, and there is no reason to suppose that the barn-door skate is less voracious. It bites readily on almost any bait, and is often caught on hand and long lines as well as in otter trawls, and in weirs along shore.

Little is known of the breeding habits. The yellowish or greenish brown egg cases are about 4⅞ to 5⅛ inches (124–132 mm.) long by 2⅝ to 2⅞ (68–72 mm.) inches broad, not counting the horns, and thus much larger than those of any other Gulf of Maine skate. Females containing fully formed egg capsules have been taken in December and January in Nova Scotia waters, evidence that the eggs are laid in winter. However, it seems that the young are not hatched until late spring or early summer, for we have seen one, taken on Nantucket Shoals in July, so small (about 7⅜ in. long) that it could not have been set free long before its capture.

General range.—Atlantic Coast of North America from the Banks of Newfoundland, Gulf of St. Lawrence and outer coast of Nova Scotia and the Nova Scotia Banks to North Carolina.[64] It is replaced in European seas by a very close ally, the common skate, *Raja batis*.

Occurrence in the Gulf of Maine.—This is a common fish in all parts of our Gulf, and any very large skate taken or reported there is almost certain to be a "barn-door." Following the coast around from east to west we find it reported as plentiful off the outer Nova Scotia shore; it is known from St. Mary Bay; is found very generally though not abundantly in the Bay of Fundy and in Passamaquoddy Bay; is reported from Eastport, Casco Bay, and generally along the coast of Maine; is known from various localities in Massachusetts Bay, where we have seen many caught; and its abundance on Georges Bank and on Nantucket shoals is illustrated by an average catch of about 21 per haul (about 14 percent of all the skates caught), in 42 trawl hauls by the *Eugene H*, fishing from Nantucket Lightship out into the south central part of Georges Bank in late June 1951. In short, it is to be expected anywhere within the limits of the Gulf. Like most other skates, it is often taken in shoal water in our Gulf in summer; seldom or never in winter. Huntsman tells us that it comes to Passamaquoddy Bay from May to November. We once caught one nearly 5 feet long at Cohasset in Massachusetts Bay in less than a fathom of water in midsummer; indeed, it is often stranded on the beach. This inshore migration, however, does not involve the entire stock, witness its presence in 20 to 60 fathoms on Georges Bank and off Cape Cod throughout the year, and the fact that it is reported by fishermen and has been trawled by vessels of the former Bureau of Fisheries, also by the *Atlantis*, as deep as 100 fathoms in summer. In the warmer waters off the southern coast of New England it comes inshore in spring and autumn, descending to somewhat deeper water for the summer.

Commercial value.—The barn-door skate is of no commercial value except as entering into the small landings of skates mentioned on page 61.

[64] Doubtfully reported from Florida.

Big skate *Raja ocellata* Mitchill 1815

SPOTTED SKATE; WINTER SKATE; EYED SKATE

Bigelow and Schroeder, 1953, p. 240.
Garman, 1913, p. 339, pl. 29, fig. 2, as *Raia diaphanes*.

Description.—This skate looks very much like the little skate, but it is larger and has more numerous teeth. The front angle of the disc is much blunter than a right angle, bulging opposite the eyes, and the tip of the snout is rounded. The teeth are in from 72 to 110 series in each jaw instead of 66 series, or fewer as in *erinacea*, and they are sharper in males than in females. The backs of both sexes are rough with sharp spines on the head, around the eyes, along the anterior margins of the pectorals, over the shoulders, and on the sides of the tail. The midline of the back behind the shoulders is almost always free of spines in adults. But we have one specimen, a female 18 inches long taken near Jeffreys Ledge, November 1, 1927, which bears a row of large spines along the midline of back and tail from the shoulder girdle to the first dorsal fin. Males, like those of other skates, have rows of retractile hooks on the outer parts of the pectorals. The two dorsal fins are close together; the outer corners of the pectorals are bluntly angular; the claspers of adult males reach about halfway back along the tail, which occupies about half the total length of the fish.

Color.—Light brown above with round dark brown spots. As a rule there is a large white eye spot with black center near the rear corner of the pectoral fin, and often two smaller ones

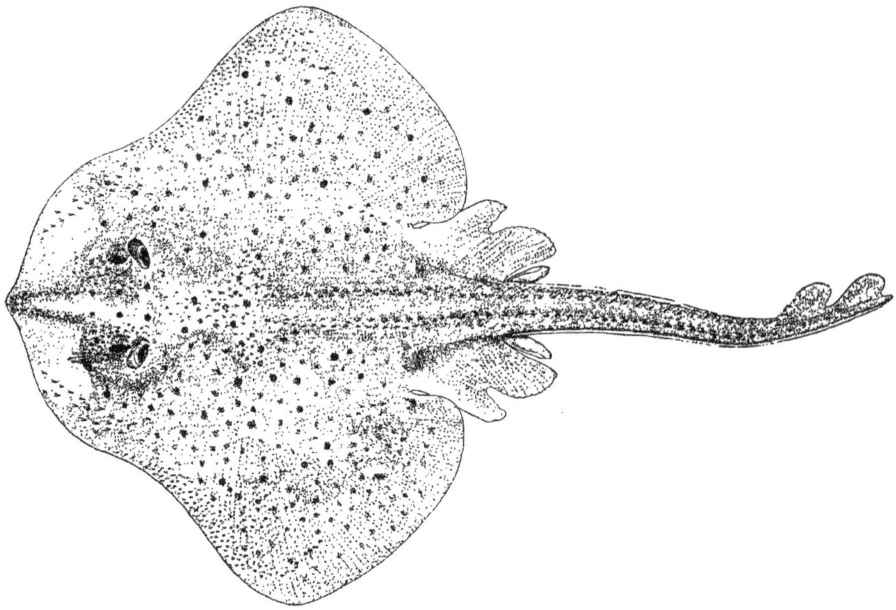

FIGURE 26.—Big skate (*Raja ocellata*), male, about 36 inches long. From Jordan and Evermann. Drawing by H. L. Todd.

close to it. And we have seen two large specimens from Georges Bank with several of these eye spots on each side of the disk. There is a translucent or white area on each side of the snout in front of the eyes and the lower surface is white.

The eye spots, if present, serve to identify this skate at a glance; sometimes, however, they are lacking, in which case half-grown specimens so closely resemble the little skate that recourse must be had to the number of teeth to tell the one from the other.

Size.—This skate does not mature until at least 25 to 26 inches long, and grows to about 3½ feet in length, commonly from 30 to 34 inches. Specimens 32 inches in length are about 20 inches wide.

Habits.—Big skates feed on the same diet as little skates do (p. 69). Rock crabs and squid are favorite prey, but they also take annelid worms, amphipods, shrimps, and razor clams, and they eat whatever small fish are readily available, the list at Woods Hole including smaller skates, eels, herring, alewives, bluebacks, menhaden, smelt, launce, chub mackerel, butterfish, cunners, sculpins, silver hake, tomcod, and hake.[65]

[65] From Vinal Edwards' and Linton's notes.

It is caught right up to the wharves in the Gulf of St. Lawrence; often comes into very shoal water on sandy beaches, and we once caught an adult male in September in only 2 or 3 feet of water in Nauset Marsh on the outer coast of Cape Cod, but few are found shoaler in our Gulf than 2 to 4 fathoms. They are much more plentiful at 25 to 35 fathoms than deeper, on the offshore grounds, as appears from average catches, of 48 per haul at 26 to 35 fathoms, but only 11 per haul at 36 to 49 fathoms, and none at 50 to 75 fathoms, in 42 trawl hauls by the *Eugene H*, fishing from Nantucket Lightship to the south-central part of Georges Bank in late June 1951, and very few are caught deeper than about 50 fathoms anywhere.

In our Gulf they inhabit about the same range of temperature as the little skate does, i. e., from 68° or so, for those along the Massachusetts coast in summer, down to 34–36° in the coastal belt as a whole in winter, and to near 32° in the Bay of Fundy region, at least in some years. In the southern side of the Gulf of St. Lawrence they are found in the icy bottom water on the banks as well as shoaler, where temperatures rise to 61° (16° C.) or more in summer. Those living the shoalest in the southern part of their range must be exposed to temperatures as high, perhaps, as 68° to 70° at the warmest time of the year.

Off the Atlantic Coast of Nova Scotia this skate deposits its eggs from summer into autumn, and probably through the same season in the Gulf of Maine for Scattergood [66] reports females with egg capsules in Maine waters in September. And it continues to do so into December and January off southern New England. Its egg cases are larger than those of the little skate, 2½ to 2¾ inches by about 1¾ inches, not counting the horns. The length of the period of incubation is not known.

General range.—Atlantic Coast of North America from northern North Carolina to the southern side of the Gulf of St. Lawrence, and to the southern part of the Newfoundland Banks.

Occurrence in the Gulf of Maine.—This, the second in size of our skates, occurs commonly all around the Gulf of Maine from Nova Scotia to Cape Cod. There are many locality records for it for the Bay of Fundy as well as from the coasts of Maine and Massachusetts, but so closely does a half or two-thirds grown big skate resemble an adult little skate (p. 68) that it is often impossible to tell to which species published reports refer. It also makes up so considerable a proportion of the skate population on Georges Bank that about 14 percent of the catch of skates made on Georges Bank by one otter trawler in September 1929, and about 18 percent (1,116) of the skates taken in 42 trawl hauls by the *Eugene H*, late June 1951, fishing from Nantucket Lightship to the southwestern part of Georges Bank, were this species. But it has never been reported from the deeper troughs of the Gulf, nor have we taken it there.

The name "winter skate" seems appropriate enough for it along the southern coast of New England, for it is only during the cold season that big skates come close inshore near Woods Hole. And they are said to be taken in larger numbers in winter than in summer in the Massachusetts Bay region (we cannot verify this). However, this is a misnomer in the cooler waters of the northern part of the Gulf of Maine, for it is common inshore in Passamaquoddy Bay from May to November, and this probably applies to the whole coastline east of Cape Elizabeth to judge from temperature.

Big skates are taken on hook and line, in weirs, and in otter trawls, but they are of no commercial value, except as they form a part of the general landings of skates. And they are only a nuisance to anglers.

[66] Copeia 1951, No. 2, p. 169.

Brier skate *Raja eglanteria* Bosc 1802

Bigelow and Schroeder, 1953, p. 165.

Garman, 1913, pl. 23.

Description.—In the brier skate, as in the thorny skate, the midline of the back and tail is armed with a continuous row of stout thorns from the shoulders to the first dorsal fin near the tip of the tail, usually with 1 or 2 in the gap between the 2 dorsal fins. But the thorns of this row are not much larger than those along the sides of the tail (they are in the thorny skate), and there are at least 16 thorns in the midrow along the tail (not more than 9 to 10 in the thorny skate). There also are groups of large thorns opposite and behind the eyes, with 1 to 5 on each shoulder and 1 to 4 rows along either side of the tail. Elsewhere the upper surface of the disc bears only small sharp prickles (hence its name), most numerous on the forward parts of the pectorals, over head and snout, and along the middle of the back and tail among the larger thorns. Thus it is a much smoother species than the thorny skate, and its snout is more acute, its outline being about a right angle with the margins bulging less opposite the eyes than in any of the blunter-nosed skates. The outer corners of the pectorals are distinctly angular, and the dorsal fins are separated by a short gap.

Color.—Brownish to grayish above; the pectorals variously marked with darker spots and blotches and with more elongate bars; this last is a characteristic feature; there is a translucent space on each side of the snout; it is white below. It is most readily recognized by its color pattern, with short dark bars as well as spots, which is not shared by any other Gulf of Maine skate.

Size.—The brier skate ordinarily grows to a length of about 2½ feet. The largest on record was about 37 inches long.

General range.—Off the eastern coast of the United States from Massachusetts Bay to both coasts of Florida.

Occurrence in the Gulf of Maine.—This is a southern species, uncommon even as far north as Woods Hole and decidedly rare in the Gulf of Maine. It has been recorded once from Gloucester, its most northerly outpost, and also from

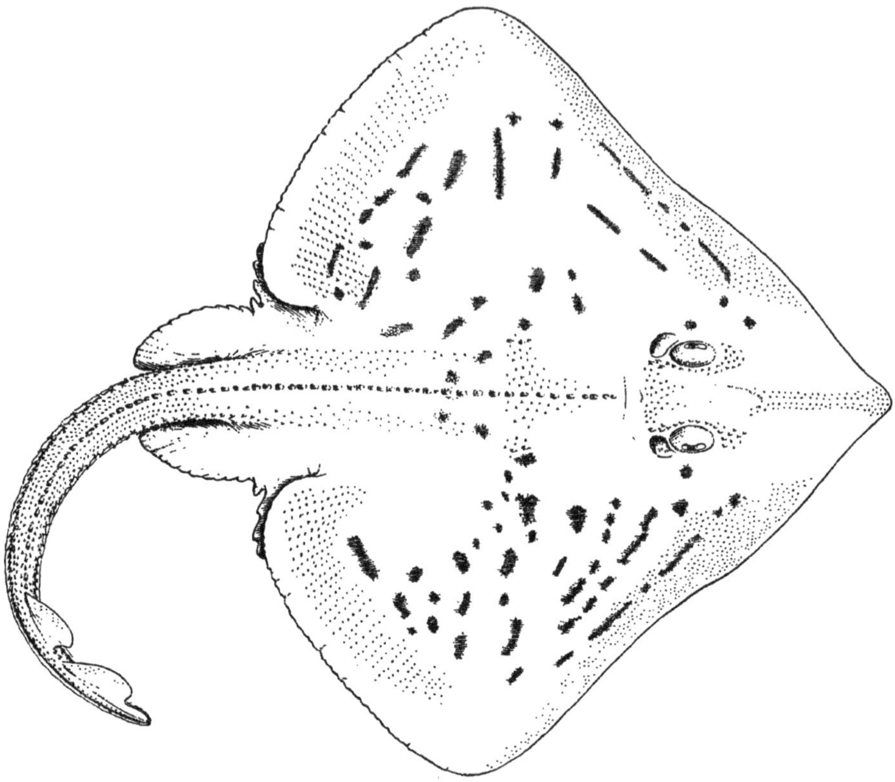

FIGURE 27.—Brier skate (*Raja eglanteria*), female, about 29 inches long, Woods Hole, Massachusetts. From Bigelow and Schroeder. Drawing by E. N. Fischer.

Provincetown. Two specimens were taken on Nantucket Shoals near Round Shoal buoy by the *Halcyon*, one in July, the other in September, in 1924.

Leopard skate *Raja garmani* Whitley 1939
ROSETTED SKATE

Bigelow and Schroeder, 1953, p. 200.

Garman, 1913, pl. 18, fig. 2.

Description.—The conspicuous dark rosettes on the upper surface make this skate recognizable at a glance, since no other skate of the western Atlantic is marked in this way. And its tail is longer relatively than that of any other Gulf of Maine skate.

The disc is considerably blunter in front than a right angle, with anterior margins bulging rather conspicuously a little anterior to the level of the eyes; the outer corners of the pectorals are very broadly rounded; the tail measured from the center of the cloaca to the tip is about 1.5 times as long as the body from tip of snout to center of cloaca; and there is a definite gap with one or two thorns between the two dorsal fins.[67]

There are thorns along the ridge of the snout; a row around the inner and posterior margins of the eyes with a few in the space between the latter; a group on each shoulder; and one row along the midbelt of the back and tail in young specimens, increasing to 2 to 6 irregular rows in large ones. In young specimens the skin of the disc, as a whole, and of the tail, is also rough with small prickles, but, most of these are lost with growth, leaving large specimens mostly naked except for the thorns. The lower surface is smooth.

There are 46 to 52 series of teeth in the upper jaw, a few less in the lower, and those of adult males are only a little sharper than those of females.

Color.—The upper side is pale buff or brown,

[67] Garman's illustration is of an abnormal specimen with three dorsal fins.

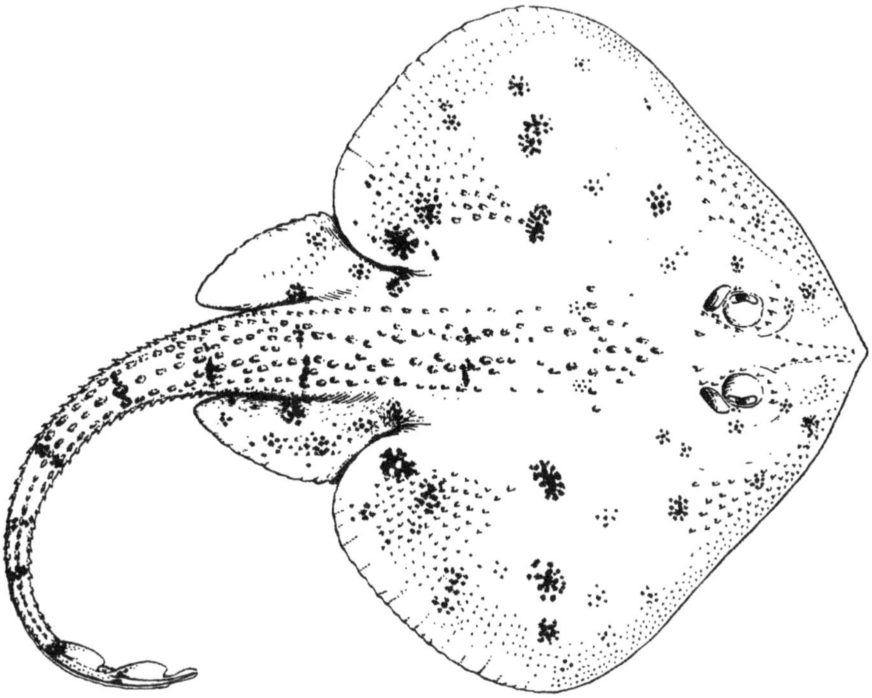

FIGURE 28.—Leopard skate (*Raja garmani*), female, 16 inches long, offing of Montauk Point, New York. From Bigelow and Schroeder. Drawing by E. N. Fischer.

freckled with small spots, darker or lighter, and conspicuously marked with dark rosettes, each consisting of a group of 6 or more dark brown or black spots surrounding a dark central spot. The lower surface is white or pale yellow.

Size.—This is one of the smaller skates, males maturing when only about 16 inches long.

General range.—Outer part of the continental shelf and upper part of the continental slope from southern Florida to the offing of Nantucket, in depths of 30 to 300 fathoms.

Occurrence in the Gulf of Maine.—Our only reason for mentioning this species is that one specimen was trawled by the *Albatross III*, May 14, 1950, at 52 fathoms southeast of Nantucket Lightship (lat. 40°05′ N., long. 69°22′ W.). And this is probably close to the eastern boundary of its range, for it has never been reported among the other skates that are trawled in abundance along the seaward slopes of Georges and of the Nova Scotia Banks. But it is one of the most plentiful of skates offshore to the westward, along southern New England.

Little skate *Raja erinacea* Mitchill 1825

COMMON SKATE; SUMMER SKATE; HEDGEHOG SKATE; TOBACCO BOX

Bigelow and Schroeder, 1953, p. 176.

Garman, 1913, pl. 20.

Description.—The most distinctive characters of grown specimens are their small size, absence of thorns along the midline of the back (except in the young) and blunt nose.

The anterior angle of the disc is blunter than a right angle and the tip of the snout is rounded, with the margins bulging opposite the eyes. The teeth are in only about 38 to 66 series. Females have thorns scattered generally over the upper surface; these are especially prominent on head, snout, shoulders, and sides of tail. Ordinarily there are no spines on the midline, back of the shoulder girdle; but we found one fish, 13½ inches long, among the many we have observed, with a median row of spines extending from the shoulder girdle to the first dorsal fin near the

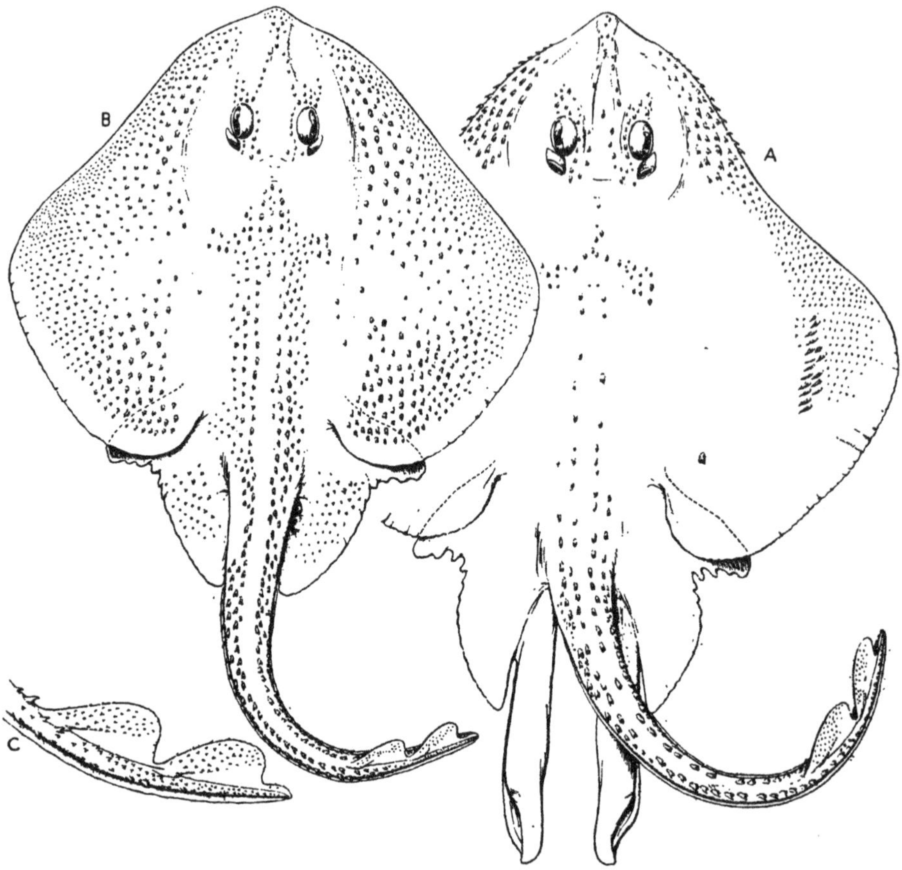

FIGURE 29.—Little skate (*Raja erinacea*). A, male, 20 inches long, Boston Harbor; B, female, 17½ inches long, Mystic Connecticut; C, side view, end of tail of same, about 0.6 times natural size. From Bigelow and Schroeder. Drawings by E. N. Fischer.

end of the tail, and this is true of newly hatched specimens in general. Males are less spiny, but the spines on tail, shoulders, and along either side of the back ridge are noticeably strong in both sexes. The two dorsal fins are close together; the tail is about half the total length. Large ones closely resemble small specimens of the big skate (*R. ocellata*, p. 63) that may chance to lack the ocellar spots with which that species usually is marked. A count of the teeth is then the only sure clue to the identity of the specimen in hand.

Color.—Grayish to dark brown above, or clouded light and dark brown, paler at the edges of the pectoral fins; usually with many small round darker spots; white or grayish below.

Size.—Ordinarily up to 16 to 20 inches long; the maximum recorded length is 21 inches (53 cm.); they weigh about ¾ to 1 pound at 16 to 17 inches and anywhere from 1⅛ to 2 pounds at 18 inches. Females mature sexually when 12½–17 inches (32–43 cm.) long, males at about 14 to 17½ inches.[68]

Habits.—It is common knowledge that this skate, like others, is most abundant on sandy or pebbly bottom; but they are likewise found on mud and we have seen them lying on ledges at times.

The usual depth range is from close to tide line down to 75 fathoms or so. Many even follow up the shelving bottoms of our beaches until they

[68] Information supplied by Dr. Daniel Merriman, Dr. Y. H. Olsen, the Misses S. B. Wheatland and L. H. Calhoun, who have made a detailed study of the little skate in southern New England waters.

strand. And the bulk of the population hold to depths of less than 40 to 50 fathoms, as appears from average catches, per haul, of 100 at 26 to 35 fathoms, and 95 at 36 to 49 fathoms, but only 12 at 50 to 75 fathoms, in 42 hauls by the *Eugene H*, fishing from Nantucket Lightship to the southcentral part of Georges Bank, in late June 1951. Fifty fathoms (off the Bay of Fundy) is, in fact, the deepest that positively identified specimens are known, in the inner parts of our Gulf; 80 fathoms off southern New England.[69]

The little skate tolerates a wide range of temperature, being found in water as warm as 68–70° in summer, while they are exposed to temperatures close to 32° in the Bay of Fundy in some winters, unless they move out, and deeper there than seems likely. In the southern side of the Gulf of St. Lawrence, writes Huntsman,[70] they are found in the intermediate zone between the icy cold waters of the banks and the surface stratum, which last warms to 61° (16° C.) or higher in summer.

They have never been reported, to our knowledge, where the water is appreciably brackish.

Hermit and other crabs, shrimps, worms, amphipods, ascidians ("sea squirts"), bivalve mollusks, squid, small fishes, and even such tiny objects as copepods have been found in their stomachs. Probably crabs loom largest in their diet, for more than 29 percent of the skates opened by Field at Woods Hole, contained them; 15 percent had bottom-dwelling shrimps (*Crago*); and 6 percent had eaten squid. In Long Island Sound, however, amphipods (*Leptocheirus*) are the dominant item in their diet, forming from one-third to one-half of the stomach contents at all seasons of the year.[71] Launce, alewives, herring, cunners, silversides, tomcod, silver hake, have all been found in their stomachs, and they bite a baited hook readily, affording amusement to vacationists.

The spawning habits of the little skate have not been followed in the Gulf of Maine. Studies, at the Bingham Oceanographic Laboratory, however, suggest that they ordinarily deposit their eggs in water not deeper than 15 fathoms and on a sandy bottom. It appears from anatomical examination of the sexual organs of the mature females that copulation may take place at any time throughout the year, and frequently. Observations, too, on skates kept in aquaria have shown that the eggs are laid in pairs at intervals of from five days to several weeks; also that they are usually buried in sand, at least partially.[72] The eggs have been taken off Southern New England, in fish traps and dredges in a few fathoms of water in abundance from July through September.

Examination of large numbers of females has shown that eggs are laid there throughout the year. And there is no reason to doubt that this is the case to the north and east of Cape Cod as well. Aquarium experiments have also shown that eggs laid in the period, May–July, hatched between the end of November and the beginning of January, i. e., after 5 to 6 months. But the incubation period is likely to be somewhat longer for spring–summer laid eggs in nature because of somewhat lower temperatures; and considerably longer for eggs laid in autumn and early winter.

The eggs measure about 1½ to 1⅞ inches by about 2⅛ to 2½ inches, not counting the horns, and the great majority of the empty skate eggs that are washed up on the beaches of our Gulf belong to this species. The young skate, which emerges through a transverse opening at the edge of the egg case at the end that has the longer pair of horns, is about 3¾ to 4 inches long at hatching; its abdomen is still swollen with yolk, and its tail terminates in a whiplash-like extension that disappears within a few days. Huntsman's observations suggest that young hatched near the head of the Bay of Fundy descend to deeper water the first winter, and this probably applies to the Gulf of Maine as a whole.

It appears from information of various sorts that a little skate 8 inches (20 cm.) long may be 1 to 1½ years old; one of 11¾ to 12 inches (30 cm.) 2 to 3 years; one of 15¾ to 16 inches (40 cm.) 3 to 4 years; one of 19¾ to 20 inches 6 to 8 years old. And the mortality rate appears to be very high after five years, for very few of those taken are longer than about 18 to 19 inches.[73]

General range.—Atlantic coast of America; southern side of the Gulf of St. Lawrence and

[69] Seventeen that we saw trawled on the *Albatross III*, May 1950.

[70] Trans. Roy. Soc. Canada, Ser. 3, vol. 12, sec., 4, 1918, p. 63.

[71] Information from Dr. Daniel Merriman, Dr. Y. H. Olsen, and the Misses S. B. Wheatland and L. H. Calhoun.

[72] This summary of the breeding habits is based on extensive information supplied by Dr. Daniel Merriman, Dr. Y. H. Olsen, and the Misses S. B. Wheatland and L. H. Calhound.

[73] Information from Dr. Daniel Merriamn, Dr. Y. H. Olsen, and the Misses S. B. Wheatland and L. H. Calhoun.

northern Nova Scotia to Virginia, in coastal waters and on the shoaler of the offshore banks.

Occurrence in the Gulf of Maine.—This, the smallest of our skates, is the commonest and the most familiar from its habit of coming up into very shoal water in summer and of stranding on the beaches, where dried skate carcasses are often to be seen. It occurs all along the coast in the southern side of the Gulf of St. Lawrence and along outer Nova Scotia; is very abundant both on the New Brunswick and on the Nova Scotia sides of the Bay of Fundy, and is to be caught everywhere and anywhere along the coasts of Maine and of Massachusetts; far more commonly, indeed, than one might suspect from the few definite records that have found their way into scientific literature.

An average catch of about 88 per haul (about 60 percent of all the skates taken) in 42 trawl hauls by the *Eugene H*, in late June 1951, fishing eastward from Nantucket Lightship suggests that this is the most plentiful skate on the southwestern part of Georges Bank and on the Nantucket grounds. But it seems to be far less numerous on the northeastern part of the Bank, if it is present there at all; at least we failed to find a single one, among 495 skates of other kinds caught there in 37 hauls by the otter trawler *Kingfisher* in September 1929. And we have never found it (nor has it been reported) in the deeper basins and troughs of our Gulf, probably because it is restricted in general to depths less than 40 to 50 fathoms (p. 69).

In our Gulf many of the little skates appear to carry out an irregular migration into shoal water in April and May, where they remain throughout the summer, autumn, and early winter, to retire again to somewhat deeper, hence, warmer water in December or January. Its migration schedule appears to be more complex in Long Island Sound waters where summer temperatures are higher; i. e., inshore in spring, offshore in mid- or late summer, inshore again in late autumn and offshore again in midwinter.[74] Doubtless little skates breed throughout the shoaler parts of the Gulf, and on the offshore banks.

They are of no commercial importance in our Gulf except as they form a part of the landings of trash fish.

[74] Information from Dr. Daniel Merriman, Dr. Y. H. Olsen, the Misses S. B. Wheatland and L. H. Calhoun of the Bingham Oceanographic Laboratory.

Smooth-tailed or prickly skate *Raja senta* Garman 1885

Bigelow and Schroeder, 1953, p. 264.

Garman, 1913, pl. 25, fig. 1.

Description.—By the time this skate has grown to one-fourth its adult size it is made recognizable by the fact that the middorsal line of thorns runs back only to about the middle of the tail, where the thorns so dwindle in size that they are not distinguishable from the tiny prickles with which the tail is clothed, generally. Newly hatched specimens in which this character is not yet established are separable from all other Gulf of Maine skates by the color pattern of the tail, which has two pale crossbars, each outlined in front and behind by a dark band or blotch.

There is a single row of 16 or more medium-sized to large thorns along the midline of the back, spaced irregularly, and usually about 20 to 30 along the anterior one-half or so of the tail; about 10 to 13 around the inner ridge of each eye; and 3 to 5 on each shoulder. Immature specimens of both sexes are also closely and uniformly roughened with small prickles over the disc as a whole, on the pelvics and on the upper side of the tail. But irregular bare areas develop on the shoulders and around the outer parts of the pelvics of females as they approach maturity while mature males lose the prickles from the central part of the disc as a whole, but develop a few thorns on the midridge of the snout besides larger thorns over a roughly triangular area on either side of the head abreast of the eyes and farther forward. They also develop two rows of the usual curved sexual spines on either side on the pectorals, about 13 to 14 per row. The lower surface of the disc is smooth, except that a few prickles develop, with growth, along the margins near the snout

The lower surface of the tail as a whole is prickly on females and on immature males, but tends to become smooth on males by the time they mature sexually.

The anterior angle of the snout is a little more obtuse than a right angle (about 110°); the tip of the snout is sharper than in either the big skate, the little skate, or the thorny skate. There are 38 to 40 series of teeth in the upper jaw, 36 to 38 series in the lower jaw; those of females are low, with only faintly indicated points, but those of mature males are longer, sharper, recurved, and

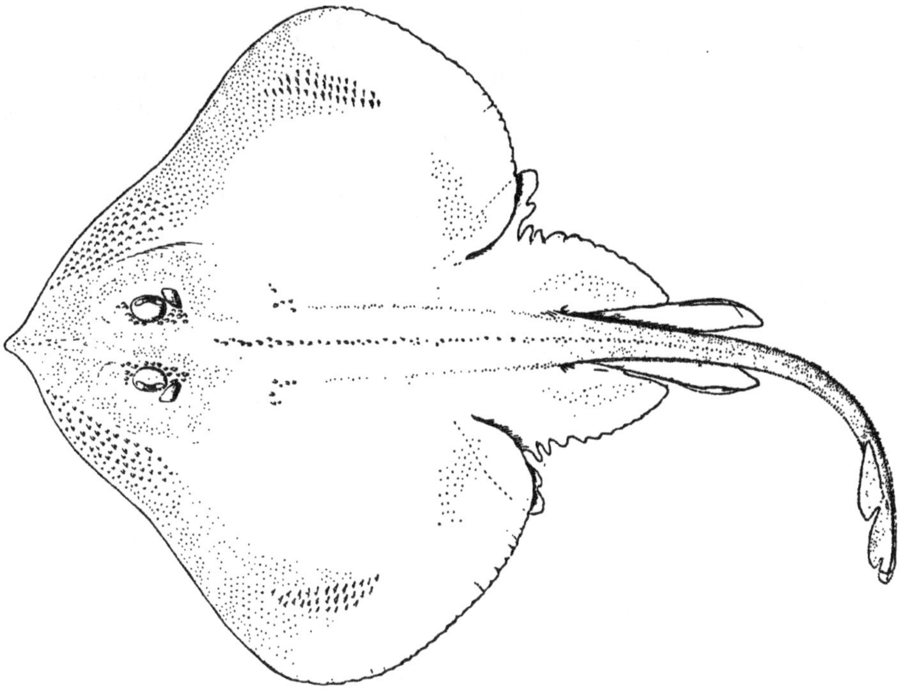

FIGURE 30.—Smooth-tailed or prickly skate (*Raja senta*), male, about 20½ inches long, Emerald Bank, Nova Scotia. From Bigelow and Schroeder. Drawing by E. N. Fischer.

spaced more loosely. There is no free space between the two dorsal fins. The disc is a little broader than long (1.2 to 1.3 times); the tail occupies about one-half of the total length.

Color.—The upper side, including the tail, is pale brown, with many obscure darker spots. Newly hatched specimens are also marked on the tail with two pale cross bars, each outlined by a darker cross bar or blotch in front of it and one behind, but these bars disappear with growth. The lower surface is white, either plain or with a few dusky blotches. Sometimes the rear part of the tail is uniformly dark below.

Size.—A male about 20 inches (515 mm.) long that we have seen seems to be sexually mature. The largest specimen of which we have record was 24 inches long.

Habits.—This skate appears equally at home on the soft mud and clay bottoms of the deeper basins of the Gulf and on the sand, broken shells, gravel and pebbles of the offshore fishing banks. Nothing is known of its diet. Egg cases, apparently of this species, have been trawled in deep water (82–164 fathoms), in the estuary of the St. Lawrence River in July and August; probably they are laid in summer in the Gulf of Maine, as well.

General range.—Atlantic shelf of North America from the offing of Charleston, S. C., to the Nova Scotia Banks and Gulf of St. Lawrence, a few reaching the southern part of the Newfoundland Banks; mostly in depths greater than about 40 to 50 fathoms. The deepest record for it is 478 fathoms off South Carolina.

Occurrence in the Gulf of Maine.—This skate, once considered a rare species, is now known to occur generally throughout the western side of the Gulf wherever the water is more than 45 to 50 fathoms deep, 25 fathoms being our shoalest record for it and on the offshore Banks as well. We have trawled it on several occasions in the deep trough west of Jeffreys Ledge; in deep water (80–100 fathoms) near Cashes Ledge; also in the basin east and southeast of Cape Cod. And, being known from the southeastern slope of Browns Bank, it is to be expected generally in the eastern side of the Gulf, as well as in the western, at the proper depth. It is widespread on Georges

Bank also, but is far less plentiful there than other skates, to judge from the fact that trawl hauls in September 1929 brought in only 37 of them, and that we counted only 8, from 42 hauls on the *Eugene H;* in June 1951, fishing from Nantucket Lightship to the south central part of Georges. We have trawled it at 50 to 250 fathoms off southern New England. To the eastward and northward, it is recorded on La Have and Emerald Banks at 50 to 100 fathoms, and in the Gulf of St. Lawrence at 82 to 178 fathoms.

Thorny skate *Raja radiata* Donovan 1807

STARRY SKATE [75]

Bigelow and Schroeder, 1953, p. 255.[75]
Garman, 1913, pl. 21, fig. 2.[76]

Description.—The thorny skate can be identified at a glance among skates of the Gulf of Maine by the fact that the row of thorns with which the midline of back and tail is armed are not only large and conspicuous, but do not number more than 19 at most from the nape back along the tail. There are also 2 or 3 large thorns on each shoulder; and one in front of each eye and one behind it; one close to the inner end of each spiracle; and other smaller thorns scattered on snout, pectoral fins, and tail. The bases of the thorns on the pectorals are star-shaped, a very distinctive character; those of the still larger thorns along the midline of the back are oval. Adult males have 2 rows of hooked, erectile thorns near the outer corners of the pectorals.

The anterior angle of the disc is considerably more obtuse than a right angle (110–140°), and the tip of the snout is blunt with the margins bulging somewhat a little in front of the level of the eyes; the outer corners of the pectorals are less broadly rounded than in either the little skate or the big skate; and the two dorsal fins may either be joined at the base or be separated by a short space. There are 36 to 46 series of teeth in each jaw, those of females and of young males with low cusps that are worn nearly smooth along the older rows; those of mature males a little sharper and spaced a little more widely.

Color.—Brown above, either uniform or slightly clouded, or spotted with darker, small specimens more conspicuously so than larger. Sometimes there is a white spot beside each eye, one on either side abreast of the nape, and another on each side on the rear part of the disc. The lower side is white, sometimes with irregular sooty or brownish blotches. Garman mentions a partial albino, white above with a few reddish brown and brown spots.

Size.—The thorny skate is about 4 inches (100 mm.) long from snout to first dorsal fin at hatching. The largest specimens so far recorded from American waters have been about 40 inches for the Nova Scotia Banks, 35¼ inches for Georges Bank, and about 31 inches for Massachusetts Bay. But some males may mature when only 21 to 22 inches long. One 32 inches long is about 23 inches wide.

Habits.—The thorny skate is a cool water fish, at home in temperatures from about 50° or so down nearly to the freezing point of salt water. It is also restricted in general to depths greater than about 10 fathoms, even in the northernmost part of its range. In the Gulf of St. Lawrence it lives indifferently on the ice cold banks and in the warmer water on the bottom of the deep Laurentian Channel. Average catches of 1 per haul at 26 to 35 fathoms, 22 per haul at 36 to 49 fathoms, and 5 per haul at 50 to 75 fathoms, in 42 trawl hauls, by the *Eugene H* fishing from Nantucket Lightship, the central part of Georges Bank, June 1951, suggest a rather definite preference for the intermediate depth zone, perhaps because of the food supply. But thorny skates have been taken at many stations, also, down to 336 fathoms off the American coast, and as deep as 459 fathoms near Spitzbergen.

The stomachs of thorny skates caught on Georges Bank contained shrimps, spider crabs, anemones, hydroids, and fish digested past identification.

The egg cases vary considerably in size, probably depending on the size of the parent fish. One from a fish 32 inches long, taken on Georges Bank, measured 3 by 2¼ inches exclusive of the horns. Others that have been measured from the Nova Scotia Banks ranged from 3 to 3½ inches in length. They are flat on one side, strongly convex

[75] When the first edition of this book appeared, it was an open question whether the thorny skate of American waters (named *R. scabrata* by Garman 1913) was identical with the thorny skate of northern Europe (*R. radiata* Donovan, 1807). Our subsequent comparison of American specimens with others from Greenland and Norway has convinced us that they all belong to the one species, which must be called by the older of the two scientific names.

[76] Figure 1 of Garman's plate 21 is not of a thorny skate, as it is named in the accompanying caption, but is of a small specimen of the big skate that we have examined.

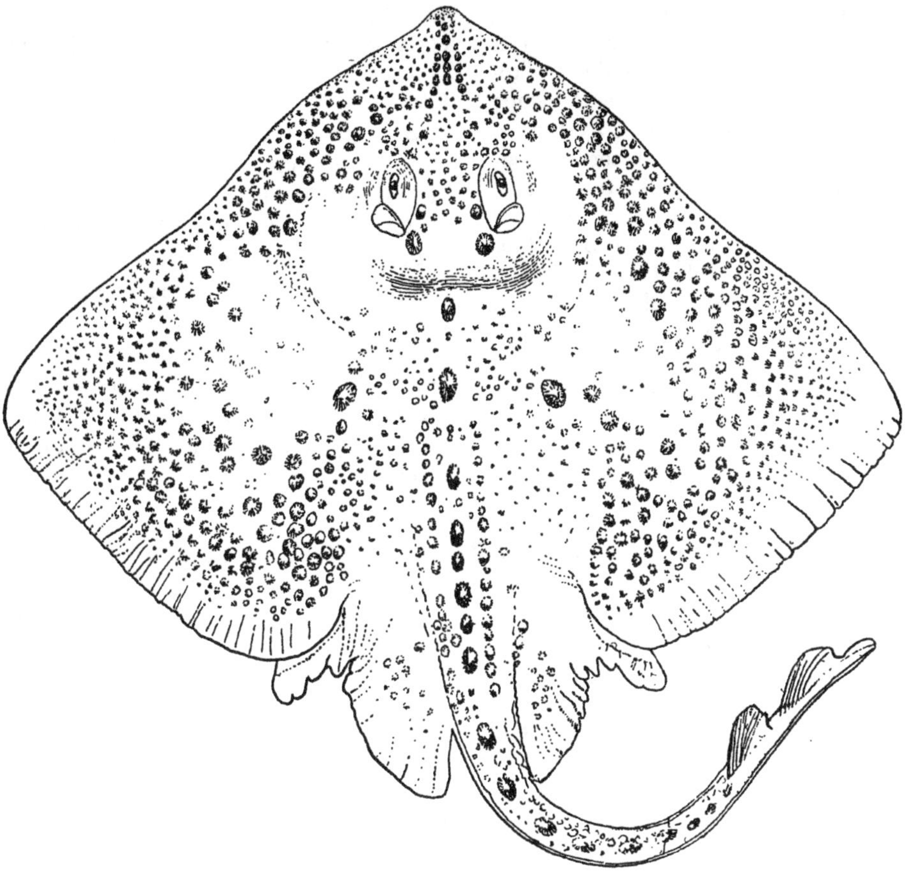

FIGURE 31.—Thorny skate (*Raja radiata*), female, about 31¼ inches long. After Garman.

on the other, and are rough with narrow cross-ridges. A mass of delicate fibrils, matted together, extends along each of the longer sides and partly over the surfaces also. And each horn ends in a slender fibril.

General range.—The thorny skate is known on both sides of the northern Atlantic. In the east its range extends from the White Sea and Barents Sea to the North Sea, Dutch coast, and western part of the Baltic;[77] in the west from West Greenland, Hudson Bay, Atlantic coast of Labrador, east and south coasts of Newfoundland, Grand Banks, Gulf of St. Lawrence and outer coast of Nova Scotia with the off-lying fishing grounds, to the Gulf of Maine, and thence westward and southward along the edge of the continental shelf to the offing of New York; and as a stray to the offing of Charleston, S. C.[78]

Occurrence in the Gulf of Maine.—The thorny skate is not often seen close inshore along our coast, being restricted in general to moderately deep water (p. 72). But it is now known to be generally distributed in the deeper waters of the Gulf. Thus it is frequently taken on the New Brunswick side of the Bay of Fundy in depths of 10 fathoms or deeper, in 20 to 30 fathoms in St. Mary Bay on the Nova Scotia side. It has been recorded from Casco Bay; from Ipswich Bay, off Gloucester, Salem and Nahant, and off Provincetown; and we have taken it ourselves in numerous places in the Gulf at 14 fathoms and deeper, including the

[77] Doubtfully reported from Belgium and the Bay of Biscay.

[78] One taken in lat. 33°10′ N., long. 77°25′ W., in 74 fathoms, by the *Albatross III* is in the Museum of Comparative Zoology.

vicinity of Mount Desert; Platts Bank; and in the bottoms of the deep troughs. It has also been trawled at many stations on Georges Bank, likewise along the upper part of the continental slope off southern New England, down to 336 fathoms.

There is nothing in the available record to suggest that it carries out any regular migrations, whether in or offshore, or along the coast. And it is more catholic in respect to its choice of bottom than some other skates, for while it is most plentiful on the good fishing grounds of sand, gravel, and broken shells, we have taken it at many stations in the Gulf on soft mud. And it is one of the most plentiful of Gulf of Maine skates at appropriate depths. Thus 325 were caught in 37 trawl hauls on the northeastern part of Georges Bank on one trip in 1929; again, in June 1951, we counted 432, from 42 trawl hauls (7 percent of the total catch of skates), on the *Eugene H* fishing from Nantucket Lightship to the south central part of Georges. We once caught 12 in the western side of the Gulf in a beam trawl only 8 feet across the mouth in 30 minutes; and we have taken 1 to 100 of them in 26 hauls with larger trawls, between Mount Desert Island and Massachusetts Bay.

Females containing eggs about ready to be laid, and deposited eggs in various stages of incubation, have been taken in Nova Scotian waters or in the Gulf of Maine, in April, June, July, and September, and they are to be expected there in winter as well, having been reported in January and February off Norway, and from February to June in Scottish waters.

THE WHIP-TAILED STING RAYS. FAMILY DASYATIDAE

The whip-tailed sting rays, like the skates, are disc-like in form, very thin toward the outer edges, with the anterior parts of the pectoral fins fused with the sides of the head, and with the eyes and spiracles on the upper surface. Their pelvic fins, however, have convex outer edges, not concave as are those of the skates. They have no dorsal fin. Their tails are long and whiplash-like toward the tip and armed, in most of them with one to several sawedged, venomous spines on the upper side. Their teeth are small and in many series, closely crowded in bands along the jaws. The upper surface of disc and tail is smooth in some of them, variously roughened with tubercles, thorns or prickles in others. They do not lay eggs as the skates do, but bear "living" young (p. 57). And the young resemble their parents closely when born. Four species are known along the Middle and South Atlantic States, but only one of them reaches the Gulf of Maine, and then only as a stray. Should any long-tailed sting ray be picked up within the limits of the Gulf that does not fit the following description, its captor is referred to Bigelow and Schroeder, 1953,[79] for its identification.

[79] Fishes Western North Atlantic, Pt. 2, Mem. 1, Sears Foundation, 1953.

Sting ray *Dasyatis centroura* (Mitchill) 1815[80]

STINGAREE; CLAM CRACKER

Bigelow and Schroeder, 1953, p. 352.

Garman, 1913, pl. 33, figs. 1, 2, as *Dasybatus marinus*.

Description.—The most distinctive features of this sting ray, among other Gulf of Maine fishes, are its very long, whiplash-like tail without dorsal fins, and the long, sawedged spine or spines with which the upper side of its tail is armed. The disc is rhomboid, about 1¼ to 1⅓ times as broad as it is long; the anterior angle is much blunter than a right angle (130–140°); and the tip of the snout projects very little if at all. The anterior margins of the disc are nearly straight, the posterior margins are only slightly convex, and the posterior corners are abruptly rounded or even angular. The tail, measured from the center of the cloaca, is about 2½ times as long as the body from cloaca to snout. The lower side of the tail has a narrow fold of skin extending rearward from below the origin of the tail spine for a distance about as

[80] This ray was mentioned as *Dasybatus marinus* and as *D. hastatus* in the first edition of this book. But the specimens in question all belong to one species, the correct scientific name for which is *Dasyatis centroura*, proposed by Mitchill in 1815, as *Raja centroura*.

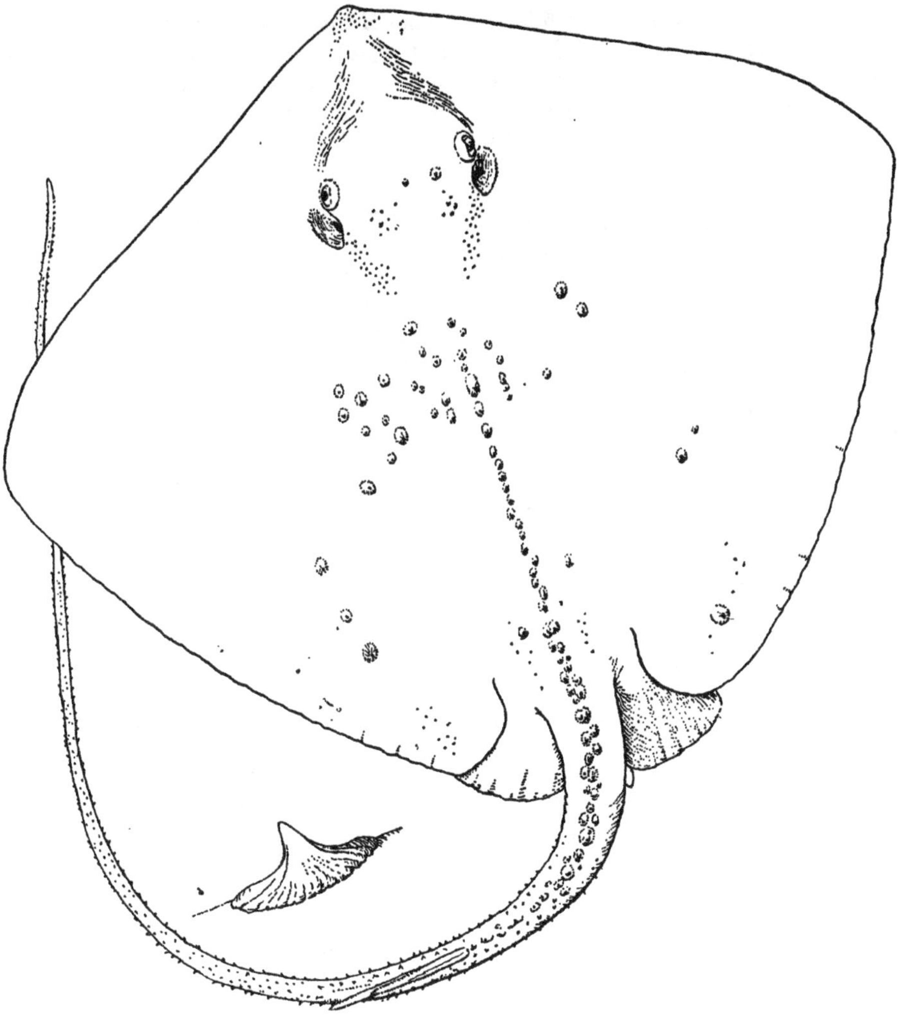

FIGURE 32.—Sting ray (*Dasyatis centroura*), male, about 55½ inches wide, Woods Hole, Massachusetts, and tubercle from tail, about 0.7 times natural size. From Bigelow and Schroeder. Drawings by E. N. Fischer.

long as from its own origin to the cloaca; the upper side of the tail is rounded, except for a low ridge close behind the spine (or spines). The spear-pointed tail spines, of which there are 1, 2, or sometimes 3, are situated well back on the tail. One spine that we examined had about 40 saw teeth on each edge.

Young ones are smooth skinned (except for the tail spines). Larger ones develop 1 to 3 irregular rows of conical tubercles along the midline of the disc, with others on the shoulders as well as on the outer posterior part of the disc, and the tail becomes very rough finally, with conspicuous thorns along its whole length on its upper side, and rearward from abreast of the tail spines on its lower side. The lower side of the disc is smooth.

Large specimens are easily distinguishable from other sting rays of our Atlantic coast by their very thorny tails and by the large tubercles on the outer parts of their discs. Small ones on which the thorns and tubercles have not yet developed, are recognizable by the shape of the disc, combined with the presence of a skin fold on the lower side of the tail but none on the upper side.

Size.—This appears to be the largest sting ray of the western North Atlantic. The greatest measured width definitely recorded for it is 5 feet, the greatest measured length 10 feet 3 inches. But some certainly grow considerably larger, for a New Jersey specimen has been reported as nearly 7 feet across; the corresponding length would be 13 to 14 feet, if the tail were intact.

Color.—Fresh caught specimens seen by us at Woods Hole have been dark brownish above with the tail black from the spine rearward; white below.

General range.—Coastal waters of the western Atlantic, from the latitude of Cape Hatteras to Cape Cod; most common from Delaware Bay to the Woods Hole region.

Occurrence in the Gulf of Maine.—The only claim of this sting ray to mention here is that one was reported from Chatham on the outer shore of Cape Cod many years ago, and that it is said to have been seen on the shoaler parts of Georges Bank. It has no real status as a Gulf of Maine fish, appearing there only as a summer straggler from the south, though it is so common near Woods Hole that the fish traps there catch some 400 to 500 of them in ordinary summers.[81]

Beware of handling any skate-like fish with a long, whip-like tail, lest it prove to be a sting ray. The tail spine (brought into action as the tail is lashed to and fro) is a dangerous weapon; and the wounds made by it cause excruciating pain.

THE COW-NOSED RAYS. FAMILY RHINOPTERIDAE

The cow-nosed rays, like the whip-tailed rays, have a very long tail armed with one or more poisonous sawedged spines; a very flat broad disc; and pelvic fins with convex outer margins. But their pectoral fins are interrupted on each side of the head, so that the forward portions form a separate two-lobed fin extending forward from the lower side in front of the mouth and nostrils; their crowns are high-domed; their eyes and spiracles are on the sides of the head instead of on its upper surface; and they have a small dorsal fin on the upper side of the tail in front of the tail spines. Their teeth have the form of large, flat grinding plates, fitting close together in mosaic arrangement; and there are only 7 to 9 series of them in each jaw.

Cow-nosed ray

Rhinoptera bonasus (Mitchill) 1815

Bigelow and Schroeder, 1953, p. 469.

Garman, 1913, pl. 37, as *Rhinoptera quadriloba.*

Description.—The cow-nosed rays with all their close relatives[82] have such characteristic outlines, the shape of their heads is so peculiar with the eyes and spiracles on the sides, and their large, flat, plate-like teeth are so different from those of any other Gulf of Maine skates or rays that they are not apt to be mistaken for anything else. The species in question is characterized among its confreres by the indented contour of the front of its head, and by the conspicuously bilobed outline of the short so-called "subrostral" fin that projects forward from the lower side of the latter. The outer corners of the pectorals are pointed, and their posterior margins distinctly concave. The pelvic fins are small, reaching but a short distance back of the posterior corners of the pectorals. The dorsal fin is rounded above, originating about opposite the rear ends of the bases of the pelvics. The tail measured from the center of the cloaca is about twice as long as the body from cloaca to front of head on adults if not damaged, nearly 3 times on small specimens. The tail spines (1 or 2) are close behind the rear limits of the pelvic fins, and thus much further forward on the tail than those of the sting rays (p. 74). There usually are 7 series of teeth in each jaw, with up to 11 to 13 rows exposed, and in function simultaneously.

Size.—The cow-nosed ray has been said to grow to a breadth of 7 feet. But the largest specimen the width of which has either been actually measured or can be calculated from some other dimension, was only about 38 inches wide.[83]

Color.—Brownish above, white or yellowish white below. Some of them are marked both above and below with many narrow faint dark lines radiating out from the center of the disc.

General range.—Western Atlantic coast from middle Brazil to southern New England.

[81] This estimate is based on our own observations near Woods Hole.

[82] The eagle rays, family Myliobatidae, and butterfly rays, family Gymnuridae, are close allies of the cow-nosed rays; none of them has yet been encountered in our Gulf.

[83] Calculated from the dimensions of the head of one from Rio de Janeiro.

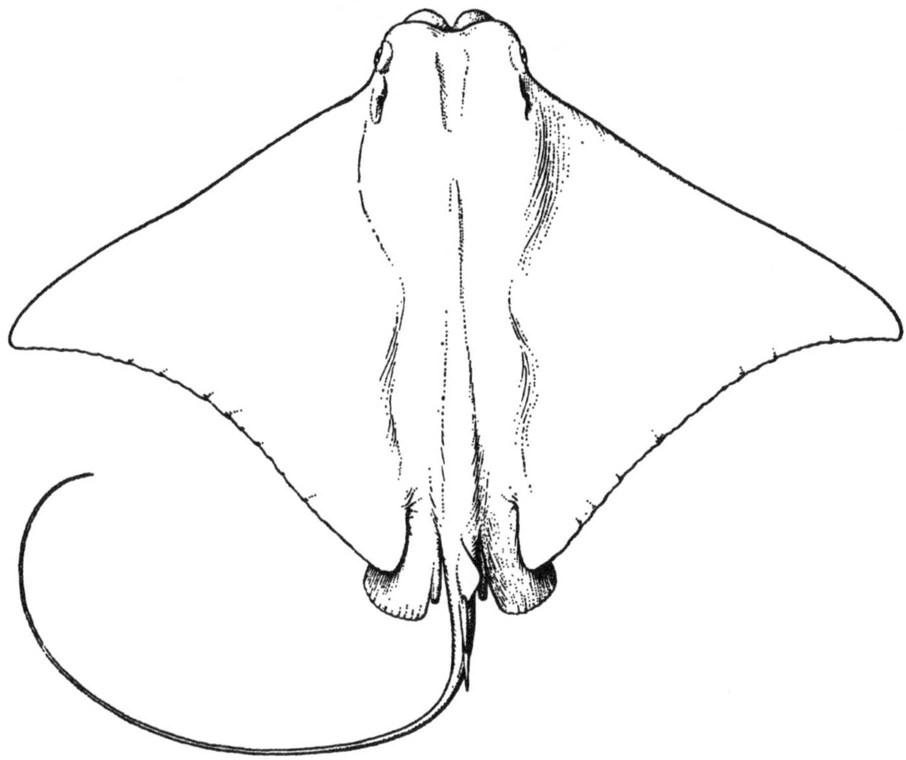

FIGURE 33.—Cow-nosed ray (*Rhinoptera bonasus*), about 22 inches wide, Newport, Rhode Island. From Bigelow and Schroeder. Drawing by E. N. Fischer.

Occurrence in the Gulf of Maine.—The cow-nosed ray has even less claim than the sting ray to be called a Gulf of Maine fish, for while schools of them appear occasionally near Woods Hole where 145 of them were taken in the fish traps in one day on one occasion, and while it is recorded from Nantucket, it has never been seen, actually, east or north of the elbow of Cape Cod.

DEVIL RAYS. FAMILY MOBULIDAE

The devil rays, like the sting rays (p. 74) and cow-nosed rays (p. 76) have the pectoral fins interrupted along the sides of the head close behind the eyes. But they differ very noticeably from the others mentioned above in the shape of the anterior parts of the pectorals, for these are in the form of two separate narrow ear-like fins, set vertically and curving forward from the front of the head. They are further unique among skates and rays in the fact that they feed on small pelagic animals, which they sift, by a complex sieve-like modification of their gill arches, out of the water that is gulped in by the mouth and passed out via the gill clefts. Some of them are the largest of the rays and among the largest of fishes. Being tropical-subtropical in nature they have no real place in the fish fauna of our Gulf, but Manta, the largest of them all, has been known to reach Georges Bank as a stray from warmer latitudes.

Devil ray *Manta birostris* (Donndorff) 1798

Bigelow and Schroeder, 1953, p. 502.

Description.—The so-called cephalic fins of the devil ray, pointing forward, give it so distinctive an appearance that it could not be confused with any other fish, except for some other member of its own family. And it is marked off from all others of these that are known in the Atlantic by the

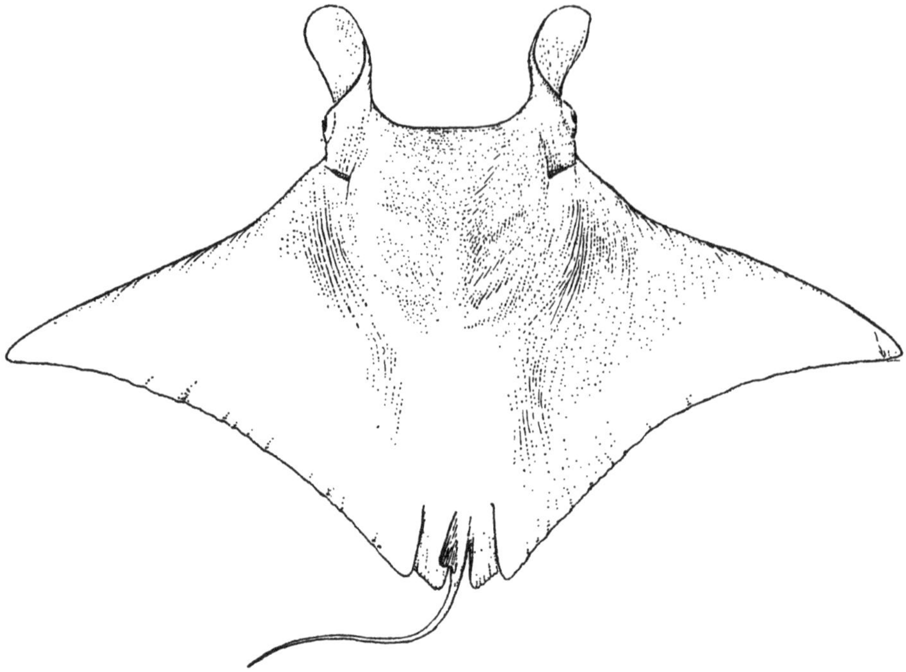

FIGURE 34.—Devil ray (*Manta birostris*), juvenile male, 11 feet 5½ inches wide, Bimini, Bahamas. From Bigelow and Schroeder. Drawing by E. N. Fischer.

position of its mouth, which extends across the front of its head instead of being on the under side. Its cephalic fins are about one-half as broad at the base as they are long, with thin lower edges and thick fleshy upper edges and rounded tips, and each arises nearly vertical from the side of the head. When the owner is swimming they point directly forward, but when the ray is feeding, they can be curved inward, one toward the other until their tips nearly meet in front of the mouth. The disc (not counting the cephalic fins) is a little more than twice as broad as long, with tapering outer corners. The tail measured from the cloaca is at least as long as the body from cloaca to front of head and perhaps longer still if not damaged. And it bears a small rounded dorsal fin on its base. Some specimens have been described as having one or two small tail spines close behind the dorsal fins. However, those that we have seen have had none, but a rounded knob in its place, supported by a mass of bony tissue with a minute pointed spur on the upper side that can be felt but does not break the skin. The skin of disc, pelvic fin, and tail is roughened with small tubercles, below as well as above. The mouth is very wide, extending across a little more than one-half the whole breadth of the front of the head. And the teeth, the lower jaw only, are minute and very numerous; we counted about 270 series in about 12 to 18 rows or a total of about 4,500 in one specimen about 11½ feet wide. The gill openings are noticeably long.

Color.—The upper side varies from reddish or olive brown to bluish slate colored or black, either plain or with various white markings. The lower side is white toward the center of the disc but gray around the margins, and there may be various dark blotches in the region of the gills and on the abdomen. The rear part of the tail is gray.

Size.—This giant ray matures when about 14 to 15 feet wide. They commonly grow to 18 feet or so, and there are recent records of measured specimens 19 feet 8 inches, 21 feet 2 inches, and 22 feet wide. One 14 feet wide weighed 1,686 pounds, one from the Galapagos Islands, 18 feet wide, 2,310 pounds; and one of 20 feet taken long ago off Venezuela weighed 3,502 pounds.

General range.—Manta is known in the Atlantic from middle Brazil to the Carolinas and as a rare stray to southern New England and Georges Bank; from Bermuda; from Madeira; and from tropical West Africa. Mantas are also widespread in the tropical-subtropical belt of the Pacific and Indian Oceans, but it is not yet known whether they are identical with the Atlantic species or not.

Occurrence in the Gulf of Maine.—The only reason for mentioning this giant ray here is that a pair, judged to be 18 to 19 feet wide, were encountered on the southeast part of Georges Bank late in August 1949, by Capt. Henry W. Klimm, while out after swordfish, and so close at hand that their cephalic fins and purplish color were noted. The nearest record to the westward and southward is of one 19 feet wide, weighing 1,686 pounds, harpooned by a sword fisherman a few miles off Block Island and landed there in August 1921.[84]

CHIMAEROIDS. SUBCLASS HOLOCEPHALI

The chimaeroids, being cartilaginous fishes, are allied to the sharks, skates and rays, but are separated from them by many important anatomic characters. Most obvious of these externally are that they have no spiracle; that they have only one external gill opening on either side; that their tails are symmetrical; and that their gill filaments are free at the tips like those of bony fishes. The chimaeroids remotely suggest the grenadiers in general body form (p. 243), but are easily separable from them at a glance; first of all by the softness of their bodies and by their naked skins, also by the location of the pelvic fins which are set far back under or behind the tips of the pectorals, and by the large size of the pectoral fins, to list only the most obvious differences. There is no danger of confusing them with any other Gulf of Maine fishes, so curious is their appearance.

They lay eggs that are astonishingly large for the size of the parent fish, and enclosed in brown horny capsules which are elliptical, spindle-shaped or tadpole-shaped in different species. But fertilization is internal.

The Chimaeras. Order Chimaerae

FAMILY CHIMAERIDAE

Chimaera *Hydrolagus affinis* (Brito Capello) 1868

Bigelow and Schroeder, 1953, p. 539

Description.—This species of chimaeroid, the only one known from within the geographic limits of the Gulf, is deepest (one-sixth to one-seventh as deep as long) just behind the gills, tapers gradually backward to a weak slender tail, and is very soft-bodied. The head is short, its dorsal profile oblique, the snout conical with a blunt tip. The forehead of the male bears a curious cartilaginous hook, armed with recurved prickles on its lower surface, which probably serves to clasp the female. The mouth, on the lower side of the head, is small, with thick fleshy lips; the upper jaw is armed with 4 flat plates in place of teeth,

[84] Reported by Gudger (Science, N. Ser., vol. 55, 1922, p. 339). There are photographs of this specimen in the American Museum of Natural History in New York.

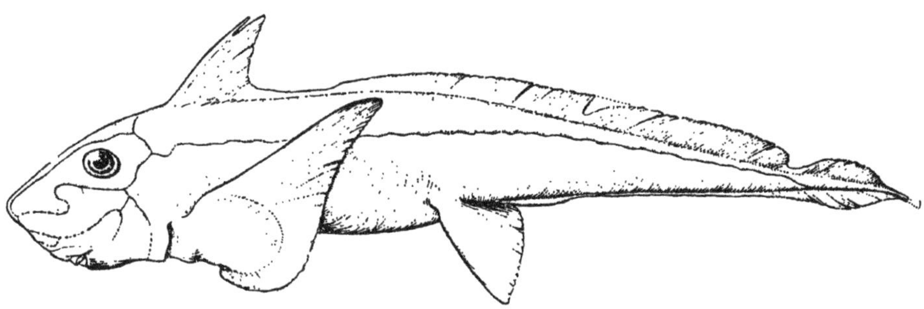

FIGURE 35.—Chimaera (*Hydrolagus affinis*), female, about 31¼ inches long, Banquereau Bank. From Bigelow and Schroeder. Drawing by E. N. Fischer.

the marginal pair set edgewise, the lower jaw with a pair of marginal plates set edgewise. The gill openings are vertical, set very low down on the sides of the neck, and each is covered with a flap of skin, paralleling the gill cover of bony fishes.

There are two distinct dorsal fins. The first of these originates about over the gill openings, is triangular, about as high as long, and supported at its anterior margin by a stout spine that is free along the terminal part, with the rear surface of the free part double saw-edged. The second dorsal is separated from the first by a space variable in length, and is less than one-third as high as the first, with straight margin. The small caudal fin, marked off from the second dorsal by a deep notch, is lanceolate in shape, ending as a short, whiplike filament; and it extends a short distance forward along the ventral surface of the trunk, there being no separate anal fin. The pelvics and pectorals both have pointed tips, the latter being much the larger and reaching back nearly to the point of origin of the pelvics. The male has a trifid copulatory organ arising from the base of each pelvic fin on the inner side, and also a supplementary bladelike clasping organ close in front of each pelvic fin, its margin armed with 4 or 5 hooks, and lying in a pocket from which it can be protruded. The skin is smooth; the lateral-line system is well developed and ramifies over the head in several branches.

This species is a close ally of the well known chimaera of north European seas (*Chimaera monstrosa*), but is distinguishable from it by the fact that it has no separate anal fin; that there is a considerable free space between its two dorsal fins; that the outline of the second dorsal fin is straight; that its caudal filament is much shorter; and that its pectorals hardly reach back to the pelvics.

Color.—Lead color, tan-brown or dark sepia below as well as above, except paler on the throat and grayish on the snout. The margin of the first dorsal, the rear and inner margins of the pelvics, and the rear margins of the pectorals are dark.

Size.—The largest specimen yet reported, taken 85 miles off Cape Sable, Nova Scotia, at a depth of between 400 and 500 fathoms, was 49 inches long and weighed 17½ pounds dressed.

General range.—Not uncommon on the continental slope of North America from the latitude of Cape Cod northeastward, along the Nova Scotia Banks, to the Grand Banks, in 160 fathoms to more than 1,200 fathoms; also in the eastern side of the Atlantic off the coast of Portugal.

Occurrence in the Gulf of Maine.—Our only reason for mentioning this chimaera is that it is (or was) so plentiful along the offshore slopes of the Banks off the eastern part of the Gulf and off Nova Scotia that many were brought in for a few years subsequent to 1875, when fishermen long lining for halibut extended their operations down to 300 fathoms or so. Only one seems to have been reported during the past 25 years, caught off Browns Bank, 85 miles southwest of Cape Sable, between 400 and 500 fathoms on October 15, 1930.[85] But perhaps it would be found no less plentiful now than of old, if sought at the proper depth. The shoalest capture of which we found record was at 160 fathoms. Nothing is known of its way of life nor have its egg cases been seen.

THE BONY FISHES. CLASS OSTEICHTHYES

THE STURGEONS. FAMILY ACIPENSERIDAE

The sturgeons, like the sharks, have an uneven ("heterocercal") tail with the vertebral column extending out along the upper lobe. But there is no danger of mistaking a sturgeon for a shark for it has only one gill opening on each side, while the gills are enclosed by bony gill covers. And the combination of gills of this kind with sharklike tail and with the fact that the head is covered by bony plates united by sutures, sets the sturgeons off from all other Gulf of Maine members of their own class. Two species of sturgeons are known from the Gulf, one of which once was rather common there; the other is extremely scarce everywhere.

[85] Reported by Firth, Bull. Boston Soc. Nat. Hist., 61, 1931, p. 9. It was 49 inches long and weighed 17½ pounds dressed.

KEY TO GULF OF MAINE STURGEONS

1. The successive bucklers in the dorsal row touch each other or even overlap; the space between the dorsal row of bucklers and the uppermost of the two lateral rows is thickly set with coarse prickles............ Sea sturgeon, p. **81.**
2. The successive bucklers in the dorsal row are separated one from the next by spaces up to ½ as long as the bucklers; the space between the dorsal row of bucklers and the uppermost of the two lateral rows is only sparsely strewn with fine prickles.. Short-nosed sturgeon, p. **84.**

Sea sturgeon *Acipenser sturio* Linnaeus, 1758[86]

Jordan and Evermann, 1896–1900, p. 105

Description.—The skin of the sturgeon is armored with a row of large bony shields or bucklers along the middle of its back (the successive bucklers touching or even overlapping) with a second row of smaller bucklers high up along each side of the body; and with a third row, also smaller, lower down, along the line of transition from side to belly. Each buckler has a longitudinal keel with a spur, which is so sharp on small fish that these are hard to handle, lower and blunter on large. On the average there are 10 or 11 (10—16) bucklers in the mid-dorsal row; 28 or 29 (26—34) in each upper lateral row; and 9 to 14 in each of the lower lateral rows. The dorsal row runs from above the gill covers back to the dorsal fin, and each of the dorsal shields reaches to the next shield or even overlaps it. The upper lateral rows run from the gill openings back to the root of the tail fin; the lower lateral rows from close behind the pectoral fin to the pelvic fin, also from the pelvic fin back as far as the anal fin. And each shield in each of the two lateral rows is separated from the next shield by a space up to one-half as long as the shields.

The body is rather slender and rendered more or less pentagonal in cross section by the five rows of shields, instead of rounded as it is in the majority of bony fishes. The snout is narrow in young sturgeons less than 2 to 2½ feet long, depressed below the level of the forehead, nearly flat below, and longer (from the eyes forward) than the distance is from the eyes rearward to the upper corners of the gill openings. But it changes shape as the fish grows, becoming blunter, straight in dorsal profile, and considerably shorter relatively. The mouth, situated on the under side of the head, is small, toothless (except in larval stages), with protractile lobed lips, and there are four pointed barbels in a row across the lower side of the snout in front of the mouth. The single rather small triangular dorsal fin stands far back, with its rear edge over that of the still smaller anal fin. The ventral fins are likewise far back. The pectorals are set almost as low as the plane of the belly.[87]

Color.—Olive greenish or bluish gray above, gradually fading on the sides and changing rather abruptly below the upper lateral rows of shields to the white of the belly.

Size.—The sea sturgeon is a very large fish. In the Delaware River where sturgeon persisted until recently in larger numbers than in New England, ripe males are up to about 6 to 7 feet in length, averaging 65 pounds in weight; the spawning females (which are larger), up to about 10 feet and to about 250 pounds,[88] with a larger one taken from time to time. And the general run was about the same in the Kennebec, to judge from an average weight of 120 pounds for males and females together, during the years when a fishery was carried on there. But some still grow considerably larger in Gulf of Maine waters. Thus 9 weighing between 350 pounds and 600

[86] It still is an open question, that we cannot answer, whether the sea sturgeon of eastern North America is identical with the European sea sturgeon, is a recognizable race of the latter, or is a separate species; if the last, its scientific name is *Acipenser oxyrinchus* Mitchill, 1815.

[87] Vladykov and Beaulieu (Natural. Canad., vol. 73, 1946, pp. 143–204), give a detailed account of the characters that separate the sea sturgeon from the lake sturgeon (*Acipenser fulvescens* Rafinesque, 1817).

[88] According to Cobb, Rept. U. S. Fish Comm. (1899), 1900, p. 277.

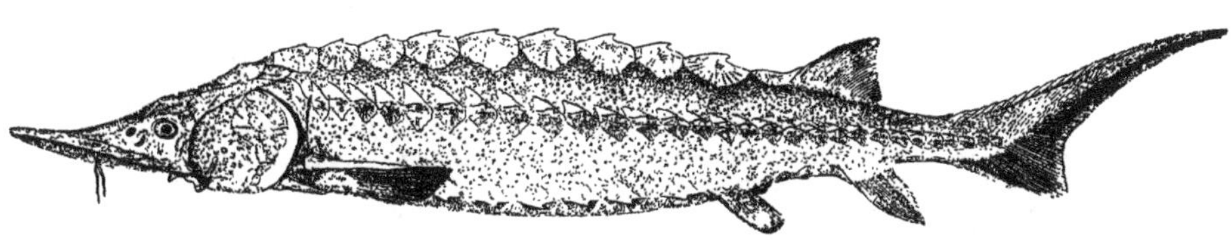

FIGURE 36.—Sturgeon (*Acipenser sturio*), Potomac River specimen. From Goode. Drawing by H. L. Todd.

pounds were landed in Portland, Maine, from the South Channel, Georges Bank, Browns Bank, and Western Bank off Nova Scotia during the period 1927–1935.[89] About 12 feet is perhaps the greatest length to be expected today. But 18 feet, reported for New England many years ago, may not have been an exaggeration, for sturgeon as long as that have been reported from Europe also. The heaviest Gulf of Maine sturgeon reliably reported (to our knowledge) was one of 600 pounds, landed in Portland by the steam trawler *Fabia* from Georges Bank, December 21, 1932.

The following relationship between length and weight, for sea sturgeons up to 7½ feet long, taken in the lower St. Lawrence River,[90] would probably apply to Gulf of Maine fish, equally: 7 to 9 pounds at 30 inches (to fork of tail); 15 to 18 pounds at 40 inches; about 35 pounds at 50 inches; 55 to 57 pounds at 5 feet; about 100 pounds at 6 feet; and about 190 pounds at 7½ feet.

Habits.—The sturgeon makes most of its growth in salt water but enters fresh-water rivers to spawn, as do the salmon, the shad, and the alewife. The large adult fish enter (or once entered) the Gulf of Maine rivers late in the spring, working their way slowly upstream beyond tidewater before depositing their eggs. So far as known, spawning takes place in our rivers in May, June, and perhaps as late as July. It has been suggested that some may spawn in brackish water from the fact that females with large eggs have been taken near Woods Hole in June and July (i. e., in the spawning season). Spawning leaves the spent "cows" in very poor condition. In the Delaware, however, and presumably in Gulf of Maine rivers, they "become again quite plump, acquiring considerable additional weight" [91] before they go down stream again, which some of them do not do until September, according to observations in the Delaware. But we do not know how many years in succession a given fish may spawn.

A single female may produce as many as 2,400,000 eggs which hatch in about a week after they are fertilized.[92] Judging from European observations on artificially reared sea sturgeon, the larvae may be expected to grow to 12 mm. in length within 5 days after hatching; to 16–17 mm. in 2 weeks; to 20 mm. in 4 weeks; and to 4–5½ inches in 2 months.

Some young sturgeon may live several years in the lower tidal reaches of the rivers in which they are spawned, until they have grown to a length of 2½ to 3 feet, as appears to be the case in the Hudson.[93] And it seems that they pass their entire growth period in the salt estuary of the St. Lawrence River, for sturgeons are taken there of all sizes from a few inches long up to 7–8 feet or longer.[94] But others may descend during their first year, for sperlets only 5 to 6 inches long have been found at the mouth of the Delaware River and of the Elbe in Europe.[95]

Some Gulf of Maine sturgeon have taken to the sea by the time they have grown to 3 feet or so, as proved by the capture of sturgeons of that size at various points around the coasts of the Gulf, and off southern New England. And recent observations in the Hudson by Greeley make it likely that all the sturgeon that are spawned in rivers emptying into the Gulf of Maine go to sea sooner or later to complete their growth.[96]

Sturgeon grow rather slowly at first while still in their parent streams. Four, for example, that were tagged in the lower St. Lawrence when 29 to 33 inches long, and recaptured nearby 2 to 3½ years later, had gained only about 2 to 5 inches in length per year.[97] Very slow growth is also indicated by ages of 5 to 6 years at 24 to 28 inches; 7 years at 25 to 31 inches; and 8 years at 32 to 34 inches, for sturgeon from the tidal waters of the lower Hudson, as estimated from the markings on their otoliths.[98] It also seems that sturgeon, like many other fish, make most of their growth during the warm season in such situations for one marked fish in the Elbe did not grow at all between November and the following February, whereas a second grew from 17 cm. (6¾ in.) to 38 cm. (15 in.) in length between January 17 and

[89] Records collected by the late Walter H. Rich of the U. S. Bureau of Fisheries.

[90] According to measurements and weights of 1,592 sturgeons by Vladykov, Rapp. Gen. Ministr. Chasse, Pêch., Quebec (1948–1949), 1949, pp. 43–54.

[91] Ryder, Bull. U. S. Fish Comm., vol. 8, 1890, p. 266.

[92] Ryder (Bull. U. S. Fish Comm., vol. 8, 1890, p. 231) describes the spawning and early development of the sturgeon in the Delaware River.

[93] See Greeley (Supp. 26 Ann. Rept. Conserv. Dept. New York, 1937, pp. 68, 78–82, 89) for a study of the sturgeon in the Hudson River.

[94] A series of 1,592 sea sturgeons from the lower St. Lawrence River, studied by Vladykov (Rapp. Gen. Minstr. Chasse, Pêch. Quebec (1948–1949) 1949, pp. 53–56) included a good representative of sizes from about 4 inches up to 90 inches.

[95] Prince reports a 6-inch sturgeon from Hudson Bay (Rept. Sixty-seventh Meeting, British Assoc. Adv. Sci., Toronto, 1897, p. 687).

[96] Greeley, Suppl. 26 Ann. Rept. Conserv. Dept. New York, 1937, p. 82.

[97] Vladykov (Rapp. Gen. Ministr., Chasse, Pêch. Quebec, 1948–1949, pp. 61–63, 66, table 19).

[98] Greeley, Supp. 26 Ann. Rept. Conserv. Dept. New York, 1937, p. 68, table 10.

the following April, and a third from 43½ cm. (17¼ in.) to 64 cm. (25¼ in.) from April 9 to the following December. But sturgeon grow much more rapidly after they go to sea, if ages (estimated from otoliths) of 11 years for a 75-inch sturgeon, and 12 years for two others of 88 and 100 inches are anywhere near the truth.[99]

The sturgeon is a bottom feeder, rooting in the sand or mud with its snout like a pig (the barbels serving as organs of touch) as it noses up the worms and mollusks on which it feeds and which it sucks into its toothless mouth with considerable amounts of mud. It also consumes small fishes, particularly sand launce. Small ones, while living in estuaries and around river mouths, subsist largely on amphipod and isopod crustacea. Sturgeon, like salmon, eat little or nothing while traveling up river to spawn.

When at ease sturgeon swim slowly to and fro, seeming very sluggish. But they are capable of darting ahead like an arrow on occasion, and they often come to the surface to jump clear of the water. Though they usually offer no resistance when netted, large ones are very strong.

General range.—Coastal waters from the St. Lawrence River to the Gulf of Mexico, running up into rivers to spawn; reported from Hudson Bay, also Scandinavia to the Mediterranean, if the American and European sea sturgeons belong to the same species.

Occurrence in the Gulf of Maine.—The sea sturgeon is (or was) well known in the St. John, Penobscot, Kennebec, and Merrimac Rivers, and has even been taken some distance from the mouths of streams no larger than the Charles River and the Parker River in Essex County, Mass.,[1] where some are still seen jumping in July and one is taken occasionally. In fact, sturgeon once entered practically every stream of any size emptying into the Gulf of Maine. Wood, writing of Massachusetts in 1634,[1a] described them as "all over the country, but best catching of them be upon the shoales of Cape Code and in the river of Merrimacke, where much is taken, pickled and brought for England, some of these be 12, 14 and 18 foote long." In fact, an odd sturgeon still enters the mouth of the Merrimac, witness one of 230 pounds netted there on September 14, 1938 and landed in Newburyport.[2]

Sturgeons may be expected anywhere off the coasts of the Gulf of Maine during their sojourn in salt water. There is definite record of them at sundry localities on both sides of the Bay of Fundy; off Mt. Desert Island; in Penobscot Bay; in Casco Bay; at the mouth of the Piscataqua River; on the Boars Head–Isles of Shoals fishing ground, where several 3 to 4 feet long were taken in gill nets during April and May 1913; at the mouths of the Essex and Ipswich Rivers, where jumping sturgeon have been reported recently in the daily press;[3] at the mouth of Gloucester Harbor, where an angler reports catching one of about 12 pounds while fishing for tautog; inside and outside Boston Harbor; at Provincetown; off Truro, Cape Cod; and at Nantucket, as well as along the southern New England coast to the westward. Some also extend their wanderings to the offshore fishing banks as they grow. Thirty, for example, ranging in weight from 120 to 600 pounds were landed in Portland and Boston by otter trawlers from Nantucket Shoals, from South Channel, and from Georges and Browns Banks, during the years 1927–1936.[4] Probably all of these were on bottom when caught, to judge from their diet (p. 83), and from the fact that sturgeon have been hooked on cod and haddock lines as deep as 25 fathoms in Scandinavian waters. Nothing beyond this is known of their movements in our Gulf.

Importance.—It is only the scarcity of the sea sturgeon in the Gulf of Maine that limits its commercial importance there and in the tributary rivers. The few taken are picked up accidentally in traps or weirs, in drift nets, or by the otter trawlers.

In former years, when our streams were less obstructed and sturgeons more plentiful, the catch was of considerable value in some of the larger rivers. It is interesting, for instance, to read that sturgeon, doubtless from the Kennebec River and cured near what is now Brunswick, Maine, were shipped to Europe as early as 1628; and that large quantities were also shipped to Europe from near Ipswich, Mass., in 1635. In the Kennebec, where an intermittent fishery had long been maintained.

[99] See footnote 98.

[1] Two sturgeon 44 and 45½ inches long, netted in the Parker River at Newbury, Mass., July 23, 1933, are (or were) in the collection of the Boston Society of Natural History, now the New England Science Museum (Bull. Boston Soc. Nat. Hist., 69, Oct. 1933, p. 8).

[1a] New England's Prospect, 1634, p. 37.

[2] Reported in the Boston Globe, Sept. 15, 1938.

[3] The Boston Herald, June 1950.

[4] Reports collected by the late Walter H. Rich, U. S. Bureau of Fisheries, and notices in the daily press.

the catch was about 250 fish in 1880, yielding 12,500 pounds of meat, and not much less in 1898 (10,875 pounds). But the yearly landings were only about one fourth as great there (2,777 pounds) by 1919. And the reported landings of sturgeon from the entire coastline of Maine (including what few were brought in from offshore) had fallen to only 300 pounds in 1940, and 400 pounds in 1947. Reported landings in Massachusetts of 5,300 pounds in 1940 (all by otter trawlers) and of 6,600 pounds (5,000 pounds by otter trawlers, from off shore), corresponding to some 50 to 70 fish, if they weighed as little as 100 pounds each, will further illustrate their present-day scarcity.

We have never heard of a large sturgeon hooked by an angler in the Gulf of Maine. But we hear from time to time of a small one caught in this way, as already remarked (p. 83).[5] And the skill of a woman angler [6] who foul-hooked a sturgeon about 6 feet long, and beached it on surf-casting tackle after a long fight, fishing alone at Wasque Point, Marthas Vineyard, on July 15, 1950, was widely heralded in the daily press.

Short-nosed sturgeon *Acipenser brevirostrum* LeSueur 1818

LITTLE STURGEON

Jordan and Evermann, 1896–1900, p. 106.

Description.—The little sturgeon resembles the sea sturgeon so closely in general appearance that we need note only the most obvious differences. These are that the shields in its dorsal row are relatively smaller, and that each is separated from the next by a space up to ½ as long as the shields themselves (successive dorsal shields in contact or overlapping in the sea sturgeon); that the space between its dorsal row of shields and the upper lateral row on each side is only sparsely set with fine prickles (closely set with coarse prickles in the sea sturgeon); and that its viscera are blackish (pale in the sea sturgeon); also the number of rays in the anal fin averages smaller in the little sturgeon (19–22) than in the sea sturgeon (23–30). The snout, too, is considerably shorter relatively, as well as broader, than it is in young sea sturgeons of equal size. And while the snout is about as long, relatively, in the one species as in the other when they are full grown, sea sturgeons are then so much the larger that there is no danger of confusing the one kind with the other.

Color.—Described as blackish above, tinged with olive above the upper lateral line of shields, marked with alternate black and pale bands; sides, below the upper lateral row of shields, reddish mixed with violet; abdomen white.[7]

Size.—This is a much smaller fish than the sea sturgeon. Males may mature when only 19–20 inches long and most of them do by the time they pass 21 inches; most of the females at about 24 inches. The largest so far recorded is one of about 36 inches, in the Museum of Comparative Zoology. One about 31 inches long weighed 7 pounds 4 ounces.[8]

Habits.—Nothing is known of the habits of the little sturgeon except that it spawns in rivers and that it does so late in April in the lower Hudson. The fact that fair sized specimens are taken there in summer and also in winter, suggests that it may not be as regularly migratory as the sea sturgeon is.[9] But the places of capture of the Gulf of Maine specimens mentioned below show that some certainly go out into the open sea and wander for some distance from their parent stream.

General range.—So far as we know, the only

[5] We once saw one small one about 23 inches (575 mm.) long foul hooked in the side off South Beach, New York, December 21, 1923, and heard of a similar experience by the same angler a year later.

[6] Mrs. George T. Rice. About 30 others were seen by her at the same time in a deep slew formed by a new bar.

[7] We have not seen a fresh-caught specimen.

[8] For further details as to size, see Ryder, Bull. U. S. Fish Comm., vol.8, 1890, p. 238; and Greeley, Suppl. 26 Ann. Rept. Conserv. Dept. New York, 1937, p. 69, table 11, pp. 82, 90.

[9] Greeley, Suppl. to 26 Ann. Rept. Conserv. Dept. New York, 1937, p. 90, makes this suggestion.

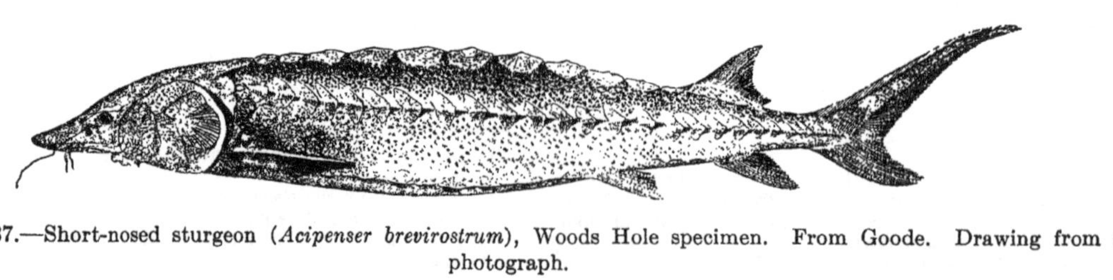

FIGURE 37.—Short-nosed sturgeon (*Acipenser brevirostrum*), Woods Hole specimen. From Goode. Drawing from a photograph.

locality records definitely belonging to this species, not to young sea sturgeons, are from Provincetown and Waquoit, Mass.; from the Hudson River, N. Y.; from Delaware Bay and River; and from Charleston, S. C.

Occurrence in the Gulf of Maine.—The only recent record of the little sturgeon in the Gulf is of one about 23 inches long, taken at Provincetown about 1907 and now mounted in the Museum of Comparative Zoology.[10] The Museum of the Essex Institute, Salem, also has—or had—a stuffed sturgeon from Rockport, Mass., identified as this species by Goode and Bean.[11] Evidently the sturgeon is now very scarce in our Gulf and there is no reason to think that it ever has been more plentiful there.

The Herring and Tarpon Tribes

FAMILIES CLUPEIDAE, DUSSUMIERIIDAE, AND ELOPIDAE

The true herrings (Clupeidae) are soft-finned fishes wholly lacking spines, with one short dorsal fin, deeply forked tails, ventral fins situated on the abdomen far behind the pectorals, teeth small or lacking in adults, deep bodies flattened sidewise, and large scales that slip off at a touch. They are, perhaps, the most familiar of northern sea fishes and certainly are the most abundant in number of individuals. Seven species of herring occur in the Gulf of Maine—the hickory shad (not very common), the sea herring, the alewife, the blueback, and the shad (regular and plentiful), thread herring (scarce), and the menhaden (irregular in its occurrence). The shad, menhaden, sea herring, and thread herring are easily named; but the alewife and the blueback resemble one another so closely that they are often confused, even by the fishermen who handle them constantly. The round herrings (Family Dussumieriidae) differ from the true herrings chiefly in their rounded bellies and less deep bodies. The members of the Tarpon Tribe (Family Elopidae) are very closely allied to the true herrings (Clupeidae), from which they differ in having a bony plate on the throat between the branches of the lower jaw. There are only about five species, all of them tropical. Two are known from the Gulf, as strays.

[10] This Museum also has another of about 36 inches from Waquoit, on the southern shore of Massachusetts.

[11] Bull. Essex Inst., vol. 11, 1879, p. 27. A sturgeon was reported as *brevirostris* from Boston Harbor many years ago, but there is no way now of checking the identification.

KEY TO GULF OF MAINE HERRINGS AND TARPONS

1. Last dorsal fin ray prolonged ---- 2
   Last dorsal ray not prolonged ---- 3
2. Dorsal fin originates in advance of the ventrals; scales only moderately large ---- Thread herring, p. 112
   Dorsal fin originates behind the ventrals; scales very large ---- Tarpon, p. 87
3. Belly rounded ---- 4
   Belly sharp edged ---- 5
4. Scales very small; mouth very large with upper jaw-bone extending considerably beyond the rear edge of the eye; point of origin of dorsal fin about over that of the ventral fins ---- Ten pounder; p. 86
   Scales large; mouth small, with upper jaw-bone extending rearward only about as far as the front edge of the eye; point of origin of dorsal fin well in advance of that of the ventral fins ---- Round Herring, p. 87
5. Head (tip of snout to edge of gill cover) very large, occupying about one-third the total length of the body to base of the central rays of the caudal fin; free edges of scales fluted, not rounded ---- Menhaden, p. 113
   Head about one-fourth the total length of the body; free edges of the scales rounded ---- 6
6. Distance from point of origin of dorsal fin to tip of lower jaw (mouth closed) about as long as from origin of dorsal fin to base of central rays of caudal fin; edge of belly hardly saw-toothed, though sharp; general form comparatively shallow; there is a cluster of teeth on the roof of the mouth ---- Sea herring, p. 88
   Distance from point of origin of dorsal fin to tip of lower jaw (mouth closed) considerably shorter than from point of origin of dorsal fin to origin of central rays of caudal fin; edge of belly more or less strongly saw-toothed, especially in space between the ventral and anal fins; general form deep; there are no teeth on the roof of the mouth ---- 7
7. The tip of the lower jaw extends noticeably beyond the upper when mouth is closed ---- Hickory shad, p. 100
   The tip of the jaw does not extend appreciably beyond the upper when mouth is closed ---- 8

8. The upper outline of the forward part of the lower jaw (visible if mouth is opened) is nearly straight, and does not show a pronounced angle; the upper jaw extends back about level with the rear edge of the eye ---------- Shad, p. 108
   The upper outline of the forward part of the lower jaw is concave with a pronounced angle; the upper jaw reaches back only about to the level of the center of the eye ---------- 9
9. Breadth of eye is greater than distance from front of eye to tip of snout; back distinctly grey green; lining of belly cavity pale grey ---------- Alewife, p. 101
   Breadth of eye is only about as great as distance from front of eye to tip of snout; back distinctly blue green; lining of belly cavity sooty or black ---------- Blue back, p. 106

Ten pounder *Elops saurus* Linnaeus 1766

Jordan and Evermann, 1896–1900, p. 410.

Description.—The ten pounder is herring-like in the arrangement of its fins, with the single and soft-rayed dorsal fin originating about midway along its back; in having no adipose fin; in the position of its ventral fins about midway between tip of snout and fork of tail; and in its forked-tail fin. But its scales are very much smaller relatively than those of any of our herrings, and its mouth is much larger, with the upper jawbone extending rearward considerably beyond the rear edge of the eye. Being about one-sixth as deep as it is long, it is a much more slender fish than any of our herrings except the round herring, and its belly is rounded like that of the latter. But its trunk is more flattened sidewise than that of the round herring, its dorsal fin-origin is over the ventrals (well in advance of the ventrals in the round herring), and its tail fin is much wider relatively than that of any herring, and more deeply forked.

A more important structural character is that its throat is stiffened between the branches of its lower jaw by a long bony plate, which it shares with the tarpon, but which no member of the herring tribe has. Its closest affinity among fishes yet known from our Gulf is with the tarpon. But its scales are very much smaller than those of the latter, nor does its dorsal fin have the prolonged ray characteristic of the tarpon.

Color.—Silvery all over, with the back bluish, the lower parts of the sides and the lower surface yellowish; the dorsal and caudal fins dusky yellowish and silvery; the ventral and pectoral fins yellowish speckled and dusky.

Size.—The ten pounder is said to grow to a length of 3 feet,[12] but few of those caught are longer than about 20 inches.

General range.—Atlantic coast of America, from Brazil northward; commonly to North Carolina, in small numbers and less regularly to southern New England, and perhaps straying around the elbow of Cape Cod on rare occasions. The ten pounder of our Atlantic coast is represented in tropical-warm temperate seas in various other parts of the world by relatives so close that they may all finally prove to represent only the one wide-ranging species.[13] Our only reason for mentioning this southern fish is that one reported as from Chatham, Mass., may have been taken on the Gulf of Maine shore of Cape Cod.[14] Ten pounders are taken from time to time near Woods Hole.

[12] Jordan and Evermann, Bull. 47, U. S. Nat. Mus., Pt. 1, 1896, p. 410.
[13] Smith (Sea Fishes Southern Africa, 1949, p. 86) considers this probable.
[14] This specimen, taken on October 19, 1888, and reported by Bigelow and Schroeder (Copeia, 1940, p. 139) is in the Museum of Comparative Zoology.

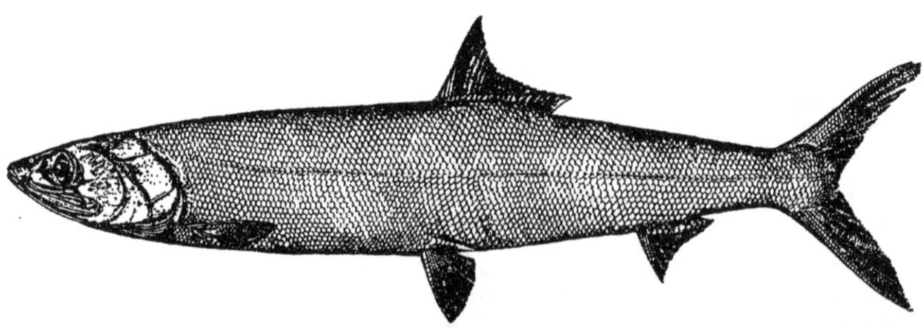

FIGURE 38.—Ten pounder (*Elops saurus*), Massachusetts. From Goode. Drawing by H. L. Todd.

Tarpon *Tarpon atlanticus* (Cuvier and Valenciennes) 1846

Jordan and Evermann, 1896–1900, p. 409, fig. 177.

Description.—The tarpon is herring-like in general form and appearance, but it is made easily recognizable by the fact that the last ray of the dorsal fin is greatly prolonged, its free portion being as long as the fin is high or longer, and by the presence of the bony plate on the throat mentioned above (p. 85) in the characterization of the family to which it belongs. Furthermore, the anal fin of the tarpon is deeply falcate; that of all Gulf of Maine herring-fishes rhomboid in outline. The ventral fins, which are situated under or behind the dorsal fin in herrings, alewives, shad, and menhaden, are considerably in front of the dorsal fin in the tarpon, while the lower jaw of the latter projects relatively further; its scales are relatively larger; and its caudal fin is relatively wider.

Color.—Bright silvery all over, the back darker than the belly.

Size.—Tarpon grow to a length of 6 to 8 feet; the longest recorded was 8 feet 2 inches; the heaviest taken on rod and reel weighed 247 pounds.[15]

General range.—Tropical and subtropical coasts of America, from Brazil to Long Island, casually to Cape Cod, and to Nova Scotia, where it has been recorded off Isaacs Harbor and in Harrigan Cove.[16] Its chief center of abundance is in the West Indies, about Florida, and in the Gulf of Mexico.

Occurrence in the Gulf of Maine.—A specimen 5½ feet long, taken at Provincetown on July 25, 1915,[17] is the only record of the tarpon in the Gulf of Maine, which it reaches only as an accidental straggler from the south.

Round herring *Etrumeus sadina* (Mitchill) 1815

Jordan and Evermann, 1896–1900, p. 420.

Description.—The most distinctive feature of this fish, among herrings, is that its belly is rounded, not sharp edged. It is, furthermore, the most slender of our herrings, its body being only

[15] Taken on rod and reel in the Panuco River, Mexico, Mar. 24, 1938, by H. W. Sedgewick.

[16] Halkett, Check List, Fishes Canada, Newfoundland, 1913, p. 45.

[17] Radcliffe, Copeia, No. 26, 1916, p. 3.

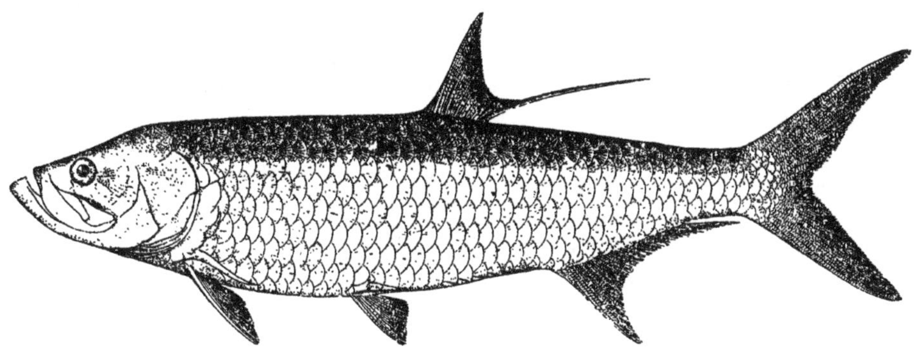

FIGURE 39.—Tarpon (*Tarpon atlanticus*), New Jersey. From Goode. Drawing by H. L. Todd.

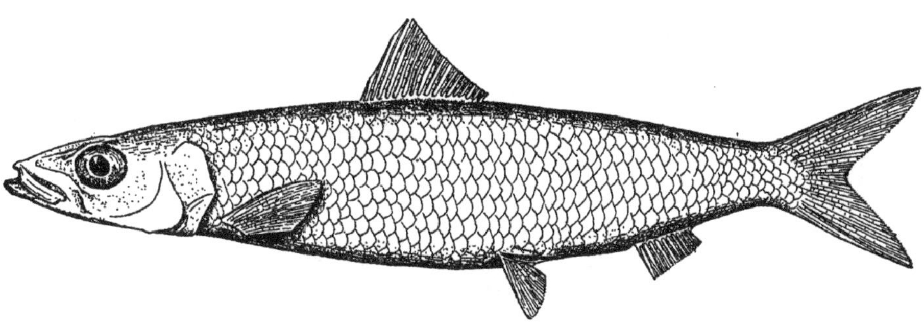

FIGURE 40.—Round herring (*Etrumeus sadina*).

one-sixth as deep as long, thus suggesting a smelt in its general outline. Its dorsal fin, too, stands wholly in front of the ventrals instead of over the latter, as in herring, alewives, and shad; and there are fewer anal fin rays (only about 13, whereas the herring has about 17, the alewife about 19, and the shad about 21) than any of the latter.

Color.—Olive green above with silvery sides and belly.

Size.—Eight to ten inches long when adult.

General range.—Atlantic and Gulf of Mexico coasts of the United States; occasionally common as far north as Woods Hole; sometimes straying past Cape Cod, to the mouth of the Bay of Fundy.

Occurrence in the Gulf of Maine.—This southern fish has been taken at Provincetown, Mass., whence the Museum of Comparative Zoology has two specimens; one was taken in the Yarmouth River which empties into Casco Bay, and one in the bay itself on September 15, 1924;[18] it has been reported from Jonesport, Maine; also from Eastport, Maine, in 1908.[19] And a number of them were taken at Campobello Island, at the mouth of Passamaquoddy Bay in September 1937.[20]

Herring *Clupea harengus* Linnaeus 1758

SEA HERRING; LABRADOR HERRING; SARDINE; SPERLING; BRIT

Jordan and Evermann, 1896–1900, p. 421.

Description.—The sea herring is typical of its family in form, with body so flattened that it is much deeper than thick; moderately pointed nose; large mouth situated at the tip of the snout and lower jaw projecting a little beyond the upper when the mouth is closed; sharp-edged belly; and deeply forked tail. The dorsal fin stands over the much smaller ventrals, its origin about midway the length of the body. The scales are large, their rear margins rounded, and so loosely attached that they slip off at a touch. There is no adipose fin, and its absence at once distinguishes all the herrings from any of the salmon tribe. The chief anatomical character separating the sea herring from the shad and from the several alewives (genus *Pomolobus*) is that it has an oval patch of small teeth on the vomer bone in the center of the roof of the mouth. Conspicuous field marks separating herring from shad, hickory shad, and alewife are that the point of origin of its dorsal fin is about midway of the length of its trunk (considerably farther forward in the others); its body is not so deep, a difference shown better in the illustrations; and the sharp midline of its belly is only very weakly saw-toothed but is usually strongly so in the others, especially along the space between ventral and anal fins.

Color.—Deep steel blue or greenish blue on the back with green reflections; the sides and belly silvery; the change from dark belly to pale sides often marked by a greenish band. The gill covers sometimes glisten with a golden or brassy gloss; indeed, fish just out of the water are iridescent all over with different hues of blue, green, and violet; but these colors soon fade, leaving only the dark back and silvery sides. The ventral and anal fins are translucent white; the pectorals, however, are dark at the base and along the upper edge; the caudal and dorsal fins are dark grayish or shading into green or blue.

Size.—Herring grow to a length of about 17 inches and to a weight of about 1½ pounds.

Habits.—The herring is a fish of open waters,

[18] Reported to us by the late Walter H. Rich of the U. S. Bureau of Fisheries.

[19] Reported in the newspapers.

[20] Reported by Leim, Progr. Rept. 21, Atlantic Biol. Sta. Fish. Res. Bd. Canada, 1937, p. 5; and by McKenzie, Proc. Nova Scotia Inst. Sci., vol. 20, 1939, p. 15.

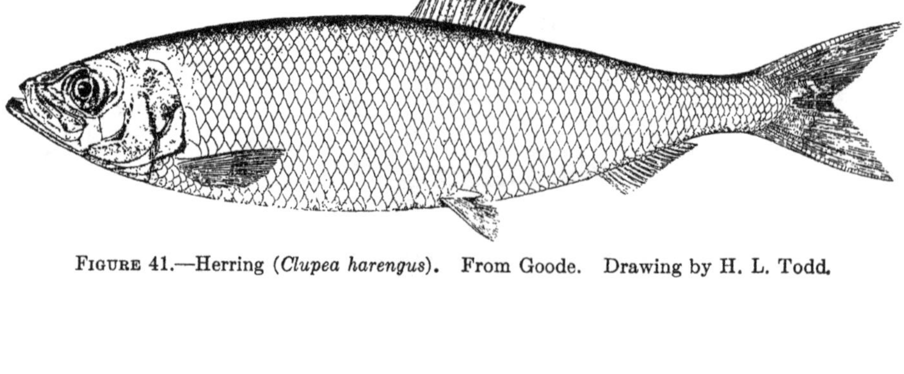

FIGURE 41.—Herring (*Clupea harengus*). From Goode. Drawing by H. L. Todd.

traveling as a rule in schools of hundreds or thousands; single fish are seldom seen, or even small companies of a few dozen. As a rule all the individual members of a school are about the same size, whether large or small. It is not known how long any given school may preserve its identity as such. Fridriksson and Aasen,[21] it is true, found that herring tagged and released together might be recaptured from widely separated localities, suggesting that schools are more or less temporary formations. But this may not apply to schools that have assembled under natural conditions.

When a school is at the surface, as often happens on a calm day, its presence is betrayed by a fine rippling of the water, but we have never seen herring "finning" or lifting their noses above the surface as menhaden often do (p. 114). They come to the surface most often by night, when their presence is betrayed by their luminous trails, if the water is "firing," as we have often seen. A school is likely to be more or less stationary when feeding, its members swimming slowly to and fro and drifting as a whole with the current.[22] But at other times schools are seen traveling with individual fish swimming side by side, rank below rank, as far down in the water as the eye can see from a boat, all heading in one direction apparently with some purposeful intent. We have often watched schools of "sardine" size streaming close past a certain rocky headland in the southern side of Massachusetts Bay, seemingly in unending procession.

As Dr. Huntsman points out, "There is no indication that herring swim against the current unless the water is somewhat turbulent." [23] If they do so under such conditions, it depends on the relationship between their rate of swimming and the strength of the current whether they actually make headway against it or lose ground, tail first.

We might also add that schools of herring, like schools of menhaden, are not so easily frightened by the approach of a boat, as mackerel often are, and striped bass. Herring do not jump unless frightened. But the smaller sizes are often seen jumping when pursued from below by larger predatory fishes, such as silver hake or striped bass, a common spectacle. Fridriksson and Aasen found that herring, held in live-nets, swam constantly at a rate of about 0.2 to 0.25 sea miles per hour (6–8 meters per minute) when not disturbed. And it is certain that they are capable of long journeys, for a number of herring tagged on the northeast coast of Iceland have been recaught in southern Norway, and some vice-versa.[24]

The activity of the herring is controlled in great part by the temperature of the water. In Passamaquoddy Bay, for example, they are "observed to move very sluggishly when the water is coldest in February and March," [25] and probably this applies all around the periphery of our Gulf, for the upper 20 fathoms ordinarily cools to about 33 to 36° F. during those months, with the surface often chilling to the freezing point of salt water in bays and harbors. The herring become active again when the water has warmed to about 40 to 43°.

Food.—The herring is a plankton feeder. When first hatched, and before the disappearance of the yolk sac, the larvae (European) feed on larval snails and crustaceans, on diatoms, and on peridinians, but they soon begin taking copepods, and depend exclusively on these for a time after they get to be 12 mm. long, especially on the little *Pseudocalanus elongatus*.[26] As they grow older they feed more and more on the larger copepods and amphipods, pelagic shrimps, and decapod crustacean larvae. Examination of 1,500 stomachs[27] showed that adult herring near Eastport were living solely on copepods and on pelagic euphausiid shrimps (*Meganyctiphanes norwegica*), fish less than 4 inches long depending on the former alone, while the larger herring were eating both.

When feeding on euphausiids, we have often seen them pursuing the individual shrimps, which frequently leap clear of the water in their efforts to escape. Even in winter, when shrimp are rarely seen at the surface, Moore found them an important article in the diet of the Eastport herring. And it is likely that the local appear-

[21] Rept., Norwegian Fish. Mar. Invest. Skrifter, vol. 9, No. 11, Rept. 1, 1950, p. 22.

[22] Huntsman (James Johnstone Memorial Vol., 1934, p. 83) gives an interesting account of the movements of herring schools in Passamaquoddy Bay.

[23] James Johnstone Memorial Vol., 1934, p. 84.

[24] Fridriksson and Aasen, Rept. Norwegian Fish. Mar. Invest., Skrifter, vol. 9, No. 11, Rept. 1, 1950, pp. 26–27.

[25] Huntsman, James Johnstone Memorial vol., 1934, p. 83.

[26] The diet of herring, young and old, in the English Channel and in the North Sea has been described by Lebour in a series of papers (see especially Jour. Mar. Biol. Assoc. United Kingdom, vol. 12, 1921, pp. 458–467), by Hardy (British Fisheries Invest., Ser. 3, vol. 7, No. 3, 1924), and by Jesperson (Medd. Komm. Havund. Ser. Plankton, vol. 2, No. 2, 1928, Copenhagen).

[27] Moore, Rept. U. S. Comm. Fish. (1896), 1898, p. 402.

ancės and disappearances of schools of large fish in the open Gulf are connected with the presence or absence of euphausiid shrimp of one species or another. A few of the larger fish, however, as well as the smaller ones, will usually be found full of copepods, even when both shrimp and copepods abound, and copepods are the chief dependence of all our herring, large and small, in the absence of shrimp. The amphipod genus *Euthemisto* also is an important food for herring in European seas; hence the absence of *Euthemisto* from the herring stomachs examined by Moore and by us has doubtless been due to the comparative scarcity of this large active crustacean in the coastwise waters of the Gulf of Maine.

The particular species of copepods on which Gulf of Maine and Woods Hole herring depend have not been identified, but we might guess that *Calanus* predominates, with *Pseudocalanus*, *Acartia*, and *Centropages*, and *Temora* also, at its times of abundance, while *Euchaeta* offers a rich food supply when the schools seek the deep waters of the basin frequented by these mammoth copepods.

In default of an abundant supply of Crustacea, and sometimes even when these are plentiful, herring feed on whatever molluscan larvae, fish eggs, Sagittae, pteropods, annelids that the water contains, even on microscopic objects as small as tintinnids and Halosphaera. But the smaller microscopic plants, either diatom or peridinian, are never found in the stomachs of herring more than 15 to 20 mm. long, probably because their gill rakers are not fine enough to retain them.

Although herring normally are not fish eaters, small launce, silversides, and the young of their own species have been found in them at Woods Hole. And Templeman[28] reports them as consuming quantities of small capelin, in winter, in Newfoundland waters.

Herring ordinarily pick up their food objects individually by a "definite act of capture" as Battle expresses it,[29] while she found that herring in the aquarium at St. Andrews did not feed in complete darkness, though they did in faint light. But it seems that when feeding on very small objects they may strain these out with their branchial sieves as the manhaden does (p. 114), for Moore, a very accurate observer, described them as swimming open mouthed when feeding on minute crustaceans, crossing and recrossing on their tracks.[30]

Doubtless it is because of their feeding habits that herring seldom take a baited hook, if they ever do. But we think it likely that large ones when feeding on shrimp would take an artificial fly, as spent and hungry alewives will (p. 104) on their return to salt water, and as shad will on their way upstream (p. 109).

Enemies.—The herring is the best of all bait fishes in our Gulf, where it is preyed upon by all kinds of predaceous fish, especially by cod, pollock, haddock, silver hake, striped bass, mackerel, tuna, salmon, and dogfish, and by the mackerel sharks. Silver hake, in particular, often drive schools of herring up on our beaches, where pursued and pursuers alike strand on the shoaling bottom. We once saw this happen at Cohasset in Massachusetts Bay many years ago, on an October morning, when hake and herring were so intermingled in shallow water at the height of the carnage that we soon filled our dory with the two, with our bare hands. The finback whales also devour herring in great quantities. The short-finned squid (*Ilex*) likewise destroys multitudes of the young sardines. On one occasion near Provincetown, in June 1925, we watched packs of perhaps 10 to 50 squids circling around a school of 2- to 4-inch herring, bunching them into a compact mass. Individual squids then darted in, seized one or two herring, ate only a small part, then darted back for more. A silvery streak of fragments of dead herring remaining along the beach bore witness to the carnage.

Breeding habits, development and growth.—Much attention has been devoted to the breeding habits and growth of the herring by European zoologists, by Moore, and by Huntsman in our own Gulf, and by Lea [31] in more northern Canadian waters.

Herring may spawn in spring, in summer or autumn, according to locality, or both in spring and autumn (for further information on this matter, see p. 98). They do so chiefly on rocky, pebbly, or gravelly bottoms, on clay to some extent, probably never on soft mud. Spawning in the Gulf of Maine (including the Bay of Fundy), takes place chiefly from 2 or 3 fathoms down to about 30 fathoms; perhaps never in the littoral

[28] Bull. Newfoundland Government Lab., No. 17, 1948, p. 133.

[29] Ann. Rept. Biol. Board Canada (1933), 1934, p. 14–15.

[30] Rept. U. S. Comm. Fish. (1896), 1898, p. 402.

[31] Age and growth of the herrings in Canadian waters. Canad. Fish. Exped., 1914–15 (1919), pp. 75–164.

zone, nor has herring spawn ever been reported as cast up by the surf on the beaches of New England, a fate that often overtakes it in the Gulf of St. Lawrence. Occasionally they spawn as deep as 100 fathoms in Scandanavian waters, perhaps also in the eastern basin of our Gulf where the sea floor is hard, not soft and muddy as it is in the basins in the western side. During the act of reproduction as observed by Moore at Cross Island and at Machias, Maine, "the fish were darting rapidly about, and those who have opportunity to see the fish spawning in more shallow water where observation is more favorable, state that both males and females are in constant motion, rubbing against one another and upon the bottom, apparently by pressure aiding in the discharge of the eggs and milt." [32]

A female herring may deposit from 20,000 to upwards of 40,000 eggs, according to her age and size, averaging about 30,000. In sexually mature herrings, the genital organs are so large just before spawning commences that they make up about one-fifth the total weight of the fish.

The eggs sink to the bottom, where they stick in layers or clumps to the sand or clay, to seaweeds, or to stones, by means of their coating of mucus, or to any other objects on which they chance to settle. They are often found massed on net warps, anchors, and anchor ropes. The individual eggs are 1 to 1.4 mm. in diameter, depending on the size of the parent fish and also, perhaps, on the local race of fish involved. The period of incubation is governed by temperature; European students tell us that it requires as long as 40 days at 38–39°, 15 days at 44–46° and 11 days at 50–51° F.; while experiments on the Massachusetts coast by the U. S. Bureau of Fisheries gave 10 to 12 days in the temperature prevailing there in autumn. And MacFarland [33] found that all of the eggs kept at Grand Manan at about 59° (15° C.) hatched, but that none hatched at 32–41° (0–5° C.), and that all died that were warmed to 68° (20° C.). Ten to fifteen days might be stated as an average incubation period for the Gulf of Maine, under existing temperatures.

The larvae of the herring family are very slender and can easily be distinguished from all other young Gulf of Maine fish of similar form (e. g., launce, smelt, or rock eel) by the location of the

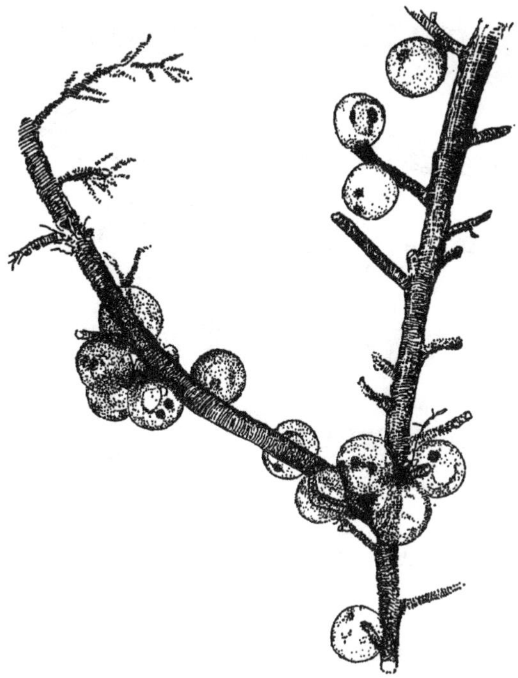

FIGURE 42.—Eggs of the herring (*Clupea harengus*), attached to seaweed (European). After Ehrenbaum.

vent, which is so far back that it lies close to the base of the tail. But it requires critical examination to distinguish our several clupeoids one from another in their early stages.

The sea herring is about 5 to 6 mm. long at hatching, with a small yolk sac that is absorbed by the time a length of about 10 mm. is reached. The dorsal fin is formed at 15 to 17 mm.; the anal at about 30 mm.; the ventrals are visible and the tail well forked at 30 to 35 mm.; and at about 40 mm. (1½ in.), the little fish begins to look like a herring.

According to Huntsman's observations, fry produced on the Grand Manan spawning grounds in late summer and early autumn grow to a length of 17 to 20 mm. by the end of November or first of December; they are 26 to 50 mm. (1–2 in.) long in March and April and 50 to 60 mm. (2–2⅜ in.) long by June when fry of this size are abundant in the St. Andrews region. This is in line with our own observations that fry of 2 to 2½ inches (50–65 mm.) predominate among the young herring at Provincetown at the end of June, and fry of 2⅛ to 4 inches (54 to 100 mm.) on Nantucket

[32] Moore, Rept. U. S. Comm. Fish. (1896), 1898, p. 412.

[33] Rept. Biol. Board Canada (1930), 1931, p. 23.

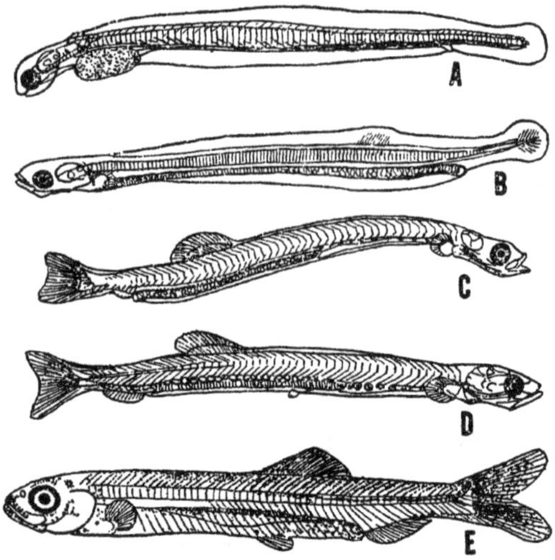

FIGURE 43.—Larval stages of the herring (*Clupea harengus*) European. After Ehrenbaum. A, newly hatched, 7 mm.; B, 10 mm.; C, 19 mm.; D. 29 mm.; E, young fry, 41 mm.

Shoals in mid-July. They grow to about 3½ to near 5 inches (90–125 mm.) by the end of their first year of life; fish of that size, presumably of the previous autumn's hatch, are abundant in the fall in the Bay of Fundy, and at Boothbay, Maine. The growth rate is about the same at Woods Hole, where herring spawned in October and early November are 3 to 5 inches (76–125 mm.) long by the following autumn. The Norwegian herring, also, average about 5 inches (125 mm.) long at the end of their first year, according to Hjort, and North Sea herring are about 4 inches (100 mm.) long then.[34]

Subsequent growth.—The herring has proved a particularly favorable object for growth studies based on the structure of the scales.[35] Without pursuing this subject, which would lead us far afield, we may point out that herring not only grow at different rates at different times of year, with the contrast between the rapid growth of summer and the slow growth of the winter greater or less in different seas, but that they grow rapidly when young and slowly thereafter in some localities, whereas they may grow slowly at first in other localities, but sustain a more even growth to old age.

The Dogger Bank herring, for example, in the North Sea approximate 4 inches in length at the end of the first year, 8½ to 9 inches at the end of the third year, 10½ at the end of the sixth, and 11½ to 12 inches at the end of the ninth, though with considerable variation. The Norwegian herring, however, spawned in the year 1899, averaged only 7¼ inches when 3 years old, but were as large as the Dogger Bank fish of equal ages by their sixth year and subsequently.[36] Newfoundland herring grow more slowly at first than those in the southern side of the Gulf of St. Lawrence, but catch up with them as they grow older.

Huntsman credits the Bay of Fundy herring with about 10 inches at the end of their third year; i. e., when 4 years old, which agrees closely with an average growth of 9⅞ inches at 4 years as calculated by Lea for Gulf of St. Lawrence fish. The average growth rate of the older Bay of Fundy fish probably falls between that of the Gulf of St. Lawrence fish and that of the herring of outer Nova Scotia which grow a little faster; i. e., to between 10¼ and 11⅝ inches at 5 years; between 11 and 12½ inches at 7 years; and between 12½ and 13⅛ inches at 9 years.[37] Bay of Fundy herring make most of their growth from May to September. In the southern parts of our Gulf, where the growth period probably continues a month later into the fall, they may grow as fast as they do along outer Nova Scotia.

When the little herring have reached an age of about 2 years and a length of 7½ to 8 inches (190–200 mm.) they accumulate large amounts of fat among the body tissues and viscera during the warm months of the year when growing rapidly, but lose this fat in winter and also at the approach of sexual maturity. We can bear witness and the fact is well known to fishermen that this "fat" stage is as characteristic of American waters as of European, where "fat" herring are the objects of extensive fisheries.

According to Moore, who examined thousands

[34] Huntsman (Canad. Fish. Exped. (1914–1915), 1919, pp. 168–169) believed he could recognize spring as well as autumn-spawned herring fry in the Bay of Fundy, and credits the former with a length of about 90 mm. by the first, and 150 mm. by the second, winter. But this seems to call for confirmation, it being unlikely that any herring now spawn there in spring (p. 98).

[35] See Lea (Canad. Fish. Exped. (1914–15), 1919, pp. 75–164) for an account of age determination by analysis of the scales, as applied to the herring.

[36] Rapp. and Proc. Verb., Cons. Internat. Explor. Mer, vol. 20, 1944.

[37] As scaled from Lea's diagrams (Canad. Fish. Exped. 1914–1915 (1929), figs. 40 and 41). It has been found that the Norwegian herring grow from April to September only, remaining practically stationary in length from October until March; see Lea (Pub. de Circ., Cons. Perm. Internat. Explor. Mer, No. 61, 1911, pp. 35–57) and Hjort (Rapp. Proc. Verb., Cons. Perm. Internat. Explor. Mer, vol. 20, 1914).

of fish about Eastport, herring rarely spawn when less than 9½ inches long; usually not until they are 10 to 10½ inches; and most of the spawners are 12 to 13 inches long. This means that some few spawn when only 3 years old, if the growth schedule outlined above is correct, but most of them not until 4 years or older, to continue spawning annually thereafter as long as they live. In Norwegian waters, too, a few spawn at 3 years, many at 4 years, and the majority at 5 years; some few not until 6 years old. Herring have been seen as old as 20 years, and they may live even longer.

Success of reproduction.—The relative abundance of any species of sea fish from year to year depends less on how many individuals spawn in any locality than it does on how many of the resultant fry survive. And the many age analyses that have been made of herring in European waters have proved that while a very large crop of young may be produced in some years, hardly any are in others, even in favorable nurseries. Apparently this applies more to the northern breeding grounds than to the southern (to some extent, however, to all) the result being that the herring spawned in some one favorable breeding season may dominate the schools over large areas for many years, or until another successful breeding year comes, producing another large crop. In Norwegian waters, for example, the herring produced in 1904 was dominant in the catches for the next six years, at least; this is a classic instance. Lea found, similarly, that herring hatched that same year (1904) dominated the catches on the west coast of Newfoundland as long afterwards as 1914 and 1915. And while precise information is not available for our Gulf, no doubt the same rule governs there.

One case, at least, is well documented of a particular body of Bay of Fundy herring that received no important recruitment for something like 10 years, when the few still remaining seem to have disappeared, from old age (p. 99).

Various explanations have been proposed to account for this, such as abundance or scarcity of microscopic plankton, favorable or unfavorable temperature, salinity, or other factors, all of which may enter in. And while it is during the first few weeks of life that the herring is most vulnerable, it is also possible that the conditions under which the parent fish lived for the year preceding spawning may influence the fate of the fry. Whatever the explanation, the fact that such fluctuations do occur from year to year, in the numbers of fry reared is of the greatest practical interest to all concerned with the sea fisheries, as evidence that variations existing in the stock of herring, and consequently in the catch, may be due more to the success or failure of reproduction than to any effect the fishery may have on the stock.

General range.—Both sides of the North Atlantic. Off the European coast the herring ranges north to Norway, Iceland, Spitzbergen, and the White Sea; south to the Straits of Gibraltar. It is known on the American coast as far north as northern Labrador and the west coast of Greenland; regularly and commonly as far south as Cape Cod and Block Island; and it is occasionally seen in small numbers as far south as Cape Hatteras in winter. It is replaced by a close ally (*C. pallasii*) in the North Pacific.

Occurrence in the Gulf of Maine.—To list the localities where herring have been recorded would be to mention every hamlet along our coasts whence fishing boats put out, for more or less herring, large or small, appear at one season or another around the entire coast line of the Gulf, and on the offshore fishing banks as well. They also enter bays and estuaries freely, but they have never been reported in our Gulf from water that is appreciably brackish; perhaps 2.8 percent salinity [38] may be set at about their lower limit.

The distribution of commercial catches, plotted by Needler (fig. 44) [39] shows that herring are far more plentiful from Casco Bay eastward along the coast of Maine, and especially in the Passamaquoddy Bay–Grand Manan region than they are along the western shores of the Gulf on the one hand, or up the Bay of Fundy on the other, or along western Nova Scotia. Thus the landings per unit length of coast averaged 3 times as great for the Passamaquoddy–Grand Manan region and for the coast of Maine to Mount Desert, as for the coast sector from Mount Desert past Penobscot Bay; about 4 times as great as for the Maine coast as a whole, westward and southward

[38] Surface, in Bay of Fundy in May.

[39] A reliable index, for the herring is a valuable fish.

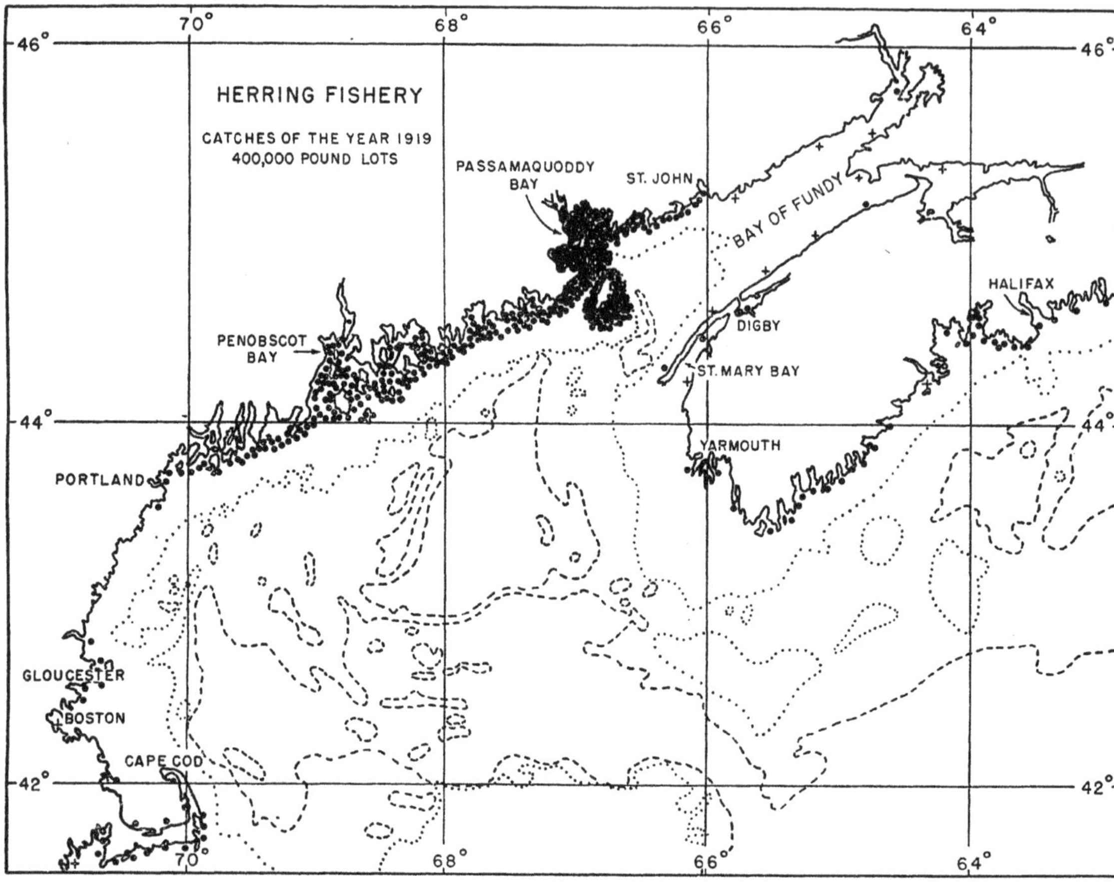

FIGURE 44.—Catches of herring for the year 1919. Each dot represents 400,000 pounds. After Needler.

from Penobscot Bay; and 13 times as great as for the coast of Massachusetts,[40] for the years 1919, 1928, 1929, and 1930.[41]

Present day landings of upwards of 30 million pounds of sardines alone, for Charlotte County, New Brunswick, even in poor years, up to something like 100 million pounds in good years, plus some 9–14 million pounds of larger herring, contrasted with a maximum of only about 17 to 18 million pounds reported for 1947 for any sector of the Maine coast of comparable length,[42] show that the Passamaquoddy–Grand Manan region has not lost its preeminence as a herring center. The abundance of little herring there is, in fact, the outstanding feature of the distribution of fishes in the Gulf of Maine. A catch of about 2,400,000 pounds for Massachusetts in 1947, contrasted with some 11,300,000 pounds for the Penobscot Bay region alone in that year, illustrates how much less rich in herring the southwestern side coast line of the Gulf is than the sector that happens to be the least productive part of the northern coast line of the Gulf.

Fishermen tell us, too, that herring are much more regular in their occurrence from year to year in the Passamaquoddy–Grand Manan region than they are either off western Nova Scotia in the one direction, or along the coast of Maine in the other. And this is borne out by such statistics as are conveniently available. Thus only one-fourth to one-fifth as many pounds of herring were caught in the Penobscot Bay region [43] in 1947 as either eastward

[40] Omitting the landings for Suffolk County, Mass., since these represent fish discharged at Boston by the vessel fisheries from offshore.

[41] Graham, Jour. Biol. Board Canada, vol. 2, No. 2, 1936, p. 129, table 2.

[42] Scattergood has given an interesting analysis, regional and seasonal, of the 1947 catch of herring for the coast of Maine.

[43] Scattergood's statistical areas 11–14.

to Lubec Narrows on the one hand, or westward past Casco Bay to Cape Elizabeth on the other,[44] whereas the catches for 1919 were rather evenly distributed along the northern and eastern Maine coast as a whole.

We find herring even more and more sporadic in their appearances and disappearances, both from place to place, from week to week, and from year to year, passing southward around the western periphery of the Gulf. Very few, for example, are seen on the southern side of Massachusetts Bay in some years (as in 1950 and 1951); many schools in others. And herring are such wandering fish in general, here today and gone tomorrow even in their centers of abundance, that the successful location of the weirs depends largely on intimate local knowledge and on close observation of the movements of the schools.

Herring appear, also, to be far less plentiful on the offshore banks and less regular in their occurrences there than they are in their inshore center of abundance in the northeastern part of the Gulf. Trawlers, it is true, occasionally pick up schools on Georges Bank and on Browns Bank, as in 1931, when catches of 3,000 pounds were reported on the northern edge of Georges and of 2,800 pounds on the southwestern part in October. Schools, too, are occasionally reported as seen at the surface, by *Albatross III* for instance, in April–May 1950. Fishermen used sometimes to set drift-nets on Georges for herring for bait in the days of the long line fishery, and small numbers up to 130–160 per haul, were trawled by *Albatross III*, widespread on the western part in depths of 20 to 50 fathoms in May 1950, as well as off southern New England.[45] But it is more usual for trawlers operating on Georges to pick up only odd fish or none. Thus the maximum catch on any trip during the otter trawl investigation of 1913 was only a dozen or two; 42 hauls by the *Eugene H*, in late June 1951, yielded only one herring, fishing from Nantucket Lightship out onto the south-central part of Georges; and the stomachs of cod caught on Georges seldom contain herring, if they ever do.[46]

The appearance of schools of large herring or of small is distinctly a seasonal event off most parts of our coast, and the picture is made still more complex by differences in the behavior of sardine-size, "fat," and spawning herring, the reasons for which are not yet well understood.

The newly spawned fry, less than ¾ of an inch (9–11 mm.) long, have been taken in September in the lower part of the Bay of Fundy, a product, doubtless, of the Grand Manan and West Nova Scotia spawning; also in October in Gloucester Harbor where one tow-net haul yielded us a great number on the 24th, in 1916. And they are to be expected wherever herring spawn in numbers in any particular year. It seems likely that most of them remain near their birth place during their first autumn and winter, when the circulation of the Gulf is in its least active stage. But they become widely distributed during the spring (March–May), when 1½ to 2 inches (30–50 mm.) long, both in the lower Bay of Fundy, around the entire periphery of the open Gulf, east as well as west, out over the basin, and on the northern and eastern parts of Georges Bank.[47]

Little seems to be known in detail about the movements of herring during their first year, but those that find their way into enclosed waters where mid-summer temperatures are high, such as Duxbury and Plymouth Bays and Provincetown Harbor, appear to move out during the early part of the summer, being reported as far less plentiful there in June than they are in April and May. Sardine-size herring, 4 to 8 inches long including 1- and 2-year-olds, are to be expected in abundance all summer east of Penobscot Bay, and particularly in the Passamaquoddy Bay region, where they support the sardine fishery for which the latter is famous, and where they are present throughout the year.

It is probable, however, though not proved, that the 1- to 2-year-olds (fish in their second and third years) do not appear along the southwestern coasts of the Gulf until several months later in the season than the little fish of ½ to 2 inches do, that were hatched the preceding autumn. Thus it usually is not until late June, July, or August that "sperling" of 4 to 7 inches are reported in numbers off the Massachusetts coast, or that we

[44] Coast sectors of comparable length.

[45] Average catches per haul about 56 fish at 22 to 40 fathoms, and 28 at 41 to 50 fathoms, but only 6 at 51 to 60 fathoms.

[46] W. F. Clapp found no herring in many cod and haddock stomachs examined by him on Georges Bank.

[47] During March and April 1920 we took them near Cashes Ledge, on the northern and eastern parts of Georges Bank, off Seal Island; off Yarmouth, Nova Scotia; near Machias, Maine, and over the basin in the offing; near Boothbay; and near the Isles of Shoals. Graham (Jour. Biol. Board Canada, vol. 2, No. 2, 1936, p. 112, fig. 8) found them equally widespread in the open Gulf in May 1932, also in the lower Bay of Fundy (none, however, at the head of the Bay).

have seen them there. They are even more erratic, too, in their appearances and disappearances in Massachusetts Bay and along Cape Cod than they are to the eastward of Mount Desert. At Cohasset, for example, on the southern shore of the Bay where we have had many years' experience, schools of sperling are here today in summer and early autumn, but gone tomorrow. It is also our impression that the sperling, like the larger herring, are not only far less concentrated in favorable localities around the southwestern shore of the Gulf than they are to the north and east, but far less numerous on the whole.[48]

These first two year classes (the fish in their first year having grown to a length of 3 or 4 inches by September; those in their second year to 7 to 9 inches) begin to thin out from the shore waters of the open Gulf after the middle of October as the water cools, and few "sardines" are taken there after early December.

The corresponding ebb and flow, so to speak, for the sardine is suggested in a striking way by the average monthly catches of sardines by the weirs in Charlotte Co., New Brunswick (Passamaquoddy Bay, Campobello, and Grand Manan) for the year 1920, which are equally illustrative of conditions today:

Month	Pounds	Month	Pounds
January	11,000	July	3,315,000
February	None	August	6,475,000
March	56,000	September	6,730,000
April	1,049,000	October	6,012,000
May	3,036,000	November	1,325,000
June	2,542,000	December	147,000

Here, however, the seasonal variation (as Dr. Huntsman informs us) is simply a matter of local availability, for sardines remain in Passamaquoddy Bay all winter, but do not move about much then. Probably the sardines winter mostly on the bottom. And there is no reason to suppose that the bulk of them travel far in any part of the Gulf.

Very little is known about the Gulf of Maine herring during their third summer, when they have passed the "sardine" or sperling stage and have not yet reached spawning age. In some years these "fat" herring, as they are often called, or "summer" herring, weighing up to about one pound (they are called "spawn" herring locally, but this is an error), are taken in the traps at Provincetown for a week or so about mid-April; they are taken at about the same time off Gloucester (in 1915 they were reported 8 to 15 miles off Cape Ann on the 17th), and they are said by the fishermen to "show" first off Seguin Island in May and June, off Mount Desert late in summer. Doubtless they form a large part (just what proportion is not known) of the catches of herring larger than sardines that are made in the Passamaquoddy Bay region, also around Grand Manan. As a rule few of them are taken inside the inner islands elsewhere, though they came into the harbor of Boothbay about May 14 in 1914.

When a mackerel seiner picks up a school of herring out in the open Gulf in summer,[49] or when an otter trawler makes a catch of herring on Georges Bank (p. 95), most of them are very fat and show no signs of approaching sexual maturity. Thus it seems that they tend to keep farther offshore than do either the younger herring or the still older mature herring.

The peak season for herring larger than "sardines" inshore in the northeastern part of the Gulf is ordinarily from July through October; i. e., some 2 months less than that for the sardines (see p. 96). But a greater proportion of the larger fish continue available there through the cold months than of the younger fish, to judge from the fact that considerably larger catches are made of big herring in winter than of sardines, whereas the total local catch is much larger for the latter than for the former.

A report [50] on the average monthly landings of large herring for Charlotte County, for the period 1920–1931, to the nearest 1,000 pounds, follows:

Month	Pounds	Month	Pounds
January	132,000	July	1,065,000
February	164,000	August	4,334,000
March	275,000	September	7,098,000
April	312,000	October	2,817,000
May	306,000	November	646,000
June	284,000	December	268,000

Large herrings, yearly average	17,701,000
"Sardines", yearly average	30,698,000

[48] No particular attention is paid to sperling around Massachusetts Bay, for they are too small to be in demand for bait, and they are not plentiful enough (or not concentrated enough) to support a sardine fishery there.

[49] Many events of this sort have been reported. For example, a large catch of fat summer herring was made on Georges Bank and reported to the Massachusetts Commissioners in the mid-summer of 1901.

[50] From Graham, Jour. Biol. Board Canada, vol. 2, No. 2, 1936, p. 130, table 3.

Large herring (Dr. Hunstman tells us) are also present there throughout late winter and spring, though few find their way then into the weirs.

In the southeastern part of the Gulf, as typified by Cape Cod Bay, large herring appear inshore in greatest numbers to about June and again in the autumn, with very few (and not many sardines) in June or July. This is illustrated by the largest and smallest catches made in 8 traps at North Truro for different months during the years 1946 to 1952. The following data are contributed by the Pond Village Cold Storage Company:

Month	Minimum (in pounds)	Maximum (in pounds)
April	0	117, 375
May	221	523, 550
June	0	88, 657
July	0	0
August	0	1, 000
September	0	57, 287
October	0	9, 526
November	0	176, 435

The earliest catch of sardines there in those years, or in 1935, 1938, or 1943 was sometime in May, the latest November 16 to 17; the earliest catch of large herring was made between April 20 and 30, the latest on December 10th.

In most years the large herring vanish from the Massachusetts coast at some time in December. In 1950, for example, they vanished about December 4th from Ipswich Bay, where considerable catches had been made for some time previous by about 15 boats.[51]

Nothing is known, definitely, as to their seasonal appearances and disappearances over the offshore banks.

About all that is known of the movements of the large mature herring (in their fourth summer and older) is that they are encountered in numbers only for the brief period before, during, and after the spawning season, when they are seen schooling at the surface, and are caught along shore. Fishermen report that they show about the off-lying islands some time before they make their way up the bays; two or three weeks earlier, for instance, at Grand Manan, Jonesport, and about Mount Desert Island than within Machias Bay. They are said to appear some time after the middle of July at Isle au Haut at the eastern entrance of Penobscot Bay, and at Castine within the Bay, though not until the end of that month or the first of August at Matinicus Island. Such of them as visit the Massachusetts Bay region are not expected there until the last week in September. But they are in full force on all the spawning areas along the shores of the Gulf by October, from Grand Manan to Cape Cod; they are equally widespread, if less abundant, inshore in November, and they are reported in December occasionally, and even later. It is probable that as the fish spawn out most of them move out promptly from the spawning grounds into deeper water, for fish recently spent are not often reported as taken in the weirs.

Probably the spawners merely descend into deeper water to winter, as is the case in European waters. How deep the great body of them go is not known. But is has been proved that herring of all ages remain in the open Bay of Fundy throughout the cold season; also in the passages between the inner and outer divisions of Passamaquoddy Bay, even when water temperatures there are as low as 32° F.[52] And the abundance of pelagic euphausiid shrimps (a favorite herring food) in the deeper water layers of the northeast corner of the Gulf suggests this as a rich winter pasture for them.

Studies carried out from the Atlantic Biological Station at St. Andrews, chiefly under Dr. A. G. Huntsman's[53] leadership, and by the International Passamaquoddy Fisheries Commission[54] during the early 1930's seem to us to have proved that the factor chiefly responsible for the great concentration of young herring in the Passamaquoddy region, and for their availability to the weir fishery there, is the differential circulation of the shoaler and deeper water layers that is set in motion by the inflow of fresh water from the tributary streams combined with superficial currents set up temporarily by the wind. In other words, the sardine-sized herring acts as does any planktonic animal such as the euphausiid shrimps and the copepod crustaceans, on which it feeds, as it swims to and fro, i. e., it drifts with the current. In technical language, it is "denatant."

The case is not so clear for the larger herring, not because there is any reason to suppose they can direct their journeys more intelligently, and because any directive swimming they may carry

[51] This happening was reported in the daily papers.

[52] Huntsman. James Johnstone Memorial Vol., 1934, p. 82.

[53] For summary, see Huntsman, James Johnstone Memorial Vol., 1934, pp. 95–96.

[54] See Graham, Jour. Biol. Board Canada, vol. 2, 1936, No. 2, pp. 95–140.

out is far more effective because more rapid; but because so little is known as to journeys any individual school actually makes as season follows season, whether of fat herring or of spawners. Perhaps the most interesting question of all, and one as yet unsolved, is how and why the spawning fish seek their spawning grounds year after year, when their sex organs mature.

Spawning grounds and season.—It appears that the most productive spawning ground for our Gulf formerly was and still is at the mouth of the Bay of Fundy, particularly on the shoals southwest of Grand Manan. The Trinity Ledges off western Nova Scotia are another important ground; and herring are reported as spawning commonly, though irregularly, in Machias Bay; about Jonesport; at Mount Desert; in Frenchmans Bay; among the islands at the mouth of Penobscot Bay (Swans, Isle au Haut, and Matinicus); in Casco Bay; also about Wood Island a few miles south of Cape Elizabeth, which has long been known as the resort of breeding schools. Herring have also been found spawning off the beaches along the western shore of the Gulf, Ipswich Bay, for example; about Cape Ann; in Massachusetts Bay; about Provincetown; along outer Cape Cod; in the Woods Hole region; near No Mans Land; and about Block Island which is the southern breeding limit. But whatever spawning does take place either southward from the vicinity of Cape Elizabeth on the one hand, or in the inner parts of the Bay of Fundy on the other, is trifling as compared with the production along the eastern coast of Maine and in the Grand Manan region.

Spawning takes place both along shore in our Gulf and on the various shoals and ledges that lie for 5 to 25 miles off the coast, a habit betrayed by the eggs that are found sticking to the anchor ropes of fishing vessels. But we find no definite record of herring spawning on Browns or Georges Banks, nor are young fry known there, a fact that was commented upon by Storer long ago.

Spawning season.—Both spring-spawning schools and summer-fall spawning schools of herring were reported formerly, in the Bay of Fundy, the spring spawners visiting the south (Nova Scotia) side of the bay from Bier Island at the mouth in as far as Digby Gut, also the Parrsboro region on the New Brunswick shore near the head of the bay, spawning during April and May. But they seem never to have been very numerous, and it is not known whether any spawn now in the bay before summer. Spring-spawning as well as autumn-spawning herring have also been reported to us by fishermen along the west coast of Nova Scotia, though we have not been able to verify this. Other than this, spring spawners are neither recorded nor rumored anywhere in the Gulf of Maine.

Around Grand Manan and in Machias Bay nearby, the heaviest runs of summer-autumn spawners usually come in July, August, and September, the spawning season continuing until late in the fall in some years,[55] but not commencing until early August, and ending by early October in others.[56] Passing westward we find the breeding period progressively later and shorter; mid-August for example until October around Petit Manan and near Mount Desert, while the few herring that spawn farther south do so chiefly during October in Ipswich[57] and Massachusetts Bays; in late October and early November in the vicinity of Woods Hole.

So many observations have been taken in the Gulf from the vessels of the Bureau of Fisheries, and in the Bay of Fundy by the Biological Board of Canada, that it is possible to establish the temperatures rather closely at which herring spawn in our waters. Around Grand Manan and in the northern part of the Gulf generally, practically all spawning is carried out in water of about 46–52° F. But such herring as spawn in the southern part of Massachusetts Bay and along the shores of Cape Cod, where autumnal cooling of the surface waters is not so rapid as it is farther north, may do so in slightly warmer water, say up to 53° or 55°. The Gulf of Maine herring spawn in rather low salinities (such characterize the coastal zone as a whole as compared with the North and Norwegian Seas), the most saline water in which it is known to spawn within our limits being not saltier than 33 per mille, the freshest probably about 31.9 per mille. They never spawn in brackish water within the limits of the Gulf, although known to do so at the mouths of certain European rivers in water that is nearly fresh.

Destruction by natural causes.—The herring is a very "tender" fish, prone to wholesale destruction

[55] So described by Moore, Rept. U. S. Comm. Fish. (1896), 1898, p. 408.
[56] Dr. Huntsman informs us that this was the case in 1917.
[57] Allen, Mem. Boston Soc. Nat. Hist., vol. 8, No. 2, 1916, p. 201.

both by stranding on beaches during storms, and by pollution of the water. Many instances of this kind have been reported. Allen,[58] for example, saw young herring in windrows for miles on the strand at Rye Beach, N. H., in August 1911. A slaughter of herring (still more instructive because the exact course of events was followed) took place at Cohasset, on the south shore of Massachusetts Bay, in October 1920. On the 5th of that month a large school of "sperling," 4 to 5 inches in length, ran up the harbor (which is nearly landlocked), probably driven in by silver hake (at least so local fishermen said); were trapped there by the falling tide, and stranded on the mud. So numerous were they that the flats were entirely covered with them and it was estimated that 20,000 barrels of fish perished. During the next few days the fish (alternately covered and uncovered by the tide) decayed, and despite the tidal circulation, so fouled the water that lobsters impounded in floating cars died. On the 10th there was a second smaller run of herring, and on the 15th a third run came as numerous as the first, the newcomers dying soon after they entered the harbor. Altogether, it was estimated that 50,000 barrels of fish perished, of which more than 90 percent were "sperling," 5 to 10 percent were large adults, and a few were small mackerel and silver hake, besides large numbers of smelt. The flats were silvery with herring scales at low tide by the last half of October, when we saw them, and the residents about the harbor found the stench almost unbearable. But the fish decomposed and the water purified itself during the winter months.

Mass destructions of young herring have also been reported in other Gulf of Maine harbors. Thus, Dr. Austin H. Clark reported that early in August 1925 the mud flats in Manchester Harbor, on the north side of Massachusetts Bay, were white with stranded herring 3 to 5 inches long, packed several deep at low tide along the sides of the little drains and hollows. Another such destruction took place in the same harbor in the summer of 1928. Vast quantities of herring spawn are likewise cast up on the beaches every year to perish in north European waters; this also happens to some extent in the Gulf of St. Lawrence.

Numerical abundance and importance.—Moore (1898), who sifted many sources of information concluded (we believe rightly) that no general decrease had taken place in the abundance of young herring at the mouth of the Bay of Fundy up to that time. But it is common knowledge among fishermen that both the numbers visiting any given locality on our coast and the duration of their stay varies widely, not only from year to year, but over longer periods. Local spawning grounds, too, may be abandoned for a term of years—a common occurrence.[59]

The best documented case of local disappearances from a previously productive ground took place, as Dr. Huntsman writes us,[60] from the shoals southwest of Grand Manan, whence large herring (previously very plentiful) withdrew in 1877, to reappear in 1881 on the Nova Scotia coast between Cape Sable and Digby. Dr. Huntsman has suggested that they had circled the Gulf offshore, for their exodus from the Grand Manan shoals was not accompanied by any coincident increase in the catch along the eastern part of the coast of Maine, but rather by the reverse.[61] They persisted on the Nova Scotia shore until 1890, when they gave out, probably from old age, for the large herring that remained in the Quoddy region also dwindled in numbers as shown by the collapse of the winter fishery there, evidence that this particular body of herring did not receive any significant recruitment after about 1880–1881. It remains to be seen whether large herring will ever reappear in their former plenty on the Grand Manan ground, as they did about 1857 in Massachusetts Bay, where the stock had been at a low ebb since 1837; or whether the yearly drain on the population of young herring by the sardine fishery (well started by about 1880) is too great.

The largest reported catch of herring for the Gulf as a whole for any year since 1928 for which statistics are readily available was 219,131,500 pounds taken in 1946, divided as follows: Massachusetts, 2,049,000 pounds; Maine, 80,107,400 pounds; and the Canadian shores of the Gulf, 136,975,100 pounds. The smallest catch was 70,519,886 pounds in 1932, divided 5,687,254 pounds, 31,988,132 pounds, and 32,844,500 pounds,

[58] Mem. Boston Soc. Nat. Hist., vol. 8, No. 2, 1916, p. 202.

[59] Moore, Rept. U. S. Comm. Fish. (1896), 1898, p. 430.

[60] Based on Canadian fishery statistics.

[61] Earll (Fisheries and Fish. Ind. U. S., sect. 5, vol. 1, 1887, pp. 423, 424) states that the fishery declined near Bois Bubert Island from 1875 to 1880, and that the catch was "considerably below average" at Matinicus during the 10 years previous to 1879.

respectively. In 1947, incidentally, the herring catch of the Gulf was topped only by the catches of haddock and of rosefish.

It is not clear to what extent this range in the catch from year to year is due to fluctuations in the supply of fish; to differences in their availability; or to the sundry economic factors that enter in. What is certain is that with some 80 percent of the catch consisting of sardine-sized fish weighing only about one-half ounce, the toll taken cannot have been less than 1½ billion fish in the poorest of recent years, 5 to 6 billion in the year when the yield was greatest, i. e., numbers far greater than that for any other Gulf of Maine fish. Come good year then or bad, Capt. John Smith's account of the herring of our Gulf thus applies equally well today: "The savages compare the store in the sea with the hair of their heads, and surely there are an incredible abundance upon this coast." [62]

The sardine catch of the Bay of Fundy is made almost wholly in weirs, 347 of which were operated on the New Brunswick shore in 1947. On the Maine coast, as a whole, a little less than two-thirds the catch of herring, large and small, is made in weirs and in purse seines, combined, a little more than one-third nowadays in stop seines (about 44,500,000 pounds in 1947). These are used "to prevent the exit of the herring school from a cove or inlet. . . . The seine is stretched around the school with the ends of the net made fast to the shore." And stop seines are used mostly at night, when the presence of fish is betrayed by their luminous trails, if the water is firing, or by the noise they make as they "flip" at the surface.[63] And some are still caught in floating traps (about 2 million pounds in 1947) which we have often seen used in the harbors of Mount Desert.

In 1947, seemingly a representative year, purse seines yielded the Maine fishermen nearly as much herring (about 36,100,000 pounds) as the weirs. Their presence is detected, Scattergood tells us, either by the firing of the water if by night, by echo sounding apparatus, or by the use of a thin wire suspended in the water, the vibrations of which indicate the presence of fish that strike it. In 1947 eleven purse seiners were active in the fall fishery for Maine herring. How many were engaged in the New Brunswick and Maine winter fishery is not known.

Large catches of herring when on bottom also are made by special otter trawls in European waters; and of the closely allied herring of British Columbia of late. But the possibility of developing an otter-trawl fishery for herring in the Gulf of Maine has not yet been explored.

Finally, we may remark that herring fresh from the water are among the most delicious of our fishes, especially the small sizes. Their only drawback is that they do not keep well, being rich-meated and oily, and in the larger sized fish the many hair-like bones are troublesome.

Hickory shad *Pomolobus mediocris* (Mitchill) 1815

Fall herring; Shad herring

Jordan and Evermann, 1896–1900, p. 425.

[62] General Historie of Virginia, New England, and the Summer Isles, 1616, reprinted in 1819 from London edition of 1629, p. 188.

[63] See Scattergood, U. S. Fish and Wildlife Service, Sp. Sci. Rept. No. 67, 1949, p. 8, for further details.

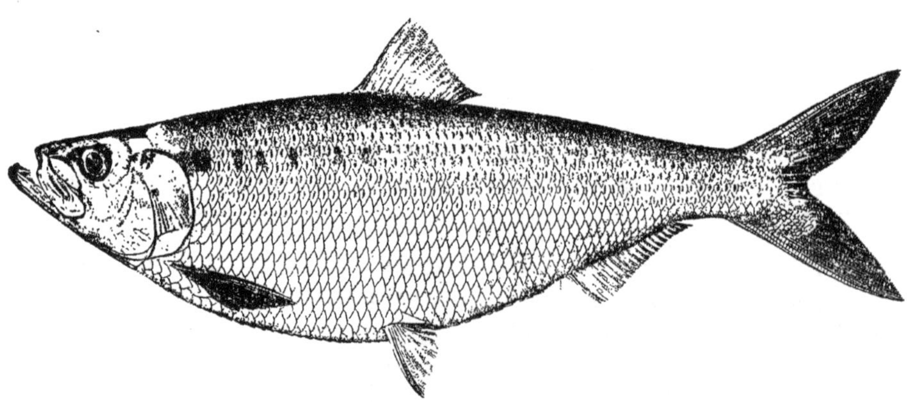

Figure 45.—Hickory shad (*Pomolobus mediocris*), Chesapeake Bay region specimen. From Goode. Drawing by H. L. Todd.

Description.—The hickory shad differs rather noticeably from the sea herring in that the point of origin of its dorsal fin is considerably in front of the mid-length of its trunk; in its deep belly (a hickory shad 13½ in. long is about 4 in. deep but a herring of that length is only 3 in. deep); in the fact that its outline tapers toward both snout and tail in side view (fig. 15); and in that its lower jaw projects farther beyond the upper when its mouth is closed; also, by the saw-toothed edge of its belly. Also, it lacks the cluster of teeth on the roof of the mouth that is characteristic of the herring. One is more likely to confuse a hickory shad with a shad or with the alewives, which it resembles in the position of its dorsal fin, in the great depth of its body, in its saw-toothed belly and in the lack of teeth on the roof of the mouth. But it is marked off from all of these by its projecting lower jaw. There is also a small difference in outline, its head tapering more to the snout, as seen in side view (fig. 45). It has only about half as many gill rakers (19 to 21 on the lower limb of the first gill arch) as either the alewife or the blueback; and its upper jaw, reaching back only about as far as opposite the center of its eye, is shorter than that of the shad in which it reaches as far as the rear edge of the eye.

Under favorable circumstances its color, also, is characteristic, for it is faintly marked on the sides with dusky longitudinal stripes, and the tip of its snout is dusky.

Size.—This is the largest of our anadromous herrings next to the shad, growing to a length of 2 feet. A fish about 15 inches long weighs a pound, one of 18 inches, 2 pounds.

Habits.—Nothing is known of the habits of the hickory shad in the sea to differentiate it from its close relatives of the herring tribe except that it is more of a fish eater. Launce, anchovies, cunners, herring, scup, silversides, and other small fish, squid, fish eggs, and even small crabs have been found in the stomachs of hickory shad at Woods Hole, as well as sundry pelagic Crustacea. It will strike a small spinner or other artifical lure, and it gives a good fight when hooked. In the southern parts of its range it is described as running up fresh streams, with the alewives in late winter and early spring to spawn.[64] But it appears not to do so in the streams tributary to Chesapeake Bay, though it is found in practically all of them. This opens the interesting possibility that the "green" fish found in Chesapeake Bay, leave the Bay, perhaps to spawn in salt water.[65]

General range.—Atlantic coast of North America from the Bay of Fundy to Florida.

Occurrence in the Gulf of Maine.—The hickory shad is a southern fish, with the Gulf of Maine as the extreme northern limit to its range. It is recorded in scientific literature only at North Truro; at Provincetown; at Brewster; in Boston Harbor; off Portland; in Casco Bay; and from the mouth of the Bay of Fundy (Huntsman doubts this record), and it usually is so uncommon within our limits that we have seen none in the Gulf ourselves. But in 1932 anglers, trolling for striped bass and mackerel off the Merrimac River, met a run of hickory shad.[66]

It is much more plentiful west of Cape Cod, being common from spring throughout summer and early autumn at Woods Hole, where as many as 3,500 have been taken at a single lift of one trap. In 1919 the Massachusetts catch of hickory shad, practically all from the south coast, amounted to 12,800 pounds, and none are listed for Massachusetts for any subsequent year.

Alewife *Pomolobus pseudoharengus* (Wilson) 1811 [approximate date]

GASPEREAU; SAWBELLY; KYAK; BRANCH HERRING; FRESH-WATER HERRING; GRAYBACK

Jordan and Evermann, 1896–1900, p. 426.

Description.—The alewife is distinguishable at a glance from the sea herring by the greater depth of its body, which is three and one-third times as long as deep (an alewife of 13½ inches is about 4 inches deep; a herring that long has a depth of only 3 inches) also by the position of its dorsal fin, the point of origin of which is considerably nearer to the tip of the snout than to the point of origin of the central rays of the tail fin. Furthermore, the alewife is much more heavily built forward than the herring, and the serrations on the midline of its belly are much stronger and sharper (hence the local name "sawbelly"), so much so that a practiced hand can separate

[64] Smith (N. C. Geol. Econ. Surv; vol. 2, 1897, p. 121) describes it as doing so in the streams tributary to Pamlico Sound, N. C., where it is plentiful.

[65] Hildebrand and Schroeder, Bull., U. S. Bur. Fish., vol. 43, 1928, p. 84.

[66] The Museum of Comparative Zoology received one from this run from Dr. J. C. Phillips, caught by him off the northern end of Plum Island, October 2, 1932.

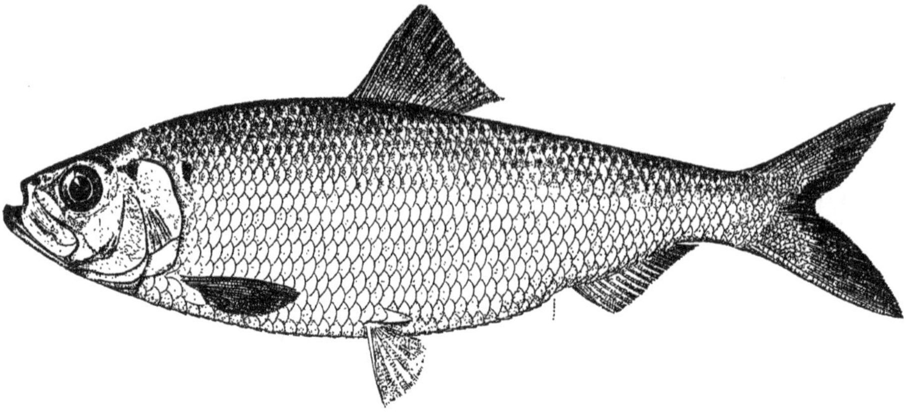

FIGURE 46.—Alewife (*Pomolobus pseudoharengus*), Chesapeake Bay region specimen. From Goode. Drawing by H. L. Todd.

herring from alewives in the dark. The most useful distinctions between the alewife and the blueback are that in the former the eye is broader than the distance from its forward edge to the tip of its snout and the back grayish green, while in the latter the eye is only about as wide as the distance from front of eye to tip of snout, and the back is dark blue (p. 107). Also the lining of the abdominal cavity is pale grayish or pinkish white in the alewife, but is usually dusky or blackish in the blueback. But this distinction may not hold in all cases.

Alewives are distinguishable from young shad by their smaller mouths with shorter upper jaws; also by the fact that the lower jaw of the alewife projects slightly beyond the upper when the mouth is closed, and by the outline of the edge of the lower jaw, the forward part of which is deeply concave in the alewife but nearly straight in the shad. The lack of teeth on the roof of the mouth distinguishes the alewife, with its brethren the hickory shad (p. 100) and blueback (p. 106) from the sea herring, anatomically.

Color.—The alewife, like the herring, is grayish green above, darkest on the back, paler and silvery on sides and belly. Usually there is a dusky spot on either side just behind the margin of the gill cover (lacking in the herring) and the upper side may be faintly striped with dark longitudinal lines in large fish. The sides are iridescent in life, with shades of green and violet. The colors change, to some extent, in shade from darker to paler, or vice versa, to match the bottom below, as the fish run up stream in shallow water.

Size.—The alewife grows to a length of about 15 inches, but adults average only about 10 to 11 inches long and about 8 to 9 ounces in weight; 16,400,000 fish taken in New England in 1898 weighed about 8,800,000 pounds.

Habits.—The alewife, like the shad and the salmon makes its growth in the sea, but enters fresh water streams to spawn. This "anadromous" habit, as it is called, forced itself on the attention of the early settlers on our coasts. In the words of an eyewitness, "experience hath taught them at New Plymouth that in April there is a fish much like a herring that comes up into the small brooks to spawn, and when the water is not knee deep they will presse up through your hands, yea, thow you beat at them with cudgels, and in such abundance as is incredible."[67] And they are no less persevering in their struggles upstream today. Numbers of them are to be seen in many streams, any spring, alternately swimming ahead; resting in the eddy behind some irregularity of the bottom; then moving ahead again, between one's feet if one happens to be standing in midstream. And they are much more successful than the shad in surmounting fishways of suitable design. During the early runs sometimes one sex predominates, sometimes the other, but the late runs consist chiefly of males, as a rule, and these are said to outnumber the females greatly on the spawning grounds. We have no firsthand observations to contribute on this score.

Alewives are decidedly general in their choice

[67] Capt. Charles Whitborne, in "The True Travels of Capt. John Smith," 1616, vol. 2, p. 250.

of streams, running indifferently up rivers as large as the St. John, Merrimac and Potomac, or streams so small that one can almost leap across, and only a few inches deep. In large rivers they run far upstream—how far they may do so we do not know—or their journey may be one of only a few yards, as it is in the artificial cuts that are kept open through barrier beaches to allow the fish access to fresh water ponds behind the latter.

The alewife spawns in ponds, including those back of barrier beaches (if there are openings to the sea, natural or artificial) and in sluggish stretches of streams, never in swift water, each female depositing from 60,000 to 100,000 eggs or more, according to her size.[68] Spawning lasts only a few days for each group of fish.

The spent fish run down stream again so soon after spawning that many of them pass others coming up, as we have often seen; fish on their return journey to salt water are familiar sights in every alewife stream.

The adults, when entering streams to spawn, make the change from salt water to fresh within a short time without damage; this is equally true of the spent fish on their return to the estuaries. But Dr. Huntsman informs us that they appear unable to endure repeated changes between salt water and fresh, and that great numbers are killed in this way in the estuaries under certain conditions of tide. The strain of spawning leaves them very thin, but they recover rapidly after they reach salt water. We have seen spent alewives that had already put on considerable fat, taken from a trap at Provincetown as early in the season as July 16 (in 1915).

Spawning ordinarily takes place at temperatures of about 55 to 60°. The eggs are about 0.05 inches in diameter, pink like those of the sea herring, and they stick to brush, stones, or anything else they may settle upon.[69] Incubation occupies about 6 days at 60°. The young alewives, which are about 5 mm. long when hatched, growing to 15 mm. when a month old, soon begin to work their way downstream. They have been seen descending as early as June 15 in the more southerly of Gulf of Maine streams; successive companies of fry move out of the pond and down with the current throughout the summer; and by autumn the young alewives have all found their way down to salt water when 2 to 4 inches long. We have seined young alewives as long as 4 to 4½ inches (102–115 mm.) in salt water near Seguin Island, Maine, at the end of July, but others, only 3 to 3⅝ inches long (78–92 mm.), near Mt. Desert Island as late as the first of October. Thenceforth the alewife lives in salt water until sexual maturity.

Hildebrand and Schroeder[70] found that little alewives in Chesapeake Bay had grown to about 4½ to 5 inches long by the time they were 1 year old.

The rate of growth of the older alewives, in salt water, has not been traced. But experiments in planting adult alewives in ponds in which there were none before, led, long ago, to the conclusion that they became sexually mature at 3 or 4 years of age, for none of their progeny returned until 3 or 4 years after the original plant. Specific instances, cited by Belding [71] are:

(1) Three years after a large number of alewives were hatched in Keene's Pond, Maine, tributary to the Calais River, from a "plant" of mature fish, a run of adult fish entered Keene's Pond stream where none had ever been seen before; this case was reported by the U. S. Bureau of Fisheries. (2) The establishment of a fishery, in the same way at Plymouth, Mass., in 4 years after restocking in 1865; and (3) G. M. Besse obtained results in 3 years in ponds in Wareham, Mass.

The fact that alewives have been known to return, for spawning, to streams in which their parents had been planted, lends support to the "parent stream" theory; i. e., that alewives, like shad, tend to spawn in the stream system in which they were hatched. But a much more intensive study is needed of this interesting question before any categorical statement can be made, as to how generally this is true; and to what extent their return depends on their never having wandered far afield.

Food.—The alewife is chiefly a plankton feeder like the herring; copepods, amphipods, shrimps, and appendicularians were the chief diet of specimens examined by Vinal Edwards and by Linton

[68] The average number of eggs in 644 females taken in the Potomac was 102,800 (Smith, N. C. Geol. and Econ. Survey, vol. 2, 1907. p. 123).

[69] The development of the eggs, larval stages, and young fry are described by Ryder (Report, U. S. Comm. of Fish. (1885), 1887, p. 505) and by Prince Contr. Canad. Biol. (1902–1905), 1907, p. 95).

[70] Bull. U. S. Bur. Fish., vol. 43, 1928, p. 91.

[71] Rept. Alewife Fish. Mass., Mass. Dept. Conservation, Div. Fish. and Game, 1921, p. 18.

at Woods Hole. However, they also take small fish, such as herring, eels, launce, cunners, and their own species, as well as fish eggs. Unlike herring, alewives often contain diatoms even when adult. Alewives fast when they are running upstream to spawn, but when the spent fish reach brackish water on their return they feed ravenously on the shrimp that abound in the tidal estuaries and which they can be seen pursuing. We have often hooked alewives on an artificial fly at such times.

Movements at sea.—The alewife is as gregarious as the herring, fish of a size congregating in schools of thousands of individuals (we find record of 40,000 fish caught in one seine haul in Boston Harbor) and apparently a given school holds together during most of its sojourn in salt water. But they are sometimes caught mixed with menhaden, or with herring. Alewives, immature and adult, are often picked up in abundance in weirs here and there along the coast,[72] and it is likely that the majority remain in the general vicnity of the fresh water influence of the stream-mouths and estuaries from which they have emerged, to judge from the success of attempts to strengthen or restore the runs of alewives in various streams, mentioned above. But it is certain that some of them wander far afield, for catches of up to 3,000 to 4,000 pounds per haul were made by otter trawlers some 80 miles offshore, off Emerald Bank, Nova Scotia (lat. about 43° 15′ N., long. about 63° W.) at 60 to 80 fathoms, in March 1936.[73]

Odd alewives were reported from Georges Bank and the South Channel in March, June, August, and November of 1913. Some (up to 78 per haul) were trawled by *Albatross III* about 25 to 60 miles out off southern New England in May 1950; also 18 adults, 10 to 11 inches long, 70 odd miles off Barnegat, N. J., on March 5, 1931; and we saw 60 alewives trawled at the 25-fathom line off Marthas Vineyard[74] in late June, 1951 by the *Eugene H.* Where these wanderers come to shore to spawn, if they succeed in doing so at all, is an interesting question.

It seems likely from various lines of evidence that alewives tend to keep near the surface for their first year or so in salt water, and while they are inshore when older. But practically nothing is known as to the depths to which they may descend if (or when) they move offshore, there being no assurance that those taken by trawlers were not picked up, while the trawls were being lowered or hauled up again.

General range.—Gulf of St. Lawrence and northern Nova Scotia south to North Carolina, running up into fresh water to spawn; landlocked races also exist in Lake Ontario, in the Finger Lakes of New York, and in certain other fresh-water lakes.[75]

Occurrence in the Gulf of Maine.—When the white man crossed the Atlantic probably there was no stream from Cape Sable to Cape Cod but saw its annual run of alewives unless they were barred by impassable falls near the mouth.

And while its numbers have declined during the past two centuries and its range has been restricted, both by actual extirpation from certain streams by overfishing, by the pollution of the river waters by manufacturing wastes, and by the erection of dams that it cannot pass, the alewife is a familiar fish still, all along around our coast[76] and yields an abundant catch in many of our streams. Alewives are taken commonly about Yarmouth, Nova Scotia; in the Annapolis Basin; in Minas Channel; and farther still, up the Bay. Alewives still run in most of the streams tributary to the Bay of Fundy, many in the St. John. A few are taken in the weirs in Passamaquoddy Bay; while young ones have been taken around Campobello Island; as deep as 50 fathoms. They enter the large river systems all along the coasts of Maine and New Hampshire, likewise many small streams, the requirements being that these shall lead to ponds or have deadwaters of sufficient extent along their courses, and no dams or falls that the alewives can not surmount. At Boothbay Harbor, for instance, a considerable number of alewives annually run, or did run, up to spawn in Campbell's Pond, a small body of water that is dammed off from the harbor, and reached by a fishway only 15 feet long. This is the shortest alewife stream of which we know.

In 1896, when the alewife fishery was the subject of inquiry by the Bureau of Fisheries,[77] catches

[72] Huntsman (Contr. Canad. Biol., [1921] 1922, p. 58) reports its young at Campobello Island, Bay of Fundy, in December and March.

[73] Reported by Vladykov, Copeia, 1936, No. 3, p. 168. One vessel brought in about 10,000 pounds.

[74] At lat. 40° 58′ N.; long. 70° 32′ W.

[75] Such a race has been reported in Cobbett Pond, Rockingham Co., N. H. by Kendall (Occ. Pap. Boston Soc. Nat. Hist., vol. 7, No. 8, 1908, p. 38) and by Bailey (Biological Survey Merrimac Watershed, New Hampshire Fish and Game Dept., 1938, p. 162).

[76] Belding (Rept. Alewife Fish. Massachusetts, Mass. Dept. Conserv., 1921) has given a very instructive report on the alewife in Massachusetts.

[77] Smith, Rept. U. S. Comm. Fish. (1898) 1899, pp. 31-43.

large enough to be worth special notice were reported from the mouths of the St. Croix, Dennys, Machias, Medomak, Penobscot, St. George, Pemaquid, Damariscotta, and Kennebec Rivers; from Casco Bay; and from sundry other shore localities in Maine; from the Piscataqua River system in New Hampshire; from the mouth of the Merrimac, and from Cape Cod Bay. Few alewives enter the Merrimac, now, so polluted is it, and so obstructed by dams.[78] And Belding found them running in only about 9 or 10 streams on the Gulf of Maine coast of Massachusetts in 1920, out of 27 streams there that had formerly supported considerable alewife fisheries.[79]

At present, we learn from John B. Burns, of the Massachusetts Division of Marine Fisheries, only a few alewives manage to run up the Merrimac past the fish ladder at Lowell; there is a small but regular run in the Parker River; a few in the Ipswich; a good run in the Essex; a few in the Saugus; perhaps some in Weymouth Back River;[80] a small run in Wier River, Hingham (really a brook); a few in Bound Brook, Cohasset; a large run in Herring Brook, Pembroke (tributary to North River) yielding about 1,000 barrels yearly; increasing numbers in Jones River, Kingston, which had been restocked previous to 1938 when a fish ladder was installed; several thousand run yearly up Barnstable Mill Pond Brook; an improving run in Stony Brook, Brewster, where a ladder was built in 1945, and a good run in Herring River (really only a brook) in Wellfleet, Cape Cod.

The first alewives ordinarily appear early in April in the few streams tributary to Massachusetts Bay that they still frequent, and equally early (March or April) in the St. John River, New Brunswick, according to McKenzie;[81] but their date of arrival varies considerably from stream to stream, according to local conditions. Thus few are seen in the streams of Maine until late April or early May; the first alewives appeared in 1915, for example, in Campbell's Creek, Booth Bay Harbor, on April 20. And the earliest good runs on the Nova Scotia shores of the open Gulf and of the Bay of Fundy may come as early as April (streams of Yarmouth, Annapolis, Hants, and Colchester Counties), in May (Digby and King's County streams), or not until June (Cumberland County).[82] Successive runs follow thereafter, all around the Gulf, until well into June, the later runs, going up, passing the earlier spawners coming down. In 1915, we saw this happening in Campbell's Creek, Boothbay, on May 20. And alewives have been seen, descending, as late as August 20, in Massachusetts streams.

The extreme range of temperature within which eggs are spawned, in Gulf of Maine tributaries, is not known; probably the bulk of production takes place between about 55° and about 60°.

Numerical abundance.—In 1896[83] reported catches were 2,677,972 individual alewives (1,356,755 lb.) for Cape Cod Bay and for the Merrimac River combined; 526,500 (293,671 lb.) for New Hampshire streams; and 5,832,900 (3,388,326 lb.) from the rivers and streams and coast of Maine. The reported catch was 5,843,000 pounds[84] for the New Brunswick shore of the Bay of Fundy that year; 1,609,400 pounds for the Nova Scotia side and for the west coast of Nova Scotia, or about 10,510,000 and about 2,895,000 individual fish, respectively, assuming that the average weight was about the same as that for the alewives of Maine. We thus arrive at a total catch for the Gulf of Maine of something like 22 million individual fish at that time and actually somewhat more, for the canvass certainly was not 100 percent complete.

The run was much greater then in the St. John River system than in any other Gulf of Maine river and doubtless is still. The Damariscotta River, ranking second, was about one-third as productive as the St. John; the Merrimac, St. George, and Penobscot Rivers only something like one-tenth as productive each. Casco Bay yielded about one-sixth as many alewives as the St. John River, the shore line of Cape Cod Bay about one-fifth as many.[85] And the catch of the St. John River system (including Kennebecasis Bay) still was about five times as great in 1931 as that for any of the other counties of New Bruns-

[78] Fishways recently constructed now allow a few to ascend beyond Lowell' Massachusetts.

[79] See his report on the Alewife Fishery of Mass. (Mass. Dept. of Conservation, Div. Fish. and Game, 1921) which gives much information as to the status of the alewife in Massachusetts streams.

[80] Stocked with 28,000 adult fish in 1949, and fish ladders under construction.

[81] Rept. Biol. Board Canada (1931) 1932, p. 34.

[82] According to McKenzie, Rept. Biol. Board Canada (1931) 1932, p. 34.

[83] A special study of the alewife fishery was made for that year, see Smith, Rept. U. S. Comm. of Fish. (1896) 1899, pp. 33–43.

[84] The Canadian catches for the year were reported in barrels; the conversion factor used is 200 pounds per barrel.

[85] Reported catches for 1896 were about 4,234,000 pounds for the St. John River system; 1,390,612 pounds for the Damariscotta River, 385,804 pounds for the St. George River, 308,844 pounds for the Penobscot, 472,500 pounds for the Merrimac, 701,287 pounds for Casco Bay, and 884,255 pounds for Cape Cod Bay.

wick or of Nova Scotia that border on the Bay of Fundy or on the open Gulf.[86]

The alewife population of the Gulf is much smaller, today, than it was half a century ago. Thus the catch was only about one-half as great for the Bay of Fundy in 1945 and 1946 [87] as it had been in 1896, and about one-third as great for Maine (1,224,600 lb.) while the Merrimac River, yielding 472,500 pounds in 1896, yielded less than 3,000 pounds in 1945.[88] And though alewives may seem almost incredibly numerous when crowding into some stream, they made but a sparse population, even in their days of greatest plenty, when spread over the coastal waters of our Gulf, as compared to the sea herring.

Importance.—Alewives are excellent food fish and they are marketed both fresh and salted, and are preferred by many to the sea herring. They are good bait for cod, haddock, and pollock; and their scales commanded a high price for use in the manufacture of artificial pearls for a brief period during the first world war and for a few years afterward.[89] By far the greater part of the catch of alewives is made in the lower reaches of the streams that they enter to spawn, in weirs, in dip nets or in haul seines according to locality. Most of those taken in outside waters (as in Casco and Cape Cod Bays) are either gill netted or are picked up in the fish traps.

[86] McKenzie, Rept. Biol. Board Canada (1931) 1932, p. 34.
[87] 5,051,100 pounds and 4,517,500 pounds, respectively.
[88] The reported catch for Essex County, Massachusetts, in that year was 2,700 pounds, only a part of which was from the region of the Merrimac.
[89] For details, see Report, Division of Fish and Game, Mass. (1920) 1921, p. 140.

Blueback *Pomolobus aestivalis* (Mitchill) 1815

GLUT HERRING; SUMMER HERRING; BLACKBELLY; KYACK

Jordan and Evermann, 1896–1900, p. 426.

Description.—Bluebacks and alewives are difficult to distinguish; experienced fishermen who recognize the existence of the two separate fish cannot always tell them apart, so closely do they resemble one another in general appearance. The most obvious external difference between them is that the back of the blueback is definitely blue green, that of the alewife gray green. But this applies only to fresh-caught fish; preserved specimens do not differ much in color, or fish that have been on ice for more than a short time. Another external difference is that the eye of the blueback is only about as broad as the distance from front of eye to tip of snout (or slightly broader), but is appreciably broader than that in the alewife; the blueback, too, with body about 3½ times as long as deep, is a slightly more slender fish (on the average) than the alewife, and its fins are a little lower, but the two species probably intergrade in both these respects.

The most dependable distinction between the two (though requiring the use of a knife) is that the lining of the belly cavity is sooty or blackish in the blueback, but pearl gray or pinkish gray in the alewife. We have yet to see a specimen that could not be named as the one or the other on this basis alone, unless so poorly preserved that the original shade of the cavity could no longer be determined.

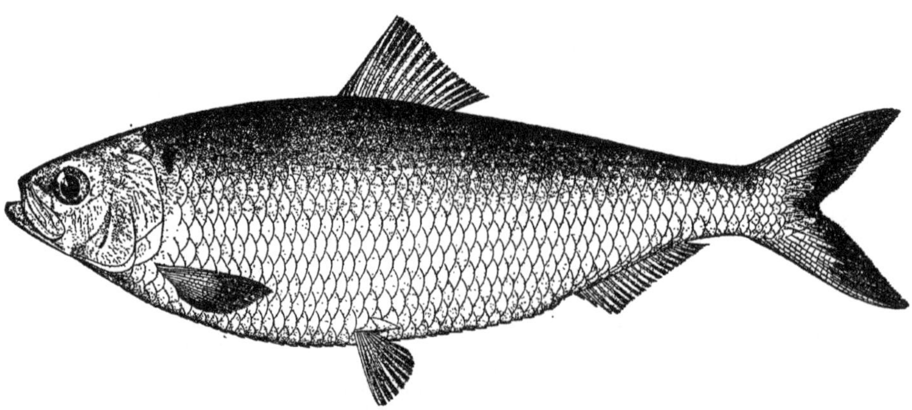

FIGURE 47.—Blueback (*Pomolobus aestivalis*), Chesapeake Bay region specimen. From Goode. Drawing by H. L. Todd.

Color.—Dark blue or bluish gray above, the sides and belly silvery, with coppery reflections at least in some waters; lining of the belly sooty or blackish.

Size.—The blueback attains about the same size as the alewife, i. e., 15 inches; the adults average about 11 inches in length and about 7 ounces in weight.

Habits.—The blueback, like the alewife, makes its growth in salt water, but runs up into fresh water to spawn. And its breeding habits do not differ in any important particular from those of the alewife, except that it "runs" later in the season, does not run up as far above tidewater, and does not spawn until the water is much warmer, 70° to 75° instead of 55° to 60°.[90] The eggs, about 1 mm. in diameter, sink like those of the alewife, and stick to anything they may chance to touch. Incubation occupies only about 50 hours at a temperature of 72°. The young are 30 to 50 mm. long within a month and already show most of the diagnostic characters of the adult. Evidently they soon find their way down to the sea, for bluebacks of 50 mm. have been seined in abundance in Rhode Island waters late in July.[91] Nothing whatever is known of their later rate of growth. The spent fish, return to sea shortly after spawning as do alewives. Practically nothing is known of their movements in the sea, except that they are schooling fishes. The fact, however, that 7 were trawled by the *Albatross II* on March 5, 1931 about 100 miles off Cape May, N. J., suggests that the blueback moves out from land and passes the cold season near the bottom.

We need only note further that the blueback is as gregarious as the herring or alewife; that it is equally a plankton feeder, subsisting chiefly on copepods and pelagic shrimp, as well as on young launce and, no doubt, on other small fish fry.

General range.—This is a more southern fish than the alewife, occurring along the American coast as far south as northern Florida; as far north as southern New England in abundance, perhaps less regularly in the Gulf of Maine though widespread there, and known definitely as far north as Cape Breton, Nova Scotia:[92] it spends most of its life in salt water but runs up into fresh water to spawn.

Occurrence in the Gulf of Maine.—Although fishermen have recognized the existence of two distinct species of alewives at least since 1816, it is difficult to arrive at a just idea of the status and migrations of this fish in our Gulf, because fish reported as "bluebacks" at sea sometimes turn out to be alewives, while the late runs of alewives are often referred to as "bluebacks." It seems, however, that schools of bluebacks are to be expected anywhere between Cape Sable and Cape Cod. Thus we have seen "gaspereau" fresh caught at Yarmouth, Nova Scotia, that appeared to be bluebacks.[93] Huntsman had specimens from St. John Harbor and Shubenacadie River; they are reported, at least by name, from the St. Croix River; from Dennys River, Eastport; Bucksport; Casco Bay; Small Point; Freeport; and sundry other localities along the coast of Maine, as well as from the shores of Massachusetts Bay, including Cape Cod.

L. W. Scattergood of the U. S. Fish and Wildlife Service has sent us about 40 typical bluebacks, about 3¾ to 5½ inches (92–124 mm.) long taken at Hodgdon Island, Sheepscot River, Maine, June 14, 1951; and we once saw thousands of fish taken from a trap near Gloucester, most of which we judged to be bluebacks from their color.[94] A few fish were reported as "bluebacks" from Georges Bank during the investigation of 1913, and while there is no way, now, of checking whether these actually were bluebacks or alewives, the fact that we saw 10 bluebacks about 1 foot long, trawled by *Albatross III* at the 45 fathom line off southern New England, in mid-May, 1950,[95] shows that they may spread as far offshore as alewives.

No definite information is at hand as to how regularly alewives run into our Gulf of Maine streams, for spawning; or what streams they enter at all.

No distinction is made, commercially, on our coast between the blueback and the more abundant alewife; it is equally useful for bait and for food.

[90] The early development and larval stages of the blueback are described by Kuntz and Radcliffe (Bull. U. S. Bur. Fish., vol. 35, 1918, pp. 87–134).

[91] In Chesapeake Bay, Hildebrand and Schroeder (Bull. U. S. Bur. Fish., vol. 43, 1928, p. 88) found that while most of the young bluebacks pass out to sea during the summer and fall, some remain in the deeper holes over the winter. By the following March when about a year old these are about 3½ to 4 inches long; those in the sea may grow faster than this.

[92] Dr. A. H. Leim has sent us four typical bluebacks about 12 inches long, taken at Cape Breton, Nova Scotia, in 1950.

[93] We had no chance to examine them critically.

[94] We did not then appreciate the desirability of positive identification.

[95] Lat. 40° 06′; long. 71° 38′ W.

Shad *Alosa sapidissima* (Wilson) 1811 [Approximate date]

Jordan and Evermann, 1896–1900, p. 427.

Description.—The shad is a typical member of the herring tribe in all respects with soft-rayed dorsal and anal fins of moderate size, the former situated above the ventrals and well forward of the middle of the body. It has a deeply forked tail and large scales that are loosened easily. Unlike the sea herring, the shad has no teeth on the roof of the mouth; adults have no teeth at all, although young shad have small ones in the jaws which may persist until the fish is a foot or so long. The shad agrees with the hickory shad, alewife, and blueback, in its deep body and sharp saw-edged belly. But it differs rather noticeably from the hickory shad in its longer mouth, with upper jaw reaching to below the rear edge of the eye, and in the fact that the tip of its lower jaw is entirely enclosed within the tip of the upper when its mouth is closed. The most clear cut character distinguishing shad from alewife and blueback is that the upper outline of the shad's lower jaw is very slightly concave, without a sharp angle, the outline of theirs deeply concave with a pronounced angle. Furthermore the lining of the shad's belly is very pale.

Color.—Dark bluish or greenish above, white and silvery low on sides and on belly, with a dusky spot close behind the rear edge of the gill cover, and usually with one or two longitudinal rows of indistinct dusky spots behind it.

Size.—The shad is the largest of the herrings that regularly visit our Gulf, growing to a length of 2½ feet. In the Bay of Fundy, according to Leim [96] shad weigh about ⅛ pound at 8 inches; about ⅜ pound at 12 inches; about 1⅛ pounds at 15 to 16 inches; about 2½ pounds at about 20 inches; and about 4½ pounds at 23 to 24 inches, though with variations according to their condition. Adult males weigh from 1½ to 6 pounds; females from 3½ to 8 pounds. Shad are occasionally reported to 12 pounds, and the older writers mention shad of 14 pounds, but none so large has been credibly reported in the Gulf of late years.

Habits.—The shad, like the alewife, spends most of its life at sea, and makes most of its growth there, but runs up into fresh rivers to spawn, the spent fish soon returning to salt water, and its fry soon running down also. During their stay in the sea shad are schooling fish, often in thousands, and they never reenter fresh water until they return to spawn, though they sometimes do appear in brackish estuaries. Schools of shad are often seen at the surface in spring, summer, and autumn. In winter they disappear from sight. Probably the shad of the year winter near the mouths of their parent streams; the larger sizes somewhat farther out and deeper. The most direct evidence as to the depths to which they may descend is that shad have been trawled at about 50 fathoms off Nova Scotia in March (see footnote 22, p. 112), and at 26 to 68 fathoms off southern New England in May (footnote 23, p. 112).

Food.—The shad, like other herrings, is primarily a plankton feeder. We have found shad taken in the Gulf of Maine in summer full of copepods (chiefly *Calanus*), and the stomach con-

[96] Contrib. Canad. Biol., N. Ser., vol. 2, 1924, p. 245, fig. 41.

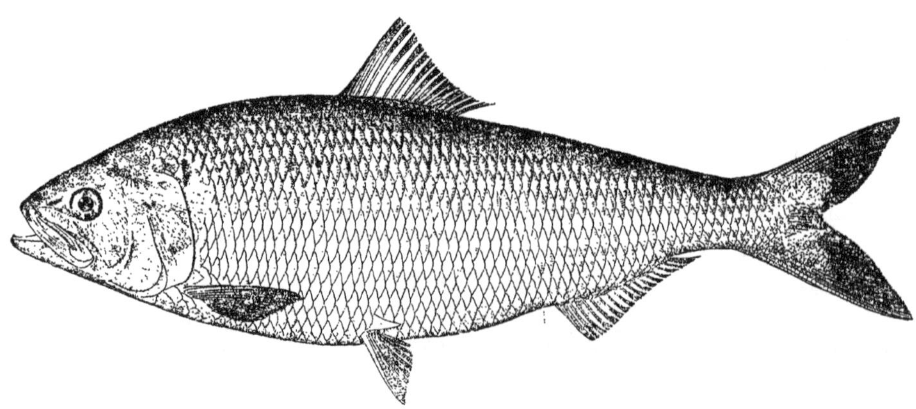

FIGURE 48.—Shad (*Alosa sapidissima*), Chesapeake Bay specimen. From Goode. Drawing by H. L. Todd.

tents of fish from the Nova Scotia Coast of the Bay of Fundy examined by Willey [97] consisted chiefly of the copepod genera *Arcatia* and *Temora* with other smaller ones, of mysid shrimps and of the larval stages of barnacles; while Leim [98] found that the shad in the open Bay of Fundy feed chiefly on copepods and mysids. Shad are also known to feed as greedily on the pelagic euphausiid shrimps as herring do, on fish eggs, and even on bottom dwelling amphipods, showing that they forage near the ground at times.

Occasionally they eat small fish, but these are only a minor item in their general diet.[99] Shad, it appears, take little or no food just prior to spawning. But they will often take an artificial fly, or a live minnow when running upstream to spawn.[1] During the past few years, crowds of anglers have caught many on flies in the Connecticut River, and doubtless could in the few Gulf of Maine streams to which shad still repair (p.110).

Reproduction and growth.[2]—The sexually mature fish enter the streams in spring or early summer when the river water has warmed to 50° to 55°. Consequently the shad run correspondingly later in the year passing from south to north along the coast, commencing in Georgia in January; in March in the waters tributary to Pamlico and Albemarle Sounds; in April in the Potomac; and in May and June in northern streams generally from the Delaware to Canada. In the Kennebec, according to Atkins,[3] the first shad appear (or did) late in April, with the main run in May and June; the first ripe females are caught the last week in May and they begin to spawn about June 1, most of them doing so during that month, a few in July, and possibly an occasional fish as late as August. Probably these dates applied equally to the Merrimac in the good old days when shad were plentiful there, but the season is somewhat later in the St. John, also in the Shubenacadie as might be expected; i.e., from mid-May until the end of June.[4]

[97] Contrib. Canad. Biol., N. Ser., vol. 1, 1923, p. 316.

[98] Contrib. Canad. Biol., N. Ser., vol. 2, 1924, p. 231.

[99] Leidy (Proc. Acad. Nat. Sci. Philadelphia, Ser., 2, 1868, p. 228) found 30 small sand eels (*Ammodytes*) in the stomach of a shad, probably caught in Delaware Bay.

[1] Bean (Bull. 60, Zool., vol. 9, New York State Mus., 1903, p. 207) commented on this long ago.

[2] Accounts of the breeding habits of the shad have been given by Ryder, Rept. U. S. Comm. Fish. (1885) 1887, p. 523; by Prince (Supp. 6, Rept. Dept. Marine Fish. Canada, Fish. Branch. 1907, pp. 95–110; in the Manual of Fish Culture, published by the U. S. Bur. of Fish., 1887; and more recently by Leim (Contrib. Canadian Biol. N. Ser. vol. 2, 1924, pp. 184–202).

[3] Fish. Ind. U. S., Sect. 5, vol. 1, 1887, pp. 683–684.

[4] Leim, Contrib. Canad. Biol., N. Ser., vol. 2, No. 11, 1924, p. 182.

In large rivers they run far upstream. In the St. John River, New Brunswick, they ascend about 200 miles to the grand falls even today according to Leim, and they still run up 300 miles (or did recently) in the Altamaha in Georgia; for 375 miles in the St. Johns River, Florida. But they could run up only about 35 miles at present in the Penobscot, where they formerly ascended some 90 miles, or 44 miles (to Augusta) in the Kennebec, which they formerly ascended 108 miles (to Carratunk Falls), though none enter either of these rivers now, so far as we know. And the dams at Lawrence, only 20-odd miles upstream, now stop any stray shad that may still enter the Merrimac, which they formerly ascended for 125 miles to Lake Winnepesaukee.[5]

In the Shubenacadie, shad spawn mostly in temperatures higher than about 54°, and spawning is interrupted if the water chills below that, temporarily.

The fish select sandy or pebbly shallows for spawning grounds, and deposit their eggs mostly between sundown and midnight. Females produce about 30,000 eggs on the average, though as many as 156,000 have been estimated in very large fish. The spent fish, now very emaciated, begin their return journey to the sea immediately after spawning. In the Kennebec they were first seen on their way down about June 20 and constantly thereafter throughout July; in the St. John spent fish are running down in July and August. According to Atkins they begin feeding before reaching salt water and recover a good deal of fat before moving out to sea.

The eggs are transparent, pale pink or amber, and being semi-buoyant and not sticky like those of other river herrings they roll about on the bottom with the current. The eggs hatch in 12 to 15 days at 52° (12° C.), in 6 to 8 days at 63° (17° C.), which covers the range characteristic of Maine and Bay of Fundy rivers during the season of incubation. And Leim has made the interesting discovery that larval development is more successful in brackish than in pure fresh water, with about 7.5 parts of salt per thousand as about the most favorable salinity.

The larvae are about 9 to 10 mm. long at the time of hatching, growing to about 20 mm., at 21

[5] Stevenson (Rept. U. S. Comm. Fish., (1898) 1899, p. 111) has given a table of the distances to which shad ascended various rivers then, and formerly from the Penobscot in Maine to the St. Johns in Florida.

to 28 days. Shad larvae resemble alewife larvae, being extremely slender with the vent almost as far back as the base of the tail.[6] The young shad remain in the rivers until fall, when they move down to salt water; they are now 1½ to 4½ inches long, resembling their parents in appearance.

According to Leim's investigation, based on scale studies and length frequencies, shad in the upper Bay of Fundy, average about 5 to 6 inches long when one year old; 9 to 10 inches long at 2 years; 13 to 14 inches at 3 years; 15 to 16 inches at 4 years; and 18 to 19 inches at 5 years. The two largest he examined, about 24½ inches (62 cm. and 63 cm.) long, appeared to be 7 and 6 years old, respectively. They may grow somewhat faster in the open Gulf of Maine, to judge from the greater abundance of pelagic crustaceans on which they feed (p. 109). Most of the spawning fish are 5 years old in the Shubenacadie, and presumably in other Gulf of Maine rivers; the oldest 8 or 9 years old.

General range.—Atlantic coast of North America from the southeastern coast of Newfoundland,[7] which shad have been known to reach as strays, and the estuary of the St. Lawrence River, where there is a considerable population of them,[8] to the St. Johns River in Florida; also represented in the Gulf of Mexico by a closely related species. The shad has been successfully introduced on the Pacific coast of the United States. It runs up rivers into fresh water to spawn.

Occurrence in the Gulf of Maine.—When the first settlers arrived in New England they found seemingly inexhaustible multitudes of shad annually running up all the larger rivers and many of the smaller streams, with the tributaries of the Gulf of Maine hardly less productive than the Hudson or Delaware. But one stream after another was rendered impassable by the construction of dams near the mouth, for shad cannot or will not run up through fishways that are readily used by alewives. Indeed, they have been practically wiped out in the Merrimac River, as appears from the following compilation:[9]

Year	Number of shad caught, reported, or estimated	Year	Number of shad caught, reported, or estimated
1789	830,000	1888	None
1805	540,000	1889	18
1835	365,000	1890–1892	None
1865	50,000	1893	2,020
1871–1873 (average)	1,942	1894	2,750
1880	2,139	1895	94
1885	130	1896	7

The Gulf of Maine rivers to which shad are known to resort regularly today are the Annapolis, Petit Codiac, Shubenecadie, and St. John, tributary to the Bay of Fundy; perhaps the St. Croix;[10] the only Maine rivers that see regular runs of a few shad are the Nonesuch and the Sheepscot.[11]

A few shad may enter other Gulf of Maine streams in some years if not yearly, and bright spots in the shad picture are that a considerable number of adult shad ran up the South River in Marshfield, Massachusetts, on the southern shore of Massachusetts Bay in 1950, and that there has been a run of something like 2,000 shad yearly in Mill Creek, Sandwich, Mass., for the past four years.[12] How successfully they may have spawned in either of these streams is not known.

It appears that most of the shad hatched in the rivers tributary to the Bay of Fundy, and the spent fish from there, remain in or near the estuaries where they take to salt water; and that most of the adults that survive the strain of spawning return to the parent stream to spawn again. Thus it is only in St. Marys Bay, in Annapolis Basin, in Cobequid Bay and Minas Basin, in Chignecto Bay and at the mouth of the St. John as well as for a few miles westward, that large Fundian shad are caught in any numbers.[13] The fact, on which Leim [13a] comments that "there is not a single record of a shad ever having been taken" at Grand Manan island, although this "lies almost directly in the path of any body of

[6] Leim (Contr. Canad. Biol., N. Ser., vol. 2, No. 11, 1924, p. 195) gives a detailed comparison of shad with alewife larvae.

[7] The most northerly record of a shad, on which we have chanced, is one taken in Bull's Bay, near St. Johns, Newfoundland.

[8] See Vladykov (Contr. Dept. Fish., Quebec, No. 30, 1950, pp. 121-135, and Natural. Canad., vol. 77, 1950, pp. 121-135) for a study of the movements of the shad in the St. Lawrence estuary.

[9] From Stevenson, Rept. U. S. Comm. Fish. (1898) 1899, p. 262.

[10] The St. Croix once had a large run of shad. None were seen there for 8 or 9 years prior to 1915, but they were there in some numbers in 1915 and 1916, according to investigations by H. F. Taylor of the U. S. Bureau of Fisheries; their present status there is not known. They have been entirely extirpated from the Saco, where they were abundant formerly, probably from the Penobscot and Kennebec, and certainly from the Merrimac, as noted above.

[11] Information from Dr. C. E. Atkinson, U. S. Fish and Wildlife Service.

[12] Reported to us by John B. Burns of the Massachusetts Division of Fisheries and Game.

[13] Leim (Contr. Canad. Biol. N. Ser., vol. 2, No. 11, 1924, fig. 2) gives a chart showing the location of shad catches for the Bay of Fundy.

[13a] Contr. Canad. Biol. N. Ser., vol. 2, No. 11, 1924, p. 173.

fish going in or out of the Bay of Fundy," is especially significant as emphasizing the localization of the St. John shad near the parent river.

The behavior of the St. John river shad raises an interesting question, as to the source of the young fish that sometimes congregate in the Bays and among the islands along the coast of Maine (Casco Bay especially), for there seem to be too many of them, in some years, to be credited to the small runs that still frequent the rivers of Maine (unless runs may have been overlooked of late in other rivers there).

Immature shad, up to 2 to 2½ pounds in weight are observed more or less commonly in Cape Cod Bay near Provincetown in summer or autumn and in the inner parts of Massachusetts Bay (sometimes taken in the traps at Beverly or Manchester), and off Cape Ann.[14] Spent shad up to 10 pounds in weight (averaging about 5 pounds), are sometimes reported by fishermen off the coast of Maine west of Penobscot Bay; near the Isles of Shoals; off York Beach, and off Cape Ann, in summer, autumn, and even in December.[15]

The few mature shad with ripening sexual organs that are picked up by the haddock netters between Cape Ann and Portland in April and May, most often about the Isles of Shoals and Boon Island,[16] probably are headed for the rivers of Maine.

Larger numbers of fish are seined in September and October, in the neighborhood of Mount Desert Island, where they have been the object of a frozen fish industry in some years.[17] These, like the green fish mentioned above, seem far too numerous to be accounted for by the small production that still takes place in the rivers of Maine. Some few of them, it seems, are Bay of Fundy fish, for one of a batch tagged near Mount Desert Rock in August 1947, was recaptured in Kings County, New Brunswick (St. John River system) the following June, and a second in the Petitcodiac River that July, while a third, tagged farther west on the coast of Maine in August or September 1948 was recaptured in the St. John River in May 1950. But it seems established that most of the medium-sized shad and larger now found in our Gulf are immigrants from the south, growing and fattening on the rich supply of plankton they find there, but returning to the rivers west and south of Cape Cod to spawn.

Direct evidence of this is that one tagged in Chesapeake Bay was recaught at Race Point, at the tip of Cape Cod, 39 days later; [18] one also was recaptured near Gloucester and another near Portland that had been tagged in the Hudson River, while 3 out of 1,380 tagged in New York Bay were recaptured in the Bay of Fundy after 37 days, 75 days, and 85 days, respectively, and one tagged off Fire Island, N. Y., was recaught at St. John, New Brunswick, after 39 days.[19] On the other hand, 18 shad, from a batch of 236 that were tagged near Mount Desert Rock in August 1947 were recaptured the next spring scattered along in different stream systems from the Connecticut to the Altamaha in Georgia. Others, from this same batch, were recaptured in the Connecticut, in the Hudson, on the coast of New Jersey, and in the Pamlico River, N. C., during the next two springs. And three others, from a batch of 431 tagged farther west along the coast of Maine in the summer and autumn of 1948, were recaptured in the Hudson River; three in Chesapeake Bay, and one in the Pamlico River, N. C.[20]

The shad that take part in this intermigration must winter somewhere between their northern feeding grounds whence they have vanished wholly by mid-autumn, and their southern breeding streams near which they do not appear until spring. But it is not yet known where they pass the cold months, how deep down they go, how far offshore, or how active they are then.

Still other shad are known to make very long journeys that can hardly be fitted into any regular migratory pattern, and from which they may never find their way back. Thus one that was tagged in the lower St. Lawrence River was recaught on Brown's Bank 258 days later; a second, from that same batch, was recaught in Cumberland basin, near Amherst, Nova Scotia, at the head of the Bay of Fundy after 322 days; a third at Province-

[14] 502 barrels (about 100,400 lb.) were taken in one set of mackerel pounds at Provincetown in June 1910; the traps picked up numbers of shad of about 14 inches from June 20 to July 6, 1921, at Magnolia and Beverly, where the catch was 10,300 pounds in 1945; and 14 shad 11 to 15¾ inches long were taken in one set of traps at Barnstable, on Cape Cod Bay, October 3, 1950.

[15] 135,000 pounds of these large spent fish were caught near Gloucester in the autumn of 1915; 125 barrels of 2- to 5-pound shad, some spent, near Seguin Island, July 19, 1925.

[16] A series of shad from that region, examined by the late W. W. Welsh in April and May 1913, averaged 5 pounds, all with well-developed sex organs.

[17] About 250,000 pounds were brought in to the local freezers yearly in 1913, 1914, and 1915.

[18] Vladykov, Trans. Amer. Fish. Soc., vol. 67, 1938, p. 64.

[19] Information supplied by C. E. Atkinson, U. S. Fish and Wildlife Service.

[20] Information supplied by E. H. Hollis of the U. S. Fish and Wildlife Service.

town at the tip of Cape Cod, some 1,200 miles away from where it had been tagged 444 days previous.[21] And one, from a batch of weir-caught fish tagged on the coast of Maine, August–September, 1948, was recaught in the Medway River, outer coast of Nova Scotia, a second, in the Miramichi River, tributary to the southern side of the Gulf of St. Lawrence in 1950, and a third, off Tor Bay, eastern Nova Scotia in 1951.

To what extent the seasonal journeys of the shad are passive with the dominant circulatory movements of the water, and to what extent (if any) they are self-directed is perhaps the most interesting question that now faces us in our studies of the shad of the Gulf of Maine.

Shad have been trawled 50 to 60 miles out, off eastern Nova Scotia;[22] have often been reported 40 to 50 miles out off the coast of Maine; also 25 to 90 miles out, off southern New England,[23] and we saw one trawled by the *Eugene H* in late June, 1951, on the southern part of Georges Bank (lat. 40°52′N., long. 67°40′W.), about 110 miles from the nearest land. Evidently they may wander as far offshore as alewives do; perhaps even as far as herring.

Shad reared in different regions may, perhaps, prove to differ enough in racial characters for recognition when taken at sea, but this is a question for the future.[24]

Abundance.—The stock of shad in the Gulf is but a shadow in comparison with that of colonial days.

In 1896, the only year for which detailed information is available as to the numbers taken in different streams, 290,122 shad were reported as caught in the Kennebec system, 9,000 in the Pleasant River, about 3,000 in the Harrington River, only 114 in the Penobscot and 12 in the St. Croix; 100 in the Piscataqua and 7 in the Merrimac.[25] In that same year the catch was about 1,059,000 pounds for the Nova Scotia shore of the open Gulf and for the Bay of Fundy;[26] 1,404,477 pounds for the rivers and coast of Maine; about 122,932 pounds (32,782 fish) for the Gulf of Maine coast of Massachusetts, or a total of about 2,586,400 pounds for the Gulf as a whole. With shad averaging about 3¾ pounds in weight,[27] this corresponds to about 690,000 fish.

But the yearly catch was only about one-third as great for the period 1916–1919 as it had been in 1896, whether for the United States shores of the Gulf or for the Canadian.[28] And it was of about that same order of magnitude in 1931, i. e., 677,540 pounds for the Gulf as a whole (157,763 pounds for Maine, 147,277 pounds for Massachusetts, 237,200 pounds for the Bay of Fundy and West Nova Scotia region). Since that time, the catches have ranged between 10,400 pounds and 306,000 pounds for the Massachusetts coast of the Gulf and between 9,300 pounds and 1,106,800 pounds for Maine, a fluctuation so extreme (no regional correlation appearing) as to suggest that market conditions were the chief governing factor. On the other hand the catches for the Canadian shores of the Gulf increased rather consistently from 1931 to a total of 1,287,600 pounds in 1939 then declined to around 780,000 pounds for 1944 and 1946, a rise and fall regular enough to suggest a corresponding fluctuation in the actual abundance of the shad. The average yearly catch for the period 1944–1946 combined, was about 20,000 pounds for Massachusetts, about 224,050 pounds for Maine, and about 780,000 pounds for the Bay of Fundy and western Nova Scotia.

Thread herring *Opisthonema oglinum* (LeSueur) 1817

Jordan and Evermann, 1896–1900, p. 432.

Description.—The thread herring is distinguishable at a glance from all the herrings that regularly inhabit the Gulf of Maine by the prolonged last ray (usually about as long as the body is deep) of its dorsal fin. It resembles the gizzard shad of

[21] See Vladykov, Nat. Canad., vol. 77, 1950, p. 121, for a detailed account of his tagging experiments on St. Lawrence River shad.

[22] Vladykov, Copeia, 1936, No. 2, p. 168, reports between 25 and 30 shad of 4–6 pounds, taken per haul, by otter trawlers in March, 1935, southwest of Middle Ground. about lat. 44°25′ N., long. 61°05′ W., at about 50 fathoms.

[23] Two shad were trawled by *Albatross III* on the eastern part of Nantucket Shoals at 68 fathoms, and 46 others at 9 stations distributed thence westward to the offing to Montauk Point (long. 71°52′ W.) at 26–64 fathoms, May 11–18, 1950.

[24] Vladykov and Wallace (Trans. Amer. Fish. Soc., vol. 67, 1937–1938, pp. 52–66) believe that Shubenacadie, Delaware River and Chesapeake Bay shad differ significantly in average number of vertebrae, of mid ventral scales and of pectoral fin rays. But Warfel and Olsen (Copeia, 1947, pp. 177–183) doubt whether any distinction can be drawn between shad in various streams along our North Atlantic coast, at least as far as average number of vertebrae goes.

[25] Stevenson, Rept. U. S. Comm. Fish (1898) 1899 pp. 265–269.

[26] These catches were reported as "barrels" presumably of 200 pounds each.

[27] Stevenson, Rept. U. S. Comm. Fish (1898) 1899, p. 121.

[28] About 460,000 pounds for the United States coast of the Gulf and about 374,000 pounds for the Bay of Fundy and in western Nova Scotia combined in 1916–17.

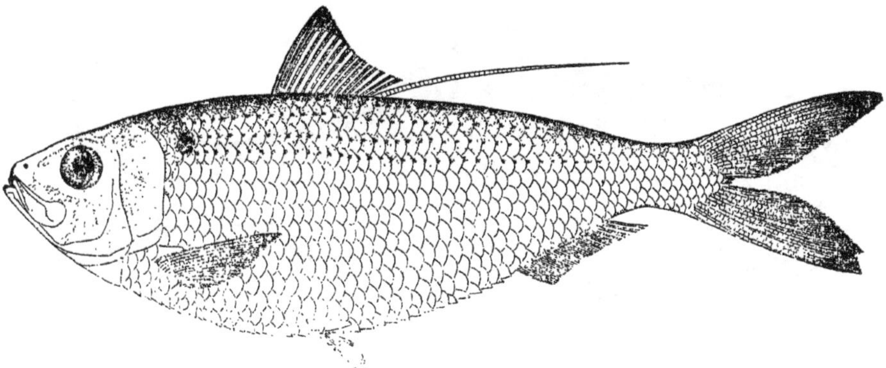

FIGURE 49.—Thread herring (*Opisthonema oglinum*). Drawing by Louella E. Cable.

fresh and brackish waters farther south in this respect, but the two differ rather conspicuously in various details. In the thread herring, the upper edge of the tail fin is about 1½ times as long as the head (only about as long as the head in the gizzard shad); the point of origin of the dorsal fin is a little in front of the origin of the ventral fins (a little behind in the gizzard shad); the distance from the origin of the ventrals to the origin of the anal fin is at least 1½ times as long as the base of the anal (only about ¾ to ⅘ in the gizzard shad); and the anal fin is very low, with its first few rays a little shorter than the eye (about 1½ times as long as the eye in the "gizzard"). There is no danger of confusing a thread herring with a young tarpon with which it shares the prolonged dorsal ray, for its dorsal fin originates in front of the ventrals, while the two fish are far apart in general appearance. This is a rather thin fish, its body about 2½ to 3 times as long (to the base of the tail) as deep; the belly is sharp and saw edged; the tail deeply forked as in our other herrings. There are 18 to 19 rays in the dorsal fin, 22 to 24 in the anal.

Color.—Bluish above, silvery on sides and belly. The scales along the back have dark centers, forming longitudinal streaks, and there is a faint dark spot just behind the upper margin of the gill cover; the dorsal and caudal fins have black tips.

Size.—Maximum length about 12 inches.

General range.—Atlantic coast of America in tropical and subtropical latitudes, south to Brazil, straying northward to Chesapeake Bay, and occasionally as far as southern Massachusetts.

Occurrence in the Gulf of Maine.—A thread herring is caught off southern New England occasionally; they were even reported as rather common in Buzzards Bay and in Vineyard Sound during the summer of 1885. But there is only one record of it within the Gulf of Maine, a single specimen 7 inches long, taken off Monomoy Point, at the southern angle of Cape Cod, in August 1931.[29] Being a tropical fish, it is not apt to reach the Gulf except as the rarest of strays.

Menhaden *Brevoortia tyrannus* (Latrobe) 1802

POGY; MOSSBUNKER; FAT BACK

Jordan and Evermann, 1896–1900, p. 433.

Description.—This fish is universally called "pogy" in the Gulf of Maine but no less than 30 common names are in use south of Cape Cod. It is flattened sidewise like all our other herrings, has a sharp-edged belly, and is as deep proportionally as the shad (body about 3 times as deep as long), though the general form is altered when the fish are fat. The very large scaleless head, which occupies nearly one-third of the total length of the body, gives the menhaden an appearance so distinctive that it is not apt to be mistaken for any other Gulf of Maine fish. It is likewise distinguishable from all its local relatives by the fact that the rear margins of the scales are nearly vertical (not rounded), and are edged with long comblike teeth instead of being smooth. The dorsal fin originates over the ventrals or very slightly behind them. We need only point out further that the pogy is toothless, its tail deeply forked, its ventral fins very small, its dorsal and anal of moderate size, its mouth large and gaping back as far as the hind margin of the eye, and that the tip of its lower jaw projects beyond the upper.

[29] Reported by MacCoy, Bull. Boston Soc. Nat. Hist., No. 61, 1931, p. 21.

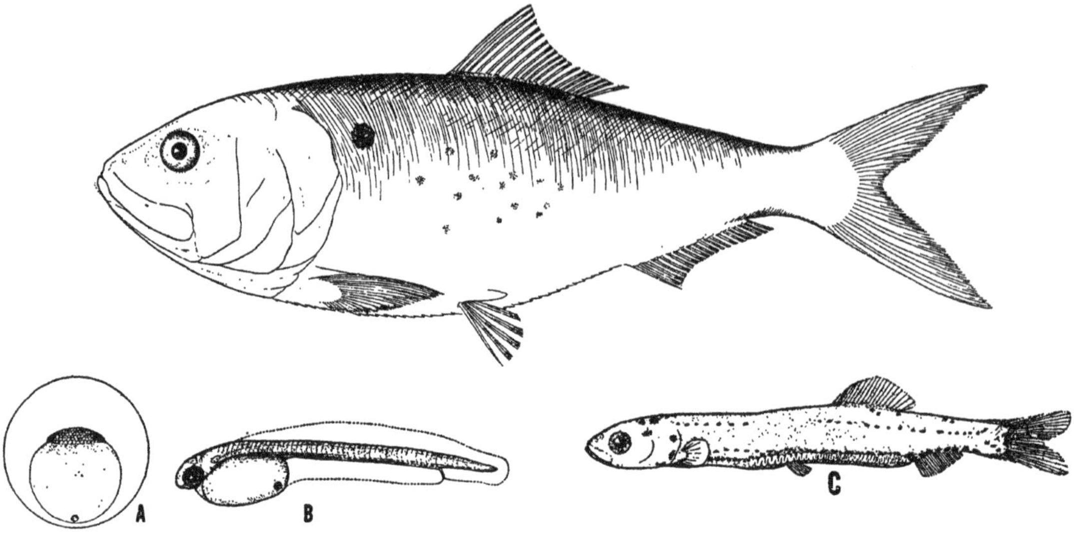

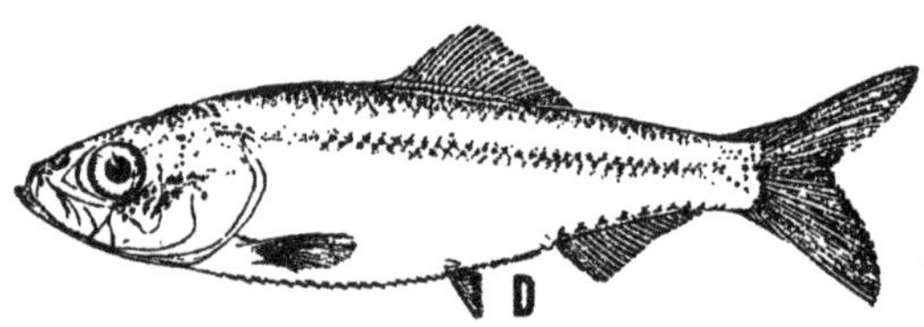

FIGURE 50.—Menhaden (*Brevoortia tyrannus*). A, egg; B, larva, newly hatched, 4.5 mm.; C, larva, 23 mm.; D, young fry, 33 mm. A–D, after Kuntz, and Radcliffe.

Color.—Dark blue, green, blue gray, or blue brown above, with silvery sides, belly, and fins, and with a strong yellow or brassy luster. There is a conspicuous dusky spot on each side close behind the gill opening, with a varying number of smaller dark spots farther back, arranged in irregular rows.

Size.—Adult menhaden average 12 to 15 inches in length, and from two-thirds to one pound in weight. One 18 inches long was taken at Woods Hole in 1876, and a fish 20 inches long has been reported. The heaviest of which we have heard was one of 1 pound 13 ounces, taken at Orient, N. Y.

Habits.—The menhaden, like the herring, almost invariably travels in schools of hundreds or thousands of individuals, swimming closely side by side and tier above tier. In calm weather they often come to the surface where their identity can be recognized by the ripple they make, for pogies, like herring, make a much more compact disturbance than mackerel do, and "a much bluer and heavier commotion than herring, which hardly make more of a ripple than does a light breeze passing over the water," as W. F. Clapp has stated to us. Also, pogies as they feed frequently lift their snouts out of water, which we have never seen herring do, while they break the water with their dorsal fins, also with their tails. And the brassy hue of their sides catches the eye (as we have often seen), if one rows close to a school in calm weather.

It is chiefly on warm, still, sunny days that the menhaden come to the surface, sinking in bad weather; and they are said to come up more often on the flood tide than on the ebb. It is also said (this we cannot vouch for) that the fish work inshore on the flood tide and offshore on the ebb.

Food.—The menhaden, formerly thought to subsist on mud, is now known to feed chiefly on microscopic plants (particularly diatoms) and on the smallest Crustacea.[30] It sifts these out of the water with a straining apparatus in the shape of successive layers of comb-like gill rakers as efficient as our finest tow nets. No other Gulf of Maine fish has a filtering apparatus comparable to that of the pogy, nor has it any rival in the

[30] For a detailed account of the food and of the branchial sieve of the menhaden, see Peck (Bull., U. S. Fish Comm., vol. 13, 1894, pp. 113–124, pls. 1–8.

Gulf in its utilization of the planktonic vegetable pasture. Menhaden feed, as Peck described, by swimming with the mouth open and the gill openings spread. We have often seen specimens in the aquarium at Woods Hole doing this.[31] And we have watched small ones in Chesapeake Bay, swimming downward as they feed, then turning upward, to break the surface with their snouts, still with open mouths.

The mouth and pharyngeal sieve act exactly as a tow net, retaining whatever is large enough to enmesh, with no voluntary selection of particular plankton units. The prey thus captured (as appears from the stomach contents) includes small annelid worms, various minute Crustacea, schizopod and decapod larvae, and rotifers, but these are greatly outnumbered as a rule by the sundry unicellular plants, particularly by diatoms and by peridinians. And the food eaten at a given locality parallels the general plankton content of the water, except that none of the larger animals appear in the stomachs of the fish on the one hand, nor the very smallest organisms (infusoria, and certain others such as the coccolithophorids) on the other. The menhaden, in short, parallels the whalebone whales, the basking shark, and the giant devil rays in its mode of feeding, except that its diet is finer because its filter is closer meshed.

Peck has calculated from observations on the living fish that an adult menhaden is capable of filtering between 6 and 7 gallons (about 24 to 28 liters) of water per minute, and while the fish do not feed continuously this will give some measure of the tremendous amount of water sifted and of plankton required to maintain the hordes in which these fish congregate. The abundance of microscopic plants in the water of bays and estuaries, and along the coast has often been invoked to explain the concentration of menhaden close to shore.

Enemies.—No wonder the fat oily menhaden, swimming in schools of closely ranked individuals, helpless to protect itself, is the prey of every predaceous animal. Whales and porpoises devour them in large numbers; sharks are often seen following the pogy schools; pollock, cod, silver hake, and swordfish all take their toll in the Gulf of Maine, as do weakfish south of Cape Cod. Tuna also kill great numbers. But the worst enemy of all is the bluefish, and this is true even in the Gulf of Maine during periods when both bluefish and menhaden are plentiful there (p. 384). Not only do these pirates devour millions of menhaden every summer, but they kill far more than they eat. Besides the toll taken by these natural enemies, menhaden often strand in myriads in shoal water, either in their attempt to escape their enemies or for other reasons, to perish and pollute the air for weeks with the stench of their decaying carcasses.

Breeding and growth.—Very little is known about the breeding habits of the menhaden, except that it spawns at sea and that the chief production of eggs takes place south of our limits. According to observations at Woods Hole,[32] the main body of the fish off southern New England spawn in June, continuing through July and August; even into October as in 1915, when the *Grampus* collected eggs and larvae in Nantucket Sound and westward from Martha's Vineyard in that month. And reports of spent fish in the Gulf of Maine in July and August, with others approaching maturity, suggest that the menhaden is a summer spawner there also. We have found no eggs in our townettings north of Cape Cod (young fry were taken in abundance in Casco Bay in October 1900), probably because our work there was carried on during a series of poor menhaden seasons. From Chesapeake Bay southward the spawning season appears to be late in the autumn, and in early winter.

Menhaden eggs are buoyant and resemble those of the European pilchard (*Clupea pilchardus*), but are easily distinguished from the eggs of any other Gulf of Maine fish by their large size (1.5 to 1.8 mm. in diameter), broad perivitelline space, small oil globule (0.15 to 0.17 mm.), and very long embryo. Incubation is rapid (less than 48 hours), as Welsh found by experiment. The newly hatched larvae are 4.5 mm. in length, growing to 5.7 mm. in 4 days after hatching. The dorsal and caudal fins first become visible at a length of 9 mm.; at 23 mm. all the fins are well developed; scales are present at 33 mm.; and at 41 mm. the fry show most of the characters of the adult, except that their eyes are much larger, proportionately. The youngest larvae much resemble young herring, but the fins are formed, the tail becomes forked, and

[31] Apparently Ehrenbaum (as quoted by Bullen, Jour., Mar. Biol. Assoc. United Kingdom, vol. 9, 1910–13, pp. 394–403) was not acquainted with the habits of menhaden when he wrote to the effect that no fish eat plankton indiscriminately, or swim about habitually with open mouth when feeding.

[32] By Kuntz and Radcliffe, Bull. U. S. Bur. Fish., vol. 35, 1918, p. 119, who describe the eggs and larvae.

the body deepens at a much smaller size, a menhaden of 20 mm. being as far advanced in development as a herring of 35 mm., which makes it easy to distinguish the older larvae of the two fish.

Welsh concluded from examination of great numbers of fry and from measurements and scale studies of fish of various ages that menhaden hatched in summer (which would apply to any fry that might be produced in the Gulf of Maine) are 2¼ to 3¼ inches (6 to 8 cm.) long by their first winter; and average about 6⅜ inches (16 cm.) by their second winter; fall-hatched fish are 1¼ inches (3 cm.) and about 5 inches (about 13 cm.) long, in their first and second winters, with every gradation between the two depending on the precise season when the fish are spawned.[33] Apparently sexual maturity is attained in the season following the third winter, and a few of the older fish that Welsh examined showed as many as 9 to 10 winter wings on their scales.

General range.—Coastal waters along the Atlantic coast of America from Nova Scotia to eastern Florida; represented in the Gulf of Mexico, and southward to northern Argentina, by a series of named forms that differ from our northern menhaden in ways that would not be apparent to any one but to a trained student of fishes.[34]

Occurrence in the Gulf of Maine.—The Gulf of Maine is the northerly limit for the menhaden; St. Mary Bay on the west coast of Nova Scotia is its most easterly known outpost. Prior to about 1850 the pogy seems to have been common at the mouth of the Bay of Fundy; it was, indeed, reported by Perley as far up the bay as St. John, and fishermen spoke of it as abundant near Eastport up to 1845–1850. But it seems to have abandoned Fundian waters altogether [35] since then except for an occasional straggler, and very few menhaden have been noticed east of Mount Desert and Jonesport of late years.

Perhaps the most interesting aspect of the occurrence of the menhaden in the Gulf of Maine is that it fluctuates tremendously in abundance there from year to year, periods of great plenty alternating with periods of scarcity or entire absence from our waters. Thus they were extremely abundant off the coasts of Massachusetts and Maine, every summer, for some years prior to 1875, when a considerable fishery developed for them in Maine. Very few, however, were taken in the Gulf during the cold summer of 1877 until September and October, when they were reported as about as abundant as normal; practically none appeared north of Cape Cod in the year 1879; and they were so scarce along the coast of Maine for the next six years that it caused comment when an occasional one was caught. In 1883, for instance, a few were reported to the U. S. Fish Commission though no schools were seen and many people thought they had gone permanently. But they were once more reported abundant off Maine and Massachusetts in 1886; they were so plentiful as far east as Frenchman Bay in 1888 that the menhaden fisheries were revived; they were as plentiful in Maine waters in 1889 as they had ever been (more than 10 million pounds taken there) and they were still so numerous in 1890 that four fertilizer factories were established, and nearly 90 million fish were taken during that season. But this period of abundance was short-lived, less than half as many fish being caught in Maine waters (about 41 million) in 1891 as the year before, while few menhaden were taken or seen north of Cape Cod in 1892. They were plentiful enough, however, in 1894, for a single steamer to seine about a million fish off the Kennebec during that summer, while 582,131 fish were taken in Boston Harbor in 10 days' fishing during the last half of that August.

Menhaden were scarce again in the Gulf during the period 1895–1897 but abundant again in 1898, when about 7 million pounds were taken along the Maine coast. They were scarce in 1902 (Maine catch about 300,000 lb.); reported as abundant again north of Cape Cod, in 1903, especially in Boston Harbor; rare north of Cape Cod from 1904 to 1921, when odd schools were seined along the Massachusetts and Maine coasts in some summers, while few or none were seen in others. They reappeared, however, in such abundance again in the southwest part of the Gulf in the summer of 1922 that 18 steamers fished for them successfully for some weeks in Massachusetts Bay, when upwards of 1,500,000 pounds were landed by the larger fishing vessels, besides what the small boats brought in. And they were so plentiful at

[33] Young menhaden that we collected at Woods Hole on September 23, 1942, were 3½ to 4 inches (91–99 mm.) long; others taken in Salt Pond, Falmouth, Mass., on November 24, 1949, were 4⅝ to 5 inches long.

[34] See Hildebrand (Smithsonian Misc. Coll., vol. 107, 1948, No. 18 for a revision of the genus *Brevoortia*). One named species, *B. brevicaudata* Goode 1878, is known only from Noank, Conn.; we doubt its validity.

[35] According to Huntsman (Contr. Canad. Biol., (1921) 1922, p. 59) one was taken in St. John Harbor in August, 1919.

least as far north as Boothbay Harbor, that about 2,500 barrels were frozen there, though no large schools were reported east of that point.

The appearance of menhaden in such abundance in the Gulf after so many years' absence prompted the Bureau of Fisheries to send the steamer *Halcyon* to Massachusetts Bay that August, and her towings indicated the presence of much greater quantities of diatoms than is usual at that season, evidence that the fish found a better pasture in Massachusetts Bay than in any summer since 1912. But we hesitate to assert that it was an unusually rich food supply that attracted them past Cape Cod.

However this may have been, there were not enough menhaden in the Gulf to be of any commercial importance from the middle 1920's to the middle 1940's. But so many visited Massachusetts Bay, in 1946 and 1947 that local boards of health were forced to clean some of the bathing beaches of the fish that drifted ashore from schools netted for lobster bait. There were a good many in Maine waters in 1948 (reported catch 145,000 pounds);[36] more still in 1949, when more than 5,000,000 pounds were taken there; and about 8,000,000 pounds off Gloucester,[37] and when small fry, 2–3¾ inches (52–95 mm.) were taken in the Sheepscot River, December 5–11, suggesting that some had been reared in the Gulf that year. But this peak of abundance lasted no longer than the peak had in the early 1920's, for there seem to have been far fewer menhaden in Maine waters in 1950 than in 1949, as there certainly were in Massachusetts Bay, where we did not chance to sight a single school, and very few were reported.

In the years when menhaden come, they appear in Massachusetts Bay about mid-May; off the Maine coast during the last half of May or first part of June. They are most abundant during July, August, and early September, and most of them depart from the coast of Maine by the middle of October, from the Massachusetts Bay region by early November; and it is unusual to find a single menhaden along these shores after the middle of that month, although small ones have been taken in the Sheepscot River as late as the first third of December.

The universal belief among fishermen, that the seasonal appearances and disappearances of menhaden in the Gulf of Maine result from a definite migration from the south around Cape Cod in the spring and a return journey in the autumn, probably is well founded.

The brevity of the peaks of abundance, the fact that they come at such long intervals, and especially the great local scarcity of young fish, are arguments against the possibility that menhaden are permanent inhabitants of our gulf, though a few fry may be produced there in favorable summers, as happened in 1949 (p. 117).

Menhaden are warm water fish, and our studies of the temperatures of the Gulf of Maine corroborate earlier observations to the effect that they never appear in spring until the coastwise water has warmed to 50° or more, or in abundance until the temperature is several degrees higher, which is in accord with Bean's[38] experience that menhaden will not survive in an aquarium if the water chills below 50°. No doubt, it is the falling temperature of autumn that forces the menhaden to leave the coasts of northern New England.

In menhaden years the fish occur all along the shores of the Gulf of Maine from Cape Cod to Penobscot Bay, even to Mount Desert. Their chief centers of abundance always lie in Massachusetts Bay within a mile or so of land, partic, ularly off Barnstable and in the mouths of Boston and Salem Harbors; in Casco Bay; and among the islands, thence to Penobscot Bay. But we have never heard of them entering water that is appreciable brackish, and in some years they may congregate as much as 40 to 50 miles offshore, as happened in 1878, for instance. But we have heard no report of menhaden in the central part of the Gulf or on the off shore Banks. The menhaden are thin when they arrive on our coasts in spring, but they put on fat so rapidly that while the average yield of oil per thousand Gulf of Maine fish was about 12 gallons for the whole summer season of 1894, it rose to 14½ gallons for Boston Harbor fish in August, and to 16 or 18 gallons in September. It is generally accepted, furthermore, that fish taken on the New England coast, south or north, always average larger and fatter than those caught farther south.

Commercial importance.—The menhaden is one of the most important, commercially, of the fishes of the Atlantic Coast of the United States, being used for the manufacture of oil, fertilizer and fish

[36] Reported by Scattergood, and Trefethen, Copeia, 1951, pp. 93–94.

[37] Reported by Scattergood, Trefethen, and Coffin, Copeia, 1951, p. 298.

[38] Rept. New York State Mus., 60, Zool. 9, 1903, p. 213.

scrap.[39] In 1946, when the catch for the Gulf of Maine was only about 20,000 pounds, the total catch for the Atlantic and Gulf States was 851,129,000 pounds; the value of the catch to the fishermen was $7,439,573; the value of the products made from menhaden was $18,196,573. Considerable numbers are used locally on the Middle Atlantic coast for bait. But the menhaden is so oily that it is unlikely to become popular as a food fish. Practically the entire catch of menhaden is taken by purse seines and in pound nets; they never bite a baited hook.

THE ANCHOVIES. FAMILY ENGRAULIDAE [40]

The anchovies are small herring-like fishes; but they are easily distinguishable from the herrings by the fact that their mouths are not only very much larger and gape much farther back, but are on the lower side of the head, and are overhung by the upper jaw, which projects like a short piglike snout in some species. Two anchovies are known to occur in the Gulf of Maine; both are stragglers from the south.

KEY TO GULF OF MAINE ANCHOVIES

1. Anal fin originates under the front of the dorsal; the silvery lateral band is diffuse; 24 to 27 anal fin rays ________________ Anchovy, p. 118.
   Anal fin originates under the rear rays of the dorsal; silvery lateral band bright and well defined; 20 or 21 anal fin rays ________________ Striped anchovy, p. 119.

Anchovy *Anchoa mitchilli* (Cuvier and Valenciennes) 1848

WHITEBAIT

Jordan and Evermann (*Stolephorus mitchilli*), 1896–1900, p. 446.

Description.—The only Gulf of Maine fishes with which one might confuse an anchovy are young herring, smelt, or silversides, but it is easily distinguished from the former by the wide mouth, as just noted; by its much larger eye; by the relative positions of the fins with the dorsal wholly behind the ventrals instead of over them and with the latter originating close behind the tips of the pectorals when these are laid back against the body; and by its much longer anal fin. The lack of an adipose fin behind the dorsal is sufficient to separate anchovy from smelt at a glance, while the silversides (*Menidia*) have two dorsal fins instead of one. The anchovy has large, thin, easily detached scales and a deeply forked tail. This species may be distinguished from the striped anchovy by the fact that its anal fin originates under the front of the dorsal; by its more or less diffuse lateral band of silver; by its more numerous anal fin rays (24 to 27 contrasted with 20 or 21 for the striped anchovy), and by its relatively small size, for it seldom exceeds 3 inches in length. The body is about 4 to 5 times as long as deep in both anchovies.

Color.—This is a whitish silvery, translucent little fish, its most characteristic marking being an ill-defined silvery band scarcely wider than the pupil of the eye, running from the gill opening back to the caudal fin. There are also many dark dots on body and fins.

Size.—Seldom more than 3½ inches long.

General range.—Coast of the United States from Maine to Texas, chiefly west and south of Cape Cod.

[39] For an account of the menhaden industry, see Harrison, Inv. Rept. No. 1, U. S. Bureau of Fisheries, 1931.

[40] For a recent review of the American anchovies see Hildebrand, Bull. Bingham Oceanographic Coll., vol. 8, art. 2, 1943.

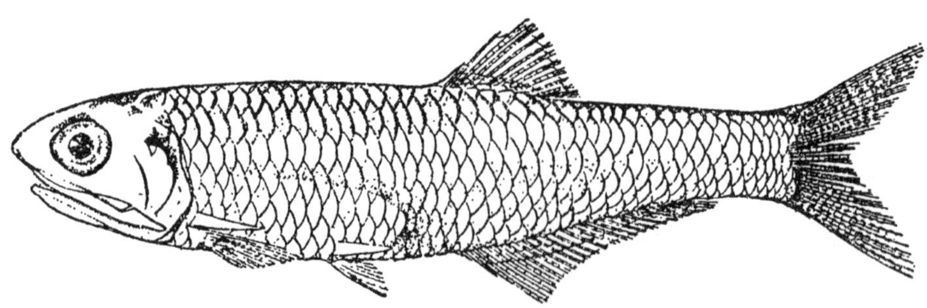

FIGURE 51.—Anchovy (*Anchoa mitchilli*).

Occurrence in the Gulf of Maine.—We mention the anchovy because it has been taken in Casco Bay and at Provincetown. It has no real place in the Gulf of Maine fauna, seldom straying past Cape Cod, though it is abundant about Woods Hole and thence westward and southward. Stragglers may be expected most often in the Gulf in midsummer for it appears from May to October in southern New England waters. Sandy beaches and the mouths of rivers are its chief resorts. An account of its embryology and larval development is given by Kuntz.[41]

Striped anchovy *Anchoa hepsetus* (Linnaeus) 1758

Jordan and Evermann (*Stolephorus brownii*), 1896–1900, p. 443.

Description.—This anchovy resembles the preceding species closely, but its anal fin is shorter (20 or 21 rays) and originates under the last rays of the dorsal, and it has a very bright and well defined silvery band along each side. It is a larger and more robust fish than the other anchovy, often more than 4 inches long.

Color.—The bright silvery lateral band, already mentioned, is the most prominent marking on this fish. Fresh specimens are pale gray and iridescent, the upper surface of the head with some green and yellow; and the back has dusky dots. The dorsal and caudal fins are more or less dusky on some specimens.

Size.—Commonly 4 to 5 inches long, maximum length about 6 inches.

General range.—Abundant from Chesapeake Bay to the West Indies, and south to Uruguay; north as a stray to Maine and to the outer coast of Nova Scotia;[42] a more southerly fish than the other anchovy.

Occurrence in the Gulf of Maine.—The claim of this species for mention in the Gulf of Maine is based on one record off the mouth of the Penobscot River, near Portland, October 8, 1930.[43] One specimen was saved and identified, and the herring fishermen who brought it in stated that there were "lots of them" on that date. It is not likely that the striped anchovy is other than a straggler to the Gulf, else it would have been found there before this. As it is a gregarious fish, nearly always traveling in small schools, it is not astonishing that they may be found together in some numbers, on occasion, even out of their usual range.

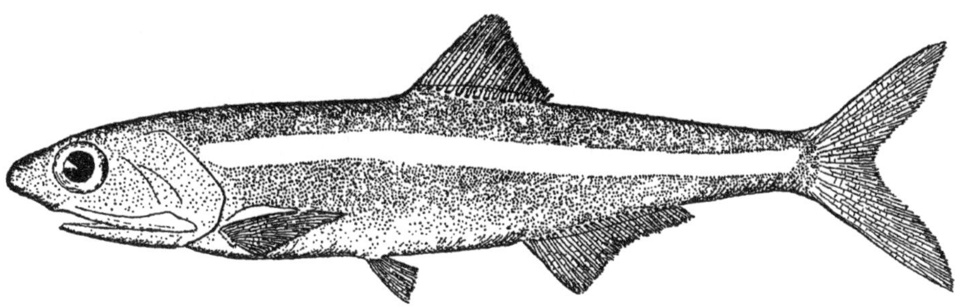

FIGURE 52.—Striped anchovy (*Anchoa hepsetus*), Somers Point, N. J., specimen 100 mm. long.

THE SALMONS. FAMILY SALMONIDAE

The salmons are soft-rayed fishes with no spines in any of the fins, with the ventrals situated on the abdomen far behind the pectorals, and with a fleshy rayless "adipose" fin on the back behind the rayed dorsal fin. The presence of this adipose fin, and its situation, separates them from all other Gulf of Maine fishes except for the smelt, capelin and the argentine, the pearlsides (p. 144), and some of the lantern, viper, and lancet fish tribes (p. 141).[44] The blunt noses, stout bodies, and nearly square tails of the salmons distinguish them at a glance from the sharper-nosed, slender, forked-tailed smelts, their large mouths and smaller eyes from the argentine; the absence of luminescent organs distinguishes them from the

[41] Bulletin, U. S. Bur. of Fish., vol. 33, 1915, p. 13.

[42] Five were taken in Bedford Basin, Halifax, Nova Scotia, on September 29, 1931 (Vladykov, Proc. Nova Scotia Inst. Sci., vol. 19, 1935, p. 3).

[43] Kendall, Bull. Boston Soc. Nat. Hist., No. 58, 1931.

[44] Sundry other deep-sea fishes have adipose fins.

pearlsides, while the lantern, viper, and lancet fishes are of different general aspect.

Four salmons [45] occur in the Gulf of Maine, or have recently, one of which, the sea trout, resorts to tidal estuaries at the mouths of a few of our streams; a second and a third—the humpback salmon and the silver salmon—were introduced from the Pacific coast, leaving the Atlantic salmon as a characteristic inhabitant of the open waters of the Gulf of Maine.

KEY TO GULF OF MAINE SALMONS

1. Scales so small that they are hardly visible; back with vermiculate markings; teeth on roof of mouth confined to a group in front..........Brook trout, p. 120
   Scales large enough to be easily visible; back without vermiculate markings; a row of teeth runs back along the mid line of the roof of the mouth..............2
2. Anal fin with only 8–10 rays..........Salmon, p. 121
   Anal fin with 12 rays or more......................3
3. Back and lower half of tail fin, as well as its upper half, conspicuously marked with large black spots Humpback salmon, p. 131
   Back with very small black spots or none at all; no black spots on lower half of tail fin..Silver salmon, p. 133

Brook trout *Salvelinus fontinalis* (Mitchill) 1815

SEA TROUT; SALTER

Jordan and Evermann, 1896–1900, p. 506.

Description.—Although brook trout vary widely in general form in different streams, they are usually salmon-like in shape when taken in salt water, that is, about one-fourth as deep as long, tapering gracefully to a small head. The nose of a trout, however, is blunter than that of a salmon, and its head is longer in proportion, the total length of the fish (not counting the caudal fin) being about four and one-half times that of the head,[46] while its mouth (gaping back of the eye) is relatively larger. The general arrangement of the fins, including the "adipose," parallels that of the salmon, but the ventral fins stand under the middle of the dorsal, thus farther forward in relation to the latter than in its larger relative. All the fins, too, are relatively larger, particularly the ventrals; as a rule the anal has one less ray in the trout (usually 8) than the salmon, but the number of dorsal rays (about 11) is the same. The tail of the sea trout is less forked than that of a young salmon of equal size.

Examination of the scales and of the teeth is the most positive means of distinguishing brook trout (in European terminology this is a "charr") from young salmon, for the teeth on the roof of the mouth of the trout are confined to a cluster near the front, instead of extending backward in a row along its midline as in the salmon; and the scales of the trout are so tiny as hardly to be visible whereas those of the salmon are large and easily seen.

Color.—Trout living in salt water almost wholly lack the yellow and red tints so conspicuous on their freshwater relatives. They are steel blue or bottle green on the back, with cheeks and sides silvery like a salmon and with a white belly. The sides above the lateral line are more or less dotted with pale yellow spots, but the dark vermiculate markings so characteristic of the fresh-water brook trout are rarely seen on the trunk of sea run fish, though evident as wavy crossbars on the dorsal

[45] A specimen of one of the whitefishes (probably *Coregonus quadrilateralis* Richardson) was taken in the mouth of the Sissibou River, St. Mary Bay, Nova Scotia, September 1919 [Huntsman, Contr. Canad. Biol., (1921) 1922, p. 59] straying down from fresh water. Whitefish have an adipose fin, like the true salmons, but have a very small mouth, and are flattened sidewise, and herring-like in appearance, rather than salmon-like.

[46] Some trout are longer headed.

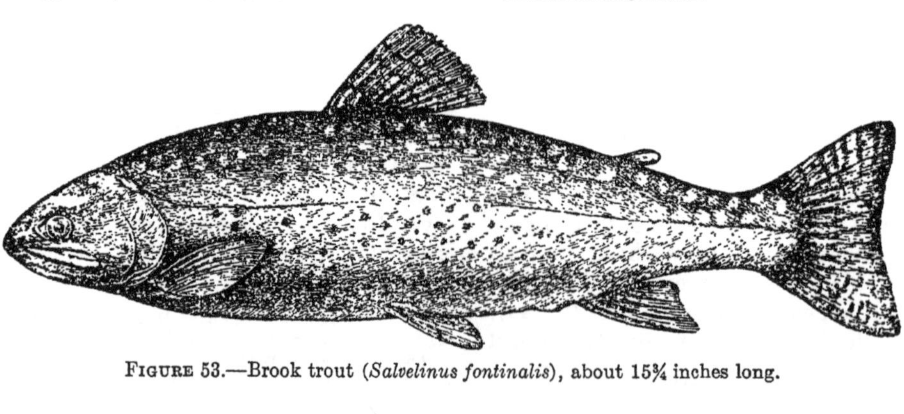

FIGURE 53.—Brook trout (*Salvelinus fontinalis*), about 15¾ inches long.

fin and on the corners of the caudal fin. The sides and flanks below the level of the lateral line usually are strewn with small pale vermillion dots, but the ventral fins are often plain white; at most, the pink edging so conspicuous in trout caught in fresh water is faint on fish in salt water.

General range.—Eastern North America, north to the outer coast of Labrador, west to Minnesota, and southward to Georgia along the Allegheny Mountains.

Occurrence in the Gulf of Maine.—Brook trout are plentiful in many of the river systems and smaller streams that empty into the Gulf of Maine. Some of the trout in some of these seek salt water after the breeding season, to remain there over the winter. This applies particularly to the brooks that flow through the sands of Cape Cod, several of those on its southern slope being famous for their sea-trout fishing. These, however, lie outside our present province, and only a couple of small streams on the Massachusetts Bay side of the Cape still support a race of trout that run down to the sea regularly. One or two small brooks tributary to Ipswich Bay, and the Merriland River, emptying between Wells and Kennebunkport, Maine, are the only places between Cape Ann and Cape Elizabeth where we have heard of sea run trout.

We cannot say how generally sea trout may now exist in the streams of eastern Maine, but according to Evermann [47] trout once inhabited the tidal portions of many of the brooks that empty into Casco Bay, and they still may. Some of good size are caught also in the Belfast River waters, tributary to upper Penobscot Bay.[48] Huntsman found no definite evidence of trout in salt or brackish water on the New Brunswick side of the Bay of Fundy, but local inquiry has elicited the information that there are fish of this habit in a few streams (notably in Salmon River) on the north and west coasts of Nova Scotia, where many streams formerly held sea run trout that have been fished out long since.

The "sea trout" are indistinguishable from the ordinary brook trout anatomically.[49] They are simply fish that have the habit of running down to salt water, and most of the trout never leave fresh water, even in streams offering free access to the sea, cold enough throughout their lengths, and harboring these "salters" (as they are called on Cape Cod). All who have given special attention to our sea trout are agreed on this. It is still an open question whether the habit is hereditary or whether it is acquired independently by each individual fish. We incline to the first view, chiefly because sea trout are slow in reestablishing themselves in any stream where they have been brought to a low ebb by hard fishing. The trout that follow this habit grow much more rapidly on the abundant rations the salt estuaries provide than do most of their relatives that remain in the brook. Sea fish weigh from 1 to 3 pounds in streams where few of the fresh-water trout exceed half a pound.

On Cape Cod the sea trout go down to salt water in November immediately after spawning, to winter there. They begin to run again in April, and all of them are in brackish or fresh water by mid-May. But it is said that they do not appear until later in the Nova Scotia streams tributary to the Bay of Fundy (we cannot vouch for this).

While in salt water (at least along Cape Cod) the trout feed chiefly on shrimps or on gammarid Crustacea, on mummichogs (*Fundulus*), and on other small fish. Trout never stray far from the stream mouths; hence they have no place [50] in the fish fauna of the open Gulf.

[47] Rept. U. S. Comm. Fish., (1904) 1905, p. 105.

[48] Towne, Striped Bass Survey, Maine Development Comm. and Dept. Sea and Shore Fisheries, 1940, p. 21.

[49] There is another species of sea trout (*Salvelinus alpinus*) in northern Canadian waters which is very plentiful along the coast of northern Labrador.

Salmon *Salmo salar* Linnaeus 1758

ATLANTIC SALMON; SEA SALMON; SILVER SALMON; BLACK SALMON; PARR; SMOLT; GRILSE; KELT

Jordan and Evermann, 1896–1900, p. 486.

Description.—The Atlantic salmon is a graceful fish, about one-fourth as deep as long, deepest below the dorsal fin, whence it tapers toward both head and tail; and oval in cross section. Its head is small (about one-fifth, or less of the fish's length, not counting the caudal fin), its nose is blunt, eye rather small, and its mouth gapes back to below the eye. The dorsal fin (about 11 rays) stands about midway between tip of snout and base of tail fin; the ventrals are under the rear end of the dorsal. The anal is similar in form to the dorsal but has only about 9 rays (7 to 10 have

[50] Trout are taken about Woods Hole, occasionally, in winter.

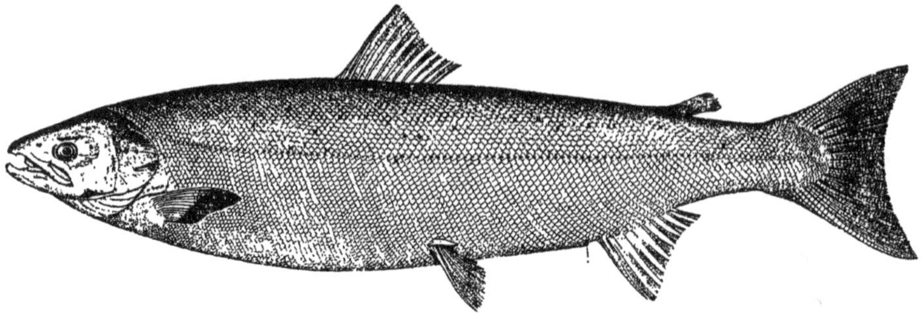

FIGURE 54.—Salmon (*Salmo salar*). Drawing by H. L. Todd.

been recorded), whereas the humpback has 14 anal rays or more. The tail is only very slightly emarginate in adults, and is almost square in large fish, but is more forked in fish that have been at sea for not more than one year ("smolts" and "grilse").

Color.—The salmon is silvery all over while in the sea, with brownish back and with numerous small black crosses and spots on head, body (chiefly above the lateral line), and fins. The young fish (or "parr") are conspicuously marked with 10 or 11 dark crossbars while in fresh water, alternating with bright red spots, much like young trout. Fish that have been at sea for only one year (grilse) are marked with a larger number of black spots than the older fish.

Size.—The largest salmon we find mentioned was an English fish of 83 pounds. None even approaching this size is recorded from our side of the Atlantic, where a 50-pounder is unusual, though fish of 40 pounds are not uncommon in some of the larger rivers emptying into the Gulf of St. Lawrence. Very few fish reach 40 pounds in the Penobscot or St. John Rivers, and 30-pounders are unusual there, the usual run being 10 to 12 pounds. Taking one river with another, large and small, 10 pounds may be set as a fair average of the mature Gulf of Maine fish. A 2-foot fish will weigh about 6 pounds, one of 3 feet, 16 to 20 pounds, with allowance for individual and seasonal variation.

Remarks.—The teeth and the scales afford the most certain distinction between small salmon and the New England sea trout (p. 120). In the salmon the roof of the mouth is armed both with a cluster in front and with a row of stout conical teeth running back along the mid-line, easily felt with the finger, whereas the sea trout has the forward group only. The scales of the salmon are so large that they are seen easily, whereas those of the trout are so minute that they are hardly visible. Old salmon sometimes lose the teeth on the roof of the mouth, but large size and large scales identify them at a glance.

It should also be easy to tell an Atlantic salmon from a humpback (should any of the latter still exist in our Gulf) for the black spots on the upper part of the body of the humpback and on its tail fin are more close set and much larger and conspicuous than the dark markings on a salmon. A more precise difference is that an Atlantic salmon never has more than 10 rays in its anal fin, whereas the humpback always has at least as many as 12, while most of them have 13 to 17.

The danger will be greater of confusing smallish Atlantic salmon with silver salmon, if the attempts now in progress to establish the latter in our Gulf should succeed, for the two fish look much alike. A reliable criterion is, again, the number of rays in the anal fin, for the silver salmon always has as many as 13 of these, an Atlantic salmon never more than 10.

Life history.[51]—It is no wonder that the life of the salmon has been the subject of much scientific study and that a whole literature has grown up about it. As everybody knows, the salmon lives the greater part of its life in the sea and makes most of its growth there but spawns in fresh water.

The salmon are silvery and very fat when they enter fresh-water on the spawning journey, but

[51] Huntsman (Bull. Biol. Board Canada, 21, 1931) has published an extensive study of the life history of the salmon of the Maritime Provinces of Canada, from which we have drawn freely in the following account. See also Huntsman and others (Migration and Conserv. of Salmon, Pub. No. 8, Amer. Assoc. Adv. Sci., 1939) for discussions of the movements of the salmon in Canadian and Newfoundland waters; also Lindsay and Thompson (Rept. Newfoundland Fish. Res. Comm., vol. 1, No. 2, 1932) for an account of the biology of the salmon in the rivers and around the coasts of Newfoundland.

they lose condition gradually as they work upstream, for they feed very little in fresh water, if at all; they make no attempt, for example, to capture the parr they meet. Most anglers believe that they may occasionally snap up a small fish or other tempting morsel. Many are caught on artificial flies, while every salmon angler knows that they will sometimes take a hook baited with angleworms or with prawns. It has been suggested that salmon recover the feeding habits of their youth to some extent after they have spent some time in the river, for they often rise to floating insects. But the stomachs of salmon caught in fresh water never contain anything but a little yellowish green fluid. And the fact that they keep better with bellies intact than if opened and gutted suggests that the secretion of effective digestive juices has ceased.

The maturing salmon of both sexes lose their silvery sheen in fresh water during the summer months, to take on a dull brownish or reddish hue, while the belly suffuses with some tint of red, large black spots develop, and the male not only becomes variously mottled and spotted with red or orange, but his jaws elongate, the lower becoming so hooked that only the tips come together. His body becomes slab-sided, his fins thicken, and his skin is covered with slime, until altogether he is but a caricature of the beautiful silvery creature that came in from the sea.

In small streams salmon may spawn only a short distance above the head of tide; but they may run upstream for more than 200 miles in large rivers that are not obstructed, as they do in the St. John system in New Brunswick. In Gulf of Maine rivers they spawn in October and early November, on sandy or gravelly bottom, the females smoothing a shallow trough or redd and covering the eggs with gravel.

As it is with the life of the salmon in the sea that we are concerned here, the reader is referred to Belding [52] and to Kendall [53] for recent accounts of the mating actions of the males and females. The spent fish, known as "kelts," "slinks," or "black salmon," are thin, weak, and so exhausted that many of them die. Most of those that survive in small rivers drop down at once to the sea after spawning. But many of them linger over the winter in large rivers, improving somewhat in condition and becoming more silvery, though they take little food. If they survive the winter (which many do not, for spawning leaves them thin and exhausted) they drop downstream to salt water the following spring.[54]

The large (6 to 7 mm.) thick-shelled eggs lie loose on the bottom and develop so slowly in the low temperature of winter that hatching does not take place until late in the following April or early in May. The newly hatched larvae are 15 to 18 mm. (0.6–0.7-inch) long, and carry a very large yolk sac for about 6 weeks, hiding among the pebbles of the spawning bed and taking no food. When the yolk sac is absorbed the little fish, now known as "parr," begin to swim and feed.

Parr live in fresh water for longer or shorter periods according to locality or to other factors not well understood. In the St. John,[55] and in the rivers of Minas Basin, most of them remain for 2 summers and 2 winters, running down to the sea the third summer. But Huntsman has found that some few stay in the Fundian rivers for 3 years. Most of the salmon of the Penobscot spend 2 years as parr, a few 3 years, according to Kendall. It is even possible that some may linger in Gulf of Maine rivers for 4, 5, or even 6 years, as is known to happen in Norway. And Dr. Huntsman informs us that some of the male parr in the rivers of the Chignecto Peninsula become sexually mature before visiting the sea.

Parr may be moving downstream any time from late spring to autumn, but most of them probably make the journey in June and July in Gulf of Maine streams, when they are 5 to 6 inches long. They put off their barred and spotted pattern as they near tidewater, to assume the silvery coat worn by the salmon during his sojourn in the sea. They are now known as "smolts."

Salmon, small or large, are voracious while in salt water, feeding altogether on live bait, chiefly on fish and on crustaceans. Among fishes available to them in this side of the Atlantic, launce, herring, alewives, smelt, capelin, small mackerel, haddock, small sculpins, and even flatfish have all been reported as entering into their diet in one place or another. Salmon caught off Norway are sometimes packed full of herring, and a hook

[52] Trans. Amer. Fish. Soc., vol. 24, 1934, p. 211.

[53] Mem. Boston Soc. Nat. Hist., vol. 9, No. 1, 1935, p. 65-68.

[54] They are voracious now, and fly-fishing for these "black salmon" as they are called, is a favorite sport nowadays, especially in Miramichi waters tributary to the Gulf of St. Lawrence.

[55] Huntsman, Bull 21, Biol. Bd. Canada, 1931, p. 31, based on studies by Kerr and by Blair.

and line fishery is carried on for salmon in the Baltic, with herring for bait, while in British waters salmon are sometimes caught on hooks baited with launce and with pieces of mackerel. Launce and capelin had been the chief diet of thousands of salmon opened by Comeau[56] in the northern side of the Gulf of St. Lawrence. And it is probable that the salmon of the Bay of Fundy and open Gulf of Maine feed chiefly on herring (herring up to 5 inches long have been found in salmon stomachs near Eastport) and on launce, taking alewives or any other small fish as occasion offers, including smelts and mummichogs (*Fundulus*), when they first enter the estuaries.[57]

Salmon also feed greedily on euphausiid shrimps (fish entering the Penobscot have been found full of "shrimp," probably euphausiids); to some extent on pelagic amphipods (*Euthemisto*), while sand fleas (gammarid crustacean) are described as ranking with launce and herring as salmon food in the North and Baltic Seas. Salmon are also credited with eating crabs.[58]

Smolts, on the other hand, fall prey to any large predaceous fish (they have been found in the stomachs of pollock), but salmon are so heavy and strong after one or two years' sojourn in salt water that only fish as large as tuna, swordfish, or the larger sharks can menace them. Their worst enemy is the harbor seal, which is a common inhabitant of the northeastern coasts of the Gulf of Maine and of the Bay of Fundy.

The young smolts grow so rapidly on the abundant diet the sea affords that they usually reach a length of at least 16 inches and a weight of anywhere from ½ to 7 pounds after one year at sea. They are now known as "grilse." And older salmon continue to put on length and weight very fast, as long as they remain in salt water. Thus, several St. John fish which were tagged and released in the river in the autumn after spawning and which were recaptured the following summer after wintering in the sea had gained 2 to 8 pounds in weight, one of them more than 6 inches in length. Others which spent two uninterrupted years in the sea (as shown by their scales) averaged about 10 pounds heavier and 6 inches longer when recaptured.[59] But they grow much less rapidly in winter than in summer. And they hardly grow at all during the years when they spawn if they enter the river early in season, though they continue growing until later if they enter late. Hence the size of a salmon depends more on the number of times it has spawned and on the date when it enters its river than on its age.

Most of the exceptionally large fish of 40 to 50 pounds are virgin females entering fresh water for the first time, but some are fish that have already spawned once. An interesting case is that of a 45-pound 2-ounce fish, caught in the Moisie River, on the north shore of the Gulf of St. Lawrence, June 1950, by E. E. Steedman, the life history of which had been as follows:[60] hatched spring 1942; went to sea June 1945; returned to river and spawned there in 1948; returned to sea autumn 1948; remained there until June 1950; then returned to the river, to be caught on a "Lady Amherst" fly; age 8 years.

Some salmon become "river mature" and return to spawn after only one year at sea; these, known as grilse, are distinguishable from the older fish by more forked tail, more slender body, thinner scales, and more numerous spots that are blue rather than black.[61] Some spawn 2 or 3 years in succession, and hence never grow large; others spawn twice in alternate years; a few three times, very few oftener. It follows from this that large salmon are to be found in the sea throughout the year, though fewer of them in summer when the spawning fish are in the rivers, than in winter when the whole stock is in salt water except for the "parr," a few immature grilse (p. 129), and such of the spent fish as winter in the rivers. Some spawn only once after 3, 4, or even 5 years at sea, growing to a great size meantime. But very few salmon live to be more than 8 or 9 years old, including the time spent in fresh water as parr.

Our ignorance of the way of life of the salmon in the sea has recently been characterized as abyssmal. Certainly they are swift swimmers, and the nature of the catches suggests that they

56 Life and Sport on the North Shore, 1909, Quebec.

57 Kendall (Mem. Boston Soc. Nat. Hist., vol. 9, No. 1, 1935, p. 34) found smelts in Penobscot salmon, alewives in salmon from the St. John.

58 See Kendall (Mem. Boston Soc. Nat. Hist., vol. 9, No. 1, 1935, pp. 33–34) for a recent survey of the diet of salmon in general; the Gulf of Maine fish in particular, with references. Eichelbaum (Cons. Perm. Internat. Explor. Mer, Rapports et Proc. Verb., vol. 21, 1916, p. 84) records the contents of many salmon from the Baltic and from the North Sea.

59 Huntsman (Bull. Biol. Board Canada, No. 21, 1931) gives an interesting account of these tagging experiments, from which this summary is drawn.

60 As worked out from its scales by Dr. D. L. Belding, and reported in Field and Stream, August 1951, p. 10.

61 It is commonly stated that this applies chiefly to the males. But Huntsman (Bull. Biol. Board Canada, No. 21, 1931, pp. 18–19) has found that grilse of both sexes spawn in the small rivers at the head of the Bay of Fundy.

live scattered for the most part. But at least one case has come to our notice of a school seen, and some of them netted.[62] While salmon often leap in the esturaries on their return journey and in the rivers, we have never heard of one doing so at sea. And they keep so constantly to the mid-depths that they are seldom seen at the surface, except in the estuaries. But this rule has its exceptions, for the school mentioned above was sighted at the surface, where they were mistaken for pollock. On the other hand, there is no reason to suppose that many of the Gulf of Maine salmon descend to any great depth, winter or summer. The weirs, gill nets, and other gear that yield so many in various regions, are all operated in rather shoal water (the Baltic hook-and-line-fishery is carried on at about 1½ fathoms). Dr. Huntsman informs us that salmon are taken on hand lines in mid-winter in the Bay of Fundy. They are caught occasionally on long lines in the Gulf, and otter trawlers get stray salmon on the offshore Banks (p. 126), proof that at least some may go as deep as 50 fathoms or so, while diet (p. 124) proves that they sometimes feed near bottom if not actually on it.

General range.—Coastal waters of both sides of the North Atlantic, entering rivers to spawn. On the European side its range extends northward well within the Arctic Circle; southward to the Miño River, at the boundary between Spain and Northern Portugal, perhaps with a few reaching the Duero River, midway of Portugal.[63] It occurs in a few rivers in western Greenland.[64] On the American side salmon ran up all suitable rivers, formerly, from northeastern Labrador to the Housatonic emptying into Long Island Sound; perhaps the Hudson also. The northern limit of the commercial fishery for it on the American side is only about latitude 54° N. (Indian Harbor, north shore of Hamilton Inlet). And while it is known to range to Hudson Strait,[65] reports of it from stream mouths northward from Hamilton Inlet seem often to have been based on the sea run form of the Arctic charr *Salvelinus alpinus*, which also grows large in the sea.[66]

Occurrence in the Gulf of Maine.—When the white man first came to New England and to the Maritime Provinces, he found salmon in every large stream not barred by impassable falls, from Cape Sable to Cape Cod; i. e., in all the Nova Scotia and New Brunswick Rivers, tributary either to the open Gulf of Maine or to the Bay of Fundy, and in the following rivers in New England: St. Croix, Dennys, Orange, East Machias, Pleasant, Narraguagus, Union, Penobscot, St. George, Medomak, Sheepscot, Androscoggin, Kennebec, Royal, Presumpscot, Saco, Mausam, Piscataqua, and Merrimac.[67] One New England river, however, after another was so obstructed by dams after the beginning of the past century, that salmon regularly entered only the St. Croix, Dennys, East Machias, Machias, Penobscot, Sheepscot, Kennebec, and Androscoggin by the 1880's. The Kennebec was still an important salmon river as late as 1895. But by 1925 the Dennys and the Penobscot alone, of the rivers of Maine, saw regular runs, with a few fish in the St. Croix where pollution by sawdust was not as bad then as it had been, perhaps with an occasional fish in other streams.

The fate of the salmon in the Merrimac[68] typifies its history in the rivers from which it is now barred. Salmon spawned plentifully in the upper tributaries, especially in the Pemigewasset, as late as 1793 (in 1790 the run was so abundant in the lower river that 60 to 100 a day was the usual catch with a 90-yard seine near the mouth at Amesbury), but the completion of the dam at Lawrence in 1847 completely barred the upper reaches of the river. For some years thereafter salmon congregated below the Lawrence dam in spring and summer, vainly endeavoring to ascend, but there has been no run of salmon in the upper Merrimac since 1859 or 1860, when the last salmon hatched above the dam had lived its span of life, nor have any spawned there since then with the possible exception of a few that have been lifted over the dam by hand.

[62] Kendall. Mem. Boston Soc. Nat. Hist., vol. 9, No. 1, 1935, p. 32.

[63] This is the southern European limit given by Berg (Zoogeographica. vol. 1, Pt. 2, 1932, p. 112.

[64] Jensen, Fauna of Greenland, vol. 1, Pt. 3, Fishes, 1928, pp. 3 and 4, Copenhagen.

[65] Vladykov (Contrib. Canad. Biol., N. Ser., vol. 8, No. 2, 1933, p. 18, fig. 1) shows a locality record near Fort Chimo, and there are salmon in the rivers of the eastern part of Ungava Bay.

[66] Blair (Res. Bull. 12, Dept. Nat. Resources Newfoundland, 1943, pp. 5-17) gives a detailed account of the salmon rivers of the outer Labrador coast, Strait of Belle Isle to Hamilton Inlet.

[67] Atkins (1887, Fish. Ind. U. S., Sect. 5, vol. 1, p. 679) has collected much information on the local history of salmon in northern New England.

[68] Lyman and Reed, Rept. Comm. Fish. Massachusetts (1865) 1866, Senate Doc. 8, pp. 36-41.

Enough salmon to yield a supply of eggs for artificial hatching continued to enter the lower Merrimac up to 1893 and there seems to have been what almost might be described as a run there in 1896, when salmon were seen leaping below the Lawrence dam nearly every day from June 10th to July 25th, often 10 or 20 at a time, and a few were lifted over. But we have not learned of a single sea-run salmon seen in the Merrimac since 1901, though watch has been kept for them by the wardens of the Massachusetts Division of Fisheries and Game,[69] and it is not likely that salmon would still run in the Penobscot were it not for the artificial propagation that is carried on there by the U. S. Fish and Wildlife Service. But the salmon situation now shows signs of improvement, for the run in the St. Croix has increased; salmon have reestablished themselves in the Narraguagus and provide sport there for many anglers since one obstructing dam has washed out and another opened. Enough salmon run regularly in the Dennys to attract anglers and a few also in the Machias and Pleasant Rivers. The Fisheries Commission of the State of Maine, and the U. S. Fish and Wildlife Service, are now concerned with the possibility of improving the runs in these streams, and of reestablishing runs of salmon in other Maine rivers.

Along the Canadian shores of our Gulf a few salmon still run in the Tusket, Salmon, and Annapolis Rivers; many in the Shubenacadie River in Nova Scotia, some in the Petitcodiac, and great numbers in the St. John River in New Brunswick, which still is a famous salmon river.

Movements in the Gulf.—After the smolts reach salt water they are found for a time in the river mouths and about estuaries. No doubt the little salmon (too small to sell) that are caught in summer and autumn in weirs at Matinicus Island have come from the Penobscot a month or two previous. They drop out of sight in winter, as do the older and larger salmon as well. But there is no reason to suppose that many of them go far out to sea in the Gulf. Odd salmon stray, it is true, as much as 90 to 100 miles seaward off the outer coast of Nova Scotia,[70] while otter trawlers pick up odd salmon from time to time in the South Channel, and even on Georges Bank up to 160 miles or more at sea from Cape Cod.[71] But the great majority of the salmon that are caught in the Gulf are taken within 25 miles of the land.

The Gulf of Maine salmon also appear to continue rather closely localized as a whole, not only within the coastal belt, but within the zone of fresh-water influence from the particular rivers or river systems from which they come. So few, for example, are caught near Cape Sable that there can be no general movement around the Cape by the fish that spawn in the rivers of the outer coast of Nova Scotia. Most of the fish that go to sea via Minas Channel from the Shubenacadie, and the few from smaller streams that discharge into Minas Basin, seem to remain along the Nova Scotia shore within a distance of 30 to 40 miles to the westward. And while tagging experiments have proved that some of them scatter more widely; i. e. to Cobequid Bay, to the estuary of the St. John River, to the Annapolis Basin, and to St. Mary's Bay, few of them leave the Bay of Fundy[72] (for some that did, see p. 127).

The much more numerous salmon from the St. John appear to hold rather closely to the tongue of low salinity that extends westerly from the mouth of the river, keeping out from the shore, for hardly any salmon are caught either on the New Brunswick shore to the eastward, except for a few near the head of the Bay (doubtless the product of the Chignecto Bay river system) or farther west than Point Lepreau, or around Grand Manan Island which stands directly in the route of any fish moving westward out of the northern side of the Bay of Fundy. Thus it appears that a radius of, say, 40 to 50 miles would enclose the wanderings of most of the St. John River fish.

The evident failure of salmon from the St. John to follow the myriads of sardine sized herring into Passamaquoddy Bay is especially interesting. The weirs there pick up a few salmon, the presence of which can be credited to the small run in the St. Croix River. And the numbers of salmon that are caught thence westward along the coast of Maine[73]

[69] A few small "salmon" reported of late in the Merrimac probably were the landlocked form, running down from tributaries stocked with this fish.

[70] Three reports of salmon caught on Western Bank have appeared in the daily press since 1925 to our knowledge, and Kendall (Mem. Boston Soc. Nat. Hist., vol. 9, No. 1, 1935, p. 33) reports one caught on La Have Bank 100 miles from Halifax, and another 60 miles off Cape Sable.

[71] Kendall (Mem. Boston Soc. Nat. Hist., vol. 9, No. 1, 1935, pp. 31–33) lists a number of such cases.

[72] Huntsman, Ann. Rept. Fish. Res. Board Canada, (1947) 1948, p. 37, and unpublished notes.

[73] The average was only 3,000 pounds (perhaps 300 fish) for the years 1939, 1940, 1943, 1944. Statistics are not readily available for 1941 and 1942.

are not larger than can be credited to such of the Maine rivers as still have runs of salmon.

It seems certain, also, that only odd salmon from the Penobscot and from the rivers farther east ordinarily disperse westward and southward beyond Casco Bay, for while the average catch for the coast of Maine east of that point has averaged about 12,000 pounds (some 1,200 fish) for the 10 most recent years of record[74] combined, the corresponding 10-year average for the whole western side of the Gulf from Cape Elizabeth to the elbow of Cape Cod was only 600 to 700 pounds, or some 60 to 80 fish at most, with more than 100 pounds reported in only 5 of the 10 years and none in 3 of the years. Further evidence of a more general kind that Gulf of Maine salmon do not scatter far as a rule is that they appear about the river mouths in spring so soon after the ice goes out that they cannot have come from any great distance.

A few do stray as far as Cape Cod Bay in most years; witness catches of one to 5 or 6 fish (10–55 pounds) in 14 out of 16 years by 8 traps, at North Truro, Cape Cod, during the period 1935 to 1950, in the months of May, June, July, September, and November.[75]

A year comes from time to time when a considerable number are taken off the coast of Massachusetts. The most recently recorded instance of this sort fell in 1937, when floating traps along the North Shore of Massachusetts Bay picked up 4,400 pounds of salmon. All of these were taken close inshore. But the 1,600 or so salmon (16,050 lb.) that were reported for Massachusetts in 1928 (the big year next previous) seemingly were farther out at sea, for all of them either hooked on long lines (10,134 lb.), or were taken in otter trawls. These must have come from as far as the Penobscot, if not from the Bay of Fundy, which is equally true of the salmon that are caught around Marthas Vineyard from time to time.[76] One, however, of about 10 pounds, reported in the North River, Marshfield, in the summer of 1938, and a few seen jumping in the Parker River (also in Massachusetts) in the summer of 1951, may have been the product of attempts to stock these streams. Occasional salmon that have been taken along the New Jersey coast and off Delaware[77] may have been the product of attempts to stock the Hudson.

Salmon, also, of 25 to 50 pounds that are sometimes caught in Minas Channel at the head of the Bay of Fundy, must come from afar, as Dr. Huntsman points out,[78] probably from the Gulf of St. Lawrence, there being no run of fish so heavy in any Bay of Fundy river or in any Maine river.

It is not astonishing that some salmon should stray far afield in Gulf of Maine waters, for marked salmon have been known to make much longer journeys, elsewhere. Thus fish marked in the southern side of the Gulf of St. Lawrence have been recaught on the north shore of the Gulf; in Newfoundland; and in the Strait of Belle Isle.[79] One marked at Bonavista on the east coast of Newfoundland was retaken 98 days later in the Margaree River, Cape Breton Island, Nova Scotia, 550 miles away[80] by the shortest possible route. One marked in Minas Channel at the head of the Bay of Fundy went out around Nova Scotia to Chedabucto Bay on the northeast, near the Gut of Canso, a journey of at least 440 miles.[81] Five, tagged in the Annapolis River system, were recaught on the east coast of Newfoundland, a minimum distance of 900 miles, while a sixth, from the same lot, was taken at Ramah on the outer coast of Labrador, more than 1,000 miles still farther away to the northward.[82] This last is the most spectacular case of wandering yet reported for any Gulf of Maine or Gulf of St. Lawrence salmon.

What is chiefly interesting about the large catches that are sometimes made off Massachusetts is their demonstration that so many fish may occasionally wander so far afield. And this applies not only to large salmon but to smolts in their first year at sea, for salmon so small that they must have run down to salt water but a few months previous have been taken in Cape Cod Bay in October.

It is not likely that these wandering salmon return at all to their home rivers; probably they

[74] 1933, 1935, 1937, 1938, 1939, 1940, 1943, 1944, 1945, and 1946.

[75] Information contributed by the Pond Village Cold Storage Co.

[76] In the spring of 1915 about 75 (including fish up to 35 lb.) were taken at Gay Head and in the neighborhood of Woods Hole

[77] Smith (Bull. U. S. Fish. Comm., vol. 14, 1895, p. 99) reports salmon seined among some mackerel off Delaware in 1893.

[78] Bull. 51, Biol. Board of Canada, 1936, p. 9.

[79] See Huntsman, Pub. Amer. Assoc. Adv. Sci. 8, 1932, p. 35, for summary of these records.

[80] Huntsman, Science, vol. 95, 1947, p. 381.

[81] Huntsman, Ann. Rept. Fish. Res. Bd. Canada, (1947) 1948, p. 37.

[82] Huntsman, Science, vol. 85, 1937, p. 314; Pub. 8, Amer. Assoc. Adv. Sci., 1939, p. 35.

are lost permanently from the breeding population. But the much greater numbers that remain localized not very far from their parent streams are believed to follow about the same routes on their return journeys that they followed when they went to sea. Thus, only a few are caught on the Nova Scotia shore between the entrance to St. Mary's Bay and Digby Gut, but fish en route to the Shubenacadie River system are taken in some numbers as they follow the shore of Annapolis and Kings Counties (the Annapolis River also yields a few salmon in its lower course, and some are taken in the Annapolis Basin). Similarly, salmon approaching the St. John River strike the coast about Point Lepreau (about 23 miles to the west) and support an important fishery from there to the mouth of the river.

A question closely bound to the movements of salmon to the sea is: what proportion of them return to spawn in the very rivers in which they were hatched? It seems demonstrated by a variety of evidence, especially by the recapture of tagged fish, that the majority do return. Huntsman, for example, reports[83] an extraordinary instance, of a kelt taken from the Sackville River on the outer coast of Nova Scotia that was tagged and released in the Shubenacadie River system at the head of the Bay of Fundy, and then found its way out of the Bay, around the Nova Scotia coast, and back again to the Sackville, where it was recaptured. We can only speculate how it directed its course, and why it did not turn in to the mouth of any of the other salmon rivers it passed en route. On the other hand, marked fish are sometimes caught in strange rivers. Fish, for instance, that were tagged in Minas Channel have been caught later in the St. John River.[84] And odd fish appear from time to time in rivers where no salmon have been hatched for many years (in the Merrimac for instance).

In short, the parent-stream theory does not always hold. Probably the truth is that while most of the fish never stray far away and do return to the home stream, wanderers that chance, in the spring, to be in the physical state leading to maturity may enter any unpolluted stream they encounter, no matter how far from home.

Dr. Huntsman's studies, carried on through many years, make it increasingly probable that the journeyings of our salmon in salt water are not the result of purposeful swimming in a definite direction, but that they tend to drift with the current as herring do (p. 97), so that the direction in which they travel depends chiefly on the depth at which they happen to be, in relation to the differential circulation of the water at different levels. If so, the St. John River fish tend to drift out with the river water as they scatter. And most of them do appear to remain more or less concentrated in the mid-depths where the principal mixing takes place between the river discharge and the water of the open Bay of Fundy, some 20 to 30 miles from St. John Harbor, living where they find an abundance of herring of various sizes as food. Here Dr. Huntsman[85] calculates the space for them is so great that no two of the approximately 50,000 fish that comprise the total yearly catch need be closer to each other than three-quarters of a mile in a layer of water 5 feet thick; so there is no crowding. But the tagging experiments have shown that the fish that go to sea from Minas Channel, where the outflow is not so definitely localized, scatter more widely, some of them drifting right around the Bay of Fundy with the anti-clockwise circulation.[86]

The situation is not so clear for the coast of Maine, partly because of the paucity of present-day information, partly because the several rivers there that once had runs of salmon are so closely spaced along the coast that it is not possible to evaluate their individual contributions to the yearly catches.

With the relationship between salmon journeys and water movements so extremely complex, all we dare say in this regard is that the inshore drift of the deeper layers (characteristic of circulation of the estuarine type) and the slackening of the offshore drift of the fresher surface water that is to be expected as the spring freshets diminish, may be the cause, at least in part, for bringing the salmon into the estuaries, and close inshore elsewhere, in spring. But the nature of the stimulus that impels a salmon to enter fresh water, and then fight his or her way upstream, remains a mystery.

It is not known whether all the salmon move inshore in spring, or only those that are destined

[83] Ann. Rept. Fish. Res. Bd. Canada (1947) 1948, p. 33.

[84] Huntsman, Ann. Rept. Fish. Res. Bd. Canada (1948) 1949, p. 40.

[85] Bulletin 21, Biol. Bd. Canada, 1931, p. 96.

[86] This was shown by Huntsman, Ann. Rep. Fish. Res. Bd. Canada (1947) 1948, p. 37.

to spawn that year, plus a certain number of immature grilse that have passed 1 year at sea. And Dr. Huntsman[87] has pointed out that the movement of the salmon riverward may be very slow; thus the salmon may take as much as a month to cover the 20-odd miles to the head of tide in the Petitcodiac River, while some of those that enter the estuary of the St. John River in autumn pass the winter there (probably in a lethargic state) before moving up to the head of tide 80 miles distant. In any case, only such fish as are approaching sexual maturity (irrespective of age), and some immature female grilse, run far up into the rivers; all the others remain in salt water, or at most they do not run above the head of tide, as has often been remarked.

The majority of the Gulf of Maine salmon become "river-mature" as it is called, long before the spawning season, for while none of them spawn before October, some of them enter fresh water as early as March and April. But the chief runs come later, varying in date, not only from river to river, but from year to year in a given river. In the Penobscot, some fish may enter in March; they are to be expected in the lower reaches after the first week in April; more come in May, perhaps two-thirds in June, with a few fish entering later still. Available information is to the effect that few enter the Narraguagus and Dennys Rivers until well into May, the chief runs there coming in June, with some entering as late as September. We have not been able to obtain definite dates for the spring and early summer runs in the St. John River. But it seems that salmon continue to enter the latter until well into the autumn, judging from catches of fish so fat that they must have come in recently from the sea. Salmon enter other streams tributary to the Bay of Fundy from May on. As a rule the large salmon come earliest, the grilse not until later, probably because it is not until later that the latter have reached the degree of fatness associated with river maturity. Accordingly, the heaviest runs in the Shubenacadie, mostly grilse (p. 130), are said to come from August until late in the autumn.

Every salmon fisherman is familiar with the fact that salmon enter in "runs" that are spaced irregularly in time, and that vary in date from year to year, depending on the height of water in the river and on the strength of the current. Freshets tend to bring them in; if the current becomes too strong they simply hold position, to breast it again as the flow slackens. The fish that are in the estuary remain there during the periods between freshets, waiting, as it were, for the message from upstream that starts them on their way. And the salmon within the river are similarly quiescent during periods of low water and weak current. This is the chief reason why salmon angling is so uncertain a sport, even in the best of rivers.

A good deal of discussion has centered about the question whether the earliest fish stay in fresh water from then until spawning time (a matter of 6 months) or whether there is more or less movement in and out of the river mouths at the beginning of the season. The latter view may be correct for the small streams, but it seems safe to say that after the run is well under way in late May or early June no fish return from fresh to salt water until autumn. Tagging experiments carried out in Canadian rivers have also yielded the very interesting information that no matter when a salmon runs upstream in one year, it may do so either early or late in the next.[88]

It is a matter of common knowledge that salmon average larger in some rivers than in others, and growth studies based on the scales have shown that these differences are due chiefly to the average ages of the fish that enter. In the St. John, as Huntsman has pointed out,[89] there are three principal groups of salmon: (a) male grilse, averaging about 6 pounds, that are mature and fated to breed that same autumn; (b) the ordinary spawners that have passed two years or more at sea; these average 10 to 15 pounds in weight and enter from May to August, the late comers running heavier than the early comers; most of them are virgins, but some of them have already spawned once or twice; (c) immature female grilse, averaging about 9 pounds, that enter from November to January. Few, however, return to spawn in the rivers of Maine until they have passed 2 years at sea; not more than 3 or 4 grilse to 70 adults were taken in the St. Croix, for example, when there still was a good run there, and not more than 1

[87] Progress Report, Atlantic stations, Biol. Bd. Canada, 8, 1933, p. 6; and unpublished notes.

[88] Fifty-fifth Annual Report of the Fisheries Branch, Department of Marine and Fisheries, Canada, (1921-22) 1922, p. 19.

[89] Nature, vol. 141, 1938, p. 421; Pub. 8, Amer. Assoc. Adv. Sci., 1939, p. 34.

grilse to 500 adults in the Dennys and Penobscot Rivers.[90]

The average weight of the salmon caught in the Penobscot was about 11½ pounds in 1905 (6,378 fish), 9 to 10 pounds in 1919 and 1920 (3,920 fish),[91] or a little less than in the St. John. The heaviest Penobscot fish of which we found definite record of late years weighed a little more than 35 pounds.[92] The fish in the rivers flowing into the head of the Bay of Fundy run much smaller, as Perley pointed out long ago, and recent studies show that most of them spawn first as grilse, i. e., after only one year at sea; a few, having spawned after one year at sea, return to spawn again a year later; and the percentage of larger and older fish is very small there. This, Huntsman points out,[93] contrasts with the prevalent 6-year-old fish in the Miramichi, which discharges into the southern side of the Gulf of St. Lawrence, and with 7- or 8-year-old fish in the Grand Cascapedia, tributary to the Bay of Chaleur. Various explanations have been advanced to account for these differences from river to river, none of them convincing in our opinion.

It also appears to be true (as often stated) that a larger proportion of the salmon are annual spawners in small streams, where most of the spent fish drop downstream again soon after spawning, than in large rivers where many of these "kelts" remain in fresh water over the winter. A plausible explanation is that kelts that return to the sea immediately after spawning have less opportunity to grow (though they recover condition sufficiently to spawn again the following summer) than such as await the spring to go downstream, and that spend a whole year at sea instead of one winter only between two successive spawnings. This, however, does not account for the fact that it is almost invariably the large rivers that yield the very large maiden fish that have spent 4 years at sea, or more.

Abundance.—The early extirpation of salmon from the Merrimac, Saco, Kennebec system, and various rivers to the eastward naturally resulted in a great decrease in the abundance of salmon in the open Gulf, clearly reflected in the catches. Data are not available for early years when all the rivers still offered free access. But the yearly catch had been reduced to about 100 to 500 fish in the St. Croix by about 1887; 200 to 1,000 each in the Dennys and in the Kennebec, and 5,000 to 15,000 in the Penobscot. The catch along the Maine coast, which had been a little more than 150,000 pounds in 1889 (more than seven-eighths of this in or about the approaches to the Penobscot), was only about 86,000 pounds in 1905 (of this 74,000 lb., or 6,378 fish from the Penobscot); was about 20,000 pounds in 1919 (13,557 lb. or 1,322 fish from the Penobscot); and was only 14,744 pounds (12,700 lb. or 1,221 fish from the Penobscot) in 1928. As 70 to 90 percent of the Maine catch comes from Penobscot River or Bay, the following table of salmon caught there in certain years from 1896 to 1928 is pertinent:

Year	Number of fish	Pounds	Year	Number of fish	Pounds
1896	6, 404	80, 225	1918	1, 653	17, 212
1898	3, 225	42, 560	1919	1, 322	13, 557
1901	6, 821	86, 055	1920	1, 598	15, 135
1903	4, 859	67, 470	1928	1, 221	12, 700
1905	6, 378	74, 158			

The Maine catch then increased again to about 88,000 pounds in 1930 and to about 70,000 in 1931, suggesting a better run in the Penobscot, and varied between 16,000 and 40,000 pounds through the period 1932–1938.[94] But the average reported catch for Maine for the period 1939 to 1947[95] was only about 3,600 pounds (maximum 9,300, minimum 600), the average Massachusetts catch for the same period only about 100 pounds (maximum 400, minimum 0). Thus the output of salmon from the rivers of Maine (none from the rivers of Massachusetts) has been only about one-fiftieth as great during the past few years as it was some 60 years ago.

The numbers of salmon have held up much better in the Canadian waters of the Gulf, thanks to wise measures of conservation such as limiting netting at the mouths of the rivers, and keeping the streams free for access by fishways at the dams. The average yearly catches, from 1870 to 1946,

[90] See Kendall (Mem. Boston Soc. Nat. Hist., vol. 9, No. 1, 1935, pp. 58–60) for age determinations of Penobscot salmon.

[91] Radcliffe. Rept. U. S. Comm. Fish. (1921) 1922, p. 146.

[92] Kendall. Mem. Boston Soc. Nat. Hist., vol. 9, No. 1, 1935, p. 32.

[93] Bull. 21, Biol. Board Canada, 1931, p. 19.

[94] No data are available for 1934 or 1936.

[95] No data for 1941.

were as follows for the west coast of Nova Scotia and for the Bay of Fundy combined:

Years	Pounds	Years	Pounds
1870–1879	655, 200	1910–1919	540, 000
1880–1889	292, 700	1920–1929	470, 300
1890–1899	634, 000	1930–1939	424, 000
1900–1909	576, 800	1940–1946	278, 000

The Canadian catch in the open Gulf and in the Bay of Fundy may be expected to run about 400,000 to 600,000 pounds at the present time, taking one year with another, or something like 40,000 to 60,000 fish, which is perhaps 100 times as great as that for the entire coastline of Maine and of Massachusetts. And the distribution of the catches shows that the St. John River contributes something like four-fifths of this, or a yearly average of some 50,000 fish,[96] contrasting with only a few hundred fish for the Penobscot in a poor year, and perhaps up to 8,000 in a good.

Salmon anglers are only too familiar with the fact that the number of fish that enter even the best of salmon rivers is much smaller in some years than in others. During the 16-year period, 1931–1946, the commercial catches reported for St. John Harbor and St. John River (best salmon river tributary to the Gulf of Maine) were good in 1931 (164,000 lbs.); in 1935 (149,300 lbs.); in 1936 (148,600 lbs.); in 1937 (172,700 lbs.); and in 1943 (157,500 lbs.); but were poor in 1939 (48,500 lbs.); in 1945 (60,000 lbs.) and in 1946 (54,500 lbs.). The yearly average for this period was 116,000 pounds.

[96] Huntsman (Bull. 21, Biol. Board Canada, 1931) has made a very interesting analysis of catches for the Bay of Fundy as a whole, as well as for the St. John, for the Chignecto system, and for the Minas system, separately.

In the Minas system the fishery produced as much as 383,800 pounds in 1907, 283,400 pounds in 1917, and 226,500 pounds in 1918; but since then, up to 1946, the best catches have been only 160,700 pounds in 1919, 165,100 pounds in 1923, and 143,300 pounds in 1925, while the poorest were 28,100 pounds in 1938 and 26,600 pounds in 1945. The average yearly catch from 1917 to 1930 was 133,000 pounds, and from 1931 to 1946, 48,000 pounds.

The reader will notice at once that the big years have not been the same for these two bodies of salmon. It seems sufficiently established that yearly and regional differences, such as these, result in the main from corresponding differences in the numbers of smolts that reach salt water in any given year. And recent investigations in Canadian waters make it likely that the factor chiefly responsible is the height of the water from summer to summer, or over periods of several summers, which of course reflects the yearly variations in rainfall. If the water is high the parr are protected from the birds that prey upon them and are more easily able to escape the trout, so that many survive to descend to the sea and to return one, two, or three years later. If the water in the river is low the parr are more at the mercy of kingfishers, mergansers, and trout, so that fewer of them live to reach salt water, and there are fewer of them to return as grilse or as older fish.

Humpback salmon *Oncorhynchus gorbuscha* (Walbaum) 1792

Jordan and Evermann, 1896–1900, p. 478.

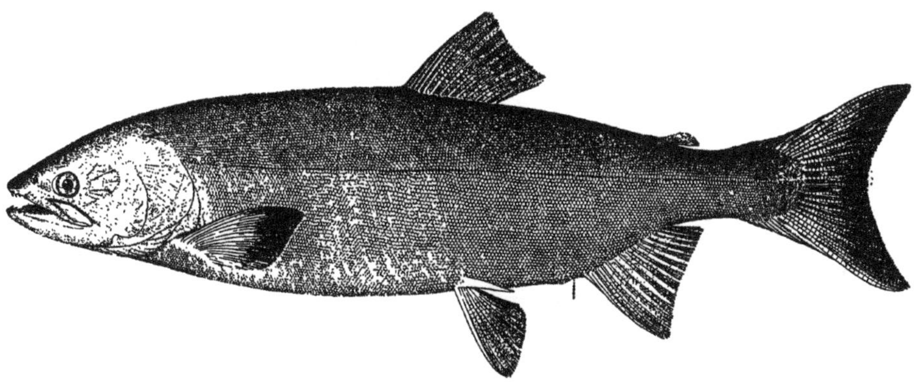

FIGURE 55.—Humpback salmon (*Oncorhynchus gorbuscha*).

Description.—The humpback is of the familiar salmon outline while living in the sea, the body being deeper than thick, with rounded belly. The head is naked but the body is covered with scales large enough to be seen easily. The dorsal fin stands about midway of the body above the ventrals, and the flaplike adipose fin is over the rear end of the anal fin. It agrees so closely with the Atlantic salmon in all this that the one might easily be taken for the other, were it not that the anal fin of the humpback invariably has 14 rays or more, whereas that of the Atlantic salmon has only about 9 rays. Also, the humpback is a stouter-bodied fish than the Atlantic salmon. The male humpback (like all the Pacific salmons, and the Atlantic salmon to a lesser degree) undergoes a very noticeable change in form in the spawning season, with the body deepening and developing a prominent hump in front of the dorsal fin; the jaws elongating and becoming hooked at the tip and the teeth increasing in size.

Color.—The back and tail of the humpback are bottle green with poorly defined black spots, while it is in the sea. These spots are particularly conspicuous on the tail, where they are oval in outline and as much as a third of an inch in longest diameter. These spots are one of the distinctive marks, whereby the humpback can be distinguished from all other salmons. The sides and belly are silvery, with a faint pinkish tinge. Young humpbacks are unique among salmon in being of practically adult coloration without "parr" marks (p. 122).

Size.—The humpback is the smallest of the Pacific salmons and much smaller than the Atlantic salmon, adults averaging only about 5½ pounds in weight and 20 to 25 inches in length. Males weigh to about 11 pounds and females to about 7½ pounds.

General range.—Pacific coast of North America and of northern Asia, from Oregon northward on the American side. This is the most abundant salmon in Alaska. It runs up fresh rivers to spawn, which it does but once and then dies. It has been introduced in the rivers of Maine.

Occurrence in the Gulf of Maine.—The history of the introduction of this west coast salmon to New England waters is as follows:

Humpback salmon eggs seem first to have been planted in Maine rivers in 1906. In the autumn of 1913 a large consignment of humpback eggs was shipped to the Craig Brook and Green Lake (Maine) hatcheries, and the approximately 7,000,000 fingerlings hatched therefrom were distributed in the Penobscot, Androscoggin, Damariscotta, Dennys, Pleasant, Union, Medomak, Georges, and St. Croix Rivers. A year later some 5,000,000 young fish were liberated. A third plant was made in 1915; a fourth of 6,235,808 fingerlings in 1916; and a fifth of about 1,000,000 in the Dennys and Pembroke Rivers in 1917.[97]

The results of this attempt at acclimatization were first seen in the summer and fall of 1915 when fishermen reported large numbers of mature humpbacks along the Maine coast, and when humpbacks ran in the Dennys River (where many were caught) from August 15 until September 24, some probably spawning there, for the bodies of spent fish were seen drifting downstream. Humpbacks again entered the Pembroke and Dennys Rivers during August, September, and October of 1917 with a few reported from the Penobscot, St. Georges, Medomak, and St. Croix, the result of the plant of 1915. And at least 2,000 mature fish were seen that summer in the Dennys alone, where many were caught averaging about 5 pounds, and one as heavy as 10 pounds 9 ounces. Definite information is lacking for 1918. But even larger numbers entered the Dennys and Pembroke Rivers in the autumn of 1919 than in 1917, with smaller runs in the Penobscot, Machias, St. Croix, and Medomak Rivers. Enough spawned that year in the Dennys and Pembroke Rivers for the fish-culturists of the Bureau of Fisheries to artificially fertilize half a million eggs there. And humpbacks were caught in the weirs in Passamoquoddy and Cobscook Bays during that season.

Adult fish were taken again in the weirs in 1920,[98] and one fish was caught in a weir as far from its native river as Lanesville, Mass. (near Cape Ann)[99] at some time during the summer of 1921.

Large numbers of eggs were collected again from wild fish between 1922 and 1926, the resultant fry being returned to the Dennys and other rivers nearby. Artificial propagation was abandoned then, for it seemed that the species was estab-

[97] More detailed accounts of these and successive plantings will be found in the annual reports of the U. S. Commissioner of Fisheries for the years 1914 to 1928.

[98] Reported catch, Washington County, Maine, 1920, 310 pounds.

[99] It was forwarded to the Massachusetts Commissioners as reported by C. E. Grant of Gloucester.

lished.[1] But natural reproduction seems not to have been successful enough for the humpback to maintain itself in the few Maine rivers open to it, much less to increase in numbers, for very few have been reported since about 1926 or 1927, and none that we have heard of for some years past.

Silver salmon *Oncorhynchus kisutch* (Walbaum) 1792

COHO SALMON

Jordan and Evermann, 1896–1900, p. 480.

Description.—The silver salmon resembles a rather stout Atlantic salmon closely in its general shape, also in the relative size and position of its fins, and in their shapes. But a safe morphological criterion for distinguishing the one from the other is that the silver always has at least 12 rays in its anal fin, and some of them have as many as 17, whereas most of the Atlantic salmons have only 8 or 9 anal rays, and never more than 10. The color is a help also, in this connection, for while a silver is silvery down its sides, like an Atlantic salmon, it is more closely sprinkled with small black spots along its back and on the upper part of its tail fin than is an Atlantic salmon. These spots, too, are always roundish or oval in a silver, never in the form of crosses. On the other hand, the black spots are much smaller and much less conspicuous on a silver salmon than on a humpback, and the lower half of the tail fin, which is as conspicuously spotted as the upper half on a humpback, usually has no spots on a silver salmon.

Size.—Up to 3 feet in length.

General range, habits, and occurrence in the Gulf of Maine.—The native range of the silver salmon is from northern California to northwestern Alaska, where it is an important food fish, and where anglers take many of them, both by trolling and by fly fishing. Like other Pacific salmons, it runs up into fresh streams to spawn, dying thereafter. Most of the young remain about one year in fresh water, but a few do not move out to sea until they are in their third year. Most of them return to fresh water at the end of the third summer at sea, a few, however, by the end of the second summer in salt water, a few others not until the fourth summer.

Our only reason for mentioning the silver salmon is that a plant of its fry and fingerlings that was made in the Duck Trap stream, tributary to the western side of Penobscot Bay, near Lincolnville, Maine, resulted in the return of 150 mature fish to Duck Trap stream in 1944, and perhaps of more of them. But nothing more was heard of them thereafter, and no returns have been reported up to this writing (Nov. 1, 1951) from other plants that were made in Maine waters[2] in 1948.

[1] Rept. of U. S. Comm. Fish. (1928), Pt. 1, 1929, p. 379.

THE SMELTS. FAMILY OSMERIDAE

The smelts are small salmons in all essential respects, except that their stomach has few pyloric caecae, or none, whereas there are large numbers of such caecae in their larger relatives of the salmon family. However, it is not necessary to look so deeply to learn whether a fish be smelt or very young salmon, for the former all have pointed noses and are very slender, whereas the young of our four salt-water salmons—humpback, silver Atlantic, and sea trout—are much stouter, with blunt noses. In most cases, too, the shape of the tail would suffice of itself to separate smelt from salmon smolt, for it is never as deeply forked in the latter as in the smelts.

Two smelt fishes occur in the Gulf of Maine: the smelt (very common), and the capelin (a sporadic visitor from the north). The argentine (p. 139) is so closely related to the smelts that it is included in the following key.

[2] In Tunk stream, Duck Trap stream, Chandler River, and Bald Hill Cove Brook.

KEY TO GULF OF MAINE SMELTS AND ARGENTINES

1. The dorsal fin originates over the tips of the pectorals; the mouth is very small .... Argentine, p. 139
   The dorsal fin situated far behind the pectorals; the mouth is large 2
2. Upper jaw almost as long as lower; teeth large; there is a group of strong fangs on the tongue; the pectoral fins have 12 rays or fewer .... Smelt, p. 135
   Lower jaw much longer than upper; teeth so small as hardly to be visible; no fangs on tongue; the pectoral fins have 15 to 20 rays .... Capelin, p. 134

Capelin *Mallotus villosus* (Müller) 1777

Jordan and Evermann, 1896–1900, p. 520.

Description.—The capelin is an even slenderer fish than the smelt, its body being only about one-sixth to one-seventh as deep and about one-twelfth as thick as it is long, and of nearly uniform depth from gill cover to anal fin (except in the case of females when their abdomens are distended with spawn), whereas the smelt is usually deepest about its mid-length (at least if the fish is fat), which gives the two species characteristically different aspects. The head of the capelin is pointed like that of the smelt, the mouth gaping back to below the center of the very large eye with the tip of the lower jaw projecting noticeably beyond the upper. The scales are minute, much smaller than those of the smelt and more numerous (about 200 per row on the sides of the body); the teeth so small as to be hardly visible to the naked eye, and the tongue fangs, so characteristic of the smelt (p. 135), are lacking here. The outline of the adipose fin likewise helps separate capelin from smelt, for it is low in the former and about half as long as the anal, but short and high in the latter. The pectoral of the capelin is broader also, usually with 15 or more rays.

The capelin exhibits a pronounced sexual dimorphism; the male has much the longer pectoral fins; and the base of his anal is elevated on a pronounced hump, whereas it follows the general outline of the belly in the female. In males, too, the scales in one of the longitudinal rows immediately above the lateral line, and in another row along each side of the belly, are pointed, distinctly larger than the other scales, and become longer still at spawning time when each pushes up the skin as a finger-like process; they form four ridges that are very evident when the fish is held in the hand.

Color.—The capelin is transparent olive to bottle green above, like a smelt, but its sides are uniformly silvery below the lateral line and the scales are dotted at the margins with minute dusky specks (in the smelt there is a distinct silvery band on each side); the belly is white. Back and head darken at spawning time.

Size.—Few capelin are more than 6½ to 7½ inches long.

Habits.[3]—Capelin are most in evidence during the spawning season, when they come inshore in multitudes along arctic-subarctic coasts. They spawn on gravel or pebbly bottom, chiefly close below tideline, many of them in the wash of the waves in the beach; many are stranded then on the beach between waves. But eggs have also been reported from as deep as 35 to 40 fathoms. Each female while spawning is accompanied by two males that crowd her between them; but she may have only one companion.[4] Spawning takes place chiefly at temperatures of 43° to 50° F. (6°–10° C.) and more actively by night than by day.

The eggs are reddish, about 1/25-inch (1 mm.) in diameter, and so sticky that they cling to each other like herring eggs, and to the gravel and pebbles with which they are intermingled by the swash of the waves. They hatch in about 15 days at a temperature of 50° F. (10° C.). And they will tolerate a salinity as low as 7 per mille,

[3] Interesting accounts of the habits of the capelin and of its rate of growth in Newfoundland waters have been given recently by Jeffers (Ann. Rept. Biol. Board Canada (1930), 1931, pp. 7-18); by Sleggs (Rept. Newfoundland Fish. Res. Comm., 1, No. 3, 1933); and by Templeman (Bull. Newfoundland Government Lab., 17 (Research), 1948.

[4] According to Lanman, Rept. U. S. Comm. Fish. (1872-1873) 1874, p. 225.

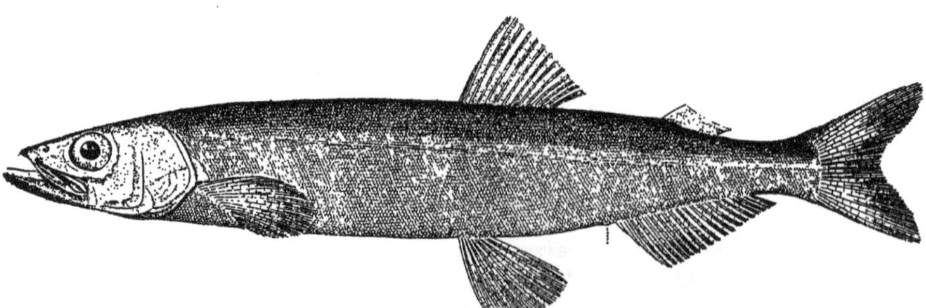

FIGURE 56.—Capelin (*Mallotus villosus*), Grosswater Bay specimen. From Goode. Drawing by H. L. Todd.

as Dr. Jeffers writes us. The larvae, described as 5 to 7 mm. long at hatching, are very slender and resemble those of smelt, herring, and launce so closely that identification is a matter for the expert. In any case, capelin are encountered so seldom in our Gulf that their larvae are not apt to be seen there.[5]

Along the coasts of Newfoundland, capelin spawn chiefly in June and July, and we have found them doing so in multitudes along the outer Labrador coast in July. Probably any spawning that may take place within the limits of our Gulf would fall in May at latest, to judge from water temperatures.

The capelin so seldom appears in the Gulf of Maine that we need only add that it is a fish of the high seas frequently encountered far out from land, coming inshore only to spawn and then as a rule moving out again; that it travels in vast schools at spawning time (when it often strands on the beach in countless multitudes). It is the chief bait fish of Arctic seas, preyed upon by whales and by every predaceous fish, particularly by cod, which are often seen pursuing the capelin at the surface in northern waters. Capelin themselves feed chiefly on small crustaceans, particularly on copepods, on euphausiid shrimps, and on amphipods. It is also known to devour its own eggs. We can bear witness that the capelin is a delicious little fish on the table.

General range.—Boreal-Arctic seas, south to the coast of Maine[6] on the Atlantic coast of America.

Occurence in the Gulf of Maine.—The capelin is a sub-Arctic fish that visits the Gulf of Maine occasionally; chiefly the eastern side as might be expected since it comes from the north.

Dr. Hunstman writes:[7]

> In the Gulf of St. Lawrence it occurs abundantly in limited areas, which shift somewhat from year to year. It occurs periodically in similar limited areas farther south. The southeastern corner of Cape Breton is the center of such an area, where large quantities were taken in 1917. Halifax is the center of another area, where, however, it is more rare. In 1916 it was abundant at Sambro, near Halifax. The next area is in the Bay of Fundy, where they have, exceptionally, been taken in large quantities at long intervals.

[5] Templeman (Bull. Newfoundland Government Lab., 17, Res., 1948, figs. 18–20) gives a series of excellent illustrations of capelin larvae at different stages of growth.

[6] According to Jordan and Evermann the capelin finds its southern limit at Cape Cod, but we find no actual records of its occurrence farther south than is mentioned.

[7] Quoted from a letter.

Apparently a period of this sort occurred about the middle of the past century, for Perley, writing in 1852, reported it from a number of points in the neighborhood of St. John, New Brunswick. It seems then to have disappeared from the Gulf of Maine, not to reappear until 1903 when it was common in the Bay of Fundy in May. A few were again taken off Passamaquoddy Bay in that same month of 1915.[8] And this was the prelude to a period of local abundance, for capelin were noticed among the herring in the weirs of the Passamaquoddy Bay region in October 1916, becoming so plentiful by the end of November that one catch of 3,000 pounds of fish consisted of 2,000 pounds of capelin and only 1,000 of herring. They were also reported at various localities along the New Brunswick coast at that time. Probably they persisted locally in the Bay of Fundy throughout the winter of 1916–1917, for large numbers of capelin appeared in Minas Basin in the following May and June. We find no record of capelin within the limits of the bay in 1918, but they were taken again in 1919 in 50 fathoms of water off Passamaquoddy Bay in January, February, and March, and they appeared with smelts a month later as far west as the Penobscot River, penetrating far inland. None, however, have been seen in the Gulf of Maine since then, so far as we have been able to learn.

Smelt *Osmerus mordax* (Mitchill) 1815

SALT-WATER SMELT

Jordan and Evermann, 1896–1900, p. 523.

Description.—The smelt is distinguishable from all other fish common in our waters by its slender form, combined with a long pointed head, large mouth, a small but evident adipose fin standing above the rear part of the anal, and a deeply forked tail. The location of its dorsal fin above the ventrals instead of in front of them, and its much larger mouth and small eye separate it from the argentine. The large, fang-like teeth on the smelt's tongue, its larger scales (of which there are about 75 along each row on the sides, all alike in the two sexes), its shorter adipose fin, its narrower pectoral fins, that its lower jaw projects only slightly beyond the upper and its scales slip off very easily, obviate any danger of confusing

[8] Huntsman (Contrib. Canadian Biol., (1921) 1922, p. 50) and Kendall (Copeia, No. 42, 1917, pp. 28–30; and Copeia, No. 73, 1919, pp. 70–71) give details.

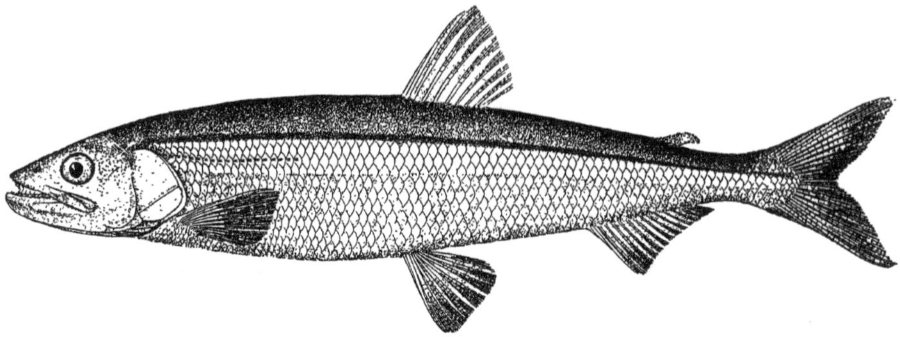

FIGURE 57.—Smelt (*Osmerus mordax*), adult, Woods Hole. From Jordan and Evermann. Drawing by H. L. Todd.

it with the capelin. The body of the smelt is only about one-fifth as deep as long (exclusive of caudal fin), with broadly rounded back but flattened enough sidewise to be egg-shaped in cross section. It is deepest about its mid-length, tapering thence toward the head and toward the tail (at least in fat fish), whereas the capelin is of nearly uniform depth from gill opening to anal fin (p. 134). Its mouth gapes back of the eye.

Printed accounts of the smelt usually credit it with a peculiar "cucumber" odor, and smelt fishermen often speak of a trace of this, but it is so faint that we have never noticed it though we have caught and handled many.[9]

Color.—Transparent olive to bottle green above, the sides are of paler cast of the same hue but each with a broad longitudinal silvery band. The belly is silvery, while the fins and body are more or less flecked with tiny dusky dots. This color pattern is shared by another slender little fish, the silverside (p. 302), but the latter has two large dorsal fins, so there is no danger of confusing the smelt with it.

Size.—Smelt grow to a maximum length of about 13 or 14 inches. Few, however, are more than a foot long, and adults run only about 7 to 9 inches. Smelt weigh from 1 to 6 ounces according to size and fatness.

Habits.[10]— The smelt is an inshore fish, confined to so narrow a zone along the coast that none has ever been reported more than a mile or so out from the land, or more than two or three fathoms in depth, while many spend the whole year in estuarine situations.

[9] The European smelt (*O. eperlanus*) smells so strong that it is not held in very high esteem as a food-fish.

[10] Kendall (Bull. U. S. Bur. Fish., vol. 42, 1927, p. 244) has given a detailed account of the habits, distribution, and catches of the smelt of the New England coast, also of the landlocked populations.

Young smelts certainly, and old ones probably, travel in schools, which are mostly composed of fish of a size, hence probably are the product of one year's hatching, and they live pelagic, not on the bottom, though confined to shoal water.

Most authorities describe the smelt as feeding on small crustaceans, which is correct as far as it goes, for shrimp (both decapods and mysids) and gammarids are probably its favorite food, and shrimp were long considered the best smelt bait. But it has been found that pieces of "sea worms" (*Nereis*) are more attractive to the larger smelt (at least in Massachusetts Bay). Small fish also form an important item in the diet. We have, for example, found smelts taken in the Sheepscot River in May packed full of young herring, and have caught many with small mummichogs (*Fundulus*) as bait; while cunners, anchovies, launce, sticklebacks, silversides, and alewives have been identified from smelt stomachs at Woods Hole. The Woods Hole diet list also includes shellfish, squid, annelid worms (*Nereis*), and crabs, but even as greedy a fish as the smelt ceases to feed during its spawning visits to fresh water. Young smelt depend chiefly on copepods and on other minute pelagic crustaceans. Smelt fishermen are familiar with the fact that a smelt approaches a bait slowly, then stops, and appears to suck it in.[11] If the smelt take their living prey in this same way, it is somewhat of a mystery how they succeed in capturing animals as active as shrimps and small fish.

Smelt, like alewives, shad, and salmon, make their growth in salt water, but run up into fresh water to spawn.

The summer habitat of the smelt varies off

[11] This method of feeding seems first to have been described in print by "Grif" (Forest and Stream, vol. 54, No. 8, Feb. 24, 1900, p. 151).

different parts of the coast of the Gulf, depending on the summer temperature of the water and perhaps on the food supply. Most of them desert the harbors and estuaries of the Massachusetts Bay region and along the southern coast of Maine during the warmest season. But it is probable that they move out only far enough to find cooler water at a slightly greater depth, and a few may be found in harbors through the summer. Smelt, for instance, are caught then in Cohasset Harbor in some years, but not in others; and east of Penobscot Bay, where the surface temperature does not rise so high as off Massachusetts, smelt are to be found in the harbors, bays, and river mouths all summer, and are sometimes taken in numbers then in the weirs.[12]

Adult smelt gather in harbors and brackish estuaries early in autumn, where smelt fishing with hook and line is in full swing by October. The schools then tend to move into the smaller harbors on the flood tide, and out again on the ebb, especially if the tidal current is strong, as it is in Cohasset, a locality with which we are familiar. But some smelt remain over the ebb in the deeper basins. And some of them have run as far as the head of tide by the time the first ice forms in December. Most of them winter between the harbor mouths and the brackish water farther up; the maturing fish commence their spawning migration into fresh water as early in the spring as the ice goes out of the streams and the water warms to the required degree.

Temperature observations by the Massachusetts Commission show that the first smelt appear on the spawning beds in Weir River, a stream emptying into Boston Harbor, when the temperature of the water rises to about 40–42°.[13] This may take place as early as the first week in March or as late as the last, about Massachusetts Bay, depending on the forwardness of the season and on the particular stream. The chief production of eggs takes place in temperatures of 50–57°, and spawning is completed in Massachusetts waters by about the 10th or 15th of May, year in and year out. East of Portland, smelt seldom commence to run before April, and continue through May. In the colder streams on the southern shores of the Gulf of St. Lawrence they do not spawn until June. On the other hand, they may commence spawning as early as February along the southern New England coast west of Cape Cod.

As a rule smelt do not journey far upstream; many, indeed, go only a few hundred yards above tidewater, whether the stream be small or large. Thus Dr. Huntsman informs us that the smelt that enter the estuary of the Stewiacke River, Nova Scotia (a tributary of the lower Shubenacadie, near the head of the Bay of Fundy) spawn only in the tidal part. And some spawn in slightly brackish water in certain ponds back of barrier beaches (e. g., Straits Pond, Cohasset, Mass.). But flooding with salt water, which sometimes happens, kills the eggs.

The adult smelts return to salt water immediately after spawning to spend the summer either in the estuary into which the stream in which they spawn empties or in the sea close by. On the Massachusetts coast north of Cape Cod all the spent fish have left fresh water by the middle of May, earlier in some years. On the Maine coast, too, a good proportion of the spent fish are in salt water by the first weeks in May; thus we have seen a bushel of large smelt taken in a weir at Cutler (near the mouth of the Grand Manan Channel) as early as May 4.

The eggs average about 1.2 mm. (0.05-inch) in diameter and they sink to the bottom, where they stick in clusters to pebbles, to each other, or to any stick, root, grass, or water weed they chance to touch. According to the Manual of Fish Culture a female weighing as little as 2 ounces will produce between 40,000 and 50,000 eggs;[14] The eggs of the closely allied European smelt (*Osmerus eperlanus*) hatch in 8 to 27 days, according to temperature, and the incubation period of the American fish is the same, probably, for smelt eggs are reported as hatching in 13 days at the Palmer (Mass.) hatchery.

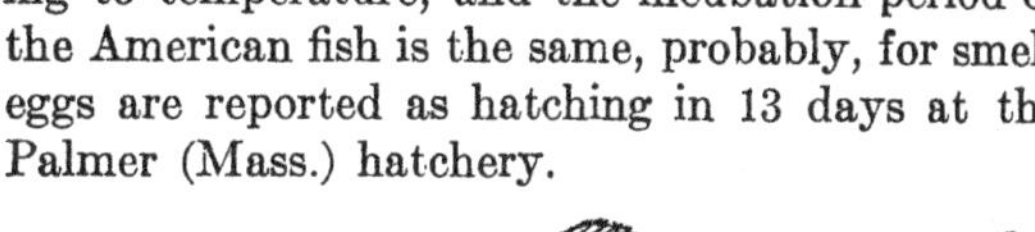

FIGURE 58.—Smelt larva, 26 mm.

The smelt has proved a favorable fish for artificial hatching and large numbers of fry are so produced yearly in Massachusetts, the eggs being

[12] Atkins (Fish. Ind. U. S., sect. 5, vol. 1, 1887, pp. 690-693) gives much information on the smelt in Maine.

[13] Kendall (Bull. U. S. Bur. Fish., vol. 42, 1927, pp. 231-233) summarizes these observations and gives additional information for streams on the coast of Maine.

[14] Rept. U. S. Fish Comm., 1897, p. 188.

taken in Weir River, just mentioned, and it has proved possible to re-establish smelt by introducing the eggs or fry into streams from which it has been extirpated. For example, good smelt fishing was reported in "Poorhouse Brook," Saugus, a tributary of Boston Harbor, three years after the stream was stocked with eggs, and attempts have been similarly successful on Long Island, N. Y. Maintenance of the stock is a question either of providing accessible spawning grounds of sufficient extent, or of making up for lack of such by artificial propagation.

The precise season when young smelt go down to the sea in the Gulf of Maine streams is yet to be learned; probably early in summer. We seined several hundred fry, 1¼ to 1⅜ inches long, October 1, 1924, on a beach of Mount Desert Island, evidence that the rate of growth is about the same for our smelt during its first summer and autumn as for the European, i. e., to a length of 1¾ to 2¾ inches.

Most of the smelt evidently do not spawn until they have passed an autumn, a winter, a summer, and a second winter in salt water.

General range.—East coast of North America from eastern Labrador, Strait of Belle Isle, and the Gulf of St. Lawrence southward regularly to New Jersey, and reported to Virginia; running up streams and rivers to spawn. Smelt, also, are landlocked naturally in many lakes and ponds in New Hampshire and in Maine, also in Lake Champlain, and in various Canadian lakes.[15]

Occurrence in the Gulf of Maine.—The smelt is a familiar little fish around the entire coast of the Gulf of Maine, but varies greatly in abundance from place to place according to the accessibility of streams suitable for spawning, from which it seldom wanders far alongshore. Smelt are plentiful, still, all around the inner parts of Massachusetts Bay and its tributary harbors, though many of the local streams are barred to them now; thence northward and eastward all along the coast of Maine; tolerably so in the region of Passamaquoddy Bay (catch for Charlotte County, New Brunswick, 7,400 pounds in 1945), and more so along the western shore of Nova Scotia (60,100 pounds for Yarmouth County in 1945). But they are less plentiful passing inward along the Nova Scotia shore of the Bay of Fundy, as illustrated by catches in 1945 of 20,100 pounds for Digby County, but only 7,600 pounds for Kings County, 2,000 pounds for Hants, and 1,800 for Colchester (covering the Minas Basin region). So few smelt exist along the New Brunswick side of the Bay, inward from the Passamaquoddy region, that none at all were reported for that stretch of coast in any year during the period 1939–1945. Doubtless this scarcity up the Bay is "due to absence of streams suitable for spawning, and the general turbidity of the water," as Jeffers has remarked.[16]

Abundance.—Smelt once were so plentiful in the Back Bay at Boston (now mostly filled in) that "distinguished merchants of lower Beacon Street might be seen, at early hours, eagerly catching their breakfast from their back doors."[17] Those happy days, however, are long since past, and smelt certainly are not so numerous as they were even 50 years ago,[18] around the Massachusetts shoreline of our Gulf, where various streams either have been closed to them, or have been rendered uninhabitable by pollution. But enough still remain to provide sport for thousands of anglers,[19] and we still hear of an occasional catch there of many dozens by some one lucky enough to hit a run of fish at the right time and tide.

In 1938, when a special effort seems to have been made to gather smelt statistics, the reported catch for the inner part of Massachusetts Bay and northward to the New Hampshire line was 25,900 pounds, or some 300,000 fish, if they ran about a dozen to the pound. The yearly catch reported for the coast of Maine, added to that of the Passamaquoddy area (which form one faunal unit so far as the smelt is concerned) averaged about 644,000 pounds during the period 1937 to 1946,[20] or perhaps some 8,000,000 fish; about 61,000 pounds for Digby and Yarmouth Counties, Nova Scotia, combined, which covers most of the catch for the Gulf, north and east of New Hampshire.

The catches of smelt that are made along the coasts of Maine, New Brunswick, and Nova Scotia may seem impressive if taken by them-

[15] The European smelt is landlocked in many lakes in northern Europe.

[16] Ann. Rept. Biol. Board Canada, (1931) 1932, p. 27.

[17] Mass. Rept. for 1870, p. 23.

[18] Kendall (Bull., U. S. Bur. Fish., vol. 42, 1927, pp. 244–249) gives many interesting details as to catches in Massachusetts.

[19] Smelt fishing has long been restricted to hook and line along this part of the coast.

[20] Maximum 675,700 pounds in 1945, minimum 316,400 pounds in 1939. No data are available for Maine for the years 1941 or 1942.

selves. But Miramichi Bay, alone, on the southern shore of the Gulf of St. Lawrence yields yearly between three and four times as much smelt as does the entire coastline of the Gulf of Maine.[21]

Catch records do not suggest any striking alteration in the abundance of smelts during the past 10 years or so for Maine or for the Canadian shores of the Gulf. But they seem to have been somewhat more plentiful along the Maine coast previous to the early 1900's, for catches of 1,125,268 to 1,279,550 pounds there in 1887, 1888, and 1902 have not been equaled since then, the nearest approach being 968,300 pounds in 1945.

We are often asked what effect the disappearance of the eel grass (*Zostera*) from our coasts has had on the abundance of the smelt. Unfortunately, the statistics of the yearly catch do not yield any clear answer. Neither can we offer any convincing explanation for the violent fluctuations that take place from year to year in the abundance (or availability?) of smelts at one point or another. Fishermen report, for example, that they were far less plentiful in Massachusetts Bay and in the Great Bay region, N. H., in 1950 than they were in either of the two previous years.

[21] Average reported catch for Northumberland County, 1937-1946, was 2,258,030 pounds.

The smelt also has a great recreational value, smelt fishing being a favorite pastime for home consumption. As many as 2,326 people, for instance, have been counted fishing at one time about Houghs Neck in Boston Harbor, and this same sort of thing is to be seen up and down the Massachusetts coast in harbors and stream mouths in autumn. Many smelt are caught in Great Bay, N. H., in good years, through the ice for the most part. And this applies equally to many localities along the coast of Maine. So plentiful are the fish on occasion and so greedily do they bite, especially on the flood tide, that it is usual to number the catch about Massachusetts Bay by the dozens rather than by the individual fish. Sea worms (*Nereis*) are generally considered the best bait, especially for the larger smelt, shrimp the second best, small minnows or clams a poor third. Smelt have also been taken with a small red artificial fly in the Gulf of St. Lawrence, and perhaps elsewhere.

THE ARGENTINES

FAMILY ARGENTINIDAE

The argentines resemble the smelts in most of their external characters. But their mouths are much smaller, with the upper jawbone reaching back only about even with the front of the eye, and the entire base of their rayed dorsal fin is in front of the ventral fins.

Argentine *Argentina silus* Ascanius 1763
HERRING SMELT

Jordan and Evermann, 1896–1900, p. 526.

Description.—The argentine has a pointed nose, deeply forked tail, and slender, compressed body, but it has much larger eyes than either smelt or capelin, a character no doubt associated with its deep-water home; its mouth is much smaller, not gaping back even as far as the eye; and its dorsal fin stands wholly in front of the ventrals, instead of above them as it does in both the smelt and the capelin.[22]

[22] The anatomy of *Argentina silus* is described, and records along the American coast are given by Kendall and Crawford (Jour. Washington Acad. Sci., vol. 12, No. 1, January 1922, pp. 8-19).

The body of the argentine (about one-fifth as deep as long) tapers toward both head and tail, but its sides are so flat, and its back and belly so broad, that it is nearly rectangular in cross section instead of oval. Its scales, too, are larger than those of the smelt, there being only 60 to 70 rows along the lateral line. Its adipose fin is very small and its jaws are toothless, though its palate and tongue are armed with small teeth.

Color.—The color of the adult is variously described by different authors. All agree, however, that the back is brownish or olivaceous, the sides silvery or with iridescent golden or brassy luster, and the belly white. The adipose fin is yellowish.

Size.—The argentine is a larger fish than the smelt or the capelin, growing to a length of about 18 inches.

Habits.—Nothing is known of the life of the herring smelt in our Gulf, and little enough is known of it in Scandinavian waters, where it is sometimes caught on deep set-lines baited with herring or mussels, and where it is occasionally

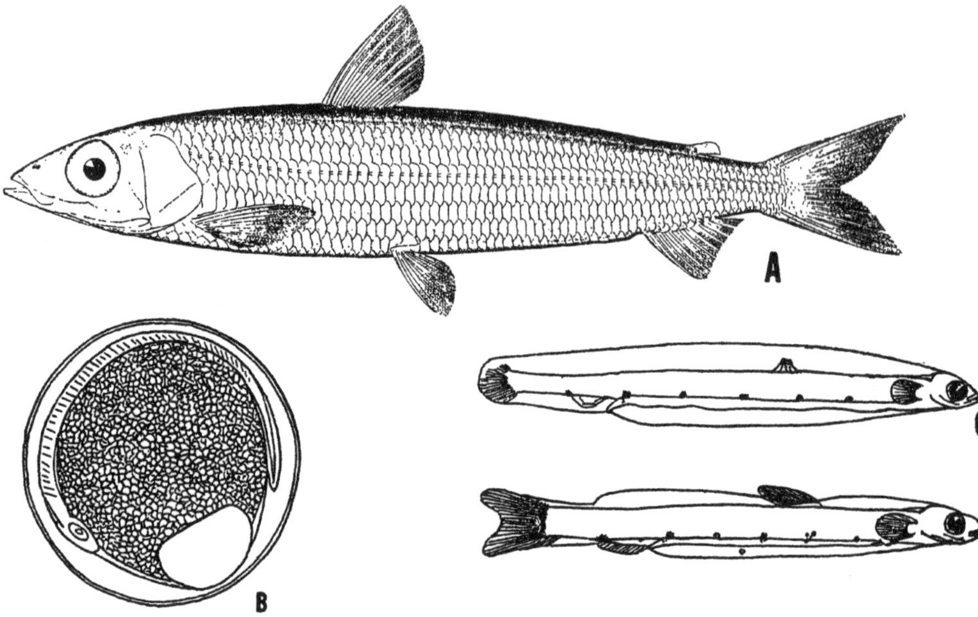

FIGURE 59.—Argentine (*Argentina silus*). A, adult, Biddeford Pool, Maine; from Goode and Bean, drawing by H. L. Todd; B, egg; C, larva, 28 mm.; D, larva, 45 mm. B–D, European, after Schmidt.

swept up to the surface like other deep-sea fishes by some upwelling of the water, to drift there helplessly. Its eggs float chiefly in the deeper water layers, seldom rising to the surface, and they are among the largest of buoyant fish eggs (3 to 3.5 mm. in diameter), with flat oil globule (0.95 to 1.16 mm.) and vacuolated yolk. Newly hatched larvae are about 7.5 mm. long and have a large yolk sac, but this has been absorbed when they have grown to a length of 12 mm. and a line of spots has appeared along the belly. The fin rays are formed by the time the little fish has reached 45 mm., the anus has moved forward, and the forked outline of the tail is apparent, but the ventral fins do not appear until the larva is about 50 mm. long.

General range.—North Atlantic, usually in water as deep as 80 to 300 fathoms; known from northern Norway south to the northern part of the North Sea on the European side, from the Nova Scotia Banks to the offing of southern New England on the American side.[23]

Occurrence in the Gulf of Maine.—The argentine was considered rare in our waters until recently. Some specimens have been brought in from widely scattered localities around the coast, namely, Belfast, Biddeford Pool, and Fletchers Neck, Maine; and from Hampton Beach, N. H. It has proved, with the development of otter trawling, that argentines are fairly common all around the edges of Georges Bank and off Cape Cod in moderately deep water. It is not unusual for one haul of the trawl to bring in from one to a dozen from depths of 30 to 100 fathoms, with much larger numbers taken occasionally; one vessel, for example, trawled 15,000 pounds on the northeastern edge of Georges Bank in about 100 fathoms during a week in mid-September 1929. Evidently there are at least a few argentines in the deep trough of the Gulf also. Firth[24] reports that ten were taken at 90 fathoms on the northwestern slope of Georges Bank on June 18; and the *Albatross II* trawled one at 115 fathoms off Mount Desert Rock. They spawn to some extent in the Gulf, for on April 17, 1920, a townet haul on the *Albatross I* from 109 fathoms in the southeastern part of the Gulf basin yielded 43 eggs, unmistakably of argentine parentage, while we have taken a scattering of argentine fry at localities as widely separated as the offing of Mount Desert Rock and the northwestern edge of Browns Bank.

[23] For recent records of argentines off Nova Scotia, see McKenzie and Homans, Proc. Nova Scotia Inst. Sci., vol. 19, No. 3, 1938, p. 277 and McKenzie, Proc. Nova Scotia Inst. Sci., 20, 1939, p. 15.

[24] Firth, Bull. Boston Soc. Nat. Hist., 61, 1940, p. 10.

LUMINESCENT FISHES

FAMILIES MYCTOPHIDAE, MAUROLICIDAE, CHAULIODONTIDAE, GONOSTOMIDAE, STOMIATIDAE, AND STERNOPTYCHIDAE

These families include a heterogeneous assemblage of small oceanic fishes, that are primitive in some respects, but are highly specialized in others for existence in mid-depths, on the high seas. They all have light-producing organs, which no other Gulf of Maine fish has; this is the only reason why we group them together here.

KEY TO GULF OF MAINE LUMINESCENT FISHES

1. Trunk at least ½ as deep as it is long from tip of snout to base of tail fin; front part of rayed dorsal fin is a hard triangular plate, supported by 7 or 8 spines ---- Hatchet fish, p. 149
   Trunk less than ¼ as deep as it is long from tip of snout to base of tail fin; rayed dorsal fin does not commence with a hard plate or hard spines ---- 2
2. Mouth does not gape back as far as the eye ---- Pearl sides, p. 144
   Mouth gapes back beyond the eye ---- 3
3. No barbel on the chin; the ventral fins are about mid-way of the trunk; origin of rayed dorsal fin either in front of mid-length of trunk or at least not much behind it ---- 4
   There is a long fleshy barbel on the chin; the ventral fins are considerably behind the mid-length of the trunk; the rayed dorsal fin is far back, close to the tail fin ---- 7
4. The rayed dorsal fin is far in advance of the ventrals; the jaws are armed with long and conspicuous fangs ---- Viperfish, p. 145
   The rayed dorsal fin is about over the ventral fins (it may be a little in front of them or a little behind); the teeth are small ---- 5
5. Eyes very small; no adipose fin behind the rayed dorsal fin; anal fin reaches nearly to the base of the caudal ---- Cyclothone, p. 146
   Eyes very large; there is an adipose fin behind the rayed dorsal; there is a considerable interspace between the rear end of the anal fin and the origin of the tail fin ---- 6
6. There are 3 or 4 separate luminescent dots at the base of the caudal fin; the Gulf of Maine species has a large luminescent patch on the snout ---- Headlight fish, p. 142
   There are only 2 separate luminous dots at the base of the caudal fin; the snout does not have a large luminescent patch ---- Lanternfish, p. 143
7. The point of origin of the anal fin is in advance of the origin of the rayed dorsal fin by a distance about as long as the diameter of the eye; the tip of the chin barbel is distinctly swollen as well as bearing several filaments ---- Stomioides, p. 147
   The point of origin of the anal fin is not in advance of the rayed dorsal fin; the tip of the chin barbel is not swollen ---- 8
8. Each side has only about 68 luminescent spots; there is a large luminescent patch crossing the top of the cheek, behind the eye; the point of origin of rayed dorsal fin is in advance of origin of anal fin by a distance about as long as the diameter of the eye; the tip of the lower jaw does not enclose the tip of the upper jaw when the mouth is closed ---- Trigonolampa, p. 148
   Each side has about 85 luminescent spots; the side of the cheek behind the eye does not have a large luminescent patch; the point of origin of rayed dorsal fin is about over origin of anal fin; the tip of the lower jaw encloses the tip of the upper jaw when the mouth is closed ---- Stomias, p. 147

LANTERN FISHES. FAMILY MYCTOPHIDAE

The most distinctive external characters of the lanternfishes are their large eyes (situated close to the tip of the blunt snout), wide mouths gaping back beyond the eye, one soft-rayed dorsal fin, a deeply forked tail, and the presence of a series of luminous organs as conspicuous pale spots along the sides. Some of them have an adipose fin on the back behind the dorsal fin, but others lack this. When present, this fin is so small and fragile that it is apt to be destroyed by the rough treatment the fish receive in the tow net in which they are taken. They most nearly resemble the anchovy (p. 118), the pearlsides (p. 144), and the cyclothone (p. 146) among Gulf of Maine fishes; but they are readily distinguished from the first of these by the presence of luminous organs and by the fact that the snout does not project beyond the mouth; from the second by their much wider mouths; and from the third by their much larger eyes.

They are among the most numerous fishes on the high seas, where they live at a considerable depth by day but often rise to the surface at night. Only two species of the group, representing as many genera (*Diaphus* and *Myctophum*), have been recorded within the Gulf of Maine. But each of these genera includes a considerable number of species that are common along the continental slope abreast of the Gulf, hence are as likely to stray into the latter as are the two that have actually been found there. And this applies equally to various other genera of lanternfishes.

The species of *Diaphus* and of *Myctophum* all resemble one another in general appearance, in having a short dorsal fin, with an adipose fin behind it; a deeply forked tail; large eyes; wide, oblique mouth; and numerous luminous organs along the sides; all, too, are blackishsilvery in color. The members of each genus are separable only by differences in the arrangement of the luminous organs. Hence, positive identification of a given specimen calls for the services of a specialist in the group. Should a lanternfish be taken in the Gulf in which the arrangement of luminous organs does not agree precisely with the two described here, we suggest that it be submitted to the U. S. Fish and Wildlife Service to be named.[25]

[25] Parr (Bull. Bingham Oceanog. Coll., vol. 3, art. 3, 1928), and Tåning (Vidensk. Meddel., Dansk Naturhist. Forening, vol. 86, p. 49, 1928) have recently published critical synopses of the lanternfishes.

Headlight fish *Diaphus effulgens* (Goode and Bean) 1895

Jordan and Evermann, 1896–1900, p. 566.

Description.—This curious little fish is separable from the lanternfish (p. 143) and from the pearlsides (p. 144) at a glance, by the large and very noticeable luminescent patch that covers the entire tip of its snout (including the anterior margin of the orbit) and that extends down over the edge of the upper jaw, a structure that has no parallel in any other fish regularly inhabiting the Gulf of Maine. It also differs from the pearlsides in its much more deeply cleft mouth, its even larger eyes, in the more convex dorsal profile of its head, and in lacking the regular horizontal row of luminescent spots along each side about at the level of the pectoral fin, that are conspicuous on the pearlsides.[26]

The four separate luminescent spots at the base of the tail (besides the organ on its snout) separate it from its close relatives of the genus *Myctophum* (p. 143). The arrangement of the fins (all of which are soft, the dorsal with about 15 rays, the anal with about 16), is essentially the same as in the latter, and in the pearlsides; the caudal fin is more deeply forked than in the pearlsides, the adipose fin proportionately shorter.

Color.—The color has not been described. Probably it is black, overlaid more or less with

[26] The structures along the lateral line shown here on the illustration of the headlight fish are large scales, not luminescent organs.

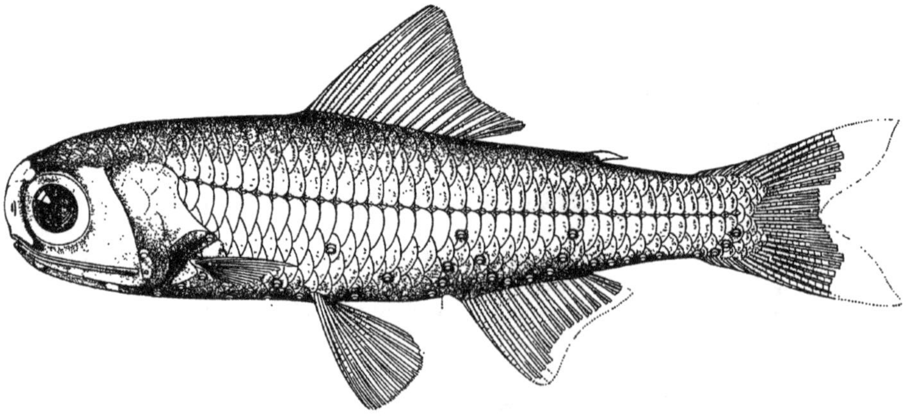

FIGURE 60.—Headlight fish (*Diaphus effulgens*), Browns Bank. From Goode and Bean. Drawing by A. H. Baldwin.

silver, with the luminescent organs pale blue or green.

Size.—The specimens from which this species was originally described seem to have been about 7 inches long.[27]

General range and occurrence in the Gulf of Maine.—This oceanic species is only a stray within the limits of the Gulf. One specimen has been found in the stomach of a cod caught on Browns Bank,[28] and another, also from a cod stomach, has been reported on Western Bank off the outer coast of Nova Scotia.[29]

Lanternfish *Myctophum affine* (Lütken) 1892

Jordan and Evermann, 1896–1900, p. 570.

Description.—The most noticeable features of this little oceanic fish are its silvery black color, the luminous dots along its sides, its enormous eye situated close to the tip of the snout, its very deep oblique mouth, and its deeply forked tail. The anal fin is mostly or wholly behind the short, soft dorsal, and there is an adipose fin behind the latter, as in the headlightfish (p. 142). The longer snout and smaller mouth of *Myctophum*, with the fact that the luminous organs on its snout are in the form of small dots instead of a large patch covering the entire tip of the snout, are the readiest field marks to distinguish it from the latter. The dorsal profile of the head is much arched, the body moderately flattened sidewise, tapering gently backward to the rather deep caudal peduncle. The location of the luminescent spots is shown in the drawing (fig. 61).

Color.—This lanternfish is silvery when alive, the silver underlain on the back with deep brownish black, the sides below the lateral line, and the belly varying (below the silver) from dark brown to dusky gray, or even to white finely dotted with gray. The luminescent organs are pale green or blue.

Size.—All members of the genus *Myctophum* are small; a little more than 3½ inches (89 mm.) is the maximum length recorded for this particular species.

General range.—All the species of this genus are oceanic, occurring only as strays inside the edge of the continent.

Occurrence in the Gulf of Maine.—Goode and Bean[30] report the capture of this lanternfish over the southeast slope of Browns Bank (lat. 42° 21′ N., long. 65° 07′ W.) at 104 fathoms, which still remains the only record for it in the Gulf of Maine,[31] or for any *Myctophum* for that matter.

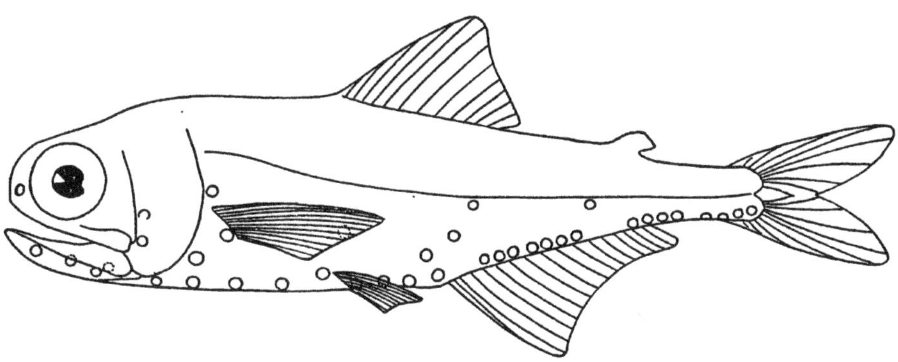

FIGURE 61.—Lanternfish (*Myctophum affine*). After Parr.

PEARLSIDES. FAMILY MAUROLICIDAE

The Pearlsides resembles the lanternfishes (p. 141) in shape of body, but it has a shorter rayed dorsal fin, a longer adipose fin, a longer anal, and a much smaller mouth.

[27] The illustration (Goode and Bean, Smithsonian Contrib. Knowl., vol. 31, 1895, fig. 103), about 6 inches long, is characterized in the legend as "slightly reduced."

[28] Reported by Goode and Bean (Smithsonian Contrib. Knowl., vol. 30, 1895, p. 88) as *Aethoprora effulgens*.

[29] Vladykov, Proc. Nova Scotia Inst. Sci., vol. 19, 1935, p. 2.

[30] Smithsonian Contrib. Knowl. (vol. 30, 1895, p. 72) as *M. opalinum*.

[31] It is likely that *Myctophum glaciale* will be found in the Gulf of Maine sooner or later, judging from its widespread distribution in the boreal belt of the Atlantic and from the fact that it has often been caught at the surface. It resembles *M. affine* very closely in appearance, and in the general arrangement of the luminous organs, but differs from it in that one of the luminescent spots above the base of the ventral fin is elevated above the others.

Pearlsides *Maurolicus pennanti* (Walbaum) 1792

PEARLFISH

Jordan and Everman, 1896-1900, P. 577.

Description.—The presence of an adipose fin between the dorsal and caudal fins, together with luminous organs, distinguishes the pearlsides from all other fishes that occur regularly in the Gulf of Maine. It agrees in both these respects with the lanternfish (p. 143) and with the headlightfish (p. 142), but it has a much smaller mouth and a longer adipose fin than the first of these, and it lacks the large luminous patch on the snout that is so striking a feature of the second. Also, the pearlsides, with its herring-like coloration (p. 88) differs strikingly from the lanternfish, which has a black back overlaid with silver; and probably the headlight fish as well.

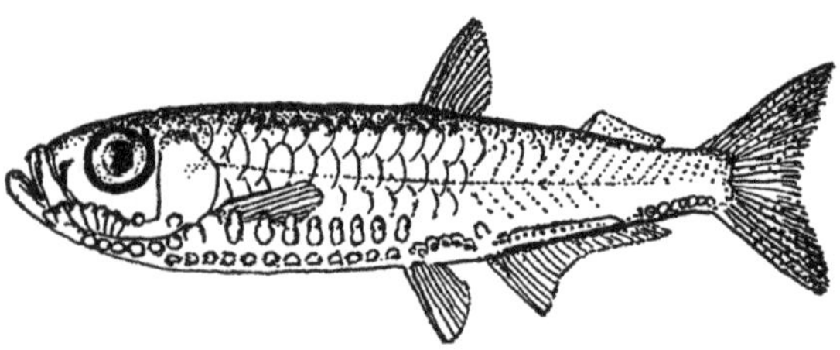

FIGURE 62.—Pearlsides (*Maurolicus pennanti*). After Smitt.

The pearlsides is a flat-sided, large-headed little fish, its body (about one-fifth as deep as long, excluding caudal fin) deepest forward of the ventral and dorsal fins; its eye very large; its lower jaw projecting; its mouth oblique; and both its jaws armed with minute teeth. The dorsal fin (about 11 or 12 rays) stands above the space between the ventrals and the anal; the anal is longer than the dorsal. The adipose fin (both of Woods Hole [32] and of Norwegian [33] examples) is low and long, much as it is in the capelin.[34] The caudal fin is broad and slightly forked.

The pearlsides has been described as without scales, but this is not correct, for both Scandinavian and Woods Hole specimens have been found to be clothed with large but extremely thin transparent scales. There is no definite lateral line.

The most interesting and diagnostic feature of the pearlsides is the presence of a series of luminescent dots situated as follows: [35] First, 12 pairs along the belly between the pectoral and the ventral fins, followed by 5 or 6 from the ventral fins to the anal fin, and, after a gap, by 24 or 25 between the center of the anal fin and the base of the caudal fin; all these together form a practically continuous row on each side of the belly from throat to tail. Second, there is a row of larger spots a little higher up on each side, 6 from chin to pectoral fin, and 9 thence backward to the ventrals. Third, there is a group of 6 low down on each side of the cheek and throat; there is likewise a spot in front of the base of each pectoral fin and 2 on the chin.

Color.—The pearlsides is colored much like a herring, with dark bluish or greenish back and lustrous silvery-white sides and belly. The luminescent spots are described as black rimmed, their centers as pale blue in life but turning yellow in alcohol; and there is a narrow black band along the base of the anal fin and from there to the base of the caudal, the latter being barred with a similar black band.

Size.—Only 1 to 2½ inches long.

Habits.—The relatives of the pearlsides are oceanic, living in the mid-depths mostly below 150 fathoms, but the pearlsides itself has been found so often in the stomachs of cod and of herring (fish that do not descend to any great depth) that there is no reason to regard it as a "deep-sea" stray, nor has it ever been taken far from land so far as we can learn. It probably spawns in early spring, females with large eggs having been taken in Scottish waters in winter.

General range.—The pearlsides (there are several other species closely allied to it) ranges widely in the open Atlantic, occurring at times in shoals on the coasts of Norway and in British waters. It is especially common off the coast of Scotland, but has not been recorded often on the American side of the Atlantic.

Occurrence in the Gulf of Maine.—The known occurrences of the pearlsides in the Gulf have been few. Storer [36] (1867) records one found alive on the beach at Nahant, Mass., in December, 1837; another taken from the stomach of a cod at

[32] Sumner, Osburn, and Cole, Bull. U. S. Bur. Fish., vol. 31, Pt. 2, 1913, p. 743.

[33] Smitt, Scandinavian Fishes, vol. 1, 1892, p. 933, pl. 44, fig. 3.

[34] Goode and Bean (Smithsonian Contrib. Knowl., vol. 30, 1895 p. 96) describe it as "very small," but probably their specimens were damaged.

[35] This account is based chiefly on Smitt's description and plate, the specimens we have seen being in poor condition.

[36] Fishes of Mass., 1867, p. 150, as *Scopelus humboldtii.*

Provincetown; a third picked up alive there in July, 1865 (pictured by Storer on pl. 25, fig. 5); and five others found on the Provincetown beach soon afterward. We have seen one specimen 41 mm. long taken from the stomach of a cod, on Platts Bank, July 27, 1924; one 43 mm. long, also from a cod's stomach, on Cashes Ledge, August 16, 1928; and four, 32 to 39 mm. long, taken from the stomachs of two pollock that we caught in 20 fathoms, 7 miles southeast of Bakers Island, Mount Desert, Maine, July 24, 1930. It has been found twice at Grand Manan,[37] and specimens were picked up on the beach at Campobello Island at the mouth of the Bay of Fundy in July 1914,[38] while others were taken from the stomach of a pollock caught near by. It has also been recorded twice near Woods Hole.

These locality records are distributed widely enough to show that it is to be expected anywhere in our Gulf. And we suspect that the pearlsides is not as scarce there as the paucity of actual records for it might suggest (in fact, Storer tells us that a Nahant fisherman reported finding them repeatedly in the stomachs of haddock many years ago), but that it keeps out of sight, being an inhabitant of the deeper water layers as its luminescent organs would suggest, coming up to the surface chiefly at night.

VIPER FISHES. FAMILY CHAULIODONTIDAE

The viper fishes have slender bodies, bulldog-like faces with long fangs; the first dorsal very far forward, the anal far back; and no barbel on the chin.

Viperfish *Chauliodus sloani* Bloch and Schneider 1801

Jordan and Evermann, 1896–1900, p. 585.

Description.—The viperfish not only has luminescent organs, but it is very different in general appearance from all the fishes that are regular inhabitants of the Gulf of Maine. Most obvious of its characteristics is its bulldog-like mouth. It shares this with its fellow strays, *Stomias* (p. 147), *Stomioides* (p. 147) and *Trigonolampa* (p. 148) and the general form is much alike in the three. But there is no danger of confusing it with any one of these if one looks closely, for the viperfish has an adipose fin and its rayed dorsal fin is far forward, whereas *Stomias*, *Stomioides*, and *Trigonolampa* have no adipose fin and their rayed dorsal fin stands far rearward.

In the viperfish the lower jaw is longer than the upper, the upper is armed with four long fangs on each side, while the lower has a series of pointed teeth set far apart, those in front very elongate and all of them so long that they project when the mouth is closed. Furthermore, the snout is so short that the very wide mouth gapes far back of the eye. The body is about seven times as long as deep, flattened sidewise, deepest close behind the head, and tapering evenly to the tail. The very short dorsal fin (6 or 7 rays) stands far forward and its first ray is separate, very slender, and about half as long as the fish when not broken off, as it usually is. The ventrals are about midway between the snout and the origin of the anal fin, variously pictured as either larger or smaller than the dorsal. The small anal is close to the caudal, with the adipose fin over it. The

[37] Cox (Bull. Nat. Hist. Soc. New Brunswick, 14, 1896, append., p. 55) reported one found dead there, on the shore.

[38] Huntsman (Contrib. Canadian Biol., (1921) 1922, p. 61.)

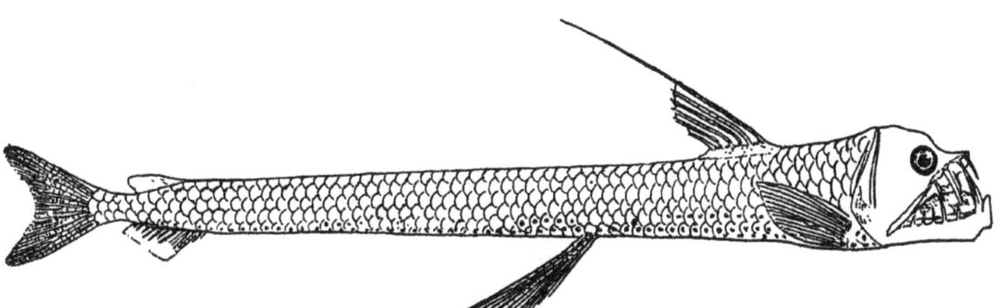

FIGURE 63.—Viperfish (*Chauliodus sloani*), southern slope of Browns Bank. After Goode and Bean.

body is clothed with large but very thin scales. There are several longitudinal rows of small luminescent spots on the ventral surface, running from throat to tail; several more such spots on each side of the head; and many tiny unpigmented dots scattered over the trunk.[39]

Color.—Greenish above, the sides with metallic gloss; blackish below.

Size.—Up to about one foot long.

Habits.—Nothing is known of its habits except that it is an inhabitant of the mid-depths of the Atlantic Basin and that it probably does not rise closer to the surface than 150 or 200 fathoms except, perhaps, during its larval stages. Its teeth suggest a rapacious habit but there is no actual record of its diet.

Occurrence in the Gulf of Maine.—The only definite Gulf of Maine records are of one specimen found in the stomach of a cod caught on Georges Bank in 1874, and of a second found in the stomach of a swordfish that was harpooned in the gully between Browns and Georges Banks in 1931.[40] But the viperfish may be expected on the offshore banks as a stray at any time, for several have been taken off the continental slope abreast of southern New England[41] in deep water.

THE STOMIATIDS. FAMILIES GONOSTOMIDAE AND STOMIATIDAE

The stomiatids include many soft-rayed fishes of the mid-depths, of most diverse appearance, all of them with well developed luminescent organs, with large eyes, large mouths, and teeth in both jaws. Some have and others lack the adipose fin, but the ventrals are inserted more than one-third of the way back on the abdomen in all of them. They differ from the herrings and salmons in the structure of the skull. Four species have been taken in our Gulf, as strays from offshore.

Cyclothone *Cyclothone signata* Garman 1899

Garman, Mem. Mus. Compar. Zoology, vol. 24. 1899, p. 246, pl. J, fig. 3.

Description.—The general aspect of cyclothone is extremely characteristic, the somewhat compressed body being deepest at the gill opening with the upper surface of the head concave in profile, the mouth so large that it gapes back of the eye, the lower jaw projecting, the eye very small, and the gill openings very long. The dorsal fin stands over the anal (the latter is much the longer of the two), both originating close behind the middle of the body. The caudal fin is deeply forked and there is no adipose fin.

The luminescent spots are arranged as follows: One on the head; 1 close below the eye and in front of it; 2 on each gill cover; 9 or 10 between the branchiostegal rays; 2 longitudinal rows along each side of the body, a lower row of 13 from throat to ventral fins, 4 from ventrals to anal fin, and 13 from anal to caudal, and an upper row of 7 reaching about as far back as the ventrals.

Color.—*Cyclothone signata* is colorless or pale gray, except that the blackish, dark silvery lining of the abdominal cavity shows through, that the luminous organs are black rimmed and silver centered, and that there are the following black markings: a Y-shaped mark on the forehead; a series of spots or short transverse stripes on the flank; spots between the bases of the dorsal and anal fin rays; one or two transverse streaks across the bases of the caudal fin rays; and a number of

[39] Brauer, Tiefsee-Fische Wissensch. Ergeb. Deutschen Tiefsee-Exped., (1898-1899) 1906, vol. 15, Pt. I, p. 40.

[40] Reported to us by Walter H. Rich.

[41] Goode and Bean (Smithsonian Contrib. Knowl., 1895, p. 97) list these captures.

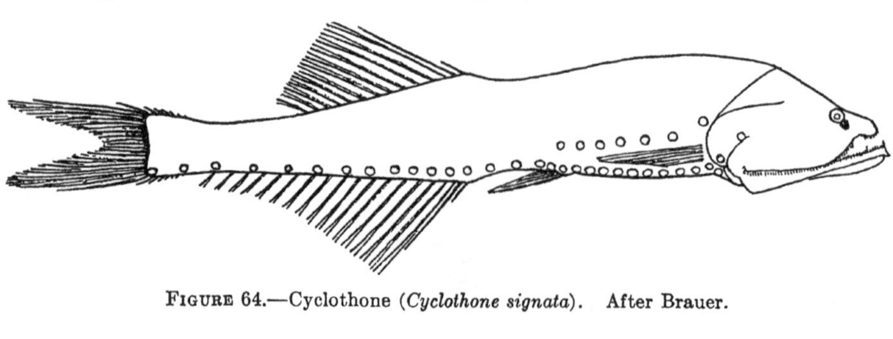

FIGURE 64.—Cyclothone (*Cyclothone signata*). After Brauer.

irregular flecks and dots along the back and on the gill covers.[42]

General range.—This is an oceanic fish, very abundant in temperate latitudes in the Atlantic where it lives pelagic from about 100 fathoms down to 250 fathoms; hundreds have often been taken in a single haul. It is also known from the Pacific.

Occurrence in the Gulf of Maine.—Cyclothone appears within our limits only as a stray from the Atlantic Basin; one 23 mm. long that we took in a haul from 30 fathoms on Browns Bank, June 24, 1915, and a second mutilated specimen probably of this species from the Fundy Deep (haul from 90 fathoms), March 22, 1920, are the only definite records of it within our limits.

Stomias *Stomias ferox* Reinhardt 1842

Jordan and Evermann, 1896–1900, p. 588.

Description.—The members of this genus (there are several), resemble the viperfish in their bull-dog-shaped heads, with large mouth and long, fanglike teeth. But they do not have an adipose fin; the dorsal fin and the anal fin both stand far rearward close to the tail fin; the dorsal fin is even with the anal fin; and the first dorsal fin ray is not prolonged as it is in the viperfish. The chin bears a fleshy barbel nearly as long as the head and ending in a group of about three simple filaments. The sides of the body are clothed with about 6 rows of large, thin, somewhat irregular, hexagonal scales, and there is one row of luminescent spots low down along each side and two rows along the belly; also one small, circular light organ below each eye.

The tip of the lower jaw overlaps and encloses the tip of the upper jaw when the mouth is closed in the only member of the genus that has been reported from the Gulf of Maine (or is likely to be found there); the slender body is about 17 times as long as it is high; the ventral fins are only about as long as the head; the dorsal fin is of about the same size and shape as the anal fin, over which it stands; and there are about 85–86 light organs in each of the ventral rows, about 60 light organs in each of the lateral rows.

Color.—Black below as well as above, the sides with metallic iridescence.

General range and occurrence in the Gulf of Maine.—This oceanic fish is so widespread in the northern North Atlantic that it was taken at almost all the stations that the *Michael Sars* occupied there in 1910,[43] mostly between the 75 to 80 fathom level and the 410 fathom (750 meter) level, most plentifully at about 275 fathoms (500 meters). The early cruises of the *Blake* and *Albatross I* took it at many localities also, along the continental slope of North America between the southeastern slope of the Newfoundland Banks and the Bahama Channel.[44] Our only reason for mentioning it is that one specimen about 12 inches long (tip of snout to base of tail fin) was taken by a trawler on the northeastern part of Georges Bank (lat. 42°10′ N., long. 67°05′ W.), at about 100 fathoms, on January 20, 1936.[45]

Stomioides nicholsi Parr 1933

Parr, Copeia, 1933, No. 4, p. 177.

Description.—The chief anatomical character separating *Stomioides* from *Stomias* is the structure of the chin barbel. In *Stomias* this terminates in three simple filaments. But in *Stomioides* it not only has these barbels, but the main trunk is swollen at the tip and has two additional filaments on one side a little inward from its tip. Another difference is that the point of origin of the anal fin is in advance of the origin of the dorsal fin by a distance about as great as the diameter of the eye in *Stomioides*, whereas the point of origin of the anal fin is about even with that of the dorsal in

[42] For detailed accounts and colored illustrations see Garman (Mem. Mus. Comp. Zool., vol. 24, 1899, p. 246, pl. J, fig. 3), Brauer (Wissensch. Ergeb. Deutschen Tiefsee-Exped. (1898–1899), 1906, vol. 15, Pt. 1, p. 77, pl. 6, fig. 6), Murray and Hjort (Depths of the Ocean, 1912, pl. I).

[43] Murray and Hjort, Depths of the Ocean, 1912, pp. 603, 611, 629.

[44] For a list of these stations, see Goode and Bean, Smithsonian Contrib. Knowl., vol. 30, 1895, p. 107.

[45] This specimen is now in the Museum of Comparative Zoology.

FIGURE 65.—Stomias (*Stomias ferox*), Banquereau Bank. From Goode and Bean. Drawing by H. L. Todd.

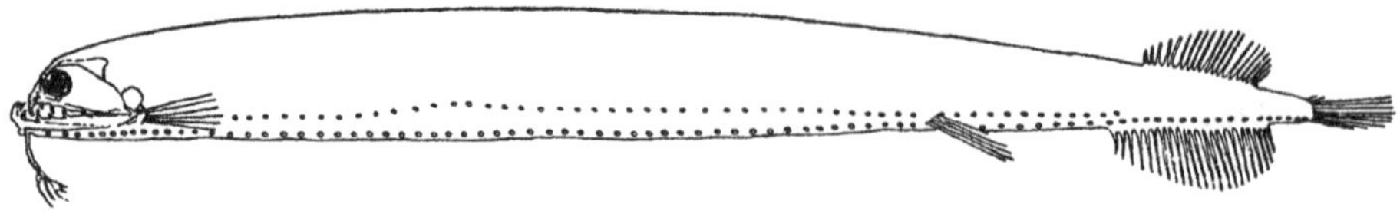

FIGURE 66.—*Stomioides nicholsi*, Browns Bank. Drawing by Myvanwy M. Dick.

Stomias. *Stomioides* resembles *Stomias* in all other respects so closely that should a specimen of either be taken, that is not easily identified, we suggest forwarding it to the U. S. Fish and Wildlife Service for naming.

Color.—Black below as well as above, the luminescent organs showing as whitish dots.

Size.—The only specimen yet seen is about 10½ inches long, from tip of snout to base of tail fin.

Range and occurrence in the Gulf of Maine.—The only known specimen of this species was taken from the stomach of a swordfish harpooned from the schooner *Barbara*, Capt. C. A. Turner, on the southeastern edge of Browns Bank,[46] over the 250 fathom line, August 3, 1932. Presumably it had strayed from the mid-depths offshore.

Trigonolampa miriceps Regan and Trewavas 1930

Regan and Trewavas, Danish *Dana* expeds. 1920–1922, No. 6, 1930, p. 55, pl. 1, fig. 1.

Trigonolampa resembles *Stomias* in general appearance, in the relative sizes and locations of the fins, and in having a long fleshy barbel on its chin. But it not only has a small light organ below the eye (as in *Stomias*), but also has a small luminescent patch close behind it, and likewise a larger triangular patch extending from close behind the eye back across the top of the cheek; these are its most distinctive characters. The one species of the genus yet known differs further both from *Stomias* (p. 147) and from *Stomioides* (p. 147) in a considerably deeper body (cf. fig. 67 with figs. 65, 66); also in that the tip of its lower jaw does not enclose the tip of its upper jaw when the mouth is closed; that the point of origin of its dorsal fin is in advance of its anal fin by a distance about as great as the diameter of the eye; and that it has only about 68 light organs in each of its ventral rows, as against 85 or 86 in *Stomias* (p. 147).

Color.—Not known, but probably black or very dark brown.[47]

Size.—The largest specimen yet seen (in the Museum of Comparative Zoology) is about 9 inches (230 mm.) long to the base of the caudal fin.

Range and occurrence in the Gulf of Maine.—Only three specimens have been seen yet. The first was taken in the eastern Atlantic by the Danish research vessel *Thor* in 1906 at a depth of about 600 fathoms; a second was found by Capt. John Toothaker in the stomach of a swordfish harpooned on the southern edge of Georges Bank in the summer of 1922,[48] and a third, now in the Museum of Comparative Zoology, was recorded simply as taken on Georges Bank about 1913. It reaches the slope of our outer Banks only as a stray from the mid-depths offshore.

[46] This specimen, described by Parr (Copeia, 1933, p. 177), is now in the Museum of Comparative Zoology.

[47] One that we have seen is brown below as well as above wherever the skin is intact, with the light organs showing as darker dots.

[48] Parr (Copeia, 1933, No. 4, p. 178) has given a detailed description of this specimen, which is now in the Museum of Comparative Zoology.

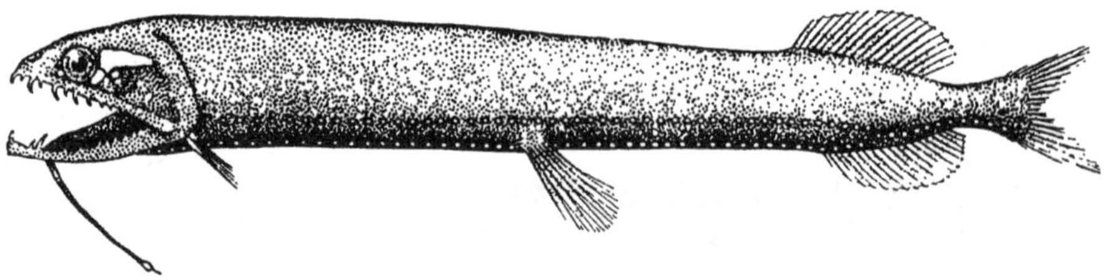

FIGURE 67.—*Trigonolampa miriceps*. After Regan and Trewavas.

HATCHET FISHES. FAMILY STERNOPTYCHIDAE

These are deep, thin, flat-sided little fishes, with various spiny projections, large oblique mouths with small teeth, large eyes which are directed upwards in some of them but sidewise in others, and ventral fins placed far back. Some of them have an adipose fin behind the rayed dorsal, but others do not. All of them are silvery, and all of them have series of large and conspicuous luminescent organs on the lower part of the body. They are to be found in the mid-depths in all oceans, sometimes in great abundance.

Silver hatchetfish *Argyropelecus aculeatus* Cuvier and Valenciennes 1849

Jordan and Evermann, 1896–1900, p. 604, as *A. olfersi* Cuvier, 1829.

FIGURE 68.—Silver hatchetfish (*Argyropelecus aculeatus*). After Brauer.

Description.—This little fish is of so bizarre an appearance that once seen it could hardly be mistaken for any other species yet known from our Gulf, or for any that is likely to stray thither, unless one of its own tribe.[49] Its body is very thin sidewise, with its forward part a little less than three-fourths (70 percent) as deep as it is long from snout to base of tail fin, but with the ventral contour bending upward abruptly about midway of its length in characteristic contour, so that the rear half is much less deep than the forward half. This break in the ventral contour is marked by two short bony spurs, which are outgrowths from the pubic bones, and there is a short single spur (outgrowth from the pectoral arch) in front of them in the midventral line.

The eyes are large, so high up that the space between them on the top of the head is very narrow, and they are directed more upward than sidewise. The mouth is noticeably large, with wide gape, and it is so strongly oblique that the upper jaw is nearly vertical.[50] The tips of the two jaws are about even one with the other when the mouth is closed, and both jaws are armed with a large number of tiny sharp teeth. The dorsal fin is short, about midway of the fish, and of two parts, separated by a deep but short notch. The forward subdivision is in the form of a hard, triangular plate (apex rearward) supported by 8 or 9 hard spines, the rearmost of which is the stoutest and longest. The rearward subdivision is supported by 9 soft rays, that are bifid toward their tips. The adipose fin is long and low. The pectorals are as long as about two-fifths the greatest height of the body. The ventral fins, each with 6 soft rays, stand close behind the break in the ventral contour of the body, and they are connected with the anal fin by a thin transparent ridge. The anal, commencing about under the rear end of the base of the soft rayed part of the dorsal, is notched midway of its length; its forward part is supported by 7 rays close together, the rear part by 5 shorter rays spaced more widely. The caudal fin is forked. A noticeable feature is that the ventral edge of the deep forward part of the body, from the pectoral spur to the pubic spines, is sharp, with a series of 12 hard, plate-like scales or scutes, that extend for some distance up the sides, each slightly overlapping the next rearward, and the profile is saw-edged between the ventral and anal fins.

The hatchetfishes are provided with a complex system of conspicuous light-producing spots. The species *aculeatus* has one row of 12 very low down along each side of the deep forward part of the body; also, a second row higher up consisting of 6 in front of each pectoral fin, 2 along the base of the pectoral, 6 between pectoral and ventral fins, 4 between the ventral and the anal fins, 6 along the anal, and 4 very small ones between anal and tail fins. There is also one light-organ a little below and behind

[49] The several known species of *Argyropelecus* resemble one another so closely that their identification calls for a specialist. They have been reviewed by Schultz, Proc. U. S. Nat. Mus., vol. 86, 1938, pp. 147–153. The most detailed description of this particular hatchetfish, with the best illustration (copied here as fig. 68) is by Brauer, Wiss. Ergeb. Deutschen *Valdivia* Tiefsee-Exped., vol. 15, Pt. 1, 1908, p. 110, fig. 47.

[50] Most of the published illustrations of *Argyropelecus* fail to show this.

each eye, 2 on the lower part of the gill cover on each side and about 5 on the lower jaw on each side.

Color.—Perhaps the most conspicuous feature of the hatchetfishes, as taken from the water, is that their entire bodies are glistening silvery. On some specimens the silver is underlaid with velvet black over the trunk as a whole; on others the black under pigment is confined to a marginal band, broader or narrower. The luminescent spots are pale yellow or white.

Size.—Maximum length probably not more than 3 inches or so.

Range and occurrence in the Gulf of Maine.—All the members of this genus are oceanic, and inhabit the mid-depths. Localities listed by Schultz [51] for this species include the Grand Banks, between Georges and Browns Banks, and the offing of New Jersey and Virginia in the western Atlantic; the Gulf of Mexico; West Indies; off the South African coast; the Indian Ocean; and the Philippines. Our only reason for mentioning the hatchetfish is that one specimen was taken on August 31, 1883, by the *Albatross* between Georges and Browns Banks where the depth was 144 fathoms.[52]

THE EELS. FAMILIES ANGUILLIDAE, CONGRIDAE, SIMENCHELYIDAE, SYNAPHOBRANCHIDAE, NEMICHTHYIDAE, AND OPHICHTHYIDAE

Eels have no ventral fins; either they have no scales or these are so small as to be hardly visible; their fins are soft, without spines; the gill openings are very small; the vertebrae extend in a straight line to the tip of the tail; and a single fin runs over the back, around the tail and forward on the belly with no separation into dorsal, caudal, and ventral portions. All the species of eels known from the Gulf of Maine have pectoral fins, but most of the morays of warmer seas are without pectorals. There are several other fishes of eel-like form in the Gulf of Maine, viz., the hag and the lampreys the rock eel (*Pholis*); the snake blenny (*Lumpenus*); the wrymouth (*Cryptacanthodes*); the eel pout (*Macrozoarces*); and the sand eel (*Ammodytes*). But the jawless, sucker-like mouth of the first two separates them, at a glance, from the true eels, while there either is a well-marked separation between anal and caudal fins in all the rest; or they have ventral fins (large or small), or the dorsal fin is spiny, not soft.

Only five true eels are known from the Gulf of Maine: the common eel (p. 151), the slime eel (p. 157), the conger (p. 154), the snipe eel (p. 159), and the snake eel (p. 159), which fall into five different families according to American usage. A sixth species, the long-nosed eel (a deep-water form p. 158) is to be expected in the deepest parts of the Gulf though it has not actually been recorded there as yet. The group likewise includes the morays of warm seas and sundry deep-sea forms, some of them exceedingly bizarre in appearance.

Common, conger, slime, and long nosed eels look much alike in general form, but are separated from one another by the size of the mouth and by the relative lengths of the fins. In the snipe eels the two jaws are prolonged into a very long slender beak, recalling that of a silver gar, the tail is whiplike, the neck noticeably slimmer than the head, and the general form extremely slender, while the snake eel is very slender with a hard pointed tail.

[51] Bull. U. S. Nat. Mus., vol. 86, 1938, pp. 146–147 and pp. 151–152.

[52] *Albatross* station 2063, lat. 42° 23′ N., long. 66° 23′ W. This specimen was recorded by Goode and Bean (Smithsonian Contrib. Knowl., vol. 30, 1895, p. 127) as *A. olfersii*. But Schultz (Proc. U. S. Nat. Mus., vol. 86, 1938, p. 151) has found, on reexamination, that it is an *aculeatus*.

KEY TO GULF OF MAINE EELS

1. Both jaws are prolonged into a long slender bill ---------- Snipe eel, p. 159
   The jaws are not bill-like ---------- 2
2. The anal fin originates well in front of the point of origin of the dorsal fin ---------- Long-nosed eel, p. 158
   The anal fin originates well behind the point of origin of the dorsal fin ---------- 3
3. The dorsal fin originates far behind the tips of the pectorals ---------- Eel, p. 151
   The dorsal fin originates close behind the tips of the pectorals ---------- 4
4. Mouth very small, its gape not reaching back as far as the eye; body very soft ---------- Slime eel, p. 157
   Mouth large, gaping back as far as the middle of the eye; body firm ---------- 5
5. Mouth gaping back only about as far as the middle or rear edge of eye; body moderately stout; tip of tail soft, rounded ---------- Conger, p. 154
   Mouth gaping back considerably beyond eye; body very slender, tip of tail hard and pointed ---------- Snake eel, p. 159

Eel *Anguilla rostrata* (LeSueur) 1817

American eel; Silver eel; Fresh-water eel; Elver (young)

Jordan and Evermann 1896–1900, p. 348, *A. chrysypa* Rafinesque 1817.

Description.—In the common American eel the dorsal fin originates far behind the pectorals, this character is enough to distinguish it from the conger, from which it also differs in that the lower jaw projects beyond the upper or at least equals it in length, and its eyes are small and round. Furthermore, it develops scales as it grows, though these are so small that they might be overlooked. The eel, however, has a pointed snout, like the conger, a large mouth gaping back as far as the middle of the eye or past it; and its gill slits are set vertically on the sides of the neck, their upper corners abreast of the center of the base of the pectoral fin. It is very closely related to the European eel (*Anguilla vulgaris*), but has fewer vertebrae (average about 107 as compared with about 114 or 115 in the European species).

Color.—The colors of eels vary widely with the bottom on which they live. As a rule they are dark muddy brown or olive-brown above, more or less tinged with yellow on the sides; the lower surface paler brown and yellower, with dirty yellowish-white belly. It is common knowledge that eels are dark if living on dark mud but much paler on pale sand. And Parker [53] has found that they can change from pale to dark in about 1½ hours and from dark to pale in a little more than 3 hours, if moved from a white background to a black or vice versa, under a strong light.

Size.—Eels are said to grow to 4 feet in length and to 16½ pounds in weight. Full-grown females average only about 2 to 3½ feet, however, and males are smaller. Any eel more than 18 inches long would probably be a female, and one more than 24 inches in length would certainly be one. The smallest mature males are about 11 to 12 inches long, females about 18 inches.

Habits.—The life history of the eel remained a mystery until very recently. It has been common knowledge for centuries that young elvers run up into fresh water in spring, and adults journey downstream in autumn. A host of myths grew up to explain the utter absence of ripe eels of either sex, either in fresh water or along the seacoast. But it was only a few years ago that the breeding places of the European and American eels were discovered and the history of their larvae

[53] Jour. of Exper. Zool., vol. 98, 1945, No. 3, pp. 211–234.

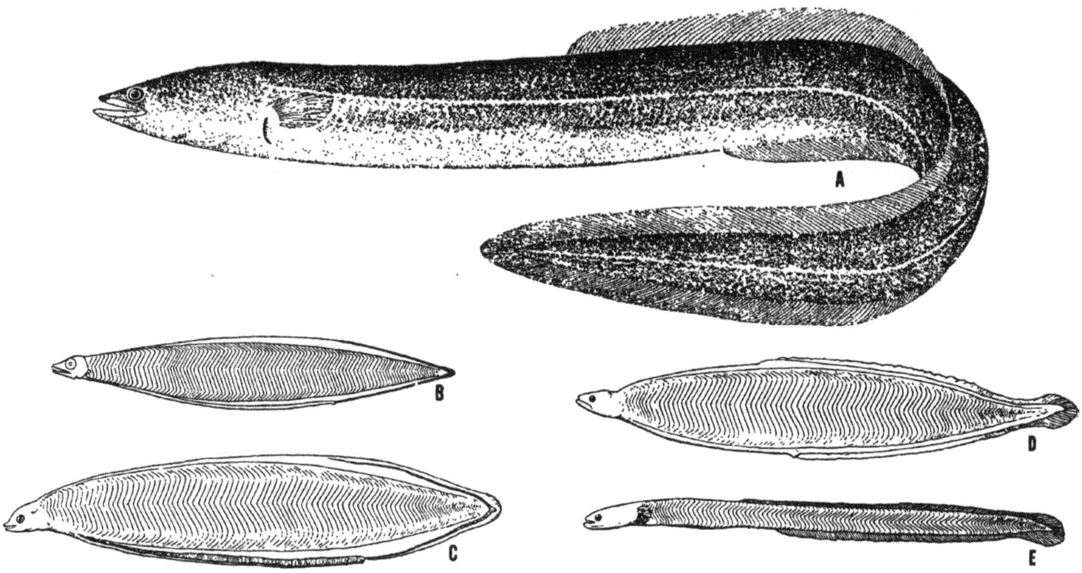

Figure 69.—Eel (*Anguilla rostrata*). A, adult, Connecticut River, Massachusetts; from Goode, drawing by H. L. Todd; B, "Leptocephalus" stage, 49 mm.; C, "Leptocephalus" stage, 55 mm.; D, "Leptocephalus" stage, 58 mm.; E, transformation stage, 61 mm. B–E, after Schmidt.

traced, chiefly by the persevering researches of the Danish scientist, Johannes Schmidt.[54] Now we know that the life history of the eel is just the antithesis of that of the salmon, shad, and alewife, for eels breed far out at sea, but make their growth either in estuarine situations or in fresh water.

The young elvers, averaging from 2 to 3½ inches in length, appear along our shores in spring. As yet we have few data on the exact date of their arrival on the Gulf of Maine coast. They appear as early as March at Woods Hole; by mid- or late April both in Narragansett Bay and in Passamaquoddy Bay at the mouth of the Bay of Fundy, while Welsh encountered a tremendous run in Little River, near Gloucester, on May 5, 1913, suggesting that they may be expected in the mouths of most Gulf of Maine streams during that month. And they are found ascending streams in the Bay of Fundy region during the summer. A run may last for a month or more in one stream, only for a few days in another. And there is a noticeable segregation even at this early stage, some of the elvers remaining in tidal marshes, in harbors, in bays back of barrier beaches, and in other similar situations, some even along the open coast, especially where there are beds of eel grass (*Zostera*); while others go into fresh water, some of them ascending the larger rivers for tremendous distances.[55]

It is now generally believed that most of the eels that are caught in fresh water are females. But some of the females remain in salt marshes and harbors, to judge from the large size of many of the eels that are caught there. And nothing is known as to what preference the males of the American eel may show in this respect.

It is no wonder that the ability of the elvers to surmount obstacles as they run upstream is proverbial, for they clamber over falls, dams, and other obstructions, even working their way up over damp rocks as Welsh saw them doing in Little River, where they were so plentiful on May 5 and 7, 1913, that he caught 1,500 in one scoop of a small dip net and 545 with a few grabs of his bare hand. Elvers in equal multitudes have often been described in other streams, American as well as European. Eels can live out of water so long as to give rise to the story that they often travel overland. There is no positive evidence for this. But Sella [56] has proved, by experiments with European eels marked so as to be recognizable if recaught, that they can carry out journeys as long as 31 miles (50 kilometers) along underground waterways. Doubtless it is this ability that explains the presence of eels in certain ponds that have no visible outlet nor inlet, a fact often attested.

It is true in a general way that eels seek muddy bottom and still water, as has been said so commonly. But this is not always so whether in salt water or in fresh. Thus the rocky pool at the outer end of the outlet from Little Harbor, Cohasset, on the south shore of Massachusetts Bay, is a good place to catch eels; and large ones are only too common in swiftflowing, sandy trout streams on Cape Cod; we have had one follow and nibble at the trout we were dragging behind us on a line. The fact is, they can live and thrive wherever food is to be had, which applies to them in estuarine situations and in fresh water.

No animal food, living or dead is refused, and the diet of the eels in any locality depends less on choice than on what is available. Small fish of many varieties, shrimps, crabs, lobsters, and smaller crustacea, together with refuse of any kind (for they are scavengers) make up the bulk of the diet in salt, estuarine, and brackish water. Being very greedy, any bait will do to catch an eel. They are chiefly nocturnal in habit, as every fisherman knows, usually lying buried in the mud by day to venture abroad by night. But eels, large and small, are so often seen swimming about, and so often bite by day that this cannot be laid down as a general rule.

Eels tolerate a wide range of temperature. But it is common knowledge that those inhabiting the salt marshes and estuaries of our Gulf, and its tributary streams, mostly lie inactive in the mud during the winter.

Eels grow slowly. Hildebrand and Schroeder [57] concluded from a series of measurements taken at different seasons in lower Chesapeake Bay that those

[54] The life history of the eel is presented in more detail than is possible here by Schmidt (Philos. Trans. Roy. Soc. London, Series B, vol. 211 (1922) 1923, pp. 179–208, summarized in Nature, vol. 110, 1922, p. 716), and by Cunningham (Nature, vol. 113, 1924, p. 199). See also Schmidt (Rapp. et Proc.-Verb. Cons. Perm. Internat. Explor. Mer, vol. 5, No. 4, 1906, pp. 137–204, pls. 7–13); for a popular account see Smith (Nat. Geog. Mag., vol. 24, No. 10, October 1913, p. 1140).

[55] Eels are native in Lake Ontario which they reach by way of the St. Lawrence River; and up the Mississippi drainage systems even as far as North Dakota, Wisconsin, Ohio, and western Pennsylvania.

[56] Mem. R. Comit. Talassogr. Ital., vol. 158, 1929.

[57] Bull. U. S. Bur. Fish., vol. 43, 1928, p. 114.

2½ inches long in April are about 5 inches long a year later, or about 2 years after their transformation. The winter rings on the scales have shown that full grown adults of the European species are from 5 to 20 years old, depending on food supply, and other conditions; this is corroborated for the American species by the fact that Dr. Hugh M. Smith, former Commissioner of the United States Bureau of Fisheries, found that a female, on the way down the Potomac, was in her twelfth year.

At the approach of sexual maturity, which takes place in the fall, the eels that are in fresh water drop downstream, traveling mostly at night. They now cease feeding, as do those that have been living in the river mouths, bays, and estuaries; the color of the back changes from olive to almost black, the ventral side turns silvery, and the eyes of the males grow to twice their previous size. Both males and females then move out to sea, and it is not until after they reach salt water that the ovaries mature. In fact, no perfectly ripe female eel has ever been seen, and only one ripe male (of the European species).

So little is the life history of the eel understood by our fishermen that we again emphasize the undoubted fact that no eel ever spawns in fresh water.

The eels drop wholly out of sight when once they leave the shore;[58] no one knows how deep they swim, but they certainly journey out beyond the continental slope into the oceanic basin before depositing their eggs. Schmidt has been able to outline the chief spawning center of the American species (from the captures of its youngest larvae) as between latitudes 20° and 30° N. and between longitudes 60° and 78° W.; i. e., east of Florida and of the Bahamas south of Bermuda. But it may also spawn (always in deep water) farther north as well.[59]

The American eel spawns in midwinter, thus occupying one to two months in its journey from the coast to the spawning ground, for Schmidt found very young larvae (7 to 8 mm.) in February. Eels, like Pacific salmon, die after spawning, the evidence of this being that no spent eels have ever been seen and that large eels have never been known to run upstream again. Smith suggests that they probably "jellify" and disintegrate, as the conger does.

Eels (European) are among the most prolific fish, ordinary females averaging 5 to 10 million eggs and the largest ones certainly 15 to 20 million. It is doubtful whether eggs laid by the American eel have been seen, or of the European either, for that matter.[60] But it is generally supposed that they float in the upper or intermediate water layers until hatching. The larval, so-called "leptocephalus" stage, like that of all the true eels, is very different in appearance from the adult, being ribbon-like and perfectly transparent, with small pointed head; and it has very large teeth, though it is generally believed to take no food until the time of its metamorphosis. These leptocephali of our eel, living near the surface, have been found off our coasts as far north as the Grand Banks, but never east of longitude 50° W.

Inasmuch as the breeding areas of the American and European eels overlap, not the least interesting phase of the lives of the two is that the larvae of the American species should work so consistently to the western side of the Atlantic, and those of the European to the eastern side that no specimen of the former has ever been taken in Europe or of the latter in America.

The American eel takes only about one-third as long as the European to pass through its larval stage; i. e., hardly a year, as against 2 to 3 years. The leptocephali reach their full length of 60 to 65 mm. by December or January, when metamorphosis takes place to the "elver"; the most obvious changes being a shrinkage in the depth and length of the body but an increase in its thickness to cylindrical form, loss of the larval teeth, and total alteration in the aspect of head and jaws, while the digestive tract becomes functional.

It is not until they approach our shores, however, that the adult pigmentation develops or that the elver begins to feed, a change that is accompanied by a second decrease in size. How such feeble swimmers as the leptocephali find

[58] Large eels, on their seaward journey, have occasionally been caught by otter trawlers in the western part of the British Channel, but we know of no such occurrence on this side of the Atlantic.

[59] See Schmidt (Ann. Rep. Smithsonian Inst., (1924) 1925, pp. 279–314) for a readable account of the investigations which enabled him to chart the breeding places and seasons of the American and European eels.

[60] Four eggs taken on the *Arcturus* expedition near Bermuda in 1925 were provisionally identified as those of the American eel by Fish who has pictured them and the larvae hatched from one of them (Zoologica, New York Zool. Soc., vol. 28, 1927, pp. 290–293, figs. 103–107). But the date at which they were taken (July 15–17) makes it more likely that they belonged to some other member of the eel tribe.

their way into the neighborhood of the land remains a mystery. It seems certain, however, that all the young eels bound for the Gulf of Maine complete the major part of their metamorphosis while they still are far offshore. Thus we have never taken one in the leptocephalus stage in the Gulf of Maine in all our tow-nettings, whereas (more significant) the *Albatross* towed three young eels in the so-called "glass-eel" stage, 54 to 59 mm. long, of practically adult form but still transparent, during her spring cruise in 1920, one of them on Georges Bank, March 11; a second on Browns Bank, April 16; and one in the western basin of the Gulf off Cape Ann, February 23. Evidently they were intercepted on their way in to land. And since all three were on the surface, we may take it that glass eels, like leptocephali, keep to the uppermost water layers during their journey.

General range.—Coasts and streams of West Greenland,[61] eastern Newfoundland,[62] Strait of Belle Isle, and northern side of the Gulf of St. Lawrence south to the Gulf of Mexico, Panama, West Indies, and (rarely) to the northern coast of South America; also Bermuda; running up into fresh water but going out to sea to spawn p. 153.

Occurrence in the Gulf of Maine.—The occurrence of the eel around the periphery of our Gulf can be described in the one word "universal." There is, we believe, no harbor, stream mouth, muddy estuary, or tidal marsh from Cape Sable on the east to the elbow of Cape Cod on the west but supports eels in some numbers, and they run up every Gulf of Maine stream, large or small, from which they eventually find their way into the ponds at the headwaters unless barred by insurmountable barriers such as very high falls. Examples of long journeys by eels upstream, in New England rivers, are to the Connecticut Lakes, New Hampshire, at the head of the Connecticut River; to the Rangeley Lakes at the head of the Androscoggin, and to Matagamon Lake, at the head of the East Branch of the Penobscot. Eels are even caught in certain ponds without outlets, as noted above (p. 152). On the other hand, we have seen a few (and some large ones) along the open coast, at Cohasset, for example, but always close in to the shore line and in only a few feet of water, where flounder fishermen catch them from time to time.

Importance.—Schmidt has suggested that the American eel is not as plentiful in actual numbers as the European, arguing from the facts that its larvae have not proven so common on the high seas, and that the American catch of eels (about 2,000 tons yearly) was but a fraction as large as the European catch (about 10,000 tons annually). But it is not safe to draw any conclusions from the statistics because the American catch is limited more by the fact that eels are not much in demand, than by the available supply. And the local demand is less for them today than it was 30 years ago, as is reflected in a decrease in the reported landings from about 305,000 pounds for Maine and about 240,000 pounds for Massachusetts in 1919 to about 19,000 pounds for Maine and about 32,000 pounds for Massachusetts in 1947. The yearly landings of eels along the Canadian shore of our Gulf and from the tributary fresh waters are 30,000–40,000 pounds nowadays.

Practically the entire coastwise catch is made in salt marshes, estuaries and stream mouths; the numbers captured up stream are negligible of recent years, except in New Brunswick where 16,000 pounds were caught in the lower sections of the St. John River System in 1950.[63] In Germany, however, where the demand for eels is much greater, the yearly catch is nearly four times as great for rivers and other fresh waters as it is for the coast. And many millions of elvers were transplanted, during the 1930's, from British rivers (the Severn in particular) to landlocked bodies of water in Central Europe which the young eels could not reach naturally.

The greater part of the catch is made in nets and eelpots; and some are speared, mostly in late autumn and winter, often through the ice.

American conger *Conger oceanica* (Mitchill) 1818 [64]

Sea eel

Jordan and Evermann, *Leptocephalus conger* (Linnaeus 1758), 1896–1900, p. 354.

[61] Jensen (Invest. of the *Dana* in West Greenland Waters, 1925, Extr. Rapp. et Proc. -Verb Cons. Internat. Expl. Mer, vol. 39, 1926, p. 101) records the American eel as one of the four fresh-water fishes known from the west coast of Greenland.

[62] Reported by Dr. G. W. Jeffers as common.

[63] Information from A. H. Leim.

[64] The American conger had long been considered identical with the European. But Schmidt (Nature, vol. 128, 1931, p. 602) has recently shown that it is a distinct species, characterized by having fewer vertebrae; a relationship paralleling that between the American and European eels of the genus *Anguilla*.

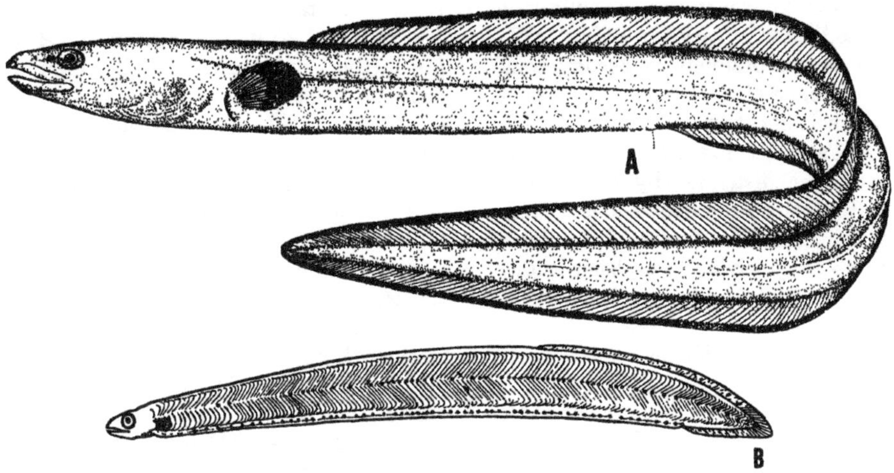

FIGURE 70.—American conger (*Conger oceanica*). A, adult, Connecticut; from Goode, drawing by H. L. Todd; B, "Leptocephalus" stage, 84 mm., Chesapeake Bay.

Description.—The readiest characters by which to distinguish the conger from other eels are noted in the key (p. 150); notably the origin of the dorsal fin above or only very slightly behind the tip of the pectoral when the latter is laid back, the rather long-pointed snout, the large mouth cleft back at least as far as the middle of the eye, and the scaleless skin. The conger has many more vertebrae than the common eel and there are other skeletal differences.[65] The conformation of the tip of the snout likewise helps to identify the conger, for its upper jaw usually projects beyond the lower, whereas in the common eel the reverse is true, or at least the lower equals the upper. Furthermore, the eyes of the conger are oval and larger than the round eyes of the common eel.

To give an idea of the proportions of the conger, we need only add that the distance from tip of snout to dorsal fin is about one-fifth of the total length; the length of the snout is one-fourth that of the head; the length of the pectorals is equal to one-third to one-fourth of the distance from dorsal fin to tip of snout; and that the body is of the snake-like form characteristic of eels in general.

Color.—Bluish gray or grayish brown above, sometimes of a reddish tinge, sometimes almost black; paler on the sides; dingy white below.

Size. - This is a much larger fish than the common eel. The larger ones taken off southern New England and New Jersey are said to measure 4 feet up to 7 feet in length. The general run of those caught weigh 4 to 12 pounds, the heaviest we have seen weighed about 22 pounds. But the North American species never attains the enormous size reached by the European species; the largest European conger reliably reported, of which we have read, was 9 feet long, and weighed 160 pounds.[66]

Habits.—The depth range of the conger is from close to the coastline (they are caught from the dock at Woods Hole) out to the edge of the continental shelf, the deepest record for it being for one that we trawled at 142 fathoms off southern New England, on the *Albatross III*, in May 1950. It feeds chiefly on fish: butterfish, herring, and eels have been found in their stomachs at Woods Hole. They also prey on shrimps and small mollusks at times. And we have caught them (and have seen them caught) on crabs, on soft clams (*Mya*), on sea clams (*Mactra*) and on cut fish bait.

It is now well established that the European species (hence no doubt the American also) breeds but once during its life and then perishes like the common eel. Ripe congers are never caught on hook and line, for they cease to feed, hence to bite, for some time previous. But the males of the European species, kept in aquaria,

[65] For an account of these, see Smitt (Scandinavian Fishes, vol. 2, 1895 pp. 1015–1017, 1037.

[66] Jenkins, Fishes of the British Isles, 1925, p. 275; see also Day, Fishes of Great Britain, vol. 2 ,1884, p. 253, for large European congers.

have repeatedly been known to become fully ripe, females nearly so,[67] then invariably dying. The ripening of the sexual products is accompanied by changes in the shape of the head; in the loss of the teeth; and in a jellification of the bones, while the eyes of the males become enormous and the females become much distended by the ovaries. It is probable that the American conger ripens off the coast of southern New England in summer; European congers in captivity have been known to do so every month in the year except October and November.

It seems that the conger, like the common eel, moves out from the coast to spawn, for its young larvae have never been taken inshore, and Dr. Johannes Schmidt's[68] discovery of very young larvae in the West Indian region, but nowhere else, points to this as the chief spawning ground of the American conger, if not the only one.

The congers are extremely prolific fish, the number of eggs a European female may produce having been estimated as high as 3 to 6 millions. American conger eggs have never been identified, for although eggs taken over the tilefish grounds 30 miles south of Nantucket lightship in July 1900[69] have been credited to this species, there is no certainty that this was their true parentage.

It has long been known that the congers, like the common eels, pass through a peculiar ribbon-like larval stage (the so-called "leptocephalus") very broad and thin and perfectly transparent, with a very small head.[70] In fact the first leptocephalus ever seen (about 1763) was the larval European conger. But its identity was not established definitely until 1886, when the famous French zoologist, Delage,[71] reared one through its metamorphosis at the biological station at Roscoff.

The leptocephalus stage of the conger is relatively more slender than that of the common eel, it grows larger (to a length of 150–160 mm.), and its vertebrae and muscle segments are far more numerous (140–149 in the American conger, 154–163 in the European) than in the common eels (about 107 in the American eel and about 114 in the European). But the number of body segments (visible only under a lens) is not of itself a safe clue to identity, for there are as many or more in the long-nosed eel (p. 158) which has been reported in the Gulf; also in the morays, and in various other members of the eel tribe.[72]

The duration of the larval period of the conger is not known. The process of metamorphosis consists essentially in a thickening and narrowing of the body, an enlargement of the head, the formation of the swim bladder and permanent teeth, and the development of pigment in the skin, a change that occupied about two months (May to July) in the case of Delage's European specimen. His young conger was 9.3 centimeters (3.6 inches) long at its completion.[73]

General range.—Continental shelf of eastern America: adults are known north to the tip of Cape Cod; larval stages to eastern Maine. Its southern boundary cannot be stated until the congers of the coasts of North and of South America have been critically compared. It is represented by a closely allied species (*Conger conger*) in the eastern North Atlantic.

Occurrence in the Gulf of Maine.—The only records for adult congers within the limits set here for the Gulf of Maine are of one taken at North Truro, Cape Cod; a second specimen trawled close to Provincetown in Cape Cod Bay, July 5, 1951, by Capt. Herman Tasha; and a third, trawled south of Nantucket shoals by *Albatross III*, in mid May 1950.[74] But the conger must be much more plentiful at times off the shoals than the foregoing would suggest for Capt. Henry Klim of the dragger *Eugene H* reports trawling 1,400 pounds of them there, at 76 fathoms, March 25–30, 1951.[75] And its curious band-like "leptocephalus" larvae have been found within the Gulf on several occasions. Thus, half a dozen specimens were picked up on the beach at Cherryfield and Old Orchard, Maine, and at Nahant, Mass.,

[67] Cunningham (Jour. Mar. Biol. Assoc. United Kingdom, N. Ser., vol. 2, 1891–92, pp. 16–42) gives an interesting account of this and other phases of the life history of the conger.

[68] See Nature, vol. 128, 1931, p. 602, for a discussion of this question by Dr. Schmidt.

[69] Eigenmann, Bull. U. S. Bur. Fish., vol. 21, 1902, p. 37.

[70] For photographs of the leptocephalus stage of the European conger, see Schmidt, Rapp. et Proc. Verb. Cons. Perm. Internat. Explor. Mer, vol. 5, No. 4, 1906, pl. 9, figs. 8, 9; and Meddelelser Komm. Havundersøgelser, Ser. Fiskeri, vol. 3, No. 6, pl. 1, figs. 1–3.

[71] Conptes Rendus Acad. Sci. Paris, vol. 103, 1886, p. 698.

[72] Fish (Zoologica, New York Zool. Soc., vol. 8, 1927, pp. 307–308) gives a table of the numbers of body segmentsfor various eels and for "leptocephalus" larvae of known and unknown parentage.

[73] Schmidtlein (Mittleil, Zool. Stat. Neapel, vol. I, 1879, p. 135) speaks of young "congers" at Naples in April as hardly one-third as long as this, a discrepancy suggesting that these may actually have belonged to one of the Muraenoid eels.

[74] Local reports of congers do not necessarily relate to the true conger, for the eel pout (p. 510), which is common in the Gulf, is often misnamed thus.

[75] At lat. 40° N., long. 69° 50′ W.

more than a half century ago. Two specimens, also picked up on the beach, were sent up from Newburyport, Mass., in November 1929; and A. H. Clark, of the U. S. National Museum, informs us that he has found many larvae of the leptocephalus type at Manchester, Mass., which probably were congers to judge from their size.

The conger occurs regularly and commonly to the west and south of Cape Cod, being taken near Woods Hole from July into the autumn, and about Block Island from August until November. Very little is known about their movements. But we suspect that they shift offshore into deeper and warmer water for the winter, judging from their absence then in shoal water, contrasted with the large offshore catch in March mentioned above (p. 156) and with the fact that we saw several trawled at 50 to 142 fathoms off southern New England on the *Albatross III*, in May in 1950.

Slime eel *Simenchelys parasiticus* Gill 1879

SNUB-NOSED EEL

Jordan and Evermann, 1896–1900, p. 349.

Description.—The most distinctive characters of the slime eel, its eel-like form, snub nose, long dorsal fin, and soft and slimy body, have been mentioned already (p. 150). It is stouter and more sway-bellied than the common eel, very soft, and with a more tapering tail. The dorsal fin originates a short distance behind the tips of the pectorals when the latter are laid back against the body, and the anal runs forward on the lower surface almost to the vent, which is situated about midway of the body. The head is much shorter than in either the common eel or the conger; the mouth is small, gaping back only about half way to the forward edge of the eye, with upper and lower jaws of equal length and each armed with a single series of small, close-set cutting teeth. The gill openings are small, and instead of being vertical and on the sides of the neck as they are in the common eel, they are longitudinal and lower down on the throat.

Color.—Dark brown, with the belly only a little paler than the back, though usually more or less silvery.

Size.—About 2 feet long.

Habits.—It is partly parasitic in habit, burrowing into the bodies of halibut and other large fish, circumstances under which a considerable number of specimens have been brought in by fishermen. Very likely it was common inshore in the old days when halibut were plentiful there. It also lives independently on the bottom. Nothing is known of its manner of life beyond this, nor of its breeding habits. We may add from experience that it is as slimy as a hag and drips with sheets of mucus when drawn out of the water.

General range.—The continental slope, and the slopes of the offshore banks, from abreast of the eastern end of Long Island to the Newfoundland Banks, in depths ranging from 200 to more than 900 fathoms; also in deep water about the Azores, and represented in Japanese waters by an extremely close relative, if, indeed, it is separable at all from the Atlantic slime eel.[76]

Occurrence in the Gulf of Maine.—There is no definite record of the snub-nosed eel actually within the southern rim of the Gulf so far as we can learn, and our only first-hand experience with it was on the slope south of Nantucket lightship, where we captured 21 in a Monaco deep-sea trap

[76] The Japanese slime eel, described first as a distinct species (*leptosomus*) by Tanaka in 1908, has been classed more recently by him (Fishes of Japan, vol. 42, 1928, p. 810, pl. 173, fig. 476) as identical with the Atlantic *parasiticus*.

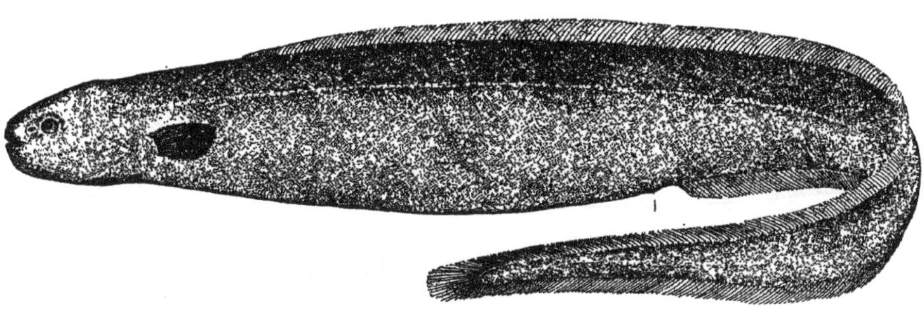

FIGURE 71.—Slime eel (*Simenchelys parasiticus*), off Sable Island Bank. From Goode and Bean. Drawing by H. L. Todd.

in 455 fathoms, on the *Grampus* in July 1908. It must be extremely abundant along that zone, however, for so many to find their way into the trap in as short a set as two hours. And it has been recorded so often in water as shoal as 200 fathoms that it may be expected in the bottom of the Eastern Channel and in the southeastern deeps of the Gulf of Maine.

Long-nosed eel *Synaphobranchus pinnatus* (Gronow) 1854

Jordan and Evermann, 1896–1900, p. 351.

Description.—This deep-sea species, a typical eel in general appearance, is readily identifiable among its tribe by the fact that while its dorsal fin originates about as far back as in the common eel (p. 151), relative to the length of the fish, its point of origin is considerably behind the vent instead of in front of the latter, and that its anal fin originates considerably in front of the dorsal fin instead of behind it as is the case in all other Gulf of Maine eels. Furthermore, its mouth is much wider, gaping far back of the eye, and its snout is pointed. The most interesting anatomic characteristic of this eel is that its gill openings, opening longitudinally on the lower side of the throat, join together in front, apparently as a single V-shaped aperture, though actually they are separate within.

Color.—Grayish, darkest below, with the vertical fins darker behind but pale-edged in front, and with the inside of the mouth blue black.

Size.—The largest of 89 specimens measured by Goode and Bean was nearly 22 inches (545 mm.) long, the smallest about 9 inches (221 mm.) The largest we trawled on the *Caryn*, in June 1949, was 24 inches (605 mm.) long. Collett [77] mentions one 26½ inches (675 mm.) long from the Azores.

[77] Result. des Camp. Sci. Prince de Monaco, Pt. 10, 1896, p. 154.

Habits.—Nothing is known of its habits except that it is a ground fish; that the readiness with which it bites a baited hook proves it predaceous; and that specimens in spawning condition have been taken in summer.[78] On June 17, 1949 in lat. 42° 38′ N., long. 64° 04′ W., in 400-460 fathoms, we trawled many on the *Caryn*, both males and females, 18½ to 24 inches (470-605 mm.) long that had well developed gonads, one female having already spawned. The ripe eggs are orange in color and about 1 mm. in diameter. In its development it passes through a leptocephalus stage even more slender than that of the American conger (p. 156), and its body segments (144-149) overlap those of the American conger (140–149) in number.

General range.—This deep-water species has a wide distribution. In the western side of the North Atlantic it has been taken at many localities along the continental slope from the offing of South Carolina to the Grand Banks; it is known in the east from the Cape Verdes; off Morocco; from the Canaries; from the Azores; near Madeira; also from the Faroe Bank and Faroe-Shetland Channel. And its leptocephalan larvae have been taken in such numbers from north of Spain to south of Iceland that it must be one of the most plentiful of deep-water fishes there.[79] It is also recorded off Brazil in the South Atlantic; likewise in the Arabian Sea; about the Philippines; and in Japanese waters, or is represented there by a very close relative. Most of the captures have been from depths of 300 to about 2,000 fathoms, but it has been taken as shoal as 129 fathoms.

[78] The "leptocephalus" larvae of the long-nosed eel are described, with photographs by Schmidt (Rapp. et Proc. Verb. Cons. Perm. Internat. Explor. Mer, vol. 5, No. 4, 1906, p. 191, pl. 9, figs. 4-6; and Meddel. Komm-Havundersøgelser, Ser, Fiskeri, vol. 3, No. 6, 1909, p. 7).

[79] This fact is commented on by Schmidt (Rapp. Cons. Perm. Internat. Explor. Mer, vol. 5, No. 4, 1906, p. 191). For further details as to its distribution see Koefoed, Rept. *Michael Sars* North Atlantic Exped., (1910), vol. 4, Pt. 1, 1927, pp. 11, 14.

FIGURE 72.—Long-nosed eel (*Synaphobranchus pinnatus*), La Have Bank. From Goode and Bean. Drawing by H. L. Todd.

Occurrence in the Gulf of Maine.—This eel has not actually been reported within the geographic limits of the Gulf. But it is to be expected in the eastern channel and possibly above 150 fathoms along the slopes of Georges Bank, for fishermen have caught them in water as shallow as that off La Have Bank, while they have been trawled in 168 fathoms and 129 fathoms off southern New England by the *Fish Hawk* and *Albatross*. So many of them have been brought in by fishermen [80] from deep water off the fishing banks to the eastward of longitude 65°, and so many have been trawled along the continental slope thence westward,[81] that this eel must be one of the commonest of fishes below 150 to 200 fathoms, all the way from the Grand Banks to abreast of New York.

Snake eel *Omochelys cruentifer* (Goode and Bean) 1895

Jordan and Evermann, 1896–1900, p. 377, *Pisodonophis cruentifer.*

Description.—The most striking feature of the snake eel and one that distinguishes it from all other Gulf of Maine eels is that the tip of its tail is hard and pointed. Other distinctive features are that it is only about one thirty-seventh to one thirty-eighth as deep as it is long; that its dorsal fin originates only a short distance behind the tips of the pectorals when these are laid back; that its anal fin originates far behind its dorsal fin; that its snout is bluntly pointed; and that its mouth gapes rearward considerably beyond its eyes (but not so far back as in the long-nosed eel, p. 158). The dorsal and anal fins end a little in front of the tip of the tail. The gill openings are short new-moon-shaped slits, close in front of the bases of the pectoral fins. Its "peculiar and savage physiognomy" was stressed by its describers.[82]

[80] Many such instances are listed in the Rept. U. S. Comm. Fish. (1879) 1882, p. 787.

[81] Goode and Bean, Smithsonian Contrib. Knowl., vol. 30, 1895, pp. 143-144.

Color.—Originally described as uniform brownish yellow. But those that we have seen have been uniform light brown below as well as above, large ones darker than small ones. A young one about 2½ inches (6½ cm.) long was pale with dark speckles.

Size.—The largest yet seen was 16⅝ inches long.

Habits.—The original account of the snake eel includes the information that specimens had been received that had been taken from the bodies of other fish, evidence that it is a parasitic-boring form. Nothing else is known of its habits.

General range.—Western side of the Gulf of Maine to the offing of Cape Henry, Va.

Occurrence in the Gulf of Maine.—The snake eel was originally described in 1895 from 6 specimens trawled off Nantucket by the *Fish Hawk*, and a number have been taken thence southward to the latitude of Cape Henry, Va., by the *Albatross II*, in depths of 24 to 245 fathoms. The only report of it within the Gulf of Maine is by its describers of specimens taken by fishermen on Jeffreys Bank many years ago.

Snipe eel *Nemichthys scolopaceus* Richardson 1848

Jordan and Evermann, 1896–1900, p. 369.

Description.—The snipe eel is easily recognizable by its extremely slender body (the fish may be 75 times as long as deep), with its tail tapering to a thread, and by its elongate, slender, bill-like jaws, one as long as the other, the upper one curving upward, but the lower more nearly

[82] Goode and Bean, Smithsonian Contrib. Knowl., vol. 30, 1895, p. 147.

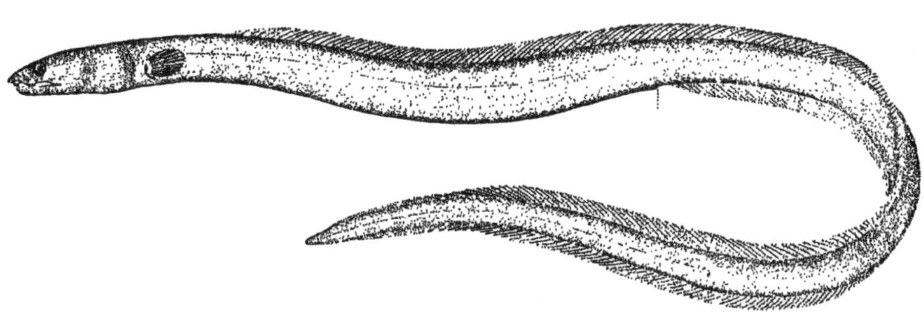

FIGURE 73.—Snake eel (*Omochelys cruentifer*), continental slope south of Nantucket Shoals. From Goode and Bean. Drawing by H. L. Todd.

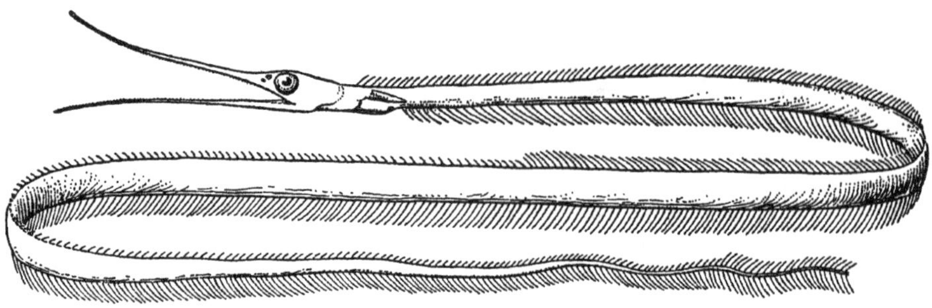

FIGURE 74.—Snipe eel (*Nemichthys scolopaceus*). Drawing by E. N. Fischer, from Bigelow and Welsh.

straight. The head is much deeper than the neck, with large eyes. The dorsal fin originates in front of the pectorals, the anal about abreast of the tip of the pectorals, and both dorsal and anal run back to the tip of the tail.

There has been some confusion in the published accounts and illustrations as to the dorsal and anal fins, for while Vaillant[83] shows both as about as high throughout their length as the fish is deep, Goode and Bean[84] picture the dorsal as much higher than the anal (the artist evidently having transposed the two fins), whereas Brauer[85] represents the anal as approximately twice as high as the dorsal and the latter as soft rayed in its anterior and posterior portions but composed of short thorn-like spines along its central third. The fins of two specimens taken off New England, now in the collections of the Museum of Comparative Zoology are as follows:

Dorsal, soft-rayed and nearly as high as the body is deep for its first half; back of that it consists of a series of very short, stiff rays that extend to the tip of the tail.

Anal, soft-rayed throughout its length and about as high as the body is deep, tapering to almost nothing on the tail.

The confusion has been due in part to the rather fragmentary state in which these deep-water fish usually arrive on board, but at the same time it is probable that two distinct species have been confused under the name *scolopaceus*, as Brauer suspected.

Color.—Described as pale to dark brown above with the belly and anal fin blackish after preservation. Judging from experience with other deep-sea fishes and from Brauer's plate (which, however, may be another species), we suspect that it is chocolate brown above in life and velvety black below.

Size.—Maximum length about 3 feet.

Habits.—Although commonly spoken of as a "deep-sea" fish, this species is undoubtedly an inhabitant of the mid depths, not of the bottom, and judging from the occurrence of other black fishes it probably finds its upper limit at 100 to 200 fathoms. Nothing further is known of its habits, but Mowbray's[86] capture near Bermuda of a snipe eel clinging by its jaws to the tail of a large red snapper has suggested that such may be a regular habit of this curious species.

General range.—The snipe eel has been taken in deep water at many stations off the east coast of North America between latitudes 31° and 42°N., longitudes 65° and 75°W.; also in the South Atlantic; near the Azores; near Madeira; off the Cape Verde Islands; off West Africa; and in the Pacific of New Guinea.

Occurrence in the Gulf of Maine.—One specimen taken from the stomach of a codfish caught on Georges Bank in 45 fathoms is the only Gulf of Maine record, but several have been taken in depths of from 300 to 2,000 fathoms on the seaward slope of the bank.

THE LANCET FISHES. FAMILY ALEPISAURIDAE

The lancet fishes have one long and very high dorsal fin, soft-rayed from end to end; a small adipose fin behind the dorsal fin, like that of a salmon or smelt; a deeply forked caudal fin; a short anal, most of which is behind the rear end of the dorsal; large pointed pectorals and ven-

[83] Poissons. Exped. Sci. *Travailleur* et *Talisman*, 1888, pl. 7, figs. 2 and 2a.
[84] Smithsonian Contrib. Knowl., vol. 31, 1895, pl. 46, fig. 170.
[85] Tiefsee-Fische, Wiss. Ergeb. Deutsch. Tiefsee-Exped. (1898-1899), vol. 15, Pt. I, 1900, p. 126, pl. 9, fig. 1.

[86] Copeia, No. 108, 1922, p. 49.

trals; and a very wide mouth with large teeth. Several species are known, all belonging to deep water; only one has been taken within the province covered by this report. Their closest affinities seem to lie with the lanternfishes (p. 141).

Lancetfish *Alepisaurus ferox* Lowe 1833

HANDSAWFISH

Jordan and Evermann, 1896–1900, p. 595.

Description.—The combination of a long and very high dorsal fin, soft-rayed from end to end, with the presence of an adipose fin behind it, distinguishes the lancetfish from all other Gulf of Maine fishes. The body is slender, somewhat flattened sidewise, deepest at the gill covers, and tapers back to a slender caudal peduncle. The snout is long and pointed, the mouth wide, gaping back of the eye, and each jaw has two or three large fangs, besides smaller teeth. The dorsal fin (41 to 44 rays) originates on the nape and occupies the greater length of the back, is rounded in outline, about twice as high as the fish is deep, and can be depressed in a groove along the back. The adipose fin recalls that of the smelt in form and location. The caudal is very deeply forked; its upper lobe is prolonged as a long filament, and although most of the specimens so far seen have lost this we have an excellent photograph showing it. The anal fin originates under the last dorsal ray, and is deeply concave in outline. The ventrals are about halfway between the anal and the tip of the snout, while the pectorals are considerably longer than the body is deep and are situated very low down on the sides. There are no scales and the fins are exceedingly fragile.

Color.—Sides described as metallic silvery. We have not seen a newly taken specimen.

Size.—The collection of the Boston Society of Natural History contains the cast of a specimen about 6 feet long that was taken off Nova Scotia in August 1910, and this is probably about the maximum size.

Habits.—This is an oceanic species, of the mid-depths, appearing only as a stray shoaler than 200 fathoms. Nothing is known of its habits. A Block Island specimen had eaten a small spiny dogfish.

General range.—Widely distributed in the deep waters of the Atlantic, also reported from the northeastern Pacific.[87]

Occurence in the Gulf of Maine.—A specimen brought in by a fisherman from Georges Bank[88] about 1878 or 1879 is its only claim to mention here. Goode and Bean and Vladykov and McKenzie[89] have reported other captures of lancetfishes from La Have Bank, from southeast of Emerald Bank and Banquereau. Another specimen 5½ feet long was caught alive in the surf on Block Island, R. I., March 12, 1928, and reported by Mrs. Elizabeth Dickins who sent us a photograph of it.

[87] Crawford (Copeia, No. 164, 1927, p. 66) reports several *A. ferox* from the halibut banks off the northwestern coast of British Columbia.

[88] No definite information is available as to this specimen.

[89] Proc. Nova Scotia Inst. Sci., vol. 19, 1935, p. 63.

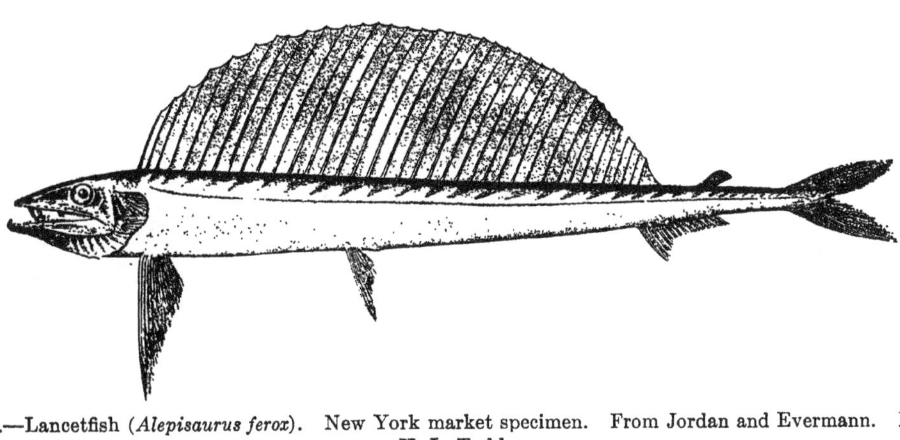

FIGURE 75.—Lancetfish (*Alepisaurus ferox*). New York market specimen. From Jordan and Evermann. Drawing by H. L. Todd.

THE MUMMICHOGS OR KILLIFISHES. FAMILY POECILIIDAE

The mummichogs are small fishes made recognizable by having only one short soft-rayed dorsal fin situated far back, and ventrals situated on the abdomen, combined with a small mouth at the tip of the snout, a very thick caudal peduncle, and a rounded tail fin. The family is represented in the Gulf of Maine by three species, two of *Fundulus* and one of *Cyprinodon*, the former slender and the latter deep in outline, a difference in body form sufficient to distinguish the one genus from the other at a glance. The teeth are likewise different in the two genera, those of *Fundulus* being sharp-pointed, whereas they are wedge-shaped in *Cyprinodon* and incisorlike. The two local species of *Fundulus* are separable by their markings, *majalis* of all ages being barred or streaked with black while the adult *heteroclitus* is not.

Common mummichog *Fundulus heteroclitus* (Linnaeus) 1766

KILLIFISH; SALT-WATER MINNOW; CHUB; MUMMY

Jordan and Evermann, 1896–1900, p. 640.

Description.—This is a stout-bodied little fish, about one-fourth as deep as long, its body thickest just back of the pectoral fins, whence it tapers to the tail. Both its back and its belly are rounded, but the top of the head is flat between the eyes. The snout, as seen from above, is blunt. The mouth is at the tip of the snout and is so small that it does not gape back to the eye. Perhaps the most striking feature of *Fundulus* is its very deep caudal peduncle and rounded caudal fin. The fins are of moderate size, the dorsal situated behind the middle of the body above the anal, the pectorals broad and rounded. Both head and body are covered with large rounded scales. On males in breeding condition the scales on the sides of the head and those on the flanks below and behind the dorsal fin develop fingerlike processes on their free edges, called "contact organs."

The mummichog shows a striking sexual dimorphism in the dorsal and anal fins, which are not only larger in the male than in the female, and the anals of a different shape,[90] but are more muscular and are used as claspers in the act of spawning.

Color.—Males and females differ in color as well as in the sizes of the fins. Out of breeding season the males are dark greenish or steel blue above, with white and yellow spots, and marked on the sides with narrow irregular silvery bars or mottlings made up a series of dots. The belly is white, pale yellow, or orange; the dorsal, anal, and caudal fins are dark green or dusky with pale mottlings; the front edges of the anal and of the ventrals are yellow. Sometimes there is a dark-edged, pale-centered eyespot on the rear part of the dorsal fin. At spawning time the pigmentation of the male is generally intensified, the back and upper sides darkening almost to black, while the yellow of the belly becomes more brilliant and the body generally takes on steel-blue reflections. The females (much paler than the males) are uni-

[90] A detailed account of the sexual differences is given by Newman (Biol. Bull., vol. 12, No. 5, 1907, pp. 314–348).

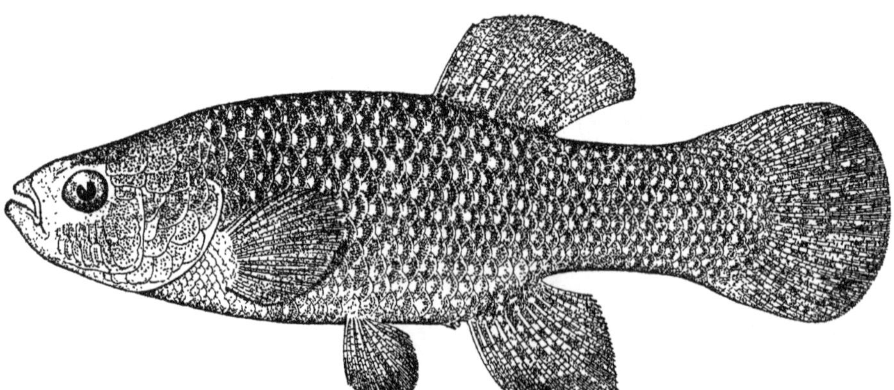

FIGURE 76.—Common mummichog (*Fundulus heteroclitus*), Maryland. From Jordan and Evermann. Drawing by A. H. Baldwin.

form olive to bottle green, darker above, lighter below, without definite markings though their sides often show faint and indefinite crossbars of a deeper tone of the same hue. Their fins are much paler than those of males. Very young fry of both sexes show dark transverse bars on the sides, but these bars are lost with growth.

Killifishes vary in shade from very pale to dark, according to the color of their surroundings. And recent experiments [91] have proved that their ability to change from light to dark depends on the sense of sight.

Size.—The maximum length is 5 to 6 inches, but adult mummichogs are seldom more than 3½ to 4 inches long in the Gulf of Maine. Several varieties of this species have been described, but they are so closely allied that it is not necessary to discuss them here.

Habits.—The home of the mummichog is along sheltered shores where the tide flows over beds of eelgrass or salt hay (*Spartina*), among which shoals of "mummies" may often be seen moving in with the flood tide. They abound in the tidal creeks that cut the salt marshes, on the shores of our harbors, and in the brackish water at the mouths of our streams and estuaries, particularly in little muddy pools, creeks, and ditches. So closely, indeed, do they hug the shore that a line drawn 100 yards out from land would probably inclose practically all the mummichogs in the Gulf of Maine. Where the shore is bold and rocky, as it is about the Bay of Fundy, the mummichog is practically restricted to brackish water, and it often goes up into fresh water. At the other extreme, it is not likely that they ever descend to a depth of more than a couple of fathoms in their journeys in and out of the creeks or along the shore.

Mummichogs are so resistant to a lack of oxygen, to the presence of carbon dioxide, and to unfavorable surroundings generally, that they can survive in very foul water. At ebb tide "mummies" are often trapped in little pools where they remain until the next tide if the water holds, often huddled together in swarms. Should the pool go dry, they work their way for the time being into the mud, where we have often found them.[92] And it is probable that they can flop overland for a few yards to some other drain as the striped mummichog does.

They winter in a more or less sluggish state on the bottoms of the deeper holes or creeks, where they have been found buried 6 or 8 inches deep in the mud,[93] and there is no evidence that they move out to sea during the cold season; in short, this is one of the most stationary of fishes.

Food.—"Mummies" are omnivorous, feeding on all sorts of edible things, vegetable as well as animal. They have been found full of diatoms, eelgrass, and other vegetable matter; foraminifera; shrimps and other small Crustacea, also small mollusks; and they sometimes have small fish in their stomachs.[94]

They soon congregate about any dead fish or other bit of carrion, to prey either upon it or upon the amphipod scavengers that gather on such dainties and they eat their own or each other's eggs at spawning time.

Spawning probably takes place at the same season in the Gulf of Maine as on the southern coast of New England; i. e., in June, July, and early August. As sexual activity approaches, the males (now brilliantly tinted) court and pursue the females, rivalry among them being very keen; those that are the most highly colored or most excited usually driving off the others. Sometimes they fight fiercely. They spawn in a few inches of water, seeking shady spots. At the moment of spawning the male clasps the female with his anal and dorsal fins just back of her anal and dorsal, usually forcing her against some stone or against the bottom, the bodies of both are bent into an S and their tails vibrate rapidly while the eggs and the milt are extruded.[95] Occasionally, pairs clasp and spawn free in the water without coming in contact with any object, and sometimes a female is seen to pursue and court a male.

The eggs, which are about 2 mm. in diameter, colorless or pale yellowish and surrounded by a firm capsule, sink and become so sticky on contact with the water that they mass together in clumps, or stick fast to sand grains or to anything else they chance to rest upon. Incubation occupies from 9 to 18 days, the exact duration probably

[91] Parker and Lanchner, Amer. Jour. of Physiol., vol. 61, 1922, p. 548.

[92] This habit is described by Mast (Jour. of Animal Behavior, vol. 5, No. 5, 1915, p. 351).

[93] Chichester, Amer. Naturalist, vol. 54, 1920, p. 551.

[94] Lists of stomach contents are given by Field (Rept. U. S. Comm. Fish., (1906) 1907, p. 29).

[95] Newman (Biol. Bull., vol. 12, No. 5, April 1907, p. 315) gives an interesting account of the courtship and spawning, from which the preceding is condensed.

depending on temperature. The larva is about 7 to 7.7 mm. long at hatching, its yolk absorbed already, and its pectoral and caudal fins fully formed. By the time the little fish has grown to 11 mm. the dorsal and anal fin rays are present in full number, and the first trace of the ventrals is to be seen. At 16 mm. the ventrals are apparent, and fry of 20 mm. resemble their parents.

General range.—Coast of North America, from the Gulf of St. Lawrence to Texas. Port au Port Bay, on the west coast of Newfoundland,[96] is the most northerly record that we have found for it.

Occurrence in the Gulf of Maine.—The "mummie" is one of the few fish which can fairly be characterized as "universal" in suitable locations around the entire coastline of the Gulf. We dare say that there is not a single bit of salt marsh, muddy creek, harbor, sheltered shore line, or brackish estuary, where they are not to be found, from the elbow of Cape Cod around to Cape Sable.

[96] Johansen (Canadian Naturalist, vol. 40, February 1926, p. 34).

Importance.—The mummichog is of some commercial value as bait, but only locally. It is also a favorite for biological experiment.

Striped mummichog *Fundulus majalis* (Walbaum) 1792

MUMMICHOG; MUMMY; KILLIFISH

Jordan and Evermann, 1896–1900, p. 637.

Description.—This fish resembles the common mummichog closely in general form, in the shape of its dorsal and anal fins, in its sexual dimorphism, and in the development of "contact organs" on the scales of breeding males. But it is more slender, its snout more pointed, its body more definitely fusiform, tapering toward both head and tail, and its caudal peduncle not so stout. But the color pattern is the most striking difference between the two, both sexes of *Fundulus majalis* being definitely barred with black at maturity as well as when young. In the male the barring is transverse throughout life, the stripes increasing

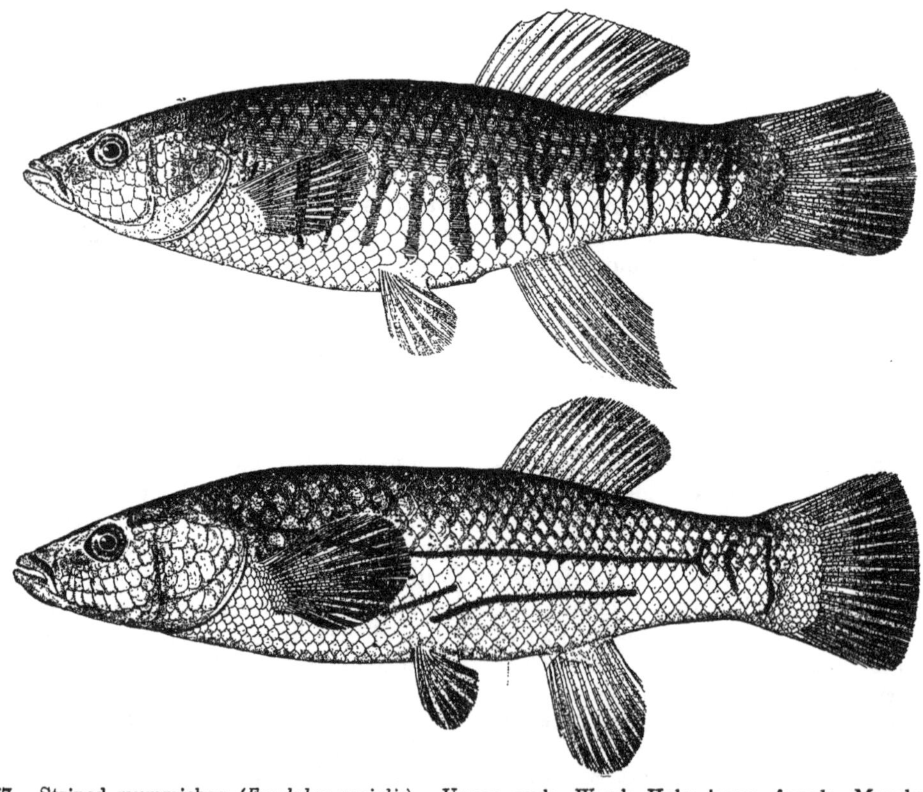

FIGURE 77.—Striped mummichog (*Fundulus majalis*). Upper, male, Woods Hole; lower, female, Maryland. From Jordan and Evermann.

in number from 7 to 12 in the young to 14 to 20 in adult fish. When the females reach a length of about two inches, however, the original 7 to 12 transverse bars are transformed with growth into two or three longitudinal stripes on each side, the upper stripe running uninterrupted from gill opening to tail, the lower stripes in two segments, the one from close behind the pectoral to above the ventral, the other thence backward to close behind the rear edge of the anal fin. One or two transverse bars persist however on the caudal peduncle, even on the oldest females.

Color.—This is a decidedly paler fish than the other "mummy." Apart from the black bars the male is dark olive green above with silvery sides, a greenish-yellow belly, and a black spot on the rear part of the dorsal fin; his pectorals and caudal are pale yellowish. The male becomes more brilliant at breeding time, the back turning almost black, the lower sides and belly changing to orange or golden, and the fins to bright yellow. The female is olive green above and white below, striped as described above.

Size.—This is a larger fish than the common mummichog, occasionally growing to a length of 7 inches and often to 6 inches.

Habits.—The striped mummie parallels the common mummie in being restricted to the immediate neighborhood of the land, and in its way of life, except that it keeps more strictly to salt water, and is found more often along open beaches. Its most interesting habit is its ability to flop back into the water if it becomes stranded with the receding tide, jumping unerringly toward the water in almost every instance, and progressing from several inches to several feet at each jump.[97] And so noted are they for this peculiar ability that a special article has been devoted to it.[98] Their diet consists of small animals including mollusks, crustaceans, fish, insects, and insect larvae. Westward and southward from Cape Cod they spawn from late spring to late summer.

General range.—Coast of the United States, from the vicinity of Boston, Mass., to Florida.

Occurrence in the Gulf of Maine.—The striped mummie is very abundant along the southern shores of New England, westward from Cape Cod. But the only published records for it in the gulf are for the vicinity of Boston and Salem, many years ago, and we had not seen it north of Cape Cod before 1937. In that autumn, however, B. Preston Clark brought in four specimens that he had taken at Cohasset, on the southern shore of Massachusetts Bay; it was reported to us as in numbers there in 1939,[99] and we have seen small schools of them in recent summers in the salt marshes at the entrance to Cohasset Harbor, as well as nearby. If this little fish actually has extended its regular range northward and if its dispersal-route has been via the Cape Cod Canal, as has been suggested,[1] it is to be expected anywhere in the marshes around Cape Cod Bay and along the southern shore of Massachusetts Bay, and we suspect that a resident population is to be found in the Nauset Marshes and in Pleasant Bay, on the outer shore of Cape Cod.

Sheepshead minnow *Cyprinodon variegatus* Lacépède 1803

Jordan and Evermann, 1896–1900, p. 671.

Description.—The sheepshead minnow is so deep bodied (its body is nearly half as deep as long, not counting the tail fin) that it is not apt to be mistaken for either of the mummichogs. And it is separated further from the mummichogs by its teeth, which are large, wedge-shaped and with tricuspid cutting edges, instead of small and pointed. It is a flat-sided little fish, with high arched back, small flat-topped head, small terminal mouth hardly gaping back to the forward edge of the eye, and it has the thick caudal peduncle characteristic of its family. Its tail is square (rounded in the mummichogs), and the fact that almost the whole of its dorsal fin is in front of the anal instead of over it affords another point of difference. The pectorals are large, reaching back past the base of the ventrals, which seem very small by contrast. Both its body and its head are covered with large rounded scales, largest on top of the head and on the cheeks, with one much larger than the others just above the pectoral fin. Young fish are proportionally more slender than old ones. The dorsal, ventral, and anal fins are higher in the males of this species than in the females, much as they

[97] Hildebrand and Schroeder, Bull. U. S. Bur. Fish. vol. 48, Pt. 1 1928 p. 141.

[98] Mast, Jour. of Anima Behavior, vol. 5, No. 5, 1915, pp. 341–350.

[99] By John W. Lowes.

[1] Schroeder, Copeia, 1937, No. 4, p. 238.

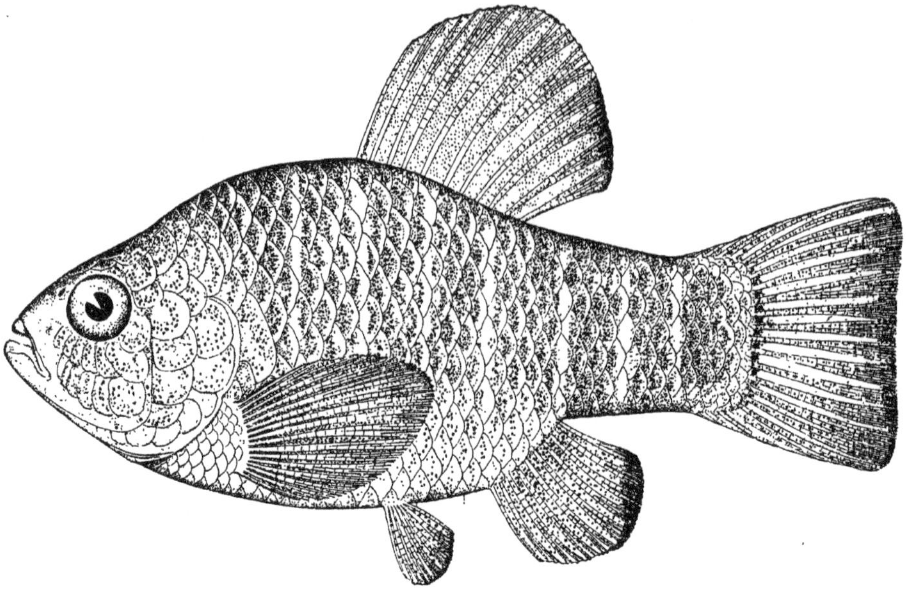

FIGURE 78.—Sheepshead minnow (*Cyprinodon variegatus*), Maryland. From Jordan and Evermann. Drawing by A. H. Baldwin.

are in the mummichogs. The male is deeper bodied and the average size is larger than that of the female.

Color.—Out of breeding season both males and females are olive above (males rather darker and greener than females) with pale yellow or yellowish-white belly, dusky dorsal fin, and pale orange pectoral, ventral, and anal fins. The young of both sexes are irregularly barred with black transverse stripes, which persist through life in the female but become obscured in adult males. Females, furthermore, have a black spot on the rear corner of the dorsal fin, which is lacking in males, while the caudal fin of the male is marked by two black cross stripes, one at the base and the other at the margin. In breeding season the male assumes a brilliant coat, his upper parts turning to steel blue in front of the dorsal fin with a greenish luster behind it, while his belly brightens to a deep salmon, his ventrals and anal change to dusky margined with orange, and the front edge of his dorsal turns orange.

Size.—The largest specimens are about 3 inches long.

Habits.—The sheepshead minnow (like the common mummichog) is confined to the shallow waters of inlets, harbors, and the heads of bays, and salt marshes, often in brackish water. Its diet is partly vegetable, partly animal. It is very pugnacious, often killing fishes larger than itself, making repeated attacks with its sharp teeth and finally devouring its victim. Its breeding habits recall those of the mummichog (p. 163), the males fighting fiercely among themselves and clasping the females just forward of the tail with dorsal and anal fin, while the eggs and milt are extruded. Spawning takes place in shallow water from April to September, the eggs maturing a few at a time, so that any given female spawns at intervals throughout the season. The eggs sink and stick together in clumps by numerous threads. They are 1.2 to 1.4 mm. in diameter, with one large oil globule and many minute ones. Incubation occupies 5 or 6 days, and even at hatching the larvae (4 mm. long) show alternate light and dark crossbands. At a length of 9 mm. all the fins are formed, and at 12 mm. the fry show most of the characters of its parents.[2]

General range.—Atlantic coast of the United States, Cape Cod to Mexico, in brackish as well as in salt water.

Occurrence in the Gulf of Maine.—This fish, like many others, finds its northern limit at Cape Cod

[2] An account of courtship and spawning is given by Newman (Biol. Bull., vol. 12, No. 5, 1907, p. 336) and of development by Kuntz (Bull., U. S. Bur. of Fish., vol. 34, (1914) 1916, p. 409).

and would not deserve mention here at all were it not recorded from the Cape by Storer. West and south of Cape Cod, however, as at Woods Hole, it is common enough in suitable situations. We have seined many of them with *Fundulus* at the head of Buzzards Bay.

THE BILLFISHES OR SILVER GARS. FAMILY BELONIDAE

The most noticeable feature of the billfishes is that both of their jaws are prolonged to form a long slim beak well armed with teeth. Their bodies are very slender, and their anal, dorsal, and ventral fins set far back. They have no finlets between the dorsal and anal fins and the caudal, the absence of these being the readiest field mark to separate the billfishes from the needlefishes (*Scomberesox*, p. 170). They are swift-swimming, predaceous fishes, represented by many species, most of them American. Only two have ever been recorded in the Gulf of Maine.[3]

KEY TO GULF OF MAINE BILLFISHES

Body as thick as it is deep; dorsal, anal, and caudal fins only moderately concave............Silver Gar, p. 167
Body less than ½ as thick as it is deep; dorsal, anal, and caudal fins deeply concave.............Garfish, p. 168

Silver gar *Tylosurus marinus* (Walbaum) 1792[4]

BILLFISH; SALT-WATER GAR; SEA PIKE; AND VARIOUS OTHER LOCAL NAMES

Jordan and Evermann, 1896–1900, p. 714.

Description.—Its long bill and slender body give the gar so peculiar an aspect that it is not likely to be confused with any other Gulf of Maine fish other than the half beak (p. 169), the needle fish (p. 170), or its own close relative *Ablennes* (p. 168). And it is easily distinguishable from the first of these by the fact that both of its jaws are prolonged instead of only the lower; from the second by lacking detached finlets on its back between its dorsal and caudal fins. The most conspicuous differences between the silver gar and *Ablennes* (p. 168) is that the body of the former is thicker than it is deep, and that its fins are only moderately concave, whereas the latter is so strongly flattened sidewise that it is less than one-half as thick as it is deep with deeply concave fins.

The head of the adult silver gar occupies nearly one-third of the total length of the fish; the upper jaw, from the eye forward, is twice as long as the rest of the head; both jaws are armed with sharp teeth; and the eyes are large. The long, slender body is only about one-twentieth as deep as long, rounded (not laterally flattened) in cross section, and thicker than deep. Both the body and the sides of the head are scaly. The dorsal fin, with 13 to 17 rays, and the anal fin, with 17 to 21 rays, are alike in outline, the anterior rays of both being much longer than those toward the rear, and the rear two-thirds of each can be depressed along the back and nearly concealed in a groove, while the forward one-third continues erect. Both fins, too, are situated far back, with the dorsal arising a little behind the forward end of the anal.

The ventral fins stand about halfway between a point below the eye and the base of the caudal. The margin of the caudal fin is only moderately concave, this fact being the readiest field mark to separate this particular gar from the only other species of its genus (*Tylosurus acus*) taken yet near the Gulf of Maine (see footnote, p. 167), for the tail of the latter is deeply forked. There is a distinct

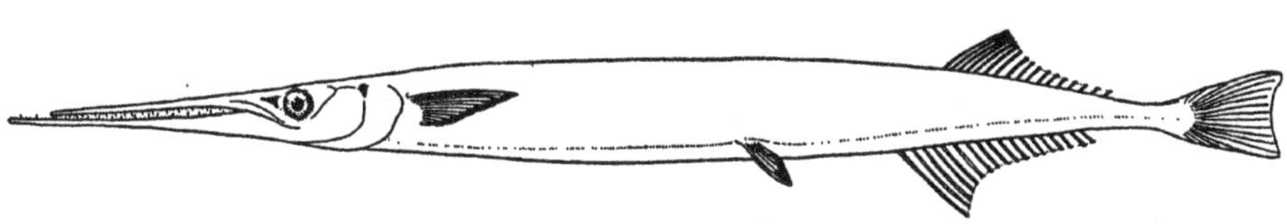

FIGURE 79.—Silver gar (*Tylosurus marinus*).

[3] The closely allied houndfish (*Tylosurus acus* Lacépède 1803) has been taken at Nantucket, but has not been found within the Gulf of Maine. Since it may appear there as a stray from the south, we may point out that it is easily distinguished from the silver gar by its deeply forked tail and by the fact that its dorsal and anal fins are much longer, the former with 23 rays, the latter with 21. The following characters in combination will serve to identify it among the several tropical gars; mouth nearly closable and upper jaw not arched; dorsal and anal fins long; beak at least twice as long as rest of head; greatest depth of body not more than two-thirds as great as length of pectoral fin; no lateral stripe.

[4] Jordan, Evermann, and Clark (Rept. U. S. Comm. Fish., (1928) Pt. 2, 1930, p. 196) place this species in the genus *Strongylura* Van Hasselt 1824.

longitudinal ridge, or low keel, along either side of the caudal peduncle.[5]

Color.—Greenish above, silvery on the sides, white below; a bluish silvery stripe along each side becoming broader and less distinct toward the tail; snout dark green; there is a blackish blotch deeper than long on the upper part of the cheek. The fins are without markings for the most part; the dorsal may be somewhat dusky, and the caudal bluish at its base.

Size.—The silver gar grows to a length of about 4 feet.

General range.—Maine to Texas; abundant along the South Atlantic and Gulf coasts of the United States, often running up fresh rivers above tide water.

Occurrence in the Gulf of Maine.—The silver gar is common enough along the southern shores of New England, e. g., in Rhode Island waters and at Woods Hole where quite a few are found from June to October. Like many other southern fishes, however, it seldom journeys eastward past Cape Cod, the only definite records of it within the Gulf of Maine being of several collected by Dr. William C. Kendall at Monomoy Island, forming the southern elbow of Cape Cod; at Wolfs Neck, Freeport, and Casco Bay, Maine, and of one found by Crane[6] in the stomach of a tuna that she examined at Portland, Maine, in July 1936. We have not seen it in the Gulf, nor have we heard even a rumor of its presence there from fishermen, good evidence that it is as rare a straggler as the few records indicate, for large silver gars are not fish to be overlooked. With so little claim to mention here, we need merely note that it is voracious, feeding on all sorts of smaller fishes, and that it runs inshore, possibly even into river mouths, to spawn. The eggs, described by Ryder,[7] are about 3.6 mm. (one-seventh of an inch) in diameter, and stick together and to any object they may touch, by long threads scattered over their surface.

[5] There are many other species of gars in tropical seas, any one of which might stray northward with the Gulf Stream and so to the Gulf of Maine. The silver gar is identifiable among them by the following combination of characters (no one character alone marks it out among its relatives): mouth capable of being nearly closed; caudal peduncle with keels; dorsal and anal fins short, the former with 13–17 rays, the latter with 17–21 rays; caudal fin only moderately concave; eyes at least one-third as broad as the head is long behind the eyes; body not excessively slender but at least one-fifth to one-sixth as deep as head (including jaws) is long; body not very strongly compressed sidewise; Jordan and Evermann (Bull. 47, U. S. Nat. Mus., Pt. 1, 1896. p. 709) gives a useful key to the species of the family.

[6] Zoologica, New York Zool. Soc., vol. 21, No. 16, 1936, p. 211.

Garfish *Ablennes hians* (Cuvier and Valenciennes) 1846

Jordan and Evermann, 1896–1900, p. 718, *Athlennes hians*.[8]

Description.—This gar resembles the silver gar (*Tylosurus*) closely in general appearance, as well as in the nature and arrangement of its fins; especially in the fact that the rear parts of its dorsal and caudal fins can be depressed and almost completely concealed in a groove, with the forward parts still remaining erect. But its body is so strongly flattened sidewise as to be less than one-half as thick as it is deep, instead of about as thick as deep, or thicker, as it is in the silver gar. The dorsal fin, also, arises farther back relative to the anal fin, than is the case in the silver gar. The tail fin is broadly forked, the dorsal and anal fins deeply concave.

Color.—Back greenish with bluish green reflections; lower part of sides bright silvery, also the abdomen; snout greenish; dorsal fin mostly greenish, but with the rays black-tipped; tail fin greenish. Some individuals have the sides plain silvery, but others are marked with dark blotches or indistinct sooty or blue crossbars.[9]

Size.—Up to 3 feet long, or more.

[7] Bull. U. S. Fish Comm., vol. 1, 1882, p. 283.

[8] The original spelling was *Athlennes* (Jordan and Fordice, Proc. U. S. Nat. Mus., vol. 9, 1886, p. 342). But Jordan and Evermann (Bull. 47, U. S. Nat. Mus., Pt. 1, 1896, p. 717, footnote) state that "*Ablennes*" was intended.

[9] Smith (Sea Fishes of Southern Africa, 1949, pl. 7, fig. 26) gives a colored illustration o fone with blue crossbars.

FIGURE 80.—Garfish (*Ablennes hians*), specimens from North Truro, Mass., and from Acapulco, Mexico. Drawing by H. B. Bigelow.

General range and occurrence in the Gulf of Maine.—Widespread in tropical seas;[10] Brazil to Chesapeake Bay in the western Atlantic, and northward as a stray to Cape Cod. A specimen of this tropical fish, about 23⅜ inches (594 mm.) long to the fork of the tail, was taken in a fish trap on the shore of Cape Cod Bay at North Truro, Mass., on August 15, 1949.[11]

THE HALFBEAKS. FAMILY HEMIRAMPHIDAE

The halfbeaks are close allies of the billfishes (*Belonidae*, p. 167), but it is only the lower jaw that is greatly prolonged while the upper jaw is short in the only species of present concern. They are largely herbivorous, feeding mainly on green algae. There are many species in warm seas, only one of which is known to reach the Gulf of Maine.

Halfbeak *Hyporhamphus unifasciatus* (Ranzani) 1842

SKIPJACK

Jordan and Evermann, 1896–1900, p. 721 (*Hyporhamphus roberti* (Cuvier and Valenciennes) 1846.

Description.—The most striking feature of this halfbeak and one which is enough of itself to mark it off from every other fish known from the Gulf of Maine, is the fact that while the lower jaw is very long, the upper jaw is short.[12]

This is a slender fish, its body only one-sixth to one-tenth as deep as long (younger fish are still more slender), tapering slightly toward head and tail. Its dorsal (14 to 16 rays) and anal (15 to 17 rays) fins are situated far back and opposite each other, as in the silver gar, and are about equal in length and alike in outline. There are no detached finlets between them and the caudal fin. The ventrals stand about midway between a point below the eye and the base of the caudal. The teeth are small and the scales are largest on the upper surface of the head. The beak is much shorter in young fish than it is in adults.

Color.—Translucent bottle green above with silvery tinge, each side with a narrow but well-defined silvery band running from the pectoral fin to the caudal fin, the sides darkest above and paler below this band. The tip of the lower jaw is crimson in life, with a short filament, and three narrow dark streaks run along the middle of the back. The forward parts of the dorsal and anal fins and the tips of the caudal fins are dusky. The lining of the belly is black.

Size.—Adults are seldom more than 1 foot long.

General range.—Tropical and subtropical on both coasts of America and in the Gulf of Mexico; abundant off the South Atlantic United States, not uncommon northward to Cape Cod, and straying to the coast of Maine.

Occurrence in the Gulf of Maine.—In our cool boreal waters the halfbeak is only a rare stray from the south, previously recorded only twice in the Gulf of Maine, i. e., from Machias and from Casco Bay, Maine, many years ago in each case. We can now add one taken in Quincy Bay, Boston Harbor, July 10, 1951, by Gordon Faust; another off Revere (also in Boston Harbor) on the 19th of the month, by John M. Hodson; a third, taken in a trap at Sandwich, September 24 of the same year,[13] and several dozen taken in a pound net at Small Point, Maine, July 14–15, reported by Leslie Scattergood.

[10] We have seen specimens from Acapulco, west coast of Mexico; Panama; Mauritius; and Zanzibar.

[11] This specimen was presented to the Museum of Comparative Zoology by John Worthington of the Pond Village Cold Storage Co., North Truro, Mass.

[12] Should a halfbeak be taken in the Gulf of Maine it would be well to consult Meek and Hildebrand (Field Museum of Natural History, Zool. Series, vol. 15, Pt. 1, p. 232, 1923) for there are several other species that might reach us as strays, either via the Gulf Stream route or from offshore. One, indeed (*Euleptorhamphus velox*), has been taken at Nantucket. Its lower jaw is even longer and more slender than that of the halfbeak, its body is more flattened sidewise, and its pectoral fins are longer.

[13] These specimens are in the Harvard Museum of Comparative Zoology.

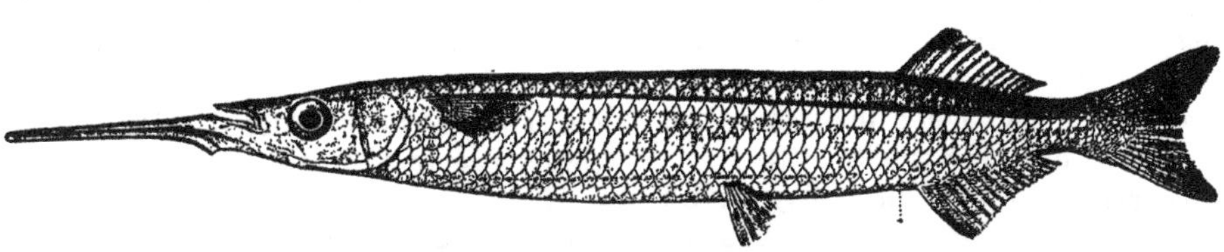

FIGURE 81.—Halfbeak (*Hyporhamphus unifasciatus*), Chesapeake Bay. From Jordan and Evermann. Drawing by H. L. Todd.

THE NEEDLEFISHES. FAMILY SCOMBERESOCIDAE

Both jaws (of the adult) are elongated to form a slender beak in the needlefishes, as in the billfishes (p. 167), and the anal, dorsal, and ventral fins are set far back. But the presence of several finlets between the dorsal and anal fins and the caudal in the needlefishes (which the billfishes lack) is a ready field mark for their identification. Furthermore, their teeth are small and weak, and their bodies only moderately slender. Four or five species are known in warm seas, one of which is not uncommon in the Gulf of Maine.

Needlefish *Scomberesox saurus* (Walbaum) 1792

BILLFISH; SKIPPER; SAURY

Jordan and Evermann, 1896–1900, p. 725.

Description.—The needlefish resembles the silver gars in its slender form and in the fact that both its upper and lower jaws are prolonged, but it differs from the gars in having a series of five or six little separate finlets, both on its back in the space between the dorsal fin and the caudal, and on its lower side between the anal fin and the caudal. Its body is about nine times as long (not counting caudal fin) as deep, flattened sidewise, tapering toward the head and tail, with slender caudal peduncle, and all its fins are small. Its dorsal fin originates slightly behind the origin of its anal; these two fins are alike in outline and stand far back. Its ventrals are situated about midway the length of the body. Its caudal is deeply forked and symmetrical, much like the tail of a mackerel. Its trunk is covered with small scales as is a patch on each gill cover. Its lower jaw projects a little beyond the upper, and its teeth are pointed but very small.

Color.—Olive green above with a silver band on each side at the level of the eye and about as broad as the latter. There is a dark green spot above the base of each pectoral; the dorsal fin is greenish; the lower parts are silvery with golden gloss. Young fry, which live in the surface waters of the open Atlantic, have dark blue backs and silvery sides.

Size.—Up to 18 inches long. Those caught along Cape Cod run a foot and more in length.

Habits.—The skipper is an oceanic fish. So far as known it always lives close to the surface; so much so indeed that in English waters, where it is plentiful in summer, few are caught in nets set as deep as a fathom or two. Its hordes are preyed upon by porpoises and by all the larger predaceous fishes; cod and pollock, for instance, feed greedily upon them, as do bluefish. When they strand on the beaches, as often happens, it is probably while they are fleeing from their enemies. At sea they attempt to escape by leaping, whole companies of them breaking the surface together as has often been described, and as we have seen them doing in Massachusetts Bay.

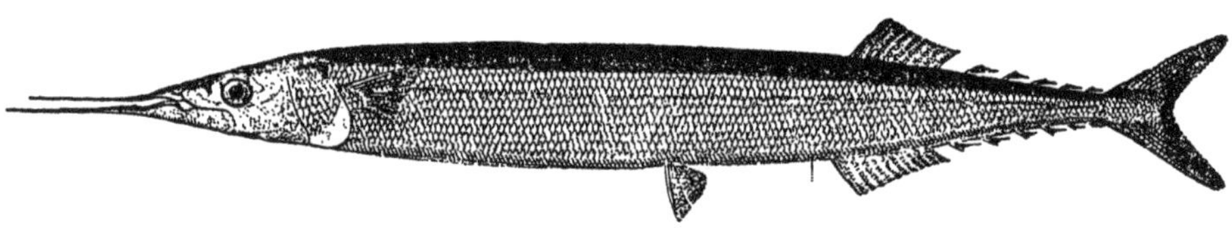

FIGURE 82.—Needlefish (*Scomberesox saurus*). Adult, Woods Hole. From Goode. Drawing by H. L. Todd.

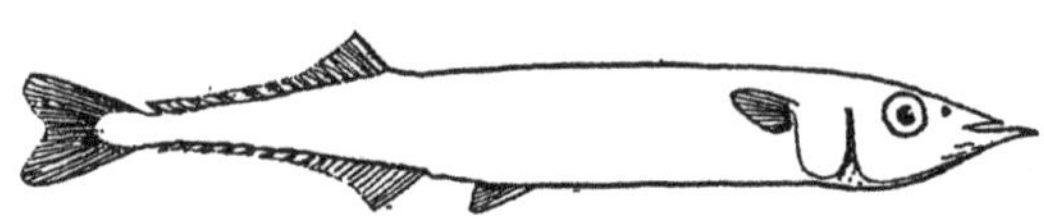

FIGURE 83.—Needlefish, young, about 2½ inches long. After Murray and Hjort.

It is not likely that they ever spawn in the cool waters of the Gulf of Maine, for we have never taken their fry in our tow nets, although they are among the most numerous of young fish in the open Atlantic between the latitudes of 11° or 12° N. and 40° N. Although their eggs are covered with filaments like those of the silver gars,[14] they are not adhesive as the latter are, but drift near the surface. The most interesting phase in the development of the skipper is that its jaws do not commence to elongate until the fry have grown

[14] Skipper eggs were so described by Haeckel (Archiv für Anatomie, Physiologie, und Wissenschaftliche Medecin, 1855, p. 23, pl. 5, fig. 15,) 75 years ago. They were not seen again until 1910 when similar eggs, 2.2 mm. in diameter, covered with filaments, were towed in the Atlantic by the *Michael Sars* (Murray and Hjort. Depths of the Ocean, 1912, p. 742, fig. 531).

to about 1⅝ inches (40 mm.), and that the lower jaw out-strips the upper at first, so that fry of 4 to 6 inches (100 to 150 mm.) look more like little halfbeaks ("Hemiramphus" stage) than like their own parents.

European students tell us that the skipper feeds on the smaller pelagic Crustacea and probably also on small fish, for it is sometimes caught on hook and line. One examined by Linton at Woods Hole contained chiefly annelid worms, fragments of fish, copepods and crustacean larvae, with some vegetable debris.

General range.—Temperate parts of the Atlantic, Pacific, and Indian Oceans, known in the open sea as far north as northern Norway off the European coast, and to southern Newfoundland and southern Nova Scotia [15] off the eastern American coast.

Occurrence in the Gulf of Maine.—While the skipper is a straggler to our Gulf from warmer waters offshore or farther south, it has been taken along the northern coasts of New England more often than have any of its relatives; specifically along Cape Cod; at Provincetown; at several locations in Massachusetts Bay where we have seen schools of them; at Annisquam a few miles north of Cape Ann; at Old Orchard (Maine); in Casco Bay; at Monhegan Island; in the central part of the Gulf; [16] among the islands at the northern entrance to the Bay of Fundy; and on the northern part of Georges Bank, where one was gaffed from the *Albatross II* on September 20, 1928. But we find no record of it along the Nova Scotia shore of the Gulf of Maine. The inner curve of Cape Cod from Provincetown to Wellfleet seems to be a regular center of abundance for it, as Storer long ago remarked, for schools of billfish are picked up in the traps along that stretch of beach almost every year, the catch occasionally amounting to hundreds of barrels, and hosts of them have been known to strand there. Its numbers fluctuate greatly from year to year, however, and it often fails to appear.[17]

They are likely to be taken any time from mid-June to October or November, the largest catches usually being made late in summer.[18] We have seen several schools skipping, as is their common habit, off the Scituate shore on the southern side of Massachusetts Bay. But skippers are so much less common farther within Massachusetts Bay that some fishermen had never heard of them there. They appear only as strays north of Cape Ann. But it would not be astonishing if a large school were to be encountered anywhere within the Gulf; witness their occasional abundance off northern Nova Scotia.[19] When skippers do invade the waters of our Gulf, they may be expected in multitudes, for they usually travel in vast schools. Day,[20] for example, mentions the capture of 100,000 in a single haul in British waters.

Commercial importance.—The skipper is not of much commercial importance, being too sporadic in its appearances. However, when large catches are made on Cape Cod they find a ready sale near by. If too many are caught for the local trade to absorb, they are sent to Boston, where they are sold for bait.

THE FLYING FISHES. FAMILY EXOCOETIDAE

The typical flying fishes have one dorsal fin and one anal fin, both of them soft rayed, both of them located far rearward, and with the anal below the dorsal. Their ventral fins are well behind their pectorals, their tails are very deeply forked with the extreme tips rounded, the lower lobe the longer, and they have small mouths and large rounded scales. Their most distinctive feature is that their pectorals are so long and so stiff that their owners can plane through the air on them, several feet above the water, which they do mostly in attempts to escape their enemies, and as has been described, time and again. Jordan and Evermann have given a popular account of this so-called "flight" (really not flight at all, for the flying fish does not flap its wings) in their

[15] Cornish (Contrib. Canadian Biol., (1902-1905) 1907, p. 83) states that large schools can often be seen at Canso, Nova Scotia, skipping over the water as they flee from the pollock.

[16] The Museum of Comparative Zoology has a specimen, taken 115 miles southeast of Portland Lightship.

[17] Blake (American Naturalist, vol. 4, 1870, p. 521) remarked that while years before he saw thousands stranded at Provincetown not one was seen in 1870. It failed in 1921, also, and no doubt in many intervening years.

[18] We are indebted for information on the local abundance of billfish on Cape Cod to Capt. L. B. Goodspeed, a fisherman of long experience and close observation.

[19] Cornish (Contrib. Canadian Biol., 1902-1905 (1907), p. 83) states that large schools can often be seen at Canso skipping over the water as they flee from the pollock.

[20] The fishes of Great Britain and Ireland, vol. 2, 1880-1884, p. 152.

Guide to the Study of Fishes, 1905, p. 157. For a more detailed study, based similarly on first-hand observations, we refer the reader to Hubbs, Papers of the Michigan Academy of Sciences, vol. 17, 1933, pp. 575–611. Voyagers in tropical seas are perhaps more familiar with flying fishes than with any other fishes. And they are often seen in the warm ultramarine-blue waters of the Gulf Stream abreast of our northeastern coast. But none of them are to be expected in the boreal waters of our Gulf except as the rarest of strays.

A flying fish could hardly be mistaken for anything else, except possible for a flying gurnard (p. 472). But a glance should be enough to tell which of them one has in hand, for the flying fishes have stiff, narrow, pointed wings, only on dorsal fin and a very deeply forked tail, whereas the so-called wings of the flying gurnard are broad, rounded, and extremely flexible; they have two dorsal fins, and a tail fin that is only weakly concave in outline.

Flying Fish *Cypselurus heterurus* (Rafinesque) 1810

Jordan and Evermann, 1896–1900, p. 2836.

Distinctive features of flying fishes of the genus *Cypselurus*, among its tribe, are pectoral fins so long that they overlap the anal fin considerably when laid back; ventrals standing far rearward and nearly or quite half as long as the pectorals; anal fin with its point of origin only a litle forward of the mid-point of the base of the dorsal fin; and second ray of the pectoral fins branched. The species *heterurus* has a plain gray dorsal fin; it has no teeth on the palatine bone in the roof of its mouth; and the pale edging of the outer margin of its pectoral fins is extremely narrow. The number of its scales is distinctive, also, as is the number of its vertebrae. But these last two characters are matters for the specialist.

Color.—Dark blue gray on the back and on the upper part of the sides, silvery lower down on the sides, and below; the dorsal fin is plain gray, the rear margin of the pectorals with a very narrow pale edging.

Size.—The species *heterurus* is one of the larger flying fishes, commonly growing to a length of about 1 foot (to the base of the tail fin).

Occurrence in the Gulf of Maine.—A flying fish, about 9¼ inches long to the fork of the tail, seemingly of this species but not in good enough condition for certain identification, was taken in a trap of the Pond Village Cold Storage Co. at North Truro, on the Massachusetts Bay shore of Cape Cod, on August 4, 1952. This is the only record of one of its tribe, from our Gulf. And the only record of a flying fish from Nova Scotian coastal waters is by Jones, of one taken at Sable Island, in 1859.

Flying fishes are taken now and then at Woods Hole, the species *heterurus* perhaps more often than any other, according to published report, but several of the kinds to be expected in the Gulf Stream off our coast resemble one another very closely indeed. So we suggest that if a flying fish should be taken in our Gulf that does not seem to fit the accompanying illustration (fig. 83A) it be forwarded either to the Fisheries Laboratory of the U. S. Fish and Wildlife Service, Woods Hole, Massachusetts; to the Department of Fishes, U. S. National Museum, Washington, D. C.; or to the Department of Fishes, Museum of Comparative Zoology, Cambridge, Massachusetts, to be named.

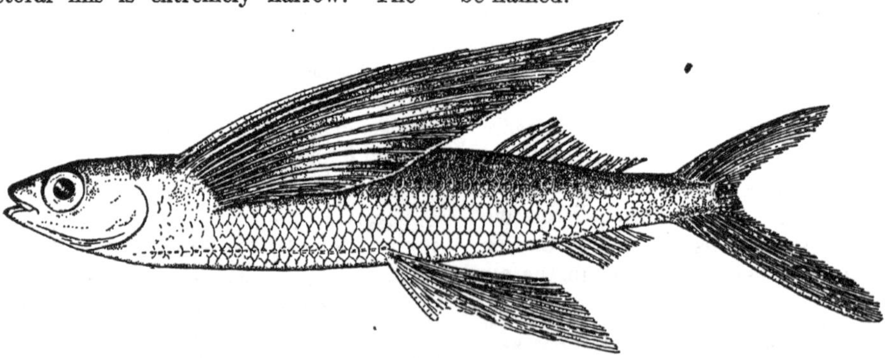

FIGURE 83A.—Flyingfish (*Cypselurus heterurus*). After Bruun and a specimen from North Truro, Mass. Drawing by Jessie Sawyer.

THE SILVER HAKE AND COD FAMILIES—FAMILIES MERLUCCIIDAE AND GADIDAE

The silver hakes and the cods are so closely allied that many European ichthyoloigsts group them in a single family. American practice, however, is to separate them because of certain differences in the structure of the skull and ribs. They are soft-finned fishes, lacking true spines at any stage in development (though in one local species, the silver hake, the basal parts of the dorsal and anal fin rays are so stiff as to feel like spines to the touch), but they are distinguishable from all other soft-rayed Gulf of Maine fishes by the fact that their large ventral fins are situated under the pectorals or in front of them, and not behind them, as they are in the herrings and salmons. They and their relatives, the grenadiers (p. 243), are separated from most of the typical spiny-rayed fishes by the structure of the skull.[21]

Key To Gulf Of Maine Hakes, Cods, And Other Species

1. There are three separate dorsal fins and two anal fins ---------- 2
   There are two separate and well developed dorsal fins ---------- 5
   There is only one well developed dorsal fin ---------- 11
2. The lateral line is black; there is a black blotch on each shoulder ---------- Haddock, p. 199
   The lateral line is pale; there is no shoulder blotch ---------- 3
3. The lower jaw projects beyond the upper; the chin barbel is very small, if there is one ---------- Pollock, p. 213
   The upper jaw projects beyond the lower; the chin barbel is large ---------- 4
4. The ventral fins are narrow, and prolonged as filamentous feelers that are as long as the rest of the fin; the eyes are small ---------- Tomcod, p. 196
   The ventral fins are broad, and their filamentous tips are less than one-third as long as the remainder of the fin; the eyes are large ---------- Cod, p. 182
5. The anal fin originates considerably in front of the point of origin of the second dorsal fin ---------- Hakeling, p. 233
   The anal fin originates under the point of origin of the second dorsal fin or behind it ---------- 6
6. The ventral fins are short and of ordinary form ---------- Silver hake, p. 173
   The ventral fins are very long and feeler-like ---------- 7
7. The first dorsal fin is hardly higher than the second dorsal, and none of its rays are prolonged or filamentous ---------- Spotted hake, p. 230
   The first dorsal fin is much higher than the second dorsal, with one or two long filamentous rays ---------- 8
8. The ventral fins reach nearly or quite as far back as the rear end of the anal fin ---------- Long-finned hake, p. 232
   The ventral fins do not reach back to the middle of the anal fin ---------- 9
9. The anal fin is so deeply notched about midway of its length as to suggest two separate fins ---------- Blue hake, p. 233
   The anal fin is of about equal height from end to end ---------- 10
10. There are about 140 rows of scales along the lateral line from gill opening to base of caudal fin; the upper jaw bone reaches back to below the rear edge of the eye ---------- White hake, p. 221
    There are only about 110 rows of scales along the lateral line; the upper jaw bone reaches back only as far as the rear edge of the pupil ---------- Squirrel hake, p. 223
11. There are no isolated rays in front of the dorsal fin, nor barbels on the top of the snout ---------- Cusk, p. 238
    The dorsal fin is preceded by a fringe of short rays and one long ray; the top of the snout bears barbels as well as the chin ---------- 12
12. There are three barbels on the top of the nose ---------- Four-bearded rockling, p. 234
    There are only two barbels on the top of the nose ---------- Three-bearded rockling, p. 237

Silver hake *Merluccius bilinearis* (Mitchill) 1814

WHITING; NEW ENGLAND HAKE

Jordan and Evermann, 1896–1900, p. 2530.

Description.—The presence of two separate and well developed dorsal fins, both of them soft-rayed, the second much longer than the first, combined with the location of the ventrals on the chest, is sufficient field mark to distinguish the silver hake from all other Gulf of Maine fishes except for the true hakes (genus *Urophycis*, p. 221). And there is no danger of confusing it with any of the latter, for it lacks the chin barbels so characteristic of them, and its ventrals are of the ordinary finlike form, whereas those of the true hakes are altered into long feelers. It is a rather slender fish, about five to six times as long as it is deep, its body rounded in front of the vent but flattened sidewise behind it, with large flat-topped head occupying

[21] The hypercoracoid bone lacks an aperture (technically a "foramen").

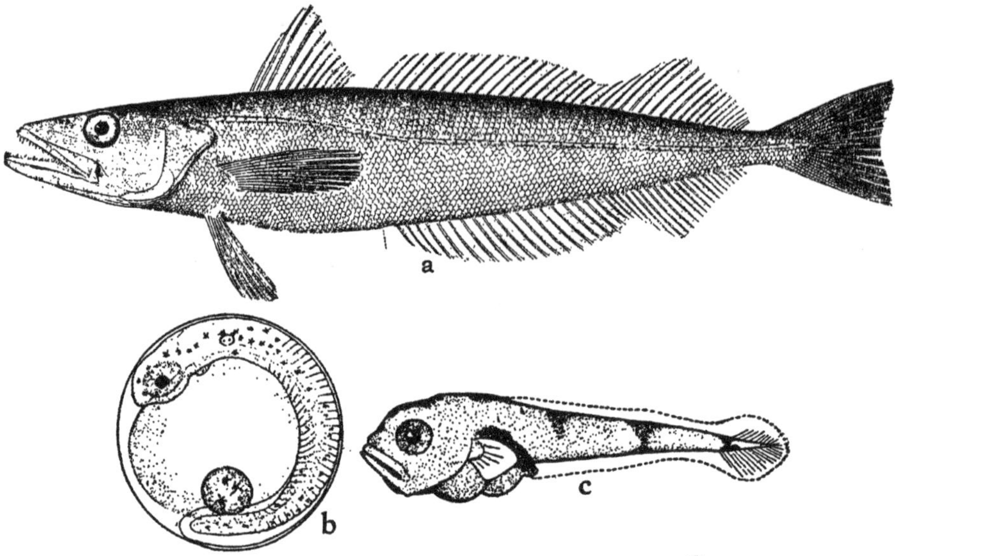

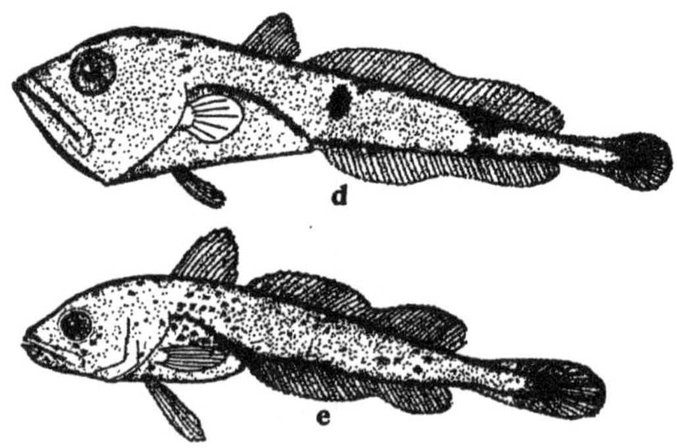

FIGURE 84.—Silver hake (*Merluccius bilinearis*). A, adult, Nova Scotia, from Goode, drawing by H. L. Todd; B, egg; C, larva, 6.5 mm.; D, larva, 11 mm.; E., young fry, 23 mm. B–E, after Kuntz and Radcliffe.

about one-fourth of the total length; large eyes; and wide mouth armed with two or more rows of sharp recurved teeth, and with the lower jaw projecting beyond the upper.

The first dorsal fin (11 to 14 rays) originates close behind the gill openings, is roughly an equilateral triangle in shape, and is separated by a short space from the second dorsal. The second dorsal (38 to 41 rays) is about four times as long as the first dorsal, but hardly more than half as high, and is of distinctive outline, being deeply emarginate two-thirds of the way back, with the rear section the higher of the two. The anal fin (38 to 41 rays) corresponds in height and in shape to the second dorsal, under which it stands. The caudal fin is square tipped when widespread, but its rear margin is weakly concave, otherwise. The pectorals are rather narrow, their tips slightly rounded, and they reach back far enough to overlap the second dorsal a little. The ventral fins, situated slightly in front of the pectorals, are perceptibly shorter than the latter, with about half as many rays (7).

Color.—The silver hake is dark gray above of brownish cast; but silvery-iridescent, as its name implies, or with golden reflections. The lower part of its sides and its belly are silvery. The inside of its mouth is dusky, the lining of its belly blackish. The fish is brightly iridescent when taken from the water, but fades soon after death.

Size.—Maximum size about 2½ feet long and about 5 pounds in weight, but adults average only about 14 inches long.

Habits.—Silver hake are strong swift swimmers, well armed and extremely voracious. They prey on herring and on any other of the smaller schooling fish, such as young mackerel, menhaden, alewives, and silversides. Probably a complete diet list would include the young of practically all the common Gulf of Maine fishes, for Vinal Edwards recorded the following from silver hake taken at Woods Hole: alewife, butterfish, cunner, herring, mackerel, menhaden, launce, scup, silversides, smelt, also the young of its own species. A 23¼-inch silver hake, taken at Orient, N. Y., had 75 herring, 3 inches long, in its stomach.[22] And it is probable that the silver hake that frequent Georges Bank feed chiefly on young haddock. They eat squid when occasion offers. The small ones in particular prey regularly on large shrimp (*Pandalus*) in the deep troughs in our Gulf, where experimental trawlings by the *Atlantis* in the summer of 1936 took about four times as many silver hake at stations where these shrimps were abundant as at stations where shrimp were scarce.[23] They sometimes take crabs, and bite freely on almost any bait, such as clams or cut fish.

Though silver hake do not school in definite bodies, multitudes of them often swim together, and such bands sometimes drive herring ashore, and strand themselves, in the pursuit. Events of this sort are oftenest reported in early autumn when the spent fish are feeding ravenously after the effort of spawning, but this may also happen at any time during the summer. Thus, Prof. A. E. Gross saw the beach at Sandy Neck, Barnstable, Mass., covered with them on several occasions in June and July 1920.[24] Doctor Huntsman informs us that spent fish frequently strand on the beaches on both sides of the Bay of Fundy in September. We once saw an army of silver hake harrying a school of small herring on a shelving beach at Cohasset, Mass. We half filled our canoe with pursuers and pursued, with our bare hands.

It is said that European silver hake rest on the bottom by day and hunt by night, and it is usually at night that the American fish run up into the shallows and enter the traps. But strandings also take place by day. Silver hake, like many other rapacious species, are wanderers, independent of depth within wide limits, and of the sea floor. Sometimes they swim close to the bottom, sometimes in the upper levels of the water, their vertical movements being governed chiefly by their pursuit of prey. Their upper limit is the tide line; at the other extreme they have been trawled repeatedly as deep as 150 to 400 fathoms on the continental slope off southern New England, and as deep as 296 fathoms off North Carolina.[25] When they are on bottom they are caught indifferently on sandy or pebbly ground, or on mud (as in the deep trough west of Jeffreys Ledge, p. 175); seldom around rocks.

The lowest temperatures in which we have known of silver hake being taken have been between 38° and 40° F. (probably), in the bottom of the deep trough west of Jeffreys Ledge, August 1936,[26] about 40° F. (4.4° C.) at 28 fathoms off New York, February 28, 1929, and about 39.5° F. (4.2° C.) at 19 fathoms in the same general region, February 5, 1930.[27] And most of the winter and early spring records for it have been where the bottom temperature was warmer than about 43° F. (6° C.).[28] At the other extreme, we have never heard of them in any numbers where the water was warmer than about 64° F. (18° C.); the monthly catches made in Cape Cod Bay (see p. 180) are especially instructive in this regard.

Breeding habits.—The silver hake is the most important summer spawner among Gulf of Maine fishes that are important commercially, just as the haddock is for spring and the pollock for autumn. The Gulf is probably its most prolific nursery, too, and it spawns over the outer part of the Nova Scotia Banks also, as far east as Sable Island, Dannevig [29] having recorded large egg catches in the offing of Halifax. But this is probably its eastern breeding limit, for the Canadian Fisheries Expedition found no silver hake eggs or fry on Banquereau or Misaine Banks; in the Laurentian Channel; or on the Newfoundland Banks. In the opposite direction, eggs in fair numbers have been taken in the tow nets off Woods Hole in July and August; the *Albatross II* has found them and the resultant larvae near shore off Long Island in June and July, with eggs as far south as the offing of Cape May; and the young fry have been caught off New York [30] from spring to autumn.

We have no evidence that silver hake commence to spawn before June, north of Cape Cod, our earliest egg record having been for the 11th of that

[22] Nichols and Breder, Zoologica, N. Y. Zool. Soc., vol. 9, 1927, p. 163.

[23] For details, see Bigelow and Schroeder, Biol. Bull., vol. 76, 1939, p. 315.

[24] The Auk, vol. 40, 1923, p. 19.

[25] Goode and Bean, Smithsonian Contrib. Knowl., vol. 30, 1895, p. 387.

[26] No temperature was taken, but 38.6° F. (3.66° C.) was recorded there at 85 fathoms, on August 15, 1914, and 39.8° F. (4.33° C.) at 72 fathoms on August 15, 1913.

[27] Specimens trawled by *Albatross II*.

[28] *Albatross II* trawled a considerable number at stations scattered along the continental slope, from the offing of southern New England to the offing of Chesapeake Bay, in February 1929 and 1930, and in April 1930.

[29] Canad. Fish. Exped. (1914–1915), 1919, p. 27.

[30] Nichols and Breder, Zoologica, New York Zool. Soc., vol. 9, 1927, p. 163.

month; in our Gulf, egg production (as evidenced by the numbers of eggs taken in our tow nets) is at its height in July and August and continues through September, though less freely, with October 22 as our latest date. Similarly, the Canadian Fisheries Expedition found no silver hake eggs in Nova Scotia waters east of Cape Sable in May, but many in July.

It is impossible to establish the exact temperature at which silver hake are spawning at any particular station without knowing at what level ripe fish are in the water, which may be anywhere between the surface and the bottom with this species. It may be definitely stated, however, that they never spawn in as cold water as cod and haddock usually do in the western Atlantic. In 1915, for example (a representative season), it was not until the entire column of water was slightly warmer than 41° F. at the locality in question that we found the first silver hake eggs in our Gulf. And if the parent fish were in the upper water layers, as they may have been, all the rich spawnings we encountered in the Gulf during that year, and during the next, took place in temperatures considerably higher still. Similarly, the silver hake eggs towed off Halifax by the Canadian Fisheries Expedition in July 1915, and off Shelburne, Nova Scotia, by the *Grampus* on September 6 of that same year may have been spawned in water warmer than 50° F., there being no need to assume that the parent fish were lying in the colder bottom stratum. As the spawning season draws to its close, in September and October, the minimum temperatures for most of our egg stations have been higher than 46°, with one (our latest record for the season) as warm as 57° F. at all depths. These data point to 41° to 45° F. as the lowest temperature limit for the spawning of the silver hake, with most of the eggs produced at 45° to 55° F.

In the case of any fish producing buoyant eggs the tendency of the latter to rise (unless counteracted by active vertical circulation of the water) insures that their development shall take place at the temperature of the upper stratum of water, not at that of the deeper levels where they were spawned. And the silver hake is no exception to this rule. While we have towed its eggs in June, when the surface was still only about 42° F., most of the egg records, and all our rich catches, were all made where the upper 5 fathoms or so were warmer than 50° and usually warmer than 55° F., with the temperature of the immediate surface 60° or higher in most cases. Similarly, silver hake eggs taken off Halifax by the Canadian Fisheries Expedition in July 1915, and off Shelburne, Nova Scotia, by the *Grampus* on September 6 of that year, may well have been in water at least as warm as 53° F., there being no reason to suppose they were far below the surface.[31] All this suggests that incubation does not proceed normally in water cooler than about 50°, and that it is most successful in temperatures as high as 55° to 60° F. This evidence that while the eggs of the silver hake may be spawned in low temperatures, a comparatively warm surface layer is necessary for their later development, offers a reasonable explanation for the failure of this fish to breed successfully along the New Brunswick shore of the Bay of Fundy, where active vertical circulation maintains surface temperatures as low as 50° to 55° F. throughout the summer, at least in most years. At the other extreme, the failure of the eggs that had been fertilized artificially to develop in the hatchery at Woods Hole in August temperatures points to 65° to 70° F. as the upper limit to successful incubation.

According to Kuntz and Radcliffe[32] only part of the eggs mature at one time, but we know of no estimate of the number of eggs a single female may produce. The eggs are buoyant, transparent, about 0.88 to 0.95 mm. in diameter, with a single yellowish or brownish oil globule of 0.19 to 0.25 mm. Incubation is rapid; Kuntz and Radcliffe assumed a duration of 48 hours at Woods Hole, but it has not been determined for the cooler waters in the Gulf of Maine. The larvae are about 2.8 mm. long at hatching, slender, with small yolk sac, and they are made recognizable by the fact that the vent is located on one side, near the base of the larval fin fold, as is the general rule in the cod family, not at its margin as in most larval fishes, and that the trunk behind the vent is marked with two black and yellow cross bars. The dorsal and anal fins and the caudal fin have all assumed their definite outlines by the time the little fish is 10 to 11 mm. long, and fry of 20 to 25 mm. begin to resemble their parents in general appearance.

[31] These catches were all made either at the surface or in oblique hauls with open nets.

[32] Kuntz and Radcliffe (Bull. U. S. Bur. Fish., vol. 35, 1918, p. 109) describe the spawning and early development.

Evidently the young silver hake take to the deeper water layers toward the end of their first summer or that autumn, when about 1 to 3 inches long, for fry as small as this have been trawled in good numbers off southern New England at 150 fathoms and deeper at that season during the early explorations of the U. S. Fish Commission,[33] by the *Albatross II* off Rockaway and off Long Beach, N. Y., in November 1928. By February they may be anywhere from 1¼ to 5 inches long, depending on how early they are hatched, on when they take to bottom, and on the feeding conditions they find there.[34] They may be anywhere from about 2 inches long to about 6½ inches long by April.[35]

The sizes of the many small silver hake that we have collected at different times of the year, both within the Gulf of Maine and southward as far as the offing of Chesapeake Bay, suggest that they reach an average length of 5½ to 7½ inches when 1 year old, and of about 9½ to 11 inches at 2 years of age, i. e., in their third summer.[36] Fish of 11 to 14 inches that dominated the pound-net catches made near Provincetown, August 1939,[37] were three-year-olds, probably. The rate of growth has not been traced for the older fish, nor can it be deduced from that of the European silver hake for the latter grows to a considerably greater length, averaging as much as 30 inches at 8 years in the extreme northerly part of its range (Iceland) and considerably larger still, in the southern part (Gulf of Gascony and off Morocco).[38] But it is reasonable to assume that the growth of the American fish varies similarly with the latitude (i. e., that it is most rapid in high temperatures) and that the American female, like the European, grows faster than the male. The European *Merluccius* matures at 2 years, which is probably true of the American species as well.

General range.—Continental shelf of eastern North America, northward to the Newfoundland Banks, southward to the offing of South Carolina;[39] most abundant between Cape Sable and New York. It is represented farther offshore and in the Gulf of Mexico by forms, the relationship of which to the *Merluccius* of our northeastern coast has not yet been determined. The silver hake is represented in Europe by a close relative, the European hake (*Merluccius merluccius*), an excellent account of the natural history and migrations of which is given by Le Danois.[40]

Occurrence in the Gulf of Maine.—Silver hake are familiar fish all around the coasts of the Gulf of Maine from Cape Cod to the Bay of Fundy and to the west coast of Nova Scotia. But it has long been a matter of common knowledge that their chief center of abundance is in the southwestern part of the Gulf. Thus in 1945 (most recent year for which detailed regional statistics are available), the reported landings were between 46 and 47 million pounds [41] from off eastern Massachusetts in general, including the shores of Cape Cod out to the western slope of the so-called South Channel, contrasting with only about 4 million pounds for the western and central coasts of Maine, and with only about 6,500 pounds for eastern Maine. Silver hake, it is true, are said to be common in the Passamaquoddy region (more so in some years than in others), also around Grand Manan at the mouth of the Bay of Fundy. And they are reported at various localities along the Nova Scotia side of the Bay and along western Nova Scotia. But they are not mentioned in the statistics of the Canadian catches for these waters, hence cannot be very plentiful there.

Silver hake are numerous over the west-central deeps of the Gulf also; in fact we found this the most plentiful fish at 70 to 90 fathoms in the basins off Cape Cod in the southwestern part of the Gulf and off Mount Desert in the northeastern, in August 1936; also in the trough west of Jeffreys Ledge, where the catches of them averaged 292 fish (maximum 840, minimum 1) as reduced to the common standard of one hour's trawling with an 82-foot shrimp trawl. And it is interesting that the catch there averaged about four times as great

[33] Goode, Fish. and Fish. Ind. U. S., Sect. 1, 1884, p. 242.

[34] Fry taken in February of different years by *Albatross II* have ranged from 1¼ inches (31 mm.) long to 4¾ inches (120 mm.).

[35] In April 1930 *Albatross II* trawled young fry ranging in length from 2 inches (54 mm.) to 6⅝ inches (163 mm.) long at a number of stations from the offing of Rhode Island to the offing of Chesapeake Bay, at 14 to 85 fathoms.

[36] For further details, see Bigelow and Schroeder (Biol. Bull., vol. 76, 1939, pp. 319–320, fig. 8).

[37] Information supplied by Wm. A. Ellison, Jr.

[38] Belloc, Notes et Memoires No. 21, Office Scientifique et Technique des Pêches Maritimes, France, 1923.

[39] The silver hake has been said, repeatedly, to range southward to the Bahamas, in deep water, following Jordan and Evermann (Bull. 47, U. S. Nat. Mus., Pt. 3, 1898, p. 2530). But the most southerly positive record we have found for it is off Charleston, S. C. (Blake Sta. 313, lat. 32° 32′ N., long. 78° 45′ W.; Goode and Bean, Smithsonian Contrib. Knowl., vol. 30, 1895, p. 387).

[40] Notes et Mem., 2, Off. Sci. Tech. Pêches Maritimes, France, 1920.

[41] "Round" and dressed fish combined.

(494 fish) at the stations where shrimp (*Pandalus*) were plentiful as at the stations where these were scarce (114 fish), evidence that silver hake congregate where feeding conditions are good.[42]

Reported landings throw little light on the numbers of silver hake that frequent the offshore rim of our Gulf, both because the otter trawls used there are so large-meshed that many pass through, and because most of those that are caught on Georges and Browns Banks are thrown overboard when the price is low.[43] Experimental trawling, however, on Georges Bank, April to September 1913, yielded about one-seventh as many silver hake on the average (about 1,800 fish) as haddock (about 14,000 fish) per trip, and the *Albatross III* caught an average of about 150 silver hake, running about one-half pound in weight, per trawl haul, in 250 hauls on various parts of Georges Bank, July, August, and September of 1948, 1949, and 1950. Thus they are moderately plentiful at least over Georges Bank as a whole, and there is no reason to doubt that this applies to Browns Bank equally.

These catches do not suggest any definite concentration on any one part of the bank, at least for summer, except that the largest that were made on its northern part were in hauls from shoaler than 30 fathoms, whereas the largest catches on the southern part were in hauls from deeper than 60 fathoms, a difference which may well have been a matter of the food supply.[44] In April, however, of 1950, the silver hake were not only more plentiful along the northern edge of the bank (average 305 per haul) than on the southern part (average 77 per haul) but so strictly confined to the deeper levels that the total yield of 66 trawl hauls at shoaler than 60 fathoms was only 11 fish, contrasting with an average catch of 232 fish per haul at 60 fathoms and deeper (25 hauls).[45]

Silver hake spawn along the entire coastal zone from Cape Cod to Grand Manan, as proved by the locations of the egg catches (fig. 85). The sloping sandy bottom around the northern extremity of Cape Cod and off the eastern slope of the Cape evidently is an important center of reproduction. Thus we found an abundance of eggs off Race Point on July 7, 1915; our tow nets yielded many eggs at two stations off the outer shore of the Cape on July 22 of the following year, when a 15-minute tow there at 20 fathoms, with a net one meter in diameter, produced approximately 25,000 larvae of 3 to 7 mm., the richest haul of young fish we have ever made in our Gulf. And the fish were still spawning there a month later, as proved by the presence of eggs.

Other occasions when we have taken silver-hake eggs in large numbers have been off Duck Island near Mount Desert on July 19 and on August 18, 1915; near Monhegan Island, August 4, 1915; off Wooden Ball Island near the mouth of Penobscot Bay on August 6, 1915; and off Rye, N. H., on July 23 of that same year. But we have never found them in any number in Massachusetts Bay though some eggs have been taken there on several occasions (fig. 85).

Unfortunately, no quantitative hauls were made at any of the more productive egg stations, hence the number of silver-hake eggs present in the water cannot be approximated. But the vertical net yielded about 190 eggs per square meter of sea surface at one station in the eastern basin.

Apparently the silver hake does not breed successfully in the northern side of the Bay of Fundy for neither its eggs nor its fry have ever been found there. But the capture of a few eggs in Petit Passage in our tow nets on June 10, 1915, suggests that it may spawn on the southern side of the bay as the cunner does (p. 478). And it may be expected to do so along the west coast of Nova Scotia, for the Canadian Fisheries Expedition found eggs at several stations off outer Nova Scotia, eastward to the longitude of Canso.

The presence of silver hake on Georges Bank throughout the summer is presumptive evidence of local spawning, though we have taken no silver hake eggs or larvae there.

The locations where we have found its eggs suggest that the silver hake, in the Gulf of Maine, spawns chiefly in water shoaler than 50 fathoms. But we have made one rich haul of its eggs in the center of the eastern basin. And the discovery of its eggs over the continental slope off Nova Scotia

[42] For further details, see Bigelow and Schroeder (Biol. Bull., vol. 76, 1939, p. 308, table 1; p. 314, table 5.

[43] Reported landings, 1945–1947, ranged between 3,000 and about 33,000 pounds for Georges Bank, between 0 and 6,000 pounds for Browns.

[44] The average catch per haul was 262 fish from shoaler than 30 fathoms and 151 fish from deeper than 60 fathoms on the northern part of the bank; 90 fish per haul from shoaler than 30 fathoms and 285 fish per haul from deeper than 60 fathoms on the southern part.

[45] Twenty-one trawl hauls at 60 fathoms and shoaler yielded none at all in March; but no hauls were made in that month deeper than 60 fathoms, where the silver hake doubtless were.

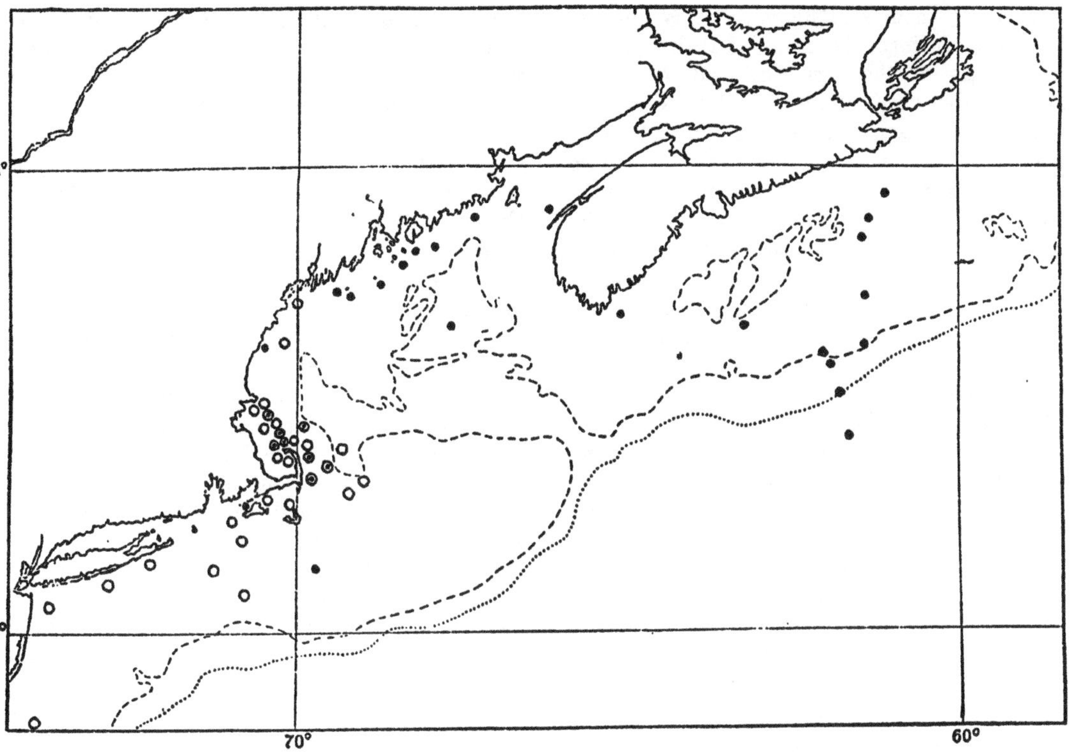

FIGURE 85.—Localities where eggs (●), and larvae (○) of silver hake, or both (⊙) have been taken.

by the Canadian Fisheries Expedition,[46] with the presence of ripe fish as well as of spent, in depths as great as 150 fathoms and more off southern New England [47] proves that it spawns over deep water also. The European silver hake usually spawns in 50 to 100 fathoms.

All our records for the free-drifting larvae of the silver hake in the Gulf, unlike those for its eggs, have been in the southwestern part. And we have towed along the coast of Maine so often in August, September, and October (when the larvae spawned from June to August might have been expected) that our failure to find them east of Cape Elizabeth seems sufficient evidence that they actually are limited, in their regular occurrence, to the southwestern part of the Gulf (they parallel the pelagic stages of the cod, the haddock, and the flatfishes in this) and to the waters westward from Cape Cod. Dannevig, too, has called attention to the absence of larvae of the silver hake in Nova Scotia waters, contrasted with the presence of their eggs there.[48]

One possible explanation for this contrast between larvae and eggs is that it may mirror the relative percentage of eggs that hatch in the regions in question. A more likely explanation we think, when taken with other similar facts of distribution, is that it results from a peripheral drift around the shores of the Gulf from northeast to southwest, in which the eggs take part first and then the resultant larvae. This type of circulation, in fact, has been established so well for our Gulf by hydrographic evidence, that some such involuntary migration is inevitable, not only for various buoyant fish eggs and larvae that are produced near the coast line, but likewise for the drifting communities of invertebrates, and of plants.

It is now known that large numbers of the silver hake that descend to the deeper water layers in the southwestern part of the Gulf during their first autumn remain there during the following

[46] Dannevig, Canadian Fish. Exped., (1914-15) 1919, p. 28.

[47] Goode, Fish. Ind. U. S., Sect. I, 1884, p. 242.

[48] Canad. Fish. Exped. (1914-1915) 1919, p. 28.

year, some of them still longer. Thus our experimental trawlings, in August 1936, yielded large numbers of the 1-year-olds at 70 to 90 fathoms in the deep basins off Cape Cod and west of Jefferys Ledge; also off Mount Desert.[49] And nearly all of the silver hake that come close enough inshore to enter the traps in the southwestern part of the Gulf, or to strand on the beaches there, are good sized individuals of 9 inches and larger. In fact, the only instance that has come to our notice of any considerable inshore catch of one-year-old fish (about 6 to 8 inches long) in the Cape Cod Bay region was near Provincetown, August 1939, when about 1,900 of them were taken during a 14-day period.[50] Huntsman, however, reports all sizes near shore from yearlings on, in the Passamaquoddy region to the northeast, and in the neighboring parts of the Bay of Fundy.

The silver hake 3 years old and older, that provide the commercial catches, sometimes appear in the Cape Cod Bay-northern Massachusetts region as early as the last week in March, regularly by May. Welsh saw some fish, for example, in Ipswich Bay in March and April in 1913 (a fairly representative season), considerable numbers in May, and an abundance in June. And this may be taken as typical for the whole coast line of the Gulf south of Portland; also for Georges Bank, where the first silver hake were taken by the otter trawlers from April 27 to 29 in 1913, and on almost every trip thereafter. We have not been able to learn how early silver hake appear on the coast of Maine east of Portland, or off western Nova Scotia, where it is only within the past few years that any attention has been paid them.

Around Cape Cod Bay, silver hake are usually the most plentiful in June; disappear more or less during August and September; and reappear in numbers in October, though far fewer then than in June, as is illustrated by the average monthly catches made by a set of eight traps at North Truro, for the years 1946–1947 and 1950:[51] June, 185,200 pounds; July, 36,700 pounds; August, 1,206 pounds; September, 1,780 pounds; and October, 10,852 pounds.

Whether their withdrawal thence in summer is a matter of food, or whether they move deeper to escape the heat of summer is a question for the future. Farther offshore in the western side of the Gulf, and to the northward, silver hake are about as plentiful in July and August as they are in June, as indicated by the vessel landings at Boston and Gloucester; somewhat less so in September and October. And what little information we have suggests that summer is the season of greatest plenty for them in the Bay of Fundy region, though there are far fewer of them there.

The great majority of the silver hake vanish from the inshore waters of the Gulf during the late autumn, November seeing the last of them in Massachusetts and Cape Cod Bays, according both to our own observations and to general report. The latest catches made on Georges Bank during the experimental trawlings of 1913 were on December 3 and 12. And though a few are brought in from the grounds off Massachusetts and Cape Cod during January, February, and March, the catches average less than 1/70 as great for those months as for the period May through October, as illustrated by the monthly landings by trawlers at Boston and Gloucester for 1947:[52]

January	1, 400	July	4, 444, 000
February	2, 255	August	4, 879, 000
March	1, 700	September	1, 974, 000
April	7, 540	October	2, 381, 000
May	860, 000	November	438, 000
June	1, 158, 000	December	207, 000

It is probable that the fish of the year and those that are only 1 year old winter in the deeper depressions near where they first took to the bottom. It is unlikely that fish as small fish as those we have trawled in these situations, in August, can travel far.

The wintering ground of the Gulf of Maine population of larger silver hake is not known. Many of them may winter near the sea floor in the deep open troughs of the Gulf,[53] where the bottom water at 75 to 100 fathoms and deeper continues warmer than 39° F. (4° C.) even at the coldest time of year. Evidence in this direction is that it is only deeper than 60 fathoms that good April catches have been reported on Georges Bank (p. 180). It is also possible that part of them move out to the shelf off southern New England to winter, or

[49] For further details, see Bigelow and Schroeder, Biol. Bull. vol. 76, 1939, pp. 308, 319–320, fig. 8.

[50] Information supplied by William A. Ellison, Jr.

[51] Information supplied by the Pond Village Cold Storage Co., North Truro, Mass.

[52] Pounds of round fish and dressed fish combined.

[53] Practically no trawling is done in winter in the deepest parts of the Gulf.

even to the continental slope as the European silver hake do. Scattered catches, in fact, of half-grown silver hake and larger are made by otter trawlers off southern New England, and off New York in January and February.[54] But it seems more likely that these are fish that either remain there throughout the year or that visit the coasts of New York and of southern New England at other times of year, than that they come from the Gulf of Maine.

Fluctuations in abundance in the Gulf of Maine.—Every shore fisherman in the Massachusetts Bay–Cape Cod region knows that silver hake vary widely in abundance from year to year. Catches by one set of six pound-nets at North Truro on Cape Cod yielded about 60,000 pounds in 1946; 237,000 pounds in 1948; 232,000 pounds in 1949; and only about 10,000 in 1944; but about 458,000 pounds in 1950. Yearly fluctuations of this sort are to be expected at any given locality, in the case of any predaceous wanderer. And there is nothing in the available record to suggest that a major alteration has taken place in the numbers of silver hake in its center of abundance in the Gulf, whether upward or downward, since it has been an important fish on the market.

Occurrence to the westward and eastward of the Gulf of Maine.—Silver hake are described as abundant from October to December as far westward as New York, sometimes in May also, though few are seen there in summer. And yearly catches of some 2 to 5 million pounds of "whiting" by pound nets [55] suggest that the beaches of New Jersey rival those of the Cape Cod Bay region in the seasonal abundance of silver hake. But we have not heard of any great numbers of them close in shore beyond Cape May, though pound nets do take a few as far south as the mouth of Chesapeake Bay. Farther out on the shelf, silver hake of all sizes are to be found at all times of the year, from the offing of southern New England, westward and southward, in numbers large enough for otter trawlers to land 3 to 5 million pounds yearly in New York and New Jersey,[56] and smaller amounts in Delaware.

Eastward from our limits we find the silver hake described as abundant [57] in outer Nova Scotian waters generally. But we have no clue as to their actual numbers there, relative to the Gulf of Maine, for they are not yet important enough commercially to be included in the Canadian fisheries statistics. The experimental cruises of the Newfoundland Fisheries Research Commission took them on Banquereau and Misaine Banks; in the northern side of Cabot Straits; on the southern part of the Grand Banks; and at Bay Bulls on the east coast of the Avalon Peninsula, which is the most northern record for them of which we chance to know. But it seems they are not known anywhere in the Gulf of St. Lawrence.[58]

Importance.—Silver hake are as sweet a fish as one could ask, if eaten fresh or if slack salted over night and used for breakfast next morning. But they soften so fast that there was no regular market demand for them of old, and most of those that were caught incidentally were thrown overboard. In fact, we can remember seeing them used locally for fertilizer. Thus only some 37,000 pounds were saved in Maine and Massachusetts combined, even as recently as 1895. But improved methods of freezing fish were followed by landings of about two million pounds by 1902; of between four and five million pounds in 1905, rising through the years of the first world war to more than 14 million pounds in 1919.[59] The yearly landings then fell off, for some reason, to only about 6 million to 9 million pounds for the period 1924 to 1933, which was far less than the potential catch. But the landings then increased again, as frozen whiting became more popular in the Middle West, to about 15 million pounds in 1935, to about 40 million pounds by 1940, with from 46 million to 74 million pounds during the 6-year period 1942 to 1947.[60]

All but a small part of the Maine and Massachusetts landings, recorded in the following table, are from within the limits of the Gulf of Maine.

The silver hake now ranks fourth or fifth among Gulf of Maine fishes in amount landed. But it

[54] *Albatross II* trawled 8 fish, 7 to 9 inches long, off New York, February 28, 1929, at 28 fathoms; and the dragger *Eugene H.*, Capt. Henry Klimm, picked up 115 of market size in a week's trip, about 80 miles off Martha's Vineyard, at 47 to 67 fathoms, January–February 1950.

[55] 1942, 5,343,300 pounds; 1945, 5,842,900 pounds; 1947, 1,784,500 pounds.

[56] Otter trawl landings of "whiting," for New York and New Jersey combined, were 3,468,200 pounds in 1942; 5,243,700 pounds in 1945; and 7,498,600 pounds in 1947. Delaware trawlers reported 203,500 pounds for 1947.

[57] Vladykov and McKenzie, Proc. Nova Scotia Inst. Sci., vol. 19, 1935, p. 72.

[58] According to Dr. Huntsman all ostensible reports of their presence in the Gulf of St. Lawrence have been based in reality on the other hakes of the genus *Urophycis* (p. 221). And it is these that are meant when "hake" are mentioned in the early publications of the U. S. Fish Commission, such as Baird's (Rept. U. S. Comm. Fish. (1886) 1889, app. A.) report on the fisheries of eastern North America.

[59] Landings for Maine and Massachusetts combined.

[60] Maine and Massachusetts combined.

ranks only about seventh in value. In 1945, the year when the catch was largest, its value was $1,736,200. Its rank is low as a sportman's fish, for while it bites greedily, it puts up only a feeble resistance when hooked.

Year	Catch to nearest 1,000 pounds	Year	Catch to nearest 1,000 pounds
1919	14,607,000	1938	24,851,000
1924	6,377,000	1939	27,539,000
1929	7,875,000	1940	39,990,000
1930	7,943,000	1942	45,900,000
1931	6,936,000	1943	48,460,000
1932	6,379,000	1944	47,373,000
1933	8,678,000	1945	73,866,000
1935	15,420,000	1946	48,844,000
1937	21,038,000	1947	58,936,000

Cod *Gadus callarias* Linnaeus 1758 [61]
ROCK COD

Jordan and Evermann, 1896–1900, p. 2541.

Description.—The most noticeable external characteristics of the cod, emphasized above in the general survey of the cod family (p. 173), are its three dorsal fins and two anal fins; its lack of fin spines; the location of its ventral fins forward of its pectoral fins, and the fact that its upper jaw protrudes beyond the lower; that its tail is usually nearly square, and that its lateral line is pale, not black.

The cod is a heavy-bodied fish, only slightly flattened sidewise, its body deepest under the first dorsal fin (cod neither very fat nor very lean are about one-fourth to one-fifth as deep as they are long), tapering to a moderately slender caudal peduncle, and with a head so large that it takes up about one-fourth of the total length of the fish. The nose is conical and blunt at the tip; the mouth wide, with the angle of the jaw reaching back as far as the anterior part of the eye; and there are many very small teeth in both jaws. The first dorsal fin usually (if not always) originates well in front of the midlength of the pectoral fins; it is the highest of the three dorsals, triangular, with rounded apex and convex margin. The second dorsal fin is nearly twice as long as the first dorsal and about twice as long as it is high, decreasing in height from front to rear with slightly convex margin. The third dorsal fin is a little longer than the first dorsal, and is similar to the second dorsal in shape.

The caudal fin is about as broad as the third dorsal fin is long (rather small for the size of the fish) and broom-shaped. The two anal fins stand below the second and third dorsals, to which they correspond in height, in length, and in shape. The number of fin rays was as follows, in a large series of Gulf of Maine cod, 23 to 37 inches long, examined by Welsh.

Number of finrays	Dorsal			Anal	
	First	Second	Third	First	Second
Least	13	19	18	20	17
Average	15	21	19	22	18
Most	16	24	21	24	22

[61] Jordan, Evermann, and Clark (Rept. U. S. Comm. Fish. [1928], Pt. 2, 1930, p. 210) use the species name *morrhua* Linnaeus 1758. But the use of *callarias* accords better with modern practice, because it preceded *morrhua* on the same page of the Systema Naturae.

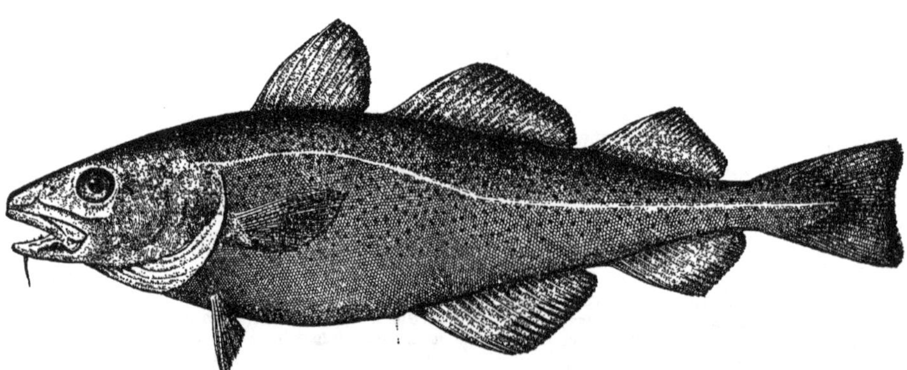

FIGURE 86.—Cod (*Gadus callarias*), Eastport, Maine. From Goode. Drawing by H. L. Todd.

As few as 12 rays have occasionally been recorded for the first dorsal, 16 for the second, 17 for the third, 17 for the first anal and 16 for the second. The pectoral fins, set high up on the sides, reach back as far as the rear end of the first dorsal. The ventral fins are nearly as long as the pectorals in young cod but are shorter, relatively in large fish, with the second ray extending beyond the general outline as a filament for a distance almost one-fourth as long as the entire fin. Both the head and the body are clothed with small scales.

Young cod are easily distinguished from large tomcod by their relatively broad ventral fins with slender filaments, by the location of the first dorsal fin, and by their larger eyes, as explained in the description of that species (p. 196). The pale lateral line readily distinguishes the cod from the haddock; and the square-tipped tail, projecting upper jaw, and spotted color pattern of a cod give it an aspect quite different from that of the pollock.

Color.—Cod vary so widely in color that sundry of its color phases have been named, but all of them fall into two main groups, the gray-green and the red. The back and upper sides of the former range from almost black through dark sooty or brownish gray, olive gray, olive brown, sepia brown, mouse gray, ashy gray, clay colored, and greenish to pale pearly (darker on the back than on the sides); the fins are of the general body tint, and the belly is whitish, usually tinged with the general ground color. The red or "rock" cod vary from dull reddish brown to orange or brick red, with white belly tinged with reddish, and with red, olive, or gray fins. In most cod the upper surface of the body, the sides of the head, and the fins and tail (but not the nose or belly) are thickly speckled with small, round, vague-edged spots. On the "gray" fish these are of a brownish or yellowish cast, darker than the general body color, while they are usually reddish brown or sometimes yellowish on the "red" fish. Occasionally one sees a spotless cod, but these are unusual. The lateral line is invariably paler than the general body tint, pearly gray or reddish according to the hue of the particular fish in question, and it stands out against the darker sides.

Size.—Cod sometimes grow to a tremendous size. A huge one of 211¼ pounds and more than 6 feet long, was caught on a long line off the Massachusetts coast in May 1895;[62] one that weighed 138 pounds dressed (hence must have weighed 180 pounds or more alive) was brought in from Georges Bank in 1838; and Goode[63] mentions several others of 100 to 160 pounds as caught off Massachusetts. But cod of a hundred pounds are exceptional, the largest New England cod of which we have heard recently being one of 90 pounds, that was taken off the coast of Maine early in July 1922. Even a 75-pound fish is a rarity, but 50 to 60 pounders are not unusual. The so-called "large" fish that are caught near shore run about 35 pounds; and "large" ones taken on Georges Bank about 25 pounds. But the shore fish, large and small together, average only between 6 and 12 pounds in weight.

The relationship between length and weight is usually about as follows for fish caught on the inshore grounds between Cape Ann and Portland, though this varies with the condition of the fish and with their state of sexual development.[64]

Females		Males	
Inches	Pounds	Inches	Pounds
19 to 20	2⅛- 3	20 to 21	3 - 3¾
21 to 22	3⅛- 4	23 to 24	4 - 5½
23 to 24	4½- 7	25 to 26	6¼- 8
25 to 26	5 - 7	27 to 28	7 - 8½
27 to 28	7 - 9	30 to 31	7 -11
30 to 31	7½-10	32 to 33	7 -13
32 to 33	9 -13	34 to 35	12 -17
34 to 35	12½-17½	36 to 37	12¾-17
36 to 36½	16 -23	38 to 39	17 -21
38 to 39	18 -22	40 to 41	19 -25
40 to 42	16 -32	43 to 45	25½-29
43 to 44	29½-32	46	43
48½ to 50½	31 -51		
52	50		
57½	54		

A 99½-pound fish recorded by Earll was 62 inches long, and one of 100 pounds caught off Wood Island, Maine, on April 9, 1883, measured 65 inches, its head 17½ inches. Any fish of 5½ to 6 feet will weigh 100 pounds or more.

Habits.—Cod in one place or another range from the surface down to 250 fathoms at least.

During the first year after the young cod take to bottom (p. 186) many of them live in very shoal water, even along the littoral zone, and many young fry have been taken at Gloucester and elsewhere along the shores of New England, while

[62] Jordan and Evermann, American Food and Game Fishes. 1902, p. 514.

[63] Fish Ind. U. S., Sec. 1, 1884, p. 220.

[64] Based chiefly on measurements given by Earll (Rept. U. S. Comm. Fish. [1878], 1880, p. 734), and on a large series of cod measured fresh from the nets by Welsh during the spring of 1913.

many small cod are caught about the rocks only a fathom or two deep even in summer. But it is certain that many cod fry take to bottom on the offshore banks also, for we have trawled young fry at many localities between Nantucket Shoals and Browns Bank. As a rule, the large cod lie deeper than 7 or 8 fathoms in summer in our latitudes. But the fishing is often good in only 3 to 5 fathoms of water in wintertime, especially in Ipswich Bay. At the other extreme, comparatively few cod are caught much deeper than 100 fathoms within the Gulf of Maine. And although fishermen sometimes do well at much greater depths on the slopes of the offshore banks, the 5- and 75-fathom contours probably include the great majority of all the cod living in the Gulf, summer or winter.

The largest catches of cod are made on rocky and pebbly grounds; on gravel; on sand, and on a particularly gritty type of clay with broken shells. They also frequent the deeper slopes of ledges along shore, where they forage among the Irish moss (*Chondrus crispus*) and among seaweeds of other kinds. Young red ones are especially common in these situations, while one sometimes catches a large rock cod as these dark brown or red fish are called. And the bottoms where cod and hake are found are so distinct that a long line set from a hard patch out over the soft surrounding ground will often catch cod at the one end, hake at the other. But fair catches are sometimes taken on mud, as off Mount Desert, where large- and medium-sized cod are regularly caught on soft ground in winter. And a few very large cod (35–60 lb.) have also been brought in from the mud bottom of the deep basin to the westward of Jeffreys Ledge (about 90 fathoms).

The cod, as appears from the foregoing, is typically a ground fish; except on some journey (a subject to be discussed later) or when following its prey, it usually lies within a fathom or so of the bottom. And large ones keep closer to the ground than small ones as a rule, so that the closer one fishes to bottom the larger the cod are likely to run. But even the large ones sometimes follow herring up to the surface; we have known of large cod gaffed from a vessel's side in Northeast Harbor, Mount Desert Island, in September, while they were chasing sardines. And they come to the surface more commonly on the Grand Banks and along the eastern coast of Labrador, when they are following capelin. Cod even strand on the Labrador beaches while harrying schools of capelin, but we have never known cod to strand anywhere around the coasts of the Gulf of Maine, as silver hake so often do (p. 175).

The adult cod is at home in any temperature from 32° to 50°–55° F.; in all but the superficial layers of the Gulf of Maine, that is, at all seasons. But experience at the Woods Hole hatchery, proves that freezing may be fatal by the formation of anchor ice. On the other hand, while large cod tend to avoid water warmer than about 50° F., except that they are abundant at times in temperatures as high as 58°–59° F. on Nantucket Shoals (the most southerly year-round cod-ground in the Atlantic). Small cod are somewhat less sensitive to heat than large, a fact reflected in the presence of greater numbers of them in shoal water in summer than of larger fish. The relationship of the spawning of the cod to temperature is discussed below (p. 194).

Food.—When the larval cod first breaks from the egg it subsists on the yolk with which its abdomen is distended (fig. 88), as do most other sea fishes. But this source of nutriment is completely absorbed by the sixth day after hatching, and the future existence of the little fish depends as much on finding a plentiful supply of food as on escaping the enemies by which it is encompassed. So far as known, the larval and post-larval cod subsist almost exclusively on copepods and on other minute Crustacea, during the several months while they are drifting in the upper layers of water.[65] And this same diet, varied with amphipods, barnacle larvae, and other small crustaceans, as well as with small worms, is the chief dependence of the little cod when they first seek the bottom[66] but as they grow larger they consume invertebrates in great variety and in enormous amount.

Mollusks, collectively, are probably the largest item in the cod's diet in the Gulf of Maine; any shellfish that a cod encounters is gobbled up, so that their stomachs are mines of information for students of mollusks. Large sea clams (*Mactra*),

[65] Bumpus, Science., N. Ser., vol. 7, 1898, p. 485.

[66] For further details on the diet of cod larvae and fry, see Brook (5 ann. Rept., Fish. Board Scotland (1886) 1887, p. 327), McIntosh and Masterman (British Marine food fishes, 1897, p. 242), Kendall (Rept. U. S. Comm. Fish. (1896) 1898, p. 179), Bumpus (Science, N. Ser., vol. 7, 1898 p. 485), and Goodchild, Graham and Carruthers (British Minist. Agric. Fish., Fish. Inv., Ser. 2, vol. 8, No. 6, [1925] 1926.

the empty shells of which are often found neatly nested in cod stomachs; cockles (*Polynices*); and sea mussels (*Modiolus*) are staples, all of which they swallow whole. Cod also eat crabs, hermit crabs, lobsters (large and small), shrimps, brittle stars (of which they are sometimes crammed full), sea urchins, sea cucumbers, and sea worms (*Nereis*). Brittle stars and small crabs, for example, had been the chief diet of the cod examined by Welsh on the Isles of Shoals-Boon Island ground in April 1913, while Wilcox [67] states that a number of 17-pound fish caught in Ipswich Bay were full of large red prawns 2 to 4 inches long (evidently the northern edible shrimp *Pandalus*). And we have found crabs (*Cancer*; *Libinia*) the chief food of the cod on Nantucket shoals.

Tunicates (sea squirts) also bulk large in their diet. Occasionally they eat hydroids, bryozoans, and algae, perhaps taking these for the amphipods that are hidden among them. And in late summer cod frequently feed on ctenophores (*Pleurobrachia pileus*). But while its diet list would probably prove almost as extensive as that of the haddock (p. 202), the cod shows so decided a preference for large shells rather than for small ones that the stomach contents of cod and haddock taken side by side differ noticeably. Nor is it likely that cod root the bottom as haddock do (p. 202), for worms.

Cod pursue and gorge on squid at every opportunity, and on various small fish, particularly on herring, on launce, and (in the north) on capelin; also on shad, mackerel, menhaden, silversides, alewives, silver hake, young haddock, and even on their own young, rising into the upper waters for this purpose when necessary (p. 184). They also pick up flounders, cunners, rock eels (*Pholis*), blennies, sculpins, sea ravens, small hake and skates from the bottom. In fact, they take any fish small enough to swallow, including the hard slim alligatorfish (p. 457) and even the sea horse (p. 315). And Welsh noted that many cod taken near the Isles of Shoals on May 1, 1913 spat up small rosefish from 4 to 6 inches long. The eggs of the longhorn sculpin [68] and of the eelpout (*Macrozoarces*) [69] also have been found in cod stomachs. Adult cod as well as small are also known to feed on pelagic shrimps in the waters around Iceland,[70] but we have never heard of them doing so in the Gulf of Maine.

Even a wild duck does not escape from a large cod now and then. Thus we have heard of several scoters found in the stomachs of large fish caught off Muskeget Island in 1897; and though sea fowl are not a normal article in their diet, the flesh of the greater shearwater (hagdon) has long been considered excellent cod bait. Objects as indigestible as pieces of wood and rope, fragments of clothing, old boots, jewelry, and other odds and ends have repeatedly been found in cod stomachs. And they often swallow stones; but probably for the anemones, hydroids, and other animals growing thereon, and not to take on ballast for a journey as the old story has it.

Although cod are so rapacious they fast generally while they are spawning; the stomachs of nearly all the ripe fish examined by Earll, and recently by Welsh, were empty.

Experiments performed on the cod in captivity,[71] combined with the general experience of fishermen, suggest that they capture moving objects by sight. But apparently cod (and for that matter other fish), can see clearly only for a few feet, and their greediness in snapping up the naked meat of clams and cockles (foods which they never find in that condition in nature), added to the fact that they bite as readily by night as by day, seems sufficient evidence that they depend largely on smell.

Enemies.—In the Gulf of Maine, large sharks and the spiny dogfish are the worst enemy of the adult cod. Formidable enemies of young cod fry are the small pollock which infest our harbors. These are so fierce that a single pollock 7 or 8 inches long will disperse a school of hundreds of cod fry, driving them to shelter among the weeds and rocks, while Earll remarks that in the aquarium a cod so fears a pollock of equal size that it will invariably hide if possible. Young cod, up to 7 to 8 inches, are also devoured in large numbers by the larger cod.

[67] Bull. U. S. Fish. Comm., vol. 6, 1887, p. 95.

[68] Warfel and Merriman, Copeia, 1944, p. 198.

[69] Olsen and Merriman, Bull. Bingham Oceanogr. Coll., vol. 9, art. 4, 1946, p. 77.

[70] Schmidt (Skrift. Komm. Havundersøgelser, No. 1, 1904, p. 70) and Paulsen (Meddelel. Kommis. Havundersøgelser, Serie Plankton, vol. 1, No. 8, 1909, p. 39).

[71] Bateson, Jour. Mar. Biol. Assoc. United Kingdom, N. Ser., vol. 1, 1889–90, p. 241.

Migrations and wanderings.—It has long been known that cod carry out extensive migrations in some regions, but that they are more nearly stationary in others. European (especially the Scandinavian) biologists have succeeded in tracing the major outlines of their movements for North European seas, and enough evidence has accumulated to show that their travels fall into the same categories in the one side of the Atlantic as in the other. These categories are: (*a*) involuntary drifts by the eggs and by the larvae before they take to the bottom; (*b*) the various journeyings by the older cod in search of food; (*c*) journeys associated with the concentrations of cod on particular spawning grounds; and (*d*) regular seasonal migrations (with return movement) between different regions that are suitable for cod during different parts of the year.

To begin with, the eggs, larvae, and young fry of the cod, like those of so many other sea fishes, drift helplessly with the current from the time they are spawned until they seek the bottom (a fact established by European observations too numerous to list).[72] The length of this period (varying in duration in different seas) depends partly on whether the fry are near land or are far out at sea, and partly on whether they are floating over deep water or over shoal. It is not likely to last for more than two months for fish that are hatched on the inshore spawning grounds in the Gulf of Maine, where the bottom is within easy reach. Even so, it is extremely unlikely that any cod fry take to the bottom near where they were spawned.

This matter is discussed further in relation to the occurrence of the cod in our Gulf (p. 190).

The journeyings of the cod that are associated with their spawning are especially extensive along the Norwegian coast, where they have been the subject of much study, leading (among other things) to the very interesting probability that their journeys up and down the coast of Norway are chiefly involuntary, for the ripe fish drifting north become so fat that they tend to be suspended in the water near the surface, whereas the spent fish become so thin that they are deeper down in the water.[73] But there is no reason to suppose that any of our Gulf of Maine cod need travel far to reach the localities where they spawn.

In the extreme northern and southern fringes of their geographic range cod are regularly "migratory" in the common understanding of the term. Thus it is only in summer and early autumn that they visit the waters of the polar current along the eastern coast of Labrador, from which they withdraw again later in the autumn, to pass the winter and spring either to the southward or in deep water. On the other hand, it is only during autumn, winter, and early spring that cod are caught off the coasts of southern New England, of New York, of New Jersey, or further south.

The fish that winter along this westerly and southerly extension of the cod's geographic range appear off southern Massachusetts in mid-October; off western Long Island and off the coast of New Jersey in November; they go back eastward again by the first part of May. And the numbers involved are large enough to support a profitable autumn-winter and early spring fishery from Nantucket to New Jersey.

Tagging experiments carried out by the U. S. Bureau of Fisheries, first at Woods Hole in the winters of 1898–1901,[74] and in various parts of our Gulf on a much larger scale from 1923–1930,[75] have shown that most of the fish that take part in this westerly movement pass the summers in the Nantucket Shoals region. But it is clear that a large part of the cod stock that summers on the Shoals fails to join this westerly mass movement in autumn, for fish tagged there in summer have been recaptured there the next winter, while many others have been recaught there the following spring. And it is established now that the great majority of the cod that live off our coasts from Cape Cod to northern Nova Scotia, in the southern part of the Gulf of St. Lawrence, and on the southern part of the Grand Banks, can fairly be termed "nonmigratory" in a broad sense.

Breeding habits.—The cod is one of the more prolific fishes. A female 39 or 40 inches long may be expected to produce about 3,000,000 eggs yearly, one of 41 inches at least 4,000,000. And Earll estimated the number in a 52½-inch fish weighing 51 pounds at 8,989,094, with 9,100,000 in

[72] In European seas young cod often live under the disks of the large red jellyfish (*Cyanea*), but they have not yet been found in this situation in the Gulf of Maine.

[73] See especially Hjort, Journal du Conseil, Cons. Perm. Internat. Explor. Mer, vol. 1, No. 1, 1926, p. 9.

[74] Smith, Rept. U. S. Comm. Fish (1901) 1902, pp. 193–208.

[75] 22,884 fish tagged in the region of Nantucket Shoals, and about 30,000 in other parts of the Gulf of Maine, including the offshore Banks, 308 fish recaptured westward from Marthas Vineyard,. For further details, see Schroeder (Bull. U. S. Bureau of Fisheries, vol. 46, 1930, pp. 1–136).

a 75-pounder. But the average production of eggs is perhaps not more than 1,000,000 for the general run of Gulf of Maine fish.

The eggs are buoyant, transparent, without oil globule, and 1.10 to 1.82 mm. in diameter. Gulf of Maine eggs, artificially fertilized and measured by Welsh, averaged about 1.46 mm. in diameter, but the size varies somewhat with the temperature of the water, being larger in cold than in warm.[76]

The period of incubation for cod eggs depends on temperature. According to experience at the hatcheries, hatching may be expected in 10 or 11 days at 47° F., in 14 or 15 days at 43° F., in 20 to 23 days at 38° to 39° F. and not for 40 days or more if the water is as cold as 32° F. Fertilization can take place and development commence in temperatures even lower than this, as proved by experiments by Krogh and Johansen.[77] But their observation that the mortality is great among eggs incubated at 32° F. (although full development can take place) corroborates the experience of the hatcheries, where it has proved impossible to hatch more than 25 to 50 percent of the eggs in water as cold as that. And the relative strength of the larvae that are hatched at different temperatures points to 41° to 47° F. as most favorable for incubation. All this suggests that extreme cold prevents the successful reproduction of the cod, not by interfering with spawning (for this can take place in the lowest temperatures to be found anywhere in the open sea, p. 195), but by its effect on the developing eggs. And it is interesting that cod in the tank at Woods Hole produced eggs in February, when the water may have cooled to 30° F. (and quite normally to judge from the fact that the eggs incubated successfully in the warmer water of the hatchery), for these same fish would have spawned naturally in temperatures at least as high as 36°–38° F. if they had been left at liberty.

Newly spawned cod eggs are indistinguishable from those of the haddock, with which they intergrade in size. But shortly before hatching, the pigment of the cod gathers in 4 or 5 distinct patches: one over the region of the pectoral fins, one above the vent, and the others equally spaced behind the latter (fig. 87); whereas in the haddock

[76] Fish (Bull. U. S. Bur. Fish., vol. 43, 1929, p. 292) found cod eggs taken in the tow net in Massachusetts Bay to average about 1.53 mm. in February, smaller (1.46 to 1.49 mm.) in December and in May.

[77] Dannevig, Canadian Fisheries Exped. (1914-1915)., 1919, p. 44.

FIGURE 87.—Egg. After Heincke and Ehrenbaum.

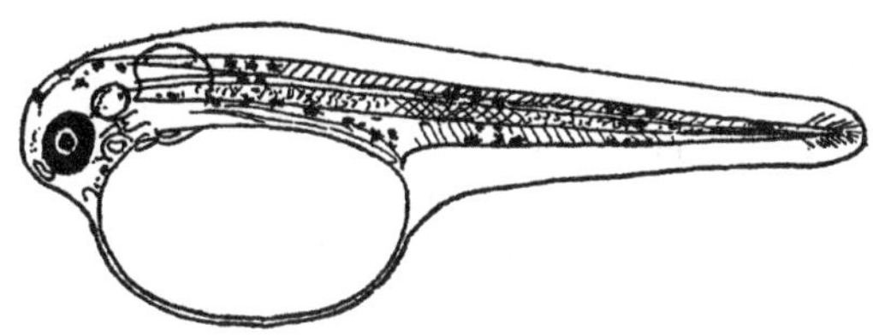

FIGURE 88.—Larva, just hatched, 4 mm. After Masterman.

FIGURE 89.—Larva, 4.5 mm. After Schmidt.

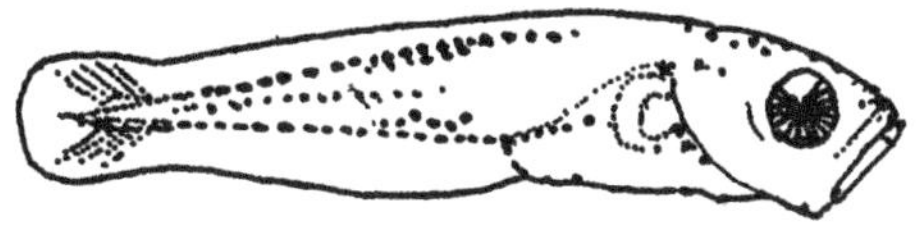

FIGURE 90.—Larva, 9 mm. After Schmidt.

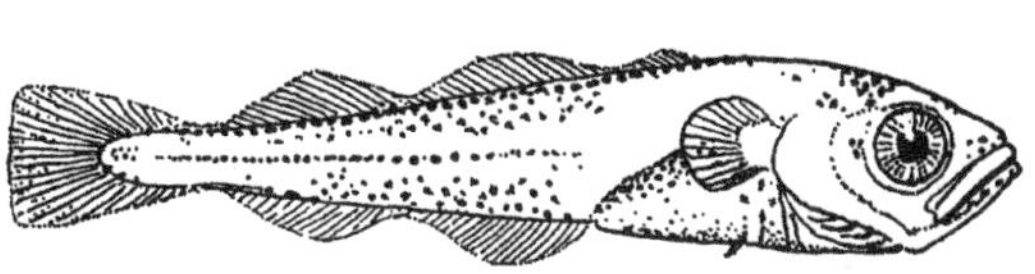

FIGURE 91.—Fry, 20 mm. After Schmidt.

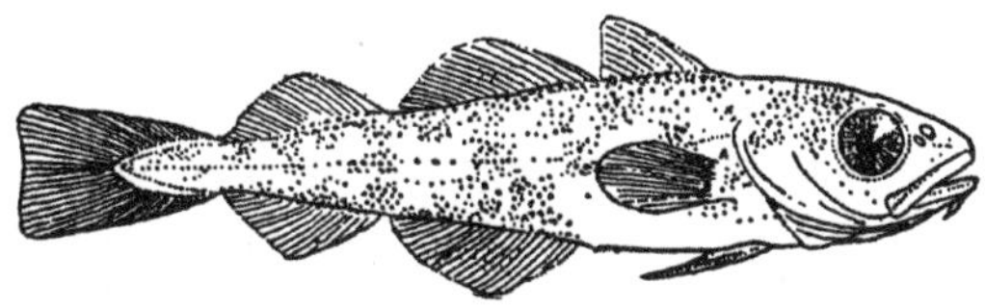

FIGURE 92.—Young, 40 mm. After Schmidt.

COD (*Gadus callarias*), developmental stages, European.

the pigment cells are arranged in a row along the ventral side of the trunk (p. 203). There is also danger of confusing newly spawned cod eggs with those of the witch flounder (p. 287), which they overlap in size; but the black pigment of the cod eggs identifies them as gadoid as soon as this appears, for the embryonic pigment of the witch is yellow. (See also haddock on p. 203.)

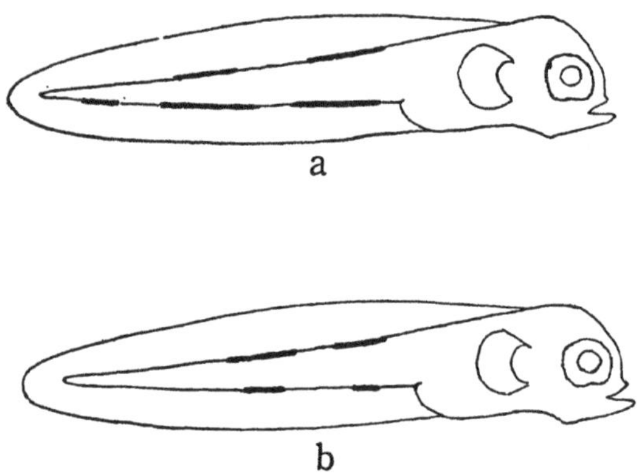

FIGURE 93.—Diagram of the pigmentation of the young larvae of the cod, A, and of the pollock, B. After Schmidt.

The larvae are about 4 mm. long at hatching with the vent (which is close behind the yolk sac) located at the base of the ventral fin fold on one side instead of at its margin, so that the intestine seems to end blindly, as is also the case with haddock and pollock larvae. At this stage young cod much resemble the latter, but are separable from them by the fact that the pigment is in two dorsal and three (rarely two) ventral bars, with the dorsal bars shorter than the ventral bars opposite them, whereas the dorsal bars are longer than the opposing ventral bars (fig. 93) in pollock larvae up to 10 mm. long. Neither is there any danger of confusing cod larvae with haddock even at this early stage, for the latter are not barred but have a continuous row of pigment cells along the ventral margin of the trunk behind the vent, besides other patches on the nape and in the lining of the abdomen.

The young cod float helplessly, when first hatched, yolk uppermost. But they assume the normal position in about 2 days; the yolk being absorbed and the mouth formed in 6 to 12 days, according to temperature, when the larvae are about 4.5 mm. long. As the little cod grows the pigment bars gradually fuse, and at 8 to 10 mm. a median band forms. Cod 10 to 20 mm. long may easily be distinguished from pollock by the fact that the pigment extends to the tail, whereas it ends abruptly some distance in front of the tail in the pollock. Haddock of this size show much less pigment (p. 203). Cod fry of 15 to 30 mm. are made recognizeable by the location of the vent under the second dorsal fin, combined with dense pigmentation. At 20 mm. the dorsal and anal fin rays have attained their final number and the separate fins are outlined, while at 30 mm. the fry begin to show the spotted color pattern so characteristic of the cod.

Rate of growth.—In 1898 a large number of newly hatched larvae were released in December at Woods Hole in the "eel pond" (a lagoon freely communicating with the harbor and with a temperature about paralleling that of the outside water), where they grew to an average length of 50 to 100 mm. by the following June.[78] The experiment was repeated in the winter of 1899[79] with similar results, as appears from the following table showing the growth of approximately 2 million freshly hatched larvae that were placed in the eel pond on January 11.

Date	Extreme lengths	Average length	Date	Extreme lengths	Average length
	mm.	*mm.*		*mm.*	*mm.*
Apr. 8	29 to 38	32.9	May 25	28 to 68	64
Apr. 25	34 to 49	40	June 6	71 to 76	75.5
May 13	35 to 51	42.8	June 20	73 to 77	75

Captures of young fry 1½ to 3 inches long in the neighborhood of Cape Ann late in June (Earll 1880), and subsequently around Woods Hole and on Nantucket Shoals, show that cod hatched from January to March in the Gulf of Maine grow at about this same rate. But fish that are hatched in the rising temperatures of spring might be expected to grow faster during their first few months. European experience[80] is to the effect that young cod are 4¾ to 8 inches long by the end of their first autumn, which probably applies equally to the Gulf of Maine.

In later life cod grow at varying rates in different seas, and even fish that are caught in the same haul

[78] Bumpus, Science N. Ser., vol. 8, 1898, p. 852.

[79] Smith, Bull. U. S. Fish Comm., vol. 19, 1901, p. 307.

[80] Damas (Rapp. et Proces-Verb., Cons. Perm. Intern. Explor. Mer, vol. 10, No. 3, 1909) gives an account of the European investigations on the life history of the cod, up to that date.

may have grown at very different rates, as shown by the structure of their scales. Consequently, the length of a fish older than a yearling is no criterion to its age within 2 or 3 years. Wodehouse's [81] studies on cod caught at the mouth of the Bay of Fundy and the Bureau of Fisheries investigations on Nantucket Shoals, suggest that cod grow more rapidly in the Gulf of Maine than in European waters, as follows:

Age, in years	Average length, in inches, Nantucket Shoals	Average length, in inches, Bay of Fundy	European (approximate average)
1	7–8	6	5
2	14–17	14	8
3	19–22	20	12
4	23–26	26	15
5	27–29	32	19
6	30–32	36	21
7	33–34	39	24
8	----------	45	27
9	----------	49	29

The fact that cod run much larger in the Gulf of Maine than in either the North Sea or the Norwegian Sea, and that those of 75 pounds and heavier, such as are brought in every year from our coastal waters are unusual on the other side of the Atlantic, tends to corroborate the American age estimates, but the desirability of further investigation along this line is self-evident.

Judging from the foregoing table the general run of mature shore cod caught in the Gulf of Maine (5 to 20 pounds) are 3 to 8 years old, but whether the very large fish have grown exceptionally rapidly or are many years old, remains to be learned.

The smallest ripe male recorded for American waters weighed about 3½ pounds; the smallest ripe female 4 pounds,[82] that is, they were in their fourth winter. Probably a considerable proportion of our cod mature when they are 5 to 6 years old; and practically all of them do so by the time they are 9 years old, as Thompson found for the cod of Newfoundland.[83]

General range.—Both sides of the North Atlantic, north to West Greenland, Davis Strait, Resolution Island, Hudson Strait in the west,[83a] south nearly if not quite to Cape Hatteras on the American coast; abundant from northern Labrador to Nantucket Shoals, and to New York and New Jersey in winter, when a few are annually caught as far south as the northern part of the North Carolina coast. The continental slope marks the offshore boundary for the cod off the North American coast. The range of the cod in the eastern Atlantic extends from Nova Zembla, Spitzbergen, and Bear Island in the north to the northern part of the Bay of Biscay in the south, and up the Baltic to Finland. The North Pacific cod, with smaller air bladder (*G. macrocephalus*) cannot be separated from the Altantic cod by external appearance.

Occurrence in the Gulf of Maine.—The cod ranks with the herring, mackerel, rosefish, haddock, pollock, and silver hake as one of the most plentiful of the important food fishes in the Gulf of Maine. Cod were the mainstay of its commercial fisheries from earliest colonial times and until the market began to welcome the haddock. We fancy there is no patch of hard bottom, rock, gravel, or sand with broken shells, from Cape Sable in the east to Cape Cod on the west, but supports more or less cod at one time or another. Cod are even caught on soft mud bottoms, though they are not common there. And while the cod are essentially fish of the open sea, they appear regularly in various river mouths in Maine and Massachusetts during the late autumn and winter. One is taken in brackish water occasionally.

The eastern half of Georges Bank has always been a most productive cod ground and one of the most famous south of the Grand Banks of Newfoundland. The next largest Gulf of Maine fares are brought in from the South Channel–Nantucket Shoals region in the southwestern part of the Gulf, and from Browns Bank in the eastern part, the latter being especially productive in winter. The broken bottom off Seal Island, Nova Scotia, the ground near Lurcher Shoal, and Grand Manan Bank are all famous cod grounds. Other well-known inshore grounds are certain hard patches off Chatham (Cape Cod); between Provincetown and Plymouth and off the latter port; Jeffreys Ledge, Ipswich Bay, Cashes Ledge, Platts Bank, and Fippenies. Small vessels likewise make good catches on the succession of hard and rocky patches that border the coast

[81] Contrib. Canadian Biol. (1914–15), 1916, p. 103.

[82] Earll, Rept. U. S. Comm. Fish. (1878) 1880, p. 717.

[83] Research Bull. No. 14, Newfoundland Dept. Nat. Resources, 1943, p. 87.

[83a] Dunbar (Kennedy, Natural History, Amer. Mus. Nat. Hist., vol. 62, No. 2, 1953, p. 78) has recently reported cod landlocked in southern Baffin Land in a so-called "lake" where the surface is fresh but the deeper water salt.

from the Isles of Shoals to the mouth of Casco Bay; on "Seguin" and "Kettle" bottoms off Seguin Island; on the "Matinicus ground" off Matinicus Island; on the "Grumpy" off Isle au Haut; in the neighborhood of Mount Desert Rock and of Mount Desert Island; and on sundry small ridges thence eastward to the mouth of the Bay of Fundy. Rich,[84] in fact, lists no less than 175 cod grounds around the inner parts of the Gulf, and many other smaller spots all up and down the coast yield a few cod to the small-boat fishermen.

The following summary of the landings of fresh cod from several of the more important Gulf of Maine grounds for 1935 [85] illustrates their relative productivity at that time, and there is no reason to suppose that the situation has altered significantly since then, so far as the numbers of cod are concerned.

Locality	Pounds	Percentage of cod in total catch of ground fish
Georges Bank	21, 598, 594	26
Browns Bank	9, 288, 806	30
South Channel	2, 993, 580	18
Cashes Ledge	602, 901	18
Stellwagen Bank	284, 265	37
Fippenies Bank	48, 865	19
Jeffreys Ledge	42, 430	21
Nantucket Shoals	26, 075	14
Platts Bank	20, 060	18

Cod, for some reason not yet explained, become scarcer passing up the Bay of Fundy, and very few are caught near the head, though there are plenty about the mouth of the Bay.

Movements of cod in the Gulf of Maine.—The young cod that are hatched within our Gulf tend to follow around the general coastline from northeast to southwest, during the period while they are adrift, as has been shown by Fish [86] very clearly for the Cape Ann—Massachusetts Bay spawning grounds. Our few captures of pelagic cod fry have, in fact, all been in the southwestern part of the Gulf, in which they agree with those of haddock, silver hake, and most of the common flatfishes. As Fish [87] pointed out, the fry from eggs that are spawned north of Cape Ann and on the Massachusetts Bay grounds have ample time to become distributed over the offshore banks before they seek the bottom (with 14 to 30 days' drift as eggs, and two months or more as pelagic larvae). They might even circle around to the coast of western Nova Scotia and so to the eastern Maine coast. And fry from the Georges Bank spawning grounds would have ample time to do this in years when they are neither held over the Bank by the local circulation nor carried out over the continental slope, to be lost, as happens in the case of the haddock in some years (p. 212). Our Gulf may also receive contributions of cod larvae and fry drifting past Cape Sable, from outer Nova Scotia waters farther east. On the other hand, the cod fry that are taken at Woods Hole in spring may have come from Nantucket Shoals. But those that we found as far south as the Capes of the Chesapeake in April 1930, probably were the product of the spawning that has long been known to take place in winter off New York and off New Jersey.

Little is known of the wanderings of the cod in the Gulf of Maine from the time they first seek the bottom when 1½ inches long or so, until they are large enough to be caught on hook and line, say 10 or 11 inches long, or 1½ to 2 years old. Young fry, however, from 2 to 4 or 5 inches long and upwards, have been trawled often enough offshore as well as inshore, and they have been found in the stomachs of older cod often enough to show that they soon become distributed all around the Gulf, including the outer part of the Bay of Fundy where it seems that none are hatched (p. 193). But they usually are much more plentiful on the rough inshore bottoms than on the smoother offshore banks. A reasonable explanation is that if young cod take to the bottom on rough, rocky grounds, or among algae, they have a fair chance of escaping their various enemies, but that they find no hiding places on the smooth bottoms that characterize extensive areas on Georges Bank and on Nantucket shoals, hence, are soon decimated.

Some of the larger Gulf of Maine cod probably travel very little out of the spawning season, except as they gradually exhaust the food supply in one spot and are therefore driven to move on over the bottom to fresh foraging grounds. Such fish usually are dark and dull colored, with large heads, a sign of scanty diet. Thus tagging experiments, involving many thousands of fish, have shown that a large percentage of the rather

[84] Rept. U. S. Comm. Fish. (1929), 1930, App. 3, table 2, pp. 85–86; table 3, p. 96.

[85] Most recent year for which catches for the smaller inshore grounds are listed separately in the published catch statistics.

[86] Bull. U. S. Bur. Fish., vol. 43, 1929, pp. 266–290.

[87] Bull. U. S. Bur. Fish., vol. 43, 1929, p. 289.

small cod that make up most of the population along the coast of Maine shift ground but little from season to season. The red fish that haunt the rocks also belong to this category, and red "rock" fish are sometimes caught as large as 10 or 20 pounds.

Other cod (and these compose the greater part of the Gulf of Maine stock), are always on the move over the bottoms of their chosen banks. Though cod can hardly be described as schooling in the same sense as herring or mackerel school, these traveling cod often hold together so closely that it is common enough for one-half of a long line to come in loaded with cod, but the other half to come in empty, and these bodies of fish often run very even in size, color, and shape, suggesting that they may hold together for considerable periods. But fishermen report them mixed as to sex, sometimes males predominating, sometimes females. It is these "school" fish, as they are called, that most often prey on fish and on squid, though they feed chiefly on shellfish as all cod do. They run slenderer and lighter colored than ground cod and have smaller heads, but it is probable that such differences are only temporary reflections of the surroundings of the individual fish, and that a cod that is a ground fish this month, may start on its travels next, turning brighter and becoming more shapely as it goes, either from a change of diet, from a change of surroundings, or from more active exercise. Furthermore, cod may flee a given locality if harassed too much by the spiny dogfish (p. 48), and no doubt other enemies drive them at times.

When cod are on their travels they often rise to the middepths (a fact proved by the levels at which they are caught in nets); netted fish are so often empty, whereas those caught on hook and line are full of food, that they are popularly (and perhaps rightly) believed to fast while they are on a journey.

It is probable that the wanderings of these schools of fish are confined to rather small areas, in most instances. Very few cod, for example, that have been tagged on one of the major Gulf of Maine grounds north or east of Cape Cod have been recaught on any other ground. But the experience of fishermen makes it probable that a certain amount of intermingling does take place between Browns Bank and Georges; also between the latter and Nantucket Shoals.

An interesting fact in this connection, and one for which we see no explanation, is that the majority of such cod as stray afield from the coast of Maine tend to travel to the eastward as a rule, as shown by tagging experiments. Thus 50 out of 76 cod that were marked near Mount Desert, and that are known to have journeyed more than a few miles afield went eastward to Petit Manan (5); to Grand Manan (6); to the west coast of Nova Scotia (20); to the outer coast of Nova Scotia as far as Scatari, Cape Breton (16); to Browns Bank (1); to La Have Bank (1); and to Sable Island Bank (1). But only 26 of them were recaptured to the southward and westward; i. e., Penobscot Bay to Cape Ann, including Cashes and Jeffreys Ledges (20); inner part of Massachusetts Bay (1); off Provincetown (1); South Channel (1); Nantucket Shoals (1) and Georges Bank (2).[88]

Canadian tagging experiments have shown a similar state for Nova Scotian cod, most of them remaining nearly stationary for long periods, some straying eastward, very few moving westward.[89] And Thompson's very extensive tagging experiments have shown that the movements of most of the cod of Newfoundland waters are confined similarly within regions where physical conditions are comparatively uniform.

Some of the cod there make long journeys, discussions of which would carry us too far afield.[90] And in two different winters, (1877–1878 and 1892–1893) hooks of a kind that are used by French fishermen on the Grand Banks of Newfoundland have been found in cod that were caught near Cape Ann,[91] evidence that cod sometimes carry out journeys from north and east to south and west along the American coast, comparable in length to the seasonal migrations that cod have long been known to make along the Norwegian coast, and between Iceland and the West Greenland Banks.[92]

[88] About 12,000 cod were tagged by us near Mount Desert, on the U. S. Bureau of Fisheries vessels *Halcyon* and *Albatross II* and from other craft, from 1924 to 1931. Recaptures nearby totaled 1,754.

[89] For details as to tagging experiments in Nova Scotian waters, see McKenzie, Contrib. Canadian Biol. and Fisheries, N. Ser., vol. 8, No. 31, 1934.

[90] See Thompson (Research Bull. 14, Newfoundland Dept. Nat. Resources, 1943, pp. 20–45, charts 1–8) for detailed discussion in relation to spawning and to racial subdivisions of the local stock.

[91] Earll, Rept. U. S. Comm. Fish. (1878) 1880, p. 706. Kendall, Rept. U. S. Comm. Fish (1896), 1898, p. 178.

[92] See especially Hjort, Journal du Conseil. Cons. Perm. Internat. Explor. Mer., vol. I, No. 1, p. 9, 1926; also Schmidt, Rapp. Proc-Verb. Conseil Perm. Intern. Explor. Mer., vol. 72, p. 37, 1931.

The only regular seasonal migrations that the cod within our Gulf are known to carry out are: (a) their concentrations on their spawning grounds, followed by their dispersal thence after they are spawned out; and (b) a tendency of the fish living closest in shore and shoalest to shift depth with the season, according to the temperature of the water. Thus the cod tend to work in shore, and shoaler, around Massachusetts Bay in autumn, to work out into deeper (hence cooler) water again for the summer. On the other hand, local fishermen report that the cod abandon the shoalest (7–10-fathom) parts of Nantucket Shoals, after the water there has been chilled by the first heavy snows, to congregate from January until April in the deeper (12–20-fathom) channels (warmer in this case.).

Spawning grounds and season.—Thanks to Earll's painstaking studies, and to the large scale on which the Bureau of Fisheries subsequently collected and hatched cod eggs at the Gloucester and Woods Hole hatcheries, the spawning season and the major spawning grounds of the cod are fairly well known for the coastal waters between Nantucket Shoals and the Bay of Fundy.

According to the reports of fishermen and to W. F. Clapp's first-hand experience, large bodies of cod spawn on the eastern part of Georges Bank east of Georges shoal, centering at about latitude 41°21′ to 41°31′, longitude 66°50′, to 67° F. in about 35 fathoms of water. Vague rumors are our only indication as to where and when cod spawn on other parts of Georges; they may do so there, wherever the water is shoaler than 35 to 40 fathoms. And there is every reason to suppose that they spawn regularly on Brown's Bank, though we have no definite record of it.

The broken bottom of Nantucket Shoals, east and south of Nantucket Island (fig. 94), has long been known as a center of abundance for ripe cod fish in late autumn and early winter.

So far as we can learn few cod, if any, spawn on the sandy bottom along the outer shores of Cape Cod. But great numbers of ripe fish congregate in Massachusetts Bay on well-defined grounds 3 to 10 miles offshore, extending from abreast of Sandwich (some 12 miles south of Plymouth) to Minots Light off Cohasset. Years ago many cod also spawned over a small area off Boston Lighthouse and thence northward toward Bakers Island. Few breeding fish have been reported there of late, however, probably because this general locality has been used as the dumping ground for the refuse from Boston, but a few still spawn on various small rocky patches off Gloucester.

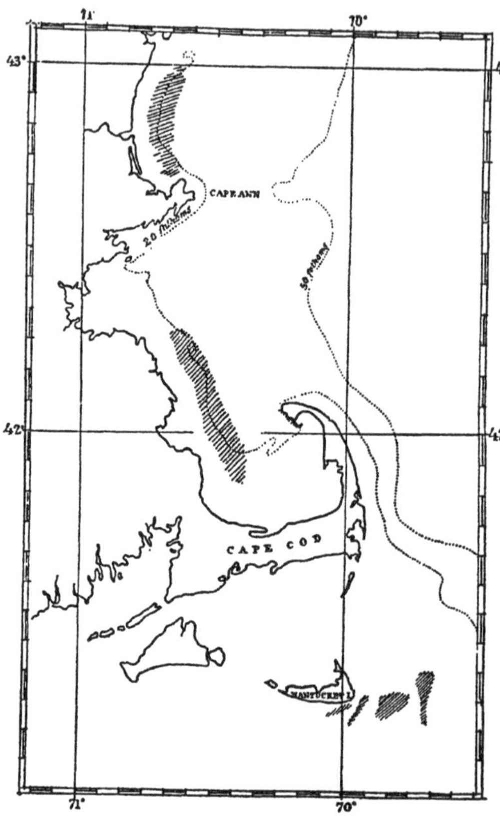

Figure 94.—Chief spawning grounds of cod in the western side of the Gulf of Maine.

The Ipswich Bay region, where large schools of ripe cod gather in winter and spring, is probably the most important center of production for the inner part of the Gulf of Maine north of Cape Ann, but this ground, like the Massachusetts Bay spawning ground, is limited to a rather small and well defined area extending only from a few miles south of the Isles of Shoals to abreast of the mouth of the Merrimac River and (less productively) to Cape Ann, chiefly within 4 to 6 miles of land. A glance at the chart (fig. 94) will show how limited the more important breeding grounds of the southwestern part of the Gulf of Maine are in extent (not more than 300 square miles in all) compared to the whole peripheral zone of this part of the Gulf within the 50-fathom curve. And ripe fish are seldom found even close by, though the fishing for green or spent fish may be good there.

One consequence of the limited extent of these spawning grounds is that the cod congregate on them at the spawning season, in great numbers. During the spring of 1879, for example, when fishing was less intensive than it is at present, and when the cod may have been correspondingly more plentiful, more than 11,000,000 pounds of cod, mostly spawning fish, were taken on the Ipswich Bay ground alone by local fishermen.

Spawning cod are caught only in small numbers, and at scattered localities in the coastal zone north and east of the Isles of Shoals, the more productive of these minor grounds being near Cape Elizabeth; off Casco Bay; off the Sheepscott River; off Boothbay; and in the neighborhood of Mount Desert Island. Very few ripe cod are reported along the Maine coast farther east. And the egg-collecting activities of the several hatcheries have been carried on over so many years that important centers of production there could hardly have been missed. Cod eggs have been taken in the Bay of Fundy but the larvae are unknown there. Neither has any definite evidence been obtained that cod breed in any abundance off the west coast of Nova Scotia. And we should emphasize that the small ledges in the western part of the Gulf, e. g., Jeffreys and Platts, are not breeding centers though they are important feeding grounds. We cannot speak for Grand Manan Bank or for German Bank. Cod, in short, are quite as local in their choice of spawning grounds in the Gulf of Maine as they are in Norwegian waters.[93]

Cod spawn at least as far south and west as New Jersey,[94] and captures, in 1930, of a considerable number of fry 1½ to 2½ inches long off New Jersey and off Virginia in April are evidence that spawning is successful at least as far south as the offing of Chesapeake Bay. But the fate of these southern-spawned cod is yet to be learned.

Following the cod eastward and northward, we learn that eggs are produced in profusion as far north as the Gulf of St. Lawrence and the Grand Banks. But it is not known how much spawning takes place along the eastern coast of Labrador, although eggs have been taken in some numbers along the west coast of Greenland as far north as latitude 66°56′ N.[95]

Cod spawn in shoaler water than haddock on the whole. In fact, we can find no record of ripe cod deeper than 50 fathoms in our waters, and most of the Gulf of Maine spawning takes place on considerably shoaler bottoms. The Georges Bank ground, for example, is about 25 to 35 fathoms deep; the Nantucket grounds are hardly anywhere deeper than 20 fathoms, and as shoal as 7 fathoms in places; the Massachusetts Bay grounds are about 12 to 25 fathoms; and the Ipswich Bay ground is only 5 to 25 fathoms deep according to the precise locality.

It has long been known that while cod spawn chiefly in winter, both in American and in European waters, the breeding season lasts much longer and is less definitely limited at either end for cod than it is for the haddock or for the pollock. And experience has shown that the season when the production of eggs is most active differs widely even within the comparatively small area now under discussion. On Nantucket Shoals, ripening fish are caught from late October on, with the cod spawning there in early November to mid-February, and occasionally until April. Corresponding to this, the brood fish taken off Nantucket that were formerly brought in to the Woods Hole pool spawned there from about the first of December until well into February and occasionally as late as March, with the major production usually from December 20 to January 7.[96] And the spawning season is about the same as this off Plymouth in Massachusetts Bay, where ripe cod of both sexes are common from November until as late as April.[97] On the north side of Cape Ann, however, only 50 miles distant, ripe fish seldom appear in any numbers until January and not until February in some years, though odd ones may be expected from November on.

Earll, for example, found that not one female in ten had commenced to throw her eggs by February, in Ipswich Bay, though spawning was then

[93] See Hjort (Rapp. Proc.-Verb., Cons. Perm. Internat. Explor. Mer., vol. 20, 1914).

[94] Smith, Rept. U. S. Comm. Fish. (1901) 1902, p. 208; Schroeder, Bull. U. S. Bur. Fish; vol 46, 1930, p. 70.

[95] Jensen (Rapp. et Proc. Verb., Conseil Internat. Explor. Mer., vol. 39, 1926, p. 85.

[96] Information from W. H. Thomas, former superintendent of the Woods Hole hatchery.

[97] Information from C. G. Corliss, former superintendent of the Gloucester hatchery.

at its height in Massachusetts Bay, nor were as many as 50 percent of the Ipswich Bay fish ripe before mid-March. Commencing to spawn later there and near Cape Ann than they do off Plymouth, they also continue later, i. e., until the end of April or even into the first part of May, as appears from the following table of cod-egg collections supplied by the Gloucester hatchery:

Season	Collecting field	Number of eggs taken	Spawning season
1911–12	Plymouth	67,032,000	Nov. 24 to Jan. 3.
1912–13	Off Rockport (Ipswich Bay).	170,840,000	Feb. 16 to Apr. 7.
1913–14	Off Gloucester	91,980,000	Feb. 1 to Apr. 15.
1914–15		82,460,000	
1915–16	In Ipswich Bay and off the New Hampshire coast.	145,630,000	Feb. 9 to Apr. 13.
1916–17		92,540,000	Feb. 27 to Apr. 13.
1917–18	Off Gloucester	119,020,000	Feb. 25 to Apr. 27.
1918–19	do	249,510,000	Feb. 27 to Apr. 30.
1919–20	do	570,740,000	Dec. 28 to Apr. 30.
1920–21	do	210,040,000	Jan. 15 to Apr. 29.

Off the western coast of Maine, according to Capt. E. E. Hahn, former superintendent of the Boothbay Harbor hatchery, cod spawn from late February or early March until the last of May, with the production of eggs at its peak in March; they spawn from March through May off the eastern Maine coast, and cod eggs (and hence spawning cod) have been recorded in spring in the Bay of Fundy.

On Georges Bank cod spawn in abundance in February,[98] March, and April.

The records of the hatcheries just summarized tell when eggs are produced in maximum abundance, but they throw little light on the limits of the spawning season, for it was only during the period when there were enough ripe fish to warrant the effort and expense that spawn taking was carried on. And occasional ripe cod of both sexes are seen long before the bulk of the fish breed, and long after. Thus Earll [99] reports the first ripe female as taken near Cape Ann on September 2 during the season of 1878–79, while we have taken cod eggs, far enough advanced in incubation for positive identification as such, off Shelburne (Nova Scotia) on September 6; near Mount Desert on September 15; and off Penobscot Bay on October 6 (all in 1915).

On the other hand Earll saw ripe fish about Cape Ann as late as June. And our tow-nettings make it likely that some may even spawn in midsummer in the coastal zone east of Cape Elizabeth, for we have occasionally found eggs identifiable as either cod or haddock by their black pigment, and probably the former, near Mount Desert Island on July 19; near Wooden Bell Island at the mouth of Penobscot Bay on August 6 and near Cape Elizabeth on September 30.

This sporadic summer breeding of cod in our Gulf is hardly comparable to the so-called "after-spawning" that has been observed off the north coast of Iceland by Schmidt,[1] in the North Sea, and in the Baltic.[2] But it is not unusual for cod to breed in summer off the outer coast of Nova Scotia where ripe fish are reported by local fishermen in June and July. Similarly, spawning cod were caught from the deck of the *Grampus* (Capt. E. E. Hahn in command) on Bradelle Bank in the Gulf of St. Lawrence late in August many years ago, while gadoid eggs (probably cod) were towed at various localities there during June, July, and August of 1915 by the Canadian Fisheries Expedition.[3]

Cod spawn chiefly if not altogether in summer on the Grand Banks where Arctic temperatures prevail during the spring.

Corresponding to the prolonged period of reproduction, spawning takes place over rather a wide range both of temperature and of salinity in our Gulf. On the Ipswich Bay grounds, for example, some are spawning late in November when the bottom water at the depth in question (p. 193) is at its warmest for the year (near 48°); they ripen regularly in temperatures of 41°–43° F. (January); spawning is at its height in the minimum temperatures of the year (35°–37.5°), and some spawning continues until the bottom water has once more warmed to 38°–41° (mid-May).

On the Massachusetts Bay ground, spawning fish appear in numbers (late November) when the bottom water is still as warm as 47°–48°; the chief production taking place in temperatures of 36°–42° (December through January), hence in warmer water than in Ipswich Bay. And the peak of the

[98] This fact has long been common knowledge, and W. F. Clapp, formerly of the Museum of Comparative Zoology, has seen many cod with eggs running, caught on Georges Bank in February and March.

[99] Rept. U. S. Comm. Fish. (1878) 1880, p. 713.

[1] Rapp. et Proc. Verb., Cons. Perm. Internat. Explor. Mer., vol. 10, 1909, pp. 21, 123.

[2] Ehrenbaum (Nordisches Plankton, vol. 1, 1905–1909, p. 225) and Fulton (Cons. Perm. l'Explor. Mer. Pub. de Circonstance, No. 8, 1904).

[3] Dannevig, Canadian Fish. Exped. (1914–15) 1919, p. 22.

spawning season has passed before the temperature drops to its winter minimum, although some cod spawn there through the coldest season (minimum temperature 33°–37°). The temperature range through which the cod breed on the offshore grounds cannot be stated so precisely, for want of data for autumn and for early winter.

In the Gulf of St. Lawrence, cod are known to spawn in water as cold as 32° F. or even slightly colder,[4] though the eggs develop at higher temperatures for they rise to the upper water layers. Around Newfoundland, the cod appear to seek temperatures of 35°–40° F. (1.5–4.4° C.) for spawning, with the chief production of eggs taking place at 37°–41° F. (3–5° C).[5]

Cod spawn in rather colder water on the whole in the Gulf of Maine (still more so in the Gulf of St. Lawrence and on the Newfoundland Banks) than they do in the other side of the North Atlantic, or about Iceland, where the chief production of eggs takes place at temperatures of 40°–45° F.

Probably no cod spawn in water fresher than about 32 per mille nor saltier than about 32.8 per mille, either on the Ipswich Bay grounds or on the Massachusetts Bay grounds. And our records (as far as they go) point to a salinity of about 32.6 per mille as typical for the spawning of the cod on Georges Bank. This is water much less saline than ripe cod seek in European seas, and necessarily so, the Gulf of Maine being decidedly fresher at all times of the year than the Norwegian Sea or the waters around Iceland.

On the Massachusetts Bay spawning ground the specific gravity of the water is high enough to insure that the eggs shall float throughout the breeding season, but in Ipswich Bay the spring freshets often so lighten the surface that late-spawned cod eggs and haddock eggs may fail to rise to the uppermost water layers, a phenomenon which hinders the operations of the hatchery but which does not militate against the successful incubation of the eggs in nature, since the eggs merely float suspended at some deeper level. This subject is discussed at greater length in connection with the haddock (p. 208).

We have yet to learn what proportion of the cod larvae that are hatched in the Gulf of Maine (doubtless a very small one) survive to grow to market size. And what few bits of evidence we have in this regard are contradictory.[6]

Importance.—In 1945, the most recent year for which detailed statistics of the catch are available for the coastlines of Massachusetts and Maine, as well as for the offshore Banks, the Gulf of Maine yielded about 62,500,000 pounds of cod to United States fishermen;[7] some 8,000,000–9,000,000 to Canadian fishermen;[8] or a grand total of some 70–71 million pounds, plus an indeterminate amount landed in small Nova Scotian harbors between the Yarmouth County line and Cape Sable. This is about the same amount as the Gulf had yielded in 1919 (about 67,000,000 pounds); nor is there anything in the catches of intervening years to suggest that any very pronounced fluctuations had taken place meantime in the abundance of cod within our Gulf.

A representative yield, in round numbers, broken down into the statistical areas now employed by the U. S. Fish and Wildlife Service, would be about 7,000,000 pounds along the western coast of Nova Scotia and along the lower Nova Scotian shore of the Bay of Fundy; about 380,000 pounds for the upper Nova Scotian shore of the Bay; about 1,600,000 pounds for the New Brunswick shore of the Bay near its mouth;[9] about 500,000 pounds for eastern Maine; about 4,500,000 pounds for central Maine; about 3,350,000 pounds along western Maine; about 600,000 pounds from the small fishing grounds in the inner-central part of the Gulf; about 5,000,000 pounds off eastern Massachusetts: a little less than 5,000,000 pounds for the grounds from Cape Cod out to the so-called South Channel; about 17,000,000 pounds for Georges Bank as a whole; about 2,000,000 pounds for the western part of Browns Bank; and about 2,200,000 pounds for Nantucket Shoals.

During the early days of the fishery, the entire Gulf of Maine catch of cod was made on hook and line; on hand lines at first, but with long or

[4] Hjort, Canadian Fish. Exped. (1914-1915) 1919, p. XXVII.

[5] Thompson, Research Bull. 14, Newfoundland Dept. Nat. Resources, 1943, p. 89.

[6] Fish (Bull. U. S. Bur. Fish., vol. 43, 1929, p. 266) caught no cod larvae in Massachusetts Bay, though eggs were abundant there, but the *Albatross II* towed several hundred little cod (4 to 9½ mm.) off the tip of Cape Cod nearby, on May 28, 1927. The paucity of our other catches of cod larvae (80 to 90 all told) for other parts of the Gulf of Maine may have been accidental.

[7] Total landings in New England ports were about 139,700,000 pounds, but something over 77,000,000 of this was taken on the grounds along outer Nova Scotia.

[8] About 9,259,900 pounds in 1944, about 8,226,000 pounds in 1945, and about 8,174,800 pounds in 1946.

[9] No cod are mentioned for the head of the Bay on the New Brunswick side in the Canadian statistics of late years.

trawl lines coming into general use about the middle of the 19th century. And it is not astonishing that a fish so nearly omnivorous as the cod should be caught on various baits. Those most in use in the Gulf of Maine are clams (*Mya arenaria*), cockles (*Polynices*), herring (fresh, frozen, or salt), and squid. General experience suggests that there is little to choose between the first two of these, while the razor clam (*Ensis directus*) is equally attractive though limited by the small supply. And tests made in the Gulf of St. Lawrence [10] proved that fresh herring and fresh squid are about as good as clams, but that frozen and salt herring are less attractive. Other kinds of fish are also used as cod bait in other parts of the world; capelin, especially, in more northern seas, and launce.

The earliest important addition to fishing methods came during the winter of 1880–1881, when gill nets, based on the Norwegian system, were introduced in the Ipswich Bay region, yielding unexpectedly large catches.[11] Since about 1908, when otter trawls came into general use in our waters, an increasing proportion of the catch has been taken by this method. Today about 80 to 85 percent of the Gulf of Maine catch is made in otter trawls; only about 10 percent on long lines; about 1 percent in gill nets; less than 1 percent in pound nets, and less than 1 percent on hand lines.

Cod still bite as greedily, however, as they ever did on clams, cockles (*Polynices*), or on pieces of squid or herring. We have even caught fair-sized cod on a pickerel spinner tipped with a bit of pork rind, over ledges in shallow water; we have heard of small cod caught on bucktail lures, also on tin-clad lures cast in the surf. And anglers fishing from small craft for pleasure or for home use catch large numbers all along the coast, though these are mostly of the smaller sizes. So far as we can learn, cod have never been jigged successfully in the Gulf of Maine, as they are in abundance in northern Labrador waters.

[10] Knight, Contrib. to Canad. Biol. (1906–1910) 1912, pp. 23–32.

[11] For account of cod fishing methods in North American waters before the introduction of the otter trawl, see Goode and Collins, Fish. Industries U. S., Sect. 5, vol. 1, 1887, pp. 123–198.

Tomcod *Microgadus tomcod* (Walbaum) 1792

FROSTFISH

Jordan and Evermann, 1896–1900, p. 2540.

Description.—The tomcod resembles a small cod so closely in its fins, in the projection of its upper jaw beyond the lower, in the presence of a barbel on its chin; and in its pale lateral line, that the one might easily be taken for the other. But the outlines of the ventral fins offer a field mark by which the two fish may be separated, for while their second rays are filamentous at the tip in both species, the ventrals of the cod are moderately broad, rounded, and with the filament occupying less than one-fourth the total length of the fin, whereas the ventrals of a tomcod are so narrow, so tapering, and with so long a filament (as long as the rest of the fin) that the whole suggests a feeler rather than a conventional fin. Furthermore, the margin of the caudal fin of a tomcod is noticeably rounded, while that of the cod is square or slightly concave; the eye of the

FIGURE 95.—Tomcod (*Microgadus tomcod*), Woods Hole. From Goode. Drawing by H. L. Todd.

tomcod is decidedly smaller than that of a cod (about one-fifth to one-sixth as long as the head in the tomcod, about one-fourth in the cod, in fish 7 to 10 inches long); and the general form of its body is more slender. A less obvious difference is that the first dorsal fin of the tomcod originates over the middle of the pectoral fins or farther back still, farther forward in the cod; and the pectoral fins reach back only a little beyond the middle of the first dorsal fin in the tomcod, but nearly to the rear end of the first dorsal on a cod.

Unfortunately, the number of fin rays varies so widely in both these fish that it is not diagnostic, there being from 11 to 15 in the first dorsal, 15 to 19 in the second dorsal, and 16 to 21 in the third dorsal of the tomcod: 12 to 21 in its first anal fin and 16 to 20 in its second anal fin. Most of the recent accounts list the position of the vent as the chief external distinction between tomcod and cod, describing it as in front of the origin of the second dorsal fin in the former and back of it in the latter. But we must caution the reader that it is only for adults of the two species (which no one could confuse in any case, cod being so very much the larger) that this distinction holds; cod as small as tomcod (that is, up to a foot long) often have the vent well in front of the second dorsal, while it may hardly be further forward than that in adult tomcod in breeding condition.

Color.—Tomcod are not so variable in color as cod. Those we have seen (a considerable number) have been olive or muddy green above, with a yellowish tinge, darkest on the back, paling on the sides, and mottled with indefinite dark spots or blotches. The lower parts of the sides usually show a decided yellowish cast in large fish; the belly is grayish or yellowish white; the dorsal and caudal fins are of the same color as the back; the anals are pale at the base but olive at the margin; and all of the fins are more or less dark mottled. The tomcod has often been described (following Storer) as thickly speckled with black dots, but we have never seen one marked in that way.

Size.—The maximum size is about 15 inches and 1¼ pounds, but few of them are more than 9 to 12 inches long.

Habits.—The tomcod is strictly an inshore fish; probably few ever descend more than two or three fathoms, or stray as much as a mile outside the outer headlands. In our Gulf they chiefly frequent the mouths of streams and the estuaries into which these empty, as well as shoal, muddy harbors like Duxbury Bay. As often as not they are in brackish water, and they run up into fresh water in winter. Dr. Huntsman, for example, writes us that they are caught in the Petit Codiac River 12 miles above the head of tide. Tomcod are less plentiful in harbors where there is no stream drainage, but now and then they are caught off open shores, off Nahant, for instance, and such fish are usually large ones. South of Cape Cod, most of them move out from the shore into slightly deeper (hence cooler) water in spring, coming in again in autumn to winter in the estuaries. But a year comes from time to time (such as 1925) when they are plentiful close inshore all summer, as far south even as New York.[12] And they do not carry out any inshore–offshore migrations of a regular sort in the cooler Gulf of Maine, so far as is known. Indeed, they are so resistant to cold that we find no record of them killed by winter chilling, a fate that sometimes overtakes other fishes that live in shoal water. And they are equally hardy toward sudden changes of salinity.

Tomcod feed chiefly on small crustaceans, especially on shrimps and amphipods, a great variety of which have been found in their stomachs; also on worms; small mollusks; squids; and fish fry, such as alewives, anchovies, cunners, mummichogs, herring, menhaden, launce, sculpins, silversides, smelt, and sticklebacks.

According to Herrick[13] tomcod are not so keen-sighted as pollock nor so active as hake, but spend most of their time quietly on the bottom in the aquarium. His experiments also proved that they are able to recognize concealed baits by the sense of smell if they chance to swim near and that they search the bottom by dragging the chin barbel and the sensitive tips of the ventral fins as they swim to and fro, either for food, or to stir up shrimps and other food items.

Tomcod spawn in the shoal waters of estuaries, in stream mouths and such places, either in salt water or in brackish, and their eggs have been hatched artificially in fresh water. The season lasts from November to February, inclusive, with the height of production in January. The eggs are about 1.5 mm. in diameter with a conspicuous oil

[12] Nichols and Breder (Zoologica, N. Y. Zool. Soc., vol. 9, 1927, p. 166) state that tomcod up to 10¾ inches long were common throughout that summer in Sandy Hook Bay.

[13] Bull. U. S. Fish Comm., vol. 22, 1904, p. 262.

globule, and (unlike those of its larger relative) they sink to the bottom where they stick together in masses, or to seaweeds, stones, or any available support. Incubation occupies about 24 days at an average temperature of 43°; 30 days at 40°. The larvae are not only somewhat larger (5 mm.) at hatching than those of the cod, but are farther advanced in development, the mouth being formed. And they differ from all other Gulf of Maine gadoids at a corresponding stage by the presence of the oil globule and by the fact that the vent opens at the margin of the ventral fin fold and not at its base at one side.[14] Although great numbers of tomcod have been hatched artificially by the State of New York, its later larval stages have not been described, nor have we seen them ourselves. The fry, which are said to remain through their first summer in the waters where they are hatched, grow to a length of 2½–3 inches by the following autumn. But nothing is known of the rate of growth of older fish.

General range.—North American coastal waters from the Gulf of St. Lawrence and northern Newfoundland to Virginia, running up into fresh water.

Occurrence in the Gulf of Maine.—The tomcod is locally common around the entire coastline of the Gulf. It is reported at Pubnico and in St. Mary Bay, for example, on the west coast of Nova Scotia; at various localities on both shores of the Bay of Fundy (e. g., Annapolis Basin and River, Minas Basin, St. John Harbor, and the St. Andrews region); at Eastport; from almost every river mouth along the Maine coast; in the vicinity of Boothbay Harbor; at sundry stations in Casco Bay; and in Portland Harbor in Maine. And it is to be found in practically every estuary around the Massachusetts Bay region.

Tomcod are caught from docks and bridges and in salt creeks in mid-summer as well as in winter. Tomcod are in the inner parts of Duxbury bay, for example, in midsummer; there are also plenty of them in a certain salt marsh creek at Cohasset at all seasons; and this applies to many similar locations all up and down the coast, including the Bay of Fundy, where tomcod are in and near the estuaries the year round, as Huntsman [15] remarks.

Westward and southward from Cape Cod, the tomcod is plentiful in suitable situations all along the coast to New Jersey, where Abbott [16] described them many years ago as a "very common" little fish, and we have often caught them while fishing from docks in lower New York Harbor.

In the opposite direction, they are common along the outer shores of Nova Scotia. They are plentiful enough around the shores of the Gulf of St. Lawrence for catches of 684,000 pounds to be reported from the New Brunswick coastline of the Gulf in 1947, 20,400 pounds from the southern shore of the estuary of the St. Lawrence River, 152,900 pounds from the north shore of the estuary and Gulf, while Jeffers [17] reports them as taken in considerable numbers through the ice in winter, on the Newfoundland side of the Strait of Belle Isle. And they are to be expected along the southern and eastern coasts of Newfoundland, though they seem not to have been reported there as yet.

Importance.—The tomcod is a delicious little fish. But it seems to have been more highly considered a century ago, when between 5,000 and 10,000 pounds were caught annually in the Charles River tributary to Boston Harbor; today, it is unusual to see any for sale in a Massachusetts fish market. And, in any case, tomcod are not plentiful enough anywhere around our Gulf to support a regular commercial fishery of any magnitude. In 1929 the reported catch was about 6,000 pounds for Massachusetts, about 16,500 pounds for Maine, and about 6,100 pounds for the Canadian shores of the Gulf. In 1942,[18] 27,500 pounds were reported for Maine, none for Massachusetts, about 10,000 pounds for the Nova Scotian shore of the Bay of Fundy. Since that time a few thousand pounds have been reported yearly from the Nova Scotia shores of the open Gulf and of the Bay of Fundy;[19] none at all, however, from its New Brunswick shore.

Most of the tomcod marketed in Maine (also most of those formerly marketed in New Brunswick) are taken in bag nets or in pocket nets set

[14] Ryder (Rept. U. S. Comm. Fish., (1885) 1887, p. 523, pl. 13, fig. 67) describes and pictures the newly hatched larva of the tomcod.

[15] Contrib. Canadian Biol., (1921) 1922, p. 67.

[16] Geol. New Jersey, 1868, p. 818.

[17] Contrib. Canadian Biol., N. Ser., vol. 7, No. 16 (Ser. A, general, No. 13), 1932, p. 7.

[18] Most recent year when tomcod were mentioned in the United States catch statistics for the Gulf of Maine coast.

[19] 35,000 pounds of tomcod were reported for Digby County in 1944, but this amount is so much larger than for preceding years, or for 1946, as to suggest some error.

in the courses of the larger rivers, a few in weirs. In the days when the commercial catch for Massachusetts was large enough to be worth reporting, most of it was taken on hook and line north of Plymouth, in weirs and traps south of Plymouth.

Besides the fish reported in catch statistics, a considerable number are caught in autumn on hook and line by smelt fishermen and by anglers fishing especially for "frost fish," all along the shores of northern New England and used for home consumption. Hence they are not reported or included in the fishery statistics.

Tomcod bite any bait greedily. Clams, shrimp, sea worms, or cut fish will serve, and they afford amusement to a larger number of anglers in harbors and stream mouths than the meager commercial catch might suggest.

Haddock *Melanogrammus aeglefinus* (Linnaeus) 1758

Jordan and Evermann, 1896–1900, p. 2542.

Description.—The most obvious ways in which the haddock differs from the cod are in its black

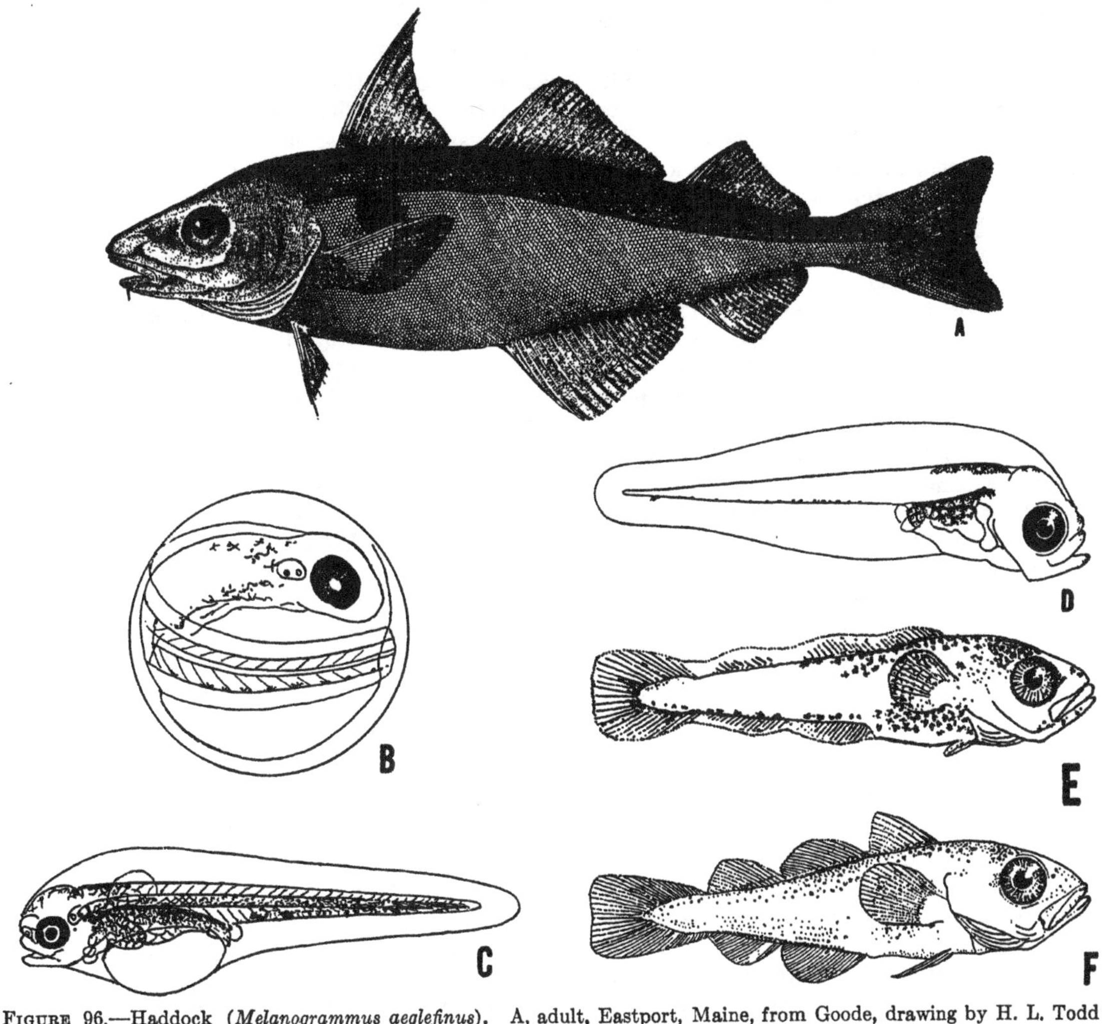

FIGURE 96.—Haddock (*Melanogrammus aeglefinus*). A, adult, Eastport, Maine, from Goode, drawing by H. L. Todd; B, egg (European); C, larva (European) just hatched; D, larva (European), 4.2 mm.; E, larva (European), 15 mm.; F, young fry (European), 25 mm. B and C, after Heincke and Ehrenbaum; D, after Ehrenbaum; E and F, after Schmidt.

lateral line (that of cod and of pollock is paler than the general ground tint) and in the presence of a dusky blotch on each side over the middle of the pectoral fin, and close below the lateral line. Furthermore the first dorsal fin of a haddock (higher than that of a cod, relatively) is considerably higher than either the second or third dorsal, more acutely triangular in outline, and with slightly concave margin. The margin of the haddock's tail is more concave than that of the cod; and its second and third dorsal fins are more angular than is usually the case with the cod, though they are similarly rhomboidal in outline.

The haddock's mouth is relatively the smaller, not gaping back to below the eye, and the lower profile of its face is straight, with the upper profile only slightly rounded, giving the nose a characteristic wedge-shaped outline in side view. The upper jaw projects further beyond the lower in the haddock than in the cod, and the snout is usually more pointed and the body more flattened sidewise. But the general arrangement of the fins is the same; there are about the same number of dorsal fin rays in haddock as in cod (14 to 17, 20 to 24, and 19 to 22, in the first, second, and third fins, respectively); and while the anal fins average one or two more rays each (21 to 25 and 20 to 24), individual cod may have more anal rays than individual haddock. Finally, the haddock is a slimmer fish than the cod and although its scales (which clothe it from nose to tail) are of about the same size relatively (about 160 rows along the side), they are scarcely visible through the mucus with which the skin is coated.[20]

Color.—When a live haddock is first taken from the water, the top of its head, back, and sides down to the lateral line are dark purplish gray, paling below the lateral line to a beautiful silvery gray with pinkish reflections, and with the black lateral line and the sooty shoulder patch (just mentioned) standing out vividly. This patch, the "devil's mark," is indefinitely outlined and varies in size and in distinctness, but only very rarely does a haddock fail to show it. The belly and lower sides of the head are white. The dorsal, pectoral, and caudal fins are dark gray; the anal fins pale like the lower part of the sides and black specked at the base; and the ventrals are white, more or less dotted with black. Haddock usually run very uniform in color, but occasionally one shows from one to four dark transverse bars or splotches in addition to the black shoulder blotch. Several of these serially striped haddock have been taken in Passamaquoddy Bay [21] and we have seen such near Mount Desert. Occasionally a haddock may be decidedly golden on the back and sides, with the lateral line golden, and such fish may lack the dark blotches.

Size.—The haddock is a smaller fish than the cod, the largest on record having been only 44 inches long, weighing about 37 pounds.[22] One of 30 pounds, caught on La Have Bank in the autumn of 1949 [23] is said to have been the heaviest ever landed at the Boston Fish Pier. The largest among 1,300 fish that were measured and weighed by Welsh near Gloucester during the spring of 1913 was 35½ inches long, weighing about 16½ pounds. Only 4 or 5 out of the more than ten thousand haddock that we have helped to tag were as long as 32 to 34 inches. And the great majority of the fish that are brought in measure from 14 to 23 inches long, and weigh from 1⅛ to 4¾ pounds. The largest among 627,996 fish measured during the period 1931–1948 was 34½ inches long.[24] The relationship between length and weight averages as follows, according to Shuck;[25] 10 inches, 7 ounces; 12 inches, 12 ounces; 14 inches, 1 pound 2 ounces; 16 inches, 1 pound 11 ounces; 18 inches, 2 pounds, 6 ounces; 20 inches, 3 pounds 3 ounces; 22 inches, 4 pounds 3 ounces; 24 inches, 5 pounds 5 ounces; 26 inches, 6 pounds 9 ounces; 28 inches, 8 pounds 3 ounces; 30 inches, 9 pounds 15 ounces.

Habits.—Haddock live deeper than cod on the whole; few are caught in less than 5 to 10 fathoms of water and most of them in 25 to 75 fathoms. In fact, they so seldom come into shoal water where young cod are so plentiful that the pound nets of Massachusetts reported only about 5,000 pounds of haddock in 1919, as compared with almost 300,000 pounds of cod. Neither do we remember hearing of a haddock of any size in any of the shoal harbors where little pollock so abound. And the difference in habitat between these closely related species holds from the time the young fry

[20] Vladykov (Canadian Field Natural., vol. 49, No. 4, 1935, p. 64) describes a haddock with 3 eyes, and includes a photograph of it.

[21] Prince, Contrib. Canadian Biol., (1915–1916) 1917, p. 86.

[22] This giant was an Icelandic fish, reported by Thompson (Rapp. et Proc. Verbaux, Conseil Internat. Perm. Explor. Mer, vol. 57, 1929, p. 29).

[23] Received by O'Hara Bros., and reported by Moore, Boston Herald, Nov. 29, 1949.

[24] Information from Howard W. Schuck.

[25] Fishery leaflet No. 198, U. S. Fish and Wildlife Service, 1947.

first seek bottom, for haddock usually do so in 20 to 50 fathoms or deeper, seldom close to the shore, and perhaps never in the littoral zone.[26] On the other hand, comparatively few haddock, are caught deeper than 100 fathoms in American waters,[27] though they have been taken as deep as 120 fathoms (220 m.) on the slopes of the Faroe Bank, and as deep as 164 fathoms (300 m.) off Iceland.[28]

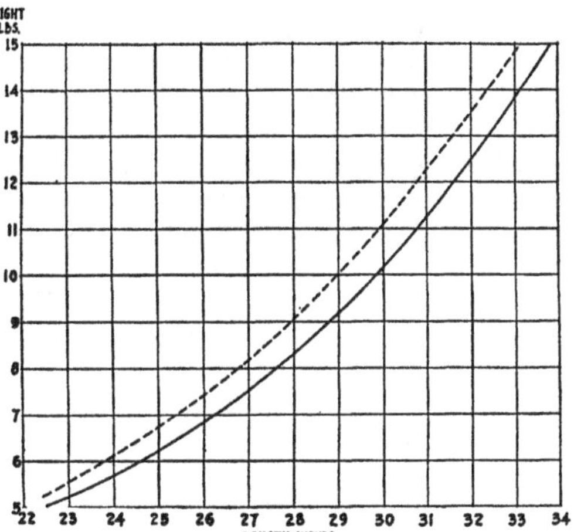

FIGURE 97.—Average weight of ripe haddock of different lengths; male (—) and female (- -) at Gloucester, Mass., March to May 1913.

The haddock, like the cod, is a cold-water fish, though it is not at home in temperatures quite as low. Thus it is almost wholly absent off Newfoundland, in the Gulf of St. Lawrence, and off Nova Scotia when the bottom water is as cold as 32° F.; few are caught there, generally speaking, where the bottom water is colder than about 35–36° F. (2° C.) though good catches are sometimes made in temperatures as low as 34°. At the opposite extreme, haddock appear to avoid water warmer than about 50–52° F. Thus Vladykov [29] reports that young haddock withdraw from Halifax Harbor if the temperature near the bottom rises above about 52°, though they can survive considerably higher temperatures for limited periods.[30] It is evident from this that the entire Gulf of Maine, at the depths frequented by the haddock, is suitable for them so far as temperature is concerned, but that the upermost stratum may be too warm from late summer through early autumn, and too cold from late winter through early spring. In exceptional years, too, such as 1926, the whole column of water may chill to a temperature too low for their comfort in the Bay of Fundy (p. 210).

The salinities at the localities and depths where haddock live in our Gulf range from about 31.5 per mille inshore to a maximum of about 34.5 per mille on the offshore edge of Georges Bank, with most of the catch made in water more saline than about 32 per mille. And while they enter the bays and reaches between the islands along the coast of Maine in some numbers (p. 210), they never run up estuaries into brackish water. Thus, haddock seem to require somewhat higher salinities than cod, which are sometimes caught in considerable numbers where the water is below 31 per mille (as in the Bras d'Or Lakes, Nova Scotia).[31]

In general, the haddock live in rather cooler and less saline waters in the American side of the Atlantic than in the European, as Thompson [32] has emphasized.

The haddock is more exclusively a ground fish than the cod and though they sometimes pursue herring and other small fish, as cod do more often, we have never heard of haddock coming to the surface when so engaged, events by no means unusual with cod, and a characteristic phase in the life of the American pollock (p. 214).

Haddock are more selective than cod in the type of bottom they frequent, being rarely caught over ledges, rocks, or kelp (where cod are so plentiful), or on the soft oozy mud to which hake resort. They are chiefly taken on broken ground, gravel, pebbles, clay, smooth hard sand, sticky sand of gritty consistency, and where there are broken shells; they are especially partial to the smooth areas between rocky patches.

Food.—During their first few months, while living pelagic near the surface, haddock fry probably depend on copepods as cod do. After

[26] The fact that haddock fry less than 1 year old have never been reported in shoal water in the Gulf or at Woods Hole corroborates European fishing experiments summarized by Damas (Rapp. et Proc.-Verb., Cons. Internat. Explor. Mer, vol. 10, 1909) and by Schmidt (*ibid.*).

[27] Thompson, Research Bull. No. 6, Newfoundland Dept. Nat. Res., 1939, p. 9.

[28] Goode and Bean (Smithsonian Contrib. Knowl., vol. 30, 1895, p. 354) list a haddock from 499 fathoms but with suspicion as to the accuracy of its label.

[29] Contrib. Canadian Biol., N. Ser., vol. 8, No. 29, 1934, p. 418.

[30] At the St. Andrews Laboratory, haddock kept at a temperature varying between about 57° and about 68° F. survived for 3 to 4 months.

[31] Needler, Contrib. Canadian Biol., N. Ser., vol. 4, No. 20, 1929, p. 10.

[32] Research Bull., No. 6, Newfoundland Dept. Nat. Resources, 1939, p. 12.

they take to the bottom they become bottom feeders like cod, devouring all kinds of invertebrates so indiscriminately that, as Baird [33] remarked long ago, "a complete list of the animals devoured by the haddock would doubtless include nearly all the species belonging to the fauna" of the particular ground on which the fish in question were living. And they begin to depend on this adult diet when they are small. Thus we have found 7- to 9-inch fish full of brittle stars, bivalve mollusks, small worms, and amphipods. The larger Crustacea, such as hermit, spider, and common crabs, shrimps, and amphipods, with gastropods and bivalve mollusks in great variety, worms, starfish, sea urchins, sand dollars, brittle stars, and sea cucumbers all enter regularly into the dietary of the haddock, according to locality.

W. F. Clapp, for instance, listed no less than 68 species of mollusks, both bivalves and gastropods, from 1,500 haddock that were caught on the northwest part of Georges Bank in 40 to 60 fathoms, and he has called our attention to the fact that haddock usually contain smaller shells than do cod, and never the very large sea clams (*Mactra*) which are so important a constituent of the diet of the latter. Neither do haddock eat crabs larger than about 2 inches across, as cod so greedily do. On the other hand, haddock depend more on worms than cod do, and they are often packed full of worm tubes when they are caught on bottoms covered with the latter (the "spaghetti bottom") as in the locality known as "Cove Clark" on the northwest face of Georges Bank (about lat. 41° 08′, long. 68° 40′). Haddock caught near Eastport, Maine, contained 8 species of annelid worms, and they must root out much of their food from the mud and sand of the sea bottom; in no other way could they obtain the burrowing worms and mollusks that their stomachs contain so often.

Haddock take squid when opportunity offers; they are said to prey on herring in Norwegian waters; on launce around Iceland; on fish, mostly launce, on the Nova Scotian banks; [34] on young eels off Cape Breton, Nova Scotia; [35] on herring near Woods Hole and, in 1931, we received reports of haddock having eaten small mackerel on Georges Bank in January. And many baby haddock about 8 inches (20 cm.) long, trawled on the southwest part of Georges Bank, August 13, 1945, were not only seen by John R. Clark of the U. S. Fish and Wildlife Service, to disgorge large numbers of small fish (apparently young silver hake) on the deck of the vessel, but had been feeding chiefly on them. They have also been accused of feeding greedily on herring spawn, perhaps without much justice. But fish ordinarily form so small a part of the diet of the haddock of our Gulf that none of those examined by Welsh near Cape Ann in 1913, nor the Georges Bank haddock opened by Clapp (about 5,000 altogether), and only two of the many that we have ourselves opened, contained fish of any kind, nor have any of the fishermen of whom we have inquired (and their practical experience is of course vastly wider than ours) described Gulf of Maine haddock as feeding to any great extent on fish. And none of the Eastport haddock that were opened by Doctor Kendall had risen to take the large pelagic shrimps (euphausiids) that are so abundant there and which are the chief food of the local pollock.

Welsh's experience with the haddock near Cape Ann during April 1913 was that they are apt to fast at spawning time; more than 95 percent of the hundreds of fish caught there in the gill nets were totally empty, while long lines set nearby were bringing in very few haddock though they were taking hake in fair numbers. But spawning haddock elsewhere "both male and female, have been found with well filled stomachs, and many spawners have been observed in the catches of line fishermen," [36] so the rule is not universal. It also seems that they feed less actively, or at least they take the hook less freely, at temperatures lower than about 36°, as it is in the coldest parts of the Gulf in winter, and the best hook and line catches are made at about 45°-50° F.

The haddock, like the cod, is a prolific fish for its size. Earll [37] estimated the number of eggs in a female weighing 2⅝ pounds and 19¼ inches long at 169,050; 634,380 in one of 4¾ pounds and 24 inches long; 1,839,581 in one 9 pounds 9 ounces and 28½ inches long. Incubation occupies 15 days at a temperature of 37°; 13 days at 41°, a fair average for the eggs that are spawned in the Gulf of Maine. The eggs are buoyant, without oil

[33] Rept. U. S. Comm. Fish (1886) 1889, p. 37.

[34] See Homans and Needler (Proc. Nova Scotian Inst. Sci., vol. 21, 1946, pp. 15-49) for a study of the haddock.

[35] Needler, Copeia, No. 171, 1929, p. 41.

[36] Needler, Contrib. Canadian Biol. and Fish., N. Ser., vol. 6, 1930, No. 10. p. 7.

[37] Rept. U. S. Comm. Fish. (1878) 1880, p. 733.

globule, and from 1.19 to 1.72 mm. in diameter; eggs taken at Gloucester in March 1913 averaged 1.57 mm., varying from 1.47 to 1.72 mm. Thus they average slightly larger than those of the cod. The haddock egg cannot be distinguished from that of the cod in early stages in its development, hence the term "cod-haddock," and when they are newly spawned there is even danger of confusing them with the eggs of one of our commonest flounders, the "witch" (p. 287), whose breeding season immediately follows that of the haddock. But the formation of black pigment soon identifies the cod-haddock egg as such (the embryonic pigment of the "witch" is yellow).

The newly hatched larva is about 4 mm. long, with the vent close behind the yolk sac and at the base of the ventral fin fold, not at the margin, so that it seems to end blind. It resembles a cod so closely that the two would be indistinguishable one from the other, were it not that the post-anal pigment granules of the haddock are arranged in a row along the ventral surface of the trunk from vent to tip of tail, and not in bands as they are in the cod (p. 188) and in the pollock (p. 216), while the dorsal wall of the body cavity of the haddock is densely pigmented. In water of 41° F. the yolk sac is absorbed in about 10 days when the little fish is about 5.5 mm. long; the dorsal and anal fins are fully formed at 16 to 20 mm.; and the young haddock begin to take on the general aspect of the adult by the time it is 30 to 40 mm. long. The arrangement of the larval pigment serves to differentiate the little haddock until it is about 12 mm. long. Larger fry are distinguishable from both cod and pollock by their pale pigmentation, and by the greater height of their first dorsal fin.

Gulf of Maine haddock average about 6 inches long (extremes, 5 to 7 inches) at the end of their first year, and investigations show that the relationship between length and age averages about as follows for larger haddock in different seas:

Age, years	Gulf of Maine	North Sea	Norway
	Length, inches	Length, inches	Length, inches
2	12	10	10.5
3	17.5	12	13
4	19	15	15.5
5	21	17.5	17.5
6	22.5	20	19.5
7	24	22	21.5
8	25	24.5	23

Thus, American haddock grow more rapidly on the whole than European haddock while they are young, but more slowly when older, so that haddock on both sides of the Atlantic appear to be of about the same size by the time they reach 7 or 8 years of age. Needler [38] has found too, that haddock also differ considerably in their rate of growth in different parts of the Gulf of Maine, St. Andrews fish growing faster than those of Browns Bank, with Nantucket Shoals fish intermediate in this respect, as is illustrated in the following table:

Age, years	Average length, inches			
	St. Andrews	Nantucket Shoals	Browns Bank	Eastern Nova Scotia
3¼	18½	18¼	16¾	16¼
4¼	20¼	20¼	18¼	19¼
5¼	22½	22	19¾	21
6¼	24	23½	20¾	22½
7¼	25½	25	21¾	24
8¼	26¾	25¾	22¾	25¼

According to Thompson [39] haddock on the Grand Banks grow more slowly than the Nova Scotian fish, averaging about 23 to 26 inches when 8 to 10 years old, while in the vicinity of Halifax Vladykov [40] gave about 12¼ inches as the length of 2+-year-old haddock and 13¼ inches for 3+-year-old, a rate of growth slower than for other parts of the western Atlantic and perhaps not typical for all years. But individual fish grow at such different rates (probably due to food supply) that a haddock of a given length may differ by 1 or 2 years in age, or even by 3 years in the case of the larger fish. Thus a Gulf of Maine haddock, 14 inches long, may be 2 to 2½ years old; one of 20 inches, 3 to 4 years; one of 28 inches, 8, 9, or 10 years old.

An illustration of this variability is that 6 out of 10 fish that were tagged by the vessels of the U. S. Bureau of Fisheries and were recaptured later had gained ¼- to ½-inch in 2 months though another had not grown at all in that period; one grew 2 inches in 9 months, but two others grew only ½- to ¾-inch in 11 months.[41] And Vladykov's

[38] Contrib. Canadian Biol. and Fish., N. Ser., vol. 4, No. 20, 1929, pp. 11–20, 275–284; N. Ser., vol. 6, No. 10, 1930, p. 54 [295], fig. 17, p. 55 [296].

[39] Research Bull. No. 6, Newfoundland Dept. Nat. Resources, 1939, p. 15, fig. 3 and table 3.

[40] Vladykov (Contrib. Canad. Biol., vol. 8 (29), 1934, p. 7) gave his lengths to the last vertebra, but we have converted these into total lengths to middle of caudal fin.

[41] Schroeder, Jour. Marine Res., vol. 5, No. 19, 1942, p. 16.

studies of the age-length relationship among young haddock of different sizes near Halifax, Nova Scotia, have shown, similarly, that their average rate of growth may differ considerably within short distances in Nova Scotia waters.[42]

The oldest haddock noted by Needler, one about 28¼ inches (72 cm.) long, taken off Ingonish, Nova Scotia, was in its 14th year. But the largest, about 30¾ inches (78 cm.) long, taken off Campobello Island at the mouth of the Bay of Fundy, was in its tenth year, only.

In general, Gulf of Maine haddock grow most rapidly in late summer and early autumn, when the temperature of the water is highest at the depths in which they live, but there is much variation in this respect from place to place and from year to year, as various authors have noted.

Shuck [43] describes the haddock of New England waters as maturing sexually at 3 or 4 years, when they weigh 2 or 3 pounds. And the smallest sexually active specimens found by Welsh among 1,300 haddock were 2 females of about 20 inches long each; i. e., about 4 years old. Most of the Nova Scotia haddock also spawn first in their fourth or fifth year, according to Needler, as some do in Icelandic waters, also. This supports Duff's [44] view that the slackening of the rate of growth at 4 or 5 years of age, which she observed, reflects the first ripening of the sexual organs. In the eastern Atlantic, mature haddock have been reported as small as 9 inches. And almost all the fish spawn there by the end of their third year.

General range.—Both sides of the North Atlantic. On the American coast haddock are the most abundant from the southern part of the Grand Bank and from the more easterly of the Nova Scotian Banks to Cape Cod. In winter they are taken southward to New York and New Jersey, and they have been recorded in deep water as far southward as the latitude of Cape Hatteras. But the species as a whole is so much more closely confined to waters east of Marthas Vineyard than is the cod, that in 1947, for example, only 158,992 pounds of haddock were caught off New York and New Jersey, contrasting with 2,962,559 pounds of cod for that part of the coast.[45] Neither does the range of the haddock extend as far north as that of the cod. Small catches are made in the southern side of the Gulf of St. Lawrence; also along its north shore both in the St. Lawrence estuary and nearing the Strait of Belle Isle, and a scattering are taken among the cod along the west coast of Newfoundland.[46] And while the experimental trawling campaigns of the Newfoundland Fishery Research Laboratory have shown that there is a distinct and extensive stock of haddock on the southern part of the Grand Banks region [47] very few are caught farther north along Newfoundland, though some fish have been reported from the Strait of Belle Isle, likewise from West Greenland.[48] And haddock are unknown in the icy waters along the outer coast of Labrador, where great quantities of cod are caught every summer.

Occurrence in the Gulf of Maine.—Haddock are very plentiful all around the open Gulf, as well as on all the offshore banks, especially on Georges where they greatly out-number the cod. This is, in fact, one of the two species that now rank at the top among Gulf of Maine fishes, from the commercial standpoint; the rosefish is the other (p. 430). Good haddock grounds, it is true, are less extensive close inshore and more scattered there than good cod grounds, haddock being confined for the most part to depths greater than 5 to 10 fathoms (p. 200), and being more selective in types of bottoms they frequent (p. 201). But the number of individual haddock that inhabit the coastal belt of the Gulf within 15 to 20 miles of the land may be as great as the number of individual cod, for while the yield of the inshore small boat fisheries has run only one-third to one-half as great in pounds for haddock as for cod, in Maine and Massachusetts, in years for which data are readily available,[49] and one-half to three-fourths as great for haddock as for cod in the Bay of Fundy,[50] this discrepancy may

[42] Contrib. Canadian Biol. Fish., N. Ser., vol. 8, No. 29, 1934, p. 415, fig. 2.

[43] Unpublished manuscript.

[44] Contr. Canadian Biol. (1914-1915) 1916, p. 39.

[45] This is exclusive of 4,110,508 pounds of haddock and 739,759 pounds of cod landed at New York City, most if not all of which were caught in waters to the east of Marthas Vineyard.

[46] For locations, see Needler, Contrib. Canadian Biol., N. Ser., vol. 6, No. 10, 1930, p. 5 [245], fig. 1.

[47] Thompson, Research Bull. No. 6, Dept. Nat. Resources Newfoundland, 1939, p. 7.

[48] Jensen and Hansen (Undersøgelser over den Grønlandske Torsk, p. 52, 1930).

[49] Between 14 and 15 million pounds of cod and about 5 million pounds of haddock in 1919; between 6 and 7 million pounds of cod and about 3 million pounds of haddock in 1924, these being the only two recent years when the yield of the small boat inshore fishery was listed separately in the published statistics of the catch.

[50] Bay of Fundy catch, about 7 million pounds of cod and about 5 million pounds of haddock in 1919; about 6 million pounds of cod and about 4 million pounds of haddock in 1946, years that seem to have been fairly representative. The inshore catches for western Nova Scotia are not separated from the offshore catches in the published statistics.

not be greater than can be accounted for by the considerably greater weights of individual cod than of individual haddock. And haddock certainly are far more numerous than cod on Georges Bank as a whole, especially on its western half.

Haddock, for example, large and small, made up 60 to 70 percent by number of all the fish caught on various parts of the bank, spring to autumn, by certain otter trawlers in 1913, cod less than 10 percent; similarly, in 1948, 1949, and 1950 haddock formed about 21 percent by number, cod less than 1 percent of the fish trawled there by the *Albatross III*.[51]

In 1945 (most recent year for which detailed statistics are available both for the New England fishery and for the Canadian), the landings were as follows, for different parts of the Gulf, to the nearest 100,000 pounds: western part of Browns Bank, 6,000,000; grounds along the Nova Scotian shore of the open Gulf, 1,000,000; Nova Scotian side of the Bay of Fundy, 3,400,000; New Brunswick side of the Bay of Fundy near the mouth, 1,100,000;[52] off eastern Maine, 200,000; off central Maine, 2,100,000; off western Maine, 900,000; off eastern Massachusetts, 5,400,000; small grounds in the inner central part of the Gulf, 400,000 to 500,000; northern part of the Gulf, not classified, 1,700,000; Cape Cod out to the so-called South Channel, 3,900,000; Nantucket Shoals, 2,200,000; Georges Bank as a whole, 53,200,000. If this proportional relationship is roughly representative, as seems likely on various grounds, the Georges Bank-South Channel area as a whole harbors perhaps two-thirds to three-fourths of the total haddock population of our Gulf, with an average yearly yield of about 94,000,000 pounds, for the period 1931–1948, equivalent to something like 37 million fish.[53] This indeed, is perhaps the greatest haddock ground for its size in the world, or has been in the past.[54]

According to the combined landings for the years 1942–1947, the northwestern [55]–northern parts of the Bank, and its central-southeastern part, are two to three times as productive each, as is the southwestern part, which agrees with fishermen's reports in general.[56] Browns Bank, much smaller in area than Georges, is perhaps equally densely populated.

The following table shows the percentages of the total catch of haddock taken on Georges Bank in each of the major statistical areas, in different years:

Year	Northwestern part	Northern edge	Central and southeastern part	Southwestern part
1942	19	39	36	6
1943	17	27	45	11
1944	20	37	35	9
1945	31	24	24	22
1946	26	35	29	11
1947	19	40	33	9
Average	22	34	34	11

Proceeding next to a more detailed survey of the inshore grounds we find that considerable numbers of haddock are caught on German Bank, and on the broken grounds off Lurcher Shoal. And while haddock are less plentiful than other ground fish on Grand Manan Bank at the mouth of the Bay of Fundy, perhaps because of the type of bottom, yearly landings of something like 3 million pounds along Digby Neck, Nova Scotia,[57] reflect a rich center of population at the mouth of the Bay of Fundy on the Nova Scotia side.[58] Haddock, like cod, diminish in numbers inward into the Bay, so much so that the counties at its head (Hants, Colchester, Cumberland, Westmoreland) report a few hundred pounds, at most, in some years, none at all in others. But they are plentiful enough on the New Brunswick side of the Bay near its mouth and within Passamaquoddy Bay to yield yearly catches about one-third as great as on the Nova Scotia side.

The most productive of the small grounds in the western side of the Gulf [59] are Cashes Ledge,

[51] Information contributed by Clyde C. Taylor of the U. S. Fish and Wildlife Service.

[52] Few haddock are landed near the head of the bay on the Nova Scotian side; none there on the New Brunswick side.

[53] Estimate by Howard W. Schuck, from Fish. Bull. 66, 1951.

[54] Herrington (Fishery Circular No. 23, U. S. Bur. Fish., 1936) so classed it.

[55] During recent years this part of the Bank has been classified as "eastern side South Channel" in the catch statistics published by the U. S. Fish and Wildlife Service.

[56] Needler's chart of haddock catches, 1917–1925 (Contrib. Canadian Biol., N. Ser., vol. 6, No. 10, 1930, p. 5 [245], fig. 1) would suggest that haddock were concentrated on the western edge of the Bank chiefly and on the neighboring parts of Nantucket Shoals. But it is probable, as he points out, that "an exaggerated impression is given of the abundance on the grounds nearest Boston, which is the most important market center."

[57] Classified in Canadian Fisheries statistics as "Digby County, from Sissiboo River to Annapolis County line."

[58] This appears clearly on Needler's (Contrib. Canadian Biol., N. Ser., vol. 6, No. 10, 1930, p. 5, fig. 1) chart of the distribution of the haddock catch, 1917–1925.

[59] Rich (Rept. U. S. Fish Comm. (1929) 1930, App. 3, pp. 51–117) gives a detailed account of the fishing grounds of the Gulf of Maine. In table 2, pp. 85–86, and table 3, p. 96, he lists 130 grounds in the inner parts of the Gulf where haddock are taken regularly.

Jeffreys Ledge north of Cape Ann, Stellwagen Bank at the mouth of Massachusetts Bay, and the several areas of "haddock bottom" off Chatham, Cape Cod. Small isolated rocky banks, such as Cashes and Platts, usually yield fewer haddock than cod, but in recent years of intensive fishing, haddock have been taken in numbers even on these so-called "cod grounds," as appears from the following table (landings to the nearest 1,000 pounds):

Locality	1919	1929	1934	1935
Platts Bank	68,000	193,000	75,000	18,000
Fippenies Bank	34,000	83,000	85,000	26,000
Cashes Ledge	[1] 320	494,000	423,000	384,000
Jeffreys Ledge	1,094,000	1,705,000	226,000	27,000
Stellwagen Bank	736,000	790,000	682,000	236,000
Off Chatham	1,373,000	1,044,000	678,000	339,000

[1] The reported landings from Cashes Ledge for 1919 were so small as to suggest some error.

Spawning grounds.—One part or another of Georges Bank appears to be the most productive spawning ground for haddock off the American coast, one of the most productive anywhere, for that matter. And Walford's detailed studies [60] have shown that haddock may spawn anywhere on the Bank eastward from Nantucket Shoals, except on Georges Shoals where the water is not deep enough. In most years there is a definite spawning center on the northeastern part of the bank, just east of Georges Shoals; Walford found this to be the case in 1931 and in 1932, corroborating our experiences on the *Albatross I* in 1920, when we found haddock eggs in great abundance [61] over an area there of at least 1,600 square miles. In 1932, there was a second spawning center in the so-called South Channel, where there seems to have been little spawning the year before. That Browns Bank, also, is a productive spawning center is proved both by Walford's studies, and by the fact that a fair proportion of the many gadoid eggs we towed there on the *Albatross I* in April 1920 were far enough advanced in development to show a haddock parentage.

Our own egg records, added to reports from the hatcheries and from local fishermen, show that haddock also spawn here and there, along the coastal belt from the entrance to the Bay of Fundy to Cape Cod, though in much smaller numbers than on Georges and Browns.

[60] Bull. U. S. Bur. Fish., vol. 49, Bull. 29, 1938, pp. 3-12.

[61] Captures of ripe fish, male and female, in the trawl established the identity of these eggs as haddock, not cod.

The more productive of the inshore spawning grounds which are neither as sharply circumscribed as those of the cod, nor as regularly occupied, are along the outer (eastern) and northern slopes of Stellwagen Bank, whence many eggs have been obtained for the Gloucester hatchery; the coastal belt between Cape Ann and Cape Elizabeth, especially off Ipswich Bay; the vicinity of the Isles of Shoals; about Boon Island; and off Wood Island, Maine.

Breeding haddock are plentiful east of Cape Elizabeth in some years and scarce or altogether absent there in other years, or for terms of years. Thus, Captain Hahn, former superintendent of the Boothbay hatchery, has informed us that spawning haddock came into Boothbay Harbor in abundance and into Linekin Bay in April and May of 1912, while gill-netters made large catches in the general vicinity, but that spawning haddock did not approach this part of the coast at any time during the next 12 years in numbers large enough either to support any extensive fishery there, or to provide the hatchery with more than a few eggs.

Spawning haddock have also been reported to us from the neighborhood of Mount Desert Island and off Cutler, Maine, while we found a few cod-haddock eggs near Petit Manan Island on April 12, 1920.[62] But there is no reason to suppose that any considerable body of haddock spawn along the Maine coast east of Mount Desert, nor on the northern side of the Bay of Fundy, where neither eggs, larvae, nor young fry have ever been seen. However, our capture of a few haddock eggs [63] and others in the younger "cod-haddock" stage (p. 203) in Petit Passage on June 10, 1915, proves that some spawn on the Nova Scotian side of the bay near its entrance; a few do so on the coastal banks along the western shores of Nova Scotia southward to Cape Sable according to general report, and we have taken a few cod or haddock eggs on German Bank in our tow nets in May.

Turning, now, southward and westward, we learn that gill-netters sometimes get good fares of ripe fish off Boston Harbor, though no great body spawns in the inner part of Massachusetts Bay, and few if any on the cod-spawning grounds off

[62] In a previous report (Bulletin, Museum of Comparative Zoology at Harvard College, vol. 59, 1917, p. 258) we recorded eggs taken along this part of the coast in June as "cod-haddock", but fresh examination of the material shows that they might equally have belonged to the witch flounder, none being sufficiently advanced in incubation to show the pigment.

[63] Far enough advanced to show the pigment in its distinctive arrangement.

Plymouth (p. 192). Some ripe haddock are caught on the shelving-sandy bottom along Cape Cod as far south as Nauset; spawning fish, too, are caught off southern New England every winter. Nearly 800 baby haddock less than 1 year old were taken off Fire Island Inlet, Long Island, and 10 miles off Ambrose Lightship, in November 1948.[64] But their presence there does not necessarily mean that they were spawned so far west, as Dr. Howard A. Shuck of the Fish and Wildlife Service has pointed out to us. Haddock may at times deposit their eggs within a couple of fathoms of the surface in our Gulf, as, for instance, in Boothbay Harbor on the occasion just noted (p. 206). But this is most unusual, 15 to 20 fathoms being the upper limit to regular spawning with the depths of the more productive Gulf of Maine spawning grounds as follows: Browns Bank, 30 to 50 fathoms and probably deeper; Georges Bank, from about 30 fathoms; Cape Cod grounds, about 40 to 70 fathoms; Stellwagen ground, 20 to 40 fathoms; grounds between Cape Ann and Cape Elizabeth, 20 to 65 fathoms.

The presence of newly spawned eggs out to the 100-fathom contour on the southeastern slope of Georges Bank at the height of the breeding season (late March 1931)[65] is evidence that the fish were spawning down nearly or to that depth. But about 100 fathoms appears to be the lower limit to any regular spawning. When eggs are found over greater depths they have drifted from shallower regions, as Walford has emphasized. The few eggs, for example, that we found over the deep basin of the Gulf, and in the Eastern Channel, in April 1920, were flotsam from the neighboring slopes or banks.

The haddock spawn rather shoaler in the Gulf of Maine on the whole than they do in the North Sea region, where the maximum production of eggs takes place at 50 to 100 fathoms. Consequently, there is less difference in this respect between haddock and cod in the western North Atlantic than in the eastern. Neither do haddock confine their spawning so definitely to smooth bottom in American seas as they do in European waters. Welsh found ripe fish chiefly on broken ground "wherever sand, gravel, mud and rocks alternate—if anything, more are taken on the mud in such localities," between Cape Ann and Cape Elizabeth.

The Gulf of Maine haddock spawn chiefly from late February until May and the following record, supplied by C. G. Corliss, former superintendent of the local hatchery, illustrates how brief the peak period of reproduction is near Cape Ann:

Year	First eggs taken	Last eggs taken	Period of greatest abundance	Total eggs collected
1917	Apr. 16	May 3		10,820,000
1918	Mar. 22	Apr. 24	Apr. 9 to Apr. 23	32,380,000
1919	Feb. 12	Apr. 30	Feb. 20 to Apr. 23	332,740,000
1920	Jan. 20	Apr. 29	Mar. 25 to Apr. 25	303,380,000
1921	Jan. 22	Apr. 25	Jan. 27 to Apr. 14	629,130,000

It appears from the hatchery records, corroborated by Welsh's experience in 1913, that the commencement of spawning varies considerably in date from year to year, with the fish breeding freely as early as the end of January in early seasons, but not until the end of March or even until the first part of April in late. But most of them are spawned out invariably by the middle or end of May at the latest.

In normal years the spawning season is about the same on Georges Bank as it is near Cape Ann. In 1920, for example, we found cod-haddock eggs in moderate numbers across its western end late in February; great numbers of them (and took ripe haddock in the trawl) on the eastern end of the Bank on March 11 and 12; and they were still plentiful there on April 16 and 17, but we found none on the western part of the bank on May 17. Similarly, Douthart, of the Bureau of Fisheries, towed haddock eggs over the north-central portion of the bank on April 14 and again on the 26 and 27th in 1913, while Walford found that spawning commenced in February, was at its peak in March and April, and had about come to an end by late May in 1931. Spawning is likewise at its height in mid-April on Browns Bank (large egg catches were observed in our tow-nets April 16, 1920).

Occasional haddock, however, may spawn long after the majority are spawned out. Thus we have towed eggs off Petit Passage, Nova Scotia, on June 10, and have caught a ripe female and a ripe male on Nantucket shoals on June 13 (in 1927). Ripe haddock have even been taken as late as the first part of July near Gloucester,[66] but this is exceptional.

The spawning season continues well into the summer in the colder water along the outer shores

[64] As reported by Arnold, Copeia, 1949, p. 239.

[65] Walford, Bull. U. S. Bur. Fish., vol. 49, Bull. 29, 1938, p. 16, fig. 7.

[66] Earll, Rept. U. S. Comm. Fish., (1878) 1880, p. 730.

of Nova Scotia and south of Newfoundland. Thus we took several unmistakable haddock eggs among numerous newly spawned cod or haddock eggs a few miles off Shelburne on June 23, 1915, while Dannevig[67] records occasional haddock larvae off Halifax on July 23; near Sable Island on July 25 and 26; and on St. Pierre Bank off Newfoundland on July 27 and 28 for that same summer.

The breeding season is about the same in European as in American seas, that is, end of January until June, with the peak of production falling as early as March and April in the North Sea region but not until June around Iceland.[68]

The Georges and Browns Bank haddock spawn in temperatures ranging from about 36.5° to about 42°–43° F., and spawning is likewise completed on the coastwise grounds between Cape Cod and Cape Elizabeth before the stratum of water in which the fish are living has warmed more than a few degrees from its coldest for the year; i. e., in temperatures of about 35° to 40°–42°. Allowing for annual variations, this gives an extreme range of from about 35° to about 44° F. for the most active spawning over the Gulf of Maine as a whole, temperatures averaging considerably lower than those in which haddock spawn the most freely in European waters (41° to 50°).

The Gulf of Maine haddock likewise spawn in less saline water than does its European congener; and necessarily so, for the more important Gulf of Maine spawning grounds are considerably less saline at all depths and seasons (about 31.5 to 33.5 per mille, mostly).

The specific gravity of the water at the temperature *in situ* (the factor that determines whether buoyant fish eggs float suspended, and develop, or sink to the bottom and die) is usually between 1.0255 and 1.0270 in our Gulf in spawning season, at the depths where the fish spawn, both along shore and on the offshore Banks. Experiments by us and by Walford have shown that these values are high enough for the flotation of the eggs. And while the water at the surface often is so light, near shore, as to interfere with the operation of the hatcheries, this layer of low specific gravity is so thin there is no reason to suppose that any of the haddock eggs produced in the Gulf fail to rise from the bottom.[69]

Populations and migrations within the Gulf of Maine.—Needler's[70] analysis of the results of tagging experiments, and of the differences in rate of growth between fish caught in different regions, and Vladykov's[71] studies of the number of vertebrae, confirmed by comparison between the growth rates of the haddock of Georges Bank and of Browns Bank by Schuck and Arnold,[72] have shown that the haddock of North American waters include three more or less self-contained populations; one (Needler's "New England population") inhabiting the Georges Bank–Nantucket shoals region and the inner waters of our Gulf from Cape Cod around to the New Brunswick shore of the Bay of Fundy; a second (Needler's "Nova Scotian") in the Nova Scotian side of the Bay of Fundy, and around Nova Scotia (including Browns Bank) to the Laurentian Channel; and a third in Newfoundland waters.

The geographic ranges of the New England and Nova Scotian populations are separated by the deep so-called "Eastern Channel" between Georges Bank and Browns, which extends inward as the "Fundian Channel" more than 100 fathoms deep, to the mouth of the Bay of Fundy. And it is probable that the depth is an actual barrier in this case, there being no evidence that haddock normally cross channels that are deeper than about 100 fathoms (at least in American waters), once they have taken to the bottom. Only within the Bay of Fundy, where there is no intervening water as deep as 100 fathoms, have tagging experiments given any evidence of a mixture between these two adult populations.[73] And the still greater depth of the Laurentian Channel probably makes it an even more effective barrier between the Nova Scotian and the Newfoundland populations.

The movements of individual fish within each of these populations fall in three groups: (a) those of the eggs and larvae while they are still adrift in the intermediate and upper water layers; (b) those of the young fry from the time they take

[67] Canadian Fish. Exped. (1914–15) 1919, p. 21.

[68] Damas, Rapp. et Proc.-Verb. Cons. Internat. Explor. Mer, vol. 10, 1909; Schmidt, *ibid.*

[69] For a discussion of the relationship between flotation of haddock eggs and the specific gravity of the water, with references to European studies, see Walford, Bull. U. S. Bureau of Fisheries, vol. 49, Bull. 29, 1938, pp. 13–15.

[70] Contrib. Canadian Biol. and Fish., N. Ser., vol. 6, No. 10, 1930.

[71] Progress Rept. Atlantic Biol. Sta. Biol. Board, Canada, No. 14, 1935.

[72] Fish. Bull. No. 67, U. S. Fish and Wildlife Service, 1951.

[73] One fish that was tagged by us near Mount Desert Island was recaptured in the Nova Scotian side of the Bay of Fundy off Digby.

to bottom until they are large enough to figure in the commercial catches, and (c) those of the larger fish.

It may be assumed that the pelagic life of the haddock lasts about as long in American waters as in European; i. e., for three months or so (we have no first-hand information) before the fry seek the bottom. Meantime the eggs and larvae, like those of many other fishes may drift for considerable distances from where they were spawned. And these involuntary drifts may be greatly extended by a habit that the very young haddock have (like those of other gadoids) of living under the bells of the larger kinds of jellyfishes. Welsh, for instance found many small haddock of 2¼ to 3 inches (60 to 77 mm.) in company with the common red jellyfish (*Cyanea*) on Georges Bank and off Nantucket Island, in late July of 1916, while Willey and Huntsman[74] found young haddock about 2 inches long under *Cyanea* in the Bay of Fundy. In fact, it is in company with *Cyanea* that young haddock in the late larval stage have been taken most often in the other side of the Atlantic.

Our few records for the pelagic larvae in the inner parts of the Gulf all have been in the southwestern part. Thus the coastal zone east of Cape Elizabeth, and the whole deep basin of the Gulf, seems to be as barren of larval haddock (so far as our catches go) as of larval cod, of larval silver hake, of larval flatfishes, and, in fact, of most other larval fishes except rosefish (p. 433) and herring. It appears from Walford's studies that in normal years, as represented by 1931, the haddock population of Georges Bank is recruited by a good supply of larvae hatched from eggs that have been spawned on the bank itself. But a large proportion of the Georges Bank eggs and larvae drift off the bank in other years, as in 1932, either to the westward and southward past Nantucket Shoals, where their mortality is too great for them to support a population of any importance, or southward out over the continental slope, to even more certain destruction,[75] with results disastrous to the ensuing brood of young fry (p. 212).

It is during their pelagic stage (whether drifting independently or with Medusae) that intermingling is the most likely to take place in significant amount between the New England and the Nova Scotian populations of haddock. All that is known in this respect is that Georges Bank seems not to have received any important recruitment from elsewhere, either in 1931 or in 1932.

In any case, hosts of young fry settle on the bottom on the offshore banks generally. Thus we have repeatedly found 10 or more little haddock 3 or 4 inches long, in the stomachs of pollock caught on Georges, while we have trawled numbers of equally small ones there as well as on the other offshore grounds. And 1- to 2-year-old fish, 6 to 12 inches long (too small to market) sometimes make up as much as 35 to 40 percent of the total catch of haddock on Georges as well as in the South Channel, while many more of them doubtless escape through the meshes of the trawls. On the other hand, very young haddock are seldom seen inshore for they are too small to be caught either on long lines or in gill nets. But it is probable that they are plentiful there, also, for yearlings are reported in the Bay of Fundy, by Huntsman.

Nothing is known about the movements of the young haddock during the first year or two after they take to the bottom. But our fishermen have long realized that the larger haddock, like the larger cod, are so constantly on the move in search of food that the fishing may be poor tomorrow where it was good today, or vice versa. And analysis of the catches that we made on Nantucket Shoals during the tagging campaigns of the U. S. Bureau of Fisheries, 1923–1931, shows that considerable changes took place in the abundance of fish within periods of a few days or weeks at the spots fished, also with occasional brief periods of unusual abundance that are most reasonably interpreted as reflecting the passage of large bodies of fish from elsewhere.[76]

The extensive tagging experiments that we have made within the Gulf of Maine on vessels of the U. S. Bureau of Fisheries,[77] and that have been made in Nova Scotian waters[78] by the Biological Board of Canada have now proved

[74] Canadian Field Natural, vol. 35, 1921, p. 2.

[75] For further details we refer the reader to Walford's very interesting study (Bull. U. S. Bur. Fish., vol. 49, Bull. No. 29, 1938).

[76] For details, see Schroeder, Jour. Marine Research, vol. 5, No. 1, 1942, p. 9, table 2.

[77] Schroeder, Jour. Marine Research, vol. 5, No. 1, 1942.

[78] Needler, Contrib. Canadian Biol. and Fish., vol. 6, No. 10, 1930.

(as was previously suspected) that most of the wanderings of the Gulf of Maine haddock are of short extent. Thus 34 fish that had been tagged on Nantucket Shoals were recaptured nearby, 16 after periods longer than 200 days; only 10 were captured at a distance. And the preponderance of relatively stationary fish is even more impressive for the vicinity of Mount Desert Island, where 114 tagged fish were recaptured within a few miles of the tagging stations after an average period of 224 days, contrasting with recaptures of only 25 of them at a distance.

The haddock of the coasts of Massachusetts and of western Maine, with the offlying banks, may be less stationary, for only two of the fish that were tagged on Stellwagen Bank and between Boone Island and Boothbay were recaptured locally; 13 of them far afield.

The tagging experiments do not suggest that such of the Gulf of Maine haddock as do wander follow any regular migratory routes. Thus some of the few Nantucket Shoals fish that are known to have strayed were recaptured to the eastward (eastern part of Georges Bank, 2); some of them to the northward (western side of Gulf and Platts Bank, 6); and some to the northeastward (northern entrance to Bay of Fundy, 2). Conversely, it was in the opposite direction, i. e., to Platts Bank, to the coasts of western Maine and of Massachusetts, to the South Channel, and to Georges Bank that wanderers are known to have strayed from the Mount Desert tagging ground. And the few fish that were recaught from those tagged at localities intermediate between Nantucket Shoals and Mount Desert, have fanned out in various directions.

An obvious reason why haddock of the New England population, that commence their adult journeyings in the northeastern part of the Gulf, should tend to stray southwestward, southward, and perhaps then eastward along Georges Bank, whereas others, commencing in the southwest should tend either eastward, or northward and then northeastward, is that these are the only routes left wide open to them within the Gulf, between the coastline on the one side and the barrier that is set for them by the 100-fathom depth line on the other side. How effective is this barrier is emphasized by the fact that only one fish, among 9,416 that we tagged off the coasts of Massachusetts and of Maine was recaptured in Nova Scotian waters (it had gone from Mount Desert to the southern side of the Bay of Fundy); and that none of the haddock that were tagged in Nova Scotian waters by the Biological Board were recaptured west of the Fundian Channel.

Very little is known as to the shifts in location and in depth that haddock may make between winter and summer, the difficulty lying in the interpretation of the differences from season to season in the amounts of haddock that are caught on neighboring grounds in the inner parts of the Gulf.[79]

In general, it appears that when the temperature of the upper 15–20 fathoms of water rises above about 50° to 52° F., as happens along the coasts of Maine and Massachusetts in July or August, the haddock tend to withdraw from the shallower grounds where they are plentiful in spring and early summer. But certain bodies of fish may linger all summer in the deeper channels among the islands of Maine, on patches of suitable bottom. In 1923, for instance, haddock were caught throughout July, August, and September, between Suttons Island and Bear Island, near Mount Desert Island, as well as at other inshore localities near by. Fishermen report them as working inshore again in autumn or early winter, as the water cools, but those that come closest inshore then are supposed to work out again, in mid and late winter, to avoid extreme chilling. Thus few or none are caught at that season in the Bay of Fundy, where the temperature may fall as low as 32° in occasional winters,[80] though it does not drop below 34° to 36° in most years.

We must caution the reader, however, that these supposed disappearances in winter from inshore localities are based on failure to catch haddock then on hook and line, which may actually result more from a reluctance on their part to bite at low temperatures (p. 202) than from seasonal scarcity of fish. Experimental trawlings at different seasons are needed to clarify this matter. At any rate, the temperatures of the open Gulf of Maine at the depths where haddock are the most plentiful never fall too low for their comfort in the winter, nor rise too high in the summer.

---

[79] Rich (Rept. U. S. Fish Comm. for 1929, 1930, App. 3) gives information in this respect.

[80] As happened in 1926 (Needler, Contrib. Canadian Biol. and Fish., N. Ser., vol. 6, No. 10, 1930, p. 19 [259]).

Accordingly, haddock are caught on all the major grounds the year around.

Except for shifts in depth, apparently associated with temperature, the haddock as a whole are year-round residents as far east as the offing of southeastern Nova Scotia (Roseway, La Have, and Sambro Banks); many of them as far east as Halifax and Sable Island Bank. But they appear only as late spring to early winter visitors farther to the east and north where they are wholly absent (as indicated by the catches) in late winter and early spring. Large catches, for example, are made in traps near Ingonish on the northeastern coast of Cape Breton Island in late May and in June. The first haddock are caught within the Gulf of St. Lawrence in June, whether on the Cape Breton shore, or westward, the largest catches are made there from July through the late autumn, and very few are taken as late as December. But catches are made again near Ingonish in December and January, of haddock, seemingly en route out of the Gulf of St. Lawrence. And it now seems established that these visitors to the Gulf of St. Lawrence pass the late winter and early spring on Sable Island Bank and farther to the westward in Nova Scotian waters.[81]

Part of the haddock in the southern part of the Grand Banks region, which form a distinct population separated from that of Nova Scotia by the deep Laurentian Channel, are described by Thompson [82] as making a summer inshore migration to the southwest coast of Newfoundland, and as clearly avoiding regions where the bottom water is colder than about 34° F.

Abundance.—The haddock and the rosefish rank next after herring in numbers, among the fishes of our Gulf that are important commercially. In good years it has not been unusual for a trawler to take 10,000 to 20,000 haddock in 5 or 6 days' fishing on the Georges Bank and South Channel grounds; a catch of 240,000 pounds of haddock (something like 70,000 fish) brought in by the trawler *Fabia* in March 1926 is one of the largest of which we have chanced to hear. One must remember, too, that this represents only the fish that are large enough to be worth saving, and that multitudes of baby haddock too small to be marketable, caught on Georges, are thrown back dying or dead; in 1947, for example, the number so wasted was estimated at almost 17 million on Georges Bank alone.[83] Howard W. Schuck informs us that the average weight of the haddock landed from Georges Bank during 1928 was about 3¼ pounds.

Fishermen have long been aware that the haddock vary widely in abundance from year to year and over periods of years, on one ground or another, independent of any effects the fishery may have had on the numbers of fish. It has been amply proved by investigations both in Europe and in North America, that these fluctuations result chiefly from differences, from year to year, in the number of young that survive and take to the bottom on the grounds in question; the Gulf of Maine is no exception. The production, for example, of young haddock at the mouth of the Bay of Fundy, on the New Brunswick side was low from 1915 to 1919, very much higher in 1920, but somewhat lower again in 1921–1923 though somewhat better than it had been in the 5 years preceding 1920.[84] But a larger number of haddock (by report) were produced near Digby, on the Nova Scotian side of the Bay in 1921 than had been in 1920.[85]

Similarly, two exceptionally successful year classes that were spawned in the Georges Bank–South Channel region during the period 1921–1924 were followed by poor year classes from 1925–1928, but then by an abundant class that was spawned in 1929.[86] Since then Georges Bank has been abundantly recruited with haddock fry in 1936, 1937, 1939, 1940, and 1945.[87] On the other hand, the crop, so to speak, was unusually scanty on the Bank in 1930, 1931, 1932, 1942, and 1947.

Perhaps a good crop comes a little more often for the Nova Scotia population, and every 3 years or so in the North Sea, "where the fry have a much better chance of being retained in the area owing to the prevailing currents." [88]

[81] For further discussion, see A. W. H. Needler, Contrib. Canadian Biol. and Fish., vol. 6, No. 10, 1930, and A. B. Needler, Bull. 25, Biol. Bd. Canada, 1931.

[82] Research Bull. No. 6, Newfoundland Dept. Nat. Resources, 1939, p. 7.

[83] Schuck, Commercial Fish. Review, vol. 10, No. 10, October 1948, p. 5.

[84] Huntsman and Needler, Contrib. Canadian Biol. and Fish., N. Ser., vol. 3, No. 18, 1927, see summary on p. 14 [436].

[85] Needler, Contrib. Canadian Biol. and Fish., N. Ser., vol. 6, 1930, No. 10, p. 44 [284].

[86] The data for 1921-1929 are summarized in the Proc. No. 2, for 1931-1933, N. Am. Council on Fishery Investigation, Ottawa, 1935, p. 13.

[87] From data supplied by Howard A. Schuck of the U. S. Fish and Wildlife Service.

[88] Thompson, Res. Bull. No. 6, Newfoundland Dept. Nat. Resources, 1939, p. 22.

Analyses from year to year of the relative proportions of fish of different ages in the catch [89] show, too, that our offshore banks may receive as much as 30 times as many fry in a good year as in the average for a run of years, and as much as 60 times as many as in the poorest years.

One essential for a good year class of haddock, perhaps the chief essential, is that large numbers of larvae shall not only be hatched and survive until old enough to take to bottom, but shall remain in the area in general, to take to bottom there, as happened in 1931, and not drift elsewhere. Conversely, a poor brood automatically ensues if the circulation of the water is such that a large proportion of the larvae drifts away, as happened in 1932, when so many of them drifted off Georges Bank altogether, to be lost permanently to the local population, that the success of that year class was seriously affected.[90] Herrington has also suggested that in years when large fish are the most plentiful the resulting competition for the supply of available food makes conditions difficult for the survival and growth of the young fry. Evidence is that the "largest spawning stocks have almost invariably yielded the leanest year classes 3 years later, and the poorer spawning stocks have done much better." [91] No doubt a combination of various other factors helps to determine whether any particular year class shall be plentiful or the reverse. But the relative importance of these factors has not yet been evaluated for our haddock.

The incidence of a good brood in any particular year, or the reverse, shows up in the commercial catch 2 years later; i. e., when the young fish first reach market size in significant numbers. And it is now well established, for both sides of the Atlantic,[92] that the differences in the numbers of fry reared in different years are the chief cause for the short term fluctuations in the catches that are so characteristic of the haddock fishery.

Our reason for emphasizing the qualification "short term" in this connection is that the situation is complicated by the unhappy fact that the haddock populations of Georges and Browns Banks have been seriously reduced by the fishery.

Commercial importance and effects of the fishery.—The haddock was once much less in favor than the cod. But the expansion of the fresh-fish trade [93] brought an increasing acceptance of haddock on the market because of their good keeping qualities and convenient size for the table. In 1919 the Gulf of Maine, inshore and offshore combined, yielded something like 85 million pounds of haddock to United States and Canadian fishermen. And the development of the filleting and packaging of fresh and frozen haddock soon brought so great an increase, both in the demand and in the intensity of the fishery, that some 206 million pounds were caught in 1929 from the New England population, with some 17 million pounds more from the Nova Scotian population on Browns Bank, off western Nova Scotia, and in the Nova Scotian side of the Bay of Fundy, making a total of at least 223–224 million pounds from the Gulf of Maine as a whole, corresponding to perhaps 60 to 70 million individual fish.

This, however, was the high point, for trawlers working on Georges during the five years, 1930–1934, "averaged scarcely one-third as much haddock per day as during the previous five years," [94] while the Gulf of Maine catch as a whole had fallen by 1934 and 1935 to only about one-quarter of what it had been in 1929.[95]

Since then, down to 1947 (most recent market year for which we have seen the returns), the yearly yield of market-size haddock from the New England population has varied between about one-third to one-half as great, and about two-thirds as great as it was in 1929, to judge from the landings in the major New England ports, which form at least 90 percent of the total take from this population.[96]

A recent estimate is that there were only about one third as many haddock on Georges Bank in

[89] From unpublished data for Georges Bank and the South Channel area supplied by Howard A. Schuck.

[90] For details, see Walford's (Bull. U. S. Bur. Fish., vol. 49, Bull. 29, 1938) very interesting study of the drift of the Georges Bank eggs and larvae in these two years.

[91] Trans. 9th North American Wildlife Conference, 1944, p. 260.

[92] See especially Thompson's studies for Iceland (Fisheries Scotland, Sci. Invest. [1928], No. 5, 1929), and Raitt's for the North Sea (Journal du Conseil, Cons. Internat. Explor. Mer, vol. 11, No. 2, 1936, p. 211).

[93] Fish that are iced at sea, not salted.

[94] Herrington, Fishery Circular No. 23, U. S. Bur. Fish., 1936, p. 9.

[95] About 78 million to 80 million pounds in 1934, judging from the landings at Portland, Boston, and Gloucester from within the Gulf (which usually run about ¾–⅘ of the total catch in the Gulf by United States and Canadian vessels combined) plus perhaps 4 million to 5 million pounds taken by Canadian fishermen off western Nova Scotia and in the Bay of Fundy.

[96] For tabulations of the total catches of haddock in the western Atlantic by Canadian and United States vessels, 1880–1927, see Needler, Contrib. No. 2, North American Council on Fish. Investigations, Ottawa, 1929, 13 pp., also Rept. U. S. Comm. Fish. (1930) 1930, App. 2, pp. 27–40.

1931 as there had been there a year or two earlier.[97] This conclusion is based on the assumption that yearly changes in the average yearly catches, per day's fishing of a standard group of the large otter trawlers, fishing consistently for haddock, over the period in question, have been proportional to the relative changes in the number of haddock on the banks. In 1939–1947 the catch statistics suggest that the total population on the banks had, on the average, increased somewhat from the relatively small population of 1931.[98]

Landings of haddock

Year	Landings by United States vessels in the major New England ports, to nearest 100,000 pounds		Canadian landings to nearest 100,000 pounds	Total Gulf of Maine
	Georges Bank, Nantucket shoals, and inner parts of Gulf of Maine	Browns Bank and off western Nova Scotia	Western Nova Scotia and southern side of Bay of Fundy	
1929	174, 700, 000	8, 200, 000	11, 500, 000	194, 400, 000
1934	45, 300, 000	14, 800, 000	6, 500, 000	66, 600, 000
1935	66, 200, 000	18, 000, 000	5, 500, 000	89, 700, 000
1936	78, 500, 000	13, 600, 000	5, 100, 000	97, 200, 000
1937	87, 500, 000	14, 900, 000	4, 700, 000	107, 100, 000
1938	83, 200, 000	22, 500, 000	8, 200, 000	113, 900, 000
1939	95, 600, 000	11, 300, 000	7, 200, 000	114, 100, 000
1940	88, 800, 000	8, 200, 000	7, 100, 000	104, 100, 000
1941	116, 400, 000	6, 100, 000	5, 600, 000	128, 100, 000
1942	101, 400, 000	5, 100, 000	5, 900, 000	112, 400, 000
1943	89, 700. 000	2, 800, 000	4, 600, 000	97, 100, 000
1944	86, 800, 000	4, 400, 000	5, 200, 000	96, 400, 000
1945	72, 500, 000	6, 000, 000	5, 400, 000	83, 900, 000
1946	99, 300, 000	5, 200, 000	5, 200, 000	109, 700, 000
1947	107, 400, 000	4, 900, 000	---	---

The yield from Browns Bank and the Nova Scotian side of the Gulf has also been significantly smaller since 1939 than it was during the few years previous, when American vessels began to fish Browns Bank more intensively than they had previously.

The persistence of poorer catches through so long a term of years in the face of sustained demand, added to continued improvement in the gear and in the general efficiency of the fishing fleet, is only too clear evidence of overfishing.

The decrease in the yield of haddock from within the Gulf of Maine has been partially offset by increased catches from the Banks along outer Nova Scotia eastward to Banquereau Bank. The landings, for example, were about 8 times as great, from east of Cape Sable in 1947 (about 26,400,000 pounds) as had been the case back in 1929 (about 3,300,000 pounds). Further discussion, however, of the fishery aspects of the matter would lead us too far from our main theme.

Previous to the general adoption of the otter trawl in American waters, haddock were caught mostly on hand lines or on long lines; some in gill nets, especially in spawning time inshore between Cape Ann and southern Maine. Today all but a very small part of the catch is made in otter trawls. In 1947, for example, nearly 97 percent of the haddock that were landed in Maine and Massachusetts had been taken in otter trawls; only 3 percent of them on long lines; and only a small fraction of 1 percent on hand lines and in gill nets.

While the haddock is of primary interest from the commercial standpoint, it deserves a word from the angler's viewpoint also, for it bites as freely as the cod does, on almost any bait, and, being a much more active fish, a haddock of fair size is likely to prove an astonishment to anybody who is lucky enough to hook one while fishing with a light sinker. A new-caught haddock is also a very beautiful object.

American pollock *Pollachius virens* (Linnaeus) 1758

POLLOCK; BOSTON BLUEFISH; COALFISH (IN GREAT BRITAIN); GREEN COD (IN GREAT BRITAIN)

Jordan and Evermann, 1896–1900, p. 2534.

Description.—The American pollock[99] has a deep, plump body (about four and one-fourth times as long as it is deep) tapering to a pointed nose and to a slender caudal peduncle. Its mouth is of moderate size. Its projecting lower jaw (giving it an undershot facial aspect); its forked, sharp-cornered tail, small ventral fins, small chin barbel (as a rule the latter is lacking altogether in large fish), and its beautiful olive green color, are ready field marks when it is caught with cod and haddock.

Its first dorsal fin (13 or 14 rays), originating slightly behind the pectoral, is triangular, and is a little the highest of the three dorsals. The second dorsal, also triangular, is the longest of the

[97] Herrington, Trans. 9th North American Wildlife Conf., 1944, p. 259. Schuck, Commercial Fish. Rev., vol. 10, Oct. 1948, p. 1.

[98] See Schuck (Biometrics, Amer. Statistical Assoc., vol. 5, No. 3, 1949, p. 215, table 1, and p. 216, fig. 2).

[99] This is the "coalfish, green cod, or saithe" of British, Scotch, and Irish fishermen. The European "pollack" is a different species (*Gadus pollachius*)

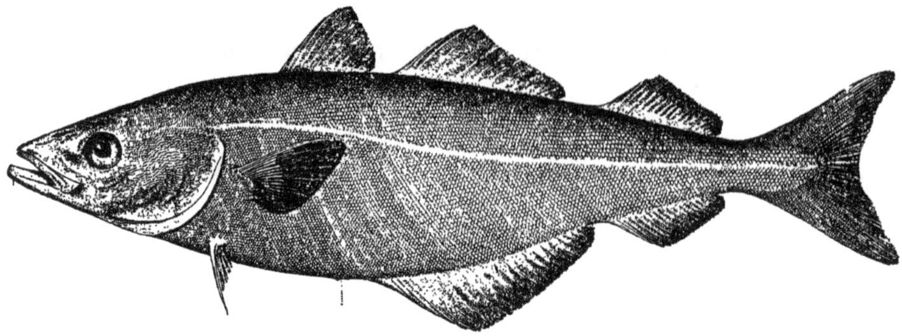

FIGURE 98.—American pollock (*Pollachius virens*), Eastport, Maine. From Goode. Drawing by H. L. Todd.

three (21 or 22 rays) and is separated by a considerable space from the third dorsal fin (19 or 20 rays) which is more rhomboid in outline. The second anal fin (20 or 21 rays) corresponds in shape and size to the third dorsal, under which it stands, but the first anal (24 to 28 rays) is considerably longer than the second dorsal though similar to the latter in shape. The ventral fins are a little in front of the pectorals, and are only about half as long as the latter. The pectorals are set high on the sides, and are longer than the first dorsal, but shorter than the second dorsal; they have rounded lower corners and bluntly pointed tips. The caudal fin is noticeably forked, with angular corners, unless it is spread to its widest when its margin becomes nearly straight.

Color.—Pollock are always of a greenish hue, usually deep rich olive green or brownish green above, paling to yellowish or to smoky gray on the sides below the lateral line, and to silvery gray on the belly. The lateral line is white or very pale gray, contrasting strongly with the dark sides. The dorsal, caudal, pectoral, and anal fins are olive, the latter pale at the base. The ventral fins are white with a reddish tinge. Young fish are darker than large ones, and many of them are more tinged with yellow on their sides.

Size.—Pollock reach a maximum length of about 3½ feet and a weight of about 35 pounds. But fish of this size are exceptional, few growing larger than 40 inches or 30 pounds, with about 2 to 3 feet and 4 to 15 pounds as the average for adults. The proportion of length to weight was as follows among fat fish measured by Welsh off Boon Island on April 22 to 25, 1913:

Length, in inches	Weight, in pounds	Length, in inches	Weight in pounds
24½	4–5½	30	8½–9½
26	4	31	10
27	7½	31½	10
27½	8½	32	10–12
28½	8	33	12
29	8–9	35	14
29½	8½–9		

Large pollock, however, of a given length vary widely in weight; for example, we have found 40-inch fish to weigh from 25 to 35 pounds; 35-inch fish, from 14 pounds to 21 pounds.

Habits.—The pollock is an active fish, living at any level between bottom and surface according to the food supply and on the season, often schooling, and sometimes gathering in bodies so large that it is on record that a purse seiner once took 60,000 fish from one school at a single set. In our Gulf their depth range is from the surface down to 100 fathoms at least,[1] while they may descend somewhat deeper in the deepest troughs. And it is the local presence or absence of prey that governs the movements of the larger pollock.

Pollock feed chiefly on small fish, and on pelagic crustaceans; among the latter most often on the large pelagic shrimp-like euphausiids. It is commonplace that pollock destroy great quantities of small herring, launce, young cod, young haddock, young hake, silver hake, and other small fish in the Gulf of Maine just as they do on the other side of the Atlantic. Pollock chasing schools of herring are a familiar sight;[2] pollock of 1 to 1½ pounds commonly run up estuaries in pursuit of smelt in autumn; and newly hatched haddock or other

[1] We have seen them trawled as deep as this on the northern slopes of Georges Bank.

[2] Sars (Rept. U. S. Comm. Fish., (1877) 1879, p. 619–620) has given a graphic account of pollock rounding up schools of launce and of young cod in Norwegian waters.

larvae that are liberated in harbors from the hatcheries are always in danger of being snapped up by the young pollock that are plentiful in such situations. When a pollock only 9 inches long is capable of eating 77 herring up to 2½ inches long at one meal,[3] "ravenous" is only mildly descriptive. However, pollock so seldom strand in pursuit of prey that we have never seen one on the beach though schools often come close in and are caught in the traps.

In the Gulf of Maine, pollock depend perhaps as much on pelagic shrimps as on fish. At Eastport, for example, where these shrimps (genera *Meganyctiphanes* and *Thysanoessa*) are very abundant all summer, Kendall [4] reports pollock of all sizes not only fattening on them but so evidently preferring them to young herring that he did not find a single "sardine" in a pollock stomach, though these were plentiful enough at the time. He adds that "if at any time the crustaceans disappeared from a place the large pollock disappeared also." And pollock, breaking the surface in pursuit of shrimp are familiar sights there, as we can bear witness with many others.

Similarly, Welsh found large pollock in schools feeding on the surface on shrimp (*Thysanoessa raschii*) off the Isles of Shoals and off Boon Island in April 1913, remarking in his field notes for the 25th that "in the last few days pollock have begun to appear in small schools of 400 to 500 fish with the appearance of large schools of feed (shrimp, 'all eyes'), the feed (shrimp) breaking water trying to get away from the pollock which are after them." He described the fish themselves as "rising and sinking at intervals; when at the surface swimming like porpoises, leaping up and over with open mouths, the feed being in dense streaks 6 inches to 1 foot down." These feeding fish were "very sluggish and tame on this feed and easily taken in the purse seines." All were "stuffed to capacity" with shrimps, and only a few contained herring.

Large pollock take morsels as small as copepods. Willey [5] for example, speaks of a fish caught near Campobello Island which contained proportionately as many of these as of euphausiid shrimps, and it is probable that the little pollock depend chiefly on copepods. Glass worms (*Sagitta*), too, have been found in pollock stomachs. Sometimes they consume considerable quantities of ctenophores; we found many pollock full of them on Cashes Ledge and on Platts Bank in August 1928; one had 105 of these watery organisms in its stomach. They also feed to a small extent on bottom-dwelling crustaceans on both sides of the Atlantic, thus crabs, and bottom-dwelling shrimp have been found in fish caught at Woods Hole and in the Gulf of Maine. They have also been reported as gorging themselves on herring spawn. They never take shelled mollusks, so far as we are aware. But they bite on clams as greedily as on fish baits. And fishermen speak of them as one of the few species that will bite, that is, feed, during the spawning period.

Experiments on fish kept in captivity at Woods Hole [6] have shown that the pollock captures its food more by its keen sight than by scent.

The pollock is a cool-water fish. We have never seen any large ones caught at the surface when the temperature there was higher than about 52° F., though there may be plenty of them a few fathoms deeper down where the water was cooler. Even the little "harbor pollock" of 8 inches or so do not appear in any great numbers at times or places where the water is warmer than perhaps 60° F. At the other extreme, pollock of all sizes from the 1 year-old fish upward must experience temperatures as low as 32° F. on the fishing grounds in the southern side of the Gulf of St. Lawrence, and on the more easterly of the Nova Scotian banks during the late winter or early spring, unless they descend then to considerably greater depths, a possible shift of which we have no direct evidence. But it is probable that the pollock's need of water as warm as about 38° F. for the incubation of its eggs (p. 216) and perhaps of temperatures a little higher than that for the maturation of its sex organs (p. 216) is the factor that sets the northern boundary to the maintenance of a permanent resident population (p. 218).

The pollock is a late autumn and early winter spawner, and the shortness of the spawning season (p. 220), with the fact that the vertical temperature gradient covers a range no greater than 3° to 5° F. down to 50 fathoms at that season, makes it easy to establish the physical conditions under which the eggs are produced

[3] Smitt, Scandinavian Fishes, vol. 1, 1892, p. 503.

[4] Rept. U. S. Comm. Fish., (1896) 1898, p. 180.

[5] Proc. Amer. Acad. Arts, Sci., vol. 56, 1921, p. 192.

[6] Herrick, Bull. U. S. Comm. Fish., vol. 22, 1904, p. 258.

and in which they develop. On the Massachusetts Bay grounds breeding commences when the whole column of water has cooled to about 47° to 49°, and is at its climax (late in December) in temperatures of 40° to 43°, while the major production of eggs takes place long before the water has cooled to its winter minimum of 35° to 36° F. at the level at which the fish lie. Thus the pollock spawns on a falling temperature, with most of the eggs produced within a comparatively narrow range and in water several degrees warmer than that in which haddock spawn most actively (p. 208). This agrees closely with the European pollock which spawns only in temperatures near 44.5°, so far as is known.

As the successful propagation of any fish depends as much upon the incubation of its eggs as on its spawning, we should note that incubation proceeds normally, and that the resultant larvae are strong and active over the whole range of temperature just outlined, that is, from about 38° to about 48° as proved by experience in the Gloucester hatchery.

The Massachusetts Bay spawning takes place in salinities ranging from as low as 32 per mille to as high as 32.8 per mille, according to precise locality, depth, and season, salinities much lower than those in which pollock breed on the other side of the Atlantic (35.14 to 35.26 per mille).

The number of eggs produced by a female pollock averages about 225,000, but more than 4 million eggs were reported in one fish of 23½ pounds. The egg is buoyant, has no oil globule, and averages about 1.15 mm. in diameter. Thus it is decidedly smaller than the egg of the cod or of the haddock. Incubation occupies 9 days at a temperature of 43°; 6 days at 49°.

The larvae are about 3.4 to 3.8 mm. long at hatching, slender, with large yolk sac, and with the vent situated on one side of the body at the base of the ventral fin fold as it is in other larval gadoids; they are sprinkled with black pigment cells. About 5 days' time is required for the entire absorption of the yolk sac and for the formation of the mouth; meantime the pigment of the post-anal section of the trunk becomes grouped in longitudinal bars, two dorsal and two ventral, the former longer than the latter. At this stage pollock closely resemble cod of the same size, but the ventral bars are longer than the dorsal bars opposite them in the cod, and usually three in number in the cod instead of two as in the pollock. These bars persist until the pollock grows to a length of about 15 mm., when the pigment becomes more scattered. The caudal fin rays appear at about 9 mm., all the dorsal and anal rays and the ventral fin rays at about 15 mm., the dorsal fins are separate from one another and also the anal fins at 20 mm. (at about 2 months), and fry of 25 to 30 mm. show most of the characters of the adult.

In European seas the young pollock lives near the surface for its first 3 months. The young fry have been taken similarly, in the tow nets near the surface at Woods Hole from January to May, and they are to be expected in Massachusetts Bay then, though we have no actual record of them there.

Rate of growth.—Thanks to the shortness of its breeding season and to the readiness with which its scales can be "read" European students[7] have found it easy to trace the rate of growth of these "saithe" or "coal fish"; and this has been done for the American pollock by Mavor,[8] also by us. Judging from scale studies and from the sizes of the fry that are caught near Woods Hole in the spring, pollock hatched in mid-winter are about 1 to 2 inches long by the following spring, growing to 3–5 inches by late summer; to 5–7 inches in their first winter, when a year old; to 12–13 inches at two years of age; to 17–18 inches at three years; to 21–22 inches at four years; to an average of about 23 inches at 4½ years; of 25 inches at 5½ years; and of about 27 inches at 6½ years. Thus the 11–12 inch and 15–16 inch pollock that appear in such numbers along the New England coast late in summer are about 1¾ and 2¾ years old, respectively. The annual rate of growth thus is about 5 or 6 inches for the first three years, 2–4 inches for the next three years, and 1–2 inches for the next few years, after which they grow still more slowly.

These sizes are somewhat larger than the averages given by Damas for European fish of corresponding ages, but the difference is so small that it is safe to apply the European figures to older Gulf of Maine fish, for which we have no data. On this basis we may expect the American pollock to average about 28 inches at 7½ years; about 29

[7] For résumé see Damas (Rapp. et Proc. Verb., Conseil Perm. Internat. Explor. Mer, vol. 10, No. 8, 1900, p. 167).

[8] Contr. Canad. Biol., (1917-1918) 1918, No. 6.

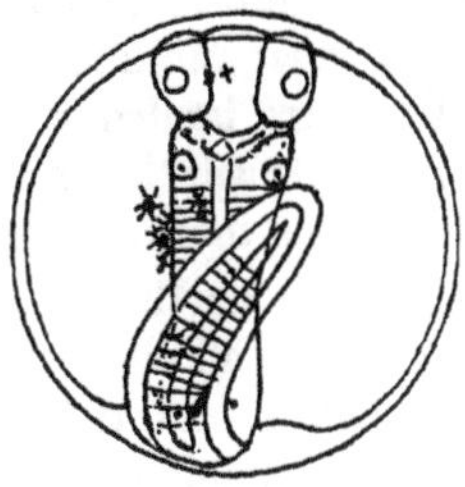

FIGURE 99.—Egg (European). After McIntosh.

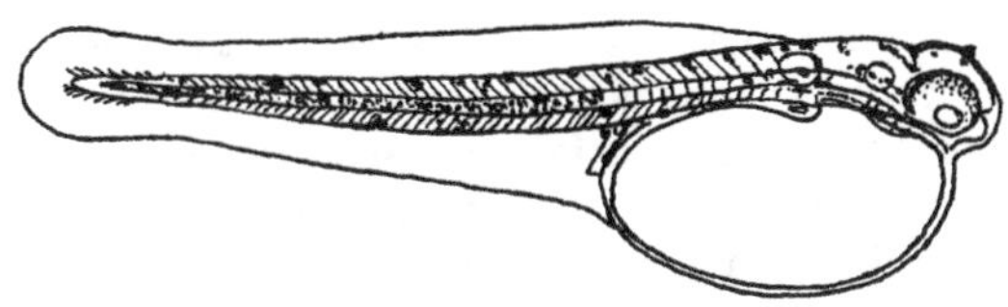

FIGURE 100.—Larva (European), 5 days old, 4.3 mm. After McIntosh.

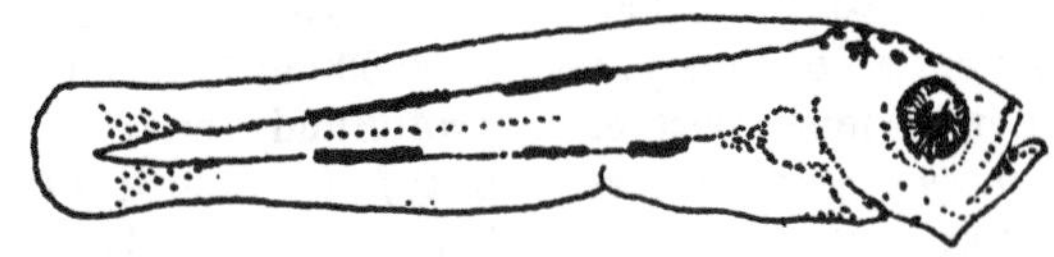

FIGURE 101.—Larva (European), 6.75 mm. After Schmidt.

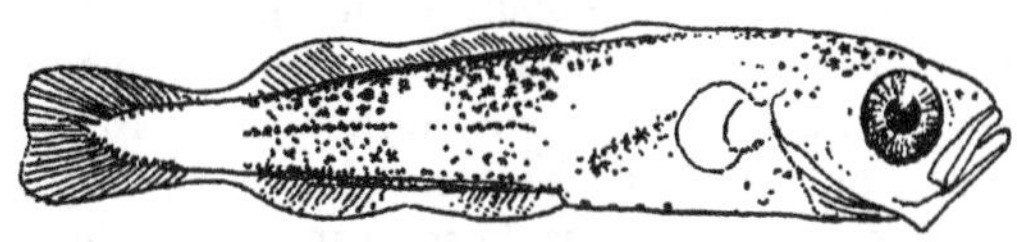

FIGURE 102.—Larva (European), 12.5 mm. After Schmidt.

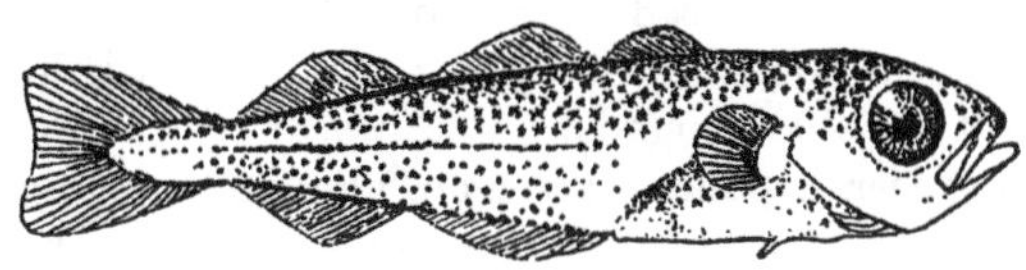

FIGURE 103.—Fry (European), 23 mm. After Schmidt.

AMERICAN POLLOCK (*Pollachius virens*).

inches at 8½ years, and about 30 inches at 9½ years. Fish of 3 feet and upward are therefore of considerable age. The oldest recorded by Damas among the thousands he examined was in its nineteenth year. In European seas pollock grow faster in the southern part of their range than in the northern, but we have yet to learn whether this applies to the American fish.

The age at which Gulf of Maine pollock first mature is not known, but this is probably at a somewhat greater size than in Norwegian waters, where most of them mature by the time they are 1½ feet long; i. e., 3 years old. All of them that are 2 feet long, or longer, in summer have spawned at least once.

General range.—Continental waters on both sides of the North Atlantic in cool temperate and boreal latitudes; regularly in the west from the southeastern part of the Gulf of St. Lawrence [9] and northeastern Nova Scotia to New Jersey; southward occasionally to Chesapeake Bay and to Cape Lookout,[10] N. C., and northward in small numbers to the southern part of the Grand Banks, to the southeastern coast of Newfoundland, and to Sandwich Bay on the southeastern coast of Atlantic Labrador;[11] West Greenland; Spitzbergen; Iceland; and the coasts of northwestern Europe south to the North Sea, English Channel, and Brittany coast of the Bay of Biscay in the eastern Atlantic; occasionally to the Gulf of Gascony (Arcachon).

Occurrence in the Gulf of Maine.—In our side of the Atlantic the pollock has its chief center of abundance in the Gulf of Maine, where it is caught in large numbers both on the offshore banks, and all around the coast line, from Nantucket Shoals and Cape Cod to Cape Sable. The only regional exception is in the inner part of the Bay of Fundy along the New Brunswick shore, where so few pollock are taken that they do not appear at all in the landings reported thence (Albert County).

The following statistics of the United States catch for 1945, combined with the Canadian catches for 1944 and 1946,[12] give a general idea of the regional abundance of pollock, on a broad scale, also of how universal they are, with the one exception just noted.

Browns Bank, about 965,000 pounds; western coast of Nova Scotia to the Annapolis County

[9] Pollock appear not to be known anywhere farther within the Gulf or in its northern side.

[10] Reported from Chesapeake Bay by Hildebrand and Schroeder (Bull. U. S. Bur. Fish., vol. 43, pt. 1, 1928, p. 156) and from Cape Lookout by Coles (Copeia, No 151, 1926, p. [105]).

[11] The pollock is listed in the Reports of the Newfoundland Fisheries Research Commission for 2 stations on the southern edge of the Grand Bank, from Bay Bulls, Newfoundland, and from Sandwich Bay, Labrador.

[12] We have not yet seen the Canadian statistics for 1945.

line, at least 7,300,000 pounds;[13] Nova Scotian shore of Bay of Fundy, about 500,000 to 1,000,000 pounds;[14] inner part of Bay of Fundy on New Brunswick side (Albert County), 0; mouth of the Bay of Fundy on New Brunswick side, about 2,000,000 to 3,500,000 pounds;[15] off eastern Maine, about 1,045,000 pounds; off central Maine, about 2,573,000 pounds; small fishing banks in west-central part of the Gulf, about 516,000 pounds; off western Maine, about 1,861.000 pounds; off eastern Massachusetts and off northern Cape Cod, about 7,347,000 pounds; Cape Cod out to the so-called South Channel, about 1,518,000 pounds; Georges Bank as a whole, about 3,184,000 pounds.

In general, pollock are more abundant around the coastal belt of the Gulf, out about to the 75 to 80 fathom line, on the isolated fishing grounds enclosed within that depth limit, and over the offshore banks than they are over the deeper central basin of the Gulf; though some are taken there also. And this has always been one of the principal fishes to be caught with hook and line on the various small banks and ledges in the inner part of the Gulf; near Lurcher Shoal for instance; on Grand Manan Bank; on Jeffreys Ledge, and on Stellwagen Bank at the entrance to Massachusetts Bay, while the neighborhoods of Boon Island and of the Isles of Shoals long have been famous pollock grounds.

Pollock are decidedly less plentiful on the Nantucket grounds in general (only about 56,000 pounds landed thence in 1947) and west of Cape Cod than they are either farther within the Gulf to the northward or on Georges Bank to the eastward. But commercial quantities are caught yearly (in season) along southern New England and New York. The landings for Rhode Island, Connecticut, and New York, combined, ranged, for example, between 787,000 pounds and 883,000 pounds for the years 1930 to 1933. And though the landings ran less than one-half as great (between 135,000 pounds and 452,000 pounds) for 1935 to 1947 we suspect that this decrease resulted from market conditions, rather than from any decrease in the numbers of pollock that are available there. A few hundreds or thousands of pounds of pollock are landed yearly[16] in New Jersey ports also. But this is the extreme southern limit for the pollock as a market fish.

To the eastward and northward, we find pollock caught in abundance all along the outer Nova Scotian coast and banks. In 1946, for example, 840,000 pounds were landed in Guysborough County, Nova Scotia, and 277,200 pounds along the Atlantic coast of Cape Breton Island, east of the Gut of Canso.[17] This, however, is the northeastern limit of our pollock as a market fish of any importance. True, a few thousand pounds were reported yearly from the southeastern side of the Gulf of St. Lawrence near the Gut of Canso during the early nineteen hundreds.[18] But the catch is so small that pollock have not been mentioned in the catch statistics for more recent years, nor anywhere else within the Gulf of St. Lawrence.

Small pollock, 8 to 10 inches long and weighing less than half a pound (1 or 2 years old) swarm inshore after early April, when we have seen thousands of them taken from the traps at Gloucester and Magnolia. In the southern part of Massachusetts Bay these "harbor pollock," as they are called locally, move out in June, probably to avoid the rising temperature, to return again in autumn. But they continue abundant all summer and auumn in the harbors and bays and among the islands along the coast northward from Cape Ann and eastward to Nova Scotia. Most of them seek slightly deeper water in winter, however, probably to avoid the cold.

The larger fish tend to keep farther offshore than the small ones; they live deeper on the whole, except when they are pursuing some particular feed (p. 214), and they are caught in more definite localities, not everywhere and anywhere along the coast as are the little fish. In the southwestern part of the Gulf, as exemplified by Massachusetts Bay and by the belt from Cape Ann to the Isles

[13] U. S. catch, 492,400 pounds, 1945; Canadian catches 7,017,000 pounds in 1944 and 6,642,000 pounds in 1946, besides an indeterminate amount landed along this part of the Shelburne County coast line.

[14] 1944, 513,000 pounds; 1946, 983,000 pounds.

[15] Charlotte and St. Johns Counties, about 2,000,000 pounds in 1944, about 3,507,000 in 1946.

[16] Maximum, 10,700 pounds, minimum 600 pounds for the years 1930–1937 and 1939–1947, 101,200 pounds were credited to New Jersey in 1938; an amount so much larger than usual as to suggest that it was because of economic reasons that the fish were landed in New Jersey rather than in New York.

[17] Richmond County, Nova Scotia, 223,600 pounds; Cape Breton County, 53,600 pounds.

[18] Yearly catch, 1,600–4,000 pounds; for 1902 to 1906 and 1909 to 1915–1916, 61,500 pounds were credited to Inverness County in 1901, but this amount is so much larger than usual as to suggest some error.

of Shoals, large pollock are taken in greatest number in late autumn and early winter when the gill-net fishery taps the spawning fish (fig. 220). Few are caught there later in the winter after they finish spawning, showing that the spent fish do not winter on particular grounds, but scatter to and fro in search of food.

Along this part of the coast they often reappear in abundance at the surface near land during April and May and even into June. In 1951, for example, we heard of schools of large pollock at various points off the tip of Cape Cod, in the northern side of Massachusetts Bay, and off the Merrimac River during the first week of that month. They tend to move out again, and deeper, as the surface warms with the advance of the season, and very few large ones are taken inshore in the Massachusetts Bay region during July and August. But it is not likely that they travel far, or sink very deep, for good fares of fish 2 to 3 feet long are brought in by line fishermen from Jeffreys Ledge throughout the summer, most of them caught some distance above bottom.

North of the Isles of Shoals, pollock are more commonly seen on the surface during the hot months. Thus, we remember one year (1922) when small boats from Cape Porpoise and from neighboring ports were doing well trolling during July and early August; in 1951 schools were reported off Baileys Island, Casco Bay, during the first week of July. And great numbers of good sized pollock are caught all summer in the tide rips at the mouth of the Bay of Fundy; in Passamaquoddy Bay; in Digby Gut on the Nova Scotian side of the Bay of Fundy; and along outer Nova Scotia.[19]

No information is available as to the relative frequency with which pollock appear at the surface over Georges Bank and other offshore fishing grounds, though they are caught all through the year at deeper levels, with no greater seasonal fluctuation in the landings than might result from the various vicissitudes of fishermen's luck, the weather, and the market.

Pollock spawn in great numbers at the mouth of Massachusetts Bay, especially on the broken bottom southeast of Gloucester and along the seaward (eastern) slope of Stellwagen Bank, where most of the eggs were taken during the years when pollock were hatched in great numbers at the Gloucester hatchery of the U. S. Bureau of Fisheries.[20]

In some years many pollock spawn (and large quantities of their eggs have been collected for the hatchery) right up to Boston Lightship in the inner part of Massachusetts Bay, though this is not a regular annual event. And gill-netters also catch an abundance of ripe fish between Cape Ann and the Isles of Shoals, where breeding pollock congregate in such abundance that they have supported a lucrative fishery in some years.

This in general seems to cover the most productive spawning area so far as the inner parts of the Gulf are concerned. Few spawning pollock are caught in the Gulf south of the Massachusetts Bay region, while we find no report of them as breeding anywhere west of Cape Cod, although fry of the winter's hatch appear at Woods Hole in spring (p. 220). On the other hand only a fewripe fish are seen along the coast of Maine, though the Boothbay hatchery has made diligent search for them east of Casco Bay; neither have we found pollock eggs anywhere north of the Isles of Shoals in our autumn or winter towings. And it seems that very few larvae are hatched at the mouth of the Bay of Fundy on the New Brunswick side, for none of their young fry have been found in the Passamaquoddy–Grand Manan region, though yearlings, half-grown fish, and adults are there in great numbers. Thus it is safe to say that no production of any importance takes place anywhere in the Gulf of Maine east of Cape Elizabeth.

We cannot offer any explanation for this regional contrast in pollock productivity. Temperature seems not the cause, for this differs by only a couple of degrees between Massachusetts Bay and Passamaquoddy Bay at the commencement of the spawning season. And while the coastal water as a whole is slightly cooler east than west of Cape Elizabeth at the height of the season, the differences from station to station have been small; and all the readings we have taken there during late December and early January have fallen well within the range at which pollock spawn freely in

[19] Near Canso good-sized fish are caught on hook and line at the surface from June to December, according to Cornish, Contributions Canadian Biology (1902–1905) 1907, p. 189.

[20] Information supplied by C. G. Corliss, former Superintendent of the Gloucester Hatchery.

Massachusetts Bay, as appears in the following table:

Water temperatures, Massachusetts Bay to Lurcher Shoal, 1920–1921

Depth in fathoms	Off Gloucester, Dec. 29, station 10489	Off Cape Elizabeth Dec. 30, station 10494	Off Mount Desert I., Jan. 1, station 10497	Off Machias, Jan. 4, station 10498	Fundy Deep, Jan. 4, station 10499	Off Lurcher Shoal, Jan. 4, station 10500
0	42	42	40.5	42	42	42.5
10	43.7	42.5	41.4	42	42.4	42.7
20	44.4	43.1	41.8	42.1	42.6	43.1
40	44.4	44.9	42.3	42.1	42.9	43.9
75	44.6	----------	----------	----------	43.5	----------

Presumably the pollock of Georges and Browns Banks and of outer Nova Scotian waters to the eastward reproduce themselves there. But we have no definite information in this regard.

A few ripe fish are caught in the Massachusetts Bay region as early as the last week in October, and the first of November to the middle of January covers the most active production there, as illustrated by the following table supplied by C. G. Corliss, former Superintendent of the hatchery, where many millions of pollock eggs were once hatched yearly.

Year	First eggs taken	Last eggs taken	Eggs most plentiful	Total eggs collected
1911–12	Nov. 10	Jan. 22		499,875,000
1912–13	Nov. 1	Jan. 31		856,680,000
1913–14	do	Feb. 6		974,240,000
1914–15	do	Feb. 9		855,020,000
1915–16	do	Feb. 17		1,713,730,000
1916–17	Nov. 7	Jan. 27	Nov. 16 to Jan. 20	2,081,400,000
1918–19	Nov. 6	Jan. 23	Nov. 20 to Jan. 8	1,110,470,000
1919–20	Nov. 10	Jan. 16	Nov. 17 to Jan. 16	954,800,000
1920–21	Nov. 15	Jan. 21	Nov. 21 to Jan. 16	650,850,000

The first week of March is the latest that the gill netters have reported any spawning fish.

The pollock spawns considerably earlier in the Gulf of Maine than in European waters, where spawning does not begin until January, is at its height in March, and continues into April.

The Gulf of Maine pollock, like the cod and haddock, spawn in comparatively shoal waters. Thus we have towed a considerable number of pollock eggs over Stellwagen Bank where the water was only 16 fathoms deep (on November 8, 1916) and most of the ripe fish that supplied the Gloucester hatchery with eggs were netted in depths of 25 to 50 fathoms. Probably few spawn deeper than 50 to 60 fathoms, and there is no evidence in egg records, in captures of ripe fish, or in fishermen's reports, that any pollock eggs are produced in the deep basins of the Gulf. In European waters, however, this fish is described as breeding only in depths greater than 75 fathoms.

The gill netters have described it to us as spawning over hard bottom chiefly, though the pollock is not a ground fish at other seasons.

The migrations of the young pollock in our Gulf, from hatching until they appear on the coast as yearlings, are of special interest because of the probability that the great majority of all the pollock that frequent the eastern coast of Maine and the Bay of Fundy region are produced elsewhere. Some of them may come from spawning grounds (as yet unmapped) off southern or western Nova Scotia; our own observations throw no direct light on this point. But what is known of the general circulation of the Gulf in spring and early summer suggests, rather, that the bulk of them come from the spawning grounds on the western side, south of Cape Elizabeth, having circled around first southward, then eastward and northeastward, and so finally to the Bay of Fundy and to the east part of the Maine coast. Others, hugging the coast more closely in their involuntary journeyings, may follow past Cape Cod and so westward, evidence of which is the presence of an abundance of pollock fry in spring at Woods Hole, for pollock are not known to spawn in quantity anywhere west of the Cape (p. 219).

Strangely enough, we have caught no pollock less than 8 or 9 inches long on the offshore banks either on hook and line or in our tow nets, nor have we seen any that had been trawled there. Whether this is because the young are too nimble to be taken in trawls, whether because they live well off bottom, or whether because they are scarce offshore, is not known.

The larger pollock of our Gulf seem to wander but little, for many that have been tagged by the U. S. Bureau of Fisheries have been recaptured within short distances of the localities where they were marked, and after long periods of time. And while a few of the marked fish are known to have made considerable journeys eastward, (one, for example, from Jeffreys Ledge to Sable Island), instances of this sort have not been numerous enough to suggest any mass movements.

Pollock appear to be similarly stationary all along the outer Nova Scotian coast, for they are caught there throughout the fishing season. But we think it likely that the few pollock that are caught within the Gulf of St. Lawrence wander in via the Gut of Canso. On the other hand, pollock are seldom caught west of Rhode Island after June,[21] and it is chiefly as cold season visitors that they appear off the coasts of Connecticut, New York, or New Jersey; the commercial catches reported thence are made mostly in winter and in early spring.

Importance.—At the time the first edition of this book appeared (1925) our Gulf was yielding about 35 million to 40 million pounds yearly. In 1946, most recent year for which we have seen the Canadian catch statistics as well as those for the United States, the total catch for the Gulf was close to 48 million pounds,[22] say 5 million to 7 million fish.

A quarter of a century ago, the gill net was regarded as the most effective apparatus for catching pollock; hand and long lines ranking next; otter trawls yielding only a few,[23] while schools that were seen at the surface were often seined, especially the smaller sizes. But the relative proportions have been reversed with the great development of the otter trawl fishery. In 1946, for example, a representative year, a little less than ¾ of the pollock landed from the Gulf of Maine by United States fishermen were caught in otter trawls; a little less than ⅙ in gill nets; with hand and long lines, traps of one sort or another, and purse seines accounting for the remainder in the order named.[24]

Some of our readers will be more interested in the fact that pollock will take an artificial lure and put up a strong resistance. Small ones up to 4 or 5 pounds will take a bright artificial fly freely (silver body with white wings of hackle or hair is good, especially with a touch of red). We have caught many fly casting from the rocks in autumn when smallish pollock are inshore after smelt or other small fish. And a pollock rises so fiercely to the fly and makes so long and strong a run when it is hooked that a small one gives fully as good sport as a trout caught on a light fly rod; a medium-sized pollock provides nearly as good sport as a salmon of equal weight. When the larger pollock are schooling at the surface near shore in May and June, many of them are taken by anglers trolling with spoons or with feather lures of one kind or another, from party boats out of Plymouth, Gloucester, Ipswich, Newburyport, Hampton, York, Casco Bay, and various other places along our coasts; also off Gay Head, Marthas Vineyard, and still farther to the westward. And pollock of all sizes bite eagerly on clams, minnows, or on bait of cut fish.

WHITE HAKE AND SQUIRREL HAKE

We are forced to discuss these two hakes together, for they are so hard to tell apart that they are often confused, while they agree so closely in habits and distribution that what is said of one applies equally to the other, except as noted below.

White hake *Urophycis tenuis* (Mitchill) 1815 [25]

BOSTON HAKE; BLACK HAKE; MUD HAKE; HAKE; LING

Jordan and Evermann, 1896–1900, p. 2555.

Description.—Although the hakes of the genus *Urophycis* (true hakes in general parlance on this side of the Atlantic) are close relatives of the cod and of the haddock, they are not at all codlike in appearance, being more slender and softer-bodied fish, tapering backward from the shoulders to a slim caudal peduncle and to a small weak tail, with much larger eyes than the cod but with smaller chin barbels. Furthermore, they have only two dorsal fins, the second much longer than the first, and only one anal fin instead of the three dorsals and the two anals of the pollock, cod, and haddock. The ventral fins are long, narrow, and feelerlike.

The body of the white hake is rounded in front of the vent, flattened sidewise behind the vent, and is about five and one-half times as long as it is deep. The mouth is so large that it gapes back to below the eyes, the upper jaw projects beyond the lower, and the chin bears a small barbel. The first dorsal fin (9 to 10 rays) originates close behind the pectorals, and is shorter

[21] Two pollock tagged on Nantucked Shoals in June and October were caught off Block Island in the following May and July respectively.

[22] 47,670,776 pounds, plus an indeterminate amount for Shelburne County, Nova Scotia, that may have been caught on the Gulf of Maine side of Cape Sable.

[23] Bigelow and Welsh, Bull. U. S. Bureau of Fisheries, vol. 40, Pt. 1, 1925, p. 406.

[24] No statistics are available as to what proportions of the Canadian catch in the Gulf are made with the different kinds of gear.

[25] The European hake is a *Merluccius* (p. 177).

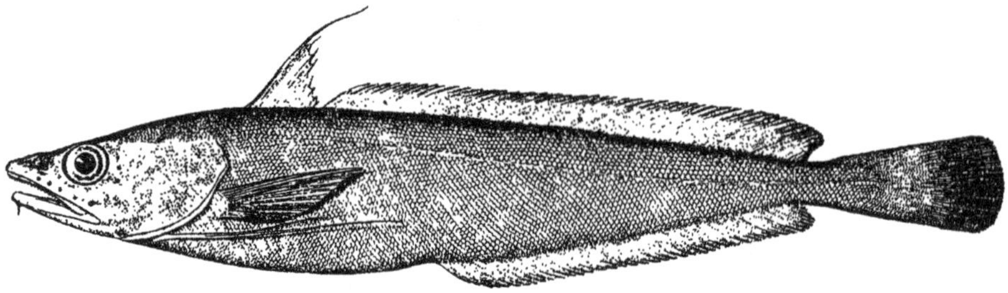

FIGURE 104.—White hake (*Urophycis tenuis*), Halifax, Nova Scotia. From Goode. Drawing by H. L. Todd.

than the latter, triangular, with the third ray prolonged as a filament that is longer than the fin proper is high. The second dorsal fin (about 54 to 57 rays) runs the whole length of the trunk from close behind the first dorsal fin to the caudal peduncle, is of about equal height from end to end, with rounded corners, and is only about half as high as the first dorsal. The anal fin is similar in outline to the second dorsal but is shorter (about 48 to 50 rays). The pectorals are rounded when spread; the ventral fins are situated considerably in front of the pectorals, and each is reduced to two very much prolonged rays (apparantly a single branched ray), with the lower (longer) ray of the two falling slightly short of the vent.

The length of the ventral fins has often been given as an alternative character separating the white hake from the squirrel hake (p. 223), in which the ventrals are usually described as reaching beyond the vent. This distinction is not to be relied on, however, for we have seen squirrel hakes in which the ventrals lacked something of reaching the vent.

The scales on both head and body are smaller than those of the closely allied squirrel hake, and their number is the most reliable distinction between the two species, there being about 140 oblique rows of scales along the lateral line between gill opening and base of tail fin in the white hake, but seldom, if ever, more than 110 rows of scales in the squirrel hake.

Another difference, which seems equally dependable, though it is less obvious, is that the upper jaw (maxillary) bone reaches as far back as the rear edge of the eye in the white hake, but only as far as the rear edge of the pupil in the squirrel hake.

Color.—Like most bottom fish, white hake vary in color. As a rule they are muddy or purple brown above, sometimes almost slaty (we saw one of this shade caught in Northeast Harbor, Maine), the sides sometimes bronzed, and the belly dirty white or yellowish white peppered with tiny black dots. The dorsal fins are of the same color as the back, the anal fin the same as the belly, and both the dorsals and the anal are edged with black. The ventral fins are pale, like the belly, but usually they are more tinged with yellow.

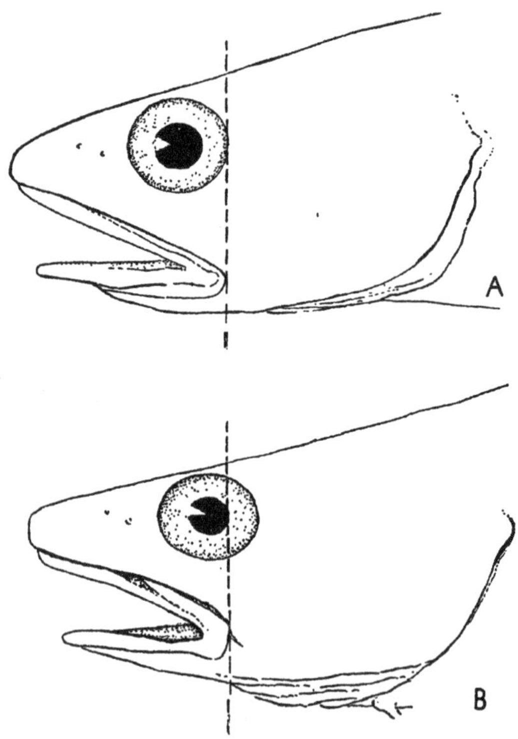

FIGURE 105.—Side view of heads of white hake, A, and of squirrel hake, B, to show the difference between the two in length of the upper jaw bone.

Size.—The maximum length is about 4 feet, the maximum weight about 40 pounds. But most of the fish that are caught weigh between 1 and 20 pounds, averaging no more than 8 pounds. A hake 28 inches long will weigh about 8½ pounds if it is in good condition; one of 30 inches, about 9 pounds; 36 to 38 inches, 13 to 16 pounds; and about 18 pounds at 40 inches, according to Welsh's experience.

Squirrel hake *Urophycis chuss* (Walbaum) 1792

RED HAKE; LING

Jordan and Evermann, 1896–1900, p. 2555.

Description.—The squirrel hake resembles its larger relative, the white hake (p. 221) so closely that the one is often taken for the other. The number of scales affords the most reliable means of identification, those of the squirrel being much larger relatively than those of the white, and arranged in only about 100 to 110 oblique cross rows along the side from gill opening to base of caudal fin, and in about 9 longitudinal rows on the upper part of the sides between lateral line and dorsal fin, as against about 140 transverse rows and about 12 longitudinal rows in the white hake (p. 222). Also, the upper jaw (maxillary bone) reaches back only as far as the rear edges of the pupil in the squirrel hake, but as far as the rear edge of the eye in the white hake (p. 222), and this difference can be relied upon, even for very small fish.

The ventral fins of the squirrel hake overlap the vent as a rule, whereas those of the white hake fall short of it, but this is not invariably the case, as already remarked (p. 222), for we have seen squirrel hakes in which the ventrals did not reach to the vent. Furthermore, the filamentous part of the third ray of the first dorsal fin is much longer (if undamaged) in the squirrel than in the white hake, i. e., three to five times as long as the rest of the fin, and the nose is blunter. The color, too, is of some value in identifying these species, for while the squirrel hake is almost always reddish brown, the white hake has a decidedly purplish lustre when fresh caught.

Color.—The squirrel hake is reddish, muddy, or olive brown on sides and back, darkest above; sometimes almost black, sometimes more or less mottled, and sometimes plain, with pale lateral line. The lower part of its sides usually are washed with yellowish, and sometimes marked with dusky dots. Its belly and the lower parts of the sides of its head are pure white, grayish, or yellowish; its dorsal, caudal, and anal fins are of the same color as the back except that the anal is pale at the base. The ventral fins are very pale pinkish or yellowish.

Size.—The squirrel hake does not grow so large as the white hake, seldom reaching a greater length than 30 inches (the largest of 780 Bay of Fundy fish measured by Craigie was about 27 inches long), or a greater weight than 6 to 7 pounds, and the average of those caught will not run above 1 to 3 pounds. In fact, a fish as heavy as 5 pounds is exceptional. Females are both longer and heavier than males of the same age (p. 226).

Habits.—These two hakes,[26] like many other sea fishes, spend their first months drifting at or near the surface, and fry of ½ to 4 inches (among which both species are no doubt represented) are often taken in summer under floating eelgrass or rockweed. On calm days we have seen them darting to and fro on the surface on many occasions (p. 224). And it is evident that the duration of this pelagic stage varies, for we have towed fry as long as 4 inches on the surface although others seek the bottom while they still are only

[26] The youngest stages of the two species are so much alike that in most cases we have been forced to list them simply as "hake," awaiting more critical examination than we have been able to give them.

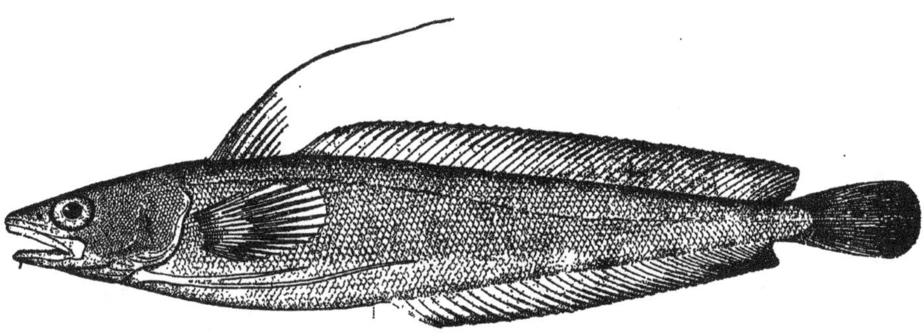

FIGURE 106.—Squirrel hake (*Urophycis chuss*), off Marthas Vineyard. From Goode Drawing by H. L. Todd.

about 2 inches long. But it is not known how far they may journey while they are at the mercy of currents. After they have taken to the bottom, they are ground fish for the remainder of their lives, only rising into the upper layers in pursuit of food. They are rather sluggish swimmers, as their body form suggests, and even a large one makes only a very feeble resistance when it is hooked.

When hake first take to bottom many of them do so in very shallow water, fry 2 to 6 inches long being common close below the tide mark among the eelgrass (*Zostera*); and fish a little larger are often caught by flounder fishermen in the harbors all around the Gulf of Maine. Others, however, seek the ground in somewhat deeper water, where they have an interesting habit of hiding within the living shells of the giant scallop (*Pecten magellanicus*). This has often been observed on the outer part of the Continental Shelf off southern New England; Nichols and Breder [27] have found little hake hiding in the mantle cavities of scallops in 20 fathoms off New York, and scallop fishermen have informed us that they often find little hake in the scallops that they dredge off the coast of Maine. Both of the common species of hake are known to use this curious refuge (they do not feed on the scallops but merely use their shells as hiding places), but most of the specimens so taken have proved to be squirrel hake. And the latter adopts this form of commensalism so commonly that Welsh records as many as 27 taken from 59 scallops in one haul of a scallop dredge, and 11 hake from 9 scallops in another haul, besides many others not counted off southern New England, New York, and New Jersey during the summer and autumn of 1913.

Slightly larger hake of both species, up to 8 to 12 inches long, are not only plentiful offshore, but are rather common close inshore in a fathom or two of water, in harbors, and even well up estuaries. The larger fish usually keep to deeper water, especially in summer, when hake of marketable sizes are most plentiful below 20 fathoms, and when only a few large ones are caught in less than 10 fathoms of water. But this rule, like most others, has its exceptions. For instance, we once saw a white hake of about 8 pounds caught from a float in Northeast Harbor, Maine, in about 10 feet of water, in July (in 1922). On the other hand, hake of both the species in question are to be caught in the deepest parts of the Gulf, and white hake have been taken down to 545 fathoms at least, on the offshore slope of Georges Bank.

Both of these hake haunt soft bottom chiefly, few being caught on the gravelly or shelly grounds that are so prolific of cod and haddock, or on rocky grounds. And it has been our experience that the whites are the more strictly mud fish of the pair.

The temperatures in which hakes of different ages are found cover the entire range proper to the Gulf except perhaps the very lowest. At the one extreme many of the youngest fry that are seen swimming at the surface in the west central part of the Gulf in summer are in water as warm as 68° to 70° F., while young hake are in still higher temperatures west and south from Cape Cod if they are at the surface. And the somewhat larger fry found on our beaches a little below tide mark may be in water as warm as 60° locally. But the great majority of the hakes living deeper are in water at least as cool as 50° throughout their later lives, most of them in temperatures lower than 45° F.

At the other extreme, all of the hakes living around the inner slopes of the Gulf at depths less than 50 fathoms experience temperatures as low as 35° to 37° F. in late winter and early spring; as low as 33° to 34° locally if they are living as shoal as 20 fathoms, which many of them do. But the fact that the bottom temperatures at the particular stations on the Grand Banks (all on the southern part) where white hake have been reported by the Newfoundland Fisheries Research Commission have all been between about 42° and about 33° F. (5.5° C. and 0.6 C.), and that they were not taken on other parts of the Bank where the bottom is colder, suggests that they tend to avoid regions where the temperature is as low as 32° F. or lower. And this finds some corroboration in the report (see p. 228) that hake tend to withdraw in autumn from Passamaquoddy Bay, where the water chills at least as low as 32° at some time during some winters.

Food.—Less is known of the diet of the hakes than of the cod, the haddock or the pollock. However, it is certain that they are not shell eaters to any extent, for it is seldom that their stomachs contain even the smaller univalve or

[27] Zoologica, N. Y. Zool. Soc., vol. 9, 1927, p. 172.

bivalve mollusks, and neither large mollusks nor echinoderms have ever been found in a hake, so far as we know. The stomach contents so far recorded[28] show that shrimps (*Pandalus*), amphipods, and other small crustacea which they find on the bottom are their chief dependence at most times and in most places. They also feed as greedily on squid as others of the cod tribe do, and a variety of small fish have been found in hake stomachs at Woods Hole,[29] such as alewives, butterfish, cunners, eels, flatfishes, tautog, herring, mackerel, menhaden, launce, silversides, silver hake, sculpins, sea robins, smelt, and tomcod.

Small white hake trawled some 75 miles south of Martha's Vineyard, in 56 fathoms, January 29, 1950, by the dragger *Eugene H* had fed on small squid, crabs (*Cancer*) and small butterfish (*Poronotus*); others trawled off Chesapeake Bay (lat. 38°13′ N., long. 73°49′ W.) in 52 fathoms by the *Albatross II*, March 2, 1931, had small mackerel, flounders, crabs, and squid in their stomachs. And we have seen squirrel hake caught off northern New Jersey with their bellies distended with launce, and with launce hanging from their mouths.

Hake of both species bite on fish bait such as herring readily; in fact, most of those that are caught on long lines (p. 230) are hooked on pieces of herring. But they also take clams on the hook greedily enough. In the northeastern part of the Gulf of Maine hake feed far enough off bottom to capture the pelagic euphausiid shrimps (*Meganyctiphanes* and *Thysanoessa*) that are so plentiful there, while the general character of their diet is sufficient evidence that they do not root in the ground like haddock.

Ever since 1616, when Capt. John Smith[30] wrote "Hake you may have when the cod failes in summer, if you will fish in the night," it has been common knowledge that they bite best after dark, from which it is fair to assume they do most of their foraging between sunset and sunrise.

[28] Goode, (Fish. Ind. U. S., Sect. 1, 1884, p. 235); Kendall, (Rept. U. S. Comm. Fish., (1896) 1898, p. 180); Linton, (Bull. U. S. Fish Comm., vol. 19, 1901, p. 478); Hansen, (Proc. U. S. Nat. Mus., vol. 48, 1915, p. 94); Breder (Zoologica, N. Y. Zool. Soc., vol. 2, No. 15, 1922, p. 350); and Vinal Edwards' notes.

[29] A large white hake taken at Woods Hole in May 1908 had a fish (*Lepophidium*) encysted in the wall of its body cavity, having no doubt penetrated the hake's stomach after it had been swallowed. (Sumner, Osburn, and Cole, Bull. U. S. Bur. of Fish., vol. 31, pt. 2, 1913, p. 768).

[30] General Historie of Virginia, New England, and the Summer Isles, 1616, ed. of 1819, vol. 2, p. 188.

Herrick[31] has given an interesting account of the perceptions of squirrel hake kept in a tank at Woods Hole, where they proved to have keen sight (though less so than pollock) and usually caught bits of meat before these had sunk. But it seems that it was only while food was in motion that the fish recognized it by sight, and that they depend chiefly on the sense of touch for their livelihood. They exercised this by swimming close to bottom with the sensitive tips of the ventral fins dragging the ground. When a hake touched a fragment of clam in this way it immediately snapped it up, but not otherwise. And they paid no attention whatever to live clams in their shells, though they often brushed over them. These observations, applied to the conditions under which hake actually live, suggests that they recognize shrimps, crabs, and other foods by their ventral feelers, and that they snap up their victims as these dart ahead, when the feelers drag over them.

No doubt the eggs of the white hake are bouyant like those of the squirrel hake (p. 225), but few wholly ripe females, no eggs naturally spawned, or young larvae have been seen yet.

We were equally ignorant of the spawning and early stages of the squirrel hake up to the summer of 1912. But we trawled squirrel hake with running spawn and milt in Ipswich Bay in that July, fertilized the eggs on board the *Grampus*, and thus identified the eggs. Since then large numbers of squirrel-hake eggs have been hatched artificially at the Gloucester hatchery.

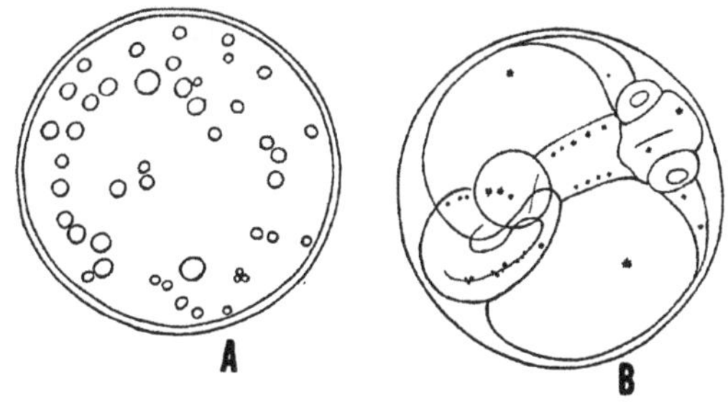

FIGURE 107.—Squirrel hake (*Urophycis chuss*), eggs, after 1 hour's incubation, A; and after 74 hours' incubation, B.

The eggs are buoyant, spherical, transparent, and 0.72 to 0.76 mm. in diameter. When first spawned they have variable numbers of small

[31] Bull. U. S. Fish Comm., vol. 22, 1904, p. 258.

colorless oil globules 0.02 to 0.07 mm. in diameter, scattered over the yolk, but most of these globules unite shortly after fertilization into one large one of 0.15 to 0.17 mm., which is sometimes single but which usually has two or three tiny ones close beside it. The embryo extends half way around the yolk sphere within two days after fertilization (at a temperature of 60° F.), and pigment has appeared, one of the most characteristic features of this species being the development of black chromatophores not only on the embryo, but over the yolk, and finally on the oil globule as well. In late stages of incubation this feature, combined with the small size of the egg and (usually) with a multiple oil globule, distinguishes eggs of the squirrel hake from all other buoyant fish eggs of known parentage that have been found in the Gulf, except for any rockling eggs that may have pigmented oil globules (p. 236). There is also some danger of confusing newly spawned eggs of the squirrel hake with those of the butterfish (of about the same diameter) for these sometimes have two oil globules (p. 364).

The newly hatched larvae have not been described. Older fry (identity established either as white hake or squirrel hake by comparison with young fry that have been reared in the hatchery by Louella E. Cable) already show the long, slender ventral fins, the short first dorsal but long second dorsal, and the tapering body form, characteristic of the adults. These little hakes, greenish blue on the back, with silvery sides, are separable from rockling fry by their more slender form, and by their scattered pigment. Older stages are separable from rocklings by their two well developed dorsal fins, while their silvery sides mark them at a glance from the dull colored fry of the cusk.[32]

Rate of growth.—The rate of growth during the first few months cannot be stated until many more young fry have been measured and identified as the one species or as the other. It is probable that two year classes are represented among the fry that are caught along shore in summer. Some of the smaller ones (2 to 3 inches long) may be from the earliest spawned eggs of that same season, but other squirrel hake of 2⅜ to 4⅛ inches (60–110 mm.) that were seined at Provincetown, on June 26, 1925, must have been about one year old, as they were taken too early in the season to have been the product of that year's spawning. And the larger ones of 6 or 7 inches are yearlings.

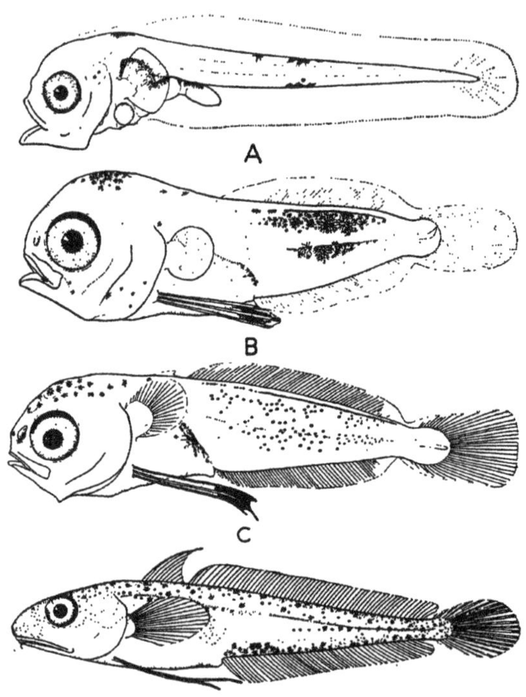

FIGURE 108.—Young stages of either white hake or squirrel hake. A, larva, 2.2 mm.; B, larva, 6.2 mm.; C, larva, 9 mm.; D, young fry, 40 mm. silvery still, and living at the surface of the water. Specimens collected off Woods Hole.

The growth of older squirrel hake in the Bay of Fundy has been studied by Craigie; his conclusions from scale studies,[33] combined with the relative abundance of fish of different size groups, being as follows.:

Age	Average length, inches — Male	Female
1 year-old	8	8
2 years-old	13	14
3 years-old	16	19

The indicated rate of growth is so uniform during the first three years of life as to suggest that spawning (an event so exhausting that it

[32] Fry figured by A. Agassiz (Proc. Amer. Acad. Arts, Sci., N. Ser., vol. 9, 1882, pl. 7, fig. 6; pl. 8, figs. 1-3), as *Motella argentata* undoubtedly were either white hake or squirrel hake. But the simple post anal pigment band, short, stocky bodies, and fan-like ventrals of the younger stages pictured by him under this same name (pl. 7, figs. 1-4) suggest that they were rockling.

[33] Contrib. Canadian Biol., (1914-1915) 1916, p. 87. Unfortunately, hake scales do not show the yearly growth zones as clearly as cod and haddock scales do.

slows the growth of any fish) probably does not take place until they have passed their third birthday. Nothing definite is known of the rate of growth of the white hake, but it is fair to assume that it grows faster than the squirrel, to attain its greater length and weight.

General range.—Both the white hake and the squirrel hake are exclusively American, occurring in continental waters from the Gulf of St. Lawrence and the southern part of the Grand Bank of Newfoundland southward to the Middle Atlantic States. The squirrel, though common as far south as Chesapeake Bay, has not been reported from farther south than Virginia. But the white hake is known off North Carolina (we have seen a 30 inch specimen that was trawled off Bodie I., North Carolina, lat. 35°52′ N., long. 74°51′ W. in 70 fathoms by the *Albatross II*, Feb. 24, 1931). And very likely the "squirrel" actually ranges as far south as the "white" does. Both of them occur from near tide mark, the squirrel down to about 175 fathoms, the white down to about 545 fathoms.

Occurrence in the Gulf of Maine.—Hake are very common fish in our Gulf, where the two species, white and squirrel, are caught side by side regularly. In the Bay of Fundy there are so few toward the head that stragglers are caught, or none at all, but they are plentiful enough toward the mouth where, for example, about 6,400,000 pounds were landed on the Nova Scotian side by Canadian fishermen in 1944, and about 8,200,000 pounds in 1946, while the yearly catch on the New Brunswick side is about 500,000 to 600,000 pounds. Other centers of abundance for them inshore are along the coast of Maine between Machias and Mount Desert Island, in Frenchman's Bay (formerly the site of an important hake fishery), the ground known locally as the "Grumpy" near Isle au Haut, and off Penobscot Bay.

Sundry small grounds outside the islands from Penobscot Bay to Cape Elizabeth and all along the western side of the Gulf, also yield good numbers of hakes, especially near Boon Island; the vicinity of the Isles of Shoals, a famous hake ground for small boat fishermen; Ipswich Bay; the lower slopes of Jeffreys and Stellwagen Banks; also the deeper parts of Massachusetts Bay, which yielded 750,000 pounds in 1919 when the demand for hake was better than it is now.

Hake, indeed, are so widespread on the lower slopes of all the banks and ledges in the inner parts of the Gulf, as well as on the mud floors between them, that Rich[34] listed 119 named grounds in the western side of the Gulf as good haking bottoms. Hake, with flounders, rosefish, and silver hake are practically the only commercially valuable fish one is likely to catch on the floors of the deep basins and channels of the Gulf; and a catch of 2,880 of them with 580 cusk, but no cod or haddock, by long-line fishing 15 miles southeast of Monhegan on June 24 to 25, 1913, will illustrate how completely they may monopolize suitable bottoms.

Hake are plentiful in the so-called South Channel also, and on the northwest slope of Georges Bank, whence about 2,000,000 pounds were landed in 1919, about 1,500,000 pounds in 1947. And it has long been known that there is an abundance of hake at depths greater than 60 to 70 fathoms all along the southern slope of Georges Bank. Long-line fishermen, too, have told us that while it was unusual to hook a hake on the shoaler parts of Georges, many were caught wherever the line was run off into deeper water on the northwest face of the bank; i. e., onto soft bottom. And this is borne out by the statistics of the catches, for the good trawling grounds on Georges Bank yield far fewer hake of marketable size than the inner parts of the Gulf do, if the year 1945 can be taken as representative.[35]

It has only been since 1944 that the landings of white hake and of red (i. e., squirrel) hake have been reported separately. Taken at their face value, these would point to the white hake as by far the more plentiful member of the pair throughout the inner parts of the Gulf as a whole, and on Georges Bank. In 1945, for example,[36] the reported landings were some 14 times as great for white as for red hake, for every one of the subdivisions into which the inner part of the Gulf is divided for statistical purposes, while only a few thousand pounds of red hake were reported from off eastern or central Maine, or from the northwestern part of Georges Bank; and none from the eastern or southwestern part of the

[34] Rept. U. S. Comm., Fish. (1929) 1930, App. 3, pp. 85–86, 96.

[35] Landings of hake in 1945 were about 414,000 pounds for Georges Bank; about 12,700,000 pounds for the inner parts of the Gulf by United States fishermen and about 9,140,000 pounds by Canadian fishermen.

[36] The only year when the landings of the two have been reported by counties for Maine and Massachusetts, besides the landings at the major ports.

Bank. But this would give a wholly false picture of the actual situation, because most of the red hake that are caught on these grounds are thrown overboard because they are too small to be worth gutting and icing under present market conditions.

Actually, we trawled 34 squirrel hake and only two other hake[37] in Ipswich Bay, in 22 fathoms on one occasion in July 1912; Welsh counted 5,450 squirrel hake to 652 white hake caught in otter trawling on the northwest slope of Georges Bank in June 1912; we counted 2,457 red hake to only 196 white hake from 29 trawl hauls at 22 to 150 fathoms on the southwestern part of Georges Bank and south of Nantucket Shoals on the *Albatross III*, May 11–18, 1950; and we saw 2,040 squirrel hake taken in 42 hauls by the *Eugene H* on the southwest part of Georges Bank, in late June 1951, but only 51 white hake. Reported landings also, in pounds, for 1945, were about 100 times as great for red as for white hake from the Nantucket grounds, whence all the little hake are brought in for the trash fish industry. And the discrepancy is greater still in numbers, for the white hake are much the heavier of the two, individually. Red hake also predominate over white among the hake landed in New York and to the southward, as is illustrated by the catch statistics for 1947.[38]

Landings, for 1947, to nearest 100,000 pounds:

	New York	New Jersey	Delaware
Red hake	1, 200, 000	5, 600, 000	200, 000
White hake	1, 000, 000	200, 000	100

On the other hand, inquiries of fishermen, corroborated by our own experience, point to the white hake as the more plentiful of the two in the basin of our Gulf at depths greater than 40 to 50 fathoms. The *Atlantis*, for example, trawled about 700 white hake in the deep basins off Cape Cod, west of Jeffreys Ledge and off Mount Desert, in August 1936, but only a scattering of squirrel hake. This appears to apply equally to the deeper holes in Massachusetts Bay at depths greater than 30 fathoms or so (both Storer and Goode and Bean spoke of the "white" as the more common of the two there), also to the Bay of Fundy region in general, including Passamaquoddy Bay, according to Huntsman. And nearly all of the hakes that have been listed by name from the more easterly of the Nova Scotian Banks, or from the southern part of the Grand Banks in the annual reports of the Newfoundland Department of Natural Resources, have been the white (*tenuis*). *Tenuis*, also, is the only member of the pair that was reported by Cornish[39] from Canso, but *chuss* alone is recorded from the Cape Breton shore of the Gulf of St. Lawrence by Cox,[40] who also records one from 205 fathoms on the Nova Scotian side of Cabot Strait.

The situation is made more confusing by Cornish's report of hake with 123 rows of scales from Prince Edward Island, and with 130 rows of scales from Canso, fish intermediate, that is, between *chuss* and *tenuis* in this regard, though favoring the latter. Perhaps the separation between the two species in number of scales, and also in other features, may not be so sharp toward the northern boundaries of their geographic ranges as we have found it farther south.

A more or less regular inshore movement of hakes of one or the other species, or of both of them, is said to take place in autumn, especially in the northeastern part of the Gulf, made evident by capture of considerable numbers in winter in the deeper, muddy harbors and bays east of Penobscot Bay, including St. John Harbor, and Kennebecasis Bay, which they reach by running up the St. John River, and where they are caught all winter through the ice. They also carry out corresponding movements in and offshore off southern New England, with goodly numbers appearing in shoal water at Woods Hole in autumn. But it is only in the spring and autumn that they are found close inshore off New York and off New Jersey. On the other hand, they are said to enter Passamaquoddy Bay in early summer, to withdraw in autumn.

Probably the explanation is that the adults, being cool water fish, are barred from the shallows in summer by high temperature along the coasts of Massachusetts and of west-central Maine, but that the low summer temperature of Passamaquoddy Bay allows large hake to summer there, as well as small. Their reported withdrawal from

[37] The latter were listed by Welsh as *U. regius*, but probably they were white hake.

[38] About 13,000 pounds of white hake were reported from Maryland in 1947, about 65,000 pounds from Virginia, and about 4,000 pounds from North Carolina, with no reds. But we suspect that reds were actually included as well as whites, and spotted hake also.

[39] Contrib. Canadian Biol. (1902–1905), 1907, p. 89.

[40] Contrib. Canadian Biol. (1918–1920), 1921, p. 113.

Passamaquoddy Bay in autumn may be in avoidance of extreme winter chilling. But we should remind the reader that failure to catch fish on hook and line in the cold season of the year (it is in this way that hakes are caught in the Passamaquoddy region) does not necessarily mean that they have departed. The hake may have stopped biting, as every fisherman knows by experience. The evidence of otter trawl catches is much more reliable in this respect, for ground fishes in general.

Except for in and offshore movements, hake are resident throughout the year in the open Gulf of Maine wherever they are found, once they have taken to the bottom. And they appear to be much more stationary than either cod or haddock.

The localities where we have found eggs, provisionally identified as squirrel hake (fig. 109), show that it spawns all around the Gulf from Cape Cod to Nova Scotia. And despite its rather deep-water habitat and preference for soft bottom, most of these egg stations have been in shoal water near the coast; a haul in the eastern basin which yielded both squirrel hake and silver hake eggs (p. 178) has been the only exception. This, of course, points to a movement from the basins into shoaler water for spawning.

It seems that the white hake spawns from late winter through spring to late summer, for we saw a

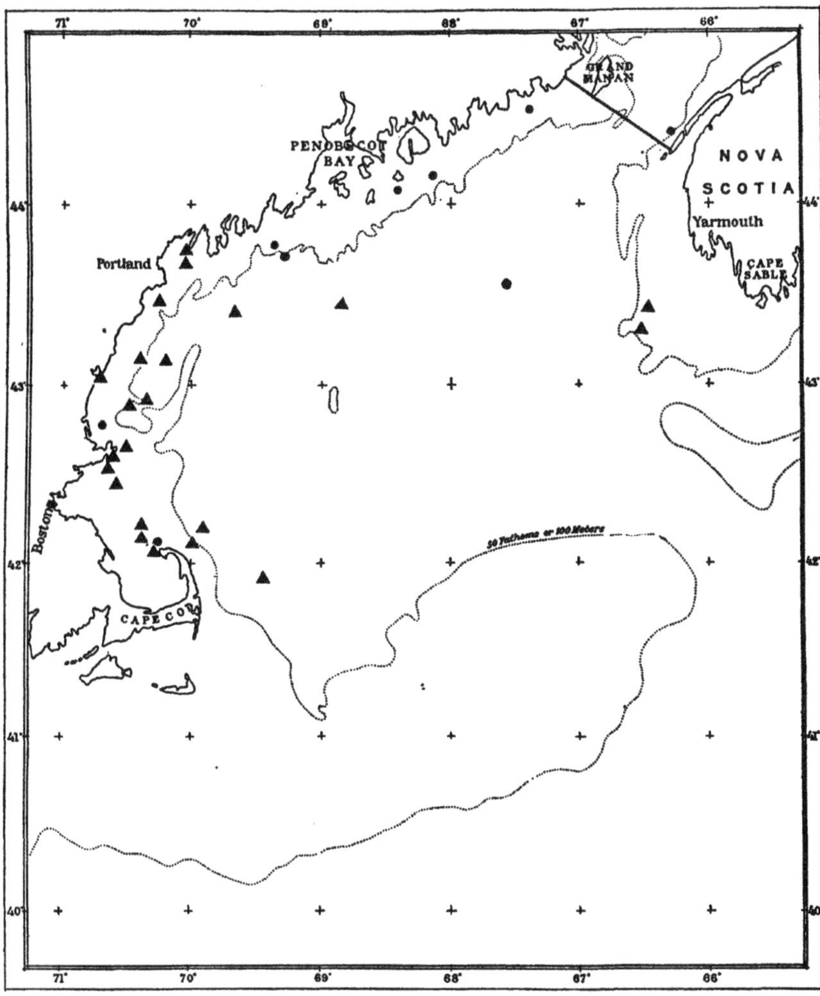

FIGURE 109.—Locality records for squirrel hake eggs (●), and for larvae of rockling (▲) in the Gulf of Maine.

30-inch female, with large roe, trawled off New Jersey (lat. 39°23′ N.) by the *Albatross II* at 88 fathoms, on February 27, 1929; Welsh saw a male with milt flowing, near Gloucester on April 22, 1913; we saw spent females, but also a male just ripening, trawled by the *Albatross III* on the southwestern part of Georges Bank on May 15, 1950; and Nichols and Breder[41] report a ripe female with ripe eggs at Woods Hole in July.

The height of the spawning season of the squirrel hake comes at least as early as June south of Cape Cod, and through early summer in the Massachusetts Bay region, judging from the size (27 to 70 mm.) of the fry mentioned above as found in scallop shells in late summer and autumn (p. 224). The extreme limits of the season are not known. But we have towed eggs of this species as early as June 10 in Petit Passage, Nova Scotia, and as late as September 20 in various parts of the Gulf, while captures of fry of 72 mm. as early as the last week in July (in Shelburne Harbor, Nova Scotia), and others as small as 36 mm. in the western part of the Gulf as late as November 1 (in 1916), point to a spawning season lasting from late spring until early autumn.

Abundance.—There is nothing in the statistical picture to suggest that hake of either species fluctuate very widely in abundance in our Gulf from year to year, for the ups and downs in the amounts caught are not greater than can be charged to market conditions. Neither has any attempt been made to estimate the periodic variations in the relative abundance of different year classes. Earlier characterizations of the numbers of the two hakes in our waters have been in relative terms, ranging from "common" to "in immense numbers."[42] More precise information is that the *Atlantis* caught an average of about 32 *tenuis* per hour's trawling with an 82-foot trawl at 65 to 125 fathoms (120–228 meters) off Cape Cod, west of Jeffreys Ledge and off Mount Desert, in August 1936,[43] while the *Albatross III* trawled about 245 *chuss* and about 18 *tenuis* per hour's hauling on the southwestern part of Georges Bank and south of Nantucket Shoals in May 1950. And we find record of 145 *tenuis* taken per hour's trawling,[44] on Sable Island Bank, off outer Nova Scotia.

[41] Zoologica, N. Y. Zool. Soc., vol. 9, 1927, p. 172.

[42] Huntsman, Contrib. Canadian Biol., (1921) 1922, p. 68.

[43] Bigelow and Schroeder, Biol. Bull., vol. 76, 1939, p. 308.

[44] Report, Newfoundland Fish. Res. Comm., vol. 1, No. 4, 1932, p. 109.

Importance.—The hakes are soft-meated and have rather poor keeping qualities, but both the white and the squirrel hake are readily absorbed by the fish markets if they are large enough, and great numbers of small squirrel hake are now used for mink and poultry feed. A quarter of a century ago the yearly catch in the Gulf ran between 20 and 35 million pounds, and it has been much the same of late years (1941–1946), with yearly landings by Canadian and United States fishermen of between 19 and 30 million pounds. In 1946, which may serve as representative, Canadian fishermen landed about 2,100,000 pounds in outer Nova Scotian ports (Cape Sable to Cape North), about 4,800,000 pounds along the southern shore of the Gulf of St. Lawrence.

At the present time (as represented by 1946 and 1947) 4 to 5 times as much hake is marketed in Maine and Massachusetts in the form of fresh and frozen fillets as is marketed there salted, some are used for fish cakes, and a very small part[45] as smoked fillets. Hake sounds (swim bladders), especially of those that are caught off Nova Scotia in deep water, are also used to make isinglass,[46] and increasing amounts of small squirrel hake brought in from Nantucket Shoals, are utilized from year to year in the trash-fish industries.[47]

Roughly two-thirds of the poundage of hake that is landed in Maine and Massachusetts are caught in otter trawls nowadays, roughly one-fifth in gill nets, and only a little more than one-eighth on long lines.[48]

The hakes are such dull and inactive fish that they are of no special interest to the angler. But a good many fair-sized ones are caught hand-lining from party boats, for they bite readily, and small hake are caught from small boats in harbors and bays, along the Maine coast especially.

Spotted hake *Urophycis regius* (Walbaum) 1792

Jordan and Evermann, 1896–1900, p. 2553.

Description.—This species is distinguishable among the hakes of the Gulf of Maine by the fact

[45] About 39,000 pounds reported for 1946, none for 1947.

[46] For further information see Tressler, Marine Products of Commerce, 1949, pp. 489–491.

[47] Landings of red hake at New Bedford from the Nantucket Shoals region, mostly used in this way, were about 5,600,000 pounds in 1947.

[48] The amounts recorded for 1947 were:

	Pounds
Otter trawls	10,399,800
Gill nets	3,380,200
Long lines	2,097,200
Hand lines	102,200

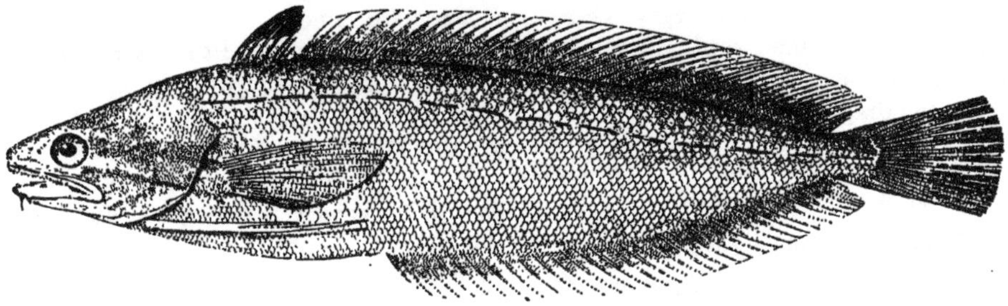

FIGURE 110.—Spotted hake (*Urophycis regius*). From Jordan and Evermann. Drawing by H. L. Todd.

that it has no prolonged rays in its first dorsal fin (which is hardly higher than the second dorsal, and has 8 or 9 rays); by the smaller number of rays in its second dorsal fin (46 to 51 as against 54 or more in the squirrel and white hakes); and by having only 90 to 95 oblique rows of scales along its sides from gill opening to caudal fin, instead of about 105 to 110 rows and about 140 rows, respectively, in the other two species. The anal fin has 43 to 49 rays, somewhat fewer, on the average, than the squirrel or the white hake (48 to 50).

Convenient field marks are that the outer half of its first dorsal fin is black with a whitish margin; that its pectoral fins reach back as far as the origin of the anal fin, whereas they fall considerably short of the latter in both the white and the squirrel hake; and that its lateral line is darker brown than the general body color, instead of paler, and is interrupted by a series of distinct whitish spots. Otherwise the spotted hake, like the commoner hakes, is dull brown, darker above than below, with dorsal and anal fins of the same color as the back. Its ventrals are whitish.

Size.—The largest of many measured by Welsh at Atlantic City, in August 1920, were about 16 inches long, and weighed between 1 and 1½ pounds; the usual length is less than 12 inches, and the longest, of about 600 taken by the *Albatross II* at 14 stations between the offing of Delaware Bay and Cape Hatteras, in late winter and spring (1930 and 1931) was 5⅛ inches (130 mm.).

Habits.—The spotted hake resembles the other hakes in its habits. It may be more of a fish eater, for Vinal Edwards noted that the few he examined at Woods Hole contained alewives, menhaden, launce, and squid. But it also feeds on the crustaceans on which the white and squirrel hakes subsist, for Hildebrand and Schroeder[49] found mysid-shrimps in most of those examined in Chesapeake Bay where small spotted hake are very common. The capture of spawning fish by the *Albatross*, off the coast of the Carolinas in December in 1919, recorded in Welsh's field notes, is evidence that it is a winter breeder.

[49] Bull. U. S. Bur. Fish., vol. 43, pt. 1, 1928, p. 161.

General range.—Coast of the United States, regularly from southern New England and New York to Cape Hatteras (including Chesapeake Bay where it is plentiful), and ranging southward as far as the offing of northern Florida in deep water.[50]

Many were trawled as far northward as the offing of Delaware Bay by the *Albatross II*, in 1930 and 1931; it is reported as rather uncommon at New York;[51] it has been taken occasionally at Woods Hole; it has been known to reach the coast of Maine as a very rare stray; it was reported more than a century ago off Halifax, Nova Scotia, by Richardson;[52] and a single specimen was reported as taken, pelagic, near Sable Island (lat. 44°10′, long. 59°45′) in August 1931.[53]

Occurrence in the Gulf of Maine.—The spotted hake strays past Cape Cod so seldom that specimens taken off Seguin Island many years ago, and four, trawled on the southwestern part of Georges Bank, by the *Albatross III* in May 1950, are the only definite records of it for the Gulf of Maine.[54] But it may well have been overlooked among the hosts of young hake of the two common species (white and squirrel) that are caught in

[50] The U. S. National Museum has specimens taken off Charleston, S. C., at 87 and 124 fathoms.

[51] Nichols and Breder, Zoologica, N. Y. Zool. Soc., vol. 9, 1927, p. 169.

[52] Fauna boreali Americana, vol. 3, 1836, p. 253. Richardson's wood cut of the specimen in question, from a sketch by Lt. Col. Hamilton Smith, shows the low first dorsal with black apex that is characteristic of the species *regius*.

[53] Report, Newfoundland Fishery Res. Comm., vol. 1, No. 4, 1932, p. 109.

[54] This species was also listed from Ipswich Bay, from Casco Bay, and off of Monhegen Island in the *Grampus* collections of 1912 (Bull. Mus. Comp. Zool. vol. 58, No. 2, 1914, p. 113), but it is probable that these specimens were white hake in reality.

the southwestern part of the Gulf every year, for 49 spotted hake were taken south of Block Island, in 47 to 67 fathoms, January 27 to February 3, 1950, by the dragger *Eugene H.*

Long-finned hake *Urophycis chesteri* (Goode and Bean) 1878

Jordan and Evermann, 1896–1900, p. 2556.

Description.—The most distinctive character of the long-finned hake is its very long ventral fins, the longest of the rays of which reach back nearly to the rear end of the anal fin (about to its fortieth ray), with the next longest ventral ray considerably overlapping the origin of the anal. The filamentous dorsal ray is longer also, than in the other Gulf of Maine hakes, reaching back to about the middle of the second dorsal fin or beyond. Furthermore, there are only about 90 rows of scales from gill opening to caudal fin along the lateral line, and the scales are relatively larger than in either the white hake or the squirrel hake; the eye is larger in the long-finned hake; the anal fin rays are more numerous (average about 56), the rear corners of the dorsal and anal fins are more rounded. The outline of the anal is slightly concave instead of straight (fig. 111); the pectoral fins are more slender and more pointed, and the caudal fin is narrower with more strongly convex margin; these differences are more clearly shown in the illustrations than verbally. The skin of the long-finned hake is curiously loose, like that of many deep-sea fishes.

Color.—Freshly preserved specimens are olive above and on the sides, with a silvery white belly. The fins are olive, with dusky markings on the dorsal filament, on the outer edge of the dorsal fins, on the caudal fin, and on the ventrals.

Size.—Specimens 14 to 15 inches (36 to 38 cm.) long, trawled by *Albatross III*, on the southwestern slope of Georges Bank and off Nantucket Shoals, in 105 to 240 fathoms, May 11–18, 1950, are the largest yet recorded.

Habits.—The long-finned hake is a bottom fish, living chiefly between 100 and 500 fathoms, the deepest record for it is from 538 fathoms. It is a summer and autumn spawner, judging from the fact that Goode and Bean saw specimens in breeding condition at that season. We have taken pelagic young of 8 to 35 mm. in our tows off Marthas Vineyard during the last week of August.[55] And captures of 3 fry, about 2¼ to 2¾ inches (57–71 mm.) long on April 26 (1931) and of 16 fish of about 3 to 4¼ inches (74–110 mm.) late that July suggests that a length of 4 to 5 inches is reached at 1 year of age.

General range.—This is a deep-water fish, occurring in great abundance on the continental slope off North America from the Laurentian Channel in Cabot Strait to abreast of Cape Lookout, N. C.

Occurrence in the Gulf of Maine.—This hake is plentiful all along the seaward slopes of Browns Bank, of Georges Bank and of Nantucket Shoals at depths greater than 100 fathoms, where it has been trawled at many stations.[56]

The *Albatross III*, for example, caught 861 in 63 half-hour trawl hauls, at 105 to 240 fathoms on the slope between the south-central part of Georges Bank (long. 67°14′ W.) and the offing of the eastern end of Long Island, New York (long. 71° 57′ W.) in May 1950. Up to 1931 the only

[55] Bigelow, Bull. Mus. Comp. Zool., vol. 59, No. 8, 1917, p. 275.

[56] For early locality records see Goode and Bean, Smithsonian Contrib. Knowl., vol. 30, p. 361.

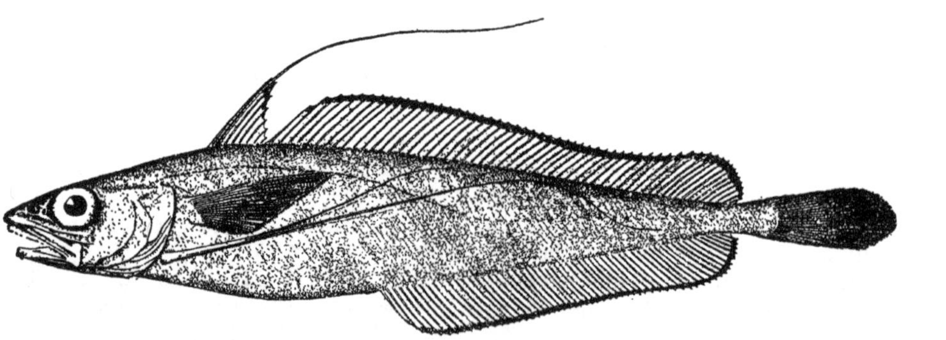

FIGURE 111.—Long-finned hake (*Urophycis chesteri*), off Cape Ann. From Goode. Drawing by H. L. Todd.

definite records of them from the inner parts of the Gulf had been of three specimens taken off Cape Ann,[57] in 110 to 140 fathoms, in 1878; and of a few others that were trawled on the northern edge of Georges Bank by the *Kingfisher*, in September 1929, in 85 to 100 fathoms. But captures of a number to the westward along the Bank and in the central basin of the Gulf in the summer of 1931, by the *Albatross II* [58] at depths of 70 to 140 fathoms, show that long-finned hakes are more numerous in the deeper parts of the Gulf than had been suspected previously.

Blue hake *Antimora rostrata* Günther 1878

Jordan and Evermann, 1896–1900, p. 2544, as *A. viola* Goode and Bean.

Description.—This species resembles the white, squirrel and spotted hakes in the form of its body and in having two separate dorsal fins, the first very short and the second very long; but it is readily distinguished from them by the fact that its anal fin is so deeply notched about midway of its length that it almost seems to have two separate anals, and that each of its ventral fins is 6-rayed, with the second ray prolonged and filamentous. The shape of the snout, which is flattened above, keeled at the sides, and blunted at the tip in some but forming an acute angle in others is distinctive, likewise its vent is situated much farther back than in the true hakes (genus *Urophycis*), and its body, in life, is deep violet, blackish brown, or blue black, below as well as above.

Size.—The longest yet measured was one of about 21½ inches (545 mm.).

Range.—The blue hake was reported at so many localities along the continental slope from the early cruises of the U. S. Bureau of Fisheries [58a] on the Nova Scotian slope; off southern New England; and southward to the offing of Cape Hatteras, North Carolina, at 350 to 1,000 fathoms that it must be one of the more plentiful of fishes there.

We have recently trawled a few at 220 to 460 fathoms, on the southeastern Nova Scotian slope, on the *Caryn* of the Woods Hole Oceanographic Institution, between longitudes 64° W. and 65° 56′ W., and halibut fishermen have occasionally brought them in. But the blue hake has not been taken within the limits of the Gulf of Maine, and it is hardly to be expected there; the shoalest capture recorded for it so far is from 220 fathoms. We mention it on the chance that vessels trawling on the slope may occasionally work deep enough to pick up a few.

The known range of this deep sea hake includes the North Atlantic from Denmark Strait to the offing of Gibralter in the east and from the Newfoundland Banks to the offing of Cape Hatteras in the west; Uruguay; the eastern Pacific, British Columbia to Panama, and the southern Indian Ocean. It has been taken as deep as 1,456 fathoms. [59]

Hakeling *Physiculus fulvus* Bean 1884

Jordan and Evermann, 1896–1900, p. 2547.

Description.—This fish is hakelike in its general appearance, also in the general arrangement of its

[57] These were the basis of Goode and Bean's original description of the species (Proc. U. S. Nat. Mus., vol. 1, 1878, p. 256).

[58] Reported by Bigelow and Schroeder, Bull. U. S. Bur. Fish., vol. 48, 1936, p. 339.

[58a] See Goode and Bean (Smithsonian Contrib. Knowl., vol. 30, 1895, pp. 374–375) for list of stations.

[59] For a recent summary, see Schroeder, Copeia, 1940, No. 4, pp. 236–237.

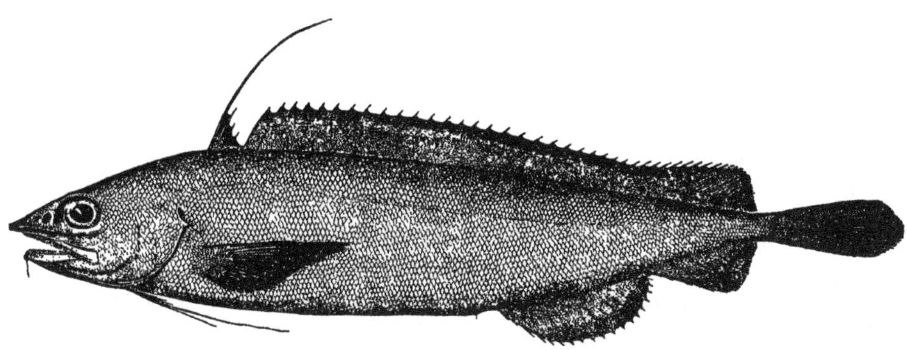

FIGURE 112.—Blue hake (*Antimora rostrata*), La Have Bank. From Goode. Drawing by H. L. Todd.

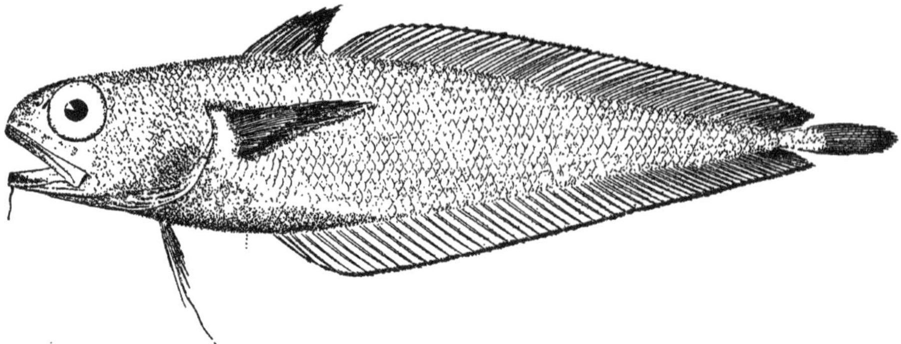

FIGURE 113.—Hakeling (*Physiculus fulvus*), outer edge of Continental Shelf off Nantucket. From Goode and Bean. Drawing by H. L. Todd.

fins, for it has two dorsals, the first (10 rays) triangular and much shorter than the second (about 49 rays) which is of nearly uniform height from end to end; one long anal fin (about 54 rays) which is similar to the second dorsal in shape; and ventral fins situated in front of the pectorals. It is separable from the white, squirrel, and long-finned hakes (genus *Urophycis*, pp. 221 and 232) in that its anal fin originates in front of the origin of the second dorsal fin instead of considerably behind the latter and that its ventral fins have 5 rays each instead of 2 and are much shorter than those of the true hakes, with the longest ray (the second, which is filamentous at the tip) hardly reaching back as far as the middle of the pectoral fins. Furthermore, the snout of the hakeling is blunter than that of any true hake; its caudal fin much smaller; its body tapers more abruptly; and none of the rays of its first dorsal fin are prolonged.

Color.—Described as light yellowish brown with the lower surface of the head, the abdomen, and the margins of the dorsal and anal fins very dark brown, and with a dark brown blotch on each cheek (on the subopercular bone). We have not seen it fresh from the water.

Size.—The maximum size is not known.

Habits.—Nothing is known of the habits of the hakeling except that it is a deep-water fish, having been taken from 79 fathoms down to 955 fathoms, where it lives on or near the bottom, to judge from its general structure.

General range and occurrence in the Gulf of Maine.—This hakeling has been taken at several localities in the Gulf of Mexico and on the continental slope off the eastern United States. The most northerly record for it is off Nantucket (lat. 40° 01′ N., long. 69° 56′ W.) in 79 fathoms, and it is on this record that the hakeling is mentioned here.[60]

Four-bearded rockling *Enchelyopus cimbrius* (Linnaeus) 1766

ROCKLING

Jordan and Evermann, 1896–1900, p. 2560.

Description.—The rocklings, of which this is the only common local representative, differ from their near relatives, the hakes (genus *Urophycis*), in the facts that their ventral fins are short, with 5 to 7 rays, and that the first section of their dorsal fin consists of only one ray, which is nearly as long as the head, and which stands over the upper corners of the gill openings, followed by a series of about 50 very short, separate, hairlike rays without connecting membrane, which can be laid down in a groove on the back. Thus there is only one well-developed dorsal fin. Rocklings differ further from all other gadoids in the presence of long barbels on the top of the nose as well as on the chin, the number of these being the most obvious specific character among the several species of rocklings. In the present species there are a pair of these barbels close in front of the nostrils, a third and somewhat shorter barbel standing alone on the tip of the snout, and there is a fourth barbel hanging from the chin.

Rocklings remind one of young hake in their slender bodies tapering back from the shoulders;

[60] Another small hakeling (*Lotella maxillaris*) has been taken off Marthas Vineyard. It is separable from the hakeling described above by the fact that its anal fin originates behind the origin of the second dorsal fin, and by its larger teeth.

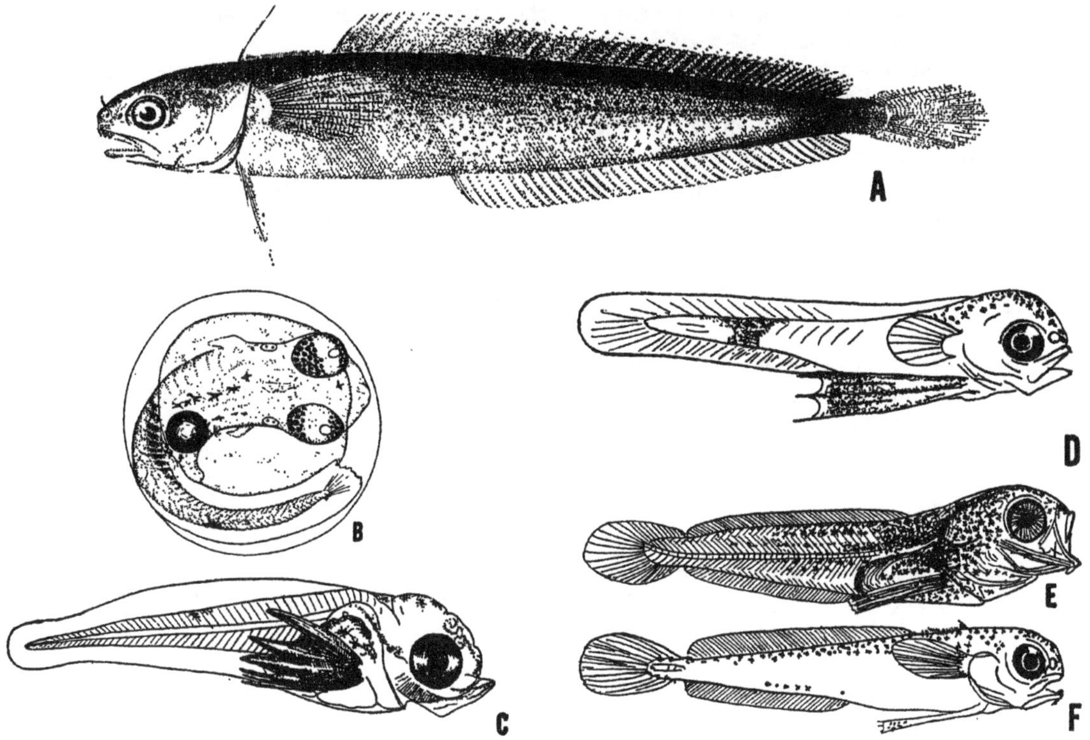

FIGURE 114.—Rockling (*Enchelyopus cimbrius*). A, adult, Bay of Chaleur, from Jordan and Evermann, drawing by H. L. Todd; B, egg; C, larva (European), 3.6 mm.; D, larva (European), 5.3 mm.; E, larva (European), 13.6 mm.; F, silvery fry (European), 17.5 mm. B, after Battle; C, after Ehrenbaum and Strodtman; D, after Brook; E, after Ehrenbaum; F, after Brook.

and (hakelike) they are rounded in front of the vent but flattened sidewise behind it. Their upper jaw is longer than the lower and their teeth are smaller than in the hakes, while their noses are shorter and blunter; their eyes are smaller, and the dorsal profile of their heads is more rounded than it is in any of the hakes. The pectorals are rounded and the narrow pointed ventrals are situated well in front of the latter. The second dorsal fin (45 to 53 rays) originates over the mid length of the pectorals, runs back nearly to the base of the caudal fin, and is equally high from end to end with a rounded rear corner. The anal fin is similar to the second dorsal in shape, but it is shorter (39 to 43 rays).[61] The caudal fin is oval when it is spread.

Color.—The color of this rockling is comparatively constant by all accounts and this is corroborated by our own experience. Its back is dark yellowish olive or dusky brown, its sides are paler, and its belly is white dotted with brown. On some individuals the sides behind the vent are more or less clouded with a darker shade of the general body hue. The first dorsal ray, the posterior edges of the second dorsal fin and of the anal fin, the lower half of the caudal fin, and the pectoral fins are sooty or bluish black. Otherwise the vertical fins are grayish or bluish brown. The ventral fins are pale, and the lining of the mouth is dark purplish or bluish.

Size.—This rockling has been described as growing to a length of 16½ inches in Scandinavian waters, but about 12 inches is the longest recorded from the Gulf of Maine, where they average only about 6 to 10 inches.

Habits.—Rocklings are bottom fish, like hake. Occasionally they have been found in very shallow water, on Nahant Beach in Massachusetts Bay, for example; in water only a few feet deep at Woods Hole; in 6 to 7 fathoms, both in St. Mary's

[61] Storer credits it with 48 rays, but subsequent students have not found so many.

Bay, Nova Scotia, and in Buzzards Bay on the south coast of Massachusetts. But they appear to be more plentiful in depths of 25 to 30 fathoms or more; there are rocklings in the deep gully off Halifax, and also in the deep trough of the Gulf of St. Lawrence.[62] They have been taken on the continental slope off southern New England to a depth of 724 fathoms.[63] And there is no reason to suppose that the adult fish ever rise far above the bottom, unless by accident.

The occasional appearance of adult rockling in very shallow water in winter near Woods Hole [64] suggests that some may work inshore and into shoal water in autumn, to work offshore again and deeper in spring, for the summer. Beyond this they seem to be year-round residents wherever they are found.

The name "rockling" is a misnomer for this fish for it is found most often on soft bottom in the Bay of Fundy, while those that we have trawled in Massachusetts Bay and in Ipswich Bay from the *Grampus* were on smooth muddy sand between the hard patches. And most of the rockling living in the deep sinks and channels in the western side of our Gulf, and on the continental slope, are on soft smooth ground.

Judging from the stomach contents of Scandinavian and British fish (their stomach contents have not been examined on this side of the water so far as we know) they feed chiefly on shrimps, isopods, and other small crustaceans, less often on fish fry. On the other hand, rockling have been found in cod stomachs in Massachusetts Bay, and no doubt all fish of prey devour them on occasion.

The eggs are buoyant, described (we have never seen them) as 0.66 to 0.98 mm. in diameter. When newly spawned the oil is in small droplets, most of which soon coalesce into one globule of 0.14 to 0.25 mm., often with one or two smaller ones close to it. The danger of confusing them with squirrel-hake eggs is discussed in the account of that fish (p. 226). And Battle has found that they develop normally at temperatures ranging from 55° to 66°.[65]

Newly hatched larvae are a little more than 2 mm. long. The yolk is absorbed at about 3.6 mm. and the later larval stages, up to about 10 mm. long, are characterized by the very large black ventral fins shown in the illustrations (fig. 114); by the presence of only one post anal band of black pigment; and by the short stocky body-form. Young hake are more slender and have scattered pigment; young cusk have two post anal bands; and all other Gulf of Maine gadoids have short ventral fins. After the rockling is 17 to 20 mm. long the structure of the first dorsal fin serves to identify it.

These larger fry are silvery, awaiting their descent to bottom before assuming the dull colors of the adult. In British waters they are sometimes called "mackerel midges" because they suggest little mackerel remotely, in their general appearance. In European waters, where there are more plentiful populations of the silvery fry of one or the other species of rockling they are often cast ashore. And one such instance is described for our Gulf by Storer [66] who writes that many were picked up on Nahant Beach during one tide in the summer of 1860; and others found in the surf at West Beach, Beverly.[67]

Rockling fry, like those of other gadoids, drift at the surface for their first few months. How long they do so in our waters is not known, but analogy with cod, haddock, and other species suggests three months at most. And it may be assumed they seek the bottom at a length of about 2 inches for our largest pelagic fry were 40 to 45 mm. long. During this pelagic stage they drift with the current like any other fish fry, and are at the mercy of mackerel and other fish. But they are not plentiful enough in the Gulf of Maine to be as important an article in the diet of the mackerel as the fry of the far commoner European

[62] Huntsman (Trans. Roy. Soc. Canada, 3, vol. 12, Sect. 4, 1918, p. 63) and further information contributed by him.

[63] Goode and Bean, (Smithsonian Contrib. Knowl., vol. 30, 1895, pp. 384–385) give a long list of locality records for the rockling on the shelf and slope between the offings of eastern Nova Scotia and of North Carolina (lat. 35°40′ N.).

[64] Sumner, Osburn, and Cole, Bull. U. S. Bur. Fish., vol. 31, Pt. 2, 1913, p. 771.

[65] Battle (Contrib. Canadian Biol., N. Ser., vol. 5, No. 6, 1930) has made a careful study of the effects of extreme temperatures and salinities on the development of the eggs of the rockling.

[66] Fishes of Massachusetts, 1867, p. 279.

[67] These fry, and one recorded at Nahant earlier by Gill (Proc. Acad. Nat. Sci., Philadelphia, (1863)1864, p. 241) were reported as an Arctic 3-bearded species (*Gaidropsarus argentatus* Reinhardt) which was described originally from Greenland and which has been found widely distributed in Denmark Strait; on the north coast of Iceland; and in the Norwegian Sea from the Faroes north to Bear Island. But there is no reason to suppose that the Nahant specimens were anything other than the fry of our common four bearded rockling. For a recent account and discussion of the species *argentatus*, with excellent illustrations, see Jensen, Spolia Zool. Mus. Hauniensis, Copenhagen, vol. 9, 1948, pp. 167–173, pl. 4, fig. 4.

rocklings are, on the other side of the Atlantic. Nothing is known of their subsequent rate of growth.

General range.—Both sides of the North Atlantic. The American range is from the northern part of the Gulf of St. Lawrence and the northeastern coast of Newfoundland (perhaps even farther north) to Narragansett Bay and Long Island Sound in coastal waters, and to the latitude of Cape Fear (N. C.) in deep water along the continental slope.[68] The Arctic three-bearded rockling (*Gaidropsarus ensis* Reinhardt), otherwise known only from Greenland, has been trawled on the lower part of the continental slope in the offings of southeastern Nova Scotia, of Cape Cod, of Martha's Vineyard, of New York and of New Jersey at depths of 858 to 1106 fathoms, by the *Fish Hawk* and *Albatross I*, but this is not shoal enough to bring it within our limits.[69]

There are several other species of rockling in north European waters, but none of them have been recorded from our side of the Atlantic.

Occurrence in the Gulf of Maine.—The little rockling is of no commercial value, and it seldom comes up into very shallow water where it would force itself on the notice of seaside visitors. But it is a common bottom fish in the deeper parts of Massachusetts Bay as Goode and Bean[70] remarked long ago, while our experience, corroborated by Huntsman for the Bay of Fundy, is that this applies to the entire Gulf. Definite Gulf of Maine records for adult rocklings are from St. Mary Bay (Nova Scotia); various localities in the Bay of Fundy including Passamaquoddy Bay; Jonesport; off Mount Desert; off Pemaquid; near Seguin Island; mouth of Casco Bay; the deep gully to the westward of Jeffreys Ledge; Ipswich Bay; Gloucester; Nahant; various stations in the deeper parts of Massachusetts Bay; Provincetown; the deep open basins of the Gulf;[71] and Georges Bank. And we have taken its young fry rather frequently in our tow nets in season.

Huntsman[72] and Battle[73] have found the eggs of this rockling in Passamaquoddy Bay throughout the summer, commencing in May and most abundantly at the time the bottom water warms to 9° or 10° C. And its breeding season probably continues from spring to early autumn in the western Atlantic as it does in the eastern,[74] for Dannevig[75] (1919) records rockling eggs (probably this species) as early as the end of May near Halifax, while we have taken rockling larvae only 5.5 mm. long as late as September and October in our tow nets in Massachusetts Bay.

It is probable that the rockling spawns all around the peripheral belt of the Gulf, with Massachusetts Bay as an important nursery, to judge from our repeated captures of its larvae there. And we have taken the pelagic fry in our tow nets at the various localities marked on the accompanying chart (fig. 109) from the first week in July until October; seldom, however, more than half a dozen in any one haul (the largest catch was 18 specimens). Huntsman, similarly, describes the fry as common in the center of the Bay of Fundy, and they have been taken in the tow nets at Woods Hole in April. But we have taken neither the eggs, the larvae, nor the pelagic fry in any of our tow nettings in the central parts of the Gulf, which perhaps justifies the assumption that the spawning grounds of the rockling within our Gulf are limited mostly to depths less than 75 fathoms, though it may spawn much deeper than that on the continental slope.

To the west of Cape Cod, the rockling is now known to occur in coastal waters as far as Narragansett Bay, and in Long Island Sound, where it was found generally in 5½ to 9 fathoms, and abundantly at 21 fathoms by the *Fish Hawk* in the summer of 1914.[76] And it has been trawled by the *Fish Hawk* and by the *Albatross I* at many stations in deeper water offshore along the shelf and slope, southward to the offing of Cape Hatteras (lat. 35° 40′ N.).[77]

---

68 A specimen trawled by the *Albatross II* in 12 fathoms off the mouth of Chesapeake Bay on February 10, 1930, is the only one recorded in shallow water so far southward.

69 Goode and Bean (Smithsonian Contrib. Knowl., vol. 30, 1895, p. 381) give a list of these localities. For a recent account of *G. ensis*, with illustrations, and list of Greenland localities, see Jensen, Spolia Zool., Mus. Hauniensis, Copenhagen, vol. 9, 1948, p. 167, pl. 4, fig. a.

70 Bull. Essex Inst., vol. 11, 1879, p. 9.

71 The *Atlantis* trawled it both in the Jeffrey bowl, and in the open basin of the Gulf, August 1936; and we trawled it in the central basin in July 1931.

72 Contrib. Canadian Biol. (1921) 1922, p. 69.

73 Contrib. Canadian Biol., Fish., N. Ser., vol. 5, No. 6, 1930, p. 13 [119].

74 It spawns from the end of January until August in the Baltic.

75 Canadian Fisheries Exped., (1914–1915) 1919, p. 53, table 1C.

76 Nichols and Breder, Zoologica, N. Y. Zool. Soc., vol. 9, 1927, p. 172.

77 For list of early stations, see Goode and Bean (Smithsonian Contrib. Knowl., vol. 30, 1895, pp. 384–385). They also report a specimen apparently of this species from the offing of Cape Fear, N. C. (lat. 34° 01′ N., long. 76° 11′ W.). But it was in poor condition, hence of doubtful identity.

Eastward and northward from our limits, the rockling is said to be rather common in Nova Scotia waters in general, coastwise as well as on the fishing banks. The *Albatross* trawled it at three stations along the continental edge between the offing of southwestern Nova Scotia and of Sable Island, at 93 to 134 fathoms; and while Huntsman[78] describes it as characteristic of the deep channels of the Gulf of St. Lawrence, Dannevig[79] points out that the stations within the Gulf of St. Lawrence where the Canadian Fisheries Expedition took rockling eggs and larvae in any number, rather generally distributed in the southern part, a few in the northeastern part, were all "close to land or above the more shallow banks." Pelagic rockling fry are listed under this name in the Reports of the Newfoundland Fishery Research Commission also, from many stations in the Grand Banks region, and around the coast of Newfoundland to the Northern Peninsula on the east and to the inner end of the Strait of Belle Isle on the west. But it would not be astonishing if the fry of the three-bearded rockling (p. 237) should prove to be represented in these collections, together with those of our four-bearded species. Dannevig, indeed, has suggested that part of the rockling eggs taken by the Canadian Fisheries Expedition in Nova Scotian waters and south of the Grand Banks in May and June belonged to some species other than *cimbrius*.

Importance.—The rockling is neither large enough nor plentiful enough to be of importance commercially, or of interest to the angler.

Cusk *Brosme brosme* (Müller) 1776

TUSK; TORSK

Jordan and Evermann, 1896–1900, p. 2561.

[78] Trans. Roy. Soc. Canada, Ser. 3, vol. 12, Sect. 4, 1918, p. 63.

[79] Canadian Fisheries Exped. (1941–1915) 1919, p. 27: charts figs. 18, 19.

Description.—The cusk is separable from all its Gulf of Maine relatives at a glance by the fact that it has only one dorsal fin. The relationship of the anal and dorsal fins to the caudal and the outline of the latter are distinctive also, for both the dorsal and the anal are continuous with the caudal at the base but are separated from it by notches so deep that they are obviously distinct. And the caudal is evenly rounded. The cusk is a more slender fish than the hakes, being only about one-fifth to one-sixth as deep as it is long, round-bodied in front of the vent but flattened sidewise behind the vent, and tapering evenly backward to the base of the caudal fin. The mouth is large, gaping back to opposite the rear third of eye, is set slightly oblique, and is armed with small, sharp, curved teeth. The snout is blunt at the tip. The upper jaw encloses the lower when the mouth is closed; the eye is of moderate size; the chin bears one barbel; and the entire head and trunk are clad with small scales. The dorsal fin (85 to 105 rays) runs the whole length of the back from the nape of the neck, and is of uniform and moderate height from end to end with rounded corners. The anal fin is similar to it in outline but is only a little more than half as long (71 to 76 rays). The pectoral fins are rounded, and about half as long as the head. The ventral fins are about as long as the pectorals, with their 5 rays free at the tips, and are situated a little (but obviously) in front of the pectorals. All the fins are so thick and fleshy at their bases that it is only near their margins that the rays are to be seen.

Color.—The cusk varies in color, no doubt conforming to the bottoms on which it lives. Its upper parts range from dark slaty to dull reddish brown or to pale yellowish, paling to grayish on the lower part of the sides and to dirty white on the belly. Old fish are plain colored, the sides of small ones, however, are often cross-barred

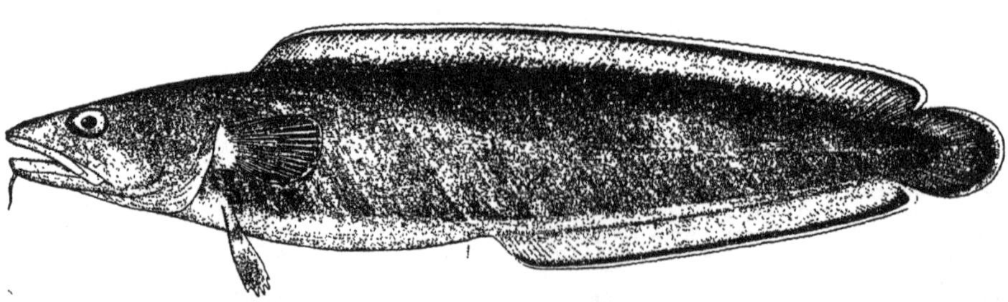

FIGURE 115.—Cusk (*Brosme brosme*). Boston market. From Goode. Drawing by H. L. Todd.

with about half a dozen yellowish bands. The pectoral and ventral fins are of the same color as the sides, and the ventral fins are sooty at their tips. The most characteristic color mark is that all three of the vertical fins (dorsal, caudal, and anal), which are of the general body tint at their bases, are black at the margin, and they are narrowly edged with white, except that the anal may lack the white edging on some individuals.

Size.—Cusk grow to a maximum length of about 3½ feet; one 40 inches long, weighing 27 pounds, trawled by the *Albatross II* in the central part of the Gulf of Maine, in 120 fathoms, was the largest that has been recorded definitely from our waters. But those caught in the Gulf of Maine average only 1½ to 2½ feet long, and from 5 to 10 pounds in weight. The relationship of weight to length, in fish we have handled recently, was as follows: 26 inches, about 5½ pounds; 33½ inches, about 14¾ pounds; 36 inches, about 20 pounds. The size at which cusk first mature sexually seems not to have been recorded.

Habits.—Once the young fry have taken to the bottom they are ground fish so exclusively that we have never heard of one swimming up to the upper waters, as cod so often do, and even hake. They are sluggish, too, and weak swimmers, but powerful of body; when a cusk is hooked it is likely to twine itself around one's line in a bothersome way.

They are more or less solitary, not so abundant anywhere as cod, haddock, or hake are, as may be illustrated by the following catches counted as they came from the water by representatives of the Bureau of Fisheries in 1913: Twenty miles east of Cape Cod Light, November 16 and 17, 1913, long line, 460 cusk to 2,150 haddock and 1,228 cod; 15 miles southeast of Monhegan Island, June 24 and 25, 1913, long lines, 580 cusk to 2,880 hake; Jeffreys Ledge, December 11 and 12, 1913, long line, 230 cusk to 470 haddock and 475 cod; northwest part of Georges Bank, October 10 to 13, 1913, otter trawl, 4 cusk and 12,473 haddock; 6 miles east of Boon Island, March 30, 1913, gill net, 5 cusk, 1,055 haddock; 51 cod, 20 pollock, and 76 dabs (*Hippoglossoides*).

It also seems that cusk move little from bank to bank. Thus the "Massachusetts fishermen tell me," wrote Goode [80] "that these fish are usually found in considerable abundance on newly discovered ledges, and that great numbers may be taken for a year or two, but that they are soon all caught. Sometimes, after a lapse of years, they may be found again abundant on a recently deserted ground." Neither is there any definite evidence that the cusk performs in or offshore migrations with the seasons, at least in our Gulf.

The cusk is so purely a fish of at least moderately deep water that we have never heard of one taken in less than 10 to 15 fathoms of water within our Gulf. On the other hand, it is safe to say that there are few cusk living below 100 fathoms or so in the deep basins of the Gulf. But they range down to 250–300 fathoms on the continental slope off southern New England, according to Goode and Bean.[81] And they have been caught down to 530 fathoms in the Faroe Channel.

Cusk are decidedly fastidious, too, in their choice of bottoms, being found chiefly on hard ground, especially where the sea floor is rough with rocks or boulders; on gravelly or pebbly grounds; occasionally on mud with hake, but seldom on smooth clean sand. In Norwegian waters they often lurk among gorgonian corals, and they may have this same habit on the parts of our offshore banks where these are plentiful.

The cusk is a fish of cool water, but not of the coldest. In the Gulf of Maine (once the fry have deserted the surface for the bottoms at their chosen depths), cusk spend their lives in water which does not warm above about 48° to 50° at the warmest season, nor cool below about 33° to 34° at the coldest. And it is probable that temperatures of 32° F. or lower are the factor that limit their American range in the north (p. 242).

Food.—Little is known of the diet of the cusk. European students describe the stomachs as usually containing crustaceans, sometimes mollusks. And crabs, with occasional mollusks, that we found in the stomachs of several cusk caught on Platts Bank in the summer of 1924, are the only record of its food of which we know, for this side of the Atlantic. But the cusk is not fastidious as to bait, accepting clams, cockles, and herring readily.

Cusk spawn in spring and early summer in both sides of the Atlantic. In European waters the season lasts only from April until June; but

[80] Fish. Ind. U. S., Sect. 1, 1884, p. 233.

[81] Smithsonian Contrib. Knowl., vol. 30, 1895, p. 385.

throughout July in the Gulf of Maine, for we have caught several nearly ripe females on Platts Bank and around Boon Island at the end of that month, though we have seen no perfectly ripe fish. In the eastern Atlantic cusk spawn chiefly deeper than 100 fathoms, to judge from the distribution of the eggs at the surface. But the chief production of eggs probably takes place in shallower water in the Gulf of Maine, since most of the stock lives in lesser depths there. And some must spawn close inshore, for we have taken cusk larvae only 6 to 13.8 mm. long off Cape Cod; in Provincetown Harbor; and near the Isles of Shoals.[82]

We owe what is known of the eggs and larvae to European students. The cusk is among the more prolific of fishes, more than 2 million eggs having been estimated in a female of medium size. Their eggs are buoyant like those of other gadoids; 1.29 to 1.51 mm. in diameter, with one oil globule of 0.23 to 0.3 mm.; and they may be recognized by the brownish or pinkish color of the oil globule, together with the fact that the entire surface of the egg is finely pitted.

The larvae are about 4 mm. long when they hatch. The vent is situated at the base of the ventral finfold as it is in other gadoids, but they are separable from all other gadoid larvae that occur in the Gulf of Maine by the pinkish oil globule at the posterior end of the yolk. The yolk is absorbed in about a week after hatching, when the larvae are about 5 mm. long. The ventral fins of the little cusk elongate as it grows, like those of young hake and of young rockling, besides becoming heavily pigmented with black. But cusk larvae are separable from those of hake and of rockling by the fact that their ventral fin rays are separate one from another, and by the presence of three patches of black pigment: one on the top of the head; a second over the gut; and a third at the tip of the tail, besides two vertical black bands which divide the trunk behind the head into three nearly equal sections. The rockling has only one band of pigment behind the vent, and neither of the hakes that are common in the Gulf of Maine has a definite cross-band of pigment.

The first traces of the vertical fin rays of the young cusk are visible at about 12.5 mm.; the dorsal and anal fins are differentiated at about 28 mm.; and it is at this stage that the ventrals are at their longest, relatively. Fry of 40 mm. and upward show most of the characters of the adult. And the relationship of their dorsal and anal fins to the caudal, and the presence of only one dorsal fin and one anal fin is sufficient to identify them from this stage on.

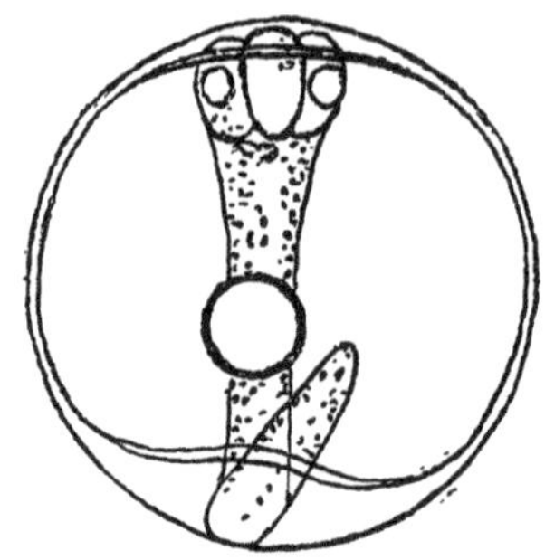

FIGURE 116.—Egg (European). After Schmidt.

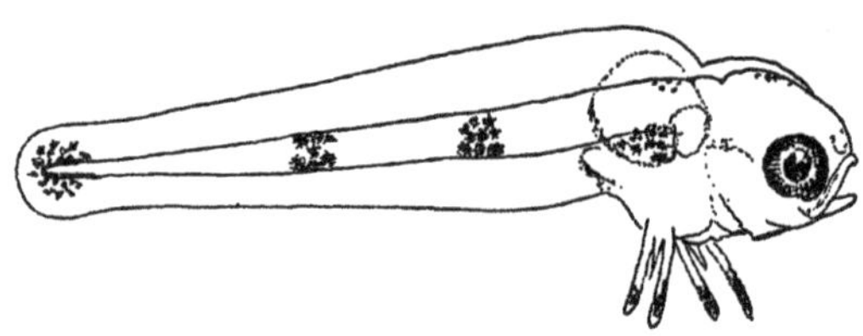

FIGURE 117.—Larva, 6.8 mm. (European). After Schmidt.

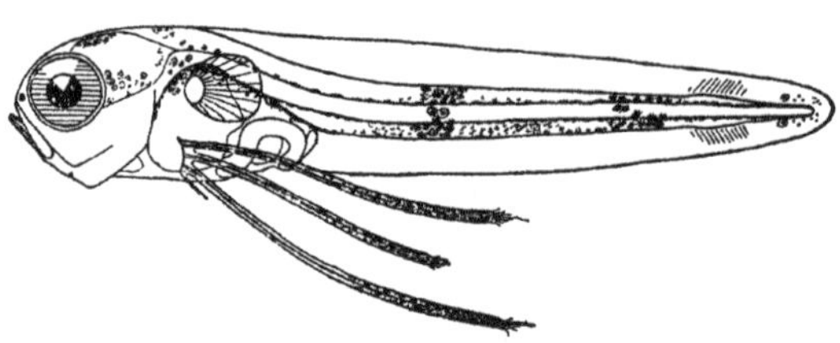

FIGURE 118.—Larva, 9.25 mm., off northern Cape Cod.

CUSK (*Brosme brosme*).

The older cusk fry, while still living at the surface, are described by Schmidt[83] as greenish yellow with blue eyes, not silvery-sided.

The young cusk drifts near the surface, as other gadoids do, until it is 2 inches long or more, and there is reason to believe that in European seas they first seek the bottom in considerable depths. But we have nothing to offer on this point for the Gulf of Maine.

[82] The records are July 22, 1912, 1 specimen; July 20, 1916, 4 specimens; and July 22, 1916, 1 specimen.

[83] Meddel. Kommis. for Havundersøgelser, Serie Fiskeri, vol. 1, No. 8, 1905, p. 7. He also describes the larval stages of the cusk.

The rate of growth of the cusk has not been studied, so far as we know.

General range.—Both sides of the North Atlantic, chiefly in moderately deep water and on hard bottoms; north on the American coast to the Newfoundland Banks, and to the Strait of Belle Isle, south regularly to Cape Cod, rarely to southern New England, and occasionally to New Jersey; northern coasts of the British Isles, Denmark (Jutland), northern part of the North Sea, and Kattegat off Bohuslan, Sweden, to Iceland and the Murman coast in the eastern Atlantic. It reaches east and west Greenland only as a rare stray from the south.

Occurrence in the Gulf of Maine.—The cusk is distributed very generally in the Gulf in water deeper than 10 to 15 fathoms, its presence or absence depending on the precise type of bottom. Because of its preference in this respect (p. 239), it varies greatly in abundance in different parts of the Gulf, and the grounds occupied by it are much less extensive than those haunted by cod, by haddock, by pollock, or by the hakes. Thus cusk are rarely taken in Cape Cod Bay or in the deeper holes in Massachusetts Bay, and we have taken none on the soft mud of the deep bowl west of Jeffreys Ledge. But considerable numbers are caught on the ledges off Chatham, Cape Cod, on Stellwagen Bank, and on the broken grounds between the latter and Cape Ann, while they are plentiful off Cape Ann and on Jeffreys Ledge, the latter being one of the most productive cusk grounds of our Gulf. The rocky slopes of Cashes Ledge, also have long been famous for cusk. In past years when more fishing was done there (as in 1902 and 1905) this ground was the chief source of supply for the cusk landed in New England. In 1935, similarly, about 30 percent of all the cusk landed in Portland, Gloucester, and Boston came from Cashes. And we have caught more cusk there than anywhere else. As might be expected, cusk are also caught on Fippenies and Platts Banks by the few vessels that fish there as is illustrated by the catches reported from these inshore grounds for the 5-year period 1931–35.[84]

The landings, 1931–1935, in Boston, Gloucester, and Portland (to nearest 1,000 pounds) follow:

Locality	*1931*	*1932*	*1933*	*1934*	*1935*
Cashes	225,000	98,000	173,000	612,000	1,023,000
Fippenies	19,000	69,000	35,000	47,000	61,000
Platts	7,000	6,000	165,000	84,000	45,000
Jeffrey Ledge	301,000	143,000	148,000	122,000	53,000
Stellwagen Bank	65,000	63,000	85,000	259,000	78,000

We are inclined to believe that the wide differences from year to year, in the catches on these small grounds reflect the number of vessels that fished there, rather than the number of cusk waiting there to be caught.

Cusk are said to be plentiful on the rather indefinite ground off Penobscot Bay that is known as Jeffreys Bank (not Ledge) or "Matinicus Sou'-sou'west." In 1921, for example, 43,545 pounds were reported thence, and considerable numbers are taken, in the aggregate, on the patches of hard bottom that skirt the coast of Maine, as appears from the approximate amounts landed in the smaller ports [85] in the different Maine counties in 1919 and in 1945: York, 9,000 pounds and 2,600 pounds; Cumberland (exclusive of vessel landings at Portland), 79,000 pounds and 182,000 pounds; Sagadahoc, 15,000 pounds and 44,000 pounds; Lincoln, 27,000 pounds and 3,000 pounds; Knox, 52,000 pounds and 109,000 pounds; Hancock, 12,000 pounds and 22,000 pounds; Washington, 4,000 pounds and 500 pounds, respectively.

Some cusk are caught at the mouth of the Bay of Fundy also, especially about Grand Manan on the New Brunswick side, and off Brier Island on the Nova Scotian side, as Doctor Huntsman informs us, though none are reported toward the head of the Bay. Small rocky patches along the west Nova Scotian shore and off Seal Island also yield some cusk; and they are taken regularly on Grand Manan Bank. German Bank and the fishing grounds off Lurcher Shoal are less productive of cusk, perhaps because they are floored, mostly, with patches of gravel and pebbles and small stones alternating with sand and clay. But large catches are taken on Browns Bank, and fair numbers on the rougher spots on Georges Bank, though its smoother expanses yield only an occasional cusk.

The only important exceptions in our Gulf to the rule that cusk hold to rocky ground are that they are at least tolerably plentiful in the co-called

[84] 1935 is the most recent year when landings were reported from these grounds, separately.

[85] Mostly by small boat fishermen.

South Channel, where the bottom is mostly smooth (see regional summary of 1945 catches, p. 242); that some are caught with hake off the coast of Maine on broken or even muddy bottom; and that we have trawled a few, on the *Atlantis*, in depths greater than 78 fathoms off Cape Cod, where the bottom is mostly a sticky sand.

One striking accompaniment of the preference of cusk for rough or stony grounds in moderately deep water, is that many more are caught around the peripheral belt of the Gulf, between, say, the 15-fathom and the 75-fathom contour lines, than are on the off-shore rim formed by Nantucket Shoals, Georges Bank, and Browns Bank. The one notable exception is that there are so few cusk, if any, in the inner parts of the Bay of Fundy that they are not mentioned in the fishery returns for the Bay, except for a few thousand pounds taken near its mouth on the Nova Scotian side.

This regional contrast is illustrated by landings by United States fishermen (1945) [86] and Canadian fishermen (1944, 1946) combined, of between 215,000 and 250,000 pounds off western Nova Scotia; [87] 1,000 to 15,000 [88] pounds at the mouth of the Bay of Fundy, Nova Scotian side; about 63,000 pounds off eastern Maine; about 333,000 pounds off central Maine; about 255,000 pounds off western Maine; about 419,000 pounds off eastern Massachusetts; about 338,000 pounds from the small grounds in west central part of the Gulf; about 68,000 pounds from the South Channel; a few hundred pounds only from Nantucket Shoals; about 25,000 pounds from the northwest part of Georges Bank; none reported from the southwestern part of Georges; about 17,000 pounds from the eastern central and northeastern parts of Georges Bank; and about 18,000 pounds from Browns Bank.

Following the cusk eastward and northward, we find that considerable quantities are caught all along the Nova Scotian Banks, from Browns to Banquereau and to the Canso grounds off Cape Breton Island (catch, in 1946, about 542,000 pounds by United States and Canadian vessels combined). Cusk were also reported from the Newfoundland Banks many years ago by Goode,[89] but there cannot be many of them there, for they are not included among the fishes reported thence from cruises of the Newfoundland Fishery Research Commission.[90] And the only report we have found of cusk anywhere in the Gulf of St. Lawrence is at Cheticamp, on the Cape Breton shore.[91] In fact, the only definite record we have found of cusk on the American coast farther north than Cabot Strait is of one that was caught in the Strait of Belle Isle at 80 fathoms many years ago.[92] And while the cusk has been credited repeatedly to Greenland, it is a rare stray there from the south, only 7 specimens having been reported there during the period 1936–46, 5 of them on the west coast, 2 on the east.[93]

Westward from Cape Cod, the cusk is said to have been "not uncommon" formerly in Vineyard Sound, but it is so rare there now (if it ever occurs there) that we have not heard of one caught anywhere in the Woods Hole region of late years. But one was caught off Newport, Rhode Island, in November 1898,[94] and two were reported from Cape May, New Jersey, many years ago.[95]

Importance.—The cusk is a good food fish and there is a ready market for all that are brought in. The landings from the Gulf of Maine by United States fishermen ranged between about 1,600,000 pounds and about 2,200,000 pounds for the years 1945–47; between about 100,000 pounds and about 200,000 pounds by Canadian fishermen for 1944 and 1946, which contrasts with 2 to 7 million pounds yearly by United States fishermen alone for the few years that preceded the publication of the first edition of this book (in 1925). We attribute this decrease to the evolution that has taken place in the fishery from long lining to otter trawling chiefly, and to the

[86] Most recent year for which landings have been published by counties, for Maine and Massachusetts, in addition to the landings at Portland, Gloucester, Boston, and New Bedford.

[87] Off western Nova Scotia, by United States fishermen, 1945, about 108,000 pounds; Yarmouth County landings, Nova Scotia, about 140,500 pounds in 1944, about 106,000 pounds in 1946.

[88] 1944, 15,000 pounds; 1946, 700 pounds.

[89] Fish. Ind. U. S., Sect. 1, 1884, p. 232.

[90] Frost (Service Bull. 8, Newfoundland Dept. Nat. Resources, 1938, p. 29) states that there is no definite record of cusk on the Newfoundland fishing grounds.

[91] Recorded by Cornish (Contrib. Canadian Biol. (1918–1920) 1921, p. 114) from fishermen's reports. W. R. Martin of the Fisheries Research Board of Canada, writes us that any fisheries reports of cusk for the Gulf of St. Lawrence actually refer to hake.

[92] Weitz, Proceedings, Boston Soc. Nat. Hist., vol. 10, 1866, p. 274; Packard, Labrador Coast, 1891, p. 819.

[93] For further details and discussion of the status of the cusk as a Greenland fish, see Jensen (Spolia Zool., Mus. Hauniensis, Copenhagen, vol. 11, 1948, p. 175).

[94] Tracy, 40 Ann. Rept. Commiss. Inland Fish. Rhode Island, 1910, p. 159.

[95] Abbott, Geol. New Jersey, 1868, p. 819.

fact that the cusk (frequenting rough bottom) is not a good trawl fish. And 80 to 90 percent as much cusk (pounds) are caught on long lines as are caught in otter trawls even today (as illustrated by 1947), although the yearly landings of fish of all kinds in Maine and Massachusetts now are 70 to 80 times as great by otter trawls as by long lines.[96]

A few cusk are caught from party boats by sportsmen hand lining for ground fish in general, but most of the cusk live too deep to be of any particular interest to anglers.

THE GRENADIERS. FAMILY MACROURIDAE

The grenadiers are characterized externally by having large heads, projecting snouts, and slender bodies that taper to whiplike tails, with no definitely demarked caudal fin. They have two dorsal fins, the first high, the second very low but occupying the greater part of the back. The anal fin is nearly as long as the second dorsal, or longer.

The grenadiers are allied to the cod family, in classification, by the structure of their skull, but they differ from the cod tribe in having one stout spine in the first dorsal fin. They are deep-sea fishes, living on the bottom, loose in texture and weak swimmers. Many species are known, but only three of them have ever been taken within the confines of the Gulf of Maine.

Besides the species described below, three others, *Coryphaenoides rupestris*, *C. carapinus* and *Nematonurus armatus*,[97] have been taken on the continental slope abreast of the Gulf and off southern New England often enough to show that they are common there below 350 fathoms. They are typical inhabitants of the deep-sea floor, never likely to rise shoal enough to come within the limits of the Gulf of Maine.[98] But fish have a way of straying, and if any grenadier should be picked up in the Gulf that proves difficult to identify, we recommend forwarding it either to the Laboratory of the Fish and Wildlife Service at Woods Hole, to the U. S. National Museum, Washington, D. C., or to the Museum of Comparative Zoology, Cambridge, Mass., to be named. Parr [99] has recently published a detailed synopsis of all the species known from the western North Atlantic and from central American seas.

[96] Otter trawlers landed about 499 million pounds of fish of all sorts in Maine and Massachusetts in 1947; long liners about 7 million pounds.

[97] According to Parr (Bull. Bingham Oceanogr. Coll., vol. 10, art. 1, 1946, p. 54) this is the correct name of the grenadier that was reported by Goode and Bean (Smithson. Contrib. Knowl., vol. 30, 1895, p. 407) as *Hymenocephalus goodei* Günther, 1887.

[98] For descriptions and lists of localities where they were taken during the early cruises by vessels of the U. S. Fish Commission, see Goode and Bean (Smithson. Contrib. Knowl., vol. 30, 1895). In June 1949 we trawled about 200 *rupestris* on the slope off southern Nova Scotia and off the southeastern face of Georges Bank, at 290–420 fathoms, from *Caryn* of the Woods Hole Oceanographic Institution.

[99] Bull. Bingham Oceanographic Coll., vol. 10, art. 1, 1946.

KEY TO GULF OF MAINE GRENADIERS

1. The dorsal spine is perfectly smooth ---------------- Long-nosed grenadier, p. 246
   The dorsal spine is serrated, with teeth which can be felt if not seen ---------------- 2
2. The vent is considerably in front of the origin of the anal fin; the skin surrounding the vent is naked and black; the dorsal fin spine is strongly serrated ---------------- Common grenadier, p. 243
   The vent is close to the origin of the anal fin; the skin around the vent is scaly and pale colored; the serrations on dorsal fin spine are so fine that they are hardly visible, though they can be felt ---------------- Rough-headed grenadier, p. 245

Common grenadier *Macrourus bairdii* Goode and Bean 1877.[1]

RAT-TAIL; MARLIN-SPIKE

Jordan and Evermann, 1896–1900, p. 2583.

Description.—This grenadier could hardly be mistaken for any other fish except for one of its own tribe, so characteristic is its slender body (flattened sidewise behind the vent and tapering to a whiplike tail with no definite caudal fin), in combination with a pointed snout that overhangs the mouth; very large eyes; and high first dorsal fin with one large spine; but very low second dorsal fin. And it has a chin barbel like a cod (not shown in the illustration). As noted above, the second ray of the first dorsal fin is a true spine, serrated along its front edge with about 15 sharp and very noticeable teeth pointing upwards.

The first dorsal fin (2 stiff rays, the first very short, and 11 softer rays) is triangular, about twice as high as it is long; and it originates over the pectorals, close behind the gill openings. The space between the two dorsal fins is about as long

[1] Parr (Bull., Bingham Oceanogr. Coll., vol. 10, art. 1, 1946, p. 37) places this grenadier in the genus *Nezumia* of Jordan, 1904. But it seems wiser to follow the older and more familiar usage here.

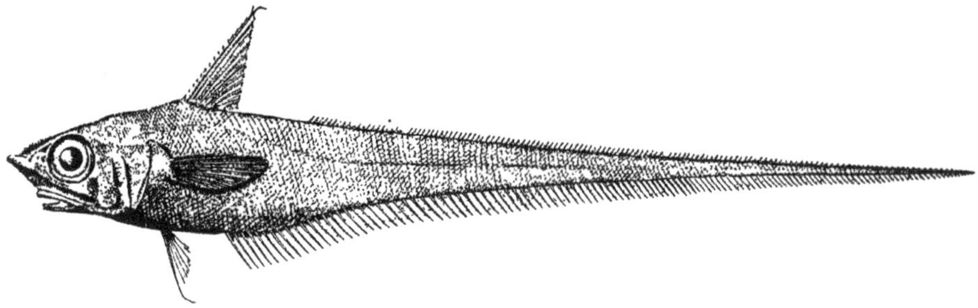

FIGURE 119.—Common grenadier (*Macrourus bairdii*), off Cape Ann. From Goode and Bean. Drawing by H. L. Todd.

as the height of the first dorsal fin. The second dorsal fin (about 137 rays) extends back to the tip of the tail, is so low that its membrane is hardly visible, and tapers to practically nothing at the rear end. The anal fin is considerably longer than the second dorsal (only about 120 rays, however) and more than twice as high as the second dorsal. The pectoral fins are rounded at the tip. The ventral fins, which stand under the pectorals or a little behind the latter, are triangular, with the first ray prolonged as a threadlike filament.

The exposed parts of the scales on the body, including the head and shoulders, are rough with minute sharp spines closely crowded together. The jaws are armed with several bands of small recurved teeth. The vent is situated a considerable distance in front of the point of origin of the anal fin, and the skin immediately surrounding it is scaleless and black.

Color.—The many we have seen have been uniform gray above and below. Also described as light brownish gray above, silvery below, with dark bluish or blackish belly. The lower surface of the snout is pink, the throat is deep violet, the first dorsal is pink with blackish spines, and the eyes are dark blue.

Size.—Usually about 1 foot long. The largest we have seen was 16 inches long.

Habits.—Grenadiers are bottom fish, usually found on soft mud, and they are very feeble swimmers. They usually live in at least 80 to 90 fathoms of water, and down to 1,000 to 1,200 fathoms (deepest record 1,255 fathoms). But one was trawled in 9 fathoms in Vineyard Sound by the *Fish Hawk* many years ago; a second was found floating near the surface at Eastport, Maine, by Dr. W. C. Kendall; and a third was taken in a weir at Lubec, Maine, as reported by Huntsman. Hansen [2] reports pelagic euphausiid shrimps (*Thysanoessa longicaudata*) in a grenadier stomach, while several examined by us from 100 fathoms on the edge of Georges Bank contained amphipods chiefly, together with an occasional worm and euphausiid shrimp.

It is probable that grenadiers spawn in summer and autumn, for the spermaries of a specimen taken in the western basin of the Gulf on August 19 were nearly ripe, while a fully ripe male has been reported from South Channel in the last week of September. The eggs of this fish have not been seen, but it is probable that they resemble other macrourid eggs described by European authors[3] in being buoyant at least for the first part of the period of incubation, with a large oil globule, wide perivitelline space, and with the surface sculptured into concave hexagonal facets. The larvae have not been seen yet. Those of other species of grenadiers have the rays of the first dorsal and ventral fins greatly prolonged.

General range.—This (normally) deep-water fish has been found at many localities along the continental slope from the West Indies northward and eastward to the Grand Banks of Newfoundland,[4] and rarely in the Gulf of St. Lawrence. It is also known from the mouth of the Laurentian Channel, on the Scotian Banks, in the Gulf of Maine, and even in Vineyard Sound. It has also been reported from the Azores.

Occurrence in the Gulf of Maine.—The common grenadier was formerly regarded as a rare stray in the inner parts of the Gulf of Maine for only two had been recorded there aside from the Eastport and Lubec specimens mentioned above, the

[2] Proc. U. S. Nat. Mus., vol. 48, 1915, p. 99.

[3] Ehrenbaum (Nordisches Plankton, vol. I, 1905–1909) summarizes what little is known of the eggs and young of this group of fishes.

[4] Rept. Newfoundland Fish. Res. Comm., (1933) 1934, p. 116.

one from the western basin in 160 fathoms, the other from off Gloucester, both of them taken many years ago. But they must be rather common on the muddy bottoms of the deeper parts of the Gulf in 85 to 125 fathoms, for we have caught more than 100 of them at various localities on recent trawling trips. No doubt it is because few vessels ever fish on these grounds, which are not productive either of cod or of haddock, that the presence of grenadiers there has been overlooked. A grenadier, too, was reported from the slope of Jeffreys Ledge, in about 50 fathoms, during March 1934.

Grenadiers, together with the long-finned hake (p. 232), are the most abundant fish on the continental slope abreast of the Gulf below 100 fathoms.[5]

Rough headed grenadier *Macrourus berglax* Lacépède 1802 [6]

RAT-TAIL; ONION-EYE

Jordan and Evermann, 1896–1900, p. 2583.

Description.—This species resembles the common grenadier (p. 243) so closely in general appearance that we need only indicate the points of difference. Most obvious of these are that its snout is shorter and blunter, with more highly arched dorsal profile; that it has 4, 5, or 6 distinct ridges on the top of its head; that its head is relatively larger (about one-fourth to one-fifth the length of the fish, only one-sixth to one-seventh in the common grenadier); that its trunk is relatively stouter (about six times as long as it is deep); that its vent is close to the point of origin of the anal fin with the skin scaly around it, and no darker colored than on the back; and that the serrations on the large spine in the first dorsal fin are so fine that they are hardly visible.

Furthermore, there are fewer (about 124) rays in the second dorsal fin, but more rays (about 148) in the anal than in the common grenadier, and its first dorsal fin is of rather different outline. The second dorsal fin, too, is relatively higher than in the common grenadier and with its membrane more developed (compare fig. 120 with fig. 119), while the filamentous prolongation of the outer ray of the ventral fins is not so long in *berglax* as it is in *bairdii*. The structure of the scales, too (visible to the naked eye), is diagnostic, for those on the head and shoulders of *berglax* are armed with either one longitudinal row of spines (10–12 rows of spines on each scale), or with up to 3 or 4 radiating ridges of spines while those farther back each have a single row of spines, which together form conspicuous longitudinal ridges along each side of the rear part of the body.

Color.—The only newly caught specimens we have seen were ash gray below as well as above; with the chest a little darker; with the rear edges of the scales on the rear part of the body still darker; with the anal fin narrowly dark edged; with the first dorsal fin and the pectoral fins sooty; and with the ventral fins sooty, except that the outermost rays are white after preservation in alcohol.

Size.—This fish is larger than the other grenadiers (p. 243). It is credited with a maximum

[5] For a list of captures on the continental slope during the early cruises of the U. S. Fish Commission vessels, see Goode and Bean (Smithsonian Contrib. Knowl., vol. 30, 1895, p. 394).

[6] Authorities disagree as to the correct scientific name of this grenadier, for while it has usually been referred to as *berglax* Lacépède, 1802, Jensen (Spolia Zool., Mus. Hauniensis, Copenhagen, vol. 9, 1948, p. 178) prefers the species-name *fabricii* that was applied to it more recently by Sundevall, 1840, on the ground that the grenadier to which Lacépède gave the name *berglax* was another fish, *Coryphaenoides rupestris* Gunnerus, 1765, which is commonly termed "berglax" on the west coast of Norway.

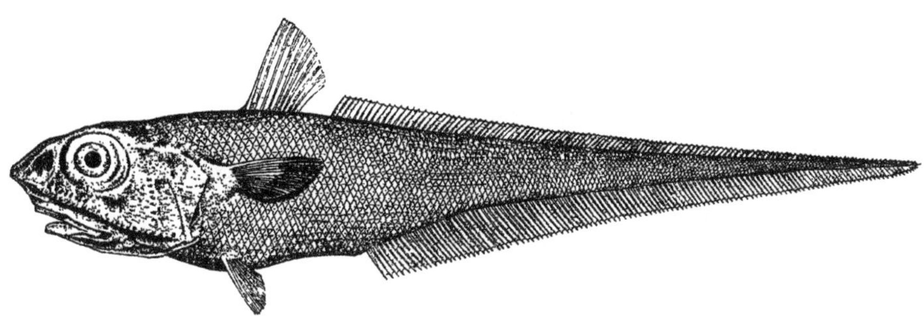

FIGURE 120.—Rough-headed grenadier (*Macrourus berglax*), Banquereau Bank. From Goode and Bean. Drawing by H. L. Todd.

length of 3 feet and a weight of 4 or 5 pounds but the largest we have seen is only 29 inches long.

General range.—This is a deep-water fish like its relative, but is more northerly in its distribution, being known off northern Norway, Spitzbergen, Iceland, southern Greenland, in Davis Strait, and southward along the continental slope of North America as far as Georges Bank. One has even been found floating dead on the surface, off New York Harbor, but it may have been thrown overboard from a fishing boat returning from the off shore banks.

Occurrence in the Gulf of Maine.—Three quarters of a century ago, when halibut were more plentiful in the Gulf of Maine than they are today, and when vessels, long-lining from Gloucester, still resorted regularly to the deep channel between Georges Bank and Browns Bank as well as to the deep gullies that interrupt the Nova Scotian banks, large grenadiers were often hooked. Fishermen described them as common enough to be a nuisance, for they stole the baits meant for other fish and were of no commercial value themselves. It was on the strength of such reports that Goode[7] characterized them as "exceedingly abundant on all of our offshore banks." A few were brought in "from off the coast of New England."[8] And our re-examination of three specimens, one taken on the outer edge of either La Have Bank or Banquereau in 1878 a second taken "off New England" in 1880, the third (probably from the Grand Banks) obtained in Boston Market by Prof. G. H. Parker in 1903 [9] has proved that earlier identifications of them as *berglax* were correct.

We have not heard of one, either from Nova Scotian waters or from the Eastern Channel since 1903;[10] not because they have vanished thence, but simply because very little long-line fishing is now done deep enough off our coasts. And there is always the chance that some vessel, fishing down the slopes of Sable Island Bank, La Have Bank, or southeastern Georges, may pick a few rough headed grenadiers at any time when least expected.

One hundred fathoms may be set as about their upper limit; most of those caught have been from 100 to 300 fathoms on both sides of the Atlantic; and they have been taken as deep as 677 fathoms by the *Albatross* off the southeast slope of Georges Bank. They are supposed to feed on small fish and on Crustacea but we find no definite record of the contents of their stomachs. Females with the roe nearly ripe have been taken off northern Norway in May, suggesting that this is a spring spawner, but nothing definite is known of its breeding habits.

Long-nosed grenadier *Coelorhynchus carminatus* (Goode) 1880

Jordan and Evermann, 1896–1900, p. 2588.

Description.—This species resembles the common grenadier (p. 243) so closely in its general appearance that there is danger of mistaking it for the latter; but it is identifiable by the facts that its dorsal spine is perfectly smooth and that its first dorsal fin is rounded instead of triangular; and that its snout not only overhangs the mouth slightly farther, but is thinner tipped.[11]

[7] Fish. Ind. U. S., Sect. 1, 1884, p. 244.

[8] Bean, Proc. U. S. Nat. Mus., vol. 3, 1881, p. 80.

[9] These three specimens, the largest 29 inches long, are in the Museum of Comparative Zoology.

[10] The most recent record with which we are acquainted is of one 16 inches long that we trawled on the southeastern slope of Georges Bank, at 500 fathoms, June 1949, on *Caryn* of the Woods Hole Oceanographic Institution.

[11] Commonly described as "sturgeon-like," but this characterization applies better to other members of the genus which have still longer snouts.

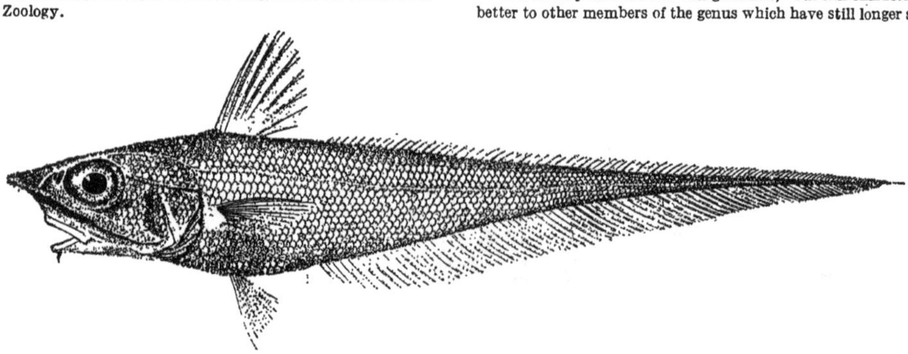

FIGURE 121.—Long-nosed grenadier (*Coelorhynchus carminatus*), continental slope off Marthas Vineyard. From Goode and Bean. Drawing by H. L. Todd.

Color.—Described as silvery gray.

Size.—About 10 inches long.

General range and occurrence in the Gulf of Maine.—This deep-water ground fish has been taken at many localities off the American coast from the West Indies and Gulf of Mexico northward along the continental slope to abreast of southern Nova Scotia, in depths of 104 to 464 fathoms. It is included here because it has been recorded once off Nantucket in 148 fathoms.

THE OPAHS. FAMILY LAMPRIDAE

For the characteristics of this family, see the following description of its unique representative, the opah.

Opah *Lampris regius* (Bonnaterre) 1788

MOONFISH; JERUSALEM HADDOCK

Jordan and Evermann, 1896–1900, p. 954, as *Lampris luna* (Gmelin) 1789.

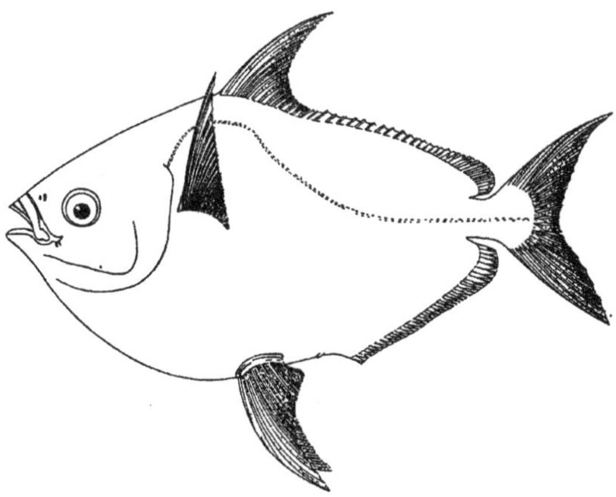

FIGURE 122.—Opah (*Lampris regius*). After Goode and Bean.

Description.—The thin, deep form of the opah (trunk less than twice as long as it is deep) with moderately slender caudal peduncle, which does not have longitudinal keels, and the rather pointed snout, might suggest an enormous butterfish, were it not provided with very long falcate ventral fins, whereas the butterfish has no ventrals. The ventrals, also, of the opah have 14 to 17 rays; none of the mackerel or pompano tribes has more than 8.

The forward part of the single dorsal fin (53 to 55 rays) is high, its outlines strongly falcate. The anal (38–41 rays) is shorter than the dorsal, and it is about equal in height to the low part of the dorsal throughout its length. Both anal and dorsal fins extend back close to the base of the caudal fin, and each of them is depressible in a groove. The tail fin is emarginate, the pectorals are conspicuously pointed, with their bases horizontal instead of vertical. The mouth is small and toothless, the scales are minute, and the lateral line is strongly arched upward above the pectoral fin, then downward toward the rear.

Color.—We have never seen this fish alive, but it is described as of a beautiful dark steel blue above, shading into green with silver, purple, gold, and lilac luster down the sides, and as rosy on the belly, with vermilion fins, while the whole body is speckled with silvery and milk-white spots.

Size.—The opah grows to a length of 3 to 6 feet; most of them are 3 to 4 feet long.

Habits.—The opah is usually spoken of as a deep-sea fish, but this is a misnomer, for it is caught on hook and line no deeper than 50 to 100 fathoms off Madeira, where it is taken in some numbers. Being so very rare off our coast, we need merely note that it feeds chiefly on squid, isopods, and small fish, as well as on seaweeds; that it is an excellent food fish; and that nothing is known of its breeding habits.

General range.—Open waters of the Atlantic and Pacific Oceans; recorded off Madeira, Scandinavia, the British Isles, Norway, Iceland, Newfoundland, Nova Scotia, Maine, Cape Cod, and Cuba in the North Atlantic; also in the Gulf of Mexico off the west coast of Florida.

Occurrence in the Gulf of Maine.—Two specimens, only, of this oceanic wanderer have been reported definitely within the limits of our Gulf, one caught on a long line on Browns Bank in the spring of 1932,[12] the other, weighing 165 pounds fresh, was taken in an otter trawl on the northeastern part of Georges Bank, in August 1947.[13]

One also was reported from Maine by Goode and Bean,[14] but this may have been based on a letter to D. S. Jordan from Everett Smith, July 19, 1888, reporting that a "Sun Fish," identified

[12] Reported by Vladykov, Proc. Nova Scotia Inst. Sci., vol. 19, 1935, p. 6.

[13] This specimen is in the Museum of Comparative Zoology.

[14] Goode and Bean, Smithsonian Contrib. Knowl., vol. 30, 1895, p. 223.

as an opah by the included description and sketch, had been landed in Portland, Maine, from the Grand Banks.[15]

One was caught off Sable Island, Nova Scotia, about 1856, and another off La Have Bank many years ago;[16] a specimen about 3 feet long was taken in July 1925 on Western Bank, southwest of Sable Island, by the schooner *Falmouth;*[17] and another of the same size stranded on the beach at Hyannis, Mass., on September 17, 1928.[18]

THE FLOUNDERS AND SOLES. FAMILIES HIPPOGLOSSIDAE, PARALICHTHYIDAE, PLEURONECTIDAE, BOTHIDAE, AND ACHIRIDAE

The flatfishes are a very homogenous tribe, so different from all other fishes that no one is likely to mistake any one of them for any other sort of fish. What strikes one first is their flatness; less obvious is the fact that they do not lie on the belly but on one side, right or left. And their skull twists in the course of development so that the eye which was originally on the side that is fated to be underneath, migrates around the head, until both the eyes finally come to lie close together, on the side that is uppermost as the fish lies on bottom. But the mouth retains its original position more nearly, so that it is often described as opening sidewise. The larval flounder swims on edge like any other fish; the migration of the eye takes place shortly before the fry take to the bottom.

All of the flatfishes have a single long fin on each edge, one the dorsal and the other the anal; they also have well-developed ventral fins (at least on the eyed side) which are either on the right-hand edge or on the left-hand edge as the fish lies. Most of the Gulf of Maine species also have pectoral fins, one on the upper side as the fish lies on the bottom, the other on the lower side. The ventral fins are in front of the pectorals or in line with them; the abdominal cavity is very short, and some species are armed with a stout anal spine.

Our several flatfishes look much alike; indeed, they are often confused. But it is not difficult to tell one from another, for the distinctive characters are rather precise, even if not obvious at first glance. Huntsman[19] has published a very useful key to the eastern Canadian species, which is expanded here to cover the Gulf of Maine.

KEY TO GULF OF MAINE FLATFISHES

1. Eyes on the left-hand side, and guts at left-hand edge, as the fish lies on bottom 2
   Eyes on the right-hand side, and guts at right-hand edge 5
2. The lateral line is straight .... *Citharichthys arctifrons*, p. 294
   The lateral line is arched over the pectoral fin[20] 3
3. The two ventrals are not alike, the left (upper) being continuous with the anal fin, the right (lower) separate from it .... Sand flounder, p. 290
   The two ventral fins are alike 4
4. The upper side is marked with four large oblong black eye-spots; there are fewer than 82 rays in the long right-hand (dorsal) fin .... Four-spotted flounder, p. 270
   The upper side is marked with many small spots; there are more than 84 rays in the long right-hand (dorsal) fin Summer flounder, p. 267
5. There is a well-developed pectoral fin on the eyed side 6
   There are no pectoral fins .... Hog choker, p. 296
6. Mouth large, gaping back as far as the eye; jaws and teeth nearly equally developed on both sides 7
   Mouth small, not gaping back as far as the eye; the jaws are nearly straight on the upper side, but curved on the lower side 9
7. Margin of tail fin rounded .... American dab, or plaice, p. 259
   Margin of tail fin slightly concave, with angular corners 8
8. Lateral line arched close behind the gill opening .... Halibut, p. 249
   Lateral line nearly straight .... Greenland halibut, p. 258

[15] We are indebted to Norman J. Wilimovsky for showing us a copy of this letter.

[16] Vladykov, Proc. Nova Scotia Inst. Sci., vol. 19, 1935, p. 6.

[17] Radcliffe, Copeia, No. 151, 1926, p. 112.

[18] Reported by Robert Goffin of the Bureau of Fisheries station at Woods Hole, Mass.

[19] Our Eastern Flat Fishes, Canadian Fisherman, vol. 5, No. 6, 1918, pp. 788–790.

[20] In all the flounders of this type so far recorded from the Gulf of Maine both of the pectoral fins are well developed. Should one be taken with no pectoral fin on the blind side it would probably be the deep-water *Monolene sessilicauda*.

9. Lower side of head with large open mucous pits; 100 or more rays in the long left-hand (dorsal) fin--Witch, p. 285
   Lower side of head lacks open mucous pits; fewer than 90 rays in the long left-hand (dorsal) fin------------- 10
10. Lateral line arched behind the gill opening--Yellow-tail, p. 271
   Lateral line nearly straight-- 11
11. Top of the head between the eyes rough with scales--Winter flounder (including the Georges Bank flounder) p. 276
   Top of the head between the eyes naked and smooth----------------------------------Smooth flounder, p. 283

Atlantic halibut *Hippoglossus hippoglossus* (Linnaeus) 1758

Jordan and Evermann, 1896–1900, p. 2661.

Description.—This is not only the largest of flatfishes, but is one of the best characterized; its most obvious distinctive characters, apart from its size, being the fact that it lies on the left side;[21] that its mouth gapes back as far as the eyes, and is armed with sharp curved teeth; that the rear edge of its tail fin is concave, not rounded; that its two ventral fins are alike; and that its lateral line is arched abreast of the pectoral fin. Furthermore it is a narrower fish, relatively, than most of our flatfishes (only about one-third as broad as it is long) but is very thick through, and its eyes are farther apart than they are in most of the other flounders.

The dorsal (long) fin (98 to 105 rays) commences abreast of the eye and runs back the whole length of the fish, broadening but slightly for the first third of its length and then abruptly, to narrow again toward the caudal peduncle. The anal fin is similar to the dorsal fin in shape but is shorter (73 to 79 rays), originates close behind the pectorals, and is preceded by a sharp spinelike extension of the post-abdominal bone, which projects in young fish but is hidden by the skin in old fish. The two pectoral fins are of different shapes, the one on the upper (eyed) side of the fish being obliquely pointed while the fin on the lower side is rounded. The rather small ventral fins, which are situated in front of the pectorals and are separated from the anal by a considerable space, are alike. Halibut, like other flatfishes are scaly on the whole head and body and they are very slimy with mucus.

[21] Left-handed halibut have been caught, occasionally.

Color.—The halibut is chocolate to olive or slaty brown on the eyed (upper) side. Young fish are paler, and are more or less mottled, while large ones are more uniform and darker, sometimes almost black. The blind (lower) side usually is pure white in small fish, but large ones are often more or less blotched or clouded below with gray (known by fishermen as "grays"). Occasionally a halibut is taken the blind side of which is marked with patches of the same color as the eyed side. And we have seen one medium-sized fish in which the rear third of the lower surface was uniform dark brown.

Size.—Only swordfish, tuna, and some of the larger sharks reach a greater size than the halibut, among Gulf of Maine fishes for while reports of specimens as large as 600 to 700 pounds have usually been looked on as exaggerations we are glad to be able to give at least one record of a Gulf of Maine halibut in this weight class. The

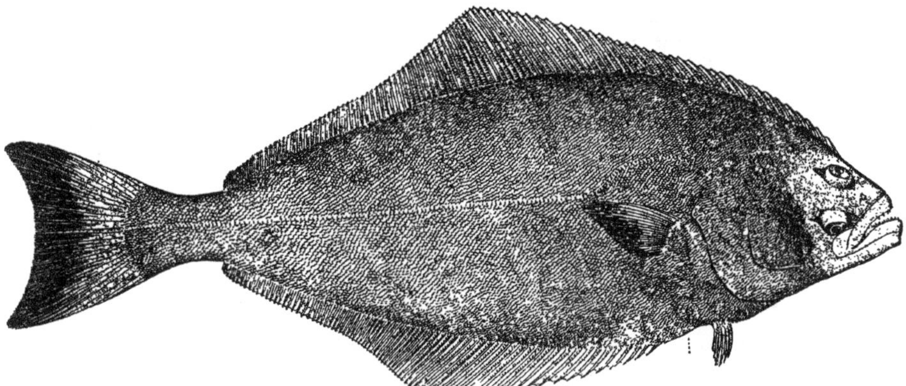

FIGURE 123.—Halibut (*Hippoglossus hippoglossus*), Eastport, Maine. From Goode. Drawing by H. L. Todd.

fish in question was taken in June 1917, by Capt. A. S. Ree, about 50 miles eastnortheast of Cape Ann, and since it weighed 615 pounds, eviscerated with the head still attached, when brought in to the Boston fish pier, it must have been as heavy as 700 pounds while alive.[22] Another halibut of 602 pounds is said to have been taken near Isle au Haut in 1902, but we cannot vouch for this one.

Halibut of 500 to 600 pounds are rumored almost every year, but the next largest of which we have definite knowledge was one of about 450 pounds caught on a hand line in the deep water between Browns and Georges Banks in 1908 by W. F. Clapp. Goode [23] likewise had records of a dozen fish of 350 to 400 pounds caught off the New England coast; the heaviest was one of 401 pounds taken near Race Point, Cape Cod, in July 1849. But a 410-pound halibut that was brought in to the Boston fish pier by the *Dawn*, March 27, 1941,[24] was spoken of as the largest that had been landed there in a "score of years," and it seems that halibut heavier than 300 pounds always were rarities anywhere in the North Atlantic.

Full-grown females average about 100 to 150 pounds. Males run smaller, and most of the "large" fish landed in New England ports weigh from 50 to 200 pounds. The largest we have caught, taken on Browns Bank, weighed exactly 100 pounds and was 5 feet long. Halibut between 7 and 8 feet long usually weigh 300 to 350 pounds, and the following table based on Icelandic fish measured by Jesperson,[25] and others from the Gulf of Maine, give the relationship of length to weight for the smaller sizes.

Iceland		Gulf of Maine	
Length in inches	Weight in pounds	Length in inches	Weight in pounds
74	215	63	[1] 120
70	168	42½	30
61	107	42	33
54 to 56	60½	41½	27½
40 to 42	29	31	12
36	11 to 12	20	2¾
30	9½	--	---
27	6½	--	---
24	5½	--	..

[1] This fish weighed 98 pounds dressed, the intestines accounting for 15 pounds and the ovaries (with immature eggs) for 7 pounds.

Habits.—The halibut, like all the flatfish tribe, is normally a ground fish, once the young fry have taken to bottom. But it comes to the surface on occasion (p. 257), and it is a very powerful fish, when hooked. Halibut caught in shallow water are very active, usually starting off at great speed when they are hauled up from the bottom, often spinning the dory around in their attempts to escape.[26] They are usually found on sand, gravel, or clay, not on soft mud or on rock bottom; 400–500 fathoms may be set as the lower boundary to their existence in any numbers,[27] but their absolute depth limit is not known.

The young halibut, like the young of so many other ground fishes, drift helplessly with the current for some months after hatching (just how long is not known); not at the surface, however, but in the mid-depths (p. 253). During this period they tend both to rise in the water as they grow, and to be carried inshore, so that when they finally take to the bottom they do so in quite shallow water (p. 254). But the fry as a whole tend to work offshore again thereafter, and deeper, so that halibut caught in deep water are larger than those caught in shallow water. This fact was noticed early on Georges Bank, where most of the fish taken on the bank in depths of 30 to 40 fathoms or less ran from 125 to 180 pounds, whereas much larger ones were caught on the deeper slope to the southeast. Fishermen have also reported catching smaller fish on the inner ends of long lines set from shoal water out into deep, and larger fish on the outer ends.[28] And this rule holds equally for the other side of the Atlantic.

The halibut is a boreal, not an arctic fish, in its relationship to temperature. Thus, large catches are (and were) made only at times and places where the water is at least as warm as 36°–38° F. (about 3° C.). In the Grand Banks region, for instance, halibut are mostly caught either far enough down the slope to be below the icy touch of the Labrador Current, or at times and places where the latter does not reach bottom, if the fish are on the bank. But the lower limit to the temperature range of the halibut is not sharp-cut. We do, in fact, find record of at least one halibut

[22] An account of this fish was published in the Boston Globe, June 12, 1917. It was bought by the Shore Fish Co.

[23] Fish. Ind. U. S., Sect. 1, 1884, p. 194.

[24] Reported in the Boston Herald, March 28, 1941.

[25] Meddelel. Kommis. Havundersøgelser, Ser-Fiskeri, vol. 5, No. 5, 1917.

[26] Goode and Collins (Fish. Ind. U. S., Sect. 5, vol. 1, 1887, p. 17) give a readable account.

[27] Capt. Baldersheim (Rapp. Cons. Internat. Explor. Mer, vol. 56, 1929, p. 25) reports good catches at that depth in Davis Strait off west Greenland.

[28] Goode, Fish. Ind. U. S., Sect. 1, 1884, p. 195.

trawled on the southern part of the Newfoundland Bank in bottom water as cold as 33° F. (0.6° C.), while others reported from Bay Bulls, on the east coast of Newfoundland and from Cut-throat Harbor on the outer coast of Labrador (p. 254) doubtless were in water equally cold.

At the opposite extreme, it has been found that only a few halibut are taken in the parts of the North Sea where the bottom water is warmer than 46°–47° (8° C.), none at all where it is warmer than 59° F. (15° C.). And there is no reason to suppose that halibut ever were plentiful anywhere in the western side of the Atlantic in temperatures much higher than about 46°–47°, for while the bottom water warms locally to 50°–52° on Georges Bank in summer, and to 52°–59° on Nantucket Shoals, it was only during the winter and spring that there ever was any regular fishing for halibut on either of these grounds.

On the other hand, the halibut that summer on banks where the bottom chills below about 36° in winter have been described repeatedly as withdrawing to deeper (i. e., to warmer) water for the coldest part of the year. Perhaps the best known example is off west Greenland. Here the halibut work in over the banks regularly in summer, from the deeper waters of Davis Strait, as the temperature rises, but work out again, and deeper, in autumn, as the water cools again.[29] Thus it was only deeper than 350 fathoms that long liners, fishing there in 1926–28 [30] found halibut in paying quantities at the beginning of June, when the bottom temperature on the banks was about 33°–37°. But good catches were made as shoal as 200 fathoms by the middle of the month when the temperature had risen to 35°–38°. And there was good fishing as shoal as 70 fathoms by mid-July, when the banks had warmed to 37°–39°, though many of the halibut were in deeper water still.

Halibut have been described as shifting ground in the same way in the coastal belt of the Gulf of Maine (p. 257) from season to season. On the other hand, we suspect that halibut finding themselves in water shoaler than 30 fathoms or so in the southernmost part of the range of the species, on the American side, at the onset of summer may withdraw to slightly deeper water for the time being, but definite information is lacking.

The seasonal movement of halibut in onto the Greenland Banks as early in the summer as temperature allows seems to be in search of food, as Jensen points out, for a much richer supply of small fish is available to them on these shoaler bottoms than deeper down the Davis Strait slope, where they must depend chiefly on large shrimps (p. 252). And we suspect that the food supply is equally important in influencing the seasonal movements of halibut in our Gulf.[31]

If the prevalent view is correct, the Atlantic halibut resort to rather definite and circumscribed ground to spawn, much as the Pacific halibut do.

Halibut have also been credited with extensive wanderings from bank to bank, for no evident reason. And recent tagging experiments carried out off Nova Scotia by the Fisheries Research Board of Canada,[32] have proved that some of them certainly do so, in American waters. Thus fish that were marked on German and Browns Banks have been recaught as far to the eastward as Western Bank and in the general vicinity of Sable Island, while one that was tagged at Anticosti was recaught at Seven Islands more than 100 miles to the westward. But most of the recaptures were made within a few miles of the places where the fish had been tagged. And available evidence as to halibut migrations in the Gulf of Maine and in Nova Scotian waters is so contradictory, and so greatly complicated by the local effects of hard fishing, that it is not worth while to attempt any further discussion here.

Food.—The halibut is very voracious, preying chiefly on other fishes, a long list of which have been reported from their stomachs, including cod, cusk, haddock, rosefish, sculpins, grenadiers, silver hake, herring, launce on which they often gorge in northern seas,[33] capelin, flounders of various sorts (these seem to be their main dependence), skates, wolffish, and mackerel. Halibut are also known to eat crabs, lobsters, clams, and mussels;

[29] Jensen (Meddelelser, Dansk Komm. Havunders., Ser. Fiskeri vol. 7, No. 7, 1925, pp. 17–18) seems to have been the first to bring this to scientific attention.

[30] Baldersheim, Rapp. Proc. Verb. Conseil Internat. Explor. Mer, vol. 56, 1929, pp. 25–28.

[31] For a further discussion of the range and movements of the halibut in relation to temperature, with references, see Thompson and VanCleve, Rept. Internat. Fish. Comm. No. 9, 1936, pp. 22–38.

[32] Martin and McCracken, Fish. Res. Board Canada, Progress Rept., Atlantic Coast Station, No. 50, 1950, pp. 3–8.

[33] Capt. Baldersheim described halibut off west Greenland as sometimes in schools, preying on launce (Rapp. Proc. Verb. Conseil Internat. Explor. Mer., vol. 56, 1929, p. 25).

even sea birds have been found in them.[34] Fishermen have reported finding in halibut the heads and backbones of cod thrown overboard, and a variety of indigestible objects such as pieces of wood or iron, and even fragments of drift ice.

The diet of the halibut in any particular locality depends chiefly on what other ground fish are most easily available. Thus they are reported as feeding chiefly on flatfish on Georges Bank, but on cod, haddock, cusk, and sculpins on other grounds.

Halibut, like other flounders, must be nearly invisible as they lie on bottom, capturing any fish that passes within reach by a sudden rush. On one occasion a halibut of about 70 pounds was seen at the surface trying to kill a small cod with blows of its tail. "We hove out a dory and two men went in her taking with them a pair of gaff hooks. They soon returned bringing not only the halibut but the cod." [35] And halibut are very destructive to smaller fish. We read, indeed, of half a bushel of flatfish taken from one halibut. And fishermen said the appearance of a school of halibut soon drove away the cod and haddock, in the days when halibut were still plentiful on the shoaler banks.

It appears that halibut do not eat many invertebrates at least in the Gulf of Maine, or in Nova Scotian waters. But a case is on record when 6 lobsters, 6 inches long, were found in the stomach of one. And Jensen found that halibut caught in deep water off west Greenland had fed chiefly on large shrimps (*Pandalus borealis*).[36]

According to fishermen who have watched them in clear shallow water, "The halibut will advance to the bait . . . then retreat 4 or 5 feet from it . . . after repeating this performance several times—generally three or four—the fish seems to make up its mind to eat the bait, and, suddenly darting toward it, swallows it down at a gulp." [37]

Halibut, in their own turn, fall prey to seals, and especially to the Greenland shark, for which they are a staple article of diet.

Large halibut are very prolific, the ovaries of an Atlantic female of about 200 pounds having been estimated as containing 2,182,773 eggs, while a female of the Pacific form of "140 pounds may have as many as 2,700,000." [38]

Very little is known about the breeding of the Atlantic halibut. In the eastern Atlantic halibut spawn chiefly in March, April, and May with the chief production of eggs in April, while a few females may ripen as early as the end of January, and some not until June.[39] Off west Greenland they spawn late in spring.[40] Off the American coast it seems that the spawning season continues through the summer, for fishermen have reported ripe fish, both male and female, in April, May, June, July, August, and early September at various localities from Georges Bank to the Grand Banks; [41] while the report that part of the eggs in the ovaries of a fish examined on Banquereau by representatives of the Bureau of Fisheries on September 13, 1878, were ripe, but others immature, is evidence (if correct) that individual halibut may spawn over a considerable period.

Presumably they spawn on bottom, like other flat-fishes, but definite information is lacking. The Pacific halibut is known to spawn at depths of 150 fathoms to about 225 fathoms: [42] and European students, generally, have believed that the Atlantic fish spawns deeper still, perhaps even outside the 400-500 fathom line; evidence is that naturally spawned eggs have been taken only where the depth was greater than about 550 fathoms (1000 meters), the drifting larvae less than 19 mm. long only over depths greater than about 220 fathoms (400 meters).[43] On the other hand, halibut spawn regularly in the aquarium at Trondhjem, Norway, where the eggs have been fertilized artificially and hatched successfully.[44] This, with fishermen reports of ripe fish, both females and males, on the slopes of all the offshore Banks east of Cape Cod and, with Cox's report of

[34] Smitt (Scandinavian Fishes, vol. 1, 1892, p. 414) speaks of a halibut that had eaten a razor-billed auk; Good and Collins (Fish. Ind. U. S., Sect. 5, vol. 1, 1887, p. 35) record an "ice bird" (probably a dovekie) as taken from a halibut caught on Georges Bank; and Scudder (Fish. Ind. U. S., Sect. 5, vol. 1, 1887, p. 119) reports finding the skeleton of a gull in the stomach of another.

[35] Goode, Fish. Ind. U. S., Sect. 1, 1884, p. 196.

[36] Meddelelser, Dansk. Komm. Havunders., Ser. Fisk., vol. 7, No. 7, 1925, p. 18.

[37] Goode and Collins, Fish. Ind. U. S., Sec. 5, vol. 1, 1887, p.. 17.

[38] Clemens and Wilby, Bull. No. 68, Fisheries Res. Board Canada, 1946, p. 312.

[39] For a general survey of available information, see Tåning, Meddelelser Komm. Danmarks Fisk. Havunders., Ser. Fisk., vol. 10, No. 4, 1939, p. 14.

[40] A larva 14.75 mm. long taken on June 19 and another 21 mm. long taken on July 7 is recorded by Jensen (Rapp. Proc. Verb. Cons. Intern. Explor. Mer, vol. 39, p. 96, 1926).

[41] Goode (Fish and Fishery Industr. U. S., Sec. 1, 1884, pp. 196-197), mentions reports to this effect.

[42] Clemens and Wilby, Bull. 68, Fish. Res. Bd. Canada, 1946, p. 312.

[43] Tåning, Meddel. Komm. Danmarks Fisk. Havunders., Ser. Fiskeri, vol. 10, No. 4, 1936, p. 8.

[44] Rollefsen, Kgl. Norske Vidensk. Selsk. Trondhjem, Forhand., vol. 7, No. 7, 1934.

two larval halibut, 20 and 21.5 mm. long, taken close in to the southern coast of Nova Scotia in shoal water,[45] suggests that the American fish may spawn at least as shoal as the Pacific fish does, and perhaps even shoaler.

The eggs are buoyant, drifting suspended in the water at depths greater than 30 to 50 fathoms, not at the surface. Usually they are 3 to 3.8 mm. in diameter, and they do not have any oil globule.[46]

The only other buoyant fish eggs equally large that are likely to be found in the Gulf of Maine are those of the Argentine (p. 140), but these have a large oil globule, so there is no danger of mistaking them for halibut eggs. The buoyant eggs of the Greenland halibut (p. 258) are larger still.[47]

In the Trondhjem aquarium the incubation of artificially fertilized eggs occupied 16 days at a temperature of about 43° (6° C.). The larvae were 6.5 to 7 mm. long at hatching, with very large yolk sac and no pigment, growing to about 8.5 mm. by the sixth day, and developing pigment by the 10th day.[48]

The smallest naturally hatched Atlantic halibut yet seen [49] was 13.5 mm. long, with the vertical fin rays appearing. The dorsal and anal fins are developed and the ventral fins are visible at about 22 mm. (fig. 125), by which time the left eye has moved upward until its margin is just visible above the contour of the head, forecasting that the fish is to be a right-handed flatfish. Fish of this size also show the large mouth characteristic of the species. Up to this stage there is little pigment. About one-fourth of the eye appears above the profile when the little halibut is about 27 mm. long, but even at 34 mm. (the largest pelagic stage yet found) the eye has not entirely completed its migration (fig. 126), though the pigmentation is stronger on the right side than on the left, and the caudal fin (previously rounded) has become square tipped.

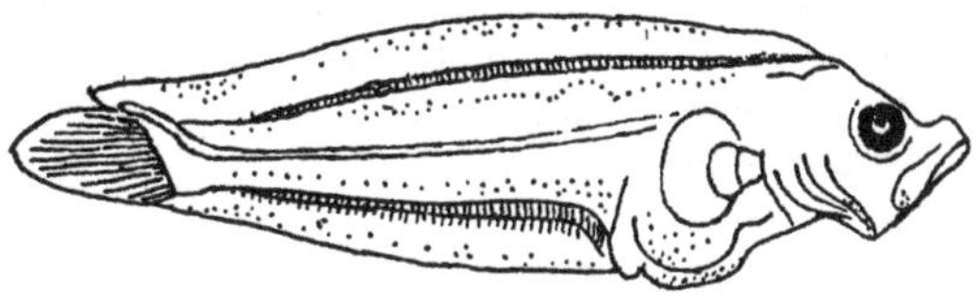

FIGURE 124.—Larva, 16.2 mm. (European). After Schmidt.

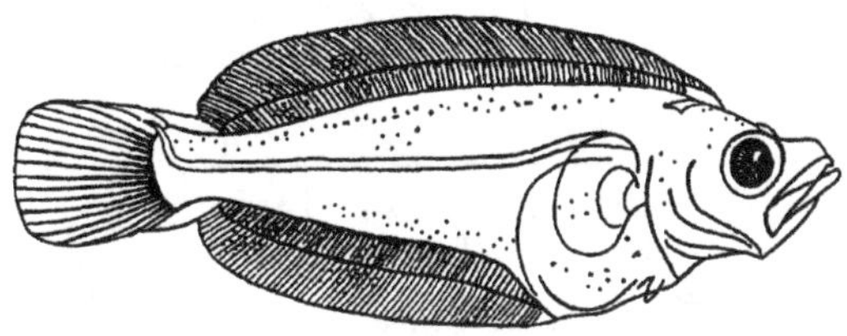

FIGURE 125.—Larva, 22 mm. (European). After Schmidt.

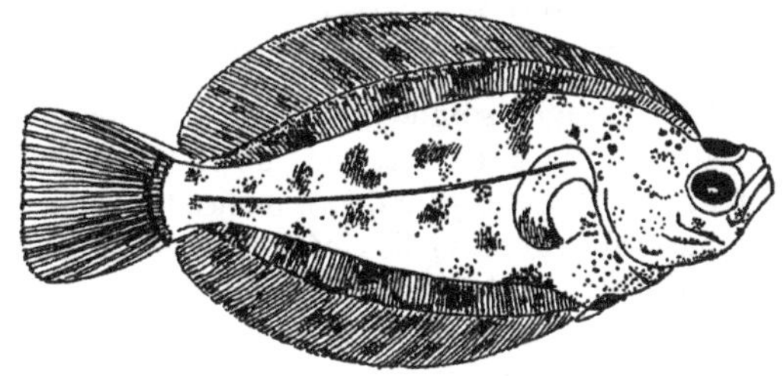

FIGURE 126.—Larva, 34 mm. (European). After Schmidt.

HALIBUT (*Hippoglossus hippoglossus*).

The younger larvae (up to about 25 mm. in length) are made recognizable as halibut by their curiously upturned snout. Older ones that are large enough to show that they belong to some right-handed large-mouthed flounder are separable from the American dab larvae (the only other common Gulf of Maine flatfish with which they agree in both these respects) by the outlines of the head and abdomen.

The early life history of the Pacific halibut has been worked out especially by Thompson and Van Cleve, who have given an excellent series of illustrations of successive stages from newly hatched larvae to young fry a little more than 1 inch long.[50] How long the young halibut lives adrift at the mercy of the currents, is not known. But the young fry, so small (47–64 mm long) that they had evidently been spawned the preceding spring or

[45] Contrib. Canadian Biol., N. Ser., vol. 1, No. 21, 1924, pp. 409–412.

[46] For description of eggs artificially fertilized in the Trondhjem aquarium, see Rollefsen, Kgl. Norske Vidensk. Selsk. Forh., vol. 7, No. 7, 1934, p. 20–23; for descriptions of naturally spawned eggs taken in tow nets in Icelandic waters, see Tåning, Meddel. Komm. Danmarks Fisk. Havunders. Ser. Fiskeri, vol. 10, No. 4, 1936, p. 5; for description of the eggs and larvae of the closely allied Pacific halibut, see Thompson and Van Cleve, Rept. No. 9, Internat. Fisheries Comm., 1936.

[47] 4 to 4.5 mm. in diameter according to Jensen, Kgl. Dansk Vidensk. Selsk. Skr. Nat., afd. 9, R. 6, 1935, p. 4.

[48] For illustration of these artificially fertilized eggs in incubation stages, and of the larvae hatched from them, see Rollefsen, Kgl. Norske Vidensk. Selsk. Forhand., vol. 7, No. 7, 1934.

[49] What little we know of the early stages of the halibut is due to European students, chiefly to Schmidt (Meddel. Komm. Havunder-søgelser, Ser. Fiskeri, vol. I, No. 3, 1904), to Jespersen (Ibid., vol. 5, No. 5, 1917), and to Tåning, (Meddel. Komm. Danmarks Fisk. Havundersøgleser, Ser. Fiskeri, vol. 10, No. 4, 1936.)

[50] Rept. No. 9, International Fisheries Commission, 1936, figs. 38–49.

summer, have been trawled off Iceland during the last week of July. And the smallest bottom stages have so far been taken only in water shoaler than about 27 fathoms (50 meters), evidence that the larvae of the Atlantic halibut tend both to rise toward the surface, and to drift inshore during their pelagic stage, as is also true of the Pacific halibut.

Fry of 3⅛ to 5⅞ inches (80–150 mm.) such as have been taken in considerable numbers in Icelandic waters in June and July probably are in their second year. The average relationship between age and size is as follows according to Jespersen: [51]

Age	Average length	Extremes of length	Age	Average length	Extremes of length
	Inches	*Inches*		*Inches*	*Inches*
1 year	3.9	3.1 to 5.9	6 years	25.6	20.9 to 34.3
2 years	9.1	7.1 to 12.6	7 years	27.6	21.7 to 40.9
3 years	13	8.3 to 18.9	8 years	29.1	22.8 to 40.6
4 years	18.5	11.8 to 24.4	9 years	33.9	26.8 to 42.1
5 years	22.4	16.1 to 28	10 years	37.4	29.5 to 55.5

Females averaged somewhat longer and heavier than males of the same age, and the fact that the oldest was a fish of 20 years, 68½ inches long, suggests that the immense fish of 400 pounds and more, and upward of 7 feet long, which are occasionally caught, may be half a century old, always assuming about the same rate of growth for the Gulf of Maine halibut as for those that are caught about Iceland.

According to Thompson [52] Pacific halibut grow at approximately the same rate for the first few years, more slowly after about the eighth year, though with wide differences in the rate of growth on different banks, probably caused by differences in the food supply.

It is probable that most of the female halibut do not mature sexually until they are 9 or 10 years old, some not until they are several years older still; males mature when they are somewhat younger.[53]

General Range.—Boreal and subarctic Atlantic, in continental waters.

The most southerly record of a halibut, in the western side of the Atlantic is of a 6-foot fish that was picked up in a pound net near Reedville, Va.[54] Stragglers have been reported off New Jersey and New York, and off Block Island. And halibut are caught in commercial quantities (or once were) from Nantucket Shoals, inner parts of the Gulf of Maine, Georges Bank, and the Nova Scotian Banks northward to the northern part of the Gulf of St. Lawrence (including the west coast of Newfoundland), the southern part of the Grand Bank, Flemish Cap and to the outer edge of the continental shelf off outer Labrador at depths of 70–90 fathoms or more. But stray specimens, only are reported in along the outer coast of Labrador, i. e., in the icy Labrador current.[55] And while the range of the halibut was said by Goode [56] to extend to Cumberland Gulf, we doubt whether there are any halibut in the icy waters along Baffins Land, for the halibut is not known off the Arctic coasts either of Asia or of America. though the Greenland side of Davis Strait supports a regular halibut fishery as far north as Disco Bay.

In the eastern Atlantic, halibut have been reported doubtfully from the Gulf of Cadiz, and definitely from the Bay of Biscay.[57] Small catches are made regularly as far south as the Irish Sea and English Channel;[58] and they are more numerous around Northern Ireland and Scotland and in the northern part of the North Sea; in the Faroe–Shetland Region; around Iceland; along the Norwegian Coast; around Spitzbergen and Bear Island; also in Barent's Sea.

The Pacific halibut, an ally so close that it is hardly to be distinguishable to the untrained eye, is one of the most important food fishes of the northeastern Pacific.

Occurrence in the Gulf of Maine.—The history of the halibut in the Gulf of Maine, like that of the salmon, must be written largely in the past tense, for their numbers have been sadly depleted there by over-fishing. In Colonial days the halibut was a familiar fish and seemingly a very abundant one on the coast of northern New England, but was considered hardly fit for food. Wood [59] for instance, writes "the plenty of better fish makes

[51] Meddelelser fra Komm. Havundersøgelser, Ser. Fiskeri, vol. 5, No. 5, 1917; based on a study of the otoliths of more than 2,000 fish caught around Iceland.

[52] Report. Comm. Fish. British Columbia, (1914) 1915, pp. 76–99.

[53] Females of the Pacific form may mature as young as 8 years, or not until as old as 16 years, with an average of 12; males considerably younger on the average.

[54] Reported by Walford, Copeia, 1946, No. 2, p. 100.

[55] One, about 20½ inches long, was reported to us by R. H. Backus as found dead in the water, in Cutthroat Harbor, August 5, 1950, by the *Blue Dolphin.*

[56] Fish. Ind. U. S., Sect. 1, 1884, p. 190.

[57] Moreau, Hist. Nat. Poissons France, vol. 3, 1881, p. 288.

[58] 6,614 pounds, for example, were landed from the English Channel in 1932. For further details as to landings from the various statistical areas in the eastern Atlantic, see Thompson and Van Cleve, Rept. 9, International Fisheries Comm., 1936, p. 21.

[59] New Englands Prospect, 1634, p. 37.

these of little esteem, except the head and finnes, which stewed or baked is very good; these hallibuts be little set by while basse is in season." They seem to have maintained their numbers there down to the first quarter of the nineteenth century, when contemporary writers described them as extremely numerous in Massachusetts Bay and along Cape Cod, in fact around the whole coast line of the Gulf of Maine. And they were discovered in abundance on Nantucket Shoals, on Georges Bank, on Browns Bank, and on the Seal Island ground as soon as fishing was regularly undertaken offshore.

The cod fishermen of those days looked upon them as a nuisance, seldom worth bringing to market. And "It was the practice of the fishermen when halibut were troublesome to string them on a line and hang them over the stem of the vessel." [60] But a demand for halibut developed in the Boston market sometime between 1820 and 1825, and they have been pursued relentlessly ever since then, first inshore and then farther and farther afield.

The Massachusetts Bay—Cape Cod region yielded large numbers of these great fish during the early years of the fishery. Four men, for instance, are reported as having caught 400 in two days off Marblehead in 1837, while a party of equal size is said to have landed 13,000 pounds off Cape Cod in three weeks. And it was discovered some time prior to 1840 that halibut congregated in winter in the 25–30 fathom gully between the tip of Cape Cod and Stellwagen Bank. However, a shrinkage in the supply had been noticed along shore even before 1839, for we find halibut described in that year (in the Gloucester Telegraph) as "formerly" caught along Cape Cod and in Barnstable Bay. And they had been so nearly fished out in the Massachusetts Bay region by about 1850 that it no longer paid small boats to go there especially for them.

Halibut held out better in the northeastern corner of the Gulf where there was not as ready a market for them as there was in Boston; Perley wrote of them as plentiful enough to be a plague to the local fishermen off Brier Island as recently as 1852. But it was not long thereafter before their numbers were greatly reduced there also.

[60] Goode and Collins, and Fish. Ind. U. S., Sect. 5, vol. 1, 1887, p. 29–30.

The offshore fishery for halibut began about 1830, when cod fishermen brought word to Gloucester of a great abundance of them on Georges Bank,[61] and they were caught there for a few years thereafter in numbers that seem almost unbelievable today. Thus we read of 250 caught in three hours; of vessels loaded in a couple of days; and of a single smack landing 20,000 pounds in a day. They were taken in great plenty on Nantucket Shoals, also, during this same period. But the supply seems to have dwindled suddenly, in the shoal waters both of Georges Bank and of Nantucket Shoals, and so permanently that few vessels went thither especially for halibut after 1850. Now forced to go further afield, the fishing fleet found that halibut were plentiful on the Seal Island ground; on Browns Bank; and in the Eastern Channel or gully that separates the latter from Georges Bank (localities which supplied the New York and Boston markets for the next decade). And in 1875 halibut fishing was extended to deeper water (100 to 200 fathoms) on the southeast slope of Georges Bank. But it was not long before all these grounds were fished out to the point where it was seldom possible to make paying trips to them for halibut alone. And for many years now, what few halibut have been caught in the Gulf of Maine have been taken incidentally.

The history, in short, of the halibut fishery leaves no doubt that this species shows the effect of hard fishing sooner than most sea fish, it being possible to catch the majority of the stock on any limited area in a few years. Long liners and otter trawlers search all the good ground-fish bottoms of the Gulf of Maine and its banks so thoroughly and constantly that the halibut never have a chance to reestablish themselves in any abundance on the shoaler grounds. They maintain their numbers better on the deeper slopes chiefly because they are subject to less intensive fishing there.

It was fortunate for the fishing industry that the depletion of the Gulf of Maine of halibut was counterbalanced by the discovery of halibut in abundance along the deeper slopes of the banks to the north and east. And halibut fisherman sailing from Gloucester had begun resorting to the Grand Banks region by 1864–1866; to the west Greenland Banks by 1866; to the Magdalens by

[61] Goode and Collins (Fish. Ind. U. S., Sect. 5, vol. 1, 1887, p. 3) have collected data on the Georges Bank halibut fishery and the former abundance of the fish there.

1873; to the northern part of the Gulf of St. Lawrence near Anticosti by 1881; to Iceland by 1884.[62] And the Gloucester vessels continued sailing to the Greenland Banks for halibut until the early 1880's. But by 1889 practically all the salt halibut that was landed in Gloucester, was being brought from Iceland. With salt fish in less and less demand, it became unprofitable, next, to sail so far afield. And it is many years, now, since any halibut fisherman from Gloucester has outfitted for Iceland.

Long liners, out of New England ports, fished especially for halibut in the northern part of the Gulf of St. Lawrence until 1938; on the Grand Banks grounds until 1940 or 1941, when competition with frozen halibut from the northwest coast had become severe, while the majority of fishermen preferred to ship on otter trawlers, for comfort and safety. Vessels continued long lining for halibut down the slopes of the Nova Scotian banks, and in the deep gullies between these, until the middle of the 1940's, but we have not heard of a vessel making a special trip from any New England port for these great fish, during the past few years.

Although there is not one halibut in our Gulf today, where there were hundreds or even thousands of them during the first quarter of the nineteenth century, the geographical range of this noble fish is as extensive there as it ever was. Thus a few halibut are still caught along Cape Cod; in Massachusetts Bay (a few "chickens," 10 pounds and upward, are brought in every summer with some larger fish); along the Maine coast; and on all the offshore grounds. Rich, writing in 1929,[63] listed 25 named inshore grounds off the coasts of Massachusetts and of Maine as still yielding a few halibut. We have enjoyed the acquaintance of several fishermen, especially interested in halibut, who treasure to themselves a hard-gained knowledge of particular spots, not too far offshore, where they are likely to catch one, in a day's pleasure fishing. And small groups of halibut accumulate occasionally on suitable patches of bottom; soon to be decimated, however, when their presence is accidentally discovered. Thus, we knew of some 25 or 30 halibut, ranging from 40 to 110 pounds in weight, being caught within 1 to 3 miles of land, near Mount Desert Island, in 1930, in 10 to 15 fathoms of water. And one of 54 pounds was caught off Boston Harbor, from the steamer *Westport*, on June 24, 1951.[64]

A catch of 9,500 pounds, or perhaps about 135 fish (assuming an average weight of 70 pounds) was reported off the coast of Maine in 1947 on hand lines, while eleven fish (largest 125 pounds) had been caught inshore, off Casco Bay, by local fishermen, up to the last week in May 1951.[65] And many other instances of this sort might be quoted, no doubt, were our knowledge sufficient.

Halibut are also caught fairly regularly still, about Grand Manan (4,700 pounds reported thence in 1947), but only occasionally about Campobello and near St. Andrews, and not at all along the north (New Brunswick) shore of the Bay of Fundy east of St. John. Small numbers occur, however, right up to the head of the bay on the Nova Scotia side.[66] And there are enough of them off Brier Island at its mouth and on the fishing grounds along western Nova Scotia to have brought the landings for Digby County and for Yarmouth County to 108,300 pounds in the year in question.

The largest catches of halibut now made within the limits of the Gulf of Maine come from the Cape Sable–Browns Bank ground, from the deeper slopes of Browns Bank, from the deep gully that separates Browns from Georges, and from the eastern part and the deeper slopes of Georges, where otter trawlers are likely to pick up anywhere from 1 to 75 fish per trip. But not many are caught now on Nantucket Shoals where they were once so plentiful.

In 1945 (most recent year for which detailed information is readily available for the coasts of Maine and Massachusetts), landings for the different parts of the Gulf, by United States and Canadian fishermen, were about as follows: off eastern Massachusetts, about 31,000 pounds; off western Maine, about 800 pounds; off central Maine, about 10,000 pounds; small banks in the inner west central part of the Gulf (Cashes, Fippenies, Platts), about 2,500 pounds; off eastern

[62] See Collins, Fish. Ind. U. S., Sect. 5, vol. 1, 1887, pp. 43–89, and Scudder, Fish. Ind. U. S., Sect. 5, vol. 1, 1887, pp. 90–119, for historical surveys of the fresh and salt halibut fisheries from New England ports.

[63] Rept. U. S. Comm. Fish. (1929) 1930, pp. 85–86, 96.

[64] Reported in the Boston Globe, June 25, 1951.

[65] Reported in Salt Water Sportsman for May 29, 1951.

[66] 11,300 pounds reported for Annapolis and King's Counties, Nova Scotia, in 1947.

Maine about 2,000 pounds; mouth of Bay of Fundy on New Brunswick side, about 700 pounds; Nova Scotian side of Bay of Fundy, about 45,000 pounds; off western Nova Scotia and Browns Bank (Canadian and United States vessels combined), about 73,000 pounds plus an indeterminate part of the landings for Shelburne County, Nova Scotia, that may have come from Browns Bank; Georges Bank, about 65,000 pounds; South Channel, about 4,000 pounds; Nantucket Shoals, about 1,400 pounds; or a total of about 235,000 pounds that can be credited definitely to the Gulf.

For some unknown reason, 1945 was a poor year; the Georges catch alone, for example, was about 110,000 pounds in 1946, about 211,000 pounds in 1947.[67] And the yearly catch for the Gulf as a whole, by United States and Canadian fishermen combined, averaged about 316,000 pounds for the 6-year period 1941–1946, plus what fish may have been landed in Shelburne, Nova Scotia, from Browns Bank. Even so, the Gulf yields only about one-tenth as much halibut by weight today as it did, say, 30 years ago.[68]

We dare not guess in what degree this continued decrease has been a result of the progressive replacement of long-line fishing by otter trawling, of market conditions, or of a continuing decrease in the numbers of halibut.

Halibut may have maintained their numbers somewhat better on the outer Nova Scotian Banks and slopes, which yielded about 3,400,000 pounds in 1934 (with Browns Bank); about 1,350,000 pounds in 1946.[69]

In the early days of the fishery, halibut were common in the Gulf of Maine in water no deeper than they were farther north; near Anticosti in the Gulf of St. Lawrence for example, or near Miquelon, south of Newfoundland, where many were caught in 5 to 10 fathoms.[70] A case is on record, for example, of a catch of 5 halibut, made in 1849, on one set of a long line with only 37 hooks, in 7 fathoms, just off the mouth of Gloucester Harbor. A good many, too, were caught in those days on the southeastern part of Stellwagen Bank, where the depth (on the fishing grounds) ranges from 15 fathoms to about 30 fathoms. And many were reported as wintering in the gullies west of Stellwagen and between the latter and the tip of Cape Cod, in depths of 30 to 50 fathoms. Similarly, the early fishery also on Georges was on the shoaler parts of the bank in depths of 15 to 30 fathoms. And the early visitors to this ground describe the halibut, not only as schooling at the surface in pursuit of herring and launce (not an uncommon event in the Gulf of St. Lawrence and off Newfoundland when they are chasing capelin), but as often following their hooked companions up to the top of the water, so that more than one vessel made a good part of her fare by gaffing them alongside. The Nantucket Shoals halibut of old were likewise in less than 30 fathoms depth, and when the fleet first repaired to Browns Bank and to the Seal Island grounds they found halibut very plentiful in water but little deeper than that.

In fact, it was not until 1874 or 1875 that the presence of this fish was suspected in the deeper gullies or on the offshore slopes of the banks below 100 fathoms. But it did not require many years of hard fishing to catch most of the halibut that were living in very shallow water, and so thoroughly that very few are now taken shoaler than 25 to 40 fathoms in our Gulf, while most of the halibut that are caught still on the offshore banks are from water deeper than 75 fathoms.

All that has come down to us as to any general movements of the halibut in the Gulf of Maine during the days of their plenty there, beyond the prevailing tendency of the larger fish to work down deeper than the smaller (p. 250), is that some of them (though not all) worked inshore into shoaler waters for the winter, to work offshore again and deeper for the summer. But this offshore movement in winter may not have extended far, or very deep, if it was to avoid low temperatures, for halibut (or any other fish for that matter), that summer inshore in shoal water where they would be most subject to winter chilling, need never move out for more than 60 miles or so off any part of the coast line of the open Gulf, nor descend deeper than about 70 to 75 fathoms, to find water permanently warmer than 38°, except in the Bay of

[67] The otter trawlers that carried on investigations for the Bureau of Fisheries in 1913 took halibut on more than half their trips to Georges. Contrast this with a catch of 570 halibut by a long-liner on a patch of rocky bottom there in one day, in the early years of the Georges fishery!

[68] The Gulf of Maine catch was nearly 3 million pounds in 1919.

[69] Catch, Cape Sable to Cape Breton in 1946, about 50,000 pounds by United States vessels, about 1,300,000 pounds by Canadian vessels. For a general survey of the catches of halibut in both sides of the Atlantic, for 1934, see Thompson and Van Cleve, Rept. No. 9, International Fisheries Commission, 1936, p. 21.

[70] Goode and Collins, Fish. Ind. U. S., Sec. 5, vol. 1, 1887, p. 17.

Fundy in the coldest winters, or in the submarine embayment between Jeffreys Ledge and the coast.

We have nothing to contribute as to present-day spawning of halibut within the Gulf of Maine. Of old, ripe females were reported on Georges in May and June, and have been, repeatedly, on the deeper slopes of the Nova Scotian banks generally, to the eastward, as well as off the Grand Banks and in the Gulf of St. Lawrence. But halibut smaller than a couple of pounds are seldom caught in the inner parts of the Gulf though "chickens" of 10 to 20 pounds are not uncommon there, there being a sharp contrast in this respect between the Gulf of Maine and the waters around Iceland, where Jespersen found an abundance of little fish of 8 to 10 inches.

This, added to the fact that the inshore grounds were fished out so soon with little apparent tendency to recover when the fishery slackened, and that depletion by overfishing has not been accompanied by any corresponding decrease in the average size of the fish that are caught, suggests that the halibut population of the inner parts of our Gulf always depended more on immigration from east and north of Cape Sable for its maintenance than it did on local production. Fry may have been produced in greater numbers over the offshore slope of Georges Bank, where the *Albatross III* trawled two little halibut about 6 inches long, at 175 to 195 fathoms, on May 16, 1950.

Importance.—The halibut, because of its present-day scarcity, is of only minor importance commercially in our Gulf; in 1947 the landings in New England, including what halibut were brought in from the Nova Scotian banks eastward from Cape Sable, amounted to only about 586,000 pounds, valued at $144,680. But the demand is always so good that all that are brought in are readily salable, and (being so large) each one that is caught is well worth saving. In the year in question (representative of present-day conditions), about one-fifth of the total New England landings, were caught on long lines [71] three-fourths by otter trawlers. The small remainder (10,000 pounds) were taken on hand lines, mostly by small-boat fishermen off the coast of Maine.

We can only regret that there are not enough halibut inshore in our Gulf today to be of any general concern to anglers, for this is a very "sporting" fish as well as welcome on the table.

Greenland halibut *Reinhardtius hippoglossoides* (Walbaum) 1792

GREENLAND TURBOT; NEWFOUNDLAND TURBOT

Jordan and Evermann, 1896–1900, p. 2611.

Description.—This is a right-handed, large-mouthed flatfish (that is, it lies on its left side, with its eyes on its right side, and its abdomen at its right edge), with slightly concave tail, and symmetrical ventral fins like a halibut. In fact it so closely resembles the halibut that it might easily be taken for one were it not that its lateral line is nearly straight abreast of the pectoral fin, (arched in the halibut) and that its long fins (dorsal and anal) are of rather different shape (compare fig. 127 with fig. 123), though with about the same number of rays (about 100 dorsal and

[71] Goode and Collins (Fish. Ind. U. S., Sec. 5, vol. 1 1887, pp. 10–18) have given a readable account of the long-line fishery.

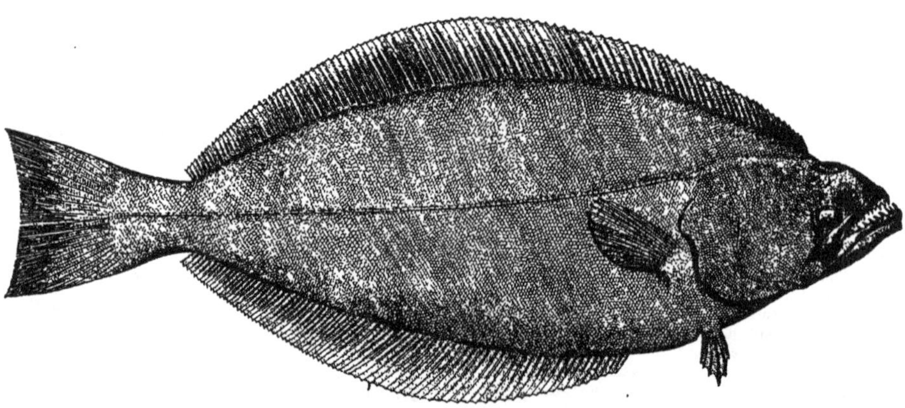

FIGURE 127.—Greenland halibut (*Reinhardtius hippoglossoides*). From Goode. Drawing by H. L. Todd.

75 anal). Its mouth, furthermore, is larger, its eyes smaller relatively and its jaw teeth stronger, though the differences in these respects are not great enough to serve as useful field marks. It is yellowish or grayish brown, paler below than above but not white.

Size.—This is one of the largest of the North Atlantic flatfishes, next to the halibut, growing to a length of about 40 inches and to a weight of 20 to 25 pounds. But fish caught about the Grand Banks weigh only from about 5 to 10 pounds.

General range and occurrence in the Gulf of Maine.—This is a fish of the Arctic and subarctic Atlantic. It is taken from northern Norway and northern Iceland to the Faroe ridge, and to southwest of Iceland [72] as a stray. It supports a fishery off west Greenland that is important for the Eskimos.[73] In the west considerable numbers are taken off the south coast of Newfoundland,[74] also on the Grand Banks, hence it is to be expected along outer Labrador, though it has not been reported thence as yet. Odd specimens are to be expected here and there in the Gulf of St. Lawrence too, for it has been taken near the Biological Station at Trois Pistoles.[75]

[72] Norman (Ann. Mag. Nat. Hist., ser. 9, vol. 13, p. 539) reports a single specimen taken southwest of Iceland.

[73] See Jensen (Meddel. Dansk. Komm. Havundersøgelser, Ser. Fiskeri, vol. 7, No. 7, 1927) for a general account of the Greenland halibut off west Greenland.

[74] According to Goode (Fish. Ind. U. S., Sect. 1, 1884, p. 198) long liners have reported it as "very abundant" in and off Fortune Bay, in 60 to 300 fathoms.

[75] Vladykov and Tremblay, Natural. Canad., vol. 62 (Ser. 3, vol. 6), 1935 p. 82.

It is described as "not uncommon" off Canso, Nova Scotia,[76] and as occasionally brought in from the more northerly of the Nova Scotian fishing banks long ago.[77]

Our only reason for mentioning it here is Goode and Bean's[78] statement that "fishermen take them frequently in the gully between La Have and Georges Bank at depths greater than 200 fathoms." This has been corroborated by the capture of 16 specimens, at 300 to 530 fathoms, along the slope of La Have Bank to the southwestern slope of Georges Bank by the *Cap'n Bill II*, in July 1952. It has no real place in the Gulf of Maine fish fauna.

American dab *Hippoglossoides platessoides* (Fabricius) 1780 [79]

CANADIAN PLAICE; LONG ROUGH DAB

Jordan and Evermann, 1896–1900, p. 2614.

Description.—The most obvious distinctive characters of the American dab are that it is right-handed and large-mouthed like the halibut, but with a rounded tail instead of concave, and with the lateral line nearly straight instead of arched; it is the only Gulf of Maine flounder in which these characters are combined. Our only other large-mouthed flat-fishes with rounded tails (the

[76] Cornish, Contrib. Canadian Biol. (1902–1905) 1907, p. 90.

[77] Jones, Proc. Trans. Nova Scotia Inst., Sci., vol. 5, Pt. 1, 1879, p. 92.

[78] Amer. Jour. Sci. Arts, Ser. 3, vol. 17, 1879, p. 40.

[79] Various other common names are applied to this fish in different seas. It is usually termed "Long rough dab" in England and is so listed in British fishery statistics. It is not the "plaice," or the "dab" of Europe.

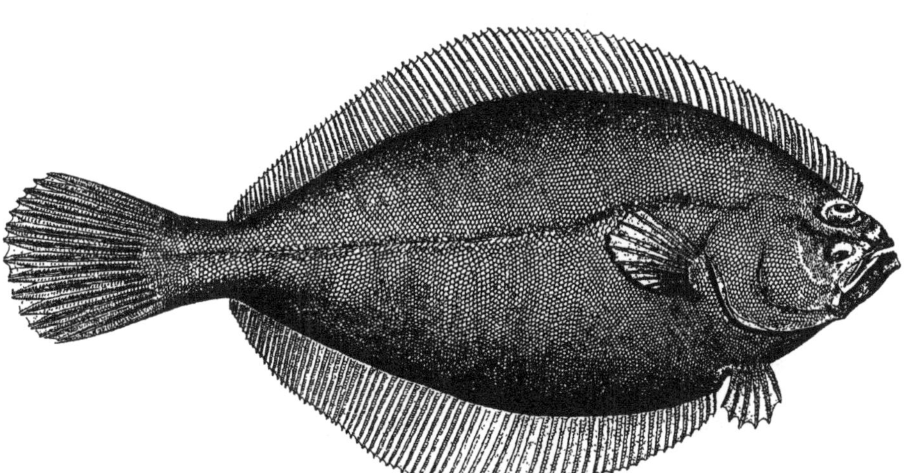

FIGURE 128.—Canadian plaice, or Dab (*Hippoglossoides plattessoides*), La Have Bank. From Goode. Drawing by H. L. Todd.

sand-, summer- and four-spotted flounders, pp. 290, 267, and 270) are left-handed, and the wide-gaping jaws readily distinguish the American dab from the various small-mouthed flounders.

It is a comparatively broad (really deep) flounder (about two and one-half times as long to base of caudal fin as it is broad), more rounded in outline than the halibut, with pointed nose, mouth gaping back to abreast of the middle of the eyes, and with one irregular row of sharp conical teeth in each jaw. The free edges of the scales on the entire eyed (upper) side of its body and of its head are serrated with sharp teeth, which give the fish a characteristic rough feeling when handled, but the scales of the blind (lower) side are smooth-edged except on the rear part of the body and along the bases of the fins. The dorsal fin (76 to 96 rays) originates in front of the middle of the left eye and the anal fin (64 to 77 rays) arises slightly in advance of the base of the pectorals. Both of these long fins taper toward the head and toward the tail, and there is a short, sharp, spine (the prolongation of the post abdominal bone) pointing forward close in front of the anal fin. The pectoral fin on the eyed side usually (not always) has one or two more rays than its fellow fin on the blind side, and is longer and more rounded, but the two ventral fins, which are close in front of the anal fin though entirely distinct from it, are alike in size, shape, and location. The margin of the caudal fin is always convex, either rounded or with its middle rays so much the longest as to form a blunt angle. The lateral line on the eyed side is more clearly evident on the dab than on most of our flatfishes, and it is straight from end to end, except for a slight arch over the pectoral fin.

Color.—Dabs run more uniform in color than most of our smaller flatfish, ranging from reddish to greyish brown (darker or paler) above and pure or bluish white below. The tips of the rays of the two long (dorsal and anal) fins are white. On one specimen we saw the right edge of the eyed side was white (like the blind side) from the gill opening to the rearmost ray of the ventral fin but this is unusual. Small fish are usually marked with three to five dark spots along each edge of the body; large ones are occasionally, though they are plain colored as a rule.

Size.—Adults measured by Welsh off Cape Ann ran from about 12 inches to 24 inches, and few of those that are caught in our Gulf are longer than 2 feet. Nova Scotian fish measured by Huntsman [80] ran from 12 to 24 inches in length, while fish caught in the colder waters off Newfoundland averaged 18 inches.[81] The largest dab recorded from American waters, taken near Sable Island, May 1939, was 32½ inches long and weighed 14 pounds.[82] The next largest, taken in 90 fathoms on the northern edge of Georges Bank, November 1951, was 29 inches long.[83]

According to Huntsman, Nova Scotian fish average about half a pound at 12 inches, 1¼ pounds at 16 inches, 1¾ pounds at 18 inches, 2¾ pounds at 20 inches, 4 pounds at 22 inches, and 6 pounds at 24 inches. Massachusetts Bay fish are about equally heavy at corresponding lengths. And a 16-inch fish from Georges Bank that we measured weighed 1 pound 5 ounces; two fish of 18¼ inches weighed 1 pound 13 ounces, and 2 pounds, respectively; one of 19½ inches weighed 2 pounds 8 ounces, and one of 29 inches weighed 9 pounds 6 ounces.

This flatfish tends to differentiate into local races in different seas. Thus the fin rays are more numerous on the average in fish from high latitudes than in those from low latitudes, while the body is relatively wider in fish caught off Greenland and off America than in those from Scandinavia and from the North Sea. But these characters vary so widely even in limited areas that the Arctic-American and European species (*platessoides* and *limandoides*) have been united by common consent long since, and we doubt whether the corresponding "varieties" still recognized by several recent authors will stand the test of time. Huntsman's statement that the dorsal rays average more numerous in dabs from Bay of Islands, Newfoundland, than in those caught on the New Brunswick shore of the Gulf of St. Lawrence, with Welsh's note of a variation of 7 in the number of dorsal rays and of 6 in the anal rays in one lot of fish caught off Gloucester, illustrates this variability. Notwithstanding the low latitude of the locality of capture (about 42° 30′ N.), this same lot contained a specimen with the largest number of fin rays yet reported (96 dorsal and 77 anal).

All we dare say until many more specimens are examined is that hereditary local races may perhaps exist off different parts of the American

---

80 Bull. No. 1, Biol. Board Canada, 1918, p. 10.

81 Frost, Res. Bull. 4, Newfoundland Dept. Nat. Resources, 1938, p. 8.

82 McKenzie, Proc. Nova Scotian Inst. Sci., vol. 20, 1940, pp. 43–44.

83 We measured this dab, taken by Capt. Arthur Nelson of the *Eugene H.* who also caught several others, 27 to 28 inches long on this same trip.

shore line, and that the growth marks on the scales, in relation to the length of the fish, may give a clue to the local origin of a given specimen, for it seems that the rate of growth is governed by the temperature of the water (p. 263).

Habits.—Dabs are bottom fish like other flatfishes. But they must rise some distance from the ground on occasion, and move about to a considerable extent to account for the capture of so many in gill nets (p. 264). We once caught one a foot long in a tow net at least 5 to 10 fathoms above the bottom off Ipswich Bay, where the water was about 50 fathoms deep.

Like some other flatfishes, they avoid rocky or hard bottom, preferring a fine, sticky but gritty mixture of sand and mud, such as floors much of the Gulf between the hard patches, from the 20-fathom contour out to the 100 fathom contour. And they are also to be caught in numbers on the soft oozy mud of the deeper basins in the western side of the Gulf, as pointed out below (p. 264).

In one part of their range or another, they are found from tide line down to as deep as about 390 fathoms (700 meters).

This is an arctic-boreal species in its temperature relations, reaching its highest development in water of 35° to 45° F.; able to live, however, in the lowest polar temperatures (29° to 30°); and finding the upper temperature limit to its regular occurrence at about 50° to 55° F.

In different seas it lives through a wide range of salinity, from 30 per mille or lower in the Baltic to upwards of 34 per mille in the open Atlantic. So far as we are aware, it is never found in water which could be described as brackish along the coasts of New England or of the Maritime Provinces. But R. H. Backus informs us that the *Blue Dolphin* found it in brackish water (salinity 23 per mille) at the west end of Lake Melville, Labrador.

Huntsman's[84] statement that it feeds on minute planktonic plants (diatoms) at first, but on copepods as it grows larger and more active is our only information as to the diet of the young fry in American waters, while they are drifting near the surface. When they first take to the bottom they eat small shrimps and other Crustacea of various sorts. But they turn (as they grow) to a diet consisting chiefly of sea urchins, sand dollars, and brittle stars, as proved by the contents of their stomachs, though they also take various shrimps, hermit and spider crabs and other crustaceans, mollusks, worms and ascidians (sea squirts), in fact, practically any bottom living animals that are small enough for them to devour. Occasionally they catch small fish.

They do not bite a baited hook as readily as various other ground fishes, partly, no doubt, because they are sluggish fish, but partly, we believe, because the clams, cockles, and herring that are usually used for bait are not their favorite food. Still, considerable numbers are caught on hand and long lines.

All the large predaceous fish that feed near bottom probably prey more or less upon them, and halibut no doubt destroyed great numbers of them in the Gulf of Maine formerly. But the adults can have no serious enemy in our Gulf today except large cod and perhaps the spiny dogfish. In more northern seas Greenland sharks prey regularly on them. Smitt and Huntsman both speak of the numbers of round worms to be found in the intestines and body cavity of the dab, and its gills are sometimes attacked by parasitic copepods.

While the young are drifting near the surface (p. 262), they share in the same involuntary journeyings as other fish fry do, that are spawned at the same place and time. But it is one of the more stationary fishes from the time it seeks bottom. It has been said to work inshore more or less in winter, though not on very definite evidence, and it may congregate on definite grounds for spawning, though this is yet to be proved. But it is certain that they are to be caught at any season of the year wherever they are plentiful. And Huntsman,[85] who has paid special attention to this fish, believes that it "remains pretty much in the same place from season to season and year to year. Perhaps in the course of years it may shift a few miles."

Individual females produce 30,000 to 60,000 eggs, according to size. The eggs are buoyant and have no oil globule, but they have a transparent (perivitelline) space around the yolk so broad that they are not likely to be confused with those of any other Gulf of Maine fish.

[84] Bull. Biol. Board Canada, No. 1, 1918, p. 15.

[85] Bull. Biol. Board Canada, No. 1, 1918, p. 18.

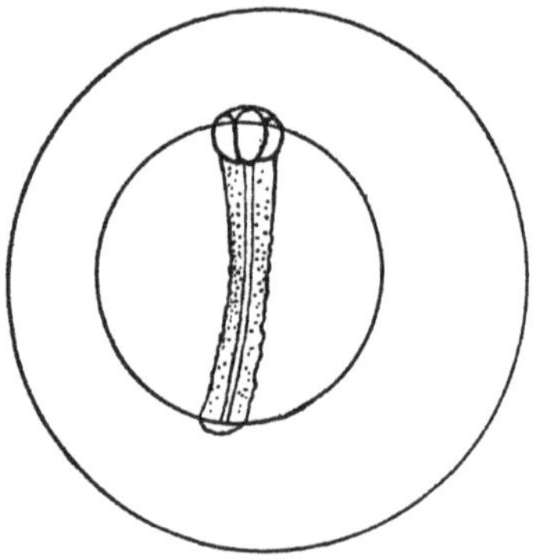

FIGURE 129.—Egg (European). After Cunningham.

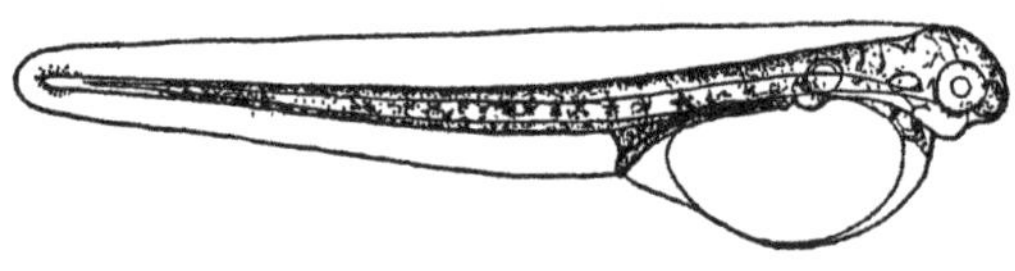

FIGURE 130.—Larva, just hatched, 4 mm. (European). After McIntosh.

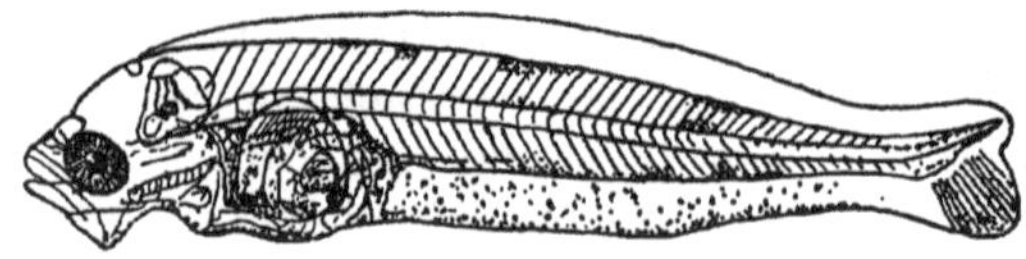

FIGURE 131.—Larva (European), 9 mm. After Ehrenbaum.

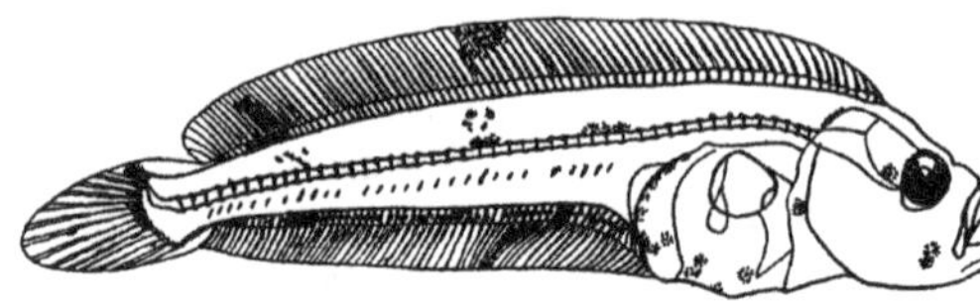

FIGURE 132.—Larva, 14.5 mm., off Massachusetts Bay.

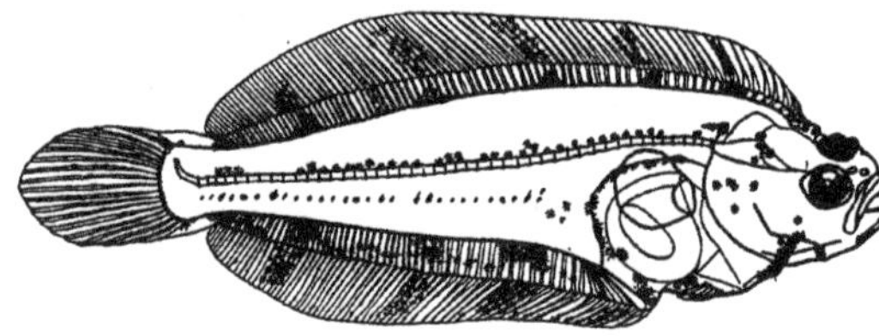

FIGURE 133.—Larva, 22.5 mm., off Cape Cod.

CANADIAN PLAICE, OR DAB (*Hippoglossoides platessoides*).

This space is formed by the entrance of water between the egg proper and its covering membrane, after the eggs are shed, and it about doubles the total diameter of the egg. The eggs we have taken in the Gulf of Maine have averaged about 2.5 mm. in diameter, but they have been reported as small as 1.38 and as large as 3.2 mm. in other seas, depending on the breadth of the perivitelline space.

Incubation occupies 11 to 14 days at a temperature of 39° F., and it seems that the eggs gain weight as development proceeds, for Huntsman found, in the Gulf of St. Lawrence, that the newly spawned eggs floated at the surface, but that eggs nearly ready to hatch drifted suspended at a depth of some 10 fathoms. We have no first-hand information to offer on this point.

During the development of the egg, minute black and yellow pigment cells are scattered over the embryo, not aggregated into any diagnostic clusters. But the pigment gathers in five definite groups very soon after hatching (which takes place when the larvae are 4 to 6 mm. long); one on the gastric region, one about the vent, and three behind the vent; a pattern similar to that of the larval witch flounder (p. 287).

The yolk is absorbed about 5 days after hatching, when the larva has grown to 6.2 to 7.5 mm. in length. The caudal rays appear shortly after this, the dorsal and anal rays at about 11 to 12 mm., and the three vertical fins are differentiated at about 15 to 18 mm. By this stage the body has begun to assume the deep but very thin form characteristic of all young flounders, while the jaws have developed sufficiently to show that the little fish belongs to one of the large-mouthed species. The left eye may commence its migration when the larva is about 20 mm. long, while Welsh found it visible above the outline of the snout in Gulf of Maine specimens of 24 mm., and almost at the dorsal edge at 34 mm. But larvae as long as 35 mm. may still be symmetrical in other seas.

The only other Gulf of Maine species for which the larval dab might be mistaken (except in its very earliest stages) are the witch flounder and the halibut; but the witch is longer at corresponding stages of development, but with the distance from snout to vent proportionately much shorter, and the outlines of throat and abdomen are sufficiently different to distinguish the dab from the halibut (p. 253).

The young dab drifts freely up to the time of its metamorphosis, as the young of most sea fishes do; close to the surface at first but sinking deeper as it grows, until it seeks the bottom finally.

Welsh's observations suggest that this takes place, in our Gulf, when the little fish are about 1½ to 1¾ inches long, with their metamorphosis already complete, their body scaly, and their eyed side densely pigmented. But there is wide variation in this respect. And European authors report that the fry may take to the bottom even before the left eye has completed its migration around the head.

The period occupied in larval growth and in metamorphosis varies with temperature. Probably it covers three to four months in the Gulf of Maine, where we have taken the pelagic larvae as early in the season as May 26 and as late as August 2.

The little fish grow to a length of 2 to 3 inches by their first winter, with their exact size then depending upon how early in the season they are hatched, and probably on the temperature in which they live. And they average about 3 inches long[86] when they are one year old. Thus it may be assumed that bottom stages 2¾ to 3⅛ inches (69–80 mm.) long that we have trawled off Cape Cod, on May 1, were about one year old; others of 3⅜ to 4⅝ inches (85–118 mm.) that we have trawled in July and August off Mount Desert, in the deep gully to the westward of Jeffreys Ledge, on Cashes Ledge, and on the edge of Stellwagen Bank were between 1¼ and 1½ years old; and that those of 8 to 10 inches were 2¼ to 2½ years old. Subsequent growth is more rapid in higher temperatures than in lower, throughout the temperature range favorable to this particular flatfish. Huntsman,[87] for example, has found that it takes only 3 to 5 years for dabs to grow to a length of 12 inches in Passamaquoddy Bay, where the bottom water at 15 to 18 fathoms warms to about 49° to 51° F. in August, but that it requires 4 to 6 years in the open Bay of Fundy, where the bottom temperature in summer is somewhat lower (45°–48°); 6 to 9 years in the cooler water (about 38°) of Chedabucto Bay, eastern Nova Scotia; and upwards of 8 years in the still lower temperatures (colder than 35°) of the Gulf of St. Lawrence.

On this basis, dabs living on the shoaler parts of Georges Bank, and as shoal as 15 fathoms or so in coastwise waters from Cape Cod to Cape Elizabeth, probably grow about as fast as the Passamaquoddy Bay fish, i. e., they may reach a length of 15 inches in 5 years or even sooner, gaining something like 4 ounces in weight yearly. Those in the eastern side of the open Gulf of Maine may be expected to grow about as fast as those in the Bay of Fundy, but somewhat more slowly there if they are living as deep as 50 fathoms, though not so slowly as in the Gulf of St. Lawrence. Some individuals may become sexually mature when only 6 inches long, probably all of them do so by their third year; and they are known to live to an age of 24–30 years, perhaps longer, at least in the Gulf of St. Lawrence.

[86] Huntsman, Bull. Biol. Board Canada, No. 1, 1918.

[87] Bull. Biol. Board Canada, No. 1, 1918, p. 23.

In general, females grow faster than males.

Huntsman has also found, in the Gulf of St. Lawrence, that a majority of the fish of 3 years and younger were males, but that females outnumbered the males among the older fish, while all of those 14 years old and older that he saw were females. We have no explanation to offer for this (apparently) higher mortality rate for the males than for the females among the older fish.

General range.—This is a very common fish on both sides of the North Atlantic, where its range parallels that of the cod, except that it does not extend as far south and west along the American seaboard. It is found in abundance along the outer coast of Labrador, southward from Hamilton Inlet, where (Frost[88] writes) they are so abundant locally that a 5-minute haul with a torn trawl yielded 50 (at lat. about 54°) in Newfoundland waters in general; on the Grand Banks, including the eastern edge;[89] in the Gulf of St. Lawrence as a whole, and thence westward and southward to Cape Cod, from close inshore out to the 100-fathom contour. Westward from Cape Cod, a few are caught in the Woods Hole region; off Marthas Vineyard; and off Narragansett Bay which marks their western limit in general. The most southerly and westerly record with which we are acquainted is of one 15½ inches long that was caught off Montauk Point, N. Y., in 112 fathoms, February 6, 1930.[90]

[88] Res. Bull. 4, Newfoundland Dept. Nat. Resources, 1938, p. 8. R. H. Backus informs us also that the *Blue Dolphin* collected them at various localities as far north as the northern shore of Hamilton Inlet (lat. 54°30′ N.), but did not take any farther north, in spite of extensive collecting.

[89] Reported in abundance down the eastern edge of the Grand Banks, in the 20th Rept. Dept. Fish. Canada (1949–1950) 1951, p. 36.

[90] We find no other credible records from New York or from New Jersey, those mentioned by DeKay being market fish which might have come from anywhere to the eastward.

It is common in west Greenland waters, as far north as Upernavik near the Arctic Circle, in latitude about 72° N.[91] And it ranges in European waters from Iceland and Spitzbergen southward to the North Sea, where it is an important commercial fish, and to the west Baltic; the English Channel is the southern boundary to its regular occurrence.

Occurrence in the Gulf of Maine.[92]—This is not as familiar a fish as are the winter and smooth flounders (pp. 276 and 283), for it is not common in water shallower than 15 to 20 fathoms. But it is probably the most abundant of all Gulf of Maine flatfishes at depths greater than 30 to 50 fathoms, except, perhaps, the witch (p. 285). Thus they are recorded from Provincetown; from Massachusetts Bay; off Cape Ann; on Stellwagen Bank, where we have hand-lined a number of them in 25 fathoms; in Ipswich Bay; near Boon Island; off Cape Porpoise; off Casco Bay; on Cashes Ledge, where we have trawled both young and adults; off Seguin; south of Monhegan (we trawled them at the last four localities on the *Grampus*); close in to Little Duck Island, off Mount Desert; in Passamaquoddy Bay; in St. Mary Bay; and right up to the head of the Bay of Fundy. In fact, they are to be caught all around the inner parts of the Gulf wherever the water is more than 15 fathoms deep or so, and where the bottom is smooth. Trawlings, too, by the *Albatross II* and by the *Atlantis* have shown that they are generally distributed throughout the basin of the Gulf down to 120 fathoms. This, indeed, was the only flatfish, other than the witch (p. 288), that was taken by the *Atlantis* on the soft mud bottoms off Cape Cod, west of Jeffreys Ledge, or off Mount Desert, at 66 to 105 fathoms during her experimental trawlings for the edible shrimp (*Pandalus*) in August 1936.[93]

Dabs are widespread on Georges Bank also, for they were reported at many localities there by representatives of the Bureau of Fisheries in 1913, while we have seen catches of up to 100 per trawl haul on the northern edge of Georges, in 60 to 100 fathoms of water. They are so plentiful along the 50–100 fathom zone on the northern edge of Georges Bank that draggers fishing there during 1951–1952 were making catches averaging about 5,000 pounds per day. A good example of their numbers there is furnished by the dragger *Eugene H* of Woods Hole which brought in catches of 10,000 to 25,000 pounds of dabs, fishing in 75 to 95 fathoms, throughout the period August 1951 to January 1952. Many of these fish were large, ranging from 4 to about 9 pounds in weight. And in this same region, in the spring, they appear to be plentiful in water much shoaler, for Capt. Arthur Nelson of Woods Hole reports a catch of 18,000 pounds taken in 25–30 fathoms in four days' fishing early in May 1952. Also, we have the definite evidence of commercial catches, as well as of newly spawned eggs taken in our tow net, that dabs are plentiful on Browns Bank also.

Huntsman has calculated from fishing experiments that they are about one-tenth as numerous as cod in the Gulf of St. Lawrence. No general estimate of this sort is yet possible for the Gulf of Maine. But catches in gill nets (gear not very well adapted for flounder fishing) of 76 dabs to 1,055 haddock, 51 cod, 20 pollock, and 39 rosefish near Boon Island on March 30; of 125 dabs to 40 other flounders, 89 cod, and 113 haddock in part of the net at the same locality on April 20; and of many dabs, but more cod and haddock, on May 3, 1913,[94] are pertinent here.

This flatfish is often found in very shoal water in colder seas. They are often seen under wharves around Newfoundland, for example, according to Frost.[95] And some are seined right on the beach[96] on the West Greenland coast. But we have never seen or heard of an adult specimen caught in less than 10 fathoms of water in the Gulf of Maine, probably because of the high summer temperatures of the shoaler waters, and they are the most plentiful in 15 to 60 fathoms there (in our experience). At the other extreme, 120 fathoms is the deepest definite record for the Gulf of Maine with which we are acquainted; hence this may be set as the lower limit to their occurrence there in any numbers, which, by report, applies to the whole American coastline, including the Scotian banks and the Grand Banks region.

This preference of the dab for moderately deep water in the southern part of its range bars it from most of the Gulf of Maine harbors and river

[91] For an account of it in west Greenland waters, see Jensen (Meddel. Dansk Komm. Havundersøgelser, vol. 7, No. 7, 1925, p. 24).

[92] Huntsman (Bull. 1, Biol. Board Canada, 1918) gives an interesting account of this fish in Canadian waters.

[93] Bigelow and Schroeder, Biol. Bull., vol. 76, 1939, p. 308.

[94] Recorded by Welsh.

[95] Research Bull. 4, Newfoundland Dept. Nat. Resources, 1939, p. 8.

[96] Jensen, Meddel. Dansk. Komm. Havundersøgelser, Ser. Fiskeri, vol. 7, No. 7, 1925, p. 24.

mouths, which are such favored haunts for the winter flounder. But it enters the deeper estuaries and passages between the islands in the northeastern part of the Gulf, those near Mount Desert, for example, Passamaquoddy Bay, and St. Mary Bay.

We hesitate to draw any definite conclusions from published statistics of the landings of "dab"[97] as to the regional abundance of this particular flatfish in our Gulf, partly because of the likelihood that other flatfish may appear under this name, or dabs under some other name, and partly because only a few of the otter trawlers fish in the deeper basins of the Gulf where dabs are known to be plentiful.

The returns for 1945, if taken at face value, show about 48,000 pounds landed from off eastern Maine; about 586,000 pounds from off central Maine; about 311,000 pounds from off western Maine; about 43,000 pounds from small grounds in the west central part of the Gulf; about 897,000 pounds from off eastern Massachusetts; about 8,000 pounds from Nantucket Shoals; about 910,000 pounds from the South Channel and Georges Bank combined; about 48,000 pounds from Browns Bank; and about 40,000 pounds from off western Nova Scotia (by United States fishermen); or a total of some 2,890,000 pounds. It was not until 1946 that the dab was listed (as "Canadian plaice") in the Canadian fisheries statistics for Nova Scotia; in that year landings for western Nova Scotia (Yarmouth County) were about 140,000 pounds, and about 41,000 pounds for the Nova Scotian side of the Bay of Fundy (Digby County).

The presence of dabs or Canadian plaice of catchable sizes in the Bay of Fundy in general, and in Passamaquoddy Bay in particular, is interesting as evidence that this is not so stationary a fish there as it seems to be elsewhere, for none are reared there so far as is known (p. 266), so that the maintenance of the local stock appears to depend on immigration from outside. Huntsman's observation is interesting, too, that large ones form a much smaller proportion of the population in Passamaquoddy Bay and in the Bay of Fundy than they do in the Gulf of St. Lawrence. And it seems, similarly, that large ones are less plentiful relatively in Passamaquoddy Bay than they are in the western side of the open Gulf of Maine. The death rate may be higher in Passamaquoddy waters, as Huntsman has suggested, or it may prove that the fish tend to work out from there into the open Gulf as they advance in age.

The dab is a spring spawner on both sides of the Atlantic, as is well known. The earliest date at which we have taken its eggs in our tow net in the Gulf of Maine has been March 4 (in 1920), off Casco Bay. We have also found the eggs on Browns Bank on the 13th, while Welsh records large female fish, half spent and with eggs exuding, as well as males with running milt, on the 14th of March, near Cape Ann, in 1913. But other fish of both sexes taken with them were unripe still, evidence that spawning is not general until the last of March or first part of April. Dab eggs have appeared regularly in our towings in April (twice in great numbers, namely off Seguin Island on the 10th and off Mount Desert Island on the 12th in 1920). Spawning continues unabated throughout May, when eggs were taken at nearly all our towing stations in 1915. And April and May similarly cover the height of the spawning season in the Bay of Fundy, according to Huntsman.[98] Our latest seasonal record has been for a single egg, on the 14th of June in 1915.

The dab spawns chiefly during May and June on the banks off Cape Breton and in the southern part of the Gulf of St. Lawrence: until the end of July on the southern part of the Newfoundland Banks (a few eggs were found by the Canadian Fisheries Expedition); until fall around the southeastern and eastern coasts of Newfoundland, and along the outer coast of Labrador, according to Frost. And the eggs are reported from May into July off West Greenland, by Jensen.

It spawns somewhat earlier in the North Sea than in American waters; i. e., from mid-January till May with the climax in March and April. Huntsman also remarks that there is a difference in the breeding season according to the depth of water, those living shoalest commencing to spawn the earliest, as the vernal warming of the water makes itself felt from above. But we have no clear evidence on this point to offer for the Gulf of Maine.

---

[97] It is only during the past few years that the landings of this particular flatfish have been reported separately, as "dab."

[98] Bull. 1, Biol. Board Canada, 1918, p. 14.

Our egg records, added to Huntsman's observations, show that the dab spawns all around the Gulf of Maine, from Cape Cod on the west to Cape Sable on the east, including the Bay of Fundy, and from close inshore out to the 50-fathom contour. It also spawns on Browns Bank (p. 265), and, while we found no eggs on Georges Bank either in February, March, April, or May, of 1920, the fish is so common there and so stationary in general that it is likely that we simply missed its eggs, either by a failure to tow over the precise spawning localities or by timing our visits between the waves of production. Dabs also spawn abundantly on Sable Island Bank (no doubt on all the other Nova Scotian Banks); off Cape Breton; in the shoaler parts of the Gulf of St. Lawrence;[99] throughout the general region of the Grand Banks; off the east coast of Newfoundland; along the outer coast of Labrador to Hamilton Inlet at least;[1] and as far north along the west coast of Greenland as the species is known to exist, as is proven by the presence of its eggs in the water there in quantities.[2]

Although the dab is rather a deep-water fish compared to most of the other flatfishes that are common in the Gulf of Maine, it is doubtful whether it ever spawns at depths much greater than 50 fathoms in the Gulf, for we have few egg records from more than a mile or two outside the 50-fathom curve, while these few have been based on only one or two eggs each. And we have trawled spawning females off Mount Desert, in 20 fathoms. This concentration of our egg catches inside the 50-fathom contour implies that the dabs that live deeper in our Gulf tend to work up into shoaler grounds to spawn. Beyond this, there is no reason to suppose that they gather in any definite localities for the purpose.

The temperatures and salinities in which the eggs are produced can be stated rather definitely for the Gulf of Maine because the dab lies close to the bottom, if not actually on it. The earliest spawning takes place at nearly the minimum temperature for the year, averaging about 37° for all the March and April stations where eggs were taken. And while the water warms to 41°–43° F. by late May and early June at the depths known to be inhabited by the ripe fish, we have not found its eggs where the bottom temperature was higher than about 40°. Thus the optimum for breeding may be set at 37°–40° for the Gulf of Maine as a whole. Dabs spawn freely in 31°–32° off Cape Breton, and even in water as cold as 29.3°–32° in the Gulf of St. Lawrence, in Newfoundland waters and northward, as well as along the West Greenland coast, proving that the lowest polar temperatures are no bar to the ripening of its sexual products. Neither does the distribution of the bottom stages suggest that warmer water is needed for the survival of the resultant larvae.

In the Gulf of Maine the dab spawns in relatively low salinities, the range there being only from about 31.8 per mille to about 32.8 per mille at the bottom at the stations where eggs were taken in any number. But it does so in considerably more saline waters in the other side of the Atlantic, generally speaking.

Although this flatfish spawns so generally throughout the whole area that it inhabits, there is evidence that different regions differ in their suitability as nurseries, either for its eggs or for the larvae. The southwestern part of the Gulf of Maine must be favorable in this respect, for we have taken larval dabs at 14 stations there, most of these off the Massachusetts Bay region. And they have also been taken at various localities off the southeast coast of Nova Scotia; on the Newfoundland Banks; in the Gulf of St. Lawrence; along the east coast of Newfoundland; in the Strait of Belle Isle; and northward for some distance along the outer coast of Labrador. But it seems that reproduction does not succeed in the Bay of Fundy, for neither the larvae nor the young fry have ever been found there, although dabs spawn there and the eggs develop, at least partially. Failure to find any dab larvae off the coast of Maine east of Penobscot Bay, though eggs are produced there in abundance may be due to the prevailing drift from northeast to southwest along this part of the coast, because of which buoyant eggs produced there are likely to hatch a considerable distance to the west of where they were spawned. The influence that this drift may have on the distribution of larval fish in the Gulf of Maine offers a fertile field for future study.

---

[99] Dannevig, Canadian Fisheries Expedition (1914–15) 1919, p. 18, figs. 11, 12, and 13.

[1] See Frost, Res. Bull. No. 4, Newfoundland Dept. Nat. Resources, 1938, chart 2, for the regional and seasonal distribution of dab eggs in Newfoundland and Labrador waters.

[2] Jensen, Meddel. Dansk. Komm. Havundersøgelser, Ser. Fiskeri, vol. 7, no. 7, 1925, p. 24.

Commercial Importance.—This is an excellent pan fish, but there is no special demand for it in New England markets, as distinguished from other flat fishes of about the same size. If the landings reported as "dab" do not include any significant proportion of other flounders, and if most of the dabs that are taken are reported under that name, the yearly catch in the Gulf by United States fishermen ranged between about 2,700,000 pounds and about 4,400,000 pounds for the period 1942 to 1947, averaging about 3,600,000 pounds. In 1946 Canadian fishermen brought in an additional 181,200 pounds from the eastern side of the Gulf and from the Bay of Fundy, plus an indeterminate amount landed in Shelburne County from Cape Sable to the Yarmouth County line.[3] We have no doubt that the catch could be increased greatly in our Gulf if any special demand were to develop for dabs.

The dab lives too far out from the land, on the whole, and too deep, and it does not bite eagerly enough for it to be of any interest to anglers along our shores.

Summer flounder *Paralichthys dentatus* (Linnaeus) 1766

FLOUNDER; FLUKE; PLAICEFISH

Jordan and Evermann, 1896–1900, p. 2629.

Description.—The summer flounder is left-handed; that is, it lies on the bottom on its right side, with its eyes on its left-hand side, and its abdomen is on its left edge as it rests on the bottom, which differentiates it at a glance from the American dab (p. 259). It is large-mouthed, like the sand flounder, which is similarly left-handed (p. 290); but its two ventral fins are alike and each of them is separated from the long anal fin by a considerable space, whereas the upper left-hand ventral fin of the sand flounder is continuous with the anal fin. The only Gulf of Maine flatfish with which the summer flounder shares its left-handedness, large mouth, and symmetrical ventral fins, is its close relative, the four-spotted flounder (p. 270), but the color pattern of the latter is distinctive (p. 270) and it has fewer fin rays. The summer flounder is one of our narrower flounders. Its dorsal fin (85 to 94 rays) originates opposite the forward margin of the eye; its anal fin has from 60 to 73 rays; the margin of its caudal is rounded, and its pectoral fins and ventral fins are smaller than those of the dab, relatively.

Color.—It has long been known that flatfishes are generally dark on a dark bottom and pale on a pale one. Perhaps the summer flounder is the most variable in color of all our local species and the one which adapts its pattern the most closely to that of the ground on which it lies. It is white below and of some shade of brown, gray, or drab above, like most flatfishes. But it can assume a wide range of tints, from nearly white on white sand through various hues of gray, blue, green,

[3] The landings for that year were 60,100 pounds for Shelburne County as a whole.

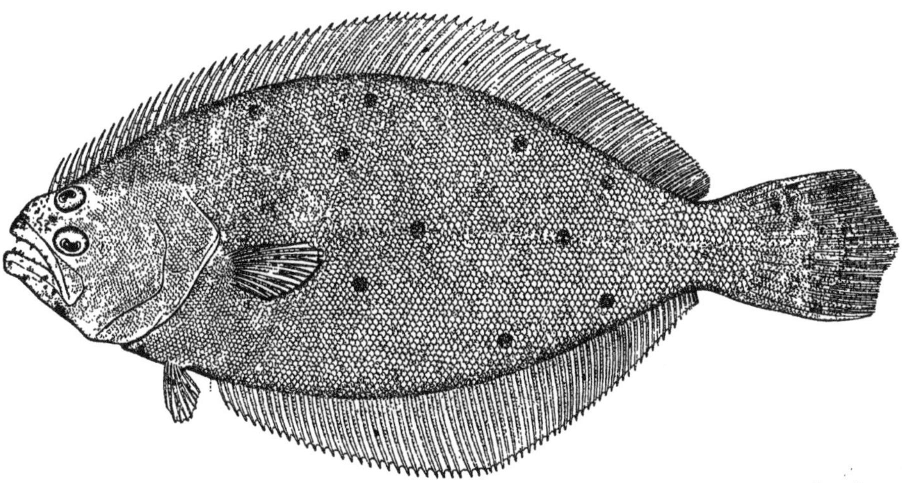

FIGURE 134.—Summer flounder (*Paralichthys dentatus*), Maryland. From Jordan and Evermann. Drawing by A. H. Baldwin.

orange, pink, and brown to almost black.[4] Its upper surface is variegated with pale and dark, as a rule, with the pattern fine or coarse according to the bottom, and it may or may not be marked with small eyespots of a darker tint of the general ground color. Mast's experiments show that it is slower in adapting its coloration to the actual colors of the bottom than to the general pattern, and also that it responds more rapidly to yellows and browns than to reds, greens, or blues, on which the adaptation may not reach its maximum for two or three months. He also observed that the skin simulates the pattern of the background, and does not reproduce the latter.

Size.—Summer flounders ordinarily grow to a maximum weight of 15 pounds or so, and to a length of 3 feet, or a little more, though one of about 30 pounds has been reported as taken off Fishers Island about 1915.[5] The largest of which we find definite record weighed 26 pounds. The largest on record, taken in sport fishing, was 37 inches long, weighing 20 pounds, caught at Oak Beach, N. Y., September 7, 1948, by F. H. Kessel, but the average size of the fish caught is only 2 to 5 pounds. The relation of length to weight is about as follows:[6]

Length	*Average weight, pounds*
15–16 inches	1 to 1¼
17–18 inches	2 to 2¾
20 inches	3 to 3½
22 inches	4
27 inches	8
30 inches	10
37 inches	20

Habits.—Many fluke come close inshore during the warm half of the year, when they are caught regularly both along open coasts and in bays and harbors, the smaller sizes often from docks and bridges, and some even run up into fresh water rivers. But the great majority of the population, especially of the larger ones, lie farther offshore even at that season, in depths of 8 to 10 fathoms and deeper, at least in the northern part of the fluke's geographic range, as illustrated by the fact that nearly 40 times as many (by weight) are landed in New Jersey and in New York by otter trawlers as from the many pound nets operating there.[7] And all of those that do come close inshore from Chesapeake Bay northward move offshore again at some time during the autumn, presumably to escape winter chilling.

The earliest landings from offshore of which we have heard for southern New England have been on October 6th, when some were brought in to Woods Hole from northwest of Nantucket Lightship, from 25 fathoms, and on the 16th of that same month, when the dragger *Eugene H* landed 6,000 pounds, taken west of Nantucket Lightship in about 25 fathoms. Corresponding to this, only a few are seen near Woods Hole after the middle of October, or after the last week of November near New York. And very few reappear near New York before the first week in May, or before about the 10th of May near Woods Hole.

It has been learned since the first edition of this book appeared that the medium sized and larger ones, at any rate, pass the winter and early spring out on the continental shelf from the 25 to 30 fathom contour about to the 80 fathom contour. Otter trawlers now make paying catches there as far north and east as the offing of southern New England, and as far south as the offing of northern North Carolina, during the part of the year when there are only a few "fluke" inshore, or none at all. In 1950 and 1951, for example, the *Eugene H*,[8] fishing in the general offing of Marthas Vineyard, brought in many fares ranging from a few hundred pounds to more than 20,000 pounds, between the first week of October and the third week of May, with the most productive fishing between early January and mid-April, from 25 to 75 fathoms. But it is doubtful whether many of them work deeper than that, for the *Albatross III* did not take any at depths greater than 80 fathoms off southern New England or New York in mid-May, 1950.

Fluke spend most of their lives on bottom, or close to it, as other flatfishes do. During their stay in shoal water they prefer sandy bottom, or mud, where they are often seen. And it takes one only an instant to bury itself to the eyes in the sand. Fluke often lurk in eel grass, or among the piling of docks; but they are swift swimmers when disturbed.

[4] Mast, Bull. U. S. Bur. Fish., vol. 34, 1916, p. 177.

[5] Nichols and Breder, Zoologica, N. Y. Zool. Soc., vol. 9, 1927, p. 177.

[6] From Goode, Fish. Ind., U. S., Sect. 1, 1884, p. 179; Hildebrand and Schroeder, Bull. U. S. Bur. Fish., vol. 43, Pt. 1, 1928, p. 167; and World Record Marine Game Fishes, Amer. Mus. Nat. Hist., New York, to 1950.

[7] 1947 landings, New York and New Jersey, about 2,300,000 pounds by otter trawlers; only about 80,000 pounds from pound nets.

[8] Information contributed by Capt. Henry Klimm.

This is a predaceous fish, like the halibut, feeding largely on smaller fish of various sorts, on squids, crabs, shrimps, and other crustaceans; on small shelled mollusks; on worms, and on sand dollars. It is very fierce and active in pursuit of prey, often following schools of small fish right up to the surface, to jump clear of the water in its dashes, actions very different from those of the sluggish dab and winter flounder.

Little is known of its breeding habits. The fact that nearly ripe females have been taken in October in Chesapeake Bay, in November and February at Beaufort, N. C., and as late as April 15, at 75 fathoms off Nantucket,[9] whereas Beaufort fish taken in March and April appeared to be spent, show that it is a late autumn, winter, and early spring spawner.[10] This implies that the flukes that spawn in the northern part of their range do so well offshore, and this may also be true of them in the southern part of their range, for fluke that were kept in aquaria at Beaufort through the winter failed to spawn.

The eggs of the summer flounder laid naturally have not been described yet. But it is likely that they are buoyant like those of the four-spotted flounder (p. 271). And their future "lefthandedness" and large mouths are foreshadowed at an early stage in the development of the larvae. Larvae either of the fluke, or of a form (*P. albiguttus* Jordan and Gilbert, 1882), so closely allied that it may prove a race of that species, resemble corresponding stages of the four-spotted flounder in their deep outlines and large heads, but the pigmentation on the rear part of their body is less dense. At a length of 16 mm. the right eye has nearly completed its migration, and the outlines of young fry 26 mm. long approach those of the adult.[11]

Young fry taken in Chesapeake Bay, had increased in length from about 0.9–2.4 inches long in May and June, to 3–5 inches in the last week of July; were 4.7–7.1 inches by December and January when one year old or a little less; about 8–10 inches long in the following October, when they were a little short of two years old; and they measured 10½ to 11 inches by their second May; i. e., when a little more than 2 years of age. The subsequent rate of growth has not been traced, so far as we know.

General range.—Continental waters of the eastern United States, from Maine to South Carolina, possibly to Florida,[12] chiefly south of Cape Cod.

Occurrence in the Gulf of Maine.—This is the most important flatfish commercially to the west and south of Rhode Island, and the one most sought after by sportsmen there. It is also plentiful offshore eastward to Nantucket Shoals and to the western part of the so-called South Channel, whence about 531,000 pounds were landed in 1947 (most recent year for which information is at hand). Trawlers also pick up a few on the southwest part of Georges Bank (about 6,000 pounds in 1947), as well as a fish here and there on other parts of the bank.[13] But there is no reason to suppose that fluke ever stray eastward and northward as far as Brown's Bank, or to outer Nova Scotian waters.

Coastwise, the angle of Cape Cod is the northern boundary to the regular range of the fluke in any great abundance. A number are caught each summer in Pleasant Bay, Chatham, Mass.,[14] where we read of one of 11⅛ pounds taken as early as the last week of May, in 1951,[15] a few in Town Cove, Orleans, some miles farther north, and a fluke is picked up occasionally by someone casting into the surf on the outer Cape Cod beach.[16] And they were so common near Provincetown and along the inner shore of Cape Cod as far as Wellfleet during the period from 1840 to 1850 that Captain Atwood carried them regularly thence to Boston, recording a catch of 2,000 pounds in a single afternoon inside Provincetown Harbor. But this is the most northerly region where fluke have ever been known to occur in commercial quantities. Even there its numbers were so reduced by a few years of hard fishing that they were described by Goode[17] in 1884 as "only

[9] Trawled by the *Eugene H*, Capt. Henry Klimm, in 1951.

[10] We dare not draw any conclusions as to spawning season from Hildebrand and Cable's table (Bull. U. S. Bur. of Fisheries, vol. 46, p. 470, table 12) of the seasonal distribution of young fry of different sizes because two species of flounders are included there.

[11] Our account of the young stages is based chiefly on Hildebrand and Cable's description (Bull. U. S. Bur. of Fisheries, vol. 46, pp. 469–475), from Beaufort, N. C.

[12] Florida is usually given as the southern limit for this flounder, but it is possible that the early records from that State (there are no recent ones) actually referred to the southern flounder (*P. lethostigmus*), a common Floridian fish.

[13] 645 pounds reported from the northwest part of Georges in 1947, 100 pounds from the northeastern edge, and 157 pounds from the central and southeastern part.

[14] This opens on the outer coast of Cape Cod.

[15] Reported in Salt Water Sportsman, June 1, 1951.

[16] There is a record of this, by Kendall, in 1896, and we have known of other cases, of late years.

[17] Fish. Ind. U. S., Sec. 1, 1884, p. 178.

occasionally taken" there. And they have never reappeared in any abundance, so far as we can learn, a fact suggesting that the local body of fish concerned was not very numerous, and that it received but few recruits from the more abundant stock to the southward.

The fluke is so rare a straggler north of Cape Cod Bay that there is only one definite record—for Casco Bay (specimens collected in 1873). We may add that we have never seen or heard of one caught in the inner part of Massachusetts Bay, and that it is unknown in the Bay of Fundy.

Importance.—This is one of the best of our flatfishes on the table, usually bringing a higher price than any other except the halibut; in 1947 it sold for 15 cents on the average in New Bedford, the halibut about 21 cents. And the landings of fluke from within the limits of the Gulf of Maine, totaling about 543,000 pounds (mostly from near Nantucket Shoals) were worth about $90,000 to fishermen that year. This is also the gamest of our flatfishes, biting freely on almost any bait, even taking artificial lures at times, while large ones put up a strong resistance when hooked. It is too bad that the fluke is not so common north of Cape Cod as it is to the south.

Four-spotted flounder *Paralichthys oblongus* 1815 (Mitchill)

Jordan and Evermann, 1896–1900, p. 2632.

Description.—This flatfish resembles the summer flounder (p. 267) so closely in its general make-up that we need mention only the points of difference. Most apparent of these are that it has fewer dorsal fin rays (72 to 81 dorsal and 60 to 67 anal rays, contrasted with 85 to 94 and 60 to 73, respectively, in the summer flounder), and that its mottled gray back is invariably marked with four large, oblong, and very conspicuous black eye spots edged with pale pinkish, two of them situated at each margin of the body, as the illustration shows (fig. 135). Incidentally, we have seen two of them on which the lower side, rearward from the gill openings was as dark as the upper side, and marked, similarly, with four eye spots; also others that were more or less dark below.[18]

This is also a much smaller fish than the summer flounder, for the adults average only about 12 inches long with 16 inches as about the maximum.

Habits.—Although this is a rather common fish about Woods Hole in May and June, and is still more numerous along the coast of New York, very little is known of its habits. It does not usually come into as shoal water as the summer flounder often does, being caught most often in 7 to 17 fathoms in Vineyard Sound, for example, near Woods Hole. And the many that have been trawled by the *Albatross II* and *Albatross III* between Georges Bank and northern North Carolina, have been generally distributed from about 23 fathoms down to at least 150 fathoms.

[18] Fish trawled by the *Eugene H* off Marthas Vineyard, Jan. 27 to Feb. 3, 1950, at 47 to 67 fathoms.

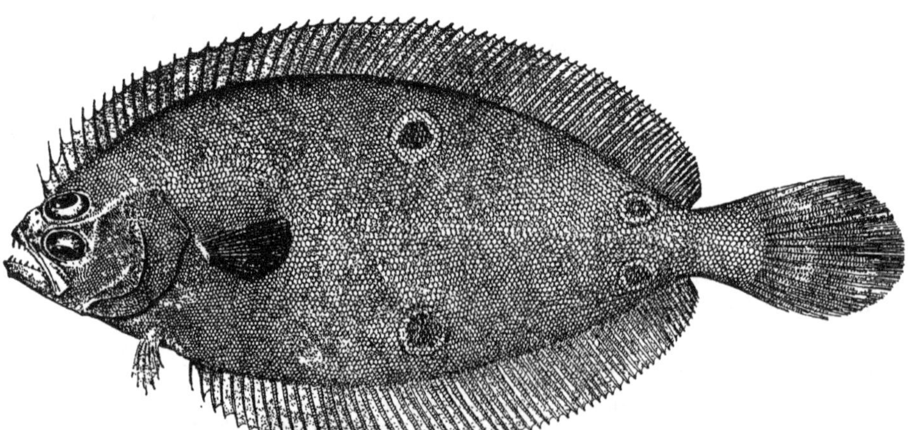

FIGURE 135.—Four-spotted flounder (*Paralichthys oblongus*), Woods Hole. From Goode. Drawing by H. L. Todd.

Its diet is much the same as that of the summer flounder; chiefly small fish and squid, with crabs, shrimps, shellfish, and worms.

It spawns from May until mid-July. The eggs are buoyant, 0.95 to 1.05 mm. in diameter, with a single oil globule of 0.16 to 0.19 mm.[19] The early stages have not been described previously, but certain large mouthed and lefthanded [20] flatfish larvae of 8 to 11 mm. that have been taken in tow nets off New Jersey, by the *Grampus* in 1913, and from Nantucket Shoals southward by the *Albatross II* subsequently, seem likely to have been young four-spots, not summer flounders, because they were taken in June and July (p. 269). If this identification is correct, an aggregation of pigment over the rear part of the trunk, combined with deep outline and a large head are distinctive for this species. Small fry of 2 to 3 inches have been taken at Woods Hole in autumn, evidence that the fry of this flounder complete their metamorphosis and take to bottom about 3 months after they are hatched.

General range.—This flounder has been taken between the eastern part of Georges Bank and the coast of South Carolina.[21] Its center of abundance appears to lie between southern New England and Delaware Bay.

Occurrence in the Gulf of Maine.—The four-spotted flounder is so plentiful along the continental shelf as far eastward as the general offing of Nantucket, and to the neighboring part of Georges Bank that we counted about 1,800 of them trawled there, by the *Eugene H* in 56 hauls at 47 to 67 fathoms, January 27 to February 3, 1950, also 968 of them on the southwestern part of Georges, in 26 to 75 fathoms, in late June 1951. And a few were trawled in 1931 by the *Albatross II* also along the southern and southeastern parts of Georges Bank. But this last is their most easterly known outpost.

The only records for the four-spot in the coastal waters of our Gulf are from Monomoy at the southern angle of Cape Cod; from the vicinity of Provincetown (where Storer saw a number of them in June 1847); and from somewhere on the northern shore of Massachusetts Bay where one was taken by the United States Fish Commission in 1878.[22]

This is a fair table fish but there is no market for it at present.

Yellowtail *Limanda ferruginea* (Storer) 1839

Rusty Flounder

Jordan and Evermann, 1896–1900, p. 2644.

Description.—The yellowtail is right-handed (that is, its eyes are on the right side and its viscera are at the right-hand edge as the fish lies on the bottom), and small-mouthed like the winter flounder, the smooth flounder and the witch. But it is easily distinguished from the first of these by its more pointed snout, thin body, arched lateral line, and more numerous fin rays; from the smooth flounder by the last two characters as well as by the concave dorsal (left hand) profile of its head and by being scaly between the eyes; and from the witch by its arched lateral line, its less numerous fin rays, concave dorsal (left) profile of the head, and especially by lacking the mucous pits on the left (white) side of its head that are conspicuous on the witch (p. 285).

The yellowtail is a comparatively wide flounder, nearly one-half as broad as it is long, with an oval body. The dorsal (left hand) outline of its head is more deeply concave than in any other Gulf of Maine flounder; its head is narrower; its snout is more pointed, and its eyes are set so close together that their rounded orbits almost touch each other. The fact that its mouth reaches scarcely as far back as the eyes, with its small teeth and thick fleshy lips, marks it off at a glance from all the large-mouthed flounders. The dorsal fin (76 to 85 rays) originates over the eyes, its middle rays are the longest. Its anal fin is similar in outline to the dorsal, but is much shorter (56 to 63 rays), and it is preceded by a short, sharp spine pointing forward. The two ventral fins are alike, and each of them is separated by a considerable space from the anal fin. But the pectoral fin on the blind side is slightly shorter than its mate on the eyed side. The scales are rough on the eyed side, but smooth on the blind side.

[19] Eggs artificially fertilized by O. E. Sette of the Bureau of Fisheries.

[20] Lefthandness foreshadowed in the larger ones by the fact that it is the right eye that had begun to migrate.

[21] The most southerly record for it is for five specimens trawled in the general offing of Charleston, S. C., lat. 33°00′ N., long. 77°44′ W., at 92 fathoms, by *Albatross III*, January 30, 1950. Another flounder, *Ancylopsetta quadrocellata* Gill, similar in appearance, for it is strikingly marked with four large spots, is found along the South Atlantic and Gulf coasts.

[22] In one paper (Am. Jour. Sci., Ser. 3, vol. 17, 1879, p. 40) Goode and Bean state that this specimen was trawled in Gloucester Harbor; in another paper (Bull. Essex Inst., vol. 11, 1879, p. 7) they credit it to the mouth of Salem Harbor.

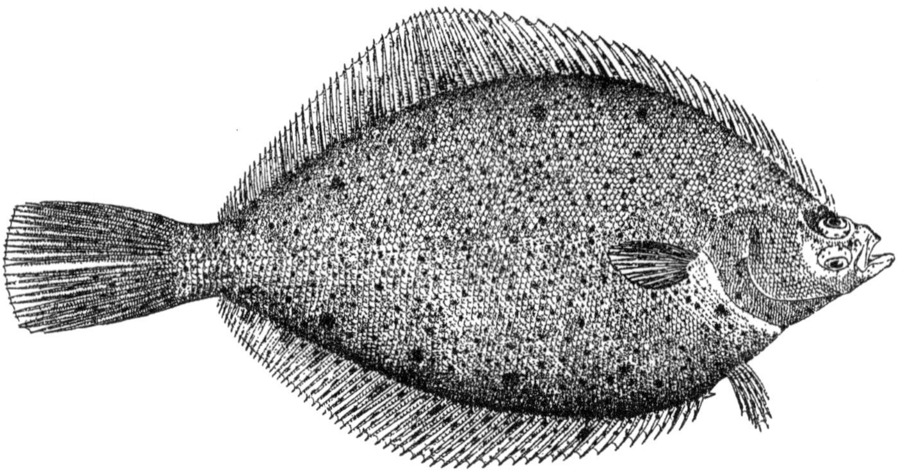

FIGURE 136.—Yellowtail (*Limanda ferruginea*), Gloucester, Mass. From Jordan and Evermann. Drawing by H. L. Todd.

Color.—The yellowtail is more constant in color than most of the other Gulf of Maine flatfishes. Its eyed side, including the fins, is brownish or slaty olive, tinged with reddish and marked with large irregular rusty red spots. The caudal fin and the margins of the two long fins are yellow, the yellow tail in particular being a very diagnostic character. The blind side is white, except for the caudal peduncle which is yellowish.

Size.—This is a medium-sized flatfish. Several hundred adults caught in gill nets between Cape Ann and Cape Elizabeth (measured by Welsh) ran as follows: Males, average length 15¾ inches, extremes 11¾ inches to 18¾ inches; females, average length 18 inches, extremes 15½ inches to 21¾ inches. This series includes the largest specimens that have ever been reported. A yellowtail 12 inches long weighs about one-half pound; one 15 inches long, about 1 pound; and one 18 inches long about 2 pounds.

Habits.—A yellowtail is caught in very shoal water now and then: We heard, for example, of several taken in Pleasant Bay, Cape Cod, in 1950. But 5 to 7 fathoms may be set as its upper limit, generally speaking. Thus it keeps to rather deeper water than either the winter flounder or the smooth flounder. On the other hand, most of those caught are at least from no deeper than 50–60 fathoms,[23] and the bulk of the catch is made shoaler than 40 fathoms. We saw many yellowtails trawled by the *Albatross III* off Marthas Vineyard and Nantucket in 20 to 40 fathoms, in May, 1950, but only 6 in 41 to 50 fathoms, and none in deeper water. Again, in late June 1951, *Eugene H* averaged about 240 yellowtails per trawl hawl, at 26 to 45 fathoms on the western part of Georges Bank, but took only three of them in deeper hauls.

Almost any sandy bottom or mixture of sand and mud suits them, and most of those that Welsh saw taken in gill nets on the Isles of Shoals-Boone Island grounds (p. 274) were over fine black sand between the hard, rocky patches. Rocks, stony ground, and very soft mud are shunned by yellowtails, as they are by most of the other flatfishes.

The yellowtail feeds chiefly on the smaller crustaceans such as amphipods, shrimps, mysids, and on the smaller shellfish, both univalves and bivalves, and on worms. It is also known to eat small fish, but it is not likely that it can catch these often. Its European relative also feeds on sea urchins, starfish, and on algae at times. And it is probable that our yellowtail would be found equally omnivorous were their stomachs examined from various localities. Fish in breeding condition are empty as a rule.

The diet of the yellowtail suggests that it is one of the more sluggish of our flatfishes, and there is no reason to suppose that it ever travels about much after it once takes to the bottom except that it has been described, in Massachusetts Bay, as "in-

[23] One was taken at 50 fathoms by *Albatross II*, September 5, 1926, on the northwestern part of Georges Bank, and two of about 10 inches at 90–95 fathoms on the northern edge of Georges Bank, by *Cap'n Bill II*, August 22, 1952.

habiting the deep water . . . in summer, and approaching the shores in winter," [24] as do various other ground fishes that tend to avoid high temperatures.

If the yellowtails are as stationary as they seem to be, they must be subject to considerable range of temperature from season to season at different depths, in one part of the Gulf or another, from a maximum of about 52°–54° to a minimum of about 33°–36°. And some of them are exposed to still lower temperatures on the Grand Banks, and in the Gulf of St. Lawrence.

The eggs of the yellowtail, artificially fertilized by Welsh in 1912, and hatched at the Gloucester hatchery, were buoyant, without oil globule, spherical, very transparent, and with a narrow perivitelline space. One hundred eggs measured by him ranged from 0.87 mm. to 0.94 in diameter, averaging about 0.9 mm. The surface of the egg is covered with very minute striations, and the germinal disk is of a very pale buff color while alive. The embryonic pigment gathers in three groups shortly before hatching (which takes place in 5 days at a temperature of 50° to 52°); one group on the head, a second group in the region of the vent, and a third group half way between the vent and the tip of the tail. Unfortunately the fish which Welsh hatched were destroyed accidentally, so we cannot describe the early larval stages. Larvae of 11 mm. are still symmetrical. But the left eye is already visible above the profile of the head at 14 mm. (fig. 139, *Grampus* specimen), all the fins are outlined, with their rays present in the final number (76 dorsal and 59 anal in the specimen illustrated). Thus, they show enough of the distinctive characters of the adult for positive identification.

The early larval stages of yellowtails and of winter flounders resemble one another closely; in fact, it is probable that some of the young flatfishes pictured by A. Agassiz [25] as winter flounders were yellowtails in reality. But the number of fin rays usually places the larvae in one species or the other after these appear. And the yellowtail does not take to bottom until upward of 14 mm. long, whereas the winter flounder completes its metamorphosis when it is only 8 to 9 mm. long.

[24] Goode and Bean, Bull. Essex Inst., vol. 11, 1879, p. 6.

[25] Agassiz, Proc. American Acad. Arts, Sci., N. Ser., vol. 6, 1879, pl. 4.

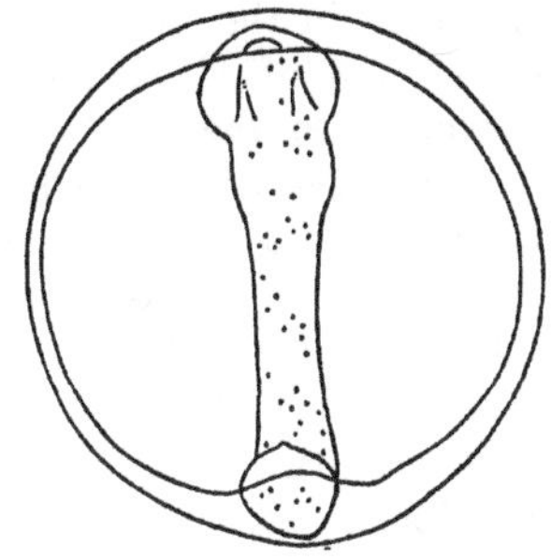

FIGURE 137.—Egg.

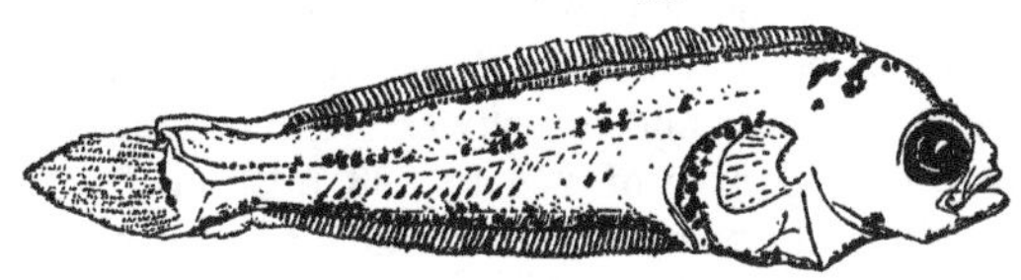

FIGURE 138.—Larva, 10.3 mm.

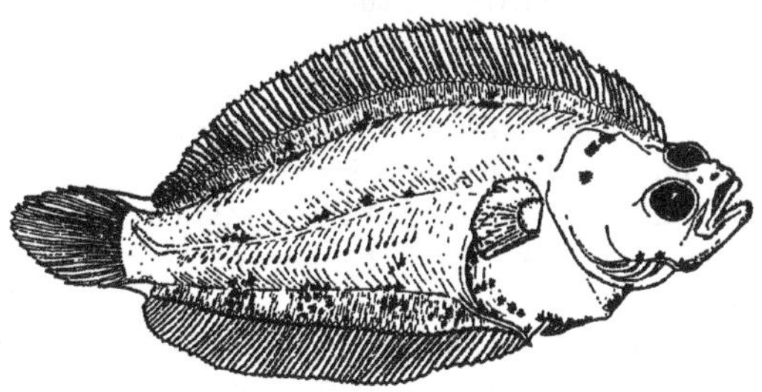

FIGURE 139.—Larva, 14 mm.

YELLOWTAIL (*Limanda ferruginea*)

Captures of young fish 2 to 4 inches long in February; 2¼ to 4½ inches long in April; 2⅛ to 5⅞ inches long in May; 3 to 5 inches long in June; and 3 to 6⅛ inches in July indicate that the yellowtail grows to an average length of about 5 inches by the time it is one year old. Its subsequent rate of growth has not been traced.

General range.—North American continental waters, from the north shore of the Gulf of St. Lawrence, the Labrador side of the Strait of Belle Isle,[26] northern Newfoundland (there are specimens from St. Anthony's in the Museum of Comparative Zoology), and the Newfoundland Banks, southward to the lower part of Chesapeake Bay.[27] It is most plentiful on the western

[26] Recent records from the Labrador side of the Strait are of one from Barge Bay, July 29, 1910 (Jeffers, Contrib. Canad. Biol., N. Ser., vol. 7, No. 16, 1932, p. 210); and of another taken at Forteau Bay, June 29, 1949, by the Blue Dolphin Expedition, reported to us by Richard H. Backus.

[27] We have records of one taken off Hog Island, Va., in lat. 37° 41′ S. (Bigelow and Schroeder, Bull. U. S. Bur. Fisheries, vol. 48, 1939, p. 340). And it was reported from the southern part of Chesapeake Bay by Uhler and Lugger (Rept. Comm. Fish., Maryland, 1876, p. 95; 2d Ed., 1876, p. 79.)

half of Georges Bank; in the western side of the inner parts of the Gulf of Maine; on the Nantucket grounds; and off southern New England.

This flatfish is represented in north European waters by the European dab, *Limanda limanda*, a close ally, from which it is distinguishable by its smaller scales, more pointed snout, more numerous fin rays, and shorter pectoral fins.

We should also mention the deep-water dab (*Limanda beanii* Goode), for while it has not been taken within the limits of the Gulf of Maine it would not be astonishing to find it on the seaward slope of Georges Bank, for it has been taken westward and southward from Marthas Vineyard in depths of 120 to 896 fathoms.[28] It differs from the rusty dab in a shorter head (occupying only two-elevenths of the total length instead of one-fourth); in the fact that the dorsal (left hand) profile of its snout is convex, not concave; in having only about 64 dorsal fin rays instead of 76 or more; in having only 88 rows of scales along its lateral line instead of 90 to 100; and in the fact that its tail fin is marked with a conspicuous black blotch on the outer rays on each side.

Occurrence in the Gulf of Maine.—Little was known of the distribution of the yellowtail in our Gulf previous to the introduction of the otter trawl there, for it is seldom seen close inshore; while its mouth is so small that one is seldom caught on hooks as large as those that are used for cod, pollock, or for haddock. But it has proved so abundant since then, in the general region of Nantucket Shoals and in the neighboring side of the so-called South Channel, that about 4,400,000 pounds were landed thence in the most recent year (1947) for which we have information. The western half of Georges Bank as a whole is good yellowtail ground also. But yellowtails seem to be less numerous on the eastern half of the Bank (though generally distributed there), and less so on Browns Bank, as is illustrated by the landings (in pounds) for 1947,[29] as follows: northwest Georges Bank, 930,000; southwest Georges Bank 1,740,000; northeast Georges Bank 210,000; central and southeast Georges Bank 540,000; and Browns Bank 40,310.

Yellowtails are so plentiful on the sandy bottoms in the eastern side of Cape Cod Bay, also, and on Stellwagen Bank, that (with winter flounders) they have long been the mainstay of the draggers that fish there; no doubt the greater part of the 1,150,000 pounds of yellowtails that were reported as taken off eastern Massachusetts in 1947 were trawled on these particular grounds. There are yellowtails in the deeper parts of Massachusetts Bay, too, as Goode and Bean[30] remarked long ago. And since Welsh saw many hundred of them taken in gill nets (not very effective gear for flatfishes) between the Isles of Shoals and Great Boars Head, during March and April of 1913, the yellowtail must be one of the most numerous of its tribe in the western side of the Gulf in general, in suitable depths.

Apparently they are less plentiful, however, around the Gulf to the north and northeast, for the reported catches for 1945 were only about 44,500 pounds for Cumberland County, Maine,[31] which covers Casco Bay and the grounds in its offing; only about 9,000 pounds for Knox and Hancock Counties combined, and none for Washington County in that particular year, though a few hundred pounds have been reported from "eastern Maine" in some other years.

We have taken no yellowtails in the deep basins of our Gulf nor have we heard of any there, probably because of the depth, for the bottom would seem hard enough for them in the eastern trough, at least, even if it is not in the western, or in the bowl west of Jeffreys Ledge. They certainly are uncommon in the Bay of Fundy, too, if not altogether lacking there. And though Huntsman did find a few in St. Mary's Bay, Nova Scotia, United States fishermen bring in only a few hundred pounds from off western Nova Scotia in some years, and none at all in others, though considerable amounts are brought in from the outer Nova Scotian grounds, as mentioned below (p. 275).

Most of the yellowtails that are caught in the inner part of the Gulf of Maine are in 10 to 30 fathoms of water, though they are reported in Shoal water at the mouth of Penobscot Bay; those caught on Georges Bank are in 20 to 45 or 50 fathoms (see p. 272).

[28] Localities are listed by Goode and Bean (Smithsonian Contrib. Knowl., vol. 30, 1895, p. 430).

[29] To nearest 10,000 pounds.

[30] Bull. Essex Inst., vol. 11, 1879, p. 6.

[31] Apart from those that were brought in to Portland by the large trawlers from more distant grounds.

Trawlers find yellowtails in even greater numbers off the southern New England coast than on Nantucket Shoals, at the proper depths, as illustrated by reported landings thence of about 17¼ million pounds in 1947.[32] And they are moderately plentiful offshore, as far as the offing of New York.[33] But southern New Jersey is about the southern limit to their regular ocurrence.[34]

Turning our attention eastward, we find the yellowtail plentiful all along the outer Nova Scotian banks, where about 2,700,000 pounds were taken in 1947 by vessels from Massachusetts,[35] besides about 2½ million pounds by Canadian vessels.

They are also reported as numerous on the southern part of the Grand Banks from experimental trawling by the Newfoundland Fishery Research Commission, but are "not in any numbers along the Newfoundland coast," [36] so far as is known, though they are recorded from as far north as the Strait of Belle Isle, as already noted (p. 273). They are also distributed generally in the Gulf of St. Lawrence, but no information is available as to their numbers there, for none are reported from the Gulf in the Canadian Fisheries statistics.

The neighborhood of the Isles of Shoals and of Boars Head, at 20 to 30 fathoms, certainly is an important spawning ground for the yellowtails; [37] so, too, the edges of Stellwagen Bank where we have caught spawning specimens. In fact, it is likely that yellowtail eggs are produced in abundance all around the western and northwestern periphery of the Gulf, between the 20 fathom and 50 fathom contours; few, however, in the eastern side, and none in the Bay of Fundy; nor have we found any of its eggs anywhere over depths greater than 50 fathoms. No doubt the yellowtail spawns as actively on the offshore Banks as it does inshore, for though we have not actually found its eggs there we have taken larvae only 7 to 11 mm. long over the western and eastern parts of Georges Bank in July,[38] as well as near Gloucester and near the tip of Cape Cod in July and August.

To the eastward and northward, yellowtail spawn on Sable Island Bank, Banquereau Bank, and the Newfoundland Banks, eggs (no doubt of this species) having been collected on these grounds by the Canadian Fisheries Expedition in 1915.[39] In the opposite direction, it certainly breeds as far westward as New Jersey, for our tow net yielded 88 of its larvae (6.5 to 19 mm. long) 11 miles off Sandy Hook on August 1, 1913; adult fish approaching ripeness have been trawled as far southward as Little Egg Inlet, N. J., in April 1930.

Spawning, Welsh found, begins on the Isles of Shoals-Boone Island ground by the middle of March; and many ripe fish were taken there during the last half of April, but the majority were still green as late in the season as May 8, though others were already spawned out. And spawning must last all summer, for we have trawled many ripe males and females in depths of 17 to 25 fathoms on the edge of Stellwagen Bank at the end of July; have taken eggs indistinguishable from those of the yellowtail in our tow nets in June, July, and August, with one even on September 11: and have taken its newly hatched larvae (6 mm. long) off Race Point as late as August 31. And the individual females evidently spawn over a considerable period of time, for Welsh found that only a small part of the eggs ripened simultaneously.

Importance.—The yellowtail is one of the most valuable of the flatfishes caught within the Gulf of Maine. It compares favorably in quality with the summer flounder and the winter flounder, but because its body is thinner it brings a lower price to the fishermen. Thus in 1947 the average price, as landed in New Bedford, was about 8 to 9 cents a pound for yellowtails; winter flounders, about 9 to 10 cents a pound; and summer flounders, about 17 to 18 cents a pound. All the yellowtails that are brought in find a ready sale and they make up a large part of the fillet of sole sold to consumers. In 1947 our Gulf yielded between 15 and 16 million pounds of them. But yellowtails live rather too deep to be of any interest to anglers.

[32] Landings in 1947 in Massachusetts ports, from grounds westward from Nantucket Shoals, about 12 million pounds; landings in Rhode Island, about 2¼ million pounds; landings in Connecticut, about 3 million pounds.

[33] About 3½ million pounds were landed in New York in 1947.

[34] *Albatross II* trawled many yellowtails as far southward as the offing of Delaware Bay (lat. 38° 32′ N., long. 74° 24′ W.) in 12 to 28 fathoms during February, April, and June, of 1929 and 1930.

[35] The Newfoundland Fishery Research Commission (Rept., vol. 1, No. 4, 1932, p. 110) reports 680 yellowtails taken per 10 hours trawling on Banquereau.

[36] Frost, Research Bull. 14, Newfoundland Dept. Nat. Resources, 1940, p. 15.

[37] Welsh obtained many ripe fish there.

[38] Station 10059, July 9, 1913; and station 10224, July 23, 1914.

[39] Dannevig (Canadian Fisheries Expedition [1914–15], 1919, p. 17) refers these provisionally to the European dab, which does not occur on our side of the Atlantic. Its egg is indistinguishable from that of the American species.

Winter flounder *Pseudopleuronectes americanus* (Walbaum) 1792

BLACKBACK; GEORGES BANK FLOUNDER; LEMON SOLE; FLOUNDER; SOLE; FLATFISH; ROUGH FLOUNDER; MUD DAB; BLACK FLOUNDER

Jordan and Evermann, 1896–1900, p. 2647.

Description.—This is a small-mouthed, right-handed species (eyes on the right side and viscera on the right). But it is easily separable from the yellowtail, which is similarly characterized, by the fact that its lateral line is nearly straight (at most only slightly bowed abreast the pectoral fin); that the dorsal profile of its head is less concave; that its nose is blunter; that its eyes are farther apart; that it has fewer fin rays; and that its fins are less tapering in outline. The most obvious differences between the winter flounder and the smooth flounder (p. 283) is that the former is rough scaled between the eyes, the latter smooth there, and that the winter flounder has the larger number of anal fin rays. On the other hand, it has only about two-thirds as many dorsal rays as the witch (p. 285); it lacks the mucous pits that are conspicuous on the left (lower) side of the head of the witch, and its tail is much larger proportionately than that of the witch. It is oval in outline, about two and one-fourth times as long to the base of the caudal fin as it is wide, thick-bodied, and with proportionately broader caudal peduncle and tail than any of our other small flatfishes.

Its dorsal fin (60 to 76 rays) originates opposite the forward edge of the eye, and is of nearly equal height throughout its length. Its anal fin (45 to 58 rays) [40] is highest about midway, and it is preceded by a short, sharp spine. Its ventral fins are alike on the two sides of the body, and both of them are separated from the long anal fin by a considerable gap. The mouth is small, not gaping back to the eye, and the lips are thick and fleshy like those of the yellowtail. The left (under) half of each jaw is armed with one series of close-set incisor-like teeth, but the right (upper) side has only a few teeth, or it may even be toothless. The scales are rough on the eyed side, including the space between the eyes, but they are smooth to the touch on the blind (white) side.

Color.—The winter flounder, like other flatfishes, varies in hue according to the bottom on which it lies, but it is the darkest of Gulf of Maine flatfishes as a rule. Large ones are usually of some shade of muddy or slightly reddish brown, olive green, or dark slate above, sometimes almost black. And they vary from plain or more or less mottled to definitely marked with smaller or larger spots of a darker shade of the general ground tone. There usually is a wide variation in this respect, among any lot of flounders. And fish caught on Georges Bank average more reddish in

[40] Perlmutter (Bull, Bingham Oceanogr. Coll., vol. 11, Art. 2, 1947, pp. 19, 20) gives a detailed tabulation, and graph of the number of dorsal and anal fin rays from upwards of 1,100 specimens including both the smaller inshore form and the larger Georges Bank form.

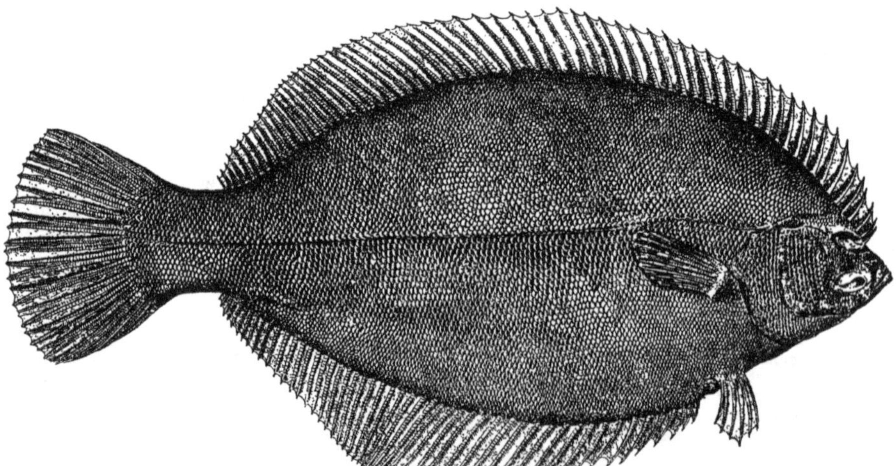

FIGURE 140.—Winter flounder (*Pseudopleuronectes americanus*). From Goode. Drawing by H. L. Todd.

hue than those caught inshore. But this rule does not always hold (see p. 277).

The blind side is white, more or less translucent toward the edge, where it is often faintly tinged with bluish, and the lower side of the caudal peduncle is yellowish on some specimens, but is pure white on others. The long fins usually are tinged with pinkish, reddish, or are yellowish on the eyed side; the ventrals and pectorals of the eyed side are of the general ground tone, but their mates on the blind side are pure white. Small fish average paler and more blotched or mottled than large ones.

Various color abnormalities have been recorded (fish, for example, that are partially white on the eyed as well as on the blind side, or with the blind side yellow-edged) and it is not uncommon to see specimens with dark blotches on the blind side. In fact, one-third of the fish caught near Providence, R. I., during the winter of 1897–98 were these "black bellies," as fishermen call them, but the commissioners of fisheries of that State estimated them as forming only 4 percent of the catch in 1900. And none (or at most only an occasional fish) has been seen since. In 1898, some fry that had been hatched artificially from eggs of black-bellied flounders were released in Waquoit Bay, southern Massachusetts, where this race had been unknown previously, and several "black bellies" 7 to 8 inches long (hence probably two years old) were taken there in 1900, probably the offspring of this planted stock.[41]

Winter flounders change color to some extent to suit their surroundings, usually being very dark on mud, and pale on bright sand bottoms. But field experience suggests that they have less control over shade and pattern than the summer flounder has.

Size.—The largest winter flounder on record caught inshore was one 22¾ inches long mentioned by Scattergood;[42] Nichols and Breder[43] report one 20 inches long, weighing 5 pounds; and Welsh saw three of about 19½ inches, weighing 3¼, 3¾, and 4 pounds, respectively, that were caught near Boon Island in April 1913. But fish longer than 18 inches or heavier than 3 pounds are unusual inshore, the general run of adults caught there being from 12 to 15 inches in length and 1½ to 2 pounds in weight. Flounders grow larger on Georges Bank, where many of 4–6 pounds are taken, and where they often are caught up to 7–8 pounds; we have handled one Georges Bank fish of 25 inches, weighing 8 pounds.

Remarks.—The winter flounder shows some tendency to break up into local races in the number of its fin rays,[44] in the size to which they grow, and perhaps in other characteristics.

The most interesting of these races, from the fisheries standpoint, is the population on Georges Bank, for the flounders tend to grow larger there than they do anywhere inshore. This fact was first brought to scientific attention in 1912, when some of these large flounders from Georges were received by the Bureau of Fisheries, to be made the basis of a new species, *Pseudopleuronectus dignabilis*, by Kendall.[45] Since that time this Georges Bank flounder has been accepted provisionally as a separate species, supposedly characterized by rather more numerous fin rays, by reddish color, and by a caudal peduncle yellow on the under side, as well as by large size. But our own comparison of specimens of the winter flounder group of various sizes, from Georges Bank, with others from the No Mans Land ground, from Nantucket Shoals, and from many localities, inshore, from Labrador to New York, leads us to conclude that it is simply a large, more rusty-brownish, local race of the winter flounder, for we find no definite regional discontinuity in the number of fin rays or of gill rakers, in the teeth, or in color (p. 277). The names "blackback" and "lemon sole," as used by fishermen, have no bearing on the case, for their choice of the one or of the other is based solely on the size of the fish in question (p. 282).[46]

Habits.—Tide mark, high or low according to the stage of the tide, is the upper limit for this flounder. It runs up into brackish water in river mouths, and we have even caught them in the Susquehanna River, tributary to Chesapeake Bay, where the water was fresh enough to drink.[47]

Its lower limit cannot be stated definitely. It is plentiful certainly at 10 to 20 fathoms in Cape

[41] Bull., U. S. Fish Comm., vol. 19, 1901, pp. 305–306.

[42] Copeia 1952, p. 206.

[43] Zoologica, New York Zool. Soc., vol. 9, 1927, p. 180.

[44] See Bumpus (American Naturalist, vol. 32, 1898, pp. 407–412) and especially Perlmutter (Bull. Bingham Oceanographic Coll., vol. 11, Art. 2, 1947, pp. 18–23) in this connection.

[45] Bull. U. S. Bur. Fish., vol. 30, 1912, p. 391, pl. 57.

[46] Perlmutter has already emphasized this point in his detailed study of the blackback (Bull. Bingham Oceanogr. Coll., vol. 11, Art. 2, 1947, p. 18).

[47] Hildebrand and Schroeder, Bull. U. S. Bur. Fish., vol. 43, Pt. 1, 1928, p. 170.

Cod Bay and on Stellwagen Bank, while the gill-netters sometimes take very large ones at about this same depth about Boon Island. According to general report, however, few, if any, are caught deeper than this in the inner parts of the Gulf except in the Bay of Fundy, where they are to be taken in winter on soft bottoms down to 30 to 50 fathoms. On Georges Bank they are taken mostly between 25 fathoms and 45 fathoms; 70 fathoms is the deepest definite record for them there of which we know. Usually the smaller fish live the shoalest and the larger ones deeper. But we have seen large flounders caught so often in only a few feet of water that no general rule can be laid down. The young fry are found chiefly in the shallows.

Most of those that are caught inshore are on muddy sand, especially where this is broken by patches of eelgrass. But winter flounders are common enough there on cleaner sand, on clay, and even on pebbly and gravelly ground. And the populations on the offshore banks are on hard bottom of one type or another. When they are on soft bottom they usually lie buried, all but the eyes, working themselves down into the mud almost instantly when they settle from swimming. And flounders that live on the flats usually lie motionless over the low tide to become more active on the flood, when they scatter in search of food. They keep near the bottom, and we have never heard of them coming up to the surface as the summer flounder so often does (p. 269). But though they spend most of their time lying motionless, they can dash for a few yards with astonishing rapidity, to snap up any luckless shrimp or other victim that comes within reach, or to snatch a bait, as any one may see, who will take the trouble to watch them on the flats on a calm day. It is in this manner that they usually feed, not by rooting in the sand. But flounders can sometimes be attracted by stirring the bottom with an oar when they are not biting, or by dragging anchor to bring up small animals from the mud, an old trick.

How close inshore they may come (how shoal) in any particular locality at any particular time depends largely on local conditions of temperature. Generally speaking, the summer temperature is low enough for their comfort close in to shore and up to within a few feet of the surface all around the open coast line of the Gulf, and among the island passages, but the winter temperatures may be uncomfortably low for them in enclosed situations locally. In Passamaquoddy Bay, for instance, where the temperature of the water falls close to the freezing point in winter, those that are closest inshore in summer work out in winter unless the year is a very mild one. Others, however, that are living at 15 fathoms or so remain there the year around, while it is only in winter that they are known to descend as deep as 30 to 50 fathoms in the Bay of Fundy.[49]

In shallow enclosed bays, however, or harbors, where extensive flats are heated by the sun at low tide in summer but are exposed to very severe chilling in winter, the flounders tend to desert the flats for the deeper channels during the heat of summer, work back again into shoal water in autumn, desert the ice-bound flats once more in winter, and then work up again in spring. Duxbury Bay is a case in point, also Barnstable Harbor, where we have speared many of them in spring, while wading on the flats.

A migration of flounders out into deeper water in the summer and back to shoal for the winter is generally characteristic south of New York, where the coastal waters are warmer, hence the common name "winter flounder." They are very scarce, for instance, in the bays of southern New Jersey in summer, but very plentiful there in winter. And many are caught in Chesapeake Bay from November to the first of June, but none are taken in shoal water there in summer or early autumn.

It has long been believed that the winter flounder is one of the most stationary of our fishes, apart from seasonal movements of the sorts just mentioned, and apart from a general tendency (recently emphasized by Perlmutter)[50] for the fry that are produced in bays and estuaries to work offshore as they grow older. This essentially stationary nature has been demonstrated recently by extensive marking experiments that have been carried out in Long Island Sound, along southern New England, and on the coast of Maine, for about 94 percent of the recaptures were made in the general areas where the fish had been tagged. Thus the population consists "of many independent localized stocks inhabiting the bays and estuaries along the coast" as Perlmutter words it,

[49] As proved by captures in shrimp trawls, as reported by Huntsman.

[50] Bull. Bingham Oceanogr. Coll., vol. 11, Art. 2, 1947, p. 17

with the fish merely tending to scatter "from population centers, a characteristic phenomenon with nonmigratory animals." [51] But some of them may stray for considerable distances. Thus winter flounders tagged at Waquoit Bay, near Woods Hole, in 1931 were recovered off Chatham, on the outer coast of Cape Cod, and on Nantucket Shoals. The case of one that was tagged near Block Island on April 17, 1941, and was recaught on the central part of Georges Bank (lat. 41°45′ N., long. 67°06′ W.) on August 27, 1945, is especially interesting,[52] as showing that some interchange does take place between the inshore and offshore populations of adult fish.

The normal distribution of the winter flounder covers a wide range of temperature at one season or another, from a minimum close to the freezing point of salt water around Newfoundland, in Nova Scotian waters, in the Gulf of St. Lawrence, and in the shoaler parts of the Gulf of Maine in late winter, to a maximum of about 64°–66° F. in shallow water in the southwestern part of the Gulf in summer, and of perhaps about 68°–70° in the southern part of its range.

They sometimes perish by the thousands in very hot spells of summer weather, if they are trapped in shallow enclosed bays, as happened in Moriches Bay, Long Island, N. Y., in 1917, between July 29 and August 4, when the air temperature rose to 82°–89°, and the temperature of the water on the very shallow flats nearly as high, probably.[53] But we have never heard of this happening in the Gulf of Maine where cooler water is always close to hand. On the other hand, they may succumb to anchor ice in winter if they are overtaken in very shoal water in a severe freeze, for dead "flounders" of one sort or another are sometimes reported in such locations after unusually severe weather. And observations at Woods Hole have shown that freezing temperatures (say 30° to 29°) drive them down into slightly warmer water.

Experience at the Boothbay and Woods Hole hatcheries, combined with the results of the trawl fishery (p. 283), proves that those living a few fathoms down are as active in winter as they are in summer, both north and south of Cape Cod. Bean, it is true, has described the winter flounder as going into "partial hibernation in the mud in winter,[54] but (as Breder[55] has pointed out) the reason the hook-and-line fishermen cannot take them in late winter or early spring may simply be that they will not bite then, this being the spawning period when winter flounders fast, as so many other fishes do.

According to Sullivan[56] diatoms are the first food taken after the yolk of the larval flounder is absorbed. A little later they begin preying on the smaller Crustacea, and Sullivan invariably found isopods in the stomachs of fry that had just passed their metamorphosis. A series of young flounders 1 to 4½ inches long from Casco Bay were found by Welsh to have fed chiefly on isopod crustaceans, with lesser amounts of copepods, amphipods, crabs, and shrimps, which together formed 36 percent of the stomach contents; worms (39 percent); mollusks (2 percent); and various unidentifiable material (22 percent). Linton[57] who examined about 398 young flounders of various sizes at Woods Hole, likewise found them feeding chiefly on amphipods and on other small Crustacea, together with annelid worms. And his tables of stomach contents show an increase in the ratio of mollusks to Crustacea as the fish grow. The adult winter flounder, like the yellowtail (p. 271), is limited by its small mouth to a diet of the smaller invertebrates and of fish fry. Sometimes they are full of shrimps, amphipods, small crabs, or other crustaceans; sometimes of ascidians, seaworms (*Nereis*), or other annelids; or of bivalve or univalve mollusks. Three hundred "seed" clams, for example, were found in an 11-inch flounder at St. Andrews, New Brunswick.[58] And it seems that they often bite off clam siphons that protude from the sand. They also eat squid, holothurians, and hydroids; occasionally they capture small fish; and they sometimes take bits of seaweed. Examination of the stomachs of adults taken at Woods Hole in February 1921 by Breder showed that they cease feeding when they are about to spawn.

In spite of its small mouth the winter flounder bites very readily on clams, pieces of seaworm, or

[51] Perlmutter, Bull. Bingham Oceanogr. Coll., vol. 11, Art. 2, 1947, pp. 26, 27.

[52] This specimen is on display at the Laboratory of the U. S. Fish and Wildlife Service at Woods Hole.

[53] This occurrence is described by Nichols (Copeia, No. 55, 1918, pp. 37–39), also by Nichols and Breder, Zoologica, N. Y. Zool. Soc., vol. 9, 1927, p. 79.

[54] Bull. 60, New York State Mus., Zool., 9, 1903, p. 778.

[55] Bull. U. S. Bur. Fish., vol. 38, 1923, p. 311.

[56] Trans. Amer. Fisheries Soc., vol. 44, 1914–15, No. 1, p. 135.

[57] App. 4, Report U. S. Comm. Fish. (1921) 1922, pp. 3–14.

[58] Fisheries Research Board of Canada, Progress Reports of the Atlantic Coast Stations, No. 52, January 1952, p. 3.

almost any other bait for that matter, provided the hook is small enough.

Breeding habits.—The winter flounder is a winter and early spring breeder, spawning from January to May (inclusive) in New England. The season is at its height during February and March south of Cape Cod and in the Massachusetts Bay region,[59] but it is somewhat later along the coast of Maine; near Boothbay spawning commences about March 1 and continues until about May 10 or 15 with the chief production of eggs usually taking place from March 30 to April 20, according to information supplied by Capt. E. E. Hahn, former superintendent of the Boothbay hatchery. Local differences of this sort in the spawning season are probably due to variations in the temperature of the water. After the severe winter of 1922–23, for example, when the vernal warming of the coastwise waters was slower than usual, Captain Hahn wrote us from Boothbay that "the fish were 10 to 15 days later in spawning than in any previous year, the first eggs being taken on March 24." On Georges Bank spawning fish have been reported in April and into May.

Thus spawning is well under way inshore while the water is still near its coldest for the year; i. e., about 32° to 35° F. in the Woods Hole region, about 32° to 37° near Gloucester, and about 31° to 35° near Boothbay, according to precise locality and depth. And the major production of eggs takes place there before the water has warmed above about 38°, with about 40° to 42° as perhaps the maximum for any extensive spawning in the inner parts of our Gulf. The picture is not so clear for Georges Bank, for we do not yet know how early in the season flounders commence spawning there. Those that spawn on the Bank in April may do so in temperatures ranging from about 38° to perhaps 42°, depending on the year, on the precise date, and on the locality.

Winter flounders spawn on sandy bottom, often in water as shoal as 1 to 3 fathoms, but as deep as 25 to 40 fathoms on George Bank, and they do so throughout the range of the fish, including the Bay of Fundy, where Huntsman found its larvae common near the mouths of estuaries. Most of the eggs are produced in salinities from about 31 to 32.3 per mille in the inner parts of the Gulf, to somewhere between 32.7 and 33 per mille on Nantucket Shoals and on Georges Bank. But those that spawn in estuaries are known to do so in brackish water, in salinities as low as 11.4 per mille near Woods Hole, for instance.

Individual females produce an average of about 500,000 eggs annually, and nearly 1,500,000 have been taken from a large one of 3⅜ pounds. They spawn at night, at least those did that were kept in the tanks at Woods Hole, where they seemed indifferent to the electric lights overhead. And Breder [60] describes the fish of both sexes as swimming in a circle, about one foot in diameter, clockwise so that the vent is outward, with the eggs from the females flowing back along the upper side of the anal fin and along the tail. After about 10 seconds of activity, they sink motionless to the bottom.

This species is peculiar among our local flatfishes in that its eggs are not buoyant but sink to the bottom, where they stick together in clusters, usually so closely massed that the individual eggs are forced into irregular outlines. They are 0.74 to 0.85 mm. in diameter, and newly shed eggs have no oil globule, but some of them (if not all) develop one as incubation proceeds.[61] Incubation occupies 15 to 18 days at a temperature of 37° to 38° F., which is about what they encounter in nature. The young larvae, which are 3 to 3.5 mm. long at hatching, are marked by a broad vertical band of pigment cells that subdivides the post anal part of the body, a characteristic feature; and the end of the gut also is heavily pigmented. In water of about 39° the larva grows to 5 mm. in length, and the yolk is absorbed (fig. 142) in 12 to 14 days. The vertical fin rays begin to appear in 5 to 6 weeks after hatching, at a length of about 7 mm., and the left eye has moved upward by then until about half of it is visible above the dorsal outline of the head, while the whole left eye shows from the right side and the fins are fully formed in larvae of 8 mm. Metamorphosis continues rapidly.[62] The left eye moves from this position to the right side of the head; the pigment fades from the blind side; the eyed side becomes uniformly pigmented; and the little fish now lies and swims with the blind side down, its metamorphosis complete when it is only 8 to 9 mm. long.

---

[59] This species was propagated artificially at the Woods Hole, Gloucester, and Boothbay hatcheries in large numbers.

[60] Copeia, No. 102, 1922, pp. 3–4.

[61] Breder, Bull. U. S. Bur. Fish., vol. 38, 1923, fig. 274g.

[62] Williams, Bull. Mus. Comp. Zool., vol. 40, 1902, No. 1, pp. 1–58, pls. 1–5. See also Sullivan (Trans. Amer. Fish. Soc., vol. 44, 1914–15, pp. 125–136, figs. 1–4) and Breder (Bull. U. S. Bur. Fish. vol. 38, 1923, p. 311).

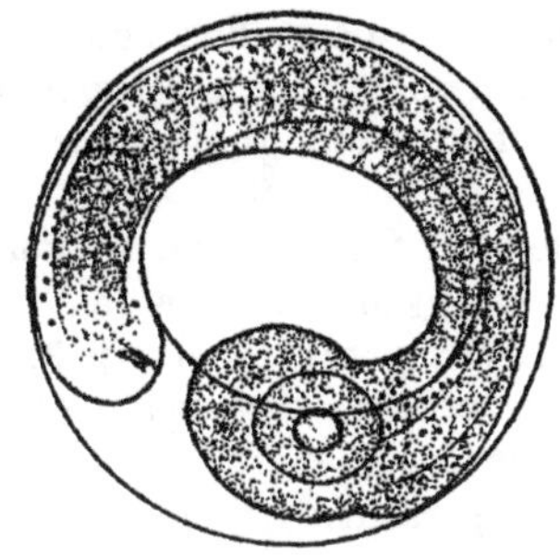

FIGURE 141.—Egg.

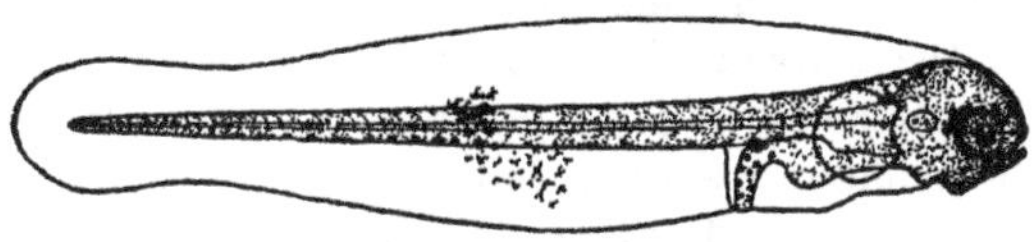

FIGURE 142.—Larva, 4.5 mm.

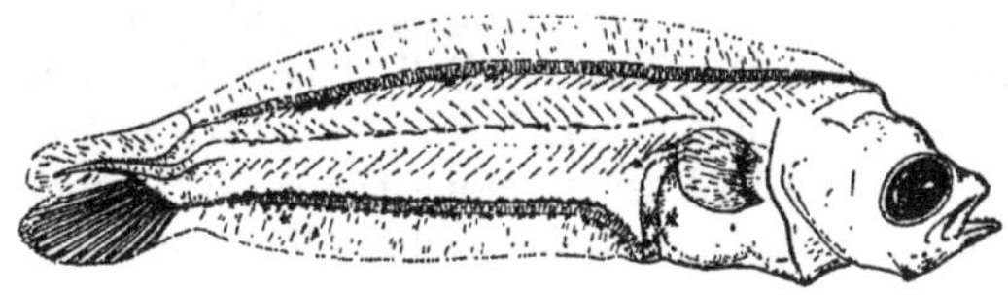

FIGURE 143.—Larva, 5 mm.

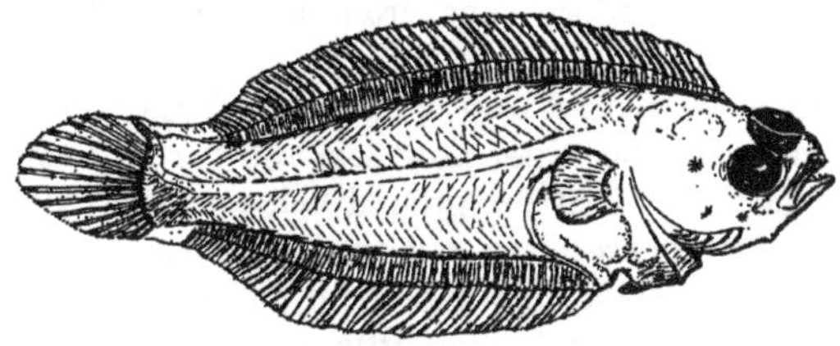

FIGURE 144.—Larva, 8 mm.

WINTER FLOUNDER (*Pseudopleuronectes americanus.*)

The youngest larval stages are made indentifiable as winter flounders by the pigment bar just mentioned. After the fin rays appear their small mouth separates them from any of the large-mouthed flounders; their short, deep body, combined with the small number of fin rays, separates them from the witch; and the number of fin rays marks them off from the yellow tail (p. 273). The winter flounder also completes its metamorphosis at a smaller size than either of these other small-mouthed flatfishes (pp. 287 and 273).

The rate of development of the larvae is governed by temperature, occupying from about 2½ to about 3½ months, according to the data available, and the larvae that are hatched later may catch up with the earlier hatched ones before their metamorphosis takes place. Larvae in their later stages have been taken in abundance in the tow nets at Woods Hole. But their habits in aquaria suggest that they are less at the mercy of the tide and current than our other flatfishes are, for they have been described as alternately swimming upward and then sinking, to lie for a time on the bottom, instead of remaining constantly adrift near the surface, as the larvae of most of the flatfishes do at a corresponding stage in their development. At any rate, we have not taken any in our towings in the open Gulf [63] that were certainly identifiable as winter flounder.

Judging from a large series from Casco Bay, measured by Welsh, and from others seen by us off near Boothbay Harbor and at Mount Desert, the fry of the previous winter grow to an average length of 1½ to 3¼ inches by August, with an occasional specimen as long as 4 inches; they are 2 to 4 inches long by the end of September; and 4 to 6 inches long off southern New England in January and February, when nearing 1 year old, which probably applies north of Cape Cod as well. They may grow somewhat faster in more southern (warmer) waters, as in Chesapeake Bay, where fish of the year are 4¼ to 7 inches long in January and February.[64]

Welsh also concluded, from measurements gathered from various sources, that the winter flounders are 5 to 7½ inches in length at 2 years of age, 7½ to 9½ inches at 3 years, and 9⅕ to 10 inches long when 4 years old, which accords with 8 to 10 inches at 2 to 3 years in New York waters as reported by Lobell [65] and by Perlmutter.[66] Probably they mature sexually at 3 years, for most of the spawners are upwards of 8 inches long. Our only information as to the rate of growth of older fish is that one tagged near Block Island, April 17, 1941, when it was 10⅞ inches long, was 17 inches long when it was recaptured on Georges Bank, 4 years and 4 months later.

[63] Three larvae taken in the Gulf in July 1912, were provisionally identified by Welsh as this species.

[64] Hildebrand and Schroeder, Bull. U. S. Bur. Fish., vol. 43, Pt. 1, 1928, p. 169.

[65] 28th Rept., New York Conserv. Dept. 1939, Sup., Pt. 1, p. 86.

[66] Bull. Bingham Oceanogr. Coll., vol. 11, Art. 2, 1947, p. 17.

General range.—Atlantic coast of North America from the coast line out to the offshore fishing banks; common from the Strait of Belle Isle,[67] the north shore of the Gulf of St. Lawrence where it has been characterized as "all along the coast,"[68] and southern and southeastern Newfoundland to Chesapeake Bay; recorded from the southern part of the Grand Banks,[69] and as far north as Ungava Bay, northern Labrador;[70] and from as far south as North Carolina and Georgia.[71]

Occurrence in the Gulf of Maine.—This is the commonest shoal water flounder, and perhaps the most familiar of all the ground fishes of the Gulf of Maine. There is no bay or harbor from Cape Cod to Cape Sable, no inter-island passage, and no stretch of open coast where it is not to be caught, unless the bottom be too smooth and hard, except, perhaps in the very turbid waters at the head of the Bay of Fundy.

As one looks down at low tide from some pier where the water is clear enough, or from a boat, drifting over the flats, one is almost sure to see a flounder here and there, lying partly buried in the sand or mud. And they often come into water so shallow that it is easy to spear them. A flounder spear used to be almost as familiar an instrument along our coasts as an eel spear.

With most of the flounder population of the inner parts of the Gulf living shoaler than 30 fathoms (20 fathoms is the deepest we have caught one there, close in to Little Duck Island, off Mount Desert), the zone occupied by them around the coast north of the elbow of Cape Cod is hardly as much as 8 to 10 miles wide, measured from the outer headlands or islands, except for Stellwagen Bank which lies a few miles farther out, and off Cape Sable, where their outer-depth limit lies something like 15 miles offshore. But their range extends out along the offshore rim of the Gulf, in somewhat deeper water, to include the Nantucket Shoals region as a whole (they must be plentiful to account for the 2 to 4 million pounds of blackbacks and lemon soles that are brought in from there yearly) and from the shoaler parts of Georges Bank.

The flounders on Georges run so much larger than they ordinarily do in-shore that they have been described as a separate species (p. 277). During the summer of 1913 these soles (as they are called now, if they weigh more than 3 pounds) constituted about 4 percent by number of all the fish of all kinds that were caught on Georges by the several otter trawlers that carried investigators from the Bureau of Fisheries. Nowadays most every otter trawling trip brings in anywhere from a few hundred to several thousand of them according to depth and precise location on the bank. About 4 million pounds of lemon soles (larger than 3 pounds) and blackbacks (smaller than 3 pounds) were brought in from Georges Bank as a whole in 1947.

They seem not to be so plentiful on Browns Bank, to judge from a catch of about 23,000 pounds of large sole and smaller blackbacks there by United States vessels in that same year. But much larger numbers are landed in the fishing ports along the outer coasts of Nova Scotia; about 420,000 pounds of flounders and soles combined, in 1946, the most recent year for which we have seen the Canadian Fisheries statistics.

Fluctuations in abundance.—Declining catches in the fyke nets that were used to take brood fish for the Booth Bay (Maine) hatchery leave no doubt that winter flounders were decidedly less abundant in that vicinity from 1934 to 1940 than they had been from 1925 to 1933. And some decrease in their abundance during the same period is indicated for the southern Cape Cod shore by the catch records of the Woods Hole hatchery; also along Connecticut and near New York, by the evidence of fishermen's logbooks.[72]

Importance.—The winter flounder, whether blackbacks or lemon soles, is the thickest and meatiest of all the flatfishes smaller than the halibut that are common on our coasts eastward and northward from the elbow of Cape Cod.

In 1946 (most recent year when statistics are available for the Canadian catch as well as for the United States catch), the inner parts of the Gulf, from the tip of Cape Cod around to Cape Sable,

[67] Jeffers (Contrib. Canadian Biol., N. ser., vol. 7, No. 16, ser. A, General, No. 13, 1932, p. 210) reports it as not uncommon at Raleigh, on the Newfoundland side of the Strait.

[68] Stearns, Proc. U. S. Nat. Mus., vol. 6, 1883, p. 125.

[69] At 2 stations, see Rept. Newfoundland Fish. Res. Lab., vol. 2, No. 3, 1938, p. 79.

[70] Reported from Fort Chimo, Labrador by Kendall (Proc. Portland Soc. Nat. Hist., vol. 2, Pt. 8, 1909, pp. 225, 233); specimen in U. S. Nat. Museum, collected in 1882 or 1883 by L. M. Turner and identified by T. H. Bean.

[71] Reported from Beaufort, N. C. (by Yarrow, Proc. Acad. Nat. Sci., Philadelphia, vol. 29, 1877, p. 205); from the Neuse River, near New Bern, N. C. (by Smith, North Carolina Geol. and Econ. Surv., vol. 2, 1907, p. 390); and from Georgia (by Hildebrand and Schroeder, Bull. U. S. Bur. Fish., vol. 43, Pt. 1, 1928, p. 170).

[72] For details, see Perlmutter, Bull. Bingham Oceanogr. Coll., vol. 11, Art. 2, 1947, pp. 6-13, who has made a special study of the blackback.

yielded not far from 4 million pounds of flounders smaller than 3 pounds (blackbacks) to New England fishermen,[73] and about 49,000 pounds of fish heavier than 3 pounds (lemon sole). Nantucket Shoals, and the neighboring side of the so-called South Channel yielded about 5 million pounds of blackbacks and 1 million of soles; Georges Bank[74] about 3 million pounds of the larger soles and about 600,000 pounds of the smaller blackbacks.

In addition to all this, Canadian fishermen caught some 4,400 pounds of flounders at the mouth of the Bay of Fundy on the New Brunswick side, 16,200 pounds of "flounders and soles" on the Nova Scotian side of the Bay, and 82,000 pounds off the west coast of Nova Scotia.

Our Gulf as a whole thus yielded something like 14 million pounds of winter flounders, large and small, in the year in question, which seems to have been a representative one.

Most of the commercial catch is made today by the otter trawlers, a small part on hook and line, or in nets of one sort or another. Years ago numbers were speared on the flats; as lately as 1919, about 7,000 pounds were reported as taken in this way on Cape Cod. But flounder spearing has gone out of fashion so completely of late that no flounders, only eels, are listed under the heading "spears" in the Massachusetts landings by gear for 1945 or for 1946.[75]

Flounder fishing, too, for amusement and for home use goes on in harbors, estuaries, and other sheltered situations all around the shores of the Gulf, from bridges, piers, and small boats. And the number taken in this way must be very large in the aggregate for flounders are easy to catch (as well as very toothsome) provided the hook is not too large (Nos. 4 to 8 are best) and the bait is on bottom. Pieces of clam, of large snails, of sea worms (*Nereis*) or of squid, shrimp, and mussels, all are good. And they will take angle worms.

Smooth flounder *Liopsetta putnami* (Gill) 1864

SMOOTHBACK FLOUNDER; EELBACK; FOOLFISH; CHRISTMAS FLOUNDER; PLAICE

Jordan and Evermann, 1896–1900, p. 2650.

Description.—This flatfish is right-handed (eyes on the right side) and small-mouthed like the winter flounder, yellow tail, and the witch. It resembles the winter flounder (with which it is often caught) closely in its general outline and in

[73] The amount cannot be stated any more precisely because of uncertainty as to how much of the 2½ million pounds of black backs reported that year from Cape Cod was caught off the Gulf of Maine coasts of the Cape, and how much off the south shore of Massachusetts.

[74] Including the statistical area classed as Eastern Side South Channel.

[75] "Spears" are not included for 1947; only "harpoons," for larger game.

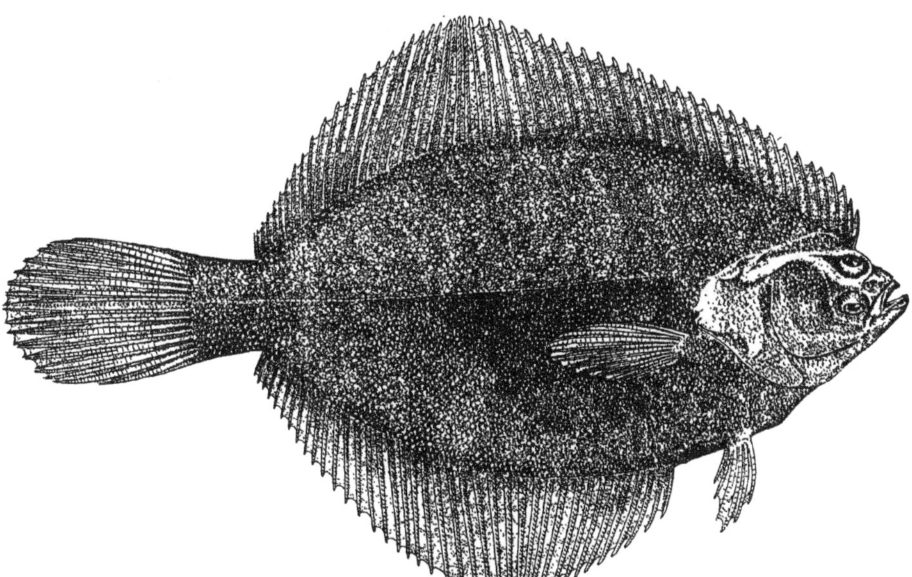

FIGURE 145.—Smooth flounder (*Liopsetta putnami*), Salem, Mass. From Goode. Drawing by H. L. Todd.

the considerable thickness of its body. But it is distinguishable from the winter flounder by the fact that the skin of its head between the eyes is smooth and scaleless. Females are more easily recognized than males, their bodies also being smooth to the touch on both sides; males are nearly as rough skinned on the eyed side (except between the eyes) as the winter flounder, but they have much longer pectoral fins than the latter. Both sexes have fewer anal fin rays (only 35 to 40) and dorsal fin rays (about 56), too, while the caudal fin of the smooth flounder is narrower and more rounded than that of the winter flounder.

The smooth flounder can always be separated from the yellowtail by the facts that its very prominent lateral line is straight, not arched, that the dorsal (left) profile of its head is straight, not concave; and that it has fewer fin rays. It has little more than half as many dorsal and anal rays as the witch, and its long fins are highest midway of the body and tapering toward the head and tail, whereas they are nearly uniform in height from end to end in the witch. It lacks the mucous pits that are so characteristic of the blind side of the head of the latter, a convenient field mark for separating these two species.

The smooth flounder is peculiar among our local flatfishes for its sexual dimorphism. Besides the difference in the scales of the two sexes noted above, the pectorals on the eyed side are longer (about four-fifths as long as the head) and more pointed on the males than they are on the females.

Color.—The smoothback varies from grayish to dark muddy or slaty brown above, or to almost black, either uniform or variously mottled with a darker shade of the same tint; the dorsal, anal, and caudal fins are of the general ground color. These fins were mottled darker or paler, in specimens we have examined, but Storer described them as black spotted. The blind side is white.

Size.—This is the smallest flatfish that is common in the Gulf of Maine, for it grows to a maximum length of only about a foot, and to a weight of about a pound and a half.

Habits.—This flatfish is confined to the close vicinity of the coast throughout its geographic range, occurring chiefly in estuaries or river mouths, and in sheltered bays and harbors; mostly on soft mud bottom. Correspondingly, it is found from tide line down to a maximum depth of perhaps 15 fathoms, with 2 to 5 fathoms as its zone of greatest abundance in our Gulf.

It prefers soft bottom to hard; so much so that a seine haul on soft mud yielded 23 smooth flounders to 4 winter flounders in St. Mary Bay, whereas another haul, only 100 yards or so distant, but on harder bottom, brought in only 3 smooth flounders to 189 winter flounders, as we learn from Dr. Huntsman's notes.

The shoal water habit of the smooth flounder exposes it to temperatures close to the freezing point of salt water in winter, and as high as 60° in summer, and perhaps higher temperatures still in some places. Little more is known of its life. But its small mouth suggests a diet similar to that of the winter flounder, and Kendall found that young fry 3 to 4 inches long from Casco Bay has been feeding chiefly on small crabs, shrimps, unidentified crustaceans, and polychaete worms.

Winter is its breeding season, females nearly ripe having been taken in Salem Harbor in December and spent fish at Bucksport, Maine, the first week in March, which corroborates fishermen's reports of more than half a century ago that it comes into Salem Harbor to breed at about Christmas time. It is not known whether the eggs sink or are buoyant, nor have its larvae been seen.

General range.—The smooth flounder is Arctic-boreal. It is definitely recorded from as far north as Ungava Bay, hence no doubt occurs along the Atlantic coast of Labrador; it is described as the most plentiful flatfish along the coasts of the Strait of Belle Isle at all seasons;[76] its young are common in Pistolet Bay on the Newfoundland side of the Strait in shallow sun-warmed pools,[77] and there are two specimens from the north shore of the Gulf of St. Lawrence in the Museum of Comparative Zoology (collected many years ago, labeled "Labrador").

Evidently it is widespread on the southern side of the Gulf of St. Lawrence, for it is the next most plentiful flatfish after the winter flounder on the Cape Breton shore and at the Magdalens, according to Cox;[78] it is reported from Prince Edward

[76] Jeffers (Contrib. Canadian Biol., N. Ser., vol. 7, No. 16 (Ser. A, No. 13 1922, p. 210). There are specimens from St. Anthonys, northern Newfoundland, in the Museum of Comparative Zoology.

[77] Rept. Newfoundland Fishery Res. Comm., vol. 1, No. 4, 1932, p. 110.

[78] Contrib. Canadian Biol. (1918-1920), 1921 p. 113.

Island,[79] also from Trois Pistoles;[80] and it has been classed by Huntsman [81] as characteristic of the estuarial transition from fresh to salt waters in the southern side of the Gulf of St. Lawrence generally. We find no record of it on the outer coast of Nova Scotia between Cape Breton and Cape Sable; but we suspect that it has been overlooked there, for it is widespread in the Gulf of Maine to Massachusetts Bay, as detailed below, and has been reported as a stray as far south as Providence, R. I.

Its range probably is continuous in the north with that of its polar relative (*L. glacialis*) of the Arctic coasts of North America and Siberia. Indeed, it is a question whether any valid distinction can be drawn between the two species.

Occurrence in the Gulf of Maine.—The smooth flounder is to be found in estuaries, river mouths and harbors, all along the shores of the Gulf, from the Bay of Fundy to the northern side of Massachusetts Bay. Localities whence it has been recorded in print, or has been definitely reported otherwise, are Annapolis basin, Minas Channel and St. Mary Bay; Grand Manan; Bucksport at the mouth of the Penobscot River; Belfast in Penobscot Bay; Casco Bay; Portland; Salem Harbor; and Boston Harbor. Apparently the latter is the southern limit to its regular occurence for while there is a specimen in the Museum of Comparative Zoology, from Provincetown at the tip of Cape Cod, it seems to be unknown in Cape Cod Bay, along the outer shore of Cape Cod, or in the Woods Hole region, though a stray individual has been caught at Providence, R. I.[82]

This flatfish (often confounded with the winter flounder) has been found so often in various markets among the winter flounders as to suggest that it is more plentiful along the coasts of northern New England, than is realized, generally.

In Casco Bay and in estuaries of the Bay of Fundy such as the mouths of the St. Croix and Annapolis Rivers it is abundant in summer, which no doubt applies equally to the intervening coast line. But it is said to run up into harbors in Massachusetts Bay in autumn and winter only;[83] nor would such a local difference be astonishing in the case of a cold-water fish, which might well be driven out into slightly deeper water by summer heat in the southern and western parts of the Gulf, but not in the northern and eastern parts.

Commercial importance.—This is an excellent table fish for its size, sweet-meated and thick-bodied like the winter flounder. But it is neither large enough, plentiful, nor widely enough distributed in the open Gulf to be of any commercial importance.

Witch flounder *Glyptocephalus cynoglossus* (Linnaeus) 1758

GRAY SOLE; CRAIG FLUKE; POLE FLOUNDER

Jordan and Evermann, 1896–1900, p. 2657.

Description.—The witch or "gray sole" as it is now named in the United States fishery statistics, is right-handed (viscera on the right hand as the fish lies) and small-mouthed like the winter flounder, the smooth flounder and the yellowtail. But there is little danger of confusing it with any of these for its fin rays are much more numerous, its body narrower relatively, its head much smaller, and the open mucous pits on the blind side of its head large and conspicuous. It is two and one-half to three times as long as it is broad (deep, in reality), elliptical in outline, very thin but with its head occupying only about one-fifth of the total body length, and it has a very small mouth. The dorsal (left-hand) profile of its head is convex. It has 100 to 115 dorsal fin rays and 87 to 100 anal rays, and the anal fin is preceded by a short, sharp spine pointing forward, which is a prolongation of the post-abdominal bone. The two long fins are of about uniform width throughout most of their lengths, except that they narrow gradually toward head and tail. The pectoral fins and the ventral fins are alike on the two sides, or nearly so, while the caudal fin is much smaller, relatively, than that of the yellowtail, of the winter flounder, or of the smooth flounder, though similarly rounded in rear outline.

The lateral line is straight, as a rule, but it is somewhat arched abreast the pectoral fin in some specimens. The teeth are small, incisorlike, and in a single series. There are about 12 open mucous pits or depressions on the blind side of the head, and less obvious ones on the eyed side also. The whole body and head (except for the tip of the

---

[79] Cornish, Contrib. Canadian Biol. (1906–1910) 1912, p. 81.

[80] Vladykov and Tremblay, Nat. Canad., vol. 62 (Ser. 3, vol. 6), 1935, p. 82; (many specimens reported).

[81] Trans. Roy. Soc. Canada, Ser. 3, vol. 12, Sect. 4, 1918, p. 63.

[82] This specimen, formerly in the Museum of Comparative Zoology, is no longer to be found.

[83] Our experience corroborates this to the extent that we have never seen it there in summer.

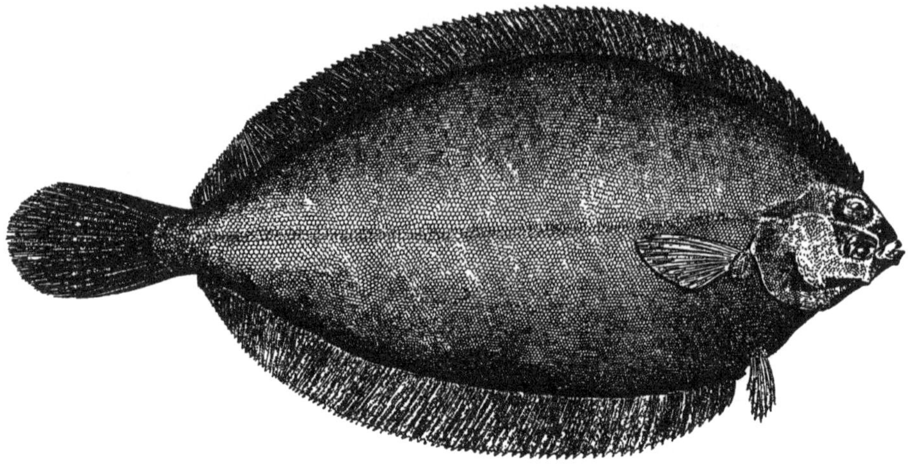

FIGURE 146.—Witch flounder (*Glyptocephalus cynoglossus*). From Goode. Drawing by H. L. Todd.

snout and the lower jaw) are scaly, but the scales are smooth to the touch, which make the witch as slippery to hold as a female smooth flounder (p. 284).

Color.—By all accounts (and the fish we have seen are in line with this) the witch is less variable in color than most of the flatfishes. Most of them are brownish or russet gray on the eyed side, either uniform or with darker transverse bars, with the vertical fins of the general body hue, tinted or tinged with violet, and either plain or spotted. The pectoral fin membrane on the eyed side is dusky or even black, a feature distinctive of this particular flatfish. The lower (blind) side is white, and more or less dotted with minute dark points. An occasional fish is colored on the under side as well as on the upper side; one of this sort, 19 inches long, was landed at the Boston Fish Pier early in March 1931.

Size.—The maximum length is about 25 inches, and fish of 23 or 24 inches, weighing about 4 pounds, are not uncommon. But the general run of those caught are only about 12 to 20 inches long.

Habits.—The witch flounder is rather a deep-water fish, seldom caught shoaler than 10 or 15 fathoms once it has taken to bottom, though taken occasionally close inshore (see footnote, p. 288). Off the American coast the best catches are made between about 60 fathoms and about 150 fathoms. Thus the *Albatross III* caught an average of about 57 witch flounders per trawl haul at 100 to 150 fathoms on the southwestern part of Georges Bank in mid-May 1950. but an average of only about one fish per haul between 31 fathoms and 80 fathoms. And they have been trawled widespread down the continental slope as deep as 858 fathoms off southern Nova Scotia; to 732 fathoms off Marthas Vineyard; to 788 fathoms off Chesapeake Bay; and to 602 fathoms off North Carolina.[84]

In Swedish waters, according to Melander,[85] the best catches are made between 80 and 140 fathoms.

They are caught most abundantly on fine muddy sand, on clay, or even on mud. They are said to frequent hard reefs in Scandinavian waters, but this does not seem to be the case in the Gulf of Maine, though they are common there on the smooth ground between rocky patches.

When the witch has once taken to the bottom it seems to be even more stationary in our gulf than some other flounders, for it is caught the year round, with no evidence that it moves in or off shore with the change of the seasons. In Swedish waters, however, it is said to work up into shoaler water in autumn, and deeper again in late winter and spring.[86]

It occurs in the Gulf of Maine in temperatures ranging from about 35°–38° F. (late winter and early spring), to 45°–48° (late summer and early autumn), according to precise locality and depth. In the Gulf of St. Lawrence it occurs in the icy

[84] Goode and Bean (Smithsonian Contr. Knowl., vol. 30, 1895, p. 433) give a long list of deep-water stations for the witch off southern New England.

[85] Pub. de Circonstance No. 85, Cons. Internat. Explor. Mer, 1925, p. 3.

[86] Melander, Pub. de Circonstance, Cons. Internat. Explor. Mer. No. 95, 1925, p. 3.

cold waters (30°–32°) on the banks as well as in the higher temperatures (40°–42°) of the deep channels.[87] Apparently it is never found in any numbers in water warmer than 50°, but we hesitate to propose high temperature as the factor barring it from shoal water because there is no evidence that it works inshore in our gulf in winter when this bar would not operate.

Food.—It feeds on invertebrates, like other small-mouthed flatfishes; European experience points to small crustaceans, starfish, small mollusks, and worms, as its chief diet.[88] It is not known to eat fish and does not take a bait often.

Breeding as it does through a long season, over many degrees of latitude, and in both sides of the Atlantic, the witch spawns in temperatures ranging from close to the freezing point of salt water up to 48°–50° F. (p. 288). And experiments, added to captures of eggs naturally spawned, and of newly hatched larvae, have shown that incubation proceeds normally in water at least as cold as 45°–46° F., and as warm as 50°–55° F.

The eggs are buoyant, spherical, transparent, with narrow perivitelline space (the perivitelline space is broad in the eggs of the dab or Canadian plaice, which overlap them in dimensions), without oil globule, and 1.07 to 1.25 mm. in diameter. As noted (pp. 288 and 203), there is danger of confusing newly spawned witch eggs with those of the cod and haddock, for they overlap these in size and in season. But identity is easily recognizable after a few days' incubation, for black pigment is to be seen in the gadoid eggs soon after the embryo is visible as such, but does not appear in the witch-flounder eggs until after hatching.

Incubation occupies 7 to 8 days at temperatures varying from 46° to 49° F., and the newly hatched larvae are about 4.9 mm. long, with a larger yolk sac than those of our other flatfishes. The yellow and black pigment becomes aggregated into five transverse bands on body, yolk (now much reduced in size), and fin folds within a few days after hatching, when the larva is 5 to 6 mm. long. One of these bands is at the region of the pectoral fin, one at the vent, and three of them on the trunk rearward from the vent. The yolk is entirely absorbed in about 10 days after hatching, the caudal rays have begun to appear at a length of 15 mm., the rays of the vertical fins are well advanced at 21 mm. and they are complete in their final number at about 30 mm. The eyes are still symmetrical, or nearly so, up to this stage. But the left eye has moved to the dorsal surface of the head in larvae of about 40 mm. And the migration of the eye is complete at a length of 40 to 50 mm., when the young fish takes to the bottom.

The witch is perhaps the most easily recognizable of Gulf of Maine flatfishes throughout its

[87] According to Huntsman, Trans. Roy. Soc. Canada, Ser. 3, vol. 12, Sect. 4, 1918, p. 63.

[88] No witch-flounder stomachs have been examined in the Gulf of Maine, so far as we know.

Figure 147.—Egg (European). After Cunningham.

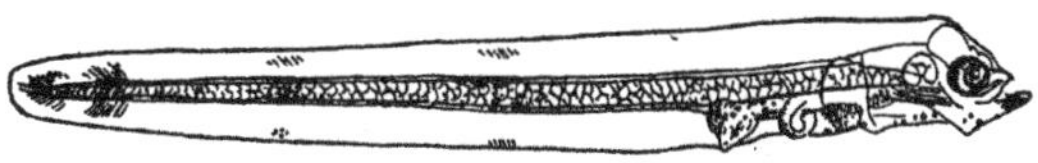

Figure 148.—Larva (European), 10 days old, 5.6 mm. After Holt.

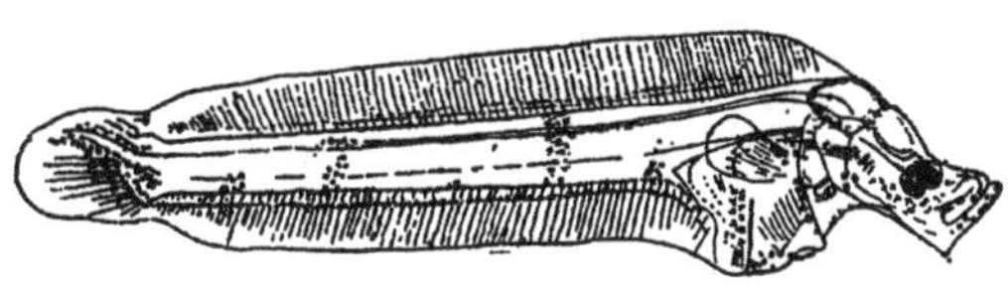

Figure 149.—Larva (European), 16 mm. After Kyle.

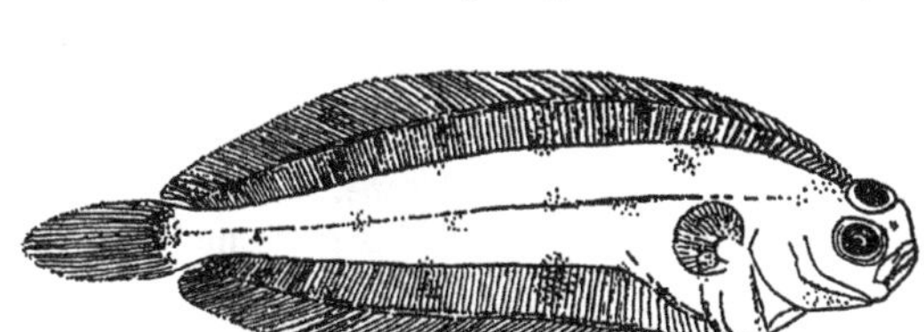

Figure 150.—Smallest bottom stage (European), 42 mm. After Petersen.

Witch flounder (*Glyptocephalus cynoglossus*)

larval stage. The transverse pigment bars are diagnostic prior to the appearance of the caudal rays, while the curiously concave ventral profile of the throat region with the comparatively long slender trunk are equally so, thereafter. And the great number of dorsal and anal fin rays, coupled with the small mouth, make identification easy after the fins are formed. The witch also grows to a larger size before it completes its metamorphosis than does any other of the right-handed, small-mouthed flatfishes that are found in the Gulf of Maine.

Measurements of the young (American as well as European), suggest that the free-drifting stage may last as long as 4 to 6 months for the witch, which is much longer than for any of our other flatfishes.

Fry of 2½ to 4¼ inches, and of 3½ to 4⅝ inches, such as we have trawled in July and August, respectively, probably are in their second summer, their sizes depending on how early in the season they were hatched the year before. The subsequent rate of growth has not been traced for American fish. If Molander's [89] estimate for European fish is correct, the size group centering at 6¾ to 8 inches that was prominent in our August catches of 1936 were in their third summer.[90] And subsequent growth is very slow.

General range.—Moderately deep water in both sides of the North Atlantic. Its European range is from northern Norway and Iceland south to the west coast of France. In American waters its free-drifting larvae are reported from as far north as the Strait of Belle Isle, around the coasts of Newfoundland, and over the Grand Banks region in general.[91] The adult is known from the Gulf of St. Lawrence; the south coast of Newfoundland; the southern part of the Grand Banks; in Cabot Strait; along outer Nova Scotia and the Scotian Banks; throughout the Gulf of Maine; and thence westward and southward along the continental shelf and slope as far as the offing of northern Virginia (lat. 37° 50′ N.) in moderate depths, to the offing of Cape Hatteras in deep water.[92]

Occurrence in the Gulf of Maine.—The distribution of this flatfish in our Gulf is governed by the fact that it is a fish of at least moderately deep water, seldom caught as shoal as 10 fathoms.[93] In fact, its very existence remained unsuspected by Massachusetts fishermen until 1877, when the United States Fish Commission caught numbers of them while trawling in the deeper parts of Massachusetts Bay. Since that time it has been reported (or we have trawled it, or both) from St. Mary Bay on the Scotian side of the Gulf; in the Bay of Fundy and its tributaries (where Huntsman describes it as taken very generally below 15 fathoms, if not in any great numbers); at Eastport; off Mount Desert, where we have trawled it as shallow as 10 fathoms; near Monhegan Island; off Seguin Island; off Cape Porpoise; near the Isles of Shoals (where Welsh saw a few taken from the gill nets set in about 25 fathoms in April 1913); in the deep trough to the westward of Jeffreys Ledge; in Ipswich Bay; near Gloucester; off Boston Harbor; at various localities in the deeper parts of Massachusetts Bay; and in both branches of the deep trough of the Gulf west and east down to a depth of 140 fathoms; in the deep channel, between Browns Bank and Georges Bank, and on the slope to the southeast.

Trawlers bring them in regularly from Browns Bank, also from Georges, where Welsh found them widespread, and from Nantucket Shoals.

This is enough to show that the witch is to be expected anywhere in our Gulf where the water is deeper than 15 to 20 fathoms, if the bottom is suitable.

The largest catches are made on the so-called South Channel grounds which include the slopes that lead down from the offing of Cape Cod on the one side and from Georges Bank on the other, into the southwestern part of the basin; farther north off eastern Massachusetts; and off western Maine. And the published statistics suggest that gray soles are about as plentiful as the American dabs are on the various grounds where the trawlers work regularly.

Reported landings of gray soles by New England vessels in 1947 were as follows for the several statistical areas:[94] Browns Bank, 44,000 pounds; off western Nova Scotia, 2,000 pounds; off eastern Maine, 17,000 pounds; off central

[89] Based on the structure of the otoliths; Pub. de Circonstance, No. 85, Cons. Perm. Internat. Explor. Mer. 1925, pp. 12–14.

[90] Bigelow and Schroeder, Biol. Bull., vol. 76, 1939, pp. 318–319.

[91] See Frost, Res. Bull. No. 4, Newfoundland Dept. Nat. Resources, 1938, Chart 6, for Newfoundland localities.

[92] Goode and Bean (Smithsonian Contrib. Knowl., vol. 30, 1895, p. 433) list it from lat. 34° 39′ N., 603 fathoms.

[93] A stray specimen, picked up in a pound net at Eastport, Maine, many years ago, was reported by Gill (Proc. Acad. Nat. Sci., Philadelphia, 1873, p. 360) as a new species, *Glyptocephalus acadianus*.

[94] To the nearest 1,000 pounds.

Maine, 12,000 pounds; off western Maine, 630,000 pounds; small grounds in west central part of Gulf, 77,000 pounds; off eastern Massachusetts, 582,000 pounds; South Channel grounds, east and west, 629,000 pounds; other parts of Georges Bank, 94,000 pounds; Nantucket Shoals region, 16,000 pounds.[95]

More precise evidence as to their local numbers on suitable bottoms in the appropriate depths is that as much as 500 pounds have been taken in a 15- to 20-minute haul with a small beam trawl in Massachusetts Bay, and that we caught 48 of them in Ipswich Bay in 22 fathoms, in a short haul with an 8-foot beam trawl on July 16, 1912. We also saw 519 of them, 10 to 22 inches long, trawled on the southwestern part of Georges Banks by the *Eugene H*, in 41 hauls at 26 to 65 fathoms in late June 1951, and learned that this dragger caught 9,000 pounds on the northeastern edge of Georges Bank, in 85 to 95 fathoms, October 12–18, 1951.

Neither the witch flounder nor the American dab is as plentiful as the yellowtail on good flounder bottoms, or the flatfishes of the winter flounder group (blackbacks plus lemon soles). And *Atlantis* took only 156 witch to 279 dabs on soft bottom at 90 to 103 fathoms during experimental trawling in the deeper parts of the Gulf in August 1936.

Gray soles are at least moderately plentiful off southern New England. The *Albatross III*, for example, took 90 there in one trawl haul at 101 to 150 fathoms in mid-May 1950, a few as shoal as 31 to 40 fathoms. And a few thousand pounds are landed yearly in New York and in New Jersey ports.[96] But records of the witch from farther south than New Jersey are of an occasional fish only.

Reported landings suggest that gray soles are about as plentiful all along the Nova Scotian banks as they are in the Gulf of Maine region. In 1947, for example, New England vessels landed about 555,000 pounds of them from the various grounds from the eastern part of Browns Bank to Banquereau, about half of which came from the Horseshoe ground between Halifax and Sable Island.[97] And they seem to be moderately plentiful in the southern part of the Gulf of St. Lawrence, for Cox [98] wrote of many (large and small) as taken off the Cape Breton shore, and in Cabot Strait off Cape North. But no information is available as to their numbers elsewhere in the Gulf of St. Lawrence, or on the Grand Banks.

It seems that the witch does not breed successfully in the Bay of Fundy; at least its eggs have never been found there, nor have its larvae. But probably it does so in other parts of the Gulf in general, including the offshore Banks, though our only positive egg records for it have been off Penobscot Bay, and at the mouth of Massachusetts Bay. And there is no reason to doubt that the more northerly populations are equally self supporting, for the pelagic larvae have been taken at many localities on the more easterly of the Nova Scotian Banks; in the Gulf of St. Lawrence; over the Grand Banks; and along the south and east coasts of Newfoundland, by the Canadian Fisheries Expedition of 1915, and during the cruises of the Newfoundland Fishery Research Commission more recently.[99] But there is no evidence that the witch spawns to any extent to the west of Cape Cod.

Captures of eggs, certainly of this species, in our tow nets in July and August, with larvae up to 20 to 23 mm. long as early as the first week of July, but others as small as 9 to 10 mm. as late as mid-October, show that the witch is a late spring and summer spawner in the Gulf of Maine as it is in European waters also, with the peak of production probably falling in July and August. Thus its spawning season overlaps that of the haddock. (p. 207).

Its eggs are shed in the Gulf of Maine in temperatures ranging from 39° to 41° F. at the beginning of the season, to 43° to 48° in midsummer. But (being buoyant) the temperature may be considerably higher at the level where their development takes place than deeper down where the spawning fish lie. In fact, it is doubtful if any eggs develop in our Gulf in water as cold as 42° to 43°. Neither is there any reason to suppose that witch eggs develop in water any colder than this in the Gulf of St. Lawrence, or off Newfoundland, for the surface stratum to which they rise after

[95] An additional 182,000 pounds were landed in Cape Cod fishing ports, source not stated.

[96] About 19,000 pounds in New York in 1947, about 28,000 pounds in New Jersey.

[97] Only a few thousand pounds are reported yearly from Nova Scotia, in the Canadian fishery statistics.

[98] Contrib. Canad. Biol. (1918-1920) 1921, p. 113.

[99] See Frost, Research Bull. 4, Newfoundland Dept. Nat. Resources, 1938 Chart 6, for Newfoundland localities.

they are shed is comparatively warm (upward of 45°) in these seas also, during the spawning season.

At the other extreme, our captures of eggs and of newly hatched larvae near the surface in July prove that the latter may be hatched in the Gulf in water at least as warm as 50° to 55°. But the upper limit to normal development cannot be stated from the evidence yet in hand, for with a temperature gradient as steep as it is over most of the Gulf of Maine in summer a difference of only a few fathoms in the depth at which the eggs or young larvae are suspended may mean a difference of several degrees of temperature.

One result of the protracted spawning season, combined with the long period occupied by larval development, is that witch larvae of various sizes are to be taken in tow nets throughout the summer and early autumn, as appears from the following table of our catches on the *Grampus*.

Date	Number of larvae	Length in millimeters	Date	Number of larvae	Length in millimeters
July 7, 1915	109	8 to 23.5.	Aug. 15, 1912	3	18.5 to 37.5.
July 8, 1913	19	8.5 to 21.5.	Aug. 24, 1912	6	10 to 18.
July 9, 1913	1	14.	Aug. 25, 1914	19	10 to 19.
July 19, 1916	100+	5 to 19.	Aug. 26, 1913	2	8 and 14.
July 22, 1912	1	9.5.	Aug. 29, 1916	100+	5 to 19.
July 24, 1912	2	8.5 and 16.5	Aug. 31, 1912	20+	9 to 16.5.
Aug. 5, 1913	27	5.5 to 12.5.	Sept. 29, 1915	22	10 to 14.
Aug. 9, 1913	7	10 to 23.	Oct. 18, 1915	1	9.5.
Aug. 14, 1912	1	18.5.	Nov. 1, 1916	20+	29.5 to 50.

All of these catches, like those for other larval flatfishes, and for larval gadoids, have been concentrated in the southwestern part of the Gulf, which must be an important nursery for the witch also. And we may note in passing that the presence of young fry at all stages from immediately after their metamorphosis (that is, 4 to 6 months old) in the Bay of Fundy, where few or none are hatched, points to an immigration of the late larvae, or of the youngest fry, into the Bay, either just before they take to the bottom or soon after they have done so.

Importance.—The witch was of no commercial importance in our Gulf a quarter of a century ago; few fishermen distinguished it from other flounders then, and no record was kept of the catch. It is an excellent table fish; and the bases of its fins are provided with astonishingly large amounts of gelatinous fat for so thin a flounder, of the sort for which the European turbot is famed.

It is now in such demand that it brings about as high a price as either the yellowtail or the American dab. In 1947, for instance, the average price at Massachusetts ports was about 7 cents for gray sole, about 8 cents for yellowtail, about 7 cents for dab, about 9 to 12 cents, according to size, for flatfish of the winter-flounder type (blackbacks and lemon sole), about 17 cents for summer flounders (fluke), and about 25 cents for halibut. The Gulf yielded between 2 million and 2½ million pounds of gray sole both in 1946 and in 1947, corresponding to something like 1 to 1¼ million individual fish.

The otter trawl is the only gear now in use in our waters that is adapted to the capture of witch flounders on a commercial scale. They live too deep, and their mouths are too small for them to be of any concern to small-boat fishermen.

Sand flounder *Lophopsetta maculata* (Mitchill) 1814

WINDOWPANE; SPOTTED FLOUNDER; NEW YORK PLAICE; SAND DAB; SPOTTED TURBOT

Jordan and Evermann, 1896–1900, p. 2660.

Description.—This is the closest North American relative of the European turbot and brill. It is left-handed (eyes and viscera at the left-hand side) and large-mouthed, like the summer and four-spotted flounders, but it is readily separable from both of these by the outlines of its ventral fins. In all other Gulf of Maine flatfish (except for the hogchoker, p. 296) these are narrow at the base and widen toward the tip, but the ventrals of the sand flounder are as wide at the base as they are at the tip, each simulating a detached segment of the anal fin. Furthermore the two ventral fins are not alike either in location or in size, the left-hand (upper) fin, which is the longer of the pair, being practically a continuation of the anal fin so far as its appearance goes, whereas the right-hand (lower) ventral fin is situated a short distance up the right-hand side of the throat. The general appearance of the dorsal fin is no less diagnostic, for its first 10 or 12 rays are not only free from the fin membrane along the outer half of their lengths, but they are branched toward their tips, so that they form a conspicuous fringe which is without parallel among Gulf of Maine flatfishes. Furthermore, the sand flounder is more nearly round in outline than any of our other local flatfishes (it is only about one and one-half times as long as it is broad), and so thin through that its body is translucent when it is held up against the light. Its pectoral fins, too, are longer than in our other left-handed flatfishes; its caudal fin is more rounded; and its teeth smaller although the gape of the mouth is wide.

The dorsal (right) fin (63 to 69 rays) tapers toward the tail; the anal (left) fin (46 to 52 rays) tapers toward head and tail, while both of these fins are noticeably thick and fleshy at the base; and there is no free anal spine. The pectoral fin on the eyed side is longer and more pointed than its mate on the blind side; the scales are smooth to the touch; and the lateral line is bowed abreast of the pectoral fin.

Color.—The sand flounder varies less in color than most shoal-water flatfishes do, the general ground tint of its eyed side (both as described by previous authors and in those we have seen) being of a pale and rather translucent greenish olive or slightly reddish or light slaty brown more or less mottled with darker and paler, and usually (if not always) dotted with many small brown spots of irregular shapes. Some fish are also marked on the body and on the bases of the dorsal, anal, and caudal fins with white spots that vary in number and in size from fish to fish. But others lack these spots. The dorsal, anal, and caudal fins are of the general body tint, more or less mottled with darker, while the pectoral of the eyed side is dark crossbarred or speckled. The blind side is white in most of them, but specimens have been seen on which it was irregularly dark-blotched.[1]

Size.—The sand flounder is said to grow to a maximum length of 18 inches and to a weight of 2 pounds. But the largest we have seen (from Waquoit on the southern shore of Massachusetts), were about 15 inches long. And adult fish run only about 10 or 11 to 12 inches in length. Sand flounders from southern New England measured by Moore[2] averaged about ¼ pound at 8 inches: about ½ pound at 10 inches; about ¾ pound at 12 inches; and a little more than 1 pound at 14 inches.

Habits.—The sand flounder is a shoal-water fish. Its upper limit is close below the tide mark, and the 20 to 25 fathom line probably marks its lower limit, in general, in the coastal zone north of Cape Cod. But Moore reports it as occurring regularly down to 27 fathoms off Connecticut, and

[1] Moore, Bull. Bingham Oceanogr. Coll., vol. 11, art. 3, 1947, p. 20.
[2] Bull. Bingham Oceanogr. Coll., vol. 11, art. 3, 1947, p. 53, fig. 12.

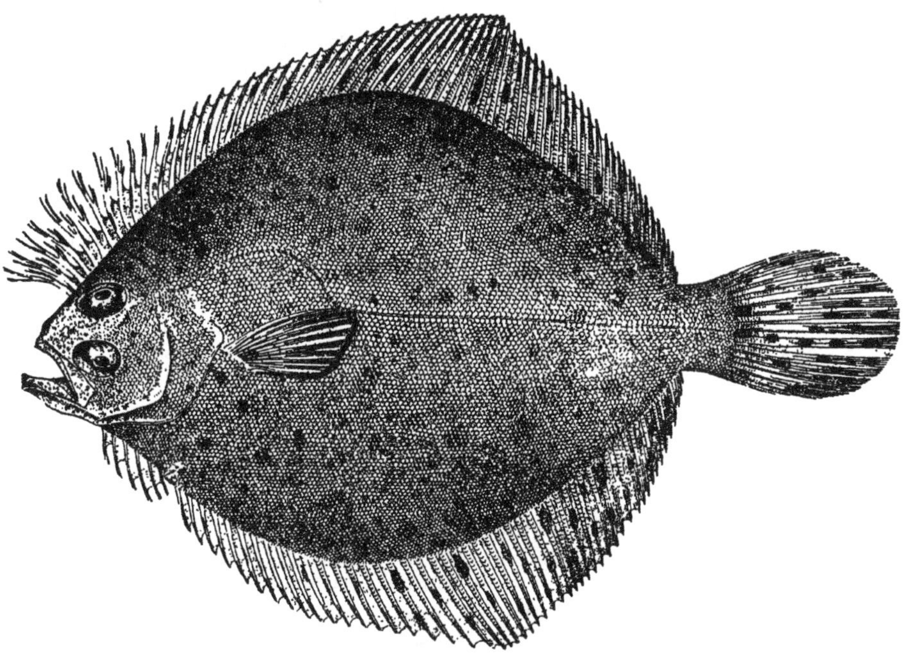

FIGURE 151.—Sand flounder (*Lophopsetta maculata*). From Jordan and Evermann. Drawing by H. L. Todd.

Welsh saw it taken on Georges Bank down to 30 or 40 fathoms, while the *Albatross III* trawled a few on the southwest part of the Bank along this same depth zone in May 1950.

It is caught chiefly on sand bottom off southern New England and southward, as its name implies, but its comparative abundance in Casco Bay and in Minas Channel shows that it also frequents softer and muddier grounds in the Gulf of Maine.

The sand flounder is a year-round resident off the southern New England coast, and probably this applies to it in the Gulf of Maine also, there being no evidence that the adults carry out any migrations inshore or offshore, with the change of the seasons. But such of the young fry as settle to bottom in shallow water inshore tend to work offshore as they grow, and deeper, while tagging experiments off southern New England have shown that individual sand flounders may wander along the coast for considerable distances, or across open water, much as winter flounders do (p. 279). Some of them went as far as 80 miles in 3 months.[3] And it is probable that the wanderings of the adults play an important part in the intermingling of local populations.

The adult sand flounder is necessarily attuned to a wide temperature, occurring as it does over many degrees of latitude, and in shallow waters where it is exposed to the extremes of winter chilling and of summer warming. Such of them as winter in shoal bays experience winter temperatures close to the freezing point of salt water in winter, not only in the northern part of their range, but even as far south as the Connecticut shore.[4] And it is probable that the entire population in the Gulf of Maine winter in water colder than 36° F. But these same fish summer in temperatures of 50° to 70°, according to locality and depth. And some sand flounders summer in still higher temperatures farther south. Nevertheless, it seems that temperature is the factor that governs the northerly range of the species and its local abundance, for it is only where the surface waters warm to 55° or higher in summer, as happens in Massachusetts Bay, in Casco Bay, in Minas Channel, and over the southern shallows of the Gulf of St. Lawrence, that the sand flounder is able to maintain itself in any numbers. Apparently either its eggs or its young larvae, or both, fail to develop in lower temperatures (p. 293). And these isolated breeding centers are not productive enough to stock the intervening stretches of shoreline in the case of a fish as stationary as the sand flounder. Thus its distribution is somewhat analagous to that of the oyster.

The large mouth of the sand flounder suggests that it feeds on active prey. Welsh, in his field notes, remarked, in fact, that sand flounders caught off Atlantic City, N. J., were full of "schizopod shrimps" (mysids) and of these alone, and mysid shrimps (*Neomysis americana*) had similarly been the predominant item in all months of the year, for 654 Long Island Sound fish examined by Moore,[5] with shrimps of other kinds ranking second. Moore also concluded that the few fishes included in their diet were not enough to class the sand flounder as a fish eater. But hake, herring, launce, and silversides have been found in their stomachs at Woods Hole, while North Carolina specimens had eaten fish, also crabs and shrimps.[6] And we suspect that they seize small fish whenever they can, for we once hooked a sand flounder only about 12 inches long on a 2½-ounce metal jig, while we were casting for striped bass in the surf on Orleans Beach, Cape Cod.

A variety of small invertebrates other than shrimp have also been found in their stomachs; Vinal Edwards noted annelid worms, crabs, squid, small mollusks, ascidians and even seaweed, to which Moore adds gammarids and other small Crustacea, worm tubes, sea cucumbers (holothurians), glass worms (*Sagitta*), and sand. A larval sand flounder 11.5 mm. long examined by Moore[7] contained minute copepods (*Temora* and *Centropages*) and amphipods (*Unciola* and *Leptocheirus*).

The sand flounder is a late spring and summer spawner in the northern part of its range; thus Welsh found them spawning late in June at Gloucester, and ripe fish are taken at Woods Hole in May and June, while Moore reports sand flounders ripe in Long Island Sound from early May to August, with some still incompletely spawned out there in September. And it seems

[3] For details, see Moore's (Bull. Bingham Oceanogr. Coll., vol. 11, art.3 1947, pp. 58-63) detailed study of the sand flounder in southern New England waters.

[4] Warfel and Merriman (Bull. Bingham Oceanogr. Coll., vol. 9, art. 2, 1944, pp. 61-62) give details of temperature and salinity for Connecticut waters, with references.

[5] Bull. Bingham Oceanogr. Coll., vol. 11, art. 3, 1947, pp. 54-58.

[6] Smith, North Carolina Econ. Geol. Survey, vol. 2, 1897, p. 392.

[7] Bull. Bingham Oceanogr. Coll., vol. 11, art. 3, 1947, pp. 26-27

that they commence spawning still earlier in the season to the westward and southward, for Nichols and Breder[8] report young fry 20 mm. long in Sandy Hook Bay by May, while the sizes of the young fry taken in winter in Chesapeake Bay suggest that they are hatched there as early as March or April.[9]

It is not yet possible to state the extremes of temperature within which the sand flounder spawns. But 50° to 60° F. has proved favorable for hatching artificially fertilized eggs at Woods Hole, with even 70° not too warm for successful incubation. The eggs are spherical, transparent, buoyant, 1 to 2 mm. in diameter (measurements taken at Gloucester by Welsh), with a single colorless or pale-lemon oil globule of 0.15 to 0.28 mm. And the surface of the egg shows faint irregular markings. Incubation occupies about 8 days at 51°–56°; its duration has not been recorded for higher temperatures. The sand flounder, like the winter flounder, completes its metamorphosis while it is smaller than either the yellowtail (p. 273) or the witch (p. 287). Thus the dorsal and anal fin rays were complete and the ventral fins had formed in one only 8½ mm. long (fig. 153), and its right-hand eye had already moved around to the back-line of the head, while the migration of the eye is completed, and they are ready to take to bottom by the time they have grown to 10 mm. long.[10]

Rate of Growth.—It seems that the sand flounder passes through its larval stage more rapidly than most flatfishes do, for many of its fry with the migration of the eye completed have been taken at Woods Hole only 1 to 2 months after spawning commences there. One that was kept in an aquarium there by Williams[11] grew from 10 mm. to 22 mm. in length in 11 days; and Nichols and Breder's[12] observation that fry of the year in Sandy Hook Bay grew from an average length of about ¾-inch (to the base of the tail fin) in May, to about 2 to 2⅜ inches by late September, is in line with Tracy's statement[13] that the fry are 2 to 3 inches long in July in Rhode Island waters, growing to 4 inches and upwards in December. Fry only 1 to 2 inches long reported by Nichols and Breder at Orient, N. Y., in December, seem to have been from a late-hatched brood.

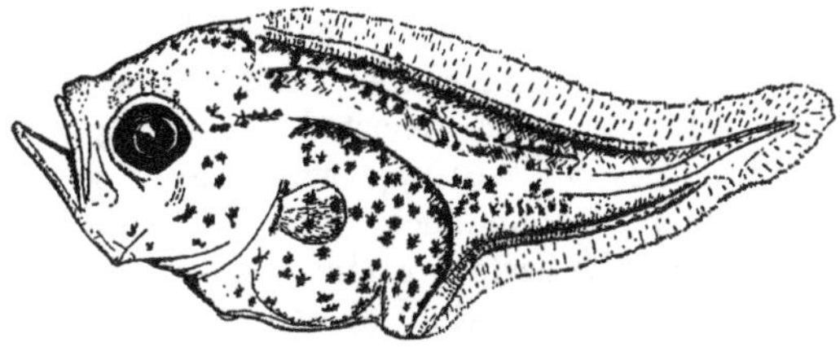

FIGURE 152.—Larva, 5.5 mm.

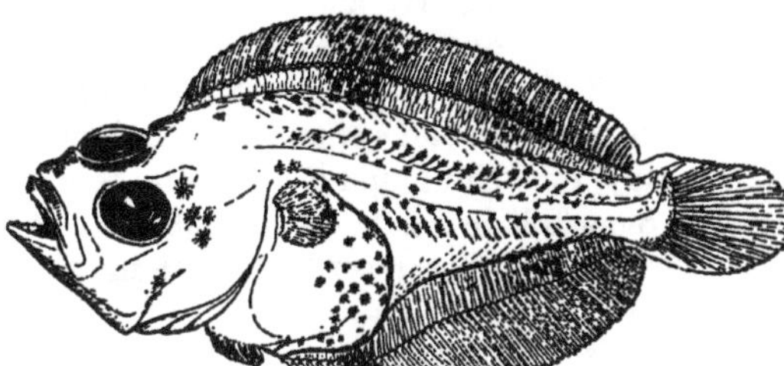

FIGURE 153.—Larva, 8 mm.

SAND FLOUNDER (*Lophopsetta maculata*).

Moore concludes, from her very detailed study of the growth zones on scales and otoliths, that sand flounders in Long Island Sound average about 4½ inches long when they are 2 years old (i. e., at the beginning of their third summer); about 7½ inches at 3 years; about 9 to 10 inches at 4 years; about 11 inches at 5 years; about 11½ inches at 6 years; and about 12 inches at 7 years.[14] And Gulf of Maine fish probably grow at about this same rate. They mature at 9 to 10 inches; i. e., in the third or fourth year, according to the foregoing schedule.

General range.—Coastal waters of eastern North America, from the Gulf of St. Lawrence to South Carolina; most abundant west and south of Cape Cod, north and east of which it is confined to favorable localities.

Occurrence in the Gulf of Maine.—This flounder is not common in the Gulf of Maine, except locally. Dr. W. C. Kendall found it at Monomoy; we have caught one (p. 292) on the outer shore of Cape Cod; Storer found it at Provincetown, where he saw a considerable number in shoal water; it is reported from North Truro; from Gloucester

[8] Zoologica, New York Zool. Soc., vol. 9, 1927, pp. 181–182.

[9] Hildebrand and Schroeder, Bull. U. S. Bur. Fish., vol. 43, Pt. 1, 1928, p. 172.

[10] Williams (Bull. Mus. Comp. Zool., vol. 40, 1902, No. 2) has given a brief account of the anatomical changes that take place during the passage of the eye in the sand flounder. and a more detailed account for the winter flounder. For photographs of larvae and small fry, see Moore (Bull. Bingham Oceanogr. Coll., vol. 11, art. 3, 1947, fig. 3).

[11] Bull. Mus. Comp. Zool., vol. 40, 1902, p. 3.

[12] Zoologica, New York Zool. Soc., vol. 9, 1927, pp. 181–182.

[13] Rept. 40, Comm. Inland Fish. Rhode Island, 1910, p. 166.

[14] Bull. Bingham Oceanogr. Coll., vol. 11, art. 3, 1947, pp. 47–51.

Harbor, where a considerable number were collected in 1878 (Welsh found it there in 1916), and at Milk Island nearby. But we have not learned of it anywhere else in the Massachusetts Bay region, and it has never been recorded between Cape Ann and Cape Elizabeth, nor did Welsh see it taken there by the gill-netters during the spring of 1913. It has been reported repeatedly at several localities in Casco Bay, which seems to be a local center of abundance. But it cannot be common along the eastern Maine coast or on the New Brunswick side of the Bay of Fundy, for the only records from this stretch of coastline are from Bucksport, from Eastport, and from Passamaquoddy Bay where one was taken in 1880 and another in 1912. Minas Channel on the Scotian side seems to be a second center of abundance, like Casco Bay, for Leim found it common there.[15] Huntsman reports it in St. Mary Bay also. But we have found no other record of it along the western coast of Nova Scotia.

Welsh saw it taken by the otter trawlers on Georges Bank in June 1913, and we have seen it there on four recent trawling trips, including about a dozen specimens trawled by the *Albatross III* on the southwest part of the bank and off Nantucket in 22 to 39 fathoms, in mid-May 1950, and 132 taken by the *Eugene H* in that same general region, in 36 hauls at 25 to 45 fathoms, in late June 1951. Beyond this, nothing is known of it on the offshore fishing grounds.

The evidence of the Gloucester specimens mentioned above proves that it breeds in the Massachusetts Bay region to some extent, while its local abundance suggests the same for Casco Bay, as does the capture of its larvae for Minas Channel. It may also breed at the heads of the warmer and shoaler bays between Casco Bay and Grand Manan. Seemingly it does not do so in any of the estuaries on the New Brunswick side of the Bay of Fundy for no larvae have ever been found in Passamaquoddy Bay, a fairly representative situation, probably because of low temperature. But we have no doubt that the local stocks in the Gulf of St. Lawrence (p. 294) are self-sustaining.

The sand flounder is much more plentiful west of Cape Cod than it is anywhere in the Gulf of Maine, southward at least to Chesapeake Bay, where it is very generally distributed in depths down to 25 fathoms, especially in the southern part. And it is reported as common at Beaufort, N. C.[16]

The sand flounder is known only here and there to the eastward and northward of our Gulf. Its pelagic larvae have been reported on Middle Ground off Halifax and near Sable Island; [17] a few adults have been taken in Chedabucto Bay, eastern Nova Scotia; [18] Cox [19] states that it is "by no means uncommon" around the Magdalen Islands, in the southern side of the Gulf of St. Lawrence, where Huntsman [20] classes it as characteristic of the warm surface stratum inshore; and it has been taken off Port-au-Port on the west coast of Newfoundland.[21]

Importance.—Sand flounders are so small and so thin bodied, and so few of them are caught in the Gulf of Maine that they are of no commercial importance there, nor likely to be. However, a market developed for them during the war years in New York, where a much larger supply was near at hand, culminating in landings of about 340,000 pounds in 1944, and about 360,000 pounds in 1945. But as Moore has pointed out,[22] the demand fell off during 1945, as the war drew to its close. And now the sand flounder is a neglected fish again.

Gulf Stream flounder *Citharichthys arctifrons* Goode 1880

Jordan and Evermann, 1896–1900, p. 2683.

Description.—This little flatfish is left-handed (eyes on the left-hand side and viscera at the left-hand edge as the fish lies), with a wide mouth gaping back as far as the forward edge of the eye; with a nearly straight lateral line; and with both of its pectoral fins well developed, though the one on the eyed side is considerably larger than its mate on the blind side. Its left-hand ventral fin stands on the midline of the body, but the right-hand ventral fin is a short distance above it on the blind side, and while the two ventral fins are alike in females, the one on the blind side is much the

[15] Huntsman, Contrib. Canadian Biol. (1921), No. 2, 1922, p. 70.

[16] Smith, North Carolina Geol. and Econ. Survey, vol. 2, 1897, p. 392.

[17] Report 1, No. 4, Newfoundland Fishery Res. Comm., 1932, p. 110.

[18] Cornish, Contrib. Canadian Biol., (1902–1905) 1907, p. 90.

[19] Contrib. Canadian Biol., (1918–1920) 1921, p. 113.

[20] Trans. Roy. Soc. Canada, Ser. 3, vol. 12, Sect. 4, 1918, p. 63.

[21] Rept., Newfoundland Fish. Res. Comm., vol. 2, No. 1, 1933, p. 127.

[22] See Moore (Bull. Bingham Oceanogr. Coll., vol. 11, art. 3, 1947, p. 71) for detailed tabulation of the New York landings, 1943–1945, from the Daily Market News Service, Division of Fishery Industries, U. S. Fish and Wildlife Service. The sand flounder is not included in the general fisheries statistics published yearly by the Fish and Wildlife Service.

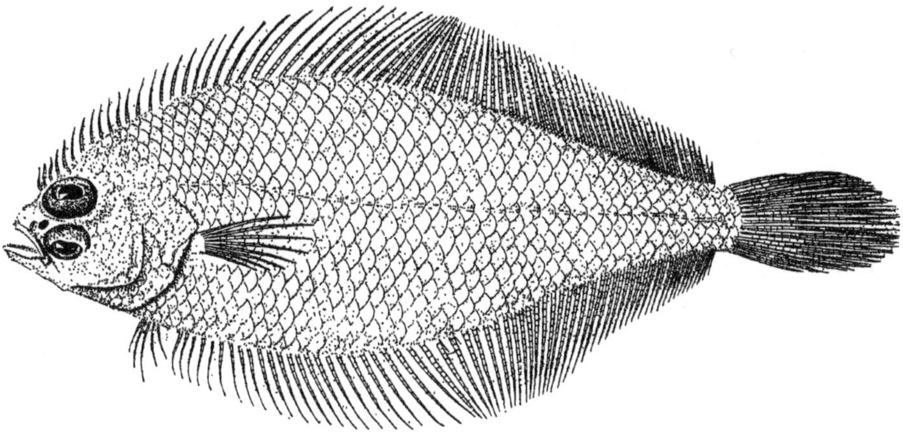

FIGURE 154.—Gulf Stream flounder (*Citharichthys arctifrons*), off Rhode Island. From Goode and Bean. Drawing by H. L. Todd.

longer of the pair in males. The body is ovate in outline and very thin. The long (ventral and dorsal) fins are of moderate breadth, with the dorsal fin (78 to 83 rays) originating over the forward margin of the eye, the anal (left-hand edge) fin (61 to 67 rays) originating a little in advance of the pectorals; and the caudal fin rounded. The scales are so large that there are only about 40 rows of them along the lateral line.

Fish living at different depths vary so widely in the number of fin rays that future studies may reveal the existence of distinct races, if not of species.[23]

Color.—Light brown above, with the scales usually more or less outlined with darker brown; brownish white below.

Size.—Maximum length about 7 inches.

Remarks.—This little flatfish parallels the summer, four-spotted, and sand flounders (the latter its closest Gulf of Maine ally) in its left-handedness. But it is distinguishable from all of these by its nearly straight lateral line; by the great disparity in size between its two pectoral fins; and by its very large scales. Its narrow shape and the fact that none of its dorsal fin rays are branched are further points of distinction between it and the sand flounder; also it is much smaller at maturity than any of the flatfishes that are common in the inner parts of the Gulf of Maine.

Habits.—Little is known of its habits. It is found chiefly in water deeper than 40 fathoms. but it has been trawled as shoal as 12 fathoms. Apparently it spawns from spring through summer, for we have found females with well-developed ovaries in February, while Goode had ripe ones in September. It is not large enough to be of commercial value, but we can witness that it is excellent on the table.

General range.—Eastern coast of America, along the outer part of the continental shelf from the southwestern part of George's Bank to the offing of Charleston, S. C., where the *Blake* took it many years ago,[24] usually at depths of 40 to 200 fathoms, but occasionally as shoal as 12 to 18 fathoms.

A fish occupying this geographic province is misnamed when it is called "Gulf Stream," but this is the only English name by which it has been known.

Occurrence in the Gulf of Maine.—This little flatfish has never been reported from the inner

[23] A second species of this genus (*C. unicornis* Goode 1880) may be expected on the outer slope of Georges Bank in depths of 100 fathoms and more, since it has been taken off Marthas Vineyard in 115 to 150 fathoms. The male is separable from *C. arctifrons* by the fact that there are several short spines on the eyed side of the head above the upper lip (the head of *arctifrons* is spineless although old fish may have a bony protuberance on the snout). Further points of distinction are that *unicornis* has fewer fin rays (only about 74 to 77 dorsal rays and 60 anal rays) and that its body is broader (actually higher) Parr (Bulletin of the Bingham Oceanographic Collection, vol. 4, art. 1, 1931) has published a revision of the genus *Citharichthys* of the western Atlantic.

We have towed the pelagic larvae of still a third small deep-water flounder (*Monolene sessilicauda* Goode 1880) off the seaward slope of Georges Bank (Bull. Mus. Comp. Zool., vol. 41, No. 8, 1917, p. 277), while the adults have been trawled in depths of 100 fathoms and more off Marthas Vineyard and thence westward and southward along the continental slope. It is left-handed like the summer, four-spotted, and sand flounders, with arched lateral line, but it has no pectoral fin on the blind side. For a detailed description of it see Goode and Bean, Smithsonian Contrib. Knowl., vol. 30, 1895, p. 452).

[24] Goode and Bean (Smithsonian Contrib. Knowl., vol. 30, 1895, pp. 443–444) give a long list of localities where it has been trawled, along the continental shelf from the offing of Nantucket to the offing of Charleston, S. C.

parts of the Gulf, nor is it to be expected there, to judge from its general distribution. But the *Albatross I* took one in a tow net over the southwestern part of Georges Bank at about the 82 fathom (150 meters) contour line.[25] And subsequent captures of scattered specimens in that general neighborhood in 1931 by the *Albatross II;* at 8 stations (30 specimens) between the offing of Nantucket and longitude about 67° 10′ W., in 41 to 150 fathoms, by the *Albatross III* in May 1950; and on those same general grounds in 39 to 65 fathoms by the *Eugene H* in late June 1951, show that its regular range extends eastward far enough to include not only the slope of Nantucket Shoals, but the southwestern sector of Georges Bank arc as well, at the appropriate depth. And it must be considerably more plentiful on the outer part of the shelf off southern New England, for the *Albatross III* has trawled a considerable number of them there, including one catch of 100 off Montauk Point, in February 1950, and another of as many more off Rhode Island on May 13, 1950, at 41 to 50 fathoms.

[25] Station 20045, lat. 40° 18′ N., long. 68° 09′ W., February 22, 1920.

Hogchoker *Achirus fasciatus* Lacépède, 1803

American sole

Jordan and Evermann, 1896–1900, p. 2700.

Description.—This fish is the closest relative, in northeastern American waters, of the famous sole of Europe. It is right-handed and small-mouthed, and it can be told at a glance from all other Gulf of Maine flatfishes by the fact that it has no pectoral fin on either side. Its mouth gapes along the general fore-and-aft line as the fish lies, with the upper jaw projecting beyond the lower, whereas the gape is oblique in all other local flatfishes, and it is their lower jaw that projects. Furthermore, the rounded outline of the head of the hogchoker, and the lack of a definite snout, gives it an aspect very different from that of any other Gulf of Maine flatfish.

Equally diagnostic among right-handed species is that its right-hand ventral fin is continuous with the anal fin; its long fins are highest toward their rear ends; its dorsal (left-hand) fin originates at the very tip of the nose (thus, further forward than in our commoner flounders); and its small eyes are set flat instead of in prominent orbits. Other characters worth mentioning are that the

Figure 155.—Hogchoker (*Achirus fasciatus*), Woods Hole. After Jordan and Evermann. Original drawing by H. L. Todd.

gape of its mouth is shorter and much more crooked on the blind side than it is on the eyed side (an asymmetry that has been emphasized in most of the descriptions of this species); that it is evenly oval in outline without a definite caudal peduncle; and that there are 50 to 56 dorsal-fin rays and 36 to 42 anal-fin rays, but no pre-anal spine. The scales are very rough on both sides, those of the upper part of the head and chin on the eyed side and on the whole head on the blind side are larger than the body scales, and its skin is slimy with mucus.

Color.—Dusky or slaty olive to dark brown on the eyed side, barred transversely with a varying number (usually 7 or 8) of indistinct darker stripes, with a dark longitudinal stripe along the lateral line, and sometimes with pale mottling. The dorsal, caudal and anal fins are of the general body tint, variously dark clouded. The blind side is dirty white, usually marked with dark round spots which vary in size and number from fish to fish. But some specimens lack these spots.

Size.—Eight inches is about the maximum length.

Habits.—The hogchoker is confined to the immediate vicinity of the coast, is most common in bays and estuaries where the water is more or less brackish, and sometimes runs up into fresh water. It is a late spring and summer spawner. At Woods Hole fish apparently ripe have been taken in May, while in Chesapeake Bay ripe or nearly ripe fish have been collected in June, July, and August. One female, 6½ inches long, contained about 54,000 eggs about 0.3 mm. in diameter, whether buoyant or not is not known. It reaches a length of 2 to 3 inches at one year of age, and matures when about 4½ inches long. It feeds chiefly on annelid worms and on small crustaceans.[26] Fragments of algae also have been found in hogchoker stomachs, but these probably were swallowed with its animal prey.

General Range.—Off the Atlantic and Gulf coasts of North America, from Massachusetts Bay to the Atlantic coast of Panama. The hogchoker is abundant in Chesapeake Bay and to the southward, and moderately common as far north as southern New England, but it is rare north of Cape Cod.

Occurrence in the Gulf of Maine.—This little flatfish has been reported from Provincetown (where Captain Atwood spoke of it as plentiful); from Boston Harbor, whence the Museum of Comparative Zoology has several, all caught long ago; from the mouth of the Charles River (two specimens reported in 1847); and from Nahant (one taken in 1840). But it is more than three-quarters of a century since it has been brought to scientific attention anywhere to the north of Cape Cod; if it is caught there from time to time, as it doubtless is, it has not been recognized. It is not known north or east of Cape Ann, nor on the offshore banks.

Importance.—The hogchoker is said to be delicious eating. But it is so small that it is of no commercial value even in Chesapeake Bay where it is plentiful. Incidentally, the rumored origin of the name "hogchoker" is that hogs that "feed on fish discarded on the beaches, have great difficulty in swallowing this sole, because of the extremely hard, rough scales."[27]

THE JOHN DORIES. FAMILY ZEIDAE

American John Dory[28] *Zenopsis ocellata* (Storer) 1858

Jordan and Evermann, 1896–1900, p. 1660.

Description.—The John Dory is easily distinguishable from all other Gulf of Maine fishes of similar body form by its long dorsal fin spines, bony armor, tiny tail fin, and the curious profile of its head. Like the butterfish it is very deep (only about one and three-fourths to twice as long as it is deep) and very much flattened sidewise. Its body is rounded in side view, with the dorsal profile of its head noticeably concave, its large mouth is set very obliquely, and its caudal peduncle is very slender. Its dorsal fin is in two parts, spiny and soft rayed; the former, originating over the upper corner of the gill covers, has 9 to 10 spines; the first, second, and third spines very long, the others graduated. And all the spines are filamentous toward the tip. The soft dorsal fin

[26] Hildebrand and Schroeder, Bull. U. S. Bur. Fish., vol. 43, pt. 1, 1928, p. 176.

[27] Hildebrand and Schroeder, Bull. U. S. Bur. Fish., vol. 43, Pt. 1, 1928, p. 177.

[28] Separable from the common John Dory of Europe by having three anal spines instead of four, and by a greater development of the bony plates.

FIGURE 156.—American John Dory (*Zenopsis ocellata*), Provincetown. From Goode. Drawing by H. L. Todd.

(25 to 27 rays) is somewhat longer than the spiny dorsal fin, but less than half as high, and its anterior rays are only about half as high as the posterior ones. The two dorsal fins, together, occupy the entire length of the back of the fish from nape of neck to caudal peduncle.

The anal fin (24 to 26 rays preceded by 3 short stout spines) corresponds to the soft dorsal in location, height, and outline. The very small caudal fin is brush shaped, the ventral fins are very long, with the rays free at their tips, and they are situated in front of the pectorals. The pectorals are short and rounded. The skin is naked except for a series of bony bucklers, each with a hooked thorn or double thorn; two or three of them along the base of the spiny dorsal fin and four along the base of the soft dorsal; two in front of the ventral fins; one in the midline behind the ventrals, followed by six pairs along the belly to the anal fin; and five along the base of the anal fin.

Color.—Silvery all over. Specimens that we have seen up to about 10 inches long are marked on either side with about 12–24 vaguely outlined dark spots, irregularly arranged, and fish up to about 15 inches long retain some of the spots.[29] But it seems that the spots tend to fade out with growth, for larger specimens that we have at hand, 16–20 inches long, have only one vague blotch on each side, a short distance behind the gill opening.

Size.—The largest four specimens yet seen measured 18¼ and 18½ inches;[30] 19 inches, weigh-

[29] This agrees with the original account of the species (Storer, Proc. Boston Soc. Nat. Hist., vol. 6, 1858, p. 386) and with a photograph of one about 3½ inches long, from Campobello, New Brunswick, sent us by Dr. A. H. Leim.

[30] Taken off Long Island, N. Y., by the *Albatross III*, lat. 39°39′ N., long. 72°08′ W., May 12, 1950.

ing 3 pounds;[31] and 20 inches, weighing 4½ pounds and 24 inches, weighing 7 pounds.[32]

Remarks.—The presence of plates along the base of its first (spiny) dorsal fin, as well as along the bases of its second (soft) dorsal and anal fins, and of only three anal spines marks our fish off from its close counterpart, the European John Dory (*Zeus faber*), which has four stout anal spines and lacks plates along the first dorsal fin. Other structural differences are that the plates are much larger in our species than in the European, but the thorns smaller and less conspicuous;[33] that the base of each of the dorsal fin spines (except for the first and last one or two) is armed in the European species with a stout thorn (not in the American); and that the upper profile of the head is much the more deeply concave in the American species.

Habits.—All that is known of the habits of our John Dory is that we found two butterfish 6 to 7 inches long and one squid in the stomach of a large one (of about 18½ in.) trawled by the *Albatross III* about 74 miles off Long Island, N. Y., May 12, 1950, at 72 fathoms; and that the ovaries were well developed with orange colored eggs 1.2 to 1.4 mm. in diameter, in a 20-inch female that we saw trawled between January 27 and February 2 on the outer part of the shelf off Marthas Vineyard.[34]

General range.—Outer part of the continental shelf from the latitude of Chesapeake Bay to the vicinity of Sable Island, Nova Scotia, and perhaps to the Laurentian Channel that separates the Nova Scotian Banks from the Newfoundland Banks. It reaches the inner parts of the Gulf of Maine now and then as a stray.

Occurrence in the Gulf of Maine.—Only four specimens are known to have been taken in the inner parts of the Gulf of Maine. One (the specimen from which the species was described) was found at Provincetown, at the tip of Cape Cod many years ago; one found in a herring weir at Campobello Island, New Brunswick, at the mouth of the Bay of Fundy in 1942;[35] one trawled 25 miles off Cape Ann in 75 fathoms, January 1948.[36] One also was trawled on the northeastern edge of Georges Bank in the summer of 1941,[37] and one taken in Cape Cod Bay, July 7, 1952, by the dragger *Santina.*

It is to be expected anywhere along the seaward slope of the offshore rim of the Gulf, for the dragger *Eugene H* took them in nearly every trawl haul on the southwest slope of Georges Bank, near Veatch Canyon, at about the 75-fathom contour line, in late March 1951, some hauls bringing in several hundred (estimated) specimens. Other specimens[38] have been trawled recently on the outer part of the continental shelf southeast of Cape Henry, Va., from between 28 and 50 fathoms;[39] off Long Island, New York, in 72 fathoms and from between 145 and 200 fathoms; off Marthas Vineyard in 55 to 68 fathoms; off Nantucket in 66 to 75 fathoms; on Emerald Bank off Halifax, Nova Scotia, in 70 fathoms; and west of Sable Island, Nova Scotia, at 62 fathoms.

GRAMMICOLEPID FISHES. FAMILY GRAMMICOLEPIDAE[40]

Grammicolepid *Xenolepidichthys americanus* Nichols and Firth 1939

Nichols and Firth, Proc., Biol. Soc., Washington, vol. 52, 1939, pp. 85–88.

Description.—This curious little oceanic fish resembles its near relative the John Dory (p. 297) in the arrangement of its fins, and in general shape, with body so strongly flattened sidewise as to be as thin as a pancake, and with a slender caudal peduncle. But it has a much smaller mouth than the John Dory, its scales are linear in shape with their long axis dorso-ventral, so that the sides of the trunk are cross marked with a large number of narrow lines, closely crowded together, and the series of bony plates that arm the dorsal and ventral edge of the body of the John Dory are replaced in the Grammicolepids by a double series of short thorns that embrace the bases of the dorsal and ventral fins. Each side of the trunk of the only species known from our waters is

[31] Caught on the northeast edge of Georges Bank in the summer of 1941 and reported in the Boston Traveler for September 9 of that year.

[32] Taken 85 miles off Marthas Vineyard by the dragger *Eugene H*, May 15, 1950.

[33] Double and sometimes triple in the European *Z. faber*.

[34] Trawled by the dragger *Eugene H* from between 55–68 fathoms.

[35] Reported to us, with a photograph, by Dr. A. H. Leim.

[36] This specimen, trawled by the *Agatha and Patricia*, is in the Museum of Comparative Zoology.

[37] Reported in the Boston Traveler for September 9, 1941.

[38] Specimens seen by us or reliably reported.

[39] Reported by Firth, Copeia, 1931, p. 162.

[40] For a recent account of this family see Myers, Proc., U. S. National Museum, vol. 84, 1937, pp. 145–156.

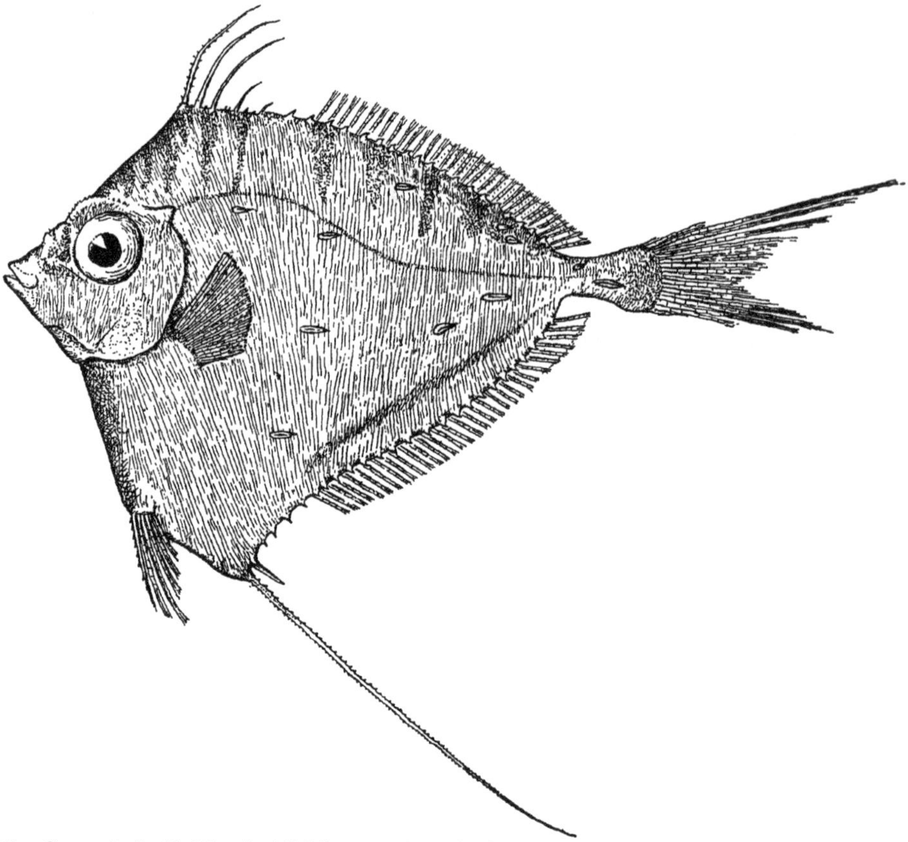

FIGURE 157.—Grammicolepid (*Xenolepidichthys americanus*), Georges Bank. Drawing by H. B. Bigelow. Tail fin after the original illustration by Firth and Nichols.

armed with about 11 or 12 conspicuous, horizontally flattened spines, pointing rearward.

Size.—The only specimen yet seen is about 4 inches (100 mm.) long, to the base of its tail fin.

We need only add, further, of our species, that the forward division of the dorsal fin consists of 5 spines, the forward edge of the first saw-edged, and all of them filamentous toward the tip; that the second dorsal fin, of 33 soft rays (separated from the first by a considerable gap), is about as high as two-thirds the diameter of the eye; is of about equal height from end to end, and reaches back to the caudal peduncle; that the tail fin is deeply forked, its tips pointed, and its upper lobe longer than the lower (unless this is the result of mutilation); that the soft-rayed anal fin, corresponding to the second (soft) dorsal fin, is preceded, after a considerable gap, first by a short, smooth spine, then by another very long spine, saw-toothed along both its front margin and its rear margin for most of its length, but filamentous toward its tip; and that the ventral fins, of 1 stout, saw-edged spine followed by 6 soft rays, stand a little in advance of the brush-shaped pectorals.

Color.—After preservation in alcohol, the color is "pale, with a series of dark marks on the midline of the back, and about 10 narrow dark bands extending downward from these to the level of the top of the eye . . . the flattened spines, scattered over the body are blackish. Base of anal with a series of dusky blotches, and posterior part of caudal dusky".[41]

Range and occurrence in the Gulf of Maine.—So far known only from Georges Bank, where the only specimen yet seen was picked up, in a bucket, from the Sword Fisherman *America*. A closely related species, *X. dalgleishi* Gilchrist 1922, is known from the Caribbean, South Africa, and the Philippines.

[41] Quoted from the original account by Nichols and Firth.

SNIPE FISHES. FAMILY MACRORHAMPHOSIDAE

Snipe fish *Macrorhamphosus scolopax* (Linnaeus) 1758

Jordan and Evermann, 1896–1900, p. 759.

Description.—The snipe fish is given so unusual an appearance by a long tubular snout with small toothless mouth at the tip, combined with a very long, stout dorsal fin spine that is saw-toothed along the rear edge that it could hardly be mistaken for any other Gulf of Maine fish.

Its body is about two-fifths as deep as long, measured from front of eye to base of caudal fin, so flattened sidewise that it is only about one-third to three-eighths as thick through as it is deep; the snout, measured from the front of the eye is about 1⅓ times as long as the depth of the body. The eye is noticeably large. The two dorsal fins stand far behind the mid-length of the trunk. The first dorsal is of 5 to 7 spines and very short, the second, far the longer, is pointed, with about 11 to 13 soft rays; and the two dorsal fins are separated by an interspace nearly as long as the base of the first dorsal. The anal, with 19 to 20 rays, is much longer than the second dorsal, but lower; the caudal is square-tipped, of moderate size. The very small ventrals are located considerably behind the pectorals. The snout, head, and sides are clothed with small rough scales. And the body is further stiffened with bony plates, of which there are 2 longitudinal rows of 4 each, high up on each side behind the gill opening; also 3 longitudinal series of 6 each along the lower breast and belly in front of the ventral fins, followed by 3 pairs behind the latter and finally by a single plate close in front of the anal fin, these last forming a sharp keel.

Color.—Pinkish or reddish on sides above, fading to silvery white below. Described as sometimes golden above.

Size.—Maximum reported size about 6¼ inches (16 cm.);[42] the few we have seen were about 4 inches long.

General range.—Widespread in warm seas. Eastern Atlantic, from the coast and Banks of Morocco, where it is sometimes taken in numbers, and the Mediterranean, northward to southern England (Cornwall, Devonshire); so far known in the western Atlantic only from the offing of Nantucket and from Massachusetts Bay.

Occurrence in the Gulf of Maine.—Oddly enough, the few records of this eastern Atlantic fish in our side of the Atlantic have all been within the limits of the Gulf of Maine; namely, one reported from Massachusetts Bay;[43] a second trawled south of Nantucket, at the 130-fathom contour line,[44] both many years ago; and eight specimens trawled in that same general vicinity (lat. 39°59′ N., long. 69°47′ W.) at 80 fathoms, by the *Albatross III* on May 14, 1950. Evidently it reaches the inner parts of the Gulf only as a stray, and at long intervals although it is taken from time to time by otter trawlers along the southwestern edge of Georges Bank in 75 to 85 fathoms.

[42] One of this size is pictured by Murray and Hjort (Depths of the Ocean, 1912, p. 397, fig. 268).

[43] Goode and Bean (Smithsonian Contrib. Knowl., vol. 30, 1895, p. 483), without further details.

[44] Original of Goode and Bean's illustration (Smithsonian Contrib. Knowl., vol. 31, 1895, pl. 127, fig. 396).

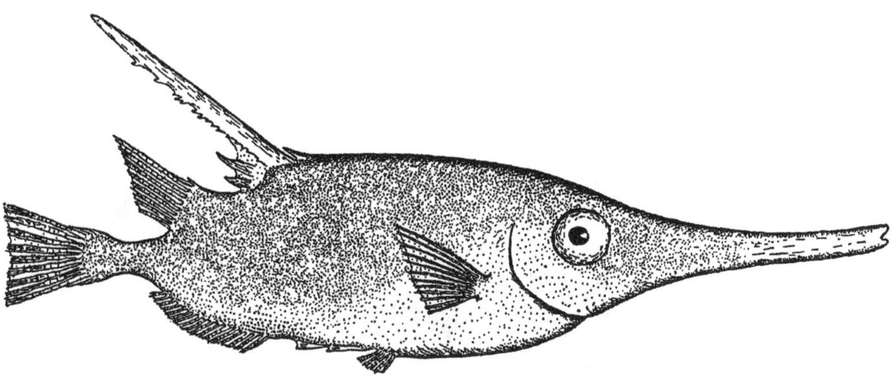

FIGURE 158.—Snipefish (*Macrorhamphosus scolopax*), off Nantucket.

THE SILVERSIDES. FAMILY ATHERINIDAE

These are small fishes, smelt-like in appearance, except that they have a spiny dorsal fin as well as a soft dorsal fin; do not have the adipose fin, and have much smaller mouths than the smelt. Two species are known from the Gulf of Maine.

KEY TO GULF OF MAINE SILVERSIDES

1. About 24 rays in the anal fin Common silverside, p. 302.
   Only about 15 or 16 rays in the anal fin Waxen silverside, p. 304.

Silverside *Menidia menidia* (Linnaeus) 1766

GREEN SMELT; SAND SMELT; WHITE-BAIT; CAPELIN; SPERLING; SHINER

Jordan and Evermann, 1896–1900, pp. 800, 2840.

Description.—This silvery little fish is often confused with the young smelt, but it does not require very close examination to tell them apart for the adipose fin characteristic of the smelt is lacking in the silverside, while the latter has a spiny dorsal fin as well as a soft dorsal fin instead of the one soft dorsal only, as the smelt does; this last character distinguishes it equally from young herrings; its anal fin too is much longer than that of the smelt.

It is a slender fish, about one-sixth as deep as long, not counting caudal fin; thin-bodied but with rounded (not sharp-edged) belly; with short head; large eye; and small mouth, gaping hardly as far back as the front of the eye, and set very obliquely. Both head and body are clothed with large scales. The first dorsal fin (3 to 7 spines) is smaller than the second and originates about midway between the tip of the snout and the base of the caudal fin; the second dorsal has 7 to 10 soft rays and originates over the middle of the anal. The anal (of 23 to 26 rays, the first stiff and the others soft) is falcate in outline. The caudal peduncle is slender, the tail moderately forked.[45]

Color.—Translucent bottle green above, with top of head, nose, and chin dusky. The upper parts of the sides are thickly speckled with dark brown, and there is a silver band outlined above by a narrow black streak, running along each side from close behind the pectoral fin to the base of the caudal fin. The belly is white.

Size.—The silverside grows to a length of about 5½ inches, adults usually running 4 to 4½ inches long.

Habits.—Silversides tend to congregate in schools usually made up of even-sized individuals. They frequent sandy or gravelly shores chiefly, and there is no reason to suppose that they ever venture out to sea. At high tide they are often seen among the sedge grass (*Spartina*), where it grows sparsely between tide marks, particularly about the inner bays and in river mouths where they follow the tide up and down the beach within a few yards of the water's edge. They also run up into brackish water; near St. Andrews, in fact, they are chiefly found in brackish situations though more generally distributed on the New Brunswick shore further up the Bay of Fundy and on the Nova Scotian side as a whole. They do not ordinarily descend deeper than a fathom or so in summer. But some of them, at any rate, sink

[45] The common silverside is represented on the coasts of the eastern United States by two races, a southern and a northern, not, however, very distinct and connected by such various intergradations that they hardly deserve two names, subspecies *menidia* for the southern and subspecies *notata* for the northern. The southern form has fewer scales than the northern, only 4 instead of 5 spines in the first dorsal, and is rather a stouter-bodied fish. Kendall has given an account of the genus (Report, U. S. Comm. Fish (1901) 1902, p. 241). For a recent discussion see Bayliffe (Publ. 90, Chesapeake Biol. Lab., Maryland Dept. Nat. Res., 1950, p. 5).

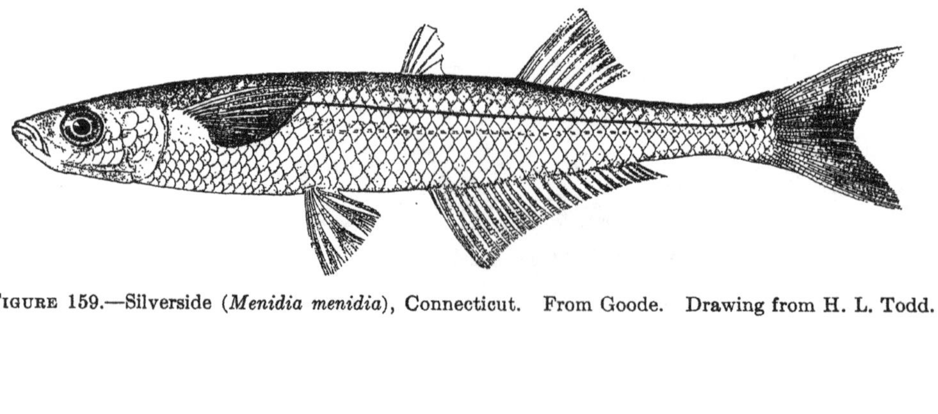

FIGURE 159.—Silverside (*Menidia menidia*), Connecticut. From Goode. Drawing from H. L. Todd.

deeper in winter, probably to avoid low temperature, for Hildebrand and Schroeder [46] found them at depths of 5 to 27 fathoms at that season in Chesapeake Bay. But this is not their universal habit, for they are taken in winter through the ice, as well as in summer, in the southern side of the Gulf of St. Lawrence. And they are resident throughout the year wherever found, generally speaking. Such, at least, is the case along southern New England.

They are omnivorous, feeding chiefly on copepods, mysids, shrimps, small decapod shrimps, amphipods, Cladocera, fish eggs (including their own), young squid, annelid worms, and molluscan larvae. Insects, too, that fall into the water have been found in their stomachs, as have algae and diatoms mixed with sand and mud. On the other hand, they are harried by every predaceous fish that comes close inshore, especially by bluefish and by striped bass. It was interesting in this connection to find that a dolphin (rare in the Gulf), taken at Sandwich, in Cape Cod Bay, in July 1951 (p. 361) was packed full of silversides.

They spawn in May, June, and early July [47] on the southern New England coast. Spawning may begin a little later in the Gulf of Maine, corresponding to more tardy vernal warming, while Leim writes [48] that they do so in June at Prince Edward Island.

The gap in the presence of silversides in abundance that seems to exist along the cool-water stretch from the western side of the Gulf of Maine to the southern side of the Gulf of St. Lawrence (p. 304) suggests that they need summer temperatures as high as 68° or so for successful reproduction. But young fry and adults alike are indifferent to temperatures down to a degree or two above the freezing point of salt water, witness their presence in winter in Cape Cod Bay (p. 304) and below the ice in the bays on the northern side of Prince Edward Island (p. 304).

When the silversides are spawning they gather in schools to deposit their eggs on sandy bottom, often among the sedge grass at high tide, or above low-water mark. Capt. John B. Smith has described them spawning in the sedge at the head of Buzzards Bay, June 13, 1872, rolling from side to side, some jumping clear of the water, and in such multitudes that the water was "whitened with the milt, and the grass was so full of eggs that they could be taken up by the handfull," while small fishes of various kinds were "helping themselves to the dainty repast." [49]

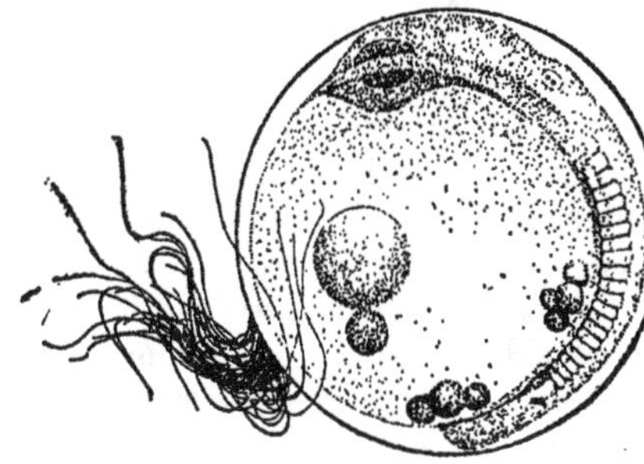

FIGURE 160.—Egg.

FIGURE 161.—Larva, 8 mm.

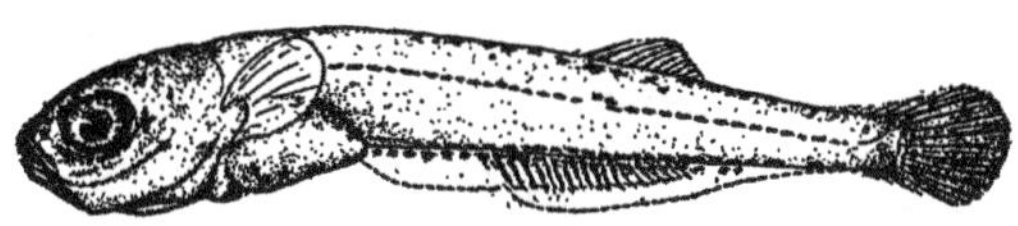

FIGURE 162.—Fry, 13 mm.

SILVERSIDE (*Menidia menidia*). After Kuntz and Radcliffe.

The eggs, 1.1 to 1.2 mm. in diameter and each bearing a bunch of sticky filaments, sink and stick fast in ropy clusters or sheets. Incubation occupied 8 or 9 days in the laboratory at Woods Hole. The yolk is absorbed before hatching, at which time the larvae are about 3.85 to 5 mm. long, and the dorsal, anal, and caudal fins are formed in larvae of 12 to 15 mm. in length. The young grew to a length of 9.3 to 11.7 mm. during the first 20 days in the aquaria. Probably they grow more rapidly at liberty, for all sizes from fry of an inch or less to adults are to be found throughout the summer. Probably the silverside attains maturity at 1 year of age.

General range.—The northern variety of this silverside is common locally from the southern side of the Gulf of St. Lawrence and the outer

[46] Bull. U. S. Bur. Fish., vol. 43, Pt. 1, 1928, p. 189.

[47] Kuntz and Radcliffe (Bull. U. S. Bur. Fish., vol. 35, 1918, p. 127) describe its development, and Hildebrand (Bull. U. S. Bur. Fish., vol. 38[1921–22] 1923 that of the southern race. For a recent account of its life history see Bayliffe, Pub. 90, Chesapeake Biol. Lab., Maryland Dept. Nat. Res. ,1950.

[48] Proc. Nova Scotian Inst. Sci., vol. 20, Pt. 2, 1940, p. 38.

[49] Goode, Fish. Ind. U. S. Sect. 1, 1884, p. 457.

Nova Scotian coast to Massachusetts Bay, and very abundant thence southward to Chesapeake Bay, south of which it gives place to the southern form or intergrades with it; the southern form has been detected as far north as Woods Hole, but never east of Cape Cod.

Occurrence in the Gulf of Maine.—The silverside is to be found all around the shores of the Gulf from Nova Scotia to Cape Cod, always, however, closely confined to the coastline. They are exceedingly plentiful around the sandy shores of Cape Cod Bay. And while we have seen them from Chelsea Beach in Boston Harbor, from Beverly and from Gloucester, many summers spent on the coast leave us with the impression that the silverside is neither as omnipresent nor as abundant from Massachusetts Bay northward, although large schools of them are often to be seen here and there along the sandy beaches on the Maine coast. Bushels, in fact, have been caught in a single haul of the seine in Casco Bay and very likely could be elsewhere.

Silversides are seldom seen along the stretches of rocky coast exposed to the open sea, which make up a large part of the northern shore line of the Gulf of Maine. In Passamaquoddy Bay Huntsman tells us[50] "they are largely restricted to brackish water and hence not very common," but they must be rather generally present in suitable situations around the shore line of the Bay of Fundy, being reported from St. John and Kennebecasis Bay, from Annapolis basin and from St. Mary Bay. Nothing is known as to their status along the Nova Scotian coast of the open Gulf of Maine, or even whether there are any silversides there at all. Halifax is the most northerly locality where they are recorded on the outer coast of Nova Scotia.[51] But Leim [52] reports them so plentiful in the shallows of the southern side of the Gulf of St. Lawrence that "hardly a seine haul has been made without catching several, and as many as 3,500 have been taken at once" in Malpeque Bay on the north shore of Prince Edward Island, where they are taken in winter through the ice, as well as in summer. Enough of them, in fact, are sometimes caught there to be worth canning.[53] Their abundance there contrasted with their evident scarcity along outer Nova Scotia suggests the presence of an isolated population (or populations) in suitable situations in the southern side of the Gulf of St. Lawrence, able to maintain itself because summer temperatures in the shallows there are high enough for its successful propagation.

Importance.—The chief function of the silverside in the economy of the sea is to feed predaceous fishes such as bluefish, mackerel, and striped bass. The silverside is of no commercial value north of Cape Cod, being too small and too soft to answer the never satisfied demand for bait for the offshore fisheries, but they are very generally used to bait eelpots on the Rhode Island coast, and they are excellent on the table, fried, as whitebait.

Waxen silverside *Menidia beryllina* (Cope) 1866

Jordan and Evermann 1896–1900, p. 797 (*Menidia gracilis* Günther).

Description.—This species resembles the common silverside so closely in general appearance that it would be apt to be overlooked among the schools of the latter were it not paler in color,

[50] Contrib. Canadian Biol. (1921) 1922, p. 61.

[51] Cornish (Contr. Canadian Biol. [1902–1905], 1907, No. 9) does not include it in his list of the fishes of Canso.

[52] Proc. Nova Scotian Inst. Sci., vol. 20, Pt. 2, 1940, p. 38.

[53] Needler, Rept. Fish. Res. Board Canada (1941) 1942, p. 11.

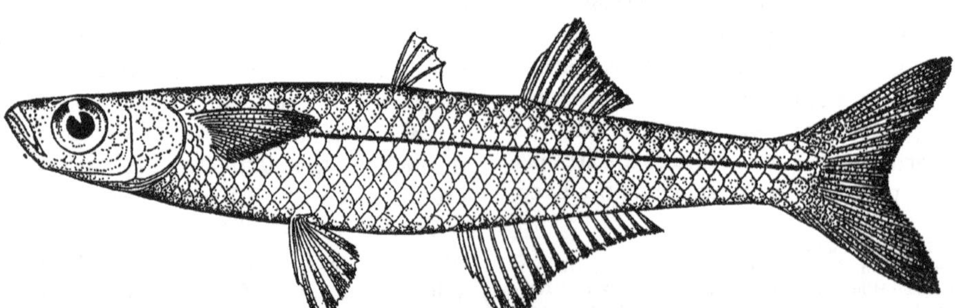

FIGURE 163.—Waxen silverside (*Menidia beryllina*). From Kendall

and stouter bodied as a rule. A more dependable difference, which will always serve to separate the two (for which neither color nor form can be relied upon) is that the anal fin is much shorter (only 15 or 16 rays) in the waxen silverside than in the common silverside.

Color.—Pale greenish on the back, silvery below; the sides with a well-defined silvery band bounded above by a dark line; scales on the back with numerous brown dots; fins without markings.

Size.—Smaller than its relative *menidia*, the maximum length being about 3 inches.

General range.—Cape Cod to South Carolina.

Occurrence in the Gulf of Maine.—Specimens reported by Kendall [54] (1902) from Truro, and from Sandwich in Cape Cod Bay, with one taken in Cohasset, on the southern shore of Massachusetts Bay in the autumn of 1939,[55] are the only records for this fish within the Gulf of Maine, where it appears only a stray from warmer waters to the west and south. At Woods Hole, where it is abundant, its habits are the same as those of the common silverside, though it spawns somewhat later (in June and July).

[54] Occ. Pap. Boston Soc. Nat. Hist., vol. 7, No. 8, 1908, p. 66, as *Menidia beryllina* subspecies *cerea* Kendall (Rept. U. S. Comm. Fish. (1901) 1902, p. 261.)

[55] This specimen is in the Museum of Comparative Zoology.

THE MULLETS. FAMILY MUGILIDAE

Mullets have two separate dorsal fins, the first spiny and the second soft-rayed. Their ventral fins are on the abdomen behind the point of insertion of the pectorals; their tails are forked and they have large scales. Their closest affinity among Gulf of Maine fishes is with the silversides, which they resemble somewhat in the relative size and locations of the fins; but they differ from the silversides in their short, broad heads, small eyes, and relatively deeper and thicker bodies, while they have only 24 vertebrae instead of 35 or more. Furthermore, they are vegetable amd mud eaters instead of carnivorous, their stomachs are thick walled and gizzard-like, the intestines long, corresponding to their food. The lining of the belly of the mullet is black while that of the silverside is pale.

There are many species of mullets. Most of them, however, are tropical, and only one has ever been known to stray within the confines of the Gulf of Maine.[56]

Mullet *Mugil cephalus* Linnaeus 1758

COMMON MULLET; STRIPED MULLET; JUMPING MULLET

Jordan and Evermann, 1896–1900, p. 811.

Description.—The common mullet, the only one of its numerous tribe (there are more than 100 species of mullets) that has been known to stray north of Cape Cod, has a spiny first dorsal and soft second dorsal fin, the two well separated as in the silverside, and its ventrals are located on the abdomen. It is a much larger fish than the silverside, however, and even very young mullets of the

[56] The so-called red mullet or goat fish (*Mullus auratus*) of more southern waters, which is not a true mullet but belongs to a different family (Mullidae), is taken from time to time near Woods Hole, and it has been reported from Halifax Harbor, Nova Scotia (by Leim, Proc. Nova Scotian Inst. Sci., vol. 17, No. 4, 1930, p. XLVI), hence it may be expected as a stray in our Gulf, though it has not actually been found there as yet. There is no danger of mistaking it for a mullet, for it is bright crimson, with a fleshy barbel on its chin, and with its ventral fins far forward, below its pectorals.

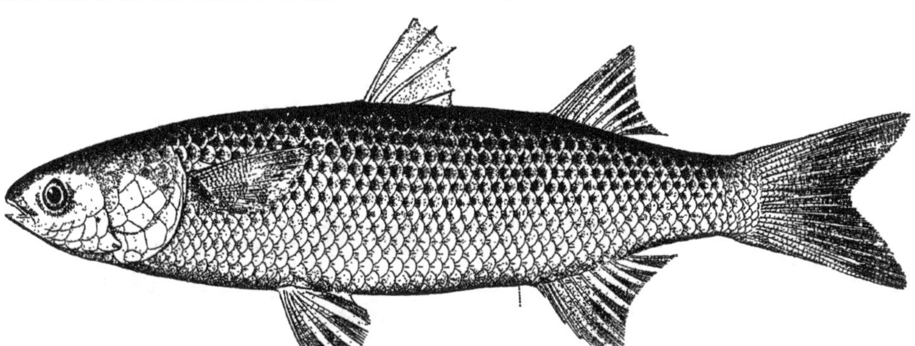

FIGURE 164.—Mullet (*Mugil cephalus*), Woods Hole. From Goode. Drawing by H. L. Todd.

size of the latter (4 to 5 inches long) are easily separable from silversides by the fact that their anal fin is only about half as long in relation to the length of the body, while the second dorsal originates over the origin of the anal instead of well behind it. Furthermore, the head of the mullet is shorter; its nose blunter; its profile quite different (compare fig. 164 with fig. 159); its eye smaller; its body stouter (about one-fourth as deep as long); and it lacks the silvery side stripes so characteristic of the common silverside. There are four spines in its first dorsal fin, 1 spine and 8 soft rays in the second dorsal, 3 spines and (usually) 8 rays in the anal. Young fish, 2 inches long or less, have only 2 spines in the anal, the first soft ray later developing into a spine.[57] The first dorsal stands over the tips of the pectorals or close behind them; and the tail is forked moderately deep. The soft dorsal fin and anal fin are almost naked (they are scaled in most of the other American mullets), but the body and head are clothed with large rounded scales.

Color.—Adults are bluish gray or greenish above, silvery on the lower part of the sides and below; the scales on the sides have dark centers which form longitudinal lines; the fins are sometimes partly dusky. Young fry are bright silvery.

Size.—The common mullet grows to a length of 2½ feet in warmer waters, but small specimens alone have been found along our northern coasts.

General range.—Both sides of the temperate Atlantic; from Brazil to Cape Cod on the American coast, and as a stray to outer Nova Scotia; also along the west coast of America from Monterey (Calif.) to Chile, and in other parts of the Pacific.

Occurrence in the Gulf of Maine.—Mullets are common as far north as New York, less so to Woods Hole, but so rarely do they stray past Cape Cod that there are only a half dozen records of them in the Gulf of Maine, viz, at Provincetown, at Essex [58] in northern Massachusetts, at Freeport, Harraseeket River, Clapboard Island, and Casco Bay in Maine, each based on an odd fish. And one has also been taken in Bedford Basin near Halifax, Nova Scotia.[59] Mullet are more likely to visit the cool waters of the Gulf in late summer or early autumn than at any other season. They have been known to winter as far north as New York, in the mud, but it is not likely that the few strays that round Cape Cod survive the cold season, nor is there any reason to suppose they ever breed in the Gulf, for immature fish alone are found at Woods Hole.

THE BARRACUDAS. FAMILY SPHYRAENIDAE

The slim bodied barracudas, with their long, pointed heads, somewhat resemble the pikes in general appearance. But they are distinguishable from the latter at a glance by having two dorsal fins. The lower jaw projects beyond the upper, and both jaws are studded with large pointed teeth of unequal sizes. The gill covers are scaly, and there is a well-developed lateral line. The first dorsal is spiny, the second soft-rayed. The anal is roughly opposite the second dorsal, the ventrals opposite the first dorsal, the pectorals short, the caudal forked.

Northern barracuda *Sphyraena borealis* DeKay 1842

Jordan and Evermann, 1896–1900, p. 825.

Description.—The combination of slender shape with long head, projecting lower jaw, a first dorsal

[57] See Jacot (Trans. Amer. Microscopical Soc., vol. 39, 1920, pp. 204–214) or a study of the growth of the mullet.

[58] There is (or was) a specimen so labeled in the collection of the Boston Society of Natural History.

[59] Reported by Vladykov, Proc. Nova Scotian Inst. Sci., vol. 19, 1935, p. 6.

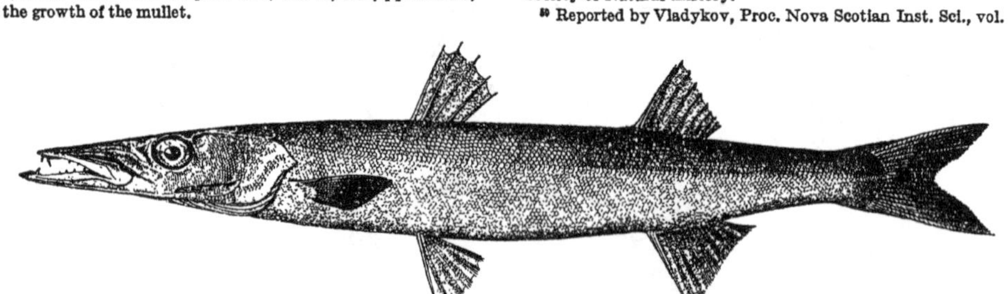

FIGURE 165.—Northern barracuda (*Sphyraena borealis*), Woods Hole. From Goode. Drawing by H. L. Todd.

situated opposite the ventrals, a second dorsal opposite the anal, and a forked tail, separates the barracuda from any other Gulf of Maine fish.

Color.—The adult is olivaceous above, silvery below. The young have dusky blotches along the back and along the lateral line.

Size.—This is the smallest of the barracudas, few growing longer than one foot.

General range.—Atlantic and Gulf coasts of America from Cape Cod to Panama.

Occurrence in the Gulf of Maine.—A specimen, about 2 inches long, found alive in the surf at Nauset Beach, Cape Cod, September 26, 1930, by the late Dr. Edward P. Richardson, is the only record for the Gulf of Maine. Young fry, a few inches long, have been taken from time to time in Vineyard Sound, however, and in Buzzards Bay on the southern coast of Massachusetts between July and December.

THE STICKLEBACKS. FAMILY GASTEROSTEIDAE

Sticklebacks are small fish, made easily recognizable by the presence of two, three or more stout free spines on the back in front of the dorsal fin (spines that they can erect or depress at will) and by the fact that each ventral fin is represented by an even larger spine with only one or two rudimentary rays. Some of them have bony plates in the scaleless skin, but others do not.

KEY TO GULF OF MAINE STICKLEBACKS

1. Seven dorsal spines or more
   Nine-spined stickleback, p. 307
   Not more than five large dorsal spines ---------- 2
2. No bony plates on the upper part of the sides, but there is a bony ridge on either side of the abdomen
   Four-spined stickleback, p. 311
   The upper part of the sides are armed with bony plates, and there is a plate in the midline of the belly, but there are no ridges on the sides of the abdomen ---------- 3
3. Many (28 or more) plates on each side
   Three-spined stickleback, p. 308
   Only 5 or 6 plates on a side
   Two-spined stickleback, p. 310

Nine-spined stickleback *Pungitius pungitius* (Linnaeus) 1758

Jordan and Evermann, 1896–1900, p. 745.

Description.—The nine-spined stickleback is a slender little fish five to 6 times as long (not counting the caudal fin) as it is deep, with a very slim caudal peduncle. The latter usually has a well-developed longitudinal keel on either side; but this keel may be very low or even wanting. There are no bony plates along the sides of the body, but only along the bases of the anal and dorsal fins and on the caudal keels. There are no true scales. The most distinctive character is that there usually are 9 spines on the midline of the back (from 7 to 12 have been counted) in a continuous row from close in front of the pectorals to the dorsal fin, set in a slightly zigzag line and leaning alternately to one side and to the other. The spines are weakly curved rearward; wider at the base than at the tip; fairly uniform in size; about one-half to one-third as long as the height of the dorsal fin; each has a small triangular fin membrane at its base; and there is a shallow groove along the back, into which the spines can be depressed. Each ventral fin is represented by a stout curved spine thicker and longer than the dorsal spines. The dorsal and anal fins (the former stands above the latter) are alike in form, tapering from front to rear, the anal preceded by a single stout recurved spine. The tail fin is weakly rounded.

Size.—Large adults are seldom more than 3 inches long, more commonly 2 to 2½.

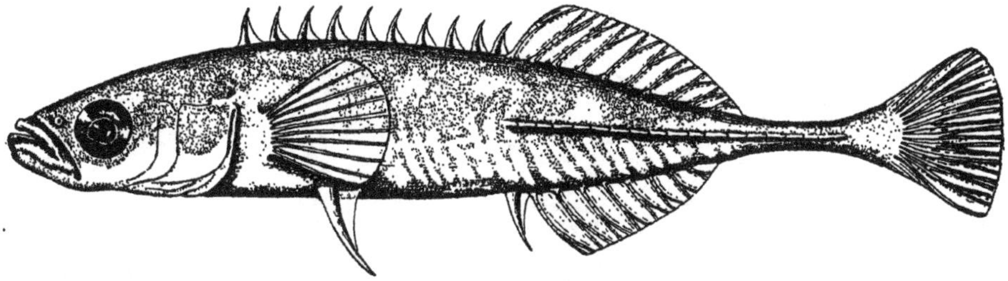

FIGURE 166.—Nine-spined stickleback (*Pungitius pungitius*). From Bigelow and Welsh.

Color.—Usually dull olive brown above, the upper part of the sides faintly barred or blotched darker; the belly silvery; the pubic and thoracic regions often black. The color varies, however, with the season of the year, with the state of sexual maturity, and with the color of the bottom on which the fish is living, those on dark mud being darker and those on bright sand paler. All become more brilliant during the breeding season when reddish tints appear under the head, the belly turns greenish, and black dots develop here and there over the entire body. The male has also been described as assuming a rosy tint beneath.

Habits.—Since the range of the nine-spined stickleback hardly touches the open waters of our Gulf, we need only note that its mode of life is much the same as those of its three-spined relative next to be considered (p. 308); that it is similarly destructive to the spawn and young of other fish, and similarly pugnacious. Probably it spawns in summer [60] on the shores of the Gulf, for its breeding season in northern Europe covers June and July. The male often (but not always) builds a nest attached to grass or weeds which the female spawns, and he guards nest and eggs until the latter hatch, which occurs in about 12 days.

General range.—This is one of the most widely ranging of northern fishes, occurring both in fresh water and in salt in the northern parts of both hemispheres; from northern Scandinavia to France, the western Mediterranean and the Black Sea on the European coast; from Arctic seas south to New York along the American, and westward to Saskatchewan and Alaska.

Occurrence in the Gulf of Maine.—This stickleback is to be found all around the shores of the Gulf of Maine from Nova Scotia and the Bay of Fundy to Cape Cod, but it is chiefly restricted there to harbors and the creeks in salt marshes, where large numbers may often be taken in company with the mummichogs that swarm in such locations, and where it is to be found throughout the year. It is also found in fresh water. In fact, the most exposed situations around the Gulf, where we have heard of it, are Biddeford Pool, Maine,[61] Passamaquoddy Bay,[62] and St. Mary's Bay on the west coast of Nova Scotia.

Commercial importance.—This stickleback is of no commercial importance in America, but it is sometimes tried out for oil in northern Europe when enough can be caught.

Three-spined stickleback *Gasterosteus aculeatus* Linnaeus 1758

TWO-SPINED STICKLEBACK; STICKLEBACK; THORNFISH; THORNBACK

Jordan and Evermann, 1896–1900, p. 747.

Description.—The three-spined stickleback has a very slender caudal peduncle, and squarish tail fin, like its nine-spined relative, but it is a stouter fish, being about one-fourth as deep as long, and it is more flattened sidewise. Its most diagnostic characters are the number of dorsal spines, of which there are three (occasionally four and rarely five), with the first two usually much the larger, and each with a small triangular fin membrane; the small size of the anal spine (this is free in the three-spined stickleback but attached to the fin by the fin membrane in the four-spined); and especially the presence of a series of 28 to 33 bony plates on each side, besides a single ventral plate

[60] At Woods Hole it spawns in April and May.

[61] MacCoy, Bull. 74, Boston Soc. Nat. Hist. 1935, p. 16.

[62] Huntsman, Contr. Canadian Biol., (1921) 1922, p. 61.

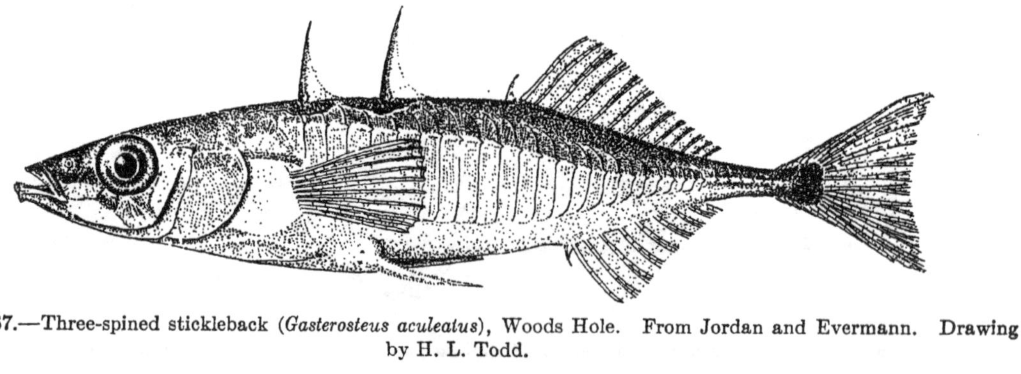

FIGURE 167.—Three-spined stickleback (*Gasterosteus aculeatus*), Woods Hole. From Jordan and Evermann. Drawing by H. L. Todd.

on the lower surface between and behind the ventral fins. The fact that the dorsal fin (1 spine, 10 to 14 rays) originates some distance in front of the anal (1 spine, 8 to 10 rays) is diagnostic also, while its ventral spines are longer and stouter than those of the nine-spined stickleback.

This is one of the most variable of fishes; Smitt,[63] lists no less than 32 named species or races based on its varieties. Thus its dorsal spines may be long or short, and they vary in number as noted above; its bony plates range from none at all to very well developed; and its caudal peduncle may be keeled or it may not. Most American authors have recognized an American species, at the least as contrasted with a European, the former supposedly with longer dorsal spines; the latter with shorter. But the long-spined, as well as the short-spined form is known to occur on the other side of the Atlantic with every possible gradation between the two. Seeing that we have found both in the Gulf of Maine among fish indistinguishable otherwise, we believe that the various forms are environmental races of the one species. And this is well established for the relative strength of the dermal armature, which is weak in fresh water, but strong in salt.

Color.—This stickleback is extremely variable in color, a fact hardly mentioned in most American accounts. They are deep grayish, or olive, or greenish-brown above, or sometimes blue; paler and often with silvery reflections on the sides; silvery on the belly. The fins are pale, except that the fin membranes often are red. In breeding season the males are described as turning reddish below from nose to vent and often up the sides. In females, the whole body except the top of the back may then be reddish; at the same time the back turns brownish with transverse bands, and the sides develop brassy reflections.

Size.—Maximum length about 4 inches, but few of them are more than 3 inches long. It matures sexually at a length of about 2 inches.

Habits.—This is distinctively a shore fish like all the sticklebacks, the great majority of them living their whole lives in estuarine situations. But it is equally at home in sea water of full salinity as in fresh water. And enough stray out to sea for it to be rather a common experience to pick up a few here and there in the tow net, far from land. On such occasions they usually hide in clumps of floating eelgrass (*Zostera*) or of rockweed (*Fucus*, *Ascophyllum*); indeed we have learned to expect a stickleback or two whenever we dip up bunches of weed of any size. These wanderers keep to the surface except, perhaps, in very rough weather.

It is a permanent all-the-year resident wherever it is found alongshore, entering creeks and the mouths of streams in the spring to spawn, and dropping down into slightly deeper water for the winter. In such situations it probably lies in schools in a more or less sluggish condition while the temperature is lowest.[64] It is a proverbially pugnacious fish, using its spines with good effect as weapons of offense and defense, even on other fishes much larger than itself. It feeds indiscriminately on the smaller invertebrates, on small fish fry, and on fish eggs, to which it is exceedingly destructive in fresh water. The diet list of specimens examined by Vinal Edwards at Woods Hole included copepods, of which they are often full, isopods, schizapod shrimps, and young squid, while some had fed on diatoms only. And it is not only omnivorous but very voracious.

This stickleback affords the classic instance of nest building and of the care of eggs among fishes, and its nesting has been described so often in popular natural histories that a bare outline will suffice here.[65] Recent studies in Europe make it likely that this stickleback spawns chiefly in brackish or fresh water, if not exclusively there, for which purpose it enters the estuaries and the mouths of streams. The spawning time is probably the same in the Gulf of Maine (May to June) as in north European waters,[66] when the fish assume the nuptial dress described above, and the males fight fiercely. It is the male that builds the nest, selecting some sheltered spot in shoal water for his purpose, or some rock pool. Here he makes a barrel-shaped mass of bits of grass, weed, and other vegetation an inch or so in diameter, cementing it together with mucous threads, which he spins from his kidneys, and

[63] Scandinavian fishes, vol. 2, 1895, p. 648.

[64] Large numbers are sometimes seined in winter in Scandinavian waters.

[65] Smitt (Scandinavian Fishes, vol. 2, 1895, pp. 653–656) and Regan (The fresh-water fishes of the British Isles, 1911, pp. 247–249) give accounts of the nest building on which the following is based.

[66] About Woods Hole it spawns from May until the last week in July.

weighting it down with pebbles. He then escorts one or a succession of females to this nest, and each of them deposits about 100 to 150 eggs in the central cavity. The male then enters the nest to fertilize the eggs, which stick in clumps to each other and to the nest. Incubation occupies 6 to 10 days, during which period the male guards the nest, driving away intruders large or small. He tears down the nest when hatching-time approaches, but he continues to guard the fry until these can shift for themselves. Many males die after spawning. Those that survive go back to sea in summer; the females, too.

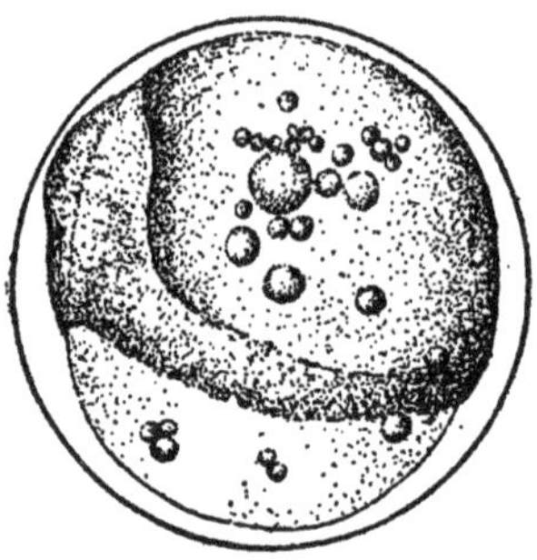

FIGURE 168.—Egg.

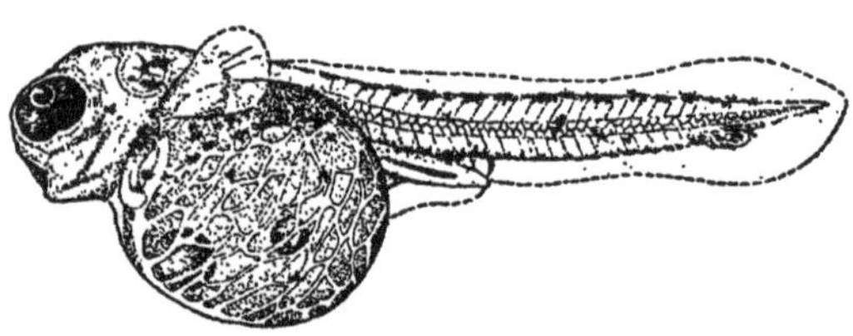

FIGURE 169.—Larva, newly hatched, 4.3 mm.

FIGURE 170.—Larva, 6.3 mm.

Three-spined stickleback (*Gasterosteus aculeatus*). After Kuntz and Radcliffe.

The young fish are 4.25 to 4.5 mm. long when hatched. The yolk sac is absorbed in three or four days; when a week old they are almost 8 mm. long; and the fry are of adult form with fins and spines fully formed when 6 weeks old, and 14 to 16 mm. long.[67] They are 1¾ to 2 inches (40–50 mm.) long when 2 years old, 2 to 2⅛ (50–55 mm.) at 3 years, according to European studies.

General range.—Coasts and fresh waters of the northern hemisphere; from Labrador, the Strait of Belle Isle and northern Newfoundland to lower Chesapeake Bay on the eastern coast of America and represented on the northwestern coast by a form (*Gasterosteus cataphractus* Pallas 1811) that probably is identical with the Atlantic species. Its European range is from northern Norway and Iceland to Spain, the Mediterranean, and the Black Sea.

Occurrence in the Gulf of Maine.—This stickleback is very plentiful all around the shores of the Gulf from Nova Scotia to Cape Cod, living indifferently in brackish water and in salt. The ditches and creeks of the tidal marshes, brackish ponds and lagoons, rock pools, and weedy shores in shallow water are its favorite habitats. It may be found practically anywhere in such places, often in great numbers and in company with other sticklebacks, for it is the commonest of its tribe in the Gulf, as it is about Woods Hole. And so many of them drift out to sea around the shores of the open Gulf that we have taken them on the eastern part of Georges Bank; over German Bank; in the western basin off Cape Cod; near the Isles of Shoals; off Seguin Island; and off Matinicus Island. In the Bay of Fundy, however, they are known only close to land and in the mouths of estuaries.

Importance.—This little fish is of no commercial value in America. In Scandinavia, however, it is sometimes seined in such quantities that it is worth boiling down for oil.

Two-spined stickleback *Gasterosteus wheatlandi* Putnam 1867 [68]

Jordan and Evermann, 1896–1900, as *Gasterosteus gladiunculus* Kendall, p. 2836.

Description.—This stickleback is said to differ from the three-spined stickleback in having a deeper body, fewer fin rays (9 or 10 dorsal and 7 or 8 anal); fewer dermal plates (5 or 6 as against 28 to 33); a caudal peduncle without keels; and a strong cusp both above and below at the base of

[67] Figures of stages in development of this fish are given by Kuntz and Radcliffe (Bull. U. S. Bur. Fish., vol. 35, 1918, p. 131); A. Agassiz (Proc. Amer. Acad. Arts Sci., vol. 17, 1882, p. 288, plate 9), and by Ehrenbaum (Nordisches Plankton, vol. I, 1905–1909, p. 319).

[68] This is the *Gasterosteus biaculeatus* of Mitchill 1815 and Storer 1867; *bispinosus* of Walbaum 1792; *gladiunculus* of Kendall 1896, but not the *G. bispinosus* of Jordan and Evermann 1896, which is a variety of *G. aculeatus*. For the reason for using the specific name *wheatlandi*, see Hubbs, Occasional Papers, Museum of Zoology, University of Michigan, No. 200, 1925.

the ventral spine. Dr. Kendall writes[69] that careful examination of large series has convinced him that this is actually a distinct species, not a race of the extremely variable three-spined stickleback, although he saw one specimen apparently intermediate between the two.

Color.—Grass-green above in life, mottled and finely speckled with black on the top of the head and back; sides of head and body golden with dark blotches; breast silvery; ventral fins scarlet.

Habits.—Its mode of life is the same as that of the three-spined species so far as known, and sticklebacks of this type have been described as building nests with bits of straw on sandy bottom in New York waters,[70] but the two species or races have been confused so often that nothing more definite can be said of its habits.

General range.—Newfoundland to New York.

Occurrence in the Gulf of Maine.—Sticklebacks of this type are common in company with the three-spined sticklebacks in Passamaquoddy and St. Mary Bays[71] and in the Bay of Fundy. They may be expected anywhere on the Maine coast, being recorded at Winter Harbor; off Monhegan Island; off Seguin Island; from Casco Bay and its tributaries in both salt and brackish water; and from Kittery. They have also been taken at Swampscott, in Massachusetts Bay, and they are fairly common in summer at Woods Hole. We have taken them in our tow-nets, also, off Cape Porpoise; on Platts Bank; in the Western Basin of the Gulf of Maine; and on German Bank.

[69] Proc. U. S. Nat. Mus., vol. 18, 1896, p. 624.
[70] Bean, Bull. 60, New York State Mus., Zool. 9, 1903, p. 341.
[71] Huntsman, Contrib. Canadian Biol. (1921) 1922, p. 61.

Four-spined stickleback *Apeltes quadracus* (Mitchill) 1815

Bloody Stickleback

Jordan and Evermann, 1896–1900, p. 752.

Description.—The four-spined stickleback has no bony plates in its scaleless skin, but it does have a bony ridge on each side of the abdomen, making it triangular in cross section, with flat belly and sharp back; this gives it an aspect very different from the other sticklebacks. It is fusiform in side view, tapering to the rather pointed nose and to the slim caudal peduncle. There are two to four free dorsal spines standing close one behind the other, inclining alternately to one side or the other, and another spine is attached to the dorsal fin by the fin membrane. The anal fin is similarly preceded by an attached spine, and each ventral fin is represented by a stouter curved spine, strongly saw-edged, followed by about two slender rays. The dorsal fin stands over the anal as in the nine-spined species, but both these fins are more rounded in outline, while the caudal fin is relatively longer and narrower than in any of our other sticklebacks.

Color.—Brownish olive or greenish brown above, with dark mottlings that alternate below the lateral line with the silvery white of the belly. The fin membrane of the ventrals is red. Males are much darker than females.

Size.—One and one-half to two and one-half inches long.

Habits.—This is a common little fish in the salt marshes, where it consorts with other sticklebacks and with mummichogs. Like the three-

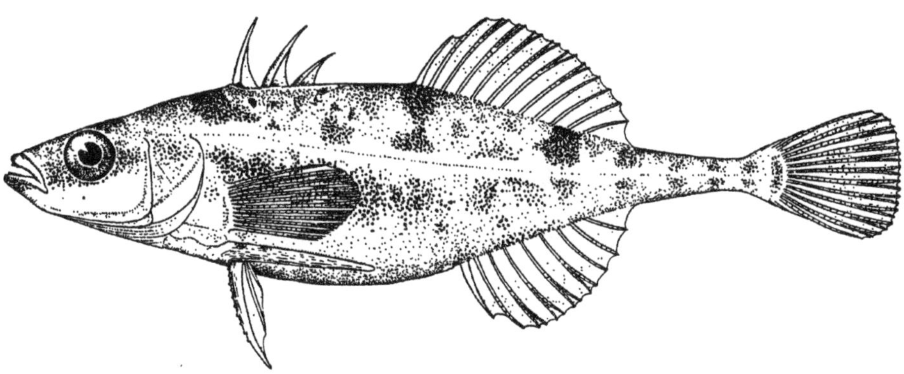

FIGURE 171.—Four-spined stickleback (*Apeltes quadracus*), Woods Hole. From Jordan and Evermann. Drawing by A. H. Baldwin.

spined stickleback it often runs up into freshwater, though it is primarily a salt and brackish water fish. And it is never found far in from the coast or out at sea.

In the Woods Hole region this stickleback spawns as early as May and as late as the last week of July, after which spent females are found; but the onset of spawning may be somewhat later in the cooler waters of the Gulf. The male builds a nest of plant fragments which it brings in its mouth, cemented with mucous threads that he spins out of a pore near his vent; a small rudimentary affair, however, compared with that of the three-spined stickleback, described by Ryder[72] as less than 1 inch in diameter, conical, with an opening at the top. Finally, the male stickleback picks up the eggs that have been laid by the female and deposits them in the hollow at the top of the nest, guarding them, presumably, during incubation. The eggs are yellow, approximately 1.66 mm. in diameter; they sink like those of the other sticklebacks and stick together in clumps. Incubation occupies six days or thereabouts at laboratory temperature (about 70°). Newly hatched larvae are about 4.5 mm. long and similar in appearance to those of the three-spined species but more densely pigmented.[73]

General range.—This is an American fish, known along the coast from the southern side of the Gulf of St. Lawrence and Nova Scotia to Virginia; at home both in salt water and in brackish, and running up into fresh water.

Occurrence in the Gulf of Maine.—This stickleback is common all around the shores of the Gulf on the Nova Scotian side as well as the New England side. We have taken it at Yarmouth; Huntsman[74] records it from St. Mary Bay and along the New Brunswick shore well within the Bay of Fundy (Maine has usually been given as its northern limit), and there are many locality records for the coasts of Maine and Massachusetts. But it is so much more closely restricted to estuarine situations than is its three-spined relative (p. 310) that we have never taken it in our tow nets nor do we find a single record of it in the open sea. On the south shore of New England it is a year-round resident. Probably this is equally true in the Gulf, where it may be expected to gather in the bottoms of the deeper creeks in winter, as it is known to do in Chesapeake Bay. It resembles the three-spined stickleback in its feeding habits so far as known (copepods and other small crustaceans being its chief diet) and in its general mode of life.

THE PIPEFISHES. FAMILY SYNGNATHIDAE

The forward portion of the head has the form of a long tubular snout in the pipefishes, with the small mouth situated at its tip; the skin is armed with rings of bony plates; there is only one dorsal fin (soft-rayed), the body is very slender, and there are no ventrals. The snout recalls that of the trumpetfishes (p. 316), but pipefishes differ from them and from most other bony fishes in the structure of their gills, which form tufts of small rounded lobes, instead of the familiar filaments. Their general affinity in this respect is with the group of which the sticklebacks are the most familiar exponents. There are many species of pipefishes in warm seas, but only one inhabits the Gulf of Maine regularly, while a second has been recorded there—a stray from the south.

KEY TO GULF OF MAINE PIPEFISHES

1. Dorsal fin with 35 to 41 rays; 18 to 20 bony plates in front of the vent and 36 to 42 behind it — Common pipefish, p. 312

   Dorsal fin with 29 to 31 rays; 16 or 17 bony plates in front of the vent and 31 to 35 behind it — Pelagic pipefish, p. 314

Common pipefish *Syngnathus fuscus* Storer[75] 1839

Jordan and Evermann, 1896–1900, p. 770 as *Siphostoma fuscum* (Storer).

Description.—This is a very slender little fish, particularly so behind the vent, males being about 35 times as long as they are deep and females about 30 times. The head occupies one-eighth to one-

[72] Bull. U. S. Fish. Comm., vol. I, 1882, p. 24.

[73] The early development is described by Ryder (Bull. U. S. Fish Comm., vol. I, 1882, p. 24) and by Kuntz and Radcliffe (Bull. U. S. Bur. Fish., vol. 35, 1918, p. 132).

[74] Contrib. Canadian Biol. (1921) 1922, p. 61.

[75] This is the only pipefish that occurs regularly on our northern coasts. Jordan, Evermann, and Clark (Rept. U. S. Comm. Fish. [1928], 1930, Pt. 2, p. 242) refer it to the genus *Syrictes* Jordan and Evermann, 1927. For a synopsis of the various other species of the genus see Jordan and Evermann, Bull. 47, U. S. Nat. Mus., Pt. 1, 1896, p. 961.

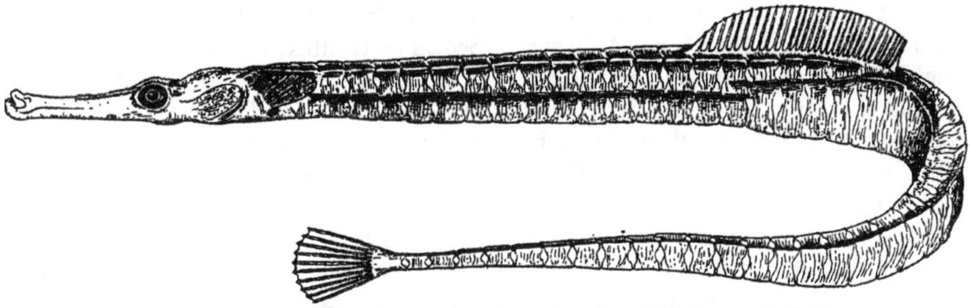

FIGURE 172.—Pipefish (*Syngnathus fuscus*). After Bigelow and Welsh.

ninth of the total length (in the trumpetfish it is nearly one-third); the snout is tubelike, blunt ended, with the small toothless mouth at its tip. The gill openings are very small. The body is hexagonal in cross section in front of the vent but is four-sided behind the dorsal fin and it is clothed in an armor of bony plates connected in rings, of which there are 18 to 20 in front of the vent and 36 to 42 behind the latter. The abdomen of the male is wider just back of the vent than elsewhere, with two lateral flaps that meet along the midline to form the so-called "marsupial" or brood pouch. The female lacks these. The dorsal fin (35 to 41 rays and 5 or 6 times as long as it is high) covers 4 or 5 of the bony rings in front of the vent and as many behind it. The caudal fin is rounded, its middle rays the longest. The anal is very small, close behind the vent; the pectorals are of moderate size; there are no ventral fins.

Color.—Greenish, brownish, or olive above, cross-barred and mottled with darker. The lower parts of the gill covers are silvery. The lower parts of the sides are sprinkled with many tiny white dots, and the longitudinal angles separating sides from abdomen are marked by longitudinal brown bars. The lower surface is colorless on the snout, but pale to golden yellow thence back to the vent, with the marsupial flaps flesh-colored. The dorsal and pectoral fins are pale, but the caudal is brown.[76] Pipefishes change color according to the color of their surroundings. We have seen them of various shades of olive and brown; and red ones have been described.

Size.—Usually 4 to 8 inches long; occasionally up to 12 inches.

Habits.—The chief home of this pipefish is among eelgrass or seaweeds, both in salt marshes, harbors, and river mouths, where it often goes up into brackish water, and on more open shores as well. In such locations it is caught as often today by boys dipping up mummichogs for bait as it was when Storer wrote of it, nearly a century ago. The pipefish, like the three-spined stickleback, sometimes strays out to sea on the surface, and while we have never taken it in our tow nets, Kendall [77] has often found it under floating rockweed along the Maine coast. But they are so seldom taken at any distance out from the land that the capture of four specimens at a depth of 19 fathoms south of No Mans Land, February 5, 1930, is of present interest, though outside the limits of our Gulf. There is no reason to suppose pipefish are at all migratory, for they are resident in the eelgrass (*Zostera*) at Woods Hole throughout the year.

They usually propel themselves by the dorsal fin, but they can travel swiftly when alarmed, with eel-like strokes of the tail from side to side. And they are able to roll their eyeballs separately, an interesting habit described many years ago by Lyman.[78]

They feed chiefly on minute Crustacea (copepods especially and amphipods), also to some extent on fish eggs, on very small fish fry, and no doubt indiscriminately for that matter on any small marine animals. And their snouts are so distensible that they can swallow larger prey than one might expect. In capturing its prey, the pipefish has been described as expelling the water from the snout and pharynx by muscular action, depending on the return rush to sweep in its victims. Pipefishes have few enemies so far as known.

[76] Colors based on Storer's (Fishes of Massachusetts, 1867, p. 412) account and on the specimens we have examined.

[77] Bull. U. S. Fish Comm., vol. 18, 1896, p. 623.

[78] Proc. Boston Soc. Nat. Hist., vol. 7, 1861, pp. 75–76.

Breeding.[79]—On the southern shores of New England pipefish breed from March to August; probably through this same period on the shores of the Gulf of Maine. Male pipefishes nurse the eggs in the brood pouch (p. 313), the flaps of which ordinarily lie flat but are swollen and have their edges cemented together during the breeding season. The protruding oviduct of the female is inserted into the opening of the pouch of the male and a dozen or more eggs are passed over. This occurs several times in succession, with intervals of rest, until the pouch is filled, the male working the eggs down toward the rear end of his pouch by body contortions. Fertilization is supposed to take place during the transference of the eggs from one parent to the other. The eggs become embedded in the lining of the brood pouch, and it has been established for the European pipefish (probably this applies equally to our North American species) that the embryos within the eggs are nourished by the epithelial lining layer of the pouch, so that the latter functions as a placenta.[80]

Incubation occupies about 10 days, according to Gudger, and the young are retained in the brood pouch until they are 8 or 9 mm. long, when the yolk sac has been absorbed. The young pipefish are then ready for independent existence, and once they leave the pouch they never return to it, as young sea horses (*Hippocampus*) are said to do (p. 315). Several observers agree on this, among them Miss Marie Poland (now Mrs. C. J. Fish), who kept pipefish under observation at the laboratory of the United States Bureau of Fisheries at Woods Hole during the summer of 1922.

Pipefish fry kept in aquaria have been found to grow from about ⅜-inch (10 mm.) to about 2¾-inches (70 mm.) in length within about 2 months after hatching.[81] It is probable that they mature when about 1 year old.

General range.—Coast of eastern North America, in salt and brackish water, from the southern side of the Gulf of St. Lawrence [82] and outer Nova Scotia at Halifax, to South Carolina.[83]

Occurrence in the Gulf of Maine.—The pipefish has been recorded from so many localities along Maine and Massachusetts that it is evidently to be expected anywhere there, in suitable situations; it is not uncommon in the Bay of Fundy; it has been reported from outer Nova Scotian waters and is common locally in the southern side of the Gulf of St. Lawrence, as noted in the preceding paragraph. It is probable also that they breed in every favorable locality all around the shores of the Gulf, but there are local differences in this respect, for while St. Mary Bay, Annapolis Basin, and Cobequid Bay, on the Nova Scotian shore of the Bay of Fundy, are breeding centers according to Huntsman, large specimens alone are known about Passamaquoddy Bay on the New Brunswick side. No doubt the estuarine waters from the Massachusetts Bay region to Penobscot Bay are favorable nurseries.

Importance.—The pipefish is of no commercial importance.

Pelagic pipefish *Syngnathus pelagicus* Linnaeus 1758

Jordan and Evermann, *Siphostoma pelagicum* (Osbeck), 1896–1900, p. 767.

Description.—Most of the species of pipefishes resemble one another so closely that they can be named only by critical examination. The pelagic pipefish differs from its common shore relative of New England (*S. fuscus*) by having fewer dorsal rays (29 to 31) and fewer rings of bony plates, of which only 16 or 17 are in front of the vent, and 31 to 35 behind the vent.

General range.—Tropical Atlantic, northward with the Gulf Stream; also the Mediterranean, and the Southern Pacific and Tropical Indian Oceans.

Occurrence in the Gulf of Maine.—A single specimen, 3½ inches (89 mm.) long, taken on Georges Bank (Lat. 42° 09′ N.; Long. 66° 41′ W.), September 20, 1927, by the *Albratross II* is the only Gulf of Maine record. This specimen was dipped up with a mass of gulf weed (*Sargassum*), and was the only one found in a large amount of weed that was examined.

[79] For a historical survey and a general account of the breeding of the closely allied *Siphostoma floridae* see Gudger (Proc. U. S. Nat. Mus., vol. 29, 1906, pp. 447–500, pls. 5–11).

[80] For detailed (if somewhat divergent) accounts of this interesting phenomenon see Huot (Annales des Sciences Naturelles, Ser. 8, Zoologie, vol. 14, 1902, pp. 197–288) and Cohn (Anatomischer Anzeiger, Centralblatt für die gesamte wissenschaftliche Anatomie, vol. 24, 1904, pp. 192–199).

[81] Tracy, 40th Rept., Rhode Island Comm. Inland Fish., 1910, p. 93.

[82] Leim (Proc. Nova Scotian Inst. Sci., vol. 20, 1940, p. 38) found them common at Prince Edward Island.

[83] There is a specimen from Charleston, S. C., in the Museum of Comparative Zoology.

THE SEA HORSES. FAMILY HIPPOCAMPIDAE

Sea horse *Hippocampus hudsonius* DeKay 1842

Jordan and Evermann, 1896–1900, p. 777.

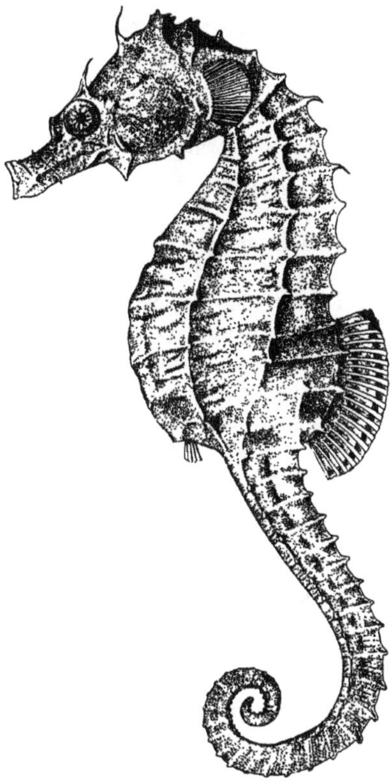

FIGURE 173.—Sea-horse (*Hippocampus hudsonius*), Virginia. From Goode. Drawing by H. L. Todd.

Description.—The sea horse grotesquely resembles the "knight" in an ordinary set of wooden chessmen in its sidewise flattened body, in its deep convex belly, in its curved neck and in its curious horse-like head carried at right angles to the general axis of the body. The head is surmounted by a pentagonal star-shaped "coronet," and the snout is tubular with the small oblique mouth at its tip, like that of its relative the pipefish. It has a sharp spine on each side above the eye and one behind it, a third over the gill cover, and a fourth on the side of the throat, which sometimes terminate in short fleshy filaments; also a blunt horn between the nostrils. Its neck, body, and tail are covered with rings of bony plates, 12 rings on the trunk, 32 to 35 on the tail, and each body ring is armed with four blunt spines. The body tapers suddenly behind the anal fin to a long tail, which is four-cornered in cross section, curled inward, and strongly prehensile. In the male the lower surface of the fore part of the tail bears the brood pouch, opening by a slit in front. The dorsal fin (about 19 rays) originates about midway of the length of the fish, opposite the vent, and runs backward over three and one-half rings to within half a ring of the commencement of the tail sector of the trunk. The very small anal fin stands opposite the rear part of the dorsal fin. The pectorals are of moderate size, broad based and round tipped; it has no ventral fins and no caudal fin.

Color.—Light brown or dusky to ashen gray or yellow, variously mottled and blotched with paler and darker, sometimes spangled with silver dots, sometimes plain colored. European sea horses change color according to their surroundings, tints of red, yellow, brown, and white all being within their capabilities, and it is probable that the American species is equally adaptable.

Size.—Adults usually are 3 to 6 inches long; one of 7¼ inches is the largest on record.[84]

Habits.—Sea horses dwell chiefly among eelgrass and seaweed,[85] where they cling with their prehensile tails, monkeylike, to some stalk. They usually swim in a vertical position by undulations of the dorsal fin, not with the tail, the trunk being too stiff for much sidewise motion.

Sea horses feed on minute Crustacea and on various larvae, in fact on any animal small enough, sucking in their prey as the pipefish does (p. 313.)

They breed in summer[86] and the breeding habits resemble those of the pipefish, the male nursing the eggs in his brood pouch where they are deposited a few at a time by the female in repeated pairings. The young, of which there may be as many as 150, are about 10 to 12 mm. long at hatching. When the yolk sac is absorbed the father squeezes them out of the brood sac, and they already resemble the adult in general appearance within a few days after they are set free. According to some students they swim out and in at will, but this calls for verification.

[84] Bull. New York Zool. Soc., vol. 16, 1913, p. 972.

[85] Gill (Proc. U. S. Nat. Mus., vol. 28, 1905, pp. 805–814) has given an excellent account of the habits and life history of the sea horse.

[86] Ryder (Bull. U. S. Fish Comm., vol. I, 1882, pp. 191–199) describes its development.

General range.—Atlantic Coast of North America, occurring regularly from South Carolina to Cape Cod, and to Nova Scotia as a stray.

Occurrence in the Gulf of Maine.—The sea horse is not common much beyond New York. Only a few are found each year about Woods Hole, chiefly in July, August, and September, and they so rarely stray past the elbow of Cape Cod that we have found only one definite (Provincetown) and one dubious (Massachusetts Bay) record of its capture in the inner parts of the Gulf of Maine, dead or alive; and one record for Georges Bank. Three specimens of the sea horse were also reported from Nova Scotia more than ¾ of a century ago;[87] and Vladykov and McKenzie have reported one, picked up in Terrance Bay, on the outer Nova Scotian coast, Sept. 18, 1934, by V. Slaunhite.[88]

Commercial importance.—The sea horse is of no commercial value, but it is an object of constant interest to visitors to marine aquaria.

THE TRUMPETFISHES. FAMILY FISTULARIIDAE

The trumpetfishes are characterized by their slender bodies and tremendously long heads, as well as by the fact that the anterior bones of the skull are prolonged in a very long tube with the small mouth at its tip. The only other Gulf of Maine species with which they could possibly be confused is the pipefish (p. 312). In the latter, however, the tubular snout occupies only about one-eighteenth of the length of the fish whereas in trumpetfishes it is nearly one-fourth. Furthermore, the pipefishes lack ventral fins which the trumpetfishes have, while the caudal fin of the trumpetfishes is forked, but that of the pipefishes is rounded.

Trumpetfish *Fistularia tabacaria* Linnaeus 1758

CORNETFISH

Jordan and Evermann, 1896–1900, p. 757.

Description.—The slender body and very long tubular snout of this fish are mentioned above. The body (to base of caudal fin) is about 30 to 35 times as long as it is deep and only about two-thirds as deep as it is thick. The head occupies almost one-third and the snout about one-fourth of the body length. The bones of the snout are so loosely united that the snout is very distensible. The mouth is small, situated somewhat obliquely at the tip of the snout, and the lower jaw projects a little beyond the upper. The caudal fin is deeply forked and its middle rays are prolonged in a filament about as long as the snout, but which is likely to be broken off. Both the dorsal and the anal fins are triangular, higher than long, the former standing exactly above the latter, about three-fourths of the distance back from eye toward base of caudal fin. The ventrals are very small, and are considerably nearer to the eye than to the rear end of the body (about one-third of the way from eye toward the base of caudal fin). The skin is without scales but with a row of embedded bony plates or shields along either side, conspicuous rearward.

Color.—Greenish brown above, the back and sides marked with many large, oblong, pale blue spots and with about 10 dark cross bars; the lower surface is pale and silvery; the caudal filament deep blue.

Size.—Said to reach a length of 6 feet, but the few specimens that stray northward are much smaller.

General range.—Tropical, southward to middle Brazil, and common among the West Indies; rarely

[87] By Knight (Catal. Fishes Nova Scotia, 1866, p. 9), as *H. brevirostris* Storer (1839); later by Jones, (Proc. Nova Scotian Inst. Sci., vol. 5, Part 1, 1882, p. 95) as *H. antiquorum* Leach 1814.

[88] Proc. Nova Scotian Inst. Sci., vol. 19, 1935, p. 5.

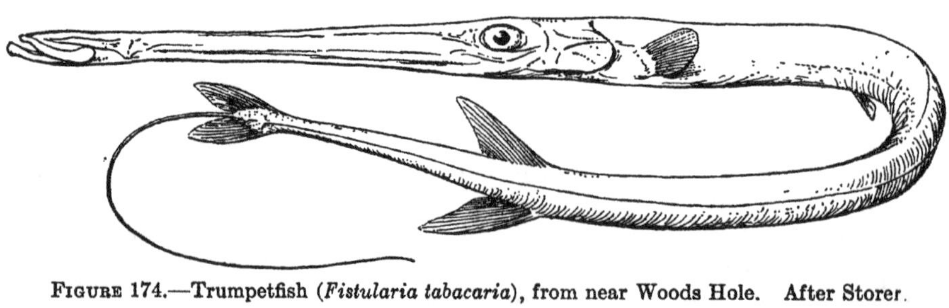

FIGURE 174.—Trumpetfish (*Fistularia tabacaria*), from near Woods Hole. After Storer.

wandering northward as far as the Massachusetts Bay region, and straying to Nova Scotia.[89]

Occurrence in the Gulf of Maine.—There are only two records of the trumpetfish from the Gulf of Maine: a specimen taken at Rockport, Mass. (north side of Cape Ann) in September 1865, preserved in the collection of the Essex Institute, where it was examined and identified by Goode and Bean [90] and a second taken on the northern edge of Georges Bank by the trawler *Flying Cloud* on October 6, 1947, in a haul at 70 fathoms.[91] Like other tropical fishes, however, it is not so rare west of Cape Cod, and a few small ones are taken at Woods Hole almost every year.

THE MACKERELS. FAMILY SCOMBRIDAE

The mackerels are a very homogeneous group, all of them agreeing in having a spiny dorsal as well as a soft dorsal fin, several small finlets behind the latter and behind the anal, a very slender caudal peduncle, a deeply forked or lunate caudal fin, a very shapely form tapering both to snout and to tail, and velvety skin with very small scales. All, too, are predaceous, swift swimmers, and powerfully muscled, while all are fish of the open sea and more or less migratory.

In the following key we mention all species so far actually recorded from within the limits of the Gulf of Maine, but it would not be astonishing if still others were to stray in from the open Atlantic from time to time.[92]

KEY TO GULF OF MAINE MACKERELS [93]

1. The two dorsal fins are separated by a space at least as long as the length of the first dorsal ---------- 2
   The two dorsal fins adjoin each other or are separated by a space much shorter than the length of the first dorsal -- 3
2. The sides below the mid line are silvery, not spotted ---------- Mackerel, p. 188, 317
   The sides below the mid line are mottled with dusky blotches ---------- Chub Mackerel, p. 209, 333
3. Body scaleless, except along the lateral line and in the region of the shoulders (the so-called "corselet") ---------- 4
   Entire body covered with scales ---------- 5
4. The lower part of the sides, below the lateral line, is marked with dark longitudinal bands, but there are no definite dark markings on the back ---------- Striped Bonito, p. 335
   There are no dark markings on the lower side below the lateral line, but the back has dark markings
   False Albacore, p. 336
5. The anal fin is about twice as high as long; the corselet of large scales is obvious ---------- Tuna, p. 338
   The anal fin only is about as high as long; there is no corselet of large scales ---------- 6
6. Second dorsal fin noticeably lower than the first dorsal ---------- Common Bonito, p. 337
   Second dorsal fin at least as high as first dorsal ---------- 7
7. Lateral line with an abrupt downward curve under second dorsal fin ---------- Cavalla p. 349
   Lateral line descending gradually ---------- 8
8. Sides with a few rows of oval bronze or yellowish spots and with one or two longitudinal dark streaks
   King Mackerel p. 348
   Sides with bronze spots but without longitudinal dark streaks ---------- Spanish Mackerel p. 347

Mackerel *Scomber scombrus* Linnaeus 1758

Jordan and Evermann, 1896–1900, p. 866

Description.—The mackerel is fusiform in outline, tapering rearward to a very slim caudal peduncle and forward to a pointed nose. Its body is about four and one-half to five and one-half times as long as it is deep, oval in section, thick and firm-muscled as are all its tribe. Its head is long (one-fourth of length to caudal) and its mouth large, gaping back to the middle of the eye (the premaxillaries are not protractile), while the jaws, which are of equal length, are armed with small, sharp, slender teeth. The eye is large, and the hollows in front of and behind it are filled with the so-called "adipose eyelid," a transparent,

[89] Dr. A. H. Leim reports the capture of a specimen at Port Mouton, Nova Scotia, on September 10, 1931; the specimen was recorded later by Vladykov (Proc. Nova Scotian Inst. Sci., vol. 19, 1935, p. 5) as *Fistularia serrata*.

[90] Bull. Essex Inst., vol. XI, 1879, p. 4.

[91] This specimen is in the Museum of Comparative Zoology.

[92] Fraser-Brunner, Ann. Mag. Nat. Hist. Ser. 12, vol. 3, No. 26, 1950, pp. 131–163, has recently given a synopsis of the mackerels, with useful keys and excellent illustrations for all known species. We follow him in uniting them all in the old and inclusive family Scombridae rather than Jordan, Evermann and Clark (Rept. U. S. Comm. Fish. (1928) Pt. 2, 1930) who have distributed them among four families, Cybiidae, Katsuwonidae, Scombridae, and Thunnidae.

[93] The long finned Albacore (*Thunnus alalunga* Bonnaterre, 1788) has been taken at Woods Hole, also on Banquereau Bank, off eastern Nova Scotia (Goode and Bean, Bull. Essex Inst., vol. XI, 1879, p. 15), so is likely to show up in the Gulf of Maine sooner or later. It is made easily recognizable among North Atlantic mackerel fishes by its very long pectoral fins which reach back past its second dorsal fin.

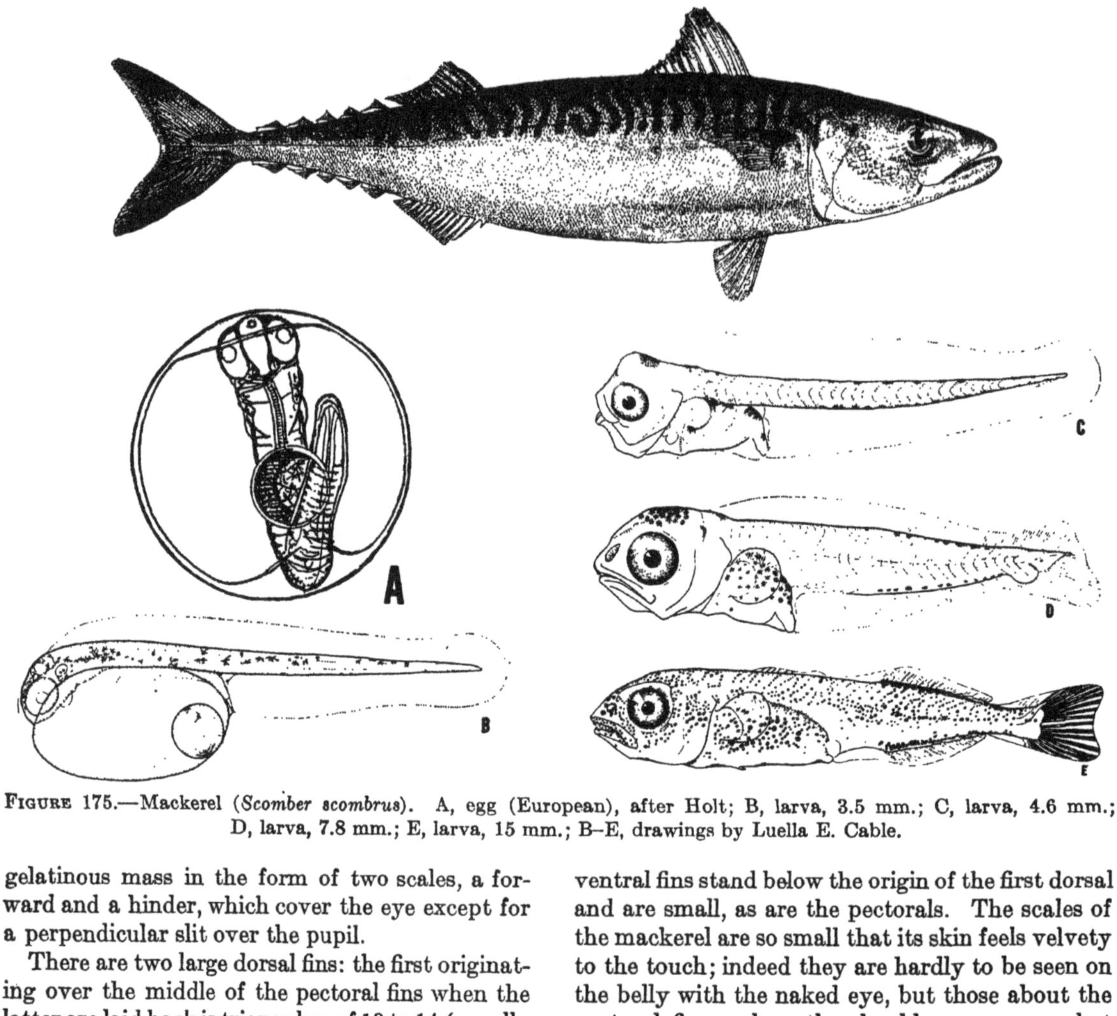

FIGURE 175.—Mackerel (*Scomber scombrus*). A, egg (European), after Holt; B, larva, 3.5 mm.; C, larva, 4.6 mm.; D, larva, 7.8 mm.; E, larva, 15 mm.; B–E, drawings by Luella E. Cable.

gelatinous mass in the form of two scales, a forward and a hinder, which cover the eye except for a perpendicular slit over the pupil.

There are two large dorsal fins: the first originating over the middle of the pectoral fins when the latter are laid back is triangular, of 10 to 14 (usually 11, 12, or 13) rather weak spines that can be laid down along the midline of the back in a deep groove; the second dorsal, separated from the first by an interspace longer than the length of the latter, is smaller (9 to 15 rays, usually 12) and is followed by several small finlets, of which there are usually 5, but sometimes 4 or 6. The anal fin is similar to the second dorsal in shape and size, originates slightly behind it, and is similarly succeeded by 5 small finlets that correspond to the dorsal finlets in size and shape. The caudal fin is broad, but short and deeply forked. The caudal peduncle bears two small longitudinal keels on either side but no median lateral keel, the absence of the latter being a distinctive character. The ventral fins stand below the origin of the first dorsal and are small, as are the pectorals. The scales of the mackerel are so small that its skin feels velvety to the touch; indeed they are hardly to be seen on the belly with the naked eye, but those about the pectoral fins and on the shoulders are somewhat larger.

Color.—The upper surface is dark steely to greenish blue, often almost blue-black on the head. The body is barred with 23 to 33 (usually 27 to 30) dark transverse bands [94] that run down in an irregular wavy course nearly to the mid-level of the body, below which there is a narrow dark streak running along each side from pectoral to tail fin. The pectorals are black or dusky at the base, the dorsals and caudal are gray or dusky. The jaws and gill covers are silvery. The lower parts of the sides are white with silvery, coppery, or brassy reflections and iridescence; the belly silvery white.

[94] Hunt (Copeia, No. 117, pp. 53–59, April, 1923) describes the variations in these stripes among young mackerel.

But the iridescent colors fade so rapidly after death that a dead fish gives little idea of the brilliance of a living one.

Size.—Most of the grown fish are between 14 and 18 inches long; a few reach a length close to 22 inches. Fourteen-inch fish weigh about 1 pound in the spring and about 1¼ pounds in the fall when they are fat; 18-inch fish weigh about 2 to 2½ pounds; a 22-inch mackerel will likely weigh 4 pounds. An unusually large mackerel is taken occasionally; in 1925, for example, the schooner *Henrietta* brought in one weighing 7½ pounds.

Habits.—Mackerel are a swift-moving fish, swimming with very short sidewise movements of the rear part of the body and of the powerful caudal fin. When caught they beat a rapid tattoo with their tails on the bottom of the boat until exhausted. And they require so much oxygen for their vital processes that when the water is warm (hence its oxygen content low) they must keep swimming constantly, to bring sufficient flow of water to their gill filaments, or else they die.[95]

Despite their great activity, they do not leap above the surface, as various others of their tribe do, unless perhaps to escape some larger fish.

The mackerel, like the herring, has the habit of gathering in dense schools of many thousands. It is not known how long these schools hold together; it would be especially interesting to know whether they do so through the winter when our mackerel are in deep water, but the general opinion of fishermen is that they do so throughout the migrations at least. Although the mackerel may scatter and the schools mix more or less, especially when they are feeding on the larger and more active members of the free-floating fauna as is said to happen in British waters, the members of any given school usually are all of about the same size, i. e., of the same age. Fish of the year almost always school separately from the others as Sette[96] has pointed out; he has also pointed out that this tendency of the fish to separate according to size is probably due to the fact that the larger ones swim faster than the smaller ones.

Mackerel school by themselves, as a rule. But sometimes they are found mingled with herring, alewives, or shad, as Kendall[97] described. We have yet to learn how mackerel schools hold together, whether by sight or by some other sense. And various explanations have been proposed to account for the schooling habit, such as that it is advantageous for feeding, that it is a concomitant of spawning (this would not explain its persistence out of the spawning season, however, or the fact that any given school is apt to contain green, and spent as well as ripe fish even at spawning time), or that it affords protection from enemies. But when all is said, the instinct prompting it remains a mystery. At any rate, schooling is not a necessity, though usual. When mackerel are at all plentiful, and even when they are not, numbers of single wandering fish are often hooked by persons trolling for them, and by flounder and cunner fishermen.

Schools of mackerel are often seen at the surface. In the daytime they can be recognized by the appearance of the ripple they make, for this is less compact than that made either by herring or by menhaden. Neither do mackerel ordinarily "fin" or raise their noses above the surface, as is the common habit of the menhaden (p. 114). An observer at masthead height can perhaps see a school of mackerel as deep as 8 to 10 fathoms by day, if the water is calm, and the sun behind him. On dark nights the schools are likely to be betrayed by the "firing" of the water, caused by the luminescence of the tiny organisms that they disturb in their progress. Sette[98] reports one case of a school recognized by its firing as deep as 25 fathoms; but the water is seldom (if ever) clear enough in the Gulf of Maine for a submerged light to be visible from above, more than 15 fathoms down.[99] The trail of bluish light left behind by individual fish as they dart to one side or the other, while one rows or sails through a school on a moonless, overcast night when the water is firing, is the most beautiful spectacle that our coastal waters afford, and one with which every mackerel fisherman is familiar.

No one knows how greatly the movements of the mackerel, from day to day, result from involuntary drifting with the circulatory movements of the water, which are different at different depths, and how greatly they depend on the directive swimming of the mackerel themselves. Our only

[95] This interesting fact seems first to have been reported by Hall (Amer. Jour. Physiol., vol. 93, 1930, pp. 417–421), and we have observed the same thing in the aquaria at the Woods Hole Oceanographic Institution.

[96] Fish. Bull. U. S. Fish and Wildlife Service, vol. 51, Bull. 49, 1950, p. 264.

[97] Bull. U. S. Bur. Fish., vol. 28, 1910, Pt. I, p. 287.

[98] Fish. Bull. U. S. Fish and Wildlife Service, vol. 51, Bull. 49, 1950, p. 267.

[99] For observations on the visual transparency of the water of our Gulf, see Bigelow, Bull. U. S. Bur. Fish., vol. 40, Pt. 2, 1927, p. 822.

contribution in this regard is that we once were able to follow on foot beside a school that was advancing along the Scituate shore at a rate of about 3 to 4 miles an hour, against a tidal current of about one-half knot, until the fish swung offshore and out of our sight.

The speed at which a school travels when it is not disturbed depends, it seems, on the size of the fish of which it is composed. It has been observed by Sette that mackerel less than one year old swim at about 6 sea miles per hour (10 ft. a second) while circling inside a live car; yearlings at a rate of about 11½ sea miles per hour (19 ft. a second), or nearly twice as fast. We find no definite observations on the normal speed of the larger fish, and no one knows how rapidly a mackerel may swim for a short distance, if it is disturbed. Mackerel seen during the warmer months of the year are always swimming, but this rule may not apply in winter, when the water holds more dissolved oxygen because colder, and when it is probable that their demand upon it is lower.

The mackerel is a fish of the open sea; while numbers of them, small ones especially, often enter estuaries and harbors in search of food, they never run up into fresh water. Neither are they directly dependent either on the coastline or on the bottom in any way at any stage in their lives. They are often encountered far out over the outer part of the shelf of the continent. But they are most numerous within the inner half of the continental shelf during the fishing season, and their normal range seems not to extend oceanward beyond the upper part of the continental slope, in which they contrast with their relatives the tunas, the bonitos, and the albacores.

The depth-range of the mackerel is from the surface down to perhaps 100 fathoms at one season or another. (We recur to this in discussing the occurrence of mackerel in the Gulf of Maine, page 325.) From spring through summer and well into the autumn, the mackerel are in the upper water layers; shoaler, mostly, than some 25 to 30 fathoms, and schools of all sizes come to the surface more or less frequently then. But they frequently disappear from the surface, often for considerable periods. And it seems, from fishermen's reports, that the larger sizes tend to swim deeper than the smaller ones, on the whole, especially in mid and late summer.[1] It is probable, also, that their vertical movements during the warmer part of the year, when they are feeding actively, are governed chiefly by the level at which food is most abundant, which for the most part is shoaler than about 50 fathoms, at least on our side of the Atlantic.

The highest temperature in which mackerel are commonly seen is about 68° F. (20° C.). At the opposite extreme they are sometimes found in abundance in water of 46°–47° (8° C.); and commercial catches are sometimes made in water as cold as 44°–45° (7° C.), but odd mackerel only have been taken in temperatures lower than that[2] in American waters. Large catches of mackerel are made, however, by trawlers in the North Sea in winter in water as cold as 43°–45° (6°–7° C.). But as Sette has emphasized, the European mackerel differs racially from the American, and may differ in its temperature relations as well.

Food.—We may assume that the diet of the young mackerel is at first much the same in the Gulf of Maine as it is in the English Channel,[3] namely, copepod larvae and eggs, the smaller adult copepods, various other minute pelagic Crustacea, and small fish larvae. But the young fish depend more and more upon larger prey as they grow. Our Gulf of Maine mackerel have repeatedly been seen packed full of *Calanus*, the "red feed" or "cayenne" of fishermen, as well as with other copepods (we have examined many in this condition). They also feed greedily, as do herring, on euphausiid shrimps (p. 89), especially in the northeastern part of the Gulf where these crustaceans come to the surface in abundance. Various other planktonic animals also enter regularly into the dietary of the mackerel. Thus, Doctor Kendall writes in his field notes that some of the fish caught on the northern part of Georges Bank in August 1896, were packed with crab larvae, others were full of Sagittae, others, again, of Sagittae and amphipods (*Euthemisto*), of small copepods (*Temora*), or of red feed (*Calanus*), so that even fish from the same school had selected the various members of the drifting community in varying proportion.

[1] See Sette, Fish. Bull. U. S. Fish and Wildlife Service, vol. 51, Bull. 49, 1950, p. 267, for further discussion of this point.

[2] Sette (Fish. Bull, U. S. Fish and Wildlife Service, vol. 51, Bull. 49, 1950, p. 257) mentions one winter record from about 40° (4.5° C.) on Georges Bank.

[3] Lebour (Jour. Mar. Biol. Assoc. United Kingdom, vol. 12, N. Ser., No. 2, 1920, p. 305) gives diet lists for 90 larval mackerel ranging from 5 mm. to 13.5 mm. in length, taken in the English Channel.

Similarly, 1,000 mackerel caught near Woods Hole from June to August contained pelagic amphipods (*Euthemisto*), copepods, squid, and launce;[4] others taken off No Man's Land have been found full of shelled pteropods (*Limacina*). And a series of small fish examined by Vinal Edwards contained copepods, shrimps, crustacean and molluscan larvae, annelid worms, appendicularians, squid, fish eggs, and fish fry such as herring, silversides, and launce. In short, practically all the floating animals, not too large or too small, regularly serve for the nourishment of mackerel except the Medusae and ctenophores, and a diet list for any given locality would include all the local pelagic Crustacea and their larvae.

Mackerel have often been seen to bite the centers out of large Medusae, but, as Nilsson suggests, they probably do this for the amphipods (*Hyperia*) that live commensal within the cavities of the jellyfish, not for the sake of the latter.

Mackerel also eat all kinds of small fish, to a greater or less extent according to circumstances. In the Gulf of Maine they devour large numbers of small herring, launce, and even smaller mackerel. They likewise feed on pelagic fish eggs when available, oftenest on those of their own species. And they bite greedily on almost any bait, especially if it moves, such as a bit of mackerel belly skin, a piece of clam, a piece of sea worm (*Nereis*), a shining jig, spoon or spinner of appropriate size, or an artificial fly, white, red, or silver-bodied. Side by side with these comparatively large objects mackerel are also known to take various microscopic organisms, chiefly the commoner peridinians and diatoms, but they never feed extensively on these as menhaden do (p. 114). And copepods are so plentiful in the Gulf of Maine, and the vegetable plankton that swarms in April has so largely disappeared over most of the Gulf before the mackerel appear later in the spring, that we doubt if they are ever reduced to a vegetable diet there or anywhere in American waters.

Mackerel are also known to feed on bottom animals to a small extent. Nilsson, for example, reports various worms and hydroids and even small stones from their stomachs, but our experience in the Gulf of Maine is to the effect that this would be exceptional there, if it happens at all.

[4] Nilsson (Publ. de Circ., Conseil Perm. Internat. Explor. Mer, No. 69, 1914) gives a similar list for Swedish waters

Most authors describe the mackerel as feeding by two methods: either by filtering out the smaller pelagic organisms from the water by their gill rakers [5] or by selecting the individual animals by sight. A good deal of discussion has centered about the relative serviceability of these two methods of feeding. Probably the truth is that when forced to subsist on the smaller objects in its dietary it must do so by sifting them out of the water, but that it selects the more desirable whenever opportunity offers to exercise its sight. It is not yet known how small objects the fish is able to pick out. It takes fish individually of course, and such large Crustacea as euphausiid shrimps and amphipods, just as the herring does, which evidently applies to the larger copepods, to judge from the fact that mackerel stomachs are often full of *Calanus* or of one or two other sorts in localities where indiscriminate feeding would yield them a variety. Whether they select the smaller copepods and crustacean larvae is not so clear. Captain Damant,[6] whose experience in deep-sea diving has given him an exceptional opportunity to observe mackerel feeding under natural conditions, describes fish among which he was at work 20 to 40 feet deep in Lough Swilly (Ireland), as "feeding on plankton, not by steadily pumping the water through the gill filters but snatching gulps from different directions and making little jumps here and there."

It has been a commonplace from the earliest days of the mackerel fishery that the fish are fat when last seen in the autumn, but that most of them are thin when they reappear in spring, obviously suggesting that they feed little during the winter. This is corroborated by the fact that the mackerel taken on bottom by British and French trawlers between December and March usually are empty, and that a few mackerel taken by the *Albatross II* along the continental edge off Chesapeake Bay in February 1931 were very emaciated. But mackerel taken in winter sometimes have food in their stomachs; some of them even are fat.[7]

[5] The mackeral has long rakers with spines on the foremost gill arch only, and these are not fine enough to retain the smallest organisms. See Bigelow, Bull. U. S. Bur. Fish, vol. 40, Pt. 2, 1926, fig. 42 C, D for photographs of the gill rakers.

[6] Nature, vol. 108, 1921, pp. 12-13.

[7] Sette (Fish. Bull. U. S. Fish and Wildlife Service, vol. 51, Bul'. 49, 1950, pp. 259, 262) reports some fat mackerel in winter.

The immature fish feed and fatten from the time they appear in spring. And it also seems that the schools of older fish destined to spawn late in the season feed until the actual ripening of their sexual products commences, for large catches of the maturing fish were regularly made on hook and line in June in the Gulf of St. Lawrence where spawning takes place in July (until the eggs began to run, in fact). But these large mackerel would not bite after that until they had spawned out (last half of July or first part of August). And available evidence, American as well as European, is to the effect that fish destined to spawn soon after their vernal appearance inshore continue their winter fast until they have spawned, when they commence feeding greedily.

Enemies.—The mackerel falls easy prey to all the larger predaceous sea animals. Whales, porpoises, mackerel sharks, threshers, dogfish, tuna, bonito, bluefish, and striped bass take heavy toll in particular. Cod often eat small mackerel; squid destroy great numbers of young fish less than 4 or 5 inches long, and sea birds of various kinds follow and prey upon the schools when these are at the surface. A considerable list of parasitic worms, both round and trematode, are known to infest the digestive tract of mackerel. But they seem more immune to danger from sudden unfavorable changes in their environment than the herring are, for they are never known to be killed by cold, and they seldom strand, except when small ones are driven ashore by larger fish.

Breeding.—Mackerel spawn off the American coast from the latitude of Cape Hatteras to the southern side of the Gulf of St. Lawrence. The spawning area covers almost the entire breadth of the continental shelf southward from Cape Cod, but it is confined more closely to the vicinity of the coast thence northward. Available data point to the oceanic bight between Chesapeake Bay and southern New England as the most productive area, the Gulf of St. Lawrence as considerably less so, and the Gulf of Maine and coast of outer Nova Scotia as ranking third.[8] Mackerel do not resort to any particular breeding grounds, but shed their eggs wherever their wandering habits have chanced to lead them when the sexual products ripen. It follows from this, and from the fact that mackerel vary so widely in abundance over periods of years that the precise localities of greatest egg production may be expected to vary from year to year, depending on the local concentrations of the fish.

The mackerel spawns in spring and early summer. As it does not commence to do so until the water has warmed to about 46° F. (8° C.), with the chief production of eggs taking place in temperatures of, say, 48° to 57°, the spawning season is progressively later, following the coast from south to north. Thus the chief production takes place as early as mid-April off Chesapeake Bay; during May off New Jersey; in June off southern Massachusetts and in the region of Massachusetts Bay; through June off outer Nova Scotia; and from late June through early July in the southern side of the Gulf of St. Lawrence, where eggs have been taken from early June to mid-August.[9] Mackerel have never been found spawning in autumn, so far as we can learn, though a considerable number of eggs that we towed in Massachusetts Bay early in November of 1916 resembled mackerel eggs from the hatchery so closely that we would not have hesitated to identify them as such, had they been taken in summer. They may have been the product of a belated fish, but more likely of some other Scombroid.

The mackerel is a moderately prolific fish; females of medium size may produce as many as 400,000 to 500,000 eggs in the aggregate, according to various estimates,[10] with 546,000 reported for one weighing 1⅞ pounds. But it is seldom that as many as 50,000 are set free at any one time, and often many fewer, for the members of a given school spawn over a considerable period. And recent observations [11] have shown that our earlier statement that they spawn chiefly at night was not correct.

The eggs are 0.97 to 1.38 mm. in diameter, with one large oil globule,[12] and drift suspended in the water, chiefly shoaler than the 5-fathom level. The rate of development is governed by the

[8] See Sette (Fish. Bull. U. S. Fish and Wildlife Service, vol. 50, Bull. 38, 1943, pp. 158–164, and especially fig. 3) for discussion of spawning seasons and temperatures, and the relative importance of different spawning areas.

[9] See Sette (Fish. Bull. U. S. Fish and Wildlife Service, vol. 50, Bull. 38, 1943, pp. 158–163) for a more detailed statement.

[10] Brice, Manual of Fish Culture, 1898, p. 212; Moore, Rept. U. S. Comm. Fish. (1898) 1899, p. 5; Bigelow and Welsh, Bull. U. S. Bur. Fish., vol. 40, Pt. 1, 1925, p. 208.

[11] Sette, Fish. Bull. U. S. Fish and Wildlife Service, vol. 50, Bull. 38, 1943, p. 165.

[12] A series of Gulf of Maine eggs measured by Welsh were about 1.1–1.2 mm. in diameter, with an oil globule of 0.3 mm.

temperature of the water. Recent experiments by Worley [13] (which corroborate early hatchery experience) have shown that incubation takes about 150 hours at 54°; 115–95 hours at 57°–61°; about 70 hours at 64°–65°; and about 50 hours at 70°; with normal development limited to temperatures between about 52° (11° C.) and 70° (21° C.).

Newly hatched living larvae are 3.1 to 3.3 mm. long [14] with large yolk sac, and with numerous black pigment cells scattered over head, trunk, and oil globule which give them a characteristic appearance. The yolk is absorbed and the mouth formed, the teeth are visible, and the first traces of the caudal fin rays have formed by the time the larva is about 6 mm. long. The rays of the second dorsal and anal fins and of the ventrals appear at about 9 mm. (to end of caudal fin); the first dorsal when the total length of the larva is about 14 to 15 mm. The dorsal and anal finlets are distinguishable as such in fry of 22 mm., and the tail fin has begun to assume its lunate shape, but the head and eyes still are much larger than in the adult, the nose blunter, and the teeth longer. At 50 mm. the little mackerel resemble their parents so closely that their identity is evident.

Rate of growth.—The sizes of the mackerel fry taken during the mackerel survey carried out by the U. S. Bureau of Fisheries in 1932,[15] added to other available evidence show that our mackerel grow to a length of about 2 inches during the first 1 to 2 months after they are hatched, a rate about the same as in British and Norwegian waters.[16]

This size is reached earlier or later in the season, depending on the date when any particular lot of fry was hatched. Thus mackerel fry of 1¼ to 2½ inches obviously spawned that spring, have been taken at Woods Hole, both in the first half of June [17] and in the last 10 days of July,[18] fry of 2½ to 5 inches in the first half of August, and fish of about 6½ inches at the end of that month. Similarly, Captain Atwood found fry of 2 inches and shorter in July in the Massachusetts Bay region, i. e., about a month after the local mackerel schools had spawned out.

Fry of 3½ to 4½ inches (obviously of the same season's crop because too small for yearlings) have been taken at Gloucester in August, and Captain Atwood reports them as 6½ to 7 inches long, near Provincetown by October. Many of these little fish, up to 7 or 8 inches long (now large enough to be caught in the fish traps and known as tacks or spikes) are caught along the western shores of the Gulf of Maine and along southern New England during the fall. And measurements of thousands of young mackerel from the Gulf and from southern New England, compiled by the U. S. Bureau of Fisheries, have shown that the fry of the year average 8 to 9 inches, or longer, by the end of their first autumn, before they leave the coast for the winter. But broods produced in different years may grow at different rates, probably depending on feeding conditions, as well as on the dates when they are hatched. Thus fry spawned in the spring of 1927 averaged 8¾ inches in November, but those spawned in 1928 averaged 9¾ inches then.

Our mackerel run about 10 to 11 inches long in spring and early summer of their second year of growth (they are known now as tinkers), which agrees closely with Stevens' [19] estimate for mackerel of the English Channel, based on studies of scales and otoliths. They grow to about 12 to 13 inches by that autumn, or to 14 inches in years of especially rapid growth,[20] and the yearlings usually are a little longer in the Gulf of Maine than at Woods Hole, and longer at Woods Hole than off Long Island, N. Y. It remains to be seen whether these differences are due to temperature, to the varying richness of the food supply, or perhaps to crowding. It is also a question for the future whether the differences persist into later life. The brood of 1923, which may perhaps be taken as typical, averaged almost 14¼ inches in their third autumn, about 15⅛ inches in their fourth, about 15½ inches in their fifth, about 16 inches in their sixth, 16⅛ inches in their seventh, and about 16¾ inches in their eighth years. Thus the American mackerel, like the European, grows very slowly after its third

[13] Jour. Gen. Physiol., vol. 16, 1933, pp. 841–857.

[14] They shrink somewhat when preserved.

[15] See Sette (Fish. Bull. U. S. Fish and Wildlife Service, vol. 50, Bull. 38, 1943, pp. 173–178) for detailed statistical analysis of these.

[16] See Ehrenbaum (Rapp. et Proces Verb., Conseil Perm. Internat. Explor. Mer. vol. 30, 1923, pp. 21, 25) for a discussion of the early growth rate of the European mackerel.

[17] Bigelow and Welsh, Bull. U. S. Bur. Fish. vol. 40, 1925, p. 204.

[18] Sette, Bull. U. S. Fish and Wildlife Service, vol. 50, Fish. Bull. 38, 1943, p. 178, fig. 8.

[19] Jour. Marine Biol. Assoc. United Kingdom, vol. 30, No. 3, 1952, pp. 549–568.

[20] Fry spawned in 1927 averaged about 13¾ inches but those spawned in 1928 averaged only about 12¼ inches in their second November according to Sette.

summer, although it is long lived. The two sexes grow about equally fast.

Nilsson's studies [21] point to a slightly slower rate of growth for the North European mackerel. But American mackerel have been found to vary so widely in this respect that the reported difference may have been only an accident of observation.

A few fish of both sexes may mature sexually in their second year; about ⅛ of the males and ⅝ of the females spawn in the third year; and practically all of them do so in their fourth year, i. e., when three full years old.[22] This coincides with the transition from fast growth to slow, as might be expected, the ripening of the sexual products being so great a strain that the adult fish do little more than recover before winter. Once a mackerel has matured sexually, it no doubt spawns yearly throughout life, as most other sea fishes do.

Proportions of the sexes.—In American waters males have been described as predominating largely over females.[23] But more recent observations have shown that there are about as many of the one sex as of the other, as there are in Sweden also.[24]

General range.—Both sides of the North Atlantic; Norway to Spain off the European coast[25]; from the northern side of the Gulf of St. Lawrence and Strait of Belle Isle [26] to Cape Lookout, N. C.[27] off the American coast.

Migrations, and occurrence in the Gulf of Maine.—The occurrence of the mackerel in the Gulf of Maine is closely bound up with the seasonal movements of the species as a whole, for this is a migratory fish wherever it occurs, appearing at the surface and near our coasts in spring, to vanish thence late in the autumn. The directions and extent of the journeys which it carries out have been the subject of much discussion ever since the fishery first assumed importance, because of their intrinsic interest, because of their bearing on the prosecution of the fishery, and because this fish has been the subject of much international dispute. The point chiefly at issue has been whether the main bodies of mackerel merely sink when they leave the coast in autumn and move directly out to the nearest deep water, or whether they combine their offshore and onshore journeys with the extensive north and south migrations in which most fishermen have long believed.[28]

The great majority of the mackerel have withdrawn from the coast by the end of December, not only from the Gulf of St. Lawrence, but from the entire inshore belt as a whole, not to be seen there again until the following spring or early summer, and it is not yet known definitely where the bulk of them go, though the subject has been widely discussed. Mackerel, it is true, have been caught, and have been found in the stomachs of cod and pollock in January, February, and early March at various localities on and around the outer Nova Scotian banks westward from Sable Island Bank; on the southern and northwestern parts of Georges Bank; in the deeper water between the latter and Nantucket Shoals; on Nantucket Shoals; and along the middle and outer parts of the continental shelf off southern New England, off New York, off New Jersey, off Delaware Bay, off Virginia, and off northern North Carolina. Most of these winter records have been along the 30–70 fathom contour zone, but sometimes as shoal as 4–5 fathoms off Nova Scotia, and as shoal as about 10–20 fathoms (near Ambrose Lightship) off New York,[29] as deep as 90 fathoms off Chesapeake Bay.[30]

Most of these winter records have been based on odd fish only, i. e., not enough to suggest the presence of any great concentration of mackerel.[31] But there were enough of them off New York in January, February, and March of 1949 for commercial fisheries to bring in what Gordon [32] has

[21] Publ. de Circ., No. 69, Cons. Perm. Internat. Explor. Mer, 1914.

[22] Sette, Fish. Bull. U. S. Fish and Wildlife Service, vol. 50, Bull. 38, 1943, p. 156.

[23] Smith, Report U. S. Comm. Fish (1900) 1901, p. 128.

[24] Nilsson, Pub. de Circ. No. 69, Cons. Perm. Internat. Explor. Mer, 1914.

[25] There is a fairly constant racial difference between American and British mackerel (Garstang, Jour. Mar. Biol. Assoc. United Kingdon, New Ser., vol. 5, No. 3, 1898, pp. 235–295), the latter showing a larger number of transverse bars, being more often spotted between them, and more often having 6 dorsal finlets instead of 5.

[26] Jeffers (Contr. Canad. Biol., N. Ser., vol. 7, No. 16 [ser. A, General No. 13], p. 207) reports that several mackerel were caught in 1929 at Raleigh, on the Newfoundland coast of the Strait of Belle Isle, where none had been seen in recent years.

[27] Coles, Copeia, No. 151, February 1926, pp. 105–106 records a three-quarter pound mackerel taken at Cape Lookout in February 1925.

[28] The literature dealing with this subject is very extensive. See especially Goode, Collins, Earll, and Clark (Rept. U. S. Comm. Fish. [1881] 1884, p. 91); Tracy (37th Annual Report, Rhode Island Commissioners of Inland Fisheries, 1907, p. 43); and Sette (Fishery Bull. U. S. Fish and Wildlife Service, vol. 51, Bull. 49, 1950, pp. 268–313) for the American mackerel.

[29] Gordon, Marine Life, Occ. Pap., vol. 1, No. 8, March, 1950, p. 39.

[30] Sette (Fish. Bull. U. S. Fish and Wildlife Service, vol. 50, Bull. 49, 1950, pp. 260–261, table 1) lists several such instances besides those cited previously by Bigelow and Welsh (Bull. U. S. Bur. Fish. vol. 40, Pt. 1, 1925, p. 196).

[31] Three hundred pounds seems to be the largest winter catch definitely reported up to 1951.

[32] Marine Life, vol. 1, No. 8, 1950, p. 39.

characterized as "huge amounts." He also reports "a large body of fish" off Montauk in mid-February of 1950. Schools of "mackerel" have also been reported as sighted at the surface on several occasions in winter, but none of these seem to have been brought in.

Direct evidence carries us only this far. But the indirect evidence of temperature is suggestive. Thus, the Gulf of St. Lawrence (where ice sometimes forms), outer Nova Scotian waters, and the upper 50 fathoms or so within the Gulf of Maine which chill to 35°–39° F. (2°–4° C.) or colder, are all too cold by late winter for mackerel, which are never encountered in commercial quantities in temperatures lower than about 45° F. (7° C.). In most years this applies equally to the inner part of the continental shelf as a whole, southward as far as northern Virginia, for the water usually cools there to 37°–40° F. (3°–4° C.) at the time of the winter minimum. But the mackerel need only move out to the so-called warm zone at the outer edge of the shelf to find a more suitable environment, for the bottom water there is warmer than 44°–46° F. (7°–8° C.) the year round as far north and east as the central part of Georges Bank, and about 41° F. (5° C.) along outer Nova Scotia.

Available evidence thus supports Sette's[33] conclusion that the bulk of the American mackerel winter on the outer edge of the continental shelf from the offing of northern North Carolina to the mid-length of Georges Bank, 30 to 100 miles off shore according to location, in depths of perhaps 50 to 100 fathoms. The few that are caught closer to land and in shoaler water in winter either represent the inshore fringe of the main population, or they are strays. Perhaps some winter off Nova Scotia as far east as Sable Island Bank. And it would not be astonishing should it prove that some winter in the deep eastern trough of the Gulf of Maine, where the temperature of the bottom water, at depths greater than 75 fathoms or so, does not fall below about 41° F. (5° C.). A few mackerel have, in fact, been caught on cod lines in deep water off Grand Manan in winter,[34] while two were found among kelp near Yarmouth, Nova Scotia, on December 28, in 1878.[35]

[33] Fish. Bull. 49, U. S. Fish and Wildlife Service, vol. 51, Fish. Bull. 49, 1950, p. 261.

[34] Collins, Rept. U. S. Comm. Fish (1882) 1883, p. 273.

[35] Goode, Collins, Earll, and Clark, Rept. U. S. Comm. Fish. (1881) 1884, p. 98; cited from the Yarmouth, Nova Scotia, *Herald*, January 2, 1879.

Sette[36] has pointed out, however, that some other factor besides temperature must have to do with the wintering habits of the mackerel, for they disappear as completely from the surface and from inshore in the southern part of their range as they do in the northern even in very warm years such as 1932, when the water (surface to bottom) was warmer than 45°–46° F. (7°–8° C.), from New Jersey southward, even at the end of the winter. On the other hand, the event (probably abnormally low temperature) that was so destructive to the tilefish in March, 1882 (p. 429), did not affect such of the mackerel as were wintering on the tilefish grounds, for they reappeared that summer in normal numbers, a point to which Sette[37] has called attention already.

Two additional facts which support the view that our mackerel do not travel very far in winter are (*a*) no mackerel, young or old, have ever been taken outside the edge of the continent, or anywhere on the high seas far from land for that matter; (*b*) their reappearance in spring takes place so nearly simultaneously along some hundreds of miles of coastline that they can hardly have come from any great distance.

Thus time and increased knowledge have corroborated the view of Captain Atwood and of Perley, of more than half a century ago that mackerel winter offshore in deep water and northward from the latitude of Virginia, not in the far south nor out in the surface waters of the warm parts of the Atlantic.

The winter home of the American mackerel appears to correspond rather closely to that of the mackerel of British seas, some of which winter on the deep northern slope of the North Sea, some in the deeper parts of the English Channel, and many on the outer edge of the continental shelf southwest of Ireland, mostly deeper than 60 fathoms.[38]

The failure of the otter trawlers to take commercial quantities of mackerel off Chesapeake Bay in winter when they fish there intensively, leads Sette[39] to conclude that our mackerel

[36] Fish. Bull. U. S. Fish and Wildlife Service, vol. 51, Fish. Bull. 49, 1950, p. 527.

[37] Sette, Fish. Bull. U. S. Fish and Wildlife Service, vol. 51, Bull. 49, 1950, p. 257, Footnote 3.

[38] Ehrenbaum (Rapp. et Proc.-Verb. Cons. Perm. Internat. Explor. Mer, vol. 18, 1914) summarizes what was known of the life history of the European mackerel up to that time. And Steven (Jour. Marine Biol. Assoc. United Kingdom, vol. 27, 1948, pp. 517–539) has recently outlined the chief wintering grounds.

[39] Fish. Bull. U. S. Fish and Wildlife Service, vol. 51, Bull. 49, 1950, p. 261.

winter in the mid-depths, not concentrated on the bottom. This, however, would imply that the wintering mackerel manage to hold position for two or three months in some way without drifting far with the movements of the water. Another possibility is that they do keep on bottom, or near it, but somewhat deeper down the continental slope than the trawlers ordinarily fish,[40] perhaps concentrated in the many gullies, large and small, with which the upper part of the slope is seamed all along from the offing of Chesapeake Bay to Georges Bank, much as the mackerel of the Celtic Sea and English Channel winter "on the sea floor, densely packed in places where its level is interrupted by banks and gullies."[41]

Whichever of these alternatives is the correct one, the oft repeated assertion that the adipose eyelids of the mackerel become opaque in winter has no foundation. And they certainly do not hibernate in thousands along the coasts of Greenland and Hudson Bay,[42] and of Newfoundland, with heads in the mud and tails protruding as a vice admiral, no less, has described them; a wholly imaginary tale, we need hardly add.[43] They may winter in a more or less sluggish state. But the presence of food in the stomachs of some of the winter-caught fish, added to the fact that some of them are fat though others are thin, shows that they move about more or less even then, and feed more or less.[44]

Most American students have looked on the vernal warming of the surface water to about 45° F. as the stimulus causing the mackerel to quit their winter quarters. European studies, however, have shown that the date of their reappearance in spring is not closely associated with any particular temperature. And if the mackerel winter on bottom along the edge of the continent, vernal changes in the temperature of the surface water nearer to land would be wholly outside their ken.

The European mackerel usually keep to the bottom on their spring migration until close in to the land before rising to the surface. But this generalization does not apply to the American fish, for while some may swim deep (so, only can we account for the fact that the first schools often show as early in Massachusetts Bay as on Georges Bank or off Nantucket) mackerel in great numbers are first sighted 30 to 50 miles offshore, and this all the way from the latitude of Cape Hatteras to the mouth of the Gulf of Maine. The first mackerel "show" off the Cape Hatteras region at any time between about March 20 and April 25, usually early in April, and by the middle of April off Delaware Bay. As the water warms they spread northward and shoreward, being joined, it seems, by additional contingents from offshore. They reach the offing of southern New England some time in May, and they are plentiful on Nantucket Shoals by the first week of that month, as a rule.

The date when they are first sighted off Cape Cod in the southwestern part of the Gulf of Maine varies from the last of April or first of May (April 29 in 1901, May 2, in 1898) to the first of June, with May 10 about the average. The earliest dates of commercial catches, for example, made in one particular set of traps near Provincetown have varied between May 14 and June 19. And the fish are plentiful in the western side of the Gulf of Maine as a whole by the end of the first week in June at the latest, if it is fated to be a good mackerel year. Mackerel (usually in smaller numbers) also appear on the Nova Scotian side of the Gulf about as early as they do in its western side; thus they were reported almost simultaneously off Yarmouth, Nova Scotia, and off Chatham on Cape Cod in 1898; in 1922 they were sighted off Yarmouth on May 7th, and off Cape Sable on the 11th. And they may appear even earlier in the season at Cape Breton, and as early well within the Gulf of St. Lawrence and in the eastern side of our Gulf. In 1894, for example, mackerel were first reported off Cape Breton on May 5 and at Gaspé on May 12, but not until May 16[45] at Yarmouth on the Gulf of Maine coast of Nova Scotia. But few of them show along the coast of Maine or in the Bay of Fundy until toward the end of June.

Sette[46] has made the very interesting discovery that two distinct populations are represented among the American mackerel, a southern and a

[40] The southern trawl fishery is mostly shoaler than 70 fathoms.

[41] Steven, Jour. Marine Biol. Assoc. United Kingdom, vol. 27, 1948, p. 537.

[42] Mackerel do not range that far north.

[43] Cited from Lacépède, Hist. Nat. Poissons, vol. 3, in Buffon, Hist. Naturelle, 1802, p. 32.

[44] Ehrenbaum (Rapp. et Proces Verbaux, Cons. Perm. Internat. Explor. Mer, vol. 18, 1914, p. 13), whose studies of the fish entitle his view to great weight, thinks that the mackerel of northern Europe probably are torpid during part of their stay on the bottom.

[45] Huntsman, Canadian Fisherman, vol. 9, no. 5, 1922, pp. 88–89.

[46] Fish. Bull. U. S. Fish and Wildlife Service, vol. 51, Bull. 49, 1950.

northern, with rather different migratory habits, and differing also in the relative success of reproduction in different years.[47] The nature of these two contingents is not known, whether genetic or environmental.

It is probable (though not proved) that the southern contingent tend to winter in the southern part of the wintering zone. The main bodies of mackerel that appear in spring along the middle Atlantic coast belong to this contingent, also most of those taken off southern New England. They summer for the most part over Nantucket Shoals; on the western part of Georges Bank; and in the western and northwestern parts of the Gulf of Maine, which they enter in the western side around Cape Cod. And they do not journey farther east than the coast of Maine. On the other hand, it seems the mackerel that appear early in the season along the Nova Scotian shore of the Gulf, to spread later to Maine, belong to the northern contingent, and also a scattering of those that enter the western side of the Gulf. These appear to winter mostly eastward from the Hudson Gorge, and their vernal migration carries most of them past our Gulf, to pass the summer along outer Nova Scotia, and in the southern side of the Gulf of St. Lawrence.[48]

A few mackerel (mostly small) from the southern contingent remain all summer in the coastwise belt from Long Island to Nantucket. Apart from these, however, the whole body of American mackerel have deserted the southern grounds altogether by the early summer, to spend the later summer either in the region of our Gulf, off Nova Scotia, or in the Gulf of St. Lawrence.[49]

If the view now held is correct as to their migratory routes, some of the mackerel that summer in our Gulf may come from as far as the offing of North Carolina; others from as nearby as the offing of New York or of southern New England. The vernal journey of the Gulf of St. Lawrence mackerel may be anywhere between, say 300 to 350 miles, and 700 miles, depending on whether they have wintered off outer Nova Scotia or as far west as the western slope of Georges Bank.

It seems certain that some of the mackerel that are first sighted on Nantucket Shoals and on Georges Bank in May remain on these offshore grounds all summer, both spawning and feeding there, for they provide good fishing there any time from June to September or October, in some years. The farther advance of such of them as continue northward into the Gulf of Maine covers a period of some weeks, with the first-comers followed by other schools later. And it seems certain (as just remarked) that fish resorting to our Gulf, do so summer after summer, never visiting the outer coast of Nova Scotia, much less a region as far afield as the Gulf of St. Lawrence. But it is an interesting question for the future, whether a given school returns to the same part of the Gulf, year after year.

Many of the mackerel that summer in our Gulf have already spawned farther south (p. 322). Others, however, are still hard, but they are soon taken there with eggs or milt running. Spawning in the Gulf of Maine is at its peak in June in most years, with the proportion of spent fish increasing through July, and only an occasional ripe fish as late as the first of August. But a year comes occasionally, such as 1882, when spawning is not at its height in the Gulf until July, with ripe fish continuing plentiful until August. And our towings there have yielded a few mackerel eggs as early as May 6, as late as September 1.[50]

The spawning season is at its height in the Gulf of St. Lawrence during the last half of June and the first half of July, continuing into August, a fact well known by the hook-and-line fishermen of half a century ago, because the ripe fish will not bite at that time, and more recently corroborated by the egg catches of the Canadian Fisheries Expedition.[51]

It seems from the relative numbers of eggs taken from place to place, that Cape Cod Bay is the only subdivision of our Gulf that has rivaled the more southern spawning grounds in egg production during the particular years when intensive studies

---

[47] This conclusion, seemingly conclusive, is based on analysis of the size (i. e. age) composition of the mackerel population at various times and places, with some evidence from tagging experiments. The data are too extensive for discussion here.

[48] For further information as to migrations of the northern contingent, see Sette, Fish. Bull. U. S. Fish and Wildlife Service, vol. 51, Bull. 49, 1950, pp. 269, 285.

[49] We have found no positive record of mackerel taken in late summer anywhere south of Delaware Bay, although they are plentiful off this part of the coast in spring. Bell and Nichols, it is true, speak of "mackerel" as found in tiger-shark stomachs off North Carolina (Copeia, No. 92, 1921, pp. 18-19), but Dr. Nichols writes us that these were "just Scombroids and probably not *Scomber scombrus*."

[50] See Bigelow and Welsh (Bull. U. S. Bur. Fish., vol. 40, Pt. 1, 1925, p. 206), for details.

[51] Dannevig, Canadian Fish. Exped. (1914-1915); 1919, p. 8.

have been made.[52] Mackerel also spawn to some extent thence northward, as far as Casco Bay, but we believe very few do so farther east than that along the coast of Maine. Neither is it likely that mackerel breed successfully in the northern side of the Bay of Fundy for neither eggs nor larvae have been taken there though some production may take place on the Nova Scotian side for Huntsman reports eggs at the mouth of the Annapolis River. And while a moderate amount of spawning takes place along the outer coast of Nova Scotia,[53] it seems that the eggs do not hatch in the low temperatures prevailing there, for no larvae have been found. But the southern side of the Gulf of St. Lawrence, where the surface waters warm to a high temperature in summer, is an extremely productive spawning ground (p. 322).

Since the large adult mackerel tend to keep farther offshore than the small ones (p. 328), such of them as spawn in our Gulf do so at least a few miles out. Very few eggs, for example, were found in 1897 (a year of plenty) in the inshore parts of Casco Bay,[54] though this was formerly thought to be a productive spawning ground.

Once the mackerel have entered our Gulf, schools are to be expected anywhere around its coastal belt, at any time during the summer; also on Nantucket Shoals, on the western part of Georges Bank, and on Browns Bank, as just noted (p. 327). And while adult fish seldom venture within the outer islands or headlands, good catches of them have been made well up Penobscot Bay, and young ones 6 to 10 inches long often swarm right up to the docks in various harbors in summers of plenty.[55]

Mackerel are proverbially unpredictable in their appearances and disappearances at any particular place, hence the common saying that "mackerel are where and when you find them." This is partly because the schools are constantly on the move, but partly because it is only while they are schooling at the surface or near it that they are seen. When they sink to lower levels in the water, as they often do, they drop out of sight entirely, unless some of them chance to be picked up by drift netters. Large mackerel are more prone to disappear in this way than small ones, especially in late summer or early autumn. In 1906, for example, the schools of large fish vanished from the Massachusetts Bay region in June, to reappear the 27th of July, on which date 28 seiners made catches ranging from 18 to 250 barrels each. And in 1892, a year of abundance, they disappeared (that is, sank) in August, not to appear again in any abundance anywhere in the Gulf of Maine until October.

The view has grown that when this happens the mackerel have deserted the Gulf for the time being. But it was common knowledge in the days before the introduction of the purse seine, when it was the regular practice to lure the fish to the surface by throwing out ground bait, that large mackerel summer as regularly in the Gulf as small, and that good hook-and-line catches of large fish could be made in one or another part of the Gulf through the season from June to October, even when none showed at the surface.

Their disappearances in summer merely mean that the fish have sought lower levels in the water; that they have wandered to some other part of the Gulf; or perhaps that the schools have dispersed more or less. When they sink in summer in our Gulf, it is not likely that they descend very deep. In the first place the water deeper than about 40 to 50 fathoms is colder than 46°–47° F. (8° C.), i. e., than they seem to prefer; in the second place the planktonic animals on which they feed are more concentrated above the 50-fathom level than deeper. And a year comes, now and then, when mackerel of all sizes school at the surface all summer long.[56]

Sette's [57] painstaking analysis of the relative frequency with which schools are seined in different localities has shown that mackerel are seen far the most often in the southwestern part of the Gulf and out along the western part of Georges Bank, with the chief concentrations in one part or another of Massachusetts Bay and off the outer shore of Cape Cod to Nantucket Shoals, though great numbers are also caught along the Maine coast, close inshore.

[52] Subsequent information, and especially the result of tow nettings on the southern grounds in 1929, 1930, 1931, and 1932 (Sette, Fish. Bull. U. S. Fish and Wildlife Service, vol. 50, Bull. 38, 1943) have shown that the Gulf of Maine as a whole is much less productive than the more southern spawning grounds, not more so as Bigelow and Welsh (Bull. U. S. Bur. Fish., vol. 4, Pt. 1, 1925, p. 206) believed.

[53] Sparks, Contrib. Canadian Biol. and Fish., N. Ser., vol. 4, No. 28, 1929.

[54] Moore, Rept. U. S. Comm. Fish. (1898) 1899, p. 16.

[55] Sette (Fish. Bull. U. S. Fish and Wildlife Service, vol. 51, Bull. 49, 1950, p. 297) discusses this point further.

[56] 1882 was an example of this.

[57] Fish. Bull. U .S. Fish and Wildlife Service, vol. 51, Bull. 49, 1950, p. 297, fig. 17.

Mackerel contrast in an interesting way with herring in this respect, the latter being caught in by far the greatest numbers in the northeastern corner of the Gulf, i. e., just where there usually are fewest mackerel. But there is much variation from year to year in their relative abundance from place to place as appears from the following table of catches, made in two successive years when the total landings from the Gulf, as a whole, did not differ greatly (landings at Boston, Gloucester, and Portland by the vessel fishery, stated in pounds).

Fishing grounds	1916	1917
Georges Bank	3,701,597	624,086
South Channel	77,157	13,600
Nantucket Shoals	2,516,414	6,277,830
Off Chatham	2,017,753	3,938,452
Off Race Point	99,250	621,751
Stellwagen Bank	1,559,972	519,550

In some years few mackerel are seen at the surface in the Gulf eastward of the Isles of Shoals, 1926, 1927, 1933, 1934, and 1935 were examples. In other years, however (e. g., in 1928, 1929, 1930, 1931, and 1932) many schools are sighted and seined along the coast of Maine as far eastward as the vicinity of Mount Desert Rock (see fig. 176, based on Sette's painstaking analysis). But the experiences of the old time hook-and-line fishermen suggest that the mackerel tend to move northward and eastward in general from the Massachusetts Bay region, for they made their best late-summer and early-fall catches between Cape Elizabeth and Mount Desert Rock in most years, notably about Monhegan Island. And the results of hook-and-line fishing are a far better clue to the presence or absence of mackerel than the seine catches are, since they draw from the fish that are deep down, as well as from those that may chance to be at the surface.

The Nova Scotian side of the Bay of Fundy has been a profitable mackerel ground, occasionally, but only for short periods and at long intervals. Thus good catches were made there for some years previous to 1876, but this fishery was abandoned a few years later for want of mackerel. There were enough fish there again in the early 1900's to yield about 7 million pounds in the 6-year period 1901 to 1906.[58] But we have not heard of any large catches made anywhere in the Bay of Fundy since that time, so events of the sort must be out of the ordinary. And very few mackerel are ever reported along the New Brunswick side of the Bay.

In most years, mackerel are few over the central deeps of the Gulf (fig. 176), but a year comes now and then when they are plentiful there, as happened in 1882 (a year of great abundance), when great numbers were caught between Georges Bank, Browns Bank, and Cashes Ledge, and thence northward to within 40 miles or so of the Maine coast. Most of the early season catch, in fact, was made in this deep water region that year, and in the weirs along the west coast of Nova Scotia. But the fish disappeared thence later in the season. And large catches have never been reported from the eastern part of Georges Bank to our knowledge.

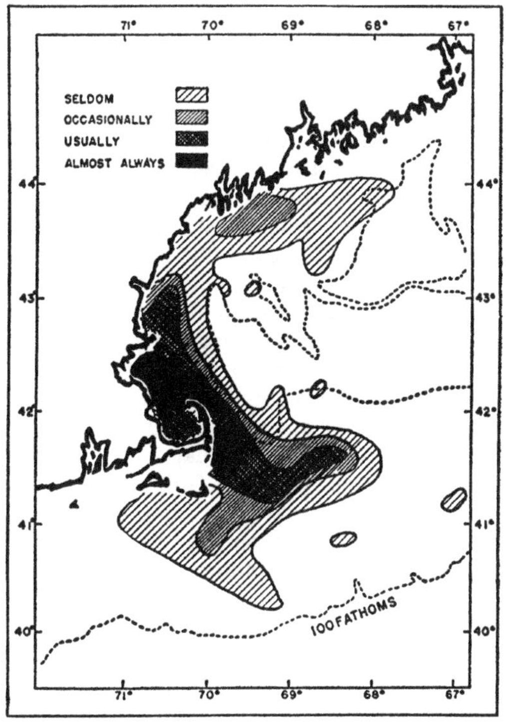

FIGURE 176.—Average distribution of mackerel in the Gulf of Maine, July through September, based on relative frequencies of catches recorded for each 10-mile rectangle, 1926 through 1935. After Sette.

As a rule, the schools tend to stay nearer the coast in years when small (i. e., young) fish dominate the population. The entire Gulf of Maine catch, for example, was taken within 45 miles of

[58] Sette and Needler, Inv. Rept. 19, U. S. Bur. Fish., 1934, pp. 1–48.

land in 1926, when the stock was dominated by fish hatched in 1923, i. e., were in their third year. In years of this sort, anglers fishing in harbors, or going out in charter boats for the day, do well, catching the smaller sizes chiefly. But in 1928, when the same year class dominated as had in 1926 (i. e., fish now in their fifth year), only about two-thirds of the catch was made that close in, with about one-third of the catch taken more than 45 miles out at sea. Nineteen twenty-nine may serve as another example, with more than one-half (57 percent) of the large fish caught more than 45 miles out, but less than 1 percent of the small ones, and a few large ones, taken as far out as 80 miles. But even the fully grown fish do sometimes come close inshore; we have ourselves caught mackerel within a few yards of the beach in the southern side of Massachusetts Bay, as large as any that we have seen taken anywhere.

Fishermen have long realized that mackerel are most likely to be found where there is a good supply of "red feed" (copepods) or other small animal life in the water. A relationship has, in fact, been found to hold in the English Channel between the catches of mackerel and the numbers of copepods present.[59] And while no attempt has been made yet to relate the local abundance of mackerel in our Gulf, or the depths at which they swim with the supply of food on a statistical basis, the mere fact that they do fatten in our waters is evidence enough that they manage in some way to congregate where food is plentiful. But it appears that their vernal journey, from their wintering grounds to the Gulf and to Nova Scotian waters, is directed by some impulse to migration more definite than the mere search for food. Thus while a large proportion of the mackerel did travel along the zone of abundant plankton in the only year (1930) when their advance along the coast has been compared with the quantitative distribution of the animals on which they prey,[60] they deserted the waters south of New England that year while the food still was abundant there, for regions (Gulf of Maine and eastward) where there is no reason to suppose that feeding conditions were any better at the time.[61]

As autumn draws on, the fish that summer along the Maine coast (chiefly belonging to the southern contingent) seem to work back southwestward toward Cape Cod, for catches were made successively off Portland, near Boon Island, and off Cape Ann, in the days when mackerel were caught on hook and line. It is probable, too, that such of the fish from the northern contingent as have entered the Gulf in the eastern side join in this general autumnal movement around the coast to the westward and southward, rather than that they leave by the route along which they enter, for schools have often been reported, and actually followed, swimming southward at the surface across Massachusetts Bay. And while reports of this sort are likely to be based on misconception,[62] they are corroborated in this instance by the fact that the latest catches are always made either in or off Massachusetts Bay, along the outer shore of Cape Cod, or on the neighboring parts of Nantucket Shoals, never either on Georges Bank, which would be on the direct route of any fish swimming westward from Nova Scotia, or in the inner parts of the Gulf of Maine.

Sette's studies indicate that the bulk, at least, of the mackerel of the southern contingent have moved out of the Gulf around Cape Cod and past Nantucket Shoals by late September or October in most years. But many of the fish of the northern contingent coming from Nova Scotia, and perhaps even from the Gulf of St. Lawrence, usually provide good fishing off Cape Ann and southward through October and late into November,[63] with good commercial catches until mid-December in some years. In 1913, for example, 1,200 mackerel were caught off Gloucester on December 10; 3,000 off Chatham, Cape Cod, a day or two earlier; and nearly 1,000 barrels (200,000 pounds) were seined off the Massachusetts coast during the early part of that month in 1922. In mild winters schools of mackerel are sometimes reported and even caught off the outer coast of Nova Scotia as late as Christmas time; i. e., somewhat later than off Cape Cod. But the

[59] Bullen, Jour. Marine Biol. Assoc. United Kingdom, vol. 8, 1908, pp. 269, 302.

[60] Bigelow and Sears, Mem. Mus. Comp. Zool., vol. 54, No. 4, 1939, pp. 259–261.

[61] See Sette (Fish. Bull. U. S. Fish and Wildlife Service, vol. 51, Bull. 49, 1950, p. 302) for a further discussion of the influence of feeding conditions on the movements of the American mackerel.

[62] The successive approach of one school after another to the coast often suggests a long-shore movement of the fish. Thus Kendall (Bull. U. S. Bur. Fish., vol. 28, Pt. 1, 1910, p. 287) tells of an instance when seiners reported "following" the schools continuously eastward along outer Nova Scotia, although the fish taken off Liverpool proved to be of quite different sizes from the catch made later about Cape Breton.

[63] In 1922, for example (Gloucester Times of April 26, 1923), mackerel netters fishing near Cape Ann did well right through November, with a catch of about 1,200,000 pounds (6,000 barrels) for the month.

whole body of Gulf of Maine, Nova Scotian, and Gulf of St. Lawrence mackerel have withdrawn thence by the end of December at the latest, except for odd stragglers. And when they do depart, they must sink at once to lower levels in the water, for schools are never sighted on their journey offshore and southward; they simply drop out of sight.

Abundance.—It has been common knowledge since early colonial days that mackerel fluctuate widely in abundance in our Gulf from year to year, perhaps more widely than any of our other important food fishes, with periods of great abundance alternating with terms of scarcity, or of almost total absence. In good years the fish may appear in almost unbelievable numbers; schools or associations of schools, miles in length, are reported; and it is common to see 50 or more separate bodies of fish from the masthead at one time. Mackerel, in short, seem to be everywhere, and a tremendous catch is made. But perhaps only an odd school will be seen here and there the next year, and the fishery will be a flat failure.

The period from 1825 to 1835 was one of abundance. In 1831, for example, more than 380 thousand barrels (76 million pounds) of salt mackerel (in those days most of them were salted) were landed in Massachusetts ports. But mackerel were scarce for the next 8 years (1837–45), only 50,000 barrels being landed in Massachusetts in 1840. The Massachusetts catch then fluctuated violently from 1851, when the landings rose once more to 348,000 barrels, down to 1879. The fleet brought in something like 294 million fish from Nova Scotian and United States waters combined in 1880. And this introduced a period of extraordinary abundance, culminating in 1885 when the catch reached the enormous total of 500,000 barrels (100,000,000 pounds). But this was followed in its turn by a decline so extreme, so widespread, so calamitous to the fishing interests, and so long continued, that the catch was only about 3,400 barrels (equivalent to 582,800 pounds of fresh fish) for the entire coast of the United States in 1910 (when the stock of mackerel fell to its lowest ebb) with almost none reported in Massachusetts Bay or along the Maine coast.

Mackerel then increased again in numbers; slowly at first, then more rapidly, as appears from the fact that the catch for the Gulf of Maine and for the banks at its mouth was about four times as great in 1911 (about 2½ million pounds) as it had been the year before, rising to about 4⅓ million pounds in 1912, 5 million in 1913, 7½ million in 1914, to something more than 11 million in 1915, and 16 to 16⅓ million each for 1916 and 1917. But this period of multiplication fell far short of equaling the banner years of the 1880's. And the catches fell off again so rapidly after 1917 that the Gulf of Maine yield for 1919 [64] was only about one-quarter as great as it had been in 1917. Although 1920 saw some slight recovery, 1921 (with a local catch of only about 1 million pounds) proved the worst mackerel season for our Gulf since 1910. The stock then built up enough (following the familiar seesaw pattern) for the Gulf to yield about 25 million pounds of mackerel in 1925. Since that time down to 1946, the Gulf of Maine catch has ranged between a low of about 20 million pounds (1937) and a high of about 59 million (1932). Thus the catch of mackerel in our Gulf may be 50 to 100 times as great in a good year as in a poor. The average Gulf of Maine catch for the period 1933–1946 was about 37 million pounds, yearly.

Various far-fetched explanations have been proposed for these astounding ups and downs in the catches from year to year, such as that the fish have gone across to Europe; have sunk; or have been driven away or killed off by the use of the purse seine. Actually, these changes reflect the ups and downs in the numbers of the fish that are in existence from year to year. Mackerel, in short, were extremely plentiful in 1885, very scarce in 1910, moderately plentiful in 1916 and 1917, very scarce again in 1921, and they have been moderately plentiful since about 1925, but probably not so plentiful as they were in the 1880's.[65]

It has long been known for the herring and for some other species that the prime factor in determining the abundance of the fish is the comparative success of reproduction from year to year, years favorable to the production and survival of larvae presaging several seasons of abundance, or vice-versa. And comparison of the relative proportions of mackerel of different sizes (that is, of different ages) in the total catches from year to year has shown that this is equally true of the

[64] In 1919, 4,091,345 pounds.

[65] See especially, Sette, U. S. Bur. Fish., Fishery Circular No. 4, 1931.

mackerel.[66] When there is an abundant crop of young mackerel, the fishing is good during the next several years, but the catches then fall off, if another good brood does not soon appear upon the scene. The course of events since the low point in 1910 may then be reconstructed about as follows:

In 1910, when the stock of mackerel was at its lowest, most of the fish caught were large, suggesting that few young had survived for several years past. Unfortunately, no information is available as to the composition of the population from the point of view of size for the next three years, when the catch was progressively somewhat larger, but great numbers of small fish, (apparently yearlings), were reported in 1912, pointing to a good breeding season in 1910, in 1911, or in both. In 1914 fish smaller than 1½ pounds again formed nearly 60 percent, by weight, of the catch in and off the Gulf of Maine, and approximately 80 percent in 1915, with an even greater preponderance in actual numbers between small (young) fish and large (old). These little fish, hatched during the period 1910 to 1912 or 1913, were responsible, as they grew, for the fairly good catches made in the Gulf in 1916 and 1917.[67] But the production of fry must have been very poor in 1916 and 1917, for the Gulf of Maine catch was only about one-seventh as great in 1919 as it had been in 1916. And reproduction must have practically failed in 1918 or in 1919, for the mackerel caught in 1920 ran very large, both south of New York that spring, and in our Gulf during that summer and autumn.

The population was now back again in about the same state as it had been in 1910, the cycle having run through a period of 10 years. The parallel goes further, too, for 1921 must have seen a wave of production to account for the swarms of small fish that appeared along the New England coast from Woods Hole to Mount Desert during the summer of 1922. This again presaged a great increase in the catches of mackerel for the next few years to come (more than 11 million pounds were taken in the Gulf of Maine in 1923). And Sette's studies show that 1923 was another productive year, resulting in a catch more than twice as great in 1925 as it had been in 1923, and about 3 times as great in 1926.[68] The very large catch of about 59 million pounds in the Gulf in 1932 was preceded similarly by the presence of great numbers of yearlings in 1929, evidence of successful reproduction in 1928.

Thus, it seems that the proportion of fish of different ages in the catch in any one year may be used as a basis for predicting the success or failure of the run of mackerel for the next year; such predictions have in fact been attempted by Sette [69] with fair success.

No record has been kept, so far as we know, of the relative numbers of mackerel of different ages, of late years. But a failure of reproduction, followed by a slump in the catch, may come at any time, for history has a way of repeating itself, especially where fishes are concerned.

Nothing definite is known as to what determines the success or failure of reproduction of the mackerel in any given year. Towings by the U. S. Bureau of Fisheries make it likely that the actual production of eggs is usually sufficient. But the vitality of the eggs spawned in any given year goes back to the physiological condition of the parents. And studies of the composition of the stock of fish in periods of high production and of low suggest that there is some correlation between the number of adult mackerel existing in the sea at any time, and the success with which they breed, for it seems that years when great numbers of fry survive always fall when the parent fish are scarce, average large, and also average very fat (by general report).

One hypothesis is that the mackerel tend to grow fast when there are only a few of them and go into the winter in excellent condition, hence are able to produce eggs of high vitality and in abundance; but they do not fare so well individually when plentiful, hence, do not emerge from their winter quarters in as good physiological condition in spring, so that fertilization and incubation may be less successful, and such larvae as hatch may be less strong. On the other hand, all this may be insignificant as compared with the success or failure of the larvae in surviving the dangers and difficulties of subsistence that confront them. Onslaughts by enemies,

[66] See especially Sette, U. S. Bur. Fish., Fishery Circular No. 4, 1931.

[67] Gulf of Maine catch, 16,391,095 pounds in 1916; 16,021,619 pounds in 1917.

[68] Gulf of Maine catch, 11,007,676 pounds in 1923; 25,475,876 pounds in 1925; 33,152,766 pounds in 1926.

[69] U. S. Bur. Fish., Fish. Circ., No. 4, 1931; No. 10, 1932; No. 14, 1933; No. 17, 1934; Fishing Gazette, vol. 50, No. 5, 1933, pp. 9 and 21.

abundance and ready availability of food, temperature, and salinity of the sea water all act upon the young fish to make their existence precarious; a favorable environment depends on a happy combination of all these.

Importance.—The mackerel is a delicious fish, but it does not keep so well as some other fishes that have less oil in their tissues. When mackerel are rather plentiful they are one of the four most valuable fishes of our Gulf commercially, surpassed in dollar value only by the haddock, cod, and rosefish, as appears from the following table of landings in New England for the years 1943–1947.

When the fishery fails, as it does periodically through periods of several years (p. 331), the value of the catch decreases accordingly, and to a point where it is only a negligible fraction of the total yield and value of the Gulf of Maine fishery.

Species	1943	1944
Haddock	$8,650,000 (1.1)	$7,550,000 (1.2)
Cod	4,000,000 (1.2)	3,500,000 (1.5)
Rosefish	4,350,000 (1.1)	4,300,000 (1.1)
Mackerel	3,180,000 (1.1)	2,400,000 (1.3)

Species	1945	1946	1947
Haddock	$7,000,000 (1.4)	$8,800,000 (1.2)	$8,900,000 (1.3)
Cod	4,280,000 (2.3)	3,940,000 (1.4)	2,780,000 (1.4)
Rosefish	3,840,000 (1.3)	4,750,000 (1.7)	4,200,000 (1 4)
Mackerel	3,160,000 (1.3)	2,340,000 (1.1)	2,000,000 (1.2)

NOTE.—The total value of the catch landed in Maine, New Hampshire, and Massachusetts, including fish from grounds outside the Gulf of Maine, as well as from within the Gulf, may be determined by multiplying the values by the figures in parentheses.

Most of the mackerel were caught formerly with hook and line, ground bait being thrown out to lure the fish close enough to the vessel.[70] But this way of fishing was gradually given up about 1870, when the use of the purse seine became general. And practically the entire catch of mackerel of the past 70 years has been made with purse seines, with pound nets, weirs and floating traps coming second, and gill nets a poor third. In 1943, for example, when the total Gulf of Maine catch was between 53 and 54 million pounds, about 80 percent was taken in purse seines; between 12 and 13 percent in pound nets, weirs, and floating traps; and between 3 and 4 percent (between 1 and 2 million pounds) in gill nets (anchored or drifting), but only 1,700 pounds on hand lines. Otter trawlers, too, bring in scattering mackerel from the offshore banks: 2,400 pounds, for example, in the year in question.

Many anglers, also, troll or bait-fish for mackerel all along the coast from Cape Cod to Penobscot Bay; as far as Mount Desert if mackerel are on the coast that far east. In good years it is not unusual for 3 or 4 anglers fishing from a party boat to bring in one or two hundred fish. And in summers when young tinkers are plentiful inshore many of them are caught from the wharves in various harbors. If one chooses to troll, an ordinary pickerel spinner, No. 3, serves well, especially if tipped with a small piece of pork rind or with mackerel skin; a small metal jig similarly adorned, or any small bright spoon. Mackerel will also take a bright artificial fly, and bite greedily on a white piece of clam, a piece of mackerel belly, or on a sea worm (*Nereis*), especially if attracted by ground bait.

Chub mackerel *Pneumatophorus colias* (Gmelin) 1789 [71]

HARDHEAD; BULLSEYE

Jordan and Evermann, 1896–1900, p. 866, *Scomber colias* Gmelin.

Description.—The hardhead (by which name it is commonly known to fishermen) resembles the common mackerel so closely that we need mention only the points of difference. Most important of these, anatomically, is the fact that the hardhead has a well-developed swim bladder connected with the esophagus, which the mackerel lacks. But it is not necessary to open the fish to identify it for there is a characteristic color difference between the two, the mackerel being silvery-sided below the mid line. whereas the lower part of the sides of the hardhead (otherwise colored somewhat like the mackerel) are mottled with small dusky blotches, and the chub has a larger eye than the mackerel. Less obvious differences are that the dorsal fins are closer together in the chub and that there are only 9 or 10 spines in its first dorsal fin instead of 11 or more, which is the usual count in the mackerel.

[70] See Goode and Collins, Fish. Ind. U. S., Sect. 5, vol. 1, 1887, pp. 275–294, for an excellent account of the hook and line fishery.

[71] This genus is separated from *Scomber* by having a well developed swim bladder which the true mackerel lacks (see Starks, Science, N. Ser., vol. 54 1921, p. 223).

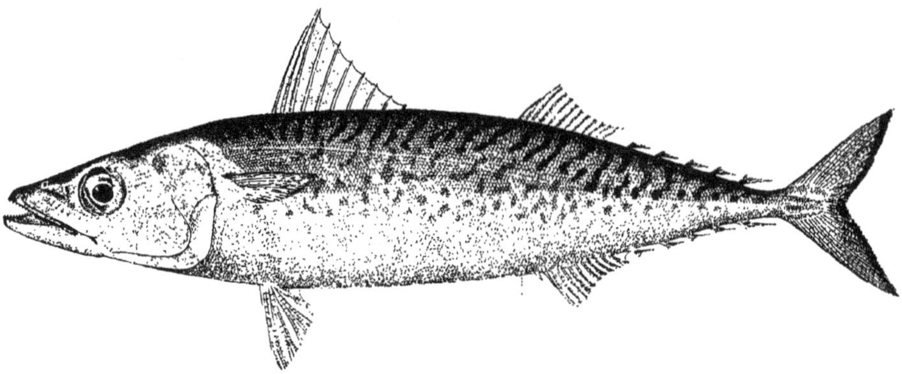

FIGURE 177.—Chub mackerel (*Pneumatophorus colias*), Provincetown, Mass. From Goode. Drawing by H. L. Todd.

Size.—This is a smaller fish than its better known relative, growing to a length of about 8 to 14 inches only.

Habits.—Hardheads school like mackerel, and their feeding habits are much the same, for Doctor Kendall found fish on Georges Bank in August 1896, full of the same species of pelagic Crustacea and Sagittae that the mackerel had taken at the same time and place, while specimens taken at Woods Hole had dieted chiefly on copepods, to a less extent on amphipods, Salpae, appendicularians, and young herring. They follow thrown bait as readily and bite quite as greedily as mackerel do. Their breeding habits have not been studied.

General range.—Temperate Atlantic Ocean, north to outer Nova Scotia and to the Gulf of St. Lawrence in the west,[72] to England in the east. It is represented in the Pacific by a close ally, *Pneumatophorus japonicus*. It is a more southerly fish than the mackerel.

Occurrence in the Gulf of Maine.—Goode,[73] long ago summarized the early history of the chub mackerel in our waters, which briefly was as follows:

It was tremendously abundant during the last of the eighteenth century and early years of the nineteenth, down to 1820–1830. Thus Capt. E. E. Merchant, an experienced and observant fisherman, described them as so plentiful off Provincetown from 1812 to 1820 that three men and a boy could catch 3,000 in a day on hook and line. But it practically disappeared from the United States coast some time between 1840 and 1850. It is interesting to note, as Captain Atwood pointed out, that destructive methods of fishing had nothing to do with the case, for its disappearance antedated the introduction of traps, pounds, or purse seines; it also antedated the reappearance of the bluefish (p. 386); hence cannot be blamed on these sea pirates. So completely did the hardheads vanish that the Smithsonian Institution tried in vain for 10 years prior to 1879 to obtain a single specimen. But a school was taken in the summer of 1879 in a trap at Provincetown (where representatives of the U. S. Fish Commission were stationed at the time), and though none were seen in 1880 there were some off the coast of New York in 1886.

We find no definite record of the status of the hardhead during the next decade. But Bean[74] describes them as abounding off New York in 1896, swimming up little creeks in such numbers that they could be dipped in boat loads. And hardheads were taken singly and in schools by the mackerel fleet on Georges Bank during that same August,[75] while many were caught on hook and line from the *Grampus* in Block Island Sound during the first week of that September.

Kendall found them at Monomoy, the southerly elbow of Cape Cod in 1898, and they were sufficiently restablished by then for Smith[76] to describe them as uncommon to abundant at Woods Hole. They then dropped out of the published record again (they are not separated from the

[72] It is reported from St. Margaret Bay and Halifax by Vladykov (Proc. Nova Scotian Inst. Sci., vol. 19, 1935, p. 7), and Schmitt (Monographie de l'Isle de Anticosti, 1904, p. 285, Paris) credits it with "apparitions irrégulières" at Anticosti.

[73] Fish. Ind., U. S., Sect. 1, 1884, p. 303.

[74] Bull. 60, New York State Mus., Zool. 9, 1903, p. 383.

[75] Field notes supplied by Dr. W. C. Kendall.

[76] Bull. U. S. Fish. Comm., vol. 17, 1898, p. 95.

common mackerel in the fishery returns) until 1900, when they were found in the Casco Bay region. There is no reason to suppose that they appeared in any numbers anywhere on our coasts during the period 1900 to 1906, but in the latter year many were taken in the traps near Woods Hole, also in 1908. And the mackerel fleet found great schools of hardheads on Georges Bank in 1909, when vessels brought in fares of 50,000 to 100,000 of them during the first week of July,[77] their small size (500 to 700 to the barrel) suggesting that there had been a great production of hardheads a year or two previous. Fishermen speak of catching a few from time to time since then, but no great numbers. We caught one at Cohasset on the south shore of Massachusetts Bay in September 1942.

In its years of plenty, which fall at long intervals, however, the chub mackerel is likely to appear wherever mackerel do off the Massachusetts coast, especially about Provincetown. Thus 13,420 pounds were taken in traps at North Truro, in 1952, between August 11 and October 5. Other definite Gulf of Maine records are mostly[78] about Casco Bay and one from Johns Bay, Maine. We found no record of it farther east along the coast of Maine; it is unknown in the Bay of Fundy, nor does it seem to reach the west Nova Scotian coast. But in good "hardhead" years, it is to be expected all along Georges Bank and on Browns as well, to judge from its occasional visits to the outer coast of Nova Scotia.

[77] Boston Herald, July 9, 1919.

[78] Scattergood, Trefethen, and Coffin (Copeia, 1951, No. 4, p. 298), report one caught in August 1949.

Importance.—The chub mackerel is as choice a table fish as the mackerel, and no distinction is made between them in the market, other than the size of the individual fish.

Striped bonito *Euthynnus pelamis* (Linnaeus) 1758

Oceanic bonito

Jordan and Evermann, 1896–1900, p. 868, *Gymnosarda pelamis.*

Description. — The various fishes commonly called bonitos, albacores, and tuna, are fusiform in shape like all their family, tapering to a pointed nose and to an extremely slender caudal peduncle. But they are much stouter-bodied than mackerel or chub mackerel, and their second dorsal fin originates close to the rear end of the first dorsal, instead of being separated from the latter by a long interspace. The present species is about one-fourth as deep as it is long; its caudal peduncle has one prominent median longitudinal keel on either side, with a smaller keel above it, and another below at the base of the tail. The very deeply concave contour of its first dorsal fin (fig. 178) is enough to separate it at a glance from the common bonito (fig. 180), or from a young tuna (fig. 181), and from the Spanish and king mackerels (figs. 182, 183). The fact that its sides have dark markings below the lateral line, but not above the latter, is the readiest field mark by which to distinguish it from its close relative the false albacore (p. 336), in which the reverse is true. Also, its anal fin originates

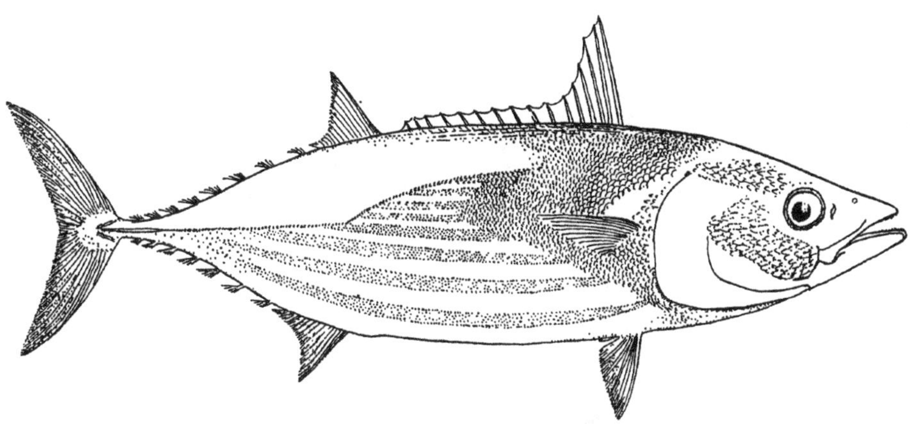

Figure 178.—Striped bonito (*Euthynnus pelamis*). After Smitt.

farther forward than in the false albacore; i. e., under the middle of the second dorsal fin, instead of under the first dorsal finlet.

Another distinctive character (shared, however, by the false albacore, p. 335) is that it has no body scales except along the lateral line, and covering a very prominent corselet on the forward and upper part of the trunk, which is outlined in the illustration (fig. 178). Its lateral line curves downward suddenly below the second dorsal which is not the case in its relative *alleteratus* (p. 336).

The first dorsal fin (about 15 spines) is not only much longer, relatively, than that of the mackerel, but its upper edge is abruptly concave behind the second spine, with the last 9 or 10 spines much shorter. The second dorsal is triangular, with concave rear edge; almost the whole of it stands in front of the anal; the anal is as large as the second dorsal and of about the same shape. There are about 8 little finlets behind the second dorsal, and about 7 finlets behind the anal. The pectorals are of moderate size, reaching back only about midway of the first dorsal. The tail fin is very short but broad and lunate in outline.

Color.—Deep steel blue above, with the lower part of the sides, the throat and the belly shining white. Each side is barred behind the corselet with 4 to 6 longitudinal blue or brown stripes, the upper ones terminating at their intersection with the lateral line, the lower 3 or 4 fading out as they near the caudal peduncle.[79]

[79] The number of stripes is different in different geographic regions; American fish usually show only 4; 7 have been described for Japanese specimens; there usually are 4, and sometimes 5 or 6, on each side in the European bonito.

Size.—This bonito grows to a length of about 30 inches.

General range.—Warmer parts of all the great oceans, the Atlantic, Pacific, and Indian.

Occurrence in the Gulf of Maine.—A specimen obtained at Provincetown in 1880 by J. Henry Blake is the only record for this oceanic fish in the Gulf, but it sometimes appears in numbers about Woods Hole, where 2,000 to 3,000 were taken in 1878, but where it did not show again until October 1905.

False albacore *Euthynnus alleteratus* (Rafinesque) 1810

LITTLE TUNNY; BONITO

Jordan and Evermann, 1896–1900, p. 258.

This bonito resembles the striped bonito (p. 335) very closely in body form, in size and arrangement of its fins, and in the fact that its body has no scales except on the forward and upper part of the trunk, the corselet, and along the lateral line. But it is distinguishable from the striped bonito by its color pattern, for it is above its lateral line that its sides bear dark markings, not below. And its lateral line does not bend downward appreciably below the second dorsal fin.[80] Also, its anal fin originates relatively farther back than in the striped bonito, i. e., under the first dorsal finlet instead of under the middle of the second dorsal fin.

[80] For further differences between the species of *Euthynnus*, see Fraser Brunner, Ann. Mag. Nat. Hist., Ser. 12, vol. 3, 1950, p. 150.

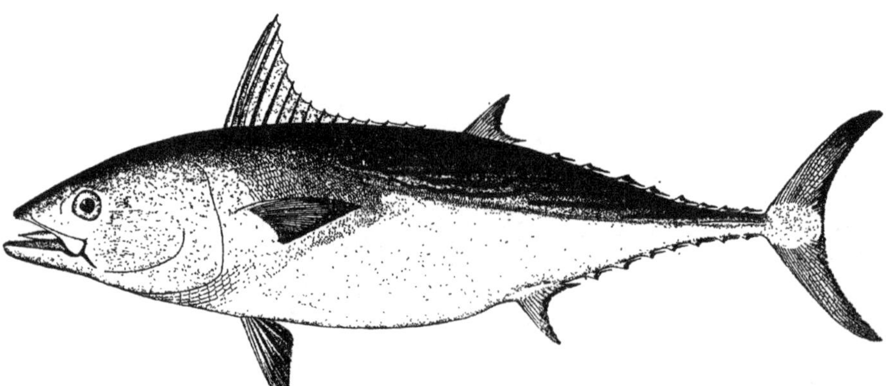

FIGURE 179.—False albacore (*Euthynnus alleteratus*), Woods Hole. From Jordan and Evermann. Drawing by H. L. Todd.

Color.—Steel blue above, glistening white lower down on the sides and on the belly. The sides are without markings below the lateral line, except for a few dark spots below the pectoral fin, but are marked above the lateral line with dark wavy bands, in various patterns.

Size.—About the same as *E. pelamis*, i. e. growing to about 2½ feet.

General range.—This, like its relative *pelamis* is a tropical-oceanic fish, widespread on the high seas, in all the great oceans.

Occurrence in the Gulf of Maine.—False albacores are picked up from time to time near Woods Hole, in July or August. But the only records of them within our Gulf are of 200 to 300 taken in a trap at Barnstable, in the autumn of 1948,[81] and of 28 taken in another trap in Cape Cod Bay, near Sandwich, on September 11, 1949.[82] Like various other tropical fishes they come our way only as strays from warmer seas; they are likely to be in schools whenever they reach our Gulf.

Common bonito *Sarda sarda* (Bloch) 1793

BONITO; SKIPJACK; HORSE MACKEREL

Jordan and Evermann, 1896–1900, p. 872.

Description.—This bonito is shaped much like a small tuna, being thick and stout bodied, about one-fourth as deep as it is long (not counting the caudal fin), and similarly tapering to a pointed snout and very slender caudal peduncle. It is tuna-like also, in that its body is scaled all over, that its caudal peduncle has median longitudinal keels, and that its two dorsal fins are so close together that they are practically confluent. But the shape of its fins distinguishes it at a glance from a small tuna, the only regular member of the Gulf of Maine fish fauna, with which it is apt to be confused,[83] its first dorsal being relatively much longer than that of the tuna (about one-third as long as the body, not counting the caudal, and with about 21 spines), and its second dorsal considerably longer than high, whereas the second dorsal is at least as high as it is long in the tuna.

The mouth, too, of the common bonito is relatively larger than that of the tuna, gaping back as far as the hind margin of the eye, and its jaw teeth are larger, with the two to four in the front of the lower jaw noticeably larger than the others. The shape of its first dorsal, with nearly straight upper margin marks it off from the oceanic bonito (p. 335), also from the false albacore (p. 336), in both of which this fin is very deeply concave in outline; the uniform scaliness of its body, also, is diagnostic, as contrasted with them.

We need only note further that its first dorsal fin is triangular, tapering regularly backward, with only slightly concave upper edge; that the margins of the second dorsal and anal fins are deeply concave; that it has 7 or 8 dorsal finlets and 7 anal finlets; that its tail fin is lunate, much broader than long; and that its lateral line is not deeply bowed below the second dorsal, but is only wavy.

Color.—The color of this bonito is so distinctive as to be a ready field mark to its identity, for while it is steely blue above with silvery lower part of the sides and abdomen, like most of the mackerel tribe, the upper part of the sides are barred with 7 to 20 narrow dark bluish bands running obliquely downward and forward across the lateral line. While young its back is transversely barred

[81] Reported to us by Frank Mather of the Woods Hole Oceangraphic Institution. All of these, weighing 2,498 pounds, were caught on September 16 in the trap of John Vetorino.

[82] Schuck, Copeia, 1951, p. 98.

[83] No one should take a bonito for a large mackerel, its dorsal fins being close together, while those of the mackerel are far apart.

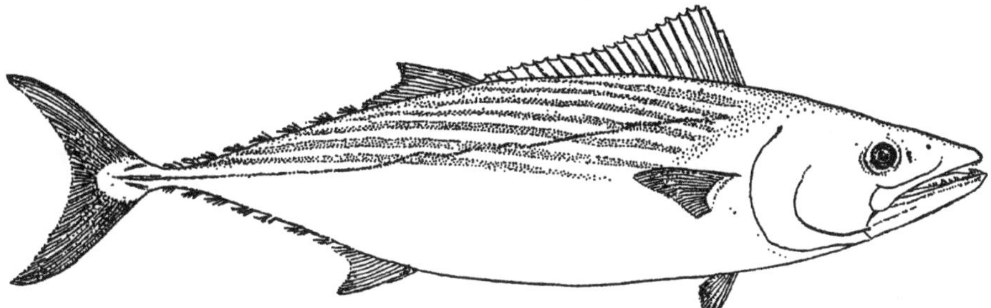

FIGURE 180.—Common bonito (*Sarda sarda*). After Smitt.

with 10 to 12 dark-blue stripes, but these dark cross-bars usually disappear before maturity.

Size.—This bonito grows to a length of about 3 feet and to a weight of 10 to 12 pounds.

Habits.—The bonito is a strong, swift, predaceous inhabitant of the open sea and like all its tribe travels in schools. When they visit our northern waters they prey upon mackerel, alewives, menhaden, and other smaller fish such as launce and silversides; also upon squid. They are very likely to be noticed, for they jump a great deal when in pursuit of their prey.

Further to the southward the bonito spawns in June; but it is not likely to spawn in the Gulf of Maine, nor does it do so in the northern part of its European range. Presumably its eggs are buoyant like those of other scombroids. Young 5 to 6 inches long have been reported as common off Orient, N. Y., early in September.[85] But nothing is known of its rate of growth.

General range.—Warmer parts of the Atlantic, including the Mediterranean; north to outer Nova Scotia,[86] on the American coast and to Scandinavia on the European coast.

Occurrence in the Gulf of Maine.—Cape Ann is the northern limit to the usual occurrence of the bonito within our Gulf. It has been taken occasionally in Casco Bay, while one was recorded from the mouth of the Kennebec River in September 1930 and two more in July 1932.[87] But we find no definite record of it east of this on the coast of Maine, or in the Bay of Fundy, although the young have been reported from Halifax on the outer coast of Nova Scotia. Its usual limitation to the southern half of the Gulf appears clearly in the location of the commercial catches.

In 1919 [88] for example, pound nets, traps, and other gear, accounted for almost 34,000 pounds in Cape Cod Bay, but only 90 pounds about Cape Ann, while the entire catch landed in the fishing ports of Maine during that year was only half a dozen fish (44 pounds). And there have been so few of them in Maine waters of late that none at all were mentioned in the fisheries statistics for that State of late years.

Bonito have been known to reach Cape Ann in larger numbers in the past, as happened in 1876, when 73 were taken in one August day in a weir near Gloucester. And probably they are far more plentiful every year out at sea in the southern part of the Gulf than these meager returns would suggest, for fishermen often mention schools of them. Capt. Solomon Jacobs reported them as very plentiful, in August 1896, for instance, in the deep water to the northward of Georges Bank. And we have seen schools of large scombroids, (probably bonito) splashing and jumping off Cape Cod more than once in August.

Apparently bonito visit New England shores only in the summer and fall. Thus the earliest catch made by a certain set of pound nets at Provincetown over a period of about 10 years was in July (1915), the latest on October 4 (1919).

The bonito is more regular in its occurrence west and south of the Cape, being common in some years at Woods Hole and especially off Marthas Vineyard, whence about 57,000 pounds were marketed in 1945. And party-boat captains have described Buzzards Bay and the waters around the Vineyard and Nantucket as full of them in some recent summers.

Importance.—The bonito is a good food fish. It readily bites a bait trolled from a moving boat, once one has the lure that it will strike on the particular occasion. A good many are caught in this way off southern New England, and we can assure the reader that a bonito is one of the strongest fish that swims, weight for weight, and one of the swiftest. Bonito are picked up now and then in Cape Cod Bay by anglers trolling for other fish; we heard of two taken in this way off Wellfleet, on August 29, 1950. But they are never abundant enough in the Gulf of Maine to be worth fishing for there with hook and line.

Tuna *Thunnus thynnus* (Linnaeus) 1758

BLUE FIN TUNA; HORSE MACKEREL; GREAT ALBACORE; TUNNY; ALBACORE [89]

Jordan and Evermann, 1896–1900, p. 870.

Description.—The two dorsal fins of the tuna are practically continuous, a character (with the

[85] Nichols and Breder (Zoologica, New York Zool. Soc., vol. 9, 1927, p. 123).

[86] "Fair numbers" have been taken in St. Margarets Bay, also some in mackerel traps near Lunenberg, and one was taken at Cape Breton, Nova Scotia, in October 1937 (McKenzie, Proc. Nova Scotian Inst. Sci., vol. 20, 1939, p. 16). It is also reported from the mouth of Halifax harbor (Jones, Proc. and Trans. Nova Scotian Inst. Sci., vol. 5, pt. 1, 1882, p. 88). One specimen, 276 mm. long, was taken off Centre East Pubnico, September 12, 1951 (reported to us by A. H. Seim).

[87] Reported by Walter H. Rich.

[88] Nineteen nineteen is the most recent year, the published statistics for which mention bonito in the regional breakdown of the total Massachusetts catch. And there is nothing in the published fishery statistics to suggest that the status of the bonito has changed since then.

[89] A comprehensive list of publications dealing with the tunas is given by Corwin, Division Fish and Game of California, Fish Bull. No. 22, 1930.

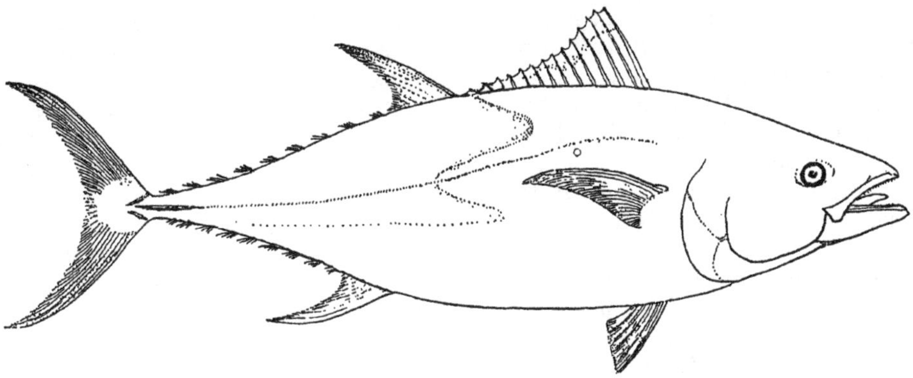

FIGURE 181.—Tuna (*Thunnus thynnus*). After Smitt.

large number of finlets) sufficient in itself to separate a very young one from either of our true mackerels. A small one is readily separable from the striped bonito and from the false albacore by the fact that the entire trunk of the tuna, including the belly, is scaly, the upper outline of its first dorsal fin only very slightly concave; and from the common bonito (p. 337) by a second dorsal that is considerably higher than it is long by the shape of its anal with only weakly concave margin, by the small size of its jaw teeth; and by the midline of the roof of its mouth armed with hairlike teeth. The plain coloration of the tuna, without dark markings, is still another convenient field mark for separating small ones from any of the bonito tribe that have been reported from our Gulf.

The tuna is shaped like a bonito rather than like a mackerel, with robust body, about one-fourth to one-sixth as deep as long, tapering to a pointed nose and to a very slender caudal peduncle which bears a strong median longitudinal keel on either side. The first dorsal fin (13 or 14 spines), originating close behind the axil of the pectoral, is triangular, its upper edge weakly concave, tapering backward from its first spine, and with the last spine very short indeed. And it can be laid down, flush, in a groove along the back. The second dorsal (about 13 rays, not depressible) is almost confluent with the first (a little lower than the latter in young fish and a little higher in old) is at least as high as it is long or higher, deeply concave behind, and with sharp-pointed apex. The anal fin originates under the rear end of the second dorsal to which it is similar in outline and size (about 12 rays). Usually there are 9 or 10 dorsal finlets and 8 or 9 anal finlets, behind the second dorsal fin and the anal fin, respectively. The tail fin is much broader than long, its margin evenly lunate, its two lobes sharp pointed, much as it is in the bonitos. The pectoral and ventral fins are of moderate size, the former scimitar-shaped and much longer than broad.[90]

Color.—The back is dark lustrous steel blue or nearly black, with gray or green reflections; the cheeks silvery; the sides and belly silvery gray, often with large silvery spots and bands, and iridescent with pink. The first dorsal is dusky to blackish; the second dusky to reddish brown; the dorsal finlets yellow with dark edgings. The anal fin is silvery gray; the anal finlets the same, or yellow; the caudal dusky but more or less silvery; the ventrals and pectorals blackish above and silvery gray below.[91]

Size.—This is the largest Gulf of Maine fish, except for some sharks; a length of 14 feet or more, and a weight of 1,600 pounds being rumored, with fish of 1,000 pounds not rare. The heaviest Rhode Island fish on record, taken about 1913, weighed 1,225 pounds, while 4 or 5 fish have been brought into Boston that weighed approximately

[90] The tunas and their allies are discussed by Jordan and Evermann (Occas. Papers, Cal. Acad. Sci. vol. 12, 1926); Fraser-Brunner (Annals and Magazine Nat. Hist., Ser. 12, vol. 3, 1950, pp. 142–146) has recently given a convenient key to all known species of tunas, with excellent illustrations; and Godsil and Holmberg have recently discussed the relationships of the bluefin tunas of New England, Australia, and California (Fish. Bull. 77, California Dept. Nat. Resources, 1950).

[91] The foregoing description of the color is based on accounts of freshly caught tuna by Storer (Fishes of Massachusetts, 1867, p. 65) and by Nichols (Copeia, No. 111, 1922, pp. 73–74); and on fish we have seen.

1,200 pounds each, and one in 1924 that is said to have reached 1,300 pounds; and Sella [92] mentions a "fairly well authenticated instance" of one caught 60 to 70 years ago off Narragansett Pier, R. I., that weighed in the neighborhood of 1,500 pounds, was divided among the various hotels, and fed 1,000 people. The largest caught so far on rod and reel weighed 977 pounds and was 9 feet 9 inches long.[93] One of 932 pounds, taken at Wedgeport, Nova Scotia, by H. E. Teller, in September 1951, is the largest that has been caught on rod and reel in the Gulf of Maine.[94] Another of 864 pounds [95] was 9 feet 4 inches long and 88 inches in girth.

Large tuna of the same length and caught the same day may vary as much as 100 pounds or more in weight, depending on their condition, as pointed out by Crane.[96] Lengths and weights of tuna, before being dressed, caught in Massachusetts Bay and off Ipswich in July and August 1951 were as follows: 28 inches, 17 pounds; 34 inches, 30 pounds; 42 inches, 56 pounds; 60 inches, 144 pounds; 63 inches, 172 pounds; 66 inches, 188 pounds; 68 inches, 200 pounds; 88 inches, 516 pounds; 93 inches, 587 pounds. Off Bimini, in May and June, 1950, two 88-inch tuna averaged 415 pounds and three 93-inch fish averaged 450 pounds indicating that they are much thinner in the spring in their more southern habitat than they are in summer to the northward.[97]

In the western side of the Mediterranean, where tuna run smaller than in our Gulf, a 500-pound fish is very large and this is equally true off the California coast. But tuna weighing as much as 1,595 pounds (725 kilograms), if the stated weights are reliable, have been reported from the eastern parts of the Mediterranean and from the Bosphorus near Constantinople.[98]

Habits.—The tuna is a strong, swift fish and an oceanic wanderer like all its tribe. Probably its chief reason for holding to continental waters along our coasts during the warm seasons is that its prey are more concentrated there and hence more easily caught than over the ocean basin.

The small, medium, and fairly large-sized fish, up to 350–500 pounds or so, commonly travel in small schools of half a dozen to 30 or 40 fish, but sometimes in much larger schools, and each school is usually composed of fish of about the same size: we have never heard of large and small tuna schooling together. And it seems that the very large fish usually are solitary.[99]

When tuna are at the surface, as they often are, they are proverbial for their habit of jumping, either singly or in schools; they may do this when swimming about, or harrying smaller fishes, or less often, when traveling in a definite direction, in which case all that are jumping do so in the same direction.

Frank Mather, for instance, reports seeing a school of 200-pounders, jumping in unison, 2 or 3 feet clear of the water. When large tuna jump, they sometimes fall flat, making a great splash, but they reenter the water a little head-first as a rule, though they do not make as complete and graceful an arc in the air as the various oceanic kinds of porpoises usually do. When schools, at the surface, are not jumping, they often splash a good deal and they are conspicuous then. We remember, for instance, sighting a large school so employed, off the Cohasset shore at a distance of about 3 miles, on one occasion. Even if they are neither jumping nor splashing, as is more commonly the case, the wakes that large ones leave behind them betray their presence, if the sea is smooth.

They sometimes cut the surface with the sickle-shaped second dorsal fin and with the tip of the caudal fin, on calm days, and they have been photographed while so doing.[1] But we have not seen this and experienced tuna fishermen have told us that tuna are not often seen finning. In any case, it seems that the first dorsal fin is laid back, when they do fin; at least we have never heard of a tuna as showing both of its dorsals above the surface, except after it had been hooked.[2]

---

[92] Internat. Rev. Gesamten Hydrobiol. Hydrogr., vol. 25, Pt. 1-2, 1931, p. 60.

[93] Caught by Comm. D. W. Hodson at Cape Breton, Nova Scotia, September 4, 1950.

[94] Reported in Salt Water Sportsman, for Oct. 1, 1951.

[95] Caught near Jordan Ferry, Nova Scotia, by Alfred Kenny in 1950.

[96] Zoologica, New York Zool. Soc., vol. 21, No. 16, 1936, p. 207.

[97] These records are from unpublished data furnished by Frank Mather of the Woods Hole Oceanographic Institution and Howard Schuck of the U. S. Fish and Wildlife Service, who have given us much first-hand information on the habits of the tuna.

[98] Heldt, 10 Rapp. Comm. Internat. Explor. Medit., vol. 11, 1938, p. 343

[99] Crane (Zoologica, New York Zool. Soc., vol. 21, 1936, pp. 207–211) has given a readable account of the tuna off Casco Bay, which we cannot better, and with which our own sightings of tuna agree.

[1] See Farrington (Fishing the Atlantic, 1950 [approximate date], upper photo facing p. 421), for an excellent photograph of a tuna finning.

[2] See Farrington (Fishing the Atlantic, 1950 [approximate date], lower photo facing p. 421), for an excellent photograph of a hooked tuna showing the first dorsal fin as well as the second dorsal.

Tuna often break the surface when striking a bait, or they may even leap clear then. But for some reason they do not jump ordinarily after they are hooked, but first make one or more swift shallow runs and then tend to bore deep unless in very shallow water.

Tuna prey on smaller fishes, especially those of the schooling kinds, the particular species depending on the local supply. In the Gulf of Maine they destroy great numbers of herring, large and small; also mackerel of which they are often full. They have been reported as pursuing silver hake; 26 tuna contained these, out of 30, that were examined by Crane at Portland, Maine, in July 1936. She also reports a rosefish (*Sebastes*) in one. No doubt they take whatever small fishes are available locally, and a tuna has been known to swallow a whole dogfish as large as 8 pounds. Southward from Cape Cod they prey on menhaden, as predaceous fishes do in general. They also eat squid: Crane found squid, in two, at Portland, and quantities of euphausiid shrimps (*Meganyctiphanes*) in two others. It is not unusual for tuna to strand in pursuit of prey. But this is a timid fish and easily frightened though so voracious.

Tuna have no serious enemies in the Gulf of Maine, but killer whales take toll of them in Newfoundland waters where, writes Wulff [3] "one or more times annually, usually in September, orcas will ravage the tuna schools in the bays they frequent most."

The tuna is a fish of at least moderately warm seas. The smaller sizes seem rather closely restricted to regions where the surface layer is warmer than 60°–62°, and while large ones are regular visitors in summer to the eastern side of our Gulf where the water warms only to about 50°–54°, this, seemingly, is about the lower limit to the thermal range they favor.[4] Few tuna, for example, whether large or small, are seen in the Passamaquoddy region in most summers (p. 343) though the multitudes of small herring there would seem to offer ideal feeding conditions, but where the temperature rises only to about 52°–54° even by August, when it is highest. And seasonal chilling is generally accepted as the factor that drives them from our northern coasts in the autumn.

Tuna tolerate a wide range of salinity, and they run well up into bays, and even into harbors in pursuit of herring; the bays on the outer Nova Scotian coast for example; Bras D'or "lake," Cape Breton; Bonne Bay on the west coast of Newfoundland; and Trinity and Conception Bays on the southeastern coast of Newfoundland. But we have never heard of one entering brackish water.

Tuna are as definitely migratory as the mackerel is, those that visit our coasts working northward in spring, to drop out of sight again late in the autumn.[5] They are said to be around Jamaica throughout the year, but most plentiful there in March and April [6] Ordinarily they appear earliest on the Bahaman side of the Straits of Florida in the first or second week in May; next off New Jersey, off Long Island, off southern New England, and in Cape Cod Bay in June. But they have been reported well within the Gulf of Maine by the last week of May (p. 342), or nearly as early as in Bahaman waters. This, with the added fact that they are not known to approach the American coast anywhere between the Bahama Channel and North Carolina or Virginia [7] suggests that we may have two separate populations, a southern and a northern.

They usually arrive in Bonne Bay, on the Gulf of St. Lawrence coast of Newfoundland in late June or in early July, and a week or two later in Trinity and Conception Bays, on the southeastern part of the Newfoundland coast.[8]

Finally, we should point out that it is not known yet whether the tuna populations of the two sides of the Atlantic are entirely separate, one from the other, or whether more or less interchange takes place between them.

The vertical range of the tuna is from the surface down to an indeterminate depth; the only barriers likely to limit their descent are the

[3] Internat. Game Fish Assoc. Yearbook, 1945, p. 65.

[4] The tuna that visit the west coast of Newfoundland find summer tempertures as high as 59°-60° along the south coast of Newfoundland, and 55°-57° in Trinity and Conception Bays on the southeastern part of the Newfoundland coast.

[5] See Heldt (Bull. No. 5, Station Oceanographique de Salambo, 1926), and Sella (Int. Rev. Hydrobiol., Hydrogr., vol. 24, 1930, p. 446) for accounts of the migration and food of tuna in the Mediterranean and eastern Atlantic.

[6] Information contributed by Capts. Eddie Wall and Walter Whiteman, for which we are indebted to Frank Mather of the Woods Hole Oceanographic Institution.

[7] Frank Mather of the Woods Hole Oceanographic Institution informs us that a 600-pound tuna has been taken in a trap 200 miles south of Chincoteague, Md., and that small ones are taken off Chincoteague. "Tuna" are reported from time to time off North Carolina, also. But it is not yet certain whether these actually are "bluefins."

[8] Wulff, Internat. Game Fish Assoc. Yearbook, 1943, p. 65.

low temperatures they would encounter in regions where there is a strong thermal gradient (the Gulf of Maine is an example, p. 344), the increasing scarcity of prey, and, perhaps, darkness.

The breeding habits of the tuna remained a mystery until recently. And while it is now known that those that visit the Mediterranean spawn in June and July, both the spawning grounds of our American tuna and their spawning season are yet to be learned.

The eggs (Mediterranean) are buoyant, small for so large a fish (1.05–1.12 mm. in diameter) with one oil globule of about 0.27 mm.

The larval stages have also been recorded in the Mediterranean in abundance; and the characters determined by which they may be distinguished from allied species.[9] Tuna fry of 3¼ inches (81 mm.) and about 6 inches (152 mm.) have also been pictured and are described from the Gulf of Mexico by Fowler.[10]

Rate of growth.—The rate of growth of so large a fish is naturally a matter of much interest. Young fry grow so rapidly that fish hatched in June in the Mediterranean reach a weight of a little less than ¾ pound to a little more than 1 pound (300–500 grams) by September. According to studies by Sella, based on the number of concentric rings in the vertebrae[11] for 1,500 individuals, Mediterranean tuna average about 10 pounds at 1 year of age, about 21 pounds at 2 years, about 35 to 36 pounds at 3 years, about 56 pounds at 4 years, about 88 pounds at 5 years, about 128 pounds at 6 years, about 170 pounds at 7 years, about 214 pounds at 8 years, about 265 pounds at 9 years, about 320 pounds at 10 years, about 375 pounds at 11 years, about 440 pounds at 12 years, about 517 pounds at 13 years, and 616 to 660 pounds at 14 years of age.

Average lengths of 20 to 24 inches in their second summer of growth, 27 to 34 inches in the third, 35 to 40 inches in the fourth, and 42 to 46 inches in the fifth, reported by Westman and Gilbert [12] suggest about the same growth rate for the American tuna. Thus the giants of 800 pounds and heavier have reached a very respectable age. According to Sella [13] Mediterranean tuna weighing only 35 pounds may already be sexually mature. But nothing definite is known about the American fish in this regard.

General range.—Warmer parts of the Atlantic (including the Mediterranean), Pacific and Indian Oceans; [14] north regularly to the western, southern and southeast coasts of Newfoundland,[15] on the western side of the Atlantic; to Iceland and northern Norway (Lofoten Islands) on the European side.

Occurrence in the Gulf of Maine.—The tuna is a yearly visitor to our Gulf. Every fisherman knows the tuna or horse mackerel, as it used to be called, and this great fish visits all parts of the Gulf of Maine, but we do not understand its comings and goings much better now than when Storer called attention to its abundance about Provincetown nearly a century ago. Scarcity is not to blame for this (it is common enough) but the fact that little attention was paid to it until recently for want of market value. And while a demand for tuna has developed of late, as is reflected in the catches (p. 346), and while many anglers now fish for them (p. 347), most of the resulting information is confined to the few inshore localities where they either seem to be the most plentiful, or where they are caught most easily from small craft, or incidentally in the fish traps.

It is now known that tuna are to be found all around the shores of the Gulf from Cape Cod to eastern Maine; in the Bay of Fundy; also along the west coast of Nova Scotia. And fishermen often report them on Nantucket Shoals, Georges Bank, and Browns. In ordinary years the first of them are likely to be seen as early in the season between Cape Ann and the Maine State line as they are off Cape Cod. In 1950, for example, the earliest report of them was off Hampton, N. H., May 26; the next off Plum Island, Mass., on June 9; and it was not until about June 16 that word came of one hooked in Cape Cod Bay, and of the first fish (one of 462 pounds) harpooned off Plum

[9] See Sella (Atti Reale accad. Lincei, Roma, Ser. 5, vol. 33, Fasc. 7–8, semestr. 1, 1924, p. 300) and Sanzo (R. Comit. Talass. Ital. Mem., No. 189, 1932) for description of the larvae; Heldt (Bulls. 5 and 18, Station Oceanographique Salambo, 1926 and 1930) for summaries of all previous observations on the breeding habits and larval stages.

[10] Monogr. 6, Acad. Nat. Sci. Philadelphia, 1944, pp. 261, 373.

[11] Memoria No. 166, R. Comitato Thalassografico Italiano, 1929, p. 10.

[12] Copeia, 1941, pp. 70–72, based on length frequencies for those up to 3 years of age and on scale studies for the older ones.

[13] Memoria No. 156, R. Comitato Thalassographico Italiano, 1929, p. 6.

[14] Sella's recent studies (Internat. Rev. Ges. Hydrobiol., Hydrogr., vol. 25, 1931, pp. 48–50) showed no characteristic differences between the bluefin tuna of the two sides of the Atlantic, and those of different oceans appear, at most, to represent races of a single wide-ranging species.

[15] Vesey-Fitzgerald and Lamonte (Game Fishes of the World, 1949, p. 183) report tuna from Hamilton Inlet.

Island. This may have been an early year. But tuna are to be expected throughout the western side of the Gulf generally by the middle or end of June, which is about as early as they ordinarily appear in any numbers off southern New England; and they appear on the Nova Scotian side of the Gulf by the first of July if not earlier. In 1950, for example, upwards of 450 had been landed from Ipswich Bay by July 31, the largest weighing 734 pounds.[16] The peak season usually is from about the middle or end of July to the middle of September off Massachusetts; July and August off Casco Bay; through August and September along western Nova Scotia.

The vicinity of Provincetown, with Cape Cod Bay, has long been known as a center of abundance for tuna. Other well known centers are from Cape Ann north to Boon Island and from the Ipswich Bay–Plum Island shore out to Jeffreys Ledge some 30 miles off shore; off the mouth of Casco Bay and for some distance thence eastward; and the vicinity of Wedgeport, on the west coast of Nova Scotia, where the International tuna matches are held. Fewer are seen along the eastern coast of Maine, though we are told that a fishery for tuna has developed during the current summer off Southwest Harbor, Mount Desert Island,[17] and in the New Brunswick side of the Bay of Fundy.

It is expecially interesting that there are so few tuna in the Passamaquoddy region in most years that the capture of even an occasional fish in the local weirs causes comment, for the astounding abundance of small herring there would seem to offer them an inexhaustible supply of food. But a summer comes now and then when they are far more plentiful there than usual; thus Passamaquoddy waters are said to have "teemed with tuna" in the summer of 1937 [18] when as many as 7 were taken at Campobello in a single seining; and several were reported again and a few caught in Passamaquoddy Bay in the summer of 1945.[19]

Dr. Huntsman writes us that "schools" were reported there in the summer of 1951, when the water was warmer than usual. And Leslie Scattergood reports 22, ranging from 113 to 161 pounds, caught in a herring weir at Grand Manan during that October.

The regional contrasts in local abundance within our Gulf may be illustrated for a representative year by the reported catches of tuna by counties around the coast from southwest to northeast, for 1945.

	Pounds
Massachusetts:	
Barnstable (chiefly Cape Cod Bay)	301, 900
Plymouth	600
Essex	50, 300
Maine:	
York	45, 300
Cumberland, Sagadahoc } vicinity of Casco Bay	815, 300
Lincoln	900
Knox (Penobscot Bay)	0
Hancock	0
Washington	0
Nova Scotia:	
Annapolis	0
Yarmouth	35, 800
Shelburne to Cape Sable	0

In most years the tuna that are seen and caught near Provincetown at the tip of Cape Cod, and in Cape Cod Bay, are small (so-called "school fish" weighing less than 200 pounds with many as small as 30 to 70 pounds; and few of those caught there in most years are large. The smallest reported in the inner part of the Gulf of Maine was a run of 20- to 26-pound fish (2-year-olds) taken in Cape Cod Bay in October 1950.[20] And good catches of "school" fish of 30–70 pounds, but few larger, if any, are being made again off the tip of Cape Cod around the shores of Cape Cod Bay at this writing (August 5, 1951), and have been for several weeks past. Large numbers of even smaller tuna, averaging about 11 pounds, have been encountered on the southwestern part of Georges Bank (p. 344), and many of these little ones (from 8 pounds or so upwards) are caught off southern New England every summer and autumn, especially near Block Island.[21] On the other hand, most of those found northward from Cape Ann, and in the Nova Scotian side of the Gulf are large, few of them as small as 100 pounds. Thus, the average live weights of 1,641 tuna that were landed at Portland, Maine, during the period 1926 to 1935, varied between 495 pounds

[16] Reported by Henry Moore in the *Boston Herald*, July 31, 1950.

[17] Information supplied by Frank Mather, of the Woods Hole Oceanographic Institution.

[18] Atlantic Fisherman, vol. 18, No. 9, October 1937, p. 28.

[19] Atlantic Fisherman, vol. 26, No. 8, September 1945, p. 52.

[20] Reported by Frank Mather of the Woods Hole Oceanographic Institution.

[21] Frank Mather, of the Woods Hole Oceanographic Institution, reports a catch of 110 of them, weighing about 10 pounds, off No Mans Land, on September 16, 1951.

and about 630 pounds yearly, as appears from the following table.[22]

Year	Number	Largest	Smallest	Average
1926	90	945	416	515
1928	176	858	74	510
1929	152	905	447	630
1930	172	967	423	620
1931	107	840	420	565
1932	91	800	409	500
1933	162	909	93	495
1934	268	955	68	530
1935	423	913	33	525

Thirty-two fish caught at the mouth of Casco Bay in 1950 averaged 468 pounds, the heaviest 643 pounds;[23] the smallest among 34 measured by Crane,[24] at Portland, Maine, weighed 65 pounds, the heaviest 860 pounds. And many fish are taken of 700 pounds and heavier. Similarly, 23 tuna caught during the international match at Wedgeport, Nova Scotia, in the second week of August 1950, weighed from 362 pounds to 744 pounds, and 72 taken there during the match of the previous year averaged about 360 pounds, the largest weighing 857 pounds. Also, most of the tuna caught in the Gulf of St. Lawrence are rather large.

The reason for this regional segregation of tuna of different sizes is not known, or for the variation therein from year to year. We suspect that temperature is chiefly responsible; i. e., that the larger fish are more tolerant than the small of the lower temperatures prevailing in the northern and northeastern parts of the Gulf, and in more northerly regions. Especially suggestive in this connection is the fact that the tuna run so large off Wedgeport, western Nova Scotia, where the abundant herring offer excellent feeding conditions, but where the water does not ordinarily warm above about 54° F. along the open coast, though to a somewhat higher figure locally, in enclosed situations.

So many tuna come so very close inshore in Cape Cod Bay that nearly all of the commercial catch made there is taken in the traps; large schools have even been sighted within Provincetown Harbor (on October 11, 1950, for example[25]) and occasionally a tuna comes into the surf either to strip the reel of some surf fisherman or to be landed (p. 347). The tuna that are taken north of Cape Ann are farther out; all of them, however, are caught within 30 miles or so of the land, at farthest. And while a great concentration of tuna was encountered by the *Albatross III* on the southwestern part of Georges Bank, on September 18, 1950, when 25 were hooked and landed, all very small, about 11 pounds apiece, it is unusual to see any large number on the offshore banks.

The tuna that are seen or caught in our Gulf all are near the surface, or at least where the water is not more than 35 to 40 fathoms deep. How deep down they might be found is not known. But it is likely that they tend to keep within 50 fathoms or so of the surface, for the deeper water in the Gulf is colder than tuna appear to like (p. 341).

In some years the tuna appear to remain fairly stationary in whatever part of the Gulf they visit, for weeks at a time, as is indicated in the consistency of catches, or the sightings reported, which is equally true of them in Newfoundland waters, according to Wulff.[26] In other years they may disappear suddenly from one locality or another, after a brief stay, and without any apparent reason. In 1926, for example, when about 70 fish were taken in July off Casco Bay, only 17 were caught there in August, 3 in September, and only 1 in October (the 4th).[27] In 1950 they deserted the Ipswich Bay-Plum Island region during the last week of August, not to reappear there in any numbers that season, though they continued plentiful enough off the Maine coast farther north to be worth fishing for until the end of September, with some in the Cape Cod Bay–Provincetown region until early October.

There are tuna in good numbers along the outer Nova Scotia coast, off Shelburne, the vicinity of Liverpool at the mouth of the Mersey River, the mouth of the La Have River, Mahone Bay, and St. Margaret Bay being centers of abundance as appear from landings of 258,000 pounds in Lunenburg County and 201,000 pounds in Halifax County in 1950. A few, also, are seen and caught around Cape Breton. It was here that the record size fish was taken with rod and reel (p. 344).

Catch records suggest that only a few visit the southern side of the Gulf of St. Lawrence; 400 pounds were reported from the Gulf shore of Cape Breton in 1944, none in 1946, and it was only in one year (1925) that any were reported (975

[22] Data gathered by the late W. H. Rich of the U. S. Bureau of Fisheries.

[23] Caught by Capt. Earl Larrabee; reported in Saltwater Sportsman for Aug. 25, 1950.

[24] Zoologica, New York Zool. Soc., vol. 21, No. 16, 1936, p. 207.

[25] Reported in Cape Cod Standard Times, October 11, 1950.

[26] International Game Fish Assoc., Yearbook, 1943, p. 65.

[27] Data from Walter H. Rich of Portland, Maine.

pounds)[28] from Prince Edward Island during the period 1917 to 1928. They may visit the west coast of Newfoundland more regularly; for Wulff speaks of them as common and gives a photograph of tuna finning at the surface in Bonne Bay,[29] but we have not heard anything to suggest that they are anywhere near so plentiful there as they are in Nova Scotian waters or southward. Wulff writes of them as "few" on the southern Newfoundland coast, at present; but they appear to be regular visitors to Conception and Trinity Bays on the southeast coast; Mr. Tibbetts informed us that he once saw an abundance of tuna in Notre Dame Bay, midway of the east coast of Newfoundland; and they are reported from Hamilton Inlet, Labrador, their most northerly known outpost on the American Coast of the Atlantic.

Most of the tuna disappear from the coasts of Maine and of northern Massachusetts by the end of September, or by the first part of October at the latest, depending on whether the season is an early one or a late.[30] But considerable numbers remain in Cape Cod Bay and around the tip of Cape Cod until well into October, or even into November in some years. Thus in 1950 large schools were seen in Provincetown Harbor, and more than 5,000 pounds of small fish, averaging about 75 pounds, were landed there on October 11,[31] while in 1949 about 2,000 pounds were caught nearby between November 1 and 14.

The dates of the earliest and latest catches, made by a set of 8 traps, at North Truro, Cape Cod Bay, during the period 1943 to 1952[32] are illustrative.

Year	Earliest catch	Latest catch
1943	July 8	Oct. 6
1944	June 29	----------
1945	June 25	Oct. 9
1946	June 15	Oct. 26
1947	June 21	Oct. 28
1948	June 11	Oct. 28
1949	June 7	Nov. 14
1951	July 7	Oct. 15
1952	June 24	Oct. 24

The monthly catches, by these same traps, mark July and August as the most productive months. The number of pounds of tuna (dressed weight) follows:

Month	Largest catch	Smallest catch
June	17, 520	0
July	148, 139	12, 255
August	185, 305	5, 029
September	70, 125	[33] 2, 365
October	43, 603	330
November [34]	2, 197	0

Catches have also been reported along western Nova Scotia as late as the third week in October, and Wulff writes of tuna lingering through the month in the bays of Newfoundland, which is as late as they remain in any part of our Gulf.

Tuna are never reported as seen moving southward on their way out of the Gulf to their winter quarters; they drop just out of sight.

The wintering grounds of the particular bodies of tuna that summer in the Gulf of Maine, and of those that go farther east and north, are not known. Small (20–50 lb.) fish, it is true, have been caught occasionally in coastal waters off southern New England from January to March; off Block Island, for example, in 1928.[35] But the bulk of the northern contingents certainly travel farther. It is probable that they winter in deep water as the Mediterranean tuna do, perhaps along the continental slope off our Middle Atlantic coast, perhaps so much farther south that some of the tuna seen (and caught) in spring in the Straits of Florida are our Gulf of Maine and Nova Scotian fish, on their way north again.

We are equally in the dark as to the spawning grounds of the American tuna, for although the Gulf of Maine fish are of breeding age, no ripe ones have ever been seen off the New England or Canadian coasts, or even fish approaching ripeness.

Abundance.—We dare not guess how many tuna are in our Gulf in any summer, there being no way to estimate how large a proportion of them the yearly catch represents. We suspect that they are fewer than reports would suggest, for being so large, a few hundred of them make a great show if they are at the surface, whereas an equal number of mackerel, for instance, would never be noticed. Neither is any definite information available as to their annual fluctuations in

[28] Sella, Internat. Rev. Ges. Hydrobiol., Hydrogr., vol. 25, 1931, p. 50.

[29] Internat. Game Fish Assoc. Yearbook, 1943, p. 66.

[30] In 1950 seven tuna of 200–300 pounds were caught off Boars Head, Maine, during the first week of October.

[31] Reported in Cape Cod Standard Times, October 11, 1950.

[32] Information contributed by the Pond Village Cold Storage Co.

[33] The year 1944 is omitted from the calculation for September–November, because the traps were not fished after September 14th that year.

[34] The only catch recorded for November was 2,197 pounds in 1949.

[35] Sella, Internat. Rev. Gesamten Hydrobiol., Hydrogr., vol. 25, 1931, p. 62.

abundance, though fishermen are well aware that their numbers in any part of the Gulf do vary widely from year to year. Thus it is on record that they were scarce in the Massachusetts Bay region for two or three years prior to 1904, but were plentiful that summer. Commercial landings suggest that they were scarce again in 1943, when the landings came to only about 380,000 pounds for Maine and Massachusetts combined.[36]

But they appear to have been much more plentiful again off the Maine coast in 1945 (catch there about 850,000 lb.); more plentiful than they have been since, if the commercial landings are a reliable index to the ups and downs of the tuna population, which they may not be. The following catch statistics of tuna landings (in pounds) suggest that the stock built up more slowly, from the 1943 low in Massachusetts waters, to a peak in 1948, which was a big year on the Ipswich Bay grounds (p. 343), as well as in Cape Cod Bay.

Year	Maine	Massachusetts
1943	25,000	129,500
1944	463,500	272,900
1945	859,500	356,400
1946	421,800	571,100
1947	186,600	726,400
1948	229,100	1,627,000

In 1949, the catch by traps in Cape Cod Bay alone was 811,160 pounds, suggesting a total of more than a million pounds from the Gulf of Maine coast of Massachusetts.

During that banner season 2,164 large tuna were taken on hand lines where the draggers work, off Ipswich Bay; while 806 fish (305,300 lb.) were taken off Wedgeport, Nova Scotia, in 1948, 1,760 fish (449,362 lb.) in 1949.

But they were scarce in 1950, to judge from reports coming in from all along our coast: a week's fishing, for instance, by the same number of rods and at about the same date, yielded only about half as many on the famous Soldiers Reef off Wedgeport during the International Match that summer as it had the year before. Many fewer were caught by anglers in Cape Cod Bay in 1950 than in 1949, and the tuna disappeared from the Ipswich Bay region unusually early that year, as noted above (p. 344). It is too early (August 5) to forecast how the 1951 season may develop.

[36] The weights given in the Fisheries statistics are for the dressed fish, and represent about 80 percent of the live weight.

The largest Gulf of Maine catches of which we have heard were of 336 fish, weighing about 75,000 pounds taken at one lift of 3 traps set for mackerel on the Barnstable shore of Cape Cod Bay, Aug. 5, 1948;[37] and of 120,000 pounds of fish ranging from 25 to 30 pounds seined some 50 miles east of Cape Cod on September 18, 1951, by the *Western Explorer*, chartered by the U. S. Fish and Wildlife Service for experimental fishing for tuna.

Importance.—Horse mackerel were formerly regarded as a nuisance on the Atlantic coast, for bands of them made trouble for fishermen by following herring or mackerel into the traps and pounds, to tear their way out again through the net unless harpooned. Many years ago, when fish oil was more valuable than now, a few were sometimes harpooned for oil, which was tried out of the heads and bellies, but there was no sale for their meat. The tuna, however, has been highly valued as a food fish for many years, not only in the Mediterranean, but on the west coast of the United States. And a local demand has developed on our coast, supplied chiefly by local fisheries off Casco Bay, in the Cape Ann-Boone Island region, and in the Cape Cod Bay region.

With this increasing demand, the reported landings on the Maine and Massachusetts coasts have risen from about 94,000 pounds in 1919, to around 250,000 pounds yearly in the early 1930's, and to about 1 to nearly 2 million pounds for the years 1945 to 1948, this last representing around 3,000–6,000 fish, if they averaged 300–400 pounds in weight (see table, p. 346). The average value to the fisherman in 1946 was about 7–9 cents per pound and all that are caught now sell readily. The annual catches off the entire coast of Nova Scotia ranged from 152,000 pounds to about 1,550,000 pounds during the period 1917 to 1928; from 402,000 to 1,820,000 pounds for the 5 years 1942–46.

The commercial catch off the coasts of Maine is made mostly by harpoon; that off northern Massachusetts by hook and line and by harpoon; that off the Cape Cod Bay region mostly in the traps.[38]

[37] A photograph of part of the catch was published in the Boston Herald, August 6, 1948.

[38] In 1945 about 60 percent of the catch reported for Maine was by harpoon, almost all the remainder on hand lines; in 1946 about 98 percent was harpooned. About 86 percent of the Massachusetts catch was taken in traps of one sort or another in 1945, about 90 percent in 1946.

But experiments are in progress, by the U. S. Fish and Wildlife Service, to find whether a profitable seine fishery or long line fishery can be developed for these great fish in our Gulf, with the hope of providing a more dependable supply, and through a longer season.

The sporting qualities of the tuna in our Gulf deserve a word, though an extended account would be out of place here. Encouraged by the famous tuna fishing off the coast of southern California, and by the knowledge that tuna run much larger in the Gulf of Maine than they do on the Pacific coast, several anglers had caught these huge fish with light tackle at various points in New England and Nova Scotia by 1925, when the first edition of this book appeared. Small tuna will often take artificial lures especially if trolled at high speed and close to the vessel's stern, while large ones will take a hook baited with herring, mackerel, or other fish. And tuna fishing has now grown to be so popular and successful a sport that many party boats go out regularly off Provincetown, in Cape Cod Bay, to the Ipswich Bay–Isles of Shoals–Boone Island region, off Casco Bay, and off Wedgeport on the Nova Scotian side.[39]

To date, the largest tuna that has been landed on rod and reel in the Gulf of Maine was one of 932 pounds, caught by H. E. Teller at Wedgeport, Nova Scotia, September 11, 1951 (p. 340). Another of 927 pounds was caught in Ipswich Bay, August 4, 1940, by Dr. J. B. Vernaglia. We have heard of one of 180 pounds landed with ordinary surf-casting tackle on the beach at Plum Island, a notable feat.[40] Even a small tuna, such as a thirty-pounder that was caught on a black plug by Wm. Lakaitis, surf casting at North Truro on the night of July 28, 1951, is a far more stronger adversary in the surf than a striped bass of equal size.

[39] Farrington (Field and Stream magazine for August 1950, p. 84) has recently given an interesting account of the methods employed by rod and reel anglers, in these localities. Crane (Zoologica, N. Y. Zool. Soc., vol. 21, No. 16, 1936, p. 210) describes in a readable way the small boat harpoon fishing for tuna off Casco Bay, Maine.

[40] Landed on August 12, 1950, by M. L. Insleyn.

Spanish mackerel *Scomberomorus maculatus* (Mitchill) 1815

Jordan and Evermann, 1896–1900, p. 874.

Description.—The Spanish mackerel has the outline of the slender mackerel rather than of the stout bonito, its body being nearly 4½ to 5 times as long as it is deep. But there is no danger of confusing it with either of the true mackerels, first, because its two dorsal fins (like those of the bonitos) are hardly separated, and second, because of its color pattern, Its high second dorsal, slender form, and spotted sides mark it off at first glance from our bonitos, while its color, slender form, long first-dorsal fin, and the outline of its second dorsal distinguish it from a small tuna. The most clear-cut distinction between the Spanish and its close relatives the king mackerel and the cavalla, is that the pectoral fins of the Spanish are naked but those of the last two are mostly covered with scales. The ventral fins, also, originate definitely behind the origin of the first dorsal in the Spanish, under it or only a very little rearward in the king; and the color differs.

The most distinctive anatomic character of the Spanish, king, and cavalla among our local scombroids is the large conical jaw teeth. The caudal peduncle of the spanish mackerel is keeled; its lateral line wavy; its first dorsal fin (17 to 18

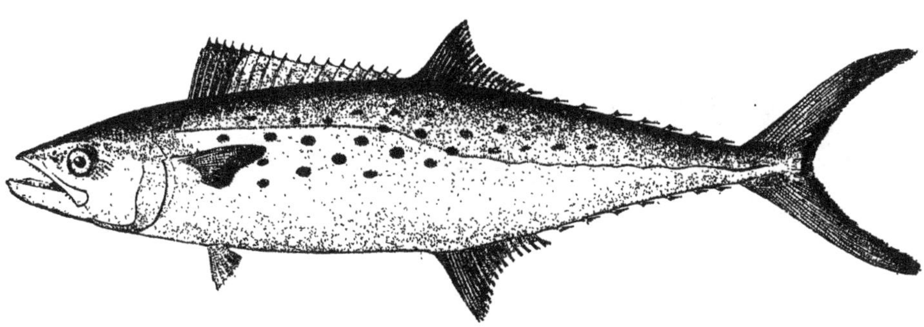

FIGURE 182.—Spanish mackerel (*Scomberomorus maculatus*). From Goode. Drawing by H. L. Todd.

spined) is triangular; its second dorsal (14 to 18 rays) is concave and originates a short distance in front of the anal, which is similar to it in form and size. It has 8 or 9 dorsal and as many anal finlets. Its pectorals are naked; its caudal is deeply lunate, with the outer rays decidedly longer than those of the common mackerel. It has 32 teeth, or fewer, in each jaw.

Color.—The Spanish mackerel is dark bluish or blue green above, pale below, like all scombroids, and silvery, its sides marked with many small, oblong-oval, dull orange or yellowish, spots, both above the lateral line and below, these spots being a very diagnostic character. The fact that the membrane of the front one-third of its first dorsal fin is black (blue in the king, p. 348), whereas its rear part is greenish white, is an equally useful field mark. The second dorsal and pectoral fins are pale yellowish with dusky edges; the anal and the ventrals are white.

Size.—The maximum weight is about 9 or 10 pounds,[41] maximum length about 36 inches, but the fish average less than 3 pounds as caught.

General range.—Both coasts of North America, north commonly as far as Chesapeake Bay in the Atlantic, and to Maine as a stray; south to Brazil.

Occurrence in the Gulf of Maine.—The Spanish mackerel is a summer visitor all along the Atlantic coast of the United States, as far north as New York; less regularly along the southern coasts of New England, though a few are taken during most summers at Woods Hole. But it is only a stray in the colder waters of the Gulf of Maine, where occasional fish are taken in Cape Cod Bay every year or two. In 1896 the local catch rose to 37 fish (Provincetown and Truro traps), and there is record of it at Lynn, Mass. But Spanish mackerel are so rare north of this point that Monhegan Island is the only locality-record for Maine, and the most northerly known outpost for the species.

[41] A weight of 25 pounds is recorded by Smith (North Carolina Geol. Econ. Surv., vol. 2, 1907, p. 191) for a specimen observed in a Washington, D. C., fish market. If the identity was correct and this was not the closely related king mackerel *Scomberomorus regalis* it must be considered a case of giantism.

King mackerel *Scomberomorus regalis* (Bloch) 1793

KINGFISH

Jordan and Evermann, 1896–1900, p. 875.

Description.—The king mackerel resembles the Spanish mackerel closely in general appearance, but its pectoral fins are mostly covered with scales; its ventrals are below the first dorsal, instead of definitely behind the origin of the latter; its head is relatively longer, its nose more pointed, its teeth more numerous (about 40 in each jaw), triangular and very sharp pointed; and the upper half of its first dorsal is deep blue. Furthermore, the king mackerel is marked by a narrow brownish stripe running from close behind each pectoral fin to the base of the caudal, crossing the lateral line as the latter bows downward below the second dorsal fin. Its side spots, too, are mostly below the lateral line and arranged in rows, whereas the spots of the Spanish mackerel are irregularly scattered, with about as many above the lateral line as below it.

Size.—Said to reach 35 pounds, but the average weight is between 5 and 10 pounds.

General range.—Atlantic Coast of North America, Cape Cod to Brazil, abundant among the West Indies and around southern Florida.

Occurrence in the Gulf of Maine.—This southern fish is recorded by Dr. W. C. Kendall at Monomoy, at the southern elbow of Cape Cod, but it has not been taken elsewhere in the Gulf of Maine.

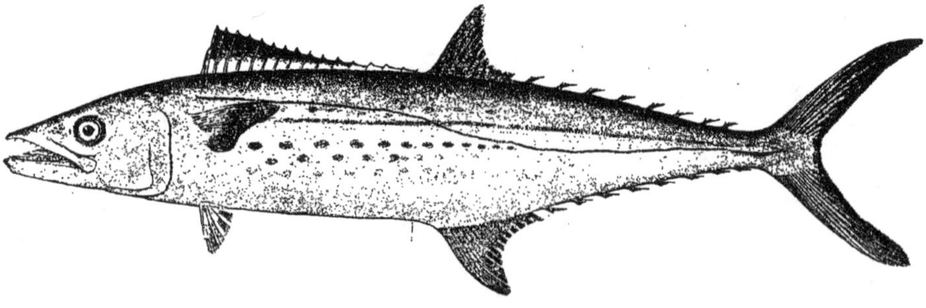

FIGURE 183.—King mackerel (*Scomberomorus regalis*), Key West, Florida. From Goode. Drawing by H. L. Todd.

Cavalla *Scomberomorus cavalla* (Cuvier) 1829 [42]

CERO

Jordan and Evermann, 1896–1900, p. 875.

Description.—The pectorals of the cavalla are scaly, and its anal fin, like that of the king mackerel originates about under the origin of its second dorsal, in which it differs from the Spanish mackerel (p. 347). In fact, it resembles the king mackerel so closely in general appearance that the one might easily be taken for the other by anybody not used to handling the two fish as southern fisherman are. But the lateral line (very conspicuous in both) is a sure clue to identity, for this dips downward abruptly in the cavalla under the forepart of the second dorsal fin, but slopes down only gradually there in the king mackerel. Other points of difference are that the outline of the first dorsal fin is concave in the cavalla (nearly straight in the king); that the cavalla has a large number of teeth (about 40 in each jaw, as against about 30); that its body is more slender (about one-sixth as deep as it is long); and that the upper forepart of its first dorsal fin is not noticeably darker than the remainder of the fin.

Color.—Iron gray above, silvery lower down on the sides and on the belly; the sides marked with darker gray or yellowish spots, which tend to disappear in large fish.

Size.—Said to reach a length of a little more than 5 feet, and a weight of about 100 pounds. The rod and reel record is 73½ pounds, for one taken off Bimini, Bahamas, February 1935, by L. B. Harrison.

General range.—Warm parts of the Atlantic; south to Brazil in the western side; north regularly to North Carolina (June-November); [43] occasionally to southern Massachusetts; and as a stray to the southern part of the Gulf of Maine.

Occurrence in the Gulf of Maine.—The only Gulf of Maine record of which we know is of one 20½ inches long (to base of caudal fin) taken in a trap at North Truro, Cape Cod, in August 1949.[44]

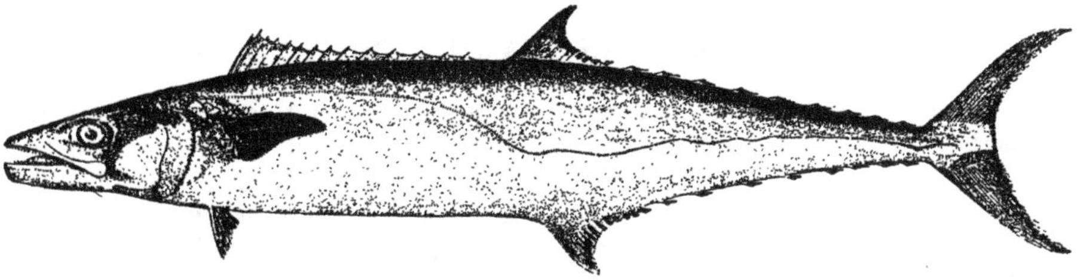

FIGURE 184.—Cavalla (*Scomberomorus cavalla*), Woods Hole. From Goode. Drawing by H. L. Todd.

THE ESCOLARS. FAMILY GEMPYLIDAE

These fishes are closely allied to the true mackerels, the most obvious differences being that they lack the keels on the sides of the caudal peduncle so characteristic of the mackerels.

Escolar *Ruvettus pretiosus* Cocco 1829

OILFISH; SCOURFISH; PLAINTAIL

Jordan and Evermann, 1896–1900, p. 879.

Description.—This fish suggests the mackerel family in its slender fusiform shape and in the general arrangement of its fins. And its first (spiny) dorsal (13 to 15 spines), like that of the Spanish mackerel, is much longer than the second dorsal (18 soft rays). But it is separable at a glance from all Gulf of Maine mackerels by the facts that it has only 2 dorsal finlets and 2 anal finlets, and that its skin is set with bony plates armed with short spines instead of being velvety with small scales, as it is in the case of the mackerel tribe. The caudal fin is deeply forked. The first dorsal is much lower than the second, and the anal is situated below the second dorsal, which it parallels in its outlines.

Color.—Described as purplish brown, darkest above, with blackish patches, and with the inside of the mouth dusky.

[42] Fowler (Proc. Acad. Nat. Sci. Philadelphia, vol. 56, 1905, p. 766) refers this species to a new subgenus *Sierra*.

[43] Taylor (Survey Marine Fisheries North Carolina: University of North Carolina Press, 1951, pp. 261–265) has given an interesting survey of the seasonal presence of various southern game fishes off the North Carolina coast.

[44] This specimen was received through the kindness of John Worthington and the Pond Village Cold Storage Co.

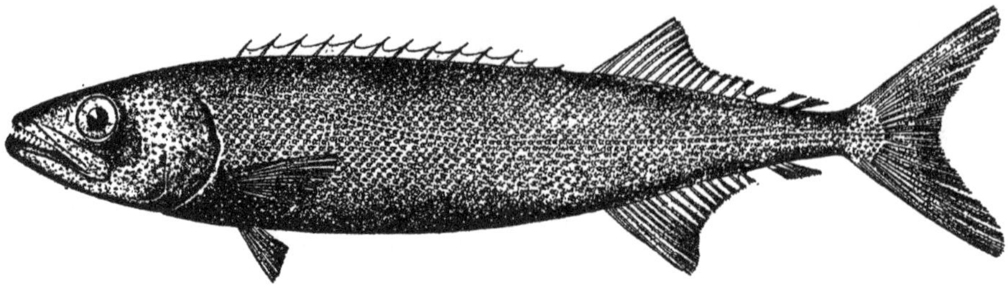

FIGURE 185.—Escolar (*Ruvettus pretiosus*), Georges Bank. From Goode and Bean. Drawing by J. C. van Hook.

Size.—The escolar grows to a weight of at least 100 pounds.

General range.—Tropical parts of the Atlantic, and the Mediterranean, in moderately deep water (usually 100 to 400 fathoms); also widespread in warm latitudes in the North Pacific [45] and in East Indian waters. It is plentiful around Cuba though not reported from Puerto Rico; is known from Bermuda; and it has been taken as a stray as far north as the Bay of Biscay in the east and to the Grand Banks of Newfoundland in the west. There are regular fisheries for it off Cuba and about the Canaries; also in the Pacific.[45]

Occurrence in the Gulf of Maine.—Two escolars, respectively, 49 inches long and 6 feet long, were brought in to the United States Fish Commission from Georges Bank during the autumn of 1891.[46] It has not been seen in the Gulf of Maine region since then. The nearest record of it to the southward, with which we are acquainted, is of two trawled about 92 miles off Cape May, N. J., in January 1950.[47]

THE CUTLASSFISHES. FAMILY TRICHIURIDAE

The cutlassfishes are characterized by a scaleless, band-shaped body tapering to a slim pointed tail, with one dorsal fin extending the whole length of the body; the anal is also long but is very low. The ventrals are absent or rudimentary, and there is no distinct caudal fin. Their large mouth is armed with strong teeth of various sizes. They inhabit the surface waters of tropical seas.

Cutlassfish *Trichiurus lepturus* Linnaeus 1758

HAIRTAIL; SCABBARDFISH; SILVER EEL; RIBBANDFISH

Jordan and Evermann, 1896–1900, p. 889.

Description.—The most striking characteristics of the cutlassfish are its band-like form tapering to a pointed whiplike tail without caudal fin; its single long dorsal fin (about 135 rays) originating close behind the eyes, about two-thirds as high at its midlength as the body is deep, and diminishing to nothing some distance in front of the tip of the tail; its long anal composed of very low detached spines pointing backward; and its long barbed fangs in the front of the mouth, four in the upper and two in the lower jaw. The depth of the body equals about one-thirteenth to one-fifteenth of its total length, about one-seventh to one eighth of which is occupied by the head. The snout is pointed, the mouth gapes back to below the eye and the lower jaw projects beyond the upper. Each of the jaws is armed with 7 to 10 smaller teeth behind the fangs. The anal fin is reduced to a series of short inconspicuous spines, about 100 to 110 in number, without connecting fin membrane, running back from the vent nearly to the tip of the tail. The small pectorals are situated a little in advance of the rear corners of the gill covers. There are no ventral fins and the skin is scaleless.

Color.—Plain silvery all over. The dorsal fin is plain yellowish or dusky green in life, dark edged or speckled along the margin with black; the tips of the jaws dusky.

Size.—Maximum length about 5 feet.

General range.—All warm seas; abundant in the West Indies and Gulf of Mexico; not rare along the

[45] Gudger (American Naturalist, vol. 62, 1928, p. 467) and Nordhoff (Natural History, vol. 28, 1928, p. 40) give accounts of the geographic distribution of the escolar, and of the fisheries for it in tropical waters.

[46] Approximate location 41° 40′ N., 67° 44′ W. See Goode and Bean, Smithsonian Contrib. Knowl., vol. 30, 1895, p. 197.

[47] LaMonte, Marine Life, vol. 1, No. 8, 1950, p. 40.

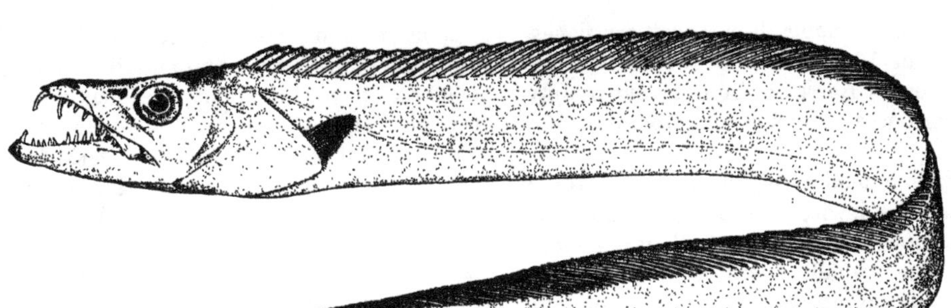

FIGURE 186.—Cutlassfish (*Trichiurus lepturus*), Florida. From Goode. Drawing by H. L. Todd.

South Atlantic coast of the United States, occasionally straying north as far as Massachusetts Bay.

Occurrence in the Gulf of Maine.—The cutlassfish is only an accidental straggler north of Cape Cod. One was taken at Wellfleet in the summer of 1845, and one in Salem Harbor also many years ago, and it is recorded from Lynn by Kendall.[48] There is no report of it farther north in the Gulf of Maine, or for Canadian waters.

THE SWORDFISHES. FAMILY XIPHIIDAE

The upper jaw and snout of the swordfish (there is only one species) is greatly prolonged, forming a flat, sharp-edged sword. It has a very high first-dorsal fin and a very small second dorsal, both of them soft rayed; a broad lunate tail; two separate anal fins, the second very small; and a strong longitudinal keel on either side of the caudal peduncle. It has no ventral fins, and the adults have neither teeth nor scales. The spearfish family (p. 357) is the only other group represented in the Gulf of Maine fauna which at all resembles the swordfish, but spearfish have ventral fins and minute teeth; their swords are round edged, and either there is one long continuous dorsal fin or, if there are two, the first is several times as long, relatively, as it is in the swordfish.

Swordfish *Xiphias gladius* Linnaeus 1758

BROADBILL

Jordan and Evermann, 1896–1900, p. 894.

Description.—The salient feature of the swordfish is the prolongation of its upper jaw into a long, flattened, sharp-edged [49] and pointed "sword" occupying nearly one-third the total length of the fish. This sword is of itself enough to identify

[48] The Massachusetts Bay and Provincetown records listed by Kendal (Occ. Pap. Boston Soc. Nat. Hist., vol. 7, No. 8, 1908, p. 76) are based on the Wellfleet specimen. He also credits it to Monhegan I., Maine, quoting Storer as his authority, but Storer stated in his latest mention of the species that only two had come to his notice; the Wellfleet specimen just mentioned, and one taken at the head of Buzzards Bay.

[49] In its tropical relatives, the sailfish and spearfish, the sword is round edged, spearlike, and relatively shorter.

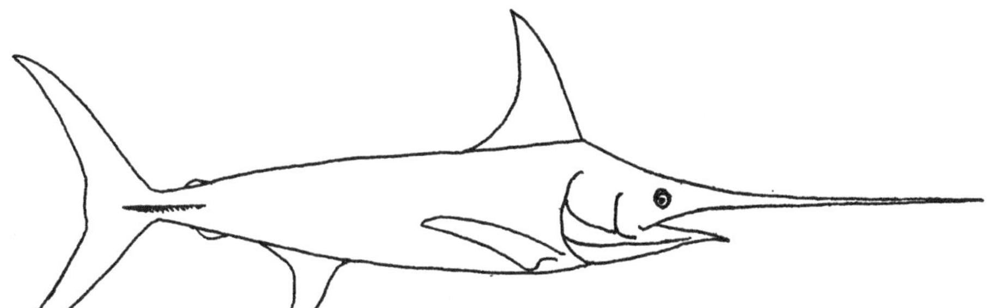

FIGURE 187.—Swordfish (*Xiphias gladius*). After California Fish and Game Commission.

the fish at a glance among all our northern fishes. On a fish 10 feet 10 inches long, which we harpooned on Georges Bank on the *Grampus* in July 1916, the sword was 42 inches long from its tip to the eyes.

The swordfish is moderately stout of body, only slightly flattened sidewise, deepest just behind the gill openings, and tapering rearward to a slender caudal peduncle, which bears a single strong longitudinal keel on either side. Apart from the sword the head is short; the lower jaw is pointed, and the mouth so wide that it gapes far back of the very large eyes, which are set close to the base of the sword. Swordfish (except young fry) are both toothless and scaleless. The first dorsal fin originates over the upper angle of the gill openings and is much higher than long (about 39 to 40 rays), with deeply concave rear margin. The second dorsal is very small and set far back on the caudal peduncle. There are two anals likewise. The second is as small as the second dorsal and located below the latter, while the first is similar to the first dorsal in outline but shorter, and located well behind it, close to the second anal. The pectorals are narrow, very long, scythe-shaped, and set very low down on the sides below the first dorsal. The caudal fin is short, but as broad as half the length of the fish from top of lower jaw to base of caudal fin, with deeply lunate margin and pointed tips. There are no ventral fins.[50]

Color.—While all swordfish are dark above and whitish with silvery sheen below, the upper surface varies from purplish to a dull leaden blue or even to black. The eye has been described as blue. Very young swordfish, like very young tuna, are transversely barred, but none small enough to show this pattern has ever been found within the limits of the Gulf. The colors fade soon after death.

Size.—Swordfish grow to a great size. The heaviest definitely recorded from the Gulf of Maine was one caught on Georges Bank in the summer of 1921 by Capt. Irving King and landed at the Boston Fish Pier, that weighed 915 pounds dressed, hence, upwards of 1,100 pounds alive.[51] This specimen was not measured, but the sword was more than 5 feet long, so that the total length of the fish must have approximated 15 feet, and 16 feet seems to be about the maximum length, though fish as long as this are very unusual. The heaviest landed in Massachusetts during 1922 weighed 637 pounds dressed; that is, upward of 750 pounds live weight,[52] while the largest taken in 1931 weighed 644 pounds dressed and was 13 feet long including its sword, which measured 44 inches. One that weighed 925 pounds before it was dressed was landed in 1932; also one weighing 650 pounds dressed, which must have weighed 800 pounds alive; while one of 850 pounds (dressed?), brought in to Halifax, Nova Scotia, was said to have been the largest ever landed in that port. And several, weighing more than 500 pounds, dressed, are reported almost every year.[53]

But the general run are much smaller. Thus the average dressed weights of sundry fares of fish landed in Portland, Boston, and Gloucester in the years 1883–1884, and 1893–1895 were between 200 pounds and 310 pounds, falling to 114–186 pounds for the years 1917, 1919, 1926, and 1929–1930. And general report has it that Block Island fish run smaller than Georges Bank and Cape Breton fish. A 7-foot fish weighs about 120 pounds; 10- to 11-foot fish about 250 pounds; fish of 13 to 13½ feet, about 600 to 700 pounds, as taken from the water.

The rod and reel record is 860 pounds, for one 13 feet 9 inches long caught off Tocapila, Chile, April 28, 1940, by W. E. S. Tuker.

Swordfish fry are quite different in appearance from their parents, having only one long dorsal fin and one long anal fin, a rounded tail, both jaws equally prolonged and toothed, and the skin covered with rough spiny plates and scales. But fish of half a pound weight such as are caught in abundance in the Mediterranean already resemble the adults, except that they have minute scales until 30 inches long.

Habits.[54]—The swordfish is oceanic, not dependent in any way either on the coast (except as this offers a supply of food), or on the bottom; and it is a warm-water fish, most plentiful in localities and at depths where the temperature is higher than about 60°. But occasional captures

[50] In the sailfishes and spearfishes the body is scaly, the jaws are toothed, ventral fins are present, and the first dorsal fin is much longer than that of the swordfish.

[51] Fishing Gazette for September 1921, p. 13.

[52] Gloucester Times, April 26, 1923.

[53] See Rich (Proc. Portland Soc. Nat. Hist., vol. 4, Pt. 2, 1947, pp. 34–37) for these and other large fish landed from year to year.

[54] Rich (Proc. Portland Soc. Nat. Hist., vol. 4, pt. 2, 1947) has recently given an extended account of the occurrence and habits of our swordfish.

of swordfish on halibut lines set near bottom as deep as 200 fathoms, together with the fact that swordfish are by no means rare on the Newfoundland Banks, whence several fish were brought back by the American cod fleet in 1920, proves that temperatures as low as 50° to 55° do not bar it, at least for a brief stay.

Although swordfish may gather in certain localities they do not school, but are always seen scattered, either singly or at most two fish swimming together. Earlier published accounts, statements by fishermen, and our own rather limited experience all agree on this point. On calm days they often lie quiet on the surface, or loaf along with both the high first dorsal fin and the tip of the caudal fin above water, and they are easily harpooned while so employed, often allowing a vessel to approach until the pulpit projecting from her bow comes directly above the fish. When a swordfish is swimming at the surface, its first dorsal fin and the upper part of its tail fin both show above the water whereas a marlin shows its caudal only. One can tell a surfacing swordfish from a shark by its sharp-pointed dorsal (that of a shark is more broadly triangular) and by the fact that its tail fin seems to cut the water in a direct line, not wobbling from side to side as the tips of the tails of most sharks do (other than the mackerel shark tribe), if they show above the water at all.

When swordfish are at the surface, they jump a good deal, perhaps in vain attempts to shake off the remoras that so often cling to them. We saw one leap clear of the water four or five times in rapid succession close to the *Grampus*, off Shelburne, Nova Scotia, on July 28, 1914. Reports by fishermen, and our own experience, are to the effect that they surface only during the hours of daylight.

The swordfish is a fish-eater. During its stay in American waters it feeds on mackerel, menhaden, bluefish, silver hake, butterfish, herring, argentines, rattails (*Macrourus bairdii*), and indeed on any smaller fish, buckets of which have been taken from swordfish stomachs. Squid, too, are often found in them and may be their chief diet at times. And the jaws of one of the giant squids (genus perhaps *Architeuthis*), taken from the stomach of a swordfish harpooned on the northern edge of Georges Bank,[55] was an especially interesting find. One that we examined on Georges Bank, July 24, 1916, was full of silver hake, one taken there in August 1929 contained 5 large haddock (p. 199), while another harpooned off Halifax contained a squid (*Ilex*) and fragments of silver hake. They have often been described as rising through schools of mackerel, menhaden, and other fishes, striking right and left with their swords, then turning to gobble the dead or mangled fish. And we have seen them so employed on more than one occasion, to judge from the commotion.

It is not unusual for swordfish taken on the offshore banks to contain deep-sea fishes of one kind or another in their stomachs; many such instances have been recorded,[56] sometimes swallowed so recently that they are still in good condition when the swordfish is opened.[57] And since these so-called "black fishes" live outside the edge of the continent, mostly below 150 fathoms, this is good evidence that the swordfish found on the banks that front our Gulf do some of their foraging at considerable depths farther out at sea. It also seems that they sometimes strip lines set for halibut and tilefish of the fish already caught, for they are sometimes brought up tangled in the line.

It was not out of the ordinary for swordfish to be hooked on long lines set for halibut in the days when this fishery flourished (p. 255). Goode[58] cites a number of cases, including one when 13 swordfish were caught in this way on one halibut trip. And fishermen have told us of more recent instances. Swordfish have often been hooked and landed on hand lines, also. A case is on record of 7 taken in this way on one trip, in the South Channel, in 15 to 25 fathoms of water, the bait being whole mackerel; evidence that swordfish seemingly do not insist on live food. We also read that of old, fishermen from Marthas Vineyard and Nantucket sometimes took them while trolling with some sort of silvery fish as bait, forecasting the big game anglers of today.

Many tales are current of swordfish attacking slow moving vessels without any provocation, and driving their swords through the planking, either

[55] Rich, Proc. Portland Soc. Nat. Hist., vol. 4, pt. 2, 1947, p. 23.

[56] The late Walter H. Rich of the U. S. Bureau of Fisheries reports the following genera as taken from swordfish on Georges and Browns Banks: *Alepisaurus*, *Chauliodus*, *Chiasmodon*, *Lampadena*, *Macrostoma*, *Myctophum*, *Notoscopelus*, and *Stomias*.

[57] Kingsley (Science, N. Ser., vol. 56, 1922, pp. 225-226) reports two freshly swallowed stomiatids (*Echiostoma barbatum*) being taken from the stomach of a swordfish harpooned on the offshore slope of Georges Bank.

[58] Rept. U. S. Comm. Fish., (1880) 1883, pp. 353-354.

in "fits of temporary insanity," as Goode[59] expressed it, or more likely, while pursuing dolphins or other fish. Most of the attacks of this sort reported from tropical seas seem actually to have been by spearfishes (p. 357) but some in northern waters seem almost certainly to have been by the broadbill. A case in point is that of the schooner *Volunteer,* out of Gloucester, which received a strong blow near Block Island, August 7, 1887, apparently from a 300-pound swordfish that was seen swimming alongside, and which proved to have lost its entire sword when it was harpooned and brought on board.[60] We can only add that we have never heard of a swordfish making an unprovoked attack on any of the fishing vessels that pursue them every summer, or on any of the other craft, large or small, that cruise off our coasts. But fish that have been harpooned often turn on their pursuers, and it is a common event for one to pierce the thin bottom of a dory. We have, indeed, known several fishermen of our acquaintance to be wounded in the leg in this way, but always after the fish had been struck with the harpoon. Under these circumstances swordfish have been known to drive their swords right through the planking of a fishing vessel.[61]

Stories of swordfish attacking whales are time-honored traditions of the sea, mostly with no more foundation than the myth that they ally themselves with the harmless thresher shark for the purpose. As a matter of fact swordfish are easily frightened, and they will not often allow a small boat to come within striking range, which made harpooning from dories difficult in the old days.[62] But for some occult reason they will allow themselves to be almost run down by a larger vessel without paying the least attention to its approach until aroused by its shadow, or by the swirl of water under its forefoot. But we have never heard of a swordfish actually being struck by a vessel; they always sound or turn aside in time.

Swordfish fight gamely on the surface or below when harpooned. Storer long ago wrote that they sometimes sound with such speed and force as to drive the sword into the bottom, which fishermen say is by no means uncommon; and we saw this off Halifax in August 1914, when a fish more than 10 feet long, which we had harpooned from the *Grampus,* plunged with such force that it buried itself in the mud beyond its eyes in 56 fathoms of water. When finally hauled alongside it brought up enough mud plastered to its head to yield a good sample of the bottom.

Full-grown swordfish are so active, so powerful, and so well armed that they have few enemies. Sperm and killer whales and the larger sharks alone could menace them. And while we can find no evidence that swordfish ever fall prey to the first two, Captain Atwood found a good-sized swordfish in the stomach of a Mako shark. A swordfisherman described seeing two large sharks bite or tear off the tail of a 350-pound swordfish, which he afterwards harpooned. A 120-pound swordfish, nearly intact with sword still attached, was found in the stomach of a 730-pound Mako taken near Bimini, Bahamas, while another Mako of about 800 pounds, harpooned off Montauk, Long Island, was seen attacking a swordfish, and was found to have about 150 pounds of the flesh of the latter in its stomach when it was landed (p. 24). And Rich[63] mentions that other like cases have been reported. Young swordfish would, of course, be preyed upon by any of the larger predaceous fishes.

Swordfish are infested with many parasites besides the remoras, several of which are often found clinging to one fish. No less than 12 species of worms and 6 of copepods have been identified from fish taken off Woods Hole.

The eggs of the swordfish have not been seen, or have not been identified if seen; probably they are buoyant. Neither is anything definitely known of the rate of growth of the swordfish. It has been supposed that the young fish of half a pound to 12 pounds that are taken in winter in the Mediterranean are the product of the past spring's spawning, which would call for unusually rapid growth. But the very large size to which swordfish grow may equally be the result of long life, as it is in the case of the tuna (p. 342).

General range.—Both sides of the Atlantic Ocean; north to northern Norway, southern and

[59] Fish. Ind. U. S., Sect. 1, 1884, p. 345.

[60] Related by Rich (Proc. Portland Soc. Nat. Hist., vol. 4, Pt. 2, 1947, pp. 48–49).

[61] Many cases of this sort are mentioned in the rather extensive literature dealing with the swordfish.

[62] Rich (Proc. Portland Soc. Nat. Hist., vol. 4, Pt. 2, 1947, p. 71) so informs us.

[63] See Farrington (Field and Stream magazine, vol. 47, February 1943) and Rich (Proc. Portland Soc. Nat. Hist., vol. 4, Pt. 2, 1947, p. 44) for more detailed accounts.

western parts of the Gulf of St. Lawrence, south coast of Newfoundland and Grand Banks, south to latitude about 35° south. Also in the Mediterranean and Red Seas; about the Cape of Good Hope; and widespread in the Indian Ocean and in the Pacific Ocean, both north and south of the equator.

Occurrence in the Gulf of Maine.—The swordfish seems to have attracted little attention in the Gulf in colonial days, and though it has long supported a lucrative fishery off New England, we know little more of its life there today than in 1883 when Goode [64] published his Materials for the History of the Swordfish. The outer half of the continental shelf off Block Island and southern Massachusetts; the offshore parts of the Nantucket Shoals region; Georges Bank; the deep channel between Georges and Browns; Browns Bank and La Have; and the banks off the outer coast of Cape Breton are its chief centers of abundance off our coasts.

On these grounds 25 or more are often sighted in a day, sometimes that many are in view at one time; in fact, "one skipper reports counting 47 fish in sight at one time, after a week-long breeze had died out to a flat calm," [65] and some 10 to 20 thousand of them are harpooned every summer off the New England coast, with as many more off eastern Nova Scotia.[66] An occasional swordfish is seen off Massachusetts Bay also, and along the Maine coast nearly every year. During some summers, of which 1884 was one, numbers of them appear there, and on such occasions some are taken in the Gulf from Cape Cod to Browns Bank, with Jeffreys Ledge and a zone about 10 to 12 miles off the coast from Boon Island to Cape Elizabeth perhaps their favorite inshore resort. But the great majority keep strictly to the offshore banks during most years, and they are seldom seen in the Bay of Fundy. Thus we find only 2,500 pounds (10 or 12 fish) brought in by the shore fishermen of Cumberland County, and 3 or 4 fish (800 lbs.) landed in York County in 1919, while none was reported as caught off the coast of Maine in 1945, though 193,000 pounds were landed on the Nova Scotian side of the Gulf (Yarmouth Co.) in that year and the offshore catch was considerable.

Swordfish seem to be less plentiful along the outer Nova Scotian coast from Cape Sable to the Gut of Canso than on Georges Bank or on Browns, though a few are brought in from the various fishing banks every summer (p. 357). But the amounts reported from the outer (Atlantic) coasts of Cape Breton are so large as to show that they are likely to be as numerous there as they are anywhere abreast of the Gulf of Maine, or off southern New England, and perhaps more concentrated. These regional variations may be illustrated by the landings for 1945, which were as follows for United States and Canadian vessels combined: offing of southern New England, westward from Nantucket Shoals, about 242,000 pounds;[67] near coast of eastern Massachusetts, probably one fish;[68] coast of Maine, 400 pounds; Bay of Fundy (including both shores), 0; Nantucket grounds and Georges Bank region (including South Channel grounds), about 800,000 pounds; off west coast of Nova Scotia and on western part of Browns Bank, about 671,000 pounds; Nova Scotian coast and banks from eastern part of Browns to offing of Cape Canso, at the entrance to the Gut of Canso, about 219,000 pounds; outer (Atlantic) coast of Cape Breton,[69] about 2,059,000 pounds.

A few are harpooned on the Gulf of St. Lawrence shore of Cape Breton also; 600 pounds were reported there in 1936, 200 pounds (one fish?) in 1943, and 1,000 pounds (4 or 5 fish) in 1946. The only other definite report of swordfish in the Gulf of St. Lawrence that has come to our notice is from Bonne Bay, on the west coast of Newfoundland, where Wulff [70] saw one, and had a strike from it (he did not hook the fish). But some few are seen and harpooned on the Grand Banks, and also along the south coast of Newfoundland, most often along the stretch between Port au Basque, on Cabot Strait, and Hermitage Bay. Here, writes Wulff, they sometimes come so close inshore that they "have been harpooned from the small wharfs, from shore, and from dories in the almost landlocked harbors," which we have never known to happen in the Gulf of Maine.

---

[64] Rept. U. S. Comm. Fish. (1880) 1883, pp. 298-394, pls. 1-24.

[65] Rich, Proc. Portland Soc. Nat. Hist., vol. 4, Pt. 2, 1947, p. 71.

[66] See Rich (Proc. Portland Soc. Nat. Hist., vol. 4, Pt. 2, 1947, pp. 55-62) for a more detailed survey of the distribution of swordfish on Georges, Browns, and the Nova Scotian Banks.

[67] The weights given are dressed; live-weights would be about 1¼ times as great.

[68] Forty pounds reported, but this probably is an error, for it is not likely that a swordfish that small was taken there.

[69] Victoria, Cape Breton and Richmond Counties, Nova Scotia.

[70] Internat. Game Fish Assoc. Yearbook, 1943, p. 66.

Swordfish are summer fish on the North American coast like the tuna, and their presence (often reported) in the blue water[72] between the outer edge of the continent and the inner edge of the Gulf Stream proper, off southern New England and the Gulf of Maine, added to the fact that few are seen along the coast south of New York, makes it likely that they come in from offshore, direct.

They appear about simultaneously off New York, off Block Island, on Nantucket Shoals, and on Georges Bank, sometime between the 25th of May and the 20th of June, but seldom on the Scotian Banks until somewhat later, or in the inner parts of the Gulf of Maine before July. They are most numerous in July and August, and they vanish at the approach of cold weather. None have ever been reported east of Cape Cod after the first half of November, so far as we can learn (in 1875 one was taken on Georges in November in a snowstorm)[73] and most of them are gone by the last week in October, though some fish have been taken off New York and New Jersey in December and even in January. A case in point is that 13 were entangled in long lines set for tilefish in 95 to 125 fathoms off Long Island between December 20, 1921, and January 1, 1922.[74]

General report has it that the fish caught early in the season average not only thinner but considerably smaller than those caught later, a phenomenon still awaiting satisfactory explanation.[75]

Nearly all the swordfish that visit us weigh upwards of 50 to 60 pounds. One of 6 pounds 7 ounces, taken by the schooner *Anna* on Georges Bank, August 9, 1922 (now or formerly to be seen at the Boston Fish Bureau); a second of 7 pounds, taken by the schooner *Courtney* on a long line, on Browns Bank in 1931; a third of 7½ pounds, taken on a long line by the *Dacia* on Western Bank, early in September 1931; and a fourth 28¼ inches long to tip of lower jaw (its sword was broken off short), weighing 5¾ pounds, caught by the trawler *Winchester*, August 15, 1951, on the southeast part of Georges Bank in a haul which fished at 46 fatboms[76] are the smallest Gulf of Maine and Nova Scotian specimens of which we have heard. Goode[77] also reported a sword, only 3½ inches long, found sticking in the nostril of a mackerel shark caught at Gloucester, probably picked up somewhere off southern New England for this particular shark does not ordinarily range farther south than that.[78] In the Mediterranean, however, young fry as small as half a pound are often brought to market.

It is generally believed that swordfish come directly in from the open seas when they appear on the offshore banks in spring; a few to enter the Gulf of Maine, but the majority to work slowly eastward along the outer part of the continental shelf. When they depart in autumn it is to return to the open Atlantic, but they are never seen on their journey offshore, or southward; they simply drop out of sight as the tuna do.

No ripe fish, male or female, have ever been seen off our coast. The ovaries and spermaries of most of those examined[79] have shown no signs of approaching maturity; most of the fishermen, too, of whom we have inquired have assured us that they have never seen "spawn" in swordfish, though they had dressed hundreds. And while the captures of 3 fish with ovaries containing eggs in early stages of development have been reported, one brought into Provincetown in September 1909, a second with ovaries weighing 15 pounds, brought to New Bedford on June 25, 1922,[80] and a third of about 150 pounds killed off Marthas Vineyard in July 1924, events of this sort are so unusual that they cause wide comment.

Evidently the swordfish that summer off our coasts spawn during the part of the year when they are elsewhere; probably in the subtropical parts of the Atlantic basin, for Lütken [81] found swordfish fry as small as 10 mm. (evidently hatched only a short time previous) between the latitudes of 20° and 39° N. The fact that the fish are thin when they return to us in spring, but fatten during the summer stay, is further evidence that they are spent before they appear off our coasts.

Abundance.—Our only clue to the numbers of swordfish that visit our waters is the poundage

[72] This is often spoken of as the "Gulf Stream"; its more accurate name is the "slope water."

[73] Rich, Proc. Portland Soc. Nat. Hist., vol. 4, Part 2, 1947, p. 58.

[74] Townsend, Science, N. Ser., vol. 56, 1922, pp. 18–19.

[75] Rich (Proc. Portland Soc. Nat. Hist., vol. 4, Pt. 2, 1947, p. 43) is "inclined to think" that there are two or more "distinct year-schools" in our waters.

[76] Reported to us by George Kelley of the U. S. Fish and Wildlife Service. The specimen is in its collection in Woods Hole.

[77] Fish. Ind. U. S., Sect. 1, 1884, p. 348.

[78] See Rich (Proc. Portland Soc. Nat. Hist., vol. 4, Pt. 2, 1947, pp. 37–39) for additional records of small Gulf of Maine swordfish.

[79] Many have been opened with this point in mind; some by us.

[80] Townsend, Science, N. Ser., vol. 56, 1922, pp. 18–19.

[81] Spolia Atlantica, in Kong. Danske Vidensk. Selsk. Skrift, Ser. 5, Nat. Math. Sect., vol. 12, No. 6, 1880, pp. 444–445.

landed yearly. The smallest year's catch reported as landed at Portland, Gloucester, and Boston, within the period 1904 to 1929 was 883,000 pounds (in 1919), the largest 4,593,000 pounds (in 1929), the average about 2,000,000 pounds, or anywhere between 4,000 and 18,000 fish per year. And the landings in New England ports ran from 1,715,000 to 5,070,000 pounds during the decade 1930 to 1939 for southern New England and the Gulf of Maine. The interruption of swordfishing by German submarines and by the diversion of manpower was reflected in much lower landings during the first two years of the war, as was to be expected.[82] But swordfishing picked up again after the war, to landings of about 1,250,000 pounds for southern New England and the Gulf of Maine, including western Browns Bank, in 1944 (New England and Canadian landings combined), to about 2,850,000 pounds in 1945, to about 2,500,000 pounds in 1946,[83] and to something like 2,000,000 pounds in 1947.[84]

A catch of somewhere between 2 million and 3 million pounds would be a reasonable expectation for southern New England and the Gulf of Maine combined in average years. The catch off Cape Breton, eastern Nova Scotia, has run between 1½ and 3 million pounds of late years (1939–1946), averaging a little more than 2 million until in 1947, when it fell to about 770,000 pounds.[85] The Nova Scotian catches were not lessened by the submarine menace during the war years.

It is not known what percentage of the total number of swordfish off our coasts is represented by the catches. But, at least, they do not suggest that any extreme ups and downs took place prior to 1947.

Importance.—Appreciation of the swordfish as a food fish is of rather recent growth. Down to the middle of the past century it was unsalable in Boston and brought a very low price in New York, but of late years the demand would have taken care of a much greater supply than has been brought in. In 1919, the price to the fishermen averaged about 24 cents; in 1928, 22 cents; and 18 cents per pound in 1929 when a large catch was made. In 1945 it brought between 40 and 42 cents; and it rose to about 60 cents in 1946, but fell again to about 40 cents in 1947.

Practically all the swordfish brought in to market are harpooned; we have never heard of one caught in net or seine, nor is it likely that any net now in use would hold a large one. Swordfish have also been taken from time to time on hand lines and on long lines baited for cod or halibut with mackerel or other fish (p. 353). But the numbers caught in these ways have never been large enough to figure to any extent in the total catches, and are not likely to be.[86] Occasional swordfish have been caught by anglers of late years, on rod and reel, and sport fishermen would agree that a good-sized broadbill is the premier prize of the sea.

THE SPEARFISHES OR MARLINS AND THE SAILFISHES. FAMILY ISTIOPHORIDAE

The spearfishes and sailfishes, like the swordfish, have a sword formed by the prolongation of the snout and upper jaw. But their sword is rounded toward the tip, not flattened, and narrower than that of the swordfish. Their bodies, too, are closely clothed with narrow lanceolate scales, pointing rearward in general and embedded in the skin, either wholly or with their sharp tips projecting slightly (fig. 188), and their first dorsal fin is much longer, occupying the greater part of the back behind the nape, and it can be depressed into a groove along the back. They fall in two groups, sailfishes with very large, sail-like dorsal fin, and spearfishes or marlins with lower dorsal.

The sailfish (*Istiophorus americanus* Cuvier and Valenciennes 1831), so common in the warmer parts of the Atlantic, is included in the following Key because it has been taken at Woods Hole on several occasions, though not yet recorded from the Gulf of Maine. It is readily recognizable by the fact that the first dorsal fin is much higher than that of the marlins while the ventral fins of the sailfish are 2- or 3-rayed instead of being reduced to a single spine, as in the marlins. The two dorsal fins of the sailfish have usually been described as connected even in the adult. This,

[82] Landings were only about 545,000 pounds in Massachusetts and 7,000 pounds in Yarmouth County, Nova Scotia in 1942; about 479,000 pounds in Massachusetts and about 17,300 pounds in Yarmouth County in 1943.

[83] Most recent year for which the landings have been published for the Canadian coast of the Gulf of Maine and for the ports in New England.

[84] The Canadian catch statistics for 1947 have not reached us yet.

[85] Information from Dr. A. H. Leim of the Fisheries Research Board of Canada.

[86] Rich (Proc. Portland Soc. Nat. Hist., vol. 4, Pt, 2, 1947, pp. 67–82) gives an interesting account of the methods of the New England swordfishery.

in fact, is given as the chief distinction between it and the marlins by Goode[87] by Jordan and Evermann,[88] and by Boulenger.[89] But there is actually a considerable gap between the two fins in large specimens as Bean[90] remarks and as appears on Goode's own illustrations of a sailfish taken at Newport, and of a skeleton.

Two species of marlins, the blue and the white are known off the middle and north Atlantic Coasts of the United States. But it is not yet clear whether the enormous marlins, with violet cross-stripes on the sides, that are caught off the North Coast of Cuba[91] are simply very large blue marlin, a separate subspecies, or even a species. And the marlins of more southern waters still await critical study.

KEY TO SPEARFISHES OR MARLINS, AND SAILFISHES

Middle Atlantic and North Atlantic Coast of United States

1. First dorsal fin much higher than the body is deep and sail-like; ventrals of 3 rays each..Sailfish, p. 357
   The first dorsal fin is not higher than the body is deep; ventrals reduced to one spine each...... 2
2. Apex of first dorsal fin and tips of pectorals pointed. Blue marlin, p. 358
   Apex of first dorsal and tips of pectorals rounded.... White marlin, p. 360

Blue marlin *Makaira ampla* (Poey) 1860.

SKILLIGALEE

Jordan and Evermann, 1896–1900, p. 892, *Tetrapterus amplus*.[92]

Description.—The marlin parallels the swordfish in the prolongation of the bones of the upper jaw to form a sword. But that of the marlin is slender and rounded above, not broad and flattened as in the swordfish, also only about half as long relatively. The two fish differ widely, too, in the relative sizes of the first dorsal fin,[93] which extends along fully two-thirds of the length of the trunk from the nape backward in the marlin and is, futhermore, of characteristic falcate outline. But more important systematically, if less apparent, is the fact that the adult marlin has ventral fins which the swordfish lacks, though they are reduced, it is true, to one long spine each (actually 5 spines fused together).

Futhermore, the second dorsal fin and the second anal fin of the marlin are relatively larger, and the pectorals smaller than those of the swordfish, while there are two small longitudinal keels on either side of its caudal peduncle instead of one broad one; its body is more slender; and its head is relatively shorter. Careful examination would show that the spearfish is not naked but has small scales imbedded in the skin and that there are small teeth in its jaws and on the roof of its mouth.

The blue marlin differs from its close relative the white marlin in the shape of the apex of its first dorsal fin and of the tips of its pectorals, both of which are pointed; also in the much darker color of its sides and belly; and in the fact that it grows much larger.

The blue marlin is deepest abreast the pectorals, (about 6¼ times as long, not counting the caudal fin, as it is deep), tapering evenly to the caudal peduncle, and its upper jaw in front of the eye (including the sword) is about twice as long as the length of its head behind the eye.[94]

The first dorsal fin, with 47 to 48 stiff rays is separated from the second dorsal by a space about as long as the latter in one fish seen by us; by a shorter space in another. The first anal fin (2 spines and about 12 or 13 rays), situated below the rear part of the first dorsal, is triangular, its long first rays forming a sharp angle. The short second anal is similar to the second dorsal fin and originates a little in front of the latter. The ventrals stand below the pectorals; the caudal resembles that of the swordfish in its lunate outline.

Color.—Dark dull blue on the back and on the sides down about to the level of the eyes, washed with coppery reflections, also on the bill, with

[87] Rept. U. S. Comm. Fish., 1880, p. 296.

[88] Bull. 47, U. S. Nat. Mus., Pt. 1, 1896, p. 890.

[89] Cambridge Natural History, vol. 7, 1904, p. 680.

[90] Bull. New York State Mus., 60, Zool. 9, 1903, p. 404.

[91] Hemmingway (in Vesey-Fitzgerald and Lamonte, Game Fishes of the World, 1949, p. 158) reports these striped marlins weighing up to 1,250 pounds off northern Cuba.

[92] Jordan and Evermann in their general review of the giant mackerel-like fishes, tunnies, spearfishes, and swordfishes (Occ. Papers, Calif. Acad. of Sci., XII, p. 28, 1926) separate the spearfishes into two genera *Tetrapterus* with the front of the first dorsal fin little if any higher than the median part of the fin and *Makaira*, with the front part of the first dorsal higher than the median part of the dorsal.

[93] Very young marlins have only one continuous dorsal fin, but this separates later into two.

[94] See Shapiro (Amer. Mus. Novitat., No. 995, 1938) for a study of the changes in proportional dimensions that take place with growth, in the blue marlin.

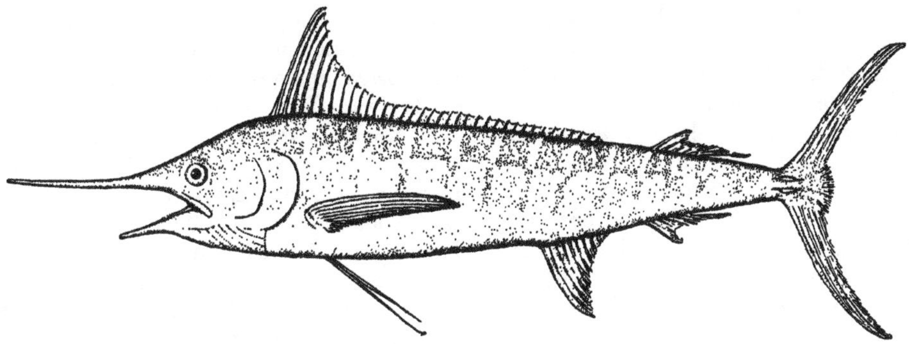

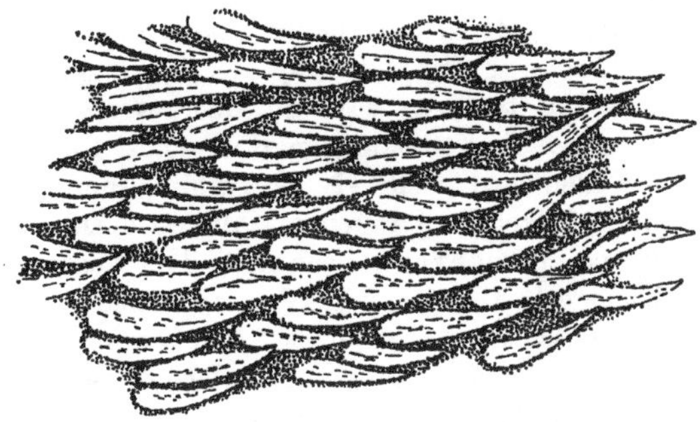

FIGURE 188.—Blue marlin (*Makaira ampla*). Drawing by Jessie Sawyer, based on Bahama and Marthas Vineyard specimens. Left, a piece of skin from the upper part of the side of the Marthas Vineyard specimen, with epidermis scraped off to show scales, about twice natural size.

rather abrupt transition to much paler gray-blue lower down the sides and on the lower surface, the belly being as dark as the lower part of the sides; the sides cross-marked with about 13 indistinct violet-blue stripes, about 1 to 1½ inches wide on a fish 8 feet long, showing pale against the dark blue of the upper parts of the body, but dark against the paler blue of the lower part of the sides. First and second dorsal fins, pectoral and ventral fins, and first anal fin dark rather vivid blue. Caudal fin of about the same color as upper part of trunk; second anal fin of same pale gray-blue as the belly.[95]

Size.—Blue marlins run fully as large as swordfish. Reports are current of fish of 1,000 pounds being harpooned; the rod and reel record is 742 pounds.[96] Many weighing more than 500 pounds are caught off the north coast of Cuba and on the Bahamas side of the Straits of Florida every year,[97] and one taken on the southern part of Browns Bank, weighed 575 pounds dressed, when landed, or about 700 pounds alive. A very large one may measure as much as 15 feet,[98] but the rod and reel record fish, mentioned above, was only 12 feet 10½ inches long. Another fish caught in the Bahamas weighed 650 pounds (not dressed), and measured 12 feet 1 inch; a third, of 621 pounds was 12 feet 3 inches long.[99]

General range—Warm parts of the northwestern Atlantic, straying northward to the Gulf of Maine. It has been reported near Sable Island, but the very small specimen in question may have been a white marlin (p. 360).

Occurrence in the Gulf of Maine.—This southern warm-water fish was reported from the South Channel, between Georges Bank and Nantucket Shoals, between 1877 and 1880, by the fishing schooner *Phoenix*. No other marlins that we can be sure were blues were reported within the limits of the Gulf of Maine until September 5, 1930, when a small one 6 feet 10 inches long,[1] was harpooned on the southern part of Browns Bank. And a very large one was caught in that same vicinity by the *Col. Lindbergh* the following July,

[95] Description based on a "blue" about 8 feet long from tip of bill to fork of tail, and weighing 169 pounds, fish taken near Bimini, Bahamas, June 1941, by R. W. Foster, mounted by the well-known fish taxidermist, H. Pfleuger of Miami, Fla., and now in the Museum of Comparative Zoology.

[96] Caught at Bimini, Bahamas, June 19, 1949, by Aksel Wichfeld.

[97] See Farrington (in Vesey-Fitzgerald and Lamonte, Game Fish of the World, 1949, p. 154) for a readable account of the blue marlin of Bahaman waters as a game fish.

[98] The blue marlin is said to reach 26 feet, but we think this much exaggerated.

[99] Reported to us by Frank Mather, of the Woods Hole Oceanographic Institution.

[1] This specimen is in the Museum of Comparative Zoology.

and brought into the Boston Fish Pier. A marlin about 5 feet long was taken on Georges Bank by the schooner *Ethel Merriam*, on August 5, 1925, but this may have been a white (p. 359).

Blue marlins are sighted at long intervals off Marthas Vineyard. And fishermen report them now and then along the southern edge of Georges (any very large marlin is a blue) but do not harpoon them, for they have no market value. They are game fish par excellence, and much sought after off Cuba and in the Bahaman side of the Straits of Florida. They also support a considerable commercial fishery off the north coast of Cuba.[2]

White marlin *Makaira albida* (Poey) 1860

Jordan and Evermann, 1896–1900, p. 892, *Tetrapterus imperator* (Bloch and Schneider) 1801, in part.

Description.—The white marlin differs from its larger relative the blue marlin in its rounded first dorsal and pectoral fins, in the pale color of the lower part of its sides, and in its white belly (p. 359); and in its smaller size. Few grow larger than 125 pounds; the rod and reel record stands at 161 pounds.[3] This fish was 8 feet 8 inches long.

General range.—Western North Atlantic; common in Cuban and Bahaman waters, and off southern Florida; north regularly in summer to the offing of Delaware Bay in abundance, and to southern New England waters in lesser numbers.

Occurrence in the Gulf of Maine.—So many white marlin come northward, as far as New York waters that about 500 were taken off Montauk, Long Island, on rod and reel during the 11 years 1925–1936, and more than 150 in 1935 alone.[4] And a few are caught off the southern Massachusetts Islands in most summers.

But their usual turning point is west of Nantucket. True, Farrington[5] speaks of "great quantities" of them as seen on Georges Bank; but we cannot find that any marlin caught there has been identified positively as a white, though one about 5 feet long taken on August 5, 1925 (p. 359) may perhaps have been one. The meager record suggests that they may stray oftener to outer Nova Scotian waters, for a 5-foot fish weighing 21 pounds, caught on Sable Island Bank, August 18, 1931, probably was a white marlin, while Farrington reports one harpooned off Glace Bay near Sidney, in 1945, and others sighted off Halifax that same year.

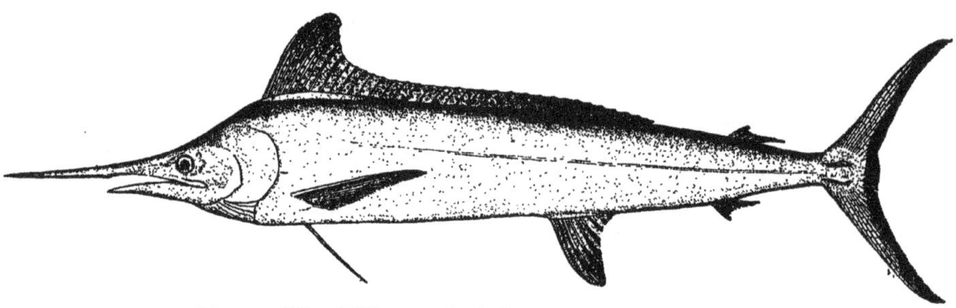

FIGURE 189.—White marlin (*Makaira albida*). From Goode.

THE DOLPHINS. FAMILY CORYPHAENIDAE

The dolphins (two species are known) are moderately slender and flattened sidewise, with slightly projecting lower jaw, a massive blunt head, a long, rather high dorsal fin without spines, extending from close behind the head to near the base of the caudal fin, an anal similar to the dorsal in shape but shorter, and a widely forked tail. They have small comb-like teeth in the jaws and on the roof of the mouth.

Common dolphin *Coryphaena hippurus* Linnaeus 1758

Jordan and Evermann, 1896–1900, p. 952.

Description.—The dolphin differs from related fishes in that its long tapering body is most massive and deepest close behind the head, and that its dorsal fin, originating over the gill cover, extends back nearly to the base of its deeply

[2] Farrington (in Vesey-Fitzgerald and Lamonte, Game Fishes of the World, 1949, p. 153) gives an interesting account of this fishery.

[3] One caught off Miami, Fla., Mar. 20, 1938, by L. F. Hooper.

[4] Information supplied by Frank Mather of the Woods Hole Oceanographic Institution.

[5] In Vesey-Fitzgerald and Lamonte, Game Fish of the World, 1949, p. 155.

FIGURE 190.—Dolphin (*Coryphaena hippurus*). From Goode. Drawing by H. L. Todd.

forked tail fin. These characters, with its sidewise flattened form, notably steep forehead, deeply forked tail, and large ventral fins, separate it at a glance from the few other Gulf of Maine fishes which have long dorsal fins with bodies that are deepest forward. Its anal fin, 26 to 30 rays, originating about midway of its body, is about half as long and half as high as the dorsal which has 55 to 65 rays. The lobes of its deeply forked tail are long and slender. Its moderately long ventrals and pectorals are situated the one below the other.

Color.—The dolphin is famous for its brilliant hues and for the vivid waves of color that flash across it when first taken from the water. Alive, in the sea, its sides are largely vivid blue, variously mottled and washed with gold; its tail largely golden yellow.

Size.—Maximum length about 6 feet.

Habits.—The dolphin, despite its blunt snout, is one of the swiftest of fishes. Voyagers on tropic seas often see them leaping in pursuit of small fry, or when pursued themselves by larger fishes. In sailing ship days dolphins were often caught by trolling from the stern. Offshore, they feed largely on flying fish; the Sandwich specimen mentioned later had some silversides in its stomach.

General range.—Cosmopolitan in warm seas; northward along our Atlantic Coast to southern New England, where it is rare inshore, occasionally straying as far as the outer coast of Nova Scotia.

Occurrence in the Gulf of Maine.—A dolphin about 3½ feet long (now in the collection of the Boston Society of Natural History) and weighing 23 pounds, taken 60 miles south-southwest of Cape Sable, in the deep gully between Browns and Georges Banks by the trawler *Natalie Hammond*, August 15, 1930, was the first record for the Gulf of Maine; a second was taken in a trap at North Truro on Cape Cod Bay, in August 1949[6] (a season when many were taken off Marthas Vineyard); a third at Sandwich, on the southern shore of Cape Cod Bay in mid-July 1951.[7]

THE SEA BREAMS OR POMFRETS. FAMILY BRAMIDAE

The sea breams are usually considered the most nearly related to the dolphins. But they rather suggest the butterfishes (Family Stomateidae, p. 363) in general appearance, with single, long, falcate soft-rayed dorsal fin; anal similar to the dorsal; lunate tails, very small ventrals; and deep, sidewise flattened bodies. They are to be expected only as strays in the Gulf of Maine.

Johnson's Sea Bream *Taractes princeps* (Johnson) 1863

Johnson, Proc. Zool. Soc. London, 1863, p. 36.

Description.—This sea bream is unique among Gulf of Maine fishes in its general appearance.

[6] Reported by Schuck, Copeia 1951, p. 171.

[7] We saw this specimen in the collection of the U. S. Fish and Wildlife Service at Woods Hole.

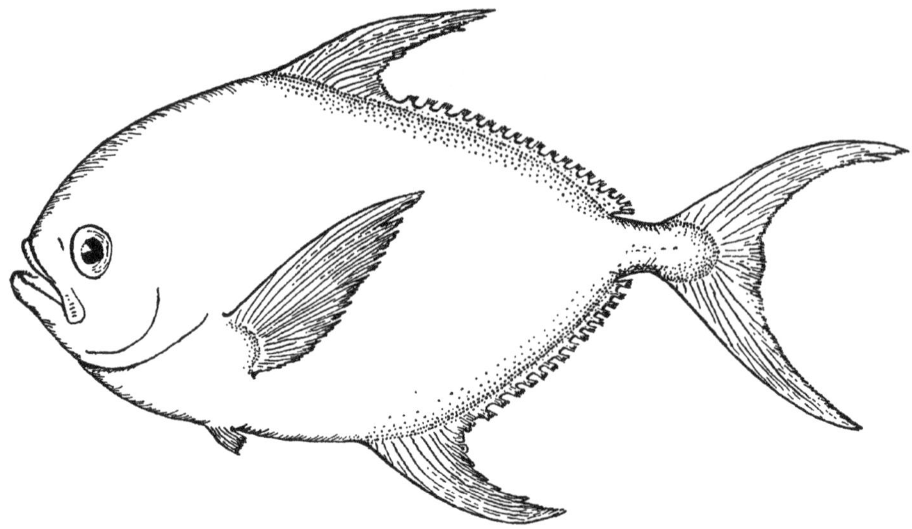

FIGURE 191.—Johnsons sea bream (*Taractes princeps*), Browns Bank. After Bigelow and Schroeder.

In the adult the body is massive, flattened sidewise, about half as deep as it is long to the base of the tail fin. The dorsal and anal fins are long, scythe-shaped in front, each followed by a row of low rays that are detached along their outer parts, but are joined by scaly skin along their bases, giving the fins the outline shown in figure 191. The tail fin is deeply lunate; the long pectorals originate a little in advance of the dorsal, and the very small ventral fins a little in front of the pectorals. The eyes are large, oval, with vertical diameter longer than the horizontal diameter. There is no lateral line. A striking character is that the scales which clothe the vertical fins and the body (about 43 scales along the median longitudinal row) vary greatly in size, being largest along the sides, smallest on the back, breast, and fins. They vary also in shape, their exposed margins being either concave, convex, notched, or straight.

This species is separated from *Brama raii* Bloch 1781,[8] the only one of its relatives yet recorded from our North Atlantic coast, by its larger scales (*Brama raii* has 80 or more in the median longitudinal row) and by the fact that its ventral fins originate slightly, but distinctly, in front of the pectorals.

Color.—The body and head of a specimen, three days after death, were blackish, tinged with salmon on the gill covers and along the sides; the dorsal and anal fins were dusky, with the free ends of the short rays pale; the caudal was black but with its concave margin white; the pectorals were gray.

Size.—Maximum length about 3 feet.

General range.—Known only from Madeira in the eastern Atlantic, and from Browns Bank in the western.

Occurrence in the Gulf of Maine.—This fish is mentioned here on the basis of one specimen, about 33 inches long, caught on a long line on Browns Bank, off Cape Sable, in January 1928, by the schooner *Wanderer*. A detailed account and comparison with allied species is given by Bigelow and Schroeder.[9]

It seems certain that *Taractes* is very rare in American waters, at least in the depths in which commercial fishermen operate, for so conspicuous a fish would almost certainly be reported, if caught. Nothing is known of its habits except that it seems to be common around Madeira in deep water.

[8] *Brama raii* has been taken at Woods Hole and on the Grand Banks.

[9] Bull. Mus. Comp. Zool., vol. 69, 1929, pp. 39–50, 1 pl.

THE BUTTERFISHES. FAMILY STROMATEIDAE

The members of this family are deep bodied and very much flattened sidewise, with one long dorsal fin that is soft rayed except for a few short weak spines at its forward end, an anal fin of corresponding size and shape, a deeply forked caudal fin, a blunt nose, and a small mouth. The two species that occur on the east coast of North America lack ventral fins, but the extremity of the pelvic bone projects through the skin as a spine but this is so short that it is likely to be overlooked unless felt for.

Two species occur in the Gulf of Maine: one (the butterfish) a common summer visitor, the other (the harvestfish) a rare stray from the south.

KEY TO GULF OF MAINE BUTTERFISHES

1. The forward one-fourth of the anal fin is only about 2 or 3 times as high as the rear portion of the fin. The margins of the anal and dorsal fins are only slightly concave in outline.......Butterfish, p. 363

   The forward one-fourth of the anal fin is at least seven times as high as the rear portion of the fin. The anal and dorsal fins are both very deeply concave in outline...................Harvestfish, p. 368

Butterfish *Poronotus triacanthus* (Peck) 1800

DOLLARFISH; SHINER; SKIPJACK; SHEEPSHEAD; HARVESTFISH

Jordan and Evermann, 1896–1900, p. 967, as *Rhombus triacanthus*.

Description.—The most distinctive characters of the butterfish are its very thin deep body, like a flounder on edge; the fish is only about twice as long as it is deep to the base of its tail fin (the only common Gulf of Maine species of this shape), combined with a single, long, soft-rayed dorsal fin, an anal fin almost equally long, and a deeply forked tail, but no ventral fins. The absence of ventral fins separates it from the pompanos; the dorsal without obvious spines from the scup (p. 411) and John Dory (p. 297); the lack of detached dorsal spines from the triggerfishes, which are, furthermore, very different in general aspect (p. 520). And it is easily distinguishable from its relative, the harvestfish (p. 194), which is rare in northern waters, by its much lower dorsal and anal fin (compare fig. 192 with fig. 194). The dorsal fin (about 45 rays) originates close behind the axils of the pectorals and tapers at first abruptly and then gradually backward, while the anal (about 40 rays) narrows evenly from front to rear. There is a forward-pointing spine close in front of the dorsal fin, so short as hardly to be visible though it can be felt; also 3 very short spines in front of the anal, almost wholly embedded in the skin, the first of which points forward. Both the dorsal fin and the anal extend rearward almost to the base of the caudal fin.

Distinctive, also, are the long pointed pectoral fins, short head, blunt snout, small mouth, weak teeth, and the short and slender caudal peduncle, which does not have longitudinal keels. The scales are very small, and are easily detached when the fish is handled, and there is a row of very conspicuous mucous pores below the forward half of the dorsal fin.

Color.—Leaden bluish above, pale on the sides, with numerous irregular dark spots which fade after death. The belly is silvery.

Size.—The largest are about 12 inches long; the general run are about 6 to 9 inches long. The weight runs about 1¾ ounces at 6 inches, 4 to 4½ ounces at 8 inches; about 1 pound at 11 inches (if fat). The largest weigh about 1¼ pounds.

Habits.—Astonishingly little is known of the manner of life of the butterfish considering how familiar and valuable it is. As a rule they travel in small bands or loose schools; and draggers report catching several times as many by night as by day, suggesting that they are active enough to dodge a trawl, except during the hours of darkness. They often come close inshore, into sheltered bays and estuaries, hence their frequent capture in pound nets. And it shows so decided a preference for sandy bottoms rather than for rocky or muddy, that few are taken in traps on muddy ground while other traps along the sandy beach nearby may yield considerable numbers. General experience is that the butterfish keeps chiefly near the surface during its stay near the coast, and schools are often to be seen. At Cohasset (on the south side of Massachusetts Bay), for instance, schools of butterfish, fifty to a few hundred, are often to be seen where the flats are covered by only 4 or 5 feet

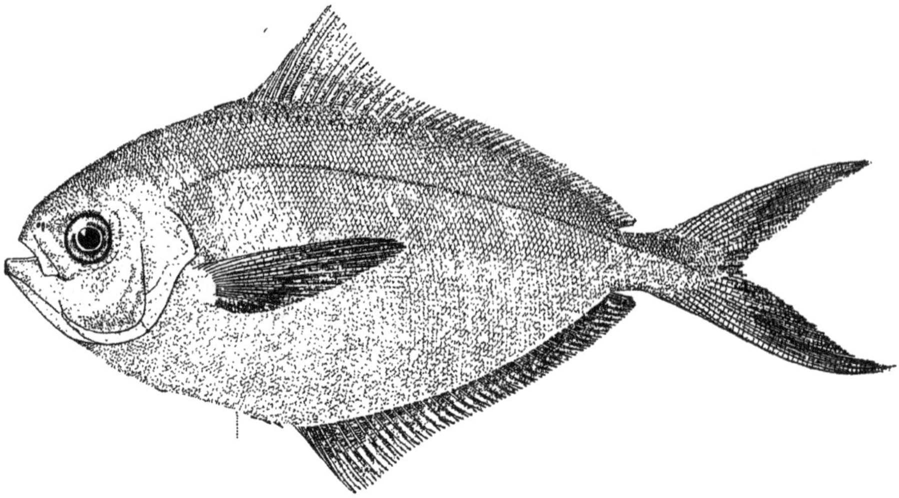

FIGURE 192.—Butterfish (*Poronotus triacanthus*), New Jersey. From Goode. Drawing by H. L. Todd.

of water. Though definite evidence is lacking, we believe butterfish seldom descend deeper than 15 to 30 fathoms during the summer, and that most of the fish caught by the otter trawlers on the Nantucket grounds and on Georges Bank in summer are picked up by the trawl on its way up or down, not while dragging on bottom. In fact, mackerel fishermen often take a few butterfish on Georges in their purse seines. But such evidence as is at hand is to the effect that they spend the winter and early spring near bottom, and in depths down to 100–115 fathoms (p. 367).

Food.—The butterfish feeds on small fish, squid, Crustacea such as amphipods and shrimp, and annelid worms. And ctenophores have been found in butterfish stomachs at Woods Hole, though these watery objects are not a regular item in its diet.

Breeding habits.—Butterfish begin spawning in the Gulf of Maine in June, soon after their arrival. The height of the reproductive season is in July and their eggs have been taken throughout August. Observations at Woods Hole suggest that butterfish spawn some few miles out at sea, returning to the coastwise waters when they are spent.[10] We have taken its eggs in our tow nets at several stations in Massachusetts Bay, and it would not be astonishing to find them anywhere off the New England and western Nova Scotian coasts or on the Scotian side of the Bay of Fundy, Huntsman having found large spawning individuals in St. Mary Bay in July. But despite the considerable number of butterfish eggs that are produced in the Gulf of Maine, we doubt whether the latter is a favorable nursery for this fish, for we have taken its larvae only twice there (off Cape Cod on August 16 and on Georges Bank on July 23, 1916) a total of only 3 specimens, 5 to 30 mm. long, although we have made hundreds of hauls widely distributed inshore as well as offshore at the season when they might be expected. Neither have young butterfish been reported from the Bay of Fundy. Butterfish fry are very plentiful, however, along the shores of southern New England.

The eggs are buoyant, transparent, spherical, 0.7 to 0.8 mm. in diameter, usually with a single oil globule of about 0.17 to 0.2 mm. In newly spawned eggs, however, there may be two globules, which coalesce as development advances.[11] At a temperature of 65° F. (about the summer state of the surface of Massachusetts Bay) incubation occupies less than 48 hours. And it is probable that development can only proceed in comparatively warm water, though the lower temperature limit to successful reproduction is not known. The larvae are about 2 mm. long at hatching and they are characterized shortly after by their short deep form, by their 30 muscle-segments, and by the row of black spots along the ventral edge in the

[10] Kuntz and Radcliffe, Bull. U. S. Bur. Fish., vol. 35, 1918, p. 112.

[11] A large series of butterfish eggs artificially fertilized at the Gloucester hatchery have been available for comparison with the pelagic eggs taken in the tow nets.

post anal region.[12] The dorsal, anal, and caudal fin rays are visible in larvae of 6 mm., when the body has already begun to assume the deep thin form so characteristic of the adult butterfish. At a length of 15 mm. the caudal fin is deeply forked, the dorsal and anal fins are formed, and the little fish resembles the adults sufficiently for ready identification.

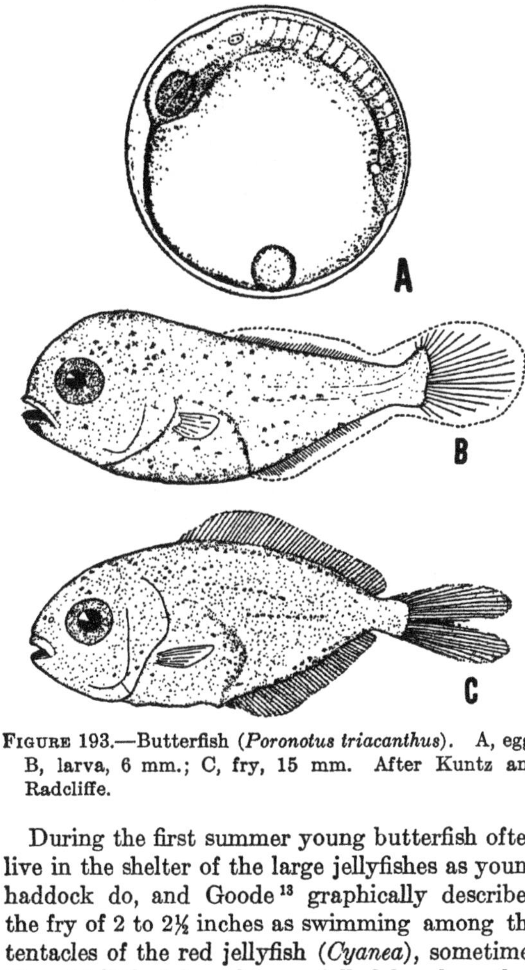

FIGURE 193.—Butterfish (*Poronotus triacanthus*). A, egg; B, larva, 6 mm.; C, fry, 15 mm. After Kuntz and Radcliffe.

During the first summer young butterfish often live in the shelter of the large jellyfishes as young haddock do, and Goode[13] graphically described the fry of 2 to 2½ inches as swimming among the tentacles of the red jellyfish (*Cyanea*), sometimes 10 or 15 little fish under one jellyfish, where they find protection from larger fish, but to which they sometimes fall prey. This association, however, is not essential to their welfare, for fry are often seen living independently at the surface, particularly in sheltered bays west and south of Cape Cod. On one occasion in late August 1925, on Nantucket Shoals, we observed numbers of young butterfish 1–1½ inches (26 to 39 mm.) long swimming free in the upper stratum of water. And we have seldom found young butterfish with the many *Cyanea* that we have captured in the Gulf of Maine.

It seems that the fry hatched earliest in the season grow to a length of 3 to 4 inches by autumn, great numbers of that size having been taken in Rhode Island waters in October. But late-hatched fish probably are not more than 2 to 3 inches long at the beginning of winter, and they can grow little during the cold season, for little fish of 3 to 5 inches are seen again in the spring. A series of measurements made by Welsh at Atlantic City, N. J., in August 1921, throws some light on the subsequent rate of growth. The fish fell into two groups: one ranging from 4 to 5¼ inches (averaging about 4¾) and the other from 7½ to 10½ inches. Very likely those of the first group (which were much the more numerous) were in their second summer, for Hildebrand and Schroeder[14] record a growth of from 4 inches to 5¼ inches from May to October in Chesapeake Bay; those of the second size group were in their third summer, some perhaps in their fourth. It is probable that the butterfish matures when 2 years old, and upward of 7 inches long.

General range.—Atlantic coast of North America from the offing of South Carolina and from coastal North Carolina waters to the outer coast of Nova Scotia and Cape Breton; northward as a stray to the Gulf of St. Lawrence[15] and to the south and east coasts of Newfoundland;[16] southward to Florida in deep water.

Occurrence in the Gulf of Maine.—This is a regular summer visitor to the Gulf of Maine, locally abundant along the shores of Massachusetts, less common along the coast of Maine. Butterfish are common also in some years along the Nova Scotian coast of the Gulf; great numbers were caught in

[12] Information furnished by O. E. Sette. The illustrations of larvae 2.1 mm. and 3.4 mm. long credited by Kuntz and Radcliffe (Bull. U. S. Bur. Fish., vol. 35, 1918, figs. 63 and 64) to the butterfish and reproduced in the previous edition of this book (Bigelow and Welsh, Bull. U. S. Bur. Fish. vol. 40, Pt. 1, 1925, fig. 116, c and d) have since been proved to belong to one of the hakes (*Urophycis*).

[13] American Fishes, 1888, p. 222.

[14] Bull. U. S. Bur. Fish. vol. 43, Pt. 1, 1928, p. 214.

[15] Hoar (Copeia, 1937, p. 238) records two large ones from Margaree Harbor on the Gulf of St. Lawrence shore of Cape Breton, and cites an earlier record for the coast of Quebec.

[16] It is reported from Rose Blanche on the south coast of Newfoundland, and from Bulls Bay and Ferryland on the east coast of the Avalon Peninsula (Rep. Newfoundland Fish. Res. Commission, vol. 1, No. 4, 1932, p. 108, and vol. 2, No. 1, 1933, p. 125.

St. Mary's Bay, for example, in 1910–1913 and again in 1938, though few were taken during the intervening years.[17] But they appear only irregularly and in small numbers on the New Brunswick shore of the Bay of Fundy, though they have been taken repeatedly in Passamaquoddy Bay.

The diminution in the numbers of butterfish, following from south and west to east and north around the coast line of the Gulf may be illustrated by catches for 1938 a fairly representative year [18] when catches in pound nets and floating traps around the shores of Barnstable, Plymouth, and Essex Counties, plus those landed in Boston and Gloucester by seiners and trawlers fishing offshore, amounted to 943,500 pounds, whereas only about 18,000 pounds were reported from the entire coast from the Massachusetts line to and including the region of Casco Bay, and none at all from farther east than that along the coast of Maine.

Butterfish also appear in the Nantucket Shoals region and on Georges Bank in summer, often in good numbers. About 1,000 fish, for example, were caught on Georges during one trawling trip in 1913; and otter trawlers accounted for nearly two-thirds of the total landings for Massachusetts in 1938, about one-half of those for 1945, most of which probably came from these offshore grounds. We have heard no rumor of them on Browns Bank but doubtless they occur there, for "fair quantities" usually visit Halifax Harbor in summer and autumn, according to McKenzie,[19] in fact, he cites one instance when about 1,500 of them were taken from two traps there in one day. And they are said to be common eastward as far as Canso.[20] But this appears to be the normal limit to their range, for strays, only, have been taken in the Gulf of St. Lawrence (p. 365), or on the Newfoundland coast (p. 365).

Season.—Butterfish are warm season fish along our coasts; we refer of course to the temperature of the water, not to that of the air. They may appear off Rhode Island by the last half of April and about Woods Hole by the middle of May, though they are not plentiful in the Woods Hole region until in June. And it is likely that these early comers move in across the shelf from offshore, rather than that they have followed along the coast, for from April 8 to 12, 1953, the *Eugene H* trawled 22,000 pounds of butterfish, close to bottom, in 85 fathoms south to Martha's Vineyard, and in 1950 the *Albatross III* trawled 10 to 723 butterfish per haul, May 11 to 18, along the 40–80 fathom zone off southern New England, where small commercial catches were also being made at the time. During the season of 1913 [21] the first butterfish were reported on Georges Bank June 5 to 8. But it is not until the end of that month or early in July that they are plentiful anywhere north of the elbow of Cape Cod. The earliest catches, for example, in one set of traps off North Truro, on Cape Cod Bay, were not made until June 26–28th in 1947, or until July 29th in 1948, but on May 29, 1951. From that time on there are butterfish in the inner parts of the Gulf throughout the late summer and autumn, also on Georges Bank.

The following tabulation of the catches made in one set of 8 traps at North Truro, on the eastern shore of Cape Cod Bay,[22] suggests that butterfish are likely to be the most numerous there in August, at least in good years, and rather more numerous in September and in October than in July. But they are exceedingly irregular and unpredictable in their appearances and their disappearances. Thus the traps just mentioned yielded butterfish on only one day in July, 2 days in August, 3 days in September, and 3 days in October in the years 1948 and 1949 combined, though catches as great as 2,856 to 7,490 pounds were made on three of these occasions. The approximate catches, in pounds, for the years 1946 through 1950 follow:

	Maximum	Minimum	Average	Total
July	5, 900	0	1, 760	8, 810
August	53, 101	0	11, 450	57, 260
September	15, 100	90	5, 850	29, 250
October	26, 440	120	8, 425	42, 130

In some years the peak for this locality may not come until October, as in 1947, when the catch by this set of traps was between five times and six times as great during that month (about 14,500 pounds) as during the next most productive month (July, about 2,300 pounds; August, about 2,500 pounds). Similarly, in 1950 the October catch

[17] McKenzie, Proc. Nova Scotian Inst. Sci., vol. 20. 1939, p. 14.

[18] This is the most recent year for which butterfish have been mentioned in the statistical breakdown by counties for Maine.

[19] Proc. Nova Scotian Inst. Sci., vol. 20, 1939, p. 17.

[20] Cornish, Contributions to Canadian Biology (1902–5) 1907, p. 85.

[21] This is the only year for which lists are available of the number of fish of all species taken on Georges Bank by certain trawlers.

[22] Information supplied by the Pond Village Cold Storage Co. of North Truro, Mass.

of these traps was about 26,400 pounds following a peak in August (about 53,000 pounds). And they linger in numbers until well into November in the Cape Cod Bay region in some years; also on Georges Bank. Thus four or five traps at Provincetown yielded some 30,000 pounds during that month in 1915, while 2 traps at Barnstable, on the southern shore of Cape Cod Bay took 4,275 pounds of butterfish on November 17, in 1950.[23]

They may linger equally late into the season along the outer Nova Scotian coast in some years, as in 1938, when two traps at Halifax yielded about 1,500 fish on November 12th.[24] They have been caught on Georges Bank until the end of that month; and in 1928 several hundred pounds were reported from Nantucket Shoals as late as the last week in December.[25] But they all vanish from the coast by the end of December at latest, and usually earlier than that, not only from our Gulf but along the more southerly part of their range as well.

It seems that the southern contingents simply move out to the outer edge of the continent into deeper and warmer water to winter, as the mackerel do also, for they are often caught by otter trawlers working out on the shelf between the latitudes of Chesapeake Bay and of Cape Hatteras in winter. The *Albatross III* trawled from 1 to 202 butterfish at a number of localities at depths of about 20 fathoms to at least 115 fathoms, between the offings of Charleston, S. C., and of Cape Hatteras in January and February of 1950.

The case is not so clear for those that summer off southern New England and farther north and east. Butterfish, it is true, have been trawled in February near the 90-fathom line abreast of the eastern part of Long Island, N. Y.;[26] also late in March on the southwestern slope of Georges Bank (where the dragger *Eugene H* had the unusually large catch of about 15,000 pounds in 1951 in the last week of that month) and in April and in May off southern New England (p. 366). These, however, may not have wintered in the vicinity, but may have been following along the outer part of the shelf northward, before turning shoreward toward their summer homes.

[23] Information from John E. Vettorino, who operates these traps.

[24] McKenzie, Proc. Nova Scotian Inst. Sci., vol. 20, 1939, p. 17.

[25] See Hildebrand and Schroeder, Bull. U. S. Bur. Fish., vol. 43, 1928, p. 215, for details as to their seasonal occurence in Chesapeake Bay.

[26] Three fish taken by *Albatross II*, February 27, 1929.

Abundance.—During the period 1928 to 1947[27] the reported catch of butterfish for Massachusetts ranged between 279,000 pounds and 2,250,000 pounds. Low points were in 1928 (about 580,000 pounds) and in 1946 (about 279,000 pounds); high, in 1932 (about 1,479,000 pounds), and during the period 1937–1940 (from about 1,226,000 pounds to about 2,250,000 pounds). And while this includes landings for the southern shore of the State as well as for the Gulf of Maine shore, the fluctuations that are indicated from year to year probably were paralleled north of Cape Cod. But the catch may be poor at any particular locality even in a good year, or vice versa. Thus the North Truro traps mentioned (p. 366) took only 1,230 pounds of butterfish in 1948, though this was a better-than average year for the Massachusetts coast as a whole.[28]

If the fish caught average about one-half pound each, the Massachusetts fishery may thus be expected to take somewhere between 560,000 and 4½ million individual fish. But it is not known what proportion this may be of the total population of butterfish in the Gulf of Maine.

Importance.—This is one of our best table fish, fat, oily, and of delicious flavor. Experience with many fresh from the net as well as on the table proves the old tale to be a myth that butterfish have a peculiar odor. However, they were often used to enrich land in planting during the first half of the past century, and appreciation of the fact that they are too good for this use is of recent growth. Even today the demand for butterfish in Boston is uncertain and the price widely variable. As late as 1938, 1,500 fish taken in traps at Halifax, Nova Scotia, were dumped for want of a market.[29]

The commercial catch is made mostly in pound nets, floating traps, purse seines, and otter trawls, and it was thought of old that they would never take a hook. But anglers have recently discovered that butterfish will sometimes bite a very small hook greedily, if baited with a bit of clam or with a small piece of a sea worm (*Nereis*). And 1,100 pounds were reported in 1945 as caught along the Massachusetts coast on hand lines.

[27] Statistics are not available for 1929, 1934, 1936, or 1941.

[28] Massachusetts catch, about 676,000 pounds.

[29] McKenzie, Proc. Nova Scotian Inst. Sci., vol. 20, 1939, p. 17.

Harvestfish *Peprilus alepidotus* (Linnaeus) 1766

STARFISH

Jordan and Evermann, 1896–1900, p. 996, as *Rhombus paru* (Linnaeus) in part.

Description.—The body of the harvestfish (not including the caudal peduncle) is almost as deep as it is long, and ovate in outline; its nose is rounded, mouth very small, and head very short. The outlines of its dorsal and anal fins afford the readiest field mark to separate it from its relative, the butterfish; both of these being very high and falcate in front, and continuing nearly straight, thence rearward (compare fig. 194 with fig. 192). The mucous pores, conspicuous in the butterfish, are lacking in the harvestfish. There is also a color difference between the two, the harvestfish being greenish silvery above, silvery sometimes tinged with yellow on its sides and belly, while the fins of some specimens are slightly dusky or yellowish. In all other respects (including size) it closely resembles the butterfish.

General range.—From Florida northward along the middle Atlantic Coast of North America; rarely straying north to Cape Cod, Mass., and to Cape Elizabeth, Maine; represented by a close ally (*Peprilus paru* Linnaeus 1758) in West Indian-Brazilian waters.[30]

Occurrence in the Gulf of Maine.—This fish, although abundant to the southward, rarely strays as far north as the outer Cape Cod coast. A specimen taken at Monomoy Point by Dr. W. C. Kendall in 1896; 5 or 6 caught in floating traps at Richmond Island, off Cape Elizabeth, Maine, in July 1929;[31] one from the Damariscotta River, Maine, in August 1933;[32] and one taken at Race Point at the tip of Cape Cod, in October 1949,[33] are the only Gulf of Maine records of which we know.

[30] See Meek and Hildebrand (Field Mus. Nat. Hist., Zool. series, vol. 15, Pt. 2, p. 411, 1925) for discussion.

[31] Reported to us by the late Walter H. Rich of the U. S. Bur. Fish.

[32] MacCoy, Bull. 69, Boston Soc. Nat. Hist., 1933, p. 9.

[33] Reported to us by Edgar Arnold.

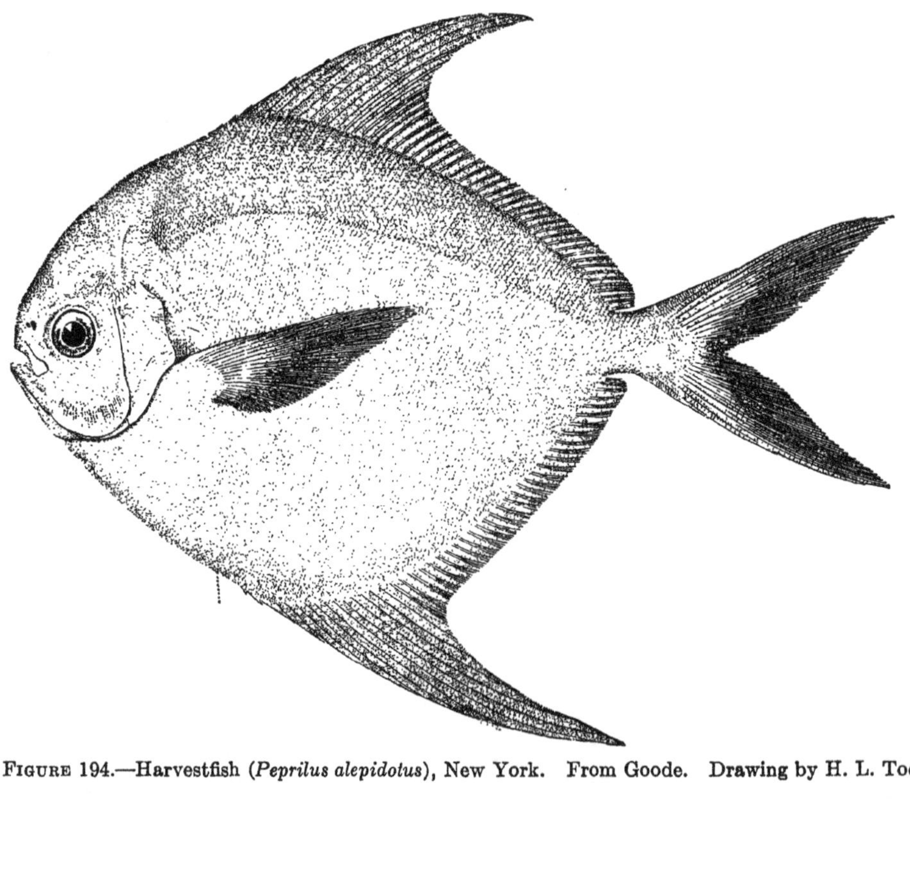

FIGURE 194.—Harvestfish (*Peprilus alepidotus*), New York. From Goode. Drawing by H. L. Todd.

THE RUDDERFISHES. FAMILY CENTROLOPHIDAE

The closest affinities of the rudderfishes are with the mackerel-like fishes. They have moderately stout bodies, short blunt snouts with convex profiles, and a moderately deep caudal peduncle without longitudinal keels. The single dorsal fin extends from over the pectorals to the caudal peduncle; the front part of the dorsal is spiny, either reduced to a few flexible spines covered over by the skin so that it is hard to find them, or represented by several detached spines so short that they might be overlooked, and preceding the much longer soft-rayed part of the dorsal. The tail fin is only slightly emarginate; the anal fin is similar to the dorsal in shape but much shorter; the ventrals are below the pectorals, and are smaller than the latter. The mouth is small, with small teeth in the jaws. Only two species are known off the Atlantic coast of the United States.

KEY TO GULF OF MAINE SPECIES

1. The single dorsal fin is preceded by 6–8 short detached spines; the sides of the head are scaly........................Barrelfish, p. 369
2. The dorsal fin is not preceded by any detached spines; there are no scales on the sides of the head........................Black ruff, p. 370

Barrelfish *Palinurichthys perciformis* (Mitchill) 1818

LOGFISH; RUDDERFISH; BLACK PILOT

Jordan and Evermann, 1896–1900, p. 964.

Description.—The reduction of the spiny portion of the dorsal fin of the barrelfish to 6 to 8 short detached spines, with very small triangular fin membranes, closely followed by a long soft-rayed dorsal fin, marks the barrelfish from all other Gulf of Maine fishes, except for certain of the pompano tribe. The caudal fin of the barrelfish is only slightly emarginate instead of deeply forked and its caudal peduncle moderately stout and without keels instead of very slender. It suggest a tautog remotely in general appearance, especially in its rather stout body (about two-fifths as deep as long, not including the caudal fin), very bluntly rounded nose, convex forehead, and small mouth. But its rudimentary spiny dorsal fin and forked caudal fin are ready field marks to distinguish it. The soft dorsal fin (20 to 22 rays) arises about mid-way from tip of snout toward base of caudal fin; the anal (16 or 17 rays) somewhat farther back. Both these fins are moderately high and they taper slightly from front to rear. The anal is preceded by three short spines so nearly imbedded in the skin as to be hardly visible. Both the ventrals and the pectorals are large with rounded tips. The top of the head is scaleless but the sides of its head and the body are clothed with small rounded scales.

The presence of the dorsal fin-spines and the scaliness of the sides of its head distinguish it from its close relative the black ruff (fig. 196).

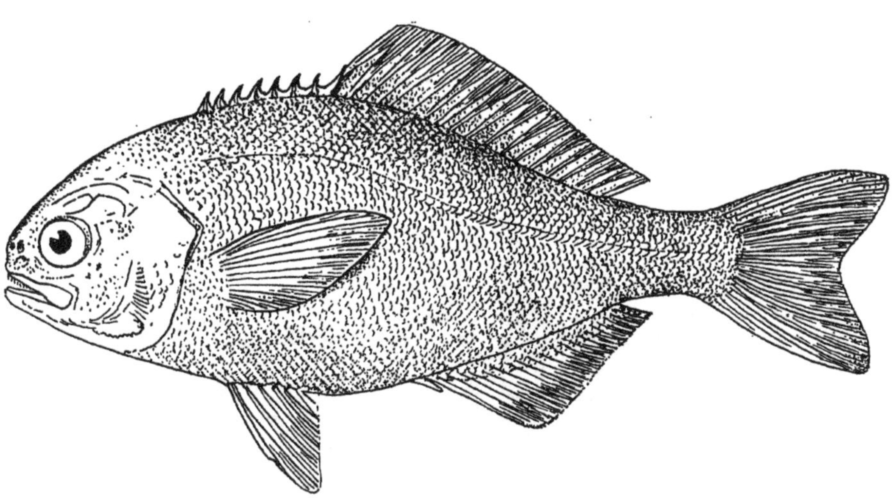

FIGURE 195.—Barrelfish (*Palinurichthys perciformis*). After DeKay.

Color.—Described as varying from blackish to green in life, and as either as dark below as above, or paling to bluish white on the belly, the latter variously mottled with darker dots and bars. It is said to change color to accord with its surroundings.

Size.—Maximum length 12 to 14 inches and about 1¼ pounds in weight, but most of those seen are smaller.

Habits.—The barrelfish owes its common name to its habit of congregating about floating spars and planks or any drifting wreckage, or inside of barrels or boxes, where it is easy to catch one in a dip net. Off southern New England they are often found under gulfweed, or under any other raft of drifting seaweed or eel grass (*Zostera*). And they sometimes gather about slow-moving vessels. Merriman [34] thinks its proper home is in the mid-depths offshore, but this is a question for the future.

It feeds on the sundry small crustaceans, barnacles, hydroids, young squids, small mollusks, and salpae, which it finds near or attached to its floating homes; on ctenophores; likewise on fish fry, the diet lists of specimens taken at Woods Hole including herring, mackerel, menhaden, launce, scup, and silversides.[35] Sometimes they contain seaweed, but we suspect that this is eaten for the animals attached to it, and not from a vegetarian taste.

Nothing is known of its breeding habits.

General range.—Atlantic Coast of North America, Cape Hatteras to outer Nova Scotia;[36] most plentiful south of Cape Cod. Probably it is oceanic, as Merriman [37] suggests, and more widely distributed than the foregoing would suggest, for one was found in a drifting packing case off Penzance Harbor, Cornwall.

Occurrence in the Gulf of Maine.—The barrelfish is caught in some numbers in the traps near Woods Hole and to the westward, or is found drifting under mats of seaweed. They were unusually plentiful in Vineyard Sound, for example, in 1920.[38] But it is so rare a fish within the Gulf of Maine that we have never seen it there,[39] nor did Doctor Kendall find it on his various collecting trips along the Maine coast. In fact, the only published Gulf of Maine records for it that we have been able to find are one from Boston Harbor; one from Salem; one from Annisquam; one from Gloucester;[40] and one vaguely described as brought in from the fishing banks off the coast of Maine. We can now add one taken on the northern edge of Georges Bank by the trawler *Squall* on September 10, 1947.[41]

Black Ruff *Centrolophus niger* (Gmelin) 1789

Jordan and Evermann, 1896–1900, p. 963.

Description.—The black ruff resembles the pilot fish (p. 372) in its general body form more than it does its closer relative the barrel fish (p. 369), being moderately slender (a little more than ¼ as deep as it is long to base of tail fin), with very blunt snout, strongly convex forehead, and small mouth. But its body (about 2½ times as

[34] Trans. Connecticut Acad. Arts Sci., vol. 36, 1945, pp. 842–843.

[35] Notes by Vinal Edwards.

[36] According to Vladykov and McKenzie (Proc Nova Scotian Inst. Sci., vol. 19, 1935, p. 87) occasional specimens are caught off outer Nova Scotia in most summers. Recent records there are of one at Halifax, October 1924, and of another there September 1927 (Vladykov, Proc. Nova Scotian Inst. Sci., vol. 19, 1935, p. 6).

[37] Trans. Connecticut Acad. Arts Sci., vol. 36, 1945, pp. 842–843.

[38] Smith, Copeia, 1921, No. 91, pp. 9–10.

[39] Our own experience with this fish is limited to a single occasion, south of Nantucket, when several were seen about a drifting box.

[40] Reported by MacCoy, Bull. 67, Boston Soc. Nat. Hist., 1933, p. 9.

[41] This specimen now in the Museum of Comparative Zoology, and was received through the kindness of J. Miggins of the U. S. Fish and Wildlife Service.

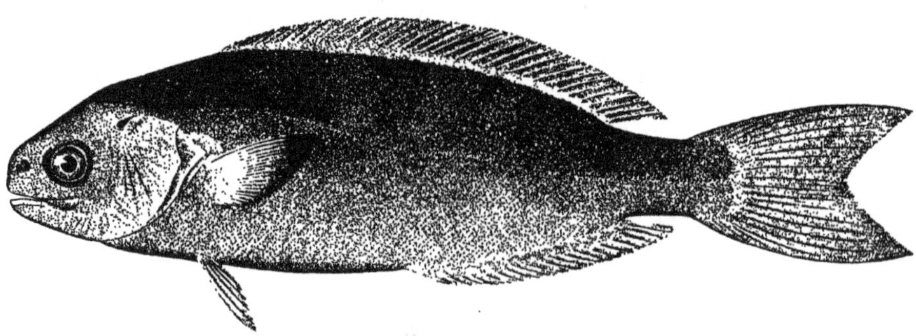

FIGURE 196.—Black ruff (*Centrolophus niger*), Dennis, Mass. From Goode and Bean. Drawing by S. F. Denton.

high as it is thick) is more flattened sidewise than that of the pilot fish; its caudal peduncle is much deeper, and has no lateral keels; its dorsal fin is considerably longer than that of the pilot fish, and there are no detached spines in front of its dorsal fin. The single dorsal fin of 3 flexible spines entirely covered over by skin and 35–38 soft rays reaches from over the pectoral fin to the caudal peduncle; the anal (3 spines concealed by skin and 20–22 soft rays) originates about under the midpoint of the dorsal and runs equally far back; both dorsal and anal fins are evenly graduated in outline from front to rear; and both are fleshy and scaly along their bases. Its ventral fins are about under the pectorals; pectorals and anals are both small; and the caudal is moderately forked.

Color.—Those we have seen (after a few weeks preservation in alcohol) are dark leaden-brown on back and sides, with the margins of the scales darkest, in so fine a pattern (because of the small size of the scales) that the general effect is sooty; the fins are darker, even, than the back; and the belly only a little paler. Other specimens have been described [42] as brownish pink all over, or brown, darkest above, some with irregular and obscure markings, either yellowish or dark blue.

Size.—Grows to about 2 feet in length.

General range.—Oceanic, and widespread in low and mid latitudes in the eastern North Atlantic; Madeira, the Azores, and the coasts of Spain north to the entrance to the English Channel, the Celtic Sea and southern Norway; also in the Mediterranean; and reported as a stray from Massachusetts and from Georges Bank.

Occurrence in the Gulf of Maine.—One specimen of this wanderer from Europe, about 12¾ inches long, was taken in a trap at North Truro, on Cape Cod Bay, September 6, 1890; [43] a second of 21½ inches was brought in from the northern edge of Georges Bank by the trawler *Thomas Whalen* in September 1936; [44] and a third of about 13 inches (330 mm.) to the fork of the tail was taken in a trap at North Truro June 23, 1951.[45]

Another about 9 inches long was taken in 1888[46] at Dennis, Mass. But it is not known whether this record should be credited to our Gulf or to the southern coast of Massachusetts, since that township fronts both on Cape Cod Bay and on Nantucket Sound.

THE POMPANOS AND JACKS. FAMILY CARANGIDAE

The pompanos are allied to the mackerels; both have deeply forked tails, very slender caudal peduncles, and ventrals situated below the pectorals. And, mackerel-like, most of them have two dorsal fins, the first hard-spined, the second soft-rayed. But they are readily separable from the mackerels by the fact that their first (spiny) dorsal, if they have one, is much shorter than the second (soft rayed) while it may be reduced to a series of very short spines, or even lost altogether in old age. And, except for the leather jacket, they either lack the dorsal and anal finlets so characteristic of the mackerel tribe, or have only one of each at most. They differ further from the mackerels in the number of vertebrae (only 24 as against upward of 30), and in that their premaxillary bones (fixed in the mackerels) are protractile (except in adult *Oligoplites*), while their anal fin is preceded by two free spines that may either take the form of a permanent finlet or may be lost in old age. Warm seas support a host of species, but none of them is more than an accidental stray to the Gulf of Maine.

KEY TO GULF OF MAINE POMPANOS

1. Rear parts of soft dorsal fin, from 7th ray backward, and of anal fin from 6th ray backward are so deeply indented between every two rays as to form a series of 12 to 14 nearly separate low finlets.......... Leather jacket, p. 380
   Rear parts of soft dorsal and of anal fins even-edged, not as series of finlets.......... 2
2. Body very much flattened, sidewise; nearly or quite half as deep as it is long to base of caudal fin.......... 3
   Body moderately stout, less than two-fifths as deep as it is long to base of caudal fin.......... 6
3. Back and belly rounded; pectoral fins reach not more than one-third the way back toward the base of the caudal fin.......... True pompanos (genus *Trachinotus*) [47]
   Back and belly sharp-edged; pectoral fins reach at least half-way back toward the base of the caudal fin.......... 4

[42] Day, Fishes Great Britain, 1880–1884, vol. 1, p. 110.

[43] Reported by Bean, Proc. U. S. Nat. Mus., vol. 21, 1898, p. 639 and now in the U. S. National Museum.

[44] Reported by Bigelow and Schroeder, Copeia, 1937, p. 61.

[45] Received through the kindness of John Worthington of the Pond Village Cold Storage Co.

[46] Goode and Bean, Smithsonian Contrib. Knowl., vol. 30, 1895, p. 214.

[47] None of these southern fish have yet been reported within our Gulf.

4. Second dorsal and anal fins are conspicuously falcate in shape, very high in front, tapering abruptly toward the rear; there are no enlarged bony plates along the lateral line on the caudal peduncle.............. Lookdown, p. 379
   Second dorsal and anal fins only moderately high in front, tapering rearward gradually; caudal peduncle with weak bony plates along the lateral line.......... 5
5. Upper anterior profile of head concave; ventral fins very small; anterior rays of soft dorsal and of anal not elongate.......... Moonfish, p. 378
   Upper anterior profile of head convex; ventral fins as long as head or longer; anterior rays of soft dorsal and anal fins elongate, threadlike.......... Thread fin (probably the young of the Cuban jack), p. 381
6. There is only one well-developed dorsal fin (the soft rayed), the first (spiny) dorsal being reduced to a few short spines, without separate fin membranes.......... Pilot fish, p. 372
   There are two well-developed dorsal fins though the first (spiny) is smaller than the second.......... 7
7. There is a detached finlet behind the dorsal fin and one behind the anal fin.......... Mackerel scad, p. 374
   There are no finlets behind the dorsal and anal fins.......... 8
8. There is no finlet in front of the anal fin; and the anal is only about one-half as long as the soft dorsal.......... Rudderfish, p. 373
   There is a finlet of 2 short spines in front of the anal fin, and the anal fin is nearly or quite as long as the soft dorsal.......... 9
9. The forward part of the lateral line is scarcely arched.......... Goggle-eyed scad, p. 377
   The forward part of the lateral line is strongly arched.......... 10
10. The breast is naked, except for a small patch of scales in front of the ventral fins.......... Crevalle, p. 375
    The breast is covered with scales.......... 11
11. The body (to base of tail) is not more than 3 times as long as it is deep; the soft dorsal fin has only 23 to 25 rays.......... Hardtail, p. 376
    The body to base of tail is more than 3 times as long as it is deep; the soft dorsal fin has 30 to 35 rays.......... Saurel, p. 377

Pilotfish *Naucrates ductor* (Linnaeus) 1758

Rudderfish; Shark pilot

Jordan and Evermann, 1896–1900, p. 900.

Description.—The pilotfish is one of the more slender carangids (body about one-fourth as deep as it is long), round-sided, about two-thirds as thick as it is deep, and somewhat mackerel-like in appearance though with a blunter, more rounded nose and smaller mouth, while its caudal peduncle is conspicuously keeled on either side like that of a bonito. But its long second dorsal fin separates it from all the mackerel tribe. The first dorsal fin is reduced to three or four short inconspicuous spines, which are connected by a membrane in young fish but this membrane is lost with growth. The second dorsal (26 or 27 soft rays) is weakly concave in outline and originates midway between tip of snout and base of caudal fin. The anal fin is similar to the second dorsal in form, but is only about half as long (16 or 17 rays), and is preceded by two very short spines. It resembles the rudderfish in this but the first dorsal of the latter is well developed and has 7 spines instead of only 3 or 4. The ventrals, situated far forward under the pectorals, are about as large as the latter. The caudal is large and deeply forked. The edge of

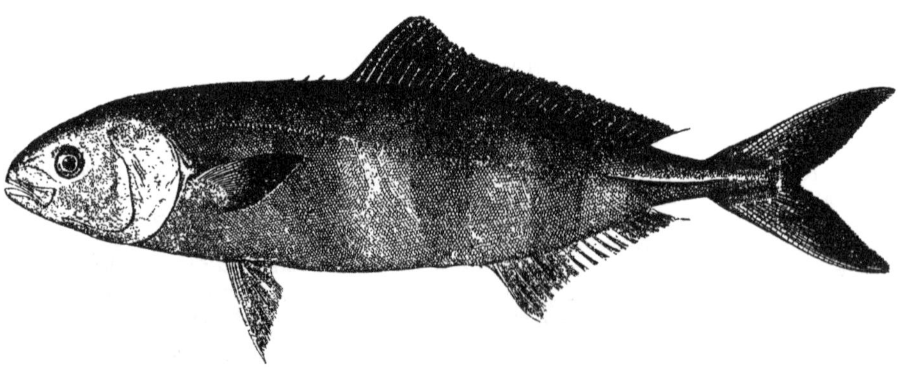

Figure 197.—Pilotfish (*Naucrates ductor*), about 13 inches long, New Bedford, Mass. After Goode. Drawing by H. L. Todd.

the gill cover is rounded in the adult but it bears a spine in young fry.

Color.—Bluish, cross-barred with 5 to 7 dark bands, 2 or 3 of which run up on the dorsal fin and down on the anal. The outer margins of caudal, ventral, and pectoral fins are nearly black The caudal is white-tipped.

Size.—Maximum length about 2 feet.

General range.—A tropical fish of the high seas, rarely straying as far north as outer Nova Scotia.[48]

Occurrence in the Gulf of Maine.—The only records of this species from within the Gulf are of one taken in a mackerel net in Provincetown Harbor in October 1858, the fish probably having followed a whale ship that arrived a few days previous; one caught near Seguin Island in 1906; one off Portland in September 1921; one taken from a mackerel net at Provincetown in August 1924; three in 1929; one of them from the northern edge of Georges Bank in October, the other two from the South Channel to the southeast of Cape Cod (one in August, one in November); one off Portland, July 1931; and one picked up in a trawl on the northern slope of Georges Bank (lat. 42°10′ N., long. 66°32′ W.) October 10, 1933.[49] We need only add that this is the fish that so commonly attends sharks in tropic seas, either picking up a living from the scraps left by the latter, or feeding on the parasites with which their protectors are infested. They often follow sailing vessels, also.

[48] Vladykov (Proc. Nova Scotian Inst. Sci., vol. 19, 1935, p. 6), reports two specimens taken on Sable Island Bank, and one from Sambro near Halifax, during the period 1932–34.

[49] Reported to us by W. C. Neville of the U. S. Bur. Fish.

Rudderfish *Seriola zonata* (Mitchill) 1815.[50]

AMBERJACK; PILOTFISH

Jordan and Evermann, 1896–1900, p. 902.

Description.—The rudderfish is deeper bodied, relatively, than the pilotfish (body about three and one-half times as long as deep), so much flattened sidewise that it is almost as thin as a butterfish (p. 363), and with a pointed nose. Its first (spiny) dorsal fin is well developed, with 7 spines. There are 36 to 38 rays in the second dorsal fin (only 26 or 27 in the pilotfish) and the ventrals are relatively much longer than in the pilot. In young fry of 2 to 3 inches the second dorsal originates a little in front of the tips of the pectorals, but it originates slightly behind the tips of the pectorals by the time the fish has grown to 8 or 9 inches, and still farther back in larger specimens.[51]

The anal fin (20 or 21 rays) is a little more than half as long as the second dorsal in the rudderfish, as it is in the pilotfish also. And in young fish it is preceded by one or two short spines which adults lack.

The ventrals are a little longer than the pectorals, and more pointed in large fish than in small; the caudal is deeply forked, its slender peduncle with a longitudinal keel on each side; the mouth gapes back to the forward margin of the eye and is armed with broad bands of hairlike teeth. The body is clad with small scales.

[50] The interrelationships of the several *Seriolas* that have been described from our South Atlantic coast still remain in doubt.

[51] We have examined specimens ranging from 3 to 9 inches in length taken in Cape Cod Bay, at Woods Hole, New Bedford, and other localities. Storer's illustration, reproduced here, was of a 2-inch fish.

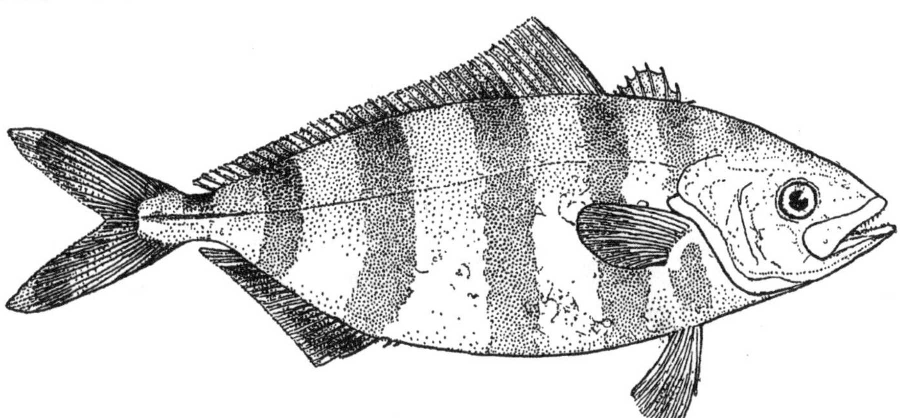

FIGURE 198.—Rudderfish (*Seriola zonata*), young, in striped stage, Wellfleet, Mass. After Storer.

Color.—Bluish or silvery brown above, paler on the sides, and white below. In young fish (no large ones have been reported from within our limits) the sides are conspicuously crossbarred with 5 or 6 broad dark blue or brown bands, the last 4 run up on the dorsal fin and the last 2 or 3 down on the anal fin. There also is a dark band running obliquely from the first dorsal to the eye in some cases. All of these bands fade with growth, however, to disappear in large fish. The first dorsal is black, the anal white at the base, the ventrals black above, pale below, and the caudal dusky green, with white tips.[52]

Size.—Maximum length about 3 feet.

General range.—Atlantic Coast of America, Halifax, Nova Scotia,[53] to Gulf of Mexico.

Occurrence in the Gulf of Maine.—The rudderfish is ordinarily a rare visitor to the Gulf of Maine, and most of those that have been seen there have been small, made conspicuous by their crossbarred pattern. Two were taken at Wellfleet in 1844 and 1849 (mentioned by Storer); another at Beverly in May 1866; one five inches long at Provincetown in 1870; and one at Salem sometime prior to 1879. A gap then follows in the record until September 1921, when one was caught by an angler fishing for smelt at a wharf in Portland Harbor.[54] Another, of 5½ inches was caught on September 22, 1929, also by an angler fishing for smelt; one of 6¼ inches was taken on Nantucket Shoals August 1, 1930;[55] several were reported in 1949 at Boothbay Harbor, the Sheepscot River, and at Gloucester.[56] However, in the summer and fall of the years 1949–51 large numbers of them were caught or observed in and around the traps at Barnstable, Cape Cod Bay, and one day's record catch by one set of pound nets, within this period, amounted to two barrels [57] indicating that, in some years, large schools of rudderfish are sometimes present in the latter region.

Small fry 1½ to 7 inches long are regular summer visitors at Woods Hole.

[52] We have no color notes from life.

[53] Reported by Leim, Proc. Nova Scotian Inst. Sci., vol. 17, 1930, No. IV, p. xlvi, as *S. dumerili*.

[54] Reported to us by Walter H. Rich.

[55] Reported by Firth, Bull. 61, Boston Soc. Nat. Hist.; 1931, p. 12.

[56] Reported by Scattergood, Trefethen, and Coffin, Copeia, 1951, p. 298.

Mackerel scad *Decapterus macarellus* (Cuvier and Valenciennes) 1833

Jordan and Evermann, 1896–1900, p. 909.

Description.—This scad is easily recognized among such of its tribe as are known from our Gulf by the presence of a small detached finlet between the second dorsal and the base of the caudal fin with another similar to it behind the anal.[58] Furthermore, it is more slender than most of the other pompanos; its body is only about one-fifth as deep as it is long, and fusiform like the mackerel. But the great length of the second dorsal fin and the fact that there is only one dorsal finlet and one anal finlet would separate a mackerel scad from a mackerel at a glance. The mouth of the scad is smaller, and its premaxillary bones are protractile. Its triangular first dorsal fin (8 spines) originates over the middle of the pectorals. Its second dorsal (about 34 rays) is

[57] Information supplied by Frank Mather who was informed of the 1949–1951 catches at Barnstable by Capt. John Vetorino in whose traps many of these rudderfish were caught.

[58] A second scad, the round robin (*Decapterus punctatus*), similarly characterized, is known as far north as the Woods Hole region. It has 40 or more scutes or shieldlike scales along the lateral line, instead of only about 30 or 31; its jaws are toothed, and it is spotted along the lateral line, characters that separate it from the mackerel scad.

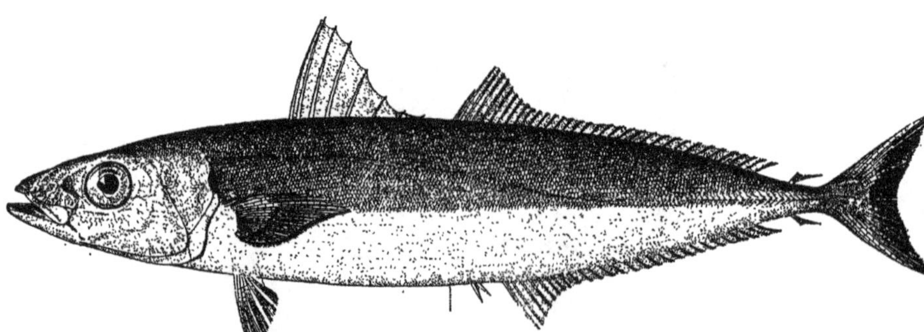

FIGURE 199.—Mackerel scad (*Decapterus macarellus*), Woods Hole. After Goode. Drawing by H. L. Todd.

separated from the first dorsal only by a very short space and extends back nearly to the base of the caudal. Its anal fin is similar to its second dorsal in shape but is shorter (about 28 rays), originates about under the seventh or eighth ray of the second dorsal, and is preceded by 2 short stout spines. The ventrals are shorter than the pectorals and situated under them.

The tail of the scad is less deeply forked than in most of the pompanos. In place of fleshy keels on the caudal peduncle, the rear half of its lateral line is armed with a series of 31 keeled shields, largest on the peduncle, and all of them much larger than the ordinary scales, a very noticeable character.

Color.—Described as slate blue or leaden above, silvery below, with a small black spot on the margin of the gill cover and with the axil of the pectoral black. We have not seen it alive.

Size.—Maximum length about 1 foot.

General range.—Warm parts of the Atlantic, rarely straying northward to the Gulf of Maine and to Nova Scotia.

Occurrence in the Gulf of Maine.—A specimen caught with smelt in Casco Bay, Maine, in October 1920, and another, 7 inches long, taken in a trap at Richmond Island, off Cape Elizabeth in September 1931, are the only Gulf of Maine records, though it has been taken at Canso and at Port Mouton Bay, Nova Scotia.[63] But being common in the autumn about Woods Hole, where as many as 10 barrels have been taken from one trap haul, it would not be surprising to find it north of Cape Cod any summer.

[63] This last fish, a 2½-inch specimen, caught October 10, 1928, was recorded by Leim (Proc. Nova Scotian Inst. Sci., vol. 17, No. 4, 1930, p. xlvi).

Crevalle *Caranx hippos* (Linnaeus) 1766.

Jack

Jordan and Evermann, 1896–1900, p. 920

Description.—The presence of a well-developed first dorsal fin (8 spines) combined with an anal (about 17 rays, preceded by 2 short detached spines) nearly as long as the second dorsal (about 20 rays), but no detached finlets, separates this particular jack from all other pompanos known from the Gulf, except the goggle-eyed scad (p. 377), hardtail (p. 376), and the saurel (p. 377). Its arched lateral line and the presence of (usually) two pairs of small but plainly visible canine teeth in the lower jaw distinguish it from the goggle eye; its naked breast and its canine teeth from the hardtail and saurel. The dorsal profile, too, of the head of the crevalle (fig. 200) is characteristic, and the long scimitar-shaped pectoral fins are a convenient field mark to separate it and other members of its immediate tribe,[64] from the pilotfish, rudderfish, and mackerel scad, in which the pectorals are short and blunter. We need only call attention further to its deeply forked tail; to the row of keeled shields along either side of its caudal peduncle; to its flattened oblong form (body

[64] The yellow tail (*Chloroscombrus chrysurus*), another species in this group straggles northward at times and, sometime, may be taken within the Gulf of Maine. It may be distinguished from the crevalle, hardtail, saurel, and big-eyed scad by the fact that its lateral line is wholly unarmed, whereas in these species it is armed with bony plates, along part of its length at least.

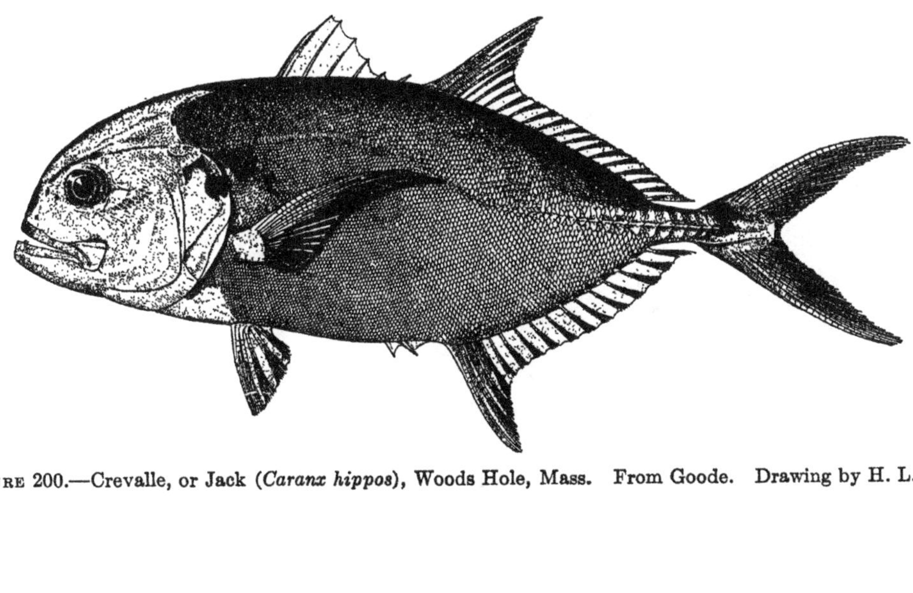

FIGURE 200.—Crevalle, or Jack (*Caranx hippos*), Woods Hole, Mass. From Goode. Drawing by H. L. Todd.

only about two and one-half times as long as deep, but with caudal peduncle as slender as that of a mackerel), and to its blunt head.

Color.—Greenish or greenish bronze above with golden sides; silvery below, sometimes with yellow blotches. There is a large black blotch on the gill cover, a fainter dark spot on the lower rays of the pectorals (in adults), and a black blotch in their axils. The fins are more or less yellowish; the edge of the dorsals is black. Very young fish have 5 or 6 dark cross-bars.

Size.—Maximum recorded weight 36 pounds.

General range.—Warm seas; abundant on both coasts of America; northward as a stray to the outer coast of Nova Scotia; [65] also among the East Indies.

Occurrence in the Gulf of Maine.—We know of only two records of this southern fish from our Gulf, one specimen picked up on Lynn Beach on the shore of Massachusetts Bay during the summer of 1847, and a second taken at Provincetown in 1933.[66] But it is a regular summer visitor at Woods Hole though it is not common there.

Commercial importance.—A famous game fish, but of minor commercial importance.

Hardtail *Caranx crysos* (Mitchill) 1815

YELLOW JACK; RUNNER; YELLOW MACKEREL

Jordan and Evermann, 1896–1900, p. 921.

Description.—The hardtail resembles the crevalle, saurel, and goggle-eyed scad in the relative sizes and arrangement of its fins, in its deeply forked tail, in its slender caudal peduncle and in the presence of a row of bony shields along at least the rear part of its lateral line. But its scaly breast, the lack of canine teeth in its lower jaw, and the lack of a black spot on the pectoral fin separates it from the first of these; the fact that the bony plates increase in size, passing rearward along the lateral line, marks it off from the saurel, and its strongly arched lateral line from the goggle eye. Its first dorsal fin has 8 spines, its second, one spine followed by 23 to 25 rays, while its anal consists of a finlet of 2 short spines followed, after a distinct gap, by the soft portion with 19 to 21 rays.

Color.—Greenish bronze above, golden or silvery below. The fins may show dusky cloudings, and there usually is a dark spot on the gill cover, near the margin, but none on the pectoral fin. Young fry are more or less distinctly cross-barred on the sides, but these bars disappear with growth.

Size.—Maximum weight about 4 pounds and length about 22 inches. Northern examples are seldom more than a foot long.

General range.—Atlantic coast of America, Brazil to Rhode Island and to Nantucket Sound regularly, and as far northward as outer Nova Scotia as a stray; represented by a closely allied species in the Pacific.

Occurrence in the Gulf of Maine.—The fact that this fish has been reported at Chatham on Cape Cod in 1933,[67] at Provincetown, in Boston Harbor,

[65] Reported near Halifax, Nova Scotia, by Vladykov (Proc. Nova Scotian Inst. Sci., vol. 19, 1935, p. 4).

[66] Reported by MacCoy, Bull. 70, Boston Soc. Nat. History, 1934, p. 6.

[67] MacCoy, Bull. 70, Boston Soc. Nat. History, 1934, p. 6.

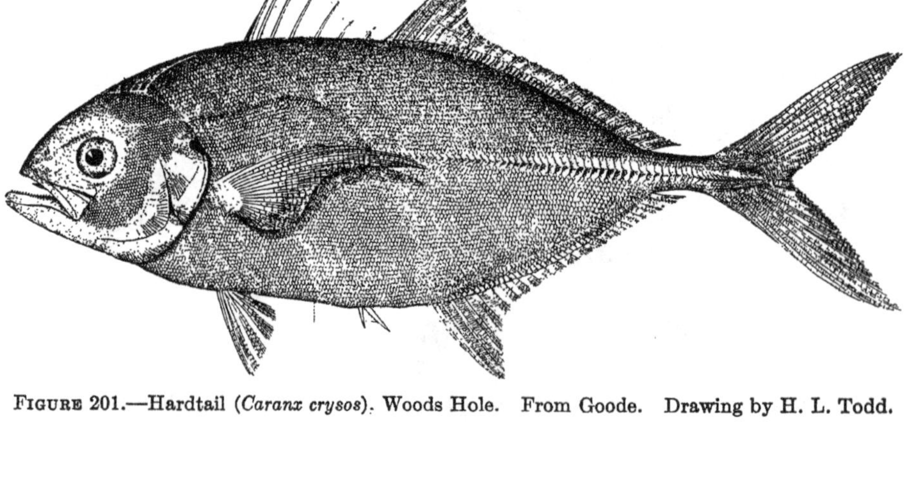

FIGURE 201.—Hardtail (*Caranx crysos*). Woods Hole. From Goode. Drawing by H. L. Todd.

off Gloucester,[68] and in Ipswich Bay,[69] and that 11 specimens, about 6 to 8 inches long, were taken in a fish trap at Barnstable on the shore of Cape Cod Bay on September 6, 1950, shows that it is more likely to round Cape Cod than is the crevalle. It is also reported from outer Nova Scotia.[70] Young fish are not rare about Woods Hole and thence westward from July to November.

Saurel *Trachurus trachurus* (Linnaeus) 1758

ROUGH SCAD

Jordan and Evermann, 1899–1900, p. 910.[71]

Description.—The saurel is distinguishable from all allied species yet known from New England waters by having about 75 bony plates along its lateral lines, as contrasted with about 30 or fewer in other Gulf of Maine carangids. It is a somewhat deeper fish than the mackerel scad but more slender than the hardtail or the crevalle, its body (to the base of tail) being about 3½ times as long as it is deep. Its first dorsal fin, of 8 spines, is closely followed by the long second dorsal of 25 to 30 soft rays. Its soft anal, opposite the second dorsal, has 24 to 26 rays, and is preceded by two small detached spines. The tail is deeply forked

[68] One netted September 18, 1878.

[69] Specimen now in the collection of the Boston Society of Natural History.

[70] A 5-inch specimen is reported from Port Mouton by Leim (Proc. Nova Scotian Inst. Sci., vol. 17, 1930, No. 4, p. xlvi), and small ones from Pubnico, and near Halifax, by Vladykov (Proc. Nova Scotian Inst. Sci., vol. 19, 1935, p. 4).

[71] Nichols (Bull. Amer. Mus. Nat. Hist., vol. 42, 1920, p. 479) considers the western Atlantic saurel distinct from the eastern Atlantic saurel and has proposed the name *lathami* for it. But this separation has not been adopted generally.

Color.—Described as bluish green above, silvery below, with a black spot on the edge of the gill cover above its rear angle.

Size.—Length about one foot.

General range.—Known from nearly all warm and temperate seas, sometimes common off the Florida Keys. A few have been recorded from the vicinity of New York, one from Newport, R. I., and three from the Gulf of Maine.

Occurrence in the Gulf of Maine.—One specimen of this rare fish was taken in Casco Bay on August 12, a second at Castine, Maine, on October 15, 1930,[72] and a third at Sandwich, Mass., on Cape Cod Bay in the summer of 1950.[73]

Goggle-eyed scad *Trachurops crumenophthalmus* (Bloch) 1793

GOGGLER; GOGGLE EYE JACK

Jordan and Evermann, 1896–1900, p. 911.

Description.—This scad resembles the mackerel scad (p. 374) in general appearance, but it has larger eyes and lacks the detached finlets behind the dorsal and anal fins. Its high first dorsal fin separates it readily from the pilotfish (p. 372), while the fact that the forward half of its lateral line is only slightly arched instead of strongly so obviates any danger of confusing it with the crevalle (p. 375), the hardtail (p. 376), or the saurel (p. 377). Its first dorsal fin has 8 spines, its second dorsal is of 1 spine followed by 23 to 26 soft rays;

[72] Kendall, Bull. No. 58, Boston Soc. of Nat. Hist., 1931, p. 11.

[73] We received this specimen from Capt. Benjamin Morrow and it is now in the Museum of Comparative Zoology.

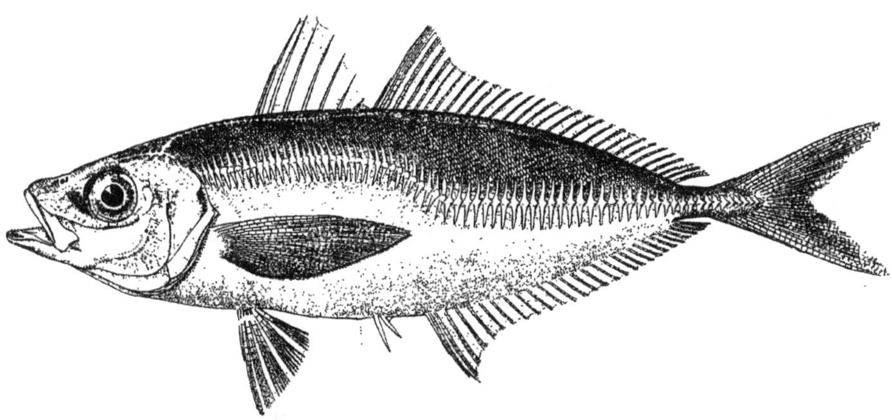

FIGURE 202.—Saurel (*Trachurus trachurus*), Rhode Island. From Goode. Drawing by H. L. Todd.

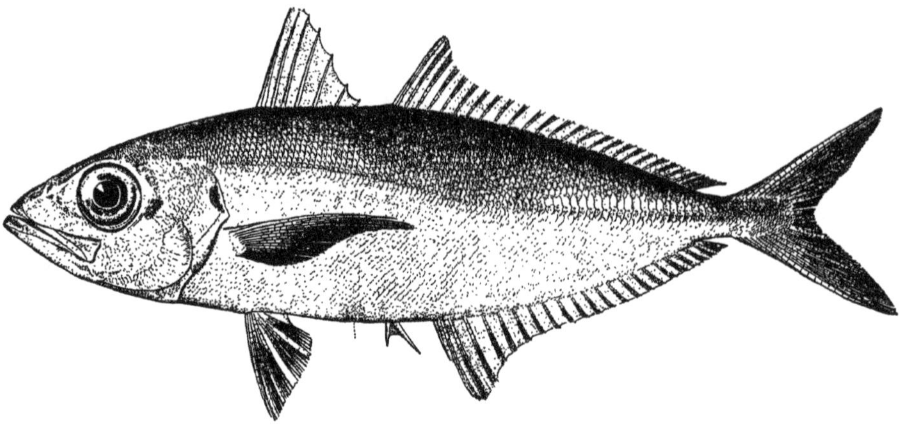

FIGURE 203.—Goggle-eyed scad (*Trachurops crumenophthalmus*), Woods Hole. From Goode. Drawing by H. L. Todd.

its anal fin has 1 spine and 20 to 23 rays, and is preceded by 2 stout detached spines. Its caudal is forked. Its ventrals originate a little behind the pectorals which reach nearly or quite as far back as its vent. Its entire breast is scaly, as are parts of its head.

Color.—Bluish above, silvery below. The fins, snout, and tip of the lower jaw have dusky markings.

Size.—Grows to a length of about 2 feet.

General range.—Cosmopolitan in warm seas, straying as far northward on our Atlantic Coast as Cape Breton, Nova Scotia.[74]

Occurrence in the Gulf of Maine.—The only positive records of this species in our Gulf are of one 5¾ inches long taken in a trap at Provincetown on August 27, 1930;[75] a second 8 miles off Chatham, Cape Cod;[76] and a third from Sandwich, on Cape Cod Bay, in the summer of 1950.[77] It may be expected to round Cape Cod from time to time, for it is taken in summer and fall as far northward and eastward as Woods Hole.

Moonfish *Vomer setapinnis* (Mitchill) 1815

SHINER; HORSEFISH; BLUNTNOSE; DOLLARFISH

Jordan and Evermann, 1896–1900, p. 934.

Description.—The very deep, thin, sharp-edged body of the moonfish (adults are scarcely twice as long as deep, and young fry even deeper, relatively), tapering to a slender caudal peduncle, and the concave upper anterior profile of its head, are enough to separate it at a glance from pilotfish, scad, crevalle, hardtail, saurel, or goggle eye; its very low dorsal and anal fins distinguish it from the lookdown (p. 379), which is of something the same shape (cf. fig. 204 with fig. 205). Its minute ventral fins, soft dorsal fin and anal fin which are nearly even in height from end to end, separate it from the threadfin (p. 381), and from the Cuban jack (*Hynnis cubensis*), now thought to be the adult of the threadfin (p. 381).

The first dorsal of the adult moonfish is reduced to 8 very short, inconspicuous, detached spines, but the first two of these are prolonged and filamentous in young fry. Its second dorsal fin (21 to 23 rays) and its anal fin (17 to 19 rays) are about equal in length, both of them low and tapering very slightly from front to rear. In very small fish the second to fourth rays of the second dorsal fin are more or less prolonged, and the anal fin is preceded by 3 or 4 short detached spines which are not to be seen in the adult. The ventrals are so small that they are likely to be overlooked except in young fry, in which the ventral rays are more or less filamentous as are the dorsal spines. The pectorals are scythe shaped. The scales along the lateral line are not large enough to be conspicuous, and the teeth are very small. There are no detached finlets, dorsal or anal.

Color.—Bluish green above, bright silvery on the sides. The second dorsal fin is plain pale greyish, sometimes light yellow at its base; the

[74] Reported from Canso by Cornish, Contrib. Canadian Biol. (1903-1905) 1907, p. 85. A. H. Leim advises us that a specimen 145 mm. long was taken off Centre East Pubnico, Nova Scotia, September 12, 1951.

[75] Firth, Bull. 61 Boston Soc. Nat. Hist., 1931, p. 11.

[76] MacCoy, Bull. 70, Boston Soc. Nat. Hist., 1934, p. 6.

[77] This specimen was received from Capt. Benjamin Morrow and is now in the Museum of Comparative Zoology.

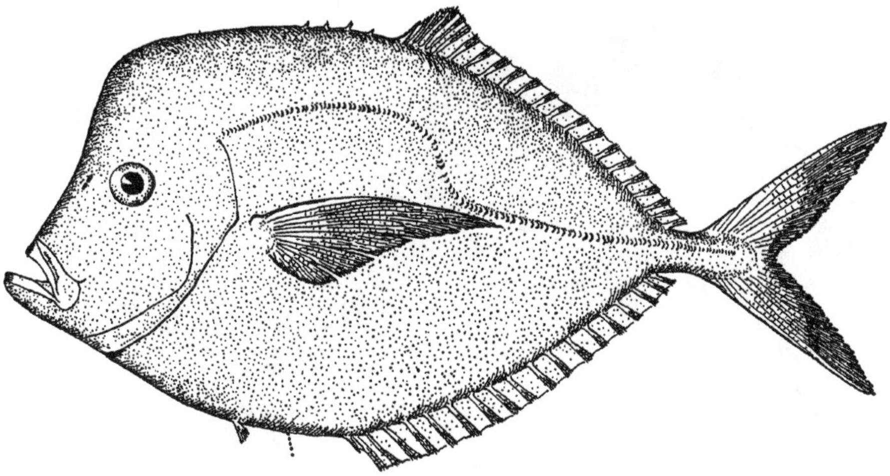

FIGURE 204.—Moonfish (*Vomer setapinnis*). After Goode. Original drawing by H. L. Todd.

caudal is greenish yellow; the pectorals light yellow or dusky greenish.

Size.—Said to reach 1 foot in length, but most of them are less than 9 inches long.

General range.—Warm seas off the east coast of America from Uruguay to Cape Cod, straying to Nova Scotia; common from Chesapeake Bay southward.

Occurrence in the Gulf of Maine.—This waif from warmer waters has been recorded from the South Channel off Cape Cod (one specimen 1¾ inches long); off Cape Cod (60 miles south by east from Highland Light), where one was taken in a mackerel seine, August 23, 1929;[78] from Gloucester (several specimens); from Magnolia, Danvers, Salem, and South Boston (a specimen 2 inches long) around Massachusetts Bay; from Saco Beach (fry of about 1 to 3 inches); and from Casco Bay in Maine. It has even been reported once or twice as far east as Liverpool and Halifax, Nova Scotia.[79] Thus it appears to reach our Gulf rather more often than any of its relatives do; not often enough, however, for most of the fishermen of whom we have inquired to know it north of Cape Cod. It appears more often (if irregularly) at Woods Hole, where young fish are sometimes common in August and September.

[78] Reported by Firth, Bull. 61, Boston Society Natural History, 1931, p. 12.

[79] Leim (Proc. Nova Scotian Inst. Sci., vol. 17, No. 4, 1950, p. xlvi); Vladykov (Proc. Nova Scotian Inst. Sci., vol. 19, 1935, p. 8).

Lookdown *Selene vomer* (Linnaeus) 1758

HORSEHEAD; MOONFISH

Jordan and Evermann, 1895–1900, p. 935.

Description.—The very high second dorsal (about 22 rays) and anal fins (about 20 rays) of the lookdown, and their peculiar falcate outline with the second ray much the longest and the next 4 or 5 rays successively shorter make distinction easy between it and the moonfish. And its peculiar form is hardly less characteristic, for it shares with the moonfish a deep, rhomboid, but very thin flat body (the trunk is only about one and one-quarter times as long as deep), abruptly truncate in front, with slightly concave upper anterior profile, and tapering rearward to a slender caudal peduncle. The mouth is set so low and the eye so high that the expression of its face is very characteristic. When adult the first dorsal is reduced to 7 or 8 short inconspicuous spines, only the first 3 of which are connected by a membrane, and the ventrals are very small; but some of the spines of the first dorsal are very long in fry up to 4 or 5 inches in length, the ventrals are much longer than in the adults, and the anal fin is preceded by two short detached spines that disappear with growth. The caudal fin is deeply forked like that of other pompanos, and the pectorals are sharp pointed and falciform, reaching back behind the middle of the second dorsal fin.

Color.—Small specimens, and northern strays usually are small, are silvery above as well as

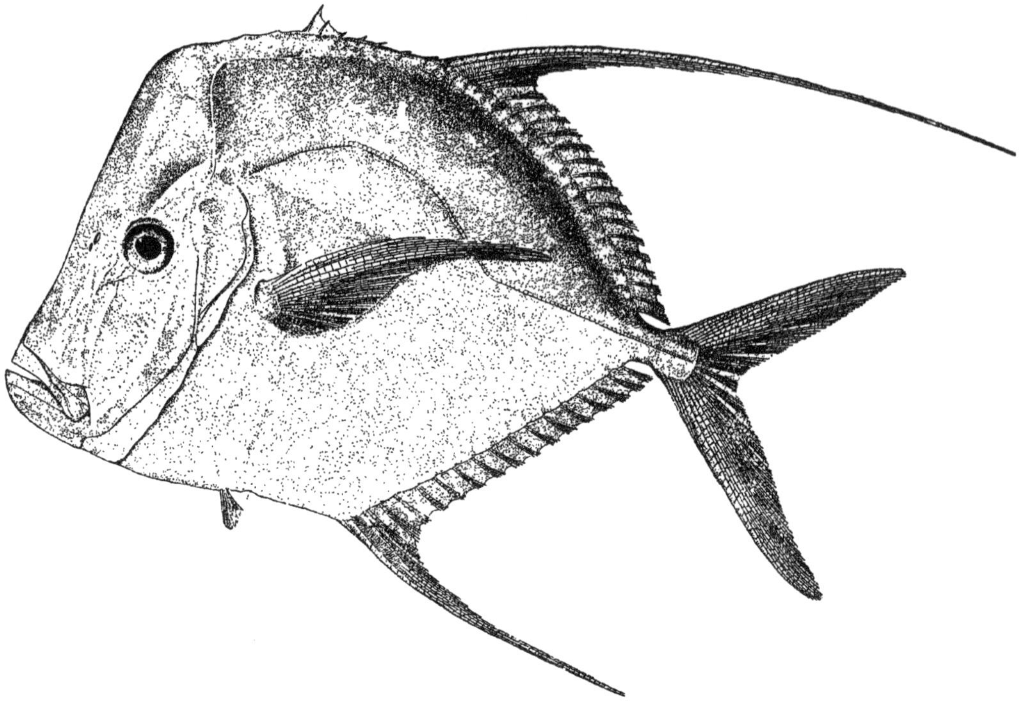

FIGURE 205.—Lookdown (*Selene vomer*). From Goode. Drawing by H. L. Todd.

below, with the ground tint of the back leaden; the sides are barred with several crossbands, variously described as dark or golden. But these bands fade out with growth.

Size.—Reaches a weight of about 2 pounds.

General range.—Warm waters on the east and west coasts of America, north rarely to Cape Cod, straying to the Gulf of Maine and to Nova Scotia;[80] common from Chesapeake Bay southward.

Occurrence in the Gulf of Maine.—There were only three records for the lookdown in our Gulf up to 1933; two of them for Casco Bay, the third for Boston Harbor (Dorchester). But many small ones were reported from the traps at the mouth of Casco Bay during that autumn, one from Beverly on the north shore of Massachusetts Bay, and one from North Truro on Cape Cod. Evidently this was an unusual incursion, for no one would be apt to overlook so bizarre a straggler from the south.

[80] Jones (Proc. and Trans. Nova Scotian Inst. Sci., vol. 5, App., 1879, p. 89) and Honeyman (Trans. Nova Scotian Inst. Sci., vol. 6, 1886, p. 328) report young fry as occasionally found in the shore waters of Nova Scotia, presumably along the outer coast, for tropical fishes are taken oftener there than along the Gulf of Maine shore of the Province.

Leatherjacket *Oligoplites saurus* (Bloch and Schneider) 1801

Jordan and Evermann, 1896–1900, p. 898.

Description.—The most interesting character of the leather jacket, and one which places it at a glance, is that the rear part of its soft dorsal fin back from the 7th ray, and also its anal fin back from the 5th ray, is broken, as it were, into a series of 12 low nearly separate finlets, the ray in each of which is subdivided at the tip like the hairs of a little brush. We need only note further that its body is about 3½ times as long as it is deep, very strongly flattened sidewise, and thin, being only about one-third as thick as it is deep; its upper jaw bone reaches back about as far as the rear edge of the eye; its snout is moderately pointed; its caudal peduncle very slender, with a low, inconspicuous keel on either side. Its first dorsal fin is reduced to about 5 separate spines, each with small fin membrane and its second dorsal has about 20 rays; its soft anal fin, also of about 20 rays, is preceded by two stout and conspicuous spines, forming, together, a separate finlet. Its lateral line is nearly straight, and its

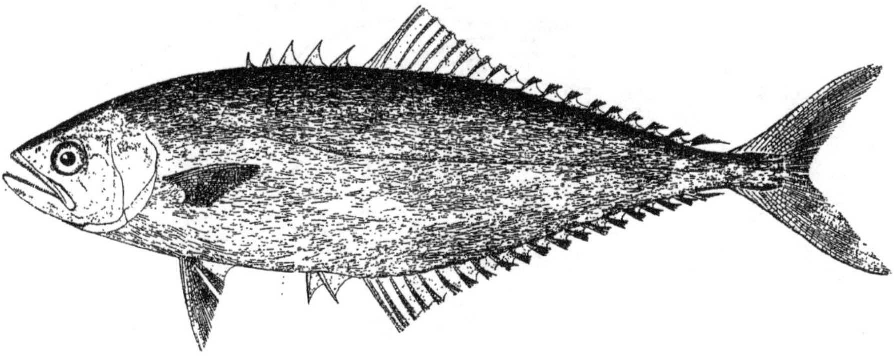

FIGURE 206.—Leatherjacket (*Oligoplites saurus*), Marthas Vineyard. From Jordan and Evermann. Drawing by H. L. Todd.

scales are very small, and imbedded in the skin, which is corrugated with a great number of short, fine, longitudinal ridges, giving it a leathery appearance, hence its common name.

Color.—Bluish above, silvery below, with yellow fins.

Size.—The largest are about 12 inches long.

General range.—Common on both coasts of tropical America; northward to New York and southern Massachusetts (Woods Hole), reaching the southwestern part of the Gulf of Maine as a stray.

Occurrence in the Gulf of Maine.—The only record of this southern fish within the Gulf is of one taken in a trap off the outer beach at Chatham, Cape Cod.

Threadfin *Alectis crinitus* (Mitchill) 1826 [81]

Jordan and Evermann, 1896–1900, p. 931, as *Alectis ciliaris* (Bloch 1788).

Description.—The combination of a head strongly convex in dorsal profile, with the fact that the first few rays of its soft dorsal fin, and of its anal fin also, are extremely long, and threadlike, places the threadfin at a glance among the carangoids of our northeastern coast. On small fish these threadlike rays are considerably longer than trunk and tail combined, but they shorten with age, probably to be entirely lost. The trunk of the threadfin is nearly as high as it is long (to the caudal peduncle), the dorsal profile of its head is strongly convex and it is strongly flattened sidewise. It has one dorsal fin of 1 stiff ray and about 19 soft rays, preceded by 6 short, separate, inconspicuous spines; the anal has 1 stiff ray and 16 soft rays, and is preceded by 2 spines so short that they are likely to be overlooked. Its lateral line is strongly arched over the pectoral; and the rear part armed with a series of bony platelike scales; the tail fin is deeply forked and the ventrals are larger than in most other carangoids. The pectorals are about as long as the head.

Color.—Upper surface bluish, the sides silvery, with traces of darker bars and blotches that tend to disappear with age; the prolonged parts of the dorsal and anal fins are bluish black; ventral fins mostly black; the fins otherwise more or less yellowish.

Size.—Specimens with the long threadlike fin rays have been reported up to about 7 inches long, in West Indian and Atlantic United States waters. But it now seems very probable that these are the young of the Cuban Jack (*Hynnis cubensis* Poey 1860). Their transformation consists chiefly in losing the filamentous fin rays; in a decrease in the depth of the body relative to its length; in a very considerable decrease in the relative size of the ventral fins; and in the assumption of a more falcate shape by the pectorals.

General range.—The threadfin (or threadfin stage of the Cuban Jack), is known on both coasts of tropical America; it strays northward on the Atlantic coast to southern Massachusetts, and it has been reported once from the Gulf of Maine. The adult Cuban Jack has not been reported north of southern Florida.[82]

[81] We follow Smith (Copeia, 1938, p. 146) in using the name *crinitus* Mitchill 1826, proposed for a specimen taken near Block Island, R. I., rather than *ciliaris* Bloch 1788 (type locality East Indies), awaiting final decision as to the true relationship between the threadfins of different oceans.

[82] We have seen one taken at Key West, and there is one from the east coast of Florida in the Museum of Comparative Zoology.

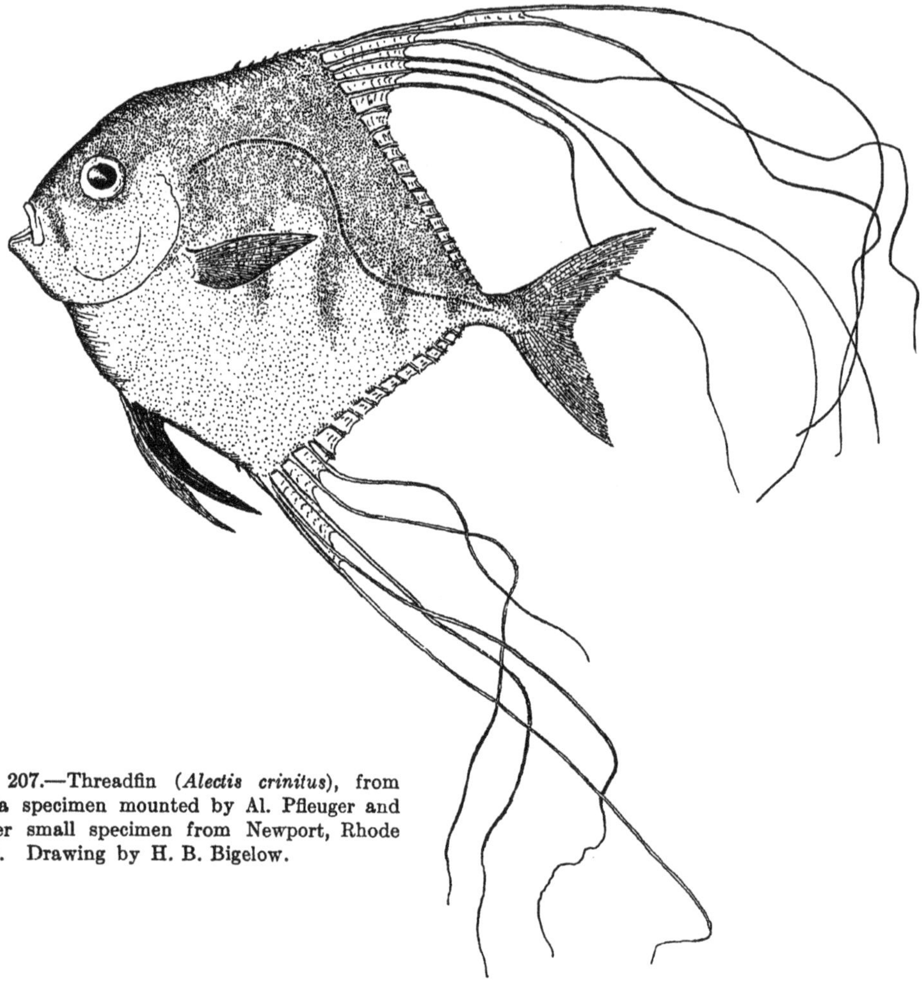

FIGURE 207.—Threadfin (*Alectis crinitus*), from Florida specimen mounted by Al. Pfleuger and another small specimen from Newport, Rhode Island. Drawing by H. B. Bigelow.

Occurrence in the Gulf of Maine.—The only records of this tropical fish for the Gulf of Maine are of one about 3¼ inches (85 mm.) long taken in a trap at Sagamore, on the southern shore of Cape Cod Bay, September 1, 1937, and another taken in a trap at North Truro, Mass., August 16, 1951. They may have come through the Cape Cod Canal as suggested by Smith.[83]

THE BLUEFISHES. FAMILY POMATOMIDAE

The bluefish (the only member of its family) resembles the pompano family in the general structure and arrangement of its fins, there being two dorsals, the first spiny and the second soft, with the ventrals well forward under the pectorals. But it lacks the free spines in front of the anal fin which are characteristic of most pompanos; its caudal peduncle is deeper; its tail is less deeply forked; and its teeth are much larger. It bears a superficial resemblance to certain of the weakfish family (p. 417) in its general body form and in the arrangement of its fins. But it is readily separable from any of the latter by the fact that its anal fin is nearly as long as its soft (second) dorsal, and from the sea-bass family in that its first (spiny) dorsal is much lower than the second. Most American ichthyologists look upon the bluefish family as closely allied to the pompanos, but

[83] Copeia, 1938, p. 146.

it should be grouped with the sea-bass tribe according to another view because of skeletal characters.

Bluefish *Pomatomus saltatrix* (Linnaeus) 1758 [84]

Snapper (Young)

Jordan and Evermann, 1896–1900, p. 946.

Description.—According to Jordan and Evermann, and to most of their successors, the bluefish is separable from its closest allies, the pompanos (Carangidae), by a tail "not deeply forked" and by larger scales, statements that may easily be misleading, for while the bluefish certainly has a less deeply forked tail than the pompanos, anyone, we think, would describe it as deeply forked as compared with any square-tailed fish. And while its scales are larger than those of most pompanos there is not much difference in this respect between a bluefish and a large crevalle (p. 375). There is, however, one positive point of difference. The jaws of the bluefish, upper as well as lower, are armed all around with a single series of stout, conical, canine teeth (one-eighth to one-fourth of an inch long in a fish of about 10 pounds), whereas the crevalle alone of northern pompanos has canines, and only two of them. Furthermore, the caudal peduncle of the bluefish is stouter than that of any pompano. It is sharply differentiated from all mackerels by the absence of dorsal or ventral finlets.

The bluefish is moderately stout bodied (large ones are about one-fourth as deep as long); its belly is flat-sided but blunt-edged below; its caudal peduncle moderately stout (slimmer, however, than in many other fish, e. g., striped bass); its head deep; its nose moderately pointed; and its mouth large and oblique, with projecting lower jaw, and with prominent canines. "Snappers," as small bluefish are called, are relatively deeper and more flattened sidewise than larger fish. The first dorsal fin (7 or 8 stout spines), originating over the middle of the pectorals, is low, rounded, depressible in a groove. It is separated by only a very short interval from the second dorsal, which is more than twice as long as the first (about 23 to 26 soft rays) and about twice as high, tapering backward with slightly concave margin. The anal fin (25 to 27 rays) is similar in form to the second dorsal though with a somewhat less concave outer margin; it originates somewhat farther back and is preceded by two very short detached spines that are often hidden in the skin. The caudal is broad and forked, moderately or deeply according to the other fish with which it is compared. The ventrals and pectorals are both of moderate size. The body, most of the head, and also the second dorsal and anal fins are clothed with medium-sized scales. There are no shields or keeled scales along the lateral line nor is the caudal peduncle keeled.

Color.—Sea-green above; silvery below. The second dorsal, caudal, and pectoral fins are of the general body tint, the latter with a black blotch at the base.

Size.—Maximum length about 3½ feet. The heaviest American fish of which we find definite record within recent years was 3 feet 9 inches long, weighing 27 pounds,[85] caught off Nantucket in 1903. One of 20 pounds was taken off Montauk, N. Y., in August 1951.[86] It is said that fish of 30 or even 50 pounds were not unheard of during the

[84] This fish has been known by various vernacular names along the middle and southern coasts of the United States. But it is the "bluefish" in the Gulf of Maine.

[85] Smith, Forest and Stream, vol. 61, October 10, 1903, p. 283.

[86] Reported in Salt Water Sportsman, August 17, 1951.

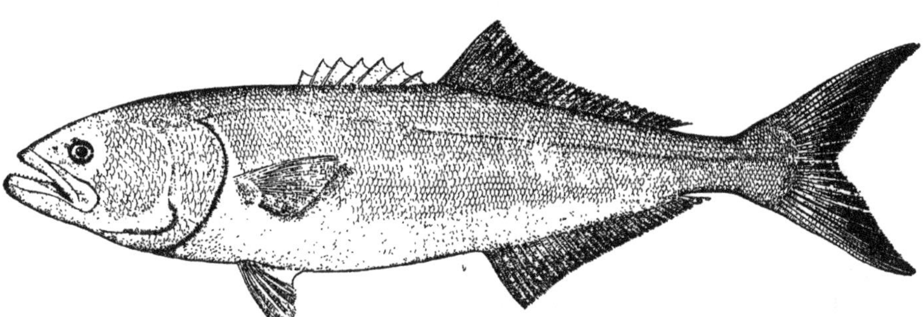

Figure 208.—Bluefish (*Pomatomus saltatrix*). From Jordan and Evermann. Drawing by H. L. Todd, from a cast.

last half of the eighteenth century, but these huge fish may not have been weighed. And the general run of the largest fish that are caught off the American coast is only 10 to 15 pounds. But they run larger off the African coast where 20-pounders are not unusual and where one of 45 pounds has been reported.[87] A 1-pound fish is about 14 inches long; a 2-pounder about 17 inches; a 3-pounder about 20 to 21 inches; a 4-pounder, about 2 feet; and an 8-pounder about 28 to 29 inches long. Fish weighing from 10 to 12 pounds are about 30 inches long.[88]

Habits.—The bluefish is oceanic in nature, found indifferently inshore, offshore, and in many parts of the ocean (p. 385). It usually travels in schools, sometimes including many thousands; in 1901, for example, a school 4 or 5 miles long was reported as seen in Narragansett Bay. And it is perhaps the most ferocious and bloodthirsty fish in the sea, leaving in its wake a trail of dead and mangled mackerel, menhaden, herring, alewives, and other species on which it preys. Goode [89] wrote long ago, the bluefish, "not content with what they eat, which is itself of enormous quantity, rush ravenously through the closely crowded schools, cutting and tearing the living fish as they go, and leaving in their wake the mangled fragments." It is not only the schooling fish that fall prey to them, but scup, squeteague, hake, butterfish, cunners, and small fish of all kinds, besides squid. Baird writing in the 1870's, when bluefish were at the height of their abundance, estimated that they annually destroyed at least twelve hundred million millions of fish during the four summer months off southern New England; and while this calculation surely was wildly exaggerated it will help give the reader a graphic realization of the havoc that they wreak during their periods of plenty. They are also known to eat various Crustacea and even marine worms on occasion. And the young "snappers," 6 to 8 inches long, feed largely on copepods, or crustacean and on molluskan larvae, as well as on fish fry smaller than themselves.

Bluefish are creatures of warm water, never found in any numbers in temperatures lower than about 58° to 60° (at least in summer); and they appear along the United States coast as warm-season migrants only. "Bluefish," writes Lyman,[90] "appear off the southern coast of Florida in midwinter," and by "late March anglers take them off the Florida coast in good quantities." "Large schools pass the Carolinas during March and April, appear off Delaware during April, and are first taken off New Jersey and Long Island, N. Y., during April and May," by commercial fishermen working well offshore. The earliest commercial catches are reported off southern Massachusetts in late May. But it is not until about a month later that they work inshore in numbers.

When they do come inshore, multitudes of little ones, known as snappers, run up into harbors and estuaries all along the coast, from Delaware Bay to Cape Cod. The larger ones, arriving somewhat later, also often come close enough in to the beach, west and south of Cape Cod, for many to be caught by anglers casting in the surf. But it is only in good years that this last holds true in our Gulf, even in the southern part.[91] When they "first appear offshore, in any locality, almost always they will be feeding deep, at or near the bottom. This means that surface lines and baits are practically worthless."[92] Later in the season schools are often seen at the surface, harrying other fish; and if they are deep, they can often be lured to the surface by throwing out ground bait.

Except for an occasional belated fish (p. 388), the bluefish disappear wholly from the entire coast northward from the Carolinas by early November. The winter home of this northern contingent has long been the subject of speculation. But the fact that we saw one trawled in 55 fathoms off Marthas Vineyard in mid-January in 1950 by the *Eugene H*, and that several hauls of 175 to 1,400 pounds per trip were brought in from the region of the Hudson Gorge by otter trawlers early that same February, makes it probable that most of the members of the northern contingent merely move offshore on bottom, to the warm zone along the outer edge of the continent, to pass the winter there. It is certain, however, that some migrate far southward (as has often been suggested for the stock as a whole) for one that was tagged off New York in August 1936 was recaptured off Matanzas,

[87] By Lt. Commander Henry Lyman (Bluefishing, 1950, p. 9) who also saw a 22-pounder weighed off northwest Africa, with still larger ones that were not weighed.

[88] Goode, Fish. Ind. U. S., Sect. 1, 1884, p. 442.

[89] Fish. Ind. U. S., Sect. 1, 1884, p. 544.

[90] Bluefishing, 1950, pp. 10, 11.

[91] We refer the reader to Lt. Comdr. Lyman's recent book (Bluefishing, 1950, pp. 34-49) for an interesting survey of the more-productive bluefishing grounds, Gulf of Mexico and Florida to Cape Cod.

[92] Quoted from Lyman, Bluefishing, 1950, p. 11.

Cuba, in January 1939.[93] Whether wanderers such as this ever return to the north is unknown.

A few bluefish are caught in winter on both coasts of Florida, southward from Cape Canaveral in the east, from Tampa Bay on the west; and enough are taken near Key West between December 15 and February 15, to yield commercial catches of 10,000 to 15,000 pounds in most years.[94] Some, also, are caught around Cuba by commercial fishermen in January and February. But these Florida fish, presumably the Cuban also, vanish at the end of the winter, not to reappear until early the next. What their relationship may be to the northern stock is not known. There are bluefish, too, off the northern coast of the Gulf of Mexico and off Texas, but nothing definite is known about their seasonal movements.

It is not likely that any interchange ordinarily takes place between the bluefish populations of the two sides of the Atlantic.

Females with large ova approaching ripeness are taken off North Carolina in spring, and off various parts of the coast farther north in summer;[95] ripe males have even been taken inside Chesapeake Bay in June and July,[96] from which it appears that they spawn from late spring through July and perhaps into August. But bluefish have never been reported actually spawning, though watch has been kept for them, which makes it likely either that they interrupt their inshore visit to move offshore for the purpose, perhaps sinking deep, or that most of them have spawned out before they appear along our northern coasts. In either case, the regular presence of "snappers" in numbers inshore, and the occasional captures of smaller fry in Chesapeake Bay[97] and in the Gulf of Maine (p. 388) make it likely that the spawning grounds of our northern bluefish are not far distant.

The eggs have not been identified with certainty. But the possibility is still open that the buoyant eggs with segmented yolk and large oil globule from Newport, R. I., provisionally referred to the bluefish by Agassiz and Whitman[98] were actually those of this species. And while the identity of their "bluefish" larvae has likewise been questioned, we believe that their identification of the oldest (9 mm. long[99]) was correct, though the younger ones may have belonged to some one of the mackerel tribe.

At this stage the second dorsal fin is formed, the first, however, still represented by the rudiments of the future spines. The anal fin is visible, also, and the tail is slightly forked. These larvae, like those of the mackerel (which they much resemble), have large blue eyes and large projecting teeth, but they are as far advanced in development as mackerel twice as large, and as ferocious in proportion to their size as the adult bluefish, devouring all other small animals in the tank with them.

The bluefish fry of three-fourths of an inch to 3 inches long, which have often been taken along shore in summer not only south of Cape Cod but even in the Gulf of Maine in some years (p. 388), are presumably the product of that season's spawning. And it seems that they grow to a length of 4 to 9 inches by autumn, fish of that size being common in October, while general experience suggests a length of 8 to 12 inches by the following spring. Nothing definite is known of the rate of growth of the older fish,[1] except that one that weighed about 1 pound when it was tagged off the coast of New York on August 10, 1936, was reported as weighing about 9 pounds when it was recaptured off Matanzas, Cuba, two years and five months later (January 15, 1939), which (if not exaggerated)[2] points to unexpectedly rapid growth.

The age at which the bluefish matures sexually is not known.

General range.—Widely but irregularly distributed in the warmer seas, its known range including the eastern coast of the Americas, northward regularly to Cape Cod, occasionally to outer Nova Scotia, south to Brazil and Argentina;[3] Bermuda;[4] eastern Atlantic off northwestern

[93] Reported by Lyman (Bluefishing, 1950, p. 10.)

[94] Schroeder, App. 12, Rept. U. S. Commissioner of Fisheries (1923) 1924, p. 12.

[95] Old statements to this effect are corroborated by Lyman (Bluefishing, 1950, p. 10), who reports females with roe and males with milt off North Carolina and near Nantucket early in summer.

[96] Hildebrand and Schroeder, Bull. U. S. Bur. Fish. vol. 43, Pt. 1, 1928, p. 232.

[97] Hildebrand and Schroeder (Bull. U. S. Bur. Fish., vol. 43, Pt. 1, 1928, p. 232) report fry as small as 2¾ inches in Chesapeake Bay.

[98] Mem. Mus. Comp. Zool., vol. 14, No. 1, Pt. 1, 1885, p. 13, pl. 4, figs 1–6.

[99] Mem. Mus. Comp. Zool., vol. 14, No. 1, Pt. 1, 1885, pl. 5, fig. 15.

[1] No growth studies based on the scales or on other exact methods have been undertaken for the bluefish, to our knowledge.

[2] Lyman (Bluefishing, 1950, p. 10), who reports the case, suggests that the fisherman who re-caught the fish "may have been stretching things a bit."

[3] Frozen bluefish have recently been imported from northern Argentina.

[4] See Lyman, Bluefishing, 1950, p. 12, for photo from the Bermuda News Bureau of a 15-pound bluefish caught at Bermuda, February 1949.

Africa; also Mediterranean; both coasts of southern Africa; Madagascar; eastern Indian Ocean and Malay Peninsula; southern Australia and New Zealand.

Occurrence in the Gulf of Maine.—Bluefish have been taken at one time or another wherever any information is available as to the local fishes around the western side of the Gulf. But they have seldom been seen east of Penobscot Bay (reported at Mount Desert in 1889); we have heard of only one taken in the Bay of Fundy, a fish caught in Minas Basin in July 1951,[5] and we have found no record of bluefish off the Nova Scotian coast of the open Gulf of Maine. But one was caught off Halifax in 1925, another more recently near Liverpool on the outer coast of Nova Scotia,[6] and they were reported "common" near Port Medway, Nova Scotia, in the summer of 1951.[7]

In our Gulf, too, they seem to be confined to the vicinity of the coast (they are unknown in the central basin or on Georges Bank), the small ones ("snappers") running up into brackish water, as in the Parker River, but the larger sizes (3 pounds or more) keeping to the outside waters.

The geographic distribution of the places where they have been recorded would suggest at first glance that bluefish are practically universal in the western side of the Gulf. But this is true only for brief terms of years and at long intervals, for while they have been known to swarm there for several summers in succession, they may then be so rare over periods of many years that the capture of a single fish causes remark.

Bluefish must have been common at the time of the first settlement, at least as far north as what is now southern Maine, for Josselyn, writing in 1672, referred to them as better meat than the salmon.

Bluefish were plentiful off southern New England and also about Nantucket in colonial times, but they seem to have disappeared thence about 1764, not to reappear there until about 1810. From that time on they increased in abundance west and south of Cape Cod, but none were reported north of the Cape until 1837. And since a fish as ubiquitous as the bluefish would certainly have attracted attention and its presence found its way into print, had it been abundant in the Massachusetts Bay region, it is safe to say that very few, if any, visited the Gulf of Maine during the late eighteenth century, or the first quarter of the nineteenth.

According to Storer, the first bluefish seen north of Cape Cod after their long period of absence, was one caught on October 25, 1837; Captain Atwood [8] saw them for the first time at Provincetown in 1838. According again to Storer, bluefish were taken yearly from the wharves at Boston after 1844. And by 1850 they were so plentiful about Cape Ann that fishermen complained of them as driving away most of the other schooling fish, while in 1863, which seems to have marked the culmination of this flood of bluefish, they were extremely abundant in the Massachusetts Bay region and especially at Provincetown.[9] They remained plentiful in the southern part of the Gulf of Maine for several summers after 1863, but by 1872 they were reported as much less so, and there have not been enough bluefish anywhere in the Gulf since the late 1870's to menace the local mackerel fishery.

The yearly catch for the Cape Cod Bay region had fallen to about 22,000 pounds by 1888 (93 pounds for Essex County), to only about 3,000 pounds for 1889. But some bluefish were seen as far north and east as Mount Desert in that year, and evidently more of them rounded the Cape during the next 9 seasons, for the catches for the years 1890–1898 were between about 26,000 pounds and 80,000 pounds for Cape Cod Bay; with a few hundred pounds for the Massachusetts coast north of Boston. But this period of moderate plenty was followed by a period of scarcity [10] so extreme (detailed statistics are wanting) that no catch as large as 5,000 pounds was reported again as made anywhere in our Gulf in any year for which statistics are available from 1900 [11] down to the early 1920's. In 1906, in fact, in 1910 and again in 1919, only an occasional school can have

[5] Reported to us by Dr. A. H. Leim.

[6] Leim, Proc. Nova Scotian Inst. Sci., vol. 17, Part IV, 1930, p. XLVI.

[7] Information from Dr. A. H. Leim, from report by L. R. Day, Fisheries Research Board of Canada.

[8] Proc. Boston Soc. Nat. Hist., vol. 9, 1863, p. 189.

[9] Baird (Rept. U. S. Comm. Fish (1871–1872) 1873, p. 237–240), and Goode, (Fish. Ind. U. S.; Sect. 1, 1884, p. 435–437) have collected much information about the early history of the bluefish.

[10] Reported catches for the Cape Cod Bay region by all methods were only about 3,600 pounds in 1899 and 7,659 pounds in 1900.

[11] Statistics of the pound net catches, by towns, were published in the Annual Report of the Commissioners on Fisheries and Game of Massachusetts for the years 1906, 1908, 1909, 1910, 1918, and 1919.

rounded Cape Cod,[12] while bluefish must have been practically nonexistent north of the Cape in 1918, for the entire reported catch there was only 34 pounds for that year.

We should also point out (we cannot explain this) that a larger proportion of the bluefish than usual that did round Cape Cod seem to have continued on to the northern shore of Massachusetts Bay during this period of general scarcity. Thus about as many (300 lb.) were reported for Essex County in 1906 as for the Cape Cod Bay region; about one-fourth as many in 1908, about one-half to one-third times as many in 1908, 1909, and 1910.[13]

Bluefish must have come north in greater numbers in 1927, for they were reported here and there from Cape Ann northward during that summer with small catches in the Casco Bay region,[14] and there seem to have been still more of them in the Gulf during the next two summers, as reflected in reported catches of 4,825 pounds for Essex County, Mass., and 140 pounds for the Casco Bay region, Maine, in 1928; 7,888 pounds for Essex County and 495 pounds for Casco Bay, Maine, in 1929. And so many blues invaded the southwestern part of the Gulf during the next three years that about 68,000 pounds were reported for Essex County, and 200 pounds for Casco Bay in 1930; 60,000 pounds for Essex County and 500 pounds for Casco Bay in 1931; and 1,414 pounds for the coast of Maine as a whole in 1932.[15]

Eighty pounds of bluefish were taken in a set of traps at North Truro on Cape Cod Bay on September 9 in 1936;[16] we heard of some large ones caught in the surf on the outer shore of Cape Cod that same year; we know of one caught at Cohasset on the south shore of Massachusetts Bay in 1937, and enough came north again in 1938 to provide a commercial catch of about 1,800 pounds for Essex County. But this slight upsurge was followed by 6 years, or more, of scarcity so extreme that only small catches were reported from Maine (200 pounds in 1944), or from northern Massachusetts (200 pounds for Essex County in 1945). The year 1946 was perhaps the low point for our Gulf, when the total catch including the southern coast of Massachusetts, was only about 1,200 pounds.

In 1947, however, when the total reported catch for Massachusetts was only 2,300 pounds (none reported for Maine)[17], schools of small bluefish, of a pound or so, appeared along the inner shores of Cape Cod, near Wellfleet, in August. They are said to have been more widespread in Cape Cod during the two next summers, and in 1950 bluefish of ¾ pound to 1½ pounds, with a few up to 4 or 5 pounds, rounded Cape Cod in such numbers that Cape Cod Bay was described by anglers as "loaded" with them during that August.

Many catches of 60 to 100 pounds were made from party boats, both in the Wellfleet side of the Bay and along the Sandwich shore; a set of 8 traps at North Truro made small catches (10 to 160 pounds) at intervals between July 18 and October 7 of that year, while other traps around Cape Cod Bay from Provincetown and Sandwich made various catches between June and October. Some also worked north across Massachusetts Bay; witness captures of about 6 bushels of 1 to 1½-pound bluefish in a trap off Marblehead on the north shore of Massachusetts Bay, July 21.[18] We heard of at least one taken at Hampton, N. H., also a few at Kennebunkport, Maine, early that September,[19] and, at least, one from the lower Kennebec River. They continued plentiful also in Cape Cod Bay until the first week of that September. And while few, if any, were seen there after the severe northeast gale of mid-September, some were caught along the outer shore of Cape Cod as late as the first week of that October, and perhaps until later still.

All we can say, as to the catch in 1951 up to this writing (August 15) is that the earliest report of a bluefish in Cape Cod Bay was for June 19; a small one was taken at Plum Island, northern Massachusetts, on July 6; good catches (presumably of small fish) were reported in Cape Cod Bay by the last week of July (2,545 pounds taken in one set of traps at North Truro, July 7–28); a tremendous run of small bluefish were reported

[12] The catches north of Cape Cod ran only between about 300 pounds and 600 pounds for those years.

[13] Pound net catches of 1,015 pounds reported for Essex County, 4,623 pounds for the Cape Cod Bay region in 1908; 600 pounds and 1,342 pounds, respectively, in 1909; 182 pounds and 419 pounds, respectively, in 1910.

[14] The Boston Post for July 24, 1927, reported 65 taken near Bald Head one day, and 35 the next by Capt. Charles F. Pye.

[15] No regional breakdown is available for Massachusetts for that year.

[16] Information contributed by the Pond Village Cold Storage Co.

[17] This is the most recent year for which catch statistics have been published.

[18] Reported in the Boston Herald for July 31, 1950.

[19] Reported in the Saltwater Sportsman for September 15, and October 6, 1950.

near Provincetown during the first part of August,[20] and a few large fish, among great numbers of small ones were being taken off the south shore of Massachusetts, and even in the Cape Cod Canal; some also were being caught in the rips and in the surf at the tip of Monomoy Point.

It remains to be seen whether this increasing run of bluefish in our Gulf is comparable to that of the 1860's. However this may prove, history will no doubt repeat itself sooner or later, and these sea pirates will again invade the Gulf in abundance, probably for several summers in succession.

The disappearance of the bluefish from the Gulf of Maine following the run of the 1860's was part of a general shrinkage in the bluefish population that visits the coast east of New York, and was to be expected, for the bluefish that reach our Gulf are only the northernmost fringe of the northern contingent. The increase in the numbers caught north of Cape Cod in the period 1928–31 was associated, similarly, with a corresponding rise in the yearly catches made off southern New England from about 55,000 pounds in 1928 to about 650,000–920,000 pounds for 1930–33.

The scarcity of bluefish north of Cape Cod from the early 1930's down to the early 1940's (interrupted in 1938 as noted) was clearly the result of the general decrease that took place in the abundance of bluefish over the northern part of their range as a whole, reflected in the southern New England catch which fell from nearly a million pounds in 1933 [21] to an all-time low of only 12,500 pounds in 1945. And there can be no doubt that the small bluefish that have reappeared in the Cape Cod Bay region and northward in increasing numbers during the past 2 or 3 years have been the overflow so to speak, from an increasing population to the southward, great numbers of which (mostly small) were being caught from New Jersey to Nantucket in 1950, and are being caught there with some large ones, at this writing (August 15, 1951).

We ought perhaps to add that it is only in the northern part of its range that the American bluefish falls periodically to a very low level; in 1945, for example, when the total catch for New England was only about 26,000 pounds, 223,000 pounds were taken in Chesapeake Bay and about 2½ million pounds along the Atlantic and Gulf coasts. This, again, was to be expected, for it is near the boreal boundary of its range that any warm-water fish is subject to the greatest vicissitudes.

An interesting phase of the fluctuations of the bluefish is that large numbers of very small ones have visited the southern coast of New England even in the poorest summers; some have been reported within the Gulf of Maine. Thus fry so small as evidently to have been the product of that season's hatch were taken in Casco Bay, Maine, in August 1899; slightly larger ones of 4 to 5 inches were caught off Plymouth in the summer of 1921, and "snappers" are sometimes reported at Provincetown, at Plymouth, and in the Parker River in northern Massachusetts. Almost all the fish, furthermore, that have been taken within the Gulf of Maine, and the majority of the larger catches that have been taken off southern New England during the past few years have been fish so small (mostly 1 to 2 pounds) that it is not likely they had reached sexual maturity. It is only in good bluefish years that many of the mature fish (weighing upwards of 4 or 5 pounds) appear that far north. In poor years large fish are caught in numbers only to the southward of Long Island, N. Y. Prior to August 15, 1951 a number of fish up to 7 pounds had been caught in southern New England waters, which may indicate better things to come.

In the years when bluefish pass Cape Cod in any numbers they usually appear in Cape Cod and Massachusetts Bays about the middle of June,[22] sometimes as early as the first of that month, and they are seen off and on all summer. Most of them depart late in September, but an occasional fish lingers into late autumn. Bluefish have even been caught about Provincetown as late as December.

Importance.—The bluefish is an excellent table fish, but it never has been plentiful enough to support a fishery of any magnitude in the Gulf of Maine. Nevertheless, its presence or absence there may be a matter of direct importance to the fishing interests, for it may drive away the mackerel when it swarms, if not the herring and menhaden as well. Being a favorite game fish, many

[20] Reported in Salt Water Sportsman for August 10.

[21] 920,965 pounds reported in 1933, to be exact.

[22] Along southern New England the first blues are expected during the last half of May (p. 384).

anglers troll for them in Cape Cod Bay in seasons when there are enough of them to be worth following; also many are caught in the surf in good years by anglers casting from the beach, as far northward along the coast as the outer shore of Cape Cod.[23]

THE SEA BASSES. FAMILY SERRANIDAE

The sea basses are an extremely numerous tribe of perchlike fishes, with both the spiny portion and the soft rayed portion of the dorsal fin well developed, either as separate fins, or at least divided by a deep notch. The ventral fins are under the pectorals, technically thoracic, in position. The anal fin is nearly or quite as long as the soft part of the dorsal; the caudal peduncle is deep and the tail is broad. In most of the species the anal fin is preceded by 3 stout spines; the margin of the gill cover bears one or two sharp conical spines in most, and the maxillary bone is not sheathed nor hidden by the preorbital bone when the mouth is closed. Smoother cheeks are a ready field mark to distinguish any of the sea basses from the rockfish family (p. 430); the 3 anal fin spines distinguish them from the croaker family (p. 417) which have 1 or 2 anal spines only; the spiny gill cover from the porgy family (p. 411); and the large mouth from the cunner and tautog tribe (p. 473).

KEY TO GULF OF MAINE SEA BASSES

1. There is one continuous dorsal fin, its front part spiny, its rear part soft rayed ---------------------------- 3
   There are two separate dorsal fins, the first spiny, the second soft-rayed ---------------------------------- 2
2. The two dorsal fins are separated by a distinct inter-space; the sides are distinctly striped ---- Striped bass, p. 389
   The two dorsal fins are joined at their bases; the sides are not distinctly striped --------------- White perch, p. 405
3. The scales are large; the space between the eyes is naked; no bony ridge on the gill covers --------- Sea bass, p. 407
   The scales are small; the space between the eyes is scaly; there is a bony ridge on the gill covers -- Wreckfish, p. 409

Striped bass *Roccus saxatilis* (Walbaum) 1792

STRIPER; ROCKFISH; ROCK; LINESIDES

Jordan and Evermann, 1896–1900, p. 1132, as *Roccus lineatus* (Bloch).

Description.—No one character alone characterizes the striped bass, but rather the combination of fin structure and arrangement with general outline and structure of the jaw. Its rather deep and keelless caudal peduncle, stout body, the presence of two well-developed dorsal fins (spiny and soft rayed, and the one about as long as the other), its lack of dorsal or anal finlets, and a tail only moderately forked, separate it from all the mackerel tribe, from the bluefish, and from the pompanos. The fact that its anal fin has 3 spines

[23] We refer the reader to Lyman (Bluefishing, 1950) for an excellent account of bluefishing methods and localities, also of the natural history of the bluefish.

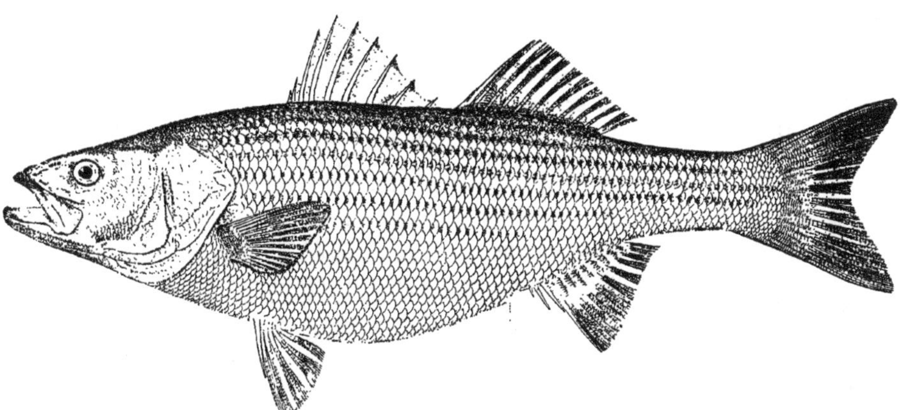

FIGURE 209.—Striped bass (*Roccus saxatilis*), Chesapeake Bay region. From Goode. Drawing by H. L. Todd.

and is almost as long as the second dorsal, also (less obvious) that its maxillary (upper jaw) bones are not sheathed by the preorbital bone, separate it from all the weakfish tribe (p. 417). Nor is there any danger of confusing it with the sea bass, cunner, tautog, or rosefish, for its two dorsal fins are entirely separate whereas in all these the spiny and soft-rayed parts are continuous, as a single fin. The white perch comes closest to it in general appearance but the two dorsal fins of the perch have no free space between them (p. 405), and its fin spines are stiffer.

The trunk of the striped bass is 3⅓ to 4 times as long (to base of caudal fin) as it is deep, thick through, its back hardly arched. It has a moderately stout caudal peduncle, a long head (almost as long as the fish is deep), two spines on the margin of each gill cover, an oblique mouth gaping back to the eye, a moderately pointed nose, and a projecting lower jaw. Young fish are more slender than old. The two dorsal fins are of about equal lengths; the first (9 or 10 stiff spines) triangular in outline, originating over the middle of the pectorals; the second (12 or 13 soft rays) regularly graduated in height from front to rear, and separated from the first by a distinct (though short) space. The anal (about 11 rays preceded by 3 spines) is of about the same size and form as the second dorsal, and originates below the middle of the latter. The caudal is moderately wide and only slightly forked. The pectorals and ventrals are of moderate size, the latter somewhat behind the former.

Color.—Dark olive green varying to bluish above, paling on the sides, and silvery on the belly, sometimes with brassy reflections. The sides are barred with 7 or 8 narrow, sooty, longitudinal stripes, which follow as many rows of scales and which may be variously interrupted. The highest stripe is the most distinct, and all of them but the lowest are above the level of the pectoral fins. The dorsal, caudal, and anal fins are somewhat dusky.

Size.—The bass grows to a great size, the heaviest of which we have found definite record being several of about 125 pounds that were taken at Edenton, N. C., in April 1891.[24] One of 112 pounds, which must have been at least 6 feet long, was caught at Orleans, Mass., many years ago. One of 100½ pounds is said to have been taken in Casco Bay, Maine,[25] and fish of 50 to 60 pounds are not exceptional. Usually bass, as caught, weigh from 3 to 35 or 40 pounds; the average weight of ones recorded in the register of the former Glades Hotel [26] at Scituate, Mass., during the period 1854 to 1858, was about 27 pounds.

Bass weigh about ¾ pound when 12 to 13 inches long; about 2¾ to 3 pounds at 18 to 20 inches; about 5 pounds at 24 inches; about 10–15 pounds at 30–32 inches; and about 18–20 pounds at 33–36 inches. Twenty-pound bass average about 36 inches in length; 30 pounders about 43 inches; 40 pounders about 47 to 48 inches.[27] On the Pacific coast 50 pounders run about 50 to 51 inches,[28] and the relationship between weight and length runs about the same for very large fish on the Atlantic coast. The record fish caught on rod and reel was one of 73 pounds, taken in Vineyard Sound in August 1913 by C. B. Church.

Females grow larger than males; probably most bass of 30 pounds and heavier are females.[29] Thus the common use of the term "bulls" for the very large ones might better be replaced by "cows."

Habits.[30]—Stripers are powerful fish; so strong in fact, that they appear to have no difficulty in handling themselves in the surf, where one is sometimes seen actually in the translucent crest of a comber just before the latter breaks. But this is not a very swift fish as compared with the mackerel tribe. Bass often swirl conspicuously at the surface or splash in pursuit of bait fish. They sometimes roll as the little northern porpoise or puffing pig (*Phocaena*) does. And we have heard of them finning (i. e., with dorsal and tail fins showing).[31] But we have never seen or heard of one leaping clear of the water as tuna and bonito so often do unless hooked in shoal water.

During the first two years they live mostly in small groups. Later they are likely to congregate in larger schools; this applies especially to those up

[24] Smith, North Carolina Geol. and Econ. Surv., vol. 2, 1907, p. 271.

[25] Atkins, Fish. Ind. U. S., Sect. 5, vol. 1, 1887, p. 694.

[26] Kindly lent by John Adams.

[27] For a detailed tabulation of the length-weight relationship for bass from ¼ pound to 47¼ pounds, see Merriman, Fish. Bull. No. 35, 1941, U. S. Fish and Wildlife Service, p. 7, vol. 50, 1950, pp. 1–77.

[28] As scaled from a graph given by Scofield, California Fish and Game, vol. 18, 1932, pp. 168–170, fig. 38.

[29] Definite information in this regard is scant.

[30] Interesting recent studies of the striped bass are by Pearson (Bull. U. S. Bur. Fish., vol. 49, 1938, pp. 825–851) and by Merriman (Fishery Bull. 35, U. S. Fish and Wildlife Service, 1941, 77 pp.).

[31] Frank Mather of the Woods Hole Oceanographic Institution reports an instance of this.

to 10 pounds or so, which are often spoken of as "school fish." The larger ones often school, but the very largest, of 30 to 40 pounds and upward, are more often found single or a few together. They are most likely to be in schools while migrating, but more scattered while feeding in one general locality.

Small fish (2 and 3 years old) in particular, tend to school densely; also they travel considerable distances without scattering but, as Merriman emphasizes,[32] it is not likely that a given school holds together for any long period, for fish of various sizes (i. e., ages) up to the very large ones often school together, showing that different ages intermingle more or less. Mixed schools running from 8 or 10 pounds to 30 or 40 pounds were reported repeatedly in 1950, for example.

The bass is very voracious, feeding on smaller fishes of whatever kind may be available, and on a wide variety of invertebrates. Lists of its stomach contents for one locality or another include alewife, anchovy, croakers, channel bass, eels, flounders, herring, menhaden, mummichogs, mullet, rock eels (*Pholis gunnellus*), launce, sculpins, shad, silver hake, silversides, smelt, tomcod, weakfish, white perch, lobsters, crabs of various kinds, shrimps, isopods, gammarid crustaceans, various worms, squid, soft clams (*Myra*) and small mussels. In our Gulf the larger bass prey chiefly on herring, smelt, sand launce, eels, and silver hake, on squid (on which they gorge when they have the opportunity), on crabs large and small, on lobsters, and on sea worms (*Nereis*); while small ones are said to feed to a considerable extent on gammarid crustaceans and on shrimps.

When bass are gorging on any one particular prey it is common knowledge among fishermen that they are likely to ignore food of other sorts for the time being. It seems also that when prey is plentiful, bass are likely to gorge, then cease feeding to digest, then to gorge again; also that all the members of a given school are likely to do this in unison, with consequent annoyance to the angler.

Bass, too, seem on the whole to be more active, and especially to feed more actively, between sunset and sunrise than while the sun is high. In estuarine situations this fits with the habits of their prey, for it is by night that the sea worms (*Nereis*) that are the chief item in their diet there emerge from their burrows to swim about. And bass fishing is often much more productive by night than by day off the open coast also, though schools of bait fish are seen at all hours (else the terns would starve), while the time when crabs, etc., are most likely to be stirred up by the surf, and are most easily caught around the rocks, depends on the stage of the tide, not on the hour of the day. So most fishermen (ourselves included) believe that it is inherent in the nature of the larger sized bass to avoid strong sunlight by sinking to the bottom. A familiar instance is the regularity with which they desert the surface soon after sunrise on bright summer days at places where large numbers are caught by trolling during the hour or two after daybreak; the eastern side of Cape Cod Bay is a local example.

It has been discovered recently that trolling deep with wire lines is often productive, irrespective of the time of day, at times and places where bass "show" only during the early morning hours. This habit, however, is not so deeply engrained but that schools of bass often rise to the surface in pursuit of bait fish at any time of day, or come within easy casting distance of the beach. We recall seeing several schools of good-sized fish (those that we landed ran up to 23 pounds) suddenly splashing all around our boat about midday, on one occasion off Wellfleet, in Cape Cod Bay, though it was only for a few hours after sunrise that the several boats fishing regularly there had taken any by top-water trolling for some time previous.

The best advice we can give the surf-caster, in this regard, is to go fishing whatever time of the day he is free to do so.

The striper is so strictly an inshore fish that we have never heard of large catches being made, or schools seen, more than 4 or 5 miles from the nearest point of land,[33] though the migrating schools doubtless pass much farther out in crossing the mouths of the larger indentations of the coast, such as Delaware Bay and Long Island Sound. And a few fish may stray far offshore in winter, for one about 18 inches long was taken in an otter trawl about 60 miles south of Marthas Vineyard, in 70 fathoms of water, in February 1949 (p. 400).[34]

[32] Fish. Bull. 35, U. S. Fish and Wildlife Service, 1941, vol. 50, p. 43.

[33] Henry Lyman informs us that bass are caught in numbers late in the autumn in the rips east of Nantucket about 4 miles out, but that verbal reports of some taken during the summer of 1950 on the offshore part of Georges Bank were actually based on two weakfish (p. 419).

[34] Reported to us by Capt. Henry W. Klimm of the dragger *Eugene H.*

On the landward side, many bass come within easy casting range of the shore; we have had a fair sized one strike our plug not 4 feet from the rock from which we were casting on the Cohasset shore. Many (especially the smaller sizes, but large ones also) run up into estuaries and into river mouths. In some rivers, good numbers (large as well as small) are caught so far upstream as to make it likely that they remain there the year round. This is notably the case in the Alabama River system where (we hear) 250 to 300 bass ranging from 5 to 40 pounds were caught near Tallasseem some 30 miles above Montgomery, which is at least 300 miles from salt water, following the river.[35] They are also known to spawn some 250 miles up the Sacramento River in California. It would be interesting to know what proportion of the bass that spawn at Weldon, N. C., 100 miles or so up the Roanoke, and that run 60 to 90 miles up the St. John, in New Brunswick,[36] ever see salt water. Bass also run up the Hudson for about 160 miles to Albany.

The great majority of the total population of bass frequent the coast line, except at breeding season. Among these, the smaller sizes, up to 15 pounds or so, are found indifferently within enclosed bays, in small marsh estuaries, in the mouths of rivers and off the open coast. But we do not often hear of fish heavier than 20 to 25 pounds caught in situations of these sorts. And the great majority of the large bass, of 30 pounds or more, hold to the open coast, except at spawning time (p. 394), and perhaps in winter (p. 400). But this is not an invariable rule; we are familiar with one narrow inlet where tides run strong, and where some lucky angler catches a very large bass now and then (p. 396).

Bass off the open coast are most likely to be found along sandy beaches, in shallow bays, along rocky stretches, over and among submerged or partially submerged rocks and boulders, and at the mouths of estuaries, the precise situations that they occupy being governed by the availability of food. Off the outer beaches they may be anywhere right to the breakers. When they are close in they frequent the troughs that are hollowed out by the surf behind off-lying bars, also the gullies through which the water rushes in and out across the bars as the rollers break, for it is in such situations that bait fish are easiest caught, and that crabs, worms, and clams are most likely to be tossed about in the wash of the breakers. When the tide is high, bass often lie on a bar, or even in the white water along the beach if there is a good surf running. When the tide falls they drop down into the troughs or move farther out, according to the precise topography. In either case, every surf fisherman knows that his chances are much better when the sea is breaking at least moderately heavy so that he can cast into white water, than when it is smooth.

They also lie under rafts of floating rockweed at times, probably to prey on the small animals they find among the weeds.

The best spots along rocky shores are in the surf generally, and in the wash of breaking waves behind offlying boulders and among them, or where a tidal current flows most swiftly past some jutting point. In the mouths of estuaries they are apt to hold to the side where the current is the strongest, and in the breakers out along the bar on that side. In shallow bays, they often pursue small fry among the submerged sedge grass when the tide is high, dropping back into the deeper channels on the ebb. And they frequent mussel beds, both in enclosed waters and on shoal grounds outside, probably because these are likely to harbor an abundance of sea worms (*Nereis*).

When bass are schooling outside they are likely to be moving along the coast in the one direction or in the other. But they may remain in the same general locality for weeks, or through the summer. Thus a body of very large fish, of 25 to 50 pounds, stayed close in to the outer beach near the tip of Cape Cod, through most of July of 1951 and into that August, yielding consistent catches to the more skillful surf-fishermen.

Bass are active over a temperature range from perhaps 70° down to about 43°–46° F. Present indications are that if the temperature falls lower they either withdraw to somewhat warmer water if off the outer coast, or lie on the bottom in a more or less sluggish state if they are in some estuary. On the other hand it is not likely that they can long survive temperatures higher than about 77°–80°, for many were found dead in

[35] Information from Henry Lyman, from an angling correspondent in Alabama. They have long been known up the Alabama as far as Montgomery (Pearson, Bull. U. S. Bur. Fish., vol. 49, 1938, p. 826).

[36] According to Adams (Field and Forest Rambles, 1873, Pt. 3, Fishes, pp. 248-249), who has given an interesting and readable account of the bass in the river.

shallow estuaries in Connecticut and in Massachusetts during the abnormally hot August of 1937.[37] They are equally at home in fresh or slightly brackish water, and in coastal salinities of 3.1 to 3.3 percent. But their usual wanderings do not take them out into waters of full oceanic salinities (3.5 percent or higher).

Migrations.—No phase of the life history of the bass arouses as much discussion among fishermen as their migrations. And the picture still remains so puzzling that we dare not attempt anything more than a brief summary of what has been learned to date.

It seems certain that stripers do not ordinarily travel far until they are 2 years old. Thus the young fish from the enormous year classes of 1934 and 1942—apparently produced in the Chesapeake Bay–Delaware Bay region chiefly—did not appear in New England waters until 2 years later. But the fact that they did appear there and in the Gulf of Maine in hordes in the summers of 1936 and 1944 shows that a bass is capable of very extensive journeys, once it has reached its third year.

It has long been known, too, that the pound nets on Long Island and along southern New England ordinarily make large catches only in the spring (peak in May), and again from early October into November;[38] also that large spring catches are made progressively later in the season, proceeding from south to north, the reverse being true in the autumn. This, of course, suggests that part at least of the bass population follows the shore line northward and eastward as far as southern New England in spring, to return westward and southward in autumn. And this is verified for bass 2 and 3 years old by the returns from tagging experiments by Merriman at the eastern end of Long Island and in Connecticut during the years 1936 to 1938,[39] for recaptures of fish that had been tagged there in May came mostly from farther east along southern New England, one from Cape Cod Bay, and another from Cohasset on the southern shore of the inner part of Massachusetts Bay. But the recaptures from fish tagged in summer were mostly from nearby (evidence of a stationary population), while those for autumn-tagged fish were scattered along the coast from the eastern end of Long Island to Chesapeake Bay, with one from Croatan Sound, one from Albemarle Sound (Stumpy Point), and one from Pamlico Sound in North Carolina.

But the picture is by no means so simple as the foregoing might suggest. To begin with, no evidence is available as to the movements of large bass, other than the successive dates when they appear or disappear off different parts of the coast.[40] And it is no less true of bass than it is of mackerel (p. 330), that successive appearances and disappearances from place to place are not conclusive evidence of along shore migration. Yet it is now certain that while some bodies of bass carry out extensive migrations north and east in spring, west and south in autumn, other bodies do not. Thus, as Merriman points out,[41] the bass of the northeastern shore of the Gulf of Mexico are completely isolated, while those of the Atlantic coast south of Cape Hatteras form another separate population, few of which (if any) ever spread farther north. The bass of the Gulf of St. Lawrence and of the lower St. Lawrence River appear to be wholly isolated also. And while some interchange may take place between the populations found in various bays and rivers around the outer coast of Nova Scotia, it is doubtful whether these have any regularly migratory association, either with the Gulf of St. Lawrence fish or with those of more southern waters, except in occasional years (p. 398).

Chesapeake Bay, however, harbors both migratory bass,[42] as proved by tagging experiments (p. 393) and other evidence (p. 393), and nonmigratory as proved by the fact that fish of all sizes are taken there both in summer and in winter, though not so many of them as in spring and fall. Similarly, some bass winter in northern waters though most of the fish appear to be migrants there; and perhaps a considerable percentage do so in the lower reaches of the Hudson River estuary.

[37] Merriman, Fishery Bulletin No. 35, U. S. Fish and Wildlife Service, 1941, vol. 50, p. 43.

[38] See Merriman, Fish. Bull. No. 35, U. S. Fish and Wildlife Service, 1941, vol. 50, pp. 33, 34, fig. 24, for details.

[39] For details we refer the reader to Merriman's original account (Fish. Bull No. 35, Fish and Wildlife Service, 1941, vol. 50, pp. 36–42, figs. 26–29; also pp. 71–73, tables 17–20), which is the most authoritative discussion of the subject that has appeared yet.

[40] The few returns so far from bass of 5 pounds and upward that have been tagged have been from nearby, and soon after they were released.

[41] Fish. Bull. No. 35, U. S. Fish and Wildlife Service, 1941, vol. 50, p. 42.

[42] Using this term to mean extensive seasonal journeys.

Merriman[43] has suggested that these northern wintering fish may be "of two types—the individuals that form the resident more or less isolated population" and others "that may have had their origin farther south but spend an occasional winter in northern waters." It may prove that a good proportion of these bass that come from the south when they are 3–4 years old may remain in the north for the rest of their lives. And there is no way for the fisherman to tell in which of these categories the bass belong, that he lands. The reader will find some further discussion of migrations in connection with the status of the bass in the Gulf of Maine (p. 395). We need only add that the existence of these nonmigratory populations and the fact that the Pacific coast bass are similarly stationary, are sufficient proof that seasonal migration is not an essential incident in the life of the striper.

Bass spawn either in brackish water at the heads of estuaries[44] (the Hudson, for example) or in fresh rivers, never off the open coast in salt water so far as is known. Those that enter fresh rivers may deposit their eggs only a short distance above the head of tide as they do in the Potomac, or they may run much farther upstream. But we have yet to learn how large a percentage of the bass that are known to spawn 100 miles up the Roanoke, near Weldon, N. C. (a major spawning ground), or still farther up the Alabama,[45] and up the Sacramento River in California, have come from salt water (p. 392).

The chief requirement for successful spawning is (it seems) a current turbulent enough to prevent the eggs from settling on bottom where they would be in danger of being silted over and smothered.

The spawning season is from late April to early May in North Carolina; in May, chiefly, in the Chesapeake Bay region; perhaps equally early in the waters of New York.[46] Any bass that may spawn in the rivers of Massachusetts, of Maine, and of the Bay of Fundy, probably do so in June; those of the southern shore of the Gulf of St. Lawrence and of the lower St. Lawrence River in June and July.

A large female during spawning may be surrounded by many small males, and the latter are described as fighting fiercely[47] with one another.

Females stripped at the Weldon, N. C. Hatchery yielded from 11,000 to 1,215,000 eggs each, during the period 1928 to 1938, with one of 4½ pounds yielding 265,000. Thus the oft-quoted estimate of 10 million fish for a really large one is within reason.[48]

The eggs average 1.1–1.35 mm. in diameter when they are deposited in the water, but the pervitelline membrane swells during the first hours after fertilization to an average diameter of about 3.6 mm. They have a large oil globule and are semibuoyant; that is, they sink in quiet water, but are swept up from the bottom by the slightest disturbance, so that they tend to drift downstream with the current. Consequently the eggs that are produced far upstream may not hatch until they have reached tidewater. The eggs are reported as hatching in about 70 to 74 hours at a temperature of 58–60°; in about 48 hours at 67°; in about 30 hours at 71–72°.

In Chesapeake Bay, the young fry of the year are about 1⅕ inches (30 mm.) long by June; 1⅘ to 2 1/12 inches (45–53 mm.) long in July; 2 to 2⅘ inches (50–70 mm.) in August; and 3¾ to 8½ inches by the following April and May; i. e., at the end of their first year.[49] According to Merriman,[50] most of the fry of the year taken in the Hudson River during their first summer are between about 1⅝ inches (40 mm.) and about 3½ inches (90 mm.) long; a few seined in the Parker River, Newbury, Mass., were from about 2¾ inches (71 mm.) to about 3⅜ inches (85 mm.) long. And this last is perhaps representative for whatever bass may now be produced in Gulf of Maine rivers, for we read that great numbers of fry of 2 to 3 inches were taken of old in winter in the rivers of Maine in bagnets set for smelt and tomcod.[51]

Two-year-old bass taken in Connecticut averaged 11 to 11½ inches (28 or 29 cm.) long in spring,

[43] Fish. Bull. No. 35, U. S. Fish and Wildlife Service, 1941, vol. 50, p. 42.

[44] See Merriman, Fishery Bulletin No. 35, U. S. Fish and Wildlife Service, 1941, vol. 50, p. 17, for precise salinites in which bass in their first summer have been taken in the Hudson River, and in the Parker River, Massachusetts. See Tresselt (Bull. Bingham Oceanogr. Coll., vol. 14, art. 1, pp. 98–110, 1952) for a survey of spawning grounds tributary to Chesapeake Bay.

[45] Pearson (Bull. U. S. Bur. Fish., vol. 49, 1938, p. 829) records a female with eggs from the Alabama River near Montgomery.

[46] Greeley (New York Conserv. Dept., Biol. Surv. Lower Hudson Watershed, 1937, p. 100) concludes that the spawning season in the Hudson "includes May."

[47] See Smith, North Carolina Geol. Econ. Survey, vol. 2, 1907, p. 272, for an eyewitness account by S. G. Worth.

[48] Merriman (Fish. Bull. No. 35, U. S. Fish and Wildlife Service, 1941, vol. 50, p. 19) gives an excellent summary of information available as to spawning, characteristics of the eggs, and period of incubation.

[49] Hildebrand and Schroeder, Bull. U. S. Bur. Fish., vol. 43, Pt. 1, 1928, pp. 248–249

[50] Fish. Bull. No. 35, U. S. Fish and Wildlife Service, 1941, vol. 50, p. 17, fig. 10.

[51] Atkins, Fish. Ind. U. S., Sect. 5, vol. 1, 1887, p. 693.

about 12 inches (30 cm.) in June, and about 14½ inches (37 cm.) in October; the 3-year-olds about 15¾ inches (40 cm.) in spring and about 18 inches (46 cm.) in October, while 4-year-olds increased in length from about 18¾ inches (48 cm.) to about 20¾ inches (53 cm.) between spring and autumn, on the average.[52] And the average rate of growth was about the same for Hudson River fish examined by Greeley.[53] But the rate at which they grow is governed largely by the food supply. Bass in captivity have been known to grow from 6 inches long to 20 inches in 11 months, while some that were kept in a certain pond in Rhode Island are described as having gained weight from 1 pound in June to 6 pounds in October.[54]

The later growth rate has not been traced for our Atlantic bass. But it is generally believed that the 35–50-pounders that were caught in considerable numbers in 1950, and are being taken in 1951, were members of the very successful year classes of 1940–1942, which fits well with the growth rate of bass on the Pacific coast, where the average age is about 7 years for 20-pound fish, 10–11 years for 30-pounders, about 14 years for 40 pounders, and 17 to 18 years for 50-pounders.[55]

On the Pacific coast females grow faster than males after the third year, which is probably true of the Atlantic bass also.[56] This certainly is a long-lived fish for one kept in the New York Aquarium lived to be 23 years old.[57]

Merriman [58] found that "approximately 25 percent of the female striped bass first spawn just as they are becoming 4 years old, that about 75 percent are mature as they reach 5 years of age, and that 95 percent have attained maturity by the time they are 6 years old," among Connecticut fish. But a large percentage of the males had matured at 2 years, probably nearly all of them by the time they were 3 years old. And it is probable that this applies equally to the Maine bass. Merriman has also made the interesting discovery that only about one-tenth of the bass of northern waters are males, but that males are nearly as numerous as females, southward from Delaware Bay.

It has been suggested that the striper may not be a regularly yearly spawner,[59] but no positive evidence is at hand as to this.

General range.—Atlantic coast of eastern North America, from the lower St. Lawrence River and the southern side of the Gulf of St. Lawrence to northern Florida; also along the northern shore of the Gulf of St. Lawrence to Alabama and Louisiana; running up into brackish or fresh water to breed.[60] In the last quarter of the 19th century it was introduced on the Pacific coast, where its range extends now from Grays Harbor, Wash.,[61] to Los Angeles County, Calif. It is now a favorite game fish there, and the yearly commercial catch since World War I ran between 500,000 and about 1,000,000 pounds there, until 1935, when commercial fishing for stripers was prohibited by the State of California.

Occurrence in the Gulf of Maine.—The range of the striper includes the coastline of our Gulf from Cape Cod to western Nova Scotia. But its distribution there in detail is determined by its very evident preference for surf-swept beaches and for particular stretches of rocky or bouldery shoreline; also for shallow bays, inlets, and estuaries. The geographic status of bass in our Gulf also depends on whether it be a good bass year (or run of years) or a poor one.

When bass are reasonably plentiful, as they have been during the past 15 years, and with a good representation of fish of different ages, the outer shore of Cape Cod provides the most productive surf casting, with Monomoy Island, the general vicinity of Nauset Inlet, and the tip of the Cape northward from Highland Light perhaps the warmest stretches, in most years. But the topography of a beach may be altered to such an extent during severe storms that a stretch that is good bass water one summer may be poor the next. Nauset beach is an example, for very few bass have been caught or seen there during the present summer (1951), though this has been one of the most productive localities on the Cape during the past few summers. Considerable numbers, mostly

[52] Merriman, Copeia, 1937, p. 23.

[53] New York Conserv. Dept., Biol. Surv. Lower Hudson Watershed, 1937, p. 62.

[54] Bean, Bull. New York State Mus., 60, Zool. 9, 1903, p. 527.

[55] Scaled from Scofield's graph (California Fish and Game, vol. 18, 1932, pp. 168–170, fig. 38).

[56] See Scofield, Fish Bull. No. 29, Div. of Fish and Game, California, 1931 for growth of bass in California.

[57] Bull. New York Zool. Soc., vol. 16, No. 60, November 1913, p. 1049.

[58] Fish. Bull. No. 35, U. S. Fish and Wildlife Service, 1941, vol. 50, p. 22.

[59] Merriman, Fish. Bull. No. 35, U. S. Fish and Wildlife Service, 1941, vol. 50, p. 16.

[60] Pearson (Bull. U. S. Bur. Fish., vol. 49, 1938, p. 827, fig. 1) charts its United States range, but does not include its Canadian range.

[61] Gerlach (Contrib. 14, Oregon Fish. Comm., 1950) gives an interesting account of the life history and distribution of the striper in Oregon waters.

of the smaller sizes, are caught in Pleasant Bay too, within Nauset Marsh, and in Town Cove, Orleans.

Considerable catches are made by boats trolling outside the surf, also, or by casting in toward the breakers along the outer Cape Cod shore, when the weather permits. But the most productive and reliable trolling grounds are along the eastern and southern sides of Cape Cod Bay in most summers, especially off the Eastham shore a few miles southward from Wellfleet, and off the mouth of Scorton Creek, Barnstable and the Sandwich shore.[62] The shores of Cape Cod and Cape Cod Bay have, in fact, been the chief center of abundance for bass within the Gulf from as far back as the record runs. Few bass are reported along the rocky stretch from the Cape Cod Canal to the entrance to Plymouth Harbor, though this would seem to be very good bass water, and schools must pass by. But many are caught in Plymouth Harbor, especially off Eel Creek, also up Duxbury Bay to the salt marsh creeks that open into its head.

Surf casters account for some along Duxbury Beach on the outside, for a few also in the boulder-strewn area at the western end of Humarock Beach.[63] The North and South Rivers in Marshfield yield considerable numbers in good years; we have seen and taken good fish there. Anglers, casting from the shore, take a few (never any great number) on boulder-strewn stretches along the Scituate shore, while Glades Point was famous for large bass in earlier periods of abundance (p. 390), when it was common practice to chum the fish by throwing out chopped lobsters, a method never likely to be revived because lobsters are far too costly nowadays. The Cohasset shoreline (with which we are familiar) yields a few yearly (mostly caught between sunset and sunrise), occasionally a very large one. In seasons when there is a good run of the smaller sizes, considerable numbers are taken at various places within the limits of Boston Harbor; Hull Gut, Weir River in Hingham, and Wollaston Beach are well known localities. And in years when there is a run of little fish, many of them are caught from the docks and from the bridges, to the head of Boston Harbor.

The north shore of Massachusetts Bay seems not to be as attractive for bass as its succession of inlets, beaches, and rocky headlands might suggest, for catches reported are small and scattering in most summers. But the beaches and enclosed waters from a few miles north of Cape Ann to and including the mouth of the Merrimac River are productive enough to rank second to the Cape Cod–Cape Cod Bay region. Bass are taken in the surf from Ipswich Beach, Cranes Beach, and along the entire length of Plum Island Beach; many more are caught by boat fishermen over the flats within the mouth of the Merrimac, as well as about the jetties at its entrance. Schools are often reported in Plum Island Sound. And the Parker River, emptying into the latter, is not only well known water for bass, especially small fish, but it holds some bass over the winter (p. 400), and it is one of the few streams along the New England shores of our Gulf where very young bass have been taken within recent years (p. 398).

Some are caught in Hampton Harbor, N. H. But the next important bass waters (moving northward) are the lower reaches of the Piscataqua River system, marking the boundary between Maine and New Hampshire; a good number, large and small, are now caught there yearly. In good years bass are to be caught in several of the streams that drain the southern part of the Maine coast, especially in the York, the Mousam, and in the Saco which is the most productive. Schools are sighted and a few are caught along the intervening beaches and some in the shallows of Biddeford Pool.

Information as to the status of bass for the coastline and streams of northern and eastern Maine, past or present, is scant, and we have come to suspect that bass may never have been as plentiful there as was supposed. A few are caught here and there around Casco Bay in good years, product perhaps of the Kennebec. But the estimated catch in the Kennebec was only about 12,760 pounds as far back as 1880;[64] and there have been far fewer bass there of late years. Our most recent information is that schools of large fish were seen in the lower Kennebec, off Popham and Reed Beaches in early October 1950 with some

[62] Many are caught by anglers casting in the Cape Cod Canal, but this is not properly a part of the Gulf of Maine.

[63] The bouldery area at the eastern end at the North River inlet is now within the limits of the military reservation; hence the only way to fish it is from a boat by casting in, toward the rocks.

[64] Atkins, Fish. Ind. U. S., Sect. 5, vol. 1, 1887, p. 675.

caught up to 26 pounds, and that a few were being taken daily, in late June 1951.[65] Nearly as many were taken in the Sheepscot, formerly, as in the Kennebec; the present condition is not known.[66] There were bass in the St. George during the period 1936–1940; doubtless there are some there still, for we heard of some in the surf near Georgetown, Maine, in August 1951.

Bass are seen in most years in Bangor Pool at the head of the estuary of the Penobscot, where some are caught by anglers casting especially for them, also by salmon fishermen. And many in the 2- to 4-pound class were reported and caught in the Belfast River and in Searsport Harbor farther down Penobscot Bay in 1938. But there have not been enough of them there during the past few years to have caused special comment. Stripers were seen in the tide rips in the narrows between Mount Desert Island and the mainland (near the Hancock–Sullivan Bridge) in August 1951, and others were reported driving squid ashore near Winter Harbor, Maine, a few miles farther east. Salmon fishermen sometimes "rise" bass in the Narraguagus, and Atkins [67] speaks of "a very few" in the St. Croix, though Huntsman found no recent record of bass in the Passamaquoddy region.

There may be an occasional bass in Maine rivers other than those we have mentioned, but there is nothing in the past record to suggest that there ever were many. In 1880, for example, the reported catch was nearly as great for the Kennebec (about 13,000 pounds) as for all the other rivers and coast of Maine combined (about 15,000 pounds). And there is no reason to suppose that the regional contrast has altered subsequently in this respect.[68] In the Bay of Fundy region, bass, as Huntsman has pointed out,[69] are confined to the large warm estuaries and the neighboring fresh water; i. e., to those of the St. John, Minas Basin–Cobequid Bay and Shubenacadie River systems, and of the Annapolis.

Available information suggests that bass always were more plentiful in St. John River waters than anywhere along the eastern part of the coast of Maine, and that they are still. Bass are occasionally caught in St. John Harbor, mostly between April and June.[70] And while they were reported as already much less numerous in St. John waters by 1884 than they had been in earlier times,[71] there still are enough of them in the St. John and its tributaries to have yielded commercial catches of 12,200 pounds in 1944, and 7,400 pounds in 1946. The most recent news that has reached us from the St. John is that salmon fishermen saw a school at the surface and caught some that weighed 3 to 11 pounds in late June or early July of 1951.[72]

Bass are well known in the Minas Basin-Cobequid region. According to local fishermen,[73] as many as 80 fish are sometimes taken in weirs there in a day, most of them in the 8- to 10-pound category, but with occasional fish reported up to 33 pounds.

The status of the bass is especially interesting in the Shubenacadie River, for they are not only caught in fresh water there and in Shubenacadie Lake where they are known to spawn, but some large fish remain throughout the year in the lake; i.e., they behave like a land-locked population.[74] A thousand or so, in fact, are caught yearly by anglers in the Lake and in the Shubenacadie River; [75] and it is said that fish as large as 50 pounds have been taken,[76] though most of them run small there.

We are informed [77] that the catch by anglers was about 620 bass (average about 4½ pounds) in the Bass River, tributary to Cobequid Bay in 1950, and that the catches for 1949 and 1950 combined were about 1,350 fish (average about 2⅓ pounds) in the Gaspereau, tributary to Minas basin; 4,650 fish (average about 5¾ pounds) in the Annapolis River; and about 125 fish (average 6 pounds) in the Bear River, tributary to Digby basin, in 1950. It is interesting, that these fish ran so small, for the bass caught in Cape Cod and northern Massachusetts waters during these same

[65] Reported in Saltwater Sportsman for October 6, 1950.

[66] Yearly catch about 1880, some 8,000 pounds in the Sheepscot according to Atkins, Fish. Ind. U. S., Sect. 5, vol. 1, 1887, p. 716.

[67] Atkins (Fish. Ind. U. S., Sect. 5, vol. 1, 1887, p. 700) reports one of 20 pounds, taken in the St. Croix in a weir in 1880.

[68] What few bass were reported from Maine in 1919 were from the Kennebec (592 pounds) and from Penobscot waters (57 pounds); bass have not been included in the fisheries statistics for Maine for any subsequent years.

[69] Contrib. Canadian Biol. (1921) 1922, p. 63.

[70] Information from Dr. A. H. Leim.

[71] Goode, Fish. Ind. U. S., Sect. 1, 1884, p. 425.

[72] Saltwater Sportsman for July 6, 1951.

[73] According to Moore, Boston Herald, August 28, 1950.

[74] Information from Dr. A. H. Leim.

[75] According to Huntsman, Ann. Rept. Fishery Board Canada, (1949) 1950, p. 41.

[76] Vladykov and McKenzie, Proc. Nova Scotia Inst. Sci., vol. 19, 1935, p. 91.

[77] Information from Maj. Howard Scott of the Fishery Division of the Nova Scotian Department of Trade and Industry, received through Henry Lyman.

years included a good number of very large fish (p. 403).

Anglers have also come to realize recently that bass are to be caught in various bays and river mouths along the western shoreline of Nova Scotia. But no definite information has reached us as to how plentiful they are there, or how large.

The regional contrasts in the abundance of bass along different sectors of the coastline of our Gulf may be illustrated more concretely by the commercial landings for 1945.[78]

Outer Cape Cod and Cape Cod Bay[79]—	perhaps about 57,000 lbs.
Cape Cod Canal to New Hampshire line	51,100 lbs.[80]
New Hampshire	9,000 lbs.
Maine	None reported.
St. John River system, New Brunswick	2,400 lbs.
Minas Basin, Cobequid Bay and Shubenacadie River region, Nova Scotia	13,800 lbs.
Annapolis County, Nova Scotia	3,100 lbs.
West coast of Nova Scotia	800 pounds in 1944; none reported in 1946.

A regional contrast of another sort, of interest to anglers, is that really large bass of (say) 30 pounds and upwards, are far more plentiful along the Massachusetts coast (especially in Cape Cod waters) than they are anywhere farther north and east in our Gulf.

Localities along the outer coast of Nova Scotia where we have heard (or read) of stripers are the head of Mahone Bay; head of Chedabucto Bay; and Mira Bay and other harbors of Cape Breton. The numbers caught there are so small that they are not included in the published statistics of the commercial catches for the counties in question. The shoal estuaries, however, of the Richibucto Bay region and also the estuary of the Miramichi River (on the southern shore of the Gulf of St. Lawrence) harbor isolated populations of bass plentiful enough to have yielded commercial catches of about 3,800 pounds and 9,000 pounds, respectively, in 1929, 4,100 and 3,000 pounds in 1931.[81] And there is also a population (or populations) below Quebec in the lower St. Lawrence River, of bass that winter in that same general region, as proved by marking experiments recently carried out by Vladykov.[82] There are enough of them, in fact, around Isle d'Orleans for bass fishing to be a favorite sport there. But the commercial catches are so small as to suggest that the stock of bass is not very large.[83]

It has been known for many years that bass spawn in the St. John River,[84] and it is probable that they also spawn in the small streams tributary to Minas Basin and Cobequid Bay at the head of the Bay of Fundy; in Grand Lake at the head of the Shubenacadie River, and probably in the Annapolis River. It is generally believed, also, that some bass spawned of old in all the larger rivers from the Penobscot westward. Great numbers, so small (2–3 inches) as evidently to be fry of the year, were caught, for example, in winter in the 1880's in the Kennebec, where ripe fish also have been reported[85] from the end of June into July. But the only Maine or Massachusetts streams where we find evidence of spawning bass in recent years are the Mousam, in Maine, where fishermen have reported taking females with ripe eggs on several occasions;[86] and the Parker, in Massachusetts, where Merriman[87] took three fry of the year 2¾ to 3¼ inches (7.1–8.5 cm. long) on August 4, 1937. Thus it seems sufficiently established that a great majority of the bass that summer in the western side of our Gulf come from spawning grounds to the west and south.

Merriman's[88] painstaking investigations show beyond reasonable doubt that most of the little bass of 2 to 5 pounds that appeared in great numbers along southern New England and to the northward in 1936 (p. 402), following a period of great scarcity of bass there, had been hatched two years previously (1934) in the region of Chesapeake Bay, perhaps some of them in the Delaware Bay region. Some of the abundant year classes of 1940 and 1942, which appeared in our Gulf in 1942, and 1944, also may have come

[78] The most recent year for which detailed statistics are readily available.

[79] Assuming that about ⅔ of the Barnstable County catch of 86,200 pounds was taken along the outer shore of Cape Cod and in Cape Cod Bay (probably an underestimate).

[80] Assuming that about ⅓ of the Plymouth County catch of 75,000 pounds was taken on the Massachusetts Bay side.

[81] This is the most recent year for which information is available for Northumberland and Kent Counties.

[82] Rapp. Gen. Ministr. Chasse et Pêch., Quebec, Pêcheries (1946-1947) 1947, p. 50.

[83] The reported catch for 1948-1949 was only about 1,800-1,900 pounds (17 quintals; See Rapp. Gen. Ministr. Chasse et Pêch., Quebec, Pêcheries (1948-1949) 1949, p. 94).

[84] Adams, Field and Forest Rambles, 1873, Pt. 3, Fishes, p. 248.

[85] Atkins, Fish. Ind. U. S., vol. 5, Sect. 1, 1887, p. 693.

[86] Towne, State of Maine Striped Bass Survey, Maine Devel. Comm. and Dept. Sea and Shore Fisheries, 1941 [approx. date], p. 14.

[87] Fishery Bulletin No. 35, U. S. Fish and Wildlife Service, 1941, vol. 50, p. 17.

[88] Fish. Bull. No. 35, U. S. Fish and Wildlife Service, 1941, pp. 46-52

from equally far away; others perhaps from intermediate spawning areas.

Since the mature bass that visit the coasts of Cape Cod and northern Massachusetts in such plenty in good years almost certainly do not spawn in any numbers in any of the Gulf of Maine rivers, we can only suppose that they repair to more southerly rivers to spawn, perhaps to the Hudson, in particular. But many of them reach northern Massachusetts so early in the season, and so little information is available as to the condition of their sexual organs when they arrive, that we still face something of a mystery, here.

In the salt estuaries and open waters of our Gulf bass are taken only from late spring, through the summer, and until late in the autumn. In years when they are plentiful enough to attract attention, they are likely to be reported about equally early in the season all along from Cape Cod to the Merrimac River. In 1950,[89] for example, bass had been reported from the outer shore of the Cape (Pleasant Bay and Orleans) by mid-May, from the North and South Rivers, at Marshfield on the southern side of Massachusetts Bay, and from the Merrimac at Amesbury by mid-May; we heard of one caught in Duxbury Bay as early as May 1 that same year; and in normally early years they are generally distributed along the Massachusetts Coast of the Gulf in May or by the first days of June. The first bass were reported in and off Hampton Harbor and in the Piscataqua River about the beginning of the second week in June (1950), and in Casco Bay about the middle of the month.

Bass are said to appear as early as the end of May in Bangor Pool at the head of the estuary of the Penobscot in some years.[90] In 1950 they were scattered all along Penobscot Bay before the end of June. And it is probable that the seasonal schedule is about the same for the bass at the head of the Bay of Fundy, but information is scant.[91]

Once the bass have appeared, they continue in evidence until well into the autumn (p. 399). During this part of the year, the bass of the coasts of Massachusetts and most of those in Maine are in salt water and in brackish, except for such as enter fresh water to spawn (p. 398). But they are caught all summer in fresh water far above the head of tide in the Shubenacadie in Nova Scotia (p. 397),[92] also in the Annapolis, and part of the stock may have here a similar habit in various of the rivers of Maine, as in the Kennebec, where they ran up as far as Waterville until they were prevented by the construction of the dam at Augusta.[93]

In rivers where bass winter, they may, of course, be taken in any month from late autumn into the spring (p. 400). As autumn approaches the bass vanish however from the open coast. What little information we have suggests that most of them have disappeared along the outer coasts of Maine by mid-October or the end of that month in most years. But they may be in evidence in Maine rivers until later in the autumn, as they were of old in the Kennebec, where Atkins[94] described them as continuing "feeding in weedy coves until November"; and in the Mousam River in southern Maine, where fishing is said to have been good until November during the period 1938–1940, when our Gulf had a spectacular run of young fish (p. 402).

Farther southward in our Gulf, they may linger equally late off the open beaches. In 1949, for example, a set of traps[95] located near Provincetown Harbor in 35–45 feet of water, took 3,705 pounds (the only large catch of the year) on November 3.

In 1950, a late season, Cape Cod Bay eastward from the Cape Cod Canal was described to us as "loaded" with bass until the third week in October, fair numbers were still being caught along the outer shore of Cape Cod at the end of the month, schools of small fish were reported on November 9, and half a dozen were landed from the surf on November 18, and one, on December 3.[96] Surf casting is likely to be much more productive along the outer Cape Cod beaches during 2 weeks or even 3 weeks of November than it is in July or August, especially for the smaller fish, and during the hours of daylight (p. 391).

And the bass in salt water may be in evidence until equally late in the season in the Minas-

[89] This is the only year for which we have detailed information.

[90] Weston, Field and Stream, March 1932, p. 69.

[91] Moore (Boston Herald, Aug. 28, 1950) reports that bass are taken in traps from July on, in the Cobequid Bay region.

[92] Huntsman, Ann. Rept. Fisheries Res. Board Canada, (1949) 1950, App. 2, pp. 41-42.

[93] Atkins, Fish. Ind. U. S., Sect. 5, vol. 1, 1887, p. 693.

[94] Fish. Ind. U. S., Sect. 5, vol. 1, 1887, p. 693.

[95] Property of the Pond Village Cold Storage Co., of North Truro, to whom we are indebted for much information.

[96] Reported by Henry Moore, Boston Herald for Dec. 7, 1950.

Cobequid Bay region, at the head of the Bay of Fundy for fishermen report taking them there through October and into November.[97]

The question where the bass that visit the different parts of the coast of the Gulf of Maine spend their winters still awaits a comprehensive answer. It has long been known that the Chesapeake Bay bass winter in the deeper channels near the head, of the bay as well as in its estuaries, and in the lower reaches of the rivers, in a more or less inactive state; also those of the New Jersey coast run up into rivers to remain until the following spring, as described more than a century ago by Mease.[98]

Knight[99] writes too, that as the weather becomes colder, the bass of the southern side of the Gulf of St. Lawrence "penetrate into the bays and arms of the sea and ascend the rivers at some distance, where they spend the winter resting on the mud in a half torpid state." The bass also, in Maine "pass the winter in quiet bays and coves of fresh water in the rivers," according to Atkins.[1] We see no reason to doubt that the Bay of Fundy bass, and also those that still frequent the Maine rivers from the Penobscot westward, still follow this habit.

It has been known, also, for many years that some bass winter in the Parker River, in northern Massachusetts. In fact, some 8,700 pounds were taken there during the financial depression of 1930 (p. 402). Local fishermen tell us also that a few bass winter in the deeper parts of the North and South Rivers in Marshfield, Mass., on the southern side of Massachusetts Bay, apparently in salt water. But these and other small streams do not seem extensive enough to provide wintering grounds for all the schools of bass that appear in summer between southern Maine and Boston Harbor in reasonably good years. Neither is there anything in the available record to suggest that the Merrimac ever was an important wintering ground. And it is hardly conceivable that the multitude of bass that sometimes frequent Cape Cod Bay and the outer shore of the Cape in good bass years can winter nearby (unless they do so offshore), there being no large rivers along this section of the coast, and no local report of bass in winter in the shallow, partially enclosed bays there, or in the salt marshes.

It was generally believed until recently that the great majority of bass that frequent the Massachusetts coasts of the Gulf (and the Cape Cod region in particular), and also those that summer off southern Massachusetts and around the offlying islands, move westward along the shore in autumn: some to contribute to the bodies of fish that are known to winter in the rivers of Connecticut and in the lower Hudson, and some to journey perhaps as far as Chesapeake Bay; i. e., to the region where many of them are hatched. The capture, however, in 1949, of an 18-inch bass some 60 miles south of Marthas Vineyard in 70 fathoms of water in February (p. 391) seems to favor the view, now gaining favor among observant anglers, that at least a part of the bass of the Cape Cod region may only move offshore to winter on bottom well out on the continental shelf in localities where the otter trawlers do not ordinarily operate, as has been found of late to be true of the summer flounder (p. 268).

If true, this would mean that some of the Chesapeake-hatched bass that spread northward to Massachusetts and Maine when 2 or 3 years old may never return to their home waters. More definite information in this regard is to be expected from tagging experiments now in progress.

Periodic fluctuations in abundance.—Nothing regarding bass is of greater interest to commercial fishermen and to anglers than the great fluctuations in its numbers that have taken place in our Gulf within historic times.

The bass was a familiar fish when New England and the Maritime Provinces were first colonized, all along the coast from Cape Cod to the Bay of Fundy; plentiful and easy to capture, because of its large size and its habit of coming into the mouths of streams and creeks; it was also an important food supply for the early settlers.

Wood,[2] for example, tells us that in what is now a part of Boston Harbor:

> The basse is one of the best fishes in the country, and though men are soon wearied with other fish, yet are they never with basse. It is a delicate, fine, fat, fast fish, having a bone in his head which contains a saucerfull of marrow sweet and good, pleasant to the pallat and wholesome to the stomach . . . Of these fishes some be three and four foote long, some bigger, some lesser; at some tides a man may catch a dozen or twenty of these in three hours . . . When they use to tide in and out of the rivers and creekes the English at the top of an high water do crosse the creekes

[97] Report by Henry Moore, Boston Herald for Aug. 28, 1950.
[98] Trans. Litt. Phil. Soc. New York, vol. 1, 1815, pp. 502-504.
[99] The River Fisheries of Nova Scotia, 1867, p. 12.
[1] Fish. Ind. U. S., Sect. 5, vol. 1, 1887, p. 693.
[2] New Englands Prospect, 1634, p. 37.

with long seanes or basse nets, which stop in the fish: and the water ebbing from them they are left on the dry ground, sometimes two or three thousand at a set, which are salted up against winter, or distributed to such as have present occasion either to spend them in their homes or use them for their grounds.

He also describes [3] how "shoales of basse have driven up shoales of mackerel from one end of the sandie beach to the other," near Salem, and mentions them in the Merrimac. In the earliest record the chief centers of abundance for them within the Gulf were Cape Cod Bay and the shore of Cape Cod; the neighborhood of Boston Bay and harbor; the vicinity of the Merrimac River; the Kennebec River and vicinity, and the larger rivers that drain into the Bay of Fundy. Inexhaustible, however, though the supply seemed, a decrease was reported as early as the last half of the eighteenth century. At first this was apparent only locally. For example, fewer were seen in the Piscataqua after about 1792. And very few were reported there from about 1880 down to 1936 or 1937, when the young fish, that were batched in southern waters in 1934, appeared in our Gulf in such numbers (p. 402).

They seem to have continued moderately plentiful in Massachusetts Bay and around Cape Cod during the first half of the nineteenth century, when bass were still being netted in abundance along the beaches between Boston and Cohasset; 300 good-sized fish were taken in one seine haul at the mouth of Barnstable Harbor in July 1829; while 700 were taken at Provincetown in a day in October 1859. Fishing for bass from the rocks with hook and line was a well recognized sport then around Massachusetts Bay. But Freeman wrote as long ago as 1862 (in his History of Cape Cod) that the bass were much less plentiful in Cape Cod waters then than they had been of old. And the catch for Cape Cod Bay and the outer shore of the Cape, combined, reached 2,000 pounds in only three of the years of record during the last quarter of the 19th century and the first quarter of the 20th.[4] Bass, in fact, had so nearly vanished soon thereafter from the Massachusetts coast north of Boston that no commercial catches were reported there for the period 1876 to 1883, though an occasional fish may have been caught.

[3] Wood, New Englands Prospect, 1634, p. 47.

[4] 1878—4,974 pounds; 1897—4,820 pounds; 1900—6,450 pounds.

There may have been a few more bass along the coast of northern Massachusetts during the next 17 years, for yearly catches ranging from none up to 441 fish (1892) were reported for Essex County between 1884 and 1900, while a number were seined in the Merrimac River in 1897. But this was again succeeded by a period of scarcity so extreme that no bass were reported for the Massachusetts coast north of Boston during the next 30 years.[5] And the capture of a single fish in the inner part of Massachusetts Bay by any method was so unusual an event then that one of 44¾ pounds, caught near Brant Rock on the southern shore of the Bay, in November 1930, was given wide publicity in the newspapers.

The bass of the coasts of Maine fared no better. They had practically gone from the Androscoggin by 1860; the reported catch for the entire coastline of Maine (already only about 27,000 pounds in 1880) had fallen to about 1,600 pounds in 1902, 4,200 pounds in 1905, to 600 pounds in 1919. And no commercial catches at all have been reported from Maine in any subsequent year,[6] except for 537 pounds in 1932.

It seems that the bass population of the St. John River system shared with those of Massachusetts and of Maine in the general decline in abundance during the first half of the nineteenth century, for they were reported as much less plentiful there by 1873 than they had been in the early 1800's when they were a familiar sight rolling and splashing at the surface in June.[7] But neither the St. John population, nor the population at the head of the Bay of Fundy, nor of the Annapolis River, seem ever to have fallen to as low an ebb as has happened along Maine and Massachusetts. In 1919, for example, when no bass were reported from the Massachusetts coastline of the Gulf (p. 401) and only 600 pounds for Maine, 2,700 pounds were reported from the Nova Scotian coastline of the Bay of Fundy,[8] 1,600 pounds for the St. John River system.

The year 1921 seems to have marked the "turning of the tide" for the bass in Cape Cod Bay waters, for 4,784 pounds were taken that year at

[5] Statistics have been published for 1903 to 1910, 1919, and 1928–1930.

[6] Statistics published for 1929–1933, and 1935–1947.

[7] See Adams (Field and Forest Rambles, 1873, pt. 3, Fishes, p. 248), who described the Indians of the Melicete Tribe as still spearing good-sized bass from their canoes in the St. John, in 1873.

[8] 2,000 pounds from the Cobequid–Shubenacadie region (Hants County), 700 pounds from Annapolis County.

Sandwich, near the Cape Cod Canal, through which they may have come, while some, that did not find their way into the official returns were taken in the inlets along the outer shore of Cape Cod during that summer or the summer before. And the commercial catches for Barnstable County rose to 8,060 pounds for 1928, to 18,665 pounds for 1929, and ran between about 27,000 pounds and about 34,000 pounds [9] for 1930, 1931 and 1932.

Surf fishermen, too, did better along the outer Cape shore from 1930 through the next couple of summers than they had for many years; (a 33-pounder was taken in the surf on Cape Cod, and one of 44¾ pounds on the south shore of Massachusetts Bay, p. 401). And it appears that the bass spread northward to the estuaries north of Cape Ann during these years (unless a small stock had persisted there through the poor period), for some were taken in the Parker River in the late 1920's while 8,700 pounds were reported thence in the winter of 1930, when net-fishing was allowed, temporarily.

This upswing was brief (the reported catches for the entire coastline of Massachusetts were only 4,500 and 5,100 pounds, respectively, for 1933 and for 1935).[10] But at least it gave a foretaste of what was to come, for the waters around Cape Cod were invaded during the summer of 1936 by countless schools of little bass, weighing about 2 to 3 pounds. These (as is now known) had been hatched in 1934 (i. e., 2 years previous) in the Chesapeake Bay–Delaware Bay region (p. 393), and it is interesting, not only that they came from so far away, but that this was the largest year's brood that has been produced in Chesapeake Bay for as far back as any record is available.[11] Unfortunately, there is no knowing in what numbers they reached the outer shore of Cape Cod and Cape Cod Bay in 1936, for no record seems to have been kept of commercial catches of them there in that year.

But they (chiefly) comprised the catches which were some 5 to 16 times as great in 1936 as in any of the 8 previous years [12] along the coast of Rhode Island. And considerable numbers of them were reported from as far north as the harbors and rivers along the southern part of the coast of Maine, where very few bass, large or small, had been caught for many years previous.

In 1937, having now grown to an average weight of about 3 to 5 pounds, they not only reappeared in such numbers that a commercial catch of something like 80,000 pounds was reported from the Gulf of Maine coast of Massachusetts,[13] but so many of them spread north past Cape Ann that the catch from the inner part of Massachusetts Bay to the New Hampshire line (about 55,000 pounds) was perhaps three times as great as that for the Cape Cod Bay–outer Cape region (in the neighborhood of 19,000 pounds). And more of these little bass were caught by anglers in the river mouths and estuaries of New Hampshire that summer, and of Maine as far as the Penobscot region, than had been the case the year before, but not enough to figure in the official statistics.

The fish of the 1934 year class averaged around 6 pounds by 1939 (many had reached 7–9 pounds); and the bass seemed so well established all along from Cape Cod to southern Maine that anglers had largely forgotten the preceding lean years. And the growth of the individual fish as they advanced in age, combined with fresh increments from the south seem to have more than balanced the death rate (natural or from fishing) for the next 5 or 6 years, for the coast of Massachusetts as a whole.[14] And a good part of the fish of the 1934 year class (still swimming in good numbers) grew meantime to 18 to 25 pounds, to the delight of the anglers.

Bass fishing improved so much in the Hampton region also, and in the Piscataqua River system that about 19,000 pounds were reported for 1943 in the commercial statistics for New Hampshire, where bass had not been mentioned in the fishery statistics for many years. But it is evident that depletion in numbers outran renewal along the coast of Maine during this same period, for there were many fewer fish there in the season 1940 than there had been in 1939 or 1938, though they ran larger, averaging about 8–10 pounds according to local reports.

---

[9] To the nearest 1,000 pounds.

[10] No data for 1934.

[11] Tiller, Publ. 85, Chesapeake Biol. Lab., 1950, p. 24.

[12] For details, see Merriman, Fish. Bull. No. 35, U. S. Fish and Wildlife Service, vol. 50, 1941, p. 10, fig. 4; p. 13, fig. 8.

[13] Assuming that about two-thirds of the catch of 28,700 pounds for Barnstable County came from the outer shore of Cape Cod and from Cape Cod Bay, probably an under estimate.

[14] Reported catches for Massachusetts as a whole were about 62,500 pounds for 1939, about 75,700 pounds for 1940, about 99,500 pounds for 1943. no data available for 1941 or 1942.

Small bass (apparently spawned in southern waters in 1940, 1941, or 1942[15] and now large enough to be included in the commercial statistics) again entered the southern part of our Gulf in 1944 in such numbers that the commercial catch for Massachusetts as a whole was nearly twice as great for that year (about 191,000 pounds) as it had been the year before (about 99,500 pounds).[16]

But it seems that very few of the fish of the 1942 year class, if any, spread northward much beyond the Merrimac River, for the reported catch for New Hampshire fell from about 19,000 pounds in 1943 to between 10,000 and 11,000 pounds in 1944, and to about 9,000 pounds in 1945, though the proportion of large fish was greater, while occasional bass, only, were reported in Maine waters in these years, or in the later 1940's.

Anglers' reports in general, and our own observations, are to the effect that few young bass appeared in the Cape Cod Bay-outer Cape region during the four years 1946–1949, or farther north along the New England coast. While this prevailing scarcity of baby bass caused widespread fears that the striper might be facing another serious decline, anglers welcomed an accompanying increase in the numbers of bass weighing upwards of 20 to 25 pounds. Thus, a larger number of fish heavier than 25 to 35 pounds were landed along the outer shore of the Cape, in Cape Cod Bay and in northern Massachusetts waters in 1950 than for many years. We saw one of 45 pounds that was caught by an angling companion in the inner part of Massachusetts Bay that September, and one of 51½ pounds, caught in the surf, was reported from the outer shore of Cape Cod that August, while others, doubtless in the 50-pound class, were taken of which we did not chance to hear. A few very large fish were reported, that summer, in the surf at Old Orchard Beach, Maine. And hundreds of bass of 25 to 45 pounds, with a few running up to 52 or 53 pounds (few smaller than 25–30 pounds) were caught in the surf on the outer beach near the tip of Cape Cod during the summer of 1951, many up to 30–45 pounds in Cape Cod Bay that July; a few as large as 20 to 30 pounds in Duxbury Bay; at least two (to our knowledge) weighing a little more than 50 pounds in the inner part of Massachusetts Bay; a few in the 30 to 45 pound class on the northern Massachusetts coast.

It is almost certain that most of these large fish belong to the abundant year classes that were spawned in the early 1940's or even previously. Hopes for the future depend, therefore, on renewed replenishments of the stock. A year ago (in 1950), prospects seemed good in this respect, for great numbers of little bass (many smaller than the legal length in Massachusetts, 16 inches from snout to fork of tail) were reported that summer and autumn from various localities along southern New England, and northward as far as Plymouth and Duxbury Bays; from the North and South Rivers in Marshfield; from Boston Harbor; from the Parker River (Plum Island Sound region); from the lower Merrimac River; and from Hampton Harbor, N. H. Many "school" fish of 2 to 8 pounds seen (and some caught) in the Saco in July and August of 1950, and a few landed every day from the York and Mousam Rivers late that June, point similarly to a fresh influx of bass to southern Maine waters, either that year or the year before, for it is not likely that these fish had been spawned in the streams along that part of the coast.

And reports that Pleasant Bay, on Cape Cod, the Massachusetts Bay end of the Cape Cod Canal, and Duxbury Bay, have all been "loaded" with small bass at times during the present summer (1951) and also of many too small to keep, off Wollaston Beach in Boston Harbor, are promising at least; so is the fact that a good many fish in the 10 to 15-pound class have been caught at various places along the coast.

An interesting aspect of the bass situation is that the young bass that invaded the water of Massachusetts and of Maine in 1936 and 1937 seem not to have spread to the St. John River system for while commercial catches ran larger there during the 1930's than they had during the 1920's, the increase may not have been greater than can be accounted for by an increasing demand for bass. And, in any case, it had taken place by 1932, i. e., two years before the fish were spawned that re-

[15] The broods of 1940-1942 were large, in Chesapeake Bay, though not as large as the brood of 1934 (Tiller, Chesapeake Biol. Lab., Pub. 85, 1950, pp. 13, 24-25).

[16] The minimum legal length for bass (snout to fork of tail) having been set in Massachusetts at 16 inches (fish 3-4 years old).

plenished the Massachusetts stock in 1936 and 1937 (p. 402).[17]

Some of these that reached Massachusetts in 1944 may have spread to Nova Scotia, also, for the average catch was larger there from 1944 to 1946, than it had been for many years. But the increase was not great enough to suggest than any large recruitment had taken place from the south.[18]

Additional evidence that the bass of the Bay of Fundy region do not intermingle to any great extent with those of the western side of the Gulf of Maine is that the fish ran so small in the Nova Scotian streams in 1949 and 1950 (p. 397) when there were so many very large ones in Massachusetts waters (p. 403). And the fact that those caught in those years ran considerably larger in the Annapolis River than in the Shubenacadie River (p. 397) emboldens us to suggest that the populations in the several Nova Scotian streams are more or less separate; and separate also from the bass of the St. John.

It would, of course, be of the greatest interest, to commercial fishermen and to anglers alike, to know what determines that an abundant brood of bass, or a poor brood is to be reared in any given year. All we dare say is that the largest brood on record (that of 1934, in Chesapeake Bay) was produced when the parent stock was at a very low ebb, which may prove a general rule, and that studies by Merriman[19] and by Tiller[20] suggest that very large broods are raised only in years when the temperature of the water is lower than normal, both before the spawning season and after it.

Importance.—Striped bass have not been plentiful enough in the Gulf of Maine at any time during the past 100 years to support a commercial fishery of any great magnitude. Even in the good years 1944–1946 the reported value of the commercial catches for Massachusetts as a whole was less than $50,000 yearly.[21] But this does not take into account bass used for home consumption, or those sold in small lots. In the seasons of 1937 and 1938, when the yield of the year class of 1934 was at its peak, about 30–40 percent of the reported commercial catch was made by hook and line, about 40–45 percent in pound nets and traps. But with the development of increasingly efficient methods of trolling with hand lines from small craft, the hook and line catch increased in importance to about 63–65 percent for the seasons of 1939, 1940, and 1943,[22] and to about 89–91 percent for 1944, 1945, and 1946, while the pound net and trap catches decreasing correspondingly.[23]

The striper is the leading game fish in its periods of plenty all along our coast, from the outer shore of Cape Cod to New Hampshire waters. The number of anglers who cast for them in the surf along the beaches of Cape Cod and northward from Cape Ann to the mouth of the Merrimac and at scattered spots elsewhere certainly is in the thousands. Many party boats troll daily for bass in Cape Cod Bay, some also along the Plum Island shore and at the mouth of the Merrimac, while many are caught by trolling, by live line fishing, and even by still fishing in the various inlets.

So far as we know, Shubenacadie River and Lake, and the Annapolis River are the only waters on the Canadian shores of our Gulf where the striper attracts attention as a game fish (p. 397); anglers visiting the St. John are far more interested in salmon.

So much has been written about the techniques of surf casting, trolling, choice of lines, and baits, that we need not delay; but it is interesting, in comparison, to read, in Wood's New England's Prospect, published in 1634 (p. 37), that "the way to catch them is with hook and line, the fisherman taking a great cod line to which he fasteneth a peece of lobster and threwes it into the sea. The fish biting at it, he pulls her to him and knockes her on the head with a sticke".

We should point out in conclusion, that the recreational value of the striper is high; its money return to the seaside communities where bass are plentiful is much greater than the price the fish bring in the market, if the amounts spent for tackle and bait, boat hire, lodging, guide service, and the patronage of service garages, and gasoline filling stations are taken into account.

[17] Maximum reported catch for St. John River system for period 1922–1943, was 21,200 pounds in 1932.

[18] Average yearly catch, Nova Scotian rivers and coast of the Bay of Fundy was about 3,355 pounds for 1922–1930; about 12,600 pounds for 1932–1943; about 18,300 pounds for 1944–1946.

[19] Fish. Bull. 35, U. S. Fish and Wildlife Service, 1941, vol. 50, p. 14.

[20] Pub. 85, Chesapeake Biol. Lab., 1950, pp. 18, 28.

[21] 1944, $29,173; 1944, $34,643; 1945, $48,748; 1946, $34,643.

[22] No data are available for 1941 or 1942.

[23] About 15–35 percent for 1939, 1940, and 1943; about 5–11 percent for 1944, 1945, and 1946.

White perch *Morone americana* (Gmelin) 1789

SEA PERCH

Jordan and Evermann, 1896–1900, p. 1134.

Description.—The white perch resembles its larger relative, the striped bass, in the number, outline, and arrangement of its fins, and in its deep caudal peduncle without longitudinal keels. But it is a deeper bodied fish, (only about 2½ to 3 times as long as deep, not counting the caudal fin); and it is more flattened sidewise (p. 390). The dorsal profile of its body is more convex than that of a bass, but its head is rather noticeably concave and its mouth is smaller. Furthermore, there is no free space between the two dorsal fins of the white perch, whereas they are separated by a short interspace in the striped bass. The perch has fewer rows of scales between gill cover and base of tail than the bass (about 48 in the white perch, 60 or more in the striped bass), and its anal spines are much stouter than those of the bass with the second and third about equal in length (graduated in the bass); also it usually has only one spine (sometimes two) at the margin of the gill cover. Finally, there is a constant difference in color.

The first dorsal fin (9 spines) of the perch is rounded in outline with its third and fourth spines longest, and although there is no free space between the two dorsal fins they are entirely separated by a deep notch. The second dorsal fin (1 spine and 12 rays) is rhomboid in outline and so short that it leaves a rather long caudal peduncle bare. The anal fin (8 to 10 rays preceded by 3 stout spines) originates under the middle of the second dorsal and is of the same shape as the latter. The ventrals originate a little way behind the pectorals and each ventral is armed with one stout spine at its forward margin. Both the pectorals and the ventrals of the perch are larger, in comparison with the size of the fish, than those of the striped bass.

Color.—The upper surface is variously olive, dark grayish green, or dark silvery gray, shading to paler olive or silvery green on the sides and to silvery white on the belly, while large fish often show a bluish luster on the head. The fins often are more or less dusky. The ventral fins and the anal fin are sometimes rose-colored at the base. The sides of young specimens are marked with pale longitudinal stripes but these usually fade out with growth.

Size.—White perch are occasionally as much as 15 inches long, 5 inches or more deep, and 2 pounds or a little more in weight; but the average is 8 to 10 inches long and 1 pound in weight, or less.

Habits.—The white perch is much more closely restricted in its seaward range than the bass, for while they are taken in undiluted sea water along southern New England, and at various other localities thence westward and southward, they are much more plentiful in ponds connected with the sea, in the brackish water of bays behind barrier

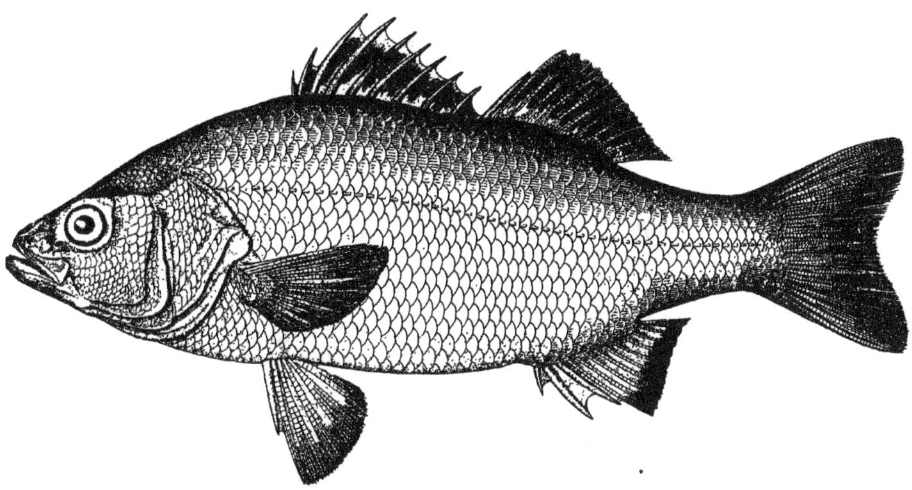

FIGURE 210.—White perch (*Morone americana*). From Goode. Drawing by H. L. Todd.

beaches, in estuaries, and in river mouths. White perch also occur landlocked in fresh-water ponds in many places.

They are ordinarily found in shallow water, usually not deeper than perhaps a fathom or two, but sometimes as deep as 10–21 fathoms in Chesapeake Bay.[24] However, they are not bottom fish (except in winter), but wander from place to place in small schools. Apart from this, they are resident throughout the year wherever found. In winter they congregate in the deeper parts of the bays and creeks, where they either hibernate, or at least pass the cold season in a sluggish condition.

When living in salt or brackish water white perch feed on small fish fry of all kinds, young squid, shrimps, crabs, and various other invertebrates, as well as on the spawn of other fish, of which they are very destructive. Swarms of young perch, for instance, have been seen following the alewives around the shores of ponds on Marthas Vineyard, eating their spawn as it was deposited. They bite freely on almost any bait, natural or artificial.

Breeding.—Along southern New England the white perch spawn in April, May, and June. Presumably the season commences a few weeks later around the Gulf of Maine, but definite data are lacking.[25] Those living in salt water run up into fresh or slightly brackish water to spawn. The eggs (about 0.73 mm. in diameter, with large oil globule) sink and stick together in masses, or to any object on which they chance to rest. Incubation occupies about 6 days at a temperature of 52°. The newly hatched larvae are about 2.3 mm. long with the vent some distance behind the yolk sac and with very little pigment. In five or six days after hatching the head begins to project forward, the yolk sac has been partly absorbed and branched pigment cells have appeared on the oil globule. The late larval and post larval stages have not been described.[26]

General range.—Atlantic coast of North America from the Gulf of St. Lawrence and Nova Scotia to South Carolina, breeding in fresh or brackish water and permanently landlocked in many fresh ponds and streams.

Occurrence in the Gulf of Maine.—The white perch inhabit salt, brackish, and fresh water indifferently along the shores of southern New England. But while this is a familiar fish in many ponds throughout northern New England, New Brunswick, and Nova Scotia, they are found regularly in only a few estuarine situations north of Cape Cod, and they hardly belong to the fish fauna of the open Gulf. Thus we have heard only vaguely of them in Duxbury Bay and in the North and South Rivers in Marshfield; and we had not been able to satisfy ourselves of their presence in the salt creeks about Cohasset, Mass. (localities apparently suited to it) until the summer of 1950, when white perch running-up stream to a pond were reported there.[27] Storer long ago described white perch as brought to Boston market from the mouths of neighboring rivers and from ponds to which the sea had access. And white perch run in salt and brackish reaches of the Parker River in northern Massachusetts, providing fishing for many small boat anglers in spring and summer.

Ordinarily white perch are so scarce along the open coast from Cape Cod northward that they did not figure in the statistics of the shore fisheries of any part of Massachusetts Bay from 1907 to 1928.[28] And ordinarily they are not common along the coast of Maine; none was reported from the shore fisheries of Maine in 1905 or 1919, and only 400 pounds in 1902; none at all of late years. But they appear locally, however, in unusual numbers on rare occasions. Thus it is probable that certain unfamiliar fish taken at Beverly on the north shore of Massachusetts Bay during the summer of 1950, and in Casco Bay, were white perch.[29] No less than 1,600 pounds of white perch were reported for the shore fisheries of the short coast line of New Hampshire in the year 1912: Casco Bay saw a run of them in the summer of 1901 when local fishermen, not knowing the fish, dubbed them "sea bass"; and they have been reported at Eastport, Maine. But apparently they do not occur around the shores of the Bay of Fundy, either in salt water or in brackish. And there is no reason to suppose that white perch were more regularly plentiful along the coast of the Gulf of Maine than they are today.

[24] Hildebrand and Schroeder (Bull. U. S. Bur. Fish., vol. 43, Pt. 1, 1928, p. 245) report ripe specimens as deep as 9½ to 21 fathoms in Chesapeake Bay.

[25] In Chesapeake Bay they spawn chiefly in April and May, but they are known to do so exceptionally in December (Hildebrand and Schroeder, Bull. U. S. Bur. Fish., vol. 43, Pt. 1, 1928, p. 245).

[26] Ryder (Rept. U. S. Comm. Fish., (1885) 1887, p. 518) describes the early development.

[27] Reported by Lenore Williams, Salt Water Sportsman for June 30, 1950.

[28] Only recent years from which detailed information is available.

[29] Moore, Boston Herald, for August 7, 1950.

In more eastern Nova Scotian waters, also, perch are "less often seen" in salt and brackish water than in fresh[30] and they appear to be restricted, in the eastern side of the Gulf of St. Lawrence, to the "estuarine transition" from salt water to fresh.[31]

Importance.—The white perch is of considerable commercial importance wherever it is abundant in tide waters. The commercial catch in Chesapeake Bay, for example, was 1,143,700 pounds for 1946, 1,851,000 pounds for 1947. And several millions of artificially hatched fry are released there yearly. It also affords good sport to many anglers wherever it is plentiful, both in brackish water or in fresh. But it is not important in the open Gulf of Maine in either of these respects.

Sea bass *Centropristes striatus* (Linnaeus) 1758

BLACK SEA BASS; BLACKFISH

Jordan and Evermann, 1896–1900, p. 1199.

Description.—The sea bass is easily distinguished from the striped bass and from the white perch by the fact that the spiny and soft-rayed portions of its dorsal fin are continuous, so that there is only one long fin instead of two short separate fins. It agrees with its nearer relative the wreck fish (p. 409) in this; also with the scup (p. 411), with the rosefish (p. 430), with the cunner (p. 473), and with the tautog (p. 478). But its general form, rounded caudal and pectoral fins, and its short but high anal fin are sufficient to separate it from the scup, its color prevents confusing it with the rosefish; and no one should take sea bass for tautog or cunner; its mouth and its pectoral fins are so much larger, its caudal of different outline (cf. fig. 211 with figs. 249, 250), and the soft portion of its dorsal as long as the spiny portion. It differs from the wreck fish (p. 409), in many respects, especially in its much larger scales; in the smoothness of its head and gill covers; and in the shape of its tail fin.

It is moderately stout-bodied, about three times as long (not counting the caudal fin) as it is deep, with rather high back but flat-topped head, moderately pointed snout, a large oblique mouth, eye set high up, and one sharp flat spine near the rear angle of each gill cover. The spiny (10 spines) and soft (11 rays) portions of its dorsal fin (which originates slightly in front of the rear corner of the gill covers) are separately rounded, the latter higher than long, with the characteristic outline shown in the illustration (fig. 211). The caudal fin is rounded. In large fish one of the upper rays is much the longest, and though the resulting outline is a trivial character and variable from fish to fish it is an extremely characteristic one that is shared by no other Gulf of Maine species except the kingfish (p. 423). The anal fin (3 short sharp spines followed by 7 soft rays) originates under or very slightly behind the origin of the soft portion of the dorsal fin, which it resembles in its rounded

[30] Vladykov and MacKenzie, Proc. Nova Scotia Inst. Sci., vol. 19, 1935, p. 91.

[31] Huntsman, Trans. Roy. Soc. Canada, Ser. 3, vol. 12, Sect. 4, 1918, p. 63.

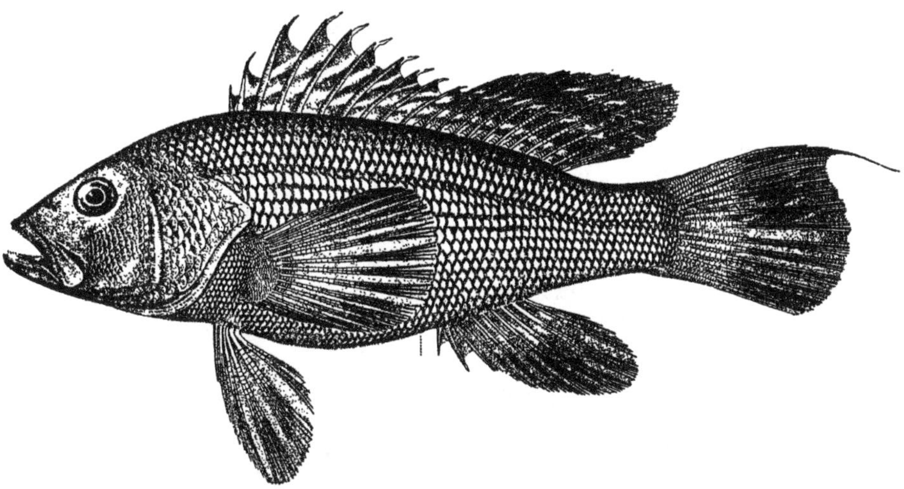

FIGURE 211.—Sea bass (*Centropristes striatus*), Connecticut. From Goode. Drawing by H. L. Todd.

outline and in being higher than long. Both the anal fin and the soft part of the dorsal are noticeably soft and flexible. The pectorals are so long that they reach back almost, if not quite to the anal, broad and round tipped; a good field mark. The ventrals, too, are larger than in any other fish with which the sea bass might be confused, and they originate in front of the pectorals, whereas they stand slightly behind the latter in scup, rosefish, cunner, and tautog. The scales are rather large, but the top of the head is naked. Adult males develop a fatty hump on the back in front of the dorsal fin.

Color.—Sea bass, like most fish that lie on rocky bottom, vary widely in color, the general ground tint ranging from smoky gray to dusky brown or blue black, usually more or less mottled. The belly is only slightly paler than the sides. On every sea bass we have seen the bases of the exposed parts of the scales are paler than their margins, giving the fish the appearance of being barred with longitudinal series of dots of a lighter tint of brown than the general hue on dark fish, but pearl gray on pale ones. The dorsal fin is marked with several series of whitish spots and bands; the other fins are mottled with dusky. Young fish 2 or 3 inches long are greenish or brownish with a dark side stripe passing from eye to caudal fin, and with dark cross bars on the sides.

Size.—Sea bass grow to a length of 2 feet or more and a few reach a weight of 7½ pounds; but northern specimens are seldom heavier than 5 pounds, and they average only about 1½ pounds. A fish a foot long weighs about one pound, one of 18 to 20 inches about 3 pounds.

Habits.—The sea bass contrasts with the striped bass in being strictly confined to salt water. Its inshore-offshore range extends from close in to the coast line, in depths of only a few feet, out about to the 70-fathom contour line, according to the season of the year. Off New Jersey, Long Island, and southern New England they appear inshore during the first or second week in May, withdrawing again late in October or early in November.

They winter offshore along the 30- to 70-fathom zone; the depth and the distance offshore being governed, it seems, by a preference for temperatures higher than about 46°–47°.[32]

It seems, too, that some of the population that summers off New Jersey and to the northward may combine this offshore movement with a southward migration, for sea bass form a considerable part of the catches that are made by the winter trawl fishery off Virginia and northern North Carolina from January to April,[33] whereas they have been taken in small numbers only (though widely dispersed) off southern New England at that time of year.[34]

During the part of the year when the sea bass are inshore they are most plentiful on hard bottom, in water less than 20 fathoms or so, often around submerged wrecks and the pilings of wharves. They are bottom feeders, subsisting chiefly on crabs, lobsters, shrimp, and various mollusks. They also eat small fish (e. g., launce and menhaden), and squid on occasion. And they take a hook readily.

The sea bass spawn in May along the North Carolina coast; from the middle of May to the end of June off New Jersey, off Long Island, and off southern New England. The eggs are buoyant.[35] The young fry are easily identifiable as sea bass by the time they have grown to a length of 2⅜ inches (60 mm.) or so.

General range.—Atlantic coastal waters of the United States, from northern Florida to Cape Cod, occasionally to Maine.

Occurrence in the Gulf of Maine.—The sea bass enters our Gulf only as a rare stray from the south, Pemaquid Point and Matinicus Island being its nothernmost known outposts. It has been taken in Casco Bay; near Gloucester (where a few have been caught in the traps); off Nahant, Salem, and Beverly in Massachusetts Bay; at North Truro and at Monomoy on Cape Cod; and 5 miles east of Pollock Rip Lightship, where a 5-pound fish was trawled in 24 fathoms, December 1930.[36] But it has never been found in any numbers north of the elbow of Cape Cod so far as we can learn. We have never seen it in the Massachusetts Bay region, nor are fishermen of whom we have inquired familiar with it there. Sea bass, it is true,

[32] Neville, Fishery Circular No. 18, U. S. Bur. Fish., 1935, p. 3-7.

[33] For quantities caught and other details, see Pearson, Investigational Report No. 10, U. S. Bur. Fish., 1932.

[34] We counted from 1 to 25 sea bass per haul in 31 out of 45 trawl hauls made by the dragger *Eugene H* off Rhode Island and off southern Massachusetts, in 46 to 67 fathoms, Jan. 27 to Feb. 3, 1950.

[35] The early development of the sea bass has been described by Wilson (Bull U. S. Fish Comm., vol. 9, 1891, p. 209).

[36] Reported by Firth, Bull. 61, Boston Soc. Nat. Hist., 1931, p. 12.

occasionally appear in the returns of the local pound nets and traps.[37]

But it is doubtful whether these records can be accepted, for when the name "sea bass" is used along the northern New England coast it usually is either striped bass (p. 389), white perch (p. 405), tautog (p. 478), or even rosefish (p. 430) that is meant.[38] No sooner do we round Cape Cod to the west, however, than we find the sea bass one of the important ground fish.

Judging from its season at Woods Hole, where it is to be caught from May to October (most abundantly in July, August, and September), sea bass are most likely to be taken in the Gulf of Maine in summer, if at all, though there is one record for December. There is no reason to suppose that they ever succeed in reproducing themselves in the Gulf or in establishing a temporary foothold even if the rare migrants should spawn there.

Importance.—Too scarce to be of any importance in the Gulf, the sea bass is a very valuable food and game fish in more southern waters.

[37] For example, 80 pounds at Provincetown for 1896; 146 pounds at Truro 1898; 101 pounds at the same locality for 1900; with occasional fish at Eastham, Barnstable, Sagamore, Manomet, and Gloucester.

[38] The 3,000 pounds of "sea bass" reported from Manchester, Mass., in 1911 certainly were not this fish.

Wreck fish *Polyprion americanus* (Bloch and Schneider) 1801

Wreck bass

Jordan and Evermann, 1896–1900, p. 1139.

Description.—The combination of a sea-bass-like body with a very rough head having a prominent ridge and strong spines on each gill cover, and a bony protuberance over the eye and on the nape, give the wreck fish an aspect so different from that of any other Gulf of Maine fish (even from its close relative the sea bass) that it should be easily recognized if caught. It is strongly flattened sidewise, about 2½ to 3 times as long as deep (to origin of tail fin), with large mouth. And the lower jaw projects considerably beyond the upper. The scales are rough, much smaller relatively than in the sea bass, and they extend over the bases of the soft-rayed fins. The first part of the dorsal fin has 11 strong spines, the second part, 11 or 12 soft rays, and the spiny part of the dorsal is continuous with the soft-rayed part. The anal fin, with 3 spines and 8 or 9 rays, is similar to the soft-rayed part of the dorsal in outline. The caudal fin is gently rounded; the pec-

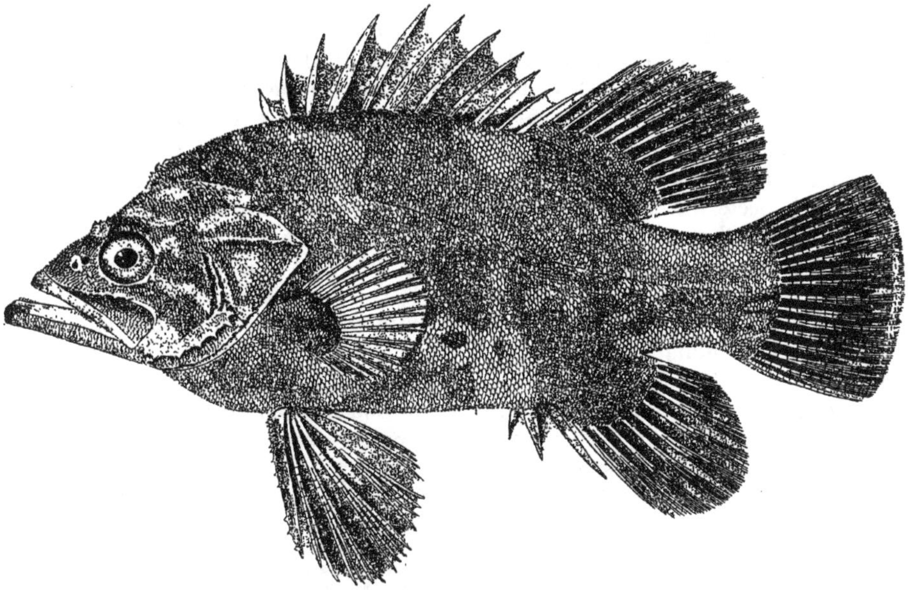

FIGURE 212.—Wreckfish (*Polyprion americanus*), Grand Bank. From Goode and Bean. Drawing by H. L. Todd.

torals (about half as long as the head) stand almost above the ventrals.

Color.—Grayish or blackish brown, the caudal fin edged with white. Young fish are mottled above with gray and cream on head and body.

Size.—Reaches a length of 4½ to 5 feet at least, and a weight of more than 100 pounds.

Habits.—Small wreck fish are most likely to be found under floating logs or wreckage, as the common name implies. When larger, they take to bottom; this, at least, is the case around Madeira and in the Mediterranean.

General range.—This is a fish of wide distribution. In the eastern Atlantic it is known as far north as Norway, as far south as the Canaries; also in the Mediterranean. It has been reported only occasionally in the western Atlantic, but its latitudinal range there extends from the Grand Banks of Newfoundland to the La Plata River. It is also known from the Cape of Good Hope and Indian Ocean.[39]

Occurrence in the Gulf of Maine.—The only report that has reached us of a wreck fish in any part of the Gulf of Maine is of one 24½ inches long, weighing 9 pounds 7 ounces (dressed), taken on the northern edge of Georges Bank, August 13, 1951, by the trawler *Winthrop*.[40] Another, 6 inches long, was caught on the surface off No Man's Land Island, near Martha's Vineyard, August 21, 1925; and two have been brought in from the Grand Banks, one of them many years ago,[41] the second in 1929.[42]

THE CATALUFAS OR BIG EYES. FAMILY PRIACANTHIDAE

The big eyes are very closely related to the sea basses (Serranidae), from which they differ chiefly by the fact that the entire head, including the snout and upper jaw, is clothed with rough scales.

Short big-eye *Pseudopriacanthus altus* (Gill) 1862

Jordan and Evermann, 1896–1900, p. 1239.

Description.—The most striking characters of this fish are its very large eyes and its brilliant red color. Apart from these, it is distinguishable from the seabass tribe by the fact that its whole head, as well as its body, is clothed with rough scales and that the anal fin is longer than the soft-rayed portion of its dorsal fin. Its sidewise flattened body, unusually stout dorsal fin spines, very large ventral fins, and small pectorals, are ready field marks to separate it from the rosefish, the only common Gulf of Maine species of similar appearance that rivals it in color. The big-eye is ovate in outline, very thin through, with rounded dorsal profile, large head, notably oblique mouth, and enormous eyes. The spiny (10 spines) and soft (11 rays) portions of its dorsal fin are continuous, and extend back from the nape nearly to the base of the caudal fin. The anal (3 stout spines and 9 or 10 rays) originates under the eighth or ninth dorsal spine and its soft portion is nearly of the same form as the soft portion of the dorsal, except that its outer angle is somewhat more rounded. The caudal is square-cornered and slightly convex. The ventrals, which originate a little in front of the pectorals, are much larger than the latter, round tipped, and each commences with a stiff spine.

Color.—Bright red in life, below and above; dorsal fin red, the spinous part edged with yellow, a few blackish dots on the soft rays; caudal fin pale, with blackish reticulations; anal red, edged with black; ventrals red at base, dusky on outer part; pectorals plain red. The iris is gold.

Size.—The largest specimen on record was 11 inches long.

General range.—Caribbean Sea, West Indies, and Gulf of Mexico in rather deep water, straying northward to the Woods Hole region and very rarely rounding Cape Cod.

Occurrence in the Gulf of Maine.—A big-eye found alive on Marblehead Beach, September 3, 1859; a second, found at Scituate, Mass., in 1932 or 1933;[43] and a third, about 1½ inches (38 mm.) long, picked up in a tide pool at Cohasset, Mass., by F. G. Bemis in September 1937,[44] are the only definite records for this southern fish within the Gulf. But since it occasionally appears in some numbers at Woods Hole in summer, it may round Cape Cod more often than this paucity of actual records suggests.

---

39 We have given a more detailed statement elsewhere (Copeia, 1930, No. 2, p. 46).

40 This specimen, which we have examined, is in the collection of the U. S. Fish and Wildlife Service at Woods Hole.

41 Goode and Bean, Smithsonian Contrib. Knowl., vol. 30, 1895, p. 238.

42 Schroeder, Copeia, June 1930, p. 46.

43 Reported by MacCoy, Bull. 67, Boston Soc. Nat. Hist., 1933, p. 9.

44 This specimen, reported by Schroeder (Copeia, 1937, p. 238) is in the Museum of Comparative Zoology.

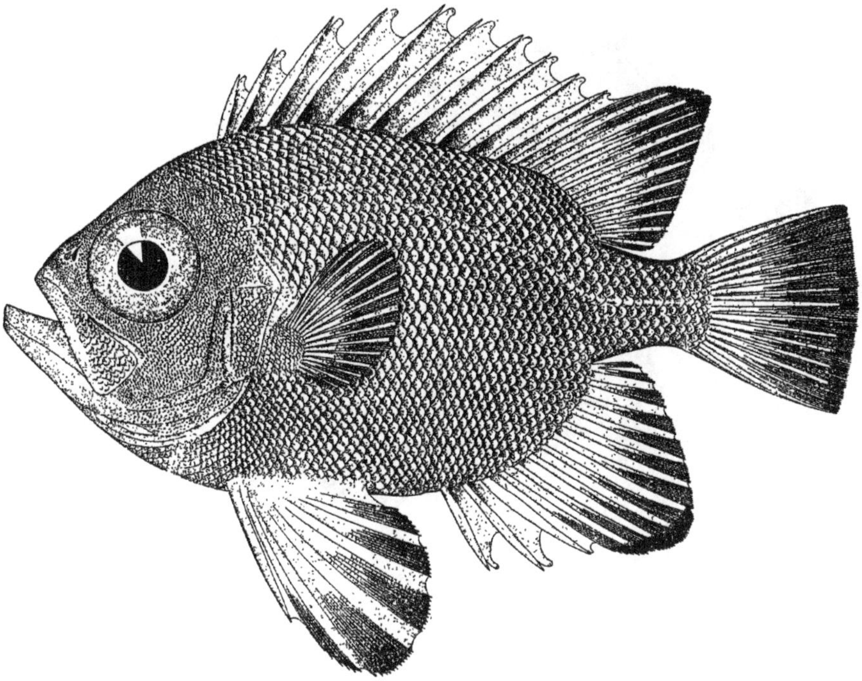

FIGURE 213.—Short big-eye (*Pseudopriacanthus altus*), Key West, Florida. From Jordan and Evermann. Drawing by H. L. Todd.

THE PORGIES. FAMILY SPARIDAE

The structure of the fins is essentially the same in this family as in the sea basses; both spiny and soft portions of the dorsal are well developed and the ventrals are situated below the pectorals. There are important anatomic differences, however, most obvious of which are that the edge of the gill cover does not end with a sharp spine in the porgies but is rounded or at most bluntly angular; and that the maxillary bone (the bone forming the margin of the upper jaw) is sheathed and hidden by the preorbital bone when the mouth is closed. Long, pointed pectoral fins are likewise characteristic of the family; the spiny and soft portions of the dorsal fin are continuous, and the soft rayed anal fin is about as long as the soft part of the dorsal.

KEY TO THE GULF OF MAINE PORGIES

1. Outline of caudal fin deeply lunate, with sharp corners .... Scup, p. 411
   Outline of caudal fin only slightly concave, with rounded corners .... Sheepshead, p. 416

Scup *Stenotomus versicolor* (Mitchill) 1815

PORGY

Jordan and Evermann, 1896–1900, p. 1346, as *Stenotomus chrysops* (Linnaeus), 1766.

Description.—Although the scup is not marked by any one outstanding character it is made easily recognizable by the fact that the spiny portion of its dorsal fin is considerably longer and higher than the soft-rayed portion, which, with its deeply lunate caudal fin, separates it from all other Gulf of Maine fishes of similarly deep and sidewise-flattened bodies. The scup is about one-half as deep as it is long (to the base of the tail fin) and very thin through, recalling a butterfish (p. 363). But the dorsal profile of its rather short head is slightly concave instead of convex, and its scales rather large, thick and firmly attached; not small, thin and easily detached as they are in the butterfish.

The mouth of the scup is small, its eyes are situated high up on the side of the head, and the margins of its gill covers are rounded. It has one

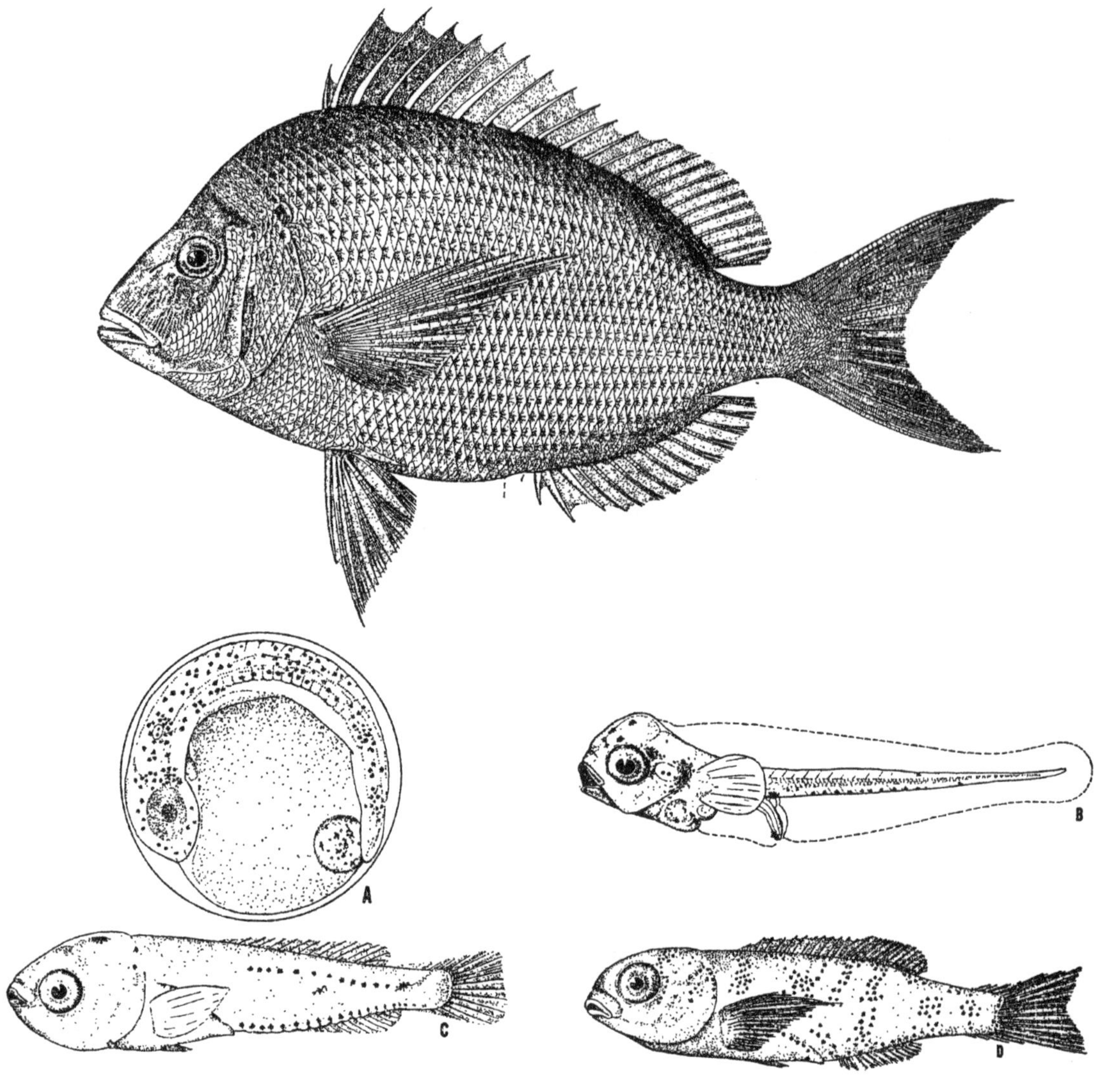

FIGURE 214.—Scup (*Stenotomus versicolor*). Adult, Woods Hole; from Goode, drawing by H. L. Todd; A, egg; B, larva, 3 days old, 2.8 mm.; C, larva, 10.5 mm.; D, larva, 25 mm. A–D, after Kuntz and Radcliffe.

long dorsal fin originating over the pectorals and preceded by a forward-pointing spine; the spiny (12 spines) and soft (12 rays) parts are continuous, forming a single fin. As a whole the dorsal fin is moderately high, its first spine much shorter than the others, its rear corner rounded, and it can be laid back in a groove along the mid line of the back. The anal (3 spines and 11 or 12 rays) is nearly as long as the soft part of the dorsal, under which it stands, and is almost even in height from front to rear, but with the first spine shorter than the others. The anal fin is depressible in a conspicuous groove, like the dorsal. The caudal is deeply concave with sharp corners. The pectorals are very long (reaching to even with the soft part of the dorsal), sharp pointed, and with slightly concave lower rear margins. The ventrals, situated below the pectorals, are of moderate size.

Color.—Dull silvery and iridescent, somewhat darker above than below; the sides and back with 12 to 15 indistinct longitudinal stripes, flecked with light blue and with a light-blue streak following the base of the dorsal fin. The head is silvery, marked with irregular dusky blotches; the belly is white. The dorsal, caudal, and anal fins are dusky, flecked with blue; the pectoral fins of a brownish tinge; the ventrals white and bluish, and very slightly dusky; the iris silvery; the pupil black.

Size.—The scup is said to reach a length of 18 inches and a weight of 3 to 4 pounds, but adults usually run only up to about 12 to 14 inches, and weigh only 1 to 2 pounds.

Habits.—Scup are inshore from early April at the mouth of Chesapeake Bay, and from early May northward to southern Massachusetts. Most of them withdraw from the coast late in October, though some few linger through November, and an occasional fish into December even as far north as the vicinity of Woods Hole.

It has been known for the past 20 years or more that many scup winter off Virginia and off northern North Carolina, in depths of 20 to 50 fathoms, where large commercial catches are made yearly by otter trawlers from January to April,[45] with a few as deep as 90 fathoms or so. And marking experiments have proved that some of the scup that summer along southern Massachusetts migrate southward in autumn as far as to the offings of Chesapeake Bay and of northern North Carolina for the winter, at least in some years, and vice versa.[46]

Scup have, however, been taken during the past few winters in depths of 45 to 70 fathoms off southern New England, in numbers large enough to show that part of the northern contingent of the species simply moves offshore in autumn, to come inshore again in spring.[47]

[45] Reported catches for 1930–1931 (the only winter for which statistics are readily available) were 9,684 pounds in December, 495,312 pounds in January, 637,595 pounds in February, 653,276 pounds in March, and 76,322 pounds in April (Pearson, Investigational Report, No. 10, U. S. Bur. Fish., 1932, p. 14, table 2). In February 1930 *Albatross II* trawled three off Chesapeake Bay in 93 fathoms.

[46] One scup, tagged in summer near Woods Hole, was recaptured in winter off northern Virginia; two off Chesapeake Bay; and one off northern North Carolina (Neville, Fishery Circular No. 18, U. S. Bur. Fish., 1935, p. 3, fig. 3). Three tagged in winter off Virginia were recaptured in summer along New Jersey.

[47] We counted from 1 to 40 scup per haul in 17 trawl hauls out of a total of 44 hauls, on the *Eugene H* off Rhode Island and southern Massachusetts, Jan. 27 to Feb. 3, 1950, at depths of 47 to 67 fathoms; a dragger that caught 7 to 30 bushels in 3 hauls nearby at the time reported catches of 2,000 to 5,000 pounds as sometimes made in the vicinity at that same season; and the *Priscilla V* reported taking 445 pounds on Jan. 12, also 1,230 pounds on Jan. 21, 1950, at 52 to 54 fathoms, some 75 to 82 miles south of No Mans Land off Marthas Vineyard. The *Eugene H* fishing near Hudson Gorge in about 62 fathoms, caught 30,000 pounds of scup on a trip April 1–6, 1953.

Differences in the locations where the largest catches are made in cool winters and in warm make it likely that a preference for water at least as warm as about 45° F. is the factor that determines how far seaward the scup move off any part of the coast in any particular winter.[48] And they are so sensitive to low temperatures that large numbers have been known to perish (both large ones and small) in sudden cold spells in shallow water.

It appears that different bodies of scup move inshore successively in spring, for in 1950 the *Albatross III* took 2,700 scup in 15 hauls at 45 to 55 fathoms, in the Hudson Gorge, on May 11–18, which is one or two weeks after the earliest scup ordinarily appear inshore near New York. And the fact that scup are more plentiful in June and July than in May points in the same direction.

It has been said that the first fish to arrive in spring are the large adults, with the immature fish following later. But there is no definite rule in this regard.

During their summer stay inshore, the scup tend to hug the coast so closely that a line drawn 5 or 6 miles beyond the outermost headlands would probably enclose the great majority of the total population at that time of year.

Scup usually congregate in schools. The young fry come close in to the land in only a few feet of water. Large fish, however, are seldom caught in summer in water shallower than 1 or 2 fathoms (occasionally at the surface), or deeper than 15 to 20 fathoms. They prefer smooth to rocky bottom, which results in a distribution so local that one trap at Manchester, on the North Shore of Massachusetts Bay, took small numbers of scup in 1885, 1886, and 1887, while another trap close by did not yield as much as one fish. They are bottom feeders in the main, seldom rising far above the ground, the adults preying on crustaceans (particularly on amphipods) as well as on annelid worms, hydroids, sand-dollars, young squid, and in fact on whatever invertebrates the particular bottom over which they live may afford. They also eat fish fry to some extent, such free-floating forms as crustacean and molluscan larvae, appendicularians, and copepods. The young feed chiefly on the latter and on other small Crustacea. Adult scup, like most other fish, cease feeding during spawning time, for which

[48] For details, see Neville, Fishery Circular No. 18, U. S. Bur. Fish., 1935.

reason few are caught then, but they bite very greedily throughout the rest of the summer on clams, bits of crab, and sea worms (*Nereis*), as do the immature fish throughout their stay.

Along southern New England scup spawn from May to August, but chiefly in June. Probably spawning both commences later and continues later for the few fish that manage to summer in Massachusetts Bay, and it may be assumed that they spawn wherever they summer.

The eggs are buoyant, transparent, spherical, rather small (about 0.9 mm. in diameter), and have one oil globule. Incubation occupies only about 40 hours at 72° (probably two to three days in the June temperatures of Massachusetts Bay) and judging from the season of spawning at Woods Hole, it is not likely that development can proceed normally in water colder than about 50° F. At hatching the larvae are about 2 mm. long, the yolk is fully absorbed within 3 days when the larva is about 2.8 mm. long, and there is then a characteristic row of black pigment spots along the ventral margin of the trunk. At 25 mm. the pectorals have assumed their pointed outline and the caudal fin is slightly forked, but the ventrals are still so small, and the body so slender, that the little fish hardly suggest their parentage until they are somewhat larger.[49]

In southern New England waters fry of 2 to 3 inches, evidently the product of that season's spawning, have been taken in abundance as early as September; they are 2½ to 3¼ inches long in October, and they may be as long as 4 inches at Woods Hole in November. Apparently young scup grow very little during the winter, for many of 4 inches are seen in the spring, probably the crop of the preceding season. According to Neville's unpublished studies,[50] scup average about 4¼ inches (11 cm.) long at one year of age (from hatching), about 6¼ inches (16 cm.) at two years, about 7⅞ inches (20 cm.) at three years, about 9 inches (23 cm.) at four years, and about 9¾ inches (25 cm.) at five years. If this age schedule is correct, the ages of the large fish of 12 to 15 inches, weighing 1½ to 2½ pounds are considerably greater than the 3 to 5 years that have been credited to them, following Baird's [51] estimate.

[49] Kuntz and Radcliffe (Bull. U. S. Bur. Fish., vol. 35, 1918, p. 106) describe the early development of the scup.

[50] Information from James A. Mason, of the U. S. Fish and Wildlife Service.

[51] Rept. U. S. Comm. Fish. (1871–1872) 1873, p. 228.

General range.—East coast of the United States, from North Carolina to Cape Cod; casual in the Gulf of Maine as far as Eastport, Maine.[52]

Occurrence in the Gulf of Maine.—Although the scup is one of the most familiar of shore fishes right up to the elbow of Cape Cod, with the southern coast of Massachusetts and its off-lying islands yielding annual catches of 1 million to 2 million pounds in good years, very few find their way past Monomoy Point into the colder waters of the Gulf of Maine.

The first definite mention of scup caught north of Cape Cod is Storer's statement that one was taken at Nahant in 1835, and another in 1836, but that it was never seen there before. Possibly these and one picked up dead at Cohasset in 1833 [53] were the survivors of a smack load that had been liberated in Boston Harbor a year or two earlier, and a similar plant was made in Plymouth Bay in 1834 or 1835. There is no reason to suppose that these planted fish established themselves. But when the practice of setting mackerel nets outside Provincetown Harbor was first adopted (about 1842) a few scup were taken in them from year to year; odd fish were caught in Cape Cod Bay yearly and between Boston and Cape Ann during the period 1860 to 1867; and a number were taken in a weir on Milk Island near Gloucester in 1878. It has been learned since (mainly from the catches of the pound nets and traps) that there were a few scup in northern Massachusetts waters in most years (or terms of years) down to the first decade or so of the present century, alternating with other years, or terms of years, when only an occasional fish was taken, or none.

In most of the years for which information is available, and when there have been any scup north of Cape Cod, the combined catches of the various traps have run from less than 100 pounds to 1 to 2 thousand pounds at most, whether for Cape Cod Bay or for the northern side of Massachusetts Bay (Essex County).[54] But Cape Cod Bay seems to have seen what might almost be called peaks of abundance in 1879 (catch, about 7,000 fish); in 1882–1885 (yearly catches 2,372–

[52] The southern scup, *Stenotomus chrysops* (Linnaeus) 1766, which was first reported from Charleston, S. C., ranges northward about to Cape Hatteras.

[53] Goode, Fish. Ind. U. S., Sect. 1, 1884, p. 387.

[54] Statistics of the shore fisheries were published by the State of Massachusetts in the Annual Reports of the Commissioner of Fisheries and Game for 1879–1911 and 1917–1919; of the Division of Fisheries and Game for 1920 and 1921.

5,354 fish); in 1887,[55] in 1890 (1,890 fish); and in 1895–1896 (14,362 and 5,083 fish, respectively); also the northern side of Massachusetts Bay in 1909–1910 (8,417 pounds [56] and 4,181 pounds); both Cape Cod Bay (6,000 pounds) and the north shore of Massachusetts Bay (3,217 pounds) in 1917.

The cataclysmic shrinkage that took place in the stock of scup off southern Massachusetts between 1896 (prior to which the annual catch there had usually run from 1 to 3 million pounds) and 1902, when it fell to only about one-tenth as much (about 200,000 pounds) appears to have involved the scup in Cape Cod Bay also, for none at all were reported there from 1907 through 1911, or in 1918–1920,[57] except that there was an unusually large run there in 1917. But 1908, 1909, and 1919 were good scup years for the north shore of Massachusetts,[58] "good," that is, for those northerly waters, suggesting that when conditions favor, a small independent population may be present there. Perhaps the fact that larger catches than usual are not always registered in both these regions in the same year may point in this same direction.

No scup were reported from Essex County for 1919, 1928, or 1930; nor were enough taken in Cape Cod Bay in those years to cause any local comment.[59] Though the fisheries statistics do not throw any light on the status of the scup north of Cape Cod subsequently,[60] there cannot have been many of them in Cape Cod Bay regularly at any time during the past 15 years or so, for the only scup that were taken in a set of 8 traps at North Truro from 1935 down through 1950 were 125 pounds taken on June 28, 1938, evidently one small school of perhaps 100–125 individuals.[61] And 33 barrels (about 4,950 pounds) taken in a trap at Sandwich on the southern shore of Cape Cod Bay on Sept. 15 or 16, 1944, after a heavy gale, were the only scup caught in this set of traps from 1944 to 1950.[62] It would be interesting to know whether they came through the Cape Cod Canal or around the Cape.

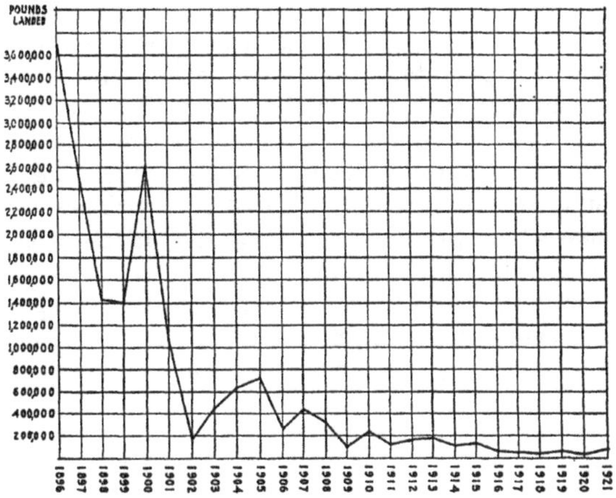

FIGURE 215.—Scup (*Stenotomus versicolor*). Annual catch of scup (pounds) in pound nets and traps in Massachusetts, from statistics published by the State Commissioner of Fisheries and Game.

Thus the presence of considerably greater numbers of scup on the southern coast of Massachusetts since about 1928 than had been there during the preceding decade [63] seems not to have been reflected in Cape Cod Bay except in sporadic cases. And we have not heard of any caught in the northern side of Massachusetts Bay during the past few summers.

In any case, Cape Ann is the northern boundary to the usual range of the scup. In 1896, a year of plenty not only in Massachusetts Bay but to the south in general, occasional specimens were taken daily in Casco Bay in the Small Point traps during the first half of July, and in July 1951, three were reported from Small Point, Maine, sporadic visits such as may be expected of any southern stray. In 1938 about 100 pounds of scup were landed in Lincoln County, Maine, probably from nearby,

[55] The reported catch for the town of Barnstable for that year was so large (69,168) as to suggest that it included scup from the south shore.

[56] If all these really were scup and not some other fish.

[57] No information is available for the years 1912–1916.

[58] Pound net catches for Essex County of 1,203 pounds, 8,417 pounds and 4,181 pounds, respectively.

[59] Catches reported for these years for Barnstable County include not only such scup as may have been taken in Cape Cod Bay, but the catches (doubtless far larger) for the southern coast, which does not fall within the limits of the Gulf of Maine.

[60] We are informed by William Royce of the U. S. Fish and Wildlife Service that catches since 1931 have been credited to the home ports of the vessels making them, wholly irrespective of where the fish were caught or landed. There is no reason to suppose that any significant part of the landings of scup reported for Essex County since then (which reached a maximum of 7,945,209 pounds for 1938) actually came from Massachusetts Bay, or from anywhere in Massachusetts waters, for that matter.

[61] Information from the Pond Village Cold Storage Co.

[62] Information from Benjamin Morrow, who operates these traps.

[63] The yearly landings of scup for Massachusetts rose from not more than 100,000 pounds for the decade 1912 to 1921 to about 1,100,000 pounds for the 5 years 1943–1947. But there is no way of knowing how large a part of the catches reported during the latter period were actually taken in Massachusetts waters and not farther west and south along the coast.

and they have been reported from Eastport.[64] But we suspect that porgies in St. Marys Bay, Nova Scotia, reported to Knight[65] were some other fish.

Probably such scup as spread north of Cape Cod in favorable summers withdraw southward again (if they survive) in autumn to the same offshore wintering grounds to which the much more numerous scup repair from the southern shores of Massachusetts. The fact that small scup, probably devoured on their way offshore, have been found in autumn in cod stomachs on Nantucket Shoals, where scup certainly are not common in summer, points in this direction. There is no reason to believe that any of these fish winter in the deep basin of the Gulf of Maine.

Importance.—Scup are never plentiful enough anywhere north of the elbow of Cape Cod to be of importance, whether commercially or to the angler. But this is an important food fish to the westward and southward where it is plentiful. Landings ran, for example, between about 3,300,000 pounds and 5,600,000 from the southern coast of New England and between about 3,300,000 and 4,300,000 pounds from New York, for the years 1945–47. The "porgy", as it is commonly called along that part of the coast, is also a favorite with anglers, for it bites greedily and is a good pan fish. Great numbers of them are caught on hook and line for home consumption.

[64] A specimen taken many years ago recorded by Kendall (Occ. Pap. Boston Soc. Nat. Hist., vol. 7, No. 8, 1908, p. 103).

[65] Descript. Cat. Fishes Nova Scotia 1866, p. 13.

Sheepshead *Archosargus probatocephalus* (Walbaum) 1792

Jordan and Evermann, 1896–1900, p. 1361.

Description.—The sheepshead resembles the scup so closely in its general organization that the family relations between the two are obvious. Like the scup it is deep bodied and much flattened sidewise, with similar profile. It has one long dorsal fin, scuplike in outline, the anterior two-thirds of which is spiny (11 or 12 spines) and the posterior one-third is soft (11 to 13 rays). Its anal fin (3 spines and 10 or 11 rays) is about as long as the soft portion of its dorsal, under which it stands, and both the dorsal fin and the anal can be depressed in a deep groove. The pectorals are long and pointed; the ventrals are situated a little behind the latter; the scales are large; and the eyes are located high on the sides of the head; in all of which the sheepshead agrees with the scup. It is readily recognized, however, by the fact that its caudal fin is not so deeply emarginate as that of the scup, and has rounded corners instead of pointed ones, while its dorsal spines are alternately stout and slender; its second anal spine is much stouter than

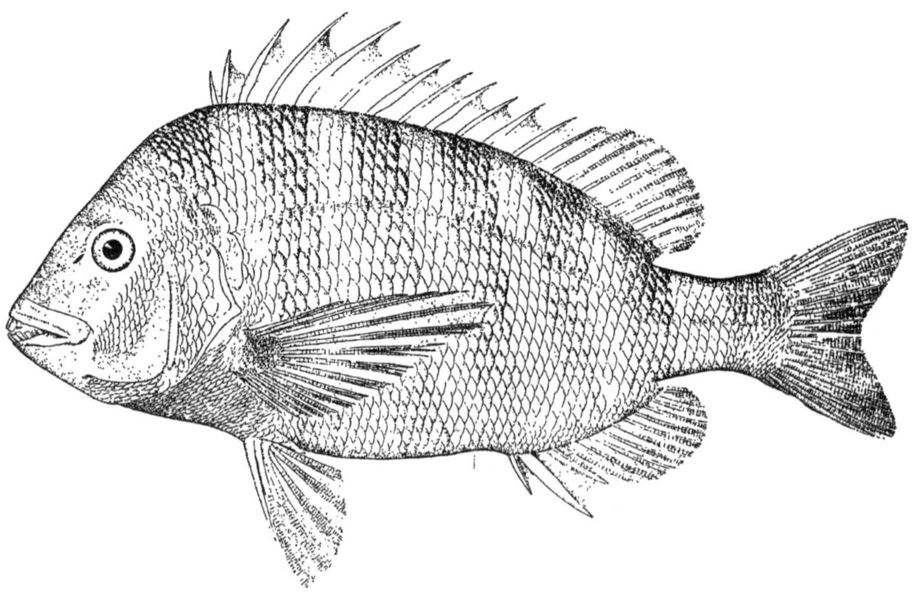

FIGURE 216.—Sheepshead (*Archosargus probatocephalus*), North Carolina. From Goode. Drawing by H. L. Todd.

that of the scup; the dorsal profile of its head is steeper; its nose is blunter; and its teeth are much broader. Furthermore, the body of the sheepshead is noticeably thicker, its back is rounded, and its sides show seven broad, dark brown or black crossbars on a gray or greenish yellow ground, instead of being plain colored like the sides of the scup.

Size.—The sheepshead grows to a length of about 30 inches and to a weight of 20 pounds.

General range.—Atlantic and Gulf of Mexico coasts of the United States from Texas to Cape Cod, and reported in the Bay of Fundy as a stray.

Occurrence in the Gulf of Maine.—The sheepshead was abundant as far north as New York formerly, and not uncommon about Woods Hole. It is common still to the southward. But it has been decidedly rare east of New York for many years past, although a number, about 6 inches long, were taken off Onset at the head of Buzzards Bay, in late August 1951.[66]

The only record of it north or east of the elbow of Cape Cod is Cox's[67] statement that it is occasional in St. John Harbor, New Brunswick. But no actual specimens are mentioned, and as it is not known ever to have strayed to Massachusetts Bay (a far more likely goal for any southern coast fish than the Bay of Fundy is), its claim to mention here is weak.

THE CROAKERS, DRUMS, AND WEAKFISHES. FAMILY SCIAENIDAE

The croakers have both the spiny portion and the soft portions of the dorsal fin well developed (either separate or as one continuous fin), and their ventrals are what is known as thoracic in position, i. e., about under the pectorals. They are readily separable from the sea basses (p. 389,) the porgies (p. 411), and the cunner tribe (p. 473) by the fact that their anal fin has only 1 or 2 spines instead of 3, and is much shorter than the soft portion of the dorsal; from the rockfishes and sculpins by their relatively smooth head; and from all the mackerels and the pompano tribe by their stout caudal peduncles and rounded or only slightly concave caudal fins. Most of them produce loud drumming sounds by rapid contractions of certain abdominal muscles against the gas-filled air bladder; hence the common names "croaker" and "drum." The kingfish (p. 423) is an exception to this rule.

KEY TO GULF OF MAINE CROAKERS AND WEAKFISHES

1. There is no barbel on the chin 2
   The chin bears one or more barbels 3
2. Body only about one-fourth as deep as it is long (to base of caudal fin); anterior profile of head sloping only moderately; snout pointed; no dark spot behind upper corner of gill opening ... Weakfish, p. 417.[68]
   Body at least one-third as deep as it is long to base of caudal fin; anterior profile of head sloping steeply; snout blunt; there is a dark spot close behind the upper corner of the gill opening .... Spot, p. 423
3. Several barbels on chin; snout ends about even with front of lower jaw; cheek smooth . Black drum, p. 425
   Only one barbel on chin; snout projects considerably beyond lower jaw; cheek with 2 short, tooth-like serrations .... Kingfish, p. 423

[66] Information from Mrs. Harold Hatch.

[67] Bull. Nat. Hist. Soc. New Brunswick, No. 13, 1895, p. 71.

[68] Jordan (Stanford Univ. Publ., Univ. Series, Biol. Sci., vol. 3, No. 2, 1923, p. 202) placed the weakfish in his new family Otolithidae, which he separated from the Sciaenidae as having a different arrangement of vertebrae. But we think it preferable (following Smith, Sea Fishes Southern Africa, 1949, p. 223) to use Sciaenidae in the older and more inclusive sense, because the only family character marking Otolithidae off from it is internal, hence requires dissection for its recognition.

Weakfish *Cynoscion regalis* (Bloch and Schneider) 1801

SQUETEAGUE; SEA TROUT; GRAY TROUT

Jordan and Evermann, 1896–1900, p. 1407.

Description.—The relative sizes and shapes of the fins of the weakfish, and its color, are such ready field marks that it is one of our most easily identified fishes. Among Gulf of Maine species with separate spiny and soft-rayed dorsal fins, it is distinguishable from the mullet by the considerable length of its dorsals as well as by many other characters; its slightly emarginate tail distinguishes it from any mackerel or pompano; this same character, combined with a short anal fin and a first dorsal fin higher than the second dorsal gives it an appearance quite different from a bluefish; and the fact that its second dorsal is much longer than the first, and that it has only 2 anal spines and a slender body obviate all possibility of confusing it with striped bass or white perch. The shape of its dorsal and caudal fins and of its head, and the absence of a chin barbel make it disinguishable at a glance from the kingfish (p.

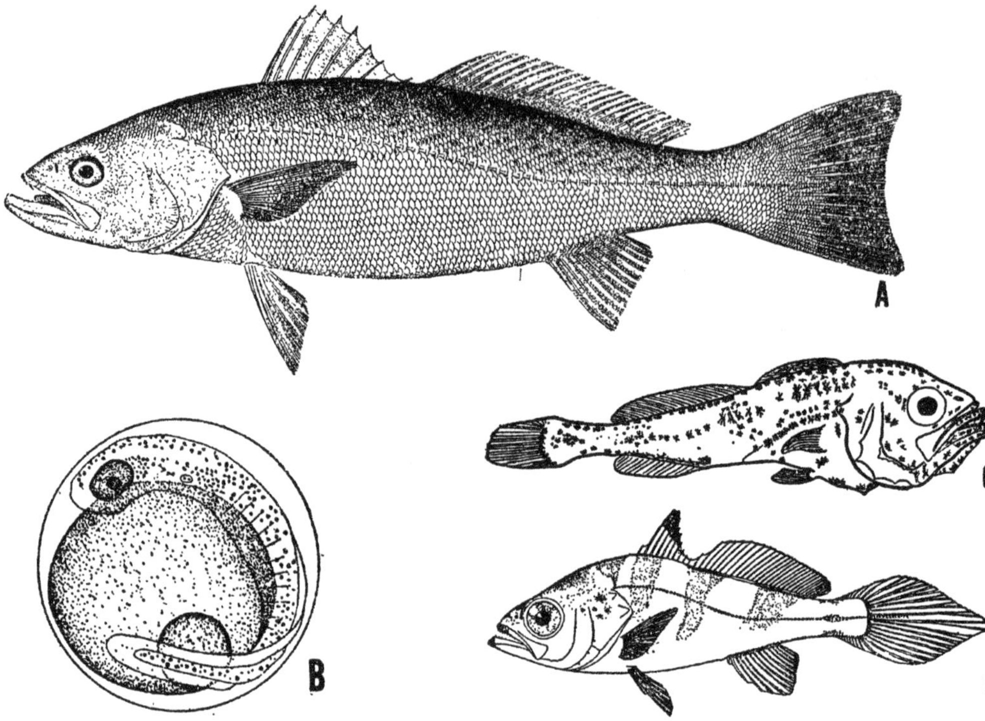

FIGURE 217.—Weakfish (*Cynoscion regalis*). A, adult; from Goode, drawing by H. L. Todd. B, egg; C, larva, 12.4 mm.; D, fry, 32 mm. B and D, from Welsh and Breder; C, after Tracy.

423), the absence of barbels on the chin separates it from a drum (p. 425); it has nothing in common with such bizarre fishes as the John Dory (p. 297), triggerfish (p. 520) or any member of the sculpin tribe.

The weakfish is a slim, shapely fish, about four times as long as deep (to the base of the caudal fin), only slightly flattened sidewise, with rather stout caudal peduncle; a head about one-third as long as body, moderately pointed snout, and large mouth. Its upper jaw is armed with two large canine teeth and its lower jaw projects beyond the upper. The first dorsal fin (10 spines), originating a little behind the pectorals, is triangular; the second dorsal (26 to 29 rays), originating close behind the first, is more than twice as long as the first and roughly rectangular. The caudal fin is moderately broad and only slightly concave in outline. The anal fin (2 very slender spines and 11 or 12 rays) is less than half as long as the second dorsal, under the rear part of which it stands. The ventrals are below the pectorals, which they resemble in their moderate size and pointed outline.

Color.—Dark olive green above with the back and sides variously burnished with purple, lavender, green, blue, golden, or coppery, and marked with a large number of small black, dark green, or bronze spots, vaguely outlined and running together more or less, especially on the back; thus forming irregular lines that run downward and forward. The spots are most numerous above the lateral line, and there are none on the lower part of the sides or on the belly. The lower surface, forward to the tip of the jaw, is white, either chalky or silvery. The dorsal fins are dusky, usually more or less tinged with yellow; the caudal is olive or dusky with its lower edge yellowish at the base; the ventrals and the anal are yellow; and the pectorals are olive on the outer side, but usually yellow on the inner side.

Size.—It is said that weakfish have been taken as heavy as 30 pounds, but the largest of which we can find authentic record in recent years was one of 17 pounds 8 ounces, taken on the New Jersey coast, on rod and reel, by A. Weisbecker, Jr., September 30, 1944. And a fish heavier than 12 pounds or longer than 3 feet is a rarity. Off southern Massachusetts the largest fish run 6 to 10 pounds in weight, while most of the larger ones taken there weigh from 1 to 6 pounds and are 14 to 26 inches long. An average of 5 pounds has been reported for Massachusetts Bay, but this is probably excessive. The average proportion between length and weight of weakfish is about as follows:

Length in inches	*Weight in pounds*
12 to 14	⅝ to 1
14 to 16	1 to 1½
16 to 18	1¼ to 1¾
18 to 20	1⅝ to 2½
22 to 23½	3½ to 4⅛
25½ to 27½	5 to 6
30 to 32	9½ to 11

The female members of a school usually run somewhat larger than the males.

Habits.—Although there are very few weakfish in the Gulf of Maine today, if any, they were for a time so plentiful in its southwestern waters (and may at any time reappear there in abundance) that their habits deserve more attention than the fish's present status would call for.

In the southern part of its range (e. g., along the Carolinas) this is said to be a resident species. But it is strictly seasonal to the northward, appearing in spring, spending the summer inshore, and withdrawing again in autumn. Within the mouth of Chesapeake Bay the fishing season usually is from the middle of April (commencing a week or two later up the bay) to the middle of November, with good catches occasionally made as late as the first of December. On the southern New England coast, as illustrated by Woods Hole, weakfish are caught from May (some years as early as April, other years not until June) until the middle of October. Probably they are not to be expected north of the elbow of Cape Cod until June (in the years when they come that far north), nor later than September or October at latest, for most of the weakfish disappear from the middle Atlantic coast before the end of October.

The lower limit to the temperature range preferred by the weakfish has not been determined. But it has long been known that they are sensitive to cold. And a case is on record (November 27, 1903) when many were benumbed by a sudden chilling of the water, near Beaufort, North Carolina.[69] Hence seasonal chilling is doubtless the event that drives them away from the middle Atlantic and New England coasts in late autumn.

The capture of weakfish in some numbers between the offings of Chesapeake Bay and of Cape Hatteras by otter trawlers during the winter months, during the past twenty-odd years,[70] has dispelled some of the mystery in which the winter home of this fish was previously shrouded. The fact that 5 small ones were picked up in the 50 to 55 fathom zone off Rhode Island by the dragger *Eugene H* in mid-January 1950, also 6 more south of Marthas Vineyard in about 55 fathoms,[71] and another 5 pounder on February 20 [72] is evidence that some of those that summer to the northward only move offshore to escape falling temperature. Others may move southward in winter for long distances, and offshore, as some of the northward-summering scup seem to do (p. 413).

Weakfish tend to hold close inshore during their summer stay on the coast; we have never heard of one on Nantucket Shoals, and only once of weakfish caught on Georges Bank.[73] They are usually found in shallow waters along open sandy shores and in the larger bays and estuaries, including salt marsh creeks. They even run up into river mouths, but never into fresh water, so far as we know.

Weakfish move in schools, often small but sometimes consisting of many thousands.[74] They have been described repeatedly as swimming near the surface, this being the general rule near New York and along the southern New England coast, where great numbers are caught on hook and line within a few feet of the top of the water. And their preference for shallow water is reflected in the large numbers caught in pound nets along the middle Atlantic coast. Probably few descend deeper than 5 to 6 fathoms during the summer, but the precise level at which they are to be caught at

[69] Smith, North Carolina Geol. and Economic Survey, vol. 2, 1907, p. 411.

[70] See Pearson, Investigational Report No. 10, U. S. Bureau of Fisheries, 1932, p. 14, table 2, for the catches for the winter of 1930–1931, by species and by months. The *Albatross III*, also, trawled 83 weakfish in 29 fathoms off Cape Hatteras, and 1 in 14 fathoms off Charleston, S. C., in late January 1950.

[71] Reported by Capt. Henry Klimm. We saw one of them.

[72] We saw this fish.

[73] Two fish were reported by an otter trawler from the offshore part of the Bank in the summer of 1950.

[74] A notable and oft-quoted instance was off Rockaway Beach, N. Y., July 1881, when a school was sighted so large that three menhaden steamers seined some 200,000 pounds of weakfish from it, averaging 1½ to 3 feet in length.

any given locality is governed by their food at the time. On open coasts they often feed on bottom right in the surf. They also feed on bottom in estuarine waters when dieting on bottom-living animals, but in the upper water layers when preying on small fish.

Weakfish feed on a wide variety of animals, including crabs, amphipods, mysid and decapod shrimps, squid, shelled mollusks, and annelid worms, but chiefly on smaller fish, such as menhaden, butterfish, herring, scup, anchovies, silversides, and mummichogs, of which they destroy vast quantities. The precise diet varies with the locality (that is, with what is most readily available), but small menhaden are probably the most important single item. The adult weakfish usually depend on fish, though occasionally they have been found feeding exclusively on crabs and shrimps. The young depend more on shrimp and on other small crustaceans than the adults.[75] Weakfish bite greedily on various kinds of bait, especially on shedder crabs, clams, shrimp, and mummichogs or other small fish. And they are often caught on artificial lures of one kind or another.

The females do not make any sounds, but the males have well-developed croaking muscles in the walls of the abdomen, with which they make a drumming noise.

Breeding habits.—On the middle Atlantic coast the weakfish spawn from May to October, with the chief production of eggs between mid-May and mid-June.[76] The eggs have been taken in tow nets at various localities in temperatures ranging from 60° to 70°, in salinities of 28.01 to 30.9 per mille. And it is probable that weakfish spawn locally around the shores of Cape Cod Bay in years when the fish are plentiful there, as they do regularly about Woods Hole, if the summer temperature of the surface is high enough. Spawning takes place chiefly in the larger estuaries or close to their mouths, usually at night. The eggs are buoyant, spherical, 0.74 to 1.1 mm. in diameter, usually with one, rarely with as many as four, oil globules that coalesce into one large one as development progresses. Incubation occupies 36 to 40 hours at a temperature of 68° to 70°, and the newly hatched larvae are 1.75 mm. long.

At 30 mm. the young weakfish have attained most of the structural characters of the adult. But they continue much deeper and more flattened sidewise until they are 6 to 8 inches long; the head and eyes are relatively larger; and their caudal fin is obtusely pointed with the center rays much the longest, instead of concave. The smaller fry (1½ to 3 inches) are marked with four dark, saddle-shaped patches extending downward on the sides to a little below the lateral line, which are not lost until a length of about 4½ inches is reached. As the young fish grow, other bands of pigment are interpolated below the lateral line, the adult coloration not being fully developed until they are 7 to 8 inches long.[77]

Rate of growth.—Weakfish fry grow at so variable a rate during the first summer that they may be anywhere between 4 inches and 6 inches long in the fall, when they are about 6 months old. The smallest fish seen in spring (no doubt yearlings) are 8 to 10 inches long. Thereafter the rate of annual growth is slower. But the variation in the length attained by the fry during their first summer and autumn, consequent on the protracted spawning season, combined with the fact that scale studies of this species have proved puzzling, make it difficult to group the older age classes by size. As far as known, a weakfish of 10 to 12 inches is likely to be about 2 years old; one of 13 inches, about 3 years; 15 inches, about 4 to 5 years; 18 inches, about 5 or 6 years; one of 22 inches about 6 to 7 years old;[78] 24 inches perhaps 9 years; and 30 inches perhaps as old as 12 years. Both males and females usually mature at 2 to 3 years of age, i. e., when 10 to 13 inches long.

General range.—Eastern coast of the United States from the east coast of Florida to Massachusetts Bay, straying northward to the Bay of Fundy, and perhaps to Nova Scotia.[79]

---

[75] For diet lists of weakfish of various sizes, see especially Welsh and Breder (Bull. U. S. Bur Fish. vol. 39, 1924, p. 159); also Peck (Bulletin U. S. Fish Comm., vol. 15, 1896, p. 352).

[76] The following account of the breeding and development of the weakfish is condensed from Welsh and Breder (Bull. U. S. Bur. Fish., vol. 39, 1924, p. 150).

[77] Tracy (Thirty-eighth Ann. Rept. Comm. Inland Fish., Rhode Island, 1908, pp. 85–91), Eigenmann (Bull. U. S. Fish Comm., vol. 21, 1902, p. 45), and Welsh and Breder (Bull. U. S. Bur. Fish., vol. 39, 1924, p. 154) describe the older larvae and fry.

[78] According to studies by Taylor (Bull. U. S. Bur. of Fish., vol. 34, 1916, p. 318); by Welsh and Breder (Bull. U. S. Bur. of Fish., vol. 39, 1924, p. 158); and by R. A. Nesbit, formerly U. S. Bur. Fish. (unpublished).

[79] It is credited indefinitely to "Maine" by Holmes (Fishes of Maine, 1862, p. 74); Goode (Fish. Ind. U. S. Sect. 1, 1884, p. 362), states that scattering individuals have been caught as far as the Bay of Fundy; and Halkett (Check List Fishes Canada, Newfoundland, 1913, p. 87) mentions one as probably caught off Nova Scotia.

Occurrence in the Gulf of Maine.—The center of abundance for the weakfish is along the coast of the middle Atlantic States from the Virginia Capes to New York. It also occurs regularly as far north and east as Cape Cod. But the stock of weakfish fluctuates widely on the southern New England coast, and it is only during periods of great abundance there that weakfish appear in any numbers in Cape Cod and Massachusetts Bays, which may be set as the extreme northern limit for its appearance except as a stray. In the years when it has passed Cape Cod in appreciable numbers it has always been far more plentiful along the inner side of the Cape and in Cape Cod Bay than north of Boston, as appears from the following statement of catches for 1906, a year of great abundance.

	Pounds
Cape Cod Bay:	
Provincetown	115, 789
Truro	202, 050
Brewster	137, 659
Sandwich	6, 221
North Shore of Massachusetts Bay:	
Nahant	[80] 369
Manchester	410

Only once, however, for a period of about 9 years, have there been many weakfish during the past century and a half, even in the Cape Cod Bay region.[81] Apparently they were plentiful off southern New England during the last part of the eighteenth century, and to judge from fishermen's reports weakfish were well known in Massachusetts Bay at that time. But they vanished so completely sometime prior to 1800 that when a stray specimen was taken at Provincetown in June 1838, it was sent to Boston for identification. And this disappearance evidently involved the whole northern part of the range of the species, for weakfish vanished similarly from the Nantucket-Marthas Vineyard region sometime between 1800 and 1837. They had reappeared, however, off southern Massachusetts by 1867; they were abundant there, once more, by 1870; and one or two were taken off Truro and Provincetown in 1884. From then on until 1895, a few were returned yearly from Truro, Provincetown, Plymouth, and even from as far north as Gloucester and Manchester, the annual catch ranging from an odd fish only (e. g., 1893 and 1894) to 700 or 800 pounds, at most, for Cape Cod Bay and for the northern part of Massachusetts Bay, combined.

The catch in the Cape Cod Bay-Massachusetts Bay region was larger for the next few years (4,892 pounds in 1896,[82] 1,006 pounds in 1897, 6,046 pounds in 1898, and 11,572 pounds in 1899), though with the catches localized chiefly on the outer side of Cape Cod and in Cape Cod Bay, as might be expected of a stray from the south. And they appeared in such numbers in Cape Cod Bay in 1900 that the catch there jumped to upward of 108,000 pounds for that year,[83] while a few were taken even as far north as Boston Harbor and Gloucester.

This marked the commencement of a period of local abundance, which was entirely unexpected (for nothing like it had been experienced since the settlement of the country), and which (with its equally sudden eclipse) is perhaps the most interesting event in the history of the local fisheries. Unfortunately definite statistics of the catches are not available for the crucial years, but weakfish were so plentiful in Cape Cod Bay in 1901 as to be a drug on the market; while in 1902 and 1903 the pound nets in Cape Cod Bay were often filled with schools of large weakfish, averaging about 5 pounds. So plentiful were they, indeed, during the summer of 1903 that the traps at North Truro alone reported 280,000 pounds.

This abundance continued through 1904, by which time it seems to have been accepted as the normal condition of affairs, and no longer worth comment. But it seems to have culminated in that summer or the next, for weakfish were reported as less plentiful in 1906. Nevertheless, the Cape Cod Bay traps (excluding Barnstable, Chatham, Yarmouth, and Dennis) reported almost half a million pounds of weakfish for that year; the North Shore of Massachusetts Bay, 20,779 pounds, which probably was not more than half or two-thirds of the actual total, for the returns were incomplete. This, however, was the last big year, for the catch north of the elbow of Cape Cod was less than one-third as great in 1907 as

[80] Twenty thousand pounds were also reported from Gloucester, but we have reason to believe that the fish were actually caught in Cape Cod Bay; and traps operated at Rockport and at Newburyport took no weakfish.

[81] There are intimations in the writings of the early historians of New England of similar disappearances and returns of the weakfish (Goode, Fish. Ind. U. S., Sect. 1, 1884, p. 363).

[82] Omitting the towns of Yarmouth, Dennis, Chatham, and Barnstable, where traps have been operated on the Vineyard Sound shore as well as on the Gulf of Maine shore line.

[83] Omitting the towns of Yarmouth, Dennis, Chatham, and Barnstable, where traps have been operated on the Vineyard Sound as well as on the Cape Cod Bay side.

it had been in 1906. And this was the beginning of the end, for only 8,249 pounds were reported there in 1908, 569 pounds in 1909, and 907 pounds in 1910.

We do not know of the capture of a single weakfish that can be credited with certainty to the outer shore of Cape Cod, to Cape Cod Bay, or to the northern shore of Massachusetts Bay from that time on, down to 1921 (most recent year for which the pound net catches were published in detail), when 21 pounds were reported for the town of Barnstable.[84] We should emphasize too that about the same number of pound nets and traps have been operated from year to year throughout this period and at about the same general localities, so that fluctuations in the catch did actually reflect similar fluctuations in the stock of fish.

There is no reason to suppose that weakfish have ever entered Cape Cod Bay in any numbers since that time. Only one, indeed (a 5-pounder), was recorded, from one set of 8 traps at North Truro during the 16 years 1935–1950;[85] another set of 2 traps at Barnstable, took only 3 weakfish during the summer of 1950;[86] and 3 other traps at Sandwich, Mass., took 2 weakfish in 1948, 1 in 1949, and none in 1950.[87]

We doubt whether any weakfish have reached the northern side of Massachusetts Bay since 1909, when 200 pounds were reported from a pound net at Gloucester. Large landings, it is true, have been reported as from the northern part of the Massachusetts coast (Essex County) in several recent years, ranging up to some 3,600,000 pounds in 1945. But there is no reason to suppose that any of them were caught north of Cape Cod for we are informed by William Royce of the Fish and Wildlife Service that all fish taken by vessels sailing out of Gloucester during these years were credited to that port, irrespective of where caught or where landed. The fish may have come from as far south as the North Carolina winter fishery. And this applies equally to a few that were credited to Maine in 1931 (45 pounds) and in 1932 (318 pounds).

We can offer no explanation for this unexpected invasion of weakfish north of Cape Cod about the turn of the present century, or for its equally sudden eclipse, the opportunity having passed long since for obtaining any information as to the sizes and ages of the fish, as to their movements, and as to the physical state of the water at the time. It was not a local event, however, but part of a corresponding fluctuation in the population as a whole existing east and north of New York. Thus the catch for the southern coast of New England was more than eight times as great in 1904 (upward of 7 million pounds) as it had been in 1889 (about 830,000 pounds), but thereafter declined so markedly that in 1908 both the commercial fishermen and the anglers of Rhode Island and of southern Massachusetts complained of the scarcity of weakfish. Less than 400,000 pounds were taken off southern New England in 1919, and the weakfish had so nearly vanished from the southern shores of Massachusetts by 1920 and 1921 that the reported catches for the pound nets of the State were only 785 and 691 pounds, respectively, for those years.[88] We should emphasize that the partial recovery that then took place off the southern Massachusetts coast, where the average catch was again nearly a quarter of a million pounds during the period 1931–1938, did not bring the weakfish back to Cape Cod Bay.

It has often been suggested that weakfish are plentiful when bluefish are scarce, and vice versa, and the argument has been advanced that the latter not only devour fry of the weakfish but its food also, and hence not only destroy many but drive others away. But no convincing evidence has been brought forward that the fluctuations of these two species of fish are mutually dependent in any way.

Importance.—At the present time the weakfish is of no importance in the Gulf of Maine, whether commercially or to the angler, though it was a very valuable addition to the shore fisheries of Cape Cod Bay during its one brief period of plenty there. However, it is one of the most important of food fishes along more southern coasts,[89] and a favorite

[84] No catch statistics are available for the years 1912-1916, and there is no knowing whether any of the weakfish reported for Barnstable Co., in 1919 (962 pounds) came from the northern (i. e., Cape Cod Bay) shore.

[85] Information from the Pond Village Cold Storage Co.

[86] Information from John E. Vetorino, who operates these traps. One hundred twenty-three pounds reported from Barnstable County in 1928, and 101 pounds in 1929, may likely have come from the Vineyard Sound shore, not from the Cape Cod Bay shore.

[87] Information from Benjamin Morrow, who operates these traps.

[88] No statistics are available for the years 1922–1929.

[89] In 1946, the reported catch of weakfish of this species was about 3,252,000 pounds for southern New England; 11,715,000 pounds for the Middle Atlantic States; 20,557,000 pounds for the Chesapeake Bay region; and 4,770,000 pounds for the South Atlantic States.

game fish which has been the subject of many accounts from the angler's standpoint.

Spot *Leiostomus xanthurus* Lacépède 1802

LAFAYETTE

Jordan and Evermann, 1896–1900, p. 1458.

Description.—The spot agrees closely with the weakfish in the arrangement and general shapes and relative sizes of its fins, and in lacking chin barbels. But it is a much deeper fish relatively (body about one-third as high as it is long, measured to base of tail fin), with blunt snout instead of pointed; it has no large canine teeth; its tail fin is more forked; and it is marked on either side with a conspicuous black spot close behind the upper corner of each gill opening.

The forward (spiny) subdivision of the dorsal fin, of 10 spines, is triangular, with rounded apex; the posterior part, of one short spine and 30–34 soft rays, is about one-half as high vertically as the spiny part. The caudal fin is moderately concave. The anal fin of two short spines and 12 or 13 soft rays, has a somewhat concave margin, and the pectorals are pointed.

Color.—Bluish gray above with golden reflections, silvery below. Medium-sized fish are marked on each side with 12–15 oblique yellowish cross bars [90] dipping obliquely forward, but these fade with age. And there is a conspicuous black spot close behind the upper corner of each gill opening. The fins are partly yellowish, partly dusky.

Size.—The spot grows to a length of about 13 to 14 inches and to a weight of 1 pound 6 ounces.[91] But adults average only about 10 to 10½ inches long, and few weigh more than three-quarters of a pound.

General range.—Inshore waters from Texas [92] to southern New England, and recorded from Massachusetts Bay as a stray.

Occurrence in the Gulf of Maine.—The spot is plentiful in some years as far north as New York, while young ones are described as common in autumn about Woods Hole. But its normal range is bounded so sharply by Cape Cod that it has been reported only once from the Gulf of Maine; a single specimen, taken in Massachusetts Bay, November 1936.[93]

Kingfish *Menticirrhus saxatilis* (Bloch and Schneider) 1801

KING WHITING; MINKFISH; WHITING

Jordan and Evermann, 1896–1900, p. 1475.

[90] Dusky on preserved specimens.

[91] These were the longest of many measured in Chesapeake Bay by Hildebrand and Schroeder (Bull. U. S. Bureau of Fisheries, vol. 43, Part 1, 1928, p. 272). The maximum length previously recorded was 11¾ inches (Nichols and Breder, Zoologica, New York Zool. Soc., vol. 9, 1927, p. 95).

[92] Once reported doubtfully from Martinque.

[93] Reported by Goffin, Copeia, 1937, No. 4, p. 236.

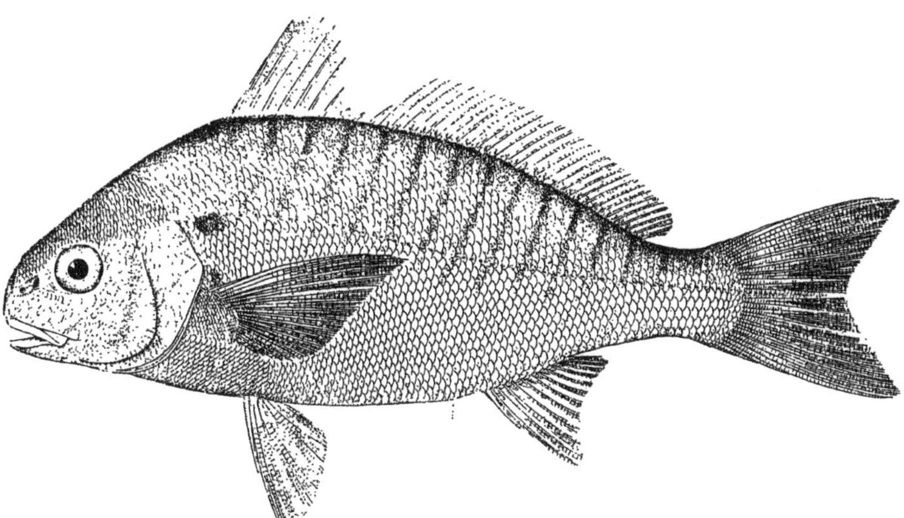

FIGURE 218.—Spot (*Leiostomus xanthurus*), Rhode Island. From Goode. Drawing by H. L. Todd.

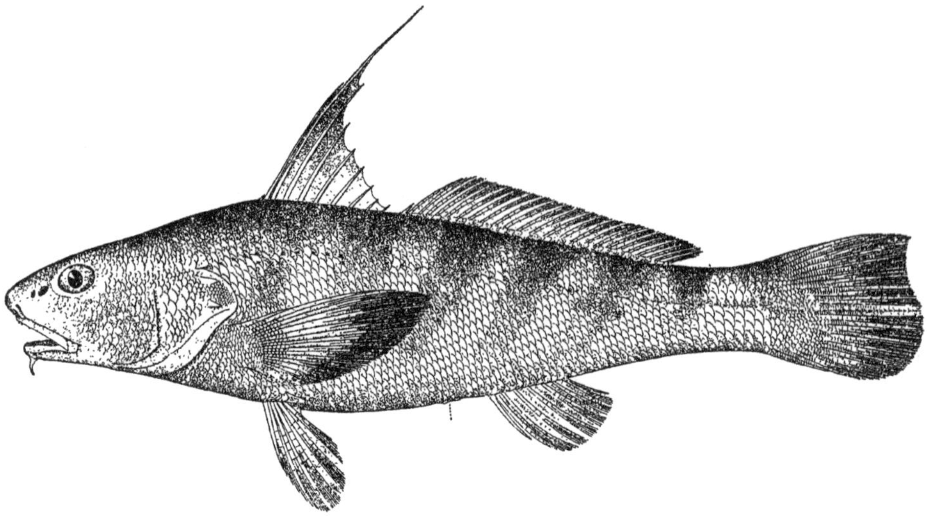

FIGURE 219.—Kingfish (*Menticirrhus saxatilis*), Pensacola, Florida. From Goode. Drawing by H. L. Todd.

Description.—The kingfish resembles the weakfish in the general arrangement and the relative sizes of its fins, the second dorsal being much longer than the first or than the anal. But its first dorsal (10 spines) is relatively much higher and more pointed than that of the weakfish, with the third spine not only much prolonged but filamentous at the tip in the adult (a noticeable character); the rather blunt nose, and snout overhanging the mouth, give the kingfish a very characteristic cast of countenance (fig. 219). Its upper jaw, furthermore, projects beyond the lower, whereas the reverse is the case in the squeteague. Its chin bears a barbel, which the weakfish lacks, its lips are fleshy, and it has no canine teeth. Its tail, too, is of very characteristic outline, with the lower half rounded but the upper half concave suggesting (though not exactly parallelling) the tail of the sea bass (p. 407). Its body is about as slender, proportionally, as that of a squeteague, but the kingfish carries its weight farther forward (it is deepest below the first dorsal fin), and it has a weak-tailed appearance remotely suggesting a hake (p. 222). We need merely note further that the filamentous spine of the first dorsal is longer in large fish than in small ones; that the second dorsal (one stout but short spine followed by 24 to 27 rays) occupies more than one-third of the length of the back and tapers slightly from front to rear; that the anal fin (one long spine and 8 rays) stands under the middle of the soft dorsal; and the pectorals are pointed and relatively much longer than those of the squeteague.

The Kingfish and its immediate relatives have no air bladder, hence makes no sounds, in which they differ from other members of their family.

Color.—Leaden or dusky gray above, (sometimes so dark as to be almost black) with silvery and metallic reflections; milky or yellowish-white below. The sides are cross marked irregularly with dark bars. These run obliquely forward and downward behind the spiny dorsal fin, but the foremost one or two bars run in the opposite direction, so that they form a V-shaped blotch or two dark V's below the fin. The pale belly is bounded above by a dark longitudinal streak on either side. The fins are dusky or blackish; the first dorsal fin anal, pectorals, and ventrals are tipped with dirty white.

Size.—Kingfish grow to a maximum length of 17 inches and a weight of about 3 pounds, but the general run are from 10 to 14 inches long, weighing ½ to 1½ pounds.

Habits.—Kingfish, like squeteague, are summer fish, appearing on the coast in May, to vanish in October. They are confined to the immediate vicinity of the coast during their stay, frequenting inclosed as well as open waters, even entering river mouths, and they are unknown on the offshore banks. They run in schools, keep close to the ground, prefer hard or sandy bottom, and feed on various shrimps (perhaps their chief diet), crabs, and other crustaceans, small mollusks, worms, and on young fish.

Breeding habits.—Kingfish spawn in bays and sounds from June until August, but it is not likely that any young that might be hatched in the Gulf of Maine from eggs laid by the occasional visitors would survive its low temperature. Welsh and Breder [94] describe the spawning and early development of this species. Young fry of ¾ to 1-inch already show most of the structural characters of the adult, including the scales, and so are readily recognizable as kingfish though they vary widely in color, ranging from the pattern of the adult to almost uniform blackish brown. Welsh and Breder found from an examination of the scales, confirmed by a large series of measurements, that kingfish are 4 to 6 inches long by the first winter, average about 10 inches the second winter, and 13¾ the third. Many males ripen when 2 years old, but few females until 3 years old.

General range.—Atlantic coasts of the United States from Florida (Pensacola, Key West) northward regularly to Cape Cod; most numerous from Chesapeake Bay to New York; known as far north as Casco Bay, Maine, as a stray.

Occurrence in the Gulf of Maine.—This excellent food and game fish reaches the Gulf of Maine only as a stray from the south. So far as we can learn the only positive records of it within our limits are as follows, south to north: Monomoy and North Truro on Cape Cod in 1896 (collected by Dr. W. C. Kendall); one taken at Provincetown, July 1847, another there in November of that same year and many small ones, apparently chilled by the cold, that appeared in that harbor in 1879; one taken at the entrance of Boston Harbor in a lobster pot some time before 1833; one at Lynn in 1840; one 8 inches long off Marblehead on October 15, 1872; one of 6½ inches at Danvers, October 28, 1874; others at Nahant (one record),[95] and in Casco Bay.

Catch statistics, if taken at face value, would suggest that kingfish reached the northern shore of Massachusetts Bay (Essex County) in unprecedented numbers during the period 1931 to 1938,[96] but we are informed by William Royce of the U. S. Fish and Wildlife Service that all the fish taken by vessels sailing out of Gloucester during these years were credited to that port, irrespective of where caught or where they were landed. There is no reason to suppose that any of these kingfish or "king whiting" actually came from as far east or north as Cape Cod, or even from anywhere in southern New England waters for that matter. And this applies equally to 466 pounds reported in 1932 from Maine. It is, in short, an unusual event for a kingfish to round the elbow of Cape Cod, or for a small school of its fry, nor have we heard of any taken anywhere in the Gulf during recent years.

Importance.—The kingfish is not plentiful enough in the Gulf to interest either commercial fishermen or anglers. It is one of the better table fishes, and a favorite with surf anglers along the coasts of New York, New Jersey, and southward, as it bites readily and fights well. In the words of a well-known angler, "no fish that swims the sea makes a better dish. Certainly no bottom living fish plays such a game for the angler's real delight." [97]

Black drum [98] *Pogonias cromis* (Linnaeus) 1766

Jordan and Evermann, 1896–1900, p. 1482.

Description.—A short deep body (less than three times as long as it is deep to the base of the caudal fin) with high-arched back but flattish belly is characteristic of the drum. The profile of the face is even more diagnostic, for the mouth is horizontal and set very low, the eye high, and the chin bears several barbels. The arrangement and sizes of the fins are essentially the same as in the weakfish, except that the second (soft-rayed) dorsal is relatively shorter, and that the anal spine is much stouter. The jaw teeth are small and

[94] Bull. U. S. Bur. Fish., vol. 39, 1924, pp. 191–194.

[95] Small amounts of "kingfish" appear in the pound-net returns published by the State of Massachusetts at various localities in Massachusetts Bay, but fishermen inform us that these were not the true kingfish but some large species of the mackerel tribe.

[96] For Essex County, Mass., 2,029 pounds reported during 1931; 34,981 pounds for 1933; 5,100–10,600 pounds for 1933, 1935, 1937 and 1938.

[97] Rhead, Bait Angling for Common Fishes, 1907, p. 145.

[98] The channel bass or red drum *Sciaenops ocellatus* (Linnaeus) 1766, a southern sciaenid uncommon east or north of New York, is represented in the collection of the Boston Society of Natural History by a mounted specimen labeled "near Portland, Me.," but as this fish was probably purchased in the market, it is likely that it had been shipped from the south than that it was actually caught nearby. Should this drum ever be taken in the Gulf of Maine, its relationship to the weakfish, kingfish, and spot would be apparent from the arrangement of its fins, especially from the shortness of the anal fin relative to the soft (second) dorsal. But it is easily distinguished from the weakfish by the fact that its upper jaw extends beyond the lower instead of vice versa; from the spot, by the barbels on its chin; and from the kingfish by having several of these barbels instead of only one; by the shape of its tail fin; and by the presence of a conspicuous black blotch (sometimes as many as 4 or more blotches) on each side at the base of the caudal fin which affords a ready field mark for its identification.

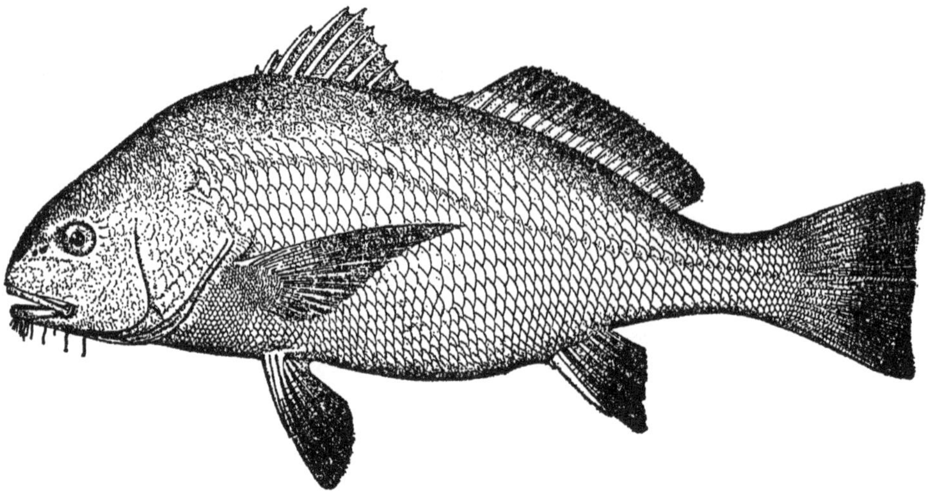

FIGURE 220.—Black drum (*Pogonias cromis*). From Goode. Drawing by H. L. Todd.

pointed, but the throat is armed with large, flat, pavement-like teeth with which the drum crushes shellfish for food, a character separating it from its allies the weakfish and the kingfish. The first dorsal fin (10 spines) is rounded-triangular; the second (1 short spine and 20 to 22 rays) oblong; the caudal is square-tipped with moderately high peduncle; the anal fin (2 spines, the first very short and the second long and stout, and 6 or 7 soft rays) is less than half as long as the soft dorsal; the pectorals are sharp pointed and relatively longer than those of the weakfish. The second anal spine is much stouter in young drums than in old ones. The eyes of the drum are comparatively small and its scales are large.

Color.—Silvery with a brassy lustre, turning to a dark gray after death. Young fish have 4 or 5 broad dark vertical bars that fade out with age. The fins are blackish. This drum occurs in two color phases, a grayish and a reddish.

Size.—Drums grow to a huge size. The largest we find positively recorded (caught in Florida) weighed 146 pounds; adults, as caught, run from 20 to 40 pounds, with 60 pounds not exceptional. The rod and reel record is 87 pounds 8 ounces, a fish 4 feet 4 inches long, caught at Cape Charles, Va., May 6, 1950, by Mrs. H. A. Bradley, Jr. A fish 40 inches long weighs about 40 pounds.

General range.—Atlantic and Gulf of Mexico coasts of America from Argentina to southern New England; common from New York southward and abundant from the Carolinas to the Rio Grande; a stray visitor as far north as Massachusetts Bay.

Occurrence in the Gulf of Maine.—This southern fish is decidedly uncommon east of New York; occasional specimens only have been reported from Woods Hole; and it is only a stray visitor to our Gulf, where 2 or 3 specimens have been taken at Provincetown, and 1 in the Mystic River, which empties into Boston Harbor.

THE TILEFISHES. FAMILY BRANCHIOSTEGIDAE

The tilefishes are sea-bass-like in appearance, but with the soft (rear) portion of the dorsal fin much longer that the spiny forward part, and the ventral fins are under the pectorals (thoracic). The only species that occurs off the northeastern United States is characterized by a large fleshy flap on the nape, suggesting (though not corresponding to) the adipose fin of salmons and smelts. But this adipose flap or fin is not shared by its relatives.

Tilefish *Lopholatilus chamaeleonticeps* Goode and Bean 1879

Jordan and Evermann, 1896–1900, p. 2278.

Description.—The presence of a thin, high, fleshy, finlike flap on the nape of the neck in front of the dorsal fin, close behind the eyes, suggesting the adipose fin of the salmon tribe in its appearance though not in its location, serves to identify the adult tilefish at a glance among Gulf of Maine

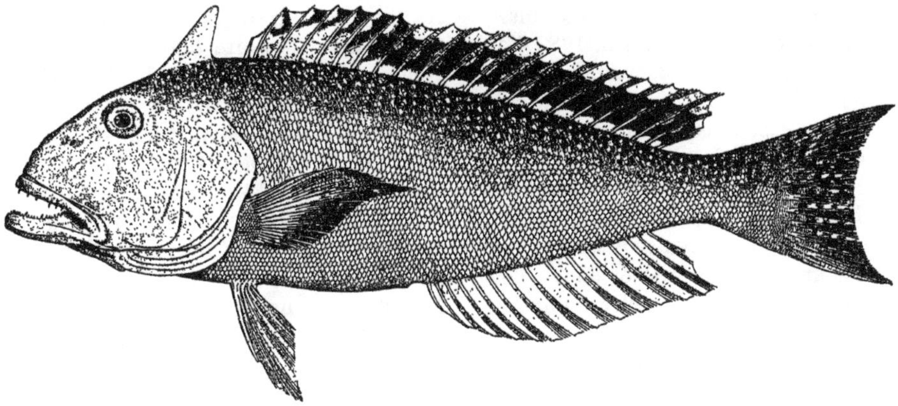

FIGURE 221.—Tilefish (*Lopholatilus chamaeleonticeps*), off Marthas Vineyard. From Goode. Drawing by H. L. Todd

fishes. In grown fish this flap is as high as the dorsal fin, higher than long, and rounded at the tip. In small fry it is relatively much lower. Equally distinctive, if less conspicuous, is a smaller fleshy flap situated on the side of the lower jaw close to the angle of the mouth, pointing backward (to be seen in the illustration, fig. 221). The large head is strongly convex in dorsal profile but nearly flat in ventral profile, with the eye high up, the mouth wide, and both the jaws are armed with an outer series of large conical teeth and inner rows of smaller teeth. The trunk (moderately flattened sidewise) is deepest close behind the head, tapering thence backward to the sidewise-flattened caudal peduncle. The spiny and soft portions of the dorsal fin are continuous, extending back from above the gill opening almost to the base of the caudal fin, as is the case in cunner, tautog, and rosefish. But the anal fin (14 or 15 rays) of the tilefish is about half as long as the dorsal fin, under the rear (soft-rayed) part of which it stands, and like the latter it is of nearly even height throughout most of its length except that its forward corner is rounded. The ventral fins are located below the pectorals, which are set low down on the sides, and both the pectorals and the ventrals are pointed. The gill covers, as well as the trunk, have moderately large scales.

Color.—This is a brilliant fish, bluish or olive green on the back and on the upper part of the sides, changing to yellow or rose lower down on the sides; its belly is of the latter tint with white midline. The head is tinged reddish on the sides; pure white below. The back and sides above the level of the pectorals are thickly dotted with small irregular yellow spots, which are particularly conspicuous below the adipose dorsal flap. The dorsal fin is dusky, marked with similar but larger yellow spots, its soft-rayed portion pale edged. The adipose flap is greenish yellow; the anal fin pale pinkish clouded with purple and with bluish iridescence; and the pectorals are pale sooty brown, with purplish reflections near their bases.

Size.—Tilefish have been reported up to 50 pounds in weight, but this is unusual. The largest fish we have seen (an unripe female) weighed 35½ pounds and was about 42 inches (108 cm.) long. Measurements taken by Bumpus[99] and more recently by us on the *Grampus* show that a 40-inch fish may be expected to weigh about 30 pounds; fish of 33 to 36 inches, 20 to 21 pounds; and 30- to 32-inch fish, 17 to 18 pounds.

Habits.—This is a bottom fish, and its depth range off our Atlantic coast is a very narrow one, none ever being taken shoaler than about 45 fathoms,[1] and very few much deeper than 100 fathoms. The deepest definite record with which we are acquainted is 170 fathoms (p. 428), and with the best fishing at 60–90 fathoms. In the Gulf of Mexico it has been caught at 90 fathoms. The thermal range to which the tilefish is exposed, normally, is very narrow also, for the temperature of the bottom water along the zone inhabited regularly by it varies only between about 47° and about 53°, in most years, summer or winter. And it appears to be very sensitive to chilling;

[99] Bull. U. S. Fish Comm., vol. 18, 1899, p. 329.

[1] The shoalest we have known any to be trawled was at 43–47 fathoms, by *Albatross III*, 35 miles southwest of Nantucket Lightship in mid-May 1950.

this having been the probable cause of a mass destruction of tilefish that took place in 1882 (for further discussion, see p. 429). It is not known whether the tilefish is equally sensitive to high temperatures, in any case it could escape such by descending to a greater depth.

Food.—A great variety of bottom-dwelling invertebrates have been taken from tilefish stomachs. Crabs, of which they are often packed full, are the most important article of diet. The list [2] also includes squid, shrimp, shelled mollusks, annelid worms, sea urchins, sea cucumbers, and sea anemones. Occasionally they catch other fish; two spiny dogfish, for instance, were found in one, and an eel (probably a conger or a slime eel) and unidentified fish bones in others.[3] The presence of pelagic amphipods (*Euthemisto*)[4] and of salpae in the stomachs of tilefish caught on long lines proves that they sometimes feed at higher levels, but they are never known to rise to the surface voluntarily, and when they are hauled up they are often "poke blown." Tilefish take any bait, perhaps menhaden best, salt herring not so readily.

Although they are strong active fish, it is probable that they suffer from the attacks of sharks, for fish caught on the long lines are often bitten in two. And we have seen numbers of sharks 7 to 8 feet long (species not determined) following them up to the surface, while the line was being hauled.

Ever since the tilefish was discovered it has been known to spawn in July, and eggs were running from 10 out of 11 females caught by the *Grampus* off New York on August 3, 1916, while the roe of the eleventh was still unripe. How early the spawning season may open is still to be learned, but August probably sees its close, for the majority of 18 females caught on the 26th of that month in 1914 were spent, only one or two still having running eggs. Among the fish that we have examined, the females have greatly predominated (only 1 male to 29 females in a total of 39 individuals).

Ripe eggs taken from a tilefish and preserved in formalin measured about 1.25 mm. in diameter.[5] As they had an oil globule of 0.2 mm., it is safe to say that they are buoyant, and tow nettings yielded eggs, indistinguishable from those stripped from the tilefish, at the station where we caught the ripe females just mentioned. But the larval stages have not been seen. The fact that a few tilefish of 2½ to 3⅛ inches were taken along the outer edge of the continental slope in April 1930, and others of 4 to 4⅛ inches in July, suggests that 4 to 5 inches is the usual length at one year of age.[6] Nothing is known of the subsequent rate of growth, nor at what age the tilefish matures sexually.

General range.—Outer part of the continental shelf and upper part of the continental edge off Nova Scotia and off the North and Middle Atlantic United States, from Banquereau Bank to the offing of Chesapeake Bay, in depths of 45 fathoms to perhaps 200 fathoms; also reported from southern Florida in more than 100 fathoms,[7] and from the Campeche Bank in the southern side of the Gulf of Mexico, whence the Museum of Comparative Zoology has received a specimen taken in 90 fathoms by the schooner *Seminole* on February 1, 1946,[8] and where local fishermen report that they have taken a number.

Occurrence off Nova Scotia and off the North and Middle United States.—The most easterly and northerly records for the tilefish are of a small one caught on Banquereau Bank (lat. 44°26′ N., long. 57°13′ W.) in 170 fathoms, December 15, 1902, from the schooner *Monitor* out of Gloucester,[9] and of another of 4½ pounds that was brought in to Boston in 1933.[10]

Its chief center of abundance is between the offings of Nantucket and of Delaware Bay. And there is some evidence that it ranges farther east in warm years than in cold. In 1908, for example, tilefish were caught off the South Channel (long. about 69°) in September, while in 1950 the *Albatross III* trawled a few at 50–80 fathoms nearly that far east (at longitudes 69°57′ to 69°35′ W.) in May, whereas the *Grampus* caught none off Martha's Vineyard (long. between 70° and 71° W.) in the very cold July of 1916, but made a fair catch off New York.

[2] Linton, Bull. U. S. Fish Comm., vol. 19, 1901, p. 47; Notes by Vinal Edwards; and our own observations.

[3] The menhaden credited to the diet of the tilefish by Sumner, Osburn, and Cole (Bull. U. S. Bur. Fish., vol. 31, Pt. 2, 1913, p. 767) were merely the pieces of bait on which the fish had been caught.

[4] Collins, Rept. U. S. Comm. Fish. (1882) 1884, p. 244.

[5] Eigenmann, Bull. U. S. Bur. Fish., vol. 21, 1902, p. 37.

[6] For details, see Schroeder, Bull. 58, Boston Soc. Nat. Hist., 1931, p. 7.

[7] One of 23 pounds, caught off Key West in more than 100 fathoms, is reported by Al Pfleuger, well-known fish taxidermist of Miami.

[8] Taken by the schooner *Seminole* on February 1, 1946. See Bigelow and Schroeder, Copeia, 1947, pp. 62–63, for details.

[9] Reported by Evermann, Rept. U. S. Comm. Fish. (1903), 1905, p. 85.

[10] Reported to us by J. Webster of the U. S. Bureau of Fisheries.

On the other hand, none have been reported alive off the Atlantic coast below lat. 37°29′ N., a few miles north, that is, of the mouth of Chesapeake Bay, which makes it likely that the tilefish of southern Florida and of the Gulf of Mexico are isolated populations.

The onshore-offshore range of the tilefish off our northern coasts, being limited in depth (p. 427), is confined to a bottom belt only some 15 to 25 miles wide—astonishingly narrow for so large a fish and one that is so plentiful. And presumably it is a year-round resident wherever it is found there, for its presence has been established northward to the offing of southern New England as early in the season as March, and as late as January, while there was no general falling off in the catches in autumn and early winter during the only year (1917–1918) for which monthly data are available.

Though the tilefish has been reported only once well within the limits of our Gulf, its history and its relationship to hydrographic factors are so interesting that it deserves more attention than its status as a Gulf of Maine fish would warrant otherwise.

It is astonishing that the very existence of so large a fish so close to our coast should have remained unsuspected until May 1879, when Captain Kirby, cod fishing in 150 fathoms of water south of Nantucket Shoals Lightship, caught the first specimens. Others were caught at 87 fathoms nearby by the schooner *Clara T. Friend* (Capt. William Dempsey) during the following July. And trips by the United States Fish Commission during the next two summers proved that the tilefish were plentiful enough to support an important new fishery. These early investigations likewise proved that it occupies a very definite environment, along the upper part of the continental slope and on the outer edge of the shelf where a narrow band of the sea floor is bathed with a belt of warm water (about 47° to 53°), varying by only a couple of degrees in temperature from season to season, and that it never ventures into the lower temperatures on the shoaling bottom nearer land, nor downward into the icy Atlantic abyss. The balance, in fact, between the physiological nature of the tilefish and its surroundings is so delicate that catastrophe overtook it within three years of its discovery. The first news of this disaster came in March 1882; throughout that month and the next vessel after vessel reported multitudes of dead tilefish floating on the surface throughout the entire zone inhabited by it north of Delaware Bay, and it has been estimated that at least a billion and a half dead tilefish were sighted.[11]

It has generally been believed that this destruction was caused by a temporary flooding of the bottom along the warm zone by abnormally cold water.[12] Consonant with this is the fact that other species of fish suffered too, and dredgings carried on during the following autumn proved that the peculiar invertebrate fauna that had been found in abundance along this warm zone in previous summers had likewise been exterminated.

The destruction of the tilefish was so nearly complete that fishing trials carried on off southern New England by the Fish Commission later in 1882; in 1883; 1884 (when a particularly careful search was made); 1885; 1886; and 1887 did not yield a single fish. But the species was not quite extinct, as the *Grampus* proved by catching 8 of them off Marthas Vineyard in 1892, and 53 in 1893. Tilefish were next heard of in 1897 when a fishing schooner caught 30 fish of 6 to 15 pounds, while long-lining for haddock south of Marthas Vineyard. And tilefish had become so numerous again by 1898 that the *Grampus* caught 363 fish, of ½ to 29 pounds, on three trips of only 1 to 3 days' duration each.

The length of the period which the fish required to reestablish itself after the mortality of 1882, together with the fact that in 1898 the catch included a considerable number of young fish, is evidence that the replenishment of the stock was chiefly the result of local reproduction, though it may have been been recruited to some extent by immigration from the southern part of the range, where destruction may not have been so complete as it was north of Delaware Bay.

The tilefish was kept in view during the next 17 years by occasional trips to the grounds by the Bureau's vessels. We caught 19, for example,

[11] Collins (Rept. U. S. Comm. Fish. [1882], 1884, pp. 237–294A) has described the event in detail, as have many subsequent authors. An account will also be found in Economic Circular No. 19 of the U. S. Bureau of Fisheries.

[12] No temperatures were taken on the tilefish ground at the season when the mortality occurred; and the bottom water was nearly as warm there by the end of the following August (48°–49°) as it usually is (about 50°–52°). The temperatures taken in this region during the early years of the Bureau of Fisheries are discussed elsewhere (Bigelow, Bull, Mus. Comp. Zool., vol. 59, 1915, pp. 238–241.)

weighing about 350 pounds, on the *Grampus* on August 26, 1914, in a set of one hour off Marthas Vineyard in 105 fathoms. In 1915, the Bureau undertook to popularize the tilefish in the market, believing it numerous enough to support an important fishery, and knowing it to be an excellent food fish. It proved so plentiful and so easily caught on long lines that the first trip stocked 38,383 pounds in 27 days. And the landings for the first 8 months after the inception of the fishery aggregated upward of 4,388,500 pounds, with a grand total of 11,641,500 pounds from July 1, 1916 to July 1, 1917. But for some reason the demand did not hold up; the catches diminished; and in 1947 (most recent year for which information is available) only 441,000 pounds were landed.[13] The tilefish continues, however, to offer a potential supply of perhaps two to three million pounds yearly, of fish that is good boiled or baked, and that is delicious for chowder. It also makes a good smoked fish, and its sounds are of value for isinglass.

THE ROCKFISHES. FAMILY SCORPAENIDAE

The rockfishes are perch-like or bass-like in general appearance. But they are related to the sculpins (p. 439) and to the sea robins (p. 467) by having a bony stay (an extension of one of the suborbital bones) stretching across the cheek, giving the latter a characteristic bony appearance. Furthermore their cheeks are spiny, and in most of the species the top of the head is marked by ridges that terminate in spines. Both the spiny portion and the soft portion of the dorsal are well developed, either as a continuous fin or subdivided by a deep notch. The ventral fins are on the chest ("thoracic"). In most of the rockfishes (including the rosefish) the eggs are retained within the mother until they hatch. There are many species, the temperate Pacific being especially rich in them. Only one, however, occurs regularly in the Gulf of Maine, though the range of another includes its offshore rim.

Key to Gulf of Maine Rockfishes

1. The lower rays of the pectoral fins, like the upper rays, are connected nearly to their tips by the fin membrane: There are 14 or 15 dorsal fin spines Rosefish, p. 430

   The lower 7–9 pectoral fin rays are free for the outer half of their length; there are only 12 dorsal fin spines ______________ Black-bellied rosefish, p. 437

Rosefish *Sebastes marinus* (Linnaeus) 1758

OCEAN PERCH; REDFISH; RED SEA PERCH; RED BREAM; NORWAY HADDOCK

Jordan and Evermann, 1896–1900, p. 1760.

Description.—The rosefish is perchlike in its general appearance, moderately flattened sidewise, about one-third as deep as it is long (to base of tail fin), with a large bony head; and its trunk tapers back from the shoulders to a moderately slender caudal peduncle. The dorsal profile of the head is concave, the mouth is large, very oblique, and gapes to below the eyes, the lower jaw projects beyond the upper, and there is a bony knob at its tip that fits into a corresponding notch in the upper jaw. Both of the jaws are armed with many small teeth. The eyes are very large and set high. The sides of the head are armed with spines, the most prominent of which are two near the rear angle of each gill cover, and a series of five confluent ones on each cheek. These, with a ridge behind and above each eye socket, give the head a bony appearance that is extremely characteristic.

The gill openings are very wide, with pointed gill covers. There is one continuous dorsal fin running from nape of neck to caudal peduncle; the spiny part (14 or 15 spines) is considerably longer than the soft part (13 to 15 rays), but the latter is higher than the former. The precise outline of the fin is easier illustrated (fig. 222) than described. The anal fin, consisting of three graduated spines and 7 or 8 longer rays, is shorter than the soft portion of the dorsal, under which it stands. The caudal fin is noticeably small, its rear edge moderately concave, and with angular corners. The pectoral fins are very large, and the smaller ventrals are situated below them. Both head and body are clad with scales of moderate size. There are about 60 to 70 oblique rows of scales from the gill opening to the origin of the caudal fin, just below the lateral line.

The rosefish agrees with the cunner, tautog, and sea bass in the union of the spiny and soft portions of its dorsal into a single long fin, and in its gen-

[13] 52,700 pounds in Massachusetts ports; 128,400 pounds in Rhode Island and Connecticut; 186,700 pounds in New York and 53,300 pounds in New Jersey.

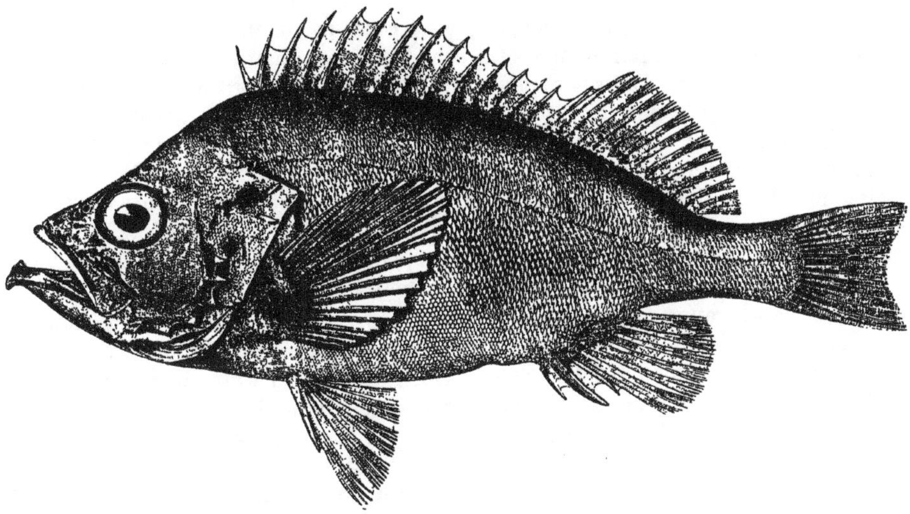

FIGURE 222.—Rosefish (*Sebastes marinus*), Eastport, Maine. From Goode. Drawing by H. L. Todd.

erally perch-like conformation. But it is separable from the first two by its much larger mouth, spiny head, large eyes, more slender caudal peduncle, and larger pectorals; and from the sea bass by its large spiny head, by the shape and small size of its caudal fin, and by the fact that its anal fin and the soft portion of its dorsal are relatively much lower. Its brilliant red color is a sufficient field mark.

Color.—Orange to flame red, occasionally grayish red or brownish red, with the belly a paler red that fades to white after death. The black eyes contrast vividly with the brightly colored body. Medium sized rosefish usually have a dusky blotch on each gill cover, and several irregularly broken dusky patches along the back. These dark markings are more conspicuous on small fish, and young fry up to 3–4 inches long are only faintly reddish, if at all so.

Size.—The rosefish matures sexually when 9 to 10 inches long, males when a little smaller than females. In the Gulf of Maine they are said to grow ordinarily to a maximum length of perhaps 2 feet. The largest measured specimen taken recently in the Gulf of Maine of which we have heard was 22 inches long, and weighed 5 pounds 11 ounces.[14] The largest we have seen measured 18¾ inches.[15] But Goode [16] reported one of about 24 inches, weighing about 14 pounds brought in to Gloucester; a 27½-inch specimen has been reported from the southern edge of the Newfoundland Bank, near the Whale Deep.[17] Another 27-inch fish, said to have weighed 13½ pounds was landed in Gloucester by the dragger *Estaela* on February 7, 1951, from somewhere off Newfoundland. And rosefish grow even larger (maximum about 31–32 inches) in the other side of the Atlantic and in Arctic Seas.[18]

The rosefish run smaller near the coast (usually 8 to 12 inches long) than on the offshore banks. In European waters, where a similar size-relationship obtains, the small inshore form represents a separate species (*Sebastes viviparus*) for it has many fewer scales than the larger, offshore form (*S. marinus*). But no racial distinctions have been found between the inshore populations and those offshore among the American rosefish.

The relationship between length and weight runs about as follows for Gulf of Maine rosefish: 9½ inches, ½ pound; 12 inches, 1 pound; 15 inches, 2 pounds; 17–18 inches, 2½–3 pounds; 20 inches, perhaps 4 pounds.

Habits.—The young rosefish drift in the upper and intermediate water layers (p. 435) until they are nearly an inch long. Fish upwards of a

[14] A fish landed in Gloucester, reported in Maine Coast Fisherman, January 1951, p. 9.

[15] One of 63 specimens trawled by *Albatross III* on the southeastern slope of Georges Bank at 175–195 fathoms, May 16, 1950.

[16] Fish. Ind. U. S. Sect. 1, 1884, p. 261.

[17] This specimen, reported by McKenzie. (Proc. Nova Scotian Inst. Sci., vol. 20, 1940, p. 44) was said to have weighed 7¼ pounds dressed, apparently an error, unless the fish was very thin.

[18] According to Saemundsson (Faune Ichthyol., Cons. Internat. Explor. Mer. 1932, plate). A length of 100 cm. (about 40 inches) has been stated, but we are inclined to doubt this.

couple of inches long tend to hold close enough to the bottom in our Gulf for great numbers of them to be caught in otter trawls. But some may also live pelagic over the deep basins as they are known to do in the Gulf of St. Lawrence; also, in the Norwegian Sea, where there is a population of all sizes living mostly at depths of about 50 to 100 fathoms, over much greater depths.[19] When they are on bottom the rosefish are chiefly on rocky or hard grounds or on mud, seldom on sand, if ever. Their depth range on the bottom is from within a few feet of tide line (p. 434) down to 350 fathoms at least; perhaps to 400 fathoms (p. 434) with the greater part of the commercial catch trawled at about 40 to 175 fathoms; and fry, living pelagic, have been taken as deep as 270 fathoms in north European waters.

Our rosefish inhabit a wide range of temperature. The maximum may be set at about 48° to 50° F., and probably it is the low temperature of parts of the Bay of Fundy, where the upper 10 fathoms or so may be as cool as 50°–52° even in midsummer that allows them to remain in shoal water there the year round (p. 435). At the other extreme they winter in Massachusetts Bay and in Passamaquoddy Bay in water as cold as 33° to 35°, and perhaps colder, though they could easily avoid these low temperatures by a short offshore migration. In fact, the rosefish has often been described as an Arctic species. But while this is true to the extent that its range extends to Arctic Seas, it is a misnomer if taken to mean that it is characteristic of Polar temperatures, for the records of its occurrence, horizontal and bathymetric, prove that the great majority of them inhabit waters warmer than 35°–36° over the greater part of their geographic range.

The distribution of the rosefish [20] in the Gulf of St. Lawrence is especially instructive in this respect, for it inhabits the comparatively warm water (39° to 42° F.) in the bottoms of the deep channels, and not the icy intermediate layer (about 32°) which, generally speaking, is so nearly an impassable barrier to its upward migration that it is seldom if ever taken on the shoal banks. And its vertical range in relation to temperature seems to be much the same as this off the southwest coast of Greenland, where rosefish are taken chiefly deeper than 90 fathoms, in water of about 37°–39°, not in the icy layer above, and where numbers of them (says Jensen) sometimes come to the surface dead in winter, apparently having succumbed to cold.[21] In the Norwegian Sea, however, rosefish of this species are caught only in the overlying layer of water of Atlantic influence at temperatures of 37°–39° or higher, never deeper in the icy cold Polar water.

Temperatures of 37°–39° are the lowest in which young rosefish are born in any numbers in our Gulf; there is no water there colder than this by the time production is well under way, say late June or early July. At the opposite extreme, practically the entire production of rosefish takes place in water colder than 46°–48°, this being the maximum to which the water warms at the 20-fathom level and deeper, except in regions of active vertical mixing where the temperature may rise a degree or two higher. In the Gulf of St. Lawrence, rosefish have been found breeding in 39°–42°. Cursory examination of station data might suggest that young are born in colder water on the Grand Banks as well as along the south and east coasts of Newfoundland, for they have been taken there in tow nets at many localities where the temperature was lower than 32°, either on the bottom or at some intermediate depth. But it is more likely that the parent fish, and the young fry also, were living above this icy layer, not in it; i. e., in water at least as warm as about 35° (1.5° C.), and warmer than about 36°–37° for the most part.

Thus the range of temperature within which American rosefish fry are produced in one place or another is from about 37° to 47° or 48°, which is about the same as for north European waters.[22] In fact it is not likely that rosefish breed successfully in temperatures lower than 35° anywhere in either side of the Atlantic.

The salinity in which rosefish breed in our Gulf is as definitely limited in one direction as is the temperature, if not in the other, for its young are produced for the most part in salinities upward of 32 per mille.

[19] For studies of the pelagic occurrence of *S. marinus* in northeastern Atlantic waters, see Murray and Hjort (Depths of the Ocean, 1912, pp. 647–648) and especially Tåning (Journal du Conseil, Cons. Internat. Explor. Mer., Vol. 16, 1949, No. 1).

[20] Huntsman, Trans. Roy. Soc. Canada Ser. 3, vol. 12, Pt. 4, 1918, p. 63.

[21] See Jensen (Vid. Meddel. Dansk Naturhist. Foren. Copenhagen, vol. 74, 1922, pp. 89-109, for an interesting study of the occurrence of the rosefish in Greenland waters.

[22] See Tåning (Journal du Conseil, Cons. Internat. Explor. Mer, vol. 16, No. 1, 1949) for a recent discussion of the thermal relationships and breeding range of *S. marinus*.

Food.—The diet of the Gulf of Maine rosefish includes a great variety of crustaceans, especially mysid, euphausiid, and decapod shrimps; small mollusks; and various other invertebrates, and small fish.[23] It bites on almost any bait. In turn, it is the prey of all the larger predaceous fish, its fry being devoured in quantity by cod, by older rosefish, and by halibut.

It has long been known that the eggs of the rosefish develop and hatch within the oviduct of the mother, and the number produced by large females may run as high as 25,000–40,000 yearly. This is a small brood compared to the numbers produced by many of the marine egg-laying fishes. But the protection offered the eggs by being retained inside the mother's body during incubation gives the young a greater chance for survival.

The larvae are about 6 mm. long at birth (fig. 223B), with the yolk mostly absorbed, the mouth already formed, and the first traces of the caudal rays already visible. At a length of 12 mm. (fig. 223D) the dorsal and anal fin rays have appeared, the ventrals are visible, and the head spines are prominent. And though the red color is not developed until the little fishes are about to take to bottom, or later, all but the very youngest larvae are recognizable as rosefish by their large spiny heads, large eyes, short tapering bodies, very short digestive tract, and by the presence of two rows of post anal pigment cells, a dorsal and a ventral row.

This is a very slow-growing fish. Available information is to the effect that they average about 2½ inches when 1 year old.[24] Studies of the scales of rosefish of different sizes [25] indicate that 5-inch fish are likely to be 4 years old; 6-inch fish, 6 years old; 7-inch fish 7 or 8 years old; 8-inch fish 8 or 9 years old; 9-inch fish 9 or 10 years old, and that many of the largest fish of 18 inches and upward may be 20 years old, or older. Thus the mature fish are 8 to 9 years old and older, with the greater part of the commercial catch 10 years old and upward. And about as slow a growth rate has been reported for the immature rosefish of this same species of Barents Sea, on the other side of the Atlantic.[26]

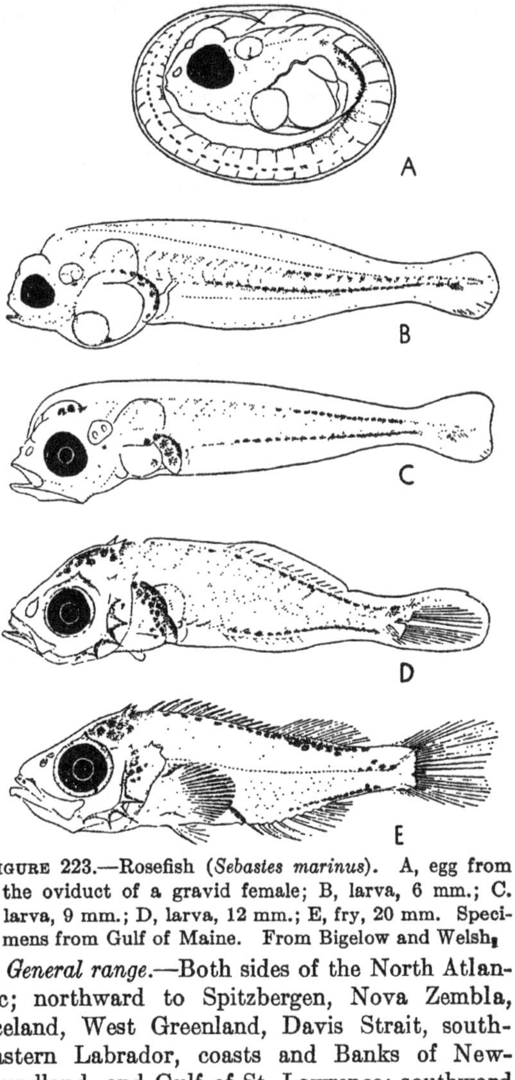

FIGURE 223.—Rosefish (*Sebastes marinus*). A, egg from the oviduct of a gravid female; B, larva, 6 mm.; C. larva, 9 mm.; D, larva, 12 mm.; E, fry, 20 mm. Specimens from Gulf of Maine. From Bigelow and Welsh.

General range.—Both sides of the North Atlantic; northward to Spitzbergen, Nova Zembla, Iceland, West Greenland, Davis Strait, southeastern Labrador, coasts and Banks of Newfoundland, and Gulf of St. Lawrence; southward to the offing of southern New England and as far as the offing of New Jersey in deep water along the American coast,[27] to the northern part

[23] Most of the rosefish that we have seen trawled had voided their stomach contents before they were brought on board.

[24] According to the sizes of young rosefish collected by us in the Gulf of Maine, May to August in various years.

[25] By Perlmutter and Clark, U. S. Fish and Wildlife Service, Fishery Bull. No. 45, 1949.

[26] By Veschezerov, in Knipovitch, Polar Sci. Inst. Sea Fisheries and Oceanogr., No. 8, 1941, pp. 236–270 (Russian).

[27] Tåning (Journal du Conseil Cons, Internat. Explor. Mer., vol. 16, 1949, p. 86) is of the opinion that the American rosefish does not belong to the same species as the European *S. marinus* hence he refers to it as *S. fasciatus*, Storer, 1854. But our own comparison of good-sized specimens from the two sides of the Atlantic has failed to show any differences that seem sufficient for specific separation, whether in number of scales, in the spines on the cheeks, in the fins, or in bodily proportions. We are much indebted to Dr. C. E. Lucas for sending us a series of rosefish of various sizes that had been landed in Aberdeen, Scotland.

of the North Sea and to the southwestern coast of Iceland along the European coast.

Occurrence in the Gulf of Maine.—This is one of the most plentiful of the commercially important fishes in all but the shoalest parts of the Gulf: on the offshore banks, in or over the deep central basin, and along shore. To list its known occurrences would be to mention practically every station where hook-and-line or otter-trawl fishing is carried on deeper than 20 fathoms. Thus considerable numbers are sometimes taken on lines or trawls in 20 to 35 fathoms or more in the Massachusetts Bay region both winter and summer, especially on or near rocky bottom, while many are caught on and near Jeffreys Ledge and at other spots between Cape Ann and Boon Island.

The fact that the *Grampus* took rosefish in 6 out of 7 hauls in 25 to 60 fathoms with a trawl only 8 feet across the mouth, between Cape Ann and Penobscot Bay in July 1912, and that *Atlantis* took 2,469 rosefish in 12 hauls with a 30-foot shrimp trawl in 66 to 96 fathoms in the mud-floored trough west of Jeffreys Ledge in August 1936, show how universal they are in the western coastal belt of the Gulf at appropriate depths; and the depth is not too great for them anywhere in the troughs of our gulf. The number of rosefish there seemed to be independent of the numbers of shrimp (*Pandalus*) on which we may assume they were feeding, the average catch per haul being almost the same (216 fish) for the group of stations where shrimp were scarce as for the group where they were plentiful.[28]

Rosefish are also caught in plenty all along the northern shores of the Gulf in depths of 25 fathoms or more; they are common in the Bay of Fundy, even in such enclosed waters as Passamaquoddy Bay; Huntsman found them in St. Mary Bay; and large commercial catches are made off the western coast of Nova Scotia (5,253,962 pounds reported in 1946).

Turning offshore, the vaguely outlined trough known as "South Channel" that separates the Cape Cod-Nantucket Shoals area from Georges Bank is one of the most productive and hardest fished of the rosefish grounds (19,016,052 pounds taken there in 1946); rosefish are generally distributed on and around Georges Bank itself, except perhaps on its shoalest parts; also on and around Browns Bank, and they range down to a depth of at least 260 fathoms on the southern slope of Georges Bank;[29] very likely down to 300 fathoms.

The relative yearly catches, from different areas, show that the inner and central parts of the Gulf in general are considerably more productive of rosefish than the offshore banks, for the poundage reported from off western Nova Scotia, from the Bay of Fundy, from the coasts of Maine and Massachusetts, and from the west-central part of the Gulf (including the Cashes Bank region and part of the deep basin) which is referred to as the "inshore grounds" in the statistical reports, were about three times as great as from the South Channel, Georges Bank, and Browns Bank combined in 1940; about 3⅓ times as great in 1946, and the regional contrast has been of this same order in other recent years of record. The South Channel in turn, has been many times as productive as the much more extensive area of Georges Bank,[30] though there were enough of them on Georges formerly for 22 successive trawl hauls to have yielded 3,887 rosefish there, September 26 to 30, 1913 (more than one-third as many as haddock).

During 1913, rosefish made up 1.8 percent of the total catch of fish of all kinds made by several trawlers operating on Georges Bank, June to December, and 5.9 percent in the South Channel.

This regional contrast between Bank and Channel emphasizes the very interesting fact that the rosefish of our Gulf, and those of outer Nova Scotian waters as well, are decidedly more plentiful in the deeper basins and depressions, and on soft bottom, than they are on the grounds that are the chief centers of abundance for cod and haddock, and for most of the commercially important flat fishes.

The statistics do not suggest any very great difference in the abundance of rosefish as between Georges Bank and the Nantucket Shoals-Nan-

[28] For further details, see Bigelow and Schroeder, Biol. Bull. vol. 76, 1939, p. 314.

[29] 63 large ones taken in one trawl haul, latitude 40°29′ N., longitude 67°10′ W., at 175-195 fathoms, by the *Albatross III*, May 16, 1950.

[30] The total reported catch for the period 1937–1946 (no report for 1942) was a little more than 158¼ million pounds for the South Channel, contrasted with a little less than 2 million pounds (1,876,000) for the whole of Georges Bank. The catches for individual years ranged between about 9½ million and about 30¼ million pounds for South Channel; between a little less than 29,000 pounds to a little more than 625,000 pounds for Georges Bank. For a chart showing the geographical limits of the statistical areas to which the catches are referred, see Fishery Statistics of the U. S., 1943, Stat. Digest, No. 18, U. S. Fish and Wildlife Service, 1943, p. 95.

tucket Lightship fishing grounds, for while the catch has averaged only about one-third as great for the latter as for the former, the statistical area in question is about one-third as extensive. But the catches of rosefish (1937–1946), made by United States vessels from southern Nova Scotia out across Browns Bank, are of the same general order of magnitude [31] as for the South Channel. And a catch of 1,400 rosefish in two sets of a line trawl on Browns Bank, April 4, 1913, will illustrate how plentiful they were there, before they were so hard-fished as they have been of late.

Large catches of rosefish are also made all along the outer Nova Scotian shelf to the eastward. There is an abundant population on the Newfoundland Banks still awaiting exploitation; some 7,000,000 pounds were taken in Hermitage Bay, on the south coast of Newfoundland from 1947 to 1950.[32] And fry have been taken along both coasts of Newfoundland; also northward from Flemish Cap, "where the Gulf Stream and the Labrador current struggle for mastery.[33] The most northerly record for the rosefish on the American coast is from the outer coast of Labrador (Camp Islands), a few miles north of the Strait of Belle Isle.[34]

It has been known for many years that there are rosefish in the deep waters of the Gulf of St. Lawrence. But we still await information as to how plentiful they may be there.

The upper limit to the vertical range of the rosefish in different parts of our Gulf is clearly correlated with temperature. Thus it is only deeper than 15 to 20 fathoms that rosefish are found during the warm half of the year in the southwestern part of the Gulf. But they have been known to run up into Gloucester Harbor in numbers in winter [35] (never in summer). Many have been taken near the surface in the spring in the drift-nets near the Isles of Shoals where it is only near bottom that they are reported in summer. We have taken them as shoal as 10 fathoms in summer off Mount Desert Island, Maine, where the water warms to about 52°–54° at that depth, and they occur in Pasamaquoddy Bay in water no deeper than 5 fathoms at that season, according to Huntsman. Verrill,[36] in fact, described them as round the wharves at Eastport, no doubt in late summer or early autumn, the season he studied the fauna there.

Apart from shifts in depth of the sort just mentioned, with the seasonal rise and fall of temperature, there is no evidence that the adult rosefish of our Gulf carry out any regular migration. But the larvae may journey for long distances while they drift helpless in the upper layers of the water (p. 436).

In 1930, we saw gravid females during the last half of April, with young nearly ready for birth, evidence that some rosefish may be born in the Gulf of Maine as early as the first part of May. Females also, with well-developed eggs, and males with well-developed milt, are taken commonly by mid-May, both within the Gulf and on Georges Bank,[37] while we have towed a few newborn fish (7 to 10 mm.) off Boothbay and off Mount Desert on May 31 and on June 14. But July 8 is the earliest that we have taken them in any numbers in our tow nets (57 larvae off Cape Cod on that date in 1913.)

Evidently the production of young continues right through July and August, for the *Albatross II* trawled many gravid females, 10 to 13½ inches long, in the central basin of the Gulf in July (1931), one of them containing about 20,000 young, 6–7 mm. long, practically ready for birth, while we have towed newly born larvae (6.5–7 mm.) in one part of the Gulf or another on July 24 and 29 and August 4, 7, 12, 14, 16, 22, and 31, and as small as 10 mm. on September 2.[38] But it is not likely that many young are produced after the first week in September.

Records for rosefish larvae and fry for late June, July, and August along the outer Nova Scotian shelf, and in the Gulf of St. Lawrence, as well as from May until into September around the Grand Banks and up the two coasts of Newfoundland, show that the season of production commences nearly as early in the season in these more northerly waters as it does in the Gulf of Maine and

[31] Yearly catches, from about 9½ million to about 27½ million pounds.

[32] Twentieth Rept. Dept. Fish. Canada (1949–50) 1951, p. 36.

[33] Tåning, Journal du Conseil, Cons. Internat. Explor. Mer., vol. 16, 1949, p. 90.

[34] See Frost, Newfoundland Dept. Nat. Resources, Res. Bull. 4, 1938, Ch. 7, for locality records of rosefish fry in Newfoundland and Labrador waters.

[35] Fish. Ind. U. S. Sect. 1, 1884, p. 262. We have not heard of them in any numbers in any other harbor south of Cape Elizabeth.

[36] American Naturalist, vol. 5, 1871, p. 400.

[37] In 1950 *Albatross III* trawled a number of large males with well-developed milt, and large females with young nearly or quite ready for birth, on the southern slope of Georges Bank on May 16, at 175–195 fathoms.

[38] For complete list, with station localities, numbers and sizes of larvae, and depths of the hauls, see Bigelow, Bull, Mus. Comp. Zool., vol. 58, 1914, p. 108; vol. 61, 1917, pp. 271–272.

that it continues equally late. In north European waters young rosefish are produced from mid-April through August, according to locality.

Seemingly the rosefish fry are ready to sink to near the bottom when they are about 25–30 mm. long, for we have not taken any larger than 27 mm. in our tow nets, while fry of 1½ inches and upwards are plentiful on bottom, both in the Bay of Fundy and in deep water off southern New England. And our failure to take any young rosefish in our tow nets off Massachusetts Bay in November or anywhere in the Gulf in winter is evidence that their descent to the bottom takes place early in their first autumn.

In north European waters such of the young rosefish as are fated to take to the bottom at all are described as continuing pelagic in the upper layers until they are 2–2½ inches (to 60 mm.) long.

Apparently rosefish never produce their young in less than 20 to 30 fathoms west or south of Penobscot Bay; and while they may perhaps do so in shoaler water about Mount Desert, and further east along the coast of Maine, Huntsman [39] reports that the spawning individuals move out into deep water. With this qualification, we have taken pelagic young in our tow nets at so many localities in the northern part of the Gulf including Southwest Harbor on Mount Desert Island, and so generally distributed, as to show that rosefish produce their young wherever they may chance to be, and do not gather on special grounds for the purpose. Rosefish (unlike most of the fishes producing buoyant eggs) also breed successfully in the Bay of Fundy, their larvae having been found both at the mouth of the bay and for some distance up the center, during the late summer.[40]

In the inner parts of the Gulf, our largest catches of its drifting young have all been located within a few miles, one side or the other, of the 50-fathom contour line. Examples are catches of several hundred off Cape Elizabeth on July 29, 1912; near Cape Sable on August 11, 1914; near Cashes Ledge on August 10, 1913, and on September 1, 1915; in the sink off Gloucester on August 9, 1913; on Platts Bank on August 7, 1912. And Goode and Bean [41] report the fry as caught "by the bushel" in the trawl by the *Fish Hawk* at 55 fathoms, presumably off Cape Cod, that being the only *Fish Hawk* station where the rosefish is listed by them. These last catches rival the swarms of young *Sebastes* that have been encountered between Iceland and the Faroes.[42]

On the other hand, most of our records for their pelagic young outside the 100-fathom contour line have been based on occasional specimens only. We have seldom taken young *Sebastes* in the western basin, though we have towed there frequently at all seasons, and never in the deep southeastern trough of the Gulf nor in the eastern channel between Georges Bank and Browns. All this suggests that the chief production of rosefish within the Gulf of Maine occurs at about 50 fathoms.

The presence of gravid females and ripe males on Georges bank (p. 435), together with the abundance of mature fish in the so-called "South Channel," shows that this general region is an important center of production. And the rosefish also breeds considerably farther west than this on the outer edge of the continental shelf, for young fry and adult females full of eggs were collected in 100 to 180 fathoms off the southern coast of New England during the early years of the United States Fish Commission.

The shelf along outer Nova Scotia (especially the depressions between the banks), the basin of the Gulf of St. Lawrence, and the waters around Newfoundland, must be productive nurseries, also, to judge from the abundance of young drifting stages that have been collected there.[43]

Importance and abundance.—The only measure of abundance of rosefish in our Gulf available before 1935 was the number taken in a few experimental trawl hauls, or on long lines (p. 434), for there was so little demand for them that nearly all of those caught incidentally were thrown back by the fishermen. Thus the reported catch for our Gulf was only 54,095 pounds in 1919, rising to a yearly average of about 209,000 pounds for the period 1931–1933. But the rosefish is a good table fish, excellent for quick freezing and filleting. The marketing of it as frozen fillets in 1935 so increased the demand that the landings from the Gulf of Maine, plus fish taken from southern Nova

[39] Contr. Canadian Biol. (1920–1921) 1922, p. 64.

[40] Huntsman, Contrib. Canadian Biol. (1920–1921), 1922, p. 64.

[41] Smithsonian Contrib. Knowl., vol. 30, 1895, pp. 260, 261.

[42] Schmidt, Skrifter, Kommiss, Havundersøgelser, No. 1, 1904, p. 9; Tåning, Journal du Conseil, Cons. Internat. Explor. Mer., vol. 16, 1949, p. 93–94.

[43] See Dannevig (Canadian Fish. Exped. (1914–1915) 1919, pp. 12–14, figs. 8–10), for records of young rosefish along outer Nova Scotia and in the Gulf of St. Lawrence; Frost (Newfoundland Dept. Nat. Resources, Res. Bull. 4, 1936, Ch. 7) for Newfoundland; also Reports, Newfoundland Fisheries Research Commission, vol. 1, No. 4, 1932; vol. 2, No. 1, 1933; vol. 2, No. 2, 1934, for details as to exact localities and dates.

Scotia out to Browns Bank rose to 17 million pounds in that year, to about 55 million pounds in 1936, about 66–89 million pounds in 1938 and in 1939, to about 106 million pounds in 1940, and to about 136 million pounds in 1941. The landings fell to about 100 million pounds in 1943, but rose again in 1945 to a peak of 151 million pounds. This corresponds to about an equal number of individual fish, a number larger than that for any other fish commercially important in our Gulf, except the herring.

It is now generally believed that this yearly drain was greater than a fish requiring 8 or 9 years to reach marketable size could withstand; the catch (Gulf of Maine and southwestern Nova Scotia) fell by about 30 percent the next year, and to only about one-fourth as much in 1949 as had been landed from these areas in 1945.[44] And this would have been calamitous for the fishery had the fleet not been able to draw on the rosefish to the eastward, along the Nova Scotian shelf, whence something like 133 million pounds were landed in New England ports in 1949, or between three and four times as much as from the Gulf of Maine.

We refer the reader to the table on page 333 for the monetary value of the catches of rosefish in recent years, as compared with cod, haddock, and mackerel.

George F. Kelly, writing in the Maine Coast Fisherman,[45] has recently emphasized the probability that the Nova Scotian catch may also be expected to decline from its present high level as soon as the accumulated stock of old fish is reduced there, as it has been in the Gulf of Maine. The fishery would then have had to depend on the annual increment of growth of a stock that has stabilized at a level considerably below its virgin state, unless operations had been extended to Newfoundland waters, where the same chain of events will eventually follow. And we must expect this increment to be far smaller for the slow-growing rosefish than it is for faster growing fishes, such as the cod or the haddock.

Finally, almost the entire commercial catch is taken in otter trawls; also while the rosefish is of such great importance to the commercial fishermen, it offers nothing to the angler; most of them live too deep to be within his reach, and any hooked would come in with very little resistance.

Black-bellied rosefish *Helicolenus dactylopterus* (De la Roche) 1809

RED BREAM; BLUE MOUTH

Jordan and Evermann, 1896–1900, p. 1837 as (*H. dactylopterus* (De la Roche) and *H. maderensis* Goode and Bean 1895).[46]

Description.—This species resembles the common rosefish closely in its general form and in the outline and arrangement of its fins. But the lower 7 to 9 rays of its pectoral fins are free from the fin membrane along the outer half to one-third of their length, and the upper margin of the pectorals is

[44] Landings of 108 million pounds for 1946; only about 36 million pounds for 1949.

[45] Vol. 5, No. 7, Jan. 1951, p. 9.

[46] We have examined some of Goode and Bean's specimens and agree with Holt and Byrne (Fisheries, Ireland, Sci. Inv. (1906), v. 1908) that the so-called *H. maderensis* is identical with *H. dactylopterus*.

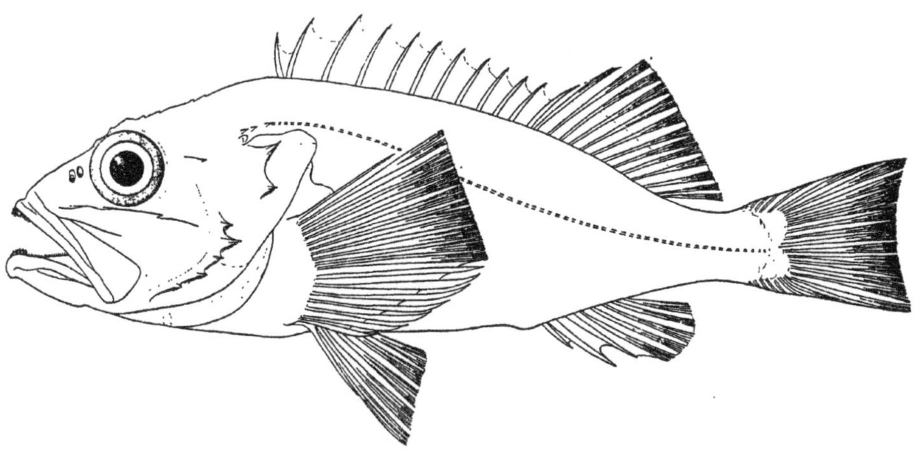

FIGURE 224.—Black-bellied rosefish (*Helicolenus dactylopterus*), off southern New England. Drawing by Louella E. Cable.

nearly straight, which gives the fins an aspect noticeably different from that of the rosefish. The space between the eyes (flat in the rosefish) is grooved in the black-bellied species; there are only 12 spines in the spiny portion of its dorsal fin (14 or 15 in the rosefish) and only 5 or 6 soft anal rays instead of 7 (in addition to 3 stiff spines).

Its caudal fin is relatively larger than that of the rosefish; its eyes closer together, the distance between them being less than one-half as great as the diameter of the eye (about two-thirds to three-quarters in the rosefish); the maximum depth of its body is somewhat less than the distance from tip of upper jaw to upper corner of gill cover; and its scales are larger relatively.[47] A more important difference anatomically is that the red bream has only 24 or 25 vertebrae, the rosefish 31.

Color.—More or less vivid reddish or pale pinkish, usually with some brown and green along the back and with irregular cross bands of darker or brighter scarlet on some specimens; the upper part of the sides marked with a sparse pattern of narrow, dusky vemiculations, roughly following the edges of the scales; and each gill cover generally has a leaden or dusky patch caused by the black inner surface shining through the bone. The lower surface is without dark markings. All the fins are pinkish, the spiny part of the dorsal mottled with white, and the soft portion of the dorsal, the ventrals, and the anal edged with white. The lining of the belly cavity is black, hence one of its common names.

Size.—Maximum length about 15 inches.

Habits.—Catch records show that the black-bellied rosefish sometimes are in the mid-depths, sometimes on bottom or close to it. Beyond this nothing is known of their daily life. Neither is it known definitely whether their eggs are hatched within the oviducts of the mother, as in the rosefish (p. 433), or whether they are set free in the water, like those of most fishes.[48]

General range.—Known from the eastern slope of Georges Bank westward and southward to Florida in depths of 68 to 373 fathoms in the western Atlantic; from Norway to the Canaries in the eastern; also in the Mediterranean.

Occurrence in the Gulf of Maine.—This fish must be generally distributed over the outer part of the continental shelf and along the upper part of the continental slope as far east as the general offing of Nantucket, for it has been reported from 27 stations between longitude 72° and a few miles east of longitude 70°,[49] including one catch of more than 100 of them, 4¾ to 11 inches long, in one haul, by the *Albatross III.*[50] One about 13 inches long was trawled on the eastern edge of Georges Bank, at 175 fathoms, October 6, 1929.[51] Subsequent records that fall within the limits set here for the Gulf of Maine, are of 24 fish, 4–10 inches long, trawled at 5 stations south of Nantucket, at 68–240 fathoms, by the *Albatross III,* May 11–18, 1950; of one brought in by the trawler *Red Jacket* from the northern slope of Georges Bank, from 120 fathoms, in 1949;[52] and of a catch of about 300 pounds of them, made in the south-eastern part of the basin of the Gulf, at 120–140 fathoms, July 24, 1948.[53]

This last catch is especially interesting, for it shows that schools of black-bellied rosefish may occasionally come in via the deep channel between Georges and Browns Banks. But they have never been reported in the inner parts of the Gulf, nor are they to be expected there unless as strays from offshore.

BOAR FISHES. FAMILY CAPROIDAE

Boar fish *Antigonia capros* Lowe 1843

Jordan and Evermann, 1896–1900, p. 1665.

Description.—This Boar Fish[54] is set apart from all other Gulf of Maine fishes by the fact that its very thin body is deeper than it is long (longer than deep in all other species yet recorded from our Gulf). It resembles the John Dory (p. 297) in the general arrangement of its fins, both the spiny portion of the dorsal and the soft portion being well developed, with the latter much the longer of the two, but lower; the soft-rayed

[47] About 42–48 oblique rows of scales from upper corner of gill opening to base of caudal fin in *Helicolenus*, 60–70 in *Sebastes*.

[48] Ehrenbaum (Nordisches Plankton, Zool., vol. 1, 1905, p. 51) thought it probable that this is an egg-laying species, and Tåning (Journal du Conseil, Conseil Internat. Explor. de la Mer, vol. 16, 1949, p. 86) so characteristizes it. But its ripe eggs have not been seen, so far as we know.

[49] For list of stations, with depths, up to 1895, see Goode and Bean, Smithsonian Contrib. Knowl., vol. 30, 1895, pp. 251–252, as *Helicolenus maderensis; Albatross III,* also, trawled one or more specimens at nine stations off southern New England in May 1950.

[50] Latitude 39°42′ N., longitude 71°57′ W., 145–210 fathoms, May 12, 1950.

[51] This specimen, reported by Firth (Bull. 61, Boston Soc. Nat. Hist., 1931, p. 13) is in the Museum of Comparative Zoology.

[52] Specimen in Museum of Comparative Zoology.

[53] Taken by the schooner *Alice M. Doughty*, Capt. Manual Silva. Six of these specimens are in the Museum of Comparative Zoology.

[54] This is the only member of the family that has been reported from the western side of the North Atlantic.

anal is about as long as the soft dorsal and is preceded by 3 spines with fin membrane. The ventrals are placed a little behind the pectorals. It lacks the bony skin plates and the filamentous prolongations of the dorsal spines so conspicuous on the John Dory; and its mouth is very small (larger in the John Dory).

Color.—Color, in life, pink and pinkish white.

Size.—Maximum reported length about 1 foot.

General range.—Tropical and subtropical Atlantic, mostly offshore.[55]

Occurrence in the Gulf of Maine.—We mention this fish because we have seen 8 specimens [56] and heard of 6 others [57] that were trawled in 55–80 fathoms, south of Nantucket Lightship in May 1950. Other records of it near the American coast are one trawled by the *Albatross III* at 50 fathoms and a second at 22 fathoms off North Carolina, in January 1950. It has also been taken near Madeira, off the Barbados, and in Cuban waters.

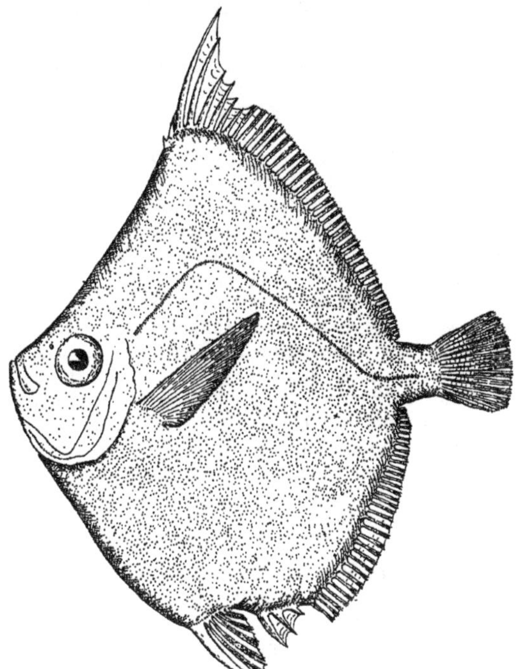

FIGURE 225.—Boarfish (*Antigonia capros*), 105 mm. specimen, south of Nantucket Lightship. Drawing by H. B. Bigelow.

[55] Reports of it from Japan, from the Kai Islands and from the Celebes Sea (Manado) may have been based on a closely allied fish. For descriptions of the species of this genus, with references, see Fraser-Brunner (Ann. Mag. Nat. Hist., Ser. 12, vol. 3, No. 32, 1950, pp. 721–724).

[56] Three trawled by *Albatross III*; five by the *Eugene H.*

[57] Reported by Capt. Henry Klimm, of the dragger *Eugene H.*

THE SCULPINS AND SEA RAVENS. FAMILIES COTTIDAE AND HEMITRIPTERIDAE

The several members of the sculpin and sea-raven tribe that are known from the Gulf of Maine are a homogeneous group, characterized by large spiny heads; very wide gill openings; very broad mouths; slender bodies; separate spiny and soft-rayed dorsal fins (united in some rare species); large fanlike pectorals but small caudals; and by ventrals that are reduced to three long rays. All of them, too, have a fashion of spreading the gill covers and of flattening the head when taken in the hand. They likewise produce grunting sounds, and some of them have the power of inflating themselves with air or water when they are molested. The only other Gulf of Maine fishes that resemble them in general form, are the sea robins (p. 467), the toadfish (p. 518), and the goosefish (p. 537). But the entire head of the sea robin is armed with bony plates, different from the soft-skinned head of a sculpin; in the toadfish the soft portion of the dorsal fin is many times as long as the spiny part (at most twice as long as the spiny part in a sculpin); and not only are the fins of the goosefish small and weak as compared with the present family, but its lower jaw projects far beyond the upper, and its mouth is full of very large pointed teeth, whereas in the sculpins the teeth are small and the upper and lower jaws are of approximately equal length.

The sculpin tribe, as a group, are egg-laying fishes.[58] Among the Arctic members of the family, including the genera *Artediellus*, *Cottunculus*, *Gymnocanthus*, and *Icelus*, the males have a long anal papilla, through which the urinary duct and the sperm ducts both pass. The supposition is that this serves as a copulating organ, fertilization taking place within the female, and the fertilized eggs being laid soon after.[59]

[58] Eggs with embryos far advanced in development have been reported within the ovaries of female short horn sculpins (*Myoxocephalus scorpius*) from Finland (Nordquist, Svensk. Fiskeri Tidskr., year 6, 1899). But it is well established that this sculpin ordinarily lays eggs, as described below (p. 447).

[59] See Jensen and Volsøe (Danske Vidensk. Selskab. Biol. Meddel., vol. 21, No. 6, 1949, p. 18) for a detailed account of the anal papilla in *Icelus*.

KEY TO GULF OF MAINE SCULPINS AND SEA RAVENS

1. There is only one dorsal fin, the spiny and soft parts being continuous, one with the other .... Arctic sculpin, p. 453
   There are two separate dorsal fins 2
2. The first dorsal fin is deeply notched between the spines; the lower jaw and the top of the head are adorned with fleshy tags .... Sea raven, p. 454
   The first dorsal is not deeply notched between the spines; there are no fleshy tags about the head 3
3. The long spine on each cheek is branched at the tip .... Staghorn sculpin, p. 452
   The long spine on each cheek is simple, not branched at the tip 4
4. The anal fin is long (25 rays); there is a series of bony plates along each side of the body .... Mailed sculpin, p. 441
   The anal fin is short (14 rays or fewer): there are no bony plates along the sides of the body 5
5. The long spine on the cheek is hooked upward .... Hook-eared sculpin, p. 440
   The long spine on the cheek is straight, not hooked 6
6. The longest (uppermost) cheek spine is four times as long as the one below it, and reaches back to the margin of the gill cover; all the head spines are very sharp .... Longhorn sculpin, p, 449
   The uppermost cheek spine is not more than twice as long as the one below it, and does not reach more than about half way to the margin of the gill cover; the head spines are blunter 7
7. Total length more than 9 inches .... Shorthorn sculpin, p. 445
   Total length less than 8 inches 8
8. Anal fin with 13 or 14 rays; the soft skin of each side of the throat is pierced by a minute pore close behind the lower part of the last gill arch .... Shorthorn sculpin, young specimens, p. 446
   Anal fin with only 10 or 11 rays; sides of throat behind last gill arch have no pore .... Grubby, p. 443

Hook-eared sculpin *Artediellus uncinatus* (Reinhardt) 1833

ARCTIC SCULPIN

Jordan and Evermann, 1896–1900, p. 1906, as *Artediellus atlanticus* Jordan and Evermann.

Description.—The most distinctive feature of this species among local sculpins is the long hooklike spine on each cheek, pointing backward and upward, plainly shown in the illustration (fig. 226). There is also a short backward-pointing spine covered by a flap of skin at the upper corner of each gill cover, two short spines on the top of the nose between the two pairs of nostrils, and a pair of blunt knobs above the eyes. Head, mouth, and tapering body are of the usual sculpin form. The skin is smooth and naked. The spiny dorsal fin is short (7 to 9 spines) and rounded in outline, the soft dorsal fin is about twice as long (13 rays), and the anal (11–12 rays) is a little shorter than the soft dorsal, which it resembles in outline and under which it stands. Each ventral fin consists of three long rays that reach back nearly to the vent; the pectorals, wide at the base and rounded in outline, reach beyond the beginning of the soft dorsal when they are laid back, and the caudal fin is narrower than it is in the commoner Gulf of Maine sculpins. The jaws and the roof of the mouth are armed with several series of small bristle-like teeth.

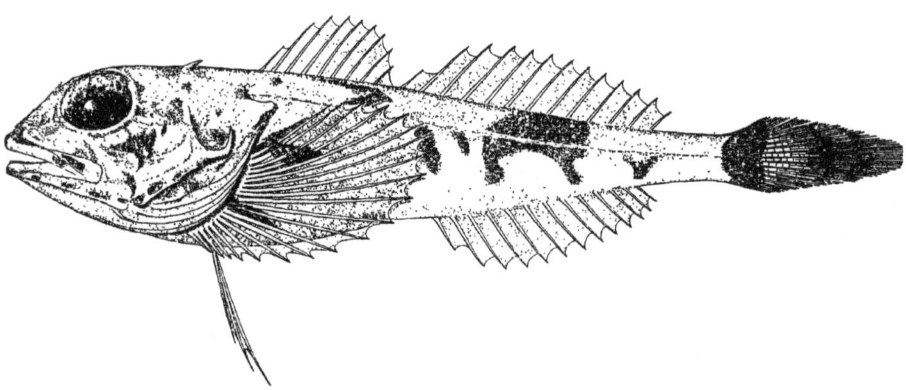

FIGURE 226.—Hook-eared sculpin (*Artediellus uncinatus*), Massachusetts Bay. From Jordan and Evermann. Drawing by H. L. Todd.

Color.—Preserved specimens are mottled with dark and pale brown, sometimes with a reddish tinge, and most of them have a blotch at the base of the caudal fin. All the fins are grayish or blackish, with oblique or vertical pale cross bands.

Size.—This is one of the smallest of sculpins, growing to a length of only about 4 inches.

General range.—This is a cold-water fish known from Labrador and the west coast of Greenland to Cape Cod in the western Atlantic; also in the littoral waters of arctic Europe, of Siberia, and of Greenland.[60]

Occurrence in the Gulf of Maine.—This sculpin, formerly thought to be rare in the Gulf of Maine, is now known to be generally distributed there in depths greater than 20 to 30 fathoms. It was dredged in numbers in the deeper parts of Massachusetts Bay many years ago. And we have since taken it repeatedly near Mount Desert; off Cape Elizabeth; in the trough between Jeffreys Ledge and the coast; around Cashes Ledge; along the northern slopes of Georges Bank; in the southeastern part of the basin of the Gulf; and at the entrance to the deep gully between Georges and Browns Banks, in depths ranging from 20 to 150 fathoms. Individual trawl hauls have yielded up to 6 or 8 specimens, both on hard bottom and on soft.

To the eastward and northward it has been taken off Cape Sable; at a number of places off the outer coast of Nova Scotia; and on the Newfoundland Banks, at depths of 50 to 190 fathoms.[61]

[60] After examining specimens from New England waters and comparing them with published drawings of European fish, we can find no significant differences between the hook-eared sculpins of the two sides of the Atlantic.

[61] For localities of record off outer Nova Scotia and on the Newfoundland Banks, see Goode and Bean, Smithson. Contrib. Knowl., vol. 30, 1895, p. 268; also Reports, Newfoundland Fishery Research Commission, Vol. 1, No. 4, 1932, p. 108; vol. 2, No. 1, 1933, p. 125.

It is common enough in the Gulf of St. Lawrence for Huntsman to have classed it as a characteristic inhabitant of the icy intermediate water layer on the Banks,[62] while Vladykov and Tremblay[63] have reported it from the estuary of the St. Lawrence River near Trois Pistoles; it has been reported from Hamilton Inlet on the outer coast of Labrador;[64] and doubtless it will be found farther north, when the fish fauna of the outer Labrador coast has been explored more thoroughly, for it is known from West Greenland.

Presumably, it is resident in small numbers wherever found, sculpins not being migratory, but nothing whatever is known of its way of life.

Mailed sculpin *Triglops ommatistius* Gilbert 1913

Jordan and Evermann, 1896–1900, p. 1923, *Triglops pingeli* (Reinhardt), 1832, in part.

Gilbert, Proceedings, U. S. National Museum, vol. 44, 1913, p. 465.

Description.—The most distinctive feature of this sculpin, apart from its very long anal fin, is that it has a row of about 45 broad plate-like scales along its lateral line on each side, with smaller spiny scales below the dorsal fins, while the skin of the sides lower down is gathered in obliquely transverse folds. The body, too, is more tapering than that of our other sculpins, the caudal peduncle more slender, and the tail fin is smaller. Furthermore, the head is smaller and smoother than in any of the sculpins that are common in the Gulf of Maine, with short spines and many prickles. The first dorsal fin (10 to 12 spines) originates over the

[62] Trans. Roy. Soc. Canada, Ser. 3, vol. 12, Sect. 4, 1918, p. 63, as *Centridermichthys uncinatus*.

[63] Natural. Canad., vol. 62, 1935, p. 79.

[64] Kendall, Proc. Portland Soc. Nat. Hist., vol. 2, 1909, p. 217.

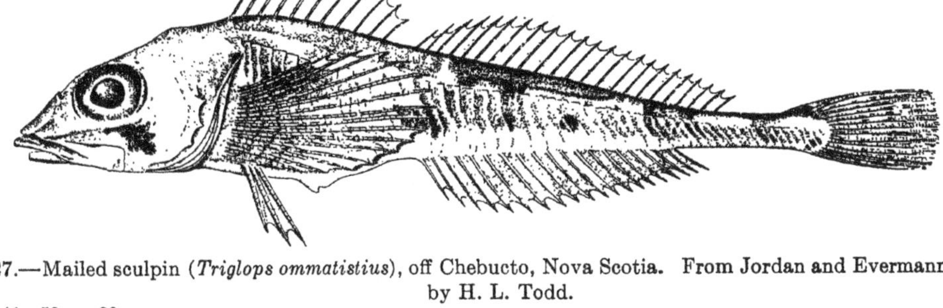

FIGURE 227.—Mailed sculpin (*Triglops ommatistius*), off Chebucto, Nova Scotia. From Jordan and Evermann. Drawing by H. L. Todd.

bases of the pectorals and is higher than the second dorsal, but only about half as long. The second dorsal has 20 to 25 rays. The anal fin is similar to the second dorsal in form and stands below it, but is a little shorter (20 to 22 rays). The ventral fins (each with 3 rays as is the rule among sculpins) reach about as far back as the rear end of the first dorsal, while the pectorals (17 rays) are of the fan-like shape usual among sculpins. The males have a very large and noticeable anal papilla.

Color.—Olive above; white, yellowish or orange below. There are four dusky blotches above the lateral line on each side, one on the caudal peduncle, one passing through the first dorsal fin, and two passing through the second dorsal fin. The fins are variously marked with yellowish and with gray-black. The first dorsal of the male has a dusky blotch between the first and second spines and another between the seventh and tenth spines; the second dorsal is marked with three horizontal olivaceous bars. Females lack the blotches on the first dorsal fin; and their second dorsal is marked with narrow lines of dots.

Size.—This is a small species, probably growing to about 8 inches, the maximum that is recorded for its European representative.[65] The largest yet recorded for the Gulf of Maine was 6 inches long.

Habits.—Little is known of its habits beyond the bare fact that it is a bottom fish, like other sculpins. Any that breed in the Gulf of Maine probably spawn in midsummer, Cox [66] having reported a ripe female at Cape Breton in July. Its eggs were pinkish, 2 mm. in diameter, with many oil globules. Presumably the eggs sink like those of other sculpins. The European mailed sculpin is known to eat worms and various crustaceans, and the diet of the American form is the same, probably.

General range.—Sculpins of this general type are circumpolar, ranging south to Cape Cod along the American coast and to the Baltic on the European side of the Atlantic, in rather deep water. But they show a tendency to split up into local races, the constancy of which is yet to be tested by a study of large series. Newfoundland specimens, for example, differ so much from typical *Triglops ommatistius* in the arrangement and number of folds of skin along the sides that Gilbert [67] has dignified them with a separate name (as the subspecies *terranovae* of species *ommatistius*); and both the eastern American forms are distinguished from the east Greenland and European mailed sculpins by the presence of the eyespot on the first dorsal fin of the male (which the European form lacks) and by slightly fewer fin rays. We do not feel convinced, however, that all these forms, together with the Bering Sea form (*Triglops beanii* Gilbert, 1895), will not finally prove to be local varieties of a single wide-ranging species.

Occurrence in the Gulf of Maine.—Judging from the scarcity of records this cold water fish is uncommon in the Gulf of Maine. Specimens have been recorded from the neighborhood of St. Andrews in the Bay of Fundy, in 15 fathoms (reported by Huntsman); a few from Massachusetts Bay and from off Race Point, Cape Cod (now or formerly in the collection of the Boston Society of Natural History); 11 others now in the United States National Museum were from Gloucester, Cape Cod, and Georges Bank; we have trawled them near Mount Desert; in Massachusetts Bay; off Cape Ann; off Cape Cod; and around the northern slope of Georges Bank, in depths of 20 to 140 fathoms in various months from spring to autumn; and two were trawled on the southeast slope of Georges [68] by the *Albatross III*, July 17, 1948, in 45 fathoms. Our most southerly record for it was about 10 miles east of Chatham, Mass.

The fact that Gilbert found differences between the Gulf of Maine and Newfoundland specimens, with others from Chebucto Head (Nova Scotia) and from Georges Bank intermediate between them, suggests that the mailed sculpin is a permanent resident of the inner parts of the Gulf, rather than that it appears there only as a wanderer, past Cape Sable, from the east and north.

Eastward and northward from our Gulf, this sculpin is described as being rather common to numerous on the outer Nova Scotian fishing grounds, and as one of the characteristic members of the fish fauna of the icy cold water on the Banks and in the Gulf of St. Lawrence.[69]

---

[65] Collett, Norske Nordhaus-Expedition, 1876-78, Zool., Fiske, 1880, p. 38.

[66] Contrib. Canad. Biol. (1918-1920) 1921, p. 111.

[67] Proc. U. S. Nat. Mus., vol. 44, 1913, p. 467.

[68] Latitute 40°48′ N., longitute 66°31′ W. (Arnold, Copeia 1949, p. 299).

[69] See Huntsman, Trans. Royal Soc. Canada, Ser. 3, vol. 12, Sect. 4, 1918, pp. 61-67, for a very interesting account of the fishes that are characteristic of the different water layers in the Gulf of St. Lawrence.

It is also reported off Bonne Bay on the west coast of Newfoundland, in the Strait of Belle Isle, and from the south coast of Newfoundland. It is so widespread on the eastern part of the Grand Banks that it was taken at 18 stations there on the cruises of the Newfoundland Fisheries Research Commission. It is also reported off the east coast of the Avalon Peninsula, and off Sandwich Bay on the outer coast of Labrador.

Grubby *Myoxocephalus aeneus* (Mitchill) 1815 [70]

LITTLE SCULPIN

Jordan and Evermann, 1896–1900, p. 1972.

Description.—The most distinctive features of the grubby, as compared with others of its tribe, are its short, simple head spines, combined with small size at maturity. It is of the typical sculpin form, though proportionately a stouter fish than either the shorthorned or the longhorned species, that is, about one-fourth as deep as it is long with smooth skin but showing the head ridges and spines typical of its genus. Most noticeable of these are a ridge with two spines running along the top of the head over each eye; a pair of spines between the nostrils; and six short spines on each side of the face between snout and gill opening. None of the cheek spines are long (p. 449). The spiny dorsal fin (9 spines), originating a little in front of the upper corner of the gill opening, is shorter (front to rear) than the second dorsal of 13 or 14 soft rays; its longest spines, measured from base to tip, are about the same length as the longest soft rays of the second dorsal; and the two fins are so close together that there is no free space between them. The anal fin (10 or 11 rays) is a little shorter than the second dorsal, under which it stands. The pectorals are of the fanlike outline characteristic of this family, while each ventral fin consists of one spine and three rays. There is no slit or pore behind the last gill arch (there is such a slit or pore in the shorthorn sculpin, at least in most specimens, p. 445).

[70] Placed in the genus *Acanthocottus* Girard, 1849, by Jordan, Evermann, and Clark, Rept. U. S. Comm. Fish (1928), Pt. 2, 1930, p. 386.

Color.—Grubbies, like other sculpins, vary in color according to the bottoms on which they lie. All that we have seen, however (this confirms the published descriptions), have been light to dark gray or greenish-gray above, with darker shadings or irregular barrings that are most evident on the sides and on the fins. The sides of the head are usually mottled light and dark; the belly is pale gray or white. According to information supplied by Dr. A. G. Huntsman, the presence of an uninterrupted pale band of considerable length along the lower sides of the caudal peduncle is a useful field character. But we have seen some specimens intermediate in this respect between the extreme condition shown in figure 228 and the variable mottlings and cloudings of the shorthorn sculpin.

Size.—This is the smallest of our common sculpins, few growing to more than 5 or 6 inches in length, and perhaps none to more than 8 inches.

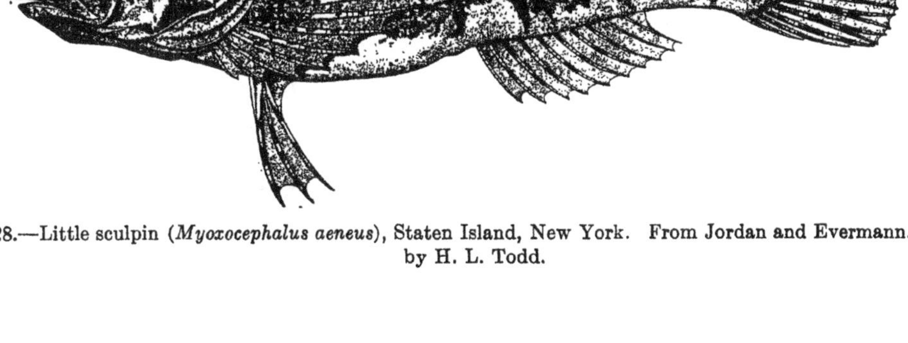

FIGURE 228.—Little sculpin (*Myoxocephalus aeneus*), Staten Island, New York. From Jordan and Evermann. Drawing by H. L. Todd.

Remarks.—The differences between the grubby and the shorthorn sculpin in number of anal rays, and in the presence or absence of a pore behind the last gill arch, seem sharp enough to forbid the possibility that the former may be a dwarf race of the latter. Determination, however, of these characters required such close examination, and grubbies resemble young shorthorns so closely in all other respects that it is not easy to tell the one from the other. We therefore suggest that any small sculpin that may prove difficult to name be forwarded for identification either to the laboratory of the U. S. Fish and Wildlife Service, Woods Hole, Mass.; to the Division of Fishes, U. S. National Museum, Washington, D. C.; or to the Department of Fishes, Museum of Comparative Zoology, Cambridge, Mass.

Habits.—On the southern shores of New England, where the grubby is not only more nearly universal than it is to the east and north of Cape Cod, but more plentiful, it is found from tide mark down to 15 fathoms or so. But we have taken it as deep as 28 fathoms in the Gulf of Maine, and Cox has reported it in the Gulf of St. Lawrence from the stomachs of cod caught in 60 to 70 fathoms. It is found on all sorts of bottoms, most abundantly among eel grass (*Zostera*) during the years when this plant was more plentiful than at present. And it is the only sculpin that summers in very shoal water along southern New England (Woods Hole and up Narragansett Bay) and near New York Harbor.

In the Gulf of St. Lawrence (e. g., around the Magdalen Islands) and on the Nova Scotian coast in general, it is found in estuaries, as in the southern part of its range, as well as outside. But it seems more restricted to the open coast in the Gulf of Maine, for Huntsman found it rare as far up Passamaquoddy Bay as St. Andrews, though common at the mouth, and more plentiful in St. Mary Bay and in Annapolis Basin than it is in Minas Basin on the Scotian side of the Bay of Fundy. Neither have we seen it in salt creeks about Massachusetts Bay.

The known distribution of the grubby in summer proves that it is certainly at home in water as warm as 69° F., and perhaps a degree or two warmer; these temperatures are several degrees higher than are preferred by its larger relative, the shorthorn. On the other hand, it survives temperatures as low as 32° in winter, if not lower, both in the Gulf of St. Lawrence, on the Nova Scotian coast, and about Woods Hole. Its presence in the inner parts of Narragansett Bay on the one hand and off open coasts on the other also proves it resistant to a wide range of salinity, but it never runs up into appreciably brackish water, as far as we can learn.

Probably the grubby breeds throughout its geographic range, certainly as far north as the southern part of the Gulf of St. Lawrence. And the Bay of Fundy appears to be the site of successful reproduction, for Huntsman found grubbies of all sizes there. The spawning season lasts all winter off southern New England and until June in the Gulf of St. Lawrence, Cox having reported a ripe female on the 18th of that month, at Amherst Island (Magdalen group). The eggs, which are described as of a beautiful green color and 1 mm. in diameter, sink like those of other sculpins and stick to seaweeds or to any other objects they chance to rest upon. Young sculpins (this species among them) have been caught in tow nets at Woods Hole from January to May.

This sculpin is omnivorous like its relatives, feeding on all sorts of small animals which it finds on the bottom, such as annelid worms, shrimps, crabs, copepods, snails, nudibranch mollusks, ascidians, and on small fish, including alewives, cunners, eels, mummichogs, launce, silversides, sticklebacks, and tomcod.[71] It also scavenges any kind of animal refuse.

General range.—North American coastal waters, from New Jersey to northern Nova Scotia and to the Gulf of St. Lawrence, both in the southern side, where it is common, and the Strait of Belle Isle.[72]

Occurrence in the Gulf of Maine.—It is probable that this little sculpin is to be found in suitable localities all around the shores of the Gulf of Maine, for it is reported as common along both shores of the Bay of Fundy (including St. Mary Bay) and at various localities in the Massachusetts Bay region, such as Cape Ann, Gloucester, Salem, Cohasset, and Provincetown, and the

[71] This list of fish fry eaten is from Vinal Edward's notes at Woods Hole.

[72] Maine has sometimes been given as its northern limit. But Doctor Huntsman writes us that in 1915 he obtained it in tide pools at Souris, Prince Edward Island; Needler (Proc. Nova Scotian Inst. Sci., vol. 20, 1940, p. 40) describes it as the common sculpin in Malpeque Bay, Prince Edward Island; Cox (Contrib. Canadian Biol. (1918–1920) 1921, p. 111) characterizes it similarly around the Magdalen Islands; Vladkykov and Tremblay (Nat. Canad., vol. 62, 1935, p. 80) report it from the estuary of the St. Lawrence, near Trois Pistoles; and Jeffers (Contrib. Canadian Biol. and Fish., N. Ser., vol. 7, No. 16, Ser. A; No. 13) (1932, p. 208) found two specimens on the beach at Raleigh, on the Newfoundland side of the Strait of Belle Isle.

Albatross II trawled 9 specimens 43–51 mm. long in 28 fathoms off the outer coast of Cape Cod abreast of Chatham, May 1, 1930. But it seems to be decidedly local in its distribution, for the only places where it has been definitely reported along the coast between Cape Ann and the Bay of Fundy is Casco Bay, nor have we caught it in any of the harbors of Maine where we have fished. In any case, it is far outnumbered in the Gulf of Maine by the two larger sculpins to be mentioned next.

Importance.—Because it is so small the grubby is of no commercial value. But wherever it is common it is something of a nuisance to anglers fishing for flounders and cunners, for it bites as greedily at any bait as do its larger relatives, and it serves as a source of food, no doubt, for more important fishes.

Shorthorn sculpin *Myoxocephalus scorpius* (Linnaeus) 1758 [73]

DADDY SCULPIN; BLACK SCULPIN; GREENLAND SCULPIN

Jordan and Evermann, 1896–1900, as *M. scorpius* and *M. groenlandicus*, p. 1974.

Description.—The shorthorn sculpin, with its large flat head, vast mouth, weak tapering body, bat-like pectorals, and insatiable appetite, typifies the sculpin race in northern seas. It has a longitudinal ridge with 3 knobs or spines running along each side of its crown; also about 6 (sometimes 5 or 7) short triangular bluntish spines on each side of the cheek between snout and gill opening, the uppermost of these less than twice as long as the one below it, and reaching not much more than halfway to the edge of the gill cover. And there is a short but sharp spine at the upper corner of each gill cover, pointing rearward and lying on a flap of skin, besides two thornlike spines on each shoulder close behind the upper corner of the gill cover.

There is a pore, or small slit, piercing the soft skin low down on each side of the throat close behind the last gill arch, easily seen on large specimens and detectable even on small ones on close examination.

The very large eyes are at least as wide as the space between them, set high up on the sides of the head with the upper edges close to the dorsal profile, and they are directed a little upward as well as outward. The two parts of the dorsal fin are entirely separated by a deep notch, but there is no gap between them. The forward part has 9 to 11 spines, the rear part about 16 or 17 (sometimes 15) soft rays, the longest of which are only a very little longer, if any, than the longest of the spines, each measured from base to tip. The anal fin, with 13 to 14 rays, is similar to the second dorsal in shape, but a little smaller; it originates about under the fourth or fifth soft

[73] Placed in the genus *Acanthocottus*, Girard, 1849, by Jordan, Evermann, and Clark, Rept. U. S. Comm. Fish. (1928), Pt. 2, 1930, p. 386.

FIGURE 229.—Shorthorn sculpin (*Myoxocephalus scorpius*), Eastport, Maine. From Goode. Drawing by H. L. Todd

dorsal ray. The caudal fin is small, its rear margin weakly rounded; the fanlike pectorals, of 17 or 18 rays, reach back about as far as the vent. On large specimens the dorsal, anal, and pectoral fins are rather noticeably thick and fleshy. There are two series of prickly platelike scales along each side of the body, one above the lateral line, the other below it.

Males and females differ in appearance, the former being the more slender, with higher fins, and the more brightly colored. Each of the scales, too, along the lateral line bears three or more prickles in males, but only one or two at the most in females, while some of the latter have no scales. Furthermore, the inner edges of the rays of the pectoral and ventral fins are armed with teeth or prickles on the males but not on the females.

Color.—The basic hue of the upper parts is usually of some shade of brown, ranging from a warm reddish tint to almost black, with the top and sides of the head marked with pale blotches and the back and sides of the body with broad dark bars on individuals on which the ground tint is pale. The lower parts of the sides are more or less spotted with yellow. The belly is whitish or yellowish in females, usually reddish orange with large round white spots in males, this being a good field mark for distinguishing the sexes. The dorsal fins are mottled dark and pale, the second dorsal often marked with 3 or 4 definite crossbars, and the caudal fin with various dark mottlings. The rays of the pectoral and anal fins are yellow with 2 or 3 irregular dark crossbars on many specimens, but they are uniformly dark in some. Males are more brightly colored than females in the breeding season, when their red and yellow tints become very brilliant, and when an intensification of the red or coppery ground color of the belly brings out the white spots more clearly than at other seasons.

Variability.—European studies have shown this to be a highly variable species, tending to break up into local races. Whether this is equally true of it on our side of the Atlantic is not known.[74]

[74] Most American ichthyologists recognize two subspecies of this fish the true "shorthorn" (*scorpius*) and the "Greenland sculpin" (*groenlandicus*) And with the prevailing tendency to call American and European fish by different names it is as the latter that our local sculpin has usually been recorded. But the differences between the two (size, relative breadth of the top of the head, and length of the dorsal fin spines) are so very slight and all of them have proved so variable, that we follow Huntsman (Contrib. Canadian Biol. (1921) 1922, p. 64) in uniting the two; the more willingly since both forms have been found on both sides of the Atlantic.

Size.—This is the largest Gulf of Maine sculpin. It has been said to grow to a length of about 3 feet, but the average run of the adults taken in our Gulf is only about 8 to 14 inches, the longest not more than 2 feet. This species increases in size from south to north, Greenland fish averaging much larger than those taken off New England or off the Maritime Provinces.

Remarks.—Young shorthorns, up to 6 or 7 inches long, resemble the little grubby (p. 443) so closely that they are likely to be confused with it. Points of difference are that the shorthorn has at least 13 or 14 rays in its anal fin, and has a pore piercing each side of the throat close behind the last gill arch, the grubby only 10 or 11 anal rays, and no such pores.

Larger specimens of the shorthorn could hardly be mistaken for any other Gulf of Maine fish, unless perhaps for the longhorn sculpin. And even a cursory look is enough to separate one of these from the other, the upper cheek spine being less than twice as long as the one below it, and not reaching more than halfway to the edge of the gill cover in the shorthorn, but about four times as long as the one below it in the longhorn, and reaching back at least as far as the edge of the gill cover.

Habits.—Bays and the vicinity of ledges that rise from comparatively smooth bottom in shoal water are the chief haunts of the shorthorn sculpin. And it is found indifferently there, on mud, sand, or pebbles, on bare bottom or among weeds. Many are also caught off piers and along our rocky shores by cunner fishermen. Off our coasts, the great majority live shoaler than 10 fathoms. And while a day's fishing on any of the ledges northward and eastward from Cape Cod is likely to yield an occasional shorthorn among other fish, few are caught on long lines set deeper than 15–20 fathoms. The deepest records for it in American waters of which we know are 50 fathoms near Campobello Island, at the mouth of the Bay of Fundy (reported by Huntsman), and 57 fathoms in the northern part of the Gulf of St. Lawrence, just within the Strait of Belle Isle.[75]

This is a cold-water fish. Even in summer it is the most plentiful at localities and at depths where the temperature is lower than 55°–60° F. In winter it endures temperatures close to the freez-

[75] Rept. Newfoundland Fish. Res. Comm., vol. 1, No. 4, 1932, p. 108, sta. 45.

ing point of salt water. We have never heard of one taken in brackish water, at least on our side of the Atlantic. It is a sluggish fish, often to be seen lying motionless, and as a rule, it hugs the bottom so closely that it is hard to tempt one to rise as much as a few feet by dangling a bait over it. Neither does it come to the surface voluntarily, though the surface may drop to the sculpin on the ebbing tide.

Sculpins usually swim slowly with undulating motion, spreading their great pectoral fins like bat's wings. As a rule, they move only a little way when disturbed, but on occasion they can dart ahead with folded "wings."

They are among the most voracious fishes, feeding chiefly on crustaceans, particularly on crabs, of which they are often full, on shrimps, sea urchins, and worms; on the fry of various other fishes; rarely on shellfish. And they are eager scavengers of any kind of refuse, congregating about fish wharves and lobster cars to feast on the debris. Like all sculpins they bite on any bait, and so greedily that we have caught one time and again, thrown it back, and seen it bite again almost as soon as fresh bait reached bottom.

The shorthorn has been described as hiding in dark crevices or among weeds by day, to emerge at night. This, however, has not been our experience, nor did Gill [76] find it doing so at Grand Manan.

This fish, like the longhorn sculpin, grunts or gurgles when drawn out of the water, particularly when handled, and it is also known to grunt in the water.

We must turn to European sources for the breeding habits of this sculpin, little attention having been paid to this phase of its life by American ichthyologists.[77] The spawning season is from November to February, both about Woods Hole and in north European waters, with the chief egg production in December, which no doubt applies equally to the Gulf of Maine. At this season the adult sculpins have been described as gathering in schools on sandy or weedy bottom, with the females greatly outnumbering the males.

Discussion has centered about the manner in which the eggs are fertilized, it being generally agreed that this takes place externally as a rule, but that they may be fertilized within the body of the mother in some parts of the Baltic Sea. In either case, the eggs sink [78] and stick together in irregular spongy masses through which the water circulates, and which retain considerable moisture even if they are left bare by the ebbing tide, as often happens. These egg masses are deposited on sandy bottoms, in pools in the rocks, among seaweeds, or in any crevice or hollow, in a tin can, for instance, or in an old shoe. Sometimes the male makes a nest of seaweed and pebbles, while he has been described as sometimes clasping the egg mass with his pectoral and ventral fins, and he has been photographed so employed.[79]

The eggs are of varying shades of red or yellow, 1.5 to 2 mm. in diameter. Incubation is so slow (occupying 4 to 12 weeks, according to temperature) that egg masses with advanced embryos have often been found as late in the spring as April or even May. Newly hatched larvae are about 7 to 8 mm. long. In a month they are 10 mm. long and the yolk sac has been absorbed. The young larvae soon rise to the surface, where quantities of them have been taken in tow nets in British waters in March, April, and May. By May and June some have grown to a length of 22 to 25 mm. They abandon their drifting life at about this size, or soon after, for the bottom, and they may be 38 mm. long by July, showing all the distinctive characters of the adult.[80] This timetable, compiled from European sources, probably applies equally to the Gulf of Maine, for larvae are found as early as February in the Bay of Fundy and thereafter throughout the spring.[81]

The subsequent rate of growth is not definitely known. But it is probable that this sculpin is 2 or 3 inches long by the end of its first summer, for we have taken a few 2-inch fish in late June on Nantucket Shoals, and 2- to 3¼-inch fish in late September off Boothbay Harbor, Maine. Most of them, it seems, do not mature sexually until they are at least 6 inches long.

General range.—One or another race of this wide ranging fish is known from Great Britain northward along the coasts of Europe; in Arctic seas

[76] Smithsonian Misc. Coll., vol. 47, 1905, p. 352.

[77] Gill (Smithsonian Misc. Coll., vol. 47, 1905, p. 352) gives a summary of its life history.

[78] Buoyant eggs taken in the tow net (Agassiz, Proc. Amer. Acad. Arts Sci., vol. 17, 1882, pl. 3) belonged to some other fish.

[79] Ehrenbaum, Wiss. Meeresuntersuchungen, Helgoland, Neue Folge, vol. 6, 1904, pl. 8.

[80] McIntosh and Masterman, Life-Histories of British Marine Food-Fishes, 1897, p. 129.

[81] Huntsman, Contrib. Canadian Biol. (1921) 1922, p. 64.

generally, including Spitzbergen, Nova Zembla, north Siberia, West Greenland, and northern Labrador; and southward along the American coast to southern New England; to New Jersey as a stray.

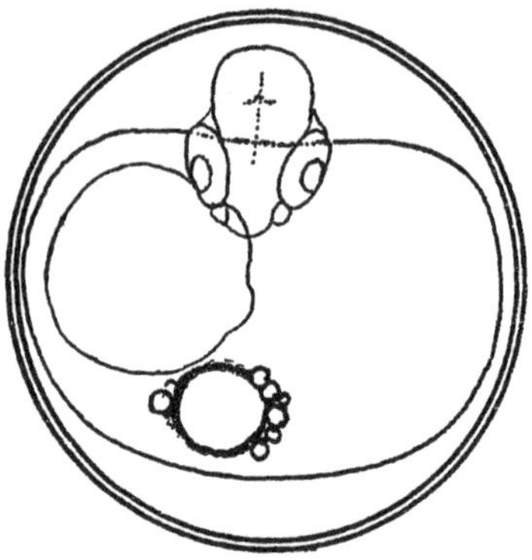

FIGURE 230.—Egg (European). After Ehrenbaum.

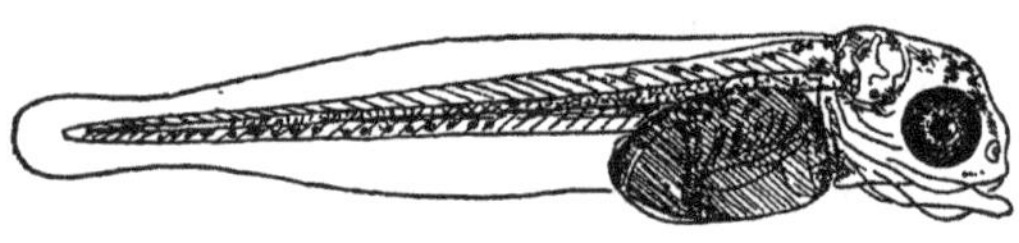

FIGURE 231.—Larva (European), 8.2 mm. After Ehrenbaum.

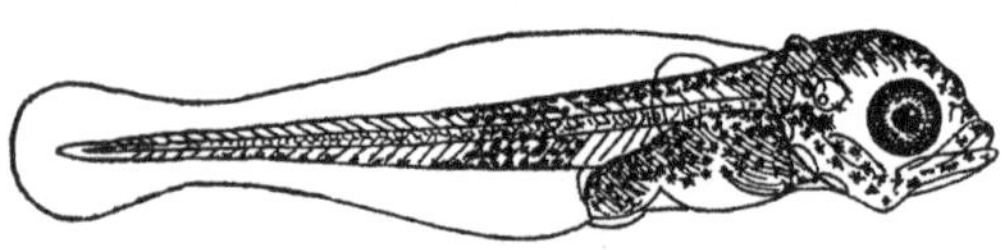

FIGURE 232.—Larva (European), 10 mm. After Ehrenbaum.

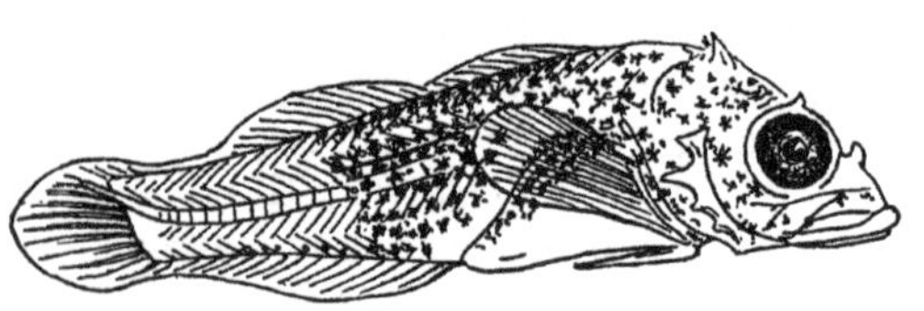

FIGURE 233.—Fry, 18 mm. After Ehrenbaum.

SHORTHORN SCULPIN (*Myoxocephalus scorpius*)

Occurrence in the Gulf of Maine.—This is one of the most familiar of our shore fishes, common all around the entire coast line of the Gulf of Maine; it is not so abundant as the longhorn sculpin (p. 449), but we doubt if there is a suitable situation, Cape Cod to Cape Sable, where some shorthorns are not to be found, except perhaps at the head of the Bay of Fundy.[82] But they do not run far up estuaries as a rule, and never into brackish water.

The shorthorn has not been reported positively either from Georges Bank or from Browns. Sculpins of some sort, it is true, are so common on the former that otter trawlers often catch from 20 to 100 per haul, and equally so on Browns Bank. But fishermen lump this and the next species together. Also the fact that the few positively identified on the banks have all proved to be longhorns, and the general predilection of the shorthorn for water shoaler than these offshore grounds, makes it doubtful whether it is to be found there in any numbers. Further evidence in this direction (if indirect) is that most of the shorthorns that were taken during the experimental trawlings of the Newfoundland Fisheries Research Commission were from hauls shoaler than 50 fathoms, only one from as deep as 57 fathoms.

Although it is more strictly confined to comparatively shoal water than the longhorn sculpin, the shorthorns are not often seen close to tide mark in summer, except in the Bay of Fundy where the temperature of even the uppermost few feet does not rise above 52°–54° in most summers, and where there are shorthorns of all sizes in very shallow water, in summer as well as at other times of the year.

On the other hand, the shorthorns living around the open shores of the Gulf show no tendency to avoid winter chilling by descending to greater depths in winter, In the Bay of Fundy, for example, where it is very common, Huntsman has characterized it as the only fish that remains near shore during the coldest part of the year; and it has been described as more plentiful along the shores of Massachusetts Bay in winter than in summer, as it certainly is south of Cape Cod. Certain shallow bays, it is true, such as Duxbury Harbor, where broad expanses of flats are exposed at low tide to heating by the sun in summer and to the formation of ice in winter, are an exception to this rule; i. e., the shorthorn sculpins tend to keep to the deeper channels through the coldest part of the winter as well as during the heat of midsummer. But we have found no evidence that they carry out any seasonal migrations more extensive than this. They are, indeed, among the most stationary of Gulf of Maine fishes.

82 Huntsman found none there.

In Scandinavian waters this fish is said to vary widely in abundance from year to year, years of plenty alternating with longer periods of scarcity, but this does not seem to be the case to any noticeable extent in the Gulf of Maine where it is always common.

To the northward and eastward, the shorthorn is common all along the outer coast of Nova Scotia, in 10–30 fathoms, and it has been taken on Banquereau Bank. It has not been reported in the Magdalen and Prince Edward Island shallows in the southern side of the Gulf of St. Lawrence, where summer temperatures are high. But it is to be expected in that side of the Gulf in slightly deeper water, for it is known all along the northern shore of the Gulf, from Anticosti to the Strait of Belle Isle. We find no record of it on the Newfoundland Banks, probably because of the depth of water; neither is any definite information available as to its status along the south coast of Newfoundland. But it is recorded off the east coast, from the trawlings of the Newfoundland Fisheries Research Commission, and along the outer coast of Labrador, at Battle Harbor (just north of the Strait of Belle Isle);[83] at Rigolet in Hamilton Inlet;[84] in the vicinity of Nain;[85] at Fort Chimo, Ungava Bay, and it is widespread in the Hudson Bay region.[86]

[83] Specimen in Museum of Comparative Zoology.

[84] See Kendall (Proc. Portland Soc. Nat. Hist., vol. 2, Pt. 8, No. 13, 1909, pp. 213, 233) for records from outer Labrador.

[85] Kendall, Proc. U. S. Nat. Mus., vol. 38, 1910, p. 509.

[86] See Vladykov (Contrib. Canadian Biol., N. Ser., vol. 8, 1933, p. 30 [No. 2, p. 18], as *groenlandicus*) for localities where it has been taken in Hudson Bay, including James Bay.

Importance.—Although this is an edible fish and accounted a good one, its appearance and habits will probably close our markets to it as long as other fish are plentiful. Nevertheless, it once was of some commercial importance, being one of the best baits for lobster pots, for which purpose great numbers were speared formerly on the Massachusetts coast in spring, and were caught along the northern coast of the Gulf on hook and line. But very little use is made of them nowadays, if any.

Longhorn sculpin *Myoxocephalus octodecimspinosus* (Mitchill),[87] 1815

GRAY SCULPIN; HACKLEHEAD; TOADFISH

Jordan and Evermann, 1896–1900, p. 1976.

Description.—This fish resembles the shorthorn sculpin so closely that the description may be confined to the points in which it differs. Chief of these is the great length of its uppermost cheek spine, which usually is about four times as long as the spine just below, and which reaches at least as far back as the edge of the gill cover. This serves equally to distinguish the young longhorn from the grubby, which is short-horned. All the head spines, too, of the longhorn are so sharp that one must be cautious in grasping one of these fish, for it turns its spines rigidly outward by spreading its gill covers. Furthermore the long spines of the long horn are naked at the tip. The number and arrangement of the head spines

[87] Placed in the genus *Acanthocottus* Girard, 1849, by Jordan, Evermann, and Clark (Rept. U. S. Comm. Fish., (1928), Pt. 2, 1930, p. 386.

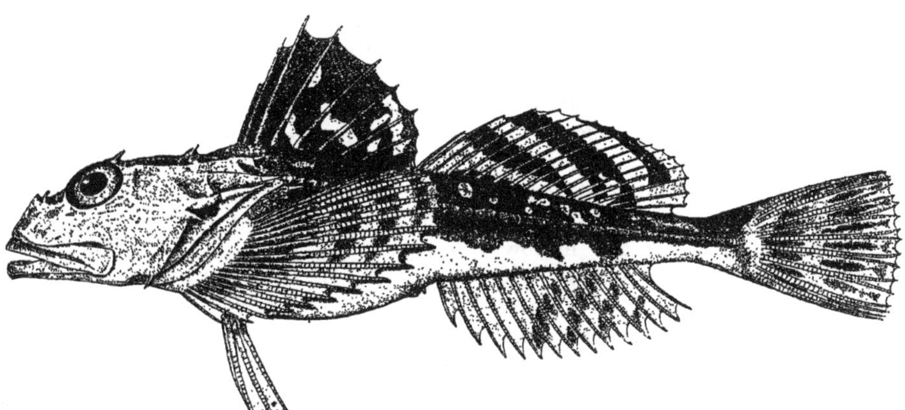

FIGURE 234.—Longhorn sculpin (*Myoxocephalus octodecimspinosus*), New Jersey. From Goode. Drawing by H. L. Todd.

is the same as in the shorthorn sculpin (p. 445), hence need not be described, and there are two thorns on each shoulder, with a larger one close above the origin of the pectoral fin. The first dorsal fin is higher than the second (in the shorthorn sculpin these two fins are of about equal heights), of rather different shape from that of the shorthorn (compare fig. 234 with fig. 229), and proportionately shorter than in the latter though with about the same number of spines (8 or 9).

The second dorsal fin and the anal have the same number of rays (15 or 16 dorsal and about 14 anal) as in the shorthorn; but the anal of the longhorn originates under the second or third ray of the second dorsal fin instead of under its fourth or fifth ray. The pectorals are of the fanlike form usual among sculpins. The lateral line of the longhorn sculpin is marked by a series of smooth cartilaginous plates instead of by prickly scales as it is in the shorthorn, a difference obvious to the touch; its body is more slender (about five and one-half times as long as it is deep); and its head is flatter.

Color.—The longhorn, like other sculpins, varies in color with its surroundings. The ground tint of the back and sides ranges from dark olive to pale greenish-yellow, greenish-brown, or pale mouse color, but is never red or black as the shorthorn so often is. As a rule it is marked with four irregular, obscure, dark crossbars, but these are often broken up into blotches and they may be indistinct. The coarseness of pattern often corresponds to that of the bottom, as does the degree of contrast between pale and dark. On mud and sand bottom this sculpin is often nearly plain colored, but when it is lying on pebbles with white corallines its back is often nearly white with dark-gray blotches, rendering it almost invisible. The first dorsal fin is pale sooty with pale and dark mottlings or spots; the second dorsal is paler olive with three irregular oblique dark crossbands; the caudal is pale gray; and the pectorals yellowish. Both caudal and pectorals are marked with 4 to 6 rather narrow but distinct dark crossbands. The anal is pale yellowish with dark mottlings; and there often is an obscure yellowish band along the lower part of the sides, marking the transition from the dark upper parts to the pure white belly.

Size.—This is a smaller fish than the shorthorn sculpin. It grows to a maximum length of about 18 inches, but only a few of them are more than 10 to 14 inches long. A 10-inch fish weighs about ½ pound, one 12 inches long about 1 pound.

Habits.[88]—Everyone who has fished along the shores of our Gulf is more or less familiar with this sculpin, for it is a nuisance to cunner and flounder fishermen. It often is bothersome to the angler to unhook when it spreads its needle-sharp spines and erects its spiny dorsal fin. It grunts when pulled out of the water and bites on any bait.

No doubt it is as omnivorous as the shorthorn. Specimens examined by Vinal Edwards at Woods Hole had fed chiefly on shrimps, crabs, amphipods, hydroids, annelid worms, mussels and sundry other mollusks, squids, ascidians, and on a considerable list of fish fry, including alewives, cunners, eels, mummichogs, herring, mackerel, menhaden, puffers, launce, scup, silversides, smelts, tomcod, silver hake, and small fry of other sculpins. Rock crabs (*Cancer irroratus*) and amphipod crustaceans (*Leptocheirus*) had been the dominant food of a large series of shorthorns in Block Island Sound, examined by Morrow; nearly all of them had eaten shrimps (*Crago*) but in small amount; a few contained small lobsters; and spider crabs (*Libinia*) were a regular article of diet in winter, but not in summer. It is interesting that these particular shorthorns had eaten only a few mollusks of any kind.[89]

The longhorn is as useful a scavenger as the shorthorn, and equally voracious, gathering about wharves, sardine factories, and under lobster cars, always keeping to the bottom. Its depth range is rather wider than that of the shorthorn. At the one extreme it is abundant in many shoal harbors and bays, where it comes up on the flats at high tide, to leave them at low; and it runs up into estuaries, salt creeks, and river mouths, though never into fresh water, so far as we know. At the other extreme it is caught in considerable numbers down to 50 fathoms or so, and it has been reported as deep as 105 fathoms.[90]

The longhorn evidently is at home in temperatures as high as about 65°–66°, for we have seen many of them in very shallow water that warm, or

[88] Morrow (Bull. Bingham Oceanographic Coll., vol. 13, Art. 2, 1951) has recently published a detailed study of this sculpin, as found off southern New England.

[89] For more extensive diet lists see Morrow, Bull. Bingham Oceanographic Coll., vol. 13, Art. 2, 1951, pp. 60–61, 88–89.

[90] In Trinity Bay, east coast of Newfoundland, Rept. Newfoundland Fish. Res. Comm., vol. 1, No. 4, 1932, p. 108, Sta. 35.

even a little warmer, in summer in the southern side of Massachusetts Bay. But in localities where the temperature of the upper few feet rises much higher than this they withdraw to somewhat deeper (i. e., cooler) water for the summer (p. 452), working inshore again in the autumn.

At the other extreme, it is subjected for the coldest part of the year to water as cold as 32°–33°, both in our Gulf, along the Nova Scotian shelf, and in the Gulf of St. Lawrence, while it has been reported from water of 31°–32° F. (−0.3° C.) in the bottom of Trinity Bay, Newfoundland. And it seems that even exposure to freezing temperature may not be fatal to it, if not too prolonged, for we find no evidence that these sculpins are ever killed by cold when they are overtaken on the flats in severe freezes, a fate that occasionally overtakes cunners and tautog (pp. 475, 480). But the fact that the geographic range of the longhorn does not reach as far north as that of the shorthorn suggests that it is not so well suited as the latter is to very low temperatures continuing throughout the year.

Off the southern New England coast the shorthorn deposits its eggs from late November through January, and perhaps into February, i. e., at the coldest time of the year, with the chief production in late December and January.[91] Presumably the spawning season is the same in the Gulf of Maine. Apparently one locality serves as well as another, nor is there any evidence that any particular depth is sought.

Ripe eggs are about 0.85 mm. in diameter before being laid, but they swell when they come in contact with the water; they are described as varying in color, from coppery green to reddish brown, orange, or purple. A 12¾-inch female, which we examined, taken near Woods Hole on November 18, 1951, contained about 8,000 chocolate brown eggs. They sink and they are so sticky when first laid that they cling together in clumps, or to anything that they may touch; and they continue to adhere during the period of incubation, but the surfaces of the eggs that are exposed to the water lose their stickiness after about 24 hours. The egg masses have been found free on the bottom, in empty clamshells or other cavities, or among the branches of the finger sponge (*Chalina*) like the eggs of the sea raven (p. 456) [92] and they are sometimes found thrown up on the beach.

The young fry have been taken in February and March off southern New England, in April on the eastern part of Georges Bank and in the channel between the latter and Browns Bank. These young stages [93] have longer cheek spines than the corresponding stages of the shorthorn sculpin (p. 447) they are more slender, and they differ further in the outline of the dorsal fin, for in the longhorns (if our identifications be correct) this is continuous from end to end, only the largest of them showing a shallow notch between spiny and soft portions, whereas in the shorthorn the two sections are separate from the time the fin first takes definite form.

Captures of many young fry 1½ to 2 inches long in September, and 3 to 3½ inches long in February suggest that the longhorn is about 2 to 2½ inches long at one year of age. According to Morrow's studies (based on the otoliths)[94] longhorns off southern New England average about 6½ to 7 inches long at 2 years of age; about 10 inches at 4 years; and 11 to 12 inches at 6 years. They are mature sexually at 3 years or older.

General range.—Coastal waters of eastern North America from eastern Newfoundland,[95] and the north shore of the Gulf of St. Lawrence, [96] south regularly to New Jersey, and reported to the Atlantic coast of Virginia.[97]

Occurrence in the Gulf of Maine.—This is our commonest sculpin, to be caught anywhere and everywhere along the entire coast line of the Gulf of Maine. We dare venture that there is not a bay, harbor, estuary, or a fishing station from Cape Sable to Cape Cod where it is not to be found. Not only is it more plentiful in most places than its short-horned relative, but it occupies a wider depth zone. It is very abundant in many shoal harbors where it comes up on the flats; it is caught

[91] For the most detailed study yet made of the breeding habits of the shorthorn, see Morrow, Bull. Bingham Oceanographic Coll., vol. 13, art. 2, 1951, p. 30–36.

[92] Warfel and Merriman (Copeia, 1944, p. 198) were the first to report this interesting habit.

[93] The smallest larva we have seen was 13 mm. long, from Georges Bank.

[94] Bull. Bingham Oceanographic Coll., vol. 13, art. 2, 1951, p. 47, table 6.

[95] Trinity Bay, Newfoundland; Rept. Newfoundland Fishery Research Commission, vol. 1, No. 4, 1932, p. 108, sta. 35.

[96] Repeated characterizations of this sculpin as ranging to "Labrador" are based on Storer's (Boston Jour. Nat. Hist., vol. 6, 1852, p. 250) report of 2 young specimens from some point not specified on the northern shore of the Gulf of St. Lawrence.

[97] *Albatross II* trawled them in small numbers at 9 stations scattered along the midbelt of the shelf, from the offing of southern Massachusetts to the offing of Delaware Bay, in the months of February, April, July, and September, at depths ranging from 11 to 50 fathoms.

also in considerable numbers down to 50 fathoms or so (p. 450). We have trawled it at 27 to 33 fathoms in Massachusetts Bay, and at 50 fathoms off Cape Elizabeth. But it has not been reported from the basin at greater depths.

It also occurs plentifully on Georges Bank, and while the composition of the sculpin population of Georges Bank is yet to be determined, the facts that this was the only sculpin (except the sea raven which it greatly outnumbered) taken there on otter trawling trips in June 1912, or in September 1929, and that the dragger *Eugene H* took 1,030 of them in 35 trawl hauls on the southwestern part of Georges, at 25–35 fathoms (but none in deeper hauls) in late June 1951, are evidence that it is the commonest member of its tribe on shoal parts of the bank. It is fair, also, to assume that this applies equally to Browns Bank, where fishermen report sculpins of one sort or another as plentiful. It is described, also, as very common along the Nova Scotian coast and banks eastward from Cape Sable, in suitable depths, and as widely but irregularly distributed around the southern shores and islands of the Gulf of St. Lawrence.[98]

The longhorn is a year-round resident, in the sense that its only periodic movements are off and on shore, and of short extent, combined with movements to and from particular grounds. Near New York it is commonest near shore from September to May, and is seen only occasionally in summer. In Long Island Sound they appear to carry out east-west journeys about which little is known; in Block Island Sound (off the mouth of Long Island Sound) they are plentiful on the productive fishing grounds from November throughApril, but mostly withdraw thence during May (either offshore, or onto more rocky grounds nearby), not to return in force until the next October.[99] These shifts do not appear to be connected with temperature.

All that is known of its movements in the Gulf of Maine is that in partially enclosed and very shallow situations where the water on the flats heats to 68°–70° in the warmest part of the season, but where ice forms in the winter (Duxbury Bay, for example), the shorthorns seek slightly deeper (i. e., cooler) water for the summer, work up again onto the flats in early autumn; move deeper again in late autumn; then work back on the flats again in early spring. They continue common, however, right up to low tide line all summer in localities where the surface does not become so warm in summer or so cold in winter. This is the general rule northward and eastward around the coast of Maine, including the Passamaquoddy region.

The presence of longhorn sculpins of all sizes, from very young fry to adult, proves that they breed all along the coasts of Massachusetts and of Maine, probably along western Nova Scotia as well. But it seems to be restricted as a breeder in the Bay of Fundy to the Scotian side. Thus it appears that the half-grown and adult fish that are plentiful along the New Brunswick shore are migrants, either from the Nova Scotian side across the bay, or from the open Gulf outside.

Importance.—The only commercial value this sculpin has had in our Gulf was as bait for lobster pots, for which they were speared formerly in some localities, and caught on hook and line in others. But very few of them are now used in this way.

Staghorn sculpin *Gymnocanthus tricuspis* (Reinhardt) 1832

Jordan and Evermann, 1896–1900, p. 2008.

Description.—This sculpin is easily distinguishable from its more common relatives by the shape of the uppermost of its three cheek spines, which is broad, flat, and with three short, sharp branches at its tip [1] instead of cylindrical and single pointed; also its anal fin (16 to 18 rays) originates well in front of its second dorsal fin instead of behind the latter, and its two dorsal fins are separated by a distinct space instead of being practically continuous at the bottom of the notch that separates them. Furthermore, the spines characteristic of the top of the head and shoulders of our other sculpins are either lacking on the staghorn or are very short, and the corners of its gill covers are rounded instead of sharp. Distinctive also, if less obvious, is the fact that the top of its head is more or less prickly or warty.

The 3-rayed ventral fins reach only about to the vent on young fry of 1¾ to 2 inches, but they

[98] Cox, Contrib. Canadian Biol. (1918–1920) 1921, p. 111; Leim, Proc. Nova Scotian Inst. Sci., vol. 20, 1940, p. 40.

[99] For further details see Morrow, Bull, Bingham Oceanographic Coll., vol. 13, art. 2, 1951, p. 54.

[1] These branches are only faintly indicated at the tips of the spatulate spines on young fry that we have seen, 1¾ to 2 inches long.

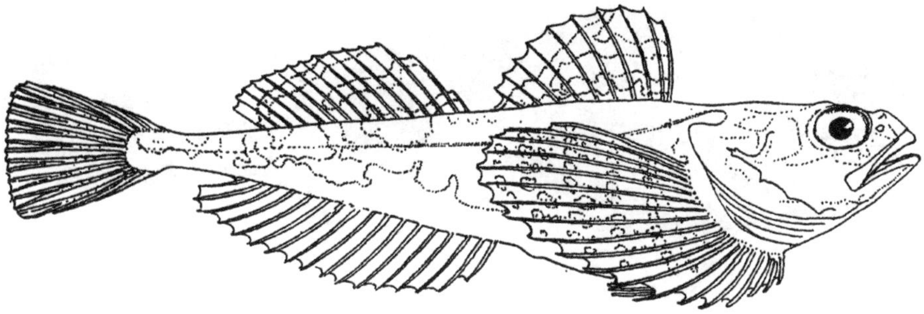

FIGURE 235.—Staghorn sculpin (*Gymnocanthus tricuspis*). After Smitt.

become relatively longer with growth until at maturity they reach considerably beyond the point of origin of the anal fin, farther in males of breeding age than in females.[2] The first dorsal fin has 11 or 12 spines; the second dorsal 15 to 17 soft rays. The caudal and pectoral fins and the general shape of the fish are of the usual sculpin type.

Color.—Described as dark brownish or gray above, the sides as marked with dark crossbands or with alternate light and dark greenish spots; the lower surface as white or yellowish with an irregular line of demarkation between dark sides and pale belly. The dorsal and pectoral fins are pale, the former with 3, and the latter with 4 or 5 irregular dark brown or black crossbands. The ventral and anal fins are yellow rayed, with membranes of the same color as the belly.

Size.—Up to about 10 inches long.

General range.—Arctic Ocean and North Atlantic, south to northern Norway on the European coast; on the American coast it ranges southward along the outer coast of Labrador[3] to the Gulf of St. Lawrence, where it is generally distributed along the north shore[4] and is characteristic of the icy water on the banks in the southern side, according to Huntsman, and it has been reported as far as Eastport, Maine, as a stray.

Occurrence in the Gulf of Maine.—The most southerly record for this Arctic sculpin, and the only one for the Gulf of Maine, is of a specimen caught at Eastport, Maine, in 1872, and now in the United States National Museum. It is only as a very rare stray from colder waters to the north that it ever reaches our Gulf.

[2] Small fry that we have examined from Saeglek Bay on the northeastern coast of Labrador, and an adult male from the estuary of the St. Lawrence River, agree with Smitt's (Scandinavian Fishes, vol. 1, 1892, p. 160 as *G. ventralis*) account of a specimen from Spitzbergen.

[3] Presumably it also occurs all around the coasts of Newfoundland; but it is not included among the species listed as taken during the trawling cruises of the Newfoundland Fisheries Research Commission.

[4] Stearns, Proc. U. S. Nat. Mus., vol. 6, 1884, p. 125; Kendall, Proc. Portland Soc. Nat. Hist., vol. 2, No. 13, 1909, p. 213–214.

Arctic sculpin *Cottunculus microps* Collett 1875

Jordan and Evermann, 1896–1900, p. 1992.

Description.—The head spines so characteristic of most sculpins, are reduced in this species to bony knobs, of which there are four on the top of the head and several on its sides. The two portions of the dorsal fin (spiny and soft) are united into one continuous fin, a feature that marks it off from all other local sculpins, while the spiny part (only 6 to 8 spines) is shorter and lower than the soft part (13 to 15 rays). But the very large bony head, wide mouth, slender tapering body, large fan-shaped pectorals, and the location of the ventrals below the pectorals, give the fish the typical sculpin aspect. The anal fin (about 10 rays) is a little shorter than the soft portion of the dorsal fin, and the caudal fin is small and rounded. The skin is roughened with small warts.

Color.—Described as pale with dusky crossbars, one on the head, two on the body and fins, and one at the base of the caudal fin. Scandinavian specimens have been reported as having still another band across the tip of the caudal, and as with the anal and pectoral fins dark mottled.[5]

Size.—Up to about 8 inches long.

General range and occurrence in the Gulf of Maine.—This is an Arctic deep-water species, known off east Greenland and about Spitzbergen in the Arctic Ocean, and from both sides of the northern Atlantic. On the eastern side it has been reported from northern Iceland, from Norwegian waters southward to the Channel, and doubtfully from the Skagerak. Off the American coast it

[5] Smitt, Scandinavian Fishes, vol. 1, 1892, p. 158.

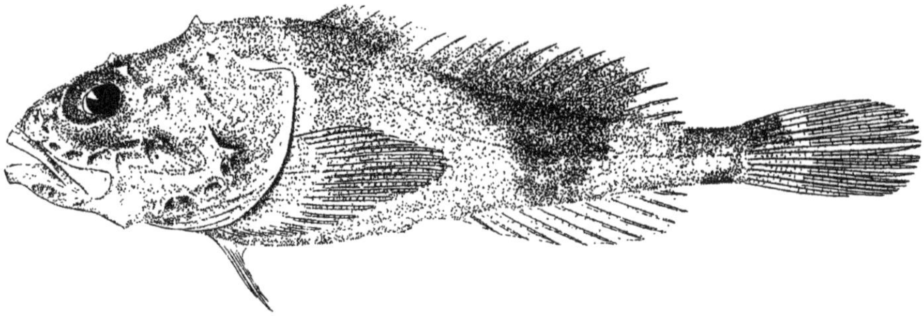

FIGURE 236.—Arctic sculpin (*Cottunculus microps*), continental slope off southern New England. From Goode and Bean. Drawing by H. L. Todd.

has been taken at numerous localities on the continental shelf and slope to abreast of New Jersey in depths of 122 to 487 fathoms. Only two of the earlier published records fall within the geographic limits covered by this report, one in the extreme southeast corner of the basin of the Gulf (latitude 42°23′, longitude 66°23′) in 141 fathoms, the other in the eastern channel between Browns and Georges Banks (latitude 42°15′, longitude 65°48′) in 122 fathoms. But we trawled one about 2 inches long, on the northern slope of Georges Bank, in 120 fathoms of water, on July 24, 1931, which (with earlier captures) shows that it is to be expected anywhere in the deep basin of our Gulf, at depths greater than 100 fathoms.[6] Nothing is known of its habits.

[6] Goode and Bean, Smithsonian Contrib. Knowl; vol. 30, 1895, p. 270, list the earlier American records.

Sea raven *Hemitripterus americanus* (Gmelin) 1789

RED SCULPIN; SEA SCULPIN; RAVEN; KING O'NORWAY

Jordan and Evermann, 1896–1900, p. 2023.

Description.—No one would be likely to confuse a sea raven with any other sculpin, for it is a "most remarkable looking fish," as Jordan and Evermann remarked.[7] It is stouter bodied than our other common sculpins, about three and three-fourths times as long as it is deep (counting caudal fin), with a very large head. Both the jaws of its wide mouth are armed with several rows of sharp teeth that are noticeably longer and stouter

[7] Bull. 47, U. S. Nat. Mus., Pt. 2, 1898, p. 2023.

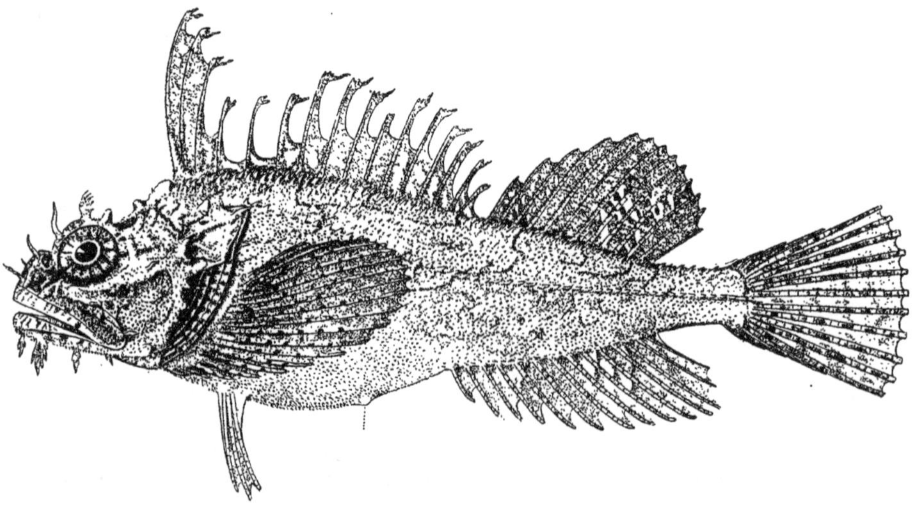

FIGURE 237.—Sea raven (*Hemitripterus americanus*), Halifax, Nova Scotia. From Goode. Drawing by H. L. Todd.

than the teeth of either the long-horned sculpin or of the short-horned sculpin. Its most distinctive features, however, which identify it at a glance, are the fleshy tabs, simple and branched, on its head; the curiously ragged outline of its first dorsal fin; and the prickly texture of its skin. There is a series of 4 to 8 of these tabs along each side of the lower jaw, three pairs on the top of the snout, and others, variable in number and size, above and in front of the eyes and along the upper jaw. There is also a short but high keel on the top of the snout with a deep hollow behind it, another high ridge above each eye, and a lower one below the eye. These ridges, with about 12 rounded knobs on the crown and several low bosses, and 2 short spines on each cheek, give the head a peculiarly bony appearance.

The first two or three spines of the first dorsal fin are the longest, and the fourth and fifth spines are shorter than those farther back, giving the fin an outline quite unlike that of any other sculpin. And the fin membrane is deeply emarginate between every two spines from the third spine backward, but expanded at the tip of each spine as an irregular flap of skin. The margin of the anal fin is similarly, but less deeply scalloped between the rays. Furthermore, the first dorsal fin originates further forward than in any other Gulf of Maine sculpin, i. e., well in front of the gill opening, and it is much longer (16 spines) than the second dorsal (1 spine and 12 rays), whereas in our other sculpins the second dorsal is longer than the first dorsal. The pectorals are fanlike, and the caudal brush-shaped, much as in other sculpins. The ventral fins are fleshy, each with 1 spine and 3 soft rays, with the first 2 rays so close that there seems to be only 1 soft ray.

The entire skin of the sea raven is prickly, belly as well as back, with the prickles largest on the back and along the lateral line; smallest, but still obvious to the touch, on the lower parts of the sides and on the belly. In all other Gulf of Maine sculpins the belly is smooth.

Color.—The sea raven varies in color from blood red to reddish purple, chocolate, or to yellowish brown, but it is invariably paler below than above, and it usually has a yellow belly. Many are plain colored. For instance, one 18 inches long, which we caught off Mount Desert recently, was uniform red chocolate on back and sides. But others are variously mottled with a paler or darker cast of the general ground tint, or even with white. The fins are variously barred with light and dark, and the pectorals and anal often are yellow-rayed.

Size.—One of the largest on record, of 25 inches and 5 pounds, is mentioned by Storer. But this specimen seems to have been in poor condition for we have caught one 22½ inches long that weighed 7 pounds. Warfel and Merriman[8] mention one about 19½ inches long that weighed 5.8 pounds inflated with water and 5.3 pounds when emptied. And many sea ravens are 18 to 20 inches long.

Habits.—The sea raven alone, among Gulf of Maine sculpins has the power of inflating its belly with water, like a bladder. If released in this condition it drifts helplessly, feebly waving its tail to and fro, and we cannot say whether it can empty itself again at will like a puffer (which, however, inflates with either air or with water, p. 526) or whether it must await the gradual escape of the water it has swallowed. Another way in which the raven differs from our other sculpins is that it can bite sharply, having larger teeth.

The raven is quite as voracious as its relatives; it takes any bait and is said to eat whatever invertebrates it finds on the bottom, such as mollusks (both bivalve and univalve), various crustaceans, sea urchins, and worms. Sea ravens also eat fish. Vinal Edwards found herring, launce, sculpins, tautog, silver hake, and both sculpin and sea-raven eggs in sea ravens taken at Woods Hole.

Our own experience, confirmed by our various inquiries, is that ravens are to be caught only on rocky ground (which is their chief haunt from Massachusetts Bay northward), pebbles, hard sand, or clay (which they frequent off Cape Cod and on the offshore Banks), never on soft sticky mud. There is no definite upper limit to their vertical wanderings other than the surface. But they are seldom caught within the smaller estuaries, perhaps never on the tidal flats at any time of year; at least we have never seen them in such situations in Massachusetts Bay, though they are not uncommon there about the off-lying ledges. The majority of them live deeper than a couple of fathoms at all times. On the other hand their usual range extends down only to about 50

[8] Copeia, 1944, p. 204.

fathoms, although one has been taken as deep as 105 fathoms.[9]

The geographic and vertical ranges of the sea raven suggest that the upper limit to its preferred temperature is about 58°–60°. At the other extreme their wide dispersal over the Magdalen Shallows in the southern side of the Gulf of St. Lawrence shows that they can winter in temperature close to the freezing point of salt water, unless they descend then into considerably deeper water, a possible shift in depth on their part of which there is no direct evidence.

Presumably the sea raven breeds throughout its geographic range. Off southern New England the eggs are deposited from early October (earliest date, October 2) on until late December; probably in autumn and early winter in the more northern part of its range as well.[10] Warfel and Merriman, who made a special study of the breeding of the sea raven, have made the very interesting discovery that it deposits its eggs chiefly at the bases of the finger-like branches of the finger sponge (*Chalina*); less often on the smaller sponge *Halichondria*, where they stick together in clusters and to the sponge. Since the eggs average only about 242 per cluster (minimum 141, maximum 478, among many clusters counted), whereas adult females contain something like 15,000 maturing eggs on the average, and occasionally as many as 40,000, it appears that a female does not lay all her eggs at one time, but deposits many clusters during each spawning season.

The eggs are large, averaging 3.9 to 4 mm. in diameter, with tough egg membrane, yellow when first spawned, but soon changing to an amber hue; so heavy that they sink; and very sticky and resistant to injury.[11] Eggs brought in to the laboratory by Warfel and Merriman hatched a few at a time, and some of those of a cluster collected on January 23 and left thereafter in a bottle fastened to a buoy (i. e., in the normal winter temperature of Long Island Sound) did not hatch until March 12.

The sizes of the few young sea ravens that have been taken in the Gulf of Maine suggest that they reach a length of 2 to 4 inches by the middle of their first summer, when 6 to 8 months old, and about 6 inches by the following April, at an age of 1½ years. Their subsequent rate of growth has not been followed.

General range.—Atlantic Coast of North America, southward to Chesapeake Bay; north to Anticosti in the northern side of the Gulf of St. Lawrence, to the Strait of Belle Isle on the Newfoundland side [12] and to the Grand Banks.[13]

Occurrence in the Gulf of Maine.—Sea ravens are to be caught all around the coastal belt of the Gulf, from a fathom or two down to about 50 fathoms, including the passages among the islands that fringe the coasts of Maine and of Nova Scotia, as well as the larger estuaries such as St. Mary, Passamaquoddy, Machias, Penobscot (it runs up the latter to the head at Bucksport), and Casco Bays; also in the deeper harbors, for example Boston, Salem, Eastport, and St. Johns. Fishermen also report them on Cashes Ledge, while the otter trawlers and long liners pick them up in small numbers over the shoaler parts of Georges Bank; likewise on Browns. But they are not known to occur on the soft mud bottoms of the deep troughs and basin of the Gulf.

Presumably the sea raven breeds in the Gulf wherever it occurs, young fry having been taken from the Bay of Fundy to Cape Cod.

Off the southern shores of New England sea ravens work inshore in autumn and out again into slightly deeper water in spring, but no seasonal movement of this sort (which means merely that shoal water is too warm for their comfort in summer) has been reported for them in the cooler waters of the Gulf of Maine.

Although they are distributed so generally in the Gulf, sea ravens are not so numerous as the shorthorn sculpins and longhorn sculpins; this is as true in the Bay of Fundy as it is in Massachusetts Bay, where one expects to catch a few ravens around any of the fishing ledges, but where it would be unusual for one person to land any considerable number in a day. Similarly, the schedules of the catches made by certain otter

[9] A specimen about 14 inches (360 mm.) long was trawled by the U. S. Fish and Wildlife vessel *Delaware* August 24, 1951, in 105 fathoms, latitude 42°06′ N., longitude 67°50′ W.

[10] Warfel and Merriman (Copeia, 1944, p. 202) probably were correct in their suggestion that a 20-inch female, containing ova of 2 sizes, caught by us off Boothbay, Maine, in April, was simply one that had failed to spawn at the proper time during the preceding winter.

[11] Described by Bean (Bull. 60, New York State Mus., Zool. 9, 1903, p. 647). The buoyant eggs referred to this species by Agassiz and Whitman (Mem. Mus. Comp. Zool., vol. 14, No. 1, 1885, p. 10) belonged to some other fish.

[12] Jeffers (Contrib. Canadian Biol. N. Ser., vol. 7, No. 16, (Ser. A, No. 13,) 1932, p. 208).

[13] Bean (Proc. U. S. Nat. Mus., vol. 3, 1881, p. 86) lists a specimen from the Grand Banks.

trawlers in 1913 and subsequently, added to our trawling experiences, show that sea ravens are much less numerous on Georges Bank than the longhorn sculpin; it is seldom that as many as a dozen are taken in a haul of the otter trawl. Thus the dragger *Eugene H* took only 46 sea ravens in 38 hawls on the southwestern part of Georges, at 26–55 fathoms in late June 1951, and none in her deeper hauls.

To the eastward and northward sea ravens are described as common all along outer Nova Scotia to Canso; they have been reported on Sable Island Bank and on Banquereau Bank in depths of about 20–30 fathoms; and they are taken here and there on the Gulf of St. Lawrence coast of Cape Breton. They are also reported from Anticosti and in the Strait of Belle Isle (footnote 12, p. 456); but there cannot be many of them in the southern side of the Gulf, for they are not mentioned in the published lists of Fishes for Prince Edward Island, or around the Magdalens. And there is only one report of a sea raven on the Grand Banks (see footnote 13, p. 456); and one for the Atlantic coast of Newfoundland, i. e., from Trinity Bay.[14]

To the westward and southward the sea raven is reported as rather common as far as New York and New Jersey; and it has been reported from Chesapeake Bay.[15]

Commercial importance.—Although the sea raven is said to be a good table fish there is no more market for it than for other sculpins in New England or Canada. But it is generally considered excellent bait for lobster pots, hence shore fishermen are likely to save what ravens they catch for this purpose.

ALLIGATORFISHES. FAMILY AGONIDAE [16]

These curious little fishes are related to the sculpins anatomically, though their general appearance gives no hint of the fact. Their most striking external feature is that the body is armed with several rows of overlapping plates. The only Gulf of Maine species somewhat suggests a pipefish in this and in its slender form, but there is no danger of confusing one with a pipefish, for its mouth is of the ordinary form; and it has ventral fins. Some agonids have a spiny dorsal fin which others lack, while the ventral fins are situated far forward (only a little rearward of the pectorals) in all of them. Twenty-six of the many species included in the family are known from the western coast of North America from Bering Sea to southern California; two are known from the eastern coast of North America.

One of the eastern American species (*Leptagonus decagonus*, Bloch and Schneider 1801), with two dorsal fins, is Arctic, ranging southward only to northern Nova Scotia;[17] the other, with only one dorsal fin (*Aspidophoroides monopterygius*) is a regular member of the Gulf of Maine fish fauna (p. 457).

[14] Rept. Newfoundland Fish. Res. Comm., vol. 1, No. 4, 1932, p. 108.

[15] Jordan, Evermann and Clark's (Rept. U. S. Comm. Fish. 1928 part 2, 1930, p. 393) reference of it to the Tortugas seems to have been an error.

[16] Some recent authors separate this group into two families, Agonidae or Sea Poachers for those with two dorsal fins, Aspidophoroididae or Alligatorfishes for those with only one dorsal fin. But it seems preferable, for the purposes of the present volume, to follow the older custom of uniting them in the one family Agonidae.

[17] A young specimen has been reported from Banquereau Bank (Rept. Newfoundland Fish. Res. Comm., vol. 2, No. 1, 1933, p. 127) as *Agonus decagonus*.

Alligatorfish *Aspidophoroides monopterygius* (Bloch) 1786

SEA POACHER

Jordan and Evermann, 1896–1900, p. 2091.

Description.—The readiest field marks for the identification of this curious little fish (so odd in appearance that it is not likely to be mistaken for any other) are that its entire head and body are clad with bony plates; that it has only one dorsal fin (the soft rayed); and that it is very slender (about 12 to 13 times as long as it is deep, not counting the caudal fin), rather broader than deep, and tapering rearward from the head to a very slender caudal peduncle. The plates are smooth, arranged in longitudinal rows. There is a double row along the back running from close behind the head to close behind the dorsal fin where it unites into a single mid-dorsal row (altogether 45 to 50 double and single plates along the back); three rows on the forward part of each side, but only two rows along the rear part; and two rows along the lower surface to just behind the anal fin, where they unite in one row. Thus the trunk is octagonal in front of the unpaired fins, but hexagonal behind

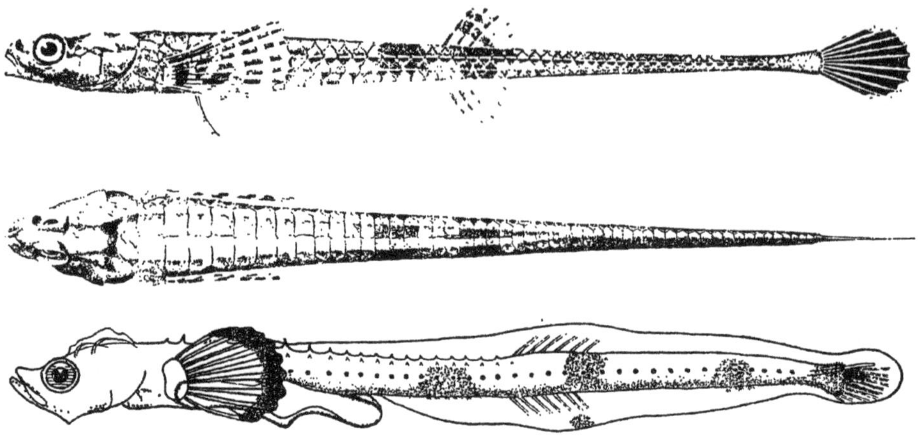

FIGURE 238.—Alligator fish (*Aspidophoroides monopterygius*). Top and center, adult, side and dorsal views, Halifax, Nova Scotia; from Jordan and Evermann, drawing by H. L. Todd. Below, larva, 11 mm. long, Gloucester, Mass.

them. There are likewise two large plates and several small ones in front of each pectoral fin. The eyes are very large, with prominent ridges above them, and there are two sharp recurved spines on the top of the nose. The mouth is small with minute teeth. The dorsal and anal fins (each of five or six rays) are fanshaped, one over the other, and are situated about midway of the trunk. The caudal fin is small and rounded, the pectorals are larger than the anal, dorsal, or caudal fins, and the ventral fins are reduced to one spine and two rays each.

Color.—The many specimens we have seen have been dark brown above, lighter brown below, with two darker cross bands between the pectoral fins and the dorsal fin; one cross band under the dorsal, and two or three cross bands between the dorsal and caudal fins. The dorsal and pectoral fins are more or less barred; the caudal is dusky.

Size.—Five to seven inches long when full grown.

Habits.—Nothing whatever is known of the life of the alligatorfish except that it is a bottom fish and that it has been repeatedly found in the stomachs of cod, haddock, and halibut although it is not "much thicker or softer than an iron spike." [18] The *Grampus* and the *Albatross II* have trawled it both on pebbly bottom, on sand and broken shells, and on soft mud. So far as known adults never stray into water shoaler than 10 to 15 fathoms, and the deepest record for it, with which we are acquainted, is from 104 fathoms.[19] Its range shows that it is a cold water fish. Its upper temperature limit is about 50°–52°; its lower limit close to the freezing point of salt water. Its breeding habits are unknown. Probably its eggs sink like those of sculpins.

The presence of its larvae in Passamaquoddy Bay, off Boothbay, and near Seal Island, Nova Scotia, from April to June, points to late autumn and early winter as the spawning season. It does not take to the bottom until of considerable size, for we have taken young ones as long as 29 mm. in our towings.[20]

General range.—From west Greenland and the east coast of Labrador southward to Cape Cod, and to northern New Jersey as a stray.

Occurrence in the Gulf of Maine.—Being of no interest to fishermen, and living too deep to strand on the beach, this fish is seldom reported. It has been taken in the Bay of Fundy in 15 to 100 fathoms; in the inner part of Passamaquoddy Bay; near Eastport; off Mount Desert in 60 fathoms; off Monhegan; near Portland; in 30 fathoms off Casco Bay; on Cashes Ledge; in Ipswich Bay; off Gloucester, Nahant, and Boston in Massachusetts Bay; off Provincetown; off Chatham; and in the South Channel to the eastward of Cape Cod. Evidently it may be expected anywhere in the Gulf in depths of 10 to 100 fathoms, and perhaps deeper.

Goode and Bean [21] described it as "abundant" in the deeper parts of Massachusetts Bay, but our

[18] Goode, Fish. Ind. U. S., Sect. 1, 1884, p. 258.

[19] Southeastern slope of Browns Bank, latitude 42°20′ N; longitude 65°08′ W; Goode and Bean Smithsonian, Contrib. Knowl., vol. 30, 1895, p. 284.

[20] Bigelow, Bull. Mus. Comp. Zool., vol. 61, 1917, p. 272.

[21] Bull. Essex Inst., vol. 11, 1879, p. 13.

experience on the *Grampus* suggests "not uncommon" as a better description, for our largest catches were 8 specimens from 32 fathoms in Ipswich Bay; 6 specimens from 27 fathoms in the inner part of Massachusetts Bay; and 14 specimens off Chatham, in May 1930, from 28 fathoms.

Other recent catches in various parts of the Gulf have been of 1 or 2 fish each. And the fact that we found it at only 4 out of our 10 trawling stations of 1912 (all in the western part of the Gulf) is in line with Huntsman's statement that it is found only occasionally in the Bay of Fundy.

It is perhaps more plentiful along the Nova Scotian shelf eastward and northward from Cape Sable.[22] It is numerous enough in the southern part of the Gulf of St. Lawrence for Huntsman [23] to describe it as "characteristic" of the ice-cold Banks water there; it has been reported in the estuary of the St. Lawrence River near Trois Pistoles;[24] and at several localities along the west coast of Newfoundland. It is so widespread over the eastern half of the Grand Banks, along eastern Newfoundland, and off southeastern Labrador that it is listed at 14 stations there from the experimental trawlings of the Newfoundland Fisheries Research Commission,[25] and is reported from Hamlton Inlet.[26]

The only records of it to the westward of Cape Cod are of the head of one that was dredged off Watch Hill, N. J., in 1874;[27] and of one that was taken off Sandy Hook, N. J., in 1864.[28]

THE LUMPFISHES. FAMILY CYCLOPTERIDAE

The lumps are characterized among Gulf of Maine fishes by their short, thick, high-arched bodies; by having a bony sucking disc on the chest with the very much reduced ventral fins at its center; and by the fact that the skin is set with tubercles.

KEY TO GULF OF MAINE LUMPFISHES

1. The body is roughly triangular in end view; the large tubercles on each side are in widely separated rows; the lower end of each gill opening is below the level of the upper edge of the base of the corresponding pectoral fin; the head (measured to the upper edge of the bases of the pectoral fins) occupies only about one-fourth of the length of the trunk; the first dorsal fin is entirely concealed within the skin after the fish is 2–3 inches long...... Common Lump Fish, p. 459
2. The body is nearly round in end view; the large tubercles on each side are close set, in irregular pattern; the lower end of each gill opening is considerably above the level of the upper edge of the base of the corresponding pectoral fin; the head (measured to the upper edge of the bases of the pectoral fins) occupies about one-third of the length of the trunk; the first dorsal fin (though more or less fleshy) continues exposed throughout life...... Spiny Lumpfish, p. 463

Lumpfish *Cyclopterus lumpus* Linnaeus 1758

LUMP; LUMP SUCKER

Jordan and Evermann, 1896–1900, p. 2096.

Description.—The lumpfish is about twice as long (counting its caudal fin) as it is deep, with a short head, and the dorsal profile of its trunk is much more arched than the ventral. There are 7 longitudinal ridges on its body. One of these ridges runs along the back as a cartilaginous flap that incloses the first dorsal fin in adults and that continues rearward as 2 ridges from the first dorsal fin to the second dorsal fin. There also are 3 ridges along each side, one of them over the eye, one close above the level of the pectoral fin, and one marking the line of transition between side and belly. Each of these ridges is marked by a line of large pointed tubercles, and the entire skin between the ridges is thickly studded with small knobs. The presence of these ridges makes the trunk of the lumpfish roughly triangular in end view, with flat belly (except when swollen by milt or roe) and sharp back, but the caudal peduncle is rounded. The profile of the head is characteristic, being concave above, convex below, with mouth at the tip of the snout. The teeth are small and the eyes and gill openings of moderate size.

The first dorsal fin (visible only on very small specimens) is of 6 to 8 spines. The second dorsal fin and the anal fin below it are alike in outline, both of them of 9 to 11 rays. The caudal is broad

[22] Vladykov and MacKenzie (Proc. Nova Scotian Inst. Sci., vol. 19, 1935, p. 97) describe it as "very common" there.
[23] Trans. Roy. Soc. Canada Ser. 3, vol. 5, sect. 4, 1918, p. 63.
[24] Vladykov and Tremblay, Natural. Canad., vol. 62, 1935, p. 80.
[25] Annual Reports, Newfoundland Fisheries Research Commission, vol. 1, No. 4, 1932, p. 108; vol. 2, No. 1, 1933, p. 126; vol. 2, No. 2, 1934, p. 115.
[26] Kendall, Proc. Portland Soc. Nat. Hist., vol. 2, No. 13, 1909, p. 214.
[27] Goode and Bean, Smithsonian Contrib. Knowl., vol. 30, 1895, p. 284.
[28] Abbot, Geology of New Jersey, 1868, p. 816.

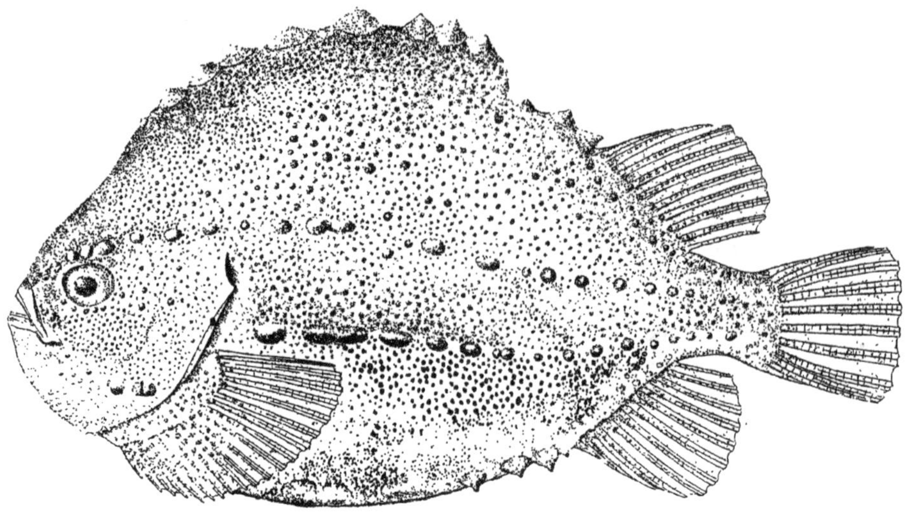

FIGURE 239.—Lumpfish (*Cyclopterus lumpus*), Eastport, Maine. From Goode. Drawing by H. L. Todd.

based and square tipped or slightly convex; the pectorals are large, rounded, and so broad based that they nearly meet on the throat, and they are larger on males than on females. The ventral fins are not visible as such, being altered into 6 pairs of fleshy knobs, surrounded by a roughly circular flap of skin. The entire disk, so formed, (sucking disk) is about as wide as the width of the head, and is situated close behind the throat.

Color.—Descriptions of this fish credit it with a great variety of tints, which we can corroborate. On adults the ground tint may be bluish gray, olive, brownish or yellow green, chocolate or kelp brown, or slaty blue, the belly usually being of a paler or more yellowish cast of the same hue, but sometimes whitish. On some specimens the back and sides are marked with dark blotches and more or less dotted with black. Others, however, are plain colored or nearly so, except that the tubercles are usually dark tipped. Young lumpfish (and it is with such that we are most familiar) often match their surroundings very closely in color, usually being mottled olive green and ochre yellow with silvery dots and stripings. Males, when mature, are more vividly colored than females, and their bellies turn red (brightest near the sucking disk) during the breeding season.

Size.—The longest lumpfish so far recorded from the American coast measured 23 inches, and weighed 13¼ pounds; the heaviest weighed 20 pounds but measured only 21½ inches (both from Orient, N. Y.), and the proportion of weight to length varies similarly in smaller fish. Few are longer than 14 to 16 inches, or heavier than 3 to 6 pounds while the largest we ourselves have seen was about 15 inches long.[29] Females average larger than males. Fulton, for example,[30] writes that 39 females taken in the Bay of Nigg (Scotland) averaged about 16 inches and 6 pounds, 30 males only 11 inches and a little less than 2 pounds.

Habits.—Although the lumpfish is ungainly, it can swim more rapidly for a short distance by vigorous tail strokes than its shape might suggest, and its young pelagic fry are very active. The adult lump is primarily a bottom fish, but is also made semipelagic by its habit of hiding in floating masses of rockweed. In European seas it ranges from tide mark down to 150–200 fathoms, but we have never heard of one taken in more than a few fathoms in the Gulf of Maine. Perhaps it is the nature of the bottom that restricts them there for the soft sticky mud that floors the deeper basins, at least in the western side,[31] can hardly be a favorable environment for them. Large lumpfish are often found hiding among rockweed or holding fast by the sucker to stones or other

[29] Smitt (Scandinavian Fishes, vol. 1, 1892, p. 294) gives 24 inches as the maximum for Scandinavian and European waters generally, apparently not accepting the enormous size (to 48 inches) credited to it by Möbius and Heincke (Vierter Bericht, Komm. wiss. Untersuch. deutschen Meere, Kiel, 1883, p. 226).

[30] Twenty-fourth Ann. Report, Fisheries Board Scotland, (1905) 1906, Pt. 3, p. 171.

[31] The eastern trough of the Gulf has a harder bottom.

objects. About Massachusetts Bay, lobster pots are favorite resorts for them when set on stony bottom. For instance, W. F. Clapp tells us that one pot in every 8 or 10 will yield a lumpfish on the broken ground off the entrance to Duxbury Harbor, and they often hold to the lower sides of lobster cars, probably for shade. Occasionally one is found clinging to one of the poles of a trap or weir, though this is a much less common event in the Gulf of Maine than it is in Scottish waters, where they are frequently caught in salmon nets set along shore. Welsh notes one entangled in a gill net set off Great Boars Head in April 1913; they have (rarely) been found clinging to floating logs or inside a floating box or barrel; sometimes they strand on the beach; and there is at least one record of a lumpfish clinging to a mackerel.

So far as known the only regular migrations carried out by the lumpfish are the involuntary drifts of its young fry at the surface, and a general movement of the adults into shoal water at spawning time followed by an offshore movement afterward.

The young fry swim at the surface, and we have taken them so often in our tow nets that we have learned to expect them wherever there are floating masses of rockweed (a refuge in which all but the smallest regularly hide, or to the fronds of which they cling).

Most of the lumpfish taken in tow nets or dipped up are less than 2 inches long, but large adults are sometimes seen at the surface; more often, perhaps, in the Bay of Fundy than elsewhere in the Gulf, their presence at the surface being determined less by the age of the individual fish than by the presence or absence of floating seaweed. However, most of the young lumpfish have left the surface by winter; indeed very few have been taken at any depth in the Gulf of Maine during the cold months,[32] but we picked up one on the surface off Lurcher Shoal on April 12, 1920, and another off Yarmouth, Nova Scotia, on January 4, 1921.

Food.—We have no first-hand information to offer as to the diet of the lumpfish. In British waters this has been found to consist chiefly of isopods, amphipods, and other small crustaceans, with various other invertebrates, including worms and soft-bodied mollusks. And its diet is much the same in the Gulf of Maine for Cox and Anderson[33] report euphausiid shrimps (*Meganyctiphanes*), fragments of jellyfish (*Aurelia*), amphipod crustaceans (*Hyperia*), caprellid crustaceans, with the remains of small fish in the stomachs of lumps from Passamaquoddy Bay. And large numbers of young clupeids have occasionally been found in their stomachs. This, too, is one of the few fish that regularly feed on ctenophores and on Medusae. Thus 25 specimens examined at Woods Hole by Vinal Edwards contained nothing but ctenophores. But lumps, like most other fishes, cease feeding during the spawning season. The lumpfish, in its turn, is said to be a favorite food of seals. Certainly it is so weak a swimmer that it would fall an easy prey to them.

In Scottish waters, where many observations have been made on the life of the lumpfish[34] spawning (and the corresponding inshore migration) takes place from February until near the end of May. And the evidence afforded by our tow nettings suggests an equally protracted spawning season in the Gulf of Maine, for while we have taken larvae 27 mm. long as early as May 10, we have taken newly hatched larvae (6 to 7 mm.) as late as June 19 in the inner parts of the Gulf and as late as July 23 on the northeast part of Georges Bank, with one only 10.5 mm. long on August 22 off Seguin Island. In the Gulf of St. Lawrence, however, where vernal warming is later than in the Gulf of Maine, lumpfish probably do not commence spawning until the middle of April, for Cox and Anderson found no larvae there until late in June, with their observations pointing to late May as the height of the breeding season and to mid-June as about its termination.[35] Presumably its period of incubation is about as long with us as it is in European waters of like temperature, i. e., 6 weeks to 2 months.

On the other side of the Atlantic the lumpfish spawns in very shallow water, chiefly close to low-tide mark. But the fact that the egg masses (more or less familiar objects on European shores)

[32] Cox and Anderson (Contrib. Canadian Biol., N. Ser., vol. 1, 1922, p. 5) state that the Canadian Research steamer *Prince* has taken only two lumpfish in the Bay of Fundy in winter, both of them small.

[33] Contrib. Canadian Biol. N. Ser., vol. 1, 1922, p. 9.

[34] McIntosh, 14th Ann. Rept., Fishery Board Scotland, (1895) 1896, Pt. 3, pp. 173–178, and Fulton, 24th Ann. Rept., Fishery Board Scotland, (1905) 1906, Pt. 3, pp. 169–178.

[35] The lumpfish spawns from late May through June on the coast of Greenland; in April and May in the Baltic; and early in the spring in Norwegian waters.

seem never to have been reported along the coast of the Gulf of Maine, although the local presence of larvae is proof that lumpfish breed all around its coast line, suggests that the eggs are deposited at least a fathom or two down with us. And our capture of recently hatched larvae over Georges Bank is evidence that the latter also serves as a spawning ground in 15 to 25 fathoms or deeper. But the lower depth limit to spawning is yet to be determined.

Large females of 18 inches may produce up to 136,000 eggs which sink and stick together in large spongy masses through which the water circulates freely. Off the coasts of northern Europe these egg masses are often found adhering to rocks or other objects, or in crannies near low-water mark, and watch should be kept for them along the rocky coast line of the Gulf of Maine.

The male lumpfish guards the eggs until they hatch; his courage and devotion has often been described.[36] Throughout the period of guardianship, which he performs fasting, he fans the egg mass constantly, keeping it free of silt and bathed in flowing water, never leaving it except to drive off some intruder. But his vigil ends as soon as the eggs are hatched, leaving him thin and exhausted. The females take no part in guarding the eggs, but are said to move out into deeper water once they have finished spawning.

The eggs are 2.2 to 2.6 mm. in diameter, pink when first laid but soon change to pale green or yellow, and deepen in tint as development progresses. The larvae are about 4 to 7.4 mm. long at hatching, shaped like a tadpole with a large head and slender tail, swimming actively, and soon able to cling to any bit of weed. At 34 mm. the tubercles begin to appear, and the fry then show most of the characters of the adult, except for the large first dorsal fin and slender form.

Lumpfish larvae and fry of all sizes are to be taken throughout the summer; the smaller ones undoubtedly are from that season's hatch, but the larger ones may be either those hatched earliest that spring, or late in the preceding summer, for the varying stage of development reached by different individuals at various sizes proves that the rate of growth varies widely. Thus Cox and Anderson describe one Cape Breton specimen that was only 33 mm. long in July, but that was so mature in outline and in its dermal armature that it must have been at least a year old, whereas they found that in the Bay of Fundy the fry of the year grow to 40 or 50 mm. by December with the yearlings averaging about 58 mm. in July and August. As they remark, the rate of growth is apparently about the same in the Bay of Fundy (probably in the Gulf of Maine as a whole) as it is in Scottish waters, while Gulf of St. Lawrence lumpfish correspond to those about Helgoland, in their slower growth.

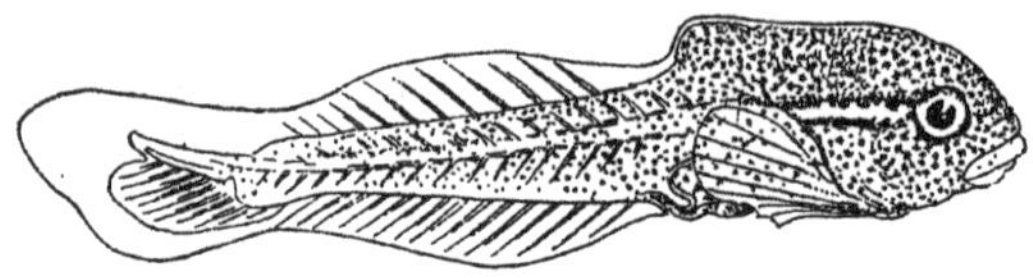

FIGURE 240.—Larva, 4–5 mm. After Garman.

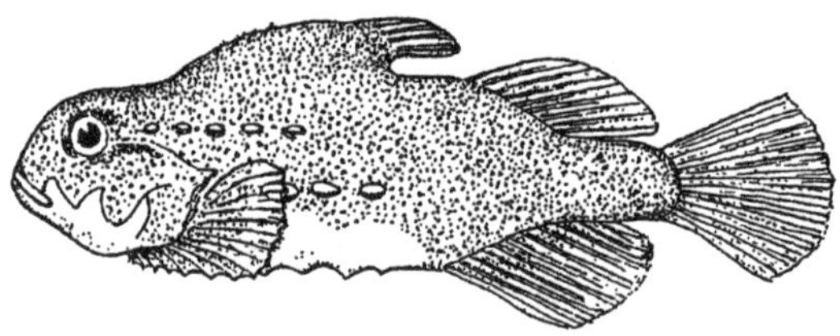

FIGURE 241.—Fry, 34 mm. After Garman.

LUMPFISH (*Cyclopterus lumpus*)

Presumably all Gulf of Maine lumpfish upward of 2½ inches long are in their second year. According to Cox and Anderson lumpfish 3¾ to 4¼ inches long are in their third year, those of about 10 inches in their fifth year.[37] But lumpfish grow much faster if fed to capacity than on the rations ordinarily available to them; one kept in the aquarium at St. Andrews increased in length from about 3.8 inches to about 12 inches in a little more than 12 months, which is as much as is to expected in 2 or 3 years under natural conditions.[38] Probably maturity is attained in the third year.

General range.—Both sides of the North Atlantic; White Sea, northern Norway and Iceland to the Bay of Biscay and occasionally to Portugal in the east (including the Baltic); northward in the west to the Gulf of St. Lawrence, Newfoundland, outer coast of Labrador, Hudson Bay, Davis Strait, and Disko (latitude about 70° N.)

[36] Fulton (24th Ann. Report, Fishery Board Scotland, (1905) 1906, Pt. 3, p. 169) gives an interesting eyewitness account of spawning and of the guardianship by the male parent over the eggs.

[37] As estimated from the structure of their vertebrae.

[38] According to McKenzie (Proc. Nova Scotian Inst. Sci., vol. 20, 1939, p. 17) this fish was kept well-fed on chopped clams and herring, but perhaps not to capacity for its appetite seemed insatiable.

in west Greenland; southward to New Jersey, and to Chesapeake Bay as a stray.

Occurrence in the Gulf of Maine.—The lumpfish is common along the outer coast of Nova Scotia and is to be found all around the shores of the Gulf of Maine. It has been reported at Yarmouth and in St. Mary Bay on the Nova Scotian side, and is abundant in all stages at various localities in the Bay of Fundy. There are many records for it along the Maine coast, including Eastport, Penobscot Bay, vicinity of Boothbay, the offing of Seguin Island, and Casco Bay; also in Massachusetts waters where it has been reported repeatedly, as at Nahant, Swampscott, Cohasset, Plymouth, Truro, along Cape Cod, and at Monomoy while we once picked up a lumpfish in the deep bowl between Jeffreys Ledge and the coast, probably as the trawl neared the surface.[39] It even enters river mouths, but it is never found where the water is appreciably brackish, so far as we can learn. According to fishermen large lumpfish are seldom seen on the offshore banks.

Importance.—The lumpfish is never eaten in the United States, but it sometimes finds its way to market as a curiosity. At one time a few were consumed locally in parts of the British Isles, and may still be.

Spiny lumpfish *Eumicrotremus spinosus* (Müller) 1777

Jordan and Evermann, 1896–1900, p. 2098.

Description.—The spiny lumpfish is easily distinguishable from its commoner relative by the fact that its skin tubercles are not only much larger relatively, and studded from the base to the tip with rough prickles, but are irregularly and closely scattered over the body and head. Furthermore, its gill openings are much shorter, while its body is not so high-arched, and is nearly round in cross section, instead of triangular. Also, its first dorsal (though fleshy in some of them) retains its fin-like appearance through life, instead of becoming entirely concealed by the skin.

Color.—Described as olivaceous to brownish.

General Range.—Arctic and northern parts of the Atlantic Ocean, south to the Gulf of Maine as a stray.

Occurrence in the Gulf of Maine.—Stray specimens of this northern fish have been reported from Eastport, Maine; from off Cape Ann; and from Salem, in the north side of Massachusetts Bay. We have also seen three small specimens, 1–1¼ inches long that were collected about 15 miles southeast of Cape Ann in 23 and 29 fathoms, by the U. S. Fish Commission in 1878 (now in the U. S. National Museum.)[40]

[39] Bigelow and Schroeder, Biol. Bull., vol. 76, 1939, p. 309.

[40] Myers and Bohlke (Stanford Ichth. Bull., vol. 3, No. 4, 1950, p. 199) have described a new species of spiny lumpfish (*E. terrae-novae*) from Newfoundland. But the specimens from off Salem and from off Eastport pictured by Goode and Bean (Smithsonian Contrib. Knowl., vol. 31, 1895, pl. 70, fig. 250) and by Garman (Mem. Mus. Comp. Zool., vol. 14, 1892, p. 71, figs. 1–3) were typical *spinosus*, as are the three small ones, from about 15 miles southeast of Cape Ann, that are mentioned above.

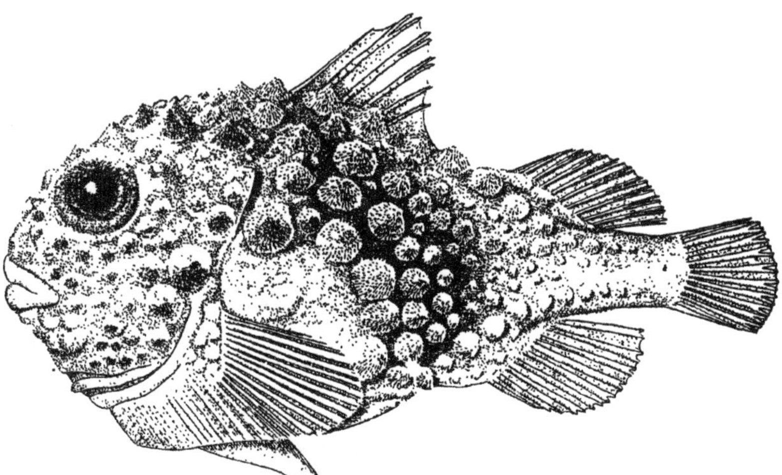

FIGURE 242.—Spiny lumpfish (*Eumicrotremus spinosus*), Massachusetts Bay specimen. From Goode and Bean. Drawing by H. L. Todd.

THE SEA SNAILS. FAMILY LIPARIDAE

The sea snails are tadpole-shaped, soft-bodied little fishes; and like the lumpfish (p. 459) most of them have a sucking disk on the chest, supported by the vestigial rays of the ventral fins.[41] But the skin is smooth, and without tubercles, and the spiny and soft parts of the dorsal fin are continuous as a single fin. The more than 115 species that are known are widely distributed in Arctic, North Temperate, and Antarctic Seas, and from the intertidal zone down to 2,000 fathoms or so.[42] The Gulf of Maine harbors two species. A third (*Careproctus ranulus* Goode and Beane, 1879) is known only from the vicinity of Halifax, Nova Scotia; from Middleground off eastern Nova Scotia,[43] from the Grand Banks, and off southeastern Newfoundland; perhaps from the estuary of the St. Lawrence River, also.[44]

We include it in the following Key, on the chance that it may be encountered in the deeper parts of our Gulf, sooner or later.

KEY TO GULF OF MAINE SEA SNAILS

1. The spiny (front) and soft (rear) portions of the dorsal fin are separated by a notch....... Sea snail, p. 464
   There is no notch between the spiny portion and the soft portion of the dorsal fin.................... 2
2. The anal fin has only about as many rays (26–29) as the pectoral fin (28–33); there are two separate nostrils; the body is opaque, variously striped or spotted.................. Striped Sea Snail, p. 466
   The anal fin has many more rays (at least 48) than the pectoral (27–28); there is only a single nostril; the body is translucent in life and colorless.....................*Careproctus ranulus*[45]

Sea snail *Neoliparis atlanticus* Jordan and Evermann, 1898

Jordan and Evermann, 1896–1900, p. 2107.

Description.—Perhaps the most noticeable character of this and of the striped sea snail (p. 466) is that it is shaped more like a tadpole than like the conventional fish, thanks to the cylindrical fore part of the trunk, together with the broad rounded snout and fat soft belly, and the abrupt flattening (sidewise) of the body close behind the vent. It is also provided with a sucking disk similar to that of the lumpfish (p. 459). In side view the body is deepest abreast the pectoral fins (about four times as long as it is deep, not counting the caudal fin), tapering evenly to a moderate caudal peduncle. The head is flat above, the mouth is at the tip of the snout and moderately wide, and the jaws are armed with many small teeth arranged in bands.

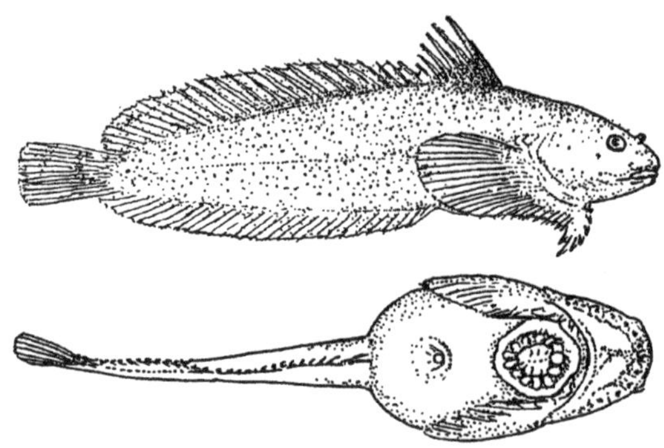

FIGURE 243.—Sea snail (*Neoliparis atlanticus*), side view (above) and ventral view (below) of adult male. After Garman.

The dorsal fin (6 spines and 26 soft rays) originates close behind the pectorals and runs continuously to the base of the caudal though it is separate from the latter. The dorsal spines are longer in males than in females, and project further beyond the membrane, giving the fin a fringed appearance. The anal fin (23 to 27 rays) originates under or a little behind the origin of the soft portion of the dorsal fin to which it corresponds in size and outline. The pectoral fins are not only very large and fanlike, but their bases run forward under the throat, where they expand into secondary lobes or wings with fringed edges. The ventral fins appear only as a circle of low knobs in the center of the sucking disk, which is situated on the throat between the pectoral fins. The skin is scaleless, and it is smooth except at spawning time, when the male becomes rough with small prickles.

[41] Some species of the genera *Paraliparis* and *Amitra* have lost the sucking disk.

[42] See V. Burke (Bull. 150, U. S. Nat. Mus., 1930) for a study of the family as a whole, giving descriptions and geographic ranges of all known species.

[43] McKenzie and Homans, Proc. Nova Scotian Inst. Sci., vol. 19, 1938, p. 278.

[44] We cannot judge whether the *Careproctus* reported by Vladykov and Tremblay (Natural. Canad., vol. 62 (ser. 3, vol. 6), 1935, p. 81) from the estuary of the St. Lawrence River as *C. longipinnis* was indeed identical with the fish from north of the Faroe Islands that was described under that name by C. V. Burke (Ann. Mag. Nat. Hist. Ser. 8, vol. 9, 1912, p. 510), or whether it is referable to *ranulus;* as seems the more likely on geographic grounds.

[45] For a detailed description, with illustrations, see Goode and Bean, Smithson. Contrib. Knowl., vol. 30, 1895, p. 275, vol. 31, pl. 70, figs. 251–251a, 251b.

Color.—Olive to reddish brown with lighter and darker cloudings and dots, the dorsal and anal fins often with crossbars. Its tints vary widely, also its markings and the strength of coloration, whether pale or dark, as is the case with so many fishes that live on the bottom.

Size.—Maximum length about 5 inches.

Remarks.—This sea snail resembles the following species (p. 466) so closely that the two are likely to be confused. The most evident difference between them is that the spiny portion of the dorsal (6 spines hardly stiffer than the soft rays) is marked off from the much longer soft part (22 to 27 rays) by a slight notch in the present species, whereas there is no such separation in the striped sea snail.

Habits.—Sea snails are inconspicuous little fish, usually found coiled up (tail to head) under stones, or attached by the sucker to some kelp stalk or other seaweed. But sometimes they swim to the upper water layers: thus the *Halcyon* specimens, for instance, mentioned below, were taken in a tow net at 8 fathoms, where the water was about 22 fathoms deep. And they seem to be indifferent to depth within moderate limits, for while they have been dredged as deep as 50 fathoms at various localities in our Gulf, they are often found clinging to lobster pots in the Bay of Fundy, (p. 465), while they have been taken in only a few feet of water in Massachusetts Bay. One was even found in a tide pool near St. Andrews, in Passamaquoddy Bay[46] so it would not be astonishing to find sea snails left in rock pools elsewhere, or on pebbly beaches by the ebbing tide, as often happens with its European representative. Young ones have been found living within the shells of the giant scallop (*Pecten magellanicus*), a curious habit that they share with the striped sea snail (p. 466) and with the hakes of the genus *Urophycis* (p. 224). Little else is known of the life of this sea snail, except that it is supposed to work inshore in winter to spawn. Presumably it feeds chiefly on small crustaceans and on small shellfish as its European relative does.

The spawning of this species has not been observed, but probably it takes place from March until midsummer in our Gulf, for Huntsman found larvae in Passamaquoddy Bay as early as April, while we towed one only 7 mm. long on German Bank as late as September 2 (1915).

The developmental stages of our sea snail have not been seen. The eggs of the European sea snail, which are about 1.1 mm. in diameter, and pale straw color to light salmon pink, sink and stick together, also to hydroids, seaweeds, and to debris of any kind. The little clusters are often brought up on long lines from 4 to 30 fathoms, but are sometimes found close below tide mark. There is no reason to suppose that the males care for the eggs, and the latter are so hardy that they do not suffer from exposure to the air for hours. Judging from the dates when newly hatched larvae have been seen, incubation of the European species occupies a month, perhaps longer in the case of the eggs that are spawned earliest in the season, at winter temperatures. The larvae are about 3.3 to 4.5 mm. long at hatching, with a small rose-red yolk sac that contains a large oil globule and that is inclosed in a net of blood vessels. Most of the characters of the adult are apparent at 11 to 12 mm. length, but the pectoral fins are brilliantly pigmented with yellow and black throughout the larval stage.[47]

General range.—Rocky shores along the North American Coast from northeastern Newfoundland, the northern part of the Gulf of St. Lawrence and the Grand Banks to southern New England.[48] It is rare west and south of Cape Cod, but has been taken at Woods Hole, on the coast of Connecticut, and off New Jersey.[49]

Occurrence in the Gulf of Maine.—This sea snail is generally distributed around the shore line of the Gulf. Thus the *Halcyon* trawled it off Yarmouth, Nova Scotia, in January 1921; it is rather common (according to Huntsman) in the Bay of Fundy and in Passamaquoddy Bay; and it has been definitely reported at Grand Manan; Eastport; Seguin Island; off Portland (where many have been collected); off Cape Elizabeth; at Kittery; and at various localities about Massachusetts Bay. It

[46] McKenzie and Homans, Proc. Nova Scotian Inst. Sci., vol. 19, Pt. 3, 1938, p. 278.

[47] McIntosh and Mastermann (Life histories of the British Marine Food-fishes, 1897 p. 191, pl. 2. fig. 9, 10) and Ehrenbaum (Nordisches Plankton, vol. 1, 1905–1909, p. 109) give descriptions of the eggs and larvae of the European sea snail *N. montagui* (as *Cyclogaster montagui*) from which the preceding is condensed.

[48] This fish is so closely allied to the north European sea snail, *N. montagui* (from which, however, it is quite distinct) that it passed under that name prior to 1898.

[49] A specimen was taken by *Albatross II*, off Atlantic City, lat. 39°24′ N., long. 74°05′ W., in 11 fathoms, in April 1930.

has been taken on Georges Bank also, and on Browns, living in scallop shells (p. 465).

Vladykov and McKenzie characterize it as "not uncommon" all around Nova Scotia;[50] it is classed by Huntsman as characteristic of the icy cold water on the banks in the southern part of the Gulf of St. Lawrence; and it has been reported from the estuary of the St. Lawrence River; from the northeastern part of the Gulf of St. Lawrence; off the south coast of Newfoundland; on the Grand Banks, and as far north on the Atlantic coast as the entrance to the Strait of Belle Isle.[51]

Striped sea snail *Liparis liparis* (Linnaeus) 1766

SEA SNAIL

Jordan and Evermann, 1896–1900, p. 2116.

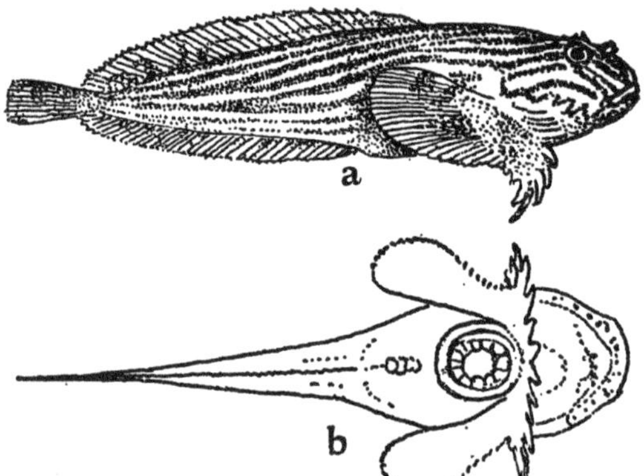

FIGURE 244.—Striped sea snail (*Liparis liparis*), side view (above) and ventral view (below) of adult. After Garman.

Description.—This little fish resembles the sea snail (p. 464) so closely, especially in its tadpole-like form, in the presence of a sucking disk on its chest, in which the rays of the ventrals (reduced to mere knobs) serve as a central support, and in the peculiar outline of the pectoral fins with secondary frilled basal lobes, that it is difficult to distinguish one from the other. The most obvious difference between the two species is that there is no indentation between the spiny and the soft parts of the dorsal fin in the striped sea snail. Furthermore, it usually has 33–35 rays in the soft portion of its dorsal fin and 26–29 rays in its anal fin, as against a maximum of 32 dorsal rays and of 27 anal rays in the other sea snail (p. 404). And the separation between the dorsal and anal fins and its caudal fin is not as definite in the striped sea snail as it is in the preceding species; in fact it is sometimes difficult to draw a sharp line between the fins. A minor character, which gives the head a rather different aspect, is that the dorsal profile is more arched in the striped sea snail.

Color.—Many color varieties of this fish have been described and named. As a rule the ground tint is of a shade of olive green, gray, or brown, variously tinged with reddish, with yellowish, or with lilac, and but little paler below than above. Red ones with pale and dark stripes have been seen among kelp in New England waters. And they are dark and pale in endless variety in varying situations, some nearly plain, some definitely striped with few or many narrow longitudinal bands, others spotted. In fact no two are alike. Usually the fins are darkly blotched or barred.

Size.—This fish grows to a length of 10 inches in Arctic seas but very few of them are more than 5 inches long in temperate latitudes.

Habits.—All that is known of its habits in our Gulf is that it lives on rocky or stony bottom, usually among the stalks and roots of kelp to which it sometimes clings fast, a habit which European writers describe as common. In British waters it is often to be found hiding in the tiny pools of water that are left under pebbles by the ebbing tide, and probably a search of the beaches would reveal it in similar situations in the Gulf of Maine also. Small ones often live inside the shells of the giant scallop (*Pecten magellanicus*), and it is our impression (though not backed by any definite evidence) that this is a more usual habit with the striped sea snail than with the preceding one (p. 465). At any rate, W. F. Clapp informs us that it is the rule to find at least one or two striped sea snails in a bushel or so of sea scallops, and fishermen have told us that sea snails of one species or the other (probably of both species) are found in scallop shells on Georges Bank.

Small crustaceans, chiefly amphipods and shrimps of various kinds, have been found in the stomachs of striped sea snails on both sides of the Atlantic; they also feed on small shellfish, and they were described by Fabricius [52] as eating small fish fry and algae.

[50] Proc. Nova Scotian Inst. Sci., vol. 19, 1935, p. 99.

[51] From the trawlings of the Newfoundland Fisheries Research Commission.

[52] Fauna Groenlandica, 1780, p. 137.

This is a winter-spring spawner; females are full of roe at Woods Hole in December and January, and the collection of the Museum of Comparative Zoology contains a female distended with eggs that was taken on April 1 many years ago. Larvae only 5.5 mm. long, which we towed near the Isles of Shoals on July 22 and in Massachusetts Bay on August 31 in 1912, must have been hatched from eggs spawned at least as late as May, if not in June.

The eggs,[53] about 1.5 mm. (0.06-inch) in diameter, sink and stick together in bunches, to hydroids, seaweeds, or other objects, like those of *Neoliparis atlanticus*, and it seems that incubation is about as long as it is with the latter, i. e., at least a month. The larvae are about 5.5 mm. long at hatching and they live adrift until they are upward of 16 mm. long, when the sucking disk is well developed.

General range.—Arctic and North Temperate Atlantic; north to the White Sea, Spitzbergen, Greenland, Davis Strait, and northern Labrador, and reported from the Kara Sea and from the Arctic Ocean north of Siberia; south to northern France and to Delaware Bay and Virginia.[54]

Occurrence in the Gulf of Maine.—The distribution of this sea snail parallels that of the preceding species in our Gulf. Thus it has been dredged not uncommonly in the Bay of Fundy region in from 5 to 100 fathoms and has been recorded from Grand Manan: from Eastport, as well as from other localities on the Maine coast; here and there about Massachusetts Bay; and on Georges Bank; also at Woods Hole.

In Nova Scotian waters it has been characterized variously as "common" [55] and as "uncommon." [56]

It has been described as "common" in the southern part of the Gulf of St. Lawrence;[57] has been reported in the estuary of the St. Lawrence River; at Anticosti; and in the northeastern part of the Gulf; also off the south coast of Newfoundland; in Conception Bay; off the eastern end of the Strait of Belle Isle; off the southeastern Labrador coast,[58] and from Fort Chimo, on Ungava Bay, in northern Labrador,[59] as well as from West Greenland.

It is of no commercial importance.

THE SEA ROBINS OR GURNARDS AND THE ARMORED SEA ROBINS

FAMILIES TRIGLIDAE AND PERISTEDIIDAE

The sea robins and their European relatives, the gurnards, suggest sculpins in their broad heads, slender bodies, large fanlike pectoral fins, in having two separate dorsal fins (a spiny and a soft rayed), and in the location of their ventral fins under the pectorals. But their entire heads are armored with rough bony and spiny plates. The Gulf of Maine is the northern limit for the family on the Atlantic Coast of America.

The armored sea robins are close relatives to the sea robins but they differ from them in four very noticeable ways: (1) the entire body is enclosed in an armor of bony plates, each plate with a spine; (2) it is only the two lower rays of the pectorals that form separate feelers; (3) each side of the front of the skull projects forward as a long flat process, so that the snout appears to be double; (4) they have 2 long barbels on the chin. They live on bottom in fairly deep water, and they are widespread in tropical to boreal seas. One species is a member of the Gulf of Maine fish fauna.

KEY TO GULF OF MAINE SEA ROBINS AND ARMORED SEA ROBINS

1. Front of snout only slightly concave as seen from above; no barbels on chin ---------------------- 2
   Front of snout so deeply concave that it seems to be double when seen from above (fig. 247); two long barbels on chin -------- Armored Sea Robin p. 471
2. Pectoral fin with 2 broad dusky blotches; there is no prominent longitudinal stripe on the side of the body ---------------- Common Sea Robin p. 467
   Pectoral fin with only 1 broad dusky blotch; there is a prominent longitudinal dark brown stripe on each side of the body --------- Striped Sea Robin p. 470

Common sea robin *Prionotus carolinus* (Linnaeus) 1771 [60]

SEA ROBIN; ROBIN, GREEN-EYE

Jordan and Evermann, 1896–1900, p. 2156.

[53] The following lines are condensed from Ehrenbaum's (Nordisches Plankton, vol. 1, 1905–1909, p. 112) account of its eggs and larvae in European waters.

[54] This sea snail, formerly known only as far south as New York, has been taken off Delaware Bay by *Albatross II*, and off Assateague, Virginia, by the *Grampus* (Welsh, Copeia, No. 18, 1915, p. 2).

[55] Jones, Proc. Nova Scotian Inst. Sci., vol. 5, pt. 1, 1882, p. 89.

[56] Vladykov and McKenzie, Proc. Nova Scotian Inst. Sci., vol. 19, 1935, p. 99.

[57] Cox, Contrib. Canadian Biol. (1918–1920) 1921, p. 112.

[58] From the cruises of the Newfoundland Research Commission.

[59] Packard Proc. Portland Soc. Nat. Hist., vol. 2, No. 13, 1909, p. 112.

[60] Jordan, Evermann, and Clark (Rept. U. S. Comm. Fish. (1928) Pt. 2, 1930, p. 407, place this species in the Genus *Merulinus* which was proposed by Jordan and Evermann in 1898 as a subgenus.

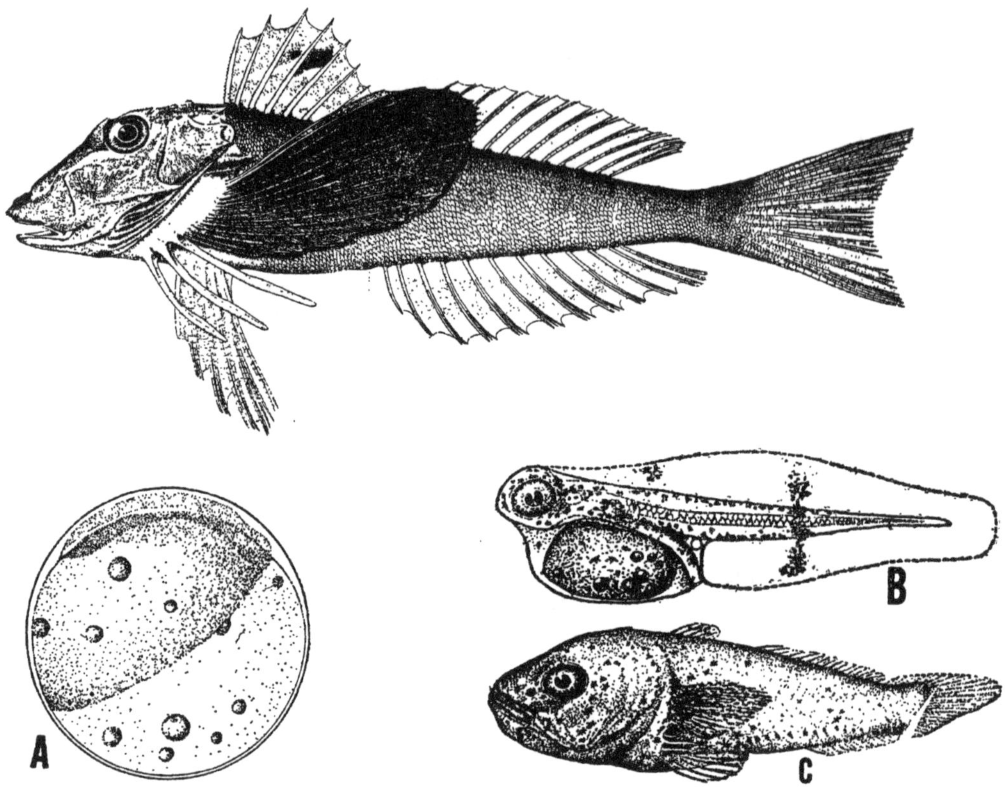

FIGURE 245.—Common sea robin (*Prionotus carolinus*). Adult, New Jersey; from Goode, drawing by H. L. Todd. A, egg; B, larva, just hatched, 2.8 mm.; C, young, 9 mm. A–C, after Kuntz and Radcliffe.

Description.—The large head, tapering body, and fanlike pectoral fins of the sea robin somewhat suggest a sculpin. But the robin is distinguished from all the sculpins by the incasement of its entire head in bony plates; by its smaller mouth; by the flat depressed dorsal profile of its snout; by its large ventral fins; and by the fact that the three lower rays of each of its pectoral fins are separate from the rest of the fin and modified into three independent feelers with slightly dilated tips, a very noticeable and distinctive feature. Furthermore, the front margin of the upper jaw is concave in outline when viewed from above, not convex as it is in most other fishes, which gives the nose of the robin a characteristic aspect. The head plates are rough and there is one sharp spine on each cheek at the angle of the gill cover; two short spines over each eye pointing backward; a spine on either side of the neck; and one on each shoulder above the base of the pectoral fin. The spiny and soft-rayed portions of the dorsal fin are separate, but they are so close together at their bases that they are almost in contact. The spiny dorsal has 10 spines, is rounded in outline, and higher than the soft dorsal (13 rays); but the soft dorsal is considerably longer than the spiny dorsal.

The caudal fin is of moderate size, its margin slightly concave. The anal fin (1 spine and 11 rays) is similar in outline to the soft dorsal, under which it stands. The pectorals (their 3 lower rays as just noted) are rounded in outline and are so large that they overlap the anal and the second (soft) dorsal when they are laid back. The ventral fins (each of 1 stiff spine and 5 rays) stand close behind the pectorals.

Color.—Usually the body is grayish or reddish brown above, with about five dark saddlelike blotches along the back, and is dirty white or pale yellow below. The dorsal fins are grayish, marked with pale spots and stripes, with a black spot between the fourth and fifth spines. The caudal fin is uniform grayish or brownish; the anal plain brown; the ventrals plain yellow to brown. The pectorals are yellow or orange, strikingly marked with two broad dusky bars, one of them crossing the middle of the fin, and the other crossing its outer third. The pectoral filaments are orange.

Size.—The maximum length is 15 to 16 inches, but few of them grow to be more than a foot long.

Habits.—Sea robins, like the sculpins, tend to keep to the bottom. But they swim actively, usually with the pectorals closed against the body. They are often hooked close to the surface; we have caught them when trolling for mackerel, and many anglers have told us of similar experiences. When on the bottom they often lie with the fanlike pectorals spread. If disturbed they bury themselves in the sand, all but the top of the head and eyes, and they are said to employ the feeler-like rays of the pectorals in stirring up the weeds and sand to rout out the small animals upon which they feed. They are usually found on smooth hard grounds, less often on mud or about rocks.

Along southern New England, where robins are far more plentiful than they are farther north, a large part of the population appear inshore in May or June, to pass the summer there; some close to tide line, but others remaining in depths of 5 to 30 or 40 fathoms, or deeper. Like many warm-water fishes, they disappear from the coast in October, to pass the cold season well out on the shelf, as recently proved by catches made at 50 to 55 fathoms off southern Massachusetts by the dragger *Eugene H* in late January,[61] in 1950, also at 21 to 93 fathoms off North Carolina, in that same month and the next, by the *Albatross III*.[62]

The fact that the *Albatross III* trawled up to 83 sea robins per haul off New York and off southern New England at 22–61 fathoms as late in the season as mid-May of that same year suggests that some of them may remain well offshore until into the summer, if not all summer.

Notwithstanding this inshore and offshore migration, some at least of the sea robins experience a temperature range of nearly 30° F. with the change of the seasons, for those that come closest inshore are in water as warm as 68°–70° at the end of the summer, while some that were trawled along the 30- to 40-fathom zone in May were in water as cold as 40.2° to 41.4°.

The sea robin is a voracious fish, feeding indifferently on shrimps, crabs of various kinds, amphipods (crustaceans are its chief diet), squids, bivalve mollusks, annelid worms, and on small fish, such as herring, menhaden, and small winter flounders. Seaweed has also been found in sea robin stomachs. They bite greedily on any bait, and are often taken with a spinner, or other artificial lure.

About Woods Hole the common sea robin spawns from June to September with July and August as the peak of the season.[63] But some may spawn earlier, for we have examined females taken at 50–55 fathoms off southern New England at the end of January with eggs so large as to suggest that they would be spawned by April or May. Unlike the sculpin tribe, the robin produces buoyant eggs, which are 0.94 to 1.15 mm. in diameter, slightly yellowish, with a variable number (10 to 25) of oil globules of various sizes, usually arranged in a more or less definite ring. Incubation occupies about 60 hours at a temperature of 72°, but any eggs that might be spawned in the cooler water of our Gulf would be slower in hatching. The newly hatched larvae are 2.5 to 2.8 mm. long, with two transverse yellow cross bands, one of these close behind the pectoral fins, the other midway between vent and tail. The yolk is absorbed, the mouth formed, and the yellow markings no longer prominent in 5 days, at a length of 3 to 3.4 mm. The dorsal and anal fin rays are visible and the lower pectoral rays have separated from the remainder of the fin at about 9 mm., and young fish of 25 to 30 mm. are darker, with transverse bands, and already show most of the distinctive characters of the adult.

General range.—Coastal waters of eastern North America from the Bay of Fundy to South Carolina; chiefly west and south from Cape Cod.

Occurrence in the Gulf of Maine.—Plentiful though the sea robin is off southern New England,[64] only a few are taken north of Cape Cod.

[61] We saw these catches which ranged from 0 up to 5,000 fish per haul, in 54 trawl hauls.

[62] One to one hundred and thirty sea robins per haul.

[63] Kuntz and Radcliffe (Bull. U. S. Bur. Fish., vol. 35, 1918, p. 105–109) give an account of its embryology and larval stages, subsequently confirmed and supplemented by Welsh.

[64] A catch of 1,000, in a day, in one trap, is recorded for Vineyard Sound, and of as many as 3,000 to 5,000, per trawl haul, at 50 to 55 fathoms off southern Massachusetts in winter. See footnote 61, p. 469.

Following the coast we find them reported at Truro, Cape Cod; Prof. A. E. Gross has informed us that he often saw as many as a dozen sea robins taken in the trap at the entrance to Barnstable Harbor in a single tide in the early summer of 1920;[65] one now in the Museum of Comparative Zoology was trawled 12 to 15 miles off Plymouth, at 30 fathoms, on November 20, 1943; the sea robin has been reported off Lynn and Salem; and in 1913 Welsh saw several in a trap at Manchester, on the North Shore of Massachusetts Bay on June 29. North of Cape Ann it has been taken at Anisquam; at Newburyport at the mouth of the Merrimac River, whence one about 1 foot long was brought in to the Museum of Comparative Zoology on August 14, 1949; also at the mouth of the Saco River. And Dr. W. C. Kendall saw more than 25 taken from the traps near Small Point, Casco Bay, between July 4 and 14 in 1896.

The only records for it for the coast east of Small Point are, however, for a single specimen caught at Campobello Island in the mouth of the Bay of Fundy in August 1911, and another in August 1949,[66] one taken in a weir in Passamaquoddy Bay at St. Andrews, October 2, 1935,[67] and of another taken in the Bay of Fundy, near Minas Channel, during the late summer or early autumn of 1951.[68]

Enough sea robins also range eastward across the South Channel for trawlers to have picked up a few (never more than a dozen or two on a trip) on Georges Bank during the summer of 1913. But it is probable that the deep channel between Georges and Browns Banks form its easterly limit, for sea robins are not known on Browns Bank or off the west coast of Nova Scotia.

It is not likely that the sea robin ever succeeds in reproducing itself in the Gulf, unless in restricted localities such as Casco Bay, where summer temperatures may be high enough. We have never taken its rather characteristic eggs in our tow nets anywhere in the Gulf, nor have its young fry ever been reported there. But when wandering fish do find their way around Cape Cod from the south, they may remain there, wintering offshore in deeper water, as they do farther south.

Importance.—The sea robin is edible, and its near relatives, the gurnards, are table fish in Europe, but it is too scarce in the Gulf of Maine to be of any potential commercial importance there. Off southern New England, where it is abundant, it is a nuisance to anglers, taking bait planned for other fishes.

Striped sea robin *Prionotus evolans* (Linnaeus) 1766

Jordan and Evermann, 1896–1900, p. 2167, as ***Prionotus strigatus*** Cuvier and Valenciennes, 1829 [69]

[65] Briefly mentioned in The Auk, vol. 40, No. 1, January 1923, p. 24.

[66] This latter fish was reported by Scattergood, Trefethen, and Coffin (Copeia, 1951, No. 4, p. 298).

[67] McGonigle and Smith, Proc. Nova Scotian Inst. Sci., vol. 19, 1936, p. 160.

[68] Reported to us by letter by Dr. Huntsman.

[69] Ginsberg (Texas Jour. Sci., vol. 2, No. 4, 1950, p. 519, 520-522) has shown that the *P. strigatus* of Cuvier and Valenciennes is not separable from the *evolans* of Linnaeus.

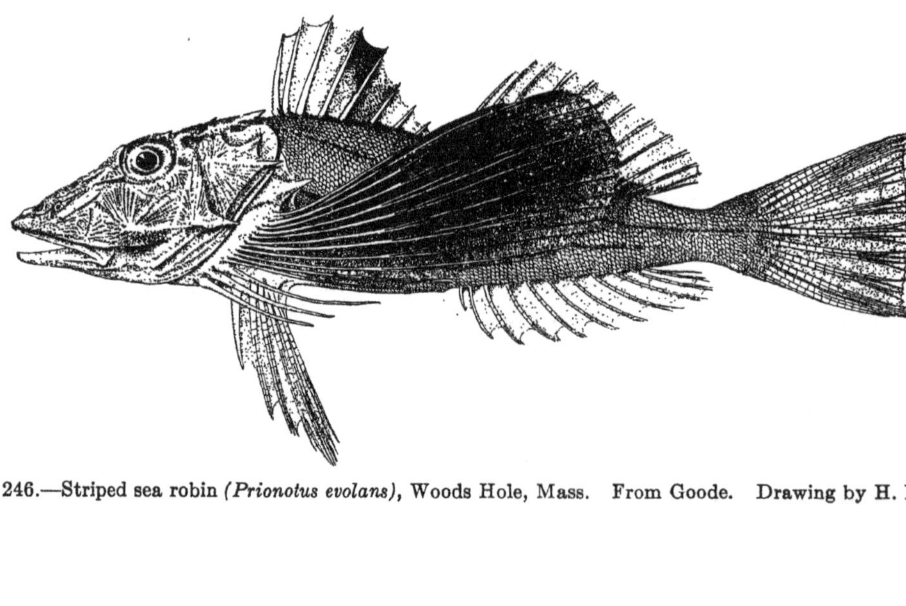

FIGURE 246.—Striped sea robin *(Prionotus evolans)*, Woods Hole, Mass. From Goode. Drawing by H. L. Todd.

Description.—The striped sea robin resembles the common sea robin so closely that one might easily be taken for the other. But its mouth is wider and gapes back almost opposite the front of the eye, with the maxillary bone more than one-third as long as the head. Its head is flatter (compare fig. 246 with fig. 245); its pectoral fins are longer relatively (reaching back to the ninth or tenth ray of the soft dorsal fin instead of only to the fifth or sixth ray); its pectoral feelers are more slender and tapering; its caudal fin is square-ended instead of concave in rear outline; and its reddish or olive-brown sides (the general ground tint varies) are marked longitudinally with a dusky or bronze-brown stripe below the lateral line. The first dorsal fin shows the same black or dusky blotch betweeen the fourth and fifth spines, so characteristic of the common robin. The pectorals are orange to brown with pale edges, their centers washed with dusky, but without the definite crossbars characteristic of the common sea robin. The pectoral filaments are pale brown or orange, marked with narrow brown bars. (The common sea robin does not show these bars.)

Size.—This is a larger fish than the common sea robin, growing to a maximum length of about 18 inches.

General range.—Shoal water along the Atlantic coast of North America from South Carolina to Cape Cod, reaching the Gulf of Maine as a stray from the south.

Occurrence in the Gulf of Maine.—This southern fish rounds Cape Cod so seldom that there are only 6 definite records for it from the Gulf of Maine: Monomoy; North Truro; Salem; Gloucester; Monhegan Island, Maine (its most northerly outpost) where one was taken in an otter trawl at 40 fathoms, November 19, 1933; and the eastern part of Georges Bank, whence one was brought in to the U. S. Fish Commission sometime between 1877 and 1880. We have never seen it north or east of the elbow of Cape Cod.

Armored sea robin *Peristedion miniatum* Goode 1880

Jordan and Evermann, 1896–1900, p. 2178.

Description.—The armored robin resembles the sea robins in general body form, and in the ar-

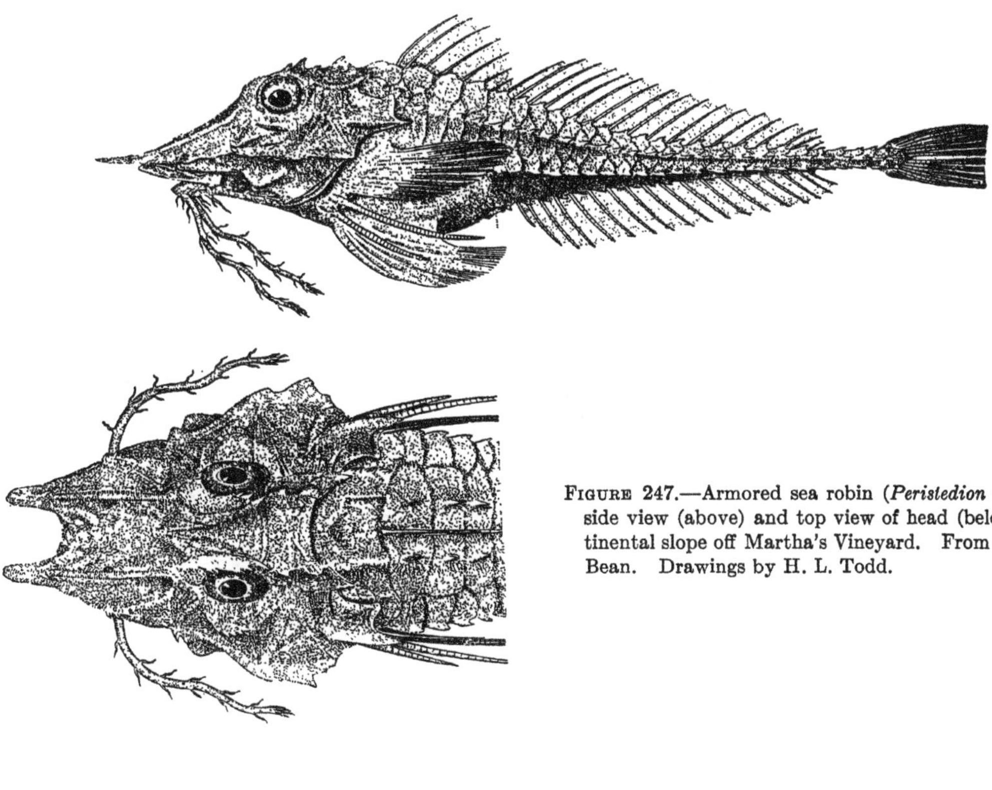

FIGURE 247.—Armored sea robin (*Peristedion miniatum*), side view (above) and top view of head (below). Continental slope off Martha's Vineyard. From Goode and Bean. Drawings by H. L. Todd.

rangement of its fins, including the fact that the two lower rays of each pectoral form separate feelers. But its body is entirely clothed with bony plates of considerable size. There are 4 rows of these on each side, from the vent rearward, with an equal number of rows of thornlike spines, the latter close set and directed rearward. The plates on the abdomen have no spines. Thanks to this armor, the trunk is very stiff. The lower jaw bears a number of short fleshy barbels; there is one long barbel with short side branches at each corner of the mouth (the sea robins have no barbels); and the front of its head is given so peculiar an appearance by the two projections from the skull (p. 467) that the armored robin could hardly be mistaken for any other fish, except for one of its own tribe.

Color.—Bright crimson, below and above.

Size.—Maximum recorded length between 13 and 14 inches (330–355 mm.).

Habits.—This is a ground fish, recorded from depths ranging from 50 fathoms down to somewhere between 200 and 235 fathoms. And it seems to be confined to the zone of warm water along the outer part of the continental shelf and upper part of the continental slope for the lowest temperature in which it has been recorded is between 44° and 45°.[70] The stomachs of those we have opened contained shrimps, stomatopods, and other small crustaceans.

General range.—Outer part of the continental shelf and upper part of the continental slope; from the southwestern face of Georges Bank to the offing of Charleston, S. C. (lat. 32°24′ N., long. 78°44′ W.).[71]

Occurrence in the Gulf of Maine.—Trawlers tell us they sometimes take these brilliant crimson fishes on the southwestern part of Georges Bank. And they must be rather common outside the 60-fathom contour, for we saw 89 specimens trawled there and south of Nantucket at depths of 66 to more than 185 fathoms, by the *Albatross III* in May 1950. But it is probable that they are barred from the more easterly parts of the bank and from the inner parts of the Gulf of Maine by low temperature.

THE FLYING GURNARDS. FAMILY DACTYLOPTERIDAE

Flying Gurnard *Dactylopterus volitans* (Linnaeus) 1758

Jordan and Evermann, 1896–1900, p. 2183, as *Cephalacanthus volitans* (Linnaeus)

Description.—The flying gurnard (only known representative of its family), built, in general, on the "sea robin" plan, is remarkable for its enormous pectorals. When it is about half grown or older, these reach nearly to the base of the caudal fin if laid back. When spread, they have the form of enormous rounded fanlike wings. Other con-

[70] We have seen it trawled off southern New England in water as shoal as 50 fathoms, and as cold as 44.4°.

[71] For list of early localities, see Goode and Bean, Smithsonian Contrib. Knowl., vol. 30, 1895, p. 471.

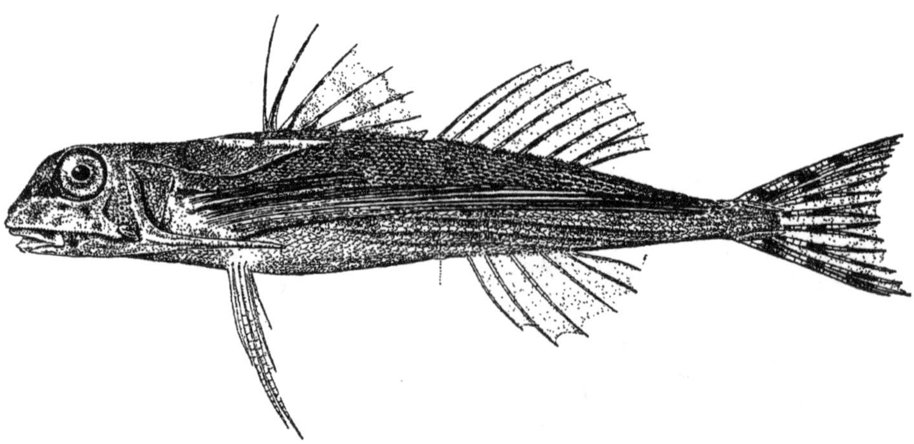

FIGURE 248.—Flying gurnard (*Dactylopterus volitans*), Key West, Florida. From Jordan and Evermann. Drawing by H. L. Todd.

spicuous respects in which the flying gurnard differs noticeably from its relatives, the sea robins, are that the first five or six rays of each pectoral, with their interconnecting membrane, form a separate fin, having no connection with the remainder of the pectoral except at the base; that the few lowermost rays of each pectoral fin are not separate, in the form of feelers, but are continuous with the remainder of the fin; that the first two spines of the dorsal fin are separate, that the bony armor covering the front and top of the head reaches rearward considerably beyond the origin of the dorsal fin on either side to end in a stout spine, that each gill cover [72] is extended rearward as a stout spine about as far as the axil of the pectoral fin, and that the scales are much larger and each armed with a short stout spine.

Color.—This is a very brilliant fish, varying widely in color; most of them are of some shade of brownish to greenish olive above, with the lower side paler, but marked irregularly with reddish salmon or salmon yellow. The winglike pectorals are variously marked with bright blue streaks near their bases, with blue spots and bars toward their tips. The caudal fin usually has about three brownish-red cross bars.

Size.—To about 12 inches.

General range.—Tropical to warm temperate latitudes of both coasts of the Atlantic; south to Brazil and north rather commonly to North Carolina on the American coast; a few to New York and the southern coast of Massachusetts in most years (in autumn [73]); recorded as a stray from Massachusetts Bay. A dried and hardened specimen that was found on the shore near Country Harbor, Nova Scotia, in September 1939, by Stanley McKinley, among the kelp and eel grass that had been washed ashore during the night, was thought by him (no doubt correctly) to have been carried north on the deck of some steamer from the south.[74]

Occurrence in the Gulf of Maine.—The only report of this warm-water fish from north or east of Cape Cod is of one said to have been taken in Massachusetts Bay.[75]

THE CUNNER TRIBE, OR WRASSES. FAMILY LABRIDAE

Members of the cunner family have a single long dorsal fin, its forward part spiny, its rear part soft rayed, with no evident demarkation between the two. The ventral fins are located under the pectorals, and the caudal peduncle is very deep. The structure of the dorsal fin is sufficient of itself to distinguish them from all Gulf of Maine fishes except the scup, sea bass, rosefish, tilefish, or certain sculpins. And there is no danger of confusing a cunner or tautog with any of these, for their rounded tails and pectorals, and their general body-forms separate them at a glance from the thin-bodied, fork-tailed scup; their small mouths and the relative sizes of their fins are obvious distinctions between them and the sea bass tribe; their smooth cheeks and broad caudal fins separate them from the spiny-headed, narrow-tailed rosefish or from any sculpin; and they do not in the least resemble the tilefish with its broad mouth, adipose "fin" on the nape of its neck, concave tail fin, and pointed pectorals. Both the roof of the mouth and the floor of the throat (pharynx) is armed with a patch of conical or knoblike teeth in the cunner tribe. It is with these that they grind the hard-shelled mollusks and crustaceans on which they feed.

KEY TO GULF OF MAINE CUNNERS

1. Gill covers scaly, snout somewhat pointed, dorsal profile of head rather flat ---------- Cunner, p. 473

   Gill covers largely naked, snout blunt, dorsal profile of head high-arched -------------- Tautog, p. 478

Cunner *Tautogolabrus adspersus* (Walbaum) 1792

Perch; Sea perch; Blue perch; Bergall; Chogset

Jordan and Evermann, 1896–1900, p. 1577.

Description.—The readiest field marks by which the cunner may be distinguished from its close relative, the tautog, are mentioned on page 479. It is moderately deep in body, moderately flattened sidewise, with a very deep caudal peduncle,

[72] Actually the preopercular bone.

[73] The most recent record from Woods Hole, of which we have heard, is of two taken there on November 24, 1948, from the deck of *Albatross III* while she was moored at the dock (Arnold, Copeia, 1949, p. 300).

[74] McKenzie, Proc. Nova Scotian Inst. Sci., vol. 20, 1940, p. 44.

[75] This specimen is now in the Museum of Comparative Zoology, to which it was transferred from the Boston Society of Natural History. There is no clue to its origin, except that it was taken many years ago.

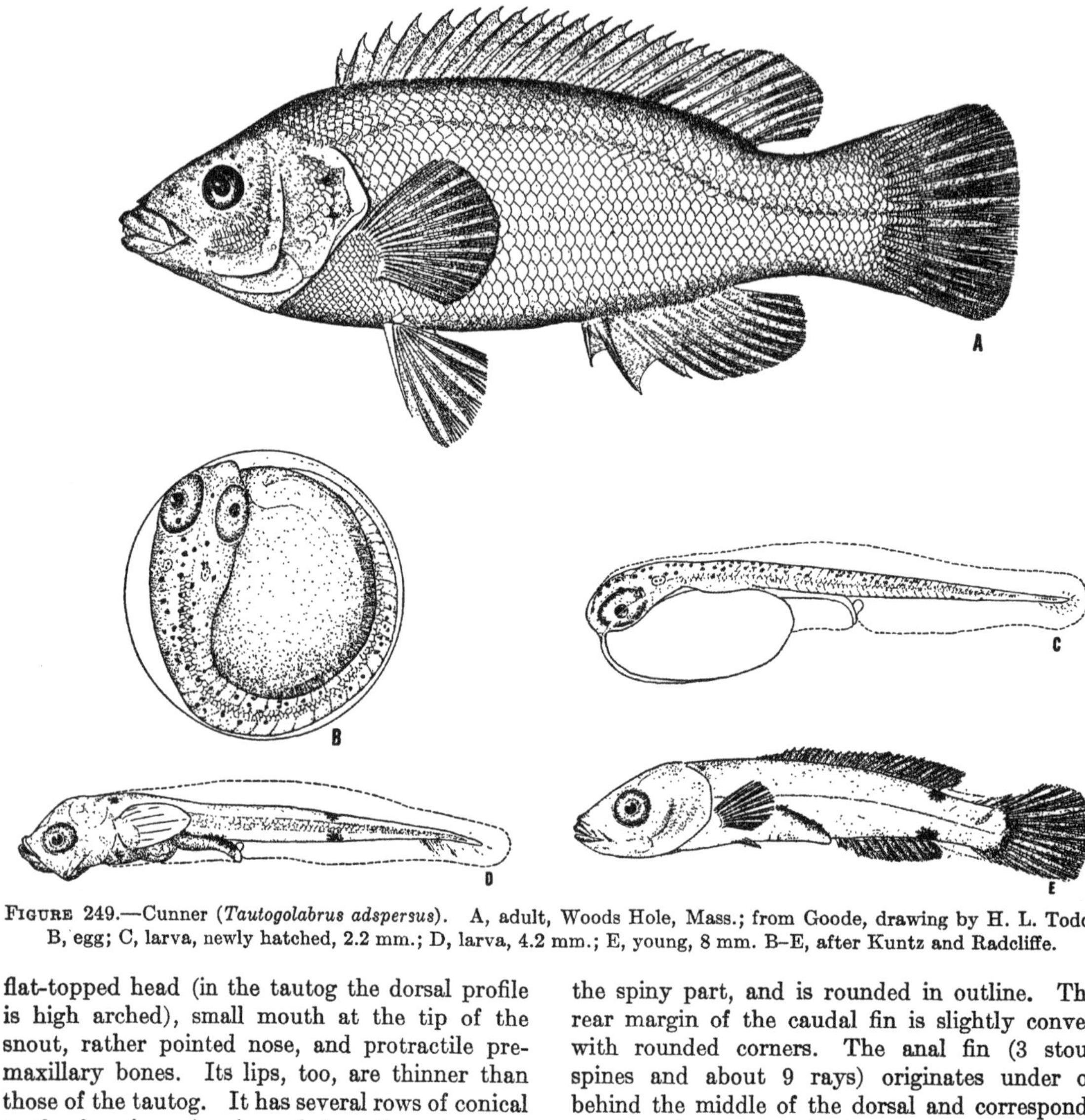

FIGURE 249.—Cunner (*Tautogolabrus adspersus*). A, adult, Woods Hole, Mass.; from Goode, drawing by H. L. Todd. B, egg; C, larva, newly hatched, 2.2 mm.; D, larva, 4.2 mm.; E, young, 8 mm. B–E, after Kuntz and Radcliffe.

flat-topped head (in the tautog the dorsal profile is high arched), small mouth at the tip of the snout, rather pointed nose, and protractile premaxillary bones. Its lips, too, are thinner than those of the tautog. It has several rows of conical teeth of various sizes in each jaw, the outer ones very stout. Its body and gill covers are covered with large scales (in the tautog there is a naked area in front of each gill opening), and its skin is so tough that the fish must be skinned before marketing. Its dorsal fin (about 18 spines and 9 or 10 soft rays) originates over the upper corner of the gill cover, i. e., a little in front of the pectoral fins, and runs back to the caudal peduncle. The first 4 or 5 rays of the dorsal fin are graduated, the others are of about equal lengths. The soft part is only a little more than one-third as long as the spiny part, and is rounded in outline. The rear margin of the caudal fin is slightly convex with rounded corners. The anal fin (3 stout spines and about 9 rays) originates under or behind the middle of the dorsal and corresponds to the soft part of the latter in outline. The ventrals stand under or a little behind the pectorals; both the ventrals and the pectorals are of moderate size, and the pectorals are rounded.

Color.—To describe the color of the cunner is to list all the colors of the bottoms on which it lives, for it is one of the most variable of fishes. As a rule the upper parts range from reddish brown (darker or paler) of a bluish cast to blue with brownish tinge, variously mottled with blue, brown, and reddish. Some fish, however, are uniform brown, while fish caught over mud bottom are often very deep sepia. In some situations they may be dull olive green mingled with blue, brown, or rust color. Some cunners are slaty, but reddish or rust tones are apt to prevail when they are living among red seaweeds about rocks. Cunners caught in deep water are often almost as red as the rosefish; on the other hand we have seen very pale ones, more or less speckled all over with blackish dots, over sandy bottom. The belly is invariably of a bluish cast, more or less vivid, sometimes whitish, sometimes dusky, sometimes little paler than the sides. Some cunners have the lips and lining of the mouth bright yellow. Young fry are more or less dark-barred and blotched.

Size.—In the Gulf of Maine adult cunners measure about 6 to 10 inches in length and weigh less than half a pound, and one a foot long is very large. But a few are caught up to 15 inches long, and as heavy as 2½ pounds.

Habits.—The cunner is chiefly a coastwise fish. In our northern waters they are the most plentiful from just below tide mark downward. They swarm among eel grass (*Zostera*) and about the piling of wharves and under floats in harbors. They also run up into the deeper salt creeks, small fish farther than larger ones, though we have never heard of one in water that is appreciably brackish; and young cunners are often found among eel grass and in rock pools. Southward, however, from New York or thereabouts, most of them keep to water at least 15 to 20 feet deep, hence somewhat farther out, depending on the topography of the coast line and of the bottom.

At the other extreme, they are common enough at 10 to 15 fathoms in the inner parts of Massachusetts Bay, and not rare as deep as 25 to 35 fathoms on the offshore ledges and banks, and we have taken them as deep as 70 fathoms on Georges Bank. But the great majority live within 5 or 6 miles of the shore. And while there are some on the offshore grounds, such as Stellwagen Bank, Jeffreys and Cashes Ledges, and even on Georges and Browns Banks where the otter trawls frequently pick up a few, we have never heard of a large catch of them made far out at sea, whether along southern New England or to the northward. Most of the cunners that are caught the deepest and the farthest offshore are large ones that have probably strayed thither, and finding good feeding, have remained.

As far as we know adult cunners never depart far from the bottom, or from the rocks about which they make their homes, nor do they school. Many, it is true, may live together, but they act quite independently of one another, simply congregating because the surroundings are attractive. Cunners, like other rockfish, spend much of the time resting quietly or swimming slowly among the bunches of Irish moss (*Chondrus*) and fronds of kelp, or in the open spaces among the eel grass (*Zostera*), wherever the latter has reestablished itself, always on the lookout for food.

Cunners are year-round residents, broadly speaking, wherever they are found. At the most, they may descend into slightly deeper water to pass the coldest months,[75] or they may desert the shoalest parts of certain enclosed bays in midsummer to escape the very high temperatures produced there as the sun strikes the flats at low tide. They have been described as hibernating in the mud during the winter, or at least as lying among eel grass or rocks in a more or less torpid state. But we find no positive evidence of this; on the contrary, practical fishermen, among them Capt. L. B. Goodspeed, to whom we are indebted for many notes, inform us that cunners are to be caught in abundance on precisely the same spots in winter as in summer. In fact a few are landed in Boston during the cold months, and the only reason more are not brought in then is that there is so little demand for them.

It has long been known that the cunner is vulnerable to very low temperatures. Hazards of this sort are more frequent south of Cape Cod, where the fish are more likely to be caught in very shoal water in a sudden freeze, than in the Gulf of Maine, where active mixing by the tide usually prevents the water from chilling to the danger point, except at the surface. However, this did take place in Massachusetts Bay in the winter of 1835, when cunners came ashore in quantities between Marblehead and Gloucester. And the failure of the cunners to produce young within the Bay of Fundy (p. 478) suggest that the lower thermal limit to their successful reproduction is about 55°–56°, though the young fry as well as the adults are at home in temperatures close to the freezing point of salt water. The upper

[75] Ambrose (Proc. and Trans., Nova Scotian Inst. Nat. Sci., vol. 2, No. 2, 1870, p. 93) describes the cunners as moving out of Saint Margaret Bay, Nova Scotia, in autumn, to return early in May.

thermal limit, for the well being of the cunner, is something like 70°–72°, to judge from the distribution of the species.

Cunners are omnivorous. As a rule they find their livelihood browsing among seaweeds, stones, or dock piles, biting off barnacles and small blue mussels, with the fragments of which they are often packed full. They devour enormous numbers of amphipods, shrimps, young lobsters, small crabs, and other small crustaceans of all kinds; also univalve mollusks and the smaller bivalves, hydroids, and annelid worms. They sometimes eat small sea urchins, bryozoans, and ascidians, and they occasionally capture small fish such as silversides, sticklebacks, pipefish, mummichogs, and the fry of larger species. Finally, eel grass is often found in cunner stomachs besides the animal food. Small cunner fry taken at Woods Hole were found by Dr. Linton to have fed chiefly on minute crustacea such as copepods, amphipods, and isopods.

The cunner is a busy scavenger in harbors, congregating about any animal refuse, to feed on the latter as well as on the amphipods and other crustaceans attracted by the same morsels. They are also said to eat fish eggs, and no doubt feed to some extent on herring spawn. Our own belief is that cunners are always hungry, no matter what the stage of the tide.

The cunner spawns chiefly from late spring through early summer. The eggs are buoyant, transparent, 0.75 to 0.85 mm. in diameter, and they do not have an oil globule. Incubation occupies about 40 hours at temperatures of 70° to 72°, but it is probable that about 3 days are required for hatching in the cooler waters of the Gulf of Maine (55° to 65°). At hatching the larvae are about 2 to 2.2 mm. long, and at 15 mm. the young cunner is of practically adult form. On newly hatched larvae the pigment cells are scattered uniformly over head and trunk, but by the 3-mm. stage they have gathered into a pair of black spots, dorsal and ventral, about halfway between the vent and the base of the caudal rays, which are characteristic of the species. And these spots persist to about the 10- to 20-mm. stage. By the time the fry have grown to about 25 mm. they are as variable in color as their parents (it is on record that Louis Agassiz had 60 colored sketches of small cunners 3 to 4 inches long, of different hues, prepared at Nahant during a single summer).[76]

Fry of 1 to 1.2 inches have often been taken in August, and young fish up to 2 inches long in September in southern New England waters. Hence we may assume that Gulf of Maine cunners (probably hatched somewhat later) may average about 2 to 2½ inches by their first autumn, and 2½ to 2¾ inches by the following June when they are one year old, which Johansen [77] found true also of the earliest hatched fry in the southern side of the Gulf of St. Lawrence. The subsequent rate of growth has not been studied for the cunners of our Gulf. But Johansen's [78] age determinations for cunners of the Gulf of St. Lawrence make it likely that Gulf of Maine cunners 3 to 4 inches long are 2 years old; those of 4 to 5 inches 2 or 3 years old; those of 5 to 6 inches 3 years old; those of 6 to 7 inches 3 or 4 years old; those of 7 to 8 inches 4 or 5 years old; those of 8 to 9 inches 5 or 6 years old; those of 9 to 10 inches about 6 years old; and those of 10 to 11 inches 6 or 7 years old. But the relationship is complicated by the fact that female cunners run larger than males, so that males may be a year older than females of the same size.

Most of the cunners mature in their third summer (i. e., when 2 full years old) when 2¾ to 3½ inches long.

General range.—Atlantic coast of North America and the offshore banks, from Conception Bay, east coast of Newfoundland, and the western and southern parts of the Gulf of St. Lawrence,[79] southward in abundance to New Jersey, and occasionally as far as the mouth of Chesapeake Bay.

Occurrence in the Gulf of Maine.—The cunner is one of our most familiar fish, to be found all around the shore line of the Gulf. The Massachusetts Bay region is perhaps their chief center of abundance, and they are so numerous there in

---

[76] The embryology and larval development and fry of the cunner have been described by Agassiz (Proc. Amer. Acad. Arts, Sci., N. Ser., vol. 9, 1882, p. 290, pls. 13 to 15); Agassiz and Whitman (Mem. Mus. Comp. Zool., vol. 14, No. 1, Pt. 1, 1885, p. 18, pls. 7–19, and Mem. Mus. Comp. Zool., vol. 40, No. 9, 1915, pls. 32–39); Kuntz and Radcliffe (Bull. U. S. Bur. Fish., vol. 35. 1918, p. 99, figs. 18–29); and more recently by Johansen (Contr. Canad. Biol., N. Ser., vol. 2, No. 17, 1925, pp. 440–450).

[77] Contrib. Canadian Biol., N. Ser., vol. 2, No. 17, 1925, p. 451.

[78] Johansen (Contrib. Canadian Biol., N. Ser., vol. 2, No. 17, 1925, pp. 451-455) worked out the age-length relationship for a large series of Gulf of St. Lawrence cunners by a study of their scales and otoliths.

[79] See Johansen, Contrib. Canadian Biol., Ser. 2, vol. 2, No. 17, 1925, pp. 5–6 [427–428]), for the distribution of the cunner in Canadian waters.

good years, along the rocky shores and around and over ledges, that no amount of fishing seems to have any effect on their numbers. Generally speaking, they are less numerous east of Casco Bay, and our experience has been that they are progressively less and less so eastward along the shore from Penobscot Bay toward the Bay of Fundy, but average larger. On the outer coast of Mount Desert, for example, it is unusual to catch one in the enclosed harbors (precisely the localities they frequent farther west and south), and most of those caught outside are very large. Thus we took many of 12 to 13 inches, averaging about 1½ pounds, near Baker's Island, off Northeast Harbor, in August 1922, and no small ones. But young fish in plenty, as well as adults, have been reported from Bluehill Bay, nearby,[80] where the water is warmer in summer.

Cunners are also taken, here and there, along the coast, eastward to the Grand Manan Channel, sometimes in numbers as in 1928, when so many were caught "about the rocks and in the coves to the south of West Quoddy," that they were reported in the press.[81] But they are so scarce ordinarily around Grand Manan and within Passamaquoddy Bay that only half a dozen large specimens had been taken there from the founding of the Biological Station at St. Andrews in 1906 down to the early 1920's.[82] And while the cunner is reported from Black River east of St. John, New Brunswick, it seems to be unknown farther in along the New Brunswick shore of the Bay of Fundy or in Chignecto Bay and Minas Basin at the head. But Annapolis Basin on the Nova Scotian side of the bay, harbors a few, while cunners of all sizes are so numerous in St. Mary Bay that this must be an important centre of reproduction and the source of the few large (i.e., old) ones that are caught farther up the Bay of Fundy. And they are reported along the western shore of Nova Scotia, as at Pubnico for example.

There are large cunners in small numbers on the offshore fishing grounds in our Gulf also, Stellwagen at the mouth of Massachusetts Bay, Cashes Ledge, and Georges and Browns Banks, as mentioned above (p. 475) in depths down to 50 fathoms or so. But it is not likely that they ever descend into the deep basins of the Gulf. Certainly our experimental trawlings have not yielded any there, 42 fathoms being the greatest depth at which we have known of a cunner taken anywhere in the inner parts of the Gulf.[83]

Extending our survey farther east and north, we find cunners reported as numerous all along the outer coast of Nova Scotia, including the many bays and inlets, also in the southern side of the Gulf of St. Lawrence from Cape Breton to the Gaspé Peninsula, including the shallow bays of Prince Edward Island and the shores of the Magdalen Islands, also up the west coast of Newfoundland as far as Bay of Islands. And they are to be expected at the heads of the bays along the south coast of Newfoundland for they have been taken in Conception Bay on the east coast. But this last is their most northerly known outpost on the Atlantic coast, and they have never been reported either from the estuary of the St. Lawrence or anywhere along the north shore of the Gulf of St. Lawrence.[84]

Cunners near Newport, Rhode Island, commence spawning by mid-May and June sees the chief production of eggs there and near Woods Hole, where most of the fish are spent after the first days of July, though eggs have been taken in abundance there until July 15, a few as late as August 15.[85] Probably spawning does not commence until June in the colder waters of our Gulf, but continues there through the later summer, for our towings have yielded many eggs, apparently of the cunner, in July and August. And the chief spawning season is about the same as this in the southern side of the Gulf of St. Lawrence, according to Johansen [86] and to Reid.[87]

Cunner eggs have been taken at our tow net stations along outer Cape Cod; near Race Point at the tip of the Cape; in Massachusetts Bay (where we have often towed them in great numbers in the tideways between the offlying ledges); and at the mouth of Penobscot Bay, as well as in sundry harbors. Blue Hill Bay inland from Mount Desert may be a breeding center, for small fry are reported there.[88] And eggs taken off

[80] Reported to us by Rear Adm. S. E. Morrison, U. S. Navy.

[81] Boston Transcript for August 29, 1928.

[82] Johansen, Contrib. Canadian Biol. N. Ser., vol. 2, No. 17, 1925, p. 5 [427].

[83] One was trawled at this depth at the mouth of Massachusetts Bay (lat. 42°28′ N., long. 70°13′ W.) by the *Albatross II*, July 28, 1931.

[84] See Johansen, Contrib. Canadian Biol., N. Ser. 2, vol. 2, No. 17, 1925, pp. 5–6 [427–428] for an account of the status of the cunner in the Gulf of St. Lawrence, and around Newfoundland.

[85] Agassiz and Whitman, Mem. Mus. Comp. Zool., vol. 14, No. 1, 1885, p. 18, Kuntz and Radcliffe, Bull. U. S. Bur. Fish., vol. 35, 1918, p. 99.

[86] Contrib. Canadian Biol., N. Ser., vol. 2, No. 17, 1925, p. 17 [439].

[87] Contrib. Canadian Biol. and Fish., N. Ser., vol. 4, No. 27, 1929.

[88] By Rear Admiral S. E. Morrison, U. S. N.

Libbey Island prove that cunners spawn in diminishing numbers eastward along the Maine coast nearly to the mouth of the Bay of Fundy. It is doubtful, however, whether eggs produced along the coast east of Mount Desert yield more than a very small proportion of fry, nor do cunners breed successfully in the cold water of the Bay of Fundy, where no small ones are ever seen. However, the Bay is simply a gap in the breeding range, for St. Mary Bay is a productive nursery. Both eggs and larvae were taken at various localities along the outer coast of Nova Scotia by the Canadian Fisheries Expedition during the summer; and the shoal inshore waters in the southern side of the Gulf of St. Lawrence are a productive spawning area.[89]

Larval cunners and small specimens generally, like their eggs, are so closely confined to the coast line that it is impossible to represent the localities where we have taken them on a general chart of the Gulf; in fact, all our catches of 100 or more have been made either in harbors or at most not a couple of miles from land.[90] There may be some successful reproduction on Casbes and Jeffreys Ledges. But we have found no evidence, whether of eggs or of young fry, that the few large cunners that wander offshore to Georges Bank produce any young there.

Variations in abundance.—No evidence is available as to how much the cunners may vary in abundance from year to year, along the coasts of our Gulf as a whole. But they may do so widely at a given locality. Thus we found very few of them in 1950 along the Cohasset shore, on the southern side of Massachusetts Bay, where they are plentiful ordinarily. And they were so scarce there during the summer of 1951, that persons raking Irish Moss (*Chondrus*) reported seeing hardly a cunner around the rocks where many are to be seen in most summers, and another acquaintance who usually baits a lobster pot or two with cunners taken in a cunner trap caught only one occasionally in that way.

Importance.—The cunner was a favorite pan fish once. During the 1870's the annual catch of the small boats fishing out of Boston was estimated as not much short of 300,000 pounds, while the fact that 104,100 pounds of cunners were reported for Maine in 1889, 148,300 pounds in 1898, and 281,500 pounds in 1905, shows that the annual harvest was still considerable to that time. But the reported catch had fallen to 30,695 pounds for Maine by 1919, and to about 10,000 pounds for the entire coast line of Massachusetts, south as well as north of Cape Cod. And Maine reported only 10,000 pounds for 1928 and 1,735 pounds for 1929, while the only cunners reported for Massachusetts were 30 pounds and 45 pounds for those 2 years, respectively. From that time down to 1947, commercial catches of cunners have been reported for Maine in only 3 years out of the 14.[91]

The landings reported for Massachusetts during this period suggests ups and downs so erratic and so extreme [92] that we hesitate to place any dependence upon them further than that landings ranging from 3,100 pounds to 18,700 pounds (average 7,450 pounds) for the years 1944–1947 show that a small demand continues for cunners. And we can witness that sizeable ones are very good pan fish.

Although not regarded as a game fish, the cunner affords amusement to thousands of vacationists near our seaside resorts. And the number caught, of which no record is kept, is so considerable that this must be classed as a useful little fish from the recreational standpoint.

Probably more cunners are caught on bits of clam than on any other bait. But they will take snails broken from their shells, bits of crab, lobster, or pieces of sea worms (*Nereis*) almost as freely. And we have even caught a few while trolling near rocks, for mackerel, with a small spinner tipped with a bit of white fish skin. The little ones are a great nuisance, often stealing the bait as fast as it is offered, and because it is a small-mouthed fish, very small hooks are best.

Tautog *Tautoga onitis* (Linnaeus) 1758

BLACKFISH; WHITE CHIN

Jordan and Evermann, 1896–1900, p. 1578.

[89] See Johansan (Contrib. Canadian Biol., New Ser., vol. 2, No. 17, 1925, p. 18 [440]; also Reid, Contrib. Canadian Biol. and Fish. N. Ser., vol. 4, No. 27, 1929.

[90] The precise records have been published elsewhere (Bull. Mus. Comp. Zool., vol. 58. 1914, p. 108, and vol. 61, 1917, p. 271).

[91] One hundred and seventy five pounds for 1933, 200 pounds for 1935, 45,300 pounds for 1938, an amount so large that we question its accuracy, especially since the entire catch was reported as made on "lines, trawl." No catch statistics are available for 1934, 1936, 1941, or 1942.

[92] Reported catches for Massachusetts jumped from 45 pounds for 1929 to 349,251 pouhds for 1931, dropped to 0 for 1932, 152 pounds for 1933 and 0 again for 1935; rose to 27,800 pounds for 1937; were 0 again in 1938; but 53,500 pounds in 1940.

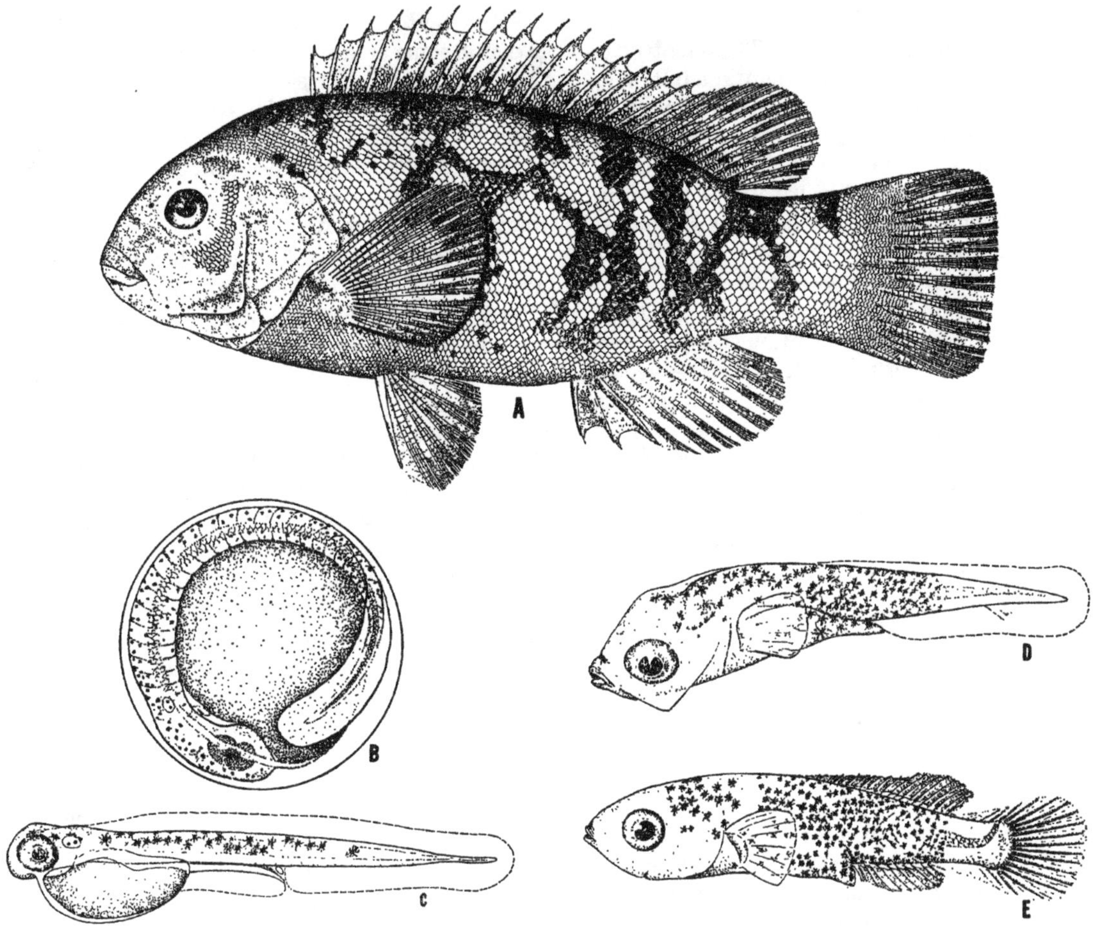

FIGURE 250.—Tautog. *Tautoga onitis*. A, adult, Woods Hole, Mass.; from Goode, drawing by H. L. Todd. B, egg; C, larva, one day old, 2.9 mm.; D, larva, 5 mm.; E, young fry, 10 mm. B–E, after Kuntz and Radcliffe.

Description.—The tautog suggests an overgrown cunner, but it is a heavier, stouter fish (about three times as long as deep, not counting the caudal fin) with caudal peduncle so broad and caudal fin so little wider than the peduncle that it is hard to hold a heavy one by the tail. The most obvious differences between the two fish are that the dorsal profile of the head of the tautog is high-arched, its nose is very blunt, and its lips are much thicker, giving it a facial aspect quite different from that of a cunner. A more precise if less obvious character is that the cheek region close in front of the gill opening (scaly in the cunner) is naked in the tautog and velvety to the touch. The fins of the tautog practically reproduce those of the cunner in relative size and location. The dorsal fin (16 to 17 spines and 10 soft rays) originates over the upper corner of the gill openings and runs back the whole length of the trunk; the anal (3 stout spines and 7 or 8 soft rays) corresponds in outline to the soft portion of the dorsal, under which it stands. The caudal fin is slightly rounded at the corners, the pectorals are large and rounded, and each of the ventrals has one stout spine. The jaw teeth of the tautog (in two series) are stout, conical, with the two or three in the front of each jaw larger than the others. The tautog has, besides, two groups of flat, rounded, crushing teeth in the rear of the mouth, as the cunner has also.

Color.—The tautog is a rather dark fish, generally mouse color, chocolate gray, deep dusky green, or dull blackish, with the sides irregularly mottled or blotched with darker. These mottlings are more evident in the young than in adults and usually they are grouped as three pairs of more or less continuous bars. Large fish are often almost plain blackish. The belly is only slightly paler than the sides, but the chin usually is white on large ones, a very conspicuous character. Tautogs, like cunners, vary greatly in color on different bottoms, and also in their markings.

Size.—Maximum length about 3 feet. The 22½ pounder, 36½ inches long, mentioned by Goode[93] as caught off New York in 1876 and preserved in the United States National Museum, still remains the heaviest fish recorded definitely. Fish of more than 14 pounds are very rare, with 12-pounders unusual. Tautog average about 2 to 4 pounds as they come to market.

Habits.—The tautog is even more strictly a coastwise fish than the cunner. Northward from Cape Cod it is unusual to catch one more than 3 or 4 miles from the land, or deeper than, say 30–60 feet; we have never heard of one caught on a long line set for cod or haddock, and they are unknown on the offshore fishing banks. But they range farther out and deeper to the southward, being one of the commoner fishes caught in 10–13 fathoms on the Cholera Bank, 10–12 sea miles offshore from Long Island, and on Seventeen Fathom Bank, 8 miles off northern New Jersey. At the other extreme, they follow the flood tide up above low-water level around ledges, to prey on the abundant supply of blue mussels along the intertidal zone, dropping back into deeper water during the ebb. We have helped to seine many small ones close along the shore in only a few feet of water, at Provincetown as well as southward, and it is not unusual for tautog to run up into brackish water, but we have never heard of them entering fresh water.

Their favorite haunts are along steep, rocky shores; around breakwaters, offlying ledges and submerged wrecks; around the piers and docks; over boulder strewn bottoms; and on mussel beds. In some places, however, good numbers are caught on smooth bottom, far from any rocks (the eastern side of Cape Cod Bay is an example, see p. 482). And young fry, 2 to 4 inches long, are often seined on sandy beaches.[94]

When tautog are not feeding they are likely to gather in some hole or cleft among the rocks, where they lie inert, on their sides, often several crowded together, until the rising tide stirs them to activity again.[95] And they are extremely local fish, perhaps more so than any other Gulf of Maine species that is interesting either to the angler or to the commercial fisherman.

While tautog are seldom seen before well into April in any part of their geographic range, or after November, they do not carry out any extensive migrations with the seasons. At most, those that find themselves in shoal water in autumn may drop off into slightly deeper water, to spend the cold season lying among eelgrass (*Zostera*), where this has reestablished itself; in crevices among rocks; or (in the case of the young ones) in empty oyster and clam shells. They move and feed little then, though they have been caught in lobster pots there and on hook and line off Rhode Island.[96]

Tautog, like cunners (p. 475), are sometimes chilled and killed if they are caught in shoal water by a sudden cold snap, as happened along Rhode Island and southern Massachusetts in 1841, 1857, 1875, 1901, and no doubt on many other occasions that have not found their way into print or into the records of the Bureau of Fisheries.

Food.—Tautog feed on invertebrates, chiefly on mollusks (both univalves and bivalves), especially on mussels which are the chief diet of the tautog living about ledges, and on barnacles that they pick off the rocks. Crabs and hermit crabs are favorite morsels. They also eat sand dollars, scallops, amphipods, shrimps, isopods, and lobsters, swallowing the smaller ones whole, but cracking the larger with their crushing teeth (p. 479). A tautog of about 2 pounds that we once caught off Cohasset, Mass., had made a meal of gammarid amphipods (sand fleas) gleaned from among the rockweed with which the ledge was clothed, though cunners caught at the same time and place were full of barnacles. We think it likely that tautog living in shallow bays (Duxbury, for example) prey

[93] American Fishes. 1888, p. 292.

[94] We have seined tautog fry in such situations in localities as far apart as Provincetown Harbor; Woods Hole, Cape Poge Bay, Marthas Vineyard, and Cape Charles Beach, Va. And good numbers of larger tautog have been reported as caught occasionally in nets in the vicinity of Provincetown; 8,700 pounds for example in 1898, and 5,800 pounds in 1899.

[95] We have often observed this habit of theirs in the large live tank at the Woods Hole Oceanographic Institution.

[96] Tautog have been described as burying in the mud, but we cannot vouch for this. And we put no credence whatever in the old myth that the vent of the tautog closes over in winter.

to a considerable extent on sea worms (*Nereis*); certainly they take these freely as bait.

Breeding habits.—About Woods Hole the tautog spawn chiefly in June, and the season for such of them as breed north of Cape Cod is probably early and midsummer. The eggs are buoyant, without oil globule and resemble those of the cunner, except that they are a little larger (0.9 to 1 mm. in diameter). At a temperature of 68° to 72° incubation occupies 42 to 45 hours, and probably 10 to 12 hours longer in the cooler water of Massachusetts Bay. The larvae [97] are about 2.2 mm. long at hatching. When 4 days old (temperature of 68°–72°) they have grown to 3.3 mm., the yolk has been absorbed, and the mouth is fully formed. Larvae of 5 mm. show the first traces of the caudal fin rays; the dorsal and anal fins are differentiated at 10 mm. and by the time the little fish are about 30 mm. long they show the fins, form, deep caudal peduncle, and blunt nose of the adult tautog. The larvae and youngest fry of the tautog and of the cunner resemble each other closely in general form, but the arrangement of the pigment offers a ready means of identification at all but the very earliest stages, for the black pigment cells remain more or less uniformly scattered over the whole trunk in the tautog, whereas they soon cluster in two definite patches in the cunner as is described elsewhere (p. 476).

Probably Tracy [98] is correct in assuming that the young tautogs of 3 to 8 inches, which may be seined in abundance along the shores of southern New England in summer, are 1 year old. Nothing definite is known of the rate of growth of older tautog, nor at what age they mature. But we suspect that large ones of 8 pounds and more may be 8 to 10 years old.

General range.—Atlantic coast of North America from the outer coast of Nova Scotia to South Carolina, chiefly south of Cape Ann; most abundant between Cape Cod and the Delaware Capes, and restricted to the immediate vicinity of the coast.

Occurrence in the Gulf of Maine.—The center of abundance of the tautog lies to the south of Cape Cod. Most of the authors, in fact, who have written of it have accepted Mitchill's [99] statement that it was not native north of Cape Cod and was introduced there shortly prior to 1814, there being no definite record of them in the Gulf of Maine prior to that date. But it seems far more likely that the anonymous writer who stated in the Gloucester Telegraph of May 5, 1860, that tautog had been plentiful there many years before, and had merely reappeared after a period of scarcity, was correct; also that this reappearance would have taken place in any event, even if none had been liberated north of Cape Cod.

Apart from Mitchill's statement that by 1814 the Boston market had a full supply (which may have come from south and not north of Cape Cod), the first positive record of any in Massachusetts Bay is of several that were caught along the Cohasset rocks in 1824,[1] which the local fishermen said was a species new to them. Tautog, however, were being caught in numbers in the inner parts of Massachusetts Bay (e. g., Lynn, Nahant, Boston Harbor) by 1839; they were more abundant then around Manomet Headland in Plymouth; and they already supported a considerable hook-and-line fishery at Wellfleet. A few years later their presence was established for the coast of Maine, and in 1851 tautog were reported as common (according to Perley) in St. John Harbor, New Brunswick, though these Bay of Fundy fish were introduced (not native). In 1876 the weirs north of Cape Cod took 2,274 pounds of tautog, and in 1879 Goode and Bean described them as abundant in many localities about Cape Ann.

At present (or within the last few years, for this fish fluctuates in abundance from year to year), the regular range of the tautog includes the whole coast line from Cape Cod around to Cape Ann, in suitable localities.

Tautog are less regular northward from Cape Ann, less abundant, and more local. But there are some tautog grounds about the Isles of Shoals, off Cape Porpoise, and about Casco Bay, where Kendall wrote of them in 1931 as having been "locally numerous" for some time previous.[2] We have also heard of tautog along the ledges near Boothbay Harbor and in Penobscot Bay. East of the latter tautog certainly are not common.

[97] Kuntz and Radcliffe (Bull. U. S. Bur. Fish., vol. 35, 1918, p. 92) describe the eggs, larvae, and fry.

[98] 40th Ann. Rept. Inland. Fish. Rhode Island, 1910, p. 137.

[99] Trans. Lit. Philos. Soc., New York, vol. 1, 1815, p. 400.

[1] Goode, Fish. Ind., U. S., Sect. 1, 1884 p. 269.

[2] According to Kendall (Bull. 58, Boston Soc. Nat. Hist., 1931, p. 10–11) the green crabs (*Carchinides maenas*) found in Casco Bay were not native there but had been introduced as tautog bait.

And it is so scarce a fish in the Passamaquoddy region (it has long since vanished from St. John Harbor) that three specimens, only, are known to have been taken there within recent years.[3]

One has been taken near the head of the Bay of Fundy on the Nova Scotian side (Scotts Bay, Kings County) one on the Nova Scotian shore of the open Gulf of Maine (Cranberry Head, Yarmouth County), and one on the outer coast of Nova Scotia near Halifax (Petpeswick Harbor, Halifax County), this last being the most northerly record for the tautog.[4]

The more productive tautog grounds north of the elbow of Cape Cod of which we chance to know are the Cape Cod Bay shore southward from Wellfleet; the Sandwich-Sagamore shore with the jetties at the mouth of the Cape Cod Canal; the bouldery ground around Manomet headland and nearby; Gurnet Point at Duxbury; the ledges off Scituate and Cohasset and especially those off Swampscott; the Nahant, Marblehead, and Magnolia Rocks; and here along the rocky shore from Gloucester Harbor around Cape Ann. The Cape Cod Bay grounds are exceptional, for the tautog caught there are on smooth bottom, not among ledges which are the usual haunts. We have also known of good-sized tautog taken inside of Nauset Inlet (where there are scattered boulders only), one in a lobster pot during the summer of 1949. And quite a number, large and small, are caught within Duxbury Bay, especially around the pilings of Powder Point Bridge.

Although tautog tend to gather in certain choice spots, they move around enough so that some idea of their relative importance along different parts of the coast line can be determined from the catches made in pound nets. Thus the average yield per pound net or trap has run from twice to 20 times as great for Cape Cod Bay as for the north shore of Massachusetts Bay in reasonably good years [5] during the periods between 1890 and 1921, when the catches for Massachusetts were reported by towns, hence can be localized.[6]

The regional discrepancy has not always been so wide in seasons when the Cape Cod Bay catch has been smaller; in 1909, for instance, when the total catch reported for Cape Cod Bay was only 635 pounds of tautog (with 27 pound nets in operation) the average catch per pound net or set of pound nets was nearly as great for the coast from Boston Harbor to Gloucester (total catch 203 pounds with 12 nets or sets of nets in operation). But the pound nets take a few tautog in Cape Cod Bay, even in years when they are so scarce north of Boston that none at all have been reported for Essex County, despite the fact that the bottom seems more suited to tautog there because rockier. The slightly lower temperature along the north shore of Massachusetts Bay may have been the contributing factor.

During the peak period 1895–1899, the chief center of abundance for tautog for Cape Cod Bay seems to have been along the Sagamore shore, where the yearly catch, per pound net, then averaged about 2¼ times as great as for the eastern shore of the Bay,[7] Brewster to Provincetown. And catches of 18,100 pounds of tautog by 2 pound nets at Sandwich in 1895 and 36,010 pounds of tautog in 12 nets in Brewster in 1898 suggest concentrations of tautog quite out of the ordinary. But the best tautog fishing has been reported from the Wellfleet region in recent years.

Catch statistics suggest, also, that not much interchange takes place between the populations of tautog of the Cape Cod Bay region, and of the the rocky coasts along the north shore of Massachusetts Bay, for the peaks of abundance (as judged from the reported landings) have fallen in different years in these two regions.

April 29 (1949) and May 1 (1950) are the earliest dates at which we have heard of tautog caught either in Massachusetts Bay or in Cape Cod Bay (Duxbury in both instances).[8] In 1950, which appears to have been an "early" season, they were reported as biting well in Cape Cod Bay by May 25 and at Duxbury by the last days of the month;

---

[3] One in Passamaquoddy Bay in 1909 or 1910 (Reported by Huntsman Contrib. Canadian Biol. (1920–1921) 1922, p. 64); a second in a tidal tributary of the St. Croix River in the summer of 1934, and another there in August 1935 (reported by McGonigle and Smith, Proc. Nova Scotian Inst. Sci., vol. 19, 1936, p. 160); all of these were taken in herring weirs.

[4] These Nova Scotian specimens are in the Provincial Museum at Halifax; see Vladykov and McKenzie (Proc. Nova Scotian Inst. Sci., vol. 19, 1935, p. 100), and Fowler (Proc. Acad. Nat. Science, Philadelphia, vol. 67, 1916, p. 517) for further details.

[5] The reported catches for 1895, 1897-1900, and 1910 were 5 to 43 times as great for Cape Cod Bay as for Essex County, made in 1.5 to 2.3 times as many pound nets or sets of pound nets.

[6] 1890–1900; 1906–1911; 1917–1921, in particular. Data are also available for earlier years.

[7] Total catch, Sagamore and Sandwich, 41,053 pounds, with 2 to 5 pound nets or sets of nets working in the different years; Brewster to Provincetown, 18,549 pounds with 14 to 24 pound nets or sets of pound nets in operation yearly.

[8] As reported in Salt Water Sportsman for May 25, 1950.

further up the Bay, however, as at Cohasset and Swampscott, very few are caught before July. In most years the best catches are made in August, September, and into October, and we have not heard of a tautog taken anywhere in the Gulf after early November at latest.

The tautog that frequent any particular ground in the Massachusetts Bay-Cape Cod region may be expected to pass the winter in a more or less inactive state close at hand, as they do farther south (p. 480). But we have no first-hand information in this respect.

Presumably the tautog spawn chiefly in June in Cape Cod and Massachusetts Bays, as they do along southern Massachusetts; perhaps into July.[9] But we have found no tautog eggs nor larvae in our towings in the Gulf of Maine, nor have any tautog less than 2 or 3 inches long been credibly reported in Cape Cod Bay or to the northward (we may have missed them as tautog spawn close in to the coast). And we have yet to learn whether the fluctuating stock north of Cape Cod is maintained wholly by local reproduction, or is reinforced from time to time by immigrants from the south. It would be especially interesting to know how many tautogs find their way from Buzzards Bay to Cape Cod Bay through the Cape Cod canal.

Fluctuations in abundance.—The pound net catches of tautog (averaged for 3-year periods) suggest that a moderate and irregular rise in abundance took place in the northern side of Massachusetts Bay from 1890–1892 (yearly average, 140 pounds) to 1899–1901 (yearly average, 1,049 pounds), followed by a corresponding decrease so extreme that none at all were reported from the pound nets of Essex County for 1917 to 1919, in the Massachusetts statistics,[10] only 42 pounds for 1920, and none for 1921. The local stock seems next to have built up again about to its former level, to continue so during the period 1928–1938.[11] Our angler-correspondents report that some tautog are caught along the Essex County rocks every summer, since then. But the fisheries statistics have not afforded information as to the tautog situation there for the past few years.

In the Cape Cod Bay region (again according to statistics of the landings) tautog seem to have been scarce for some years through 1890; then to have increased in numbers to a rather pronounced peak of abundance in 1895–1900, when the reported catch averaged about 13,190 pounds yearly (the maximum 22,264 pounds in 1895), an increase that came 5 or 6 years earlier than the upswing recorded for the north shore of Massachusetts Bay. There appear to have been fewer in 1899 (6,282 pounds recorded); perhaps not more than half as many in 1906 or in 1907 (3,168 pounds and 2,934 pounds reported) when the publication of the catches by towns was resumed, and apparently rather fewer still during the 4-year period 1908–1911.[12]

The Cape Cod Bay population seems to have been at about this same level in 1917, and tautog seem to have been more plentiful again in 1918, when the very large catch of 36,000 pounds was reported from the pound nets along the shore line of Brewster. But they fell, then, to so low an ebb that the reported yearly catches for 1919 and 1920 were only 801 and 877 pounds, respectively, and 44 pounds in 1921. Catch records tell nothing as to the status of the tautog in Cape Cod Bay since 1921.[13]

The disappearance of the eel grass (*Zostera*) about 1930–1931, must have altered their local habitat for the worse. But the stock seems to have built up again with the reappearance of eel grass here and there. And tautog have been plentiful enough around Cape Cod Bay during recent summers for party boats, hand-lining, to have made good catches there day after day. The traps at Barnstable continue to take some even though they are set on sand bottom, with their best catches in autumn when a single lift of 4 traps sometimes yields as much as 400 pounds.

According to local report, 1950 was a very good tautog season in Cape Cod waters. But the commercial fishermen took few or none smaller than one-half pound that year.[14] What this presages for the future remains to be seen.

Importance.—Tautog are not plentiful enough anywhere north of the elbow of Cape Cod to be of any great commerical importance, and never have

[9] In 1950 the "spawning run" was reported as about over in Cape Cod Bay by the end of the first week in June (Salt Water Sportsman for June 9, 1950).

[10] One hundred and fifty-eight pounds were reported for that year in the statistics of the U. S. Bureau of Fisheries.

[11] Landings, Essex County, 1928–1931, 1933, 1935, 1938, 0–803 pounds, average about 300 pounds. There is no reason to suppose that the 10,700 pounds reported for 1937 came from the Gulf of Maine. See footnotes p. 415; and p. 422.

[12] Maximum, about 3,900 pounds in 1910; minimum, 635 pounds in 1909; yearly average, about 1,400 pounds.

[13] There is no way of knowing how great a part of the catches reported in subsequent years from "Barnstable County" came from the Cape Cod Bay shore; i. e., from the Gulf of Maine.

[14] Information supplied by Henry Lyman.

been, but there is a ready sale for all that are brought to market, most people thinking this a very good table fish. And with so few fishes in the Gulf of Maine that can be classed as "game" (that is, affording sport on rod and reel), we may well wish the tautog were more plentiful there, for they put up so strong a resistance that tautog fishing is very good sport indeed.

Along the stretch from Manomet Headland, Plymouth, to Cape Ann, tautog are caught either from a boat at anchor over submerged ledges or bouldery bottom, or by casting with a long rod from dry ledges or from the rocky coast line. In either case, the fish are so local and irregular in distribution (depending on the food supply and also on the contour of the rocks) and so stationary that it is worth fishing for them only in certain spots. Even so, a few feet one way or the other may mean the difference between success and failure. In Cape Cod Bay, however, where the tautog are on smooth bottom, they lie in little openings among eel grass (whenever there is any), "with just their snouts sticking out" as an angler friend writes us,[15] "and, by lowering a fiddler or hermit crab in the clear spot in front of them, they will be caught in very shallow water."

Fishing the Cohasset rocks, we have found green crabs (*Carcinides*) the most attractive bait, whole if small enough, cut if larger; rock crabs (*Cancer*), or hermit crabs second best; large snails or cockles (*Polynices*) fairly good; lobster would perhaps be best of all, were it not so expensive. Mussels are often successful. And small whole clams are good, hooked through the "neck", (actually the siphon) with the shell cracked so as to let the juices escape, but they are next to worthless if shelled because they are stolen almost at once by the swarms of cunners. Anglers tell us that the same baits are used along the north shore of Massachusetts Bay. In Cape Cod Bay, where tautog are caught on smooth bottom (p. 480), the baits most used are hermit crabs and fiddler crabs.[16] We once had a good-sized tautog strike a sea worm (*Nereis*), behind our boat, while trolling for striped bass.

When a tautog bites, it passes the bait back to the pharyngeal teeth, to crush the shell before swallowing; in doing so he gives several distinctive jerks or twitches. This is the time to hook him; many are missed by being struck too soon by anglers not experienced in the ways of the tautog.

THE REMORAS OR SHARK SUCKERS. FAMILY ECHENEIDAE

The several remoras are easily distinguished from all other fishes by the fact that the spiny part of the dorsal fin is modified into a flat oval sucking plate, composed of a double series of cartilaginous crossplates with serrated free edges, and situated on the top of the head and neck. All the remoras, too, are slender of form with the lower jaw projecting well beyond the upper. Their mouths are armed with many small pointed teeth; their soft dorsal and anal fins are about the same in form and size, the one above the other; and their pectoral fins are set high up on the sides. The lower surface of the head is convex, the upper flat (a very conspicuous feature) with the lower surface of the body nearly as deeply colored as the upper so that the back is often mistaken for the belly. The members of this family all attach themselves to other fishes, or to sea turtles, by their sucking disk, usually clinging to the sides of their hosts, but often within the mouth or gill cavities of the larger sharks and of the giant rays.[17] They are carried about in this way, and they feed on the scraps from the meals of their transporters. All the remoras are tropical; they appear only as strays in boreal seas, usually fast to sharks or to swordfish.

We follow Sumner, Osburn, and Cole [18] in uniting under one species the shark sucker (*naucrates*), with more than 21 plates but a sucking disk less than one-fourth as long as the body, and the pilot sucker (*naucrateoides*), with only 20 or 21 plates but longer, fishes that are otherwise indistinguishable one from the other.

[15] Quoted from a letter from Henry Lyman.

[16] Not having fished there for tautog, we welcome this information from Henry Lyman.

[17] Gudger (Natural History, vol. 22, No. 3, May-June 1922, pp. 243-249) gives an interesting account of this habit.

[18] Bull. U. S. Bur. Fish. vol. 31, Pt. 2, 1913, p. 766.

KEY TO GULF OF MAINE REMORAS

1. Pectoral fins pointed; ventral fins attached to the belly for less than one-third of their length.. Shark sucker, p. 485
   Pectoral fins rounded; ventral fins attached to the belly for more than half of their length.................. 2
2. Dorsal fin of 29 rays or more; at most 17 plates in the sucker........................ Swordfish sucker, p. 486
   Dorsal fin of only about 23 rays; about 18 plates in the sucker.............................. Remora, p. 487

Shark sucker *Echeneis naucrates* Linnaeus 1758

PILOT SUCKER; WHITE-TAILED SUCKER

Jordan and Evermann, 1896–1900, pp. 2269–2270, as *Echeneis naucrates* Linnaeus 1758 and *E. naucrateoides*, Zuiew, 1789.

Description.—The most distinctive characters of the shark sucker are mentioned above. It is a very slim fish, 11 or 12 times as long as it is deep, nearly round in cross section, and tapering to a very slender caudal peduncle. The sucking plate, reaching from close behind the tip of the snout back over the nape of the neck even with the middle of the pectoral fin, is about as broad as the head, flat, oval, and with 20 or more very conspicuous transverse plates. The soft dorsal fin (32 to 41 rays) and the anal fin (31 to 38 rays) both originate about the mid length of the body, and they both extend nearly to the base of the caudal fin. Both of them taper, too, from front to rear, but the anal is more concave in form than the dorsal. The caudal fin is slightly concave in old fish but in young ones its central rays are the longest. The ventral fins are pointed like the pectorals below which they stand, and their inner rays are attached to the skin of the abdomen for only a short distance. The broad-based pectoral fins are set so high up on the sides that their upper margins are close below the overlapping edge of the sucking plate.

Color.—The general ground tint is slaty or dark brownish gray, with the belly nearly as dark as the back. Each side is marked by a broad darker brown or sooty stripe with white edges, that runs from the angle of the jaw to the base of the caudal fin but is interrupted by the eye and by the pectoral fin. The caudal fin is velvety black with white corners, a character noticeable enough to give rise to a vernacular name. The dorsal and anal fins are dark slate color or black, more or less margined with white. The pectorals and ventrals are black, either plain or more or less pale edged.

Size.—Reaches about 38 inches.

General range.—Cosmopolitan in warm seas, north as a stray to Halifax, Nova Scotia,[19] on the Atlantic Coast of North America.

Occurrence in the Gulf of Maine.—So far as we can learn no shark sucker of this species has been reported from the Gulf for many years; in fact, the only positive records of it there are for one taken from the bottom of a fishing boat in Boston Bay some time prior to 1839;[20] for a second reported by Wheatland[21] from Salem Harbor (reidentified by

[19] Leim Proc. Nova Scotian Inst. Sci., vol. 17, Pt. 4, 1930, p. XLIV.

[20] Described and illustrated by Storer (Fishes of Mass., 1867, p. 210, pl. 32. fig. 3).

[21] Jour. Essex Co., Nat. Hist. Soc., vol. 1, No. 3, 1852, p. 125.

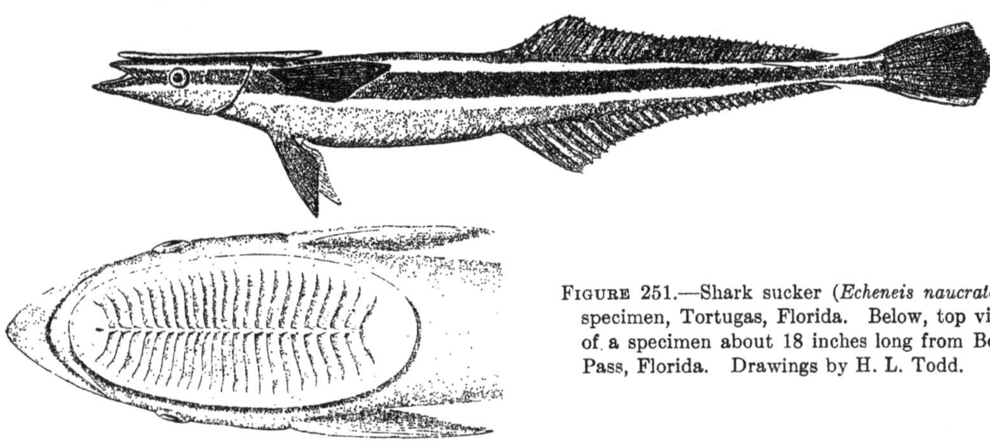

FIGURE 251.—Shark sucker (*Echeneis naucrates*), 11-inch specimen, Tortugas, Florida. Below, top view of head of a specimen about 18 inches long from Boca Grande Pass, Florida. Drawings by H. L. Todd.

Goode and Bean as *naucrateoides*); and for a third reported by Goode and Bean [22] as taken at the mouth of the Merrimac River in June 1870. And Leim [23] reports one from Halifax Harbor, Nova Scotia, during the fall of 1928. It is only as the rarest of strays that it ever wanders north of Cape Cod, clinging to some ship (for such is a common habit in its tropical home) or to some shark.

Swordfish sucker *Remora brachyptera* (Lowe) 1839 [24]

Jordan and Evermann, 1896–1900, p. 2272

Description.—This is a stouter fish than the shark sucker (p. 485), being only about seven times as long as it is deep (counting the caudal fin) and about as thick through the shoulders as it is deep, with a thicker caudal peduncle. And although the sucking plate is as long, relatively, it consists of only 14 to 17 ridges. Furthermore, the pectoral fins of the swordfish sucker are relatively shorter than those of the shark sucker, softer, and rounded instead of pointed, while the upper margins of these fins are not so close to the edge of the sucking plate. The ventral fins, too, are attached to the skin of the abdomen along their inner margins for at least one-half their length, as noted above (p. 485). The long dorsal fin (29 to 32 rays) of the swordfish sucker serves to separate it from the remora (p. 487).

[22] Bull. Essex Inst., vol. 11, 1879, p. 20.

[23] Proc. Nova Scotian Inst. Sci., vol. 17, No. 4, 1930, p. XLVI.

[24] Jordan, Evermann, and Clark (Rept. U. S. Comm. Fish. (1928), Pt. 2, 1930, p. 449) place this species in the genus *Remoropsis* Gill, 1864.

Color.—Described as light reddish brown above and as darker below, with paler dorsal and anal fins. A distinctive feature is that it lacks the side stripes and white fin edgings so characteristic of the shark sucker.

Size.—A length of 12 inches is the maximum so far recorded.

General range.—Warm and warm-temperate seas generally, probably paralleling that of the swordfish.

Occurrence in the Gulf of Maine.—Goode and Bean's [25] description of this sucker as not infrequently accompanying swordfish into Massachusetts Bay probably applies to the whole Gulf except the Bay of Fundy, for specimens have been brought in from near Matinicus Rock and near the Isles of Shoals; fishermen occasionally speak of seeing "suckers" clinging to the swordfish they harpoon on the offshore Banks; sometimes several fastened to a single swordfish. But they also report far more swordfish lacking these uninvited guests than carrying them, and this has been the case with the few fish harpooned by the *Grampus* during our cruises in the Gulf. Suckers are described by eyewitnesses as usually fast to the shoulder of the swordfish, nor have we heard of one actually within the gill cover of the latter, though very likely they refuge there, for one has been

[25] Bull. Esesex Inst., vol. 11, 1879, p. 21.

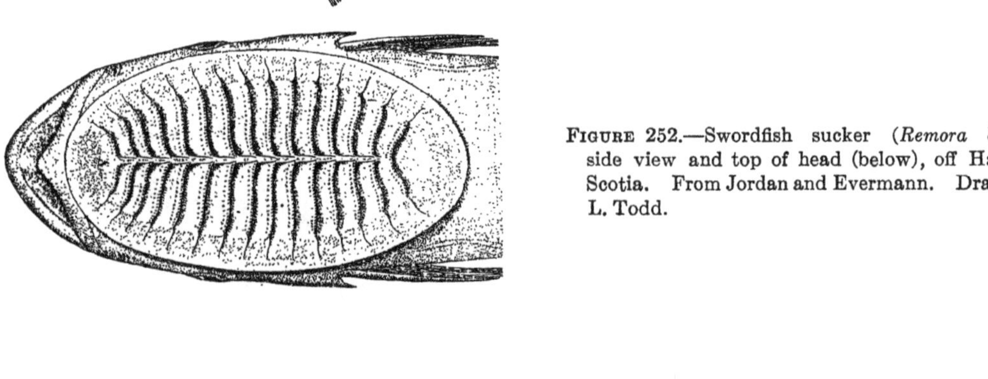

FIGURE 252.—Swordfish sucker (*Remora brachyptera*), side view and top of head (below), off Halifax, Nova Scotia. From Jordan and Evermann. Drawings by H. L. Todd.

taken from the gill cavity of a sunfish (*Mola mola*); others have been found in the mouths and gill cavities of large sharks.

Nothing beyond this is known of its way of life. Presumably it feeds on fragments of the fish killed by its host, as the shark sucker does whose actions are better known. Presumably, too, it is an active a swimmer as are its relatives. Nothing is known of its breeding habits.

Remora *Remora remora* (Linnaeus) 1758

Jordan and Evermann, 1896–1900, p. 2271.

Description.—The chief distinctions between the remora and the swordfish sucker is that it has a larger number of ridges in its sucking plate on the average (16 to 20, as against 14 to 17), and that there are only 22 to 25 rays in its dorsal fin, whereas the swordfish sucker has 29 to 32. Like the latter, it is a stouter fish than the shark sucker (p. 485), and its ventral fins are similarly attached to the skin of the abdomen along their inner edges.

Color.—Uniform brownish, blackish, or sooty, both above and below.

Size.—Maximum length about 18 inches.

Habits.—Very little is known of the life history of the remoras. The young fry of this, and of other species of *Remora* have been taken in the open Atlantic, usually in June or July which suggests a sharply limited spawning period. A remora may join a shark, or other host, when only about 1½ inches (3 to 4 cm.) long.[26] But we have yet to learn how long or how constantly one may accompany a single shark, or how often it may transfer from one host to another.

General range.—Tropical seas generally; very common in the West Indies, occasionally north to New York and to Woods Hole, and only a stray north of Cape Cod. It is usually attached to large sharks or to sea turtles.

Occurrence in the Gulf of Maine.—The only Gulf of Maine records for the remora, up to 1925, were of one taken many years ago in Salem Harbor, no doubt brought thither clinging to the bottom of some ship in from a southern voyage, as Goode and Bean [27] remarked; and of one in the Museum of Comparative Zoology that was taken at Provincetown long past. More recent records are of one found clinging to the bottom of a lobster trap in Portland Harbor in 1931, probably brought in by some West Indian schooner, several of which had recently been in the harbor;[28] of a second found sucking to the gills of a blue shark *Prionace glauca* caught on the northeast edge of Georges Bank, August 1 of that same year; and of a third fastened to a shark of the genus *Carcharhinus* that was caught at the surface over the southeast slope of Georges Bank in July 1939.[29]

Records from farther east and north along the American coast are of one taken from a blue shark, 10 miles off Cape Sable, June 1, 1933; [30] of another (also from a blue shark) west of Sable Island, September 9, 1934; and of two taken from sharks on St. Pierre Bank, south of Newfoundland, one of them on August 13, 1936,[31] the other on October 7, 1937.[32]

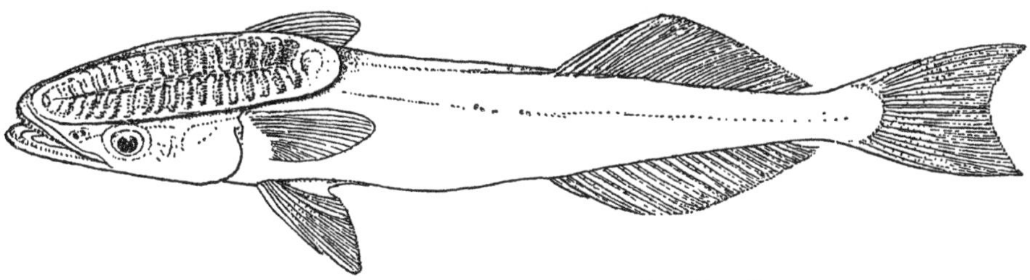

FIGURE 253.—Remora (*Remora remora*). After Day.

THE SAND LAUNCES. FAMILY AMMODYTIDAE

The slender, round-bodied sand launces suggest small eels in general appearance. Eel-like, too, they lack ventral fins, and they swim with eel-like undulations from side to side. But they are not even remote relatives of the true eels, from

[26] Tåning, Nature, vol. 20, 1927, p. 224.

[27] Bull. Essex Inst., vol. 11, 1879, p. 21.

[28] Reported to us by the late Walter H. Rich, of the U. S. Bureau of Fisheries.

[29] The last 2 are in the collection of the Museum of Comparative Zoology.

[30] Vladykov, Proc. Nova Scotian Inst. Sci., vol. 19, 1935, p. 7.

[31] McKenzie and Homans, Proc. Nova Scotian Inst. Sci., vol. 19, 1938, p. 279

[32] McKenzie, Proc. Nova Scotian Inst. Sci., vol. 20, 1939, p. 18.

which they are distinguishable at a glance by their forked caudal fin, separated by an appreciable space from both dorsal and anal fins; by their wide gill openings; and by the presence of a large bony gill cover, not to mention other anatomic characteristics equally important if less obvious. Only one species inhabits our Gulf.[33] The larger North European launce (*A. lanceolatus*), which grows to 12 inches and has 2 stout teeth on the roof of its mouth, has no representative in our side of the Atlantic.

Sand launce *Ammodytes americanus* De Kay 1842 [34]

SAND EEL; LAUNCE; LANT

Jordan and Evermann, 1896–1900, p. 833.

Description.—The sand eel is a slender fish, its body about one-tenth as deep as it is long (not counting caudal fin), with long head and sharply pointed nose, wide gill openings, and large mouth with the lower jaw projecting far beyond the upper. The jaws are toothless, and there are no teeth on the roof of the mouth. There is one long low dorsal fin, soft-rayed (59 to 64 rays; no spines), rising somewhat in front of the tips of the pectorals and running back along the whole length of the body nearly to the base of the caudal fin. The anal (28 to 32 rays), similar in outline and equally lacking spines, originates slightly behind the middle of the dorsal and runs equally far back. The tail is forked. The pointed pectorals are set very low down on the body, and there are no ventral fins. The scales are small, lying in cross series on the sides of the body between numerous skin folds that run obliquely downward and backward, and there is a low ridge of skin on either side along the belly.

The readiest field marks for the sand eel among Gulf of Maine fishes are its slender form and sharply pointed snout, coupled with long dorsal fin (separated from the caudal) and the absence of ventral fins. The only fishes with which one would be apt to confuse it are young eels, but in these the dorsal, caudal, and ventral fins are confluent, not separate, and the tail is rounded, not forked.

Color.—Authors differ in their accounts of the colors of the sand eel, probably because its iridescent luster fades at death and because it varies in shade on different bottoms. The many we have handled have been olive, brownish or bluish green above, with silvery lower sides and a duller white belly. Some have a longitudinal stripe of steel-blue iridescence [35] along each side, but others lack this.

Size.—Mature sand eels run from about 4 inches to about 6 inches in length as a rule, with a few as long as 7 inches.[36]

Habits.—Sand eels are found chiefly along sandy foreshores, also over the shoaler parts of the offshore fishing banks; they are seldom seen off rocky parts of the coast, or over muddy bottoms in deep water. They usually congregate in dense schools, often of thousands of individuals, and they swim as an eel does, by sidewise undulations that run along the body from front to rear, which makes them easy to recognize in the water.

The most interesting habit of the sand eel is the custom it has of digging itself several (4 to 6) inches deep in the sand, into which it burrows

[33] A second species of launce (the Arctic *Ammodytes dubius* Reinhardt 1838), thought to be characterized by having more fin rays (65–67 dorsals, 33–36 anals), has been reported from Boston by Günther (Catalogue Fishes British Mus., vol. 4, 1862, p. 387), and from Woods Hole by Smith (Bull. U. S. Fish. Comm., vol. 17, 1898, p. 95), but it is probable that the specimens in question were merely large *Ammodytes americanus*. In fact it is doubtful whether there is any sound distinction between the *A. dubius* of Greenland and the European *A. tobianus* on the one hand, and the American *A. americanus* on the other.

[34] Our sand eel is so closely allied to the common European launce (*Ammodytes tobianus*) that we doubt whether the distinction between the two (more slender form and longer head of *americanus*) will stand the test of time.

[35] In the European sand launce (*Ammodytes tobianus*), the sides are described as with lines of tiny brown dots and the tip of the snout as blackish, especially in the young fish.

[36] According to Bean (Bull. New York State Mus. 60, Zool. 9, 1903, p. 376) the majority are from 4⅞ to 7 inches; and the largest we measured was about 7 inches long. Storer (Fishes of Mass., 1867, p. 217) credits them with a maximum length of 12 inches; seemingly this was an error.

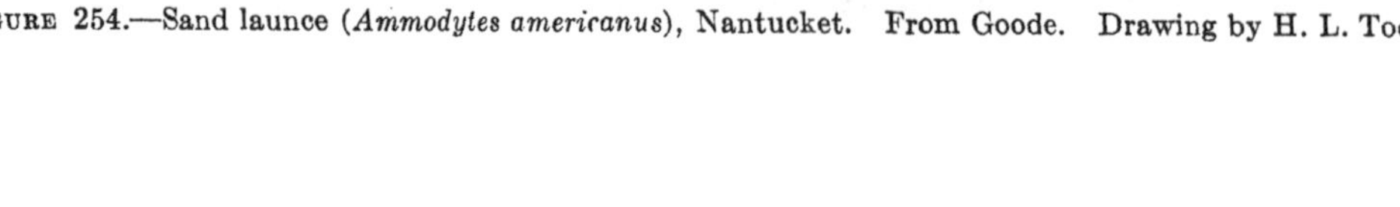

FIGURE 254.—Sand launce (*Ammodytes americanus*), Nantucket. From Goode. Drawing by H. L. Todd.

with great speed, opening the way with its sharp-pointed snout. It often does this above low-water mark at high tide to await the return of the tide. We have often seen them vanish with surprising rapidity when alarmed by clam diggers, and we cannot improve on Goode's [37] account of seeing "a great section of the beach" in Provincetown harbor become "alive with dancing forms of dozens of these agile fishes" when he stuck his clam-hoe into the sand. It has been suggested that they spend a large part of the time so buried, and that their sudden appearances and disappearances are to be explained thus, rather than as evidence of their wanderings or migrations. It is not known whether they follow this habit only in shoal water where they come under direct observation, or whether they also burrow into deeper bottoms. If the burrowing habit is for refuge, it is not always successful, for, as Smitt [38] remarks, porpoises have been seen rooting them out of the sand.

In Scandinavian waters sand eels feed on all sorts of small marine animals, but chiefly on small Crustacea, especially on copepods, and on fish fry, including their own kind. Worms have also been found commonly in the stomachs of sand eels, but it is not likely that they catch these while burrowing, as some writers have suggested.

The sand eel plays a very important role in the economy of northern seas as food for larger animals. Finback whales devour them greedily when they find them in abundance, as happened in Cape Cod Bay in June 1880, when launce appeared in swarms early in the month followed by finbacks a few days later. Porpoises, too, find them a staple article of food, and sundry predaceous fish such as cod, haddock, halibut, silver hake, salmon, mackerel, striped bass, and bluefish. When sand eels are fleeing from their pursuers especially from the silver hake, which does not hesitate to follow them up on the sand, they often strand in such numbers as to cover the flats.

Sand eels' noses are so sharp that when they are swallowed by cod, and perhaps by other fish, they sometimes work right through the stomachs and into the body cavities of their captors, to become encysted in the body wall.

[37] Fish. Ind. U. S.; Sect. 1, 1884, p. 244.

[38] Scandinavian Fishes, vol. 2, 1895, p. 579.

The spawning of the American sand eel has not been observed so far as we can learn.[39]

Ripe specimens of the European species (*tobianus*), both male and female, have been taken throughout the year, a phenomenon that has given rise to widely differing views as to its spawning season. But the chief production of its eggs takes place in autumn and early winter, at least in the southern part of the North Sea, as Ehrenbaum [40] demonstrated, both by dredging them in large numbers, and by the fact that its larvae are extremely abundant there from January to March, but have seldom been taken at other seasons.

The occurrence of larvae suggests that the season is about the same for the American form in the Gulf of Maine. Thus the *Fish Hawk* towed numbers of larval launce (identified by R. A. Goffin of the Bureau of Fisheries and by Mrs. C. J. Fish) near Provincetown and in Cape Cod Bay, during December, January, and February, 1924–1925, evidence both that this part of the Gulf of Maine (where adults are abundant) is the site of considerable reproduction, and that spawning commences as early as November there. Fry have been taken in March at Woods Hole while the *Albatross* towed a number of larvae of 11 to 17 mm. on the western part of Georges Bank on February 22, 1920. It seems that the spawning season is progressively later, however, to the northward, for we took larvae only a few days old (7 to 8 mm. long), with the yolk still showing, off Newburyport, Mass., on March 4 in 1921, while the Canadian Fisheries Expedition of 1915 obtained an abundance of slightly older stages (7 to 15 mm. long) off the southeast coast of Nova Scotia in May.

Evidently the sand eel breeds successfully throughout the more northern part of its range, for its larvae have been found, widespread, over the Nova Scotian Banks, in the Gulf of St. Lawrence northward nearly to the Strait of Belle Isle, throughout the Grand Bank region, off the east coast of Newfoundland and off the outer coast

[39] Hind (Fish. Comm. Halifax, 1877, Pt. 2, p. 7) describes the launce in the Gulf of St. Lawrence as "depositing their large reddish-colored ova on the sand between high and low water." This account, however, is widely at variance with the spawning habits of their European representative (*Ammodytes tobianus*) and with the seasonal occurrence of their larvae; and was probably borrowed from the larger European sand eel (*Ammodytes lanceolatus*).

[40] Wissenschaftliche Meeresuntersuchungen, Helgoland, Neue Folge, vol. 6, 1904, p. 184.

of Labrador, north to Sandwich Bay.[41] How far south it may do so is not known.

Sand eels were formerly thought to spawn on sandy beaches above low-water mark while burrowing in the sand, but their eggs have never been found in such a situation, and Ehrenbaum proved, by dredging them in large numbers, that those of the European species (*Ammodytes tobianus*) are actually deposited in depths of 10 fathoms or so, on sandy bottom where they stick fast to the grains of sand. His experience also suggests that they resort to very definite grounds for spawning, all of which probably applies as well to the American form as it does to the European.

The eggs of the American launce have not been seen. Those of the European *tobianus* are oval, 0.72 to 0.97 mm. in greatest diameter, with a yellow oil globule of 0.25 to 0.31 mm., and they are described as of an orange tint. The larvae are very slender, and about 7 mm. long by the time the yolk is absorbed. The dorsal and anal fin rays are visible when the larva is about 18 mm. long, but it is not until the little fish is upward of 25 mm. long that the tail begins to assume its forked outline; this is a convenient field mark for distinguishing between the launce and the herring, in which the tail is deeply forked from a much earlier stage. The early larval stages are easily recognizable by their slender form combined with the fact that the vent opens at one side, just as among the cod tribe, not at the margin of the larval fin fold, so that it apparently ends blind.

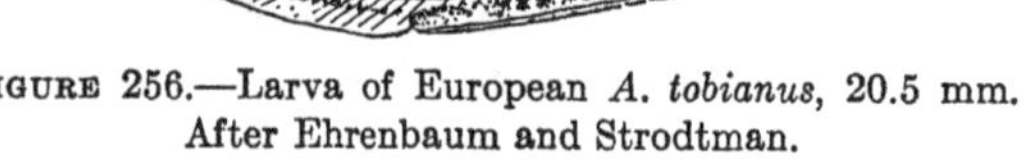

FIGURE 255.—Larva of European *A. tobianus* 6.6 mm. After Ehrenbaum and Strodtman.

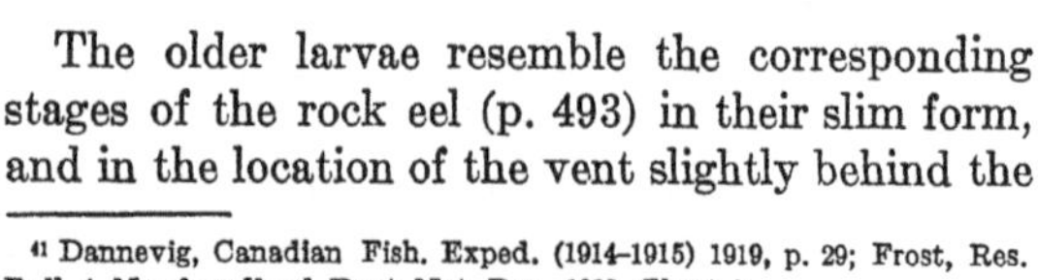

FIGURE 256.—Larva of European *A. tobianus*, 20.5 mm. After Ehrenbaum and Strodtman.

SAND LAUNCE (*Ammodytes*).

The older larvae resemble the corresponding stages of the rock eel (p. 493) in their slim form, and in the location of the vent slightly behind the middle of the trunk (it is located farther back in the similarly slender larvae of the herring tribe), but may be recognized by the row of black pigment cells along the *dorsal* side of the intestine instead of along the *ventral* side, and by their pointed noses.

The rate of growth of our launce has not been studied. But it is probable that the small ones of 3 to 4 inches, which are plentiful from July until September, are yearlings; those of 4 inches and upward probably are 2 years old, or more.

General range.—Atlantic coast of North America from Cape Hatteras to the Gulf of St. Lawrence, northern Newfoundland and northern Labrador, perhaps to Hudson Bay (p. 491). Its European relative, *A. tobianus*, occurs from Greenland, Iceland, northern Scandinavia and the White Sea south to Spain.

Occurrence in the Gulf of Maine.—The sand eel is very plentiful along the coast from Cape Cod to Cape Sable wherever there are sandy shores, but it is seldom seen off the rocky parts of the coast line. Thus it is rather scarce in the Bay of Fundy except locally, but is common on the sandy beaches that break the bold northern shores of the Gulf here and there. They swarm on the strands of Cape Cod Bay, a phase of their distribution associated with their burrowing habit. Here one may see schools of them throughout the summer in shoal water close in to tide mark, swimming with the curious undulating motion so characteristic of them; and they continue plentiful there in some years during the winter, when great numbers are sometimes cast on the beach in stormy weather. Sand eels are to be taken in shallow water on sandy flats throughout the year about Woods Hole also, but they are never so plentiful there in winter as they are in fall and spring. And since a general decrease in their numbers close inshore seems to take place during the cold months in the more northern part of their range as well, it is probable that a considerable proportion of the local stock moves out into deeper water for the winter, to return in spring, as most of the launce do in north European seas.

In Scandinavian waters this vernal inshore movement takes place in May as the coast waters warm up, and it is probable that the seasonal schedule is much the same in the Gulf of Maine, judging from its temperature. The sand eels may also be expected to leave some of the shallow-

[41] Dannevig, Canadian Fish. Exped. (1914-1915) 1919, p. 29; Frost, Res. Bull. 4, Newfoundland Dept. Nat. Res., 1938, Chart 8.

est bays of our Gulf in midsummer, when the water is at its warmest, to work in again in early autumn, such being their custom near Woods Hole.

There is no reason to suppose that they inhabit the central deeps of the Gulf of Maine regularly unless some of them repair thither in winter, though it would not be astonishing to find an odd sand eel in deep water at any time; in fact, we towed a young one about 1¼ inches long over the deep basin southeast of Grand Manan on June 10, 1915. But they must be plentiful on Nantucket Shoals, for they were found in the stomachs of cod caught there by the *Halcyon* and by the *Albatross II*, from time to time throughout the springs and summers of 1928–1930. There are also sand eels over the shallows of Georges and Browns Banks, whence they have been brought to the Bureau of Fisheries by fishermen on several occasions.[42]

Further northward, fishermen are familiar with them all along the outer coast of Nova Scotia and on the Scotian Banks; they are so common near Canso that a seine, dragged on a sandy beach there "captured hundreds in a short time";[43] they are reported from Prince Edward Island [44] and from the Magdalens in abundance,[45] here and there along the north shore of the Gulf; from the Strait of Belle Isle; also from Sandwich Bay and Sloop Harbor in southeastern Labrador; and they are to be expected all along the outer Labrador coast, for we have seen one taken at Eclipse Harbor, near Cape Chidley.[46] Sand eels have also been found on the southern side of Hudson Strait, and in the southern and western parts of Hudson Bay.[47] But these northern specimens may represent a distinct race for they have more fin rays (63–67 dorsal, 28–33 anal); Vladykov, in fact, has classed them as a new subspecies (*hudsonicus*) of the Greenland launce (*A. dubius*).

Sand eels are locally plentiful southward as far as northern New Jersey where we have seen squirrel hake gorged with them; they are reported as "common" as far as Cape May at the entrance to Delaware Bay,[48] and have been reported as far south as Cape Hatteras.[49]

Importance.—It is only for bait that sand eels are of any commercial value in our Gulf, for which purpose 67,800 pounds were landed from the traps in Massachusetts in 1919, 20,000 pounds in 1946.

BLENNY-LIKE FISHES. FAMILIES LUMPENIDAE, PHOLIDAE, AND STICHAEIDAE

The blenny tribes are characterized among Gulf of Maine fishes by the position of their ventral fins, which are under or in front of the pectorals, combined with a single dorsal fin that is spiny throughout its length and extends the whole length of the trunk, and with a slender form, eel-like in some of them. The only other Gulf of Maine fishes that resemble them are the wolffishes (p. 502) and the wrymouth (p. 500), but both of these lack ventral fins, which are present in all our blennies, though they may be very small. Furthermore, the tremendous canine tusks and molar teeth of the wolffish (p. 503) have no counterpart among the blennies, and the peculiar face of the wrymouth is equally distinctive for it (p. 501). The eelpout (*Macrozoarces*) also is somewhat blennylike in appearance, but the greater part of its dorsal fin is soft rayed, not spiny; and its anal fin is continuous with its caudal fin.

The blennies are a numerous tribe of small carnivorous shore fishes, widely distributed both in northern and in tropical seas. Five species are known in the Gulf of Maine.

KEY TO GULF OF MAINE BLENNY FISHES

1. Body very slender, about 18–20 times as long as it is high .... Snake blenny, p. 494
   Body only moderately slender, not more than 8 to 10 times as long as it is high 2
2. There is a row of conspicuous roundish black or dusky spots along the dorsal fin 3
   There is only one large and conspicuous dark spot on the dorsal fin, or none 4

[42] Rept. U. S. Comm. Fish. (1879) 1882, pp. 808, 812, 814, 817.

[43] Cornish (Contrib. Canadian Biol. [1902–1905] 1907, p. 84.)

[44] Leim, Proc. Nova Scotian Inst. Sci.; vol. 20, Pt. 2, 1940, p. 39.

[45] Cox, Contrib. Canadian Biol. (1918–1920) 1921, p. 111.

[46] This specimen about 3 inches long, collected by C. O. Iselin, now in the Museum of Comparative Zoology, appears to be a typical *americanus*, for it has 30 anal fin rays, and only 59 dorsal rays.

[47] Vladykov, Contrib. Canadian Biol., N. Ser., vol. 8, No. 2, 1933, pp. 23–25.

[48] Fowler, Rept. New Jersey State Mus. (1905) 1906, p. 411.

[49] Jordan and Evermann, Bull. 47, U. S. Nat. Mus., Pt. 1, 1896, p. 833.

3. The pectoral fins are about as long as the body is high; the ventral fins are well developed, without noticeable spines, and about as long as one-half the height of the body .... Arctic shanny, p. 497
The pectoral fins are only about one-half as long as the body is high; the ventral fins are minute (likely to be overlooked) .... Rock eel, p. 492
4. Pectoral fins evenly rounded, their middle rays the longest; dorsal fin marked on its forward part with one large and conspicuous dark blotch; only 43 or 44 dorsal fin spines .... Radiated shanny, p. 498
Pectoral fins with the lower rays longer than the upper rays and free at their tips; 58 to 61 dorsal fin spines .... Shanny, p. 497

Rock eel *Pholis gunnellus* (Linnaeus) 1758

Gunnel

Jordan and Evermann, 1896–1900, p. 2419.

Description.—The slender flexible trunk (only about one-tenth as deep as it is long and about one-half as thick as it is deep), short head, and rounded nose of this little blenny suggest an eel, but the spiny nature of its dorsal fin betrays its true relationship. The dorsal fin (73 to 86 spines) extends from the nape of the neck back along the whole length of the trunk to the base of the caudal fin, from which it is marked off by a shallow notch only; and it is of uniform height from end to end. The anal fin (2 very short spines and 37 to 44 rays) originates opposite the mid-length of the dorsal, to which it corresponds in height and outline, and runs back similarly to meet the caudal fin, from which it is marked off by a distinct notch, though there is no free space between the two fins. The caudal fin is small and rounded. The tiny ventrals, set near together close in front of or under the pectorals, are reduced to one very short spine and one rudimentary ray each. The pectoral fins, smaller than in our other blennies, are a little longer than one-half the length of the head. The skin of the trunk is clothed with very small scales, hardly visible, however, through the thick layer of slimy mucus with which the rock eel is covered. The head is naked; the mouth is small and oblique; the upper jaw is armed with several rows of conical teeth, but the lower jaw has a single row only.

Color.—A row of about 10 to 14 round, black-centered and pale-edged spots, spaced at equal distances along the middle of the back and spreading out onto the dorsal fin, are the most characteristic feature of the color pattern of this fish. The ground tint of the upper part is yellowish, olive brown, reddish, or light red, matching the seaweed or the bottom, with pale, irregularly rounded cloudings on the sides, and with an oblique streak from the eye to the angle of the jaw. The belly varies from pale gray to yellowish white. The pectoral, caudal, and anal fins are yellowish. We have seen a specimen (at Boothbay) that was brick red above and below, light and dark mottled, flecked also with tiny black dots, and with the spots on the dorsal fin dark red instead of black.

Size.—The maximum length is about 12 inches but few of those found are more than 6 to 8 inches long.

Habits.—Rock eels are often found along low tide mark, left by the ebb in little pools of water under stones, or among seaweed, where they await the return of the tide. But they are not confined to very shoal water as formerly supposed, for we have also collected them in considerable numbers both within the Gulf and on the offshore banks down to 40 fathoms, and one has been taken as deep as 100 fathoms (p. 494).

They are rather local throughout their range. In some places one is to be found under almost every stone; on others you may turn rocks in vain; their presence or absence along any particular stretch of shore probably depends on the character of the bottom immediately outside, for this fish prefers pebbly, gravelly, or stony ground, or shell beds, and not mud or eelgrass. Neither have we found them about the steep ledges so

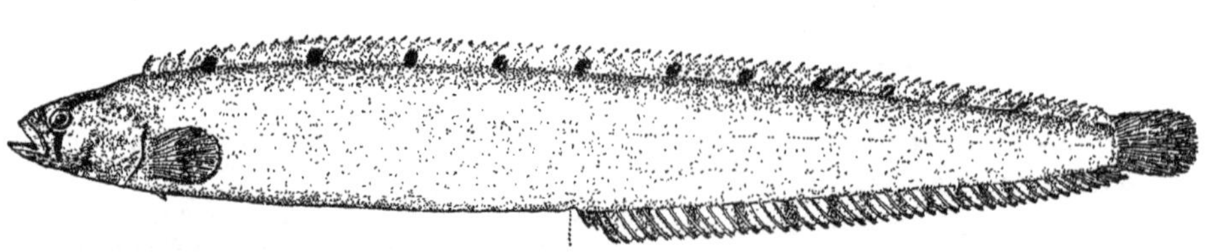

Figure 257.—Rock eel (*Pholis gunnellus*), Gloucester, Mass. From Jordan and Evermann. Drawing by H. L. Todd.

numerous along the rockbound coasts in the Gulf of Maine. In Scandinavian waters according to Smitt [50] they often take refuge inside large empty mussel shells. But as he remarks, there is no ground for the accusation that rock eels enter live bivalves of any sort to devour them. Whether they seek such places of concealment in deeper waters is not known.

When disturbed they squirm like eels. Eel-like, they swim by sidewise undulations, and they are so active and so slippery (hence the name "butterfish") that it calls for quick work to catch one by hand, even in a very small puddle.

Very little is known of the diet of the rock eel, except that it is carnivorous and that various molluscan and crustacean fragments have been found in their stomachs. Vinal Edwards records small amphipods, shrimps, and worms in the few that he examined at Woods Hole, but we have no first-hand information to offer on this point. In turn, rock eels have been found in the stomachs of various larger fishes, especially of cod, in New England waters.

So far as known the rock eel is resident throughout the year wherever it is found; at most it may move out from the beach into slightly deeper water in winter to escape chilling.

Breeding habits.—It is necessary to turn to European sources for information about its breeding habits, for its spawning has not been seen in American waters. In the eastern Atlantic and in the North Sea [51] it spawns from between tide marks down to 12 fathoms or more, from November to February or March. And its spawning season probably is the same in our Gulf, for eggs apparently belonging to the rock eel have been found off Rhode Island late in December.[52] A female from Peconic Bay, N. Y., contained 686 eggs.[53] These, by European accounts, are about 2 mm. in diameter, opaque, whitish, but iridescent on the surface, with a single globule of about 0.6 mm., and they are laid in holes or crannies. In British waters the rock eel usually chooses empty oyster shells, or the holes that are made in the limestone rocks by the boring bivalve *Pholas* for the purpose, but there are no oysters in the Gulf of Maine, except in Cape Cod Bay, and the local *Pholas* is unable to bore into the hard granite rocks of our coast line, so the rock eels must seek other nesting sites. Perhaps large mussel shells may serve them, or any crevice. The eggs are adhesive, and both the parents have been observed rolling them, by coiling around them, into balls or clumps an inch or so across, in which they stick together.

In European waters incubation occupies from 8 to 10 weeks, during which period the parent fish of both sexes have been seen lying close beside the egg clumps. But since Ehrenbaum [54] described the parent as "very negligent" in the aquarium, it seems that they merely seek the nesting holes as convenient shelters, and not that they actually guard the eggs.

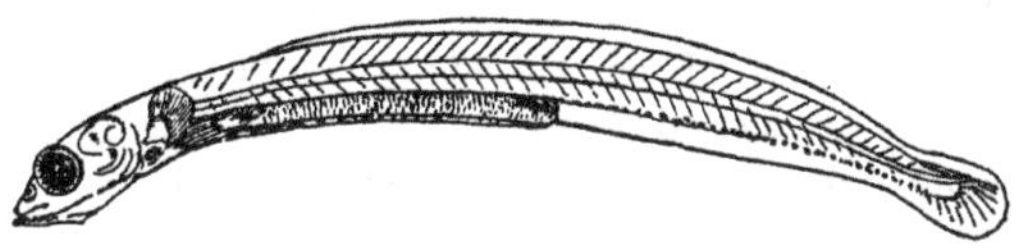

FIGURE 258.—Larva (European), 18 mm. After Ehrenbaum.

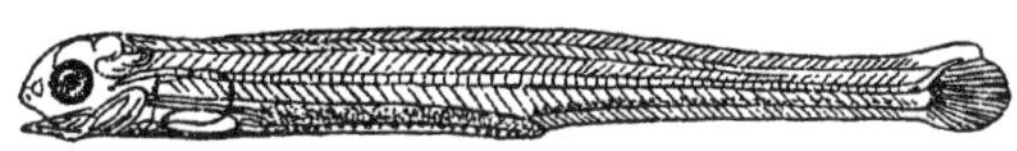

FIGURE 259.—Larva (European), 20 mm. After Ehrenbaum.

ROCK EEL (*Pholis gunnellus*).

The larvae are much larger at hatching (about 9 mm.) and further advanced in development than those of most of the fishes that lay buoyant eggs. Older larvae of the rock eel resemble corresponding stages of the launce and of the snake blenny in their extremely slender form. But they are easily distinguishable from both these species by the presence of a row of small black pigment spots below the intestine, instead of above it, and from the herring (the only other very slender larvae apt to be met in any numbers in the Gulf at the same season) by the location of the vent about midway of the body (fig. 259), and by the fact that their tails are rounded, not forked. The 12 black dorsal fin spots so characteristic of the adult are first noticeable against the transparent trunk in young fry of 25 to 30 mm. The

[50] Scandinavian Fishes, vol. 1, 1892, p. 223.

[51] MacIntosh and Masterman (Life Histories of British Marine Food-Fishes, 1897, p. 210) and Ehrenbaum (Wissenschaftliche Meeresunters., Helgoland, N. Ser., vol. 6, 1904, p. 160) give accounts of its spawning and of its larval development.

[52] Tracy, 40th Rept. Comm. Inland Fish., Rhode Island, 1910, p. 151.

[53] Nichols and Breder, Zoologica, New York Zool. Soc., vol. 9, 1927, p. 159.

[54] Wissenschaftliche Meeresuntersuchungen, Helgoland, N. Ser. vol. 6, 1904, p. 161

young fish (previously drifting near the surface) sink to the bottom when 30–40 mm. long, an event that takes place in late summer or early autumn in the Gulf of Maine. Nothing definite is known of the rate of growth of the rock eel after its first autumn.

General range.—Shoal waters on both sides of the North Atlantic; from Hudson Strait to the offing of Delaware Bay on the American coast [55] and south to France on the European coast; most numerous north of Cape Cod and north of the English Channel.

Occurrence in the Gulf of Maine.—This little fish is to be found all around the shores of the Gulf from Nova Scotia to Cape Cod. Thus it is definitely recorded at Yarmouth (Nova Scotia); at various localities on both sides of the Bay of Fundy where it occurs nearly everywhere (writes Dr. Huntsman); at half a dozen points along the Maine coast; on Cashes Ledge; at Portsmouth (N. H.), where A. H. Clark of the U. S. National Museum reports it plentiful; at Hampton (reported to us by W. F. Clapp); in Ipswich Bay; here and there along the northern shore of Massachusetts Bay: at Cohasset, on the southern shore, where we have seen them; among the stones and boulders of the Gurnet, off Plymouth, where W. F. Clapp informs us that he has found many while shore-collecting for mollusks; also at Provincetown. And doubtless it is more universal than these records indicate, for being of no consequence to the fisherman or to the angler, it comes to notice only when scientific collections are made.

We have taken its pelagic fry (20–39 mm. long) in our tow nets off Seal Island (Nova Scotia); on German Bank; near Mount Desert Island; off Matinicus Island; and off Ipswich Bay, in April, May, June, and August (a total, however, of only 6 stations), while Huntsman states that they are caught in early summer in the Bay of Fundy, evidence that it breeds all around the Gulf from early spring on into the summer.

Recent records show that rock eels also occur in considerable numbers on Nantucket Shoals, as well as on Georges and Browns Banks, where we have trawled a number, down to 40 fathoms, one as deep, even, as 100 fathoms,[56] where many of them have been found in the stomachs of cod and pollock.[57] They are described as common inshore along outer Nova Scotian waters, eastward from our limits; [58] and as "rather common" in the southern side of the Gulf of St. Lawrence,[59] where Huntsman [60] classes them as characteristic of the warm inshore waters, but where they are exposed to temperatures close to the freezing point of salt water in late winter and in spring.

They are reported from the southwest coast of Newfoundland; from the estuary of the St. Lawrence near Trois Pistoles,[61] from Anticosti; from the north shore of the Gulf of St. Lawrence; from both shores of the Strait of Belle Isle; from the southern and eastern parts of the Grand Banks:[62] from Bay Bulls, Conception Bay, and Trinity Bay on the east coast of Newfoundland; from Hamilton Inlet on the Atlantic Coast of Labrador, and from Fort Chimo, Ungava Bay, northern Labrador.

Importance.—This little fish is of no commercial interest. But it appears to be of some importance on the offshore banks as food for larger fishes.

Snake blenny *Lumpenus lumpretaeformis* (Walbaum) 1792

SERPENT BLENNY

Jordan and Evermann, 1896–1900, p. 2438, as *Lumpenus lampetraeformis*.[63]

Description.—This is a very slender little fish as its name implies, being only about one-twentieth as deep as it is long, slightly flattened sidewise, with head about one-eighth as long as body to base of caudal fin, large eyes, wide mouth, and blunt snout. It somewhat suggests a launce (p. 488) in general form, but is much more slender and eel-like. Also, its rounded tail (that of the launce is forked), its large pectoral fins, spiny dorsal fin (the launce has a soft dorsal only), and

[55] In February 1930 *Albatross II* trawled 2 rock eels, 80 and 84 mm. long, in 23 fathoms and 38 fathoms, at lat. 38°32′ N., long. 74°24′ W., and lat. 38°18 N., long. 73°51′ W.

[56] One was caught at this depth on the northeastern part of Georges Bank August 5, 1932 (Schroeder, Bull. 66, Boston Soc. Nat. Hist., 1933, pp. 5–6).

[57] We once found one alive in the mouth of a pollock caught at 34 fathoms on Browns Bank.

[58] Vladykov and McKenzie, Proc. Nova Scotian Inst. Sci., vol. 19, 1935, p. 103.

[59] Cox, Contrib. to Canadian Biol. (1918–1920) 1921, p. 112.

[60] Trans. Roy. Soc. Canada, Ser. 3, vol. 12, Sect. 4, 1918, p. 63.

[61] Vladykov and Tremblay, Natural. Canad., vol. 62 (Ser. 3, vol. 6), 1935, p. 81).

[62] The pelagic larvae.

[63] The original spelling of this name by Walbaum (*P. Artedi* Genera piscium, Ichth., Pt. 3, 1792, p. 184) was *lumpretae-formis*.

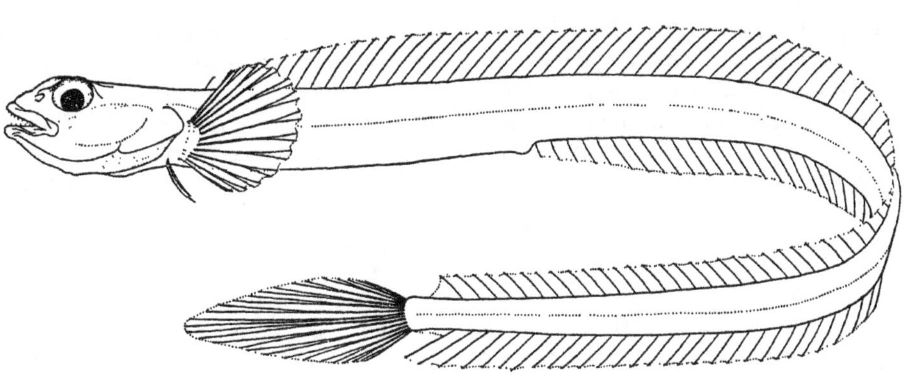

FIGURE 260.—Snake blenny (*Lumpenus lumpretaeformis*). Drawing by Louella E. Cable.

the fact that its lower jaw does not project beyond the upper, together with its color, serve to separate it from the launce at a glance.

The chief anatomic feature (apart from its slenderness) distinguishing it from the rock eel (p. 492) is that the ventral fins (each of one short spine and three longer rays) are well developed and one-third to one-half as long as the pectorals, slightly in advance of which they stand. The pectorals, too, are much larger than those of the rock eel, and its dorsal fin, with 68 to 85 spines and its anal with 49 to 62 rays are fully twice as high, relative to the depth of the body, while its anal fin originates farther forward; the separation of dorsal and anal fins from the caudal is more evident; and its eyes are noticeably larger.

The very slender body is the most obvious difference between this species and its allies the shanny (p. 497), the Arctic shanny (p. 497), and the radiated shanny (p. 498), which are rather robust. The outline of the caudal fin, which is oval (more pointed in large fish than in small), with the central rays much longer than the outer ones, is diagnostic, too.

Color.—Whitish or pale brown on the back and sides, with darker brown markings. On a 12-inch fish taken off the coast of Maine the head was pale brown, the sides of the body blotched with brown, the dorsal fin marked obliquely with 18 pale bars, the caudal marked transversely with 8 bars, the anal rays pale brown against a colorless membrane, the ventrals white, and the pectorals tinged with brown.

Size.—The largest one so far measured was 19 inches long.[64]

Remarks.—Vladykov's [65] discovery that Newfoundland specimens have a larger number of dorsal fin-spines (85) and anal fin rays (62) and a longer caudal fin (longer than the head) than others from the St. Lawrence estuary (75–79 dorsal spines, 52–56 anal rays) shows that the snake blenny tends to break up into local races. Vladykov has made the St. Lawrence race the basis of one new subspecies, *americanus*, and the Newfoundland race the basis of another, *terrae novae;* both of which have more spines and rays than have been recorded for some eastern Atlantic specimens. Gulf of Maine specimens, with 77–83 dorsal spines, and 56–59 anal rays, are intermediate between the Newfoundland and Gulf of St. Lawrence races in this respect.

Habits.—Little is known of the habits of the snake blenny on either side of the Atlantic. Although it is not found along the littoral zone, it is a fish of comparatively shoal water, never taken as deep as 100 fathoms (so far as we have been able to learn) and apparently the most common from a fathom or so below tide mark down to 40 or 50 fathoms. And as most of the specimens that have been caught in Scottish waters were picked up by the foot rope of the trawl, Sim's [66] suggestion that it burrows in mud

[64] In the Museum of Comparative Zoology, trawled about 17 miles off Cape Ann, lat. 42°28′ N., long. 70°13′ W., at 42 fathoms, in July 1931.

[65] Rept. Newfoundland Fish. Res. Comm., vol. 2, No. 3, 1935, p. 75–78.

[66] Jour. Linnaean Soc., London, Zoology, vol. 20, 1890, p. 38.

or clay bottom probably is correct. The *Albatross II* has taken it both on mud bottom and on hard.

Food.—Amphipods, copepods, and other tiny crustacea, with very small starfish, small bivalves, and holothurians have all been found in snake-blenny stomachs in British seas. These blennies are eaten in their turn by large fish, by cod and halibut, for example, in Massachusetts Bay,[67] by pollock in the Bay of Fundy; and by cod in Northumberland Strait, Gulf of St. Lawrence, as Capt. Thor Iversen informed Dr. Hunstman from his experience during the Canadian Fisheries Expedition of 1915.

The spawning season has been stated as autumn or winter in north Scandinavian seas, and it may commence by late summer there, or by early autumn, for Sim found its roe well advanced in development as early in the season as the end of April. Its drifting larvae have been taken in tow nets from February to March in the Baltic, and from March to May in the Gulf of Maine.

The eggs of this species have not been seen, but they probably sink and stick together like those of the rock eel. Apparently the larvae are of considerable size at hatching, for the smallest we have taken (the smallest on record) were about 11 mm. long, though they still lacked any trace of the dorsal- and anal-fin rays. Snake blenny larvae are very slender, resembling the corresponding stages of the rock eel and of the launce in general appearance, but they are distinguishable from both of these by the fact that the vent is situated considerably in front of the midlength of the trunk. There is no danger of confusing them with the young of the herring, the only other very slender pelagic fish larva (besides rock eel and launce) that is apt to be found in any numbers in the Gulf of Maine in spring, for the tail of the herring is forked from a very early stage and its vent is situated much farther back than that of the blenny (p. 91). Another distinctive feature of the snake blenny larvae is the presence of a large black pigment dot at the base of each pectoral fin, and of a double row of 6 to 9 black dots along the dorsal surface of the intestine with several about the vent, which are very conspicuous by contrast with the colorless body. Our largest pelagic larva (41 mm. long) showed most of the characters of the adult, although it was transparent and had the arrangement of pigment characteristic of the earlier larval stages.

General range.—Arctic and northern Atlantic Ocean; south to Scotland, the Baltic, and the southern part of the North Sea in the eastern side; to the offing of southern New England in the western side.

Occurrence in the Gulf of Maine.—It is probable that this northern fish occurs in small numbers around the coastline of the Gulf at some little depth. Thus Hunstman reports it from St. Mary Bay, Nova Scotia, in August and September; from Passamaquoddy Bay from April to August; and in the open waters of the Bay of Fundy from January on. It was recorded off Eastport in 1872; *Albatross II* trawled one specimen (about 12 inches long) 3 miles south of Great Duck Island, near Mount Desert, Maine, in 28 to 33 fathoms, April 1927; two others (8 and 8¼ inches long) 13 miles east of Boone Island, in 88 fathoms in August 1928; one off the Isles of Shoals at 72–78 fathoms in August 1926; one at 42 fathoms on the eastern slope of Stellwagen Bank, about 17 miles off Cape Ann in July 1931. And Goode and Bean [68] described it long ago as a common resident in the deeper parts of Massachusetts Bay.

Our tow nettings, also, of 1920 yielded its drifting larvae off Seguin Island; near Cape Elizabeth; over Platts Bank; near the Isles of Shoals; off Ipswich Bay; off Cape Ann; off Boston Harbor; and in the southwest basin of the Gulf off Cape Cod during March, April, and May—evidence that it breeds successfully throughout the southern part of its range. While it has not been reported on Browns or Georges banks, it is to be expected there.

It has never been reported from the banks along the outer coast of Nova Scotia, so far as we know. But it is so slender and active a fish that it can easily pass through the meshes of any of the nets that are used in our Gulf by commercial fishermen, hence is not likely to be brought in unless it is sought for especially. And the experimental trawlings by the Newfoundland Fishery Research Laboratory did take it at several stations on the Newfoundland Banks, as well as in the Gulf of St. Lawrence, and also off the southeastern coast of Labrador, while it has long been known from

[67] Goode and Bean, Bull. Essex Inst., vol. 11, 1879, p. 10.

[68] Bull. Essex Inst., vol. 11, 1879, p. 10.

as far north on the Labrador coast as Ungava Bay.[69]

Shanny *Leptoclinus maculatus* (Fries) 1837

LANGBARN

Jordan and Evermann, 1896–1900, p. 2433.

Description.—The shanny resembles the snake blenny somewhat in general appearance and in the location and shape of its dorsal and anal fins, but is not so slender (only 10 to 12 times as long as it is deep instead of about 20 times). The most important points of difference (aside from its more robust form) are that the tail of the shanny is about straight in outline instead of narrowly oval or pointed as it is in the snake blenny; that the lower rays of its pectoral are the longest and are separate at their tips; and that the shanny has only 58 to 61 dorsal fin spines, and 35 to 38 anal fin rays.

Color.—Dirty-yellowish, paler below, the back and sides marked with indistinct yellowish-brown blotches of various sizes. The dorsal fin is described as barred obliquely with about 10 rows of brownish dots and the pectorals as cross-barred with about 5 rows. These fins show no distinct markings on the several preserved specimens we have examined; the caudal fin, however, shows one or two dark crossbars, even after preservation.

Size.—Maximum length about 7 inches.

Habits and food.—In Scandinavian waters the shanny spends most of the year in deep water, probably coming up to the shallows to spawn. In the aquarium it keeps close to the bottom, with the body extended and the pectoral fins expanded, and apparently supports itself on the free lower rays of those fins.[70] Annelid worms and pelagic amphipods have been found in shanny stomachs; this is all that is known of their mode of life. The shanny is supposed to spawn in winter, but neither its eggs nor its larvae have ever been seen.

General range.—An Arctic fish, known south to Norway and Sweden in the eastern side of the Atlantic, and to Cape Cod in the western side.

Occurrence in the Gulf of Maine.—Definite records of this Arctic fish for the Gulf of Maine are of several specimens that were collected in 40 to 90 fathoms in Massachusetts Bay by the U. S. Fish Commission in 1887;[71] one that we took in a tow net near Boone Island on March 4, 1920; one from the northeast part of Georges Bank, August 1926, and four (4 to 4½ inches long) that were trawled off Chatham, Cape Cod, in 28 fathoms, May 1, 1930, by the *Albatross II*. This paucity of captures suggests that it enters the Gulf only as a chance straggler from the north, perhaps maintaining itself in small numbers in the bottoms of the deep isolated troughs where the water is coldest.

The nearest records of it to the eastward and northward are of fish taken off the Atlantic Coast of Cape Breton,[72] from the estuary of the St. Lawrence River near Trois Pistoles;[73] from St. Mary's Bay on the south coast of Newfoundland, from the eastern part of the Grand Banks, and off the east coast of the Avalon Peninsula, Newfoundland.

Arctic shanny *Stichaeus punctatus* (Fabricius) 1780

Jordan and Evermann, 1896–1900, p. 2439.

Description.—This shanny suggests the rock eel in its color pattern. But it is easily distinguished from it by having well developed ventral fins and considerably larger pectorals, but fewer dorsal fin

[69] Kendall, Proc. Portland Soc. Nat. Hist., vol. 2, No. 13, 1909, p. 224.

[70] Smitt, Scandinavian Fishes, vol. 1, 1892, p. 230.

[71] Presumably the Gulf of Maine specimens reported by Kendall (Proc. Portland Soc. Nat. Hist., vol. 3, 1914, p. 62), now in the United States National Museum, are this lot.

[72] By the Newfoundland Fisheries Research Commission; also it is listed from Nova Scotia without locality, see Vladykov and McKenzie (Proc. Nova Scotian Inst. Sci., vol. 19, 1935, p. 104).

[73] Vladykov and Tremblay, Natural. Canad. vol. 62 (Ser. 3, vol. 6), 1935, p. 81.

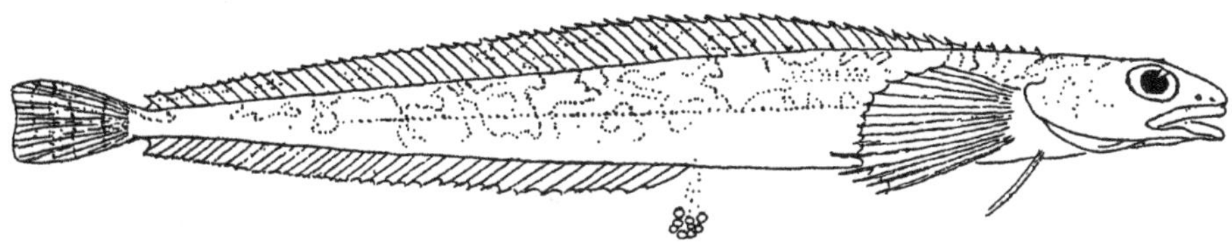

FIGURE 261.—Shanny (*Leptoclinus maculatus*). After Collett.

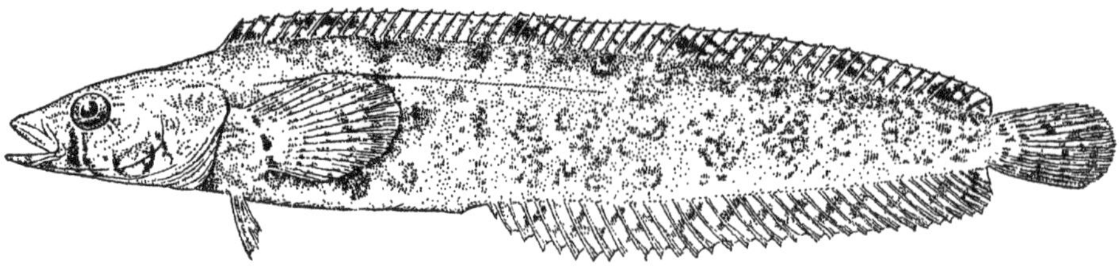

FIGURE 262.—Arctic shanny (*Stichaeus punctatus*), about 7¼ inches, Halifax, Nova Scotia, redrawn by E. N. Fischer from U. S. Fish Commission wash drawing.

spines (only 48 or 49). The presence of a series of large roundish spots on the dorsal fin separates it at a glance from the radiated shanny, which is similar to it in form, but which has only a single large blotch on its dorsal; its spotted dorsal and its evenly rounded pectorals mark it off from the shanny; and its much less slender body (only about one-seventh as deep as it is long) from the snake blenny.

The single long dorsal fin originates directly over the edge of the gill cover, and is of uniform height throughout its length, except that the first 2 or 3 spines, and the last 2 or 3, are shorter than the others. It extends backward nearly to the caudal, but the fins are separated by a conspicuous notch. The anal fin (one short spine, closely followed by about 35 rays) is about two-thirds as long as the dorsal fin, is of approximately the same height as the latter, and bears a similar relation to the caudal. The caudal fin is gently rounded in outline. The pectorals are broadly rounded, a little longer than the depth of the body; the ventral fins are somewhat less than half as long as the pectorals. The straight dorsal profile of the head, from tip of snout to origin of dorsal fin, is an outstanding feature, and the lateral line is single (double in the radiated shanny, p. 499), conspicuous, and ends at about the mid-length of the body.

Size.—Grows to a length of at least 7 inches.

Color.—The single row of 5 to 9 round black spots with pale margins on the dorsal fin, and irregular dark bars on the cheeks and chin are the most conspicuous markings of this shanny. The ground color of the body of an Alaskan specimen is described as bright scarlet,[74] but is brownish (after preservation) on a Gulf of Maine specimen, with darker cloudings extending from close behind the head to the base of the caudal fin. The lower surface of the body (except the head) is plain whitish. The anal fin is dusky, edged with white; the pectorals and caudal are crossed by pale bars, and the ventrals are plain yellow.

General range.—Arctic and circumpolar, from Greenland to northern Siberia and the Arctic Coast of America; [75] southward to Bristol Bay and Prince of Wales Island, Alaska, in the Pacific, and to Mount Desert, Maine on the Atlantic coast of America.

Occurrence in the Gulf of Maine.—A specimen 4⅓ inches (109 mm.) long, of this Arctic species, taken one-half mile off Little Duck Island near Mount Desert, Maine, from the stomach of a cod, on April 30, 1930, was in such good condition that it unquestionably had been living in the immediate vicinity.

The next most southerly records are of two specimens from Chedabucto Bay, Nova Scotia, near the Atlantic entrance to the Strait of Canso; [76] and of one taken off Cheticamp,[77] on the Nova Scotian shore, within the Gulf of St. Lawrence, where Huntsman [78] classes it as a characteristic of the icy-cold water on the banks; and of one from Battle Harbor, on the outer coast of Labrador, a few miles north of the eastern entrance to the Strait of Belle Isle.

Radiated shanny *Ulvaria subbifurcata* (Storer) 1839

Jordan and Evermann, 1896–1900, p. 2440.

Description.[79]—The most conspicuous feature of this little fish, and one that sets it apart from all the other blennies of our Gulf, is that the forward part of its dorsal fin is marked with a

[74] Jordan and Evermann, Bull. 47, U. S. Nat. Mus., Pt. III, 1898, p. 2439.

[75] Reported from Hudson Bay by Vladykov, Contrib. Canadian Biol., N. Ser., vol. 8, No. 3, 1933, p. 35.

[76] Cornish, Contrib. Canadian Biol. (1902–1905) 1907, p. 87.

[77] Cox, Contrib. Canadian Biol. (1918–1920) 1921, p. 112.

[78] Huntsman, Trans. Roy. Soc. Canada, Ser. 3, vol. 12, Sect. 4, 1918, p. 63.

[79] Based on 3 specimens from Grand Manan, the largest 5⅝ inches long.

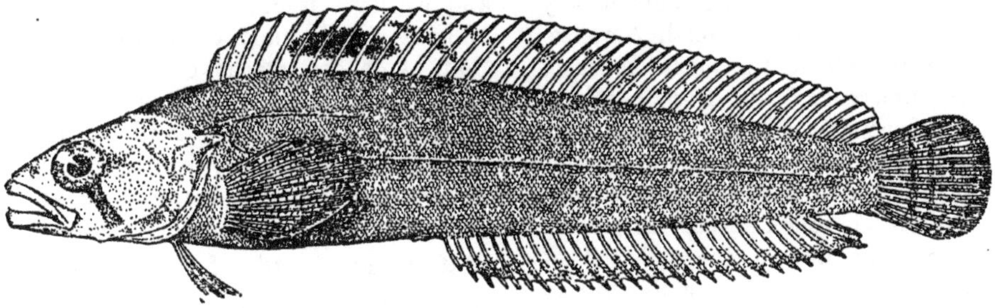

FIGURE 263.—Radiated shanny (*Ulvaria subbifurcata*). From Jordan and Evermann. Drawing by H. L. Todd.

single large dark blotch. More significant morphological characters are that it is much stouter of body (about one-sixth to one-seventh as deep as it is long) than the snake blenny; that its pectoral fins and especially its ventrals are considerably larger, relatively, than those of the rock eel (p. 492), while its ventrals are situated farther in advance of the pectorals; and that its gill openings are much wider and extend forward under the throat (confined to the sides of the neck in the rock eel). The evenly rounded outline of its pectorals distinguishes it from the shanny, in which the lower rays are the longest, and are separate at their tips. It is a stouter fish than the Arctic shanny (p. 497), and its lateral line is double, with an upper branch and a lower, whereas the lateral line of the Arctic shanny is single. The outline of the edge of its gill covers, with the upper corner terminating in a rounded fleshy flap concealing a sharp angle, is diagnostic also, for they are rounded in all the other Gulf of Maine blennies.

The dorsal profile of the head is more convex than the ventral; the lower branch of the lateral line runs the whole length of the body, but the upper branch (the more obvious of the two) reaches only about as far back as the tips of the pectoral fins. The dorsal fin, of 43 or 44 spines, is higher, relatively, than that of the rock eel, and is practically continuous with the caudal fin. The anal fin, of about 30 rays, is about half as long as the dorsal, and it is separated from the caudal by a short but definite interspace, made obvious by the abrupt rear angle of the anal. The pectoral fins, evenly rounded in outline, reach back about as far as the eighth dorsal-fin spine. The rear margin of the caudal fin is evenly rounded.

Color.—The most distinctive feature of the color pattern, one which marks this species among local blennies, is the presence of a large oval dusky blotch on the dorsal fin extending from the fifth or sixth spine to the eighth or tenth spine. The back and the upper parts of the sides are dull brown, obscurely barred or blotched alternately with paler and darker; the sides of the head are marked with a dark bar running obliquely downward and backward from the eye; and the belly is pale brownish (described also as yellowish white). The caudal fin is crossbarred with 3 or 4 series of dark dots, and the dorsal fin is marked with many tiny dark dots, besides the blotch just mentioned.

Size.—The largest one we have seen or read of is 6½ inches long, but the maximum size may well be larger.

Habits.—Nothing is known of the mode of life of this shanny except that it is a bottom fish like other blennies, living among seaweed and stones from low-tide mark down at least to 30 fathoms, and very likely much deeper. Dr. Huntsman writes in his notes, "It is found under stones near low tide mark" with the rock eel but far less abundantly than the latter and only on the more exposed shores. Cornish[80] likewise describes it as taken under stones on the beach, as well as in the dredge and trawl in 6 to 30 fathoms at Canso, Nova Scotia.

The eggs have not been seen, but the fact that we have taken larvae as small as 8 to 11 mm. in our tow nets in June, July, and October[81] points to a spawning season lasting from late spring throughout the summer (if our identification is correct).

[80] Contrib. Canad. Biol. (1902–1905) 1907, p. 87.

[81] These are listed in Bull. Mus. of Comp. Zool., vol. 58, No. 2, 1914, p. 109; and vol. 59, 1917, p. 273.

General range.—So far this fish is known only off the boreal coasts of eastern North America, from eastern Newfoundland, the north shore of the Gulf of St. Lawrence,[82] and from northern Nova Scotia to Nantucket Shoals and southern Massachusetts.[83]

Occurrence in the Gulf of Maine.—This shanny, first described by Storer from a specimen found at an unusually low tide among the seaweed at Nahant on the north shore of Massachusetts Bay in 1838, was long supposed to be rare. But it has proved to be common in the Bay of Fundy [84] among seaweed on rocky shores, a number having been taken both at Campobello Island and at Grand Manan, and one in the mouth of the St. Croix River, while the Museum of Comparative Zoology has a specimen from Grand Manan, collected by Verrill many years ago. Huntsman also reports it from St. Mary Bay on the Nova Scotia shore, and we have found several in the tide pools at Nahant, on the northern shore of Massachusetts Bay. Offshore, we have trawled it near Seguin Island; in Massachusetts Bay; also on the eastern part of Georges Bank. And we have found so many in the stomachs of cod caught on Nantucket Shoals, on Cashes Ledge, and on other offshore grounds as to show that it is widespread in the Gulf on hard bottom, from the shore down to at least 30 fathoms, while we have had one from as deep as 45 fathoms. Our tow nets also have taken its pelagic larvae near Seal Island (Nova Scotia); in the Grand Manan Channel; at the mouth of Casco Bay; near Cape Porpoise; off the Isles of Shoals; near Cape Ann; and in Massachusetts Bay.

The only importance of this little fish in the natural economy of our Gulf is that many of them are eaten by cod and by other fishes.

THE WRYMOUTHS. FAMILY CRYPTACANTHODIDAE

The wrymouths are slender, eel-like fishes, close relatives of the blennies but much larger. Like the blennies they have a long dorsal fin that is spiny throughout its length, but the demarcation between dorsal, caudal, and anal fins is so vague that they can be described as practically continuous. They have no ventral fins, and their mouth is so strongly oblique that it is nearly vertical. Only three species are known, all North American, of which the Gulf of Maine harbors one.

[82] Reparted from Trinity Bay, Newfoundland (Rept. Newfoundland Fishery Research Comm., vol. 1, No. 4, 1932, p. 109, Sta. 39), and from Canso, Nova Scotia, by Cornish (Contrib. Canadian Biol. (1902–1905) 1907, p. 87. And Dr. Vladykov writes us that he has collected a specimen at Pointe du Maurier on the north shore of the Gulf of St. Lawrence.

[83] Sumner, Osburn, and Cole (Bull. U. S. Bur. Fish., vol. 31, Pt. 2, 1913, p. 768) record 4 living specimens from Vineyard Sound and others found in the stomach of a sheldrake (*Merganser*) that was shot near Robinson Hole in the Woods Hole region.

[84] Huntsman, Contrib. Canadian Biol. (1921) 1922, p. 66, and unpublished notes.

Wrymouth *Cryptacanthodes maculatus* Storer 1839

CONGO EEL; BASTARD CUSK; GHOSTFISH

Jordan and Evermann, 1896–1900, p. 2443.

Description.—The wrymouth is eel-like in form, about thirteen times as long as it is deep but much flattened sidewise, and (eel-like) it lacks ventral fins. And (as in the eel) there is no definite demarcation between dorsal fin and the caudal fin or between the anal fin (about 47 to 50 rays) and the caudal, the one merging into the other to form a continuous fin around the tail, with no inter-

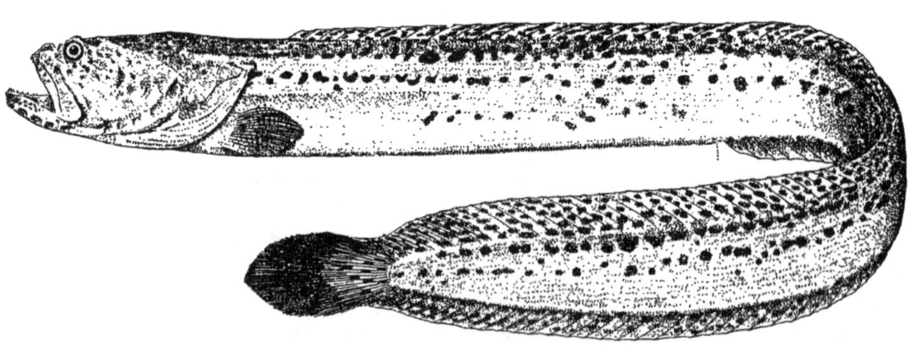

FIGURE 264.—Wrymouth (*Cryptacanthodes maculatus*). From Jordan and Evermann. Drawing by H. L. Todd.

spaces. But its dorsal fin (which extends from close behind the pectorals back to the caudal fin) is spiny (about 75 to 77 spines) for its entire length like that of its close relatives the blennies. But the absence of ventral fins separates it from all of our local blennies, and its peculiar profile is an equally useful field mark, the head being flat-topped, the eyes set high up in very prominent orbits, and the mouth strongly oblique with so heavy a lower jaw that it gives the face a bulldog-like expression when the mouth is closed. The wide gill openings, running forward under the throat, and the small size and rounded outline of the pectorals are distinctive, also, as is the fact that both the dorsal fin and the anal fin are low (less than half as high as the body is deep in large specimens, relatively higher in small), and of uniform heights throughout most of their lengths, with the anal about two-thirds as long as the dorsal. The caudal fin is oval.

Color.—Described (and the few preserved specimens we have seen correspond with this) as of varying shades of brown or reddish brown, with the upper part of the sides marked with two or three irregular rows of small darker brown spots that run from head to tail; the top of the head is thickly speckled; the dorsal and anal fins are spotted with similar but smaller dots, and the belly is grayish white. A few spotless specimens have been seen.

Size.—Maximum length about 3 feet.

Habits.—Very little was known of the habits of the wrymouth until recently, except that it is a bottom fish living from the intertidal zone down to considerable depths (where it is sometimes taken on line trawls in the Bay of Fundy). But in 1910 and again in 1920 Willey and Huntsman [85] found full-grown wrymouths living in burrows in the mud on the flats at the mouth of the Magaguadavic River, a tributary of Passamaquoddy Bay. These burrows, to quote from their account, "were found in very soft mud from the lower part of the Fucus zone downward; that is, as far up as 4 feet above low-water mark," and "each system of burrows, inhabited by only one fish, consisted of branching tunnels about 5 cm. in diameter and from 3 to 8 cm. below the surface" of the mud, originating from a more or less centrally placed mound, where the main entrance was, with other smaller openings along the tunnels and at their terminations.

It seems that the burrowing instinct is strong, for one fish kept in a tank constantly inhabited a piece of hard rubber tubing. Hence it is probable that wrymouths in other parts of the Gulf likewise live in burrows or perhaps under stones. And they seem as likely to be inshore in shoal water in winter as in summer, for one was speared in Marblehead Harbor in December many years ago.[86] Within our Gulf wrymouths have been found from a little above low water mark, as just remarked, down to about 100 fathoms; and to somewhere between 245 and 325 fathoms off New Jersey (see footnote 89, p. 502).

Huntsman and Willey found "beach fleas" or "sand-hoppers" (*Gammarus*), shrimps (*Crago*) and fragments of winter flounders in several wrymouths which they opened, and the one kept in captivity ate sand-hoppers, hermit crabs, small herring, and mollusks such as limpets, periwinkles, whelks, clams, and mussels. Apparently it located food as much by sight as by smell.[87]

Ripe wrymouths are yet to be seen; but the presence of the larvae early in spring in Passamaquoddy Bay, as reported by Huntsman, with the seasonal occurrence of the fry mentioned below (p. 502), proves it a winter spawner in the Gulf of Maine. It may breed later in the Gulf of St. Lawrence, for Dannevig [88] records a young wrymouth only 38 mm. long that was taken there as late as June 10. The localities where the young fish have been taken (see p. 502) suggest that wrymouths spawn all around the coast of the Gulf of Maine and wherever they occur on the offshore banks.

Neither the eggs nor the early larval stages are known. But by the time the young have grown to a length of 21 to 22 mm. they show the long dorsal and anal fins, and the lack of ventral fins characteristic of their parents, though they are much less slender, relatively, their caudal fins are larger and square instead of rounded and their mouths are nearly horizontal. The pigmentation of the fry is likewise extremely characteristic, the upper sides from the eye back to the caudal fin

[85] Canad. Field Natural., vol. 35, 1921, p. 4.

[86] Putnam, Bull. Essex Inst., vol. 6, 1874, pp. 11-13.

[87] Willey and Huntsman also give interesting data on its respiration and on its response to various stimuli.

[88] Canadian Fisheries Expedition, 1914-1915 (1919), p. 16. He gives an excellent figure of this specimen on pl. 2, fig. 10.

being thickly speckled with dark brown dots, which are sparser on the lower part of the sides.

General Range.—Atlantic Coast of North America, from southeastern Labrador, the coasts and banks of Newfoundland, and the Gulf of St. Lawrence to Long Island Sound and to the offing of central New Jersey.[89]

Occurrence in the Gulf of Maine.—Published records locate this fish in the Bay of Fundy; at Eastport; in Casco Bay; at Portland; in the mouth of the Piscataqua River; at Gloucester; in Marblehead Harbor; at Swampscott; Nahant; and Dorchester in Boston Harbor; and in the outer waters of Massachusetts Bay; there are specimens in the Museum of Comparative Zoology from Trenton, Maine; from outer Boston Harbor; and from near Provincetown. Two were taken in the central basin of the Gulf in July 1931 at a depth of 88–95 fathoms; one was trawled by the *Atlantis* in the deep trough west of Jeffreys Ledge at 72–78 fathoms, and another in the southwestern basin of the Gulf off Cape Cod at about 100 fathoms (183 meters), in August 1936; the *Albatross II* trawled one on the eastern slope of Nantucket Shoals[90] at 52 fathoms, in May 1950. And one of the crew of the dragger *Eugene H* reports the capture of 4 of them on the northeastern part of Georges Bank on October 12, 1951. We have also taken its late larvae and fry in tow nets (11 specimens 18 to 40 mm. long) in Massachusetts Bay off Boston Harbor; over Jefferys Bank; in the trough near the Isles of Shoals; in the western basin a few miles west of Cashes Ledge; off Penobscot Bay; near Mount Desert Island; and in the deep basin off Machias, Maine, in May 1915, and in March and April 1920. These localities are sufficiently scattered to show that it is to be found, not only all around the coasts of the Gulf, but on the offshore grounds as well.

However, it seems to be rare or at least very local, for we have caught few adults ourselves, nor have we seen it brought in by fishing boats. In fact, few of the fishermen of whom we have inquired have been aware of its existence, a fact no doubt associated with its burrowing habit. And it has not been reported as yet from Browns Bank though it is to be expected there.

Following its range to the eastward and northward, we find it described as "rather common" all along Nova Scotia (taken at 60 fathoms on Western Bank off Halifax);[91] it has been reported from a number of stations on the eastern half of the Grand Banks region, from the southern and southeastern coasts of Newfoundland, and from the outer Labrador coast some 20 miles north of the Strait of Belle Isle (see footnote 89, p. 502); also within the Gulf of St. Lawrence from the Bay of Chaleur in the southwest,[92] and from the entrance to the Strait of Belle Isle in the northeast.[93]

THE WOLFFISHES. FAMILY ANARHICHADIDAE

The wolffishes are closely allied to the blennies, and like the latter they have a single long spiny dorsal fin running the whole length of the back from the nape of the neck. But the presence of large molar teeth and canine tusks, with their total lack of ventral fins and the fact that all but the last 10 or 12 of their dorsal fin spines are soft and flexible at the tips, instead of stiff justify a separate family for them. They are much larger fish than any of our blenny tribe. Two species occur in the Gulf of Maine, one, the wolffish, commonly; the other, the spotted wolffish, only as a stray from the north. Another cold water species, *Anarhichas latifrons*, has been recorded repeatedly from Nova Scotian waters, hence it is to be expected in our Gulf sooner or later though it has not been reported there definitely. It is included in the following Key.[94]

[89] The most northern locality-record which we have found is for its drifting larvae off the outer coast of Labrador, about 20 miles north of Belle Isle (Rept. Newfoundland Fish. Res. Lab., vol. 2, No. 3, 1935, p. 79, Sta. 422); the most southern are for one trawled by the *Albatross* II off northern New Jersey, lat. 40°04′ N., long. 73°32′ W., August 1936, at 35 fathoms; and of another dredged by the *Atlantis* 30 miles farther south (lat. 39°31′ N., long. 72°10′ W.) between 245 and 325 fathoms, that same year.

[90] Lat. 40°05′ N., long. 69°22′ W.

[91] Vladykov and McKenzie, Proc. Nova Scotian Inst. Sci., vol. 19, 1935, p. 104.

[92] Halkett, Checklist Fishes Canada, 1913, p. 112, "Gaspé Bay."

[93] Rept. Newfoundland Fish. Res. Lab., vol. 2, No. 3, 1935, p. 79, Sta. 370.

[94] We think it likely that a specimen of *A. latifrons* credited by Bean (Proc. U. S. Nat. Mus., vol. 3, 1881, p. 82) to "east coast of United States" was actually brought in from Nova Scotian waters rather than that it was caught anywhere west of Cape Sable. See Bigelow and Schroeder (Proc. Boston Soc. Nat. Hist., vol. 41, No. 2, 1935, p. 15) for further remarks on this species.

KEY TO GULF OF MAINE AND NOVA SCOTIAN WOLFFISHES

1. Back and sides definitely black spotted on a pale ground; the bands of molar teeth in the roof of the mouth are all of about equal lengths. Spotted wolffish, p. 507

   Back and sides plain colored or dark barred and blotched, but not definitely black-spotted; the central band of molar teeth in the roof of the mouth is longer than the bands that flank it---------- 2
2. The central band of molar teeth originates at about the same level as the bands on either side of it, but it extends considerably farther rearward than they do; the canine teeth are very large and prominent; the flesh is firm------------------Wolffish, p. 503

   The central band of molar teeth originates considerably in advance of the bands on either side of it, but it terminates about even with them rearward; the canine teeth are small, not very prominent; the flesh is noticeably flabby when fresh-caught. Arctic wolffish, *Anarhichas latifrons*, Steenstrup and Hallgrimsson.

Wolffish *Anarhichas lupus* Linnaeus 1758

CATFISH; OCEAN WHITEFISH

Jordan and Evermann, 1896–1900, p. 2446.

Description.—The wolffish suggests a huge blenny in its general make-up, except that its dorsal fin spines are flexible at their tips instead of stiff; that it has no ventral fins; and that its mouth is armed with a set of teeth more formidable than those of any other Gulf of Maine fishes, except for its relative, the spotted wolffish (p. 507), and for some of the sharks. There is a row of about 6 very large, stout, conical canine tusks with a cluster of 5 or 6 smaller canines behind them in the upper jaw; and the roof of the mouth back of the latter is armed with three series of crushing teeth. The central series of these consists of a double row of about 4 pairs of large rounded molars that are united into a solid plate; each of the outer series consists of two alternating rows of blunt conical teeth. In the common wolffish the central series, which is the longest of the three, originates a very little in advance of the outer series, and it extends rearward noticeably farther. The lower jaw has 4 to 6 large tusks in front, behind which are two longitudinal diverging rows of rounded molars. And the throat also is armed with small scattered teeth. The great projecting tusks, rounded nose, and small eyes give the wolf a singularly savage aspect.

The body is deepest close behind the head, tapering back to a slender caudal peduncle and to a small weak tail fin. The dorsal fin (69–77 spines) is about half as high as the head is long and uniform in height from end to end except for its rounded corners, and it extends from the nape of the neck to the base of the caudal fin. The anal fin (42–48 rays) is only about half as high as the dorsal, and a little more than half as long; its rear corner is angular. The pectoral fins are large and rounded, and the caudal fin is slightly convex in outline.

Color.—Wolffishes are dull-colored, but they vary widely in tint. The upper parts and the dorsal fins of those taken off the Massachusetts coast have been described as purplish brown and we have seen them of this tint. But fish caught on Georges Bank are invariably dull olive green, according to Mr. Clapp, while they are described as purplish, brownish, or bluish gray, or slate colored in other seas. No doubt the color of the wolf, like that of many other ground fish, varies with that of its surroundings, purplish and brown tints ruling among red seaweeds and olive gray on clean bottom. Whatever its tint, its sides are transversely barred with a variable number (usually 10 or more) of irregular and broken

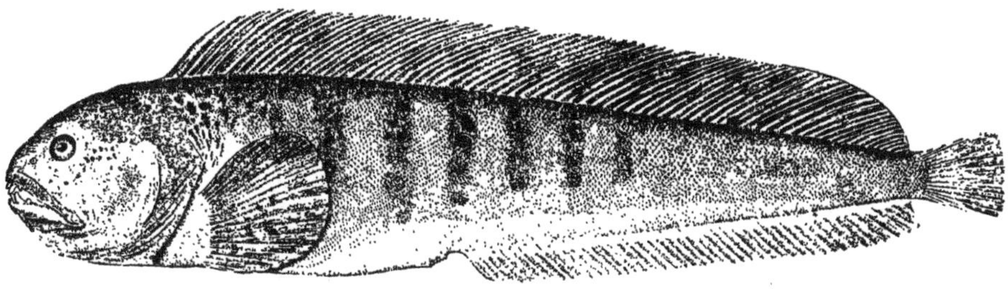

FIGURE 265.—Wolffish (*Anarhichas lupus*), Georges Bank. From Goode. Drawing by H. L. Todd.

darker bands or blotches, or scattered spots, that extend out on the dorsal fin. The throat and the belly back to the vent are dirty-white tinged with the general ground tint of the upper parts. Wolffish fade so soon after they are caught that those seen in the markets usually are much paler than they were in life.

Size.—A length of 5 feet seems about the maximum in Gulf of Maine waters; one more than 4 feet long is seldom seen, and the larger fish caught and brought in run less than 3 feet. European authors speak of wolffish of 6 feet and even longer, but they average only about 2 feet in Scandinavian waters,[95] i. e., scarcely so large as those in the Gulf of Maine. A fish 33 inches long weighs about 10 pounds, one of 37 inches between 16 and 17 pounds. The greatest weight reported for American waters so far is about 40 pounds.[96]

Habits.—The wolffish is solitary, living one here and one there, and it is not abundant anywhere, in the sense that this term can be applied to the cod, to the haddock, to the pollock, or others of our commercially important fishes. It holds close to the bottom; and it is always caught on hard ground, never on mud, a preference illustrated by the fact that our experimental trawlings on the soft bottom of the deep troughs within the Gulf did not catch one wolffish, though they did yield a variety of other fishes in plenty.[97] It is a weak swimmer, moving by sinuous side to side undulations like a blenny or an eel; and probably it spends most of its life hidden among seaweed or rocks, or nosing about such surroundings for food. There is no reason to suppose that it ever attacks other fish in its normal way of life, but when hauled out of the water it snaps like a bulldog and with good aim at anything in its way, the hands, an oar, or at other fish among which it is thrown, and it can inflict a serious bite. Goode[98] remarks that it has been known to make a furious attack on persons wading among the rock pools at Eastport, Maine.

The depth zone occupied by the wolffish at one time or another extends from a fathom or so below tide mark down to 85 fathoms at least, and very likely deeper. It has been reported in tide pools at Eastport, but we have never heard of it in such situations or at low-water mark anywhere else in the Gulf, nor does it run up estuaries, and it is probable that most of the local stock lives in depths of 10 to 50 or 60 fathoms.

The wolffish is a cool- and cold-water fish, as might be assumed from the fact that its regular geographic range extends hardly farther west than Cape Cod and Nantucket shoals. Those living in the coastal belt of our Gulf, at depths of 25 fathoms or less, regularly experience temperatures as low as 34°–36° (locally even as low as 32°) at the end of the winter, or at some time during the spring, according to locality.[99] They are in temperatures equally low or even lower, fractionally, in late spring and early summer on the fishing grounds along outer Nova Scotia, while the grounds where they are caught in the Gulf of St. Lawrence are flooded every spring and early summer, with water as cold as 32°, which they can avoid only by descending deeper into the Laurentian Trough, a movement of which we have no direct evidence. And they have been caught, widespread, on the Newfoundland Banks (p. 507) in water as cold as 30°–31°. At the other extreme, the highest temperature in which wolffish occur in any numbers is about 50°–52°, at the end of summer (again for those living shoalest) in the coastal belt of our Gulf, and also on the Nantucket shoals grounds. They are never known to run up into brackish water.

The wolffish is resident wherever it is found, to be caught throughout the year. For example, about as many are brought in from Georges Bank in one month as in another, allowance being made for the difficulties and dangers of winter fishing. And as it passes through only a brief pelagic stage when it is young (p. 506), it is a comparatively stationary fish, with much less interchange from one locality to another than is the case with cod or with haddock.

The diet of the wolffish consists wholly of hard-shelled mollusks, crustaceans, and echinoderms. So far as we can learn fish have never been found in the stomach of a wolffish. Mr. Clapp found that the 50 or 60 fish that he opened on Georges Bank had all eaten large whelks (*Buccinum*), cockles (*Polynices*, *Chrysodomus* and *Sipho*), sea clams (*Mactra*), and other shellfish, which it crushes easily in its viselike molars. Sometimes,

[95] Smitt, Scandinavian Fishes, vol. 1, 1892, p. 232.

[96] Goode, Fish. Ind. U. S. Sect. 1, 1884, p. 249.

[97] For list of species taken, see Bigelow and Schroeder, Biol. Bull. vol. 76, 1939, p. 309.

[98] Fish. Ind. U. S. Sect. 1, 1884, p. 249.

[99] For further details, see Bigelow, Bull. U. S. Bur. Fish., vol. 40, part 2, 1927, p. 542.

however, mollusks even as large as these are swallowed whole, and we have seen many caught on hooks baited with clams. In north European waters wolffish are said to subsist largely on mussels, and one examined by Vinal Edwards at Woods Hole was full of these; none, however, were found in stomachs of the fish caught on Georges Bank, although mussels are plentiful there, which points to a definite preference for the other shellfish just mentioned. The wolffish is also known to feed on large hermit crabs, on ordinary crabs, and other crustaceans, on starfish, and on sea urchins, a quart of the latter having been taken from one caught at Eastport.[1] And Mr. Clapp's observations that every one he has opened contained food of some sort is good evidence of its constant search for anything edible. With such a diet it is not surprising that wolffish are more often caught on hand lines baited with cockles or clams than on long lines, which are usually baited with herring.

Breeding habits.—The breeding habits of the wolffish have not been followed on this side of the Atlantic. In north European waters it spawns chiefly from November until January,[2] and apparently the breeding season is about the same for it in Nova Scotian waters, and in the Gulf of Maine, for McKenzie and Homans [3] report a mass of eggs dragged up on February 19, in 1937, some of them just hatching, while we have taken larvae of 20 to 22 mm. (fig. 267), that is, 2 to 3 months old from the time the eggs were deposited, as early as January 30 in 1913, and as late as March 4 in 1920.

The eggs, 5.5 to 6 mm. in diameter (among the largest fish eggs known), yellowish, opaque, and with an oil globule of 1.75 mm., are laid on the bottom in shoal water where they stick together in large loose clumps among weeds and stones. The fish have been described as making an annual shoreward journey for spawning purposes, but there is little evidence of this. The precise duration of incubation is yet to be learned; probably it is long, as it is for most of the fishes that lay their eggs on the bottom.

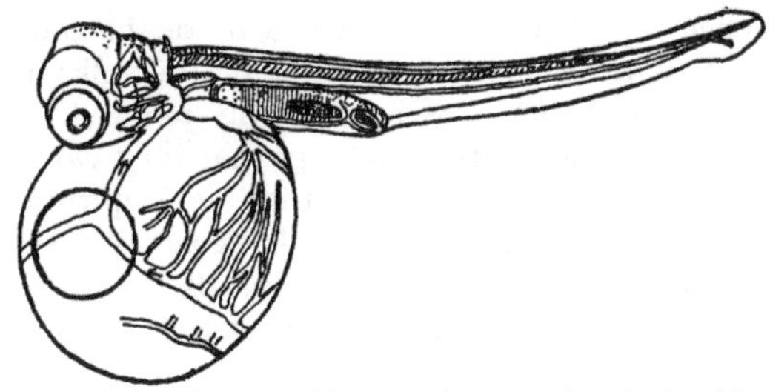

FIGURE 266.—Larva (European), just hatched. After Ehrenbaum.

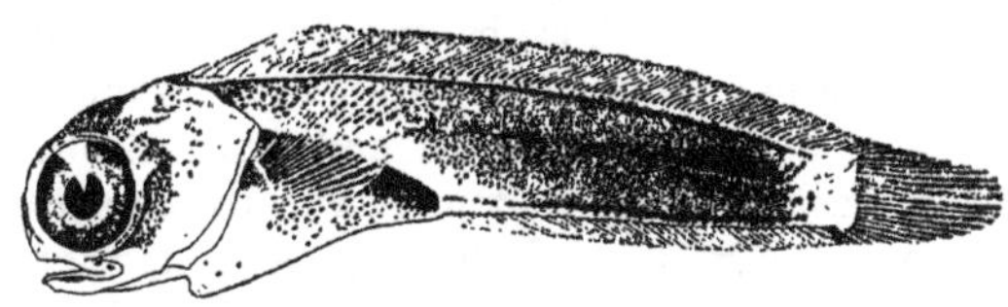

FIGURE 267.—Larva, 21.5 mm. Gulf of Maine.

WOLFFISH (*Anarhichas lupus*).

The slender transparent larvae of the wolffish of north Europe are described as about half an inch (12 mm.) long at hatching, but those that hatched from a mass of eggs dragged up off Nova Scotia were 17–18 mm. long.[4] They have an enormous baglike yolk sac enclosed in a net of highly developed blood vessels (see fig. 266, p. 505), and thus they suggest salmon or trout larvae remotely, in appearance. Wolffish hatched by McIntosh and Prince in the aquarium at St. Andrews, Scotland, did not absorb the yolk sac wholly until about 3½ months old and upward of 20 mm. long, but in natural surroundings larvae as small as 17 mm. have been found free of yolk, nor was any trace of it visible in the larvae of 21 mm. and upward which we have towed in the Gulf of Maine. Larvae of 20 to 22 mm. show the dorsal and anal fin rays in their final number, but the large head, enormous eyes, and tiny teeth, with the fact that there is no definite separation between the anal and dorsal fins and the caudal, give the young fishes an aspect very different from that of their parents until they are 1¼ to 1¾ inches long. In life the wolffish is silvery on the sides at this stage, but this metallic hue fades after preservation, leaving only the dark brown pigment granules with which the sides are thickly dotted. The largest fry we have seen (44 mm. long) are similarly pigmented but somewhat paler.

[1] Verrill, Amer. Naturalist, vol. 5, 1871, p. 400.

[2] It was formerly thought to spawn in spring, but McIntosh and Prince (Trans. Roy. Soc. Edinburgh, vol. 35, Pt. 3, No. 19, 1890), to whom we owe all that is known of its early larval development, proved it an autumn and winter spawner both by examination of its ovaries and by the discovery of its eggs.

[3] Proc. Nova Scotian Inst. Sci., vol. 19, 1938, p. 279.

[4] McKenzie and Homans, Proc. Nova Scotian Inst. Sci., vol. 19, 1938, p. 279.

When first hatched, the larvae lie on bottom like young trout or salmon, resting on the yolk; it is not until the latter is considerably reduced in size (several weeks after hatching) that they swim much. And they do no more than dart upward for a few inches and then settle back again until a month old. Thus the wolffish spends the early part of its life close to bottom instead of drifting at the mercy of tide and current, as all fishes do that produce buoyant eggs. Some of the older larvae and the young fry adopt a drifting habit for a time after the yolk is absorbed (we took some 20 specimens of various lengths from 21 to 44 mm., in our tow nets during March and April 1920). But it seems that they seldom rise to the uppermost water layers, for only 2 of the 8 hauls that took them were made at the surface, the others were at depths of 30 to 60 fathoms. And as fry no larger than this have been trawled on bottom in European waters, it seems that some do not leave the ground. It follows, then that the wolffish probably is not subject to the long involuntary migrations that are carried out by most of the members of the cod and flatfish tribes, but that it passes through its entire larval stage near where it is hatched, hence localities where the young are taken are evidence of local spawning. The brevity of the pelagic stage, if any, also implies that the stock of wolffish in any given locality depends on local reproduction for its maintenance.

In Scottish waters wolffish fry have been taken as long as 5 to 6 inches in July, and up to 7 to 8 inches in August, pointing to a rapid rate of growth for the first summer. Nothing is known of the later growth.

General range.—Both sides of the North Atlantic; north to Davis Strait in American waters; south regularly to Cape Cod; less often to the westward along southern New England, and exceptionally to New Jersey;[5] also Greenland; Iceland; and northern Europe southward to northern France.

Occurrence in the Gulf of Maine.—Although the wolffish has been recorded at only a few stations in the Gulf in scientific literature[6] it is a fairly common fish, to be caught on suitable bottom in all parts of the Gulf, though nowhere in any great numbers. Thus, something like 30,000 pounds were taken off western Nova Scotia yearly during the period 1944–1946;[7] scattering fish are caught at the mouth of the Bay of Fundy and around Grand Manan (stray fish only up the Bay, if any); on one occasion, in April 1930, we saw 35 of them, 2 to 3 feet long, caught on one set of a long line in 18 fathoms of water off Mount Desert Island; and enough are caught on the small grounds thence eastward to Eastport for 16,000 pounds to have been reported from this statistical area in 1945. The various small coastwise fishing grounds, westward from Mt. Desert, yield much greater numbers, as reflected in reported landings, for 1945, of 253,000 pounds for central Maine; about 1,400 pounds for Cashes Ledge and the neighboring patches of hard bottom, where we have caught them, as we have on Platts Bank also; about 26,000 pounds for western Maine; and about 118,000 pounds reported as taken by the vessel fishery off the coast of Massachusetts from the New Hampshire line to Cape Cod.

Recent statistical surveys have not been of a sort to localize the catches more precisely than this. But the small-boat fishermen, landed more than 37,000 pounds in Essex County, Mass., in 1905, mostly from Jeffreys Ledge, from Stellwagen Bank, and from the deeper rocky spots near Gloucester and Nahant, grounds where fishermen report them as fairly plentiful. And there is no reason to suppose that the situation in this respect has altered subsequently.

Considerable catches are also brought in from Cape Cod waters, as reflected in landings of 233,000 pounds reported for Barnstable County in 1945, about 224,000 pounds for Cape Cod in 1947.[8]

Wolffish are not taken in commercial quantities either to the westward of Nantucket Shoals, or

[5] Abbott (Geol. New Jersey, 1868, p. 818) characterized it as "not unfrequently met with" off the New Jersey coast, and Schnakenbeck (Faune Ichth.; Cons. Perm. Internat. Explor. mer. Pl. not numbered, 1933) even outlined its range as extending southward to Cape Hatteras. But we have heard of none caught to the westward and southward of Vineyard Sound at any time during the past half-century.

[6] The deep channel between Georges Bank and Browns; off Cape Sable; in St. Mary Bay; at Grand Manan; at Campobello, at the mouth of the Bay of Fundy; Eastport; Mussel Ridge Channel; Casco Bay; Ipswich Bay; Annisquam; off Gloucester; Massachusetts Bay; North Truro; Nantucket Shoals, and Georges Bank.

[7] Canadian and United States catches combined.

[8] We cannot localize these any more precisely.

on the southwestern part of Georges Bank.[9] But the so-called "south channel" grounds from Cape Cod out to the northwestern part of Georges, with the northern and eastern parts of Georges, and Browns Bank to the eastward, support so large a population that these grounds, combined, yielded about 368,000 pounds in 1945, about 840,000 pounds in 1947.[10] And it is much more likely that the differences from year to year in the catch [11] are due to the fact that wolffish are taken only incidentally, so that the catch depends on the precise grounds fished, rather than on any periodic alterations in their abundance.

In 1946 (most recent year for which we have seen statistics for the Canadian as well as for the United States catches) something like 1,571,500 pounds of wolffish were reported as caught within the limits of our Gulf, or something like 260,000 to 270,000 fish, assuming an average weight of 5 to 7 pounds. But it is anyone's guess what proportion of the total population this may be.

Wolffish appear to be about as numerous on the various fishing grounds along outer Nova Scotia (reported catch for 1949, about 800,000 pounds) as they are on Georges and Browns Banks. (p. 507). But while they are reported at several localities in the southern side of the Gulf of St. Lawrence, [12] also off the west coast of Newfoundland, and at Anticosti, they are not plentiful enough anywhere in the Gulf to yield commercial catches. [13] And this applies equally to the Newfoundland Banks, though they have been taken widespread there; also along the south and southeast coast of Newfoundland, and as far north as the offing of Hamilton Inlet on the outer Labrador coast, during the experimental trawlings carried out by the Fishery Research Board of Newfoundland.

The fact that we have taken wolffish larvae in the channel between Browns Bank and Cape Sable; near Seal Island (Nova Scotia); on German Bank and off its slope; off Lurcher Shoal; off Machias (Maine); on Jeffreys Bank (off Penobscot Bay); and in Massachusetts Bay a few miles off Gloucester, is evidence that the wolffish breeds in the Gulf wherever it is to be found, as might, indeed, be expected. And this applies, equally (it seems) to the more northern parts of the American range of the wolffish, for its pelagic young have been reported off northeastern Newfoundland; in the Strait of Belle Isle; and off Sandwich Bay on the Atlantic coast of Labrador, by the Newfoundland Fisheries Research Commission.

Commercial importance.—The market demand for wolffish is of comparatively recent growth. It is an excellent table fish, selling readily as "ocean catfish" or as "whitefish." In 1947 the average price to the fisherman was between 4 and 5 cents per pound, and the Gulf of Maine catch was worth about $70,000 to $80,000.

Spotted wolffish *Anarhichas minor* Olafsen 1774[14]

SPOTTED CATFISH

Jordan and Evermann, 1896–1900, p. 2446.

Description.—This species resembles the common wolffish closely in its general form and in the arrangement of its fins. The chief difference is that while the central ("vomerine") band of teeth on the roof of the mouth is longer than the band on either side ("palatine") in the common wolffish, these bands are of about equal lengths in the spotted wolffish, and its teeth are described as red in life, not white. Furthermore, the rear end of its dorsal fin is abruptly indented close to the base of the tail, with its last 3 to 6 spines much shorter than those further forward, while the rear end of the dorsal is evenly rounded in the common wolffish.

Color is, however, the most convenient field mark for the spotted wolffish, its pale olive or chocolate [15] upper parts together with its dorsal and caudal fins, being thickly sprinkled with blackish brown spots, of different sizes and of irregular shapes.

Size.—Notwithstanding its Latin name this is fully as large a fish as the common wolffish, said to grow to a length of 6 feet. One 37 inches long weighed 13 pounds, eviscerated.

[9] Reported landings were about 6,000 pounds for Nantucket Shoals and about 13,000 pounds for the southwestern part of Georges Bank for 1945; about 9,000 pounds and 23,000 pounds respectively for these same grounds in 1947.

[10] Weight of dressed fish.

[11] For earlier examples, see Bigelow and Welsh, Bull. U. S. Bur. Fish., vol. 40, Pt. 1, 1925, p. 373.

[12] Cheticamp; Prince Edward Island; also Gaspé Bay.

[13] They are not mentioned in the catch statistics for the Gulf of St. Lawrence coast of Nova Scotia, for New Brunswick or for Quebec.

[14] Jordan, Evermann and Clark (Rept. U. S. Comm. Fish. (1928), Pt. 2, 1930, p. 472) place this species in the genus *Lycichthys* Gill 1876; but it seems preferable to follow the older usage here.

[15] The general ground tint has been variously described.

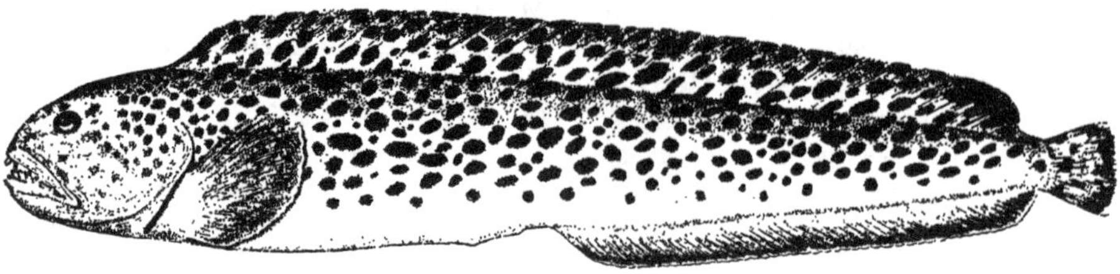

FIGURE 268.—Spotted wolffish (*Anarhichas minor*), off La Have Bank. From Goode. Drawing by H. L. Todd.

Habits.—Very little is known of its habits. Its diet is much the same as that of its more common relative. But it is said to keep to rather deeper waters, having been caught as deep as 200 to 240 fathoms off Banquereau Bank [16], to 200 fathoms in the eastern Atlantic. And its geographic range (see below) shows that it is confined to colder water than is the common wolffish.

General range.—Chiefly north of the Arctic circle; north coast of Russia, White and Barents Seas, and Iceland, south to middle Norway (vicinity of Bergen) on the European coast; Greenland; and southward occasionally to the Gulf of Maine on the American coast.

Occurrence in the Gulf of Maine.—Goode and Bean's [17] statement that "the Fish Commission has specimens from off the mouth of Gloucester Harbor and from Eastport, Maine," long remained the only notice of this northern fish for the Gulf of Maine,[18] and fishermen of whom we have inquired have either never seen it there or they have failed to distinguish it from the common wolffish, which is unlikely, so striking is its color pattern. But the late Walter Rich, of the U. S. Bureau of Fisheries, obtained a specimen that had been taken in 35 fathoms off Cape Elizabeth (now in the collection of the Portland Society of Natural History); another, weighing 3¼ pounds was caught on a long line off Portland lightship on April 23, 1927. Evidently this wolffish reaches our Gulf only as an accidental waif from its Arctic home, one to be watched for but hardly to be expected.

It appears to occur regularly off outer Nova Scotia however, though in small numbers. Thus 5 to 10 are usually landed at Boston each year from Sable Island Bank; we have records of 7 caught there in 1934;[19] one was brought in many years ago from 200 fathoms from the deep gulley between Sable Island Bank and Banquereau; and one from 250 fathoms from the northeast slope of the latter bank.[20] Another was taken on Banquereau in 50 fathoms, in 1932;[21] three were taken on Western Bank in 1937;[22] five more were definitely reported from Banquereau in that same year; and McKenzie [23] writes that half a dozen are brought in from that general region yearly.

With so many records for Nova Scotian waters it is astonishing that the spotted wolffish has not been reported at all in the Gulf of St. Lawrence so far as we can learn, nor is it mentioned among the fishes listed in Newfoundland waters or off outer Labrador from the experimental trawlings by the Fishery Research Commission of Newfoundland.

THE OCEAN POUTS AND WOLF EELS. FAMILY ZOARCIDAE

The ocean pouts and wolf eels are slender eel-like fishes with the anal fin continuous with the caudal. In most of the members of the family the dorsal fin, also, joins the caudal equally, making one continuous fin extending around the tip of the tail, but in the only common Gulf of Maine species the rear portion of the dorsal is so low that there seems to be a bare space between it and the caudal. All the members of the family, however, known definitely, either from the Gulf of Maine, or from the outer coast of Nova Scotia, are readily separable from the true eels by having ventral fins (small but unmistakable) situated a

[16] Bean, Proc. U. S. Nat. Mus., vol. 3, 1881, p. 82.

[17] Bull. Essex Inst., vol. 11, 1879, p. 11.

[18] Goode's (Fish. Ind. U. S., Sect. 1, 1884, p. 249) statement that it has been seen in the Bay of Fundy apparently refers to this Eastport record.

[19] Bigelow and Schroeder, Bull. U. S. Bur. Fish., vol. 48, 1936, p. 337.

[20] Bean, Proc. U. S. Nat. Mus., vol. 3, 1881, p. 82.

[21] Vladykov, Proc. Nova Scotian Inst. Sci., vol. 19, pt. 1, 1935, p. 3.

[22] McKenzie and Homans, Proc. Nova Scotian Inst. Sci., vol. 19, 1938, p. 279.

[23] Proc. Nova Scotian Inst. Sci. vol. 20, pt. 1, 1939, p. 18.

little in advance of the pectorals. But the green ocean pout (*Gymnelis viridis* Goode and Bean) of arctic seas has no ventrals; it ranges southward as far as the estuary of the St. Lawrence River,[24] and perhaps as far as northern Nova Scotia.[25] The closest affinities of the ocean pouts, among Gulf of Maine fishes are with the blennies (p. 491), the wolffishes (p. 500), and the wrymouths (p. 502). But they are easily separable from the blennies and wrymouths by the fact that at least the major part of the dorsal fin is soft-rayed, not spiny; and from the wolffish by their more slender form and smaller teeth.

Only two species are known definitely from our Gulf, one, the common ocean pout (p. 510) very plentiful; the other, the wolf eel (p. 515), much less so. A third, the Arctic Ocean pout (p. 516) has been reported from shoal water to the west of our Gulf as well as from the Nova Scotian Banks to the east, though not from the Gulf itself. A fourth species (*Lycodes esmarkii* Collett 1875) has been credited to the Bay of Fundy.[26] But the specimen in question was trawled by the *Albatross I*, on the southern slope of the Grand Banks in 244 fathoms of water.[27] Two others that have been reported from the Nova Scotian Banks are also included in the following key, as they are likely to be found in our Gulf sooner or later.

Still another species, *Lycodes atlanticus* Jensen 1904, has been reported from a number of stations along the continental slope from the offing of southern Nova Scotia to the offing of northern North Carolina, in depths of 543 to 1,423 fathoms.[28] But being a deep-water form, it is not to be expected either within the limits of the Gulf of Maine, or on the Nova Scotian Banks.

The various species of the genus *Lycodes* resemble one another so closely that their identification is very difficult. If one should be taken in the Gulf that does not agree with either of these that are described on the following pages, we suggest that it be sent either to the laboratory of the U. S. Fish and Wildlife Service, Woods Hole, Mass.; the Division of Fishes, U. S. National Museum, Washington, D. C.; or to the Department of Fishes, Museum of Comparative Zoology, Cambridge, Mass., to be named.[29]

KEY TO GULF OF MAINE AND NOVA SCOTIAN OCEAN POUTS AND WOLF EELS

1. The dorsal fin seems to be separated from the caudal fin by a considerable gap ---- Ocean pout, p. 510
   The dorsal, caudal and anal fins form one continuous fin ---- 2
2. The trunk is extremely slender, at least 14–16 times as long as it is deep; the dorsal fin originates over the tips of the pectoral fins ---- 3
   The trunk is stouter, less than 12 times as long as it is deep ---- 4
3. Dorsal fin with only about 92 rays and anal fin with about 88 rays; lower surface of body with only a few scales. ---- Wolf eel, p. 515
   Dorsal fin with about 118 rays and anal fin with about 110 rays; lower surface of body uniformly scaly, like upper surface ---- *Lycenchelys paxillus*, Goode and Bean 1879[30]
4. The lateral line runs along the middle of the sides; the vent is only a little nearer to the snout than to the tip of the tail; there are no scales on the belly or on the forward part of the back ---- Arctic eelpout, p. 516
   The lateral lines (or their lower branch if double) run along the lower part of the sides; the vent is considerably nearer to the snout than it is to the tip of the tail; the body is covered with scales ---- 5

---

[24] Vladykov and Tremblay, Natural. Canad., vol. 62 (Ser. 3, vol. 6), 1935, p. 82.

[25] Goode and Bean (Smithsonian Contrib. Knowl., vol. 30, 1895, p. 313) credit it to Nova Scotia but give no definite locality.

[26] By Vladykov and McKenzie, Proc. Nova Scotian Inst. Sci., vol. 19, 1935, p. 109.

[27] Reported by Goode and Bean (Smithsonian Contrib. Knowl., vol. 30, 1895, p. 305) as from lat. 44°47′ N.; long. 66°33′ W., *Albatross* Station 2470; but the actual position of this station was lat. 44°47′ N.; long. 56°33′ W. See Townsend, Rept. U. S. Comm. Fish., (1900), 1901, p. 399.

[28] Reported by Goode and Bean (Smithsonian Contrib. Knowl., vol. 30, 1895, pp. 305–306) as *L. frigidus* Collett 1875, with list of stations.

[29] Jensen's monograph of the Lycodids of Northern Europe and of Greenland (Danish *Ingolf* Exped., vol. 2, Pt. 4, 1904) includes descriptions, and beautiful illustrations of all the species of *Lycodes* that have been reported from the Gulf of Maine, or from the Nova Scotian Banks. And Vladykov and Tremblay (sta. Biol. Saint Laurent. Fauna and Flora Laurent, No. 1, 1936) have given a revision of the genus in the western Atlantic with decriptions and photographs of several new species and subspecies from the Gulf of St. Lawrence and northward.

[30] Known from several stations on the continental slope abreast of our Gulf and off southern New England at depths of 365–904 fathoms; also from the deep gully between LaHave and Sable Island Banks at 200 fathoms (see Goode and Bean 1895, p. 311 for list).

5. Lateral line single, running along lower part of side of body; forward part of dorsal fin marked with one or more sooty patches .... *Lycodes vahlii* Reinhardt 1838 [31]
Lateral line double, with the more distinct branch of the two running along the lower edge of the side of the body; the forward part of the dorsal fin is not marked with dark patches .... *Lycodes esmarkii*, Collett 1875 [32]

Ocean pout *Macrozoarces americanus* (Bloch and Schneider) 1801

EELPOUT; CONGO EEL; MUTTONFISH

Jordan and Evermann, 1896–1900, p. 2457. *Zoarces anguillaris* (Peck) 1804.

Description.—The ocean pout is blenny-like or eel-like in form, its body about 8 times as long as it is deep (10 to 11 times in young fish up to about 8 inches long), moderately flattened sidewise, noticeably sway-bellied, and tapering backward from abreast of the pectorals, where it is deepest, to a pointed tail. It is very soft, its scales are very small, and its skin as slimy as an eel. Its ventral fins are small like those of the rock eel (p. 492), and they are situated well forward of the pectorals.

The most useful field mark for the identification of the ocean pout among the several eel-like fishes with which it might be confused are its vertical fins. Its anal fin is continuous with the caudal, there being no trace of any notch between the two, as is the case with the true eels. In reality, this is also true of the dorsal fin of the eelpout. But about 16 to 24 of the dorsal rays near the rear end of the fin are so short as to be hardly visible, so that there seems to be a considerable free gap between the dorsal fin and the caudal fin. Furthermore, these short rays are spiny instead of soft as all the other dorsal rays are. The dorsal fin runs from the nape back along the whole length of the trunk, and consists of first about 95–100 soft rays; next of the short spines, then of about 17 more soft rays. The anal fin (about 105–124 soft rays) originates a little in front of the mid length of the fish. Both the dorsal fin and the anal are of nearly even height from end to end except as just noted, but the dorsal is nearly twice as high as the anal. The pectoral fins are large and rounded like those of the wolffish. The very small ventrals are on the throat, in front of the pectorals. The upper jaw projects a little beyond the lower, the soft, fleshy upper lip somewhat farther still,[33] enclosing the tip of the lower lip when the mouth is closed.

The mouth is wide, gaping back beyond the small eyes, and it is set low with thick and fleshy lips that give the profile a distinctive aspect. Both jaws are armed with two series of strong, blunt conical teeth, largest in front, but the mouth lacks the crushing teeth that are so characteristic of the wolffish tribe (p. 502). There are 131–144 vertebrae.

Color.—Although this fish has usually been described as reddish brown mottled with olive, or as salmon colored, most of those we have seen caught have been of some shade of muddy yellow, paler or darker; some tinged with brownish, some with salmon, and some with orange; a few have been pure olive green. Fishermen usually describe them as yellow, and this is evidently the prevailing hue in the offshore parts of the Gulf. Other ocean pouts we have caught inshore along the coast of Maine, however, have shown yellow only on the margins of the fins, particularly the lower edge of the pectorals, with the general ground tint of sides

[31] Originally described from Greenland; reported from Banquereau Bank, at 130–190 fathoms by Goode and Bean (Smithsonian Contrib. Knowl., vol. 30, p. 308, as *Lycodes zoarchus*. See Vladykov and McKenzie, Proc. Nova Scotian Inst. Sci., vol. 19, pt. 1, 1935, p. 109.

[32] Spitzbergen and Northern Norway; Grand Banks; LaHave Bank, and southward along the continental slope in depths of 300–420 fathoms to the offing of Rhode Island.

[33] The ocean pout has sometimes been described and pictured as with the upper lip and jaw projecting far beyond the lower; but this is contrary to our observations.

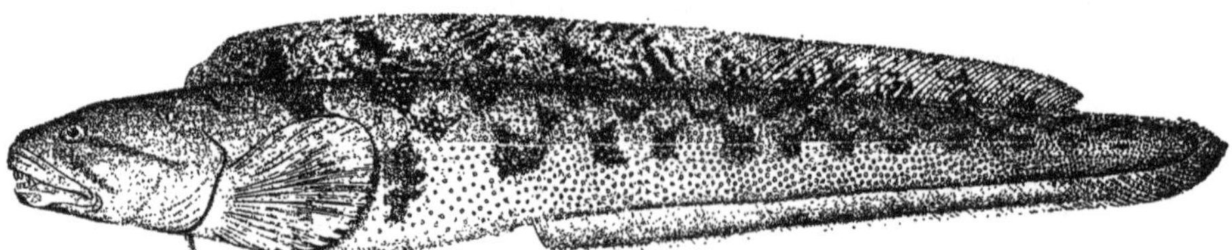

FIGURE 269.—Ocean pout (*Macrozoarces americanus*), Eastport, Maine, specimen. From Goode. Drawing by H. L. Todd.

and back ranging from pale gray (sometimes with purplish tinge) to dull brown or to dark dusky olive; the belly ranging from dirty white or yellowish or pinkish to the same dark shade as the back. One of a pair that were caught side by side in Northeast Harbor, Mount Desert, was pale grayish white below while the other was amethyst pink on the belly and on the lower side of the head. Whatever the ground tint, the sides are dotted with small dark spots clustered in irregular cross-bars, extending out on the dorsal fin. And there is a dark brown stripe running from eye to edge of gill cover.

Young fry, up to 3 or 4 inches long, are checkered along the sides and irregularly blotched on the back with light and dark brown, and they have a small but prominent black spot on the forward part of the dorsal fin until about 1 foot long, but this spot fades out with growth.

Size.—The ocean pout is said to reach a length of 3½ feet and to a weight of 12 pounds; Olsen and Merriman's largest, among some 2,500 specimens, was 38½ inches long, weighing 11¾ pounds, and we have seen one of about 3 feet among the many we have handled. But only a few grow longer than 30 inches, with 16 to 28 inches as a fair average for the general run of those that are caught.

The average weights of pouts of different sizes were about as follows for a large number taken in the southern side of the Gulf of St. Lawrence, in the Bay of Fundy, in the southwestern part of the Gulf of Maine, and off southern New England: 16 inches, ½ pound; 18 inches, 1–1⅛ pounds; 20 inches, 1½–1¾ pounds; 22 inches, 2–2¼ pounds; 24 inches, 2½–3 pounds; 26 inches, 3¼–3¾ pounds; 28 inches, 4½ pounds; 30 inches, 5½ pounds.[34]

Remarks.—The ocean pout of North Europe (*Zoarces viviparus* Linnaeus 1758), a very close relative, is distinguishable from the American eelpout, by having fewer fin rays (about 100 dorsal rays and 6 to 10 spines; 80 to 89 anal rays), fewer vertebrae (101–126), smaller head and mouth, and only a single row of teeth in the front of the jaw, while some specimens have no interruption between the dorsal fin and the caudal. Also, the European ocean pout is a smaller fish, and its eggs are retained in the ovaries of the mother until after they have hatched, hence its specific name *viviparus*.

It is well known [35] that the European ocean pout tends to break up into genetic races that are partly correlated with environmental conditions. And recent studies by Olsen and Merriman [36] make it likely that there is a slower growing race of ocean pout in the Bay of Fundy and perhaps northward, a larger, faster growing race ranging from Cape Cod southward, with each of these including minor subpopulations. This interesting subject would repay further investigation.

Habits.[37]—The ocean pout is a ground fish, as might be expected from the fact that it has no swim-bladder, as well as from its food (see p. 512). And the habits of fish kept in aquaria, where they are described as remaining coiled up in the darkest parts suggests that they spend most of their lives hiding among sea weeds and stones. They are described as moving slowly backward and forward by undulations of the fanlike pectoral fins or of swimming more rapidly by undulating motions of the rear part of the trunk and tail, with the pectorals wide spread and held horizontal, and with the dorsal and anal fins close to the body.[38] They swim actively when disturbed. And it is almost unbelievable to what a hopeless tangle of cord, fish, and slime a few ocean pouts can reduce many fathoms of long line set for other fish.

The vertical range of the ocean pout in one place or another extends at least as deep as 105 fathoms.[39] At the opposite extreme Clemens and Clemens [40] report that young ones are sometimes found around rocks and in seaweed along the shore in the Bay of Fundy during the ebb tide. They are even known to run into rivers for some distance, though always holding to the bottom, i. e., to the undercurrent of water of

[34] According to graphs by Clemens and Clemens (Contrib. Canadian Biol. (1918–1920) 1921, fig. 5, p. 79), for the Gulf of St. Lawrence and Bay of Fundy; and by Olsen and Merriman (Bull. Bingham Oceanographic Coll., vol. 9, art. 4, 1946, fig. 4, p. 43) for the southwestern part of the Gulf of Maine and for southern New England.

[35] Especially from investigations by Johannes Schmidt and by J. V. C. Smith (for list of references, see Olsen and Merriman, Bull. Bingham Oceanographic Collection, vol. 9, art. 4, 1946, p. 182).

[36] Bulletin, Bingham Oceanographic Coll., vol. 9, art. 4, 1946, p. 116–117.

[37] Olsen and Merriman (Bull. Bingham Oceanogr. Coll., vol. 9, art. 4, 1946) have recently published a detailed study of the life history of the ocean pout.

[38] Willey and Huntsman (Canadian Field Naturalist, vol. 35, 1921, p. 6), and Clemens and Clemens (Contrib. Canadian Biol. [1918–1920], 1921, p. 71) give some observations on the actions of ocean pouts kept in the aquarium at the St. Andrews Laboratory.

[39] *Albatross III* trawled 3 specimens from between 105 and 240 fathoms, on the southwestern slope of Georges Bank in May 1950.

[40] Contrib. Canadian Biol. (1918–1920) 1921, p. 72.

high salinity that tends to move inward along the bottom from offshore. But most of them live between 8–10 fathoms and perhaps 45 fathoms in the waters with which we are most immediately concerned. During the years when ocean pouts were in demand (see below), good catches were made as shoal as 10 to 12 fathoms in the southwestern part of the Gulf, also off southern New England.[41] And we have seen large numbers caught from party boats, at 8–17 fathoms along the coast of New Jersey.

We have taken ocean pouts in the Gulf of Maine on sandy mud, on sticky sand, on broken bottom, also on pebbles and gravel. They are caught in large numbers on smooth hard bottom and we have seen many more of them taken from party boats off northern New Jersey on rocky bottom, along with sea bass (p. 407), tautog, cod, and other fishes, than were taken on soft bottom when we were fishing for hake (*Urophycis*).

There is no evidence that they carry out any extensive migrations. However, information has accumulated recently to the effect that the adults congregate through the summer, autumn, and early winter on rocky bottoms where the eggs are deposited and guarded, to disperse again in midwinter (after the eggs have hatched), over the smoother grounds in the vicinity where food is more plentiful.[42] And this spawning migration appears to be complicated by an autumnal shift offshore to deeper water, with a return movement in spring, in coastal regions where the bottom water chills in winter to a temperature too low for their comfort; in the Bay of Fundy, for example (p. 514), and perhaps in the Gulf of St. Lawrence.

The ocean pout can be classed as a cool-water fish, for the great majority of them, in whatever part of their geographic range, are in temperatures at least no higher than about 62°, throughout the year. At the other extreme, they have been taken in 32° in the Bay of Fundy (p. 514); in about 31° to 32° in Trinity and Conception Bays, Newfoundland.[43] And eelpouts are exposed to temperatures as low as this, in spring, in the Gulf of St. Lawrence, unless they descend deeper into the Laurentian Channel, which they may do.

Food.—The American ocean pout feeds on a wide variety of shelled mollusks, univalve and bivalve, on crustaceans large and small, on echinoderms and on other invertebrates. The Bay of Fundy fish opened by Clemens and Clemens had dieted chiefly on the two common mussels, *Mytilus* and *Modiolaria*, on whelks (*Buccinum*), periwinkles (*Littorina*), and on scallops (*Pecten*) as well as on various other bottom-living mollusks, on sea urchins, brittle stars, and barnacles. A large specimen caught in Massachusetts Bay, January 1924, was packed full of brittle stars (ophiurans), spider crabs, and small sea scallops (*Pecten magellanicus*); a number trawled by the *Albatross III* at 42 fathoms, near Nantucket Lightship, May 17, 1950, were full of small sea scallops (*Pecten magellanicus*), as many as 100–200 per fish. Olsen and Merriman [44] write that sand dollars (*Echinarachnius*) were the chief items in the stomach contents of some 850 ocean pouts taken in the southwestern part of our Gulf and off southern New England, with crabs (*Cancer*) and isopod crustaceans (*Unicola*) as seconds; while some had eaten bivalve mollusks (*Yoldia* and *Pecten*) in large amounts; also the eggs of the longhorn sculpin (p. 451), which are often laid among the branches of the finger sponge (*Chalina*).

Ocean pouts bite on fish as greedily as they do on clams or cockles, and pouts kept in the aquarium at St. Andrews took fish as readily as clams;[45] while two of the fish opened by Clemens and Clemens,[46] and also Bay of Fundy fish examined by Olsen and Merriman,[47] contained remains of fish. But in all probability about the only fish they get are dead ones that have sunk to the bottom, or occasionally a small one that a pout may have the good luck to catch. The European representative of our ocean pout (*Zoarces viviparus*) is described [48] as taking in mouthfuls of weeds for the crustaceans and mollusks living among these, and as swallowing a considerable amount of the plant material with its animal prey. But American ocean pouts appear not to feed in

[41] Olsen and Merriman, Bull. Bingham Oceanographic Coll., vol. 9, art. 4, 1946, p. 37, 38, fig. 3.

[42] This shift of grounds has been demonstrated recently by Olsen and Merriman (Bull. Bingham Oceanog. Coll., vol. 9, art. 4, 1949, p. 40–42).

[43] Rept. Newfoundland Fish. Res. Comm., vol. 1, No. 4, 1932, p. 109.

[44] See Olsen and Merriman, (Bull. Bingham Oceanog. Coll., vol. 9, art. 4, 1946, p. 124–129) for a list of stomach contents, with discussion.

[45] As reported by Willey and Huntsman, Canadian Field Naturalist, vol. 35, 1921, pp. 6–7.

[46] Contrib. Canadian Biol. (1918–1920) 1921, p. 80, small fish, including a smelt.

[47] Bull. Bingham Oceanogr. Coll., vol. 9, art. 4, 1946, p. 129; probably herring.

[48] By Blegvad, Report Danish Biol. Stat. (1916), 1917, p. 42.

this way to any great extent, for only traces of plants have been found in their stomachs.

The eggs of the European ocean pout (*Zoarces viviparus*) are fertilized within the mother, and are retained within her oviducts until after they hatch. But the American eelpout lays eggs, as proved by the fact that the young were hatched from a mass of eggs brought up, with two eelpouts, in an old rubber boot,[49] in Blacks Harbor, Passamaquoddy Bay. And enough small specimens of 1⅛ inches and upward have been collected of late, between New Jersey and Maine, to show that the eelpout breeds successfully throughout this part of its range, at least; and probably as far north as northern Nova Scotia and the southern side of the Gulf of St. Lawrence.

Various lines of evidence [50] show that spawning takes place in September and October. And the fact that fish taken in summer differ widely in the stage of development of their sexual products suggested to Clemens and Clemens that they do not breed every year. But it seems more probable to us that this is evidence simply of a protracted breeding period, some individuals ripening early in autumn, others not until later.

Large females lay more eggs than small, the numbers of maturing eggs actually counted having ranged from 1,306 in a fish 21½ inches (55 cm.) long to 4,161 in one of about 34½ inches (87.5 cm.).

The eggs are yellow, 6–7 mm. in diameter, and are laid in masses held together by a gelatinous substance. The only egg mass so far brought in was in an old rubber boot, suggesting that they are normally deposited in crevices in rocks or among stones, which would explain the apparent tendency of the mature fish to congregate on rocky bottom as the spawning season approaches.

The fact that eggs brought up in the trawl in the Passamaquoddy region, where the spawning is supposed to take place from mid-September through October, hatched in early January, and that ocean-pout eggs taken off New York in mid-November [51] still were some weeks short of hatching, makes it likely that incubation occupies at least 2½ to 3 months. And the actions of a captive female that lay coiled around its mass of eggs, though these had not been fertilized, [52] makes it likely that the eggs are guarded by one or the other parent during this period, perhaps by both of them.

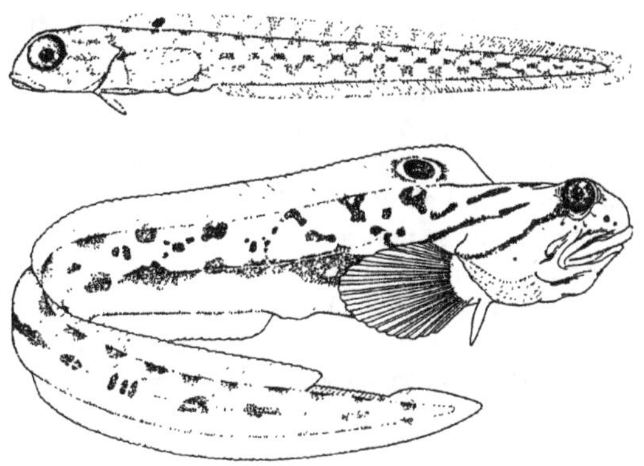

FIGURE 270.—Ocean pout (*Macrozoarces americanus*). Above, larva, 48 mm. Below, young fish, 387 mm. Drawings by Louella E. Cable.

The larvae are about 30 mm. long at hatching, i. e., much larger than those of most of our commercially important fishes, and they are so far advanced already in development that they are easily identified.[53] Being so nearly adult in form, it is probable that they hold to the bottom from the time they are hatched; all catches of immature fish recorded so far have, indeed, been on the bottom.

The sizes, of the fry in different months, show that ocean pouts reach a length of about 3 inches during their first summer, and that they are about 4 to 5 inches long when 1 year old. According to studies of otoliths by Olsen and Merriman,[54] ocean pouts in southern New England waters may be expected to reach 6 inches when between 1 and 2 years old; 12 inches at 3 years; and 24 inches when between 6 and 7 years, the very large fish of 36 inches and upward being 12 years to 16 years old. Estimates by the same method by Clemens and Clemens [55] point, however, to a considerably slower rate of growth in the colder water of the Bay of Fundy, where a 12-inch fish is likely to be nearly 5 years old, a 24-inch fish between 12 and 13 years old; and where the 8 oldest fish examined

[49] For account, see White, Jour. Fish. Res. Bd. Canada, vol. 4, pt. 5, 1939, pp. 337–338.

[50] Discussed in detail by Olsen and Merriman, Bull. Bingham Oceanogr. Coll., vol. 9, art. 4, 1946, pp. 69–77.

[51] From the stomach of a cod, see Olsen and Merriman, Bull. Bingham, Oceanog. Coll., vol. 9, art. 4, 1946, p. 76, fig. 9.

[52] See Olsen and Merriman (Bull. Bingham Oceanog. Coll., vol. 9, art. 4, 1946, p. 75, fig. 8) for a photograph of a female so employed, in the Shedd Aquarium, Chicago.

[53] For further details, see White, Jour. Res. Board Canada, vol. 4, 1939, p. 338.

[54] Bull. Bingham Oceanogr. Coll., vol. 9, art. 4, 1946, pp. 85–93.

[55] Contributions to Canadian Biology, 1918–1920 (1921), p. 74.

(16–18 years) among 190 specimens, were only 26¾ to 28½ inches long.

In one commercial catch, probably representative, made off Provincetown and analyzed by Olsen and Merriman, most of the fish were 4 to 10 years old, with only scattered fish of 11 to 16 years.

Off southern New England, according to Olsen and Merriman, a few females mature sexually when they are about 18 inches long; about half at 22–23 inches, and all of them by the time they are 24–25 inches long; males mature earlier, most of them by the time they have reached 15–16 inches; a few not until larger. The smallest females with large eggs seen in the Bay of Fundy region by Clemens and Clemens were 16–18 inches long.

General range.—Coast of North America from the Strait of Belle Isle, Gulf of St. Lawrence, and southeastern Newfoundland south to Delaware;[56] common from the southern side of the Gulf of St. Lawrence and northern Nova Scotia to New Jersey.

Occurrence in the Gulf of Maine.—The ocean pout, known more familiarly as "conger", or "congo" eel along the coast of Maine,[57] is a familiar fish in the Gulf in moderate depths of water both near shore and on the offshore banks; abundant locally off western Nova Scotia; in the Bay of Fundy;[58] all along the coasts of Maine and Massachusetts; and on Georges Bank where considerable numbers are taken both by otter trawlers and by long-line fishermen.[58] Very small ones have been collected off Chatham, Cape Cod; on Stellwagen Bank at the mouth of Massachusetts Bay; and near Mount Desert Island, Maine, by us; also in the Bay of Fundy and in Passamaquoddy Bay by Clemens and Clemens,[59] evidence that it breeds successfully all around the Gulf.

There seems to be a wide difference in the depth zone frequented by the ocean pout in different parts of the Gulf. In the Bay of Fundy some of them run up into shoal water in summer and young ones are to be found under stones and among seaweed between tide marks. Similarly, one is always apt to catch several in a half day's flounder fishing in 1 to 3 fathoms in Penobscot Bay or in Northeast Harbor, Maine, as we can bear witness. And this probably applies to bays and harbors all along the coast of Maine east of Cape Elizabeth. But we have never seen one taken in less than 10 fathoms of water in the Massachusetts Bay region, where most fishermen speak of it as a comparatively deep-water fish though it has been recorded from Gloucester Harbor. And the ocean pouts on the offshore grounds live mostly deeper than 20 to 30 fathoms. Thus *Albatross II* trawled a number in the basin of the Gulf down to 90 fathoms in July 1931, while a large number of them have been trawled on Georges Bank, at depths of 20–60 fathoms.[60] And in May 1950, the *Albatross III* trawled 3 at 105 fathoms or deeper[61] on the southwestern slope of Georges Bank.

Ocean pouts also frequent different types of bottom in different localities. In Massachusetts Bay they are seldom caught on the good fishing grounds on stony or gravelly bottoms, that is, or about ledges. But if the long line chances to run off these, the portion of it that is resting on the softer floor of the deeper parts of the bay often brings in eelpouts and nothing else except an odd hake. They are caught regularly on hard bottom, however, off Cape Cod and to the westward; we have trawled them on rather sticky sand in Ipswich Bay (22 fathoms) among good catches of hake and plaice; on broken bottom at the mouth of Casco Bay; on pebbles and mud in Penobscot Bay; they are commonly caught on stony ground farther eastward along the coast of Maine; and Huntsman describes them as taken on hard bottom in the Bay of Fundy.

In fact, the only type of bottom where we have not heard of them in our Gulf is the soft oozy mud with high organic content that floors certain of the deeper depressions, such as the trough to the west of Jeffreys Ledge.[62]

Information as to the seasonal movements of ocean pouts in different parts of the Gulf is not only scanty, but perplexing. In the open Bay of Fundy, Huntsman describes them as working inshore in spring but moving out again into deeper water in

[56] There is a doubtful record for North Carolina (Smith, North Carolina Geological and Economic Survey, vol. 2, 1907, p. 379).

[57] Years ago we heard them called "yowlers" by long-line fishermen, but we doubt that this name is still used for them anywhere.

[58] Clemens and Clemens (Contrib. Canadian Biol. (1918–1920) 1921, p. 69) give a general account of the ocean pout in the Bay of Fundy, and list the localities there whence it has been recorded.

[59] Contrib. Canadian Biol. (1918–1920), 1921, p. 77.

[60] *Albatross III*, for example, trawled 137 of them along the southern slope of Georges, at 31–60 fathoms, in May 1950; the dragger *Eugene H* trawled an average of 8 pouts per haul at 26–45 fathoms, and about 2 per haul at 46–75 fathoms on the south central part of Georges Bank, in late June 1951.

[61] The depth ranged from 105 fathoms to 240 fathoms along the strip of bottom on which the trawl was working.

[62] They were not represented among the considerable list of fishes trawled in such situations by the *Atlantis* in August 1936 (Bigelow and Schroeder, Biol. Bulletin, vol. 76, 1939, p. 309.

October or early November, which is corroborated by a report of Clemens and Clemens, that set lines made good catches in the Passamaquoddy region from early June through September, but caught none there from January to May. And their abundance in Penobscot Bay in midsummer suggests that some of them may perform a similar on and offshore migration there, too. But this may not apply to the coast south of Cape Elizabeth. Off southern New England, where they are plentiful on the commercial fishing grounds in winter and spring, only stray ocean pouts are taken there in summer and autumn. But it seems more likely that they shift, then, to regions of rockier bottom nearby, than that they move off-shore.[63]

It is not yet clear to what extent their movements depend on the local food supply, on seasonal changes in temperature, or on the habit the mature fish have of congregating on rocky grounds during the spawning season, and while guarding the eggs thereafter (p. 513). One must also bear in mind that failure to catch them on hook and line may simply mean that they are not biting at the time, not necessarily that they have moved away. This is likely to apply to the adult fish in particular during their spawning and egg-guarding season.

Importance.—Although the ocean pout has few bones and is said to be a sweet-meated fish, there was no regular market for it prior to the early 1930's; only a few, brought in by small boats, were sold on the streets of Boston, and nearly all of those that were caught incidentally by the larger vessels were thrown overboard. A small demand then developed for them resulting in landings for Massachusetts ranging between 45,600 pounds and 114,700 pounds yearly, for the period 1935 to 1942,[64] though none for Maine.

A concerted attempt was made in 1943 to market ocean pout as fillets, partly as a war measure. This was so successful that 3,943,300 pounds were reported as landed in Massachusetts ports in that year; 4,449,600 pounds in 1944, most of them caught from the tip of Cape Cod southward, and nearly all of them marketed through New York. But this popularity was short lived, for word soon spread that ocean pout are often afflicted with a protozoan parasite. Many shipments were condemned for this cause, and the demand fell off so rapidly that the landings for Massachusetts were less than one-fourth as great in 1945 (1,003,700 pounds) as they had been in 1944[65]; fell to 613,300 pounds in 1946; were 167,400 pounds in 1947; and dropped to 6,100 pounds in 1948, the most recent year for which statistics of the catch are available.

Wolf eel *Lycenchelys verrillii* (Goode and Bean) 1877.

Jordan and Evermann, 1896–1900, p. 2470.

Description.—This fish is eel-like in form and resembles the ocean pout in most other respects as well, except that it is more slender (about 14 to 16 times as long as it is deep), and that there is no separation, apparent or real, between its dorsal, caudal, and anal fins, but the three form a single continuous vertical fin running along the back, around the tail, and forward on the lower surface to the vent. The dorsal fin not only originates

[63] For discussion, see Olsen and Merriman, Bull. Bingham Oceanogr. Coll. vol. 9, art. 4, 1946, pp. 40–42.

[64] Landings for 1933 to 1940 were listed as "conger eels"; no data are available for 1934 or 1936.

[65] For a detailed history of the event, see Olsen and Merriman, Bull. Bingham Oceanogr. Coll., vol. 9, art. 4, 1946, pp. 9–10.

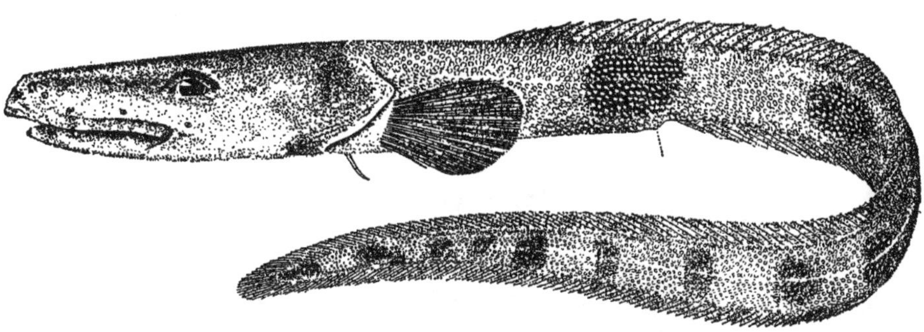

FIGURE 271.—Wolf eel (*Lycenchelys verrillii*), off Chebucto, Nova Scotia. From Jordan and Evermann. Drawing by H. L. Todd.

farther back than in the ocean pout (over the tip of the pectoral instead of in front of the base of the latter), but all the dorsal rays (about 92) are soft. Furthermore the anal fin (about 88 rays) extends relatively farther forward than in the ocean pout. The pectoral fins are rounded like those of the ocean pout, but smaller relatively, the small ventrals are similarly located well forward of the pectorals, and in small specimens the head resembles that of its relative in profile except for a somewhat wider mouth. Old males, however (fig. 271), "are transformed almost beyond specific recognition by an extraordinary development of the entire head in advance of the eyes. The snout becomes shovel-shaped, its length equal to two-fifths that of the head, while in the normal condition it is one-fourth".[66]

Color.—The sides are light brown above the lateral line, white below it, with a series of 8 to 10 irregular dark brown patches which the lateral line bisects. The belly is blue, its lining jet black.

Size.—Maximum length about 10 inches; usual length about 4 to 6 inches.

Habits.—This is a bottom fish, living on mud or sand and confined to considerable depths of water. Normally, 25 to 30 fathoms is its upper limit, but the fact that the *Grampus* specimen mentioned below was taken in a tow net, though close to bottom, proves that it sometimes rises from the ground. To the southward, on the continental slope, it has been trawled down to 603 fathoms.

Nothing is known of its way of life or of its breeding habits.

General range.—So far known only off the coasts of Nova Scotia and of New England, and southward along the continental slope to the offing of Beaufort, N. C., in rather deep water.

Occurrence in the Gulf of Maine.—The wolf eel has been trawled at many localities on the continental slope at 200 to 600 fathoms from the offing of Cape Fear, North Carolina (lat. 34°40′ N., long. 75°15′ W.) to abreast the western end of Georges Bank (long. 68°22′ W.).[67] It was formerly regarded as very rare within the Gulf of Maine, the only records for it there up to 1925 having been of a few specimens trawled off the mouth of Passamaquoddy Bay in 35 to 50 fathoms; of one 4 inches long taken off Monhegan Island by the *Grampus* on August 2, 1912, in 60 fathoms; and of several that were collected by the U. S. Fish Commission many years ago off Cape Ann in 75 to 110 fathoms, in the Western Basin in 115 fathoms, and off Cape Cod. More recent captures, however, of a number of wolf eels at about 90 fathoms in the trough west of Jeffreys Ledge by the *Albatross II* in November 1927, in August 1928, and in September 1930 (a total of 61 specimens) show that they are more plentiful in the deeper parts of the Gulf than the previous record might have suggested.

The only definite records for the wolf eel eastward from Cape Sable are of 5 specimens taken by the U. S. Fish Commission steamer *Speedwell* 3 miles off Cape Negro, Nova Scotia, in 90 fathoms, in the summer of 1877[68], and of one[69] taken 27 miles off the entrance to Halifax Harbor (Chebucto Head) in 101 fathoms.[70]

Arctic eelpout *Lycodes reticulatus* Reinhardt 1838

Jordan and Evermann, 1896–1900, p. 2465.

Description.—This fish resembles the ocean pout in its general appearance and in the arrangement of its fins. The readiest field marks for it are that the dorsal fin is not interrupted, but is continuous with the caudal fin, and that the dorsal originates behind the bases of the pectorals instead

[66] Goode and Bean, Smithsonian Contrib. Knowl., vol. 30, 1895, p. 310.

[67] Goode and Bean (Smithsonian Contrib. Knowl., vol. 30, 1895, p. 310) give a long list of localities from the early cruises of the U. S. Coast Survey and of the U. S. Fish Commission. And Beebe (Zoologica, N. Y. Zool. Soc., vol. 12, 1929, p. 18) has reported a wolf eel from the Hudson Gorge, off New York, at 69 fathoms.

[68] Reported by Goode and Bean, American Jour. Science, ser. 3, vol. 14, 1877, p. 473.

[69] This is the specimen pictured here in fig. 271.

[70] Probably these specimens were the basis for Jones' (List Fishes Nova Scotia, 1879, p. 5; Proc. Nova Scotian Inst. Sci., vol. 5, 1882, p. 91) statement that the wolf eel occurs on the Nova Scotian fishing banks.

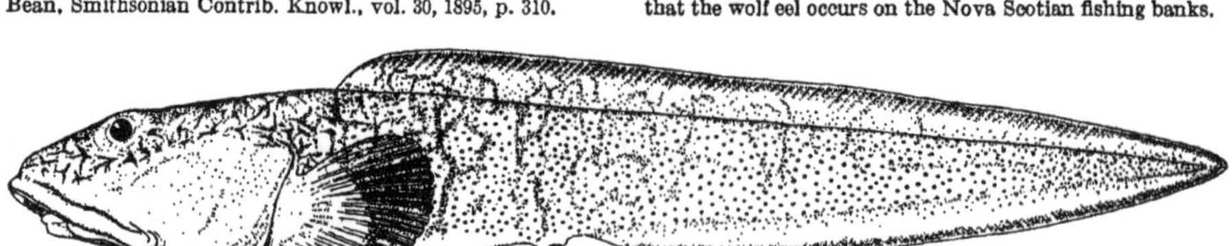

FIGURE 272.—Arctic eelpout (*Lycodes reticulatus*). Drawing by Louella E. Cable.

of in front of them, while the fanlike pectoral fins are even larger, relatively, than those of the ocean pout. Furthermore, its upper jaw projects far beyond the lower, giving it a distinctive cast of countenance (compare fig. 272 with fig. 269). The most obvious difference between this *Lycodes* and the wolf eel (p. 515) is that the former is much the stouter bodied of the two, being only about 8 times as long as it is deep (the wolf eel is 14–16 times as long as deep), and that the dorsal fin of *Lycodes reticulatus* originates farther forward, i. e., close behind the bases of the pectorals instead of over the tips of the latter.

Color.—Described as brownish, with a network of black lines on the head and with several groups of such lines or with solid dark bands on the body. The dorsal fin is dark edged. The young fry are marked with a series of large dark spots on the back and extending out on the dorsal fin.[71]

Size.—Specimens of which the measurements have been definitely recorded have ranged up to 15 inches (380 mm.) in length.

Remarks.—This lycodid tends to separate into local races; one such from northeast Greenland and Jan Mayen has, in fact, been dignified with a separate varietal name; var. *macrocephalus* by Jensen,[72] because seemingly separable from the West Greenland form. One subspecies, *hacheyi*, too, has been described subsequently from Hudson Bay by Vladykov;[73] also a second (*lavalei*) from the Gulf of St. Lawrence by Vladykov and Tremblay[74] but none of these call for consideration here.

Habits.—Little is known of its habits except that it is a ground fish, usually living in moderately deep water, and that worms, crustaceans, and small fish have been found in the stomachs of European specimens. In its turn it falls a prey to larger fishes, and frequently to Greenland sharks.[75]

General range.—Both sides of the Arctic Atlantic; reported as far south as Vineyard Sound, southern Massachusetts.

Occurrence along the Atlantic coast of North America.—This particular *Lycodes* has been reported definitely off southeastern Labrador in the offing of Sandwich Bay; in the Strait of Belle Isle; in Conception Bay, Newfoundland; on the Grand Banks; off Placentia Bay, south coast of Newfoundland; also on the Newfoundland side of Cabot Strait;[76] and on the southwest slope of Banquereau Bank at 300 fathoms;[77] while Vladykov and McKenzie report it from Nova Scotian waters in general.[78]

It has not yet been recorded from the Gulf of Maine. But it is to be expected there sooner or later, for it has not only been taken to the east and north of our limits, as just stated, but also in Vineyard Sound to the westward,[79] if the specimen in question was identified correctly.

THE CUSK EELS. FAMILY OPHIDIIDAE

The members of this family are eel-like in appearance, but they differ very obviously from the true eels in having well-developed ventral fins in the form of long forked barbel-like structures, situated on the throat. Their soft-rayed dorsal and anal fins are low but continuous around the tip of the tail; the gill openings are very wide, and the snout projects beyond the mouth. Many species are known, most of them from warm seas.

Cusk eel *Lepophidium cervinum* (Goode and Bean) 1885[80]

Jordan and Evermann, 1896–1900, p. 2484.

Description.—The cusk eel is about 12 to 13 times as long as it is deep; all its fins are soft and eel-like; and there is no separation between the dorsal, caudal, and anal fins, but the three form one continuous fin running backward along the

[71] See Smitt (Scandinavian Fishes, vol. 1, 1892, p. 605) for the coloration of the genus *Lycodes* in general; Jensen (Danish *Ingolf* Exped., vol. 2, pt. 4, pl. 2, figs. 2, 3, and pl. 8) for beautiful illustrations of this species.

[72] Danish *Ingolf* Exped., vol. 2, pt. 4, 1904, p. 66, pl. 8.

[73] Contr. Canad. Biol., N. Ser., vol. 8, No. 2, 1933, p. 25.

[74] Fauna et Flora St. Laurent., Sta. Biol. St. Laurent., No. 1, 1936, p. 34.

[75] Smitt, Scandinavian Fishes, vol. 1, 1892, p. 613.

[76] For records of it in Labrador and Newfoundland waters, see Goode and Bean (Smithsonian Contrib. Knowl., vol. 30, 1895, p. 305); also the Annual Reports of the Newfoundland Fisheries Research Commission, vol. 1, No. 4; vol. 2, Nos. 1–3, 1932–35.

[77] See Goode and Bean, Smithsonian Contrib. Knowl., vol. 31, 1895, List of plates and figures, p. 17, figs. 273, 281.

[78] They do not mention any definite locality records but write of it (Proc. Nova Scotian Inst. Sci., vol. 19, 1935, p. 109) as usually believed to be the most common *Lycodes* there.

[79] Goode and Bean (Smithsonian Contrib. Knowl., vol. 30, 1895, p. 305), *Fish-Hawk* Station 681; Sumner, Osburn, and Cole (Bull. U. S. Bur. Fish., vol. 31, Pt. 2, 1913, p. 768). Goode and Bean also report it from east of the Bahamas (*Albatross* Sta. 2652, lat. 24°13′ N., long. 77°13′ W., 140 fathoms). But this is so very far to the south of the normal range of this species that we suspect the record is an error.

[80] See Jordan and Evermann (Bull. 47, U. S. Nat. Mus., Pt. 3, 1898, p. 2482) for the nomenclatural history of the name *Lepophidium* Gill 1895.

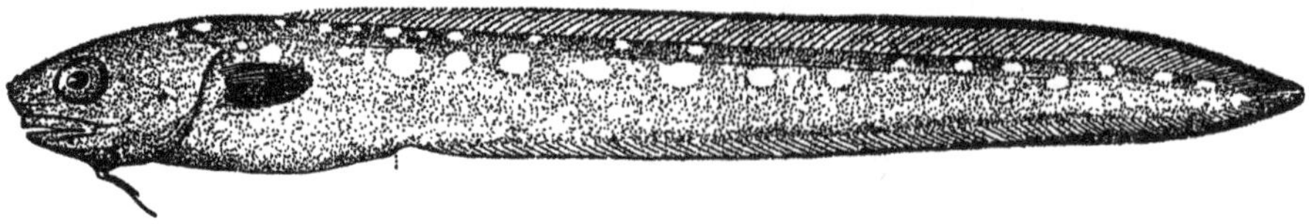

FIGURE 273.—Cusk eel (*Lepophidium cervinum*). After Goode and Bean. Drawing by H. L. Todd.

back, around the tip of the tail, and forward on the lower surface. But it is separated from all the true eels, by the presence of ventral fins, situated on the throat far in front of the pectorals, and reduced to forked barbel-like structures. The structure of the ventral fins and the uninterrupted dorsal fin separate the cusk eel from the ocean pout, its near relative among local fishes. And the presence of a short sharp spine on the top of the snout pointing forward and downward, which is easily felt if not seen (for it is nearly concealed in the skin), likewise differentiates it from such other Gulf of Maine species as it resembles in general appearance. The shape of the snout, too, is distinctive, as are its rather large scales, for the other genera of its family have naked heads, and the scales on their bodies are very small.

Color.—Brownish yellow, darker above than below, the upper part of each side marked with a row of 14 to 23 roundish white or pale brown spots. The dorsal and anal fins have narrow black or dusky margins. It seems that the adult color pattern develops late, for neither the spots nor the edgings on the fins are visible in a young specimen of 2⅝ inches.

Size.—The type specimen was about 10⅝ inches (262 mm.) long.

General range and occurrence in the Gulf of Maine.—This fish has been taken at various localities along the outer part of the continental shelf from off Florida to eastward of Nantucket, in depths of 38 to 102 fathoms. It is mentioned here because one specimen has been taken in 76 fathoms off Nantucket Shoals,[81] while two others, about 7⅞ and 8½ inches long (newly swallowed) were found in the stomach of a white hake *Urophycis tenuis*) that was trawled on the southwestern part of Georges Bank (lat. 40°31′ N., long. 68°55′ W.), at 39 fathoms, by the dragger *Eugene H.* on June 27, 1951.

THE TOADFISHES. FAMILY BATRACHOIDIDAE

The toadfishes are somewhat sculpin-like in appearance, but the resemblance is superficial, for their ventral fins are situated on the throat well in front of the pectoral fins ("jugular"), and they have only three gills and three gill arches. Both the soft and the spiny portions of the dorsal fin are well developed as separate fins, the former much longer than the latter. Most of the species belong to warm seas, only one reaching the Gulf of Maine.[82]

Toadfish *Opsanus tau* (Linnaeus) 1766

Jordan and Evermann, 1896–1900, p. 2315.

Description.—The toadfish, like the sculpins, has a large flat head, round nose, tremendous mouth, tapering body with plump belly, and fanlike pectoral fins. But it differs from all sculpins, and indeed from all other spiny-finned fishes of the Gulf of Maine except the blennies (p. 491) in the location of its ventral fins, which are under the throat well in front of the pectorals ("jugular") instead of below the latter or behind them. And no one could confuse it with any blenny, for it is not only a totally different looking fish, but its dorsal fin is mostly soft rayed while that of the blennies is spiny throughout. The presence of fleshy flaps of irregular outline on the tip of the upper jaw and along the edge of the lower jaw, on the cheeks, and over each eye, gives its head a peculiar wary appearance. Distinctive, also, is the fleshy nature of all its fins and the outline of the dorsal, the soft part of which (26 to 28 rays) is five to six times as long as the spiny part (3 spines), from which it is entirely separated by a deep notch, the two together extending the whole length of the trunk from the nape of the neck nearly to the base of the caudal fin. The anal

[81] Goode and Bean, Smithson. Contrib. Knowl., vol. 30, 1895, p. 347.

[82] See Schultz and Reid (Copeia, 1937, No. 4, p. 211) for a synopsis of the American-Atlantic species of the genus *Opsanus*.

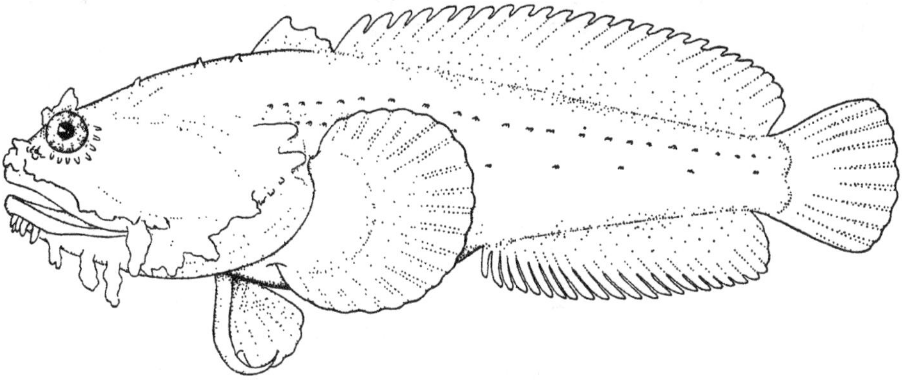

FIGURE 274.—Toadfish (*Opsanus tau*), Woods Hole. Drawing by Louella E. Cable.

fin (21 or 22 rays) is somewhat shorter than the second (soft) dorsal, orginates under about the eighth ray of the latter, and is similar to it in outline except that its rays are more or less free at their outer ends, especially in its forward half. The caudal fin is rounded; the ventrals covered by thick fleshy skin, are jagged in outline, with the first ray stouter than the others. There is a large open pit of unknown function in the axil of each pectoral fin.

We need only remark, further, that the skin is covered with a thick layer of slimy mucus. The toadfish has no scales; its teeth are large and blunt; and it has two short spines at the upper angle of each gill cover, hidden however, in the thick skin.

Color.—The general ground tint ranges from dark muddy olive green to brown or yellow, darker on back and sides, paler below, and variously and irregularly marked with darker bars and marblings, which may be restricted to head and fins or may extend over the whole fish, belly as well as back. The toadfish, like many other bottom fishes, changes color to match the bottom on which it lies.

Size.—Exceptionally 15 inches long, but few are longer than 12 inches.

Habits.—The toadfish lives in shoal water, and it is resident the year around wherever found, probable becoming torpid in winter in the northern part of its range. It is commonest on sandy or muddy bottom, hiding among eelgrass or under stones where it hollows out dens in which it lies in wait for prey. It is voracious and omnivorous, Vinal Edwards's diet list for it at Woods Hole including sea worms (*Nereis*), amphipods, shrimps, crabs, hermit crabs, a variety of mollusks both univalve and bivalve, ascidians, squid, and fish fry such as alewives, cunners, mummichogs, menhaden, puffers, sculpins, scup, silversides, smelt, and winter flounders. No doubt any small fish is acceptable.

Toadfish snap viciously when caught, and they often fight among themselves. Like some sculpins they grunt, especially at night or if handled. And despite their clumsy appearance, they can dart out of their hiding places and back again with considerable speed. They are very tenacious of life and can live out of water for an astonishingly long time.

In the northern part of its range the toadfish spawns in June and early July. The very large eggs (about 5 mm. in diameter) are laid in holes under stones, under large shells, in old tin cans, among sunken logs, or among eelgrass, where they adhere in a single layer to whatever serves as a nest, which the male guards during the 3 weeks or so that are occupied by incubation. Even after hatching the tadpole-shaped larvae remain attached to the nest by the yolk sac until the latter is absorbed at a length of 15 to 16 mm. when they break free.[83]

General range.—Shoal water along the east coast of North America from Cuba to Cape Cod, straying northward to Maine.

Occurrence in the Gulf of Maine.—The toadfish is common about Woods Hole and thence westward. But it ventures around Cape Cod so

[83] Ryder (Bull. U. S. Fish Comm., vol. 6, 1887, p. 8) and Gudger (Bull. U. S. Bur. Fish., vol. 28, 1910, pt. 2, pp. 1071–1109, pls. 107–113) describe the breeding habits, eggs, and larvae of the toadfish. For further accounts of its habits see Gill (Smithsonian Miscell. Coll., vol. 48, 1907, pp. 388–427).

seldom that none of the fishermen in Massachusetts Bay of whom we have inquired have seen or heard of it there, nor further north. In fact there are only two definite records of it in the Gulf of Maine: "Maine"[84] and Cohasset on the south shore of Massachusetts Bay, where one (now or formerly in the collection of the Boston Society of Natural History) was caught by Dr. Owen Bryant. There is also one doubtful record for Kittery, Maine.[85]

THE TRIGGERFISHES. FAMILY BALISTIDAE

The triggerfishes are very divergent from the ordinary spiny-rayed fishes anatomically, and their external appearance is so characteristic that they are not apt to be mistaken, unless for their close relatives, the filefishes (p. 521). Their most interesting external characteristics are that the first spine of the first dorsal fin is not only very much stouter than the others but it can be locked erect by the second dorsal spine, and that the large bony scales are so close set as to form a hard but flexible armor. Other distinctive features are mentioned below in the description of the Gulf of Maine species. Most of the many species of triggerfishes are purely tropical; it is only as a stray that any member of the family enters into the Gulf of Maine fauna. Most of the tropical species are more or less poisonous if eaten.

Triggerfish *Balistes carolinensis* Gmelin 1789[86]

Jordan and Evermann, 1896–1900, p. 1701.

Description.—The readiest field marks for the triggerfish are its deep, sidewise-flattened body with slender caudal peduncle; its small terminal mouth with both dorsal and ventral profiles of the nose nearly straight; its eye situated so high as to give its face a very peculiar aspect; its large projecting incisor teeth; its very short gill openings wholly above the insertions of the pectorals; the plate armor of thick scales with which its entire head and body are clad; and especially its unusually stout first dorsal spine.

The spiny dorsal fin is triangular, with three spines, the first so stout that it is more like a horn, situated close behind the eyes and with the second spine acting as a trigger to lock the first spine

[84] Storer, Mem. Amer. Acad., N. Ser., vol. 2, 1846, p. 384, gives no definite locality.

[85] Holmes (2nd Ann Rept. Nat. Hist. Geol. Maine, Pt. 1, 1862, p. 95), "noticed it while standing on the bridge which connects the Navy Yard at Kittery with one of the islands."

[86] Jordan, Evermann, and Clark (Rept. U. S. Fish Comm. (1928), Pt. 2, 1930, p. 491) prefer the name *capriscus* Gmelin, and correctly so, from the strictly nomenclatural standpoint. But *carolinensis* is preferable both because it appeared on an earlier page of the same publication, and because the great majority of references to the species have been under that name.

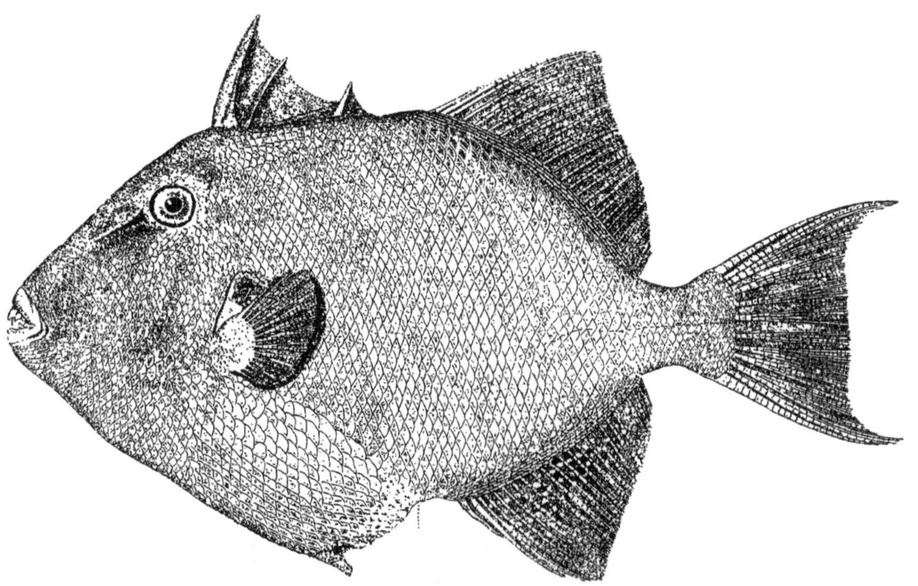

FIGURE 275.—Triggerfish (*Balistes carolinensis*), New York. From Goode. Drawing by H. L. Todd.

erect, whence the common name of the fish. The soft dorsal fin (27 rays) is separated from the first by a considerable interspace, is rhomboid in outline with the third or fourth ray longest, and tapers back to the base of the caudal peduncle. The anal fin (25 rays) corresponds to the soft dorsal in outline and in location. The caudal fin is of moderate size, its rear margin moderately concave, in a very characteristic curve, with sharp pointed, somewhat prolonged corners. The pectorals are short, rounded, and situated below the gill openings. The ventrals are reduced to one short, stout, blunt spine, mostly embedded in the skin and they are connected with the general outline of the abdomen by a sort of dewlap.

Color.—The colors of this triggerfish vary widely. A specimen 2 inches long recently taken on Georges Bank was yellowish, with many small blue-violet spots on the sides, dusky-blotched along the back, and with one broad, irregular dusky band extending from the base of the dorsal fin almost to the anal. The caudal fin was pale yellow. Other examples have been described as olive gray, marked with violet dots and with dark crossbars, the fins as variously tinted with yellow, blue, and olive.

Size.—Said to reach a weight of 4 pounds, but they average only about 1 pound.

General range.—Both sides of the tropical Atlantic, also the Mediterranean; straying north to Ireland on the European coast; to the outer coast of Nova Scotia, in the vicinity of Cape Canso [87] on the American side.

Occurrence in the Gulf of Maine.—Only one specimen of this tropical fish, taken in the Squam River at Annisquam, near Gloucester, Mass., many years ago,[88] had been reported from the Gulf of Maine previous to 1925. But it must drift in over the offshore rim of the Gulf more often than had been suspected, for two small fry of 2 to 3 inches were picked up on the northeast part of Georges Bank among Gulf weed (*Sargassum*) from the *Albatross II*, in mid-September 1927; a large one about 15 inches long was gaffed at the surface from the fishing schooner *Huntington Sanford* 14 miles southeasterly from Highland Light, Cape Cod, on July 19, 1929;[89] one, now in the Museum of Comparative Zoology, was picked up at Plymouth, Mass., on September 5, 1932, by the late C. L. Hauthaway, a well-known angler and a close observer.[90] One was reported from Casco Bay, (Small Point), Maine, in 1949, and one near Boothbay Harbor (Linekin Bay), also in 1949.[91]

THE FILEFISHES. FAMILY MONACANTHIDAE

The filefishes recall the triggerfishes in their general form, being similarly deep and flattened sidewise, with the same peculiar profiles, small terminal mouths, projecting incisor teeth, eyes set high up, very stout dorsal spines, and short gill openings; also in the fact that the ventral fins are either lacking altogether or at least are reduced to a single short blunt movable spine at the end of the very long pelvic bone, forming a keel-like continuation of the general ventral profile of the head and connected with that of the belly by a dewlap of skin. The filefishes differ from triggerfishes in having only one dorsal spine instead of three, and in the fact that their scales are so minute that the skin is velvety to the touch although very tough. Most of the species are tropical or subtropical, and none has any commercial or sporting value. Adults of the four species known from the Gulf of Maine are separable as follows:

[87] Nova Scotian records are from Halifax and from Queensport near Cape Canso (Vladykov, Proc. Nova Scotian Inst. Sci., vol. 19, 1935, p. 9; McKenzie Proc. Nova Scotian Inst. Sci., vol. 20, 1939, p. 18); also 24 miles southeasterly from Sable Island, where one was picked up by the schooner *Wanderer*, July 5, 1931 (Firth, Bull. 61, Boston Soc. Natural History, 1931, p. 13).

[88] This specimen is now in the Museum of Comparative Zoology.

[89] Reported by Firth, Bull. No. 61, Boston Soc. Nat. Hist., 1931, p. 12.

[90] Another species of trigger fish (*Balistes vetula*) is more common than *carolinensis* at Woods Hole, and is recorded from Nantucket, but it has not been taken in the Gulf of Maine as yet. It is separable from *carolinensis* by the fact that the forward rays of its soft dorsal fin and the corners of its caudal fin are elongated and filamentous; also by the presence of two blue bars on each side of its head.

[91] These last two fish were reported by Scattergood, Trefethen, and Coffin, Copeia, 1951, No. 4, p. 298).

KEY TO GULF OF MAINE FILEFISHES

1. There is a prominent external ventral spine; the gill openings are nearly vertical 2
   There is no external ventral spine; the gill openings are very oblique 3

2. Dorsal profile of head in front of the eyes is straight, or only very slightly concave; there are no thorns on the sides of the caudal peduncle .... Filefish (*Monacanthus hispidus*), p. 522
   Dorsal profile of head in front of the eyes is conspicuously concave; there are about 6 stout thorns pointing forward on each side of the caudal peduncle .... Filefish (*Monacanthus ciliatus*), p. 523
3. Dorsal fin with about 34 to 38 soft rays; anal with 36 to 41 rays; also, in small specimens, the caudal fin is more than one-half as long as the body .... Orange filefish (*Alutera schoepfii*), p. 524
   Dorsal fin with about 44 to 48 soft rays; anal with 47 to 52 rays; also, in small specimens, the caudal fin is less than one-half as long as the body .... Unicornfish (*Alutera scripta*), p. 525

Filefish *Monacanthus hispidus* (Linnaeus) 1766

Jordan and Evermann, 1896–1900, p. 1715.

Description.—In this species the rear edge of the dorsal spine, which is situated over the rear edge of the eye, is armed with a double series of barbs, but the sides of the rounded caudal peduncle do not bear any spines. The point of origin of the soft dorsal fin (32 to 34 rays) is behind the middle of the body, while the first soft dorsal ray often is much prolonged in adults and with a filamentous tip (young 1 to 2 inches long lack this filamentous ray.) Otherwise the fin is rounded in outline, narrowing from the front to the rear. The anal fin (31 to 34 rays) stands below the soft dorsal, and is of the same shape except that none of its rays are prolonged. The caudal fin is rounded. The pectorals are short, rounded, and situated lower than the gill openings, like those of triggerfishes.

Color.—Greenish, olive, or brownish. The back and sides of young fish are mottled with irregular darker blotches but adults are plain colored. The dorsal spine and the caudal fin are green. The soft dorsal fin and the anal fin are pale and translucent.

Size.—Maximum length about 10 inches.

General range.—This is a tropical species, common along the south Atlantic Coast of the United States and among the West Indies; also around the Canaries and Madeira in the eastern Atlantic, and represented in East Indian waters by a fish that does not seem to be distinguishable from it in any way. In the western side of the Atlantic it is known as far south as Brazil, has been taken from time to time as far north as Woods Hole, and has been recorded from St. Margaret Bay and from Halifax on the outer coast of Nova Scotia.[92]

[92] The only recent Nova Scotian record that has come to our notice is of one taken in Halifax Harbor in the autumn of 1928 (Leim, Proc. Nova Scotian Inst. Sci., vol. 17, Pt. 4, 1930, p. 46).

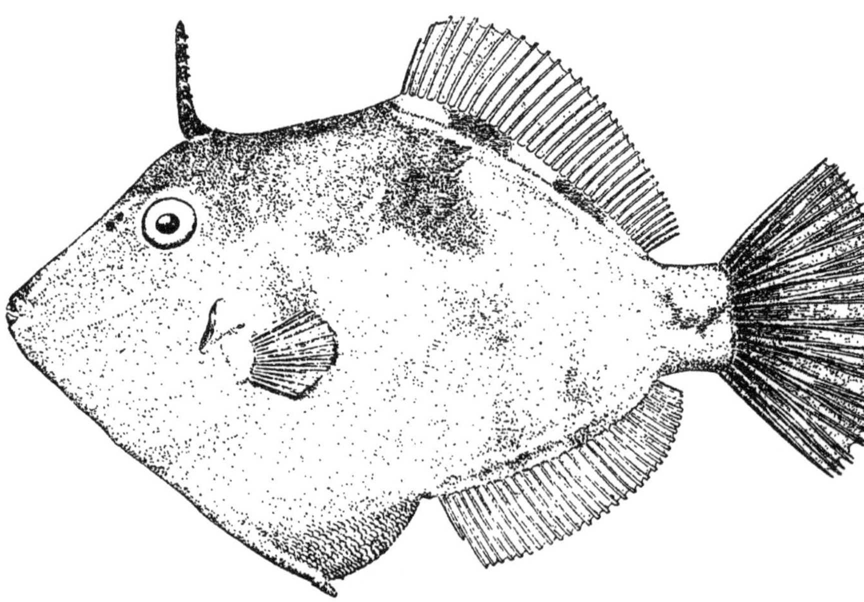

FIGURE 276.—Filefish (*Monacanthus hispidus*). From Jordan and Evermann. Drawing by W. S. Haines.

Occurrence in the Gulf of Maine.—Odd specimens of this filefish have been recorded from Hingham, Lynn, Nahant, and Boston Harbor in Massachusetts Bay; and from Cape Cod; all many years ago. More recent records of it in the Gulf are of 181 fry, 1–2 inches long, picked up from the *Albatross II* on the northeastern part of Georges Bank among floating Gulf weed (*Sargassum*), in September 1927; a larger one taken off Seguin Island, Maine, September 12, 1927;[93] one of 6 inches, at Provincetown, November 6, 1929; one picked up from the schooner *Old Glory* among floating rockweed (*Fucus* or *Ascophyllum*) and Gulf weed (*Sargassum*), on the western part of Georges Bank, September 15, 1930;[94] one taken off Portland Lightship, July 17, 1931; one taken in a trap at Provincetown, October 6, 1950; and two, about 6 inches long, taken off Wood End, Provincetown, in 17 fathoms, by the dragger *Mary Magdalyn* (Capt. Charles Santos), on October 30, 1951. It is also likely that a "filefish" taken at Beverly on the north shore of Massachusetts in 1933[95] was of this species. An occasional filefish straying from the south is thus to be expected anywhere on Georges Bank, or in the western side of the Gulf. But we find no evidence that they ever enter its eastern side, or that they ever reach the Bay of Fundy.

[93] Kendall, Bull. 58, Boston Soc. Nat. Hist., 1931, p. 11.
[94] Firth, Bull. 61, Boston Soc. Nat. Hist., 1931, p. 13.
[95] MacCoy, Bull. 67, Boston Soc. Nat. Hist., 1933, p. 9.

Filefish *Monacanthus ciliatus* (Mitchill) 1818

Jordan and Evermann, 1896–1900, p. 1714.[96]

Description.—This filefish resembles the species *hispidus* (p. 522) very closely. But its first dorsal ray is never prolonged, the ventral dewlap extends somewhat farther behind the tip of the ventral spine, and the caudal peduncle in the adult is armed with 2 or 3 pairs of strong forward-curving hooks on either side.

Color.—Described as varying from olive gray, or grass green to yellowish brown, with darker blotches or crossbands. The dorsal and anal fins are pinkish and they usually have three dark spots at the base. The ventral dewlap is edged with scarlet, and the caudal fin is greenish, mottled dark and pale.

[96] The illustration labeled "*ciliatus*" by Jordan and Evermann (Bull. 47, U. S. Nat. Mus., Pt. 4, 1900, pl. 259, fig. 634) is actually of *hispidus*, as is their figure 635.

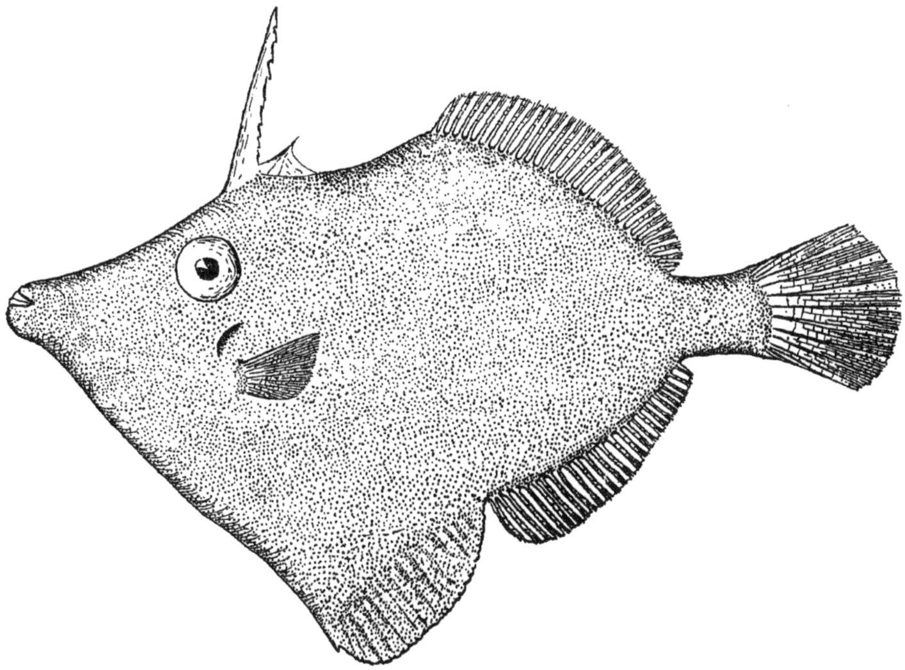

FIGURE 277.—Filefish (*Monacanthus ciliatus*), Bahamas specimen, 85 mm. long. Drawing by H. B. Bigelow.

Size.—Four to eight inches long.

General range.—Warmer parts of the Atlantic; from Brazil to Cape Cod on the American coast. A straggler has even been reported from Argentia, on the southern coast of Newfoundland, far to the north of its normal range.[97]

Occurrence in the Gulf of Maine.—The only recent report of this filefish from within the Gulf of Maine is of a 7-inch specimen taken in a trap at Provincetown, November 9, 1929.[98] We judge that earlier reports of it from Massachusetts Bay referred to in Storer's description and illustration [99] were actually based on a specimen of *hispidus.*

Orange filefish *Alutera schoepfii* (Walbaum) 1792 [1]

FILEFISH; TURBOT; HOGFISH; SUNFISH; UNICORNFISH

Jordan and Evermann, 1896–1900, p. 1718.

Description.—This fish resembles its relatives of the genus *Monacanthus* in most respects (p. 521), but while it is equally flattened sidewise, it is relatively shallower, being not over half as deep as long. The pelvic bone is as prolonged as it is in the other filefishes, but it does not project externally, nor is there a ventral dewlap, which is the readiest field mark by which to distinguish *Alutera* from *Monacanthus.* The eyes, too, are set lower down on the sides of the head, and the gill openings are relatively longer and more oblique. The dorsal spine is relatively shorter than in *Monacanthus,* and the lower jaw projects considerably beyond the upper. The soft dorsal fin (34–38 rays) originates behind the middle of the trunk and is rounded in outline, and the anal fin (36–41 rays) corresponds to the soft dorsal fin in size, shape, and position. The short rounded pectorals are situated opposite the lower half of the oblique gill openings and the tail fin is relatively narrower than in the other filefishes or triggerfishes; its longest rays are more than ½ as long as the body in small specimens, but only one-fourth to one-fifth as long as the body in half-grown fish and larger.

Color.—Described as varying from uniform olive gray to rich orange yellow or to milky white above, mottled with darker hues of the same tints; bluish white beneath. The caudal fin usually is yellowish on adults but sometimes it is dusky, edged with white.

Size.—Maximum length about 2 feet.

General range.—Atlantic and Gulf of Mexico coasts of the United States; not uncommon in summer as far north as Cape Cod; reported to Portland, Maine, and perhaps to Halifax, Nova Scotia.[2]

[97] Newfoundland Fish. Res. Comm. Rept., vol. 1, No. 1, 1933, p. 126.

[98] Firth, Bull. Boston Soc. Nat. Hist., No. 61, 1931, p. 13.

[99] The illustration of his *Monacanthus massachusettensis* shows the profile typical of *hispidus,* and neither his description nor his illustration suggests that there were any thorns on the caudal peduncle such as characterize *ciliatus.* (See Storer Mem. Amer. Acad. Arts, Sci., N. Ser., vol. 8, 1863, p. 425, pl. 34, fig. 4; also Fishes of Massachusetts, 1867, p. 231, pl. 34, fig. 4.)

[1] Jordan, Evermann, and Clark (Rept. U. S. Comm. Fish. (1928), Pt. 2 1930, p. 495) place the species in the genus *Ceratacanthus* Gill 1801.

[2] A specimen, found in Halifax Harbor, August 25, 1938, appears to have belonged to this species, but it was not in good enough condition for positive identification (McKenzie, Proc. Nova Scotian Inst. Sci., vol. 20, 1939, p. 19).

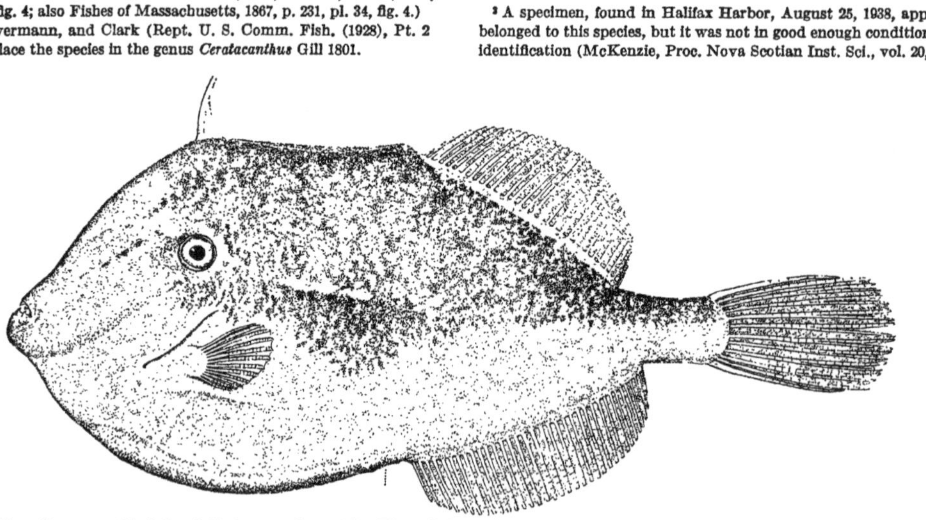

FIGURE 278.—Orange filefish (*Alutera schoepfii*), Key West, Florida. From Jordan and Evermann. Drawing by W. S. Haines.

Occurrence in the Gulf of Maine.—Although the orange filefish has been described as "rather common" at Woods Hole during the summer,[3] only 3 specimens have been reported within the limits of the Gulf of Maine: 1 from Portland, Maine, and 2 from Salem, Mass., all of them many years ago.[4] Evidently it reaches the Gulf of Maine only at long intervals, as a waif from the south.

Unicornfish *Alutera scripta* (Gmelin) 1789

Jordan and Evermann, 1896–1900, p. 1719.[5]

Description.—This species much resembles the orange filefish from which it differs chiefly in its somewhat more slender body (2 to 3 times as long as deep), longer dorsal fin (44 to 49 rays), longer anal fin (47 to 52 rays), shorter caudal fin (in small unicornfish the caudal is less than half as long as the body, while in young orange filefish it is more than half as long), and in color. The dorsal spine may be serrated in young fish, but it is smooth in adults.

Color.—The unicornfish is olive on head and body with light blue reticulations extending from the snout to the tail; in grown fish there are numerous small round black spots on the sides of the body.

Size.—Reaches a length of 3 feet.

General range.—Tropical seas; northward to South Carolina on the Atlantic Coast of America, and to Georges Bank as a stray.

Occurrence in the Gulf of Maine.—Two specimens of this fish, 5 inches and 5½ inches long, caught on the western edge of Georges Bank Sept. 15, 1930, by the schooner *Old Glory*,[6] are the only ones that have been reported from the Gulf of Maine. A third, 5 inches long, was taken by *Atlantis*, south of Sable Island (lat. 40°55′ N., long. 59°55′ W.), August 18, 1941.

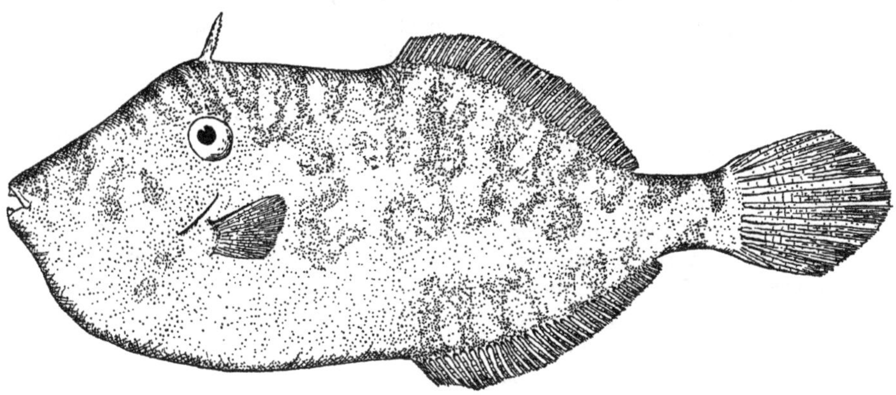

FIGURE 279.—Unicorn fish (*Alutera scripta*), Georges Bank specimen, 143 mm. long. Drawing by H. B. Bigelow.

THE PUFFERS AND PORCUPINE FISHES
FAMILIES TETRAODONTIDAE AND DIODONTIDAE

The members of these two families are so closely allied one to the other, not only anatomically but in general appearance, that they may be described as a unit. They have only one dorsal fin (the soft-rayed), the spiny dorsal being obsolete, and they have no ventral fins. Their gill openings are reduced to short slits like those of their allies, the triggerfishes and filefishes (pp. 520 and 521); their teeth are fused into cutting plates; and they have no scales. The two families are separable by the structure of the teeth, as described below in the accounts of the two species concerned, and by certain anatomical characters.

All of them are capable of inflating their bellies to balloonlike proportions with air or with water, if annoyed, and of deflating at will. And it is a matter of general interest (though not touching

[3] Sumner, Osburn, and Cole, Bull. U. S. Bur. Fish., vol. 31, Pt. 2, 1913 p. 762.

[4] There is no way to verify the identifications at this late date.

[5] Jordan and Evermann's (Bull. 47, U. S. Nat. Mus., Pt. 4, pl. 260, fig. 637) illustration labeled "*Scripta*" seems to have been based on a specimen of *schoepfi*, in an intermediate stage of development.

[6] Reported by MacCoy, Bull. Boston Soc. Nat. Hist., No. 58, 1931 p. 16.

the Gulf of Maine directly) that the flesh of some of the species of puffers, and perhaps of all of the porcupine fishes, is poisonous.

Both groups are warm-water fishes. One species of puffer reaches the southwestern part of the Gulf rather commonly; and one porcupine fish has been reported there as a stray from the south.

KEY TO GULF OF MAINE PUFFERS AND PORCUPINE FISHES

1. Skin set with large conical spines.....Burrfish, p. 527
   Skin merely prickly..................Puffer, p. 526

Puffer *Sphaeroides maculatus* (Bloch and Schneider) 1801

SWELLFISH; SWELL TOAD; BALLOONFISH; BELLOWSFISH; GLOBEFISH

Jordan and Evermann, 1896–1900, p. 1733.

Description.—When the puffer is not inflated it it moderately slender (about three times as long as deep), about as thick as it is deep, and it tapers from abreast the gill opening to a rather slender caudal peduncle in one direction, to a rounded snout in the other. Its very small mouth is situated at the tip of the snout as it is in the triggerfishes and filefishes. It has no true teeth but the bones of its upper and lower jaws form cutting edges, each divided in the middle by a suture, giving the appearance of two large incisors above and two below. The gill openings are very small and set oblique, but their obliquity is the reverse of that of the filefishes (p. 521), i. e., backward and downward. The eyes are set very high and are horizontally oval in outline. The skin has no scales, but the sides of both head and body, the back from snout to dorsal fin, and the belly as far back as the vent are rough with small, stiff, close-set prickles; those on the back are bluntish and nearly vertical while those on the sides and belly are rather sharp, pointing backward when the fish is not inflated, but erect when it is.

There is no spiny dorsal fin. The soft dorsal fin is very short (8 rays), rhomboid in outline, about twice as high as it is long, and set far back close to the caudal peduncle. The anal fin (7 rays) is similar to the dorsal in shape and size, and arises close behind it. The caudal fin is of moderate size, weakly rounded, with angular corners. The pectorals are fan-shaped, and are situated close behind the gill openings. There are no ventral fins.

The most interesting morphologic character of the puffer is its ability to inflate itself with air or with water if it is handled or at the slightest disturbance of any sort, until the skin of the belly is stretched tight as a football, and the fish is almost globular. In this condition, it floats at the surface, belly up, and apparently helpless. Leave it alone, however, and it soon deflates, discharging the air or water suddenly, and shrinks back to its normal dimensions.

Color.—Dark olive green above, sometimes ashy or dusky, the sides greenish yellow to orange, crossbarred with 6 to 8 rather indefinite dark bands or blotches. The belly is white.

Size.—The puffer is said to grow to a length of 14 inches, but few of them are more than 10 inches long. Females average larger than males.

Habits.—The puffer is an inshore fish, often coming in to the tide line. It runs up into slightly

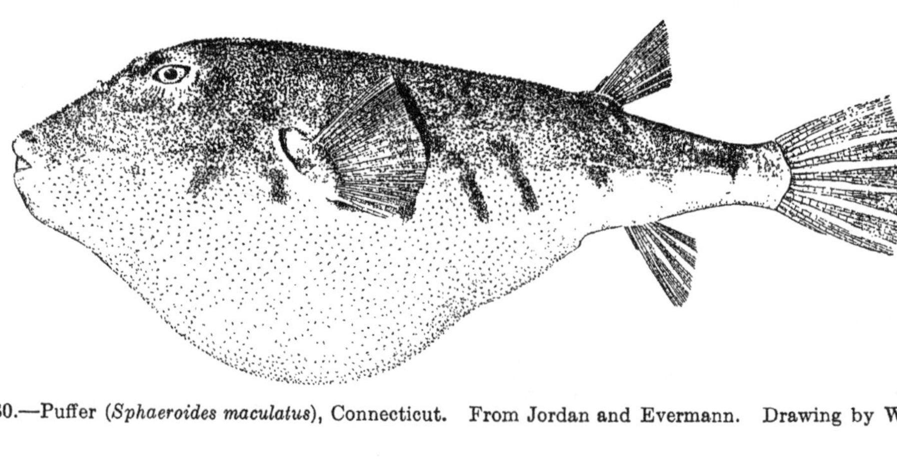

FIGURE 280.—Puffer (*Sphaeroides maculatus*), Connecticut. From Jordan and Evermann. Drawing by W. S. Haines.

brackish water in various estuaries, and seldom is caught more than a few fathoms deep, or more than a mile or two from land. Throughout the northern part of its range it belongs in the rather numerous and varied category of "summer" fishes, taken from April to November in Chesapeake Bay, from late May or early June to October or early November along southern New England. It is probable that when the puffers disappear from their usual summer haunts, with the onset of cold weather, they merely descend into somewhat deeper water nearby, to spend the winter on bottom in a more or less quiescent state.

Puffers feed on small crustaceans of all sorts especially on crabs, shrimp, isopods and amphipods, as well as on small mollusks, worms, barnacles, sea urchins, and other invertebrates, which they find on bottom. Young fry of 7 to 10 mm., examined by Dr. Linton at Woods Hole, had eaten copepods as well as crustacean and molluscan larvae. And they are only too ready to take bait, if the hook is small enough. Where they are plentiful they may be nearly as much of a nuisance in this way as the cunners.

Puffers spawn in shoal water close to the shore, from mid-May, in Chesapeake Bay, and from early June through the summer off southern Massachusetts. And they are prolific. The ovaries of a Chesapeake Bay female, 10½ inches long contained (estimated) about 176,000 ova.[7] The eggs (about 0.9 mm. in diameter, with many small oil globules) sink and stick fast to each other or to whatever they chance to touch. Incubation occupies 3½ to 5 days at a temperature of about 67°–68° F. (20° C.). The larvae are about 2.4 mm. long at hatching, and are brilliantly pigmented with red, orange, yellow, and black. In 3 days the mouth functions, and when they are 7 mm. long the young fish show most of the diagnostic characters of the adults,[8] and can inflate themselves even more, in fact, until the bulging skin entirely hides the dorsal and anal fins.

General range.—Atlantic coast of the United States from Florida to Cape Cod in abundance; to Casco Bay in small numbers, and perhaps to the Bay of Fundy as a stray.

Occurrence in the Gulf of Maine.—Anglers find the puffer only too plentiful along the southern shores of Massachusetts, but the elbow of Cape Cod marks the eastern and northern limit to their presence in any numbers. They have been reported at Monomoy, Truro, and Provincetown. Cape Cod Bay may perhaps support a small resident population, for Prof. A. E. Gross informs us that he has seen as many as four or five taken at one time in a pound net at Sandy Neck, Barnstable, at a tide, in the summer of 1920; besides others stranded there on the beach.[9] And we have heard of others there recently, or nearby. Storer described them as common at Nahant, a few miles northeast of Boston, but this seems to have been an error, for Wheatland (1852, p. 214) writing about the same period, not only spoke of them as seldom seen in Massachusetts Bay, but considered a single specimen taken in Salem Harbor in the summer of 1848 as worthy of a note. And this remained the only positive record for a puffer for Essex County until August 24, 1920, when one was caught at Gloucester.[10]

The only records of puffers north of Cape Ann that have come to our notice are of two taken in a trap in Casco Bay in 1896, and of one taken near Long Island, off Portland Harbor, Maine, on July 24, 1933. But there may be a small local population in Casco Bay, and in the vicinity of Boothbay Harbor, Maine, for L. W. Scattergood [11] writes us that the pound net fishermen have long been acquainted with them there and that he had received three specimens recently from Pemaquid Point where the fishermen report them as commonest in June. A skeleton, apparently of a puffer, has been found on the shore of Minas Basin, at the head of the Bay of Fundy on the Nova Scotian side.[12]

Burrfish *Chilomycterus schoepfii* (Walbaum) 1792 [13]

PORCUPINEFISH; RABBITFISH; OYSTERFISH

Jordan and Evermann, 1896–1900, p. 1748.

Description.—The burrfish resembles the puffer (p. 526) in the positions of its dorsal and anal fins, but its skin is armed with short, stout, triangular

[7] Hildebrand and Schroeder, Bull. U. S. Bur. Fish., vol. 43, Pt. 1, 1928, p. 348.

[8] Welsh and Breder (Zoologica, New York Zool. Soc., vol. 2, No. 12, January 1922, N. Y.) describe the early stages in the life history of the puffer.

[9] The Auk, vol. 40, 1923, p. 24.

[10] This specimen, reported by MacCoy (Bull. 67, Boston Soc. Nat. Hist., 1933, p. 9) is in the Museum of Comparative Zoology.

[11] Letter dated September 19, 1951, U. S. Fish and Wildlife Service.

[12] Reported to us (1951) by Dr. A. H. Leim of the Fisheries Research Board of Canada.

[13] Jordan, Evermann, and Clark place this species in the genus *Cyclichthys* Kaup 1855.

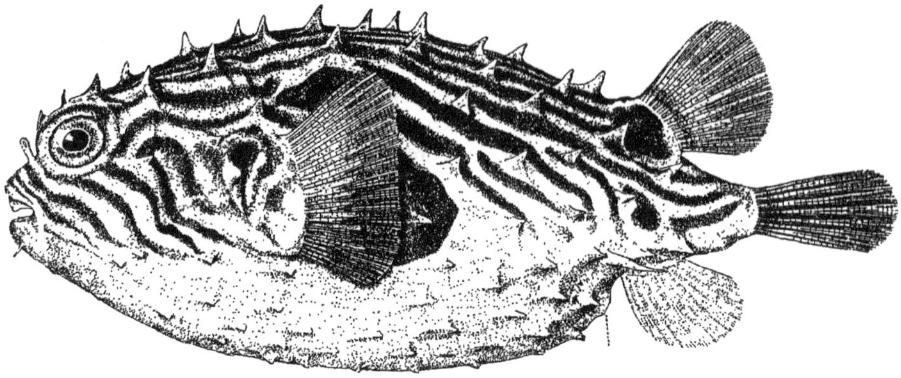

FIGURE 281.—Burrfish (*Chilomycterus schoepfii*), Connecticut. From Goode. Drawing by H. L. Todd.

spines instead of being merely prickly. These spines are sparsely scattered all over the trunk, with about 9 or 10 from nose to tail along any given line. Furthermore, the burrfish is oval in outline, not fusiform like the puffer; the openings of its nostrils are prolonged in a single tubular tentacle; its bony jaw plates are not divided by a median suture as they are in the puffer, hence each jaw apparently is armed with a single very broad incisor tooth instead of with two; the pectoral fins are not only much larger than in the puffer but their upper edge is level with the upper corner of the gill openings in the burrfish (considerable below it in the puffers); its eye is round, not oval; and its anal fin is below the dorsal, not behind the latter. We need only add that the soft dorsal and anal fins (it has no spiny dorsal) are both rounded, each has 10 to 12 rays; the caudal fin is very narrow and round-tipped; the pectorals are much broader than long, and there are no ventral fins.

Color.—The ground color varies from green to olive or brownish above, with pale belly, usually tinted with yellow or orange. The back and sides are irregularly striped with olive brown, dusky, or black lines, running downward and backward, roughly parallel one with another. There is a dark blotch on each side at the base of the dorsal fin, a smaller one between the latter and the anal fin, one above the base of each pectoral fin, and a fourth close behind the latter.

Size.—Length, to about 10 inches.

General range.—Coast of the United States, from Florida northward regularly about to New York, occasionally to Cape Cod, and straying as far as Massachusetts Bay; most plentiful from the Carolinas southward.

Occurrence in the Gulf of Maine.—The only record of this southern fish north of the elbow of Cape Cod is of one taken in Massachusetts Bay many years ago [14] and another caught at West Point, Maine, August 5, 1949.[15]

THE OCEAN SUNFISHES OR HEADFISHES. FAMILY MOLIDAE

Although the ocean sunfishes are allied anatomically to the puffers and porcupine fishes, with which they agree in the very small gill openings and in the fusion of the teeth into a sort of bony beak, they bear no resemblance whatever to them in general appearance, for they appear to consist of nothing but a "huge head to which the fins are attached," as Jordan and Evermann [16] aptly express it. They have no spiny dorsal fin; the soft dorsal and anal fins are short and very high, and they have no caudal peduncle. The caudal fin, so short that it is apparently nothing more than a flap of skin, extends all around the rear outline of the trunk. Corresponding to their extraordinary conformation the sunfishes have only 16 or 17 vertebrae.

All known members of the family are oceanic in nature, and they are widely distributed in warm seas. One (*Mola mola*, p. 529) is a rather frequent

[14] This specimen, reported by Kendall (Occas. Paper, Boston Soc. Nat. Hist., vol. 7, Pt. 8, 1908, p. 118) is (or was) in the collection of the Boston Society of Natural History.

[15] Taken in a fish trap and reported by Scattergood, Trefethen, and Coffin (Copeia, 1951, No. 4, p. 298).

[16] Bull. 47, U. S. Nat. Mus., Pt. 2, 1898, p. 1752.

visitor to our Gulf from the south; and a second (*Masturus lanceolatus*) has, perhaps, a claim to mention here, on the strength of one very young sunfish that was taken in Massachusetts Bay many years ago (p. 532).

KEY TO GULF OF MAINE SUNFISHES

1. There is no evident caudal fin--------Sunfish, p. 529
2. There is an evident caudal fin, extending horizontally across the posterior edge of the trunk, with a triangular lobelike extension a little above the midlevel of the body-----------Sharp-tailed sunfish, p. 531

Sunfish *Mola mola* (Linnaeus) 1758

Jordan and Evermann, 1896–1900, p. 1753.

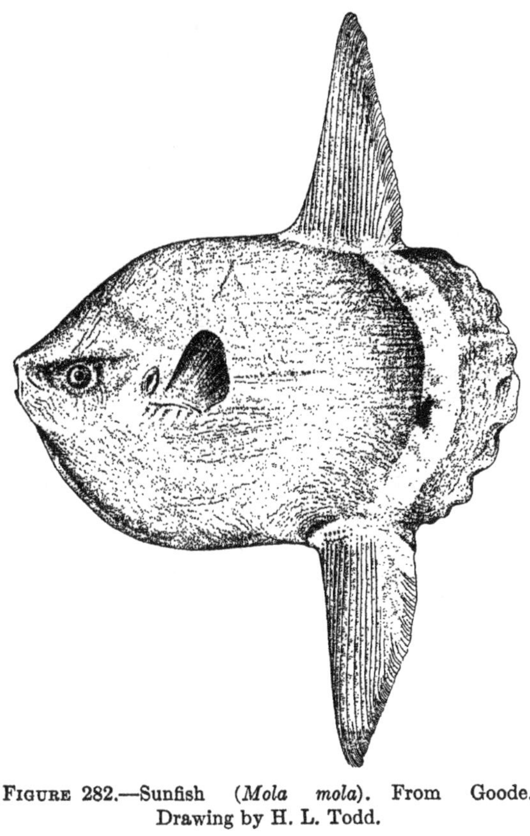

FIGURE 282.—Sunfish (*Mola mola*). From Goode. Drawing by H. L. Todd.

Description.—The oblong body of an adult sunfish (adults alone are seen regularly in the Gulf of Maine) suggests the head and fore trunk of some enormous fish cut off short, for it is truncate immediately back of the dorsal and anal fins, and has no caudal peduncle. But it tapers in front of the fins toward the snout so that the forward half of the trunk is oval in profile. The fish is less than twice as long as deep, strongly flattened sidewise (about one-fourth as thick as deep), with a very small mouth at the tip of the snout; teeth completely united in each jaw; a very small eye in line with the mouth; remarkably short gill openings, and the nose overhangs the upper jaw as a kind of rough, mobile wart or pad. The soft dorsal fin (there is no spiny dorsal) stands over the anal fin, close behind the midlength of the fish. Both these fins are very much higher than long, triangular, with sharply rounded tips, and each has 15 to 18 rays, with the seventh ray the longest.

The fins cannot be laid back, as they can in most bony fishes; and the sunfish propels itself along by waving them from side to side. The caudal fin extends around the whole posterior margin of the body. Confluent with the dorsal and anal fins in the young and hardly separated from them in the adult, it is so short and its rays so hidden by the thick opaque skin that it looks more like a fold of skin than a typical fin. Its general outline is rounded, paralleling the rear outline of the body, but its margin is scalloped, with a rounded bony prominence or knob in line with each caudal ray (11–14) and with a notch between every two of these prominences. We have counted 11 such notches on a fish 3½ feet long, and have record of 8 on one of about 4 feet.[17] The pectoral fins are small, rounded, each with 12 or 13 rays, and are situated about halfway up the body close behind the tiny gill openings. There are no ventral fins. The skin is unusually thick (about 1½ inches thick in one 47 inches long which we harpooned near La Have Bank on August 7, 1914), very tough and elastic in texture; it is crisscrossed with low ridges, and fins as well as trunk are clothed with small bony tubercles, giving the appearance of shark skin.

The sunfish is described as glowing luminescent at night in the water. We cannot verify this first hand. But we can bear witness that it grunts or groans when hauled out of the water; that its skin is covered with a thick layer of tough slime, and it is the host of a great variety of parasites, external and internal, with copepods and trematodes clinging to its skin and infesting its gills, with its muscles harboring round worms and with various

[17] Taken near Boothbay, Maine, and reported to us by Dr. Austin F. Riggs.

round worms and flat worms inhabiting its intestines.

Color.—Dark gray above, the back with a brownish cast, the sides paler with silvery reflections, the belly dusky to dirty white. Some descriptions mention a broad blackish bar along the bases of the dorsal and anal fins, but nothing of the sort was to be seen in the only example we have handled fresh from the water.

Size.—The sunfish grows to a great size. Heilner [18] describes the capture of one 10 feet 11 inches long off Avalon (Calif.), while Jordan and Evermann record another Californian specimen 8 feet 2 inches long, weighing about 1,800 pounds.

One measuring 8 feet in length and 11 feet from tip of dorsal fin to tip of anal fin was exhibited in London in 1883,[19] and an 8-foot specimen was taken off Cape Lookout (N. C.) in 1904,[20] but large ones such as this are exceptional, the general run being from 3 to 5 feet (rarely 6 feet) long, and from 175 pounds to 500 pounds in weight. A fish 4½ feet long is about 31 inches across the body and 6½ feet from the tip of the dorsal fin to the tip of the anal. One, 5 feet 3 inches long, was 4 feet 2 inches wide and 14½ inches thick.[21] A fish 4 feet 1 inch long, caught off Boston Harbor, August 14, 1922, weighed 516 pounds.[22]

Habits.—The sunfish is a wanderer of the high seas, drifting at the mercy of the ocean currents; those that are seen are at the surface (see following for an exception); how deep they may descend is not known.

When these unlucky vagrants are sighted in our cool northern waters they have usually been chilled into partial insensibility. They float awash on the surface, feebly fanning with one or the other fin, the personification of helplessness. Usually they pay no attention to the approach of a boat, but we have seen one come to life with surprising suddenness and sound swiftly, sculling with strong fin strokes, just before we came within harpoon range. When one is struck it struggles and thrashes vigorously while the tackle is being slung to hoist it aboard, suggesting that they are far more active in their native haunts than their feeble movements in fatally cold surroundings might suggest.

The sunfish lives on an unusual diet, for as a rule the contents of the stomach consists either of jellyfish, ctenophores, or salpae, or of a slimy liquid that probably represents the partially digested remains of these. This has been true of all the sunfish brought in to the Bureau of Fisheries at Woods Hole. But various crustacean, molluscan, hydroid, and serpent-star remains, even bits of algae and eelgrass (*Zostera*), have been found in sunfish stomachs in European waters, proving that at times they either feed on the bottom in shoal water, or among patches of floating weed. And their jaws certainly seem fit for harder fare than jellyfish.

The spawning habits are not known, nor have the eggs been seen; presumably these are buoyant, with many globules, as are those of the sharp-tailed sunfish *Masturus lanceolatus*. The young fry differ from their parents in being armed with 8 short stout spines on either side, and with a single median row of 4 spines along the back and 7 along the ventral margin of the body.[23]

General range.—Oceanic and cosmopolitan in tropical and temperate seas; known northward to northern Norway on the European side of the Atlantic, to the Newfoundland banks, the Gulf of St. Lawrence, and the outer coast of Nova Scotia on the American side.[24]

Occurrence in the Gulf of Maine.—The sunfish is only a stray visitor to our Gulf, which it enters now and then from the warmer and more congenial waters outside the continental slope. There are published records of its appearance in St. John Harbor, New Brunswick, near Birch Harbor; near Seguin Island; off Small Point; and off Cape Elizabeth (Maine), where it has been reported repeatedly; off Cape Ann; and from various localities in Massachusetts Bay. Sunfish have even been

[18] Bull. New York Zool. Soc., vol. 23, No. 6, November 1920, p. 126.

[19] Smitt, Scandanavian Fishes, vol. 1, 1892, p. 626.

[20] Smith, North Carolina Geological and Economic Survey, vol. 2, 1907. p. 353.

[21] As reported in the Boston American for June 24, 1930.

[22] Reported, with photograph, in the Boston Daily Post for August 14, 1922

[23] For a discussion of the young fry of the ocean sunfishes, with illustrations and references to earlier accounts, see Schmidt, Meddel. Kommission Havundersøgelser, Ser. Fiskeri, Denmark, vol. 6, 1921, No. 6.

[24] Localities where sunfish had been reported in the Gulf of St. Lawrence up to 1947 include north of Cape Breton; Bathurst, New Brunswick; Northumberland Straits; the north shore of the Gaspe Peninsula; the south shore of the Gulf opposite the Saguenay River; vicinity of Trois Pistoles; Anticosti; and Bay of Islands on the west coast of Newfoundland. See Medcoff and Schiffman (Acadian Naturalist, vol. 2, No. 7, 1947) for list with details. Dunbar (Canad. Field Naturalist, vol. 64, No. 3, 1950, p. 124) has recently reported one, 5 feet long, that was found on the beach at Metis on the southern shore of the Lower St. Lawrence River. A Gulf of St. Lawrence record that is especially interesting because so late in the season, is of one about 5 feet long that stranded late in October 1, 1926, at Curling, Bay of Islands, west coast of Newfoundland (reported in the Boston Traveler for Nov. 2, 1926).

seen in Boston Harbor, and on August 18, 1918, one 4½ feet long was killed in a narrow creek at Quincy, Mass. The *Grampus* sighted sunfish near the Isles of Shoals in 1896 (Dr. Kendall's field notes), in 1912, and in 1914, as well as one in the eastern basin of the Gulf in 1912. Seaside dwellers reported one or two near Cape Porpoise in 1921; one of 7 feet was caught off Boothbay, Maine, in August 1927; and one 5 feet 3 inches long in the northern side of Massachusetts Bay, off Bakers Island, Beverly, in 1940, an especially interesting case, for the fish in question was taken on a hand line in 20 fathoms of water.[25] And in 1950 several blundered into one of the traps at Barnstable on Cape Cod Bay.

An occasional sunfish is, in short, to be expected anywhere in the western side of the Gulf and along the coast of Maine. The only record, however, for a sunfish in the Bay of Fundy is from near its mouth at St. John Harbor.[26] Nor do we find any report of them along the Nova Scotian side of the open Gulf of Maine.

In most summers it is something of an event to see a sunfish anywhere in the inner part of the Gulf. During July and August of 1912, for example, we sighted only one from the *Grampus*, none at all in August 1913, and only one in the Gulf and another near La Have Bank during the mid and late summer of 1914. They vary, however, in numbers from year to year; 1928, for example, was a year of abundance all along the coast, while in 1950, a single trap at Barnstable on Cape Cod Bay took 26 sunfish, an astonishing number. Report also has it (we cannot verify this first-hand) that sunfish are more plentiful over and along the southern edge of Georges Bank than they are within our Gulf, as indeed might be expected from their oceanic origin.

In the inner parts of our Gulf sunfish are oftenest sighted in mid or late summer, or early in autumn. And one has been reported stranded in Bay of Islands on the west coast of Newfoundland as late as the end of October.[27] But it is not likely that any can survive the winter in our Gulf, or anywhere along the coast to the northward. Neither is there any reason to suppose that the waifs that visit our Gulf ever spawn there.[28]

[25] Reported in the Boston American, June 24, 1930.

[26] Cox, Bull. Nat. Hist. Soc. New Brunswick, No. 13, art. 2, 1896, p. 75.

[27] Reported in the Boston Traveler, November 2, 1926, from the Associated Press.

Sharp-tailed sunfish *Masturus lanceolatus* (Liénard) 1840 [29]

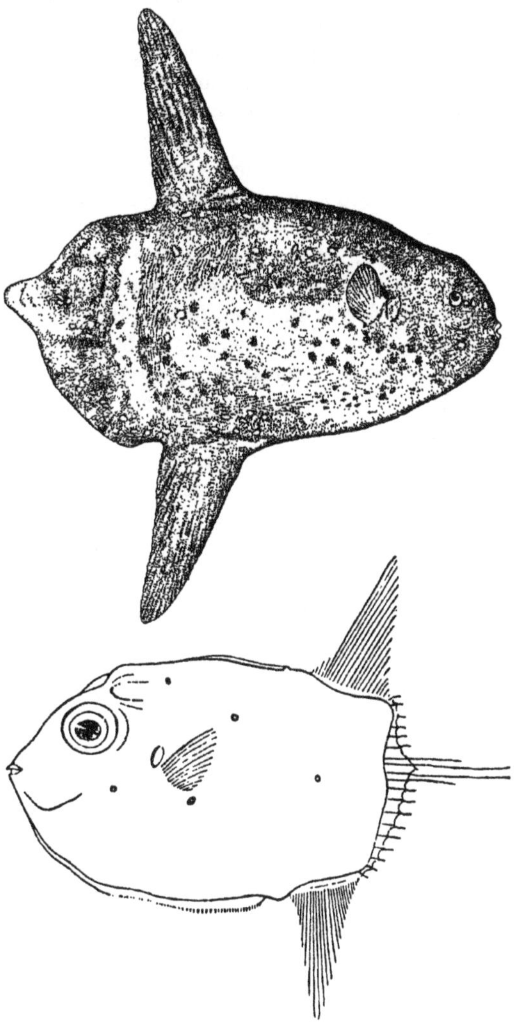

FIGURE 283.—Sharp-tailed sunfish (*Masturus lanceolatus*). Above, adult, Miami, Florida, after Gudger. Below, young, 54 mm., Massachusetts Bay, after Putnam.

[28] Sunfish fry, about 2 inches long, taken in Massachusetts Bay many years ago and reported by Putnam (Proc. Amer. Assoc. Advancement of Science, 19th Meeting (1870), 1871, pp. 255, 256, fig. 3) as this species, actually belonged to the closely allied sharptailed sunfish (p. 532), as shown by Schmidt (Meddel. Kommiss. Havundersøgelser, Denmark, Ser. Fiskeri, vol. 6, 1921, Pt. 6, p. 6), and by Gudger (Proc. Zool. Soc. London, 1937, Ser A, p. 382).

[29] It is an open question still, whether specimens with longer tail fins and others with shorter tail fins represent two separate species, or whether the differences between them are sexual ones. See Fraser-Brunner (Bull. British Mus. Nat. Hist., vol. 1, No. 6, 1951, p. 105) for a recent discussion of this subject.

Description.—The sharp-tailed sunfish differs from the common sunfish (*Mola mola*) chiefly in the fact that the rear margin of its body is edged by a short but evident caudal fin of 18–20 soft rays, that extends around from close behind the dorsal fin to close behind the anal fin, with a triangular lobelike, blunt-tipped projection a little above the midlevel of the body. Its scales, too, are much finer and less evident to the touch than those of *Mola*, and its skin is less slimy. It resembles *Mola* very closely in all other respects.

Color.—Described as with the whole trunk more or less silvery, the upper parts of the sides grayish brown to blackish, the lower parts paler; the sides either plain or variously marked with ill-defined dark spots; the dorsal and anal fins as dark slaty, the caudal fin as sometimes with pale blotches.

Size.—This sunfish appears to grow as large as the more common *Mola*, perhaps even larger. In a Florida specimen, 88 inches long (after being dried somewhat) the tail fin occupied 21 inches, the body occupied 67 inches and was 38 inches high.[30] The dimensions of a North Carolina specimen 73½ inches long were: body 54½ inches long by 37 inches high and 11 inches thick, tail fin 19 inches long, dorsal fin 27 inches high, anal fin 25 inches high.[31]

Habits.—Nothing is known of its habits to differentiate it from its more common relative.

General range.—This sunfish, like *Mola*, appears to be cosmopolitan in tropical-warm temperate latitudes, oceanic in nature but coming close inshore on occasion, and even into estuarine situations. Adults have been reported from Japan, the Hawaiian Islands, Polynesia, Amboina, and Mauritius in the Indian and Pacific Oceans; from the Red Sea; from South Africa (Table Bay); from Madeira, from near the Azores; near Habana, Cuba (7 specimens); east coast of Florida (9 specimens), and North Carolina (4 specimens) in the Atlantic. Young fry have been taken off the Azores; in the Sargasso Sea; west of the Canaries; in the Caribbean; and in the Gulf of Mexico.[32]

Occurrence in the Gulf of Maine.—The only record for the sharp-tailed sunfish for our Gulf is for 4 young fry, about 2 inches long, that were taken many years ago in Massachusetts Bay. These were originally reported by Putnam [33] as the young of the common ocean sunfish, but Schmidt [34] and Gudger [35] have shown that they were the sharp-tailed species in reality, because with projecting caudal fins. The nearest locality record for an adult of the species (to date) is for Pamlico Sound, N. C. But it would not be astonishing if one were to drift farther northward any summer, as so many stray species do from the south.

THE ANGLERS. FAMILY LOPHIIDAE

This family is the only familiar Gulf of Maine representative of the small but anatomically remarkable tribe of pediculate fishes, in which the base of the pectoral fin takes the form of an arm ("pseudo-brachium") formed by the elongation of the carpal bones ("actinosts"), which are so short in all other bony fishes that they are not noticeable externally. Coupled with this peculiar structure of the pectorals, the gill openings are reduced to small apertures in or near the axils ("armpits") of these fins. The anglers are characterized among their immediate relatives by a very large and very much flattened head; by an enormous mouth; and by the fact that they have only two bones in each pectoral "arm." The Gulf of Maine harbors one species.

[30] Hubbs and GiovannelI, Copeia, 1931, pp. 135–136.

[31] Brimley, Jour. Elisha Mitchell Sci. Soc., vol. 55, 1939, p. 295.

American goosefish *Lophius americanus* Cuvier and Valenciennes 1837

MONKFISH; ANGLER; ALLMOUTH; MOLLIGUT; FISHING FROG

Jordan and Evermann, 1896–1900, *Lophius piscatorius* Linnaeus 1766 in part, p. 2713.

Description.—The goosefish is so unlike all other Gulf of Maine fishes that there is no danger of mistaking it for any other once it is seen. It is so much flattened, dorso-ventrally, and so soft in texture that when one is left stranded on the

[32] See Gudger (Proc. Zool. Soc. London, 1937, Ser. A, p. 353), for list of locality records up to 1937 with references; Brimley (Jour. Elisha Mitchell Sci. Soc., vol. 55, 1939, p. 295) for account of North Carolina specimens.

[33] Proc. Amer. Assoc. Advancement of Science, 19th meeting, Troy, N. Y. (1870) 1871, pp. 255–256.

[34] Meddel. Komm. Havundersøgelser, Denmark, Ser. Fiskeri, vol. 6, Pt. 6, 1921, p. 6.

[35] Proc. Zool. Soc. London, 1937, Ser. A, p. 382.

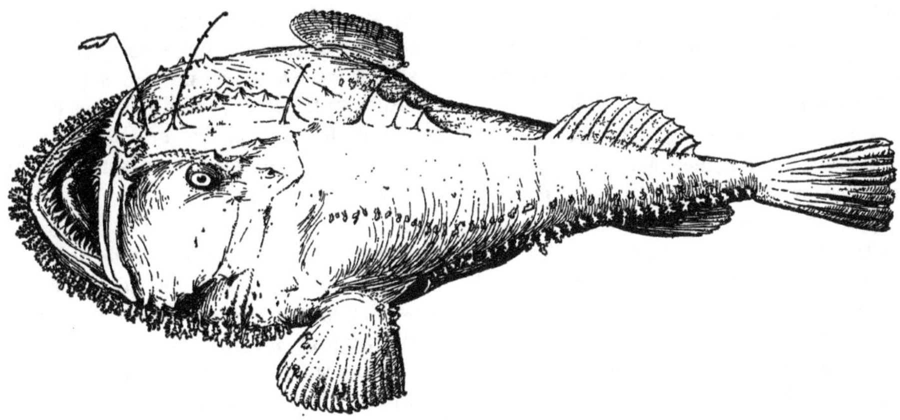

FIGURE 284.—Goosefish (*Lophius americanus*), oblique-dorsal view, Gulf of Maine specimen. From Bigelow and Welsh.

shore it collapses until it is hardly thicker than a skate. Its head is rounded as seen from above, about as broad as it is long, and enormous in comparison with the body, which is so narrow and tapering, back of the pectoral fins as to give the fish a tadpole-like appearance. The most noticeable feature is the enormous mouth, which is directed upward, with the lower jaw projecting so far beyond the upper jaw that most of the lower teeth are freely exposed even when the mouth is closed.

Both jaws are armed with long, slender, curved teeth, all alike in form but of various sizes, and very sharp, and all the teeth point inward toward the gullet. Some of them may be as much as an inch long in a large fish. The teeth in the lower jaw are in 1 to 3 rows, mostly large, while in the upper jaw the few teeth in the middle (there is a toothless space in the midline) are largest, with a single row of smaller ones flanking them. And there are several rows of thorn-like teeth on the roof of the mouth. The gill openings are behind the pectoral fins and they lack the gill covers that are to be seen in most of the bony fishes. The eyes are on the top of the head, and are directed upward.

The pectoral fins are exceedingly distinctive, for their bases have the form of thick fleshy arms as already described (p. 532) that bear the fins proper at their outer edge. The finlike parts are fanlike when spread, and so thick-skinned that the rays are hardly visible except in the scalloping of the margins.

The top of the head bears 3 stiff slender spines (representing the anterior part of the spiny dorsal fin) hardly thicker than bristles, the first standing close behind the tip of the snout, the second a little in front of the eyes, the third on the nape of the neck. And while the first and second are movable from recumbent to erect, the third slopes backward with its basal half imbedded in the skin.[36] The relative lengths of these spines vary, but the first two have been about equal in length on most of the fish we have seen, or the second a little the longer, with the third much the shortest of the three. The first spine bears an irregular leaflike flap of skin at its tip, which plays an important role in the daily life of the goosefish as a lure for its prey (p. 536) while the second and third spines have small triangular membranes at their bases, and one or both of them may be fringed with short lobes of skin. Besides these spines there are two well-developed dorsal fins, the first (of 3 spines) situated over the pectorals and the second (10 to 11 rays) on the rear part of the trunk.

The single anal fin (9 to 10 rays) stands below the second dorsal fin, and the ventral fins (about 5 thick rays) are situated on the lower surface of the head, well in front of the pectorals. The caudal fin is small and broom shaped. The dorsal fins have thin delicate membranes. But the caudal, anal, and ventral fins are thick and fleshy, like the pectorals. The skin is scaleless, very smooth and slippery to the touch, and there is a row of fleshy flaps of irregular shape running around the margin of the head and around the edge of the lower jaw, besides smaller tags that

[36] Sometimes more than one-half is so imbedded.

fringe the sides of the trunk as far back as the base of the caudal fin. Furthermore, the upper side of the head bears several low conical tubercles which vary in prominence from fish to fish.

Color.—The many goosefish we have seen (and this corroborates the published accounts) have been chocolate brown above, variously and finely mottled with pale and dark. The dorsal fins, the upper sides of the pectoral fins, and the caudal fin are of a darker shade of the same color as the back, except nearly black at the tips, while the whole lower surface of the fish is white or dirty white. Sometimes, it is said, the upper side is dotted with white spots but we have seen none that were marked in that way. Very small ones are described as mottled and speckled with green and brown. Wilson, who watched many in the aquarium at Plymouth, England,[37] writes that the European species is able to match both its color and its color pattern closely to the sand and gravel on which it lies.

Size.—Adults run from 2 to 4 feet long,[38] weighing up to 50 pounds, and heavier ones have been reported. One 38 inches long, caught at Woods Hole on July 25, 1923, weighed 32 pounds alive.

Remarks.—The goosefish of eastern North America has usually been thought identical with the widespread eastern Atlantic angler (*L. piscatorius* Linnaeus. But as Tåning [39] has pointed out, the late larval stages of our fish do not resemble those of *L. piscatorius* as closely as they do those of the angler of the Mediterranean and of neighboring parts of the Atlantic that various authors regard as a separate species, *L. budegassa*, Spinosa 1807.[40] This suggests that the goosefish of the western Atlantic is a distinct species, for which Berrill [41] has revived the old name *L. americanus* Cuvier and Valenciennes 1837.

The adults of the three forms in question certainly resemble one another so closely that we have not found any external differences that seem significant to separate Gulf of Maine fish from two specimens from northern Europe, and others from the Mediterranean (all of about the same size) with which we have compared them. But it seems wisest to retain the separate name for our form until the larval differences can be investigated further (which we are not in a position to do), and until much larger series of grown fish have been compared.

Habits.—The depth range inhabited by the goosefish extends from tide line down to at least 365 fathoms on the continental slope off southern New England, and very likely deeper still. The adults appear, for the most part, to hold to the sea floor, where many are taken by the otter trawlers. And they are found indifferently on hard sand, on pebbly bottom, on the gravel, sand, and broken shells of the good fishing grounds, and on soft mud, where we have trawled them in the deep basin of our Gulf.

Specimens of the closely allied European goosefish kept in the aquarium at Plymouth, England, spent most of the time resting quietly.[42] When they swam they did so slowly, and they used their paired fins for walking on the bottom. Wilson describes one as digging a small hollow in the bottom when it settled down, using its pelvic fins to shovel the sand and pebbles forwards-outwards, and using its pectorals, almost like webbed hands, to push the sand away to either side until its back was almost flush with the surrounding bottom. But the fact that goosefish have been known to seize and swallow hooked fish as the latter were being hauled up, and even to capture sea birds sitting on the surface, proves that while they ordinarily snap up their prey from ambush, or by a sudden short rush, they may make considerable excursions for a meal on occasion.

The American goosefish is at home through a very wide range of temperature. They have been trawled on the Newfoundland banks in water as cold as 32°,[43] and it is likely that those living shoalest in the Gulf of St. Lawrence are exposed to equally low temperatures, in late winter and in spring. But we doubt if they can survive much colder water, for many were seen floating dead in Narragansett Bay, and on the shore, during the

[37] Jour. Marine Biol. Assoc. United Kingdom, vol. 21, Pt. 2, 1937, p. 485.

[38] Rumor has it that goosefish grow to 6 feet, but we find none definitely recorded (and have seen none) longer than 4 feet.

[39] Rept. Danish Oceanogr. Expeds. (1908-1910), No. 7, vol. 2, Biol. A. 10., 1923, p. 7-16.

[40] See Regan, Ann. Mag. Nat. Hist. Ser. 7, vol. 11, 1903, p. 283 for descriptions of *L. piscatorius*, of *L. budegassa*, and of a new species, *vaillanti*, described by him from the Azores and from the Cape Verde Islands.

[41] Contrib. Canadian Biol. N. Ser., vol. 4, No. 12, 1929.

[42] Wilson (Jour. Mar. Biol. Assoc. United Kingdom, vol. 21, Pt. 2, 1937, pp. 486-490) has given a very interesting account of the habits of specimens in the aquarium at Plymouth, England, where some were kept for as long as 11 months.

[43] Rept., vol. 2, No. 1, Newfoundland Fish. Res. Comm., 1933, p. 127 sta. 97.

winter of 1904–1905, apparently killed by the unusually severe cold.[44]

At the other extreme, goosefish picked up by net fishermen near Cape Lookout, N. C., in shoal water (p. 540) are exposed to temperatures higher than 70° for part of the season, perhaps as high as 75°. But reports [45] that the inshore contingent of the goosefish population of Rhode Island waters works out (i. e., deeper) in July, to work inshore again in October suggest that they tend to avoid extreme summer heat, if they can do so by descending into deeper water.

They are tolerant to a wide range of salinity also, occurring as they do from estuarine situations out to the upper part of the continental slope. But we have never heard of one in brackish water.

The larvae of the goosefish, like those of most sea fishes, feed on various small pelagic animals such as copepods, crustacean larvae, and glass worms (*Sagitta*); and *Sagitta* is the chief diet of young goosefish in the Adriatic during the life of the latter near the surface, hence may serve this same purpose in the Gulf of Maine.

The goosefish becomes a fish eater in the main after it takes to the bottom, and the following Gulf of Maine species have been recorded from its stomach: spiny dogfish, skates of various kinds, eels, launce, herring, alewives, menhaden, smelts, mackerel, weakfish, cunners, tautog, sea bass, butterfish, puffers, various sculpins, sea ravens, sea robins, sea snails, silver hake, tomcod, cod, haddock, hake, witch flounders, American dab, yellowtail flounders, winter flounders, and various other species of flatfish unnamed, as well as its own kind. The goosefish often captures sea birds, as one of the vernacular names implies, cormorants, herring gulls, widgeons, scoters, loons, guillemots, and razor-billed auks are on its recorded dietary, while we have found grebes and other diving fowl, such as scaup ducks and mergansers, in goosefish in Pamlico Sound, N. C. It is questionable, however, whether even the largest of them would be able to master a live goose, as rumor has it, nor do the local fishermen believe it ever does so in Pamlico Sound, though the abundance of wild geese there in winter would afford it every opportunity. Goode,[46] however, tells of one which a fisherman saw struggling with a loon. Even a sea turtle has been found in one.[47]

Goosefish are also known to devour invertebrates such as lobsters, crabs of several species, hermit crabs, squids, annelid worms, shellfish, starfish, sand dollars, and even eelgrass. Linton's [48] report of one that was full of mud containing small shellfish, crustaceans, and worms is interesting. In short, nothing edible that strays within reach comes amiss to a goosefish. And examinations of stomachs have shown that the relative importance of various articles in its diet varies widely on different grounds, depending on what is available. Thus Field [49] found skates, flounders, and squid their chief dependence near Woods Hole. The 32-pounder from there, mentioned above, contained 2 menhaden, 1 spiny dogfish a foot long, and the vertebral columns of 6 others; while goosefish diet largely on hakes in the Bay of Fundy; [50] on haddock, flatfish, and on skates on Georges Bank.

The goosefish has often been cited for its remarkable appetite. We read, for instance, of one that had made a meal of 21 flounders and 1 dogfish, all of marketable size; of half a pailful of cunners, tomcod, and sea bass in another; of 75 herring in a third; and of one that had taken 7 wild ducks at one meal. In fact it is nothing unusual for one to contain at one time a mass of food half as heavy as the fish itself. And with its enormous mouth (one 3½ feet long gapes about 9 inches horizontally and 8 inches vertically) it is able to swallow fish of almost its own size. Fulton, for instance, found a codling 23 inches long in a British goosefish of only 26 inches, while Field took a winter flounder almost as big as its captor from an American specimen. One that we once gaffed at the surface, on Nantucket Shoals, contained a haddock 31 inches long, weighing 12 pounds, while Captain Atwood long ago described seeing one attempting to swallow another as large as itself. Wilson's observations, however, indicate that they are no more gluttonous than any other rapacious fish, for those that he watched in the aquarium usually refused food for 2 or 3 days

[44] Reported by Tracy, 36th Rept. Comm. Inland Fish. Rhode Island, 1906, p. 92.

[45] Tracy, 36 Rept. Comm. Inland Fish., Rhode Island 1906, p. 92.

[46] Fish. Ind. U. S., Sect. 1, 1884, p. 174.

[47] Schroeder, Copeia 1947, p. 201.

[48] Bull. U. S. Bur. Fish., vol. 19, 1901, p. 487.

[49] Rept. U. S. Comm. Fish., (1906) 1907, Doc. No. 622, p. 39.

[50] Connelly, Bull. 3, Biol. Board Canada, 1920, p. 16.

after a meal. His observation that they evidently preferred small fish is in line with their normal habits, for they feed mostly on small fish, not on large, and even the largest of them take very small fry on occasion.

In Scottish waters,[51] where the habits of the local goosefish are better known than in the Gulf of Maine, their local abundance depends on the supply of small fish. And despite their poor ability as swimmers goosefish have been found to congregate near particular shoals of herring.

Goosefish, like most fish of prey, often swallow indigestible objects. They have even been credited (on how good evidence we cannot say) with pouching lobster-pot buoys. And the story of one whose mouth made a holding ground for the anchor of a small boat has been related repeatedly.

The most interesting habit of the goosefish is that it actually does use the flap of skin at the tip of its first dorsal spine as a bait to lure small fishes within seizing distance, much as Aristotle described. W. F. Clapp (only observer who has watched the American goosefish feeding, to our knowledge) has described them to us, in Duxbury Harbor as lying motionless among the eelgrass, with the "bait" at the tip of the first dorsal ray swaying to and fro over the mouth. When a tomcod (the only fish he saw them take) chances to approach, it usually swims close up to the "bait," but never (in his observation) actually touches the latter, for the goosefish opens its vast mouth as soon as the victim comes within a few inches and closes it again, engulfing its prey instantaneously.

Further details added by observations on European anglers in aquaria at Port Erin, Isle of Man, by Chadwick,[52] and at Plymouth, England, by Wilson,[53] are that the first dorsal spine, with its terminal "bait" is held down along the top of the head, to be raised at the approach of a prospective victim; that the bait may be jerked to and fro quite actively in front of its owner's head; that the victim is usually taken in head first; that a fish swimming close enough may be snapped up without the bait being brought into play; and that some anglers use the bait often, others seldom.

[51] Fulton (Ann. Report, Fish. Board Scotland, (1902) 1903, Pt. 3, p. 195) lists the stomach contents of 541 goosefish from various localities off Scotland).

[52] Nature, vol. 124, 1929, p. 337.

[53] Jour. Marine Biol. Assoc. United Kingdom, vol. 21, pt. 2, 1937, p. 479.

Wilson also made the interesting observation that touching the "bait" does not cause a reflex snapping of the jaws, showing that the angler feeds by sight.[54]

Adult goosefish cannot have many enemies. But small ones are no doubt picked up by various predaceous fishes. And Lebour's observation [55] that goosefish larvae in aquarium jars at Plymouth, England, were devoured by the larvae of the spiny lobster (*Palinurus*), by large copepods, by ctenophores, and by hydroids when they came close enough to the walls of the jar to be seized by the latter, is an interesting illustration of the hazards that larval fishes meet during their free-drifting stages.

Goosefish spawn in spring, summer, and early autumn, according to the latitude, and through a long season. Eggs and larvae have been taken near Cape Lookout, N. C., in March and April;[56] in May off Cape Hatteras;[57] and as early as May at Woods Hole. But spawning may not commence until early summer in the Gulf of Maine, for June 24 (Passamaquoddy Bay [58]) is the earliest date when eggs have been seen north of Cape Cod. September 18 (off Seguin Island, Maine) is the latest recorded date for American waters.

The floating egg-veils of the European angler have been reported as early as February 18 in Scottish waters and as late as July 23,[59] while Tåning [60] concludes from the sizes of larvae taken at different dates that March–June is the season of chief production to the west and southwest of the British Isles in general. In the Mediterranean (with higher temperatures), anglers spawn from December and January on, as shown by the presence of larvae.[61]

The locality of spawning has been the subject of discussion, whether inshore in shoal water, or offshore in deeper. The egg veils reported from the Bay of Fundy by Connolly;[62] from Passama-

[54] Gudger (Amer. Naturalist, vol. 79, 1945, p. 542), has given an interesting and readable survey of observations, at various hands, on the use of the bait.

[55] Jour. Mar. Biol. Assoc. United Kingdom, vol. 13, 1925, p. 728.

[56] Information supplied by the late S. F. Hildebrand.

[57] Tåning, Rept. Danish Oceanogr. Expeds., 1908-1910, No. 7, vol. 2 (Biol.), A 10, 1923, p. 25.

[58] Connolly, Contrib. Canadian Biol. (1921) 1922, p. 116.

[59] See Bowman (Fish. Bd. Scotland, Sci. Invest. (1919, No. 2), 1920, p. 23) for records for angler eggs up to 1919.

[60] Danish Oceanogr. Expeds., 1908-1910, No. 7, vol. 2 (Biol.), A 10, 1923, p. 23.

[61] See Stiasny (Arbeit. Zool. Inst. Vienna, vol. 19, 1911, p. 70) for Mediterranean records, besides which an egg veil has been reported in January near Naples by Le Bianco (Mitth. Zool. Stat. Neapel., vol. 19, pt. 4, 1909, p. 725).

[62] Contrib. Canadian Biol. (1921), No. 7, 1922, p. 116.

quoddy Bay by Berrill;[63] and from Frenchman Bay near Mount Desert by Procter and others,[64] were in such early stages of incubation that they must have been spawned close at hand. And this also applies to some isolated eggs that were collected at about the 20-fathom contour line off northern North Carolina,[65] by the *Dana*. Neither is there any reason to suppose that veils farther advanced in incubation, that have been taken in the inner parts of the Gulf of Maine (p. 541); at Woods Hole; and at Newport (p. 537), had come from any great distance. Furthermore, large adult fish are present in abundance inshore throughout the spawning season, which would hardly be the case if they moved offshore or into deep water to spawn. On the other hand, veils that could not have been spawned long before have also been met with near the 1,000 to 1,100 fathom (2,000-meter) contour line over the continental slope off North Carolina [66] and at about the same relative position over the slope south of the Newfoundland Banks.[67]

It appears, in short, that the American goosefish spawns indifferently in shoal water and in deep. It differs in this respect from its European relative, which moves offshore and down the slope for the purpose, to near the 1,000-fathom contour, to judge from the localities where the newly hatched larvae have been collected in the eastern North Atlantic.[68]

The presence of egg veils off North Carolina; near Newport [69] and near Woods Hole along southern New England; in the Gulf of Maine (p. 541); and over the continental slope south of the Newfoundland Bank; with the capture of a very small (4-inch) specimen on the Grand Bank (p. 540) shows that the American goosefish breeds throughout its geographic range.

The eggs are shed in remarkable ribband-like veils of mucus, each probably the product of a single ovary, up to 25–36 feet long, and said sometimes to be as much as 2 to 3 feet broad, in which the eggs are arranged in a single layer, lying one to three or even four in separate hexagonal compartments, with the oil globule uppermost. In an egg veil found near St. Andrews, New Brunswick, between 32 and 36 feet long, about 8 inches wide, about ⅛ inch thick, and about 25 quarts (26½ liters) in volume, about 5 percent of the eggs were single, about 80 percent were in pairs, and about 5 percent were in threes, per compartment. This veil was estimated to contain about 1,320,000 eggs,[70] and Fulton estimated about the same numbers (1,345,848 and 1,317,587) in the ovaries of two in Scottish waters.[71]

The veils are light violet gray or purplish brown, made more or less blackish by the embryonic pigment of the eggs according to the stage of development attained by the latter. And they are so conspicuous when floating at the surface that fishermen have long been familiar with them, though it was not until about 1871 that Alexander Agassiz demonstrated their true parentage.[72] The eggs occasionally become isolated, perhaps when a storm shreds the mucous veil to pieces, and they float like any ordinary buoyant fish eggs when this happens. We have not actually found them in this condition in the Gulf of Maine, but Agassiz and Whitman saw isolated eggs at Newport, and Tåning has reported others from North Carolina waters.

The eggs themselves, large numbers of which have now been examined, are 1.61 to 1.84 mm. in longest diameter in the case of the American species, as they lie in their mucous compartments.[73] The yolk is straw-colored, and they have either one copper-colored or pinkish oil globule of 0.4 to 0.56 mm., or several smaller ones. Incubation proceeds normally at temperatures from as low as 41° to as high as 63°–64°, and probably in higher temperatures. The larvae, which float with the yolk uppermost at first, have been re-

[63] Contrib. Canadian Biol. Fish., N. Ser., vol. 4, No. 12, 1929, p. 145.

[64] Biol. Surv. Mount Desert Region, Pt. 2, Fishes, 1928, p. 3.

[65] Tåning, Rept. Danish Oceanogr. Expeds., 1908–1910, No. 7, vol. 2 (Biol.) A 10, 1923, p. 25.

[66] Lat. 36°16′ N., long. 74°33′ W., see Tåning, Danish Oceanogr. Expeds., 1908–1910, vol. 2 (Biol.), A 10, 1923, p. 25.

[67] Murray and Hjort, Depths of the Ocean, 1912, p. 108.

[68] For further discussion, see Bowman (Fishery Bd. Scotland Sci. Invest. [1919], No. 2, 1920, p. 21) and Tåning (Rept. Danish Oceanogr. Expeds., 1908–1910, vol. 2 (Biol). No. 7, A 10, 1923).

[69] It was at Newport that Agassiz, and Agassiz and Whitman, collected the veils and the larvae on which they based their accounts.

[70] Berrill, Contrib. Canad. Biol. and Fish., N. Ser., vol. 4, No. 12, 1929, pp. 145, 147.

[71] Sixteenth Annual Report, Fish. Bd. Scotland (1897) 1898, Pt. III, pp. 125–134, pls. 2–3.

[72] Baird, American Naturalist, vol. 5, 1871, pp. 785–786.

[73] The eggs of the European *L. piscatorius* are described as larger, averaging about 2.3 mm.

ported as from about 2.5 mm. to about 4.5 mm. long when they hatch.[74]

The first of the dorsal fin rays (which is to form the second head spine of the adult) appears within 4 days or so after hatching, as a lobe at the margin of the embryonic finfold on the nape of the neck. The pectorals are formed at about 7 days, when the larva is 5.5 mm. long, the ventral fins have now appeared as two long conical processes below and behind the pectorals (fig. 285D); and the pigment has become congregated in 3 or 4 masses behind the vent, the last being a very conspicuous feature that the larvae of the European species *L. piscatorius* do not share. The yolk has been absorbed at a length of 6–8 mm., a second dorsal ray has formed behind the first, and the ventral fins have become 2-rayed. The third and fourth dorsal rays or filaments appear while there are still only two

[74] Larval goosefish from New England, from the Bay of Fundy, and from Nova Scotian waters have been described and pictured by Agassiz (Proc. Amer. Acad. Arts, Sci., N. Ser., vol. 9, 1882, p. 280); by Agassiz and Whitman (Mem. Mus. Comp. Zool., vol. 14, No. 1, Pt. 1, 1885, p. 16, pl. 6, figs. 1-10); by Connolly (Contrib. Canadian Biol. [1921], No. 7, 1922); by Procter and others (Biol. Surv. Mt. Desert region, Pt. 2, Fishes, 1928); and by Berrill (Contrib. Canadian Biol. and Fish., N. Ser., vol. 4, No. 12, 1929, pp. 145-149). For accounts and illustrations of North European and Mediterranean *Lophius* larvae, see especially Lebour (Jour. Marine Biol. Assoc. United Kingdom, vol. 13, No. 3, 1925, pp. 721-728) who reared them from the eggs at Plymouth, England; also Bowman (Fishery Board for Scotland, Sci. Invest. [1919] 1920, No. 11), Stiasny (Arbeit, Zool. Inst. Vienna, vol. 19, 1911, p. 71), and Tåning (Report, Danish Oceanogr. Expeds., 1908-1910, No. 7, vol. 2 (Biol.), A 10, 1923).

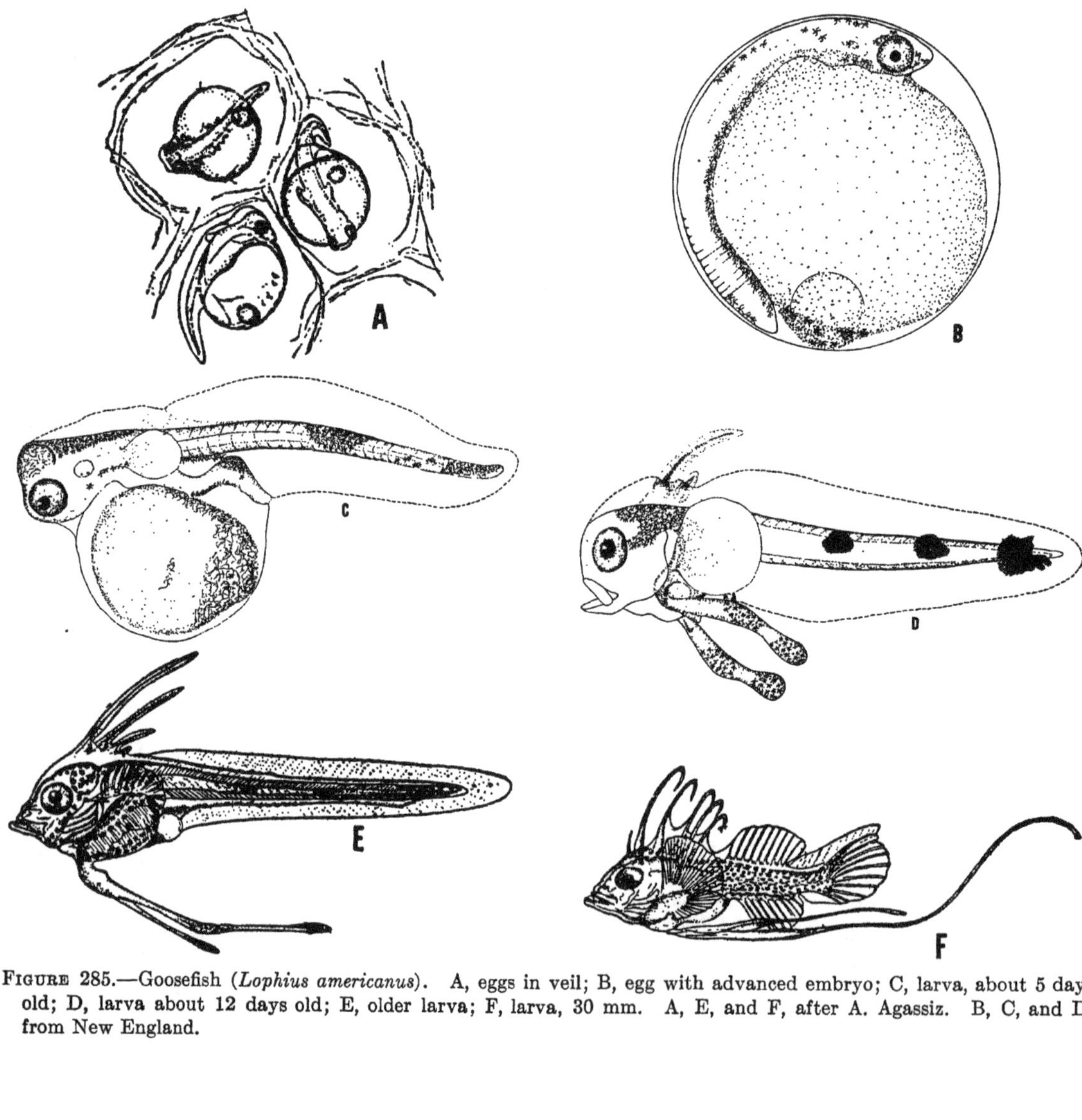

FIGURE 285.—Goosefish (*Lophius americanus*). A, eggs in veil; B, egg with advanced embryo; C, larva, about 5 days old; D, larva about 12 days old; E, older larva; F, larva, 30 mm. A, E, and F, after A. Agassiz. B, C, and D, from New England.

ventral rays in the American goosefish (fig. 285E), but they do not do so in the European species until the third rays have developed, in the ventral fins. The Mediterranean larvae so far described have agreed with the American in this respect, which makes the situation puzzling.

A fifth dorsal ray next appears behind those that have developed already, and a sixth in front of these, all of them being interconnected with membrane at their bases but free at their tips. The pectoral fins assume a great breadth and fanlike outline; the second dorsal, the anal, and the caudal fins take definite form; the ventral rays become filamentous at their tips, streaming far out behind the tail; and a complete row of teeth appears in the lower jaw, with a few in the upper. The goosefish pictured at this stage by Agassiz (fig. 285F) was 30 mm. long, and one much like it taken off Brazil Rock, described by Connolly, was 27 mm. long, but the larvae of the Mediterranean goosefish attain this stage when they are only 13 to 18 mm. long, according to Stiasny.

The older post-larval stages of the American goosefish have not been seen yet. But development no doubt follows the same course for them as it does for the Mediterranean form; i. e., the foremost dorsal ray becomes bristlelike with the flap appearing at its tip; the last three of the free rays on the nape of the neck join together as the future first dorsal fin; the lappets of skin appear around the margin of the lower jaw and along the cheeks; and the head broadens and flattens while the young fish are still living pelagic, with enormous pectoral fins and with threadlike ventrals (fig. 286).

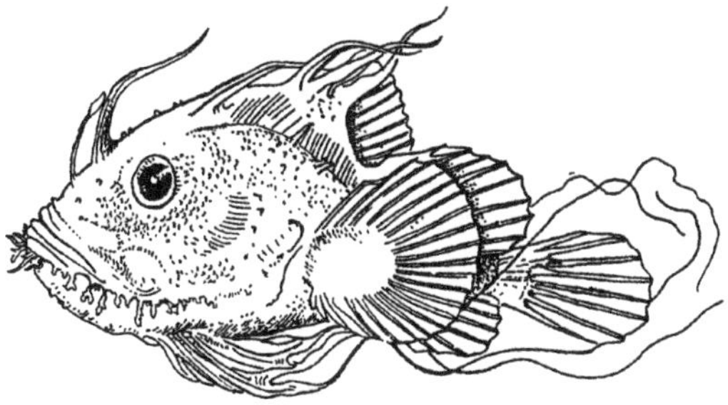

FIGURE 286.—Goosefish (*Lophius*). Larva, Mediterranean, 50 mm. After Stiasny.

The largest free-swimming Mediterranean larva seen by Stiasny was 2 inches (50 mm.) long. Probably the young take to the ground shortly after this stage, for Bowman describes European goosefish fry of about 2½ inches (65 mm.) that were trawled on the bottom, off Scotland, as of adult form, except that their pectorals were proportionately larger. To attain this state entails growth on the part of the head out of proportion to the rest of the body; enlargement of the mouth; shrinkage of all the fins (of the ventrals most of all); alteration of the second and third free dorsal rays into spines (they are soft previously); and a general flattening of the whole fish. Young of 3 inches taken at Halifax, one of 4½ inches from Campobello (both pictured by Connolly), and others as small as 4–4½ inches that we have trawled, were at about this stage in their development.

The capture of a 2½-inch specimen in October (presumably), in Halifax Harbor, and of another of 3 inches there (date not recorded),[75] suggests that the goosefish may be expected to reach about that length by the onset of their first winter in our northern waters. One 4½ inches long from Halifax, studied by Connolly, seemed, from the thickness of its otoliths, to have been in its second summer or autumn, i. e., one full year old, which probably applies to three others of 4–4½ inches, trawled in August, that we have seen. But it is not clear whether 14 others of 7½ inches in May, and of 6¾–9 inches in July, were early hatched fry in their second season, or late hatched specimens in their third season.[76]

One of the larger fish studied by Connolly showed 4 concentric rings in its vertebrae; one 31 inches long seemed to have 9 rings; one of 37 inches seemed to have 10 rings; and one of 40 inches seemed to have 12 rings. But it is not certain whether these vertebral rings are laid down regularly, one per year, or not.

Fulton's observations [77] show that the fry of the North European angler may be 5–5½ inches long by November off Scotland, where spawning commences in March or earlier; which is as large as the fry of the American species are in their second summer in our northern waters, where the first growing season is at least 3 to 4 months shorter. Fulton's measurements also point to more rapid growth by the larger Scottish fish than by the

[75] Connolly (Contrib. Canadian Biol. (1921), No. 7, 1922, pp. 119–120).

[76] Equal uncertainty applies to two of 10 inches, one of which was trawled in February, the other in April.

[77] 21st Ann. Rept. Fish. Bd. Scotland, Pt. 3 (1902) 1903, pp. 190–194.

American goosefish in the Bay of Fundy, namely to 9–16 inches at a year and a half; to 14½–18½ inches when 2½ years old; and to about 21 inches at 3 years of age.

Few goosefish mature on either side of the Atlantic until they are 30 inches long, or longer.[78]

General range.—Coast of eastern North America from the southern and eastern parts of the Grand Banks of Newfoundland, and the northern side of the Gulf of St. Lawrence [79] southward to North Carolina, in shoal and moderately deep water; also reported (as *L. piscatorius*) off the Barbadoes at 209 fathoms, on the Yucatan Bank, southern part of the Gulf of Mexico, at 84 fathoms,[80] and off Cape Frio, Brazil, in lat. 22°56′ S.,[81] if these southern specimens actually belonged to the same species.

Occurrence in the Gulf of Maine.—This is a familiar fish in the Gulf of Maine both along shore and on the outer fishing banks. It has been recorded in print from the west coast of Nova Scotia (St. Mary Bay) and from various localities in the Bay of Fundy, where, according to Huntsman, large ones are frequently taken on long lines, or found stranded on the beach. It is well known, if not abundant, all along the coast of Maine, and we once caught 8 (all large) in Ipswich Bay in one haul of a beam trawl only 8 feet wide. In Massachusetts Bay goosefish are the most common on the smooth bottom south of Boston; many enter Duxbury Bay (p. 546); and they are so numerous in Cape Cod Bay that one can hardly walk the beach for an afternoon without finding a jawbone bleaching on the sand, which applies equally all along the outer shores of Cape Cod. Fishermen speak of them as common on and about Stellwagen Bank, also. And we have trawled them in the deep basin of the Gulf.

Goosefish formed about 1 percent (in numbers of individuals) of the fishes of all kinds taken by certain otter trawlers in the South Channel and on Georges Bank in 1913. And most of the trawl hauls that we have seen made there subsequently in depths of 60 to 100 fathoms have brought in from 1 to 40 of them.

They do not show any evident preference for any particular depth zone in the inner parts of the Gulf between tide mark and 100 fathoms or so, and the *Albatross III* found them generally distributed from 22–30 fathoms down to at least as deep as 150–160 fathoms both on Georges Bank and off southern New England to the westward, in May 1950.[82] Our failure to take any in the bowl between Jeffreys Ledge and the coast suggests that they may avoid the very softest mud bottoms. And it is likely that a rather definite concentration of them in depths of 26 to 45 fathoms on the southwestern part of Georges Bank in June 1951, when the *Eugene H* caught an average of about 5 per haul there, but only 1 per haul at 46–65 fathoms, was a matter of the food supply, not of the depth.

Goosefish are said to be as common on Browns Bank as they are on Georges, also along the outer Nova Scotian coast and banks as far as Banquereau, though they may not be as common inshore there as they are in the Gulf of Maine. They must be generally distributed in the southern side of the Gulf of St. Lawrence also, to judge from the localities of record there, and they have been reported from Anse des Dunes and from near Mingan on the north shore.[83] They have also been trawled at a few localities on the southern and eastern part of the Grand Banks.[84] And a 4-inch specimen was brought back from the Grand Bank in 1856.[85] But this seems to be its northern limit in our side of the open Atlantic, for they have not been reported from the east coast of Newfoundland, or reliably from the outer coast of Labrador (see footnote 79, p. 540).

Goosefish are common westward and southward also, as far as North Carolina. We have seen many stranded in winter a few miles north of Cape Hatteras, both in Pamlico Sound and on the outside beach, and Smith [86] described it as so plentiful

[78] The smallest ripe males of the North European species seen by Fulton were 26–27 inches long, the smallest ripe females 30 inches.

[79] Pennant's (Arch. Zool., vol. 1, 1784, p. cxci) report of "the *Lophius piscatorius* or common angler" in Hudson Bay seems to have been based on a sculpin (for history of the case, see Connolly, Bull. 3, Biol. Bd. Canada, 1920, p. 7). And we think it likely that this applies also to the "*Lophius laevigatus*" reported by Weiz (Proc. Boston Soc. Nat. Hist., vol. 10, 1866, p. 269) from Okak, northern Labrador.

[80] Goode and Bean, Smithsonian Contrib. Knowl., vol. 30, 1895, p. 486.

[81] Regan, British Antarctic (*Terra Nova*) Exped. (1910), Zool., vol. 1, No. 1, 1914, p. 23.

[82] Catches of 1 to 34 per haul.

[83] See Cox (Contrib. Canadian Biol. [1902–1905], 1907, p. 90), Cornish (Contrib. Canadian Biol. [1906–1910], 1912, p. 81), and Connolly (Bull. 3, Biol. Bd. Canada, 1920, p. 7) for Nova Scotian and Gulf of St. Lawrence localities; the Annual Reports of the Newfoundland Fishery Research Commission, vol. 1, No. 4, 1932, p. 110, for additional records for the Nova Scotian Banks.

[84] Rept. Newfoundland Fish. Res. Comm., vol. 1, No. 4, 1932, p. 110, Sta. 17; vol. 2, No. 1, 1933, p. 127, Sta. 97; vol. 2, No. 2, 1935, p. 116, Sta. 204, 205, 274.

[85] Goode and Bean, Smithsonian Contrib. Knowl., vol. 30, 1895, p. 486.

[86] North Carolina Econ. Geol. Surv., vol. 2, 1907, p. 399.

near Cape Lookout that "as many as 20 large specimens are sometimes found in a sink net at one lift," though it has been seen less often of late years.[87]

Egg veils have been reported within the Gulf of Maine from Campobello Island at the entrance to the Bay of Fundy; from Passamaquoddy Bay (2 instances);[88] in Frenchman Bay, Maine;[89] about 15 miles off Seguin Island, Maine, September 18, 1925 (with eggs nearly ready to hatch, found by Capt. Greenleaf of the U. S. Bureau of Fisheries); and at Provincetown, where we found a veil within a few feet of the shore, on June 26, 1925. The captures of the pelagic larvae within the Gulf have been fewer, namely 3 taken near Brazil Rock off southwestern Nova Scotia, and two very small ones (5 and 6.5 mm. long) collected by us on the *Grampus* in Massachusetts Bay, one on July 12, 1912, the other September 29, 1915.

The question why the egg veils should have been encountered so seldom in our Gulf when they are so conspicuous in the water, and why so few larvae have been taken in our tow nets, when the parent fish are moderately plentiful and very generally distributed, is an intriguing one that we cannot answer from present information.

Importance.—No regular commercial use has been made of the goosefish in America up to the present time. But it is an excellent food fish, white-meated, free of bones, and of pleasant flavor, as Dr. Connolly assures us from personal experience. In 1948,[90] English and Scottish vessels landed about 7 million pounds of the European species, as "monk" which fetched nearly as high a price as haddock in English markets, though it brings only about one-half as high a price as haddock in Scotch ports.[91] And some were sold in retail stores during the last year.

THE SARGASSUM FISHES. FAMILY ANTENNARIIDAE

The sargassum fishes resemble their relatives the anglers (p. 532) in their peculiar armlike pectorals and in their large fleshy ventrals, as well as in the fact that the forward part of their dorsal fin is represented by a series of separate spines. But they are very different from the anglers in general appearance, for their bodies and heads are flattened sidewise instead of dorso-ventrally; their soft second dorsal fin is much longer than their anal fin; their second and third dorsal spines are clothed with fleshy skin so thick as to obscure their true nature; and their mouth is much smaller than that of the anglers.

These curious little tropical-oceanic fishes live chiefly among masses of floating seaweed, with which they sometimes drift far to the north of their normal homes.

Sargassum fish, *Histrio pictus* (Cuvier and Valenciennes) 1837

MOUSEFISH

Jordan and Evermann, 1896–1900, p. 2716, as *Pterophryne histrio* (Linnaeus) 1758.

Description.—The peculiar armlike pectoral fins; the rather long fleshy ventral fins situated on the throat; the soft flabby skin; and a body flattened sidewise (about 2¾ times as long as it is deep) distinguish this fish from any other that is known from the Gulf of Maine, or that is ever likely to be found there. The goosefish (p. 532), the only one of its close relatives which normally inhabits the Gulf, is of such different appearance with its flattened body form, and enormous mouth, that it is not necessary to enumerate the smaller differences between the two.

The head and body of the sargassum fish appear as one, for each gill opening has the form of a pore on the lower margin of the pectoral near its base, so small that it is likely to be overlooked. There are three detached dorsal spines. The first, standing over the front margin of the eye, has the form of a slender tentacle, its tip bearing a bulbous swelling, known technically as an "illicium," which is fringed at the tip. The second spine (close behind the first) and the third are much larger than the first; they are enclosed in skin so thick that they suggest conical horns in their general appearance, and they bear several tags or streamers of skin. All the other fins are also fleshy. The second (soft-rayed) dorsal fin is more than twice as long as the anal fin; and the detached

[87] Information supplied by the late S. F. Hildebrand.

[88] Connolly, Contrib. Canadian Biol. (1921) 1922, No. 7, p. 116; Berrill Contrib. Canadian Biol. Fish. N. Ser., vol. 4, No. 12, 1929, p. 145.

[89] Procter and others, Biol. Surv. Mt. Desert, Pt. 2 Fishes, 1928, p. 3.

[90] Most recent year for which the international fisheries statistics are readily available.

[91] For catches and values, see Bull. Statist., Cons. Internat. Explor. Mer., vol. 33, 1951, pp. 14, 16–18.

FIGURE 287.—Sargassum fish (*Histrio pictus*). Drawing by Louella E. Cable.

tips of both the dorsal rays and the anal rays form short fringes. The margin of the caudal fin is almost straight. The skin feels smooth to the touch; actually it is finely studded with minute granulations, and it bears variously shaped fleshy tags, as appears in the illustration (fig. 287).

Color.—Creamy white, the fins as well as the head and body mottled with pale and dark brown. The fleshy tags are yellowish.

General range.—Tropical and subtropical, living at the surface among floating seaweed; sometimes drifting far northward with the Gulf Stream.

Occurrence in the Gulf of Maine.—A specimen about 4¾ inches (12 cm.) long, that was picked up in a purse seine near the surface over the west central part of Georges Bank, by the Schooner *Old Glory* on September 15, 1930,[92] and a second of 2¼ inches, taken off the southeast slope of Georges Bank, by the sword fisherman *Leonora C*, on June 15, 1937, are the only records of this fish in the Gulf of Maine; the most northerly records, in fact, for it for continental waters in this side of the Atlantic. But it has been picked up from time to time near Woods Hole.[93] Living, as they usually do, among floating gulf weed (*Sargassum*), it is not astonishing that sargassum fish should drift in over the offshore banks, occasionally.

THE DEEP SEA ANGLERS. FAMILY CERATIIDAE

The members of this family fall with the anglers and sargassum fishes in the pediculate tribe. And the first dorsal spine bears a "bait" at its tip (known technically as an "illicium") as it does among the anglers. But the wristlike structure of the pectoral fins is not obvious in the deep sea anglers. And the members of this family, as well as those of several other families closely related to them,[94] differ from the anglers and from the sargassum fishes in lacking ventral fins. Their bodies are somewhat flattened sidewise (not dorso-ventrally as in the anglers); their dorsal and anal fins are very short (3 to 5 rays); and their central four caudal rays are branched. Their

[92] Reported by Firth, Bull. 61, Boston Soc. Nat. Hist., 1931, p. 14.

[93] For early records of it near Woods Hole, see Sumner, Osburn, and Cole, Bull. U. S. Bur. Fish., Vol. 31, Part 2, 1913, p. 774.

[94] For a synopsis of the ceratioid fishes, a numerous race, see Regan and Trewavas, Rept. 2, Danish *Dana* Exped. (1928–1930), 1932, p. 48.

mouths are oblique when closed, or even vertical. Associated with their deep-water habitat their bodies are noticeably soft and flabby. Their eyes are very small; some appear to be blind. The ceratioids, too, are unique among the vertebrates in the fact that the males of many of them (including those of the Gulf of Maine species) are dwarfs in size as compared with the females, and live parasitic, attached to the females by their heads.

They are oceanic as a group, living in the mid depths, mostly from about 200 fathoms down to perhaps 750 fathoms. And they are blackish in color as are so many other pelagic fishes of that same depth zone. One species has been taken in the Gulf of Maine as a stray.

Deep sea angler *Ceratias hölbolli* Krøyer 1844

Jordan and Evermann, 1896–1900, p. 2729 (young as *Mancalias uranoscopus* Murray); p. 2730 (young as *Mancalias shufeldti* [Gill]; Barbour and Bigelow, Proc. New England Zool. Club, vol. 23, 1944, p. 16 (adult, as *Reganichthys giganteus*, new genus and species); Clarke, Discovery Rept., vol. 26, 1950, p. 1 (adult).

Description.—This deep sea angler is so bizarre in its appearance that there is no danger of confusing it with any other Gulf of Maine fish, unless it were with some other member of its own family. In the large female, the body is strongly flattened sidewise; the eyes are very small and set high on the head; and the mouth is nearly vertical when it is closed. Perhaps their most striking external feature is the very long and extremely slender bristlelike spine or "tentacle," that is borne on the top of the head. This is jointed about two-thirds the way out along its length, and it ends in a fleshly, pear-shaped swelling ("illicium"), the tip of which is described as pierced by a small pore.[95] The illicium is supposed to be luminous,[96] and it bears 2 to 4 short filaments.[97] This head-tentacle corresponds to the whiplike head spine of the goosefish, but is situated farther back, about abreast of the eyes. It is interpreted as representing a vestige of the first dorsal fin. The basal

[95] So described by Clarke (Discovery Rept., vol. 26, 1950, p. 9) for an Antarctic specimen; the pore is not visible on the specimen we have examined.

[96] Dahlgren (Science, vol. 68, 1928, p. 65) describes the tip of the illicium of an unnamed species of *Ceratias* as with an open gland in which light is produced by bacteria.

[97] Four in the Gulf of Maine specimen described by Barbour and Bigelow (Proc. New England zool. club, vol. 33, 1944, p. 9) as *Reganichthys giganteus*; two (each bifid) in an Antarctic specimen described by Clarke.

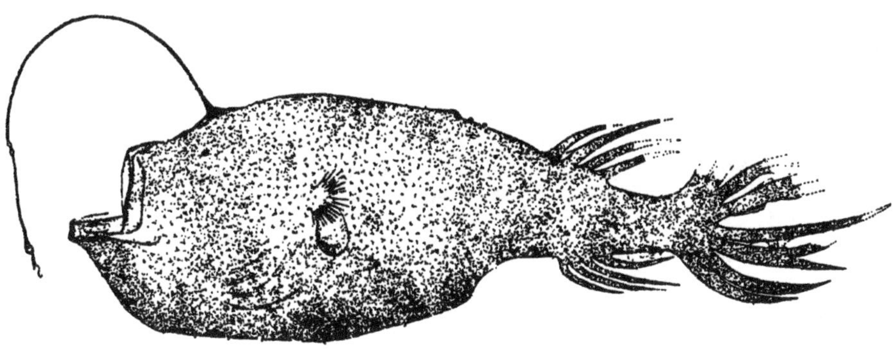

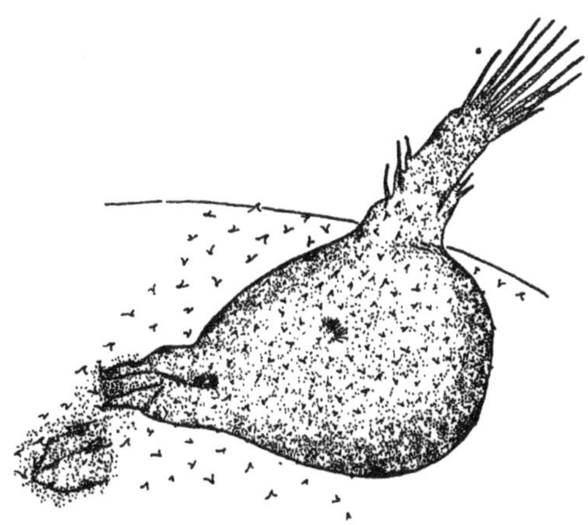

FIGURE 288.—Deep-sea angler (*Ceratias hölbolli*), adult female (above) and parasitic male (right) that was attached to her, off Mount Desert Rock. From Barbour and Bigelow.

joint of the head tentacle is provided with retractor muscles by which it can be withdrawn rearward into a tunnel-like sheath along the head and back, bringing the "bait" close to the mouth.

Bertelsen[98] has found (from dissecting a West Greenland specimen) that when this happens, the rear part of the hard axis of the head tentacle, which is enclosed in the very elastic skin, emerges from the back of the fish, about midway between the caudal fin and the base of the pectoral fins, so as to form the axis of a slender, tapering "dorsal tentacle." Thus this extraordinary and unique structure, which has been the subject of much discussion, is actually the rear end of the head tentacle which protrudes when the latter is drawn rearward.

When the cephalic tentacle is moved forward by its protractor muscles, its protruding rear end is withdrawn into the tentacular sheath, either partially, when the so-called dorsal "tentacle" appears as a short fingerlike process, or wholly, leaving simply an indentation or pore in the midline of the back, as it is in the Gulf of Maine specimen pictured in figure 288.

Close behind the so-called "dorsal tentacle" (or behind the pore representing the latter) are a pair of low, fleshy appendages or "caruncles," scarcely noticeable on large specimens, but more conspicuous on small. These have been interpreted as vestiges of the first dorsal fin, for each of them encloses a spine that can be felt if not seen. Their function is not known.[99]

The skin is strewn with small prickles on very small specimens, but is close-set with low conical, broad-based thorns on larger fish. The eyes are minute, seemingly functional on small fish, but covered over by skin and apparently blind on large ones. The gill slits are very small, C-shaped; placed below the pectorals and a little behind them. The small, slender, sharp-pointed teeth are directed into the mouth. The dorsal and anal fins each have 4 rays, thick, fleshy, and tapering, as the caudal rays are also. The central caudal rays are forked. The caudal fin has been described as occupying as much as two-fifths of the total length of the fish when it is intact.[1] But it has been much damaged in most of the specimens that have been seen, and the membranes of all the fins have been mostly torn away.

Color.—Small ones are jet black, but the dermal prickles, being colorless, show white against the black skin on large specimens, giving a granulated black and white appearance.

Size.—The largest specimen seen so far[2] was 26¾ inches (68 cm.) long to the base of the tail fin, and about 47 inches (119 cm.) long, counting the tail fin.[3]

The parasitic males are fastened to the ventral side of the female, by two outgrowths from the front of the head, that are fused at the tip. They have no teeth, no tentacle-like spine and no eyes, and the alimentary canal is vestigial; in fact, about the only important internal organ is a large testis. But their fins resemble those of their mates, as do the gill openings; their skins are prickly; and they are similarly black. Those that have been seen (1 or 2 per female) have ranged from about 3⅜ inches (85 mm.) long to about 6 inches (150 mm.) long (Gulf of Maine specimen).

General range.—Oceanic and apparently cosmopolitan, for adults have been reported from Greenland; Iceland (where it has been taken the most often); off Nova Scotia; Gulf of Maine; near the Azores; and in the sub-Antarctic. Young specimens apparently referable to this species are reported off southern New England; from the Caribbean; near the Canaries; north of the Cape Verde Islands; from the South Atlantic (lat. 52°25′ S., long. 9°50′ E.); and also widespread in Indo-Pacific waters.[4]

Occurrence in the Gulf of Maine.—A female, about 32 inches long to the base of the caudal fin, and about 40 inches counting what remained of the latter (fig. 288), with one male attached, and showing the scar of attachment of another, was taken 12 miles south of Mount Desert rock, at 125

[98] Vid. Meddel. Dansk Naturh. Foren., vol. 107, 1943, pp. 190–193; see especially his fig. 4, p. 192.

[99] See Regan and Trewavas (Rept. 2, Danish *Dana* Exped. (1928–1930), 1932, pp. 23–24) for an account of the lateral line papillae in different families of deep sea anglers.

[1] This was the case in the specimen about 47 inches (119 cm.) long described by Kröyer (Naturhist, Tiddsskr., Ser. 2, vol. 1, 1844, pp. 640–642); also in one pictured by Goode and Bean (Smithsonian Contrib. Knowl., vol. 31, 1895, pl. 117, fig. 399, after Gaimard).

[2] Kröyer's original specimen from Greenland.

[3] See Clarke (*Discovery* Rept., Vol. 26, 1950, p. 14, table 1), for measurements of several specimens.

[4] For complete list of localities for adults and young, see Clarke, *Discovery* Rept., vol. 26, 1950, pp. 23, 30.

fathoms, in October 1943, by the schooner *Dorothy and Ethel II*, Capt. Harold Paulsen.[5] A second female, about 18⅞ inches long to the base of the caudal fin, and about 24⅜ inches counting what was left of the caudal fin, trawled on the southeast part of Georges Bank, between 150 and 200 fathoms, in February 1927, appears to belong to this same species.[6] A third probable Gulf of Maine record is of a fish, about 3 feet long, and weighing about 20 pounds, that was taken by the trawler *Ebb*, in 140 fathoms, on Georges Bank, in June 1936. Photographs of it appeared in the Boston Globe and in the Boston Post for June 29 of that year.

BIBLIOGRAPHY

(Includes titles giving local references for the Gulf of Maine)

ADAMS, A. LEITH.

1869. Death of fishes in the Bay of Fundy. Amer. Naturalist, vol. 2, pp. 337–342.

1873. Field and forest rambles . . . Eastern Canada, xvi, 333 pp. London. [Fishes, pp. 201–257, 304–306.]

AGASSIZ, ALEXANDER.

1882. On the young stages of some osseous fishes. Part 3. Proc. Amer. Acad. Arts, Sciences, New Ser., vol. 9, pp. 271–303, Pls. 1–20.

AGASSIZ, ALEXANDER, and C. O. WHITMAN.

1885. Studies from the Newport Marine Laboratory. 16. The development of osseous fishes. 1. The pelagic stages of young fishes. Mem. Museum Compar. Zool., vol. 14, No. 1, Part 1, 56 pp., 19 pls.

ALEXANDER, A. B.

1905. Statistics of the fisheries of the New England States, 1902. Report U. S. Comm. Fish and Fisheries for 1904, pp. 245–325.

ALEXANDER, A. B., H. F. MOORE, and W. C. KENDALL.

1915. Otter-trawl fishery. Appendix 6, Report U. S. Comm. Fish and Fisheries for 1914, 97 pp., 2 charts, diagrams.

ALLEN, GLOVER M.

1916. The whalebone whales of New England. Mem. Boston Soc. Nat. Hist., vol. 8, No. 2, pp. 107–322, Pls. 8–16.

ANONYMOUS.

1882. List of collections made by the fishing vessels of Gloucester and other New England seaports . . . Report U. S. Comm. Fish and Fisheries for 1879, pp. 787–835.

ATKINS, CHARLES G.

1874. On the salmon of eastern North America, and its artificial culture. Report U. S. Comm. Fish and Fisheries for 1872 and 1873, pp. 226–337.

1876. The Atlantic salmon (*Salmo salar*). A. Report on the collection and distribution of Penobscot salmon in 1873–74 and 1874–75. Report U. S. Comm. Fish and Fisheries for 1873–4 and 1874–5, pp. 485–530.

ATKINS, CHARLES G.—Continued

1878. Artificial hatching of the cunner. Forest and Stream, Rod and Gun, Vol. 10, No. 26, p. 502.

1887. The river fisheries of Maine. *In* Fisheries and Fishery Industries of the U. S., Sect. 5, vol. 1, pp. 673–728, and Sect. 5, vol. 2, pls. 164–168.

ATWOOD, N. E.

1859. [Remarks on *Naucrates* at Provincetown, and on the habits of the cod.] Proc. Boston Soc. Nat. Hist., vol. 7, p. 4.

1863. [Notes on the habits of bluefish, cod, and mackerel.] Proc. Boston Soc. Nat. Hist., vol. 9, pp. 189–190.

1864. [Notes on cod.] Proc. Boston Soc. Nat. Hist., vol. 9, pp. 318–319.

1865. [On the habits of some salt-water fishes.] Proc. Boston Soc. Nat. Hist., vol. 10, pp. 65–68.

1865a. [Notes on dogfish and other sharks.] Proc. Boston Soc. Nat. Hist., vol. 10, pp. 81–82.

1866. [On the habits of the cod.] Proc. Boston Soc. Nat. Hist., vol. 10, pp. 103–104.

1866a. [On the habits of the halibut.] Proc. Boston Soc. Nat. Hist., vol. 10, pp. 182–183.

1866b. [On the habits and distribution of the haddock.] Proc. Boston Soc. Nat. Hist., vol. 10, pp. 322–323.

1866c. [Notes on the habits of several fishes.] Proc. Essex Inst., vol. 4, pp. ci–ciii.

1866d. [Note on the torpedo at Provincetown.] Proc. Essex Inst., vol. 5, pp. 14–15.

1867. [On the habits of the Gadidae.] Proc. Boston Soc. Nat. Hist., vol. 11, pp. 100–102.

1869. [Description of *Carcharias tigris* Atwood.] Proc. Boston Soc. Nat. Hist., vol. 12, pp. 268–269.

1869a. [On the natural history of a few of our edible sea fish, and particularly of the halibut and bluefish.] Proc. Boston Soc. Nat. Hist., vol. 12, pp. 402–403.

[5] This specimen, now in the Museum of Comparative Zoology, was described by Barbour and Bigelow (Proc. New England Zool. Club, vol. 23, 1944. p. 9) as "*Reganichthys giganteus*."

[6] This specimen, now in the Mus. Comp. Zool., and first described by Parr (Bull. 63, Boston Soc. Nat. Hist., 1932, pp. 12–13) as *Mancalias uranoscopus*, was later made the basis of a new genus, *Typhloceratias*, by Barbour (Proc. New England Zool. Club, vol. 21, 1940, p. 78). Its head and back have been so badly damaged that it has lost whatever tentacular structures it may have had originally.

ATWOOD, N. E.—Continued

1870. [On the habits of several marketable fishes.] Bull. Essex Inst., vol. 2, No. 3, March 1870, p. 48.

1871. [On the habits of the bluefish, *Temnodon saltator* Cuv.] Proc. Boston Soc. Nat. Hist., vol. 14, p. 92.

1871a. [Remarks on fisheries on the coast of Massachusetts.] 5th Ann. Report Comm. Inland Fisheries [of Massachusetts] for 1870, pp. 63–73.

1873. [On the fisheries on the coast of Massachusetts.] Report U. S. Comm. Fish and Fisheries for 1871–72, pp. 117–124.

BAILEY, REEVE M.

1938. The fishes of the Merrimac Watershed. Merrimac Watershed Survey, New Hampshire Fish and Game Dept., Rept. 3, pp. 149–185.

BAIRD, SPENCER F.

1873. Natural history of some of the more important food fishes of the south shore of New England. Rept. U. S. Comm. Fish and Fisheries for 1871–1872, pp. 228–252.

1874. Conclusions as to the decrease of cod fisheries on the New England coast. Rept. U. S. Comm. Fish and Fisheries for 1872–1873, pp. xi–xiv.

1879. [Pole flounder (*Glyptocephalus cynoglossus*) abundant off New England.] Report U. S. Comm. Fish and Fisheries for 1877, pp. 4*–5*.

1889. The sea fisheries of eastern North America. Report U. S. Comm. Fish and Fisheries for 1886, App. A, pp. 3–224.

BARBOUR, THOMAS

1940. More concerning ceratioid fishes. Proc. New England Zool. Club, vol. 21, pp. 77–86, pl. 9.

BARBOUR, THOMAS, and H. B. BIGELOW.

1944. A new giant ceratioid fish. Proc. New England Zool. Club, vol. 23, pp. 9–15, pls. 4–6.

BATCHELDER, J. M.

1845. [Description of a new species of Argyreiosus.] Proc. Boston Soc. Nat. Hist., vol. 2, pp. 78–79.

BATTLE, HELEN I.

1930. Effects of extreme temperatures and salinities on the development of *Enchelyopus cimbrius* (L.). Contrib. Canadian Biol., and Fish. New Ser., vol. 5, No. 6, pp. 109–192.

1930a. Spawning periodicity and embryonic death rate of *Enchelyopus cimbrius* (L.) in Passamaquoddy Bay. Contrib. Canadian Biol., and Fish., New Ser., vol. 5, No. 7, pp. 361–380.

1934. Laboratory feeding of the herring. Ann. Rept. Biol. Board Canada for 1933, pp. 14–15.

BEAN, TARLETON H.

1879. On the identity of *Euchalarodus putnami* Gill, with *Pleuronectes glaber* (Storer) Gill, with notes on the habits of the species. Proc. U. S. Nat. Museum, vol. 1, pp. 345–348.

1881. Checklist of duplicates of North American fishes distributed by the Smithsonian Institution in behalf of the United States National Museum, 1877–1880. Proc. U. S. Nat. Museum, vol. 3, pp. 75–116.

1892. Observations upon fishes and fish-culture. Bull. U. S. Fish Comm., vol. 12, pp. 49–61.

1903. Catalogue of the fishes of New York. Bull. 60, Zool. 9, New York State Museum, 784 pp.

BELDING, DAVID L.

1921. A report upon the alewife fisheries of Massachusetts. Div. Fisheries and Game; Dept. Conservation, Massachusetts, pp. 1–135, figs. 1–14.

1934. The spawning habits of the Atlantic salmon. Trans. Amer. Fish. Soc., vol. 64, pp. 211–218.

BENSLEY, B. ARTHUR.

1901. Report on the sardine industry in relation to the Canadian herring fisheries. Contrib. Canadian Biol. for 1901, pp. 59–62.

BERRILL, N. J.

1929. The validity of *Lophius americanus* Val., as a species distinct from *L. piscatorius* Linn., with notes on the rate of development. Contrib. Canadian Biol., and Fish. New Ser., vol. 4, No. 12, pp. 143–155.

BIGELOW, HENRY B.

1914. Explorations in the Gulf of Maine, July and August, 1912, by the U. S. Fisheries Schooner *Grampus*. Oceanography and notes on the plankton. Bull. Museum Compar. Zool., vol. 58, No. 2, pp. 31–134, 9 pls. [List of fishes identified by W. W. Welsh, pp. 107–114.]

1917. Explorations of the coast water between Cape Cod and Halifax in 1914 and 1915, by the U. S. Fisheries Schooner *Grampus*. Oceanography and plankton. Bull. Museum Compar. Zool., vol. 61, No. 8, pp. 163–357, 100 figs., 2 pls. [List of fishes identified by W. W. Welsh, pp. 268–277.]

BIGELOW, HENRY B., and WILLIAM C. SCHROEDER.

1927. Notes on northwest Atlantic sharks and skates. Bull. Museum Compar. Zool., vol. 68, No. 5, pp. 239–251.

1929. A rare bramid fish (*Taractes princeps* Johnson) in the northwestern Atlantic. Bull. Museum Compar. Zool., vol. 69, No. 2, pp. 41–50, 1 plate.

1936. Supplemental notes on fishes of the Gulf of Maine. Bull. U. S. Bur. Fisheries, vol. 48, pp. 319–343.

BIGELOW, HENRY B., and WILLIAM C. SCHROEDER—Con.
1937. A record of *Centrolophus niger* (Gmelin) from the western Atlantic. Copeia, p. 61.
1939. Notes on the fauna above mud bottoms in deep water in the Gulf of Maine. Biol. Bull., vol. 76, pp. 305–324.
1948. Fishes of the western North Atlantic, Pt. 1. Mem. Sears Found. Marine Research, No. 1, Chap. 2, Cyclostomes; Chap. 3, Sharks, pp. 29–576.
1953. Fishes of the western North Atlantic, Part 2. Mem. Sears Found. Marine Research, No. 2, Chap. 1, Batoidea; Chap. 2, Holocephali, xv, 562 pp., index.

BIGELOW, HENRY B., and W. W. WELSH.
1925. Fishes of the Gulf of Maine. Bull. U. S. Bur. Fish., vol. 40, Part 1, 567 pp.

BINNEY, AMOS.
1842. [On the habits of the *Orthogariscus mola*, or short sunfish.] Proc. Boston Soc. Nat. Hist., vol. 1, p. 93.

BLAKE, J. HENRY.
1870. The habits and migrations of some of the marine fishes of Massachusetts. Amer. Naturalist, vol. 4, pp. 513–521.

BOARDMAN, SAMUEL L., and CHARLES G. ATKINS.
1875. The menhaden and herring fisheries of Maine as sources of fertilization. Report to Maine Board of Agriculture, pp. 1–64, 1 pl.

BOWERS, GEORGE M.
1906. Food of dogfishes. Report U. S. Comm. Fish and Fisheries for 1905, pp. 24–25.
1907. Notes on important fisheries. Report U. S. Comm. Fish and Fisheries for 1906, pp. 16–19.

BRYANT, HENRY.
1842. [Sunfish off Massachusetts.] Proc. Boston Soc. Nat. Hist., vol. 1, p. 93.

BUMPUS, HERMON C.
1899. The reappearance of the tilefish. Bull. U. S. Fish Comm., vol. 18, pp. 321–333.

CABOT, SAMUEL.
1846. [Abundance of horse mackerel (*Temnodon saltator*) in Beverly Harbor.] Proc. Boston Soc. Nat. Hist., vol. 2, p. 179.
1849. [Capture of a nurse or sleeper shark. *Somniosus brevipinna*, at Nahant.] Proc. Boston Soc. Nat. Hist., vol. 3, p. 97.

CANADA, DEPT. MARINE and FISHERIES, FISHERIES BRANCH.
1897–1914. [Statistics of the fisheries of New Brunswick and Nova Scotia, 1897 to 1913–1914.] Ann. Reports Nos. 29 to 47.

CANADA, DEPT. NAVAL SERVICE, FISHERIES BRANCH.
1915–1918. [Statistics of the fisheries of New Brunswick and Nova Scotia, 1914–15 to 1916–1917.] Ann. Reports Nos. 48 to 51.

CANADA, DOMINION, BUREAU OF STATISTICS.
1919–1920. [Statistics of the fisheries of New Brunswick and Nova Scotia.] Yearly fisheries statistics, 1917 to 1919.

CANADA, DOMINION, BUREAU OF STATISTICS, FISHERIES DIVISION.
1921–1949. [Statistics of the fisheries of New Brunswick and Nova Scotia.] Yearly fisheries statistics, 1920 to 1946.

CARSON, RACHEL L.
1943. Food from the sea. Fish and shellfish of New England. U. S. Fish and Wildlife Service, Conserv. Bull. 33, 74 pp., 34 figs.

CLARK, A. HOWARD.
1884. Notes on the fisheries of Gloucester, Mass. Bull. U. S. Fish Comm., vol. 4, pp. 401–410.
1886. Notes on the New England fishery for swordfish during the season of 1884. Rept. U. S. Comm. Fish and Fisheries for 1884, pp. 233–239.
1887. The fisheries of Massachusetts [in 1880]. *In* Fisheries and Fishery Industries of the U. S., Sec. 2, pp. 113–280.
1887a. Historical references to the fisheries of New England. *In* Fisheries and Fishery Industries of the U. S., Sec. 2, Append., pp. 675–737.

CLEMENS, WILBERT A.
1920. Histories of new food fishes. 4. The muttonfish. Bull. Biol. Board Canada, No. 4, 12 pp.

CLEMENS, WILBERT A., and LUCY S. CLEMENS.
1921. Contribution to the biology of the muttonfish (*Zoarces anguillaris*). Contrib. Canadian Biol. for 1918–20, pp. 69–83, 1 pl.

COLLINS, JOSEPH W.
1883. Appearance of dogfish (*Squalus acanthias*) on the New England coast in winter. Bull. U. S. Fish Comm., vol. 2, p. 8.
1883a. Shad taken in mackerel gill nets. Bull. U. S. Fish Comm., vol. 2, p. 95.
1883b. Notes on the movements, habits, and captures of mackerel for the season of 1882. Bull. U. S. Fish Comm., vol. 2, pp. 273–285.
1883c. Notes on the herring fishery of Massachusetts Bay in the autumn of 1882. Bull. U. S. Fish Comm., vol. 2, pp. 287–290.
1884. Movements of mackerel in winter. Bull. U. S. Fish Comm., vol. 4, p. 15.
1884a. Note on the destruction of mackerel by dogfish. Bull. U. S. Fish Comm., vol. 4, p. 248.
1884b. History of the tilefish. Report U. S. Comm. Fish and Fisheries for 1882, pp. 237–294a, pls. 1–2.
1885. Unusual abundance of cod on Browns Bank. Bull. U. S. Fish Comm., vol. 5, p. 234.
1887. Historical notes and incidents of the fresh-halibut fishery. *In* Fisheries and Fishery Industries of the U. S., Sec. 5, vol. 1, pp. 43–89.

COLLINS, JOSEPH W.—Continued

1892. Statistical review of the coast fisheries of the United States. 3. Fisheries of the New England States [1887 and 1888]. Report U. S. Comm. Fish and Fisheries for 1888, pp. 286–322.

COLLINS, JOSEPH W., and RICHARD RATHBUN.

1887. The sea fishing-grounds of the eastern coast of North America from Greenland to Mexico. *In* Fisheries and Fishery Industries of the U. S., Sec. 3, 1887, pp. 5–78, 11 charts.

COLLINS, JOSEPH W., and HUGH M. SMITH.

1892. Report on the fisheries of the New England States. Bull. U. S. Fish Comm., vol. 10, pp. 73–176.

CONEL, J. L.

1931. Naturally deposited eggs of the Myxinoidea (Hyperotretia). Science, New Ser., vol. 75, pp. 19–20.

CONNOLLY, C. J.

1920. Histories of new food fishes. 3. The angler. Bull. Biol. Board Canada, No. 3, 17 pp.

1922. On the development of the angler (*Lophius piscatorius* L). Contrib. Canadian Biol. for 1921, No. 7, pp. 113–124.

COX, PHILIP.

1893. Observations on the distribution and habits of some New Brunswick fishes. Bull. Nat. History Soc. New Brunswick, No. 11, pp. 33–42.

1895. History and present state of the ichthyology of New Brunswick. Bull. Nat. History Soc. New Brunswick, No. 13, pp. 27–61.

1895a. Catalogue of the marine and fresh-water fishes of New Brunswick. Bull. Nat. History Soc. New Brunswick, No. 13, pp. 62–75.

1896. Report on zoology. Bull. Nat. History Soc. New Brunswick, No. 14, App., pp. 55–57.

1920. Histories of new food fishes. 2. The lumpfish. Bull. Biol. Board Canada, No. 2, 28 pp.

COX, PHILIP, and MARIAN ANDERSON.

1922. A study of the lumpfish (*Cyclopterus lumpus* L). Contrib. Canadian Biol., New Ser., vol. 1, No. 1, pp. 1–20.

CRAIGIE, E. HORNE.

1916. The life history of the hake (*Urophycis chuss* Gill) as determined from its scales. Contrib. Canadian Biol. for 1914–15, No. 8, pp. 87–94.

CRANE, JOCELYN.

1936. Notes on the biology and ecology of the giant tuna, *Thunnus thynnus* Linnaeus, observed at Portland, Maine. Zoologica, N. Y. Zool. Soc., vol. 21, pp. 207–212, 1 pl.

CRITTENDEN, A. R.

1884. Catching alewives with hooks baited with eels. Bull. U. S. Fish Comm., vol. 4, pp. 255–256.

DEAN, BASHFORD.

1900. The egg of the hagfish, *Myxine glutinosa.* Mem. New York Acad. Sciences, vol. 2, Pt. 2, Art. 2, pp. 33–46, pl. 2.

DE BROCA, P.

1876. The halibut fishery of the United States. Report U. S. Comm. Fish and Fisheries for 1873–74 and 1874–75, pp. 169–171.

DRESSLAR, FLETCHER B., and BERT FESLER.

1889. A review of the mackerels (Scombrinae) of America and Europe. Bull. U. S. Fish Comm., vol. 7, pp. 429–446, pls. 1–11.

DUFF, DOROTHY.

1916. Investigation of the haddock fishery, with special reference to the growth and maturity of the haddock (*Melanogrammus aeglefinus*). Contrib. Canadian Biol. for 1914–15, No. 9, pp. 95–102.

EARLL, R. EDWARD.

1880. A report on the history and present condition of the shore cod fisheries of Cape Ann, Mass., together with notes on the natural history and artificial propagation of the species. Report U. S. Comm. Fish and Fisheries for 1878, pp. 685–740.

1887. The coast of Maine and its fisheries in 1880. *In* Fisheries and Fishery Industries of the U. S., Sect. 2, Pt. 1, pp. 5–102.

1887a. The herring fishery and the sardine industry. *In* Fisheries and Fishery Industries of the U. S., Sect. 5, vol. 1, Pt. 6, pp. 417–524.

EATON, CYRUS.

1851. Annals of the town of Warren, in Knox County, Maine, with the early history of St. George's, Broad Bay, and the neighboring settlements on the Waldo Patent. Second edition, 1877, 679 pp.

EIGENMANN, CARL H.

1902. The egg and development of the conger eel. Bull. U. S. Fish Comm., vol. 21, pp. 37–44, figs. 1–15.

EIGENMANN, CARL H., and CLARENCE HAMILTON KENNEDY.

1902. The Leptocephalus of the American eel and other American Leptocephali. Bull. U. S. Fish Comm., vol. 21, pp. 81–92, 14 figs.

EVERMANN, BARTON WARREN.

1905. Report of the division of statistics and methods of the fisheries. Rept. U. S. Comm. Fish., for 1903. (The Salmon Fishery of Penobscot River and Bay, pp. 110–114.)

1905a. Report on inquiry respecting food fishes and the fishing grounds. Rept. U. S. Bur. Fish., 1904 (Investigations in Maine, pp. 102–106).

FIELD, IRVING A.

1907. Unutilized fishes and their relation to the fishing industries. Special Paper No. 6, Rept. U. S. Comm. Fish and Fisheries for 1906. Bur. Fish. Doc. 622, 50 pp.

FIRTH, F. E.

1931. Some marine fishes collected recently in New England waters. Bull. Boston Soc. Nat. Hist., No. 61, pp. 8–14.

FISH, CHARLES J.
1929. Production and distribution of cod eggs in Massachusetts Bay in 1924 and 1925. Bull. U. S. Bur. Fish., vol. 43, Part 2, pp. 253–296.

FOSTER, N. W., and CHARLES G. ATKINS.
1869. Second report of the Commissioners of Fisheries of the State of Maine, for 1868, 39 pp.

GARMAN, SAMUEL.
1874–75. On the skates (Rajae) of the eastern coast of the United States. Proc. Boston Soc. Nat. Hist., vol. 17, pp. 170–181.
1886. Notes and descriptions taken from selachians in the U. S. National Museum. Proc. U. S. Nat. Museum, vol. 8, pp. 39–44.
1892. The Discoboli, Cyclopteridae, Liparopsidae, and Liparidae. Mem. Museum of Compar. Zool., vol. 14, No. 2, 96 pp., 13 plates.
1913. The Plagiostomata (sharks, skates, and rays). Mem. Museum Compar. Zool., vol. 36, 515 pp. and atlas of 77 plates.

GILBERT, CHARLES H.
1913. Descriptions of two new fishes of the genus *Triglops* from the Atlantic Coast of North America. Proc. U. S. Nat. Museum, vol. 44, pp. 465–468, pl. 64.

GILL, THEODORE.
1864. Catalogue of the North American sciaenoid fishes. Proc. Acad. Natural Sciences Phila. for 1863, pp. 28–32.
1864a. Synopsis of the North American gadoid fishes. Proc. Acad. Natural Sciences Phila., 1863, pp. 229–242.
1864b. Note on some recent additions to the ichthyological fauna of Massachusetts. Proc. Acad. Natural Sciences Phila., 1863, pp. 332–333.
1864c. Note on the family of stichaeoids. Proc. Acad. Natural Sciences Phila., 1864, pp. 208–211.
1864d. Synopsis of the pleuronectoids of the eastern coast of North America. Proc. Acad. Natural Sciences Phila., 1864, pp. 214–224.
1865. Synopsis of the fishes of the Gulf of St. Lawrence and Bay of Fundy. Canadian Naturalist and Geologist, New Ser., vol. 2, pp. 244–266.
1872. [Note on *Cottus groenlandicus* Fabr.] Proc. Acad. Natural Sciences Phila., 1872, pp. 213–214.
1873. On a new American species of Pleuronectoid (*Glyptocephalus acadianus*). Proc. Acad. Natural Sciences Phila., 1873, pp. 360–362.
1873a. Catalogue of the fishes of the east coast of North America. Report U. S. Comm. Fish and Fisheries for 1871–72, pp. 779–822.
1905. The sculpin and its habits. Smithsonian Misc. Collections, vol. 47, pp. 348–359, figs. 45–55.

GILL, THEODORE—Continued
1905a. The life history of the angler. Smithsonian Misc. Collections, vol. 47, pp. 500–516, figs. 94–103, pls. 73–75.
1905b. State ichthyology of Massachusetts. Report U. S. Bur. Fish., 1904, App., pp. 163–188.
1911. Notes on the structure and habits of the wolffishes. Proc. U. S. Nat. Museum, vol. 39, pp. 157–187, pls. 17–28.

GIRARD, C. F.
1850. On the genus *Cottus* Auct. Proc. Boston Soc. Nat. Hist., vol. 3, pp. 183–190.

GODSIL, N. C., and E. K. HOLMBERG.
1950. A comparison of the bluefin tunas, genus *Thunnus*, from New England, Australia, and California. Fish Bull. 77, California Dept. Nat. Resources, 55 pp.

GOODE, GEORGE BROWN.
1875. Albino fishes. Amer. Naturalist, vol. 9, p. 517.
1879. The occurrence of *Hippocampus antiquorum*, or an allied form, on Saint George's Banks. Proc. U. S. Nat. Museum, vol. 1, pp. 45–46.
1879a. The natural and economical history of the American menhaden. Report U. S. Comm. Fish and Fisheries for 1877, pp. 1–529, pls. 1–31.
1882. The carangoid fishes of the United States . . . pompanos, crevalles, amber-fish, and others. Bull. U. S. Fish Comm., vol. 1, pp. 30–43.
1883. Materials for a history of the swordfish. Report U. S. Comm. Fish and Fisheries for 1880, pp. 289–394, pls. 1–24.
1884. Natural history of useful aquatic animals. Pt. 3, The food fishes of the U. S. *In* Fish. Ind. U. S., Sect. 1, pp. 169–549, 610–612, 629–681.
1885. A brief biography of the halibut. Amer. Naturalist, vol. 19, pp. 953–969.
1887. The swordfish fishery. *In* Fisheries and Fishery Industries of U. S., Sect. 5, vol. 1, Pt. 4, pp. 315–326.
1888. American fishes . . . xv, 496 pp., illus.

GOODE, GEORGE BROWN, and TARLETON H. BEAN.
1877. Descriptions of two new species of fishes (*Macrurus bairdii* and *Lycodes verrillii*) recently discovered by the U. S. Fish Commission, with notes on the occurrence of several unusual forms. Amer. Jour. Science and Arts, Third Ser., vol. 14, pp. 470–478.
1879. A list of the fishes of Essex County, including those of Massachusetts Bay, according to the latest results of the work of the U. S. Fish Commission. Bull., Essex Institute, vol. 11, pp. 1–38.
1879a. The Craig flounder of Europe, *Glyptocephalus cynoglossus*, on the coast of North America. Proc. U. S. Nat. Museum, vol. 1, pp. 19–23.
1879b. The oceanic bonito on the coast of the United States. Proc. U. S. Nat. Museum, vol. 1, pp. 24–26.

GOODE, GEORGE BROWN, and TARLETON H. BEAN—Con.

1879c. Descriptions of two gadoid fishes, *Phycis chesteri* and *Haloporphyrus viola*, from the deep-sea fauna of the northwestern Atlantic. Proc. U. S. Nat. Museum, vol. 1, pp. 256–260.

1879d. The identity of *Rhinonemus caudacuta* (Storer) Gill with *Gadus cimbrius* Linn. Proc. U. S. Nat. Museum, vol. 1, pp. 348–349.

1879e. Note on *Platessa ferruginea*, D. H. Storer, and *Platessa rostrata*, H. R. Storer. Proc. U. S. Nat. Museum, vol. 1, pp. 361–362.

1879f. On the identity of *Brosmius americanus*, Gill, with *Brosmius brosme* (Müller) White. Proc. U. S. Nat. Museum, vol. 1, pp. 362–363.

1879g. Discoveries of the United States Fish Commission: Notices of fifty species of east-coast fishes . . . Amer. Jour. Science and Arts, Third Ser., vol. 17, pp. 39–48.

1883. Reports on the results of dredging under the supervision of Alexander Agassiz . . . by the U. S. Coast Survey steamer *Blake* . . . Report on the fishes. Bull. Museum Compar. Zool., vol. 10, pp. 183–226.

1886. Description of *Leptophidium cervinum* and *L. marmoratum*, new fishes from deep water off the Atlantic and Gulf coasts. Proc. U. S. Nat. Museum, vol. 8, pp. 422–424.

1895. Oceanic ichthyology . . . Smithsonian Contrib. Knowl., vol. 30, xxv, 553 pp., vol. 31, 123 pls.; also Special Bull. 2, U. S. Nat. Museum, 1896; and Mem. Museum Compar. Zool., vol. 22, 1896.

GOODE, GEORGE BROWN, and A. HOWARD CLARK.

1887. The menhaden fishery. *In* Fisheries and Fishery Industries of U. S., Sect. 5, vol. 1, Pt. 5, pp. 327–415.

GOODE, GEORGE BROWN, and JOSEPH W. COLLINS.

1882. The winter haddock fishery of New England. Bull. U. S. Fish Comm., vol. 1, pp. 226–235.

1887. The fresh-halibut fishery. *In* Fisheries and Fishery Industries of U. S., Sect. 5, vol. 1, pp. 3–89.

1887a. The Banks hand-line cod fishery; the Labrador and Gulf of St. Lawrence cod fisheries; the Banks trawl line fishery; the Georges Bank cod fishery. *In* Fisheries and Fishery Industries of U. S., Sect. 5, vol. 1, pp. 123–198.

1887b. Haddock fishery of New England. *In* Fisheries and Fishery Industries of U. S., Sect. 5, vol. 1, pp. 234–241.

1887c. The hake fishery. *In* Fisheries and Fishery Industries of U. S., Sect. 5, vol. 1, pp. 241–243.

1887d. The mackerel fishery of the United States. *In* Fisheries and Fishery Industries of U. S., Sect. 5, vol. 1, pp. 247-313.

GOODE, GEORGE BROWN, JOSEPH W. COLLINS, R. E. EARLL, and A. HOWARD CLARK.

1884. Materials for a history of the mackerel fishery. Rept. U. S. Comm. Fish and Fisheries for 1881, pp. 91–531, 1 chart.

GRAHAM, MICHAEL.

1936. Investigations of the herring of Passamaquoddy and adjacent regions. Jour. Biol. Board Canada, vol. 2, No. 2, pp. 95–140.

GRANT, ISAAC H.

1883. Movements of menhaden—catch of herring. Bull. U. S. Fish Comm., vol. 3, p. 470.

GÜNTHER, ALBERT.

1862. Catalogue of the fishes in the British Museum. Vol. 4, xxi, 534 pp.

HALKETT, ANDREW.

1913. Check list of the fishes of the Dominion of Canada and Newfoundland. 138 pp., 14 pls.

HALL, ANSLEY.

1898. The herring industry of the Passamaquoddy region, Maine. Rept. U. S. Comm. Fish and Fisheries for 1896, pp. 443–487.

HASKELL, E. H.

1883. Second annual appearance of young cod hatched by the United States Fish Commission in Gloucester Harbor in the winter of 1879–80. Bull. U. S. Fish Comm., vol. 2, p. 112.

HERRINGTON, W. C.

1936. Decline in haddock abundance on Georges Bank and a practical remedy. Fishery Circular No. 23, U. S. Bur. Fisheries, 22 pp.

1944. Factors controlling population size. Trans. 9th N. American Wildlife Conf., pp. 250–263.

HILDEBRAND, SAMUEL F., and WILLIAM C. SCHROEDER.

1928. Fishes of Chesapeake Bay. Bull. U. S. Bur. Fisheries, vol. 43, Part 1, 388 pp.

HOLMES, EZEKIEL.

1862. Report on the fishes of Maine . . . Second Ann. Rept. Natural Hist. and Geol. State of Maine, Pt. 1, pp. 11–119.

HOMANS, R. E. S., and A. W. H. NEEDLER.

1946. Food of the haddock. Proc. Nova Scotian Inst. Sci., vol. 21, pts. 1 and 2, pp. 15–49.

HONEYMAN, D.

1886. Nova Scotian ichthyology. Proc. Trans. Nova Scotian Inst. Science, vol. 6, Part 4, pp. 328–330.

HUBBS, CARL L.

1923. Notes on a small collection of fishes from Monhegan Island, Maine. Copeia, No. 123, pp. 101–103.

HUNTSMAN, A. G.

1918. Histories of new food fishes. 1. The Canadian plaice. Bull. Biol. Board Canada No. 1, 22 pp.

HUNTSMAN, A. G.—Continued

1918a. The effect of the tide on the distribution of the fishes of the Canadian Atlantic coast. Trans. Royal Soc. Canada, Ser. 3, vol. 12, Sect. 4, pp. 61–67.

1918b. The growth of the scales in fishes. Trans. Royal Canadian Institute, vol. 12, pp. 61–101.

1918c. Our eastern flat-fishes. Canadian Fisherman, June 1918, vol. 5, No. 6, pp. 788–790.

1919. Growth of the young herring (so-called sardines) of the Bay of Fundy. Canadian Fisheries Exped. 1914–15, Dept. of Naval Service [Canada], pp. 165–171.

1922. Is winter mackerel fishery possible? Canadian Fisherman, May, vol. 9, No. 5, pp. 88–89.

1922a. The fishes of the Bay of Fundy. Contrib. Canadian Biol. for 1921, No. 3, pp. 49–72.

1931. The Maritime salmon of Canada. Bull. Biol. Board Canada, No. 21, 99 pp.

1933. St. John November salmon, the earliest run known. Progress Rept., No. 6, Atlantic Biol. Stations, Biol. Board Canada, pp. 7–10.

1934. Herring and water movements. James Johnstone Memorial Volume, pp. 81–96.

1936. Return of salmon from the sea. Bull. Biol. Board Canada, No. 51, 20 pp.

1937. "Migration" and "homing" of salmon. Science, New Ser., vol. 85, pp. 313–314.

1938. "Spawning urge," "homing instinct", and "waiting" in salmon return. Nature, vol. 141, pp. 421–422.

1948. Atlantic salmon investigation. Ann. Report Fish. Research Board Canada for 1947, pp. 37–38.

1949. Atlantic salmon investigations. Ann. Report Fish. Research Board Canada for 1948, pp. 40–41.

1950. Report for 1949 of fish (salmon) management. Ann. Report Fish. Research Board Canada for 1949, App. 2, pp. 41–43.

1952. The production of life in the Bay of Fundy. Trans. Roy. Soc. Canada, vol. 46, ser. 3, sect. 5, pp. 15–38.

HUNTSMAN, A. G., and A. W. H. NEEDLER.

1927. Fluctuations in the haddock fishery of the Bay of Fundy. Contrib. Canadian Biol. Fish., New Ser., vol. 3, No. 18, pp. 423–436.

HUNTSMAN, A. G., and others.

1939. Migration and conservation of salmon. Pub. No. 8, Amer. Assoc. Adv. Science, 106 pp.

JACKSON, C. T.

1852. [Capture of a basking shark off St. John, New Brunswick.] Proc. Boston Soc. Natural Hist., vol. 4, pp. 202–205,

1852a. [Capture of a thresher shark in Boston Harbor.] Proc. Boston Soc. Natural Hist., vol. 4, pp. 205–206.

JACKSON, J. B. S.

1847. Bluefish at Nahant, Mass. Proc. Boston Soc. Natural Hist., vol. 2, p. 248.

JEFFERS, A. M.

1932. The Fundy shad industry. Ann. Report Biol. Board Canada for 1931, p. 26.

1932a. The Fundy cod fishery. Ann. Rept. Biol. Board Canada for 1931, pp. 26–27.

1932b. The Fundy smelt fishery. Ann. Rept. Biol. Board Canada for 1931, p. 27.

JOHANSEN, FRITS.

1925. Natural history of the cunner (*Tautogolabrus adspersus* Walbaum). Contrib. Canadian Biol., New Ser., vol. 2, No. 17, pp. 423–468.

JONES, J. MATTHEW.

1879. List of the fishes of Nova Scotia. Proc. Trans. Nova Scotian Inst. Nat. Sci., vol. 5, Pt. 1, pp. 87–97.

JORDAN, DAVID STARR.

1891. A review of the labroid fishes of America and Europe. Report U. S. Comm. Fish and Fisheries, 1887, pp. 599–699, pls. 1–11.

JORDAN, DAVID STARR, and BRADLEY MOORE DAVIS.

1892. A preliminary review of the apodal fishes or eels inhabiting the waters of America and Europe. Report U. S. Comm. Fish and Fisheries, 1888, pp. 581–677, pls. 73–80.

JORDAN, DAVID STARR, and CHARLES L. EDWARDS.

1887. A review of the American species of Tetraodontidae. Proc. U. S. Nat. Museum, vol. 9, pp. 230–247.

JORDAN, DAVID STARR, and CARL H. EIGENMANN.

1889. A review of the Sciaenidae of America and Europe. Report U. S. Comm. Fish and Fisheries, 1886, pp. 343–451, pls. 1–4.

1890. A review of the genera and species of Serranidae found in the waters of America and Europe. Bull. U. S. Fish Comm., vol. 8, pp. 329–441, pls. 60-69.

JORDAN, DAVID STARR, and BARTON WARREN EVERMANN.

1896. A checklist of the fishes and fishlike vertebrates of North and Middle America. Report U. S. Comm. Fish and Fisheries for 1895, pp. 207–584.

1896–1900. The fishes of North and Middle America. Bull. U. S. Nat. Museum, No. 47, Parts 1–4, 3313 pp., 392 pls.

JORDAN, D. S., B. W. EVERMANN, and H. W. CLARK.

1930. Checklist of the fishes and fishlike vertebrates of North and Middle America . . . Report U. S. Comm. Fish and Fisheries, 1928, Pt. 2, pp. 1–670.

JORDAN, DAVID STARR, and CHARLES H. GILBERT.

1882. Synopsis of the fishes of North America. Bull. U. S. Nat. Museum, No. 16, lvii, 1018 pp.

1884. A review of the American Caranginae. Proc. U. S. Nat. Museum, vol. 6, pp. 188–207.

JORDAN, DAVID STARR, and DAVID KOP GOSS.
1889. A review of the flounders and soles (Pleuronectidae) of America and Europe. Rept. U. S. Comm., Fish and Fisheries, 1886, pp. 225–342, Pls. 1–9.

JORDAN, DAVID STARR, and ELIZABETH G. HUGHES.
1887. A review of the species of the genus Prionotus. Proc. U. S. Nat. Museum, vol. 9, pp. 327–338.

JOSSELYN, JOHN.
1672. New England rarities discovered in birds, beasts, fishes, serpents, and plants of that country. Printed for G. Widdowes at the Green Dragon in St. Paul's Churchyard, 1672. London. [Fishes, p. 23.] Second edition, 1675, London. Reprint of first edition, edited by Edward Tuckerman, in Trans. Amer. Antiquarian Soc., vol. 4, 1860, pp. 105–238. Another reprint by Edward Tuckerman, 1865, 134 pp., Boston. [Fishes, pp. 53–61.]
1674. An account of two voyages to New England. Printed for Giles Widdowes at the Green Dragon in St. Paul's Churchyard, 1674, London. [Fishes, pp. 103–115.] Second edition, 1675, London. Reprint of second edition in Massachusetts Historical Soc. Collections, Ser. 3, vol. 3, 1833, pp. 211–396 [Fishes, pp. 271–277]. Another reprint by William Veazie, 1865, vii, 211 pp., Boston [Fishes, pp. 83–89].

KENDALL, WILLIAM CONVERSE.
1896. Description of a new stickleback, *Gasterosteus gladiunculus*, from the coast of Maine. Proc. U. S. Nat. Museum, vol. 18, pp. 623–624.
1898. Notes on the food of four species of the cod family. Report U. S. Comm. Fish and Fisheries, 1896, pp. 177–186.
1902. Notes on the silversides of the genus Menidia of the east coast of the United States, with descriptions of two new subspecies. Rept. U. S. Comm. Fish and Fisheries for 1901, pp. 241–267.
1908. Fauna of New England. List of the Pisces. Occasional Papers, Boston Soc. Nat. Hist., vol. 7, No. 8, 152 pp.
1910. Effects of the menhaden and mackerel fisheries upon the fish supply. Proc. Fourth Internat. Fishery Congress. Bull. U. S. Bur. Fisheries, vol. 28, Pt. 1, pp. 279–293.
1912. Notes on a new species of flatfish from off the coast of New England. Bull. U. S. Bur. Fisheries, vol. 30, pp. 391–394, Pl. 57.
1914. An annotated catalogue of the fishes of Maine. Proc. Portland Soc. Nat. Hist., vol. 3, Pt. 1, pp. 1–198.
1917. The capelin (*Mallotus villosus*), with notes on its occurrence on the coast of Maine. Copeia, No. 42, pp. 28–30.

KENDALL, WILLIAM CONVERSE—Continued
1919. Second authentic record of capelin (*Mallotus villosus*) on the Maine coast. Copeia, No. 73, pp. 70–71.
1927. The smelts. Bull. U. S. Bur. Fisheries, vol. 42, pp. 217–375.
1931. Remarks on additions to the marine fauna of the coast of Maine. Bull. No. 58, Boston Soc. Nat. Hist., pp. 9–11.
1935. The fishes of New England. The salmon family, Pt. 2. The salmons. Mem. Boston Soc. Nat. Hist., vol. 9, No. 1, 166 pp., 10 pls.

KENDALL, W. C., and DONALD R. CRAWFORD.
1922. Notice of a spiral valve in the teleostean fish *Argentina silus* . . . Jour. Washington Acad. Sciences, vol. 12, No. 1, pp. 8–19.

KINGSLEY, J. S.
1922. The food habits of swordfish. Science, New Ser., vol. 56, pp. 225–226.

KINGSLEY, J. S., and H. W. CONN.
1883. Some observations on the embryology of the teleosts. Mem. Boston Soc. Nat. Hist., vol. 3, No. 6, pp. 183–212, pls. 14–16.

KNEELAND, S., JR.
1847. Dissection of *Scymnus brevipinna* (LeSueur). Boston Jour. Nat. Hist., vol. 5, No. 4, pp. 479–485.

KNIGHT, THOMAS, F.
1866. Descriptive catalog of the fishes of Nova Scotia. 55 pp. Halifax.
1867. Shore and deep-sea fisheries of Nova Scotia. vi, 113 pp. Halifax.

KUNTZ, ALBERT, and LEWIS RADCLIFFE.
1918. Notes on the embryology and larval development of twelve teleostean fishes. Bull. U. S. Bur. Fish., vol. 35, pp. 87–134.

LAIGHTON, CEDRIC.
1882. The capture of shad at Isles of Shoals, New Hampshire. Bull. U. S. Fish Comm., vol. 1, p. 421.

LANMAN, CHARLES.
1874. The Salmonidae of eastern Maine, New Brunswick, and Nova Scotia. Report U. S. Comm., Fish and Fisheries for 1872–73, pp. 219–225.
1874a. The shad and gaspereau or alewife of New Brunswick and Nova Scotia. Rept. U. S. Comm., Fish and Fisheries for 1872–73, pp. 461–462.

LEA, EINAR.
1919. Age and growth of herring in Canadian waters. Canadian Fisheries Expedition of 1914–1915, Dept. Naval Service [Canada], pp. 75–164.

LEE, L. A.
1880. Occurrence of the web-fingered sea-robin on the coast of Maine. Amer. Naturalist, vol. 14, p. 896.

LEE, L. A.—Continued
1885. The fishes of Casco Bay. Portland *Advertiser*, March 3, 1885, and Brunswick *Telegraph*, Friday, March 13, 1885.

LEIM, A. H.
1924. The life history of the shad (*Alosa sapidissima* Wilson) with special reference to the factors limiting its abundance. Contrib. Canadian Biol., New Ser., vol. 2, No. 11, pp. 163–284, 45 figs.
1930. Unusual fishes and other forms in Nova Scotian waters. Proc. Trans. Nova Scotian Inst. Science, vol. 17, Pt. 4, p. xlvi.
1937. Progress Rept. 21, Atlantic Biol. Stations, Biol. Board Canada, p. 5.

LESUEUR, C. A.
1817. A short description of five (supposed) new species of the genus Muraena . . . Jour. Acad. Nat. Sciences, Philadelphia, vol. 1, Pt. 1, pp. 81–83.
1817a. Descriptions of four new species, and two varieties, of the genus Hydrargira. Jour. Acad. Nat. Sciences, Philadelphia, vol. 1, Pt. 1, pp. 126–134.
1818. Descriptions of several new species of North American fishes. Jour. Acad. Nat. Sciences, Philadelphia, vol. 1, Pt. 2, pp. 222–235.
1821. Observations on several genera and species of fish, belonging to the natural family of the Esoces. Jour. Acad. Nat. Sciences, Philadelphia, vol. 2, Pt. 1, pp. 124–138, 2 pls.

LYMAN, HENRY.
1950. Bluefishing. 98 pp. Boston.

LYMAN, THEODORE.
1859. [Account of the habits of some animals recently observed by him at West Yarmouth, Mass.] Proc. Boston Soc. Nat. Hist., vol. 7, pp. 75–79.

LYMAN, THEODORE, and ALF. A. REED.
1866. [Concerning the obstructions to the passage of fish in the Connecticut and Merrimac Rivers.] Report of the Commissioners. Senate Massachusetts Document No. 8, Dec. 1865 (1866), pp. 3–41.

MACCOY, CLINTON V.
1929. The mackerel in New England. Bull. No. 53, Boston Soc. Nat. Hist., pp. 3–7.
1931. Museum Notes, Fishes. Bull. No. 58, Boston Soc. Nat. Hist., pp. 16–18.
1931a. Museum Notes, Fishes. Bull. No. 61, Boston Soc. Nat. Hist., p. 21.
1933. Museum Notes, Fishes. Bull. No. 67, Boston Soc. Nat. Hist., p. 9.
1933a. Museum Notes, Fishes. Bull. No. 69, Boston Soc. Nat. Hist., pp. 8–9.
1934. Museum Notes, Fishes. Bull. No. 70, Boston Soc. Nat. Hist., pp. 6–7.
1935. Fishes. Bull. No. 74, Boston Soc. Nat. Hist., p. 16.

MACDONALD, MARSHALL.
1884. The shad and the alewives. *In* Fish. Ind. U. S., Sec. 1, Text, pp. 579–609.

MACFARLAND, W. E.
1931. A study of the Bay of Fundy herring. Ann. Rept. Biol. Board Canada for 1930, pp. 23–24.

MCGONIGLE, R. H., and M. W. SMITH.
1936. Three rare fishes from Passamaquoddy Bay. Proc. Nova Scotian Inst. Sciences, vol. 19, p. 160.

MCINTIRE, EDWIN C.
1906. Numbers of dogfish. Rept. Massachusetts Comm. Fish. Game for 1905, pp. 108–109.

MCKENZIE, R. A.
1932. The Fundy haddock fishery. Ann. Rept. Biol. Board Canada for 1931, p. 34.
1932a. The Fundy alewife fishery. Ann. Rept. Biol. Board Canada for 1931, p. 34.
1934. Cod movements on the Canadian Atlantic coast. Contrib. Canadian Biol. Fish., New Ser., vol. 8, No. 31 (Series A, general, No. 46), pp. 433–459.
1939. Some marine fish and salp records. Proc. Nova Scotian Inst. Sciences, vol. 20, pp. 13–20.
1940. Some marine records from Nova Scotian fishing waters. Proc. Nova Scotian Inst. Sciences, vol. 20, pp. 42–46.

MCKENZIE, R. A., and R. E. S. HOMANS.
1938. Rare and interesting fishes and salps in the Bay of Fundy and off Nova Scotia. Proc. Nova Scotian Inst. Sciences, vol. 19, Pt. 3, pp. 277–281.

MADDOCKS, LUTHER.
1878. The menhaden fishery of Maine . . . Rept. Secretary Assoc. Menhaden Oil and Guano Manufacturers, Maine, for 1878, pp. 1–48. Portland.

MARSTON, PHILIP M.
1938. Notes on fish and early fishing in the Merrimac River System. Biol. Surv. Merrimac Water shed, Surv. Rept., No. 3, New Hampshire Fish and Game Dept., pp. 187–197.

MARTIN, S. J.
1882. First appearance of fish at Gloucester, 1881. Bull. U. S. Fish Comm., vol. 1, p. 66.
1882a. Notes on New England fisheries. Bull. U. S. Fish Comm., vol. 1, p. 133.
1882b. Notes on New England food-fishes. Bull. U. S. Fish Comm., vol. 1, p. 202.
1882c. Pollock-fishing in Boston Bay. Bull. U. S. Fish Comm., vol. 1, p. 342.
1882d. Notes on the fisheries of Gloucester, Massachusetts. Bull. U. S. Fish Comm., vol. 1, pp. 425–427.
1883. Movements and catch of mackerel. Bull. U. S. Fish Comm., vol. 2, pp. 89–90.

MARTIN, S. J.—Continued

1883a. Notes on the fisheries of Gloucester, Massachusetts. Bull. U. S. Fish Comm., vol. 2, pp. 91–93.

1883b. Notes on the fisheries of Gloucester, Massachusetts. Bull. U. S. Fish Comm., vol. 3, pp. 162–178; 297–300.

1884. Notes on the fisheries of Gloucester, Massachusetts. Bull. U. S. Fish Comm., vol. 4, pp. 89–96; 249–255; 444–448.

1885. Notes on the fisheries of Gloucester, Massachusetts. Bull. U. S. Fish Comm., vol. 5, pp. 57–62; 203–208.

MARTIN, W. R., and F. D. MCCRACKEN.

1950. Movements of halibut on the Canadian Atlantic coast. Progress Report Atlantic Coast Stations, Fish. Res. Board Canada, No. 50, pp. 3–6.

MASSACHUSETTS COMMISSIONERS OF INLAND FISHERIES.

1877–86. [Annual statistics of the pound and shore seine fisheries of Massachusetts.] Ann. Reports, for 1877 to 1885.

MASSACHUSETTS COMMISSIONERS ON INLAND FISHERIES AND GAME.

1887–1902. [Annual statistics of the pound and shore seine fisheries of Massachusetts.] Ann. Reports for 1886 to 1901.

MASSACHUSETTS COMMISSIONERS ON FISHERIES AND GAME.

1903–20. [Annual statistics of the pound and shore seine fisheries of Massachusetts.] Ann. Reports, 1902 to 1906 and 1917 to 1919.

1906. Report on the damage done by dogfish to the fisheries of Massachusetts. Ann. Report for 1905, pp. 97–169.

1907. Dogfish. Ann. Report for 1906, pp. 19–20.

MASSACHUSETTS DIVISION OF FISHERIES AND GAME.

1921–22. [Annual statistics of the pound and shore seine fisheries of Massachusetts.] Ann. Reports for 1920 and 1921.

MAVOR, JAMES W.

1918. On the age and growth of the pollock in the Bay of Fundy. Contrib. Canadian Biol. for 1917–18, No. 6, pp. 111–125.

1921. The utilization of dogfish and selachian fishes of eastern Canada. Contrib. Canadian Biol. for 1918–20, No. 13, pp. 125–133.

MERRIMAN, DANIEL.

1941. Studies on the striped bass (*Roccus saxatilis*) of the Atlantic coast. Fishery Bull. No. 35, U. S. Fish and Wildlife Service, 77 pp., vol. 50, 1950.

MITCHILL, S. L.

1815. The fishes of New York . . . Trans. Literary and Philosophical Soc. New York, vol. 1, pp. 355–492, pls. 1–6.

MOORE, HENRY FRANK.

1898. Observations on the herring and herring fisheries of the northeast coast, with special reference to the vicinity of Passamaquoddy Bay. Report U. S. Comm. Fish and Fisheries, for 1896, pp. 387–442.

MOORE, J. PERCY.

1899. Report on mackerel investigations in 1897. Report U. S. Comm. Fish and Fisheries for 1898, pp. 1–22.

MORROW, JAMES E.

1951. The biology of the longhorn sculpin . . . Bull. Bingham Oceanogr. Collection, vol. 13, Art. 2, 89 pp.

MORTON, THOMAS.

1637. New English Canaan; or New Canaan, containing an abstract of New England . . . Printed by C. Green, 1632, 125 pp. 1st edition. Printed at Amsterdam by Jacob Frederick Stam, 1637, 188 pp. Also Prince Society edition, 1883, 348 pp., Boston.

MOULTON, JAMES M.

1951. The unusual behavior of a sand launce. Copeia, p. 90.

NEEDLER, ALFREDA P.

1931. The haddock. Bull. No. 25, Biol. Board Canada, 28 pp.

NEEDLER, A. W. H.

1929. Studies on the life history of the haddock (*Melanogrammus aeglefinus*). Contrib. Canadian Biol. Fish., New Ser., vol. 4, No. 20, pp. 265–285.

1930. The migrations and the interrelationships of haddock populations in North American waters. Contrib. Canadian Biol. Fisher., New Ser., vol. 6, No. 10, pp. 241–314.

NICHOLS, J. T., and C. M. BREDER, JR.

1927. The marine fishes of New York and southern New England. Zoologica, New York Zool. Soc., vol. 9, No. 1, 192 pp.

NICHOLS, J. T. and F. E. FIRTH.

1939. Rare fishes off the Atlantic coast, including a new Grammicolepid. Proc. Biol. Soc. Washington, vol. 52, pp. 85–88.

OLSEN, Y. H., and DANIEL MERRIMAN.

1946. The biology and economic importance of the ocean pout (*Macrozoarces americanus* Bloch and Schneider). Bull. Bingham Oceanogr. Collection, vol. 9, Art. 4, pp. 1–184.

PARR, ALBERT EIDE.

1932. On a deep sea devilfish from New England . . . Bull. No. 63, Boston Soc. Nat. Hist., pp. 3–16.

1933. Two new records of deep-sea fishes from New England with description of a new genus and species. Copeia, pp. 176–179.

Peck, William D.
1804. Description of four remarkable fishes, taken near the Piscataqua in New Hampshire. Mem. Amer. Acad. Arts and Sciences, vol. 2, Part 2, pp. 46–57, pl. 2.

Perley, M. H.
1850. Statements in relation to the fisheries of the Bay of Fundy. Proc. Boston Soc. Nat. Hist., vol. 3, p. 374.
1851. Report upon the fisheries of the Bay of Fundy. v, 176 pp. [Pp. 118–159 reprinted in 1852 as Descriptive catalogue (in part) of the fishes of New Brunswick and Nova Scotia.]
1852. Descriptive catalogue (in part) of the fishes of New Brunswick and Nova Scotia. Second edition, 50 pp., Fredericton. [Pp. 118–159 of "Report upon the fisheries of the Bay of Fundy."]

Perlmutter, Alfred.
1947. The blackback flounder and its fishery in New England and New York. Bull. Bingham Oceanogr. Collection, vol. 11, Art. 2, 92 pp.

Perlmutter, Alfred, and G. M. Clark.
1949. Age and growth of immature rosefish (*Sebastes marinus*) in the Gulf of Maine and off western Nova Scotia. Fishery Bull., U. S. Fish and Wildlife Service, vol. 51, Fishery Bull. No. 45, pp. 207–228.

Piers, Harry.
1934. Accidental occurrence of the man-eater or great white shark *Carcharodon carcharias* (Linn.) in Nova Scotian waters. Proc. Nova Scotian Inst. Science, vol. 18, pp. 192–203.

Prince, Edward E.
1907. The eggs and early life-history of the herring, gaspereau, shad, and other clupeoids. Further contrib. Canadian Biol., for 1902–1905, pp. 95–110, pls. 8–10.
1913. The pearlsides. A luminous fish new to Canada. Rod and Gun in Canada, vol. 14, No. 11, pp. 1143–1145. Woodstock, Ont.
1917. On serially striped haddock in New Brunswick. Contrib. Canadian Biol. for 1915–16, pp. 86–90, pl. 9.

Prince, E. E., and A. H. MacKay.
1901. The paired fins of the mackerel shark. Contrib. Canadian Biol. for 1901, pp. 55–58, pls. 5–7.

Procter, William, and others.
1928. A contribution to the life history of the angler (*Lophius piscatorius*). Biol. Survey Mt. Desert region. Part 2, Fishes, 13 pp., 5 plates.
1933. Marine fauna. Biol. Survey Mt. Desert region. Part 5, Fishes, pp. 286–290.

Putnam, F. W.
1856. [Fishes of Essex County, Mass.] Proc. Essex Institute, vol. 1, p. 144.
1856a. [Fishes of Essex County, Mass., continued.] Proc. Essex Institute, vol. 1, p. 148.
1856b. [Fishes of Essex County, Mass., third communication.] Proc. Essex Institute, vol. 1, p. 201.
1864. [Note on the cod of Massachusetts waters.] Proc. Boston Soc. Natural Hist., vol. 9, p. 319.
1866. [Occurrence of *Leptocephalus gracilis* at Nahant, Mass., in July, 1858]. Proc. Boston Soc. Natural Hist., vol. 10, p. 373.
1866a. [Description of *Gasterosteus wheatlandi*, from Nahant, Mass.] Proc. Essex Institute, vol. 5, p. 4.
1870. [Addition of two species of fishes to the fauna of Essex County, Mass.] Bull. Essex Institute, vol. 2, No. 7, p. 111.
1870a. [On a species of Hemiramphus from Danvers, Mass.] Bull. Essex Institute, vol. 2, Nos. 11, 12 Nov. and Dec., 1870, p. 171.
1870b. On the young of *Orthagoriscus mola*. Amer. Naturalist, vol. 4, pp. 629–633. Also in Proceedings, Amer. Assoc. Advancement of Science, 19th meeting, 1871, pp. 255–260.
1873. List of fishes sent by the Museum to different institutions . . . Bull. Museum Compar. Zool., vol. 1, pp. 2–16 [*Myxine* at Grand Manan, p. 10].
1874. [Notes on several rare fishes from Essex County, Mass.] Bull. Essex Institute, vol. 6, No. 1, pp. 11–13.
1874a. [Tooth of a maneater that attacked a dory near St. Pierre Bank.] Bull. Essex Institute, vol. 6, No. 4, p. 72.
1874b. [*Chauliodus sloani* on George's Bank.] Bull. Essex Institute, vol. 6, No. 7, p. 111.
1874c. [*Liparis lineatus* and *L. montaguii* in Salem Harbor, Mass.] Proc. Boston Soc. Natural Hist., vol. 16, p. 114.
1874d. Notes on the genus *Myxine*. Proc. Boston Soc. Natural Hist., vol. 16, pp. 127–135.
1874e. Notes on Liparis, Cyclopterus and their allies. Proc. Amer. Assoc. Advancement Science, 22d meeting, pp. 335–340.

Radcliffe, Lewis.
1916. An extension of the recorded range of three species of fishes in New England waters. Copeia, No. 26, pp. 2–3.
1922. Fisheries of the New England States in 1919. App. 5, Report of the U. S. Commissioner of Fisheries for 1921, pp. 120–187.
1926. "Opah" and "Skilligalee" landed at Boston Fish Pier. Copeia, No. 151, p. 112.

RANEY, EDWARD C., and OTHERS.
1952. The striped bass. Bull. Bingham Oceanogr Coll., vol. 14, art. 1, 177 pp.

RATHBUN, R., and W. WAKEHAM.
1897. Preservation of the fisheries in the waters contiguous to the United States and Canada. Document No. 315, U. S. House of Representatives, 54th Congress, 2d session, pp. 1–178.

REID, MARGARET E.
1929. The distribution and development of the cunner (*Tautogolabrus adspersus*) along the eastern coast of Canada. Contrib. Canadian Biol. Fish., New Ser., vol. 4, No. 27, pp. 431–441.

RICH, WALTER H.
1930. Fishing grounds of the Gulf of Maine. Report U. S. Commissioner of Fisheries for 1929, App. 3, pp. 51–117, 5 maps.
1947. The swordfish and the swordsfishery of New England. Proc. Portland Soc. Natural Hist., vol. 4, pt. 2, 102 pp., 1 chart.

SCATTERGOOD, LESLIE W.
1949. The production and the fishing methods of the Maine herring industry, with notes on the 1947 season. Spec. Sci. Report No. 67, U. S. Fish and Wildlife Service, 25 pp., 7 graphs and charts.
1949a. Notes on the Maine shark fishery. Copeia, pp. 69–71.
1951. The occurrence of egg capsules in the winter skate (*Raja diaphanes*) in Maine waters. Copeia, p. 169.

SCATTERGOOD, LESLIE W., P. S. TREFETHEN, and G. W. COFFIN.
1951. Notes on the size of menhaden taken in Maine during 1949. Copeia, pp. 93–94.
1951a. Notes on Gulf of Maine fishes in 1949. Copeia, pp. 297–298.

SCHROEDER, WILLIAM C.
1930. Migrations and other phases in the life history of the cod off southern New England. Bull. U. S. Bur. Fish., vol. 46, pp. 1–136.
1930a. A record of *Polyprion americanus* (Bloch and Schneider) from the northwestern Atlantic. Copeia, pp. 46–48.
1931. Notes on certain fishes collected off the New England coast from 1924 to 1930. Bull. Boston Soc. Natural Hist., No. 58, pp. 3–8.
1933. Unique records of the brier skate and rock eel from New England. Bull. No .66, Boston Soc. Natural Hist., pp. 5–6.
1937. Records of *Pseudopriacanthus altus* Gill and *Fundulus majalis* (Walbaum) from the Gulf of Maine. Copeia, p. 238.
1939. The Provincetown "sea serpent." New England Naturalist, No. 2, pp. 1–2.
1942 Results of haddock tagging in the Gulf of Maine from 1923 to 1932. Jour. Marine Research, vol. 5, No. 1, pp. 1–19.

SCHROEDER, WILLIAM C.—Continued
1947. Notes on the diet of the goosefish *Lophius americanus*. Copeia, p. 201.

SCHUCK, HOWARD A.
1947. Haddock (*Melanogrammus aeglefinus*). Fishery Leaflet No. 198, U. S. Fish and Wildlife Service, 9 pp.
1948. Current haddock situation on Georges Bank. Commercial Fisheries Review, vol. 10, No. 10, 6 pp.
1949. Relationship of catch to changes in population size of New England haddock. Biometrics, Amer. Statistical Assoc., vol. 5, No. 3, pp. 213–231.
1951. Northern record for the little tuna *Euthynnus alletteratus*. Copeia, p. 98.
1951a. New Gulf of Maine record for occurrence of dolphin *Coryphaena hippurus*, and data on small specimens. Copeia, p. 171.

SCUDDER, NEWTON P.
1887. The salt halibut fishery . . . *In* Fisheries and Fishery Industries of the U. S.; Sect. 5, vol. 1, pp. 90–119.

SELLA, MASSIMO.
1931. The tuna (*Thunnus thynnus* L.) of the western Atlantic. An appeal to fishermen for the collection of hooks found in tuna fish. Internat. Revue gesamt. Hydrobiol. Hydrog., vol. 25, pp. 46–67.

SETTE, O. E.
1931. Outlook for the mackerel fishery in 1931. Fishery Circular No. 4, U. S. Bur. Fish., 20 pp.
1932. Outlook for the mackerel fishery in 1932. Fishery Circular No. 10, U. S. Bur. Fish., 25 pp.
1933. Outlook for the mackerel fishery in 1933. Fishery Circular No. 14, U. S. Bur. Fish., 23 pp.
1934. Outlook for the mackerel fishery, 1934. Fishery Circular No. 17, U. S. Bur. Fish., 6 pp.
1943. Biology of the Atlantic mackerel (*Scomber scombrus*) of North America, Part 1. Fishery Bull., U. S. Fish and Wildlife Service, vol. 50 (Fish. Bull. No. 38), pp. 149–227.
1950. Biology of the Atlantic mackerel (*Scomber scombrus*) of North America. Part 2. Fishery Bull., U. S. Fish and Wildlife Service, vol. 51 (Fish. Bull. No. 49), pp. 251–358.

SETTE, O. E., and A. W. H. NEEDLER.
1934. Statistics of the mackerel fishery off the east coast of the U. S., 1804 to 1930. Inv. Rept. No. 19, U. S. Bur. Fish., pp. 1–48.

SHARP, BENJAMIN.
1901. The food of the cod. Proc. Acad. Natural Sciences Philadelphia, vol. 53, p. 2.

SHARP, BENJAMIN, and HENRY W. FOWLER.
1904. The fishes of Nantucket. Proc. Acad. Natural Sciences Philadelphia, vol. 56, pp. 504–512.

SHERWOOD, GEORGE H., and VINAL N. EDWARDS.
1902. Notes on the migration, spawning, abundance, etc., of certain fishes in 1900. Bull. U. S. Fish Comm., vol. 21, pp. 27–31.

SMITH, EVERETT.
1883. A mammoth cod. Bull. U. S. Fish Comm., vol. 3, p. 443.
1889. Results of planting shad in the Kennebec River. Bull. U. S. Fish Comm., vol. 7, p. 16.

SMITH, HUGH M.
1893. [Eastward movement of menhaden.] Report U. S. Comm., Fish and Fisheries, for 1889–91, pp. 202–203.
1895. Notes on the capture of Atlantic salmon at sea and in the coast waters of the eastern States. Bull. U. S. Fish Comm., vol. 14, pp. 95–99.
1896. Notes on an investigation of the menhaden fishery in 1894 . . . Bull. U. S. Fish Comm., vol. 15, pp. 285–302.
1898. The fishes found in the vicinity of Woods Hole. Bull. U. S. Fish Comm., vol. 17, pp. 85–111.
1898a. The salmon fishery of Penobscot Bay and River in 1895 and 1896. Bull. U. S. Fish Comm., vol. 17, pp. 113–124, pl. 5, 1 map.
1899. Notes on the extent and condition of the alewife fisheries of the United States in 1896. Rept. U. S. Comm., Fish and Fisheries for 1898, pp. 31–43.
1902. Notes on the tagging of four thousand adult cod at Woods Hole, Massachusetts. Report U. S. Comm. Fish and Fisheries for 1901, pp. 193–208.
1938. *Alectis crinitus*, a fish new to Massachusetts Bay. Copeia, pp. 146–147.

SMITH, HUGH M., and R. A. GOFFIN.
1937. A fish new to Massachusetts Bay. Copeia, p. 236.

SMITH, HUGH M., and WILLIAM C. KENDALL.
1898. Notes on the extension of the recorded range of certain fishes of the United States coasts. Report U. S. Comm. Fish and Fisheries for 1896, pp. 169–176.

SMITH, JEROME V. C.
1833. A catalogue of the marine fishes taken on the Atlantic coast of Massachusetts. *In* Hitchcock, Report on the Geol., Mineral., Botany, and Zool. Massachusetts, 1833, Pt. 4, pp. 553–554.
1833a. Natural history of the fishes of Massachusetts, embracing a practical essay on angling. vii, 400 pp., 1833. Boston.
1835. A catalogue of the marine and fresh-water fishes of Massachusetts. *In* Hitchcock, Report on Geol., Mineral., Botany, and Zool. Massachusetts, 2d edition, 1835, Pt. 4, pp. 535–538.

SMITH, CAPTAINE JOHN.
1616. The generall historie of Virginia, New England, and the Summer Isles, together with the true travels, adventures, and observations and a sea grammar. 2 vols., London. Reprinted from the London edition of 1629, by the Franklin Press, Richmond, Va., in 1819.

SPARKS, M. IRVING.
1929. The spawning and development of mackerel on the outer coast of Nova Scotia. Contrib. Canadian Biol. Fish., New Ser., vol. 4, No. 28, pp. 443–452.

SPINNEY, BENJAMIN H.
1883. Return to Gloucester of young codfish hatched by United States Fish Commission in 1879. Bull. U. S. Fish Comm., vol. 3, p. 362.

STEVENSON, CHARLES H.
1899. The shad fisheries of the Atlantic coast of the United States. Report U. S. Comm. Fish and Fisheries for 1898, pp. 101–269.

STORER, DAVID HUMPHREYS.
1836. An examination of the "Catalogue of the marine and fresh water fishes of Massachusetts," by J. V. C. Smith, M. D. . . . Boston Jour. Natural Hist., vol. 1, No. 3, pp. 347–356.
1837. Description of a new species of the genus Hydrargyra; with some additions to the Catalogue of the fishes of Massachusetts in Hitchcock's Report. Boston Jour. Natural Hist., vol. 1, No. 4, pp. 416–418.
1839. Remarks on the "Natural History of the Fishes of Massachusetts, embracing a Practical Essay on Angling; by Jerome V. C. Smith, M. D." Amer. Jour. Science and Arts, vol. 36, pp. 337–349.
1839a. A report on the fishes of Massachusetts. Boston Jour. Natural Hist., vol. 2, pp. 289–558, pls. 6–8.
1839b. Reports on the fishes, reptiles, and birds of Massachusetts. Comm. Zool. and Botan. Survey of the State, 1839. Boston. [Fishes, pp. 1–202, pls. 1–3.]
1840. Supplement to the ichthyological report. Boston Jour. Natural Hist., vol. 3, pp. 267–273.
1841. Additional descriptions of, and observations on, the fishes of Massachusetts. Proc. Boston Soc. Natural Hist., vol. 1, pp. 53–54.
1842. Additional descriptions of, and observations on, the fishes of Massachusetts. Boston Jour. Natural Hist., vol. 4, pp. 175–190.
1842a. [Note on a torpedo taken on Cape Cod.] Proc. Boston Soc. Natural Hist., vol. 1, p. 94.
1843. [Description of a new species of flatfish.] Proc. Boston Soc. Natural Hist., vol. 1, pp. 130–131.

STORER, DAVID HUMPHREYS—Continued
1843a. [Notes on *Sphyraena borealis* and *Caranx chrysos*.] Proc. Boston Soc. Natural Hist., vol. 1, pp. 148–149.
1843b. Notice of the discovery of an electrical fish on our coast. Amer. Jour. Science and Arts, vol. 44, p. 213; also Ann. Mag. Natural Hist., vol. 11, p. 326.
1843c. Description of a new species of the torpedo. Amer. Jour. Science and Arts, vol. 45, pp. 165–170, pl. 3.
1844. [Notes on flat fishes, and on the sunfish, *Orthagoriscus mola*.] Proc. Boston Soc. Natural Hist., vol. 1, pp. 194–195.
1845. [Occurrence of *Prionotus tribulus* Mitch., in Massachusetts Bay.] Proc. Boston Soc. Natural Hist., vol. 2, p. 51.
1845a. [Remarks on a living torpedo taken at Provincetown.] Proc. Boston Soc. Natural Hist., vol. 2, p. 71.
1845b. [*Alutera schoepfii*, as "*Balistes aurantiacus* Mitch." at Salem, Mass., and Portland, Maine.] Proc. Boston Soc. Natural Hist., vol. 2, pp. 71–72.
1845c. [Description of a new species of Leptocephalus.] Proc. Boston Soc. Natural Hist., vol. 2, pp. 76–77.
1845d. [Description of a new species of Prionotus from Massachusetts Bay.] Proc. Boston Soc. Natural Hist., vol. 2, pp. 77–78.
1845e. [Note on a Trichiurus from Massachusetts Bay.] Proc. Boston Soc. Natural Hist., vol. 2, pp. 85–86.
1846. [Occurrence of *Palinurus perciformis* in Boston Harbor.] Proc. Boston Soc. Natural Hist., vol. 2, p. 180.
1846a. A synopsis of the fishes of North America. Mem. Amer. Acad. Arts Sciences, New Ser., vol. 2, pp. 253–550. (Reissued, repaged from 1 to 298, by Metcalf & Co., Cambridge, in 1846.)
1847. [Descriptions of new species of Alosa and Platessa.] Proc. Boston Soc. Natural Hist., vol. 2, pp. 242–243.
1848. [Description of *Motella caudacuta*.] Proc. Boston Soc. Natural Hist., vol. 3, p. 5.
1848a. [On a shark taken at Provincetown.] Proc. Boston Soc. Natural Hist., vol. 3, pp. 11–12.
1848b. [Notes on the Greenland shark, *Somniosus brevipinna* Le Sueur.] Proc. Boston Soc. Natural Hist., vol. 3, pp. 15–16.
1848c. [Occurrence of the hammerhead shark at Provincetown and at Chatham, Mass.] Proc. Boston Soc. Natural Hist., vol. 3, pp. 70–71.
1848d. [Description of *Carcharias atwoodi*.] Proc. Boston Soc. Natural Hist., vol. 3, pp. 71–72.

STORER, DAVID HUMPHREYS—Continued
1853–1867. A history of the fishes of Massachusetts. Mem. Amer. Acad. Arts and Sciences, New Ser., vol. 5, pp. 49–92, 122–168, 257–296, pls. 1–8, 9–16, 17–23. Vol. 6, pp. 309–372, pls. 24–29. Vol. 8, pp. 389–434, pls. 30–35. Vol. 9, pp. 217–263, pls. 36–39.
1854. [Description of *Sebastes fasciatus*, from Provincetown.] Proc. Boston Soc. Natural Hist., vol. 5, p. 31.
1858. [Description of *Zeus ocellatus*.] Proc. Boston Soc. Natural Hist., vol. 6, pp. 385–386.
1867. A history of the fishes of Massachusetts. 287 pp., 39 pls. Cambridge and Boston.

STORER, HORATIO ROBINSON.
1850. Observations on the fishes of Nova Scotia and Labrador, with descriptions of new species. Boston Jour. Natural Hist., vol. 6, pp. 247–270.

SULLIVAN, JAMES.
1795. The history of the District of Maine. 421 pp., Boston.

SUMNER, FRANCIS B., RAYMOND C. OSBURN, and LEON J COLE.
1913. A biological survey of the waters of Woods Hole and vicinity. Part 2, Sect. 3.—A catalogue of the marine fauna. Bull. U. S. Bur. Fisheries, vol. 31, pt. 2, pp. 549–794 [Fishes, pp. 734–774].

SWAIN, JOSEPH.
1883. A review of the species of Stolephorus found on the Atlantic coast of the United States. Bull. U. S. Fish Comm., vol. 2, pp. 55–57.

TARR, R. S.
1884. Return to Gloucester Harbor of the young codfish hatched by the U. S. Fish Commission. Bull. U. S. Fish Comm., vol. 4, pp. 57–58.

TIBBETTS, N. V.
1887. Scarcity of cod and haddock on the coast of Maine. Bull. U. S. Fish Comm., vol. 6, pp. 75–76.

TOWNE, SUMNER A.
[1942, approx. date.] State of Maine striped bass survey . . . Maine Devel. Comm. and Dept. Sea and Shore Fisheries. 30 pp.

TOWNSEND, C. H.
1901. Statistics of the fisheries of the New England States [for 1898]. Report U. S. Comm. Fish and Fisheries, for 1900, pp. 311–386.

TRACY, HENRY C.
1910. Annotated list of fishes known to inhabit the waters of Rhode Island. 40th Ann. Rept. Comm. Inland Fisheries, Rhode Island, pp. 35–176.

UNITED STATES COMMISSION OF FISH AND FISHERIES.
1873. Notices in regard to the abundance of fish on the New England coast in former times. [Excerpts reprinted from accounts by the early voyagers and settlers.] Report for 1871–72, pp. 149–172.
1882. List of collections made by the fishing vessels of Gloucester and other New England seaports for the United States Fish Commission, from 1877–1880. Report for 1879, pp. 787–835.
1894–1905. [Annual statistics of the vessel fisheries of northern New England.] Reports for 1892 and 1895 to 1903.
1898. A manual of fish culture. Report for 1897, Appendix, 340 pp., pls. 1–62, and I–XVIII.

UNITED STATES BUREAU OF FISHERIES.
1905–30. [Statistics of New England fisheries.] Reports of the Commissioner for 1904, 1905, 1909 to 1919, and 1921 to 1929.
1931–41. [New England fishery statistics] in Ann. Reports, Fishery Industries of U. S. for 1930–1939.

UNITED STATES FISH AND WILDLIFE SERVICE.
1942–52. [New England fishery statistics] in Ann. Reports, Fishery Statistics of the U. S. for 1939–1949.

VERRILL, A. E.
1871. On the food and habits of some of our marine fishes. Amer. Naturalist, vol. 5, pp. 397–400.
1872. Marine fauna of Eastport, Maine. Bull. Essex Institute, vol. 3, No. 1, pp. 2–6.
1874. Explorations of Casco Bay by the U. S. Fish Commission in 1873. Proc. Amer. Assoc. Advancement of Science, 22nd meeting, pp. 340–395, 6 pls.

VLADYKOV, V. D.
1935. Some unreported and rare fishes from the coast of Nova Scotia. Proc. Nova Scotian Inst. Science, vol. 19, pp. 1–8.
1935a. [Haddock races.] Progress Report No. 14, Atlantic Biol. Stations, Biol. Board Canada.
1950. Movements of Quebec shad (*Alosa sapidissima*) as demonstrated by tagging. Naturaliste Canadien, vol. 77, pp. 121–134, 1 chart; also as Contrib. No. 30, Dept. Fisheries, Province Quebec.

VLADYKOV, V. D., and R. A. MCKENZIE.
1935. The marine fishes of Nova Scotia. Proc. Nova Scotian Inst. Science, vol. 19, pp. 17–113.

VLADYKOV, V. D., and D. N. WALLACE.
1938. Remarks on populations of shad (*Alosa sapidissima*) along the Atlantic coast region. Trans. Amer. Fish. Soc., vol. 67, pp. 52–66.

WALFORD, L. A.
1938. Effect of currents on distribution and survival of the eggs and larvae of the haddock (*Melanogrammus aeglefinis*) on Georges Bank. Bull. U. S. Bur. Fish., vol. 49 (Bull. 29), pp. 1–73.

WASSON, SAMUEL.
1878. Survey of Hancock County. Ann. Report Agricultural Soc. Maine for 1877–1878, pp. 189–273. [Fishes, pp. 220–225.] Bound with the 22nd Ann. Report Maine Board of Agriculture for 1877.

WELSH, W. W.
1915. Note on the habits of the young of the squirrel hake and sea snail. Copeia, No. 18, pp. 2–3.

WELSH, WILLIAM W., and C. M. BREDER, JR.
1924. Contributions to life histories of Sciaenidae of the eastern United States coast. Bull. U. S. Bur. Fish., vol. 39, pp. 141–201.

WHEATLAND, R. H.
1852. Notice of several fishes of rare occurrence. Jour. Essex County Natural Hist. Soc., vol. 1, No. 3, pp. 122–125.
1862. [Sticklebacks at Nahant, Mass., and notes on color changes in fishes.] Proc. Essex Institute, vol. 2, pp. 364–366.

WHITE, H. C.
1939. The nesting and embryo of *Zoarces anguillaris*. Jour. Fish. Research Board, Canada, vol. 4, Part. 5, pp. 337–338.

WILCOX, W. A.
1883. Reappearance of young cod hatched by the United States Fish Commission. Bull. U. S. Fish Comm., vol. 3, p. 439.
1885. New England fisheries [from January 1 to August, 1885]. Bull. U. S. Fish Comm., vol. 5, pp. 162–166, 166–170, 170–174, 197–199, 332–336, 442–445.
1887. New England fisheries [from September, 1885, to December, 1886]. Bull. U. S. Fish Comm., vol. 6, pp. 33–47, 49–51, 79–80, 94–95, 106–108, 191–192, 200–202, 210–212, 241–245, 273–275, 328–331, 418–420, 423–424.
1887a. The fisheries of New Hampshire [in 1880]. *In* Fisheries and Fishery Industries of U. S., Sect. 2, Part. 2, pp. 103–112.
1889. The fisheries of Gloucester, Mass. [in January, February, March, April, May, and June 1887], with notes on those of other localities. Bull. U. S. Fish Comm., vol. 7, pp. 66–71, 73–80.

WILLARD, B. J.
1895. Captain Ben's book. A record of the things which happened to Capt. Benjamin J. Willard . . . 204 pp., illus., 1895. Portland, Maine.

WILLEY, ARTHUR.
1921. Arctic Copepoda in Passamaquoddy Bay. Proc. Amer. Acad. Arts and Sciences, vol. 56, pp. 185–196. [Food of the American pollock, p. 192.]
1923. Notes on the distribution of free-living Copepoda in Canadian waters. Contrib. Canadian Biol., New Ser., vol. 1, No. 16, pp. 305–334. [Food of shad, pp. 313–320.]

WILLEY, A., and A. G. HUNTSMAN.
1921. Faunal notes from the Atlantic Biological Station, 1920. Canadian Field Naturalist, vol. 35, No. 1, pp. 1–7.

WILLIAMSON, WILLIAM D.
1832. The history of the State of Maine . . . Vols. 1–2, 1832. Hallowell, Maine. [Fishes, vol. 1, pp. 150–164.]

WODEHOUSE, R. P.
1916. Report on the life history of the cod as determined from the scales and other data. Contrib. Canadian Biol. for 1914–15, No. 10, pp. 103–113.

WOOD, WILLIAM.
1634. New Englands Prospect . . . Printed at London by Tho. Cotes, for John Bellamie, 1634, 83 pp. Reprinted for E. M. Boynton, 1898, x, 103 pp. Boston.

WOOD, WILLIAM.
1846. Description of a shark taken off Portland, Maine. Proc. Boston Soc. Natural Hist., vol. 2, p. 174.

WYMAN, JEFFRIES.
1867. Observations on the development of *Raia batis*. Mem. Amer. Acad. Arts, Sciences, New Ser., vol. 9, Pt. 1, pp. 31–44, 1 pl.

INDEX TO COMMON NAMES

INDEX TO SCIENTIFIC NAMES

○

www.ingramcontent.com/pod-product-compliance
Lightning Source LLC
Jackson TN
JSHW061308250925
91530JS00034B/51
* 9 7 8 1 9 3 0 6 6 5 6 0 6 *